Spectral Analysis for Univariate Time Series

Spectral analysis is widely used to interpret time series collected in diverse areas such as the environmental, engineering and physical sciences. This book covers the statistical theory behind spectral analysis and provides data analysts with the tools needed to transition theory into practical applications. Actual time series from oceanography, metrology, atmospheric science and other areas are used in running examples throughout, to allow clear comparison of how the various methods address questions of interest.

All major nonparametric and parametric spectral analysis techniques are discussed, with particular emphasis on the multitaper method, both in its original formulation involving Slepian tapers and in a popular alternative using sinusoidal tapers. The authors take a unified approach to quantifying the bandwidth of nonparametric spectral estimates, allowing for meaningful comparison among different estimates. An extensive set of exercises allows readers to test their understanding of both the theory and practical analysis of time series. The time series used as examples and R language code for recreating the analyses of the series are available from the book's web site.

DONALD B. PERCIVAL is the author of 75 publications in refereed journals on a variety of topics, including analysis of environmental time series, characterization of instability of atomic clocks and forecasting inundation of coastal communities due to trans-oceanic tsunamis. He is the coauthor (with Andrew Walden) of *Spectral Analysis for Physical Applications: Multitaper and Conventional Univariate Techniques* (Cambridge University Press, 1993) and *Wavelet Methods for Time Series Analysis* (Cambridge University Press, 2000). He has taught graduate-level courses on time series analysis, spectral analysis and wavelets for over 30 years at the University of Washington.

ANDREW T. WALDEN has authored 100 refereed papers in scientific areas including statistics, signal processing, geophysics, astrophysics and neuroscience, with an emphasis on spectral analysis and time series methodology. He worked in geophysical exploration research before joining Imperial College London. He is coauthor (with Donald B. Percival) of *Spectral Analysis for Physical Applications: Multitaper and Conventional Univariate Techniques* (Cambridge University Press,1993) and *Wavelet Methods for Time Series Analysis* (Cambridge University Press, 2000). He has taught many courses including time series, spectral analysis, geophysical data analysis, applied probability and graphical modeling, primarily at Imperial College London, and also at the University of Washington.

CAMBRIDGE SERIES IN STATISTICAL AND
PROBABILISTIC MATHEMATICS

Editorial Board

Z. Ghahramani (Department of Engineering, University of Cambridge)
R. Gill (Mathematical Institute, Leiden University)
F. P. Kelly (Department of Pure Mathematics and Mathematical Statistics,
University of Cambridge)
B. D. Ripley (Department of Statistics, University of Oxford)
S. Ross (Department of Industrial and Systems Engineering,
University of Southern California)
M. Stein (Department of Statistics, University of Chicago)

This series of high-quality upper-division textbooks and expository monographs covers all aspects of stochastic applicable mathematics. The topics range from pure and applied statistics to probability theory, operations research, optimization, and mathematical programming. The books contain clear presentations of new developments in the field and also of the state of the art in classical methods. While emphasizing rigorous treatment of theoretical methods, the books also contain applications and discussions of new techniques made possible by advances in computational practice.

A complete list of books in the series can be found at www.cambridge.org/statistics. Recent titles include the following:

24. *Random Networks for Communication*, by Massimo Franceschetti and Ronald Meester
25. *Design of Comparative Experiments*, by R. A. Bailey
26. *Symmetry Studies*, by Marlos A. G. Viana
27. *Model Selection and Model Averaging*, by Gerda Claeskens and Nils Lid Hjort
28. *Bayesian Nonparametrics*, edited by Nils Lid Hjort et al.
29. *From Finite Sample to Asymptotic Methods in Statistics*, by Pranab K. Sen, Julio M. Singer and Antonio C. Pedrosa de Lima
30. *Brownian Motion*, by Peter Mörters and Yuval Peres
31. *Probability: Theory and Examples (Fourth Edition)*, by Rick Durrett
33. *Stochastic Processes*, by Richard F. Bass
34. *Regression for Categorical Data*, by Gerhard Tutz
35. *Exercises in Probability (Second Edition)*, by Loïc Chaumont and Marc Yor
36. *Statistical Principles for the Design of Experiments*, by R. Mead, S. G. Gilmour and A. Mead
37. *Quantum Stochastics*, by Mou-Hsiung Chang
38. *Nonparametric Estimation under Shape Constraints*, by Piet Groeneboom and Geurt Jongbloed
39. *Large Sample Covariance Matrices and High-Dimensional Data Analysis*, by Jianfeng Yao, Shurong Zheng and Zhidong Bai
40. *Mathematical Foundations of Infinite-Dimensional Statistical Models*, by Evarist Giné and Richard Nickl
41. *Confidence, Likelihood, Probability*, by Tore Schweder and Nils Lid Hjort
42. *Probability on Trees and Networks*, by Russell Lyons and Yuval Peres
43. *Random Graphs and Complex Networks (Volume 1)*, by Remco van der Hofstad
44. *Fundamentals of Nonparametric Bayesian Inference*, by Subhashis Ghosal and Aad van der Vaart
45. *Long-Range Dependence and Self-Similarity*, by Vladas Pipiras and Murad S. Taqqu
46. *Predictive Statistics*, by Bertrand S. Clarke and Jennifer L. Clarke
47. *High-Dimensional Probability*, by Roman Vershynin
48. *High-Dimensional Statistics*, by Martin J. Wainwright
49. *Probability: Theory and Examples (Fifth Edition)*, by Rick Durrett
50. *Statistical Model-based Clustering and Classification*, by Charles Bouveyron et al.
51. *Spectral Analysis for Univariate Time Series*, by Donald B. Percival and Andrew T. Walden

Spectral Analysis for Univariate Time Series

Donald B. Percival
University of Washington

Andrew T. Walden
Imperial College of Science, Technology and Medicine

CAMBRIDGE
UNIVERSITY PRESS

University Printing House, Cambridge CB2 8BS, United Kingdom

One Liberty Plaza, 20th Floor, New York, NY 10006, USA

477 Williamstown Road, Port Melbourne, VIC 3207, Australia

314–321, 3rd Floor, Plot 3, Splendor Forum, Jasola District Centre, New Delhi – 110025, India

79 Anson Road, #06–04/06, Singapore 079906

Cambridge University Press is part of the University of Cambridge.

It furthers the University's mission by disseminating knowledge in the pursuit of education, learning, and research at the highest international levels of excellence.

www.cambridge.org
Information on this title: www.cambridge.org/9781107028142
DOI: 10.1017/9781139235723

© Cambridge University Press 2020

This publication is in copyright. Subject to statutory exception and to the provisions of relevant collective licensing agreements, no reproduction of any part may take place without the written permission of Cambridge University Press.

First published 2020

Printed in the United Kingdom by TJ International, Padstow Cornwall

A catalogue record for this publication is available from the British Library.

Library of Congress Cataloging-in-Publication Data
Names: Percival, Donald B., author. | Walden, Andrew T., author.
Title: Spectral analysis for univariate time series / Donald B. Percival, Andrew T. Walden.
Description: Cambridge ; New York, NY : Cambridge University Press, 2020. |
Series: Cambridge series on statistical and probabilistic mathematics ;
51 | Includes bibliographical references and indexes.
Identifiers: LCCN 2019034915 (print) | LCCN 2019034916 (ebook) |
ISBN 9781107028142 (hardback) | ISBN 9781139235723 (ebook)
Subjects: LCSH: Time-series analysis. | Spectral theory (Mathematics)
Classification: LCC QA280 .P445 2020 (print) | LCC QA280 (ebook) | DDC 519.5/5–dc23
LC record available at https://lccn.loc.gov/2019034915
LC ebook record available at https://lccn.loc.gov/2019034916

ISBN 978-1-107-02814-2 Hardback

Cambridge University Press has no responsibility for the persistence or accuracy of URLs for external or third-party internet websites referred to in this publication and does not guarantee that any content on such websites is, or will remain, accurate or appropriate.

To Niko, Samuel and Evelyn (the next generation of spectral analysts)

Contents

Preface xiii
Conventions and Notation xvi
Data, Software and Ancillary Material xxiv

1 Introduction to Spectral Analysis 1

    1.0 Introduction 1
    1.1 Some Aspects of Time Series Analysis 1
            Comments and Extensions to Section 1.1 5
    1.2 Spectral Analysis for a Simple Time Series Model 5
    1.3 Nonparametric Estimation of the Spectrum from Data 11
    1.4 Parametric Estimation of the Spectrum from Data 14
    1.5 Uses of Spectral Analysis 15
    1.6 Exercises 17

2 Stationary Stochastic Processes 21

    2.0 Introduction 21
    2.1 Stochastic Processes 21
    2.2 Notation 22
    2.3 Basic Theory for Stochastic Processes 23
            Comments and Extensions to Section 2.3 25
    2.4 Real-Valued Stationary Processes 26
    2.5 Complex-Valued Stationary Processes 29
            Comments and Extensions to Section 2.5 31
    2.6 Examples of Discrete Parameter Stationary Processes 31
    2.7 Comments on Continuous Parameter Processes 38
    2.8 Use of Stationary Processes as Models for Data 38
    2.9 Exercises 41

3 Deterministic Spectral Analysis ... 47

- 3.0 Introduction ... 47
- 3.1 Fourier Theory – Continuous Time/Discrete Frequency ... 48
  - Comments and Extensions to Section 3.1 ... 52
- 3.2 Fourier Theory – Continuous Time and Frequency ... 53
  - Comments and Extensions to Section 3.2 ... 55
- 3.3 Band-Limited and Time-Limited Functions ... 57
- 3.4 Continuous/Continuous Reciprocity Relationships ... 58
- 3.5 Concentration Problem – Continuous/Continuous Case ... 62
- 3.6 Convolution Theorem – Continuous Time and Frequency ... 67
- 3.7 Autocorrelations and Widths – Continuous Time and Frequency ... 72
- 3.8 Fourier Theory – Discrete Time/Continuous Frequency ... 74
- 3.9 Aliasing Problem – Discrete Time/Continuous Frequency ... 81
  - Comments and Extensions to Section 3.9 ... 84
- 3.10 Concentration Problem – Discrete/Continuous Case ... 85
- 3.11 Fourier Theory – Discrete Time and Frequency ... 91
  - Comments and Extensions to Section 3.11 ... 93
- 3.12 Summary of Fourier Theory ... 95
- 3.13 Exercises ... 102

4 Foundations for Stochastic Spectral Analysis ... 107

- 4.0 Introduction ... 107
- 4.1 Spectral Representation of Stationary Processes ... 108
  - Comments and Extensions to Section 4.1 ... 113
- 4.2 Alternative Definitions for the Spectral Density Function ... 114
- 4.3 Basic Properties of the Spectrum ... 116
  - Comments and Extensions to Section 4.3 ... 118
- 4.4 Classification of Spectra ... 120
- 4.5 Sampling and Aliasing ... 122
  - Comments and Extensions to Section 4.5 ... 123
- 4.6 Comparison of SDFs and ACVSs as Characterizations ... 124
- 4.7 Summary of Foundations for Stochastic Spectral Analysis ... 125
- 4.8 Exercises ... 127

5 Linear Time-Invariant Filters ... 132

- 5.0 Introduction ... 132
- 5.1 Basic Theory of LTI Analog Filters ... 133
  - Comments and Extensions to Section 5.1 ... 137
- 5.2 Basic Theory of LTI Digital Filters ... 140
  - Comments and Extensions to Section 5.2 ... 142
- 5.3 Convolution as an LTI filter ... 142
- 5.4 Determination of SDFs by LTI Digital Filtering ... 144
- 5.5 Some Filter Terminology ... 145
- 5.6 Interpretation of Spectrum via Band-Pass Filtering ... 147
- 5.7 An Example of LTI Digital Filtering ... 148
  - Comments and Extensions to Section 5.7 ... 151
- 5.8 Least Squares Filter Design ... 152

5.9 Use of Slepian Sequences in Low-Pass Filter Design 155
5.10 Exercises 157

6 Periodogram and Other Direct Spectral Estimators 163

6.0 Introduction 163
6.1 Estimation of the Mean 164
    Comments and Extensions to Section 6.1 165
6.2 Estimation of the Autocovariance Sequence 166
    Comments and Extensions to Section 6.2 169
6.3 A Naive Spectral Estimator – the Periodogram 170
    Comments and Extensions to Section 6.3 179
6.4 Bias Reduction – Tapering 185
    Comments and Extensions to Section 6.4 194
6.5 Bias Reduction – Prewhitening 197
    Comments and Extensions to Section 6.5 201
6.6 Statistical Properties of Direct Spectral Estimators 201
    Comments and Extensions to Section 6.6 209
6.7 Computational Details 219
6.8 Examples of Periodogram and Other Direct Spectral Estimators 224
    Comments and Extensions to Section 6.8 230
6.9 Comments on Complex-Valued Time Series 231
6.10 Summary of Periodogram and Other Direct Spectral Estimators 232
6.11 Exercises 235

7 Lag Window Spectral Estimators 245

7.0 Introduction 245
7.1 Smoothing Direct Spectral Estimators 246
    Comments and Extensions to Section 7.1 252
7.2 First-Moment Properties of Lag Window Estimators 255
    Comments and Extensions to Section 7.2 257
7.3 Second-Moment Properties of Lag Window Estimators 258
    Comments and Extensions to Section 7.3 261
7.4 Asymptotic Distribution of Lag Window Estimators 264
7.5 Examples of Lag Windows 268
    Comments and Extensions to Section 7.5 278
7.6 Choice of Lag Window 287
    Comments and Extensions to Section 7.6 290
7.7 Choice of Lag Window Parameter 291
    Comments and Extensions to Section 7.7 296
7.8 Estimation of Spectral Bandwidth 297
7.9 Automatic Smoothing of Log Spectral Estimators 301
    Comments and Extensions to Section 7.9 306
7.10 Bandwidth Selection for Periodogram Smoothing 307
    Comments and Extensions to Section 7.10 312
7.11 Computational Details 314
7.12 Examples of Lag Window Spectral Estimators 316
    Comments and Extensions to Section 7.12 336
7.13 Summary of Lag Window Spectral Estimators 340

7.14 Exercises 343

8 Combining Direct Spectral Estimators 351

8.0 Introduction 351
8.1 Multitaper Spectral Estimators – Overview 352
      Comments and Extensions to Section 8.1 355
8.2 Slepian Multitaper Estimators 357
      Comments and Extensions to Section 8.2 366
8.3 Multitapering of Gaussian White Noise 370
8.4 Quadratic Spectral Estimators and Multitapering 374
      Comments and Extensions to Section 8.4 382
8.5 Regularization and Multitapering 382
      Comments and Extensions to Section 8.5 390
8.6 Sinusoidal Multitaper Estimators 391
      Comments and Extensions to Section 8.6 400
8.7 Improving Periodogram-Based Methodology via Multitapering 403
      Comments and Extensions to Section 8.7 412
8.8 Welch's Overlapped Segment Averaging (WOSA) 412
      Comments and Extensions to Section 8.8 419
8.9 Examples of Multitaper and WOSA Spectral Estimators 425
8.10 Summary of Combining Direct Spectral Estimators 432
8.11 Exercises 436

9 Parametric Spectral Estimators 445

9.0 Introduction 445
9.1 Notation 445
9.2 The Autoregressive Model 446
      Comments and Extensions to Section 9.2 447
9.3 The Yule–Walker Equations 449
      Comments and Extensions to Section 9.3 452
9.4 The Levinson–Durbin Recursions 452
      Comments and Extensions to Section 9.4 460
9.5 Burg's Algorithm 466
      Comments and Extensions to Section 9.5 469
9.6 The Maximum Entropy Argument 471
9.7 Least Squares Estimators 475
      Comments and Extensions to Section 9.7 478
9.8 Maximum Likelihood Estimators 480
      Comments and Extensions to Section 9.8 483
9.9 Confidence Intervals Using AR Spectral Estimators 485
      Comments and Extensions to Section 9.9 490
9.10 Prewhitened Spectral Estimators 491
9.11 Order Selection for $AR(p)$ Processes 492
      Comments and Extensions to Section 9.11 495
9.12 Examples of Parametric Spectral Estimators 496
9.13 Comments on Complex-Valued Time Series 501
9.14 Use of Other Models for Parametric SDF Estimation 503
9.15 Summary of Parametric Spectral Estimators 505

9.16 Exercises 506

10 Harmonic Analysis 511

10.0 Introduction 511
10.1 Harmonic Processes – Purely Discrete Spectra 511
10.2 Harmonic Processes with Additive White Noise – Discrete Spectra 512
      Comments and Extensions to Section 10.2 517
10.3 Spectral Representation of Discrete and Mixed Spectra 518
      Comments and Extensions to Section 10.3 519
10.4 An Example from Tidal Analysis 520
      Comments and Extensions to Section 10.4 523
10.5 A Special Case of Unknown Frequencies 523
      Comments and Extensions to Section 10.5 524
10.6 General Case of Unknown Frequencies 524
      Comments and Extensions to Section 10.6 527
10.7 An Artificial Example from Kay and Marple 530
      Comments and Extensions to Section 10.7 534
10.8 Tapering and the Identification of Frequencies 535
10.9 Tests for Periodicity – White Noise Case 538
      Comments and Extensions to Section 10.9 543
10.10 Tests for Periodicity – Colored Noise Case 544
      Comments and Extensions to Section 10.10 548
10.11 Completing a Harmonic Analysis 549
      Comments and Extensions to Section 10.11 552
10.12 A Parametric Approach to Harmonic Analysis 553
      Comments and Extensions to Section 10.12 557
10.13 Problems with the Parametric Approach 558
10.14 Singular Value Decomposition Approach 563
      Comments and Extensions to Section 10.14 567
10.15 Examples of Harmonic Analysis 567
      Comments and Extensions to Section 10.15 583
10.16 Summary of Harmonic Analysis 584
10.17 Exercises 587

11 Simulation of Time Series 593

11.0 Introduction 593
11.1 Simulation of ARMA Processes and Harmonic Processes 594
      Comments and Extensions to Section 11.1 599
11.2 Simulation of Processes with a Known Autocovariance Sequence 601
      Comments and Extensions to Section 11.2 603
11.3 Simulation of Processes with a Known Spectral Density Function 604
      Comments and Extensions to Section 11.3 609
11.4 Simulating Time Series from Nonparametric Spectral Estimates 611
      Comments and Extensions to Section 11.4 613
11.5 Simulating Time Series from Parametric Spectral Estimates 617
      Comments and Extensions to Section 11.5 618
11.6 Examples of Simulation of Time Series 619
      Comments and Extensions to Section 11.6 631

11.7 Comments on Simulation of Non-Gaussian Time Series 631
 11.8 Summary of Simulation of Time Series 637
 11.9 Exercises 638

References **643**

Author Index **661**

Subject Index **667**

Preface

Spectral analysis is one of the most widely used methods for interpreting time series and has been used in diverse areas including – but not limited to – the engineering, physical and environmental sciences. This book aims to help data analysts in applying spectral analysis to actual time series. Successful application of spectral analysis requires both an understanding of its underlying statistical theory and the ability to transition this theory into practice. To this end, we discuss the statistical theory behind all major nonparametric and parametric spectral analysis techniques, with particular emphasis on the multitaper method, both in its original formulation in Thomson (1982) involving Slepian tapers and in a popular alternative involving the sinusoidal tapers advocated in Riedel and Sidorenko (1995). We then use actual time series from oceanography, metrology, atmospheric science and other areas to provide analysts with examples of how to move from theory to practice.

This book builds upon our 1993 book *Spectral Analysis for Physical Applications: Multitaper and Conventional Univariate Techniques* (also published by Cambridge University Press). The motivations for considerably expanding upon this earlier work include the following.

[1] A quarter century of teaching classes based on the 1993 book has given us new insights into how best to introduce spectral analysis to analysts. In particular we have greatly expanded our treatment of the multitaper method. While this method was a main focus in 1993, we now describe it in a context that more readily allows comparison with one of its main competitors (Welch's overlapped segment averaging).

[2] The core material on nonparametric spectral estimation is in Chapters 6 ("Periodogram and Other Direct Spectral Estimators"), 7 ("Lag Window Spectral Estimators") and 8 ("Combining Direct Spectral Estimators"). These chapters now present these estimators in a manner that allows easier comparison of common underlying concepts such as smoothing, bandwidth and windowing.

[3] There have been significant theoretical advances in spectral analysis since 1993, some of which are of particular importance for data analysts to know about. One that we have already mentioned is a new family of multitapers (the sinusoidal tapers) that was introduced in the mid-1990s and that has much to recommend its use. Another is a new bandwidth measure that allows nonparametric spectral analysis methods to be meaningfully compared. A third is bandwidth selection for smoothing periodograms.

[4] An important topic that was not discussed in the 1993 book is computer-based simulation of time series. We devote Chapter 11 to this topic, with particular emphasis on simulating series whose statistical properties agree with those dictated by nonparametric and parametric spectral analyses of actual time series.

[5] We used software written in Common Lisp to carry out spectral analysis for all the time series used in our 1993 book. Here we have used the popular and freely available R software package to do all the data analysis and to create the content for almost all the figures and tables in the book. We do *not* discuss this software explicitly in the book, but we make it available as a supplement so that data analysts can replicate and build upon our use of spectral analysis. The website for the book gives access to the R software and to information about software in other languages – see "Data, Software and Ancillary Material" on page xx for details.

Finally, a key motivation for us to undertake an expansion has been the gratifying response to the 1993 book from its intended audience.

The following features of this book are worth noting.

[1] We provide a large number of exercises (over 300 in all), some of which are embedded within the chapters, and others, at the ends of the chapters. The embedded exercises challenge readers to verify certain theoretical results in the main text, with solutions in an Appendix that is available on the website for the book (see page xx). The exercises at the end of the chapters are suitable for use in a classroom setting (solutions are available only for instructors). These exercises both expand upon the theory presented and delve into the practical considerations behind spectral analysis.

[2] We use actual time series to illustrate various spectral analysis methods. We do so to encourage data analysts to carefully consider the link between spectral analysis and what questions this technique can address about particular series. In some instances we use the same series with different techniques to allow analysts to compare how well various methods address questions of interest.

[3] We provide a large number of "Comments and Extensions" (C&Es) to the main material. These C&Es appear at the ends of sections when appropriate and provide interesting supplements to the main material; however, readers can skip the C&Es without compromising their ability to follow the main material later on (we have set the C&Es in a slightly smaller font to help differentiate them from the main material). The C&Es cover a variety of ancillary – but valuable – topics such as the Lomb–Scargle periodogram, jackknifing of multitaper spectral estimates, the method of surrogate time series, a periodogram based upon the discrete cosine transform (and its connection to Albert Einstein!) and the degree to which windows designed for one purpose can be used for another.

[4] At the end of most chapters, we provide a comprehensive summary of that chapter. The summaries allow readers to check their understanding of the main points in a chapter and to review the content of a previous chapter when tackling a later chapter. The comprehensive subject index at the end of the book will aid in finding details of interest.

We also note that "univariate" is part of the title of the book because a volume on multivariate spectral analysis is in progress.

Books do not arise in isolation, and ours is no exception. With a book that is twenty-five years in the making, the list of editors, colleagues, students, readers of our 1993 book, friends and relatives who have influenced this book in some manner is so long that thanking a select few individuals here will only be at the price of feeling guilty both now and later on about not thanking many others. Those who are on this list know who you are. We propose to thank you with a free libation of your choice upon our first post-publication meeting (wherever this might happen – near Seattle or London or both or elsewhere!). We do, however, want to explicitly acknowledge financial support through EPSRC Mathematics Platform grant EP/I019111/1. We also thank Stan Murphy (posthumously), Bob Spindel and Jeff Simmen (three generations of directors of the Applied Physics Laboratory, University of Washington) for supplying ongoing discretionary funding without which this and the 1993 book would not exist.

Finally, despite our desire for a book needing no errata list, past experience says this will not happen. Readers are encouraged to contact us about blemishes in the book so that we can make others aware of them (our email addresses are listed with our signatures).

Don Percival
Applied Physics Laboratory
Department of Statistics
University of Washington
dbpercival@gmail.com

Andrew Walden
Department of Mathematics
Imperial College London
atwalden86@gmail.com

Conventions and Notation

- *Important conventions*

(14)	refers to the single displayed equation on page 14
(3a), (3b)	refers to different displayed equations on page 3
Figure 2	refers to the figure on page 2
Table 214	refers to the table on page 214
Exercise [8]	refers to the embedded exercise on page 8 (see the Appendix on the website for an answer)
Exercise [1.3]	refers to the third exercise at the end of Chapter 1
a, $\boldsymbol{a}$ and $\boldsymbol{A}$	refer to a scalar, a vector and a matrix/vector
$S(\cdot)$	refers to a function
$S(f)$	refers to the value of the function $S(\cdot)$ at f
$\{h_t\}$	refers to a sequence of values indexed by t
h_t	refers to a single value of a sequence
α and $\hat{\alpha}$	refer to a parameter and an estimator thereof

In the following lists, the numbers at the end of the brief descriptions are page numbers where more information about – or an example of the use of – an abbreviation or symbol can be found.

- *Abbreviations used frequently*

ACF	autocorrelation function	27
ACLS	approximate conditional least squares	549
ACS	autocorrelation sequence	27
ACVF	autocovariance function	27
ACVS	autocovariance sequence	27
AIC	Akaike's information criterion	494
AICC	AIC corrected for bias	495, 576

AR(p)	pth-order autoregressive process	33, 446
ARMA(p, q)	autoregressive moving average process of order (p, q)	35
BLS	backward least squares	477
C&Es	Comments and Extensions	24
CI	confidence interval	204
CPDF	cumulative probability distribution function	23
dB	decibels, i.e., $10 \log_{10}(\cdot)$	13
DCT–II	discrete cosine transform of type II	184, 217
DFT	discrete Fourier transform	74, 92
DPSS	discrete prolate spheroidal sequence (Slepian sequence)	87, 155
DPSWF	discrete prolate spheroidal wave function	87
ECLS	exact conditional least squares	549
EDOFs	equivalent degrees of freedom	264
EULS	exact unconditional least squares	549
FBLS	forward/backward least squares	477, 555
FFT	fast Fourier transform	92, 94
FIR	finite impulse response	147
FLS	forward least squares	476
FPE	final prediction error	493
GCV	generalized cross-validated	309
GSSM	Gaussian spectral synthesis method	605
Hz	Hertz: 1 Hz = 1 cycle per second	
IID	independent and identically distributed	31
IIR	infinite impulse response	147
LS	least squares	475
LTI	linear time-invariant	132
MA(q)	qth-order moving average process	32
MLE	maximum likelihood estimator or estimate	480
MSE	mean square error	167
MSLE	mean square log error	296
NMSE	normalized mean square error	296
OLS	ordinary least squares	409
PACS	partial autocorrelation sequence	462
PDF	probability density function	24
PSWF	prolate spheroidal wave function	64
RV	random variable	3
SDF	spectral density function	111
SS	sum of squares	467–8, 476–7
SVD	singular value decomposition	565
WOSA	Welch's (or weighted) overlapped segment averaging	414
ZMNL	zero-memory nonlinearity	634

- *Non-Greek notation used frequently*

A_l	real-valued amplitude associated with $\cos(2\pi f_l t \Delta_t)$	35, 515
$b(\cdot)$	bias	192, 239, 378
$b^{(B)}(\cdot)$	broad-band bias	378
$b^{(L)}(\cdot)$	local bias	378
$b_k(f)$	weight associated with kth eigenspectrum at frequency f	386
$b_W(\cdot)$	bias due to smoothing window only	256
B_l	real-valued amplitude associated with $\sin(2\pi f_l t \Delta_t)$	35, 515
$B_\mathcal{H}$	bandwidth of spectral window $\mathcal{H}$	194
B_S	spectral bandwidth	292, 297
B_T	bandwidth measure for $\{X_t\}$ with dominantly unimodal SDF	300
$\tilde{B}_T$	approximately unbiased estimator of B_T	300
$B_\mathcal{U}$	bandwidth of spectral window $\mathcal{U}_m$	256
B_W	Jenkins measure of smoothing window bandwidth	251
$\{c_\tau\}$	inverse Fourier transform of $C(\cdot)$ (cepstrum if properly scaled)	301
$C(\cdot)$	log spectral density function	301
$\hat{C}^{(D)}(\cdot)$	log of direct spectral estimator	301
$\hat{C}_m^{(LW)}(\cdot)$	smoothed log of direct spectral estimator	303
C_h	variance inflation factor due to tapering	259, 262
C_l	complex-valued amplitude associated with $\exp(i2\pi f_l t \Delta_t)$	108, 519
$d(\cdot,\cdot)$	Kullback–Leibler discrepancy measure (general case)	297
$dZ(\cdot)$	orthogonal increment	109
D_l	real-valued amplitude associated with $\cos(2\pi f_l t \Delta_t + \phi_l)$ or with $\exp(i[2\pi f_l t \Delta_t + \phi_l])$	35, 511, 517
$\boldsymbol{D}_N$	$N \times N$ diagonal matrix	375
$\mathcal{D}_N(\cdot)$	Dirichlet's kernel	17
$\vec{e}_t(k)$	observed forward prediction error	467
$\overleftarrow{e}_t(k)$	observed backward prediction error	467
f_k	$k/(N\Delta_t)$, member of grid of Fourier frequencies	171, 515
f_k'	$k/(N'\Delta_t)$, member of arbitrary grid of frequencies	171
$\tilde{f}_k$	$k/(2N\Delta_t)$, member of grid twice as fine as Fourier frequencies	181
f_l	frequency of a sinusoid	35, 511
$f_\mathcal{N}$	$1/(2\Delta_t)$, Nyquist frequency	82, 122, 512
$F_t(\cdot)$	cumulative probability distribution function	23
$\mathcal{F}(\cdot)$	Fejér's kernel	174, 236
g	Fisher's test statistic for simple periodicity	539
g_F	critical value for Fisher's test statistic g	540
$g(\cdot)$	real- or complex-valued function	53
$g_p(\cdot)$	periodic function	48
$\{g_u\}$	impulse response sequence of a digital filter	143
$g \star g^*(\cdot)$	autocorrelation of deterministic function $g(\cdot)$	72
$g \star h^*(\cdot)$	cross-correlation of deterministic functions $g(\cdot)$ and $h(\cdot)$	72
$\{g * h_t\}$	convolution of sequences $\{g_t\}$ and $\{h_t\}$	99

$g*h(\cdot)$	convolution of functions $g(\cdot)$ and $h(\cdot)$	67
$G(\cdot)$	Fourier transform of $g(\cdot)$ or transfer function	54, 97, 136, 141
$G_p(\cdot)$	Fourier transform of $\{g_t\}$	74, 99
$\{G_n\}$	Fourier transform of $g_p(\cdot)$	49, 96
$\{G_t\}$	stationary *Gaussian* process	201, 445
$\{h_t\}$	data taper	186
$\{h_{k,t}\}$	kth-order data taper for multitaper estimator	352, 357, 392
$H(\cdot)$	Fourier transform of data taper $\{h_t\}$	186
$\{H_t\}$	Gaussian autoregressive process	445
$\mathcal{H}(\cdot)$	spectral window of direct spectral estimator	186
$\mathcal{H}_k(\cdot)$	spectral window of kth eigenspectrum	352
$\overline{\mathcal{H}}(\cdot)$	spectral window of basic multitaper estimator	353
$\widetilde{\mathcal{H}}(\cdot)$	spectral window of weighted multitaper estimator	353
$\mathcal{HT}\{\cdot\}$	Hilbert transform	114, 562, 579
$J(\cdot)$	scaled Fourier transform of tapered time series	186, 544
$K_{\max}$	maximum number of usable Slepian multitapers	357
$\mathrm{KL}(\cdot)$	Kullback–Leibler discrepancy measure (special case)	297
$L\{x(\cdot)\}$	continuous parameter filter acting on function $x(\cdot)$	133
$L\{x_t\}$	discrete parameter filter acting on sequence $\{x_t\}$	141
$\boldsymbol{L}_N$	$N \times N$ lower triangular matrix with 1's on diagonal	464
m	parameter controlling smoothing in lag window estimator	247
N	sample size	2, 163
N'	integer typically greater than or equal to N	179, 237
N_B	number of blocks in WOSA	414
N_S	block size in WOSA	414
p	order of an autoregressive process or a proportion	33, 189, 204
$\mathbf{P}[A]$	probability that the event A will occur	23
$\mathcal{P}_k$	normalized cumulative periodogram	215
q	order of a moving average process	32, 503
$\boldsymbol{Q}$	weight matrix in quadratic spectral estimator	374
$Q_\nu(p)$	$p \times 100\%$ percentage point of χ_ν^2 distribution	265
$\{r_\tau\}$	inverse Fourier transform of $R(\cdot)$	212
R	signal-to-noise ratio	526, 532
$R(\eta)$	correlation of direct spectral estimators at f and $f+\eta$	212
$\{R_t\}$	residual process	550
$\{s_\tau\}$	autocovariance sequence (ACVS)	27, 29
$\{s_\tau^{(\mathrm{BL})}\}$	ACVS for band-limited white noise	379
$\{\hat{s}_\tau^{(\mathrm{D})}\}$	ACVS estimator, inverse Fourier transform of $\{\hat{S}^{(\mathrm{D})}(\cdot)\}$	188
$\{\hat{s}_\tau^{(\mathrm{P})}\}$	"biased" estimator of ACVS	166
$\{\hat{s}_\tau^{(\mathrm{U})}\}$	"unbiased" estimator of ACVS	166
$s(\cdot)$	autocovariance function (ACVF)	27
$S(\cdot)$	spectral density function (SDF)	111
$S_\eta(\cdot)$	SDF of (possibly) colored noise	519

$S^{(\mathrm{I})}(\cdot)$	integrated spectrum or spectral distribution function	110
$S^{(\mathrm{BL})}(\cdot)$	SDF of band-limited white noise	379
$\hat{S}^{(\mathrm{AMT})}(\cdot)$	adaptive multitaper spectral estimator	389
$\hat{S}^{(\mathrm{D})}(\cdot)$	direct spectral estimator	186
$\hat{S}^{(\mathrm{DCT})}(\cdot)$	DCT-based periodogram	217
$\hat{S}^{(\mathrm{DS})}(\cdot)$	discretely smoothed direct spectral estimator	246
$\hat{S}_m^{(\mathrm{DSP})}(\cdot)$	discretely smoothed periodogram	307
$\hat{S}_m^{(\mathrm{LW})}(\cdot)$	lag window spectral estimator	247
$\hat{S}^{(\mathrm{MT})}(\cdot)$	basic multitaper spectral estimator	352
$\hat{S}_k^{(\mathrm{MT})}(\cdot)$	kth eigenspectrum for multitaper estimator	352
$\hat{S}^{(\mathrm{P})}(\cdot)$	periodogram (special case of $\hat{S}^{(\mathrm{D})}(\cdot)$)	170, 188
$\tilde{S}^{(\mathrm{P})}(\cdot)$	periodogram of shifted time series or rescaled periodogram	184, 222, 240
$\hat{S}^{(\mathrm{PC})}(\cdot)$	postcolored spectral estimator	198
$\hat{S}^{(\mathrm{Q})}(\cdot)$	quadratic spectral estimator	374
$\hat{S}^{(\mathrm{WMT})}(\cdot)$	weighted multitaper spectral estimator	352
$\hat{S}^{(\mathrm{WOSA})}(\cdot)$	WOSA spectral estimator	414
$\hat{S}^{(\mathrm{YW})}(\cdot)$	Yule–Walker spectral estimator	451
t	actual time (continuous) or a unitless index (discrete)	22, 74
t_λ	critical value for Siegel's test statistic	541
T_λ	Siegel's test statistic	541
$U_k(\cdot; N, W)$	discrete prolate spheroidal wave function of order k	87
$\mathcal{U}_m(\cdot)$	spectral window of $\hat{S}^{(\mathrm{LW})}(\cdot)$	255
$\boldsymbol{v}_k(N, W)$	vector with portion of DPSS, order k	87
$\{v_{m,\tau}\}$	nontruncated version of lag window $\{w_{m,\tau}\}$	247
$V_m(\cdot)$	design window (Fourier transform of $\{v_{m,\tau}\}$)	247
$\{w_{m,\tau}\}$	lag window (truncated version of $\{v_{m,\tau}\}$)	247
width$_\mathrm{a}\{\cdot\}$	autocorrelation width	73
width$_\mathrm{e}\{\cdot\}$	equivalent width	58
width$_\mathrm{hp}\{\cdot\}$	half-power width	192
width$_\mathrm{v}\{\cdot\}$	variance width	60, 192
W	DPSS half-bandwidth, regularization half-bandwidth	65, 377
$W_m(\cdot)$	smoothing window	247–8
$x_0, \ldots, x_{N-1}$	time series realization or deterministic series	2
$X_0, \ldots, X_{N-1}$	sequence of random variables	3
$\{X_t\}$	real-valued discrete parameter stochastic process	22
$\{X(t)\}$	real-valued continuous parameter stochastic process	22
$\{X_{j,t}\}$	jth real-valued discrete parameter stochastic process	23
$\{X_j(t)\}$	jth real-valued continuous parameter stochastic process	23
$\overline{X}$	sample mean (arithmetic average) of $X_0, \ldots, X_{N-1}$	164
$\overrightarrow{X}_t(k)$	best (forward) linear predictor of X_t given $X_{t-1}, \ldots, X_{t-k}$	452
$\overleftarrow{X}_t(k)$	best (backward) linear predictor of X_t given $X_{t+1}, \ldots, X_{t+k}$	455
$\{Y_t\}$	real-valued discrete parameter stochastic process (refers to an AR process in Chapter 9)	23, 445

Conventions and Notation

$\{Z_t\}$	complex-valued discrete parameter stochastic process	23, 29
$\{Z(t)\}$	complex-valued continuous parameter stochastic process	23
$\{Z(f)\}$	orthogonal process	109

• *Greek notation used frequently*

α	intercept term in linear model, scalar, level of significance or exponent of power law	39, 43, 215, 327
$\alpha^2(N)$	fraction of sequence's energy lying in $0,\ldots,N-1$	85
$\alpha^2(T)$	fraction of function's energy lying in $[-T/2, T/2]$	62
β	slope term in linear model	39
$\beta^{(\text{B})}\{\cdot\}$	indicator of broad-band bias in $\hat{S}^{(\text{Q})}(\cdot)$	379
$\beta^{(\text{L})}\{\cdot\}$	indicator of magnitude of local bias in $\hat{S}^{(\text{Q})}(\cdot)$	378
β_W	Grenander's measure of smoothing window bandwidth	251
$\beta^2(W)$	fraction of function's energy lying in $[-W, W]$	62, 85
$\beta^2_{\mathcal{H}}$	indicator of bias in $\hat{S}^{(\text{D})}(\cdot)$	391
γ	quadratic term in linear model or Euler's constant	46, 210
$\Gamma(\cdot)$	gamma function	440
$\boldsymbol{\Gamma}$	covariance matrix (typically for AR(p) process)	450, 464
Δ_f	spacing in frequency	206
Δ_t	spacing in time (sampling interval)	74, 81–2, 122
$\{\epsilon_t\}$	white noise or innovation process	32, 446
$\overrightarrow{\epsilon}_t(k)$	forward prediction error: $X_t - \overrightarrow{X}_t(k)$	453
$\overleftarrow{\epsilon}_t(k)$	backward prediction error: $X_t - \overleftarrow{X}_t(k)$	455
η	equivalent degrees of freedom of a time series	298
$\{\eta_t\}$	zero mean stationary noise process (not necessarily white)	518
$\theta(\cdot)$	phase function corresponding to transfer function $G(\cdot)$	136
θ	coefficient of an MA(1) process	43
$\theta_{q,1},\ldots,\theta_{q,q}$	coefficients of an MA(q) process	32
$\vartheta_{q,0},\ldots,\vartheta_{q,q}$	coefficients of an MA(q) process ($\vartheta_{q,0}=1$ and $\vartheta_{q,j}=-\theta_{q,j}$)	594
λ	constant to define different logarithmic scales	301
$\lambda_k(c)$	eigenvalue associated with PSWF, order k	64
$\lambda_k(N,W)$	eigenvalue associated with DPSWF, order k	86
μ	expected value of a stationary process	27
ν	degrees of freedom associated with RV χ^2_ν	202, 264
$\{\rho_\tau\}$	autocorrelation sequence (ACS)	27
$\rho(\cdot)$	autocorrelation function (ACF)	27
σ^2	variance	27
σ^2_ϵ	white noise variance or innovation variance	32, 404
σ^2_η	variance of noise process $\{\eta_t\}$	519
σ^2_k	mean square linear prediction error for $\overrightarrow{X}_t(k)$ or $\overleftarrow{X}_t(k)$	453, 455
σ^2_p	innovation variance for an AR(p) process	446
$\hat{\sigma}^2_p$	estimator of σ^2_p	485
$\bar{\sigma}^2_p$	Burg estimator of σ^2_p	468

$\tilde{\sigma}_p^2$	Yule–Walker estimator of σ_p^2	451, 458
$\boldsymbol{\Sigma}$	covariance matrix	28
τ	lag value	27
ϕ_l	phase of a sinusoid	35, 511
ϕ	coefficient of an AR(1) process	44
$\phi_{p,1},\ldots,\phi_{p,p}$	coefficients of an AR(p) process	33, 446
$\hat{\phi}_{p,1},\ldots,\hat{\phi}_{p,p}$	estimators of AR(p) coefficients	485
$\bar{\phi}_{p,1},\ldots,\bar{\phi}_{p,p}$	Burg estimators of AR(p) coefficients	466, 505
$\tilde{\phi}_{p,1},\ldots,\tilde{\phi}_{p,p}$	Yule–Walker estimators of AR(p) coefficients	451, 458, 505
$\varphi_{2p,1},\ldots,\varphi_{2p,2p}$	coefficients of a pseudo-AR($2p$) process	553
$\boldsymbol{\Phi}_p, \hat{\boldsymbol{\Phi}}_p, \tilde{\boldsymbol{\Phi}}_p$	$[\phi_{p,1},\ldots,\phi_{p,p}]^T, [\hat{\phi}_{p,1},\ldots,\hat{\phi}_{p,p}]^T, [\tilde{\phi}_{p,1},\ldots,\tilde{\phi}_{p,p}]^T$	450, 485, 451
$\Phi^{-1}(p)$	$p \times 100\%$ percentage point of standard Gaussian distribution	265
χ_ν^2	chi-square RV with ν degrees of freedom	37, 202
$\psi(\cdot)$	digamma function	210, 296
$\psi'(\cdot)$	trigamma function	296
$\psi_k(\cdot;c)$	prolate spheroidal wave function (PSWF), order k	64
ω	angular frequency	8, 119

- *Standard mathematical symbols*

e	base for natural logarithm ($2.718282\cdots$)	17
i	$\sqrt{-1}$	17
$\log(\cdot), \log_{10}(\cdot)$	log base e, log base 10	
$\approx$	approximately equal to	
$\doteq$	equal at given precision, e.g., $\pi \doteq 3.1416$	
$\stackrel{\text{def}}{=}$	equal by definition	23–4
$\stackrel{\text{d}}{=}$	equal in distribution	203
$\stackrel{\text{ms}}{=}$	equal in mean square sense	49
$E\{\cdot\}$	expectation operator	24
$\text{var}\{\cdot\}$	variance operator	24–5, 27
$\text{cov}\{\cdot,\cdot\}$	covariance operator	24–5, 27
$\text{corr}\{\cdot,\cdot\}$	correlation operator	27
z^*	complex conjugate of z	25
$\Re(z)$	real part of complex-valued number z	61, 551
$\Im(z)$	imaginary part of complex-valued number z	551
$\arg(z)$	argument of complex-valued number z	54
$*$	convolution operator	67, 96, 98–9, 101
$\star$	cross-correlation operator	72, 96, 98–9, 101
$\longleftrightarrow$	Fourier transform pair relationship	69, 96–7, 99–100
$\boldsymbol{a}^T$ and $\boldsymbol{Q}^T$	transpose of vector $\boldsymbol{a}$ and of matrix $\boldsymbol{Q}$	28, 374
$\boldsymbol{Z}^H$ and $\boldsymbol{Q}^H$	Hermitian transpose of vector $\boldsymbol{Z}$ and of matrix $\boldsymbol{Q}$	374
$\boldsymbol{I}_N$	$N \times N$ identity matrix	366
$\text{tr}\{\boldsymbol{Q}\}$	trace of matrix $\boldsymbol{Q}$	376

$\lvert \boldsymbol{\Gamma}_N \rvert$	determinant of matrix $\boldsymbol{\Gamma}_N$	480
$\boldsymbol{R}^{\#}$	generalized inverse of matrix $\boldsymbol{R}$	566
$\langle \cdot, \cdot \rangle$	inner product	470
$\lVert \cdot \rVert^2$	squared norm	476
$\mathbb{R}$	set of all real-valued numbers, i.e., $\{t : -\infty < t < \infty\}$	22
$\mathbb{Z}$	set of all integers, i.e., $\{\ldots, -2, -1, 0, 1, 2, \ldots\}$	17
$L^2(\cdot)$	set of square integrable functions over specified domain	54–5
$\in, \notin$	contained in, not contained in	
δ_n	Kronecker delta function	44, 65
$\delta(\cdot)$	Dirac delta function	74, 120
$\operatorname{sinc}(\cdot)$	sinc function, i.e., $\sin(\pi t)/(\pi t)$	63
$\lfloor x \rfloor$	greatest integer $\leq x$, e.g., $\lfloor \pi \rfloor = 3$ and $\lfloor 3 \rfloor = 3$	7–8
$(a)_+$	positive part, i.e., $\max\{a, 0\}$	541
mod	modulo operator, e.g., $5 \bmod 4 = 1$	45, 101
$O(\cdot)$	$f(x) = O(g(x))$ as $x \to 0$ if $\lvert f(x)/g(x) \rvert \leq C$ for constant C	175

Data, Software and Ancillary Material

The website for this book is currently at

http://faculty.washington.edu/dbp/sauts.html

(alternatively go to www.cambridge.org/9781107028142 – this is maintained by Cambridge University Press and should have both a description of the book and a link to the current location for the book's website). The website gives access to

- an Appendix with answers to all the exercises embedded within Chapters 1 to 11;
- almost all the time series used as examples in the book;
- software in the R language for recreating the content in the bulk of the figures and tables in each chapter (and updates on the status of software in other languages);
- the current errata sheet (we encourage readers who spot errata to contact us via email at dbpercival@gmail.com and/or atwalden86@gmail.com);
- information about how to obtain a solutions guide for the exercises at the end of each chapter (this guide is *only* for instructors who are using the book as course material); and
- PDF files for the bulk of the figures and tables in the book (these are intended to help prepare course and seminar material, but please note that all figures and tables are the copyright of Cambridge University Press and must not be further distributed or used without written permission).

1

Introduction to Spectral Analysis

1.0 Introduction

This chapter provides a quick introduction to the subject of spectral analysis. Except for some later references to the exercises of Section 1.6, this material is independent of the rest of the book and can be skipped without loss of continuity. Our intent is to use some simple examples to motivate the key ideas. Since our purpose is to view the forest before we get lost in the trees, the particular analysis techniques we use here have been chosen for their simplicity rather than their appropriateness.

1.1 Some Aspects of Time Series Analysis

Spectral analysis is part of time series analysis, so the natural place to start our discussion is with the notion of a time series. The quip (attributed to R. A. Fisher) that a time series is "one damned thing after another" is not far from the truth: loosely speaking, a time series is a set of observations made sequentially in time (but "time" series are also often recorded sequentially in, e.g., distance or depth). Examples abound in the real world, and Figure 2 shows plots of small portions of four actual time series:

  (a) the speed of the wind in a certain direction, measured every 0.025 sec;
  (b) the daily record of a quantity (to be precise, the change in average daily frequency) that tells how well an atomic clock keeps time on a day-to-day basis (a constant value of zero would indicate that the clock agreed perfectly with a time scale maintained by the US Naval Observatory);
  (c) monthly average measurements related to the flow of water in the Willamette River at Salem, Oregon; and
  (d) the change in the level of ambient noise in the ocean from one second to the next.

For each of these plots, the values of the time series at 128 successive times are connected by lines to help the eye follow the variations in the series. The visual appearances of these four series are quite different.

The chief aim of time series analysis is to develop quantitative means to allow us to characterize time series, e.g., to say quantitatively how one series differs from another or how two series are related. There are two broad classes of characterizations, namely, time domain techniques and frequency domain techniques. Spectral analysis is the prime example of a frequency domain technique. Before we introduce it, we will first consider a popular time

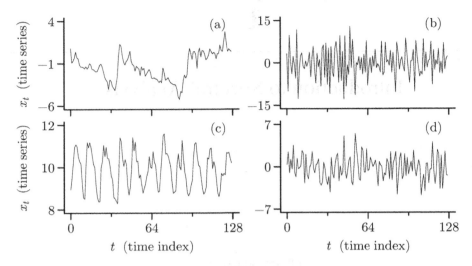

Figure 2 Plots of portions of four time series related to (a) wind speed, (b) an atomic clock, (c) the Willamette River and (d) ocean noise. For each series the vertical axis is the value of the time series (in unspecified units), while the horizontal axis is a unitless time index (the actual time between adjacent observations is 0.025 sec for the wind speed series, one day for the atomic clock data, one month for the Willamette River series and one second for the ocean noise series).

domain technique. We contend that this latter technique is not completely satisfactory and that spectral analysis is a useful and complementary alternative to it.

Let us concentrate for the moment on the wind speed and atomic clock data (top row of plots in Figure 2). How do these two series differ? In the wind speed series adjacent points of the time series tend to be close in value, while in the atomic clock series positive values tend to be followed by negative values, and vice versa. To see this effect graphically, we can plot x_{t+1} versus x_t as the time index t varies from 0 to $N - 2$, where we let $x_0, x_1, \ldots, x_{N-1}$ represent any one of our series and let N represent the sample size, i.e., the number of data points in a time series, 128 in our case. Such a plot is called a "lag 1 scatter plot," and Figure 3 shows this plot for each of our four series. We note the following:

(a) For the wind speed series, the points tend to fall about a line of positive slope. Thus a wind speed with a certain value tends to be followed by one near that same value.
(b) For the atomic clock data, the points fall loosely about a line with a negative slope.
(c) The plot for the Willamette River data resembles that of the wind speed series except that the points are more spread out.
(d) For the ocean noise data, it is not obvious that there is a tendency of the points to cluster about a line in one direction or another.

We could create a lag τ scatter plot by plotting $x_{t+\tau}$ versus x_t, but, while such plots are informative, they are unwieldy to work with. To summarize the information in scatter plots similar to those in Figure 3, note that these plots indicate a roughly linear relationship between x_{t+1} and x_t; i.e., with $\tau = 1$, we can write

$$x_{t+\tau} = \alpha_\tau + \beta_\tau x_t + \epsilon_{\tau,t}$$

for some intercept α_τ and slope β_τ (possibly equal to 0), where $\epsilon_{\tau,t}$ represents an "error" term that models deviations from strict linearity. If we make the assumption that a linear relationship holds approximately between $x_{t+\tau}$ and x_t for all τ, we can use as a summary statistic a

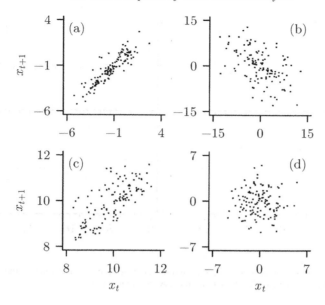

Figure 3 Lag 1 scatter plots for four time series in Figure 2. In each of plot, the value of the time series at time index $t+1$ is plotted on the vertical axis versus the value at time index t on the horizontal axis (t ranges from 0 to 126).

well-known measure of the strength of the linear association between two ordered collections of variables $\{y_t\}$ and $\{z_t\}$, namely, the Pearson product moment correlation coefficient:

$$\hat{\rho} = \frac{\Sigma(y_t - \bar{y})(z_t - \bar{z})}{[\Sigma(y_t - \bar{y})^2 \Sigma(z_t - \bar{z})^2]^{1/2}}, \qquad (3a)$$

where $\bar{y}$ and $\bar{z}$ are the sample means of the y_t and z_t terms, respectively. This coefficient can be interpreted in many ways (Rogers and Nicewander, 1988; Falk and Well, 1997; Rovine and von Eye, 1997; Nelsen, 1998). For example, if we use $\{y_t\}$ and $\{z_t\}$ to form two vectors, then $\hat{\rho}$ is the cosine of the angle between them. If we let $y_t = x_{t+\tau}$ and $z_t = x_t$, and if we adjust the summations in the denominator to make use of all available data, we are led to the lag τ sample autocorrelation for a time series:

$$\hat{\rho}_\tau = \frac{\sum_{t=0}^{N-\tau-1}(x_{t+\tau} - \bar{x})(x_t - \bar{x})}{\sum_{t=0}^{N-1}(x_t - \bar{x})^2}. \qquad (3b)$$

Note that $\hat{\rho}_0 = 1$ and that, as τ increases, the numerator is based on fewer and fewer cross products. As a sequence indexed by the lag τ, the quantity $\{\hat{\rho}_\tau\}$ is called the sample autocorrelation sequence (sample ACS) for the time series x_t. (See Exercise [1.6] for a caveat about interpreting $\hat{\rho}_\tau$ as a true correlation coefficient.)

The sample ACS up to lag 32 is plotted for our four time series in Figure 4. A careful study of these plots can reveal a lot about these series. For example, from the ACS in (c) for the Willamette River data, we see that x_t and x_{t+6} are negatively correlated, while x_t and x_{t+12} are positively correlated. This pattern is consistent with the visual evidence in Figure 2 that the river flow varies with a period of roughly 12 months.

Let us now assume that the time series $x_0, x_1, \ldots, x_{N-1}$ can be regarded as observed values (i.e., realizations) of corresponding random variables (RVs) $X_0, X_1, \ldots, X_{N-1}$. We use the term "modeling of a time series" for the procedure by which we specify the properties

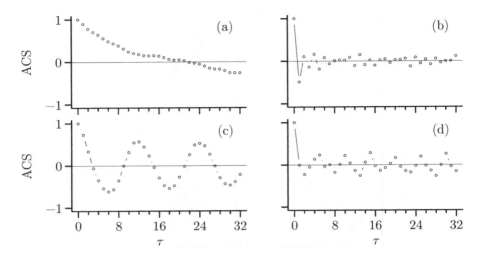

Figure 4 Sample autocorrelation sequences $\{\hat{\rho}_\tau\}$ for the time series of Figure 2. The value of the ACS at lag τ is plotted versus τ for τ ranging from 0 to 32. By definition the ACS for lag 0 is 1.

of these N RVs. For a class of models reasonable for time series such as those in Figure 2, $\hat{\rho}_\tau$ is an estimate of a corresponding population quantity called the lag τ theoretical autocorrelation, defined as

$$\rho_\tau = E\{(X_{t+\tau} - \mu)(X_t - \mu)\}/\sigma^2,$$

where $E\{W\}$ is our notation for the expectation operator applied to the RV W; $\mu = E\{X_t\}$ is the population mean of the time series; and $\sigma^2 = E\{(X_t - \mu)^2\}$ is the corresponding population variance. (Note, in particular, that ρ_τ, μ and σ^2 do not depend on t. As we shall see later, models for which this is true play a central role in spectral analysis and are called stationary.) Moreover, if we make an additional assumption, namely, that the X_t terms follow a multivariate Gaussian (normal) distribution, knowledge of the ρ_τ terms, σ^2 and μ completely specifies our model. Thus, to fit such a model to a time series, we need only estimate ρ_τ, σ^2 and μ from the available data.

As a set of parameters, the ρ_τ terms, σ^2 and μ constitute a time domain characterization of a model. Since a model is completely specified by these parameters in the Gaussian case and since these parameters can all be estimated from a time series, why would we want to consider other characterizations? There are several reasons:

[1] The parameters of a model should ideally make it easy for us to visualize typical time series that can be generated by the model. Unfortunately, it takes a fair amount of experience to be able to look at a theoretical ACS and visualize what kind of time series it corresponds to.

[2] For a lag τ that is a substantial proportion of the length N of a time series, it is often hard to get reliable estimates of ρ_τ (and even more difficult to do so for τ greater than N). This is evident from Equation (3b) since the number of cross products that are used in forming the numerator decreases as τ increases. The variance of $\hat{\rho}_\tau$ depends upon τ and the true ACS in a complicated way – typically it increases as τ increases. Moreover, for most cases of interest the estimators $\hat{\rho}_\tau$ and $\hat{\rho}_{\tau+1}$ are highly correlated. This lack of homogeneity of variance and the correlation between nearby estimators make a plot of $\hat{\rho}_\tau$ versus τ hard to interpret.

[3] Because of these sampling problems, it is difficult to devise good statistical tests based upon $\hat{\rho}_\tau$ for various hypotheses of interest. For example, suppose we entertain a hypoth-

esis that specifies values for ρ_1 and ρ_2. To evaluate to what extent the sample values $\hat\rho_1$ and $\hat\rho_2$ offer evidence for or against this hypothesis, we need to derive their statistical properties. In general this is not an easy task because it can be difficult to determine the variances of $\hat\rho_1$ and $\hat\rho_2$ and the degree to which they are correlated.

[4] Even in the rare instances where we believe we have enough data to estimate ρ_τ reliably, a second model characterization can be useful as a complementary way of viewing the properties of our data. In particular, in contrast to time domain models, the characterization behind spectral analysis makes it much easier to visualize the kinds of time series that would be generated by the model (we return to this point in Section 4.6).

Comments and Extensions to Section 1.1

[1] The reader might well ask whether the right-hand side of Equation (3b) should be multiplied by a factor of $N/(N-\tau)$ to compensate for the different number of terms in the summations in the numerator and denominator. Most time series analysts would answer "no." As discussed in Chapter 6, estimation of the ACS via Equation (3b) yields a sequence that corresponds to the ACS for some theoretical stationary process. If we introduce the factor of $N/(N-\tau)$, we cannot make a similar statement, and we could run into practical problems in using the resulting ACS estimates. For example, based upon these estimates, we could in fact obtain a negative value when attempting to compute the variance of certain linear combinations of our time series – this would obviously be nonsense since variances must always be nonnegative.

[2] The use of the term "sample autocorrelation" for the right-hand side of Equation (3b) conforms to that of the statistical literature. Unfortunately this conflicts with the engineering literature, in which sometimes either the quantity

$$\frac{1}{N}\sum_{t=0}^{N-\tau-1}(x_{t+\tau}-\bar{x})(x_t-\bar{x}) \text{ or } \frac{1}{N}\sum_{t=0}^{N-\tau-1}x_{t+\tau}x_t$$

is called the "lag τ sample autocorrelation." This latter notation can lead to unnecessary confusion between correlations and covariances and cause nonzero means to be ignored.

[3] We do not want to leave the impression that lag τ scatter plots for time series always indicate an approximately linear relationship between $x_{t+\tau}$ and x_t. As a simple counterexample, Figure 6 shows the first 24 years of a time series of monthly average temperatures at St. Paul, Minnesota, as well as the lag 6 and 9 scatter plots for the entire time series (this extends from 1820 to 1983), both of which are highly nonlinear. In these cases the summary given by the sample ACS does not give the full story about the relationship between $x_{t+\tau}$ and x_t.

[4] While it is reasonable that $\mu = E\{X_t\}$ is independent of the time index t for the wind speed, atomic clock and ocean noise series, it would seem to be an unreasonable assumption for the Willamette River data, which varies with a prominent annual pattern. A more natural assumption is that $E\{X_t\}$ is a function of which month X_t occurs in. As we shall see, the key concept of stationarity assumes that certain quantities – including $E\{X_t\}$ – are independent of the index t. It would appear at first that we cannot assume a stationary model for the Willamette River data as we have implied above. In fact, as we shall discuss later (see Sections 2.6 and 2.8), there is a mathematical trick that allows us to treat such data in the context of a stationary model (the trick involves assuming that the time origin of a periodic phenomenon can be regarded as being picked at random).

1.2 Spectral Analysis for a Simple Time Series Model

Some of the problems of estimation and interpretation that are associated with the ACS are lessened (but not completely alleviated) when we deal with a frequency domain characterization called the "spectrum." The spectrum is simply a second way of characterizing models for time series. The objective of spectral analysis is to study and estimate the spectrum.

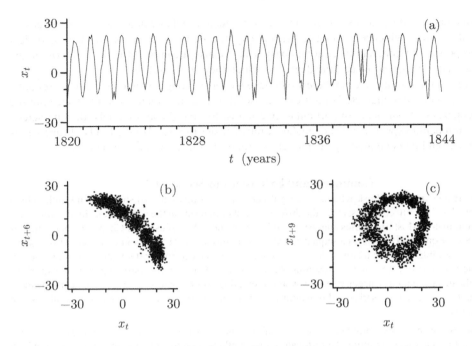

Figure 6 Plots of (a) the first 24 years of the St. Paul temperature time series, (b) the lag 6 scatter plot for the entire series and (c) the corresponding lag 9 scatter plot. The temperature series is measured in degrees centigrade. For the lag τ scatter plot ($\tau = 6, 9$), the value $x_{t+\tau}$ is plotted on the vertical axis versus x_t on the horizontal axis.

How exactly we define the spectrum depends upon what class of models we assume for a time series. A detailed definition for a useful class of models is presented in Chapter 4, but the key idea behind the spectrum is based upon a model for a time series consisting of a linear combination of cosines and sines with different frequencies; i.e.,

$$X_t = \mu + \sum_f [A(f)\cos(2\pi f t) + B(f)\sin(2\pi f t)]. \tag{6}$$

At first glance it might not seem possible to express time series such as those in Figure 2 in terms of sinusoids: these series appear to have random irregular bumps, whereas cosines and sines are deterministic regular oscillations. Figure 7 demonstrates that we can indeed get random-looking time series by combining sinusoids in a special way. The upper ten plots in the left-hand column show cosines (thick curves) and sines (thin) with equal amplitudes and with frequencies $f = (2l-1)/128$, $l = 1, 2, \ldots, 10$ (top to bottom). The bottom plot in that column shows a series $\{x_t\}$ that is equal to the sum of these twenty sinusoids; i.e.,

$$x_t = \sum_{l=1}^{10} \left[\cos\left(2\pi\tfrac{2l-1}{128}t\right) + \sin\left(2\pi\tfrac{2l-1}{128}t\right)\right], \quad t = 0, 1, \ldots, 127.$$

This artificial series is highly structured and does not particularly resemble any of our four actual series. We can, however, create series that are more random in appearance by introducing random amplitudes; i.e., we form

$$x_t = \sum_{l=1}^{10} \left[a_l \cos\left(2\pi\tfrac{2l-1}{128}t\right) + b_l \sin\left(2\pi\tfrac{2l-1}{128}t\right)\right], \quad t = 0, 1, \ldots, 127,$$

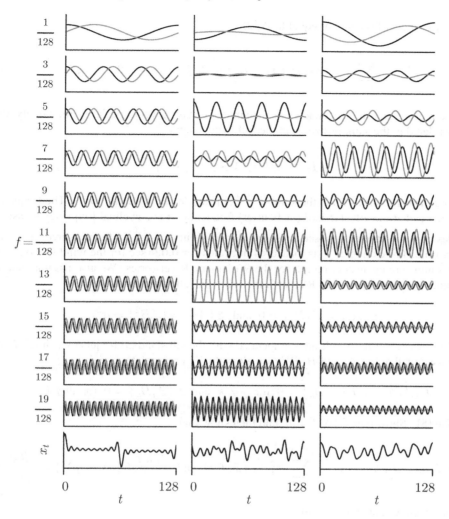

Figure 7 Sums of sinusoids with fixed and random amplitudes (see text for details).

where the a_l and b_l terms are twenty different realizations of uncorrelated Gaussian RVs with zero mean and unit variance. The upper ten plots in the middle column show $a_l \cos\left(2\pi \frac{2l-1}{128} t\right)$ (black curves) and $b_l \sin\left(2\pi \frac{2l-1}{128} t\right)$ (gray) versus t for $l = 1, 2, \ldots, 10$ (top to bottom) for one particular set of realizations, while the right-hand column shows similar plots for a second set. The bottom plots in each column show the sum x_t versus t. These artificial series are much more random in appearance, so indeed it does seem possible to construct irregular looking series out of sinusoids.

In general the summation in Equation (6) is a rather special one. To say what it means for the class of stationary processes is the subject of the spectral representation theorem (see Section 4.1). Fortunately, if we deal with a particularly simple (but unrealistic) model, we can say exactly what the summation means, define the spectrum in terms of elements involved in the summation, and thereby get an idea of what spectral analysis is all about. Let us assume that our time series can be modeled by a sum of a constant term μ and sinusoids with different fixed frequencies $\{f_j\}$ and random amplitudes $\{A_j\}$ and $\{B_j\}$ (the notation $\lfloor N/2 \rfloor$ refers to the greatest integer less than or equal to $N/2$):

$$X_t = \mu + \sum_{j=1}^{\lfloor N/2 \rfloor} [A_j \cos(2\pi f_j t) + B_j \sin(2\pi f_j t)], \quad t = 0, 1, \ldots, N-1. \quad (8a)$$

Here we require that the frequencies of the sinusoids have a very special form, namely, that they be related to the sample size N in the following way:

$$f_j \stackrel{\text{def}}{=} j/N, \quad 1 \leq j \leq \lfloor N/2 \rfloor$$

(here and throughout this book the symbol $\stackrel{\text{def}}{=}$ means "equal by definition"). The frequency f_j is often called the jth standard (or Fourier) frequency; it is a cyclical frequency measured in cycles per unit time as opposed to an angular frequency $\omega_j \stackrel{\text{def}}{=} 2\pi f_j$, measured in radians per unit time. For example, f_j is measured in cycles per 0.025 sec for the wind speed series, while its units are cycles per month for the Willamette River series. We also assume that the amplitudes $\{A_j\}$ and $\{B_j\}$ are RVs with the following stipulations: for all j

$$E\{A_j\} = E\{B_j\} = 0 \text{ and } E\{A_j^2\} = E\{B_j^2\} = \sigma_j^2.$$

Thus the variance of the amplitudes associated with the jth standard frequency is just σ_j^2. We further assume that the A_j and B_j RVs are all mutually uncorrelated; i.e.,

$$E\{A_j A_k\} = E\{B_j B_k\} = 0 \text{ for } j \neq k \text{ and } E\{A_j B_k\} = 0 \text{ for all } j, k.$$

▷ **Exercise [8]** Show that $E\{X_t\} = \mu$, and then show that

$$E\{(X_{t+\tau} - \mu)(X_t - \mu)\} = \sum_{j=1}^{\lfloor N/2 \rfloor} \sigma_j^2 \cos(2\pi f_j \tau) \text{ and } \sigma^2 \stackrel{\text{def}}{=} E\{(X_t - \mu)^2\} = \sum_{j=1}^{\lfloor N/2 \rfloor} \sigma_j^2, \quad (8b)$$

from which we can conclude that

$$\rho_\tau = \frac{\sum_{j=1}^{\lfloor N/2 \rfloor} \sigma_j^2 \cos(2\pi f_j \tau)}{\sum_{j=1}^{\lfloor N/2 \rfloor} \sigma_j^2}. \quad (8c) \triangleleft$$

(We emphasize that we are considering models defined by Equation (8a) for pedagogical purposes only. These have a number of undesirable features, not the least of which is an explicit dependence of the component frequencies f_j on the sample size N.)

For this model we *define* the spectrum by

$$S_j \stackrel{\text{def}}{=} \sigma_j^2, \quad 1 \leq j \leq \lfloor N/2 \rfloor.$$

A plot of S_j versus f_j merely shows us the variances of the RVs that determine the amplitudes of the sinusoidal terms at the standard frequencies. From Equation (8b), we have the following fundamental relationship:

$$\sum_{j=1}^{\lfloor N/2 \rfloor} S_j = \sigma^2.$$

1.2 Spectral Analysis for a Simple Time Series Model

Thus, for a time series generated by the model in Equation (8a), the population variance, σ^2, can be regarded as being composed of a sum of a number of components, each of which is associated with a different nonzero standard frequency. The contribution to the variance due to the sinusoidal terms with frequency f_j is given by S_j. A study of S_j versus f_j indicates where the variability in a time series is likely to come from. In other words, the spectrum represents an analysis of the process variance σ^2 as the sum of variances associated with the Fourier frequencies.

Equation (8c) and the definition of the spectrum tell us that we can determine the ACS and σ^2 if we know the spectrum. Conversely, it can be shown (see Exercise [1.4]) that we can determine the spectrum if we know the ACS and σ^2. The spectrum is a frequency domain characterization for a model of a time series and is fully equivalent to the time domain characterization given by the ACS and σ^2.

For a model given by Equation (8a), it is easy to simulate a typical time series: as we did earlier in creating two of the time series shown in the bottom row of Figure 7, we use a random number generator on a computer to pick values for A_j and B_j and plug these into (8a) to generate a simulated time series. To illustrate this procedure, we will generate four such series using four different spectra. This exercise will show how a spectrum can be used to tell us something about the structure of an associated time series. The four spectra that we will use are actually rough models for the four time series in Figure 2 (for the moment we ignore the question of where these models came from). Figure 10a shows the four theoretical spectra; Figure 10b shows the corresponding ACSs (calculated via Equation (8c)); and Figure 11 shows a simulated time series that corresponds to each of the four spectra (we have set μ in Equation (8a) equal to the sample mean $\bar{x}$ for the corresponding series). If a proposed spectrum is a reasonable model for a time series, the corresponding theoretical ACS should resemble the sample ACS for the series, and simulated time series from that spectrum should have roughly the same visual properties as the actual time series. Here are some specifics about our four time series and these figures.

(a) For the wind speed data, we assume that S_j is large for $j = 1$ and then tapers off rapidly as j gets large. Thus, the low frequency terms in Equation (8a) – these correspond to sinusoids with long periods – should predominate. The theoretical ACS in plot (a) of Figure 10b is positive until lag 18. This picture agrees fairly well with the corresponding sample ACS in Figure 4, which is positive until lag 22. The appearance of the simulated time series is one of rather broad swoops together with some choppiness (evidently due to the higher frequencies in Equation (8a)). The wind speed series and the corresponding simulated series appear to have the same kind of bumpiness.

(b) For the atomic clock data, we assume S_j is large for $j = \lfloor N/2 \rfloor = 64$ and then tapers off rapidly as j decreases. Thus the high frequency terms (i.e., sinusoids with short periods) should predominate. The theoretical ACS oscillates between positive and negative values with an amplitude that is close to zero after the first few lags. The sample ACS in plot (b) of Figure 4 for these data shows more variability than this theoretical ACS (particularly for the higher lags), but the discrepancy might be due to sampling variation. The appearance of the simulated time series in Figure 11 is one of choppiness as the series swings back and forth from positive to negative values. The atomic clock data and the simulated series have the same "feel" to them.

(c) For the Willamette River data, we assume a spectrum that is constant except for a spike at $j = 11$. Since $f_{11} = 11/128$, this frequency corresponds to a period of $1/f_{11} = 128/11 \approx 11.6$ months. This is the frequency with a period closest to 1 year in our model (the next closest is f_{10} with a corresponding period of 12.8 months). We would thus expect terms with about this period to be predominant in Equation (8a). The generated

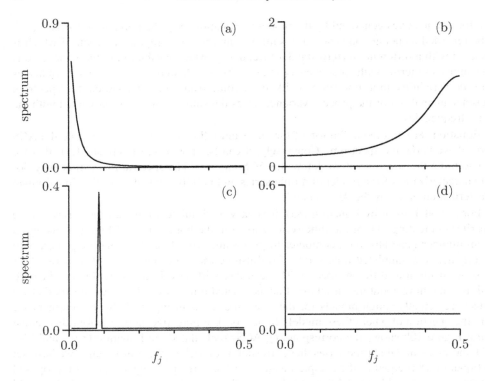

Figure 10a Plots of theoretical spectra S_j versus frequency f_j for models of four time series in Figure 2. The 64 values that determine each spectra are connected by solid lines. The horizontal axis represents frequency measured in cycles per unit time.

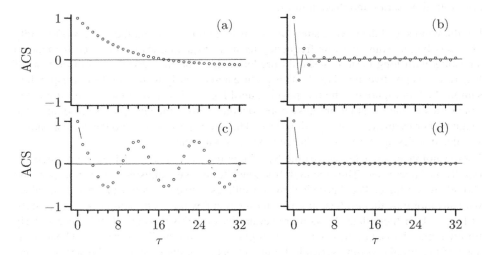

Figure 10b Plots of theoretical autocorrelation sequences of models for four time series in Figure 2 (cf. Figure 4).

time series should have a tendency to fluctuate with this period. This is roughly true for both the simulated series (Figure 11) and the Willamette River series. The sample ACS for the river flow data (Figure 4) and the theoretical ACS (Figure 10b) look fairly similar. (Here one of the limitations of our simple model is apparent: from physical considerations, it would make more sense to have a term corresponding to a frequency

1.3 Nonparametric Estimation of the Spectrum from Data

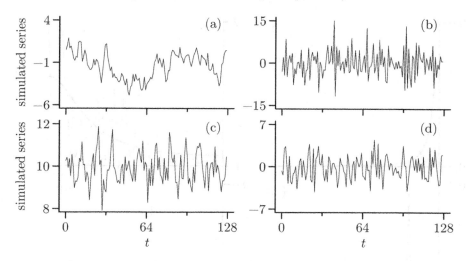

Figure 11 Plots of four simulated time series with statistical properties similar to series in Figure 2.

of one cycle per year in our model, but to include such a term would destroy some of the nice mathematical properties of our model that we will need shortly.)

(d) Finally, for the ocean noise data, we assume a spectrum S_j that is constant for all j. A time series generated from such a spectrum is often called "white noise," in analogy to white light, which is composed of equal contributions from a whole range of colors. The theoretical ACS is close to 0 for all $|\tau| > 0$ (see Exercise [1.5]). There should be no discernible patterns in a time series generated from a white noise spectrum, and indeed there appears to be little in the ocean noise data and none in its simulation. (We will return to this example in Sections 6.8 and 10.15, where we will find, using tests for white noise, evidence for rejecting the hypothesis that this time series is white noise! This is somewhat evident from its sample ACS in plot (d) of Figure 4, which shows a tendency to oscillate with a period of five time units. Nonetheless, for the purposes of this chapter, the ocean noise series is close enough to white noise for us to use it as an example of such.)

1.3 Nonparametric Estimation of the Spectrum from Data

The estimation of spectra from a given time series is a complicated subject and is the main concern of this book. For the simple model described by Equation (8a), we will give two methods for estimating spectra. These methods are representative of two broad classes of estimation techniques in use today, namely, nonparametric and parametric spectral analysis.

We begin with nonparametric spectral analysis, which also came first historically. A time series of length N that is generated by Equation (8a) depends upon the realizations of $2 \lfloor N/2 \rfloor$ RVs (the A_j and B_j terms) and the parameter μ, a total of $M \stackrel{\text{def}}{=} 2 \lfloor N/2 \rfloor + 1$ quantities in all. Now, $M = N$ for N odd, and $M = N + 1$ for N even, but in the latter case there are also actually just N quantities: $B_{N/2}$ is not used in Equation (8a) because $\sin(2\pi f_{N/2} t) = \sin(\pi t) = 0$ for all integers t. We can use the methods of linear algebra to solve for the N unknown quantities in terms of the N observable quantities $X_0, X_1, \ldots, X_{N-1}$, but it is quite easy to solve for these quantities explicitly due to some peculiar properties of our model.

▷ **Exercise [11]** Show that

$$A_j = \frac{2}{N} \sum_{t=0}^{N-1} X_t \cos(2\pi f_j t)$$

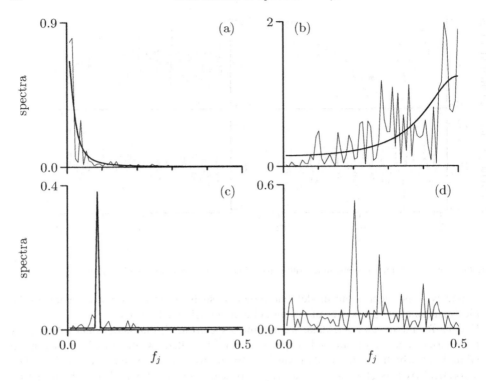

Figure 12 Comparison of theoretical and estimated spectra for four time series in Figure 2. The thick curves are the theoretical spectra (copied from Figure 10a), while the thin curves are the estimated spectra. The units of the horizontal axes are the same as those of Figure 10a.

if $1 \leq j < N/2$ and, if N is even,

$$A_{N/2} = \frac{1}{N} \sum_{t=0}^{N-1} X_t \cos\left(2\pi f_{N/2} t\right).$$

Hint: multiply both sides of Equation (8a) by $\cos\left(2\pi f_j t\right)$, sum over t and then make use of relationships stated in Exercises [1.2d] and [1.3c]. ◁

Likewise, it can be shown that for $1 \leq j < N/2$

$$B_j = \frac{2}{N} \sum_{t=0}^{N-1} X_t \sin\left(2\pi f_j t\right).$$

We now have expressions for all quantities in the model except for the parameter μ.

▷ **Exercise [12]** By appealing to Exercise [1.2d], show that

$$\overline{X} \stackrel{\text{def}}{=} \frac{1}{N} \sum_{t=0}^{N-1} X_t = \mu.$$

◁

Note that the sample mean $\overline{X}$ is *exactly* equal to the parameter μ. Perfect estimation of model parameters rarely occurs in statistics: the simple model we are using here for pedagogical purposes has some special – and implausible – properties.

1.3 Nonparametric Estimation of the Spectrum from Data

Since $E\{A_j^2\} = E\{B_j^2\} = \sigma_j^2$, for a time series $x_0, x_1, \ldots, x_{N-1}$ that is a realization of a model given by (8a), it is natural to estimate S_j by

$$\hat{S}_j \stackrel{\text{def}}{=} \frac{A_j^2 + B_j^2}{2} = \frac{2}{N^2}\left[\left(\sum_{t=0}^{N-1} x_t \cos\left(2\pi f_j t\right)\right)^2 + \left(\sum_{t=0}^{N-1} x_t \sin\left(2\pi f_j t\right)\right)^2\right] \quad (13a)$$

for $1 \leq j < N/2$ and, if N is even,

$$\hat{S}_{N/2} \stackrel{\text{def}}{=} \frac{1}{N^2}\left(\sum_{t=0}^{N-1} x_t \cos\left(2\pi f_{N/2} t\right)\right)^2. \quad (13b)$$

As examples of this estimation procedure, the thin curves in Figure 12 are graphs of $\hat{S}_j$ versus f_j for the four time series in Figure 2. We can now see some justification for the theoretical spectra for these time series given previously (shown in these figures by the thick curves): the theoretical spectra are smoothed versions of the estimated spectra. Here are some points to note about the nonparametric spectral estimates.

[1] Since $\hat{S}_j$ involves just A_j^2 and B_j^2 and since the A_j and B_j RVs are mutually uncorrelated, it can be argued that the $\hat{S}_j$ RVs should be approximately uncorrelated (the assumption of Gaussianity for A_j and B_j makes this statement true without approximation). This property should be contrasted to that of the sample ACS, which is highly correlated for values with lags close to one another.

[2] Because of this approximate uncorrelatedness, it is possible to derive good statistical tests of hypotheses by basing the tests on some form of the $\hat{S}_j$.

[3] For $1 \leq j < N/2$, $\hat{S}_j$ is only a "two degrees of freedom" estimate of σ_j^2 (i.e., it is an average of just the two values A_j^2 and B_j^2). For N even, $\hat{S}_{N/2}$ is just a "one degree of freedom" estimate since it is based solely on $A_{N/2}^2$. These facts imply that there should be considerable variability in $\hat{S}_j$ as a function of j even if the true underlying spectrum changes slowly with j. The bumpiness of the estimated spectra in Figure 12 can be attributed to this sampling variability. If we can assume that S_j varies slowly with j, we can smooth the $\hat{S}_j$ locally to come up with a less variable estimate of S_j. This is the essential idea behind many of the nonparametric estimators discussed in Chapter 7.

[4] It can be shown (see Section 6.6) that, if we consider the logarithmically transformed $\hat{S}_j$ instead of $\hat{S}_j$, the variance of the former is approximately the same for all f_j. Thus a plot of the log of $\hat{S}_j$ versus f_j is easier to interpret than plots of the sample ACS, for which there is no known "variance-stabilizing" transformation. As an example, Figure 14 shows plots of $10 \log_{10}(\hat{S}_j)$ versus f_j. In the engineering literature, the units of $10 \log_{10}(\hat{S}_j)$ are said to be in decibels (dB). A decibel scale is a convenient way of depicting a log scale (10 dB represents an order of magnitude difference, while 3 dB is approximately a change by a factor of two because $10 \log_{10}(2) \doteq 3.01$). Note that the "local" variability of the estimated spectra on the decibel scales is approximately the same across all frequencies; by contrast, in the plots of $\hat{S}_j$ in Figure 12, the local variability is high when $\hat{S}_j$ is large, and small when $\hat{S}_j$ is small. (Use of a decibel scale also allows us to see some subtle discrepancies between the theoretical and estimated spectra that are not apparent in Figure 12. In particular, plots (b) and (c) show mismatches between these spectra at, respectively, low and high frequencies.)

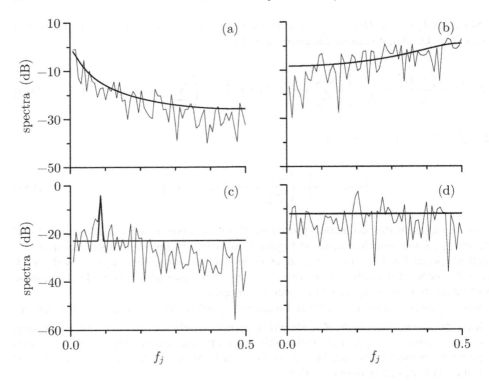

Figure 14 As in Figure 12, but now with the theoretical and estimated spectra plotted on a decibel scale (e.g., $10 \log_{10}(\hat{S}_j)$ versus f_j rather than $\hat{S}_j$ versus f_j).

1.4 Parametric Estimation of the Spectrum from Data

A second popular method of estimating spectra is called parametric spectral analysis. The basic idea is simple: first, we assume that the true underlying spectrum of interest is a function of a small number of parameters; second, we estimate these parameters from the available data somehow; and third, we estimate the spectrum by plugging these estimated parameters into the functional form for S.

As an example, suppose we assume that the true spectrum is given by

$$S_j(\alpha, \beta) = \frac{\beta}{1 + \alpha^2 - 2\alpha \cos(2\pi f_j)}, \qquad (14)$$

where α and β are the parameters we will need to estimate. (The reader might think the functional form of this spectrum is rather strange, but in Chapter 9 we shall see that (14) arises in a natural way and corresponds to the spectrum for a first-order autoregressive process.) If we estimate these parameters by $\hat{\alpha}$ and $\hat{\beta}$, say, we can then form a parametric estimate of the spectrum by

$$\hat{S}_j(\hat{\alpha}, \hat{\beta}) = \frac{\hat{\beta}}{1 + \hat{\alpha}^2 - 2\hat{\alpha} \cos(2\pi f_j)}.$$

Now it can be shown that, if a time series has an underlying spectrum given by (14), then $\rho_1 \approx \alpha$ to a good approximation so that we can estimate α by $\hat{\alpha} = \hat{\rho}_1$. By analogy to the relationship

$$\sum_{j=1}^{\lfloor N/2 \rfloor} S_j = \sigma^2,$$

we require that
$$\sum_{j=1}^{\lfloor N/2 \rfloor} \hat{S}_j(\hat{\alpha}, \hat{\beta}) = \frac{1}{N} \sum_{t=0}^{N-1} (x_t - \bar{x})^2 \stackrel{\text{def}}{=} \hat{\sigma}^2$$

and hence estimate β by

$$\hat{\beta} = \hat{\sigma}^2 \left(\sum_{j=1}^{\lfloor N/2 \rfloor} \frac{1}{1 + \hat{\alpha}^2 - 2\hat{\alpha} \cos(2\pi f_j)} \right)^{-1}.$$

In fact, we have already shown two examples of parametric spectral analysis using Equation (14): what we called the theoretical spectra for the wind speed data and the atomic clock data (the thick curves in plots (a) and (b) of Figures 12 and 14) are actually just parametric spectral estimates that use the procedure we just described. The parametric estimates of the spectra are much smoother than their corresponding nonparametric estimates, but the fact that the parametric estimates are more pleasing visually should not be taken as evidence that they are necessarily superior. For example, for the atomic clock data, there is almost an order of magnitude difference between the parametric and nonparametric estimates at the very lowest frequencies. Knowledge of how this series was formed – in combination with more data from this same clock – leads us to the conclusion that the level of the nonparametric estimate is actually more reasonable at low frequencies even though the estimate is bumpier looking.

There are two additional caveats about the parametric approach that we need to mention. One is that it is hard to say much about the statistical properties of the estimated spectrum. Without making some additional assumptions, we cannot, for example, calculate a 95% confidence interval for S_j. Also we must be careful about which class of functional forms we assume initially. For any given functional form, if there are too many parameters to be estimated (relative to the length of the time series), all the parameters cannot be reliably estimated; if there are too few parameters, the true underlying spectrum might not be well approximated by any spectrum specified by that number of parameters. The problem with too few parameters is illustrated by the poor performance at low frequencies of the spectrum of Equation (14) for the atomic clock data. An additional example is that no spectrum from this equation is a good match for the Willamette River data.

1.5 Uses of Spectral Analysis

In summary, spectral analysis is an analysis of variance technique. It is based upon our ability to represent a time series in some fashion as a sum of cosines and sines with different frequencies and amplitudes. The variance of a time series is broken down into a number of components, the totality of the components being called the spectrum. Each component is associated with a particular frequency and represents the contribution that frequency makes to the total variability of the series.

In the remaining chapters of this book we extend the basic ideas discussed here to the class of stationary stochastic processes (Chapter 2). We discuss the theory of spectral analysis for deterministic functions (Chapter 3), which is important in its own right and also serves as motivation for the corresponding theory for stationary processes. We motivate and state the spectral representation theorem for stationary processes (Chapter 4). This theorem essentially defines what Equation (6) means for stationary processes, which in turn allows us to define their spectra. We next discuss the central role that the theory of linear filters plays in spectral analysis (Chapter 5); in particular, this theory allows us to easily determine the spectra for an important class of processes. The bulk of the remaining chapters is devoted to various aspects of nonparametric and parametric spectral estimation theory for stationary processes.

We close this introductory chapter with some examples of the many practical uses for spectral analysis. The following quote from Kung and Arun (1987) indicates a few of the areas in which spectral analysis has been used.

> Some typical applications where the problem of spectrum estimation is encountered are: interference spectrometry; the design of Wiener filters for signal recovery and image restoration; the design of channel equalizers in communications systems; the determination of formant frequencies (location of spectral peaks) in speech analysis; the retrieval of hidden periodicities from noisy data (locating spectral lines); the estimation of source-energy distribution as a function of angular direction in passive underwater sonar; the estimation of the brightness distribution (of the sky) using aperture synthesis telescopy in radio astronomy; and many others.

Here we list some of the uses for spectral analysis.

[1] *Testing theories*

It is sometimes possible to derive the spectrum for a certain physical phenomenon based upon a theoretical model. One way to test the assumed theory is to collect data concerning the phenomenon, estimate the spectrum from the observed time series, and compare it with the theoretical spectrum. For example, scientists have been able to derive the spectrum for thermal noise produced by the random motion of free electrons in a conductor. The average current is zero, but the fluctuations produce a noise voltage across the conductor. It is easy to see how one could test the physical theory by an appropriate experiment. A second example concerns the wind speed data we have examined above. Some physical theories predict that there should be a well-defined peak in the spectrum in the low frequency range for wind speed data collected near coastlines. We can use spectral analysis to test this theory on actual data.

[2] *Investigating data*

We have seen that knowledge of a spectrum allows us to describe in broad outline what a time series should look like that is drawn from a process with such a spectrum. Conversely, spectral analysis may reveal certain features of a time series that are not obvious from other analyses. It has been used in this way in many of the physical sciences. Two examples are geophysics (for the study of intraplate volcanism in the Pacific Ocean basin and attenuation of compressional waves) and oceanography (for examining tidal behavior and the effect of pressure and wind stress on the ocean). As a further example, Walker (1985) in the "Amateur Scientist" column in *Scientific American* describes the efforts of researchers to answer the question "Does the rate of rainfall in a storm have a pattern or is it random?" – it evidently has a pattern that spectral analysis can discern.

[3] *Discriminating data*

Spectral analysis is also a popular means of clearly demonstrating the qualitative differences between different time series. For example, Jones et al. (1972) investigated the spectra of brain wave measurements of babies before and after they were subjected to a flashing light. The two sets of estimated spectra showed a clear difference between the "before" and "after" time series. As a second example, Jones et al. (1987) investigated electrical waves generated during the contraction of human muscles and found that waves typical of normal, myopathic and neurogenic subjects could be classified in terms of their spectral characteristics.

[4] *Performing diagnostic tests*

Spectral analysis is often used in conjunction with model fitting. For example, Box et al. (2015) describe a procedure for fitting autoregressive, integrated, moving average (ARIMA) models to time series. Once a particular ARIMA model has been fit to some data, an analyst can tell how well the model represents the data by examining what are called the estimated innovations. For ARIMA models, the estimated innovations play the same role that residuals play in linear least squares analysis. Now the estimated innovations are just another time series, and if the ARIMA model for the original data is a good one, the spectrum of the estimated innovations should approximate a white noise spectrum. Thus spectral analysis can be used to perform a goodness of fit test for ARIMA models.

[5] *Assessing the predictability of a time series*
One of the most popular uses of time series analysis is to predict future values of a time series. For example, one measure of the performance of atomic clocks is based upon how predictable they are as time keepers. Spectral analysis plays a key role in assessing the predictability of a time series (see Section 8.7).

1.6 Exercises

[1.1] Use the Euler relationship $e^{i2\pi f} = \cos(2\pi f) + i\sin(2\pi f)$, where $i \stackrel{\text{def}}{=} \sqrt{-1}$, to show that Equations (13a) and (13b) can be rewritten in the following form:

$$\hat{S}_j = \frac{k}{N^2} \left| \sum_{t=0}^{N-1} x_t e^{-i2\pi f_j t} \right|^2$$

for $1 \le j \le \lfloor N/2 \rfloor$, where $k = 2$ if $j < N/2$, and $k = 1$ if N is even and $j = N/2$.

[1.2] (a) If z is any complex number not equal to 1, show that

$$\sum_{t=0}^{N-1} z^t = \frac{1-z^N}{1-z}. \tag{17a}$$

(b) Show (using Euler's relationship in Exercise [1.1]) that

$$\cos(2\pi f) = \frac{e^{i2\pi f} + e^{-i2\pi f}}{2} \quad \text{and} \quad \sin(2\pi f) = \frac{e^{i2\pi f} - e^{-i2\pi f}}{2i}. \tag{17b}$$

(c) Show that, for $-\infty < f < \infty$,

$$\sum_{t=0}^{N-1} e^{i2\pi ft} = N e^{i(N-1)\pi f} \mathcal{D}_N(f), \quad \text{where } \mathcal{D}_N(f) \stackrel{\text{def}}{=} \begin{cases} \frac{\sin(N\pi f)}{N\sin(\pi f)}, & f \notin \mathbb{Z}; \\ (-1)^{(N-1)f}, & f \in \mathbb{Z}; \end{cases} \tag{17c}$$

here $\mathbb{Z} = \{\ldots, -2, -1, 0, 1, 2, \ldots\}$ is the set of all integers. Note that $\mathcal{D}_N(0) = 1$ and that $\mathcal{D}_N(\cdot)$ is a continuous function. This function is one form of Dirichlet's kernel, which plays a prominent role in Chapters 3 and 6. (The term "kernel" is used to denote certain functions that appear in the integrand of integral equations. For our purposes, we could just as easily refer to $\mathcal{D}_N(\cdot)$ as "Dirichlet's function," but we stick with its more traditional name.)

(d) Use part (c) to show that, for integer j such that $1 \le j < N$,

$$\sum_{t=0}^{N-1} \cos(2\pi f_j t) = 0 \quad \text{and} \quad \sum_{t=0}^{N-1} \sin(2\pi f_j t) = 0, \tag{17d}$$

where $f_j = j/N$ (for $j = 0$, the sums are, respectively, N and 0).

(e) Use part (c) to show that, for $-\infty < f < \infty$,

$$\sum_{t=1}^{N} e^{i2\pi ft} = N e^{i(N+1)\pi f} \mathcal{D}_N(f) \quad \text{and} \quad \sum_{t=-(N-1)}^{N-1} e^{i2\pi ft} = (2N-1)\mathcal{D}_{2N-1}(f). \tag{17e}$$

[1.3] The following trigonometric relationships are used in Chapters 6 and 10.
(a) Use Exercise [1.2c] and the fact that

$$e^{i2\pi(f\pm f')t} = e^{i2\pi ft}e^{\pm i2\pi f't}$$
$$= \cos(2\pi ft)\cos(2\pi f't) \mp \sin(2\pi ft)\sin(2\pi f't)$$
$$+ i\left[\sin(2\pi ft)\cos(2\pi f't) \pm \cos(2\pi ft)\sin(2\pi f't)\right]$$

to evaluate

$$\sum_{t=0}^{N-1}\cos(2\pi ft)\cos(2\pi f't),\quad \sum_{t=0}^{N-1}\cos(2\pi ft)\sin(2\pi f't) \text{ and } \sum_{t=0}^{N-1}\sin(2\pi ft)\sin(2\pi f't).$$

In particular, show that, if $f \pm f' \notin \mathbb{Z}$,

$$\sum_{t=0}^{N-1}\cos(2\pi ft)\cos(2\pi f't) = C_N(f-f') + C_N(f+f'),$$

$$\sum_{t=0}^{N-1}\cos(2\pi ft)\sin(2\pi f't) = S_N(f+f') - S_N(f-f')$$

and

$$\sum_{t=0}^{N-1}\sin(2\pi ft)\sin(2\pi f't) = C_N(f-f') - C_N(f+f'),$$

where, with $\mathcal{D}_N(\cdot)$ defined as in Exercise [1.2c],

$$C_N(f) \stackrel{\text{def}}{=} \frac{N}{2}\mathcal{D}_N(f)\cos[(N-1)\pi f] \text{ and } S_N(f) \stackrel{\text{def}}{=} \frac{N}{2}\mathcal{D}_N(f)\sin[(N-1)\pi f].$$

(b) Show that for $f \neq 0, \pm 1/2, \pm 1, \ldots,$

$$\sum_{t=0}^{N-1}\cos^2(2\pi ft) = \frac{N}{2} + \frac{\sin(N2\pi f)}{2\sin(2\pi f)}\cos[(N-1)2\pi f],$$

$$\sum_{t=0}^{N-1}\cos(2\pi ft)\sin(2\pi ft) = \frac{\sin(N2\pi f)}{2\sin(2\pi f)}\sin[(N-1)2\pi f]$$

and

$$\sum_{t=0}^{N-1}\sin^2(2\pi ft) = \frac{N}{2} - \frac{\sin(N2\pi f)}{2\sin(2\pi f)}\cos[(N-1)2\pi f].$$

(c) Show that

$$\sum_{t=0}^{N-1}\cos^2(2\pi f_j t) = \sum_{t=0}^{N-1}\sin^2(2\pi f_j t) = \frac{N}{2},$$

$$\sum_{t=0}^{N-1}\cos(2\pi f_j t)\sin(2\pi f_j t) = \sum_{t=0}^{N-1}\cos(2\pi f_j t)\sin(2\pi f_k t) = 0$$

and

$$\sum_{t=0}^{N-1} \cos\left(2\pi f_j t\right) \cos\left(2\pi f_k t\right) = \sum_{t=0}^{N-1} \sin\left(2\pi f_j t\right) \sin\left(2\pi f_k t\right) = 0,$$

where $f_j = j/N$ and $f_k = k/N$ with j and k both integers such that $j \neq k$ and $1 \leq j, k < N/2$. Show that, for even N,

$$\sum_{t=0}^{N-1} \cos^2(2\pi f_{N/2} t) = N \quad \text{and} \quad \sum_{t=0}^{N-1} \cos\left(2\pi f_{N/2} t\right) \sin\left(2\pi f_{N/2} t\right) = \sum_{t=0}^{N-1} \sin^2(2\pi f_{N/2} t) = 0.$$

[1.4] Show how σ_j^2 in Equation (8c) can be expressed in terms of $\{\rho_\tau\}$ and σ^2.

[1.5] (a) Show that the ACS for the white noise defined in Section 1.2 as a model for the ocean noise data is close to – but not exactly – zero for all lags τ such that $0 < |\tau| < N$ (see plot (d) of Figure 10b; here we do *not* assume that N is necessarily 128). This is another peculiarity of the model defined by Equation (8a): the usual definition for white noise (see Chapters 2 and 4) implies both a constant spectrum (to be precise, a constant spectral density function) and an ACS that is *exactly* zero for all $|\tau| > 0$, whereas a constant spectrum in (8a) has a corresponding ACS that is *not* exactly zero at all nonzero lags.

(b) Now consider the model

$$X_t = \mu + \sum_{j=0}^{\lfloor N/2 \rfloor} \left[A_j \cos\left(2\pi f_j t\right) + B_j \sin\left(2\pi f_j t\right) \right],$$

$t = 0, 1, \ldots, N-1$. This differs from Equation (8a) only in that the summation starts at $j = 0$ instead of $j = 1$. We assume that the statistical properties of A_j and B_j are as described below Equation (8a). Since $\cos(2\pi f_0 t) = 1$ and $\sin(2\pi f_0 t) = 0$ for all t, this modification just introduces randomness in the constant term in our model. For this new model, show that $E\{X_t\} = \mu$ and that the variance σ^2 and the ACS $\{\rho_\tau\}$ are given, respectively, by Equations (8b) and (8c) if we replace $j = 1$ with $j = 0$ in the summations. Show that, with $\nu^2 > 0$,

$$\text{if } \sigma_j^2 = \begin{cases} 2\nu^2, & 1 \leq j < N/2; \\ \nu^2, & \text{otherwise,} \end{cases} \text{ then } \rho_\tau = \begin{cases} 1, & \tau = 0; \\ 0, & 0 < |\tau| < N. \end{cases}$$

Thus, if we introduce randomness in the constant term, we can produce a model that agrees with the usual definition of white noise for $|\tau| < N$ in terms of its ACS, but we no longer have a constant spectrum.

[1.6] This exercise points out that the lag τ sample autocorrelation $\hat{\rho}_\tau$ of Equation (3b) is not a true sample correlation coefficient because it is not patterned *exactly* after the Pearson product moment correlation coefficient of Equation (3a). If we were to follow that equation, we would be led to

$$\tilde{\rho}_\tau \stackrel{\text{def}}{=} \frac{\sum_{t=0}^{N-\tau-1}(x_{t+\tau} - \bar{x}_{\tau:N-1})(x_t - \bar{x}_{0:N-\tau-1})}{\left[\sum_{t=0}^{N-\tau-1}(x_{t+\tau} - \bar{x}_{\tau:N-1})^2 \sum_{t=0}^{N-\tau-1}(x_t - \bar{x}_{0:N-\tau-1})^2\right]^{1/2}}, \quad (19)$$

where

$$\bar{x}_{j:k} \stackrel{\text{def}}{=} \frac{1}{k-j+1} \sum_{t=j}^{k} x_t.$$

The modifications to Equation (19) that lead to $\hat{\rho}_\tau$ seem reasonable: we replace $\bar{x}_{\tau:N-1}$ and $\bar{x}_{0:N-\tau-1}$ with $\bar{x}$ in the numerator, and we use the sample variance of the entire time series in the denominator rather than sample variances from two subseries. To illustrate the fact that, unlike $\tilde{\rho}_\tau$, the sample autocorrelation $\hat{\rho}_\tau$ is not a true sample correlation, consider a time series whose values all lie on a line, namely, $x_t = \alpha + \beta t$, $t = 0, 1, \ldots, N - 1$, where α and $\beta \neq 0$ are constants.

(a) Derive an expression that describes how the scatter plot of $x_{t+\tau}$ versus x_t depends upon α, β and τ. Make a plot of x_{t+70} versus x_t, $t = 0, 1, \ldots, 29$, for the specific case $\alpha = 1$, $\beta = 2$ and $N = 100$, and verify that your theoretical expression matches the plot.

(b) Derive an expression for $\hat{\rho}_\tau$ valid for the assumed x_t. Use Equation (3b) to create a plot of $\hat{\rho}_\tau$ versus $\tau = 0, 1, \ldots, 99$ for the specific case $\alpha = 1$, $\beta = 2$ and $N = 100$, and verify that your theoretical expression matches the plot. Two facts that might prove useful are $\sum_{t=1}^{M} t = M(M + 1)/2$ and $\sum_{t=1}^{M} t^2 = M(M + 1)(2M + 1)/6$.

(c) Based upon the expression derived in part (b), argue that, for large N, $\hat{\rho}_\tau$ achieves a minimum value at approximately $\tau = N/\sqrt{2}$ and that the minimum value is approximately $1 - \sqrt{2} \doteq -0.41$. How well do these approximations match up with the plot called for in part (b)?

(d) Show that $\tilde{\rho}_\tau = 1$ for $0 \leq \tau \leq N - 1$. Hint: argue that $x_{t+\tau} - \bar{x}_{\tau:N-1} = x_t - \bar{x}_{0:N-\tau-1}$.

(e) For the specific case considered in part (a), how does $\hat{\rho}_{70}$ compare to $\tilde{\rho}_{70}$? Which one is the appropriate summary of the scatter plot requested in part (a)?

2

Stationary Stochastic Processes

2.0 Introduction

Spectral analysis almost invariably deals with a class of models called stationary stochastic processes. The material in this chapter is a brief review of the theory behind such processes. The reader is referred to Brockwell and Davis (2016), Papoulis and Pillai (2002), Priestley (1981), Shumway and Stoffer (2017) or Yaglom (1987a) for complementary discussions.

2.1 Stochastic Processes

Consider the following experiment (see Figure 22): we hook up a resistor to an oscilloscope in such a way that we can examine the voltage variations across the resistor as a function of time. Every time we press a "reset" button on the oscilloscope, it displays the voltage variations for the one-second interval following the reset. Since the voltage variations are presumably caused by such factors as small temperature variations in the resistor, each time we press the reset button, we will observe a different display on the oscilloscope. Owing to the complexity of the factors that influence the display, there is no way that we can use the laws of physics to predict what will appear on the oscilloscope. However, if we repeat this experiment over and over, we soon see that, although we view a different display each time we press the reset button, the displays resemble each other: there is a characteristic "bumpiness" shared by all the displays.

We can model this experiment by considering a large box in which we have placed pictures of all the oscilloscope displays that we could possibly observe. Pushing the reset button corresponds to reaching into the box and choosing "at random" one of the pictures. Loosely speaking, we call the box of all possible pictures together with the mechanism by which we select the pictures a *stochastic process*. The one particular picture we actually draw out at a given time is called a *realization* of the stochastic process. The collection of all possible realizations is called the *ensemble*.

A more precise definition utilizes the concept of a random variable (RV), defined as a function, or mapping, from the sample space of possible outcomes of a random experiment to the real line (for a real-valued RV), the complex plane (for a complex-valued RV) or m-dimensional Euclidean space (for a vector-valued RV of dimension m). If we let the sample space be the ensemble of all possible realizations for an experiment such as the one we've just described, then for any fixed time t we can define an RV $X(t)$ that describes the outcome of the experiment at time t (this would be the vertical displacement of the picture on the

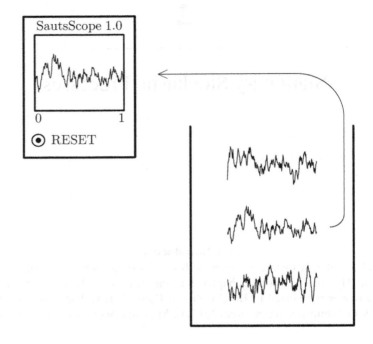

Figure 22 The oscilloscope experiment and a simplified box model. Here the ensemble contains only three possible realizations, the second of which was chosen.

oscilloscope at time t in our example). We can now give the following formal definition of a stochastic process: denoted by $\{X(t) : t \in T\}$, it is a family of RVs indexed by t, where t belongs to some given index set T. For convenience, we refer to t as representing time, although in general it need not.

In the experiment described in this section, the index set is all real numbers between zero and one. Other common index sets are the entire real axis and the set of all integers. If t takes on a continuous range of real values, the stochastic process is called a *continuous parameter* (or *continuous time*) stochastic process; if t takes on a discrete set of real values, it is called a *discrete parameter* (or *discrete time*) stochastic process.

2.2 Notation

Since it is important to be clear about the type of stochastic process we are discussing, we shall use the following notational conventions throughout this book:

[1] $\{X_t\}$ refers to a real-valued *discrete parameter* stochastic process whose tth component is X_t, while

[2] $\{X(t)\}$ refers to a real-valued *continuous parameter* stochastic process whose component at time t is $X(t)$.

[3] When the index set for a stochastic process is not explicitly stated (as is the case for both $\{X_t\}$ and $\{X(t)\}$), we shall assume that it is the set of all integers $\mathbb{Z} = \{\ldots, -2, -1, 0, 1, 2, \ldots\}$ for a discrete parameter process and the entire real axis $\mathbb{R} = \{t : -\infty < t < \infty\}$ for a continuous parameter process. Note that "t" is being used in two different ways here: the t in X_t is a unitless index (referring to the tth element of the process $\{X_t\}$), whereas the t in $X(t)$ has physically meaningful units such as seconds or days (hence $X(t)$ is the element occurring at time t of the process $\{X(t)\}$).

[4] On occasion we will need to discuss more than one stochastic process at a time. To distinguish among them, we will either introduce another symbol besides X (such as in $\{Y_t\}$) or add another index before the time index. For example, $\{X_{j,t}\}$ and $\{X_{k,t}\}$ refer to the jth and kth discrete parameter processes, while $\{X_j(t)\}$ and $\{X_k(t)\}$ refer to two continuous parameter processes. (Another way to handle multiple processes is to define a vector whose elements are stochastic processes. This approach leads to what are known as vector-valued stochastic processes or multivariate stochastic processes, which we do not deal with in this book.)

[5] We reserve the symbol Z for a complex-valued RV whose real and imaginary components are real-valued RVs. With an index added, $\{Z_t\}$ is a complex-valued discrete parameter stochastic process with a tth component formed from, say, the real-valued RVs $X_{0,t}$ and $X_{1,t}$; i.e., $Z_t = X_{0,t} + \mathrm{i} X_{1,t}$, where $\mathrm{i} \stackrel{\mathrm{def}}{=} \sqrt{-1}$ (i.e., i is "equal by definition" to $\sqrt{-1}$). Likewise, $\{Z(t)\}$ is a complex-valued continuous parameter stochastic process with a tth component formed from two real-valued RVs, say, $X_0(t)$ and $X_1(t)$; i.e., $Z(t) = X_0(t) + \mathrm{i} X_1(t)$.

2.3 Basic Theory for Stochastic Processes

Let us first consider the real-valued discrete parameter stochastic process $\{X_t\}$. Since, for t fixed, X_t is an RV, it has an associated cumulative probability distribution function (CPDF) given by
$$F_t(a) = \mathbf{P}[X_t \leq a],$$
where the notation $\mathbf{P}[A]$ indicates the probability that the event A will occur. Because $\{X_t\}$ is a stochastic process, our primary interest is in the relationships amongst the various RVs that are part of it. These are expressed by various higher-order CPDFs. For example, for any t_0 and t_1 in the index set T,
$$F_{t_0,t_1}(a_0, a_1) = \mathbf{P}[X_{t_0} \leq a_0, X_{t_1} \leq a_1]$$
gives the bivariate CPDF for X_{t_0} and X_{t_1}, where the notation $\mathbf{P}[A_0, A_1]$ refers to the probability of the intersection of the events A_0 and A_1. More generally, for any integer $N \geq 1$ and any $t_0, t_1, \ldots, t_{N-1}$ in the index set, we can define the N-dimensional CPDF by
$$F_{t_0,t_1,\ldots,t_{N-1}}(a_0, a_1, \ldots, a_{N-1}) = \mathbf{P}[X_{t_0} \leq a_0, X_{t_1} \leq a_1, \ldots, X_{t_{N-1}} \leq a_{N-1}].$$
These higher-order CPDFs completely specify the joint statistical properties of the RVs in the stochastic process.

The notions of univariate and higher dimensional CPDFs extend readily to a real-valued continuous parameter stochastic process $\{X(t)\}$ and to complex-valued discrete and continuous parameter stochastic processes. For example, for a complex-valued continuous parameter stochastic process $\{Z(t)\}$ with
$$Z(t) = X_0(t) + \mathrm{i} X_1(t),$$
we can define a univariate CPDF for $Z(t)$ in terms of
$$F_t(a, b) = \mathbf{P}[X_0(t) \leq a, X_1(t) \leq b]$$
and a bivariate CPDF for $Z(t_0)$ and $Z(t_1)$ using
$$F_{t_0,t_1}(a_0, b_0, a_1, b_1) = \mathbf{P}[X_0(t_0) \leq a_0, X_1(t_0) \leq b_0, X_0(t_1) \leq a_1, X_1(t_1) \leq b_1].$$

In practice we are often interested in summaries of CPDFs afforded by certain of their moments. If we assume that they exist, the first moment (i.e., expected value or mean) and the second central moment (i.e., variance) associated with the RV X_t having CPDF $F_t(\cdot)$ are given, respectively, by

$$\mu_t \stackrel{\text{def}}{=} E\{X_t\} = \int_{-\infty}^{\infty} x \, dF_t(x) \tag{24a}$$

and

$$\sigma_t^2 \stackrel{\text{def}}{=} \text{var}\{X_t\} = \int_{-\infty}^{\infty} (x - \mu_t)^2 \, dF_t(x),$$

where, in general, μ_t and σ_t^2 depend upon the index t. The two integrals above are examples of Riemann–Stieltjes integrals (these are briefly reviewed in the Comments and Extensions [C&Es] later in this section). If $F_t(\cdot)$ is differentiable with derivative $f_t(\cdot)$, which is called a probability density function (PDF), then the integrals reduce to the following Riemann integrals:

$$\int_{-\infty}^{\infty} x \, dF_t(x) = \int_{-\infty}^{\infty} x f_t(x) \, dx \quad \text{and} \quad \int_{-\infty}^{\infty} (x - \mu_t)^2 \, dF_t(x) = \int_{-\infty}^{\infty} (x - \mu_t)^2 f_t(x) \, dx.$$

If $F_t(\cdot)$ is a step function with steps of size p_k at locations x_k, i.e., $\mathbf{P}[X_t = x_k] = p_k$, then we have

$$\int_{-\infty}^{\infty} x \, dF_t(x) = \sum_k x_k p_k \quad \text{and} \quad \int_{-\infty}^{\infty} (x - \mu_t)^2 \, dF_t(x) = \sum_k (x_k - \mu_t)^2 p_k,$$

where $\sum_k p_k = 1$.

While μ_t and σ_t^2 summarize two aspects of univariate CPDFs, the covariance between the RVs X_{t_0} and X_{t_1} summarizes one aspect of their bivariate CPDF $F_{t_0, t_1}(\cdot, \cdot)$:

$$\text{cov}\{X_{t_0}, X_{t_1}\} \stackrel{\text{def}}{=} E\{(X_{t_0} - \mu_{t_0})(X_{t_1} - \mu_{t_1})\} \tag{24b}$$
$$= \int_{-\infty}^{\infty} \int_{-\infty}^{\infty} (x_0 - \mu_{t_0})(x_1 - \mu_{t_1}) \, d^2 F_{t_0, t_1}(x_0, x_1).$$

Note that $\text{cov}\{X_{t_0}, X_{t_0}\} = \text{var}\{X_{t_0}\}$. If $F_{t_0, t_1}(\cdot, \cdot)$ is differentiable and hence is equivalent to the bivariate PDF $f_{t_0, t_1}(\cdot, \cdot)$ in that

$$F_{t_0, t_1}(a_0, a_1) = \int_{-\infty}^{a_1} \int_{-\infty}^{a_0} f_{t_0, t_1}(x_0, x_1) \, dx_0 \, dx_1,$$

then we can write

$$\text{cov}\{X_{t_0}, X_{t_1}\} = \int_{-\infty}^{\infty} \int_{-\infty}^{\infty} (x_0 - \mu_{t_0})(x_1 - \mu_{t_1}) f_{t_0, t_1}(x_0, x_1) \, dx_0 \, dx_1.$$

On the other hand, suppose that X_{t_0} can only assume the values $x_{t_0, j}$ over some range of the index $j \in \mathbb{Z}$ and that X_{t_1} assumes only $x_{t_1, k}$ over $k \in \mathbb{Z}$. We then have $\mathbf{P}[X_{t_0} = x_{t_0, j}, X_{t_1} = x_{t_1, k}] = p_{j, k}$ with $\sum_j \sum_k p_{j, k} = 1$, and we can write

$$\text{cov}\{X_{t_0}, X_{t_1}\} = \sum_j \sum_k (x_{0, j} - \mu_{t_0})(x_{1, k} - \mu_{t_1}) p_{j, k},$$

where

$$\mu_{t_0} = \sum_j x_{t_0,j} \mathbf{P}[X_{t_0} = x_{t_0,j}] = \sum_j x_{t_0,j} \sum_k p_{j,k},$$

with a similar expression for μ_{t_1}.

The above ideas extend in an obvious manner to a real-valued continuous parameter stochastic process $\{X(t)\}$, but there is one key modification required for complex-valued stochastic processes. To set the stage, suppose that $\{Z_t\}$ is a complex-valued discrete parameter stochastic process such that $Z_t = X_{0,t} + \mathrm{i}X_{1,t}$, where $\{X_{0,t}\}$ and $\{X_{1,t}\}$ are real-valued discrete parameter stochastic processes with $E\{X_{0,t}\} = \mu_{0,t}$ and $E\{X_{1,t}\} = \mu_{1,t}$. The obvious definition for the mean value of Z_t is

$$\mu_t \stackrel{\text{def}}{=} E\{Z_t\} = E\{X_{0,t}\} + \mathrm{i}E\{X_{1,t}\} = \mu_{0,t} + \mathrm{i}\mu_{1,t},$$

where now μ_t is in general complex-valued. The key modification is in the definition of the covariance between the complex-valued RVs Z_{t_0} and Z_{t_1}:

$$\operatorname{cov}\{Z_{t_0}, Z_{t_1}\} \stackrel{\text{def}}{=} E\{(Z_{t_0} - \mu_{t_0})(Z_{t_1} - \mu_{t_1})^*\} = E\{(Z_{t_0} - \mu_{t_0})(Z_{t_1}^* - \mu_{t_1}^*)\}, \quad (25\mathrm{a})$$

where the asterisk indicates the operation of complex conjugation. In particular,

$$\operatorname{var}\{Z_t\} = \operatorname{cov}\{Z_t, Z_t\} = E\{(Z_t - \mu_t)(Z_t - \mu_t)^*\} = E\{|Z_t - \mu_t|^2\},$$

where we make use of the fact that $|z|^2 = zz^*$ for any complex-valued variable z. Note that, if we set $X_{1,t} = 0$ for all t so that $\{Z_t\}$ becomes a real-valued process, the covariance defined in Equation (25a) is consistent with the definition of covariance in Equation (24b) because $(Z_{t_1} - \mu_{t_1})^* = Z_{t_1} - \mu_{t_1}$ in this special case.

Exercise [2.1] invites the reader to prove several key properties about covariances.

Comments and Extensions to Section 2.3

[1] We state here one definition for the Riemann–Stieltjes integral and a few facts concerning it (we will need a stochastic version of this integral when we introduce the spectral representation theorem for stationary processes in Section 4.1); for more details, see section 18.9 of Taylor and Mann (1972) or section 4.2 of Greene and Knuth (1990), from which the following material is adapted. Let $g(\cdot)$ and $H(\cdot)$ be two real-valued functions defined over the interval $[L, U]$ with $L < U$, and let P_N be a partition of this interval of size $N + 1$; i.e., P_N is a set of $N + 1$ points x_j such that

$$L = x_0 < x_1 < \cdots < x_{N-1} < x_N = U.$$

Define the "mesh fineness" of the partition P_N as

$$|P_N| \stackrel{\text{def}}{=} \max\{x_1 - x_0, x_2 - x_1, \ldots, x_{N-1} - x_{N-2}, x_N - x_{N-1}\}.$$

Let x'_j be any point in the interval $[x_{j-1}, x_j]$, and consider the summation

$$\mathcal{S}(P_N) \stackrel{\text{def}}{=} \sum_{j=1}^N g(x'_j)\left[H(x_j) - H(x_{j-1})\right].$$

The Riemann–Stieltjes integral is defined as a limit involving this summation:

$$\int_L^U g(x)\,\mathrm{d}H(x) \stackrel{\text{def}}{=} \lim_{|P_N| \to 0} \mathcal{S}(P_N), \quad (25\mathrm{b})$$

provided that $\mathcal{S}(P_N)$ converges to a unique limit as the mesh fineness decreases to zero. Extension of the above to allow $L = -\infty$ and $U = \infty$ is done by the same limiting argument as is used in the case of a Riemann integral.

We now note the following facts (all of which assume that $g(\cdot)$ and $H(\cdot)$ are such that the Riemann–Stieltjes integral over $[L, U]$ exists).

[1] If $H(x) = x$, the Riemann–Stieltjes integral reduces to the ordinary Riemann integral $\int_L^U g(x)\,dx$.
[2] If $H(\cdot)$ is differentiable everywhere over the interval $[L, U]$ with derivative $h(\cdot)$, the Riemann–Stieltjes integral reduces to the Riemann integral $\int_L^U g(x)h(x)\,dx$.
[3] Suppose that b is such that $L < b < U$ and that

$$H(x) = \begin{cases} c, & L \leq x < b; \\ a + c, & b \leq x \leq U; \end{cases}$$

i.e., $H(\cdot)$ is a step function with a single step of size a at $x = b$. Then

$$\int_L^U g(x)\,dH(x) = ag(b).$$

In general, if $H(\cdot)$ is a step function with steps of sizes $a_0, a_1, \ldots, a_{N-1}$ at points $b_0, b_1, \ldots, b_{N-1}$ (all of which are distinct and satisfy $L < b_k < U$), then

$$\int_L^U g(x)\,dH(x) = \sum_{k=0}^{N-1} a_k g(b_k).$$

The last fact shows that many ordinary summations can be expressed as Riemann–Stieltjes integrals. This gives us a certain compactness in notation. For example, as noted previously, Equation (24a) handles the special cases in which $F_t(\cdot)$ is either differentiable or is a step function. This equation also handles "mixed" CPDFs, a combination of these two special cases often discussed in elementary textbooks on statistics. The Riemann–Stieltjes integral thus gives us some – but by no means all – of the advantages of the Lebesgue integral commonly used in advanced texts on probability and statistics.

2.4 Real-Valued Stationary Processes

The class of all stochastic processes is "too large" to work with in practice. In spectral analysis, we consider only a special subclass called stationary processes. Basically, stationarity requires certain properties of a stochastic process be time-invariant.

There are two common types of stationarity. The first type is *complete* stationarity (sometimes referred to as *strong* stationarity or *strict* stationarity): the process $\{X_t\}$ is said to be completely stationary if, for all $N \geq 1$, for any $t_0, t_1, \ldots, t_{N-1}$ contained in the index set, and for any τ such that $t_0 + \tau, t_1 + \tau, \ldots, t_{N-1} + \tau$ are also contained in the index set, the joint CPDF of $X_{t_0}, X_{t_1}, \ldots, X_{t_{N-1}}$ is the same as that of $X_{t_0+\tau}, X_{t_1+\tau}, \ldots, X_{t_{N-1}+\tau}$; i.e.,

$$F_{t_0,t_1,\ldots,t_{N-1}}(a_0, a_1, \ldots, a_{N-1}) = F_{t_0+\tau,t_1+\tau,\ldots,t_{N-1}+\tau}(a_0, a_1, \ldots, a_{N-1}).$$

In other words, the probabilistic structure of a completely stationary process is invariant under a shift in time.

Unfortunately, completely stationary processes are too difficult to work with as models for most time series of interest since they have to be specified by using N-dimensional CPDFs. A simplifying assumption leads to the second common type of stationarity: the process $\{X_t\}$ is said to be *second-order* stationary (sometimes called *weakly* stationary or *covariance* stationary) if, for all $N \geq 1$, for any $t_0, t_1, \ldots, t_{N-1}$ contained in the index set, and for any τ

such that $t_0+\tau, t_1+\tau, \ldots, t_{N-1}+\tau$ are also contained in the index set, all the first moments, second moments and second-order joint moments of $X_{t_0}, X_{t_1}, \ldots, X_{t_{N-1}}$ exist, are finite and are equal to the corresponding moments of $X_{t_0+\tau}, X_{t_1+\tau}, \ldots, X_{t_{N-1}+\tau}$ (second-order moments take the form $E\{X_{t_j} X_{t_k}\}$; more generally, $E\{X_{t_j}^l X_{t_k}^m\}$ is called a joint moment of order $l+m$). Immediate consequences of this definition are that

$$\mu \stackrel{\text{def}}{=} E\{X_t\} \text{ and } \mu_2' \stackrel{\text{def}}{=} E\{X_t^2\}$$

are both constants independent of t. This implies that

$$\sigma^2 \stackrel{\text{def}}{=} \text{var}\{X_t\} = \mu_2' - \mu^2$$

is also a constant independent of t. If we allow the shift $\tau = -t_1$, we see that

$$E\{X_{t_0} X_{t_1}\} = E\{X_{t_0-t_1} X_0\}$$

is a function of the difference $t_0 - t_1$ only. Actually, it is a function of the absolute difference $|t_0 - t_1|$ only since, if we now let $\tau = -t_0$, we have

$$E\{X_{t_0} X_{t_1}\} = E\{X_{t_1} X_{t_0}\} = E\{X_{t_1-t_0} X_0\}.$$

The above implies that the covariance between X_{t_0} and X_{t_1} is also a function of the absolute difference $|t_0 - t_1|$, since

$$\text{cov}\{X_{t_0}, X_{t_1}\} = E\{(X_{t_0} - \mu)(X_{t_1} - \mu)\} = E\{X_{t_0} X_{t_1}\} - \mu^2.$$

For a discrete parameter second-order stationary process $\{X_t\}$, we define the *autocovariance sequence* (ACVS) by

$$s_\tau \stackrel{\text{def}}{=} \text{cov}\{X_{t+\tau}, X_t\} = \text{cov}\{X_\tau, X_0\}. \tag{27}$$

Likewise, for a continuous parameter second-order stationary process $\{X(t)\}$, we define the *autocovariance function* (ACVF) by

$$s(\tau) \stackrel{\text{def}}{=} \text{cov}\{X(t+\tau), X(t)\} = \text{cov}\{X(\tau), X(0)\}.$$

Both s_τ and $s(\tau)$ measure the covariance between members of a process that are separated by τ units. The variable τ is called the *lag*.

Here are some further properties of the ACVS and ACVF.

[1] For a discrete parameter process with an index set given by $\mathbb{Z}$, the lag τ can assume any integer value; for a continuous parameter stationary process with an index set given by $\mathbb{R}$, the lag can assume any real value.
[2] Note that, in the discrete parameter case, $s_0 = \sigma^2$ and $s_{-\tau} = s_\tau$; likewise, in the continuous parameter case, $s(0) = \sigma^2$ and $s(-\tau) = s(\tau)$. Thus $\{s_\tau\}$ is an even sequence, and $s(\cdot)$ is an even function.
[3] Assuming the trivial condition $s_0 > 0$, we can define $\rho_\tau = s_\tau/s_0$ as the *autocorrelation sequence* (ACS) for $\{X_t\}$; likewise, we can define $\rho(\tau) = s(\tau)/s(0)$ as the *autocorrelation function* (ACF) for $\{X(t)\}$. Since we have

$$\rho_\tau = \text{corr}\{X_{t+\tau}, X_t\} \stackrel{\text{def}}{=} \frac{\text{cov}\{X_{t+\tau}, X_t\}}{(\text{var}\{X_{t+\tau}\} \text{var}\{X_t\})^{1/2}} = \frac{\text{cov}\{X_{t+\tau}, X_t\}}{\text{var}\{X_t\}},$$

ρ_τ is the correlation coefficient between pairs of RVs from the process $\{X_t\}$ that are τ units apart. There is an analogous interpretation for $\rho(\tau)$. (As already mentioned in Chapter 1, the definitions we have given for the ACVS, ACS, ACVF and ACF are standard in the statistical literature, but other definitions for these terms are sometimes used in the engineering literature.)

[4] Since ρ_τ and $\rho(\tau)$ are correlation coefficients and hence constrained to lie between -1 and 1, it follows that

$$|s_\tau| \leq s_0 \text{ and } |s(\tau)| \leq s(0) \quad \text{for all } \tau. \tag{28a}$$

[5] A necessary and sufficient condition that an even sequence $\{s_\tau\}$ be the ACVS for some stationary process is that it be *positive semidefinite*; i.e., for all $N \geq 1$, for any $t_0, t_1, \ldots, t_{N-1}$ contained in the index set, and for any set of nonzero real numbers $a_0, a_1, \ldots, a_{N-1}$,

$$\sum_{j=0}^{N-1} \sum_{k=0}^{N-1} s_{t_j - t_k} a_j a_k \geq 0 \tag{28b}$$

(if this double summation is strictly greater than 0, then $\{s_\tau\}$ is said to be *positive definite*). To see that this condition is necessary, consider the following two column vectors of length N:

$$\boldsymbol{a} \stackrel{\text{def}}{=} [a_0, a_1, \ldots, a_{N-1}]^T \text{ and } \boldsymbol{V} \stackrel{\text{def}}{=} [X_{t_0}, X_{t_1}, \ldots, X_{t_{N-1}}]^T,$$

where the superscript T denotes the operation of vector transposition. Let $\boldsymbol{\Sigma}$ be the covariance matrix for the vector $\boldsymbol{V}$. Its (j, k)th element is given by

$$E\{(X_{t_j} - \mu)(X_{t_k} - \mu)\} = s_{t_j - t_k}.$$

Define the RV

$$W = \sum_{j=0}^{N-1} a_j X_{t_j} = \boldsymbol{a}^T \boldsymbol{V}.$$

Then we have

$$0 \leq \text{var}\{W\} = \text{var}\{\boldsymbol{a}^T \boldsymbol{V}\} = \boldsymbol{a}^T \boldsymbol{\Sigma} \boldsymbol{a} = \sum_{j=0}^{N-1} \sum_{k=0}^{N-1} s_{t_j - t_k} a_j a_k$$

(see Exercise [2.2]; the sufficiency of the condition is shown in theorem 1.5.1, Brockwell and Davis, 1991). The important point here is that an ACVS cannot just be any arbitrary sequence, a fact that will become important when we discuss estimators for $\{s_\tau\}$. The same comments hold – with obvious modifications – for the function $s(\cdot)$. (For a fascinating discussion on how severely the requirement of positive semidefiniteness constrains a sequence of numbers, see Makhoul, 1990.)

[6] Let us now consider the covariance matrix $\boldsymbol{\Sigma}_N$ for a vector containing N contiguous RVs from the stationary process $\{X_t\}$, say, $[X_0, X_1, \ldots, X_{N-1}]^T$. Since the (j, k)th element of $\boldsymbol{\Sigma}_N$ is given by $\text{cov}\{X_j, X_k\} = s_{j-k}$, the elements of this matrix depend upon just the *difference* between the row and column indices. As an example, let us consider the case $N = 5$:

$$\Sigma_5 = \begin{bmatrix} s_0 & s_{-1} & s_{-2} & s_{-3} & s_{-4} \\ \underline{s_1} & s_0 & s_{-1} & s_{-2} & s_{-3} \\ s_2 & \underline{s_1} & s_0 & s_{-1} & s_{-2} \\ s_3 & s_2 & \underline{s_1} & s_0 & s_{-1} \\ s_4 & s_3 & s_2 & \underline{s_1} & s_0 \end{bmatrix} = \begin{bmatrix} s_0 & s_1 & s_2 & s_3 & s_4 \\ \underline{s_1} & s_0 & s_1 & s_2 & s_3 \\ s_2 & \underline{s_1} & s_0 & s_1 & s_2 \\ s_3 & s_2 & \underline{s_1} & s_0 & s_1 \\ s_4 & s_3 & s_2 & \underline{s_1} & s_0 \end{bmatrix}. \qquad (29a)$$

In the above we have underlined the entries such that $j - k = 1$. A two-dimensional matrix with a constant value along each of its diagonals is known as a *Toeplitz matrix*. Since $s_{j-k} = s_{k-j}$ for a discrete parameter real-valued stationary process, the covariance matrix for any N contiguous RVs from this process is a symmetric Toeplitz matrix of dimension $N \times N$.

It follows from the definitions of complete and second-order stationarity that, if $\{X_t\}$ is a completely stationary process with finite variance, then it is also second-order stationary. In general, second-order stationarity does not imply complete stationarity. An important exception is a *Gaussian process* (also called a *normal process*), defined as follows: the stochastic process $\{X_t\}$ is said to be Gaussian if, for all $N \geq 1$ and for any $t_0, t_1, \ldots, t_{N-1}$ contained in the index set, the joint CPDF of $X_{t_0}, X_{t_1}, \ldots, X_{t_{N-1}}$ is multivariate Gaussian. A second-order stationary Gaussian process is also completely stationary due to the fact that the multivariate Gaussian distribution is completely characterized by its moments of first and second order.

Hereafter the unadorned term "stationarity" will mean "second-order stationarity."

2.5 Complex-Valued Stationary Processes

We say that a discrete parameter complex-valued process $\{Z_t\}$, defined via

$$Z_t = X_{0,t} + iX_{1,t}, \qquad (29b)$$

is (second-order) stationary if its real and imaginary parts are *jointly stationary*, by which we mean that $\{X_{0,t}\}$ and $\{X_{1,t}\}$ by themselves are second-order stationary and, in addition, $\text{cov}\{X_{0,t+\tau}, X_{1,t}\}$ is a function of τ only for all t. The stationarity of the component processes $\{X_{0,t}\}$ and $\{X_{1,t}\}$ tells us that

$$\mu \stackrel{\text{def}}{=} E\{Z_t\} = \mu_0 + i\mu_1 \stackrel{\text{def}}{=} E\{X_{0,t}\} + iE\{X_{1,t}\},$$

where μ is a complex-valued constant, and μ_0 and μ_1 are real-valued constants, all three of which are independent of t.

▷ **Exercise [29]** Show that

$$\text{cov}\{Z_{t+\tau}, Z_t\} \stackrel{\text{def}}{=} E\{(Z_{t+\tau} - \mu)(Z_t - \mu)^*\} \qquad (29c)$$

is independent of t and hence defines the τth element s_τ of the ACVS $\{s_\tau\}$. Show also that the ACVS has the property $s_{-\tau} = s_\tau^*$. Finally, show that

$$r_\tau \stackrel{\text{def}}{=} E\{(Z_{t+\tau} - \mu)(Z_t - \mu)\} \qquad (29d)$$

is also independent of t, where we refer to $\{r_\tau\}$ as the *autorelation sequence*. ◁

Note that the above also holds for a real-valued stationary process – the complex conjugate of a real number is just the real number itself. For $\tau = 0$, we have

$$s_0 = E\{|Z_t - \mu|^2\} = \text{var}\{Z_t\}.$$

The ACS is defined in the same way as before, so we must have $\rho_{-\tau} = \rho_\tau^*$.

The positive semidefinite property of $\{s_\tau\}$ is now defined slightly differently: $\{s_\tau\}$ is said to be such if, for all $N \geq 1$, for any $t_0, t_1, \ldots, t_{N-1}$ contained in the index set, and for any set of *complex* numbers $c_0, c_1, \ldots, c_{N-1}$,

$$\text{var}\left\{\sum_{j=0}^{N-1} c_j Z_{t_j}\right\} = \sum_{j=0}^{N-1} \sum_{k=0}^{N-1} s_{t_j - t_k} c_j c_k^* \geq 0.$$

While the covariance matrix for a subsequence of a discrete parameter complex-valued stationary process is a Toeplitz matrix, it is not necessarily a symmetric Toeplitz matrix (as in the real-valued case) since $s_{-\tau} \neq s_\tau$ in general; however, because of the condition $s_{-\tau} = s_\tau^*$, the covariance matrix falls in the class of *Hermitian Toeplitz matrices*.

A continuous parameter complex-valued stationary process $\{Z(t)\}$ is defined in a similar way. In particular, we note that its ACVF satisfies $s(-\tau) = s^*(\tau)$ and is positive semidefinite.

We note the following definitions and results for later use.

[1] A complex-valued RV Z_t is said to have a *complex Gaussian* (or *normal*) distribution if its (real-valued) real and imaginary components $X_{0,t}$ and $X_{1,t}$ are bivariate Gaussian; likewise, a collection of N complex-valued RVs $Z_0, Z_1, \ldots, Z_{N-1}$ is said to follow a complex Gaussian distribution if all of its real and imaginary components are multivariate Gaussian (of dimension $2N$).

[2] Complex-valued processes for which the autorelation sequence $\{r_\tau\}$ is everywhere zero are termed *proper* (Schreier and Scharf, 2003, 2010). Note that Miller (1974a) gives a comprehensive discussion of complex-valued stochastic processes, but concentrates on proper processes (Miller, 1974a, p. 42). As a result, he defines $\{Z_t\}$ to be stationary if $\text{cov}\{Z_{t+\tau}, Z_t\}$ is independent of t for any given τ; this is a weaker condition than our definition, which implies that $\text{cov}\{Z_{t+\tau}, Z_t^*\}$ is independent of t for any given τ also, as shown in Exercise [29].

[3] The complex-valued process $\{Z_t\}$ is said to be complex Gaussian if, for all $N \geq 1$ and for any $t_0, t_1, \ldots, t_{N-1}$ contained in the index set, the joint CPDF of the real and imaginary components $X_{0,t_0}, X_{1,t_0}, X_{0,t_1}, X_{1,t_1}, \ldots, X_{0,t_{N-1}}$ and $X_{1,t_{N-1}}$ is multivariate Gaussian.

[4] If Z_0, Z_1, Z_2 and Z_3 are any four complex-valued Gaussian RVs with zero means, then the *Isserlis theorem* (Isserlis, 1918) states that

$$\text{cov}\{Z_0 Z_1, Z_2 Z_3\} = \text{cov}\{Z_0, Z_2\} \text{cov}\{Z_1, Z_3\} + \text{cov}\{Z_0, Z_3\} \text{cov}\{Z_1, Z_2\} \quad (30)$$

(in fact, the above also holds if we replace Z_0, Z_1, Z_2 and Z_3 with any four real-valued Gaussian RVs with zero means).

Although the main focus of this book is on the application of real-valued stationary processes to time series data, complex-valued processes arise indirectly in several contexts, including the spectral representation theorem (Section 4.1) and the statistical properties of spectral estimators (Section 6.6). Complex-valued stationary processes are used directly to model certain bivariate medical (e.g., Loupas and McDicken, 1990; Rowe, 2005), oceanographic (Chandna and Walden, 2011; Sykulski et al., 2016) and meteorological (Hayashi, 1979; Maitani, 1983) time series.

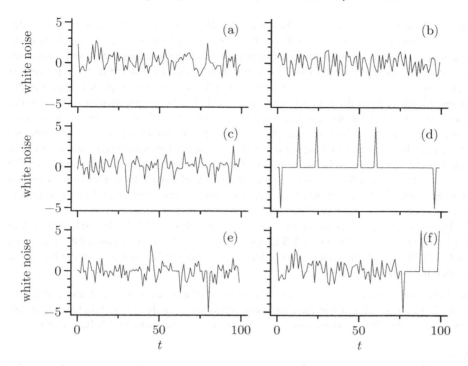

Figure 31 Realizations of length 100 from six white noise processes, all with zero mean and unit variance. The distribution for the process in plot (a) is Gaussian; (b) is uniform over the interval $[-\sqrt{3}, \sqrt{3}]$; (c) is a double exponential with PDF given by $f(x) = \exp(-|x|\sqrt{2})/\sqrt{2}$; and (d) is from a discrete distribution that assumes the values -5, 0 and 5 with probabilities 0.02, 0.96 and 0.02, respectively. The realization in (e) is, at each time index t, a random selection from one of the previous four distributions. Realization (f) was created by pasting together the first 25 values from (a), (b), (c) and (d).

Comments and Extensions to Section 2.5

[1] Picinbono and Bondon (1997) use the term *relation function* to refer to $\{r_\tau\}$ of Equation (29d), but we have added "auto" to emphasize that both RVs are from the same process, and we use "sequence" instead of "function" since we are dealing with a discrete parameter process. This quantity is also called the *complementary covariance* (Schreier and Scharf, 2003); however, note that $\{r_\tau\}$ does not have all the properties of a covariance sequence (for example, while a covariance sequence $\{s_\tau\}$ satisfies $s_{-\tau} = s_\tau^*$, this does not hold for $\{r_\tau\}$ unless it is real-valued).

2.6 Examples of Discrete Parameter Stationary Processes

White Noise Process

Let $\{X_t\}$ be a sequence of uncorrelated RVs such that $E\{X_t\} = \mu$ and $\mathrm{var}\{X_t\} = \sigma^2 > 0$ for all t, where σ^2 is a finite constant. Since uncorrelatedness means that $\mathrm{cov}\{X_{t+\tau}, X_t\} = 0$ for all t and $\tau \neq 0$, it follows that $\{X_t\}$ is stationary with ACVS

$$s_\tau = \begin{cases} \sigma^2, & \tau = 0; \\ 0, & \tau \neq 0, \end{cases} \text{ which implies } \rho_\tau = \begin{cases} 1, & \tau = 0; \\ 0, & \tau \neq 0. \end{cases} \quad (31)$$

Despite their simplicity, white noise processes play a central role since they can be manipulated to create many other stationary processes (see the examples that follow).

Six examples of realizations from white noise processes are shown in Figure 31, each with $\mu = 0$ and $\sigma^2 = 1$. The first three examples are realizations of independent and identically distributed (IID) RVs with, respectively, a Gaussian, uniform and double exponential distribution. The fourth example is also based on IID RVs, but with realizations that can only assume three distinct values (-5, 0 and 5). The fifth example was formed by selecting, at each time index t, one of the four previous distributions at random and then generating a realization from the selected distribution. While the distribution of this white noise process is complicated, it is the same for all t. The sixth example was created by pasting together the first 25 points from each of the first four examples. This final example demonstrates that the distribution of a white noise process need not be constant over time; i.e., the only requirements for a process to be white noise are that its RVs are uncorrelated and that the mean and variance for each RV are finite and the same for all t (a process that satisfies the more stringent requirement that its distribution is the same for all t is sometimes called an IID process or IID noise). Although the ACVSs for these six processes are identical, their realizations do not look the same – the characterization of a process by just its first- and second-order moments can gloss over potentially important features.

Finally we note that a sequence of complex-valued RVs $\{Z_t\}$ is deemed to be white noise if $\{Z_t\}$ is stationary with an ACVS that is the same as for real-valued white noise (see Equation (31)). The following exercise considers a special case of complex-valued white noise that has been used widely.

▷ **Exercise [32]** Show that, for *proper* complex-valued white noise, the real and imaginary parts of Z_t are uncorrelated and have the same variance, namely, $\sigma^2/2$ (this latter quantity is sometimes called the *semivariance* – see, for example, Lang and McClellan, 1980). ◁

Moving Average Process
The process $\{X_t\}$ is called a qth-order moving average process – denoted by MA(q) – if it can be expressed in the form

$$X_t = \mu + \epsilon_t - \theta_{q,1}\epsilon_{t-1} - \cdots - \theta_{q,q}\epsilon_{t-q} = \mu - \sum_{j=0}^{q}\theta_{q,j}\epsilon_{t-j}, \quad t \in \mathbb{Z}, \qquad (32a)$$

where μ and $\theta_{j,q}$ are constants ($\theta_{q,0} \stackrel{\text{def}}{=} -1$ and $\theta_{q,q} \neq 0$), and $\{\epsilon_t\}$ is a white noise process with zero mean and variance σ_ϵ^2. In other words, a moving average process at time t is the sum of a constant μ, the tth component of a white noise process and a linear combination of the same white noise process at q previous times. Note first that $E\{X_t\} = \mu$, a constant independent of t – we assume it to be zero in what follows. Second, since $E\{\epsilon_{t+\tau}\epsilon_t\} = 0$ for all $\tau \neq 0$, it follows from Equation (32a) and Exercise [2.1e] that, for $\tau \geq 0$,

$$\text{cov}\{X_{t+\tau}, X_t\} = \sum_{k=0}^{q}\sum_{j=0}^{q}\theta_{q,k}\theta_{q,j}\,\text{cov}\{\epsilon_{t+\tau-k}, \epsilon_{t-j}\} = \sigma_\epsilon^2\sum_{j=0}^{q-\tau}\theta_{q,j+\tau}\theta_{q,j}$$

depends only on the lag τ. Here we interpret the last summation to be zero when $q - \tau < 0$. Since it is easy to show that $\text{cov}\{X_{t-\tau}, X_t\} = \text{cov}\{X_{t+\tau}, X_t\}$, it follows that $\{X_t\}$ is a stationary process with ACVS given by

$$s_\tau = \begin{cases} \sigma_\epsilon^2 \sum_{j=0}^{q-|\tau|}\theta_{q,j+|\tau|}\theta_{q,j}, & |\tau| \leq q; \\ 0, & |\tau| > q. \end{cases} \qquad (32b)$$

Note that we did not need to place any restrictions on the $\theta_{q,j}$ terms to ensure stationarity. Note also that the variance of $\{X_t\}$ is given by

$$s_0 = \sigma_\epsilon^2 \sum_{j=0}^{q}\theta_{q,j}^2.$$

2.6 Examples of Discrete Parameter Stationary Processes

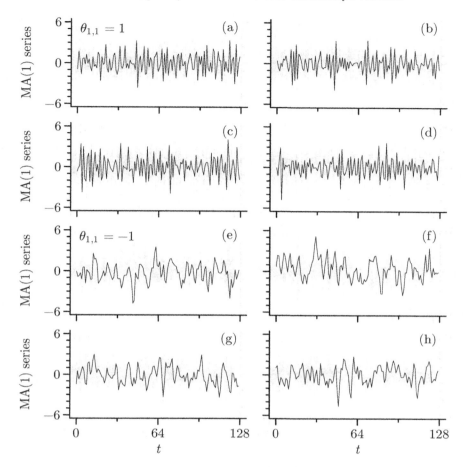

Figure 33 Realizations of length 128 of two first-order Gaussian moving average processes. The top two rows are for $\theta_{1,1} = 1$, and the bottom two, for $\theta_{1,1} = -1$.

Figure 33 shows four realizations from two different Gaussian MA(1) processes of the form $X_t = \epsilon_t - \theta_{1,1}\epsilon_{t-1}$. For the first process (top two rows), $\theta_{1,1} = 1$; for the second (bottom two), $\theta_{1,1} = -1$. Since $\rho_1 = -\theta_{1,1}/(1 + \theta_{1,1}^2)$ here, we see that adjacent values of the first process are negatively correlated ($\rho_1 = -1/2$), while adjacent values of the second process are positively correlated ($\rho_1 = 1/2$). The realizations in Figure 33 agree with this description (see Section 11.1 for a discussion on generating realizations from a moving average process).

An interesting example of a physical process that can be modeled as a moving average process is the thickness of textile slivers (an intermediate stage in converting flax fibers into yarn) as a function of displacement along a sliver (Spencer-Smith and Todd, 1941). Note that the "time" variable here is a displacement (distance) rather than physical time.

Autoregressive Process

The process $\{X_t\}$ with zero mean is called a pth-order autoregressive process – denoted by AR(p) – if it satisfies an equation such as

$$X_t = \phi_{p,1} X_{t-1} + \cdots + \phi_{p,p} X_{t-p} + \epsilon_t, \quad t \in \mathbb{Z}, \tag{33}$$

where $\phi_{p,1}, \ldots, \phi_{p,p}$ are constants (with $\phi_{p,p} \neq 0$) and $\{\epsilon_t\}$ is a white noise process with zero mean and variance σ_p^2. In other words, an AR process at time t is the sum of a white noise

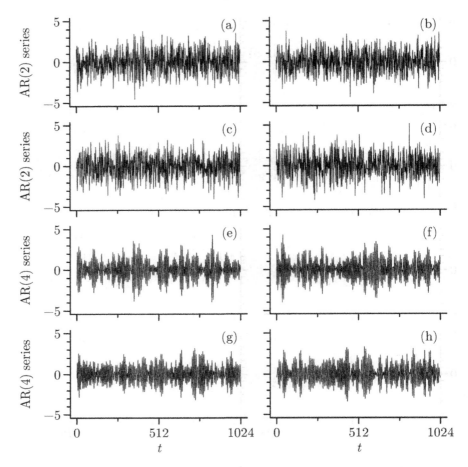

Figure 34 Realizations of length 1024 from two Gaussian autoregressive processes. The top two rows show four realizations of the AR(2) process of Equation (34), while the remaining two rows have four realizations of the AR(4) process of Equation (35a).

process at time t and a linear combination of the AR process itself at p previous times. In contrast to the parameters of an MA(q) process, the $\phi_{p,k}$ terms must satisfy certain conditions for the process $\{X_t\}$ to be stationary and "causal"; i.e., the random variable X_t just depends upon $\epsilon_t, \epsilon_{t-1}, \epsilon_{t-2}, \ldots$ and not $\epsilon_{t+1}, \epsilon_{t+2}, \epsilon_{t+3}, \ldots$ (for details, see Section 9.2 and C&E [5] for Section 9.4). For example, when $p = 1$, we must have $|\phi_{1,1}| < 1$ (see Exercise [2.17a]). When $p = 2$, stationarity and causality hold if the three conditions $\phi_{2,2} + \phi_{2,1} < 1$, $\phi_{2,2} - \phi_{2,1} < 1$ and $|\phi_{2,2}| < 1$ are true (Box et al., 2015). If a stationary AR(p) process is causal and nondeterministic (i.e., X_t cannot be perfectly predicted from $X_{t-1}, X_{t-2}, \ldots, X_{t-p}$), it can be written as an infinite-order moving average process:

$$X_t = -\sum_{j=0}^{\infty} \theta_j \epsilon_{t-j},$$

where θ_j can be determined from the $\phi_{p,k}$ terms. AR(p) processes play an important role in modern spectral analysis and are discussed in detail in Chapter 9.

Figure 34 shows realizations from two Gaussian stationary and causal AR processes; the top two rows are of the AR(2) process

$$X_t = 0.75 X_{t-1} - 0.5 X_{t-2} + \epsilon_t \text{ with } \sigma_2^2 = 1, \tag{34}$$

2.6 Examples of Discrete Parameter Stationary Processes

while the two bottom rows are of the AR(4) process

$$X_t = 2.7607 X_{t-1} - 3.8106 X_{t-2} + 2.6535 X_{t-3} - 0.9238 X_{t-4} + \epsilon_t \quad \text{with } \sigma_4^2 = 0.002 \quad (35a)$$

(in both cases $\{\epsilon_t\}$ is a Gaussian white noise process with zero mean). These two AR processes have been used extensively in the literature as test cases (see, for example, Ulrych and Bishop, 1975, and Box et al., 2015). We use the realizations in Figure 34 as examples in Chapters 6, 7, 8 and 9 (Exercises [597] and [11.1] describe how these realizations were generated).

Autoregressive Moving Average Process

The process $\{X_t\}$ with zero mean is called an autoregressive moving average process of order (p, q) – denoted by ARMA(p,q) – if it satisfies an equation such as

$$X_t = \phi_{p,1} X_{t-1} + \cdots + \phi_{p,p} X_{t-p} + \epsilon_t - \theta_{q,1}\epsilon_{t-1} - \cdots - \theta_{q,q}\epsilon_{t-q}, \quad t \in \mathbb{Z}, \quad (35b)$$

where $\phi_{p,j}$ and $\theta_{q,j}$ are constants ($\phi_{p,p} \neq 0$ and $\theta_{q,q} \neq 0$) and again $\{\epsilon_t\}$ is a white noise process with zero mean and variance σ_ϵ^2. With the process parameters appropriately chosen, Equation (35b) describes a rich class of stationary processes that can successfully model a wide range of time series (Box et al., 2015, treat ARMA processes at length).

Harmonic Process

The process $\{X_t\}$ is called a harmonic process if it can be written as

$$X_t = \mu + \sum_{l=1}^{L} A_l \cos(2\pi f_l t) + B_l \sin(2\pi f_l t), \quad t \in \mathbb{Z}, \quad (35c)$$

where μ and $f_l > 0$ are real-valued constants, L is a positive integer, and A_l and B_l are uncorrelated real-valued RVs with zero means such that var$\{A_l\}$ = var$\{B_l\}$ (Anderson, 1971; Newton, 1988). We can reexpress a harmonic process as

$$X_t = \mu + \sum_{l=1}^{L} D_l \cos(2\pi f_l t + \phi_l), \quad t \in \mathbb{Z}, \quad (35d)$$

where $D_l^2 = A_l^2 + B_l^2$, $\tan(\phi_l) = -B_l/A_l$, $A_l = D_l \cos(\phi_l)$ and $B_l = -D_l \sin(\phi_l)$ (these relationships are based in part upon the trigonometric identity $\cos(x+y) = \cos(x)\cos(y) - \sin(x)\sin(y)$). If we adopt the convention that $-\pi < \phi_l \leq \pi$, and if we draw a line from the origin of a graph to the point $(A_l, -B_l)$, then D_l is a nonnegative random amplitude describing the length of the line, and ϕ_l is a random phase representing the smaller of the two angles that the line makes with the positive part of the real axis. The RV ϕ_l is related to A_l and B_l via arctan$(-B_l/A_l)$ if we use the version of this function that pays attention to the quadrant in which $(A_l, -B_l)$ lies. (The model we considered in Equation (8a) of Chapter 1 is a special case of a harmonic process.)

Consider momentarily the case $\mu = 0$, $L = 1$ and $f_1 = 1/20$, and assume that A_1 and B_1 are Gaussian RVs with zero mean and unit variance. Three realizations of length $N = 100$ from this harmonic process are shown in the left-hand column of Figure 36. These realizations are sinusoids oscillating with a period of $1/f_1 = 20$ and differing only in their amplitudes and phases. They are quite different from the realizations of the white noise, MA and AR processes depicted in Figures 31, 33 and 34. In particular, once we have observed a small portion of a realization of this harmonic process, we know what all its other values must be. The same statement cannot be made for realizations of white noise, MA or AR processes. Despite the fact that realizations of this simple harmonic process differ markedly from ones we have seen so far for stationary processes, we have the following remarkable result.

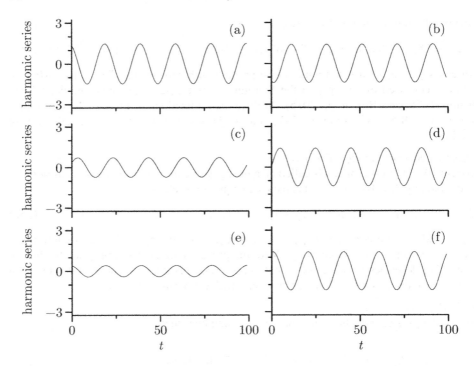

Figure 36 Realizations of length 100 from two harmonic processes for which $\mu = 0$, $L = 1$ and $f_1 = 1/20$. The realizations in the left-hand column are from a process whose distribution for each X_t is Gaussian, while those in the right-hand column have a non-Gaussian distribution that is restricted in its assumable values to the interval $[-\sqrt{2}, \sqrt{2}]$.

▷ **Exercise [36]** Letting $\sigma_l^2 \stackrel{\text{def}}{=} \text{var}\{A_l\} = \text{var}\{B_l\}$, show that the harmonic process of Equation (35c) is a stationary process with mean μ and with an ACVS given by

$$s_\tau = \sum_{l=1}^{L} \sigma_l^2 \cos(2\pi f_l \tau). \qquad (36a)$$ ◁

If we compare Equations (35d) and (36a), we see that both the harmonic process and its ACVS consist of sums of cosine waves with exactly the same frequencies, but that all of the cosine terms are "in phase" for the ACVS; i.e., the ϕ_l RVs have been replaced by zeros. Note that the sequence $\{s_\tau\}$ does not damp down to zero as τ gets large. Because

$$s_0 = \sum_{l=1}^{L} \sigma_l^2, \qquad (36b)$$

it follows that the ACS is given by

$$\rho_\tau = \sum_{l=1}^{L} \sigma_l^2 \cos(2\pi f_l \tau) \bigg/ \sum_{l=1}^{L} \sigma_l^2.$$

Let us consider two specific examples of harmonic processes. For the first example, we assume the RVs A_l and B_l have a Gaussian distribution. Since a linear combination of Gaussian RVs is also a Gaussian RV, it follows from Equation (35c) that the resulting process

2.6 Examples of Discrete Parameter Stationary Processes

$\{X_t\}$ is a Gaussian stationary process. Recall that, if $Y_0, Y_1, \ldots, Y_{\nu-1}$ are independent zero mean, unit variance Gaussian RVs, then the RV

$$\chi_\nu^2 \stackrel{\text{def}}{=} Y_0^2 + Y_1^2 + \cdots + Y_{\nu-1}^2 \tag{37a}$$

has a chi-square distribution with ν degrees of freedom. Since A_l/σ_l and B_l/σ_l are independent zero mean, unit variance Gaussian RVs, it follows that $D_l^2/\sigma_l^2 = (A_l^2 + B_l^2)/\sigma_l^2$ has a chi-square distribution with two degrees of freedom. The PDF for the RV χ_2^2 is given by

$$f_{\chi_2^2}(u) \stackrel{\text{def}}{=} \begin{cases} e^{-u/2}/2, & u \geq 0; \\ 0, & u < 0, \end{cases}$$

from which we can deduce that the PDF for D_l^2 is given by

$$f_{D_l^2}(u) \stackrel{\text{def}}{=} \begin{cases} e^{-u/(2\sigma_l^2)}/(2\sigma_l^2), & u \geq 0; \\ 0, & u < 0. \end{cases}$$

This is a special case of an exponential PDF $f(u) = \exp(-u/\lambda)/\lambda$ with a mean value of $\lambda = 2\sigma_l^2$. The random amplitude D_l is thus the square root of an exponential RV, which is said to obey a Rayleigh distribution. Anderson (1971, p. 376) notes that the symmetry of the bivariate Gaussian distribution for A_l and B_l dictates that ϕ be uniformly distributed over the interval $(-\pi, \pi]$ and be independent of D_l. Thus formulation of a Gaussian harmonic process via Equation (35d) involves Rayleigh distributed D_l and uniformly distributed ϕ_l, with all $2L$ RVs being independent of one another.

Whereas a Gaussian-distributed harmonic process consists of random amplitudes and uniformly distributed random phases, our second example shows that we can dispense with the random amplitudes and still have a harmonic process.

▷ **Exercise [37]** Suppose X_t obeys Equation (35d), but with the stipulation that the D_l terms are real-valued constants, while the ϕ_l terms are independent RVs, each having a uniform distribution on the interval $(-\pi, \pi]$. Show that the process so defined is in fact a harmonic process and that its ACVS is given by

$$s_\tau = \sum_{l=1}^{L} D_l^2 \cos(2\pi f_l \tau)/2, \tag{37b}$$ ◁

Three realizations from this harmonic process for the case of $\mu = 0$, $L = 1$, $f_1 = 1/20$ and $D_1 = \sqrt{2}$ are shown in the right-hand column of Figure 36. This choice of parameters yields an ACVS that is identical to the Gaussian-distributed harmonic process behind the realizations in the left-hand column. Note that each realization has a fixed amplitude so that X_t is restricted to the interval $[-\sqrt{2}, \sqrt{2}]$ both from one realization to the next and within a given realization. The fact that X_t is bounded implies that it cannot have a Gaussian distribution.

Our second example can be called a "random phase" harmonic process and suggests that random phases are more fundamental than random amplitudes in formulating a stationary process (see Exercise [2.19] for another connection between stationarity and a process whose realizations can also be thought of as "randomly phased" versions of a basic pattern). In fact, in place of our more general formulation via Equation (35c), some authors *define* a harmonic process as having fixed amplitudes and random phases (see, e.g., Priestley, 1981). In practical applications, however, a random phase harmonic process is often sufficient – particularly since we would need to have multiple realizations of this process in order to distinguish it from one having both random amplitudes and phases. (Exercise [2.20] explores the consequence of allowing the random phases to come from a nonuniform distribution.)

Note that any given realization from a harmonic process can be regarded as a deterministic function since the realizations of the RVs in right-hand sides of either Equation (35c) or (35d) determine X_t for all t. Assuming the frequencies f_l and mean μ are known, knowledge of any segment of a realization of X_t with a length at least as long as $2L$ is enough to fully specify the entire realization. In one sense, randomizing the phases (or amplitudes and phases) is a mathematical trick allowing us to treat models like Equation (35d) within the context of the theory of stationary processes. In another sense, however, *all* stationary processes can be written as a generalization of a harmonic process with an infinite number of terms – as we shall see, this is the essence of the spectral representation theorem to be discussed in Chapter 4 (harmonic analysis itself is studied in more detail in Chapter 10, and simulation of harmonic processes is discussed in Section 11.1).

It should be noted that the independence of the ϕ_l RVs is a sufficient condition to ensure stationarity (as demonstrated by Exercise [2.21], an assumption of just uncorrelatedness does not suffice in general). Although independence of phases is often assumed for convenience when the random phase model described by Equation (35d) is fit to actual data, this assumption can lead to subtle problems. Walden and Prescott (1983) modeled tidal elevations using the random phase harmonic process approach. The disagreement (often small) between the calculated and measured tidal PDFs is largely attributable to the lack of phase independence of the constituents of the tide (see Section 10.1 for details).

2.7 Comments on Continuous Parameter Processes

All of the stationary processes mentioned in Section 2.6 – with the important exception of harmonic processes – can be constructed by taking (possibly infinite) linear combinations of discrete parameter white noise. One might suppose that we can likewise construct numerous examples of continuous parameter stochastic processes by taking linear combinations of continuous parameter white noise. There is a fundamental technical difficulty with this approach – a continuous parameter white noise process with properties similar to those of a discrete parameter process does not exist; i.e., a continuous parameter process $\{\epsilon(t)\}$ with an ACVF given by

$$s(\tau) = \begin{cases} \sigma^2 > 0, & \tau = 0; \\ 0, & \text{otherwise,} \end{cases}$$

does not exist! It is possible to circumvent these technical problems and to deal with a "fictitious" continuous parameter white noise process to construct a wide range of continuous parameter stationary processes. This fiction is useful in much the same way as the Dirac delta function is a useful fiction (as long as certain precautions are taken). We do not deal with it directly in this book since our primary concern is analyzing time series using techniques appropriate for a digital computer. We thus do not need to construct continuous parameter stationary processes directly – though we do deal with them indirectly through the process of sampling (Section 4.5). For our purposes the spectral representation theorem for continuous parameter stationary processes will serve as the necessary foundation for future study (see C&E [3] for Section 4.1).

2.8 Use of Stationary Processes as Models for Data

The chief use of stationary processes in this book is to serve as models for various time series. Although the concept of stationarity is defined for models and not for data (see Exercise [2.19]), it is proper to ask whether certain time series can be usefully modeled by a given stationary process. We should be careful about drawing conclusions about a time series

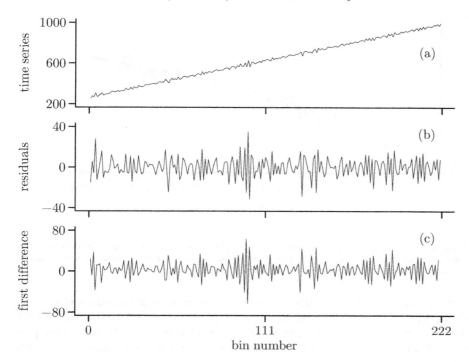

Figure 39 Spinning rotor time series. From top to bottom, the plots are (a) the original data X_t; (b) the residuals $\hat{Y}_t$ from a linear least squares fit; and (c) the first difference $X_t^{(1)}$ of the original data. The data are measured in units of microseconds. This data set was collected by S. Wood at the then National Bureau of Standards (now called the National Institute of Standards and Technology) and made available to the authors by J. Filliben.

based upon analysis techniques (such as spectral analysis) that assume stationarity when the assumed model is not likely to generate realizations that are reasonable alternatives to what was actually observed.

There are many ways in which a time series can be mismatched to a particular stationary process. Fortunately, some common violations are relatively easy to patch up. Here we examine two types of patchable nonstationarities by considering two physical time series for which use of, e.g., a Gaussian-distributed stationary AR or MA process as a model is suspect.

The first example is the time series shown in the top plot of Figure 39. These data concern a spinning rotor that is used to measure air density (i.e., pressure). During the period over which the data were collected, the rotor was always slowing down – the more molecules the rotor hit, the faster it slowed down. The data plotted are the amount of time (measured in microseconds) that it took the rotor to make 400 revolutions. The x-axis is the bin number: the first bin corresponds to the first group of 400 observed revolutions, and the tth bin, to the tth such group. (These data are a good example of a time series for which the "time index" is not actually time – however the values of the series are measured in time!)

As the spinning rotor slowed down, the time it took to make 400 revolutions necessarily increased. If we use a Gaussian-distributed stochastic process $\{X_t\}$ as a model for these data, it is unrealistic to assume that it is stationary. This requires that $E\{X_t\}$ be a constant independent of time, which from physical considerations (and Figure 39) is unrealistic. However, the time dependence of $E\{X_t\}$ appears to be linear, so a reasonable model might be

$$X_t = \alpha + \beta t + Y_t, \tag{39}$$

where α and β are unknown parameters and $\{Y_t\}$ is a stationary process with zero mean.

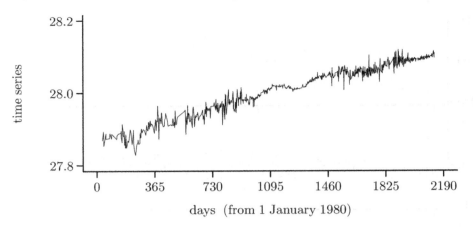

Figure 40 Standard resistor time series. The plot shows the value of a resistor measured over approximately a six-year period. This data set was collected by S. Dziuba at the then National Bureau of Standards (now called the National Institute of Standards and Technology) and made available to the authors by J. Filliben.

How we proceed now depends upon what questions we want answered about the data. For heuristic purposes, let us assume that we want to examine the covariance structure of $\{Y_t\}$. Two important ways of recovering $\{Y_t\}$ from $\{X_t\}$ are shown in the middle and bottom plots of Figure 39. The middle plot shows the residuals from an ordinary linear least squares fit to the parameters α and β; i.e.,

$$\hat{Y}_t \stackrel{\text{def}}{=} X_t - \hat{\alpha} - \hat{\beta}t,$$

where $\hat{\alpha}$ and $\hat{\beta}$ are the linear least squares estimates of α and β. The residual $\hat{Y}_t$ only approximates the unknown Y_t, but nonetheless we can analyze $\{\hat{Y}_t\}$ to get some idea of the covariance structure of $\{Y_t\}$.

A second way of analyzing $\{Y_t\}$ is to examine the first difference of the data:

$$X_t^{(1)} \stackrel{\text{def}}{=} X_t - X_{t-1} = \beta + Y_t^{(1)}, \quad \text{where } Y_t^{(1)} \stackrel{\text{def}}{=} Y_t - Y_{t-1}. \tag{40}$$

This is shown in the bottom plot of Figure 39 (the sample mean of the $X_t^{(1)}$ values is an estimator of β and is equal to 3.38 here). It can be shown (by arguments similar to those needed for Exercise [2.23]) that the first difference of a stationary process is also a stationary process – hence, if $\{Y_t\}$ is stationary, so is $\{Y_t^{(1)}\}$. Moreover, by using the theory of linear filters (see Chapter 5), it is possible to relate the covariance (or spectral) properties of $\{Y_t^{(1)}\}$ to those of $\{Y_t\}$. Because of this, differencing is often preferred over linear least squares to effectively remove a linear trend in a time series.

An extension to simple differencing for removing a trend can be applied when there is seasonality, or periodicity, in the data with period p (Box et al., 2015). If p is a multiple of the sampling interval, we could choose to examine $X_t - X_{t-p}$, which is called lag p differencing (Brockwell and Davis, 2016). Nature might choose to be less cooperative as in the case of tidal analysis where the dominant tidal period is 12.42 hours and the sampling interval is often 1 hour. Conradsen and Spliid (1981) showed that in this case the periodicity can be largely removed by using $X_t - (1-\delta)X_{t-n} - \delta X_{t-n-1}$, where $n = \lfloor p \rfloor$ and $\delta = p - n$ (these are 12 and 0.42, respectively, in the tidal example). However, as explained in Section 2.6, harmonic processes incorporating such periodicities can be examined within the framework of stationary processes if desired (see also Chapter 10).

The second example of a time series for which a Gaussian-distributed stationary process is a questionable model is shown in Figure 40. The data are the resistance of a standard resistor over approximately a six-year period. Here we see a linear or, perhaps, quadratic drift over time. If we assume only a linear drift, an appropriate model for these data might be the same stochastic process considered for the spinning rotor data (Equation (39)). However, here it might be questionable to assume that $\{Y_t\}$, the deviations from the linear drift, is a stationary process. If we compare the data from about days 1095 to 1460 with the rest of the series, it appears that the series has a variance that changes with time – a violation of the stationarity assumption. There are no simple transformations that correct this problem. One common approach is to use different stationary processes to model various chunks of the data. Here two models might work – one for the deviations about a linear drift for days 1095 to 1460, and a second for the data before and after.

2.9 Exercises

[2.1] Here we consider some basic properties of covariances. In what follows, assume for generality that all RVs and constants are complex-valued (the definition of covariance for complex-valued RVs is given in Equation (25a)). All RVs are denoted by Z (or subscripted versions thereof), while c with or without a subscript denotes a constant. Note that, as usual, Z^* denotes the complex conjugate of Z and that $E\{Z^*\} = (E\{Z\})^*$.

(a) Show that $\operatorname{cov}\{Z, c\} = 0$.
(b) Show that $\operatorname{cov}\{Z_0, Z_1\} = \left(\operatorname{cov}\{Z_1, Z_0\}\right)^*$.
(c) Show that $\operatorname{cov}\{Z_0 + c_0, Z_1 + c_1\} = \operatorname{cov}\{Z_0, Z_1\}$.
(d) Suppose that at least one of the RVs Z_0 and Z_1 has a zero mean. Show that $\operatorname{cov}\{Z_0, Z_1\} = E\{Z_0 Z_1^*\}$.
(e) Show that

$$\operatorname{cov}\left\{\sum_j c_{0,j} Z_{0,j}, \sum_k c_{1,k} Z_{1,k}\right\} = \sum_j \sum_k c_{0,j} c_{1,k}^* \operatorname{cov}\{Z_{0,j}, Z_{1,k}\},$$

where j and k range over finite sets of integers.

Since real-valued RVs and constants are special cases of complex-valued entities, the results above continue to hold when some or all of the RVs and constants in question are real-valued. In particular, when Z_0 and Z_1 are both real-valued, part (b) simplifies to $\operatorname{cov}\{Z_0, Z_1\} = \operatorname{cov}\{Z_1, Z_0\}$. Also, when Z_1 and $c_{1,k}$ in parts (d) and (e) are real-valued, we can simplify Z_1^* to Z_1, and $c_{1,k}^*$ to $c_{1,k}$.

[2.2] Show that, if $W_0, W_1, \ldots, W_{N-1}$ are any set of $N \geq 2$ real-valued RVs with finite variances, and if $a_0, a_1, \ldots, a_{N-1}$ are any corresponding set of real numbers, then

$$\operatorname{var}\left\{\sum_{j=0}^{N-1} a_j W_j\right\} = \sum_{j=0}^{N-1} a_j^2 \operatorname{var}\{W_j\} + 2 \sum_{j=0}^{N-2} \sum_{k=j+1}^{N-1} a_j a_k \operatorname{cov}\{W_j, W_k\}. \qquad (41)$$

If we further let $\boldsymbol{a} \stackrel{\text{def}}{=} [a_0, a_1, \ldots, a_{N-1}]^T$ and $\boldsymbol{W} \stackrel{\text{def}}{=} [W_0, W_1, \ldots, W_{N-1}]^T$, and if we let $\boldsymbol{\Sigma}$ be the covariance matrix for the vector $\boldsymbol{W}$ (i.e., the (j,k)th element of $\boldsymbol{\Sigma}$ is $\operatorname{cov}\{W_j, W_k\}$), show that Equation (41) implies that $\operatorname{var}\{\boldsymbol{a}^T \boldsymbol{W}\} = \boldsymbol{a}^T \boldsymbol{\Sigma} \boldsymbol{a}$.

[2.3] Let the real-valued sequence $\{x_{0,t}\}$ be defined by

$$x_{0,t} = \begin{cases} +1, & t = 0, -1, -2, \ldots; \\ -1, & t = 1, 2, 3, \ldots, \end{cases}$$

and let the real-valued sequence $\{x_{1,t}\}$ be defined by $x_{1,t} = -x_{0,t}$. Suppose we have a discrete parameter stochastic process whose ensemble consists of just $\{x_{0,t}\}$ and $\{x_{1,t}\}$. We generate

realizations of this process by picking either $\{x_{0,t}\}$ or $\{x_{1,t}\}$ at random with probability $1/2$ each. Let $\{X_t\}$ represent the process itself.
  (a) What is $E\{X_t\}$?
  (b) What is var$\{X_t\}$?
  (c) What is cov$\{X_{t+\tau}, X_t\}$?
  (d) Is $\{X_t\}$ a (second-order) stationary process?

[2.4] Suppose that the real-valued RVs X_0, X_1 and X_2 have zero means and a covariance matrix given by

$$\Sigma_3 = \begin{bmatrix} 1.00 & 0.69 & 0.43 \\ 0.69 & 1.00 & 0.30 \\ 0.43 & 0.30 & 1.00 \end{bmatrix}.$$

Is it possible for these RVs to be part of a stationary process $\{X_t\}$?

[2.5] (a) Suppose that $\{X_t\}$ and $\{Y_t\}$ are both stationary processes with zero means and with ACVSs $\{s_{X,\tau}\}$ and $\{s_{Y,\tau}\}$, respectively. If $\{X_t\}$ and $\{Y_t\}$ are uncorrelated with each other, show that the ACVS $\{s_{Z,\tau}\}$ of the complex-valued process defined by $Z_t = X_t + iY_t$ must be real-valued. Can $\{s_{X,\tau}\}$ and $\{s_{Y,\tau}\}$ be determined if only $\{s_{Z,\tau}\}$ is known?
  (b) Suppose that $Y_t = X_{t+k}$ for some integer k; i.e., the process $\{Y_t\}$ is just a shifted version of $\{X_t\}$. What is $\{s_{Z,\tau}\}$ now?

[2.6] Let $\{Z_t\}$ be a complex-valued stochastic process. For such a process we defined the notion of stationarity by making certain stipulations on the RVs associated with the real and imaginary components of $\{Z_t\}$ (see Section 2.5).
  (a) Suppose $\{Z_t\}$ takes the form $Z_t = \epsilon_t e^{it}$, where $\{\epsilon_t\}$ is a real-valued white noise process with mean zero and variance σ_ϵ^2. Find cov$\{Z_{t+\tau}, Z_t\}$, and show that $\{Z_t\}$ is *not* stationary.
  (b) Construct a complex-valued process $\{\widetilde{Z}_t\}$ that is stationary, but is such that

$$\text{cov}\{\widetilde{Z}_{t+\tau}, \widetilde{Z}_t\} = \text{cov}\{Z_{t+\tau}, Z_t\}$$

for all τ, where $\{Z_t\}$ is defined as in part (a). (This example demonstrates that there can be more than one complex-valued stochastic process with the same covariance structure.)

[2.7] (a) Let $\{X_{0,t}\}, \{X_{1,t}\}, \ldots, \{X_{m-1,t}\}$ each be stationary processes with means $\mu_0, \mu_1, \ldots, \mu_{m-1}$ and ACVSs $\{s_{0,\tau}\}, \{s_{1,\tau}\}, \ldots, \{s_{m-1,\tau}\}$, respectively; i.e., $E\{X_{j,t}\} = \mu_j$ and $s_{j,\tau} = \text{cov}\{X_{j,t+\tau}, X_{j,t}\}$. If $X_{j,t}$ and $X_{k,u}$ are uncorrelated for all t, u and $j \neq k$ (i.e., cov$\{X_{j,t}, X_{k,u}\} = 0$ when $j \neq k$) and if $c_0, c_1, \ldots, c_{m-1}$ are arbitrary real-valued variables, show that

$$X_t \stackrel{\text{def}}{=} \sum_{j=0}^{m-1} c_j X_{j,t}$$

is a stationary process, and determine its ACVS. (The procedure of forming a new process by making a linear combination of several processes is sometimes called *aggregation*.)
  (b) Let a be any real-valued constant. Consider the sequence $s_0 = 2 + a^2$, $s_1 = s_{-1} = -a$ and $s_\tau = 0$ for $|\tau| > 1$. Is this a valid ACVS? Hint: consider Equation (32b).

[2.8] Suppose that $\{Y_t\}$ is a stationary process with ACVS $\{s_{Y,\tau}\}$. We subject this process to a lag $p = 12$ seasonal differencing to obtain $X_t \stackrel{\text{def}}{=} Y_t - Y_{t-12}$, following which we form $W_t \stackrel{\text{def}}{=} X_t - X_{t-1}$, i.e., the first difference of X_t. Show that the process $\{W_t\}$ is stationary with mean zero and an ACVS given by

$$s_{W,\tau} = 4s_{Y,\tau} - 2(s_{Y,\tau+1} + s_{Y,\tau-1} + s_{Y,\tau+12} + s_{Y,\tau-12}) \qquad (42)$$
$$+ s_{Y,\tau+11} + s_{Y,\tau-11} + s_{Y,\tau+13} + s_{Y,\tau-13}.$$

(An example of a time series for which it is reasonable to apply both seasonal and first differencing is the CO_2 data discussed in Section 7.12.)

[2.9] Suppose that X_0, X_1, X_2 and X_3 are any four real-valued Gaussian RVs with zero means. Use the Isserlis theorem (Equation (30)) to show that

$$E\{X_0 X_1 X_2 X_3\} = E\{X_0 X_1\}E\{X_2 X_3\} + E\{X_0 X_2\}E\{X_1 X_3\} + E\{X_0 X_3\}E\{X_1 X_2\}. \quad (43)$$

[2.10] (a) Let $\{X_t\}$ be a real-valued stationary Gaussian process with mean zero and ACVS $\{s_{X,\tau}\}$. Show that $Y_t = X_t^2$ is a stationary process whose ACVS is given by $s_{Y,\tau} = 2s_{X,\tau}^2$. Hint: make use of the Isserlis theorem (Equation (30)).

(b) Find a real-valued stationary Gaussian process $\{X_t\}$ such that $Y_t = X_t^2$ has an ACVS given by

$$s_{Y,\tau} = \begin{cases} 200, & \tau = 0; \\ 0, & |\tau| = 1; \\ 18, & |\tau| = 2; \\ 0, & \text{otherwise.} \end{cases}$$

Hint: consider Equation (32b), recalling that $\theta_{q,0} \stackrel{\text{def}}{=} -1$.

[2.11] Let W be a real-valued random variable with mean $\mu = 0$ and finite variance $\sigma^2 > 0$, and let c be a real-valued constant. Define a discrete parameter stochastic process $\{X_t\}$ by letting $X_t = W \cos(ct)$ for all integers t.

(a) Show that $\{X_t\}$ is a stationary process if and only if $c = l\pi$ for some integer l, and, for each l, determine the corresponding ACVS $\{s_{l,\tau}\}$. How many unique ACVSs are there in all?

(b) Recall that a sequence $\{s_\tau\}$ is said to be positive semidefinite if the inequality in Equation (28b) holds. Verify directly that this defining inequality holds for each $\{s_{l,\tau}\}$. Is $\{s_{l,\tau}\}$ for any l in fact positive definite as opposed to just positive semidefinite?

(c) Suppose now that $\mu \neq 0$. Is $\{X_t\}$ a stationary process?

[2.12] Show that, if $\{s_\tau\}$ is the ACVS for a stationary process, then, for any $\alpha \geq 0$, the sequence $\{s_\tau^2 + \alpha s_\tau\}$ is the ACVS for some stationary process. (Hint: consider a product of stationary processes with different means, but with the same ACVS.) Show also that, when $\alpha < 0$, the sequence $\{s_\tau^2 + \alpha s_\tau\}$ need not be a valid ACVS.

[2.13] Suppose that $\{\epsilon_t\}$ is a white noise process with zero mean and variance σ_ϵ^2. Define the process $\{X_t\}$ via

$$X_t = \begin{cases} \epsilon_0, & t = 0; \\ X_{t-1} + \epsilon_t, & t = 1, 2, \dots. \end{cases}$$

Is $\{X_t\}$ a stationary process? (This process is one form of a *random walk*.)

[2.14] In the defining Equation (32a) for a qth-order moving average process, we set $\theta_{q,0} \stackrel{\text{def}}{=} -1$. In terms of the covariance structure for moving average processes, why do we not gain more generality by letting $\theta_{q,0}$ be any arbitrary number?

[2.15] Show that the first-order moving average process defined by $X_t = \epsilon_t - \theta \epsilon_{t-1}$ can be written in terms of previous values of the process as

$$X_t = \epsilon_t - \sum_{j=1}^{p} \theta^j X_{t-j} - \theta^{p+1} \epsilon_{t-p-1}$$

for any positive integer p. In the above, what condition on θ must hold in order that X_t can be expressed as an infinite-order AR process, i.e.,

$$X_t = \epsilon_t - \sum_{j=1}^{\infty} \theta^j X_{t-j}?$$

(A moving average process that can be written as an infinite-order AR process is called *invertible*. For details, see, e.g., Brockwell and Davis, 1991, Section 3.1.)

[2.16] Consider the following moving average process of infinite order:

$$X_t = \sum_{j=0}^{\infty} \theta_j \epsilon_{t-j}, \quad \text{where } \theta_j = \begin{cases} 1, & j = 0; \\ 1/\sqrt{j}, & j \geq 1, \end{cases}$$

and $\{\epsilon_t\}$ is a zero mean white noise process with unit variance. Is $\{X_t\}$ a stationary process?

[2.17] (a) Consider the AR(1) process $X_t = \phi X_{t-1} + \epsilon_t$, where, as usual, $\{\epsilon_t\}$ is a zero mean white noise process with variance σ_ϵ^2. Show that this process can be expressed as

$$X_t = \phi^q X_{t-q} + \sum_{j=0}^{q-1} \phi^j \epsilon_{t-j} \tag{44a}$$

for any positive integer q. If $|\phi| < 1$, it can be argued that the term $\phi^q X_{t-q}$ becomes negligible as q gets large and hence that $\{X_t\}$ is a stationary process with an infinite-order moving average representation given by

$$X_t = \sum_{j=0}^{\infty} \phi^j \epsilon_{t-j} \tag{44b}$$

(see the discussion on causality in C&E [1] for Section 9.2). Use the above representation to show that the ACVS for $\{X_t\}$ is given by

$$s_\tau = \frac{\phi^{|\tau|} \sigma_\epsilon^2}{1 - \phi^2} \tag{44c}$$

(see Exercise [9.2a] for another way to obtain this ACVS).

(b) Consider the ARMA(1,1) process $X_t = \alpha X_{t-1} + \epsilon_t + \alpha \epsilon_{t-1}$. Show that this process can be expressed as

$$X_t = \alpha^q X_{t-q} + \epsilon_t + 2 \sum_{j=1}^{q-1} \alpha^j \epsilon_{t-j} + \alpha^q \epsilon_{t-q} \tag{44d}$$

for any integer $q \geq 2$. If $|\alpha| < 1$, it can be argued that $\{X_t\}$ is a stationary process with an infinite-order moving average representation given by

$$X_t = \epsilon_t + 2 \sum_{j=1}^{\infty} \alpha^j \epsilon_{t-j}.$$

Using the above representation, show that the ACVS for $\{X_t\}$ is given by

$$s_\tau = \sigma_\epsilon^2 \left(-\delta_\tau + 2\alpha^{|\tau|} \frac{1+\alpha^2}{1-\alpha^2} \right),$$

where δ_τ is Kronecker's delta function; i.e., $\delta_\tau = 1$ when $\tau = 0$, and $\delta_\tau = 0$ when $\tau \neq 0$.

[2.18] Let $\{\epsilon_t\}$ be a white noise process with mean zero and variance σ_ϵ^2. Let c and ϕ be real-valued constants. Let $X_0 = c\epsilon_0$ and $X_t = \phi X_{t-1} + \epsilon_t$, $t = 1, 2, \ldots$. Let Σ_3 denote the covariance matrix for X_1, X_2 and X_3; i.e., the (j, k)th element of this matrix is cov$\{X_j, X_k\}$.

(a) Determine the nine elements of Σ_3 in terms of σ_ϵ^2, c and ϕ.

(b) By considering the case $c = 0$ so that $X_0 = 0$, argue that, unless $\phi = 0$, the matrix Σ_3 need not be of the form required for X_1, X_2 and X_3 to be three contiguous RVs from a stationary process.

(c) Show that, if we make the restriction $|\phi| < 1$ and if we set $c = 1/\sqrt{(1-\phi^2)}$, then Σ_3 has the required form.

[2.19] This exercise illustrates the point that the concept of stationarity is properly defined only for models and not for data. We do so by claiming that, given any observed time series $\{x_t : t \in T\}$ with $T \stackrel{\text{def}}{=} \{0, 1, \ldots, N-1\}$, we can always construct a stochastic process $\{X_t : t \in T\}$ that is stationary and has $\{x_t\}$ as one of its realizations, with the observed series having a nonzero probability of being selected. To establish this claim, define the ensemble for $\{X_t\}$ to consist of N realizations given by $\{x_t\}$ and all possible circular shifts. For example, if $N = 4$, the ensemble consists of

$$\{x_0, x_1, x_2, x_3\}, \{x_1, x_2, x_3, x_0\}, \{x_2, x_3, x_0, x_1\} \text{ and } \{x_3, x_0, x_1, x_2\}. \tag{45a}$$

If we index the realizations in the ensemble by $k = 0, 1, \ldots, N-1$, we can mathematically describe the kth realization as $\{x_{t+k \bmod N} : t \in T\}$, where, for any integer l, we define $l \bmod N$ as follows. If $0 \leq l \leq N-1$, then $l \bmod N = l$; otherwise, $l \bmod N = l + mN$, where m is the unique integer such that $0 \leq l + mN \leq N-1$. For example, when $N = 4$ and $k = 2$,

$$\{x_{t+2 \bmod 4} : t = 0, 1, 2, 3\} = \{x_{2 \bmod 4}, x_{3 \bmod 4}, x_{4 \bmod 4}, x_{5 \bmod 4}\}$$
$$= \{x_2, x_3, x_0, x_1\}.$$

To complete our definition of $\{X_t\}$, we stipulate that the probability of picking any given realization in the ensemble is $1/N$. Thus, if κ is an RV that takes on the values $0, 1, \ldots, N-1$ with equal probability, we can express the stochastic process $\{X_t\}$ as $\{x_{t+\kappa \bmod N}\}$. Show that, for all $s \in T$ and $t \in T$,

(a) $E\{X_t\}$ is a constant that does not depend on t and
(b) $\text{cov}\{X_s, X_t\}$ is a constant that depends upon just the lag $\tau = s - t$;

i.e., $\{X_t\}$ is a stationary process.

[2.20] In the random phase harmonic process $\{X_t\}$ formulated in Exercise [37], we stipulated that the ϕ_l RVs be independent and uniformly distributed over the interval $(-\pi, \pi]$. Suppose that the PDF of ϕ_l is not that of a uniformly distributed RV, but rather is of the form

$$f_{\phi_l}(u) \stackrel{\text{def}}{=} \frac{1}{2\pi}(1 + \cos(u)), \qquad u \in (-\pi, \pi].$$

A realization of ϕ_l from this PDF is more likely to be close to zero than to $\pm\pi$. With this new stipulation, is $\{X_t\}$ still a stationary process?

[2.21] Let ϕ_1 and ϕ_2 be two RVs whose joint PDF is given by

$$f_{\phi_1,\phi_2}(u,v) = \frac{3}{8\pi^3}\left(|u-v| + |u+v| - \frac{u^2}{\pi} - \frac{v^2}{\pi}\right), \quad \text{where } u, v \in (-\pi, \pi]$$

(Ferguson, 1995).

(a) Show that ϕ_1 is uniformly distributed over $(-\pi, \pi]$; i.e., its univariate (or marginal) PDF is given by

$$f_{\phi_1}(u) = \int_{-\pi}^{\pi} f_{\phi_1,\phi_2}(u,v)\,dv = \frac{1}{2\pi}, \quad u \in (-\pi, \pi].$$

Argue that the PDF $f_{\phi_2}(\cdot)$ for ϕ_2 is also uniform, from which we can conclude that ϕ_1 and ϕ_2 are not independent since $f_{\phi_1}(u)f_{\phi_2}(v) \neq f_{\phi_1,\phi_2}(u,v)$ for all u and v.

(b) Show that, even though they are not independent, the RVs ϕ_1 and ϕ_2 are uncorrelated, i.e., that $\text{cov}\{\phi_1, \phi_2\} = 0$.

(c) Define

$$X_t = D_1 \cos(2\pi f_1 t + \phi_1) + D_2 \cos(2\pi f_2 t + \phi_2), \quad t \in \mathbb{Z}, \tag{45b}$$

where D_1, D_2, f_1 and f_2 are real-valued constants satisfying $D_1 > 0$, $D_2 > 0$ and $0 < f_1 \leq f_2 < 1$. Show that, while $E\{X_t\} = 0$, we have

$$\text{var}\{X_t\} = \frac{D_1^2 + D_2^2}{2} - \frac{3D_1 D_2}{\pi^2} \cos(2\pi f_1 t) \cos(2\pi f_2 t),$$

and hence $\{X_t\}$ is *not* a stationary process.

(d) Suppose that we have a random number generator capable of producing what can be regarded as six independent realizations u_j, $j = 0, 1, \ldots, 5$, from an RV uniformly distributed over the interval $(0, 1)$. The following recipe uses these to create realizations of ϕ_1 and ϕ_2. Let $a = \log(u_0) + \log(u_1)$ and $b = \log(u_2) + \log(u_3)$, and form the ratio $c = a/(a+b)$. Let $d = u_4$. If $u_5 \leq \frac{1}{2}$, let $e = |c - d|$; if $u_5 > \frac{1}{2}$, let $e = 1 - |1 - c - d|$. The realizations of ϕ_1 and ϕ_2 are given by $\pi(2d - 1)$ and $\pi(2e - 1)$. Using this procedure, generate a large number of independent realizations of X_0, X_1, X_2 and X_3 in Equation (45b) with $D_1 = D_2 = 1$, $f_1 = \frac{1}{4}$ and $f_2 = \frac{1}{2}$. Compute the sample means and variances for these realizations, and compare them to the corresponding theoretical values derived in part (c).

[2.22] Suppose that $\{X_t\}$ is a discrete parameter stationary process with zero mean and ACVS $\{s_{X,\tau}\}$, and consider the process defined by $Y_t = X_t \cos(2\pi f_1 t + \phi)$, where $0 < f_1 < 1/2$ is a fixed frequency.

(a) If we regard ϕ as a constant, is $\{Y_t\}$ a stationary process?

(b) If we regard ϕ as an RV that is uniformly distributed over the interval $(-\pi, \pi]$ and is independent of X_t for all t, is $\{Y_t\}$ a stationary process?

[2.23] Suppose that $Y_t = \alpha + \beta t + \gamma t^2 + X_t$, where α, β and γ are nonzero constants and $\{X_t\}$ is a stationary process with ACVS $\{s_{X,\tau}\}$. Show that the first difference of the first difference of Y_t, i.e., $Y_t^{(2)} \stackrel{\text{def}}{=} Y_t^{(1)} - Y_{t-1}^{(1)}$, where $Y_t^{(1)} \stackrel{\text{def}}{=} Y_t - Y_{t-1}$, is a stationary process, and find its ACVS $\{s_{Y^{(2)},\tau}\}$ in terms of $\{s_{X,\tau}\}$.

3

Deterministic Spectral Analysis

3.0 Introduction

In Chapter 1 we modeled four different time series of length N using a stochastic model of the form

$$X_t = \mu + \sum_{j=1}^{\lfloor N/2 \rfloor} [A_j \cos(2\pi f_j t) + B_j \sin(2\pi f_j t)], \quad f_j = j/N, \qquad (47)$$

where A_j and B_j are uncorrelated random variables (RVs) with mean zero and variance σ_j^2 (this is Equation (8a)). We defined a variance spectrum $S_j = \sigma_j^2$, $1 \leq j \leq \lfloor N/2 \rfloor$, and noted that, by summing up all of its values, we could recover the variance of the process:

$$\sum_{j=1}^{\lfloor N/2 \rfloor} S_j = \text{var}\{X_t\}.$$

The variance spectrum thus decomposes (analyzes) the process variance into $\lfloor N/2 \rfloor$ components, each associated with the expected square amplitude of sinusoids of a particular frequency. This spectrum gives us a concise way of summarizing some of the important statistical properties of $\{X_t\}$.

In Chapter 2 we introduced the concept of stationary processes. The task that awaits us is to define an appropriate spectrum for these processes. We do so in Chapter 4. We devote this chapter to the definition and study of various spectra for deterministic (nonrandom) functions (or sequences) of time. Our rationale for doing so is threefold. First, since every *realization* of a stochastic process is a deterministic function of time, we shall use the material here to motivate the definition of a spectrum for stationary processes. Second, deterministic functions and sequences appear quite naturally in many aspects of the study of the spectra of stationary processes. The concepts we discuss here will be used repeatedly in subsequent chapters when we discuss, for example, the relationship between the spectrum and the ACVS for a stationary process, the relationship between a linear filter and its transfer function and the use of data tapers in the estimation of spectra. Third, just as the spectrum for stationary processes is a way of summarizing some important features of those processes, the various spectra that can be defined for deterministic functions can be used to express certain properties of these functions easily.

In model given by Equation (47) we dealt with a discrete time (or discrete parameter) process $\{X_t\}$ and defined a variance spectrum for it over a set of discrete frequencies, namely, $f_j = j/N$ with $1 \leq j \leq \lfloor N/2 \rfloor$. Now for a real- or complex-valued deterministic *function* $g(\cdot)$ defined over the entire real axis $\mathbb{R}$, it might be possible to define a reasonable spectrum over a discrete set of frequencies or a continuum of frequencies; the same remark also holds for a deterministic *sequence* $\{g_t\}$ defined over all integers $\mathbb{Z}$. There are four possible combinations of interest: continuous time with continuous frequency; continuous time with discrete frequency; discrete time with continuous frequency; and discrete time with discrete frequency. The first case arises in much of the historical theoretical framework on spectral analysis; the second provides a description for periodic functions often encountered in electrical engineering and is prevalent in the study of differential equations; the third appears in the relationship between the spectrum and ACVS of a stationary process; and the fourth is important as the route into digital computer techniques. The remaining sections in this chapter are an investigation of some of the important time/frequency relationships in each of these categories. For heuristic reasons, we begin with the case of a continuous time deterministic function with a spectrum defined over a discrete set of frequencies.

3.1 Fourier Theory – Continuous Time/Discrete Frequency

Since $\cos(2\pi n t/T)$ defines a periodic function of t with period $T > 0$ for any integer n, i.e.,

$$\cos(2\pi n[t+T]/T) = \cos(2\pi n t/T) \quad \text{for all times } t,$$

and since the same holds for $\sin(2\pi n t/T)$, it follows that

$$\tilde{g}_p(t) = \frac{a_0}{2} + \sum_{n=1}^{\infty} a_n \cos(2\pi n t/T) + b_n \sin(2\pi n t/T) \tag{48a}$$

also defines a periodic function of t with period T. Here $\{a_n\}$ and $\{b_n\}$ are arbitrary sequences of constants (either real- or complex-valued) subject only to the condition that the summation in Equation (48a) must converge for all t. We can rewrite Equation (48a) in a more compact form by expressing the cosines and sines as complex exponentials (see Exercise [1.2b]):

$$\tilde{g}_p(t) = \sum_{n=-\infty}^{\infty} G_n e^{i2\pi f_n t} \tag{48b}$$

where

$$G_n \stackrel{\text{def}}{=} \begin{cases} (a_n - ib_n)/2, & n \geq 1; \\ a_0/2, & n = 0; \\ (a_{|n|} + ib_{|n|})/2, & n \leq -1, \end{cases} \tag{48c}$$

and $f_n = n/T$. Equation (48b) involves both positive and negative frequencies. Negative frequencies are a somewhat strange concept at first sight, but they are a useful mathematical fiction that allows us to simplify the mathematical theory in this chapter considerably. If $\{a_n\}$ and $\{b_n\}$ are real-valued, then $\tilde{g}_p(\cdot)$ is also real-valued. In this case, $\{G_n\}$ is conjugate symmetric about $n = 0$ in that $G_{-n}^* = G_n$ so that $|G_{-n}| = |G_n|$. Exercise [3.1] is to express a_n and b_n in terms of the G_n (and to examine a third common way of expressing Equation (48a)).

Let $g_p(\cdot)$ be a deterministic (nonrandom) real- or complex-valued function of t that is periodic with period T and is square integrable; i.e., $g_p(t+kT) = g_p(t)$ for any integer k and

$$\int_{-T/2}^{T/2} |g_p(t)|^2 \, dt < \infty.$$

3.1 Fourier Theory – Continuous Time/Discrete Frequency

Can $g_p(\cdot)$ be represented by a series such as in Equation (48b)? The answer is "yes," in a certain sense. The exact result is as follows: define

$$g_{p,m}(t) = \sum_{n=-m}^{m} G_n e^{i2\pi f_n t}, \tag{49a}$$

where

$$G_n \stackrel{\text{def}}{=} \frac{1}{T} \int_{-T/2}^{T/2} g_p(t) e^{-i2\pi f_n t} \, dt. \tag{49b}$$

Then $g_{p,m}(\cdot)$ converges to $g_p(\cdot)$ in the mean square sense as $m \to \infty$ (Theorem 15.12, Champeney, 1987); i.e.,

$$\lim_{m \to \infty} \int_{-T/2}^{T/2} |g_p(t) - g_{p,m}(t)|^2 \, dt = 0.$$

The above says that, as m gets larger and larger, the functions $g_{p,m}(\cdot)$ and $g_p(\cdot)$ become closer and closer to one another as measured by their integrated magnitude square differences. Hence we can write

$$g_p(t) \stackrel{\text{ms}}{=} \sum_{n=-\infty}^{\infty} G_n e^{i2\pi f_n t}, \tag{49c}$$

where we define the symbol $\stackrel{\text{ms}}{=}$ to mean "equal in the mean square sense" (item [1] in the Comments and Extensions [C&Es] to this section has a discussion on the distinction between pointwise and mean square equality).

The following exercise offers some justification for Equation (49b).

▷ **Exercise [49a]** Suppose that Equation (49c) holds in a pointwise sense; i.e., we are allowed to replace $\stackrel{\text{ms}}{=}$ with $=$. By multiplying both sides of this equation by $\exp(-i2\pi f_m t)$ and integrating them from $-T/2$ to $T/2$, show that the resulting equation can be manipulated to obtain Equation (49b). ◁

Equation (49c) is called the *Fourier series representation* of a periodic function; G_n is referred to as the nth *Fourier coefficient*.

Considering the case of a real-valued function $g_p(\cdot)$ that represents, e.g., a periodic fluctuation in the voltage in a circuit, we define

$$\text{energy of } g_p(\cdot) \text{ over } [-T/2, T/2] = \int_{-T/2}^{T/2} |g_p(t)|^2 \, dt$$

and use this definition even for functions such that the above integration does not have the units of energy as defined in physics.

▷ **Exercise [49b]** Show that the energy in $g_p(\cdot)$ is preserved in its Fourier coefficients because

$$\int_{-T/2}^{T/2} |g_p(t)|^2 \, dt = T \sum_{n=-\infty}^{\infty} |G_n|^2. \tag{49d}$$ ◁

This equation is referred to as *Parseval's theorem* (or Rayleigh's theorem) for Fourier series. Since the energy in $g_p(\cdot)$ is finite, the infinite sum on the right-hand side converges, which requires that $|G_n|^2 \to 0$ as $n \to \infty$. The rate at which G_n decays toward zero is related to how "smooth" $g_p(\cdot)$ is: a faster decay rate implies a smoother function. Equation (49c) supports this interpretation because, if G_n decays rapidly, then complex exponentials with high frequencies are negligible contributors in the Fourier representation for $g_p(\cdot)$.

Because $g_p(\cdot)$ is periodic, its energy over the real axis $\mathbb{R}$ would be infinite. A concept closely related to energy is power, which is defined as energy per unit time interval:

$$\text{power over } (-T/2, T/2) = \frac{\text{energy over } (-T/2, T/2)}{T}$$

$$= \frac{1}{T} \int_{-T/2}^{T/2} |g_p(t)|^2 \, dt = \sum_{n=-\infty}^{\infty} |G_n|^2. \tag{50a}$$

Whereas the energy in $g_p(\cdot)$ over $\mathbb{R}$ is infinite, the power is finite since, for integer k,

$$\lim_{k \to \infty} \frac{1}{kT} \int_{-kT/2}^{kT/2} |g_p(t)|^2 \, dt = \sum_{n=-\infty}^{\infty} |G_n|^2.$$

Equation (50a) decomposes the power into an infinite sum of terms $|G_n|^2$, the nth one of which is the contribution from the term in the Fourier series for $g_p(\cdot)$ with frequency n/T. We can now define the *discrete power spectrum* for $g_p(\cdot)$ as the sequence whose nth element is

$$S_n \stackrel{\text{def}}{=} |G_n|^2, \quad n \in \mathbb{Z}.$$

A plot of S_n versus f_n shows us how the power is distributed over the various frequency components of $g_p(\cdot)$. Note, however, that we can recover $|G_n|$ – but not G_n itself – from S_n. This makes it impossible to reconstruct $g_p(\cdot)$ from knowledge of its discrete power spectrum alone.

As a simple example of the theory presented so far in this section, consider the 2π periodic, even and real-valued function given by

$$g_p(t) = \frac{1 - \phi^2}{1 + \phi^2 - 2\phi \cos(t)}, \tag{50b}$$

where the variable ϕ is real-valued. If $|\phi| < 1$, then $g_p(\cdot)$ is square integrable with Fourier coefficients $G_n = \phi^{|n|}$ (these results follow from section 3.616, part 7, and section 3.613, part 2, Gradshteyn and Ryzhik, 1980). Hence

$$g_{p,m}(t) = \sum_{n=-m}^{m} G_n e^{i2\pi f_n t} = 1 + 2 \sum_{n=1}^{m} \phi^n \cos(nt) \tag{50c}$$

should converge in mean square to $g_p(\cdot)$ as $m \to \infty$ (in fact, it converges pointwise – see Exercise [3.2]). Figure 51a shows plots of $g_p(\cdot)$ compared with $g_{p,m}(\cdot)$ for $m = 4, 8, 16$ and 32 and $\phi = 0.9$. For this example, the discrete power spectrum is given by $S_n = \phi^{2|n|}$. Figure 51b shows a plot of $10 \log_{10}(S_n)$ – again with $\phi = 0.9$ – versus $f_n = n/2\pi$ for $n = -32, -31, \ldots, 32$. (Exercises [3.3] through [3.6] consider four other periodic functions that can be represented by a Fourier series).

It is interesting to note that $g_{p,m}(\cdot)$ in Equation (49a) has the following least squares interpretation as an approximation to $g_p(\cdot)$. Let

$$h_{p,m}(t) \stackrel{\text{def}}{=} \sum_{n=-m}^{m} H_n e^{i2\pi f_n t}$$

3.1 Fourier Theory – Continuous Time/Discrete Frequency

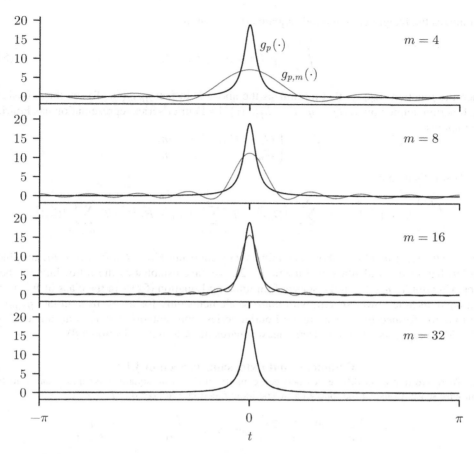

Figure 51a Approximation of a function by a truncated Fourier series. The dark curve in each plot is $g_p(\cdot)$ of Equation (50b) with ϕ set to 0.9. The light curves are $g_{p,m}(\cdot)$ of Equation (50c) for m = 4, 8, 16 and 32 – the approximation is sufficiently good in the latter case that there is little discernible difference between it and $g_p(\cdot)$.

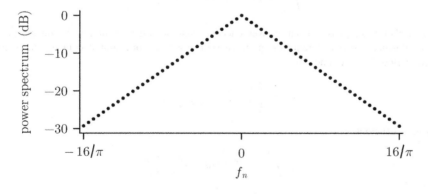

Figure 51b Discrete power spectrum $\{S_n\}$ for $g_p(\cdot)$ in Figure 51a.

be any function whose discrete power spectrum is identically zero for $|n| > m$, where we now regard m as a fixed positive integer. Suppose we want to find a function of this form that minimizes the integrated magnitude square error given by

$$\int_{-T/2}^{T/2} |g_p(t) - h_{p,m}(t)|^2 \, dt. \tag{52}$$

Then the function $h_{p,m}(\cdot)$ that minimizes the above is in fact $g_{p,m}(\cdot)$. To see this, consider the function defined by $d_p(t) = g_p(t) - h_{p,m}(t)$. Its Fourier series representation has Fourier coefficients

$$D_n = \begin{cases} G_n - H_n, & |n| \leq m; \\ G_n, & |n| > m. \end{cases}$$

By Parseval's theorem,

$$\int_{-T/2}^{T/2} |d_p(t)|^2 \, dt = T \sum_{n=-\infty}^{\infty} |D_n|^2 = T \sum_{n=-m}^{m} |G_n - H_n|^2 + T \sum_{|n|>m} |G_n|^2.$$

Because the right-hand side above is minimized by choosing $H_n = G_n$ for $|n| \leq m$, and since the left-hand side is identical to Equation (52), we have established the following: the best approximation to $g_p(\cdot)$ – in the sense of minimizing Equation (52) – in the class of functions that have discrete power spectra with nonzero elements only for $|n| \leq m$ is in fact $g_{p,m}(\cdot)$, the function formed by truncating the Fourier series representation for $g_p(\cdot)$ at indices $\pm m$. (This idea reappears when we discuss least squares filter design in Section 5.8).

Comments and Extensions to Section 3.1

[1] There are important differences between equality in the mean square sense and equality in the pointwise sense. Consider the following two periodic functions with period $2T$:

$$g_p(t) = \begin{cases} e^{-t}, & 0 < t \leq T; \\ 0, & -T < t \leq 0, \end{cases} \quad \text{and} \quad h_p(t) = \begin{cases} e^{-t}, & 0 \leq t \leq T; \\ 0, & -T < t < 0. \end{cases}$$

These functions are not equal in the pointwise sense since $g_p(0) = 0$ while $h_p(0) = 1$; however, they are equal in the mean square sense because

$$\int_{-\infty}^{\infty} |g_p(t) - h_p(t)|^2 \, dt = 0.$$

It should also be emphasized that convergence in mean square is computed by integration (or summation) and hence is very different from pointwise convergence. For example, consider the periodic function with period $T > 2$ defined by

$$g_p(t) = \begin{cases} 1, & |t| \leq 1; \\ 0, & 1 < |t| \leq T/2. \end{cases}$$

Its nth Fourier coefficient is

$$G_n = \frac{1}{T} \int_{-1}^{1} e^{-i2\pi f_n t} \, dt = \frac{\sin(2\pi f_n)}{T \pi f_n},$$

where we interpret the latter quantity to be $2/T$ when $n = 0$. Since $g_p(\cdot)$ is square integrable, evidently the sequence of functions defined by

$$g_{p,m}(t) = \sum_{n=-m}^{m} \frac{\sin(2\pi f_n)}{T \pi f_n} e^{i2\pi f_n t}$$

converges to $g_p(\cdot)$ in the mean square sense as $m \to \infty$; however, from Theorem 15.1 of Champeney (1987), it can be shown that, while we do have

$$\lim_{m \to \infty} g_{p,m}(t) = g_p(t) \text{ for all } t \neq \pm 1,$$

in fact we do *not* have pointwise convergence where $g_p(\cdot)$ is discontinuous since

$$\lim_{m \to \infty} g_{p,m}(1) = \left[g_p(1-) + g_p(1+)\right]/2 = 1/2 \neq g_p(1) = 1,$$

with a similar result for $t = -1$; here we define

$$g_p(t-) = \lim_{\substack{\epsilon \to 0 \\ \epsilon > 0}} g_p(t - \epsilon) \text{ and } g_p(t+) = \lim_{\substack{\epsilon \to 0 \\ \epsilon > 0}} g_p(t + \epsilon).$$

[2] In this chapter we use the symbol $\stackrel{\text{ms}}{=}$ whenever equality in the mean square sense is meant, but in latter chapters we follow standard mathematical practice and use = to stand for both types of equality – the distinction should be clear from the context.

3.2 Fourier Theory – Continuous Time and Frequency

Suppose now that $g(\cdot)$ is a deterministic real- or complex-valued function of t that is not periodic. Since it does not possess a periodic structure, we cannot express it in the form of a Fourier series (the right-hand side of Equation (49c)) that will be valid for all t. However, we can do the following (cf. Figure 54). For any positive real number T, define a new function $g_T(\cdot)$ with period T first by letting

$$g_T(t) = g(t) \text{ for } -T/2 < t \leq T/2$$

and then by letting

$$g_T(t + pT) = g_T(t) \text{ for } -T/2 < t \leq T/2 \text{ and } p \in \mathbb{Z}.$$

If $g_T(\cdot)$ is square integrable over $(-T/2, T/2]$, the discussion in the previous section says that we can write

$$g_T(t) \stackrel{\text{ms}}{=} \sum_{n=-\infty}^{\infty} G_{n,T} e^{i2\pi f_n t},$$

where

$$G_{n,T} = \frac{1}{T} \int_{-T/2}^{T/2} g_T(t) e^{-i2\pi f_n t} \, dt = \frac{1}{T} \int_{-T/2}^{T/2} g(t) e^{-i2\pi f_n t} \, dt,$$

since $g(t) = g_T(t)$ over the interval $(-T/2, T/2]$. Hence, for $-T/2 < t \leq T/2$, we have

$$g(t) = g_T(t) \stackrel{\text{ms}}{=} \sum_{n=-\infty}^{\infty} \left(\int_{-T/2}^{T/2} g(t) e^{-i2\pi f_n t} \, dt \right) e^{i2\pi f_n t} \Delta_{\text{f}},$$

where we take Δ_{f} to be $f_n - f_{n-1} = 1/T$. Now as $T \to \infty$, we have $\Delta_{\text{f}} \to 0$, and the summation above becomes an integral, so that (formally at least) for all t

$$g(t) \stackrel{\text{ms}}{=} \int_{-\infty}^{\infty} G(f) e^{i2\pi ft} \, df, \tag{53}$$

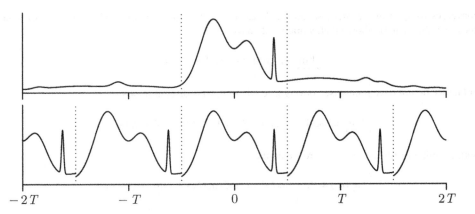

Figure 54 Periodic extension of a function. The upper plot shows a nonperiodic real-valued function of time. The bottom plot shows a new function that results from taking the values in the top plot from $(-T/2, T/2]$ – i.e., those values between the vertical dotted lines – and replicating them periodically.

where

$$G(f) \stackrel{\text{def}}{=} \int_{-\infty}^{\infty} g(t) e^{-i2\pi ft} \, dt. \qquad (54a)$$

The heuristic argument above can be made rigorous under the assumption that

$$\int_{-\infty}^{\infty} |g(t)|^2 \, dt < \infty. \qquad (54b)$$

Intuitively, this condition says that the "business part" of $g(\cdot)$ must damp down to zero as $|t|$ approaches infinity in the sense that, for every small $\epsilon > 0$, there is some finite interval of length T centered about the origin such that

$$\int_{-\infty}^{\infty} |g(t)|^2 \, dt - \epsilon < \int_{-T/2}^{T/2} |g(t)|^2 \, dt \leq \int_{-\infty}^{\infty} |g(t)|^2 \, dt.$$

The set of all functions that satisfy this property of square integrability is denoted as $L^2(\mathbb{R})$, where the superscript reflects the square, and the bounds on the integral are dictated by the set $\mathbb{R}$ within the parentheses. Equations (53) and (54a) are thus well defined for all functions in $L^2(\mathbb{R})$. The right-hand side of Equation (53) is called the *Fourier integral representation* of $g(\cdot)$. The function $G(\cdot)$ in Equation (54a) is said to be the *Fourier transform* of $g(\cdot)$. The function of t that is defined by the right-hand side of Equation (53) is called the *inverse Fourier transform* of $G(\cdot)$ (or, sometimes, the *Fourier synthesis* of $g(\cdot)$). The functions $g(\cdot)$ and $G(\cdot)$ are said to be a *Fourier transform pair* – a fact commonly denoted by

$$g(\cdot) \longleftrightarrow G(\cdot).$$

The inverse Fourier transform on the right-hand side of Equation (53) represents $g(\cdot)$ (in the mean square equality sense) as the "sum" of complex exponentials $\exp{(i2\pi ft)}$ having amplitude $|G(f)|$ and phase $\arg{(G(f))}$, where $\arg{(G(f))}$ is taken to be $\theta(f)$ when we write $G(f) = |G(f)| e^{i\theta(f)}$ in polar notation. The function $|G(\cdot)|$ is often referred to as the *amplitude spectrum* for $g(\cdot)$. Note that, if $g(\cdot)$ is real-valued, then $G^*(-f) = G(f)$; i.e., the Fourier transform is conjugate symmetric about $f = 0$ (this fact follows readily from Equation (54a)).

The energy in the function $g(\cdot)$ over the entire real axis is just the left-hand side of Equation (54b). This energy has a simple relationship to the amplitude spectrum $|G(\cdot)|$.

3.2 Fourier Theory – Continuous Time and Frequency

▷ **Exercise [55]** Show the following version of Parseval's theorem:

$$\int_{-\infty}^{\infty} |g(t)|^2 \, \mathrm{d}t = \int_{-\infty}^{\infty} |G(f)|^2 \, \mathrm{d}f. \tag{55a}$$ ◁

(For an extension, see Equation (97e), which is the subject of Exercise [3.7].) Note that Equation (55a) implies that the Fourier transform of a square integrable function is itself a square integrable function (Champeney, 1987, p. 62). The function $|G(\cdot)|^2$ is called the *energy spectral density function* for $g(\cdot)$ by analogy to, say, a probability density function (PDF). This is appropriate terminology since Parseval's relationship implies that $|G(f)|^2 \, \mathrm{d}f$ represents the contribution to the energy from those components in $g(\cdot)$ whose frequencies lie between f and $f + \mathrm{d}f$. Note that for this case the power (energy per unit time interval) is given by

$$\lim_{T \to \infty} \frac{\text{energy over } (-T/2, T/2]}{T} = 0,$$

since the energy over the entire real axis $\mathbb{R}$ is finite.

Because we can recover (at least in the mean square sense) $g(\cdot)$ if we know $G(\cdot)$ and vice versa, the time and frequency domain representations of a deterministic function contain equivalent information. A quote from Bracewell (2000, Chapter 8) is appropriate here:

> We may think of functions and their transforms as occupying two domains, sometimes referred to as the upper and the lower, as if functions circulated at ground level and their transforms in the underworld (Doetsch, 1943). There is a certain convenience in picturing a function as accompanied by a counterpart in another domain, a kind of shadow which is associated uniquely with the function through the Fourier transformation, and which changes as the function changes.

As an example of the theory presented in this section, consider the Gaussian PDF with zero mean:

$$g_\sigma(t) = \frac{1}{(2\pi\sigma^2)^{1/2}} \mathrm{e}^{-t^2/(2\sigma^2)}. \tag{55b}$$

Since the Fourier transform of the function defined by $\exp(-\pi t^2)$ is the function itself, it follows that the Fourier transform of $g_\sigma(\cdot)$ is given by

$$G_\sigma(f) = \mathrm{e}^{-2\pi^2 f^2 \sigma^2} \tag{55c}$$

(Bracewell, 2000, p. 107). Plots of portions of $g_\sigma(\cdot)$ and $G_\sigma(\cdot)$ for $\sigma = 1$, 2 and 4 are shown in Figure 56. Since the energy spectral density function is just $|G_\sigma(\cdot)|^2$, we can readily infer its properties from the plot of $G_\sigma(\cdot)$.

Comments and Extensions to Section 3.2

[1] The conventions that we have adopted in Equations (53) and (54a) for, respectively, the inverse Fourier transform (+i in the complex exponential) and the Fourier transform (−i) are by no means unique. Bracewell (2000, chapter 2) discusses some of the prevalent conventions and their advantages and disadvantages. The reader is advised to check carefully which conventions have been adopted when referring to other books on Fourier and spectral analysis.

[2] The *Schwarz inequality* is useful for $L^2(\alpha, \beta)$ functions, i.e., those that are square integrable over the interval (α, β). If $g(\cdot)$ and $h(\cdot)$ are two such functions, this inequality takes the form

$$\left| \int_\alpha^\beta g(t) h(t) \, \mathrm{d}t \right|^2 \leq \int_\alpha^\beta |g(t)|^2 \, \mathrm{d}t \int_\alpha^\beta |h(t)|^2 \, \mathrm{d}t. \tag{55d}$$

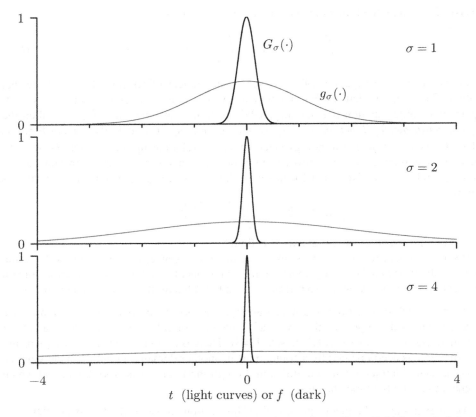

Figure 56 Gaussian PDFs $g_\sigma(\cdot)$ (light curves) and their Fourier transforms $G_\sigma(\cdot)$ (dark) for standard deviations $\sigma = 1, 2$ and 4.

The proof only requires the basic notion of the norm of a complex number and some results from algebra about quadratic equations. Define $p = \int_\alpha^\beta g(t)h(t)\,dt$. We can restrict ourselves to the case where both $\int_\alpha^\beta |g(t)|^2\,dt$ and $\int_\alpha^\beta |h(t)|^2\,dt$ are greater than zero (if not, the inequality holds trivially since then $g(t)h(t)$ must be zero almost everywhere, which implies $p = 0$). We can further assume that $p \neq 0$ because, if not, then Equation (55d) holds trivially. For any real x, we have

$$0 \leq \int_\alpha^\beta \left|g(t) + xph^*(t)\right|^2\,dt. \tag{56}$$

By expanding the integrand as $[g(t)+xph^*(t)][g^*(t)+xp^*h(t)]$ and using the fact $p^* = \int_\alpha^\beta g^*(t)h^*(t)\,dt$, the above can be rewritten as

$$0 \leq \int_\alpha^\beta |g(t)|^2\,dt + 2x\,|p|^2 + x^2\,|p|^2 \int_\alpha^\beta |h(t)|^2\,dt \stackrel{\text{def}}{=} c + bx + ax^2,$$

where a, b and c are necessarily real-valued, and $a > 0$. The roots of this quadratic expression in x are, as usual, given by $[-b \pm \sqrt{(b^2 - 4ac)}]/2a$. Because the expression is nonnegative for all x, it cannot have distinct real roots, so we must have $b^2 \leq 4ac$, implying that

$$4\,|p|^4 \leq 4\,|p|^2 \int_\alpha^\beta |h(t)|^2\,dt \int_\alpha^\beta |g(t)|^2\,dt$$

or, equivalently,

$$|p|^2 = \left| \int_\alpha^\beta g(t)h(t)\,dt \right|^2 \leq \int_\alpha^\beta |g(t)|^2\,dt \int_\alpha^\beta |h(t)|^2\,dt.$$

From Equation (56), it follows that equality in Equation (55d) holds if and only if $g(t) + xph^*(t) = 0$ for all t, i.e., if and only if $g(\cdot)$ is proportional to $h^*(\cdot)$.

The frequency domain version of Equation (55d) is obviously

$$\left| \int_\alpha^\beta G(f)H(f)\,df \right|^2 \leq \int_\alpha^\beta |G(f)|^2\,df \int_\alpha^\beta |H(f)|^2\,df,$$

provided that $G(\cdot)$ and $H(\cdot)$ are functions in $L^2(\alpha, \beta)$. If we replace (α, β) with $\mathbb{R}$, both the above and Equation (55d) hold for $g(\cdot)$, $G(\cdot)$, $h(\cdot)$ and $H(\cdot)$ such that $g(\cdot) \longleftrightarrow G(\cdot)$ and $h(\cdot) \longleftrightarrow H(\cdot)$, since in this case all these functions are in $L^2(\mathbb{R})$.

3.3 Band-Limited and Time-Limited Functions

We have just defined the amplitude spectrum and energy spectral density function for a function in $L^2(\mathbb{R})$. In practice, most $L^2(\mathbb{R})$ functions representing physical phenomena have amplitude spectra with finite support (i.e., the amplitude spectra are nonzero only over a finite range of frequencies). Slepian (1976, 1983) gives some nice illustrations. A pair of solid copper wires will not propagate electromagnetic waves at optical frequencies and hence would not be expected to contain energy at frequencies greater than, say, 10^{20} Hz (cycles per second). Recorded male speech gives an amplitude spectrum that is zero for frequencies exceeding 8000 Hz, while orchestral music has no frequencies higher than 20,000 Hz. These band limitations are all at the high-frequency end, and the cutoff frequencies are called high-cut frequencies. It is possible for measurements to be deficient in both low and high frequencies if the recording instrument has some sort of built-in filter that does so. An example is equipment used in seismic prospecting for oil and gas, which contains a low-cut (high-pass) filter to eliminate certain deleterious noise forms and a high-cut (low-pass) *antialias* filter (the necessity for which is due to effects described in Section 4.5).

If there is no energy above a frequency $|f| = W$, say, then the finite energy function $g(\cdot)$ is said to be *band-limited* to the band $[-W, W]$. In this case the function has the following Fourier integral representation:

$$g(t) \stackrel{\text{ms}}{=} \int_{-W}^{W} G(f)e^{i2\pi ft}\,df.$$

Slepian (1983) notes that band-limited functions are necessarily "smooth" in the following sense. If we replace t in the right-hand side of the above by the complex number z, the resulting function is defined over the complex plane. Its kth derivative with respect to z is given by

$$\int_{-W}^{W} G(f)(i2\pi f)^k e^{i2\pi fz}\,df.$$

The Schwarz inequality and Parseval's relationship show that the above exists and is finite for all k (see Exercise [3.8]). A complex-valued function that can be differentiated an arbitrary number of times over the entire complex plane is called an *entire function* in complex analysis. It has a Taylor series expansion about every point with an infinite radius of convergence – these properties make it "smooth" in the eyes of mathematicians.

Just as some physical functions are band-limited and hence smooth, others can be time-limited and possibly "rough": we say that $g(\cdot)$ is time-limited if, for some $T > 0$, $g(t)$ is zero for all $|t| > T/2$. Examples of time-limited functions abound in the real world. The seismic trace due to a particular earthquake is necessarily time-limited, as is the amount of gas released by a rocket.

Can a function be both band-limited and time-limited? The answer is "yes," but only in a trivial sense. Slepian (1983) notes that the only $L^2(\mathbb{R})$ function with both these properties is zero for all t. The argument for this is simple. If $g(\cdot)$ is time-limited and band-limited, we can express it as a Taylor series about any particular point. If we pick the point to be in the region where $g(\cdot)$ is zero (i.e., $|t| > T/2$), all its derivatives are necessarily 0, and hence the Taylor series representation for $g(\cdot)$ shows that it must necessarily be identically zero everywhere. In Section 3.5 we consider nontrivial band-limited functions that are as "close" as possible to also being time-limited, where "closeness" is measured by a carefully chosen criterion. This idea is then extended to sequences in Section 3.10, where we also consider time-limited sequences that are close to being band-limited. These sequences play prominent roles in Chapters 6 and 8, where they are used to form nonparametric spectral estimators with improved statistical properties.

3.4 Continuous/Continuous Reciprocity Relationships

We now explore in more detail the relative behavior of an $L^2(\mathbb{R})$ function in the time and frequency domains, i.e., the function itself $g(\cdot)$ and its Fourier transform $G(\cdot)$. We shall do this by examining three different measures of reciprocity between $g(\cdot)$ and $G(\cdot)$, the most important of which is the last one, the so-called *fundamental uncertainty relationship*.

Similarity Theorem

Exercise [3.9] shows that, if $g(\cdot)$ and $G(\cdot)$ are a Fourier transform pair and a is a nonzero real-valued number, then the functions

$$g_a(t) \stackrel{\text{def}}{=} |a|^{1/2} g(at) \quad \text{and} \quad G_a(f) \stackrel{\text{def}}{=} \frac{1}{|a|^{1/2}} G(f/a) \tag{58a}$$

form a Fourier transform pair also. Hence if one member of the transform pair contracts horizontally and expands vertically, the other member expands horizontally and contracts vertically. This is illustrated in Figure 59a for the Fourier transform pair

$$g(t) = \frac{1}{(2\pi)^{1/2}} e^{-t^2/2} \quad \text{and} \quad G(f) = e^{-2\pi^2 f^2}$$

for $a = 1, 2$ and 4.

Equivalent Width

Suppose that $g(0) \neq 0$. We can define the width of such a function as the width of the rectangle whose height is equal to $g(0)$ and whose area is the same as that under the curve of $g(\cdot)$; i.e.,

$$\text{width}_e \{g(\cdot)\} \stackrel{\text{def}}{=} \int_{-\infty}^{\infty} g(t) \, dt \Big/ g(0). \tag{58b}$$

This measure of function width makes sense for a $g(\cdot)$ that is real-valued, positive everywhere, peaked about 0 and continuous at 0 (see Figure 59b), but it is less satisfactory for other types of functions. Now

$$\int_{-\infty}^{\infty} g(t) \, dt = G(0) \quad \text{and} \quad g(0) = \int_{-\infty}^{\infty} G(f) \, df$$

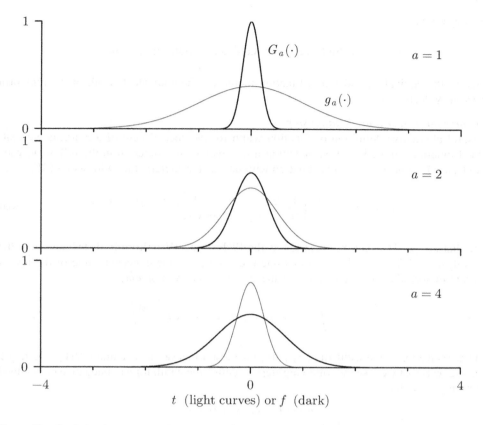

Figure 59a Similarity theorem. As $g(\cdot) = g_1(\cdot)$ (light curve in top plot) is contracted horizontally and expanded vertically by letting $a = 2$ and 4 (middle and bottom plots, respectively), its Fourier transform $G(\cdot) = G_1(\cdot)$ (dark curve in top plot) is expanded horizontally and contracted vertically.

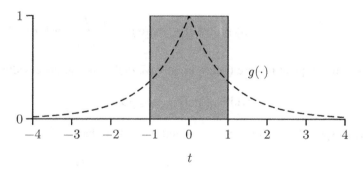

Figure 59b Equivalent width. The dashed curve shows $g(t) = \exp(-|t|)$. Since $\int_{-\infty}^{\infty} g(t)\,dt = 2$ and $g(0) = 1$, the equivalent width is 2, which is the width of the shaded rectangle.

(if $g(\cdot)$ is continuous at 0), so we have

$$\text{width}_e\{g(\cdot)\} = \int_{-\infty}^{\infty} g(t)\,dt \Big/ g(0) = G(0) \Big/ \int_{-\infty}^{\infty} G(f)\,df = 1/\text{width}_e\{G(\cdot)\}. \quad (59)$$

Note that the equivalent width of a function is the reciprocal of the equivalent width of its transform; i.e.,

$$\text{equivalent width (function)} \times \text{equivalent width (transform)} = 1.$$

Thus, as the equivalent width of a function increases, the equivalent width of its transform necessarily decreases.

Fundamental Uncertainty Relationship

A useful alternative definition of function width for the special case of a nonnegative real-valued function can be based on matching it to a rectangular function in the following way. Let $r(\cdot; \mu_r, W_r)$ be a rectangular function with unit area centered at μ_r with width $2W_r$; i.e.,

$$r(t; \mu_r, W_r) \stackrel{\text{def}}{=} \begin{cases} 1/(2W_r), & \mu_r - W_r \leq t \leq \mu_r + W_r; \\ 0, & \text{otherwise.} \end{cases} \qquad (60\text{a})$$

Note that $r(\cdot; \mu_r, W_r)$ can be regarded as the PDF of an RV uniformly distributed over the interval $[\mu_r - W_r, \mu_r + W_r]$ – in this interpretation μ_r is the expected value of the RV. A measure of spread of this PDF is its variance (second central moment):

$$\sigma_r^2 \stackrel{\text{def}}{=} \int_{-\infty}^{\infty} (t - \mu_r)^2 \, r(t; \mu_r, W_r) \, dt = \frac{W_r^2}{3}.$$

Thus we can express the width of $r(\cdot; \mu_r, W_r)$ in terms of its variance, namely, $2W_r = 2\sigma_r\sqrt{3}$. This suggests that we take the following steps to define the width for a nonnegative real-valued function $g(\cdot)$ such that

$$\int_{-\infty}^{\infty} g(t) \, dt = C \text{ with } 0 < C < \infty.$$

First, we renormalize $g(\cdot)$ by forming $\tilde{g}(t) = g(t)/C$ so that $\tilde{g}(\cdot)$ can be regarded as a PDF. Second, we calculate the second central moment

$$\sigma_{\tilde{g}}^2 \stackrel{\text{def}}{=} \int_{-\infty}^{\infty} (t - \mu_{\tilde{g}})^2 \, \tilde{g}(t) \, dt, \text{ where } \mu_{\tilde{g}} \stackrel{\text{def}}{=} \int_{-\infty}^{\infty} t \tilde{g}(t) \, dt$$

(we assume both these quantities are finite). Finally we *define* the variance width as

$$\text{width}_v \{g(\cdot)\} = 2\sigma_{\tilde{g}}\sqrt{3}. \qquad (60\text{b})$$

For the important special case in which $C = 1$ and $\mu_{\tilde{g}} = 0$, we have

$$\text{width}_v \{g(\cdot)\} = \left(12 \int_{-\infty}^{\infty} t^2 g(t) \, dt\right)^{1/2} \qquad (60\text{c})$$

(we make use of closely related formulae in Chapters 6 and 7 as one measure of the bandwidth of spectral windows and smoothing windows – see Equations (192b) and (251a)).

As we have defined it, the measure of width given by Equation (60b) is not meaningful for a complex-valued function $g(\cdot)$; however, it can be applied to its modulus $|g(\cdot)|$ or square modulus $|g(\cdot)|^2$ since both are necessarily nonnegative and real-valued. The latter case leads to a more precise statement about the relationship between the width of a function – defined now as $\text{width}_v \{|g(\cdot)|^2\}$ using Equation (60b) – and the width of its Fourier transform width$_v$ $\{|G(\cdot)|^2\}$ via a version of Heisenberg's uncertainty principle (Slepian, 1983; Chapter 8 of Bracewell, 2000; Folland and Sitaram, 1997). Let the function $g(\cdot)$ have unit energy; i.e.,

$$\int_{-\infty}^{\infty} |g(t)|^2 \, dt = \int_{-\infty}^{\infty} |G(f)|^2 \, df = 1.$$

Then both $|g(\cdot)|^2$ and $|G(\cdot)|^2$ act like PDFs, and their associated variances (a measure of spread or width) are given by

$$\sigma_g^2 = \int_{-\infty}^{\infty} (t - \mu_g)^2 |g(t)|^2 \, dt, \text{ where } \mu_g \stackrel{\text{def}}{=} \int_{-\infty}^{\infty} t \, |g(t)|^2 \, dt,$$

and

$$\sigma_G^2 = \int_{-\infty}^{\infty} (f - \mu_G)^2 |G(f)|^2 \, df, \text{ where } \mu_G \stackrel{\text{def}}{=} \int_{-\infty}^{\infty} f \, |G(f)|^2 \, df.$$

For convenience assume that $\mu_g = \mu_G = 0$. Then we have

$$\sigma_g^2 \sigma_G^2 = \int_{-\infty}^{\infty} t^2 \, |g(t)|^2 \, dt \int_{-\infty}^{\infty} f^2 \, |G(f)|^2 \, df.$$

Since

$$g(t) \stackrel{\text{ms}}{=} \int_{-\infty}^{\infty} G(f) e^{i2\pi ft} \, df,$$

it can be argued (under a regularity condition) that

$$g'(t) \stackrel{\text{def}}{=} \frac{dg(t)}{dt} \stackrel{\text{ms}}{=} \int_{-\infty}^{\infty} i2\pi f G(f) e^{i2\pi ft} \, df,$$

so that $g'(\cdot)$ and the function defined by $i2\pi f G(f)$ are a Fourier transform pair. By Parseval's theorem for a Fourier transform pair (see Equation (55a)),

$$\int_{-\infty}^{\infty} |g'(t)|^2 \, dt = 4\pi^2 \int_{-\infty}^{\infty} f^2 \, |G(f)|^2 \, df.$$

Hence

$$\sigma_g^2 \sigma_G^2 = \frac{1}{4\pi^2} \int_{-\infty}^{\infty} t^2 \, |g(t)|^2 \, dt \int_{-\infty}^{\infty} |g'(t)|^2 \, dt.$$

By the Schwarz inequality (Equation (55d)) we have

$$\int_{-\infty}^{\infty} t^2 \, |g(t)|^2 \, dt \int_{-\infty}^{\infty} |g'(t)|^2 \, dt \geq \left| \int_{-\infty}^{\infty} t g^*(t) g'(t) \, dt \right|^2$$

$$\geq \left[\Re \left(\int_{-\infty}^{\infty} t g^*(t) g'(t) \, dt \right) \right]^2$$

$$= \frac{1}{4} \left[\int_{-\infty}^{\infty} t g^*(t) g'(t) + t g(t) \left[g'(t) \right]^* \, dt \right]^2$$

(here $\Re(z)$ is the real part of the complex number z, and we use the facts that $|z|^2 \geq (\Re(z))^2$ and $\Re(z) = (z + z^*)/2$). We thus have

$$16\pi^2 \sigma_g^2 \sigma_G^2 \geq \left[\int_{-\infty}^{\infty} t g^*(t) g'(t) + t g(t) \left[g'(t) \right]^* \, dt \right]^2 ;$$

however, since

$$\frac{d|g(t)|^2}{dt} = \frac{dg(t)g^*(t)}{dt} = g^*(t)g'(t) + g(t)\frac{dg^*(t)}{dt} = g^*(t)g'(t) + g(t)[g'(t)]^*$$

(recall that the operations of complex conjugation and differentiation commute), we can conclude that

$$16\pi^2\sigma_g^2\sigma_G^2 \geq \left[\int_{-\infty}^{\infty} t\frac{d|g(t)|^2}{dt} dt\right]^2 = \left[\int_{-\infty}^{\infty} |g(t)|^2 dt\right]^2 = 1,$$

where we have used integration by parts and the fact that $g(\cdot)$ has unit energy. This is Heisenberg's uncertainty principle, namely,

$$\sigma_g^2\sigma_G^2 \geq \frac{1}{16\pi^2}. \tag{62a}$$

Note for example that, if σ_g^2 is very small, then σ_G^2 must be large to ensure that (62a) is satisfied.

Is the lower bound in Equation (62a) attainable? Consider the Gaussian-shaped function

$$g(t) = be^{-\pi a^2 t^2}, \quad \text{where } b^2 = a\sqrt{2}$$

(this condition on b forces $g(\cdot)$ to have unit energy). Then

$$\sigma_g^2 = \int_{-\infty}^{\infty} t^2 g^2(t) \, dt = a\sqrt{2}\int_{-\infty}^{\infty} t^2 e^{-2\pi a^2 t^2} \, dt = \frac{1}{4\pi a^2}.$$

The Fourier transform of $g(\cdot)$ is given by

$$G(f) = ba^{-1}e^{-\pi f^2/a^2}, \quad \text{so } |G(f)|^2 = b^2 a^{-2} e^{-2\pi f^2/a^2}.$$

Thus

$$\sigma_G^2 = b^2 a^{-2}\int_{-\infty}^{\infty} f^2 e^{-2\pi f^2/a^2} \, df = a^2/(4\pi) \quad \text{and} \quad \sigma_g^2\sigma_G^2 = 1/(16\pi^2),$$

which is the minimum value obtainable under the uncertainty relationship. (Some further mathematical insights into equivalent width and the fundamental uncertainty relationship are given by Champeney, 1987, pp. 75–6.)

3.5 Concentration Problem – Continuous/Continuous Case

Three ways of measuring time and frequency concentrations of functions belonging to $L^2(\mathbb{R})$ have been discussed in the previous section. An alternative way of measuring concentration is discussed by Slepian (1983). Concentration in time is measured by the ratio

$$\alpha^2(T) \stackrel{\text{def}}{=} \int_{-T/2}^{T/2} |g(t)|^2 \, dt \bigg/ \int_{-\infty}^{\infty} |g(t)|^2 \, dt, \tag{62b}$$

i.e., the fraction of the function's energy lying in a time interval of length T centered about 0. Analogously,

$$\beta^2(W) \stackrel{\text{def}}{=} \int_{-W}^{W} |G(f)|^2 \, df \bigg/ \int_{-\infty}^{\infty} |G(f)|^2 \, df \tag{62c}$$

3.5 Concentration Problem – Continuous/Continuous Case

is a measure of concentration in the frequency domain.

These measures of concentration have considerable intuitive appeal. They have also had a profound effect on current research in spectral analysis because an analytic solution to the following problem leads to a way of characterizing time-limited and band-limited functions. The task at hand is the following: among all functions that are band-limited to $[-W, W]$, find all those such that $\alpha^2(T)$ is as large as possible, i.e., that have the greatest concentration of their energy in the time interval $[-T/2, T/2]$ and hence are approximately time-limited to that interval. Since a nontrivial band-limited function $g_W(\cdot)$ cannot also be time-limited, we must have $1 > \alpha^2(T) > 0$. We now outline the steps that can be taken to solve this problem by following Slepian (1983).

Expressing $g_W(\cdot)$ in terms of its Fourier transform $G_W(\cdot)$ and making use of Equation (62b) with $g_W(\cdot)$ substituted for $g(\cdot)$ yield

$$\alpha^2(T, W) = \frac{\int_{-W}^{W} \int_{-W}^{W} G_W(f) K(f, f'; T) G_W^*(f') \, df' \, df}{\int_{-W}^{W} |G_W(f)|^2 \, df}, \tag{63a}$$

where

$$K(f, f'; T) \stackrel{\text{def}}{=} \frac{\sin(\pi T[f - f'])}{\pi[f - f']};$$

here we have added a second argument to $\alpha^2(\cdot)$ to emphasize its dependence on W as well as T. It follows from the calculus of variations (see, for example, Courant and Hilbert, 1953) that all functions $G_W(\cdot)$ that maximize $\alpha^2(T, W)$ must satisfy the following integral equation:

$$\int_{-W}^{W} K(f, f'; T) G_W(f') \, df' = \alpha^2(T, W) G_W(f), \quad |f| \leq W. \tag{63b}$$

An equation of this form is known in the literature as a *Fredholm equation of the second kind*; the quantity $K(\cdot, \cdot; T)$ is called the *kernel* of the integral equation. Engineers will recognize this kernel as the function defined by $T \operatorname{sinc}[T(f - f')]$, where $\operatorname{sinc}(\cdot)$ is the sinc function:

$$\operatorname{sinc}(t) \stackrel{\text{def}}{=} \frac{\sin(\pi t)}{\pi t}.$$

The unknown quantities that a solution to this equation gives us are the functions $G_W(\cdot)$ and the corresponding scalar $\alpha^2(T, W)$. The theory for integral equations such as Equation (63b) is dependent on the nature of the associated kernel. Here the kernel is such that there is a countably infinite number of solutions to Equation (63b) (this statement regards functions that differ only by a multiplicative constant – such as $G_W(\cdot)$, $2G_W(\cdot)$, $\pi G_W(\cdot)$, etc. – as being a single solution). Not all of these solutions are also solutions to our maximization problem. If we let

$$G_{W,k}(\cdot) \text{ and } \alpha_k^2(T, W), \quad k = 0, 1, \ldots,$$

represent these solutions, where $1 > \alpha_0^2(T, W) \geq \alpha_1^2(T, W) \geq \cdots$, one solution to our maximization problem is given by $G_{W,0}(\cdot)$, and the degree of concentration of energy is given by $\alpha_0^2(T, W)$.

This maximization problem that we have just described reappears in another form when we discuss filters and data tapers in Chapters 5 and 8. The set of all solutions to the integral equation of Equation (63b) is also of considerable interest. We summarize some of their properties here. We first make the following substitutions:

$$x = f/W; \quad y = f'/W; \quad \text{and} \quad c = \pi WT.$$

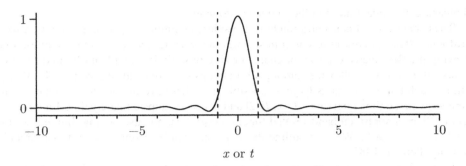

Figure 64 Prolate spheroidal wave function $\psi_0(\cdot\,; 4)$ (PSWF) and associated rescaled Fourier transform. The PSWF is the portion of the curve between the dashed vertical lines, while its rescaled Fourier transform is the entire curve.

Equation (63b) now becomes

$$\int_{-1}^{1} \frac{\sin(c[x-y])}{\pi[x-y]} G_W(Wy)\,dy = \alpha^2(T,W) G_W(Wx), \quad |x| \le 1.$$

The theory of integral equations says that the solutions to the above equation depend only upon the properties of the associated kernel. The kernel depends upon T and W only through $c = \pi W T$, so we define

$$\psi(y;c) = G_W(Wy) \text{ and } \lambda(c) = \alpha^2(T,W)$$

to reflect this dependence. Our integral equation now takes the form

$$\int_{-1}^{1} \frac{\sin(c[x-y])}{\pi[x-y]} \psi(y;c)\,dy = \lambda(c)\psi(x;c), \quad |x| \le 1. \tag{64}$$

We denote the countably infinite set of functions that solve the above by

$$\psi_0(\cdot\,;c), \psi_1(\cdot\,;c), \psi_2(\cdot\,;c), \ldots,$$

and the corresponding proportions of energy by

$$\lambda_0(c), \lambda_1(c), \lambda_2(c), \ldots.$$

The solution $\psi_k(\cdot\,;c)$ is called the kth *eigenfunction*, and $\lambda_k(c)$ is the associated *eigenvalue*. The eigenfunction corresponding to each eigenvalue is unique except for a multiplicative constant. These eigenfunctions are called *prolate spheroidal wave functions* (PSWFs). The PSWF $\psi_0(\cdot\,;4)$ is shown in Figure 64 – it is the portion of the curve between the two dashed vertical lines located at $x = \pm 1$. Here are some of the important properties of these eigenfunctions and eigenvalues.

[1] The eigenfunctions are real-valued and orthogonal on $[-1, 1]$; i.e.,

$$\int_{-1}^{1} \psi_j(x;c)\psi_k(x;c)\,dx = 0, \text{ for } j \ne k.$$

[2] The eigenvalues have the following properties:

$$\lambda_k(c) > 0 \text{ for all } k;\ 1 > \lambda_0(c) > \lambda_1(c) > \cdots;\ \text{and } \lim_{k\to\infty} \lambda_k(c) = 0.$$

3.5 Concentration Problem – Continuous/Continuous Case

[3] Equation (64) only defines $\psi(\cdot;c)$ for $|x| \leq 1$. The left-hand side of that equation, however, is well defined for all x, so we can *define*

$$\psi(x;c) = \lambda(c)^{-1} \int_{-1}^{1} \frac{\sin(c[x-y])}{\pi[x-y]} \psi(y;c)\,\mathrm{d}y, \quad |x| > 1. \tag{65}$$

With this definition, it can be shown that the eigenfunctions are orthogonal over $\mathbb{R}$ as well as on $[-1,1]$. Moreover, the Fourier transform of $\psi_k(\cdot;c)$ restricted to $|x| \leq 1$ has the same form as $\psi_k(\cdot;c)$ except for a scale change; i.e.,

$$\int_{-1}^{1} \psi_k(x;c) \mathrm{e}^{-\mathrm{i}2\pi xt}\,\mathrm{d}x \propto \psi_k(2\pi t/c;c), \quad t \in \mathbb{R}.$$

The curve in Figure 64 shows $\psi_0(\pi t/2;4)$ versus t for $|t| \leq 10$ scaled to the same values as $\psi_0(\cdot;4)$ (recall that this is shown by the segment of the curve between the dashed vertical lines). Similarly,

$$\int_{-\infty}^{\infty} \psi_k(x;c) \mathrm{e}^{-\mathrm{i}2\pi xt}\,\mathrm{d}x \propto \begin{cases} \psi_k(2\pi t/c;c), & |t| \leq c/(2\pi); \\ 0, & \text{otherwise.} \end{cases}$$

[4] If we normalize the eigenfunctions to have an energy of unity over $\mathbb{R}$, it follows that

$$\int_{-\infty}^{\infty} \psi_j(x;c)\psi_k(x;c)\,\mathrm{d}x = \delta_{j-k}$$

(here δ_n is Kronecker's delta function; i.e., $\delta_n = 1$ if $n = 0$ and $= 0$ if n is a nonzero integer) and that

$$\int_{-1}^{1} \psi_j(x;c)\psi_k(x;c)\,\mathrm{d}x = \lambda_k(c)\delta_{j-k}.$$

The PSWFs are thus orthogonal over $[-1,1]$ and orthonormal over $\mathbb{R}$. The value $\lambda_k(c)$ gives the proportion of energy in the interval $[-1,1]$.

[5] At this point we can return to the solution to the concentration problem posed in this section. Recall that $g_W(\cdot)$ is to be a function whose Fourier transform $G_W(\cdot)$ is zero outside $[-W,W]$. If we make

$$G_{W,k}(f) \propto \begin{cases} \psi_k(f/W;c), & |f| \leq W; \\ 0, & \text{otherwise,} \end{cases}$$

where $c = \pi W T$, then $G_{W,k}(\cdot)$ is zero outside $[-W,W]$, and its inverse Fourier transform $g_{W,k}(\cdot)$ is such that $g_{W,k}(t) \propto \psi_k(2t/T;c)$. Thus there is only one solution to the concentration problem, namely, the band-limited function $g_{W,0}(\cdot) = \psi_0(2\cdot/T;c)$ (if we count solutions that differ only by a multiplicative constant as just one solution). For this solution $\lambda_0(c)$ represents the degree of concentration of $g_{W,0}(\cdot)$ in $[-T/2,T/2]$. Among all functions that are band-limited to $[-W,W]$ and are orthogonal to $g_{W,0}(\cdot)$, the function $g_{W,1}(\cdot) = \psi_1(2\cdot/T;c)$ is most concentrated in $[-T/2,T/2]$ with a degree of concentration of $\lambda_1(c) < \lambda_0(c)$. This pattern continues: amongst all functions that are band-limited to $[-W,W]$ and that are orthogonal to $g_{W,0}(\cdot), g_{W,1}(\cdot), \ldots, g_{W,k-1}(\cdot)$, the function $g_{W,k}(\cdot)$ is most concentrated in $[-T/2,T/2]$ with a degree of concentration of $\lambda_k(c)$.

[6] For large c the eigenvalue series $\lambda_k(c)$ drops sharply from approximately unity to nearly zero at a value of k known as the *Shannon number*, namely,

$$k = 2c/\pi = 2WT.$$

This quantity is fundamental in electrical engineering and has a vital role in spectral analysis (see Chapter 8). Loosely speaking, if $2WT$, the so-called duration–bandwidth product is large (and hence also c), the space of functions of approximate duration T and approximate bandwidth $2W$ has approximate dimension (or complex degrees of freedom) $2WT$. Since there are no nontrivial functions that are both time-limited and band-limited, this statement is necessarily vague (Slepian, 1976, has made it mathematically precise).

[7] The eigenfunctions $\psi_k(\cdot; c)$ are even or odd as k is even or odd; $\psi_k(\cdot; c)$ has exactly k zeros in the interval $[-1, 1]$.

[8] With $c = 2\pi W$ the eigenfunctions $\psi_k(\cdot; c)$ form a complete basis for the class of all functions that are band-limited on $[-W, W]$. This means that, if $g_W(\cdot)$ is *any* $L^2(\mathbb{R})$ function in that band-limited class, it can be represented as

$$g_W(t) \stackrel{\text{ms}}{=} \sum_{k=0}^{\infty} \gamma_k \psi_k(t; c), \quad t \in \mathbb{R}, \tag{66a}$$

where

$$\gamma_k \stackrel{\text{def}}{=} \int_{-\infty}^{\infty} g_W(t) \psi_k(t; c) \, dt.$$

It follows from the symmetry between time and frequency that these eigenfunctions also form a complete basis for the class of $L^2(-1, 1)$ functions.

As an application of this theory, let us consider the problem of extrapolating a band-limited function (Slepian and Pollak, 1961). Suppose that we only know $g_W(\cdot)$ over the interval $[-1, 1]$ and that we wish to extrapolate it outside this interval. Since $g_W(\cdot)$ can be represented as in Equation (66a), we can multiply both sides of that equation by $\psi_j(\cdot; c)$ and integrate over $[-1, 1]$ to get (after exchanging the order of integration and summation on the right-hand side)

$$\int_{-1}^{1} \psi_j(t; c) g_W(t) \, dt = \sum_{k=0}^{\infty} \gamma_k \int_{-1}^{1} \psi_j(t; c) \psi_k(t; c) \, dt = \gamma_j \lambda_j(c)$$

(from property [4] earlier in this section), or, equivalently,

$$\gamma_k = \lambda_k^{-1}(c) \int_{-1}^{1} g_W(t) \psi_k(t; c) \, dt.$$

Thus, given $g_W(\cdot)$ over the interval $[-1, 1]$, we can extrapolate to values outside this interval by using

$$g_W(t) \stackrel{\text{ms}}{=} \sum_{k=0}^{\infty} \lambda_k^{-1}(c) \left[\int_{-1}^{1} g_W(t') \psi_k(t'; c) \, dt' \right] \psi_k(t; c); \tag{66b}$$

i.e., it is possible to extrapolate the band-limited function $g_W(\cdot)$ *perfectly* (at least in the mean square sense) just from knowledge of its form over the interval $[-1, 1]$. If we consider an interval of nonzero length other than $[-1, 1]$ and if we suitably rescale the time axis, we can alter the development leading to Equation (66b) to show that a band-limited function can be reconstructed perfectly (in mean square) from knowledge of its values over *any* interval of nonzero length.

Now suppose that we truncate the infinite summation in Equation (66b) at, say, $k = m$, and use

$$g_m(t) \stackrel{\text{def}}{=} \sum_{k=0}^{m} \gamma_k \psi_k(t; c)$$

to approximate $g_W(\cdot)$. The energy in the error of fit of $g_m(\cdot)$ to $g_W(\cdot)$ in $[-1, 1]$ is

$$\int_{-1}^{1} |g_W(t) - g_m(t)|^2 \, dt = \sum_{k=m+1}^{\infty} \gamma_k^2 \lambda_k(c).$$

From property [5] in this section, $\lambda_k(c)$ is close to 0 if m exceeds the Shannon number $2c/\pi$. Hence the error in $[-1, 1]$ will be small if m is chosen thus. However, the energy in the error of fit in $\mathbb{R}$ is

$$\int_{-\infty}^{\infty} |g_W(t) - g_m(t)|^2 \, dt = \sum_{k=m+1}^{\infty} \gamma_k^2,$$

which does not depend on $\lambda_k(c)$.

3.6 Convolution Theorem – Continuous Time and Frequency

We state and prove here one version of the widely used convolution theorem. This theorem is often paraphrased as "convolution in the time domain is equivalent to multiplication in the frequency domain." Let $g(\cdot)$ and $h(\cdot)$ be two real- or complex-valued functions. The convolution of $g(\cdot)$ and $h(\cdot)$ is the function of t defined by

$$\int_{-\infty}^{\infty} g(u) h(t - u) \, du,$$

provided the integral exists. Convolution involves reflecting one of the functions about the time axis, shifting it by t units, multiplying it by the corresponding coordinate of the other function and integrating this product over $\mathbb{R}$. The function just defined is conveniently denoted by $g * h(\cdot)$, and its value at t by $g * h(t)$; i.e.,

$$g * h(t) \stackrel{\text{def}}{=} \int_{-\infty}^{\infty} g(u) h(t - u) \, du. \tag{67a}$$

A change of variable in the integral shows that $h * g(\cdot) = g * h(\cdot)$. Figure 68 illustrates the construction of the convolution of rectangular functions of different widths, while Figure 69 shows convolutions involving rectangular functions we will have occasion to reference (Exercise [3.11] asks the reader to verify the contents of this figure).

The following key result says that the Fourier transform of $g * h(\cdot)$ has a particularly simple form.

▷ **Exercise [67]** By informally manipulating the order in which integrations are carried out, show that the Fourier transform of $g * h(\cdot)$ is the function defined by $G(f)H(f)$; i.e.,

$$\int_{-\infty}^{\infty} g * h(t) e^{-i2\pi f t} \, dt = G(f) H(f). \tag{67b}$$ ◁

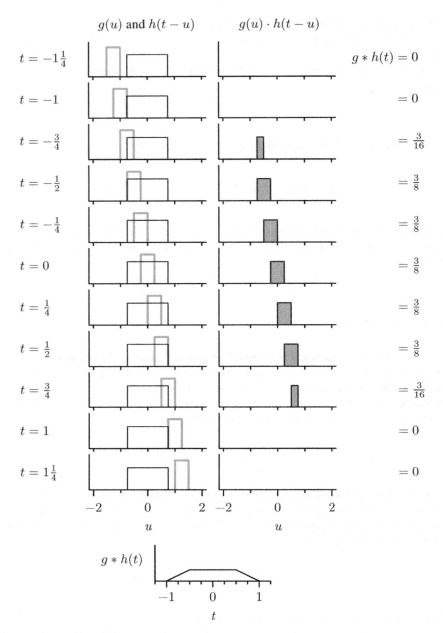

Figure 68 Convolution of functions given by

$$g(u) = \begin{cases} 3/4, & |u| < 3/4 \text{ and} \\ 0, & \text{otherwise} \end{cases} \text{ and } h(u) = \begin{cases} 1, & |u| < 1/4 \text{ and} \\ 0, & \text{otherwise.} \end{cases}$$

The black and gray rectangles in the left-hand column show, respectively, $g(\cdot)$ and the functions of u given by $h(t - u)$ for selected values of t. Each plot in the right-hand column shows the product of the two functions in the left-hand plot. The integral of this product is $g * h(t)$, the convolution of $g(\cdot)$ and $h(\cdot)$ at t. The bottom plot shows $g * h(t)$ for $t \in [-1\frac{1}{4}, 1\frac{1}{4}]$.

In other words, the Fourier transform of the convolution of $g(\cdot)$ and $h(\cdot)$ is the product of the

3.6 Convolution Theorem – Continuous Time and Frequency

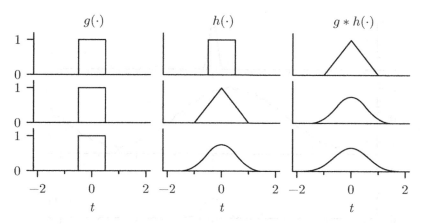

Figure 69 Examples of convolutions involving a rectangular function $g(\cdot)$ of unit height that is zero outside the interval $[-1/2, 1/2]$. The top row shows that the convolution of a rectangular function with itself yields a triangular function. The right-hand plot in the middle row shows the result of convolving a rectangular function and a triangular function. The bottom row shows what happens if we convolve a rectangular function with the function in the right-hand middle plot; i.e., the right-hand plot in the bottom row is the successive convolution of four identical rectangular functions (Exercise [3.11] is to verify mathematical expressions for these convolutions).

Fourier transforms of $g(\cdot)$ and $h(\cdot)$. This fact is written as

$$g * h(\cdot) \longleftrightarrow G(\cdot)H(\cdot).$$

There are a variety of convolution theorems that stipulate under what conditions on $g(\cdot)$ and $h(\cdot)$ the above makes sense rigorously. Exercise [3.12] concerns one of these theorems; see Champeney (1987) for others.

Convolution is often regarded as a smoothing operation (this is used extensively in Chapter 7). In this interpretation, $h(\cdot)$ is regarded as a function we desire to smooth, while $g(\cdot)$ is called a *smoothing kernel*. The convolution $g * h(\cdot)$ is then a smoothed version of $h(\cdot)$ whose degree of smoothness is determined by the properties of $g(\cdot)$. As a concrete example, let us assume that

$$h(t) = \sum_{l=1}^{L} A_l \cos\left(2\pi f_l t + \phi_l\right), \tag{69a}$$

so that our function is just a sum of sinusoids with different amplitudes, frequencies and phases. For the smoothing kernel let us pick the Gaussian PDF

$$g(t) = \frac{1}{(2\pi\sigma^2)^{1/2}} e^{-t^2/(2\sigma^2)}$$

with zero mean and standard deviation $\sigma > 0$ – here σ plays the role of an adjustable smoothing parameter (the boundedness of $h(\cdot)$ in Equation (69a) and the integrability of $g(\cdot)$ is sufficient for the convolution integral of Equation (67a) to exist). Now the Fourier transform of the function defined by $\exp\left(-\pi t^2\right)$ is the function itself (Bracewell, 2000, p. 107), a fact implying that

$$\int_{-\infty}^{\infty} e^{-\pi t^2} \cos\left(2\pi f t\right) dt = e^{-\pi f^2} \quad \text{and} \quad \int_{-\infty}^{\infty} e^{-\pi t^2} \sin\left(2\pi f t\right) dt = 0.$$

▷ **Exercise [69]** Use these two integral expressions – along with the trigonometric identity $\cos(a - b) = \cos(a)\cos(b) + \sin(a)\sin(b)$ – to show that

$$g * h(t) = \sum_{l=1}^{L} e^{-(\sigma 2\pi f_l)^2/2} A_l \cos\left(2\pi f_l t + \phi_l\right). \tag{69b}$$ ◁

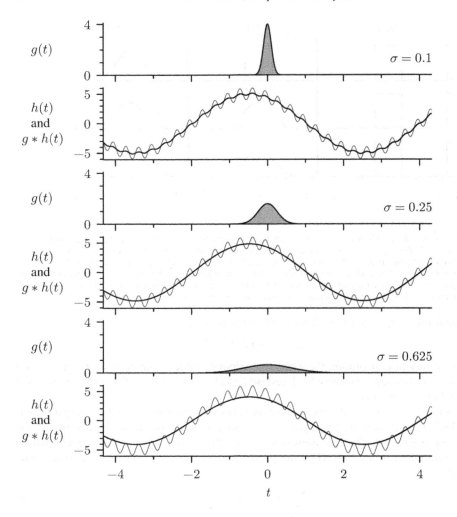

Figure 70 Convolution as a smoothing operation. Each shaded curve is a Gaussian smoothing kernel $g(\cdot)$, below which there is a plot showing $h(\cdot)$ of Equation (70) (light wiggly curve) and the convolution $g * h(\cdot)$ (dark curve), which we can regard as a smoothed version of $h(\cdot)$.

Note that the only difference between $h(\cdot)$ and $g * h(\cdot)$ is in the amplitudes assigned to the sinusoids – their frequencies and phases are unchanged. The original amplitude A_l is multiplied by the attenuation factor $\exp\left[-(\sigma 2\pi f_l)^2/2\right]$, which must lie between 0 and 1. For fixed σ, this factor is smaller for an amplitude associated with a high frequency term than for one associated with a low frequency term. This attenuation of the high frequency contributions to $h(\cdot)$ makes $g * h(\cdot)$ smoother in appearance. The degree of smoothness can be controlled by adjusting σ – the larger σ is, the more A_l is attenuated. Figure 70 illustrates this for the case

$$h(t) = 5\cos\left(2\pi f_1 t + 0.5\right) + \cos\left(2\pi f_2 t + 1.1\right), \tag{70}$$

with $f_1 = 1/6$ and $f_2 = 3$. For $\sigma = 0.1$, 0.25 and 0.625, the attenuation factors for the low frequency term f_1 are, respectively to two decimal places, 0.99, 0.97 and 0.81; the corresponding factors for the high frequency term f_2 are 0.17, 0.0 and 0.0. If we define the proper degree of smoothing here to mean that $g * h(\cdot)$ should be as close as possible to the low fre-

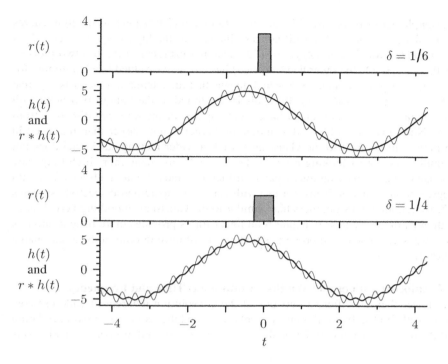

Figure 71 Convolution with a rectangular smoothing kernel $r(\cdot)$ (cf. Figure 70).

quency term in $h(\cdot)$, then Figure 70 shows that the choice of 0.25 for σ is preferable to 0.1 (undersmoothing) or 0.625 (oversmoothing).

To facilitate a discussion in Chapter 7, let us reconsider this example by looking at another widely used smoothing kernel, namely, the rectangular kernel defined by

$$r(t) = \begin{cases} 1/(2\delta), & |t| \leq \delta; \\ 0, & |t| > \delta. \end{cases}$$

The parameter δ determines the degree of smoothing. Note that

$$r * h(t) = \frac{1}{2\delta} \int_{t-\delta}^{t+\delta} h(u)\,du, \tag{71}$$

a quantity that in calculus books is called the *average value* of the function $h(\cdot)$ over the interval from $t - \delta$ to $t + \delta$.

▷ **Exercise [71]** With $h(\cdot)$ defined by Equation (69a), verify that

$$r * h(t) = \sum_{l=1}^{L} \operatorname{sinc}(2f_l\delta) A_l \cos(2\pi f_l t + \phi_l), \quad \text{where } \operatorname{sinc}(u) \stackrel{\text{def}}{=} \frac{\sin(\pi u)}{\pi u}. \quad ◁$$

It is interesting to compare this smoothed version of $h(\cdot)$ with $g * h(\cdot)$ in Equation (69b), which uses a Gaussian smoothing kernel. Note that, in the latter case, the attenuation factor $\exp[-(\sigma 2\pi f_l)^2/2]$ for the lth term in $h(\cdot)$ decreases monotonically from 1 to 0 as the smoothing parameter σ increases from 0 to ∞. This is not true for $r * h(\cdot)$ since $\operatorname{sinc}(2f_l\delta)$ oscillates about 0 with decreasing amplitude as δ increases. The effect of varying δ in $r * h(\cdot)$ is thus less transparent than varying σ in $g * h(\cdot)$.

As an example, let us reconsider $h(\cdot)$ given by Equation (70) (shown as the light curves in Figure 71). Since $\text{sinc}(u) = 0$ for all nonzero integers u, the high frequency term f_2 is completely eliminated for $\delta = u/(2f_2) = u/6$. This is illustrated in the top two plots of Figure 71 with $\delta = 1/6$. On the other hand, $\text{sinc}(u)$ has a negative-valued local minimum for u near $(4k-1)/2$ for positive integers k, which means that the attenuation factor is negative for $\delta = (4k-1)/(4f_2) = (4k-1)/12$. This is illustrated in the bottom two plots with $\delta = 1/4$, for which $r * h(\cdot)$ is seen to have small ripples in the *opposite* direction from those in $h(\cdot)$. Note that the width of the smoothing kernel is now larger than that used in the top two plots – in contrast to the Gaussian kernel, a wider rectangular kernel does not necessarily imply a smoother looking $r * h(\cdot)$. For an $h(\cdot)$ with many more high frequency components than in our simple example, it can be difficult to find a δ that yields a $r * h(\cdot)$ with a desired degree of smoothness. Thus, for use with various nonparametric spectral estimates in Chapter 7, we recommend smoothers that, similar to the Gaussian smoothing kernel, have monotonic attenuation properties. (Chapter 5 on linear filters provides more insight into this example and discusses some of the properties that $g(\cdot)$ must have in order to be a reasonable smoothing kernel.)

3.7 Autocorrelations and Widths – Continuous Time and Frequency

There are no fewer than 20 different versions of the convolution theorem in wide use (see p. 118 of Bracewell, 2000). Two important variants are called the (complex) cross-correlation theorem and the (complex) autocorrelation theorem. The cross-correlation of $g(\cdot)$ and $h(\cdot)$ is defined by

$$g \star h^*(t) = \int_{-\infty}^{\infty} g(u+t) h^*(u) \, du, \tag{72a}$$

where $\star$ is defined by

$$a \star b(t) = \int_{-\infty}^{\infty} a(u+t) b(u) \, du.$$

This is similar to – but not the same as – the convolution integral since

$$a * b(t) = \int_{-\infty}^{\infty} a(u) b(t-u) \, du = \int_{-\infty}^{\infty} a(u+t) b(-u) \, du.$$

The cross-correlation of $g(\cdot)$ and $h(\cdot)$ is in fact equal to the convolution of $g(\cdot)$ with the *time reversed* version of $h^*(\cdot)$. Whereas ordinary convolution is commutative since $g * h(t) = h * g(t)$, cross-correlation need not be because

$$h \star g^*(t) = \int_{-\infty}^{\infty} h(u+t) g^*(u) \, du = \int_{-\infty}^{\infty} g^*(u-t) h(u) \, du,$$

which in general is not equal to

$$g \star h^*(t) = \int_{-\infty}^{\infty} g(u+t) h^*(u) \, du,$$

even if both $g(\cdot)$ and $h(\cdot)$ are real-valued. Exercise [3.13a] says that

$$g \star h^*(\cdot) \longleftrightarrow G(\cdot) H^*(\cdot). \tag{72b}$$

If we put $h(\cdot) = g(\cdot)$, the cross-correlation becomes an autocorrelation:

$$g \star g^*(t) = \int_{-\infty}^{\infty} g(u+t) g^*(u) \, du, \tag{72c}$$

3.7 Autocorrelations and Widths – Continuous Time and Frequency

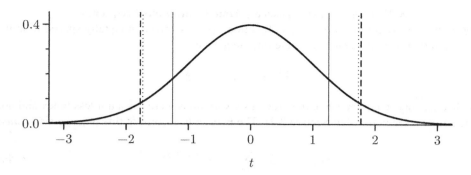

Figure 73 Standard Gaussian PDF $g_1(\cdot)$ (dark curve), along with pairs of vertical lines whose separation depicts width$_e$ $\{g_1(\cdot)\}$ (the equivalent width, solid lines), width$_v$ $\{g_1(\cdot)\}$ (the variance width, dotted) and width$_a$ $\{g_1(\cdot)\}$ (the autocorrelation width, dashed).

from which we have

$$g \star g^*(\cdot) \longleftrightarrow G(\cdot)G^*(\cdot) = |G(\cdot)|^2. \tag{73a}$$

The autocorrelation can be used to define another measure of the width of a function $g(\cdot)$. This measure is simply the equivalent width of the autocorrelation of $g(\cdot)$:

$$\text{width}_a\{g(\cdot)\} \stackrel{\text{def}}{=} \text{width}_e\{g \star g^*(\cdot)\} = \frac{\int_{-\infty}^{\infty} g \star g^*(t)\, dt}{g \star g^*(0)}. \tag{73b}$$

▷ **Exercise [73a]** Show that

$$\text{width}_a\{g(\cdot)\} = \frac{\left|\int_{-\infty}^{\infty} g(t)\, dt\right|^2}{\int_{-\infty}^{\infty} |g(t)|^2\, dt}. \tag{73c} \triangleleft$$

Note that the autocorrelation width of a function is the reciprocal of the equivalent width of its energy spectral density function since

$$\frac{\left|\int_{-\infty}^{\infty} g(t)\, dt\right|^2}{\int_{-\infty}^{\infty} |g(t)|^2\, dt} = \frac{|G(0)|^2}{\int_{-\infty}^{\infty} |G(f)|^2\, df}.$$

Let us check that the autocorrelation width makes sense in two simple examples. For the rectangular function of Equation (60a), we have

$$\int_{-\infty}^{\infty} r(t; \mu_r, W_r)\, dt = 1 \quad \text{and} \quad \int_{-\infty}^{\infty} |r(t; \mu_r, W_r)|^2\, dt = \frac{1}{2W_r}.$$

It follows from Equation (73c) that

$$\text{width}_a\{r(\cdot; \mu_r, W_r)\} = 2W_r,$$

which is intuitively reasonable. For the second example, let us consider the PDF for a Gaussian RV.

▷ **Exercise [73b]** Show that the autocorrelation width for the PDF $g_\sigma(\cdot)$ of Equation (55b) is $2\sigma\sqrt{\pi} \doteq 3.54\sigma$. ◁

The autocorrelation width is thus proportional to the standard deviation σ, which again is intuitively reasonable. By comparison, using Equations (58b) and (60b), the equivalent width and the variance width are $\sigma\sqrt{(2\pi)} \doteq 2.51\sigma$ and $2\sigma\sqrt{3} \doteq 3.46\sigma$. All three widths are depicted in Figure 73 for the case $\sigma = 1$.

We will use the autocorrelation width throughout Chapters 6, 7 and 8 as a unified way of defining the bandwidth for nonparametric estimators of the spectral density function.

3.8 Fourier Theory – Discrete Time/Continuous Frequency

Suppose that the continuous time $L^2(\mathbb{R})$ function $g(\cdot)$ is sampled at equally spaced time intervals of duration Δ_t to create a discrete time sequence

$$g_t \stackrel{\text{def}}{=} g(t\,\Delta_t), \quad t \in \mathbb{Z}$$

(note that, as opposed to its use in the previous six sections, here t is a unitless index and not physical time – see item [3] of Section 2.2). The Fourier transform of this sequence is *defined* to be

$$G_p(f) = \Delta_t \sum_{t=-\infty}^{\infty} g_t e^{-i2\pi f t \Delta_t} \tag{74a}$$

(the forthcoming Equation (74b) gives the rationale for the subscript p). The above is commonly referred to as the *discrete Fourier transform* (DFT) of an infinite sequence. We motivate this definition by considering the *continuous time* function defined by the product of Δ_t, $g(\cdot)$ and an infinite set of equally spaced Dirac delta functions; i.e.,

$$g_\delta(t) \stackrel{\text{def}}{=} \Delta_t\, g(t) \sum_{k=-\infty}^{\infty} \delta(t - k\,\Delta_t).$$

With this definition, for ϵ small enough,

$$g_t = \frac{1}{\Delta_t} \int_{t\Delta_t - \epsilon}^{t\Delta_t + \epsilon} g_\delta(u)\,\mathrm{d}u, \quad t \in \mathbb{Z}.$$

Let us regard $g_\delta(\cdot)$ as an $L^2(\mathbb{R})$ function – technically it is not, but we can manipulate it as if it were. Equation (54a) states that the Fourier transform of $g_\delta(\cdot)$ is

$$G_\delta(f) = \int_{-\infty}^{\infty} g_\delta(t) e^{-i2\pi f t}\,\mathrm{d}t = \int_{-\infty}^{\infty} \left(\Delta_t\, g(t) \sum_{k=-\infty}^{\infty} \delta(t - k\,\Delta_t) \right) e^{-i2\pi f t}\,\mathrm{d}t$$

$$= \Delta_t \sum_{k=-\infty}^{\infty} \int_{-\infty}^{\infty} g(t) \delta(t - k\,\Delta_t) e^{-i2\pi f t}\,\mathrm{d}t$$

$$= \Delta_t \sum_{k=-\infty}^{\infty} g_k e^{-i2\pi f k \Delta_t},$$

which, if we replace the dummy index k by t, is simply the right-hand side of Equation (74a).

Equation (74a) is effectively a rectangular integration approximation of Equation (54a) with the term Δ_t ensuring conservation of integrated area between the two equations as $\Delta_t \to 0$. Recalling that $e^{i2\pi t} = 1$ for all $t \in \mathbb{Z}$, we see that

$$G_p(f + \tfrac{1}{\Delta_t}) = \Delta_t \sum_{t=-\infty}^{\infty} g_t e^{-i2\pi(f + \frac{1}{\Delta_t})t \Delta_t} = \Delta_t \sum_{t=-\infty}^{\infty} g_t e^{-i2\pi f t \Delta_t} e^{-i2\pi t} = G_p(f), \tag{74b}$$

so $G_p(\cdot)$ is periodic with period $1/\Delta_t$ (hence the subscript p). Suppose for the moment that we consider $G_p(\cdot)$ to be a function of time such that

$$\int_{-1/(2\Delta_t)}^{1/(2\Delta_t)} |G_p(t)|^2\,\mathrm{d}t < \infty.$$

3.8 Fourier Theory – Discrete Time/Continuous Frequency

From our discussion of the Fourier series representation of square integrable periodic functions in Section 3.1, we know from Equation (49b) (with $T = 1/\Delta_t$) that the Fourier coefficients for $G_p(\cdot)$ are, say,

$$\tilde{g}_n = \frac{1}{T} \int_{-T/2}^{T/2} G_p(t) e^{-i2\pi f_n t} \, dt = \Delta_t \int_{-1/(2\Delta_t)}^{1/(2\Delta_t)} G_p(t) e^{-i2\pi t n \Delta_t} \, dt$$

since $f_n = n/T = n\Delta_t$ here. The Fourier synthesis of $G_p(\cdot)$ follows from Equation (49c) and tells us that (at least in the mean square sense)

$$G_p(t) = \sum_{n=-\infty}^{\infty} \tilde{g}_n e^{i2\pi f_n t} = \sum_{n=-\infty}^{\infty} \tilde{g}_n e^{i2\pi t n \Delta_t}.$$

Changing the dummy variables (i) n to t and (ii) t to f in the above yields

$$\tilde{g}_t = \Delta_t \int_{-1/(2\Delta_t)}^{1/(2\Delta_t)} G_p(f) e^{-i2\pi f t \Delta_t} \, df \text{ and } G_p(f) = \sum_{t=-\infty}^{\infty} \tilde{g}_t e^{i2\pi f t \Delta_t}.$$

Finally, letting $\tilde{g}_t = g_{-t} \Delta_t$, we obtain

$$g_t = \int_{-1/(2\Delta_t)}^{1/(2\Delta_t)} G_p(f) e^{i2\pi f t \Delta_t} \, df \tag{75a}$$

and

$$G_p(f) = \Delta_t \sum_{t=-\infty}^{\infty} g_t e^{-i2\pi f t \Delta_t},$$

which is the same as our definition of the DFT of $\{g_t\}$ in Equation (74a). We can now say that the inverse DFT of $G_p(\cdot)$ is given by Equation (75a). This development of the inverse transform for the discrete time/continuous frequency case by appealing to the continuous time/discrete frequency case emphasizes the duality between time and frequency: these two cases are in many ways complementary. For example, note that sampling in the time (or frequency) domain results in periodization in the frequency (or time domain).

We can now restate and expand upon some of the results we found in Section 3.1 in terms of our present case of discrete time/continuous frequency. If $\{g_t\}$ is real-valued, then its DFT is conjugate symmetric about $f = 0$; i.e., $G_p^*(-f) = G_p(f)$. Parseval's theorem – Equation (49d) – becomes

$$\Delta_t \sum_{t=-\infty}^{\infty} |g_t|^2 = \int_{-1/(2\Delta_t)}^{1/(2\Delta_t)} |G_p(f)|^2 \, df. \tag{75b}$$

Very often Δ_t is taken to be unity – as we shall do in the remainder of this section for simplicity – so that the Fourier transform and its inverse become

$$G_p(f) = \sum_{t=-\infty}^{\infty} g_t e^{-i2\pi f t} \text{ and } g_t = \int_{-1/2}^{1/2} G_p(f) e^{i2\pi f t} \, df.$$

If we are only given g_t for $t = -m, \ldots, m$, then the truncated Fourier series

$$G_{p,m}(f) \stackrel{\text{def}}{=} \sum_{t=-m}^{m} g_t e^{-i2\pi f t}$$

minimizes the integrated magnitude square error between $G_p(\cdot)$ and functions of the form

$$H_{p,m}(f) \stackrel{\text{def}}{=} \sum_{t=-m}^{m} h_t e^{-i2\pi ft};$$

i.e., the quantity

$$\int_{-1/2}^{1/2} |G_p(f) - H_{p,m}(f)|^2 \, df$$

is minimized by setting $h_t = g_t$ (see the discussion surrounding Equation (52)).

What other properties does $G_{p,m}(\cdot)$ possess?

▷ **Exercise [76]** Show that

$$G_{p,m}(f) = (2m+1) \int_{-1/2}^{1/2} G_p(f') \mathcal{D}_{2m+1}(f - f') \, df', \tag{76}$$

where $\mathcal{D}_{2m+1}(\cdot)$ is Dirichlet's kernel (see Equation (17c)). ◁

Thus $G_{p,m}(\cdot)$ is the convolution of $G_p(\cdot)$ and the function given by $2m+1$ times Dirichlet's kernel; note, however, that here convolution is defined differently than in Section 3.6, where we introduced it for the continuous/continuous case – since $G_p(\cdot)$ and Dirichlet's kernel are both periodic functions with unit period, the limits on the integral in the convolution are now over one complete cycle instead of from $-\infty$ to ∞. Figure 77 shows plots of $\mathcal{D}_{2m+1}(f)$ versus f for $m = 4$, 16 and 64. Since $\mathcal{D}_{2m+1}(0) = 1$ for all m, the amplitude of the central lobe of $(2m+1)\mathcal{D}_{2m+1}(\cdot)$ grows as m increases whereas, as can be seen in the plots, its width decreases; the same holds for all the sidelobes. Because

$$\int_{-1/2}^{1/2} (2m+1)\mathcal{D}_{2m+1}(f) \, df = \int_{-1/2}^{1/2} \sum_{t=-m}^{m} e^{i2\pi ft} \, df = 1,$$

the area under all the lobes of $(2m+1)\mathcal{D}_{2m+1}(\cdot)$ is constant.

Let us look in more detail at the effect of convolving a function with Dirichlet's kernel. Both the central lobe and the sidelobes of this kernel will cause $G_{p,m}(\cdot)$ to be a distorted version of $G_p(\cdot)$. The effect of the central lobe is to smooth out sharp features (peaks) in $G_p(\cdot)$ that are small compared to the width of the lobe – this is often referred to as a *loss of resolution* due to the use of a finite amount of data. This width can be measured roughly as half the distance between the two nulls of $\mathcal{D}_{2m+1}(\cdot)$ closest to 0. These occur at $f = \pm 1/(2m+1)$, so the width by this measure is $1/(2m+1)$. As an example, the light curves in Figure 78 show

$$G_p(f) = \sum_{j=1}^{2} e^{-10000(f-f_j)^2} + e^{-10000(f+f_j)^2}$$

with $f_1 = 1/4 - 1/50$ and $f_2 = 1/4 + 1/50$. This function has twin peaks whose separation in frequency is $1/25$. The dark curves show the corresponding $G_{p,m}(\cdot)$ for $m = 4$, 16 and 64 (the $G_{p,64}(\cdot)$ approximation is so good that it is indistinguishable from $G_p(\cdot)$ in the bottom plot). The corresponding widths of the central lobe of Dirichlet's kernel are, respectively, $1/9$, $1/33$ and $1/129$. The twin peaks are completely smeared together in $G_{p,4}(\cdot)$, while they are faithfully resolved in $G_{p,64}(\cdot)$. For the intermediate case $m = 16$, the width of the central lobe is just slightly less than the separation in frequency between the twin peaks. In fact, a

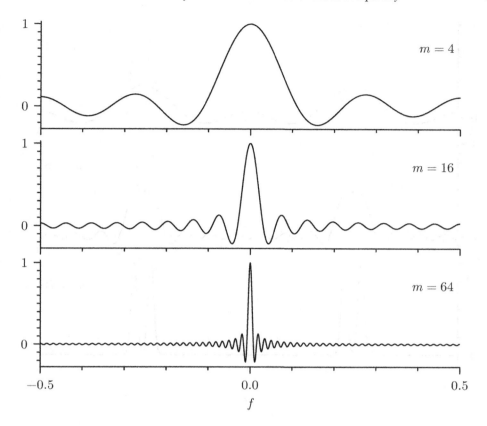

Figure 77 Dirichlet's kernel's $\mathcal{D}_{2m+1}(\cdot)$ for m = 4, 16 and 64.

magnification of $G_{p,16}(\cdot)$ around $f = \pm 1/4$ shows a slight dip so that the twin peaks are just barely resolved.

The sidelobes of Dirichlet's kernel have positive peaks at approximately $\pm 5/(4m+2)$, $\pm 9/(4m+2)$, ..., and negative peaks at approximately $\pm 3/(4m+2)$, $\pm 7/(4m+2)$, ... Thus, we can expect a significant discrepancy between $G_{p,m}(f)$ and $G_p(f)$ if $G_p(\cdot)$ happens to have large values close to, say, $f + 3/(4m+2)$. The integral in Equation (76) would then have significant contributions from parts of $G_p(\cdot)$ far away from $G_p(f)$. In the literature this phenomenon is often called *leakage*. Again, we look to Figure 78 for an example: the function $G_p(\cdot)$ there is essentially zero around $f = 0$, yet both $G_{p,4}(0)$ and $G_{p,16}(0)$ are nonzero due to leakage from the twin peaks in $G_p(\cdot)$. Leakage typically decreases as m increases – here $G_{p,64}(\cdot)$ shows little evidence of leakage.

We can investigate another phenomenon attributable to the sidelobes of Dirichlet's kernel by considering the periodic function with unit period described over $[-1/2, 1/2]$ by

$$G_p(f) = \begin{cases} 1, & |f| \leq 1/4; \\ 0, & 1/4 < |f| \leq 1/2. \end{cases} \tag{77}$$

Convolving $G_p(\cdot)$ with $(2m+1)\mathcal{D}_{2m+1}(\cdot)$ yields a $G_{p,m}(\cdot)$ that has ripples where the discontinuities in $G_p(\cdot)$ meet the lobes of the Dirichlet kernel. The result of increasing m is to make the ripples occur more frequently since the lobes become narrower but their amplitudes do not decrease. This behavior is known as the *Gibbs phenomenon* and is illustrated in Figure 79. As m increases, the maximum "overshoot" occurs closer and closer to the discontinuities at $\pm 1/4$

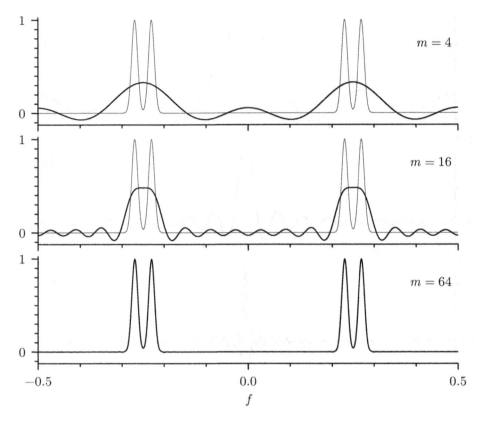

Figure 78 Illustration of loss of resolution and leakage. The function $G_p(\cdot)$ – light curve in the plots – has twin peaks and is approximated by the truncated Fourier series $G_{p,m}(\cdot)$ with $m = 4$, 16 and 64 (dark curves from top to bottom, but not visible in the bottom plot because $G_{p,64}(\cdot)$ is such a good approximation to $G_p(\cdot)$). Note the resolution of the approximation increases as m increases, while its leakage decreases as m increases.

(toward zero) while the maximum "undershoot" does likewise away from zero – each tends to about 9% of the amplitude of the discontinuity. While it is true that $G_{p,m}(f)$ converges to $G_p(f)$ for any $f \neq \pm 1/4$, the Gibbs phenomenon is still present at frequencies sufficiently close to $f = \pm 1/4$, no matter how large m gets.

There is an interesting way of reducing both leakage and the Gibbs phenomenon – at a certain cost – by approximating $G_p(\cdot)$ with a function that, like the truncated Fourier series $G_{p,m}(\cdot)$, is based upon the available subsequence $g_{-m}, \ldots, g_m$. The construction of this function is based on the concept of *Cesàro summability* (see, e.g., Titchmarsh, 1939, or Wiener, 1949). Let $\{u_t : t \in \mathbb{Z}\}$ be an infinite sequence, and, for $m = 0, 1, \ldots$, let

$$s_m \stackrel{\text{def}}{=} \sum_{t=-m}^{m} u_t \text{ and } a_m \stackrel{\text{def}}{=} \frac{1}{m} \sum_{j=0}^{m-1} s_j$$

be, respectively, its mth (two-sided) partial summation and the average of the first m partial sums. Note that

$$a_m = \frac{1}{m}\left(u_0 + \sum_{t=-1}^{1} u_t + \cdots + \sum_{t=-(m-1)}^{m-1} u_t\right) = \sum_{t=-m}^{m}\left(1 - \frac{|t|}{m}\right) u_t.$$

3.8 Fourier Theory – Discrete Time/Continuous Frequency

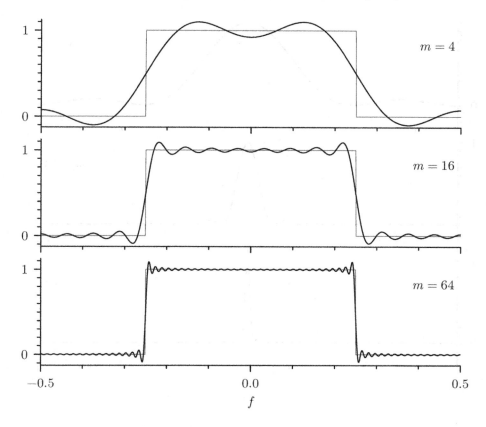

Figure 79 Illustration of the Gibbs phenomenon. The discontinuous rectangular function $G_p(\cdot)$ – light curve in the figure's plots – is approximated by the truncated Fourier series $G_{p,m}(\cdot)$ with $m = 4$, 16 and 64 (dark curves from top to bottom). Note the ripples around the discontinuities at $f = \pm 1/4$.

The Cesàro summability theorem states that,

$$\text{if } s_m = \sum_{t=-m}^{m} u_t \to s, \text{ then } a_m = \sum_{t=-m}^{m}\left(1 - \frac{|t|}{m}\right) u_t \to s \text{ also.} \tag{79}$$

For our application, let

$$s_m = \sum_{t=-m}^{m} g_t e^{-i 2\pi f t} = G_{p,m}(f) \text{ and note that } s_m \to G_p(f).$$

Then

$$G_{p,m}^{(C)}(f) \stackrel{\text{def}}{=} \sum_{t=-m}^{m} \left(1 - \frac{|t|}{m}\right) g_t e^{-i 2\pi f t} \to G_p(f)$$

also, and hence $G_{p,m}^{(C)}(\cdot)$ might be a useful approximation to $G_p(\cdot)$.

▷ **Exercise [79]** Show that

$$G_{p,m}^{(C)}(f) = m \int_{-1/2}^{1/2} G_p(f') \mathcal{D}_m^2(f - f')\, df'.$$

◁

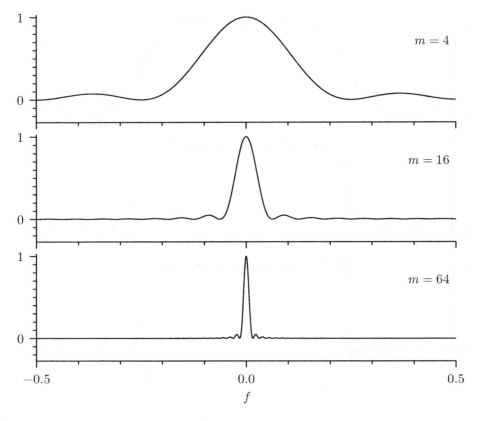

Figure 80 Square of Dirichlet's kernel, i.e., $\mathcal{D}_m^2(\cdot)$, for m = 4, 16 and 64 (cf. Figure 77).

In other words, $G_{p,m}^{(C)}(\cdot)$ is obtained by convolving $G_p(\cdot)$ with a kernel that is m times the square of the Dirichlet kernel. A rescaled version of $\mathcal{D}_m^2(\cdot)$ is called *Fejér's kernel* and is discussed in detail in Section 6.3. The quantity $\mathcal{D}_m^2(\cdot)$ is shown in Figure 80. If we compare this with the Dirichlet kernel in Figure 77, we note that $\mathcal{D}_m^2(\cdot)$ is always positive and that its sidelobes are smaller relative to its central lobe, but at the expense of a wider central lobe. If we again consider $G_p(\cdot)$ of Equation (77), we obtain the results shown in Figure 81: negative overshoots cannot occur (because of the positive nature of the kernel) and the positive overshoot can be reduced (improved "fidelity"). This is bought at the cost of a less clearly "resolved" discontinuity due to the wider central lobe.

Note that both $G_{p,m}(\cdot)$ and $G_{p,m}^{(C)}(\cdot)$ can be regarded as a summation of the form

$$\sum_{t=-m}^{m} c_t g_t e^{-i2\pi ft}$$

with an appropriate choice of c_t – "rectangular" (or uniform) for $G_{p,m}(\cdot)$ and "triangular" for $G_{p,m}^{(C)}(\cdot)$. This introduces the possibility of other definitions for the c_t terms, which are often called *convergence factors* in the literature on Fourier series and *windows* in the engineering literature (Rabiner and Gold, 1975, Section 3.8). These have been studied to manage the trade-off between fidelity and resolution. We shall meet them again in Chapter 7, where a summation similar to the above will reappear in our discussion of lag windows (see Equation (248a)). In place of resolution and fidelity, these windows manage an analogous trade-off between bias and variance in lag window spectral estimators (see Equations (256d) and (259a)).

3.9 Aliasing Problem – Discrete Time/Continuous Frequency   81

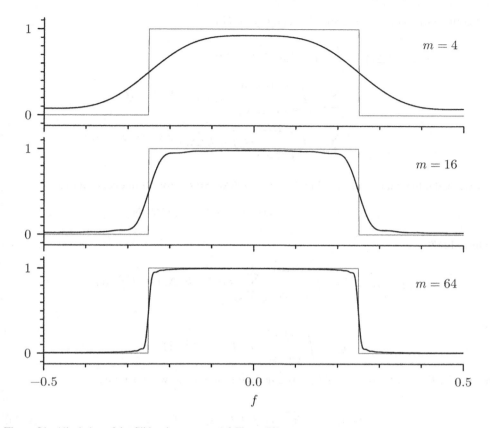

Figure 81 Alleviation of the Gibbs phenomenon (cf. Figure 79).

3.9 Aliasing Problem – Discrete Time/Continuous Frequency

In Section 3.2 we found that a continuous time $L^2(\mathbb{R})$ function $g(\cdot)$ has an amplitude spectrum defined as $|G(f)|$, where

$$G(f) \stackrel{\text{def}}{=} \int_{-\infty}^{\infty} g(t) e^{-i 2\pi f t}\, dt, \quad f \in \mathbb{R}$$

(this is Equation (54a)). The motivation for calling this an amplitude spectrum is that $g(\cdot)$ has the representation

$$g(t) \stackrel{\text{ms}}{=} \int_{-\infty}^{\infty} G(f) e^{i 2\pi f t}\, df, \qquad (81)$$

so that $|G(f)|$ represents the amplitude associated with the complex exponential with frequency f (in what follows, we assume that $g(\cdot)$ is a continuous function so that we can replace $\stackrel{\text{ms}}{=}$ with $=$ in Equation (81)).

In the previous section we found the following representation for the sequence $\{g_t\}$:

$$g_t = \int_{-1/(2\Delta_t)}^{1/(2\Delta_t)} G_p(f) e^{i 2\pi f t \Delta_t}\, df$$

(this is Equation (75a)). Again we can claim that $|G_p(f)|$ defines an amplitude spectrum. Since $\{g_t\}$ is a sequence obtained by sampling the function $g(\cdot)$, the question arises as to what connection there is between $G(\cdot)$ and $G_p(\cdot)$. Now

$$g_t = g(t\,\Delta_t) = \int_{-\infty}^{\infty} G(f')e^{i2\pi f' t\,\Delta_t}\,df'$$

$$= \sum_{k=-\infty}^{\infty} \int_{(2k-1)/(2\,\Delta_t)}^{(2k+1)/(2\,\Delta_t)} G(f')e^{i2\pi f' t\,\Delta_t}\,df'$$

$$= \sum_{k=-\infty}^{\infty} \int_{-1/(2\,\Delta_t)}^{1/(2\,\Delta_t)} G(f+k/\Delta_t)e^{i2\pi (f+k/\Delta_t)t\,\Delta_t}\,df$$

after we make the change of variable $f = f' - k/\Delta_t$. Since, for all integers t and k,

$$e^{i2\pi(f+k/\Delta_t)t\,\Delta_t} = e^{i2\pi ft\,\Delta_t}e^{i2\pi kt} = e^{i2\pi ft\,\Delta_t},$$

we have both

$$g_t = g(t\,\Delta_t) = \int_{-1/(2\,\Delta_t)}^{1/(2\,\Delta_t)} \sum_{k=-\infty}^{\infty} G(f+k/\Delta_t)e^{i2\pi ft\,\Delta_t}\,df$$

and

$$g_t = \int_{-1/(2\,\Delta_t)}^{1/(2\,\Delta_t)} G_p(f)e^{i2\pi ft\,\Delta_t}\,df.$$

The inverse Fourier transform of $\{g_t\}$ is a unique function, so we must have

$$G_p(f) = \sum_{k=-\infty}^{\infty} G(f+k/\Delta_t), \quad f \in \mathbb{R}, \tag{82a}$$

with $G_p(\cdot)$ being periodic with a period of $1/\Delta_t$. This equation tells us that $G_p(f)$ – the Fourier transform at frequency f for the sampled sequence $\{g_t\}$ – is the sum of contributions from the Fourier transform of $g(\cdot)$ at frequencies f, $f \pm 1/\Delta_t$, $f \pm 2/\Delta_t$, $f \pm 3/\Delta_t$, ... In general there is no way of exactly recovering the Fourier transform of $g(\cdot)$ given that of $\{g_t\}$. The value of $G_p(f)$ depends upon $G(\cdot)$ not only at f but also at a countably infinite set of frequencies $f + k/\Delta_t$, $k = \pm 1, \pm 2, \ldots$. This phenomenon is called *aliasing*, and the frequency f is said to be aliased with each of the frequencies $f \pm 1/\Delta_t$, $f \pm 2/\Delta_t$, $f \pm 3/\Delta_t$, ... These latter frequencies are called *aliases* of the frequency f. The highest frequency that is not an alias of a lower frequency is $1/(2\,\Delta_t)$. This frequency is often called the *Nyquist frequency* or the *folding frequency*. We shall denote it by

$$f_{\mathcal{N}} \stackrel{\text{def}}{=} \frac{1}{2\,\Delta_t}.$$

As a simple example, consider the real-valued function

$$G(f) = r(f+f_{\mathcal{N}})/4 + r(f) + r(f-1.5f_{\mathcal{N}})/2, \tag{82b}$$

where $r(f) = 1$ for $|f| \leq f_{\mathcal{N}}/8$ and 0 otherwise. This function is plotted in the top part of Figure 83a, while the bottom part is its aliased version $G_p(\cdot)$ (the vertical dashed lines indicate the Nyquist frequency and its negative). Note that $G_p(\cdot)$ is periodic with period $2f_{\mathcal{N}}$; that $G(\cdot)$ around $f = 1.5f_{\mathcal{N}}$ is aliased around $f = -f_{\mathcal{N}}/2$ in $G_p(\cdot)$; that $G(\cdot)$ for f just above

3.9 Aliasing Problem – Discrete Time/Continuous Frequency

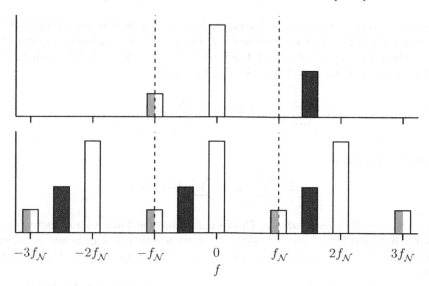

Figure 83a Example of the aliasing effect.

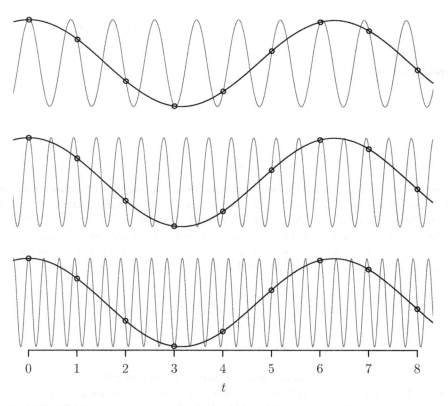

Figure 83b Illustration of the aliasing effect. The dark curves in the figure show $\cos(t)$ versus t. The light curves show $\cos([1 + 2k\pi]t)$ versus t for (from top to bottom) $k = 1, 2$ and 3. The circles highlight the common value of all four sinusoids when sampled at $t = 0, 1, \ldots, 8$.

$f = -f_\mathcal{N}$ is aliased to just below $f = f_\mathcal{N}$; and that, while $G(\cdot)$ and $G_p(\cdot)$ agree perfectly at certain frequencies (0, for example), they disagree substantially at others ($f_\mathcal{N}$).

An illustration of what causes aliasing is shown in Figure 83b. By sampling at discrete times (marked by the circles), it is impossible to know whether we are sampling from a cosine wave with frequency $1/(2\pi)$ or one of its aliases $1/(2\pi) + k$, $k = \pm 1, \pm 2, \ldots$ (here $\Delta_t = 1$).

If we compare Equations (74a) and (82a), we obtain

$$\Delta_t \sum_{t=-\infty}^{\infty} g_t e^{-i2\pi f t \Delta_t} = \sum_{k=-\infty}^{\infty} G(f + k/\Delta_t). \tag{84a}$$

If we set $f = 0$, we have

$$\Delta_t \sum_{t=-\infty}^{\infty} g_t = \sum_{k=-\infty}^{\infty} G(k/\Delta_t),$$

which is known as *Poisson's formula*. This formula (and others like it for nonzero f) make the most sense if the summations on both sides of Equation (84a) converge pointwise. Conditions sufficient to ensure this for any f are stricter than $g(\cdot)$ belonging to $L^2(\mathbb{R})$ (Champeney, 1987, p. 163).

If $g(\cdot)$ is band-limited to $W < f_\mathcal{N}$, then $G_p(f) = G(f)$ for $|f| \leq f_\mathcal{N}$. Hence

$$g(t) = \int_{-f_\mathcal{N}}^{f_\mathcal{N}} G(f) e^{i2\pi f t} \, df = \int_{-f_\mathcal{N}}^{f_\mathcal{N}} G_p(f) e^{i2\pi f t} \, df.$$

▷ **Exercise [84]** Show that

$$g(t) = \sum_{t'=-\infty}^{\infty} g_{t'} \, \text{sinc}\left(2 f_\mathcal{N}[t - t' \Delta_t]\right), \tag{84b}$$

where, as usual, $\text{sinc}(t) \stackrel{\text{def}}{=} \sin(\pi t)/(\pi t)$. ◁

Equation (84b) gives us an interpolation formula for recovering the continuous time function from its sampled values (this only works for the case of a band-limited function with W less than the Nyquist frequency). As a simple example, consider $g(t) = \text{sinc}^2(t)$. The Fourier transform of $g(\cdot)$ is band-limited to ± 1 cycles per unit time. If we sample at $\Delta_t = 1/4$, then $W = 1 < f_\mathcal{N} = 1/(2\Delta_t) = 2$. We can reconstruct, say, $\text{sinc}^2(0.1)$ from its values at $t = 0$, $\pm 1/4, \pm 1/2, \ldots$, since $g_{t'} = g(t' \Delta_t)$ so that

$$\text{sinc}^2(0.1) = \sum_{t'=-\infty}^{\infty} \text{sinc}^2(t'/4) \, \text{sinc}(0.4 - t').$$

Comments and Extensions to Section 3.9

[1] Oppenheim and Schafer (2010) cast their definition of the Nyquist frequency in terms of band-limited functions, namely, if $g(\cdot)$ is band-limited to W, the Nyquist frequency is defined to be W. Their definition is thus tied to a property of a particular class of functions rather than to the spacing between samples from any general function. They also define a *Nyquist rate* to be twice their Nyquist frequency. The inverse of the Nyquist rate is the largest sampling interval Δ_t that would allow perfect reconstruction of a band-limited function $g(\cdot)$ from samples $g_{t'} = g(t' \Delta_t)$, $t' \in \mathbb{Z}$, via Equation (84b).

3.10 Concentration Problem – Discrete/Continuous Case

Our development here is analogous to that of Section 3.5 for the continuous time/continuous frequency case. We present it because a number of details are different and because the discrete time version is most relevant to our treatment of multitaper spectral estimation in Chapter 8. The results for this case were originally derived by Slepian (1978), to whom the reader is referred for more details.

Let $\{g_t\}$ be a real- or complex-valued sequence with finite energy and a sampling interval $\Delta_t = 1$ (this yields a Nyquist frequency of $f_\mathcal{N} = 1/2$). The Fourier transform of $\{g_t\}$ is

$$G_p(f) = \sum_{t=-\infty}^{\infty} g_t e^{-i2\pi f t}$$

(Equation (74a)), and Parseval's theorem states that

$$\sum_{t=-\infty}^{\infty} |g_t|^2 = \int_{-1/2}^{1/2} |G_p(f)|^2 \, df$$

(Equation (75b)). The energy in the index range 0 to $N-1$ is just $\sum_{t=0}^{N-1} |g_t|^2$, and the fraction of the energy lying in this index range is

$$\alpha^2(N) \stackrel{\text{def}}{=} \sum_{t=0}^{N-1} |g_t|^2 \Big/ \sum_{t=-\infty}^{\infty} |g_t|^2. \tag{85a}$$

As in the continuous time/continuous frequency case of Section 3.5, the energy in the frequency range $|f| \leq W < 1/2$ is $\int_{-W}^{W} |G_p(f)|^2 \, df$, and the fraction of the energy in this range is

$$\beta^2(W) \stackrel{\text{def}}{=} \int_{-W}^{W} |G_p(f)|^2 \, df \Big/ \int_{-1/2}^{1/2} |G_p(f)|^2 \, df. \tag{85b}$$

Just as we described a function $g(\cdot)$ as being time-limited if $g(t) = 0$ for $|t| > T/2$, so we can describe a sequence $\{g_t\}$ as being *index-limited* to $t \in \{0, 1, \ldots, N-1\}$ if $g_t = 0$ for $t < 0$ or $t \geq N$. There are two concentration problems of interest here. The first is to determine how large $\alpha^2(N)$ can be for $\{g_t\}$ band-limited to $|f| \leq W < 1/2$; the second, to determine how large $\beta^2(W)$ can be when $\{g_t\}$ is index-limited.

For the first problem, we note that the concentration measure can be rewritten as

$$\alpha^2(N) = \frac{\int_{-W}^{W} \int_{-W}^{W} G_p(f) e^{i\pi(N-1)f} N \mathcal{D}_N(f' - f) G_p^*(f') e^{-i\pi(N-1)f'} \, df \, df'}{\int_{-W}^{W} |G_p(f)|^2 \, df}, \tag{85c}$$

where $\mathcal{D}_N(\cdot)$ is Dirichlet's kernel (the proof of this is Exercise [3.16]). Note that we can simplify the above somewhat by defining

$$H_p(f) = G_p(f) e^{i\pi(N-1)f}$$

to obtain

$$\alpha^2(N) = \frac{\int_{-W}^{W} \int_{-W}^{W} H_p(f) N \mathcal{D}_N(f' - f) H_p^*(f') \, df \, df'}{\int_{-W}^{W} |H_p(f)|^2 \, df}.$$

Note that $G_p(\cdot)$ and $H_p(\cdot)$ differ only in phase but not magnitude. As in Section 3.5, it can be shown that all functions $H_p(\cdot)$ that maximize $\alpha^2(N)$ must satisfy the following integral equation:

$$\int_{-W}^{W} N\mathcal{D}_N(f'-f)H_p(f)\,\mathrm{d}f = \alpha^2(N)H_p(f'), \quad |f'| \leq W \tag{86a}$$

(see Equation (63b)). This again is a Fredholm integral equation of the second kind; however, now the kernel is given by

$$N\mathcal{D}_N(f'-f) = \frac{\sin(N\pi[f'-f])}{\sin(\pi[f'-f])} \quad \text{instead of} \quad \frac{\sin(T\pi[f'-f])}{\pi(f'-f)} = T\,\mathrm{sinc}\,(T[f'-f]),$$

the kernel we encountered in the continuous time/continuous frequency case. These two kernels are both symmetric about the origin since

$$\mathcal{D}_N(-f) = \mathcal{D}_N(f) \quad \text{and} \quad \mathrm{sinc}\,(-f) = \mathrm{sinc}\,(f);$$

they are in fact close in form for small $\pi(f'-f)$ since then

$$\frac{\sin(N\pi[f'-f])}{\sin(\pi[f'-f])} \approx \frac{\sin(N\pi[f'-f])}{\pi(f'-f)} = N\,\mathrm{sinc}\,(N[f'-f]).$$

However, there is an important difference between them – whereas the kernel $N\mathcal{D}_N(\cdot)$ can be expressed as a *finite* sum of products of functions of f' and f alone, i.e.,

$$N\mathcal{D}_N(f'-f) = \sum_{t=0}^{N-1} e^{i2\pi f[t-(N-1)/2]} e^{-i2\pi f'[t-(N-1)/2]},$$

the same does not hold for the continuous time/continuous frequency kernel. This difference is summarized by calling $N\mathcal{D}_N(\cdot)$ a *degenerate* kernel. Because its kernel is degenerate, the integral equation shown in Equation (86a) possesses only a *finite* number of eigenvalues λ and eigenfunctions $U(\cdot)$, i.e., values and functions such that the equation

$$\int_{-W}^{W} N\mathcal{D}_N(f'-f)U(f)\,\mathrm{d}f = \lambda U(f') \tag{86b}$$

holds true (here – as before – we count functions that differ only by a nonzero scale factor as one function). There are in fact only N nonzero eigenvalues, say,

$$\lambda_0(N,W), \lambda_1(N,W), \ldots, \lambda_{N-1}(N,W).$$

These eigenvalues are distinct, real, positive, less than one and can be ordered such that

$$1 > \lambda_0(N,W) > \lambda_1(N,W) > \cdots > \lambda_{N-1}(N,W) > 0.$$

The first $2WN$ eigenvalues are extremely close to one, and then the eigenvalues fall off rapidly to zero. The maximum value of $\alpha^2(N)$ in our first concentration problem is just $\lambda_0(N,W)$, the largest eigenvalue of the integral equation.

For each eigenvalue $\lambda_k(N,W)$, there is an associated eigenfunction $U_k(\cdot;N,W)$ defined on the interval $[-W,W]$. As with the PSWF (see Equation (65)), the integral equation shown in Equation (86b) can be used to extend $U_k(\cdot; N, W)$ to be defined over all of $[-1/2, 1/2]$. After standardization, these eigenfunctions can be taken to be *orthonormal* over $[-1/2, 1/2]$,

$$\int_{-1/2}^{1/2} U_j(f; N, W) U_k(f; N, W) \, df = \delta_{j-k}, \tag{87a}$$

and *orthogonal* over $[-W, W]$,

$$\int_{-W}^{W} U_j(f; N, W) U_k(f; N, W) \, df = \delta_{j-k} \lambda_k(N, W).$$

The band-limited sequence that solves the first concentration problem is

$$g_t = \frac{1}{\lambda_0(N, W)} \int_{-W}^{W} U_0(f; N, W) e^{i 2\pi f [t - (N-1)/2]} \, df, \quad t \in \mathbb{Z}. \tag{87b}$$

The normalization by the term $1/\lambda_0(N, W)$ is that of Slepian (1978), but we could have used any nonzero constant (independent of t) times g_t. The function $U_k(\cdot; N, W)$ is called the kth-order *discrete prolate spheroidal wave function* (DPSWF). The sequence defined in Equation (87b) is called a zeroth-order *discrete prolate spheroidal sequence* (DPSS) or, more commonly now, a zeroth-order *Slepian sequence*.

We can be generate the kth-order Slepian sequence by substituting $U_k(\cdot; N, W)$ for $U_0(\cdot; N, W)$ in Equation (87b); however, now $(-1)^k / (\epsilon_k \lambda_k(N, W))$ is the normalizing constant instead of $1/\lambda_0(N, W)$, where ϵ_k is taken to be 1 for even k and i for odd k (again, this is done to conform to Slepian's notation). Let us now denote the kth sequence by

$$\ldots, v_{k,-1}(N, W), v_{k,0}(N, W), v_{k,1}(N, W), \ldots.$$

In this notation the band-limited sequence that solves the first concentration problem is just

$$\ldots, v_{0,-1}(N, W), v_{0,0}(N, W), v_{0,1}(N, W), \ldots.$$

There is a second way of generating the kth-order Slepian sequence. Slepian (1978) in fact *defines* this sequence via the solution to this system of equations:

$$\sum_{t'=0}^{N-1} \frac{\sin(2\pi W [t - t'])}{\pi (t - t')} v_{k,t'}(N, W) = \lambda_k(N, W) v_{k,t}(N, W), \tag{87c}$$

$t = 0, 1, \ldots, N - 1$. This is equivalent to saying that the eigenvalues $\lambda_k(N, W)$ are the eigenvalues of the $N \times N$ matrix whose (t, t')th element is

$$\frac{\sin(2\pi W [t - t'])}{\pi (t - t')}, \quad t, t' = 0, 1, \ldots, N - 1,$$

and that the N elements of the corresponding eigenvectors for this matrix – say, $\boldsymbol{v}_k(N, W)$ – are subsequences of length N of the Slepian sequences, namely, $v_{k,0}(N, W), v_{k,1}(N, W), \ldots, v_{k,N-1}(N, W)$. The subsequences are real-valued and are in fact those elements of the Slepian sequences with indices in the range 0 to $N - 1$. Slepian (1978) shows how the remaining elements of the Slepian sequences can be generated based upon these subsequences.

Remarkably the vector $\boldsymbol{v}_k(N,W)$ is also the eigenvector for a symmetric tridiagonal matrix with diagonal elements $([N-1-2t]/2)^2 \cos(2\pi W)$, $t = 0, 1, \ldots, N-1$, and off-diagonal elements $t(N-t)/2$, $t = 1, 2, \ldots, N-1$ (Slepian, 1978, p. 1379). Calculation of the Slepian sequences using this matrix is numerically stable, fast and hence the method of choice. The eigenvalues for the tridiagonal matrix are *not* the same as $\lambda_k(N,W)$ (in particular, they are not bunched near unity or zero); however, once we have determined $\boldsymbol{v}_k(N,W)$, we can use Equation (87c) to obtain $\lambda_k(N,W)$.

Let us now consider the second concentration problem, namely, to determine how large $\beta^2(W)$ in Equation (85b) can be for $\{g_t\}$ index-limited to $t \in \{0, 1, \ldots, N\}$. By making use of the relationship

$$G_p(f) = \sum_{t=-\infty}^{\infty} g_t e^{-i2\pi ft} = \sum_{t=0}^{N-1} g_t e^{-i2\pi ft}$$

(from Equation (74a) with $\Delta_t = 1$), it follows that Equation (85b) can be rewritten as

$$\beta^2(W) = \sum_{t'=0}^{N-1} \sum_{t=0}^{N-1} g_t^* \frac{\sin(2\pi W[t'-t])}{\pi(t'-t)} g_{t'} \bigg/ \sum_{t=0}^{N-1} |g_t|^2. \tag{88a}$$

The sequence $\{g_t\}$ that maximizes $\beta^2(W)$ must satisfy

$$\sum_{t'=0}^{N-1} \frac{\sin(2\pi W[t'-t])}{\pi(t'-t)} g_{t'} = \lambda_k(N,W) g_t, \quad t = 0, 1, \ldots, N-1 \tag{88b}$$

(see Exercise [3.17]). The above is equivalent to Equation (87c). We can rewrite it in the compact notation

$$\boldsymbol{A}\boldsymbol{g} = \lambda_k(N,W)\boldsymbol{g}, \tag{88c}$$

where $\boldsymbol{A}$ is a matrix of order $N \times N$ whose (t',t)th element is given by $\sin(2\pi W[t'-t])/[\pi(t'-t)]$ (the rows and columns of $\boldsymbol{A}$ are labeled from 0 to $N-1$); and $\boldsymbol{g}$ is an N-dimensional vector whose tth element (again labeled from 0 to $N-1$) is g_t. Thus the sequence that is index-limited to $t = 0, 1, \ldots, N-1$ and that has the highest concentration of energy in the frequency interval $[-W, W]$ is a vector $\boldsymbol{v}_0(N,W)$ whose elements are a finite subsequence of the zeroth-order Slepian sequence.

We note here a few important points about the solutions to Equation (88b).

[1] There are N nonzero eigenvalues that satisfy Equation (88b). These have exactly the same values and properties as the eigenvalues satisfying the first concentration problem.
[2] The N eigenvectors that are associated with these eigenvalues can be standardized such that they are orthonormal; i.e.,

$$\boldsymbol{v}_j^T(N,W)\boldsymbol{v}_k(N,W) = \sum_{t=0}^{N-1} v_{j,t}(N,W) v_{k,t}(N,W) = \delta_{j-k}.$$

With standardization, it follows from Equation (88c) that the index-limited sequence with the highest concentration of energy satisfies the condition

$$\boldsymbol{v}_0^T(N,W)\boldsymbol{A}\boldsymbol{v}_0(N,W) = \lambda_k(N,W);$$

i.e., $\boldsymbol{v}_0(N,W)$ is a solution to the maximization of the quadratic form $\boldsymbol{g}^T \boldsymbol{A} \boldsymbol{g}$, subject to the constraint $\boldsymbol{g}^T \boldsymbol{g} = 1$ (the solution $\boldsymbol{v}_0(N,W)$ is not unique since $-\boldsymbol{v}_0(N,W)$ is

3.10 Concentration Problem – Discrete/Continuous Case

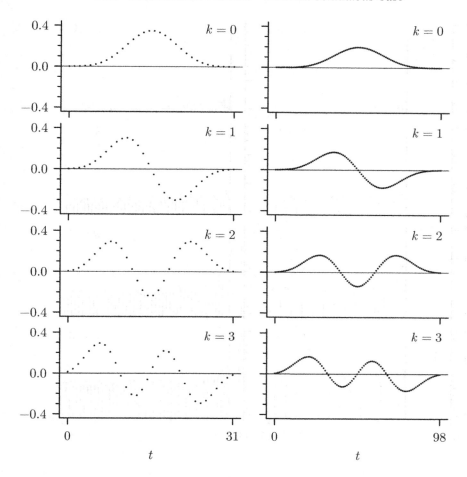

Figure 89 Plots of $v_k(N, W)$ with $N = 32$ and $W = 4/N = 1/8$ (left-hand column) and with $N = 99$ and $W = 4/N = 4/99$ (right-hand).

also a solution). With or without standardization, the vector $v_0(N, W)$ maximizes the so-called *Rayleigh quotient* $g^T A g / g^T g$ (Golub and Van Loan, 2013). Exercise [3.17] expands upon this point.

[3] These N eigenvectors form a basis for N-dimensional Euclidean space; i.e., any real-valued N-dimensional vector can be expressed as a linear combination of these eigenvectors.

[4] As we have already seen, the index-limited sequence with the highest concentration of energy in the frequency interval $[-W, W]$ is

$$v_{0,0}(N, W), v_{0,1}(N, W), \ldots, v_{0,N-1}(N, W).$$

The concentration $\beta^2(W)$ is just $\lambda_0(N, W)$. The index-limited sequence that is orthogonal to this one and has the highest concentration of energy in the frequency interval $[-W, W]$ is

$$v_{1,0}(N, W), v_{1,1}(N, W), \ldots, v_{1,N-1}(N, W).$$

The concentration in this case is $\lambda_1(N, W) < \lambda_0(N, W)$. This pattern continues in an obvious way.

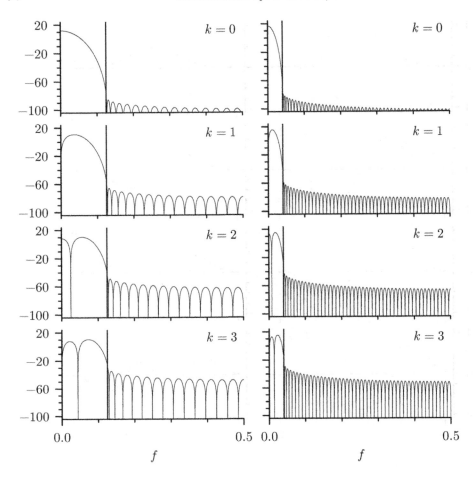

Figure 90 Plots (on a decibel scale) of the square magnitude of the Fourier transforms for each of the sequences in Figure 89. The vertical lines indicate the frequencies $W = 1/8$ (left-hand column) and $W = 4/99$ (right-hand).

As an example, the left-hand column of Figure 89 shows plots of $v_k(N, W)$ for $k = 0$, 1, 2 and 3 with $N = 32$ and $W = 4/N = 1/8$. The degree of concentration for each of these sequences is high: we have (to the first digit that differs from 9) $\lambda_0(32, 1/8) \doteq 0.999\,999\,999\,8$, $\lambda_1(32, 1/8) \doteq 0.999\,999\,98$, $\lambda_2(32, 1/8) \doteq 0.999\,999\,2$ and $\lambda_3(32, 1/8) \doteq 0.999\,98$. The square magnitude of the Fourier transforms for each of these sequences is shown in the left-hand column of Figure 90, where vertical lines indicate the frequency $W = 1/8$. We see that the energy is concentrated in the interval $[-W, W]$ (since the sequences in Figure 89 are real-valued, the square magnitude of the Fourier transform is symmetric about 0 so we only need plot it for positive frequencies). The right-hand columns of Figures 89 and 90 show similar plots for the case $N = 99$ and $W = 4/N = 4/99$.

Figure 91 shows $\lambda_k(N, W)$ plotted versus k for the two cases illustrated in the previous two figures. The upper plot is for the case $N = 32$ and $W = 1/8$, while the lower plot shows the case $N = 99$ and $W = 4/99$. In both cases the Shannon number $2NW$ is 8, so we can expect $\lambda_k(N, W)$ to be close to 1 for $k < 8$ and close to 0 for $k > 8$. The plots show this pattern.

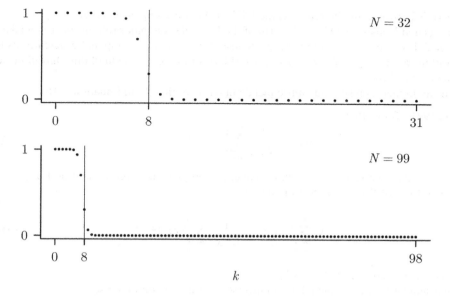

Figure 91 Plots of $\lambda_k(N, W)$ versus k for the cases shown in the left-hand columns of Figures 89 and 90 (top) and in the right-hand columns (bottom). The thin vertical lines indicate the position of $k = 2NW$, which is 8 in both cases.

3.11 Fourier Theory – Discrete Time and Frequency

Suppose now that only a finite number of samples of a continuous time real or complex-valued function is available, say, $g_0, g_1, \ldots, g_{N-1}$. In the case of equally spaced time sampling and continuous frequency, the Fourier transform is the periodic function $G_p(\cdot)$ defined in Equation (74a) as

$$G_p(f) = \Delta_t \sum_{t=-\infty}^{\infty} g_t e^{-i2\pi f t \Delta_t}.$$

We could *define* a Fourier transform $G_p(\cdot; 0, N-1)$ for our subsequence of N values by setting $g_t = 0$ for $t < 0$ and $t \geq N$ in the above:

$$G_p(f; 0, N-1) = \Delta_t \sum_{t=0}^{N-1} g_t e^{-i2\pi f t \Delta_t}. \tag{91a}$$

This definition has its uses and will appear again when we discuss the periodogram in Chapter 6; however, it involves an infinite number of frequencies. Since our subsequence has only a finite number of values, it seems plausible that we should be able to construct a representation for it that involves only a finite number of frequencies. This would be quite useful for digital computations. Accordingly, let us define a grid of N equally spaced frequencies, say,

$$f_n = \frac{n}{N \Delta_t}, \quad \text{where } n = 0, 1, \ldots, N-1.$$

This set of N frequencies is sometimes called the *Fourier frequencies* or *standard frequencies*. For this grid of frequencies Equation (91a) now becomes

$$G_n \stackrel{\text{def}}{=} G_p(f_n; 0, N-1) = \Delta_t \sum_{t=0}^{N-1} g_t e^{-i2\pi f_n t \Delta_t} = \Delta_t \sum_{t=0}^{N-1} g_t e^{-i2\pi n t/N}. \tag{91b}$$

92 *Deterministic Spectral Analysis*

The above is known in the literature as the DFT of the finite sequence $g_0, g_1, \ldots, g_{N-1}$ (see, e.g., Briggs and Henson, 1995). This form of the DFT plays a major role in spectral analysis and other fields of electrical engineering because it can often be computed efficiently using an algorithm called the *fast Fourier transform* (FFT; for the history behind this algorithm, see Heideman et al., 1984).

We can in fact represent our subsequence in terms of the G_n in Equation (91b).

▷ **Exercise [92]** Show that

$$g_t = \frac{1}{N\,\Delta_t} \sum_{n=0}^{N-1} G_n e^{i2\pi nt/N}. \tag{92a}$$ ◁

The above is the inverse DFT of a finite sequence and gives us a representation for g_t. The appropriate form of Parseval's theorem here is

$$\Delta_t \sum_{t=0}^{N-1} |g_t|^2 = \frac{1}{N\,\Delta_t} \sum_{n=0}^{N-1} |G_n|^2. \tag{92b}$$

If $\{g_t\}$ is real-valued, then $G^*_{-n} = G_n$.

Two standard forms of the DFT and inverse DFT are in common use:

(a) $\Delta_t = 1$ and (b) $\Delta_t = 1/N$.

The first choice leads to

$$g_t = \frac{1}{N} \sum_{n=0}^{N-1} G_n e^{i2\pi nt/N} \quad \text{and} \quad G_n = \sum_{t=0}^{N-1} g_t e^{-i2\pi nt/N}$$

and agrees with the definition in, e.g., Oppenheim and Schafer (2010). The second leads to

$$g_t = \sum_{n=0}^{N-1} G_n e^{i2\pi nt/N} \quad \text{and} \quad G_n = \frac{1}{N} \sum_{t=0}^{N-1} g_t e^{-i2\pi nt/N}$$

and agrees with the definition in, e.g., Bloomfield (2000). This second form also appears to be that most often used in published FFT programs.

Note that, if we use Equation (91b) to *define* G_n for all integers n instead of just for $n = 0$ to $N - 1$, the resulting infinite sequence $\{G_n\}$ is periodic with period N. Since the sequence $\{G_n \exp(i2\pi nt/N)\}$ is also periodic with period N, it follows that an alternative to Equation (92a) is

$$g_t = \frac{1}{N\,\Delta_t} \sum_{n=k}^{N+k-1} G_n e^{i2\pi nt/N},$$

where k is any integer. The frequencies f_n that are associated with the G_n in the summation range from f_k up to f_{N+k-1}. If $k = 0$ as in Equation (92a), the range is thus from 0 up to $(N-1)/(N\,\Delta_t)$, which is just less than twice the Nyquist frequency $f_\mathcal{N} = 1/(2\,\Delta_t)$; alternatively, if $k = -\lfloor (N+1)/2 \rfloor + 1$, the f_n terms all satisfy $-f_\mathcal{N} < f_n \leq f_\mathcal{N}$.

Note also that, if we use Equation (92a) to *define* g_t for $t < 0$ and $t \geq N$, the resulting infinite sequence $\{g_t\}$ is also periodic with period N. This supports the statement (sometimes found in the literature) that the DFT implies a periodic extension of the original subsequence. With this interpretation, in contrast to the discrete time/continuous frequency and continuous time/discrete frequency cases for which sampling in the time (or frequency) domain is

3.11 Fourier Theory – Discrete Time and Frequency

expressed as periodic behavior in the frequency (or time) domain, the discrete time/discrete frequency case involves sampling and periodic sequences in both domains. It is, however, also sometimes claimed that the DFT is a result of assuming g_t is zero outside the original subsequence – this alludes to the argument we used to obtain Equation (91a). Except for the trivial case of a subsequence whose terms are all zero, a periodic extension would imply a sequence such that $\sum_{t=-\infty}^{\infty} |g_t|^2 = \infty$, so the theory of Section 3.8 is not applicable. If, however, a zero extension is assumed, Equation (91b) shows that the Fourier transform of this infinite sequence is identical to the DFT of the subsequence at frequency f_n (see C&E [1]).

Suppose now, however, that our subsequence is a portion of a sequence $\{g_t\}$ whose moduli are square summable but whose terms are not necessarily zero outside the range $t = 0$ to $N - 1$. What is the relationship between the Fourier transform $G_p(\cdot)$ for $\{g_t\}$ and the G_n for the subsequence?

▷ **Exercise [93]** Show that

$$G_n = \int_{-f_\mathcal{N}}^{f_\mathcal{N}} G_p(f) P(f_n - f) \, df \stackrel{\text{def}}{=} G_p * P(f_n), \tag{93}$$

where

$$P(f) \stackrel{\text{def}}{=} \Delta_t \, e^{-i(N-1)\pi f \Delta_t} N \mathcal{D}_N(f \Delta_t). \quad \triangleleft$$

We see that $P(\cdot)$ acts as a blurring function since it is convolved with $G_p(\cdot)$. The amount of blurring is dictated by Dirichlet's kernel $\mathcal{D}_N(\cdot)$ (see Figure 77). If we are calculating G_n to obtain an approximation to $G_p(f_n)$, this blurring is obviously undesirable. We also note that Equation (93) is clearly related to Equation (76), which shows that the truncated Fourier series $G_{p,m}(\cdot)$ involving $g_{-m}, \ldots, g_m$ is proportional to the convolution of the Fourier transform $G_p(\cdot)$ of the infinite series $\{g_t\}$ with Dirichlet's kernel. On the other hand, if we consider our finite sequence as coming from the function $g(\cdot)$, i.e., $g_t = g(t \Delta_t)$, we obtain instead

$$G_n = \int_{-\infty}^{\infty} G(f) P(f_n - f) \, df.$$

Clearly, if $G(\cdot)$ is band-limited to $|W| < f_\mathcal{N}$, then

$$G_n = \int_{-W}^{W} G(f) P(f_n - f) \, df = \int_{-W}^{W} G_p(f) P(f_n - f) \, df.$$

Comments and Extensions to Section 3.11

[1] Let $\{g_t\}$ be an infinite sequence such that $g_t = 0$ for $t < 0$ and $t > N - 1$. The DFT of the subsequence $g_0, g_1, \ldots, g_{N-1}$ is

$$G_n = \Delta_t \sum_{t=0}^{N-1} g_t e^{-i2\pi nt/N} = \Delta_t \sum_{t=-\infty}^{\infty} g_t e^{-i2\pi nt/N} = G_p(f_n),$$

for $n = 0, 1, \ldots, N - 1$, where $G_p(\cdot)$ is the Fourier transform of $\{g_t\}$. Thus, as we have noted before, the DFT of the subsequence is equal to the Fourier transform of the entire sequence – at the Fourier frequencies – in the special case where $g_t = 0$ outside the subsequence. Suppose that we now consider a larger subsequence $g_0, g_1, \ldots, g_{N'-1}$ with $N' > N$. Its DFT is, say,

$$G'_n = \Delta_t \sum_{t=0}^{N'-1} g_t e^{-i2\pi nt/N'} = \Delta_t \sum_{t=-\infty}^{\infty} g_t e^{-i2\pi nt/N'} = G_p(f'_n)$$

for $n = 0, 1, \ldots, N' - 1$ and $f'_n = n/(N' \Delta_t)$. Hence, by *zero padding* our original subsequence with $N' - N$ zeros, the DFT of the extended subsequence effectively evaluates $G_p(\cdot)$ over a finer grid of frequencies $\{f'_n\}$ than that given by the Fourier frequencies $\{f_n\}$. Note that this only works because g_t is in fact zero for $t < 0$ and $t > N - 1$.

Zero padding is often used in conjunction with FFT algorithms. These can quickly compute the DFT of a finite sequence if the length of the sequence N satisfies certain restrictions (the most common restriction is that $N = 2^k$ for some positive integer k, but some FFT algorithms work for $N = N_1 \times N_2 \times \cdots \times N_q$ with each N_j being a small prime number – see Singleton, 1969). Thus, if N is not a number acceptable to an FFT algorithm but $N' > N$ is, then $N' - N$ zeros are appended to the original sequence, and the DFT of this extended sequence is computed using the FFT algorithm. This scheme works fine as long as (i) we can reasonably assume that $g_N, \ldots, g_{N'}$ are in fact zero and/or (ii) the zero padded sequence $\{G'_n\}$ is as much use or interest to us as the original sequence $\{G_n\}$. When this is not the case so that we really require the DFT of a sequence with an "unfriendly" length, we can still take advantage of FFTs to compute the DFT by use of the following *chirp transform algorithm* (Oppenheim and Schafer, 2010, section 9.6.2).

Let a_t for $t = 0, 1, \ldots, N - 1$ and b_t for $t = -(N - 1), -(N - 2), \ldots, N - 1$ be two sequences of complex or real-valued numbers of length N and $2N - 1$, respectively, and define the following sequence of length N:

$$c_n = \sum_{t=0}^{N-1} a_t b_{n-t} \text{ for } n = 0, 1, \ldots, N - 1. \tag{94a}$$

Exercise [3.20] indicates how this convolution can be computed efficiently using zero padding along with three invocations of a standard "powers of 2" FFT algorithm. Now, since $-2nt = (n - t)^2 - t^2 - n^2$ (a "trick" attributed to L. I. Bluestein), we have

$$G_n = \Delta_t \sum_{t=0}^{N-1} g_t e^{-i2\pi nt/N} = \Delta_t \sum_{t=0}^{N-1} g_t e^{i\pi\left[(n-t)^2 - t^2 - n^2\right]/N}$$

$$= \Delta_t \, e^{-i\pi n^2/N} \sum_{t=0}^{N-1} g_t e^{-i\pi t^2/N} e^{i\pi(n-t)^2/N} = \Delta_t \, b_n^* \sum_{t=0}^{N-1} a_t b_{n-t}, \tag{94b}$$

where $b_t \stackrel{\text{def}}{=} \exp(i\pi t^2/N)$ and $a_t \stackrel{\text{def}}{=} g_t b_t^*$. The right-most summation in Equation (94b) is of the same form as Equation (94a) and can hence be computed efficiently using FFTs. Bluestein's substitution thus enables us to use a standard FFT algorithm to evaluate a DFT of general length. A FORTRAN implementation of the chirp transform algorithm is given in Monro and Branch (1977).

[2] Grünbaum (1981) was the first to study the discrete time/discrete frequency concentration problem in detail. It is formulated by analogy to the cases of continuous time/continuous frequency and discrete time/continuous frequency. Thus define the following measures of energy concentration:

$$\alpha^2(N, m_1, m_2) = \sum_{t=m_1}^{m_2} |g_t|^2 \Big/ \sum_{t=0}^{N-1} |g_t|^2 \text{ and } \beta^2(N, m_1, m_2) = \sum_{n=m_1}^{m_2} |G_n|^2 \Big/ \sum_{n=0}^{N-1} |G_n|^2,$$

where m_1 and m_2 are integers such that $0 \leq m_1 \leq m_2 \leq N - 1$. Because of the obvious symmetries, we shall only study $\alpha^2(N, m_1, m_2)$. If, for simplicity, we set $\Delta_t = 1$ in Equation (92a), we can use that equation and Equation (92b) to write

$$\alpha^2(N, m_1, m_2) = \frac{\sum_{n'=0}^{N-1} \sum_{n=0}^{N-1} G_{n'} G_n^* \sum_{t=m_1}^{m_2} e^{i2\pi(n'-n)t/N}}{N \sum_{n'=0}^{N-1} |G_{n'}|^2}.$$

Now, if we define $M = m_2 - m_1 + 1$, we can write

$$\sum_{t=m_1}^{m_2} e^{i2\pi(n'-n)t/N} = \sum_{t=0}^{M-1} e^{i2\pi(n'-n)(t+m_1)/N}$$

$$= e^{i2\pi(n'-n)m_1/N} \sum_{t=0}^{M-1} e^{i2\pi(n'-n)t/N}$$

$$= e^{i\pi(n'-n)(m_1+m_2)/N} M \mathcal{D}_M\left(\frac{n'-n}{N}\right),$$

where the last line follows from Exercise [1.2c]. If we now define

$$H_n = G_n e^{i\pi n(m_1+m_2)/N},$$

we have

$$\alpha^2(N, m_1, m_2) = \frac{M \sum_{n'=0}^{N-1} \sum_{n=0}^{N-1} H_{n'} \mathcal{D}_M\left(\frac{n'-n}{N}\right) H_n^*}{N \sum_{n=0}^{N-1} |H_n|^2}.$$

By arguments similar to those for Exercise [3.17], the sequence maximizing $\alpha^2(N, m_1, m_2)$ must satisfy

$$\sum_{n'=0}^{N-1} \frac{M}{N} \mathcal{D}_M\left(\frac{n'-n}{N}\right) H_{n'} = \lambda_l H_n, \quad n = 0, 1, \ldots, N-1.$$

The highest concentration is given by the largest eigenvalue λ_0 of the corresponding matrix equation

$$\boldsymbol{A}\boldsymbol{h} = \lambda_l \boldsymbol{h},$$

where $\boldsymbol{A}$ is an $N \times N$ matrix whose (n', n)th element is

$$\frac{M}{N} \mathcal{D}_M\left(\frac{n'-n}{N}\right)$$

and $\boldsymbol{h}$ is an N-dimensional vector with nth element H_n. As also noted for the discrete time/continuous frequency case, there is an easy way to find the eigenvalues and eigenvectors of $\boldsymbol{A}$ by using an associated commuting tridiagonal matrix (Grünbaum, 1981). The sequences resulting from these calculations have been used in image processing (Wilson, 1987; Wilson and Spann, 1988).

3.12 Summary of Fourier Theory

For the four types of Fourier theory, we give (a) equations for the Fourier transform and its inverse, (b) the associated Parseval's theorems, (c) the spectral properties for the transform, (d) definitions and Fourier transform relationships for both time and frequency domain convolutions and complex cross-correlations and (e) definitions for autocorrelation widths in the time and frequency domains. We assume throughout that the functions $g(\cdot)$, $g_p(\cdot)$, $h(\cdot)$ and $h_p(\cdot)$ and sequences $\{h_t\}$ and $\{g_t\}$ are real- or complex-valued. (The subscript p is added to emphasize here that the function in question is periodic – we do not follow this convention in subsequent chapters.)

[1] *Continuous time $g_p(\cdot)$ with period T and $\int_{-T/2}^{T/2} |g_p(t)|^2 \, dt < \infty$*
    (a) Fourier representation (valid for $t \in \mathbb{R}$):

$$g_p(t) \stackrel{\text{ms}}{=} \sum_{n=-\infty}^{\infty} G_n e^{i2\pi f_n t} \quad \text{with } f_n \stackrel{\text{def}}{=} \frac{n}{T}, \qquad \text{(see (49c))}$$

where, for $n \in \mathbb{Z}$,

$$G_n \stackrel{\text{def}}{=} \frac{1}{T} \int_{-T/2}^{T/2} g_p(t) e^{-i2\pi f_n t} \, dt. \qquad \text{(see (49b))}$$

The Fourier relationship between $g_p(\cdot)$ and $\{G_n\}$ is noted as

$$g_p(\cdot) \longleftrightarrow \{G_n\}.$$

If $g_p(\cdot)$ is real-valued, then $G_{-n}^* = G_n$.

(b) Parseval's theorems: The "one-function" version says that

$$\int_{-T/2}^{T/2} |g_p(t)|^2 \, dt = T \sum_{n=-\infty}^{\infty} |G_n|^2. \qquad \text{(see (49d))}$$

Assuming $h_p(\cdot) \longleftrightarrow \{H_n\}$, the "two-functions" version says that

$$\int_{-T/2}^{T/2} g_p(t) h_p^*(t) \, dt = T \sum_{n=-\infty}^{\infty} G_n H_n^*. \qquad (96a)$$

(c) Spectral properties: $|G_n|$ defines an amplitude spectrum (in view of Equation (49c)) and $|G_n|^2$ defines a discrete power spectrum; $|G_n|^2$ is the contribution to the power due to the sinusoid with frequency f_n.

(d) Convolution and related theorems: The Fourier transform of the time domain convolution

$$g_p * h_p(t) \stackrel{\text{def}}{=} \frac{1}{T} \int_{-T/2}^{T/2} g_p(u) h_p(t-u) \, du \qquad (96b)$$

is

$$\frac{1}{T} \int_{-T/2}^{T/2} g_p * h_p(t) e^{-i2\pi f_n t} \, dt = G_n H_n, \qquad (96c)$$

so $g_p * h_p(\cdot) \longleftrightarrow \{G_n H_n\}$. The inverse Fourier transform of the frequency domain convolution

$$G * H_n \stackrel{\text{def}}{=} \sum_{m=-\infty}^{\infty} G_m H_{n-m} \qquad (96d)$$

is

$$\sum_{n=-\infty}^{\infty} G * H_n e^{i2\pi f_n t} \stackrel{\text{ms}}{=} g_p(t) h_p(t), \qquad (96e)$$

so $g_p(\cdot) h_p(\cdot) \longleftrightarrow \{G * H_n\}$. The Fourier transform of the time domain complex cross-correlation

$$g_p \star h_p^*(t) \stackrel{\text{def}}{=} \frac{1}{T} \int_{-T/2}^{T/2} g_p(u+t) h_p^*(u) \, du \qquad (96f)$$

is

$$\frac{1}{T} \int_{-T/2}^{T/2} g_p \star h_p^*(t) e^{-i2\pi f_n t} \, dt = G_n H_n^*, \qquad (96g)$$

so $g_p \star h_p^*(\cdot) \longleftrightarrow \{G_n H_n^*\}$. Letting $h_p(\cdot)$ be the same as $g_p(\cdot)$ yields an autocorrelation, for which $g_p \star g_p^*(\cdot) \longleftrightarrow \{|G_n|^2\}$. The inverse Fourier transform of the frequency domain complex cross-correlation

$$G \star H_n^* \stackrel{\text{def}}{=} \sum_{m=-\infty}^{\infty} G_{m+n} H_m^* \qquad (96h)$$

3.12 Summary of Fourier Theory

is

$$\sum_{n=-\infty}^{\infty} G \star H_n^* e^{i2\pi f_n t} \stackrel{\text{ms}}{=} g_p(t) h_p^*(t), \tag{97a}$$

so $g_p(\cdot) h_p^*(\cdot) \longleftrightarrow \{G \star H_n^*\}$. Letting $H_n = G_n$ yields an autocorrelation, for which $|g_p(\cdot)|^2 \longleftrightarrow \{G \star G_n^*\}$.

(e) Equivalent widths and autocorrelation widths: In the time and frequency domains we have

$$\text{width}_e\{g_p(\cdot)\} \stackrel{\text{def}}{=} \frac{\int_{-T/2}^{T/2} g_p(t)\,dt}{g_p(0)} = \frac{TG_0}{\sum_{n=-\infty}^{\infty} G_n} = \frac{1}{\text{width}_e\{G_n\}}. \tag{97b}$$

In the time domain we have

$$\text{width}_a\{g_p(\cdot)\} \stackrel{\text{def}}{=} \frac{\int_{-T/2}^{T/2} g_p \star g_p^*(t)\,dt}{g_p \star g_p^*(0)} = \frac{\left|\int_{-T/2}^{T/2} g_p(t)\,dt\right|^2}{\int_{-T/2}^{T/2} |g_p(t)|^2\,dt}. \tag{97c}$$

In the frequency domain we have

$$\text{width}_a\{G_n\} \stackrel{\text{def}}{=} \frac{\sum_{n=-\infty}^{\infty} G \star G_n^*}{TG \star G_0^*} = \frac{\left|\sum_{n=-\infty}^{\infty} G_n\right|^2}{T\sum_{n=-\infty}^{\infty} |G_n|^2}. \tag{97d}$$

[2] *Continuous time* $g(\cdot)$ *with* $\int_{-\infty}^{\infty} |g(t)|^2\,dt < \infty$

(a) Fourier representation (valid for $t \in \mathbb{R}$):

$$g(t) \stackrel{\text{ms}}{=} \int_{-\infty}^{\infty} G(f) e^{i2\pi ft}\,df, \qquad \text{(see (53))}$$

where, for $f \in \mathbb{R}$,

$$G(f) \stackrel{\text{def}}{=} \int_{-\infty}^{\infty} g(t) e^{-i2\pi ft}\,dt. \qquad \text{(see (54a))}$$

The Fourier relationship between $g(\cdot)$ and $G(\cdot)$ is noted as

$$g(\cdot) \longleftrightarrow G(\cdot).$$

If $g(\cdot)$ is real-valued, then $G^*(-f) = G(f)$.

(b) Parseval's theorems: The "one-function" version says that

$$\int_{-\infty}^{\infty} |g(t)|^2\,dt = \int_{-\infty}^{\infty} |G(f)|^2\,df. \qquad \text{(see (55a))}$$

Assuming $h(\cdot) \longleftrightarrow H(\cdot)$, the "two-functions" version says that

$$\int_{-\infty}^{\infty} g(t) h^*(t)\,dt = \int_{-\infty}^{\infty} G(f) H^*(f)\,df. \tag{97e}$$

(c) Spectral properties: $|G(f)|$ defines an amplitude spectrum (in view of Equation (53)), and $|G(f)|^2$ defines an energy spectral density function (in view of Equation (55a)); $|G(f)|^2\,df$ is the contribution to the energy due to sinusoids with frequencies in a small interval about f.

(d) Convolution and related theorems: The Fourier transform of the time domain convolution

$$g * h(t) \stackrel{\text{def}}{=} \int_{-\infty}^{\infty} g(u)h(t-u)\,du \qquad \text{(see (67a))}$$

is

$$\int_{-\infty}^{\infty} g * h(t) e^{-i2\pi ft}\,dt = G(f)H(f), \qquad \text{(see (67b))}$$

so $g*h(\cdot) \longleftrightarrow G(\cdot)H(\cdot)$. The inverse Fourier transform of the frequency domain convolution

$$G * H(f) \stackrel{\text{def}}{=} \int_{-\infty}^{\infty} G(f')H(f-f')\,df' \qquad (98a)$$

is

$$\int_{-\infty}^{\infty} G * H(f) e^{i2\pi ft}\,df = g(t)h(t), \qquad (98b)$$

so $g(\cdot)h(\cdot) \longleftrightarrow G * H(\cdot)$. The Fourier transform of the time domain complex cross-correlation

$$g \star h^*(t) \stackrel{\text{def}}{=} \int_{-\infty}^{\infty} g(u+t)h^*(u)\,du \qquad \text{(see (72a))}$$

is

$$\int_{-\infty}^{\infty} g \star h^*(t) e^{-i2\pi ft}\,dt = G(f)H^*(f), \qquad (98c)$$

so $g \star h^*(\cdot) \longleftrightarrow G(\cdot)H^*(\cdot)$ (this is Equation (72b)). Letting $h(\cdot)$ be the same as $g(\cdot)$ yields an autocorrelation (see Equation (72c)), for which $g \star g^*(\cdot) \longleftrightarrow |G(\cdot)|^2$. The inverse Fourier transform of the frequency domain complex cross-correlation

$$G \star H^*(f) \stackrel{\text{def}}{=} \int_{-\infty}^{\infty} G(f'+f)H^*(f')\,df' \qquad (98d)$$

is

$$\int_{-\infty}^{\infty} G \star H^*(f) e^{-i2\pi ft}\,df \stackrel{\text{ms}}{=} g(t)h^*(t), \qquad (98e)$$

so $g(\cdot)h^*(\cdot) \longleftrightarrow G \star H^*(\cdot)$. Letting $H(\cdot) = G(\cdot)$ yields an autocorrelation, for which $|g(\cdot)|^2 \longleftrightarrow G \star G^*(\cdot)$.

(e) Equivalent widths and autocorrelation widths: In the time and frequency domains we have

$$\text{width}_e \{g(\cdot)\} \stackrel{\text{def}}{=} \frac{\int_{-\infty}^{\infty} g(t)\,dt}{g(0)} = \frac{G(0)}{\int_{-\infty}^{\infty} G(f)\,df} = \frac{1}{\text{width}_e \{G(\cdot)\}}. \qquad \text{(see (59))}$$

In the time domain we have

$$\text{width}_a \{g(\cdot)\} \stackrel{\text{def}}{=} \frac{\int_{-\infty}^{\infty} g \star g^*(t)\,dt}{g \star g^*(0)} = \frac{\left|\int_{-\infty}^{\infty} g(t)\,dt\right|^2}{\int_{-\infty}^{\infty} |g(t)|^2\,dt}. \qquad \text{(see (73b) and (73c))}$$

In the frequency domain we have

$$\text{width}_a \{G(\cdot)\} \stackrel{\text{def}}{=} \frac{\int_{-\infty}^{\infty} G \star G^*(f)\,df}{G \star G^*(0)} = \frac{\left|\int_{-\infty}^{\infty} G(f)\,df\right|^2}{\int_{-\infty}^{\infty} |G(f)|^2\,df}. \qquad (98f)$$

3.12 Summary of Fourier Theory

[3] *Discrete time* $\{g_t\}$ *sampled at intervals of* Δ_t *with* $\sum_{t=-\infty}^{\infty} |g_t|^2 < \infty$

(a) Fourier representation (valid for $t \in \mathbb{Z}$):

$$g_t = \int_{-f_\mathcal{N}}^{f_\mathcal{N}} G_p(f) e^{i2\pi f t \Delta_t} \, df, \qquad \text{(see (75a))}$$

where $f_\mathcal{N} \stackrel{\text{def}}{=} 1/(2\Delta_t)$ and

$$G_p(f) \stackrel{\text{def}}{=} \Delta_t \sum_{t=-\infty}^{\infty} g_t e^{-i2\pi f t \Delta_t}. \qquad \text{(see (74a))}$$

The function $G_p(\cdot)$ is periodic, with a period of $2f_\mathcal{N}$. The Fourier relationship between $\{g_t\}$ and $G_p(\cdot)$ is noted as

$$\{g_t\} \longleftrightarrow G_p(\cdot).$$

If $\{g_t\}$ is real-valued, then $G_p^*(-f) = G_p(f)$.

(b) Parseval's theorems: The "one-sequence" version says that

$$\Delta_t \sum_{t=-\infty}^{\infty} |g_t|^2 = \int_{-f_\mathcal{N}}^{f_\mathcal{N}} |G_p(f)|^2 \, df. \qquad \text{(see (75b))}$$

Assuming $\{h_t\} \longleftrightarrow H_p(\cdot)$, the "two-sequences" version says that

$$\Delta_t \sum_{t=-\infty}^{\infty} g_t h_t^* = \int_{-f_\mathcal{N}}^{f_\mathcal{N}} G_p(f) H_p^*(f) \, df. \qquad (99a)$$

(c) Spectral properties: $|G_p(f)|$ defines an amplitude spectrum (in view of Equation (75a)), and $|G_p(f)|^2$ defines an energy spectral density function (in view of Parseval's theorem); $|G_p(f)|^2 \, df$ is the contribution to the energy due to sinusoids with frequencies in a small interval about f.

(d) Convolution and related theorems: the Fourier transform of the time domain convolution

$$g * h_t \stackrel{\text{def}}{=} \Delta_t \sum_{u=-\infty}^{\infty} g_u h_{t-u} \qquad (99b)$$

is

$$\Delta_t \sum_{t=-\infty}^{\infty} g * h_t e^{-i2\pi f t \Delta_t} = G_p(f) H_p(f), \qquad (99c)$$

so $\{g * h_t\} \longleftrightarrow G_p(\cdot) H_p(\cdot)$. The inverse Fourier transform of the frequency domain convolution

$$G_p * H_p(f) \stackrel{\text{def}}{=} \int_{-f_\mathcal{N}}^{f_\mathcal{N}} G_p(f') H_p(f - f') \, df' \qquad (99d)$$

is

$$\int_{-f_\mathcal{N}}^{f_\mathcal{N}} G_p * H_p(f) e^{i2\pi f t \Delta_t} \, df = g_t h_t, \qquad (99e)$$

so $\{g_t h_t\} \longleftrightarrow G_p * H_p(\cdot)$. The Fourier transform of the time domain complex cross-correlation

$$g \star h_t^* \stackrel{\text{def}}{=} \Delta_t \sum_{u=-\infty}^{\infty} g_{u+t} h_u^* \qquad (99f)$$

is
$$\Delta_{\mathrm{t}} \sum_{t=-\infty}^{\infty} g \star h_t^* \mathrm{e}^{-\mathrm{i}2\pi f t \Delta_{\mathrm{t}}} = G_p(f)H_p^*(f), \qquad (100\mathrm{a})$$

so $\{g \star h_t^*\} \longleftrightarrow G_p(\cdot)H_p^*(\cdot)$. Letting $\{h_t\}$ be the same as $\{g_t\}$ yields an autocorrelation, for which $\{g \star g_t^*\} \longleftrightarrow |G_p(\cdot)|^2$. The inverse Fourier transform of the frequency domain complex cross-correlation

$$G_p \star H_p^*(f) \stackrel{\mathrm{def}}{=} \int_{-f_{\mathcal{N}}}^{f_{\mathcal{N}}} G_p(f'+f)H_p^*(f')\,\mathrm{d}f' \qquad (100\mathrm{b})$$

is

$$\int_{-f_{\mathcal{N}}}^{f_{\mathcal{N}}} G_p \star H_p^*(f) \mathrm{e}^{\mathrm{i}2\pi f t \Delta_{\mathrm{t}}}\,\mathrm{d}f = g_t h_t^*, \qquad (100\mathrm{c})$$

so $\{g_t h_t^*\} \longleftrightarrow G_p \star H_p^*(\cdot)$. Letting $H_p(\cdot) = G_p(\cdot)$ yields an autocorrelation, for which $\{|g_t|^2\} \longleftrightarrow G_p \star G_p^*(\cdot)$.

(e) Equivalent widths and autocorrelation widths: In the time and frequency domains we have

$$\mathrm{width}_{\mathrm{e}}\{g_t\} \stackrel{\mathrm{def}}{=} \frac{\Delta_{\mathrm{t}} \sum_{t=-\infty}^{\infty} g_t}{g_0} = \frac{G_p(0)}{\int_{-f_{\mathcal{N}}}^{f_{\mathcal{N}}} G_p(f)\,\mathrm{d}f} = \frac{1}{\mathrm{width}_{\mathrm{e}}\{G_p(\cdot)\}}. \qquad (100\mathrm{d})$$

In the time domain we have

$$\mathrm{width}_{\mathrm{a}}\{g_t\} \stackrel{\mathrm{def}}{=} \frac{\Delta_{\mathrm{t}} \sum_{t=-\infty}^{\infty} g \star g_t^*}{g \star g_0^*} = \frac{\Delta_{\mathrm{t}} \left|\sum_{t=-\infty}^{\infty} g_t\right|^2}{\sum_{t=-\infty}^{\infty} |g_t|^2}. \qquad (100\mathrm{e})$$

In the frequency domain we have

$$\mathrm{width}_{\mathrm{a}}\{G_p(\cdot)\} \stackrel{\mathrm{def}}{=} \frac{\int_{-f_{\mathcal{N}}}^{f_{\mathcal{N}}} G_p \star G_p^*(f)\,\mathrm{d}f}{G_p \star G_p^*(0)} = \frac{\left|\int_{-f_{\mathcal{N}}}^{f_{\mathcal{N}}} G_p(f)\,\mathrm{d}f\right|^2}{\int_{-f_{\mathcal{N}}}^{f_{\mathcal{N}}} |G_p(f)|^2\,\mathrm{d}f}. \qquad (100\mathrm{f})$$

[4] *Segment of discrete time $\{g_t\}$ sampled at intervals of Δ_{t}*
   (a) Fourier representation (valid for $t = 0, 1, \ldots, N-1$):

$$g_t = \frac{1}{N \Delta_{\mathrm{t}}} \sum_{n=0}^{N-1} G_n \mathrm{e}^{\mathrm{i}2\pi nt/N}, \qquad (\text{see (92a)})$$

where, for $n = 0, 1, \ldots, N-1$,

$$G_n \stackrel{\mathrm{def}}{=} \Delta_{\mathrm{t}} \sum_{t=0}^{N-1} g_t \mathrm{e}^{-\mathrm{i}2\pi nt/N}. \qquad (\text{see (91b)})$$

G_n is associated with frequency $f_n \stackrel{\mathrm{def}}{=} n/(N \Delta_{\mathrm{t}})$. The Fourier relationship between $\{g_t\}$ and $\{G_n\}$ is noted as

$$\{g_t\} \longleftrightarrow \{G_n\}.$$

If $\{g_t\}$ is real-valued, then $G_{-n}^* = G_n$. Since both $\{g_t\}$ and $\{G_n\}$ can be considered as periodic sequences with period N, it is useful to arbitrarily shift the indices t and n; hence the notation

$$\{g_t : t = j, j+1, \ldots, N+j-1\} \longleftrightarrow \{G_n : n = k, k+1, \ldots, N+k-1\} \qquad (100\mathrm{g})$$

means that, for $t = j, j+1, \ldots, N+j-1$,

$$g_t = \frac{1}{N\Delta_t} \sum_{n=k}^{N+k-1} G_n e^{i2\pi nt/N}, \tag{101a}$$

where, for $n = k, k+1, \ldots, N+k-1$,

$$G_n = \Delta_t \sum_{t=j}^{N+j-1} g_t e^{-i2\pi nt/N}. \tag{101b}$$

(b) Parseval's theorems: The "one-sequence" version says that

$$\Delta_t \sum_{t=0}^{N-1} |g_t|^2 = \frac{1}{N\Delta_t} \sum_{n=0}^{N-1} |G_n|^2. \tag{see (92b)}$$

Assuming $\{h_t\} \longleftrightarrow \{H_n\}$, the "two-sequences" version says that

$$\Delta_t \sum_{t=0}^{N-1} g_t h_t^* = \frac{1}{N\Delta_t} \sum_{n=0}^{N-1} G_n H_n^*. \tag{101c}$$

(c) Spectral properties: $|G_n|^2$ could be used to define a discrete power spectrum (but only in a very limited sense).

(d) Convolution and related theorems: the Fourier transform of the time domain convolution

$$g * h_t \stackrel{\text{def}}{=} \Delta_t \sum_{u=0}^{N-1} g_u h_{t-u} \tag{101d}$$

(where $h_s \stackrel{\text{def}}{=} h_{s \bmod N}$ for s outside the range 0 to $N-1$, with $s \bmod N$ being defined as in Exercise [2.19]) is

$$\Delta_t \sum_{t=0}^{N-1} g * h_t e^{-i2\pi nt/N} = G_n H_n, \tag{101e}$$

so $\{g * h_t\} \longleftrightarrow \{G_n H_n\}$. This type of convolution is called *cyclic* (see Exercise [3.20]). The inverse Fourier transform of the frequency domain convolution

$$G * H_n \stackrel{\text{def}}{=} \frac{1}{N\Delta_t} \sum_{u=0}^{N-1} G_u H_{n-u} \tag{101f}$$

is

$$\frac{1}{N\Delta_t} \sum_{n=0}^{N-1} G * H_n e^{i2\pi nt/N} = g_t h_t, \tag{101g}$$

so $\{g_t h_t\} \longleftrightarrow \{G * H_n\}$. The Fourier transform of the time domain complex cross-correlation

$$g \star h_t^* \stackrel{\text{def}}{=} \Delta_t \sum_{u=0}^{N-1} g_{u+t} h_u^* \tag{101h}$$

is
$$\Delta_t \sum_{t=0}^{N-1} g \star h_t^* e^{-i2\pi nt/N} = G_n H_n^*, \qquad (102a)$$

so $\{g \star h_t^*\} \longleftrightarrow \{G_n H_n^*\}$. Letting $\{h_t\}$ be the same as $\{g_t\}$ yields an autocorrelation, for which $\{g \star g_t^*\} \longleftrightarrow \{|G_n|^2\}$. The inverse Fourier transform of the frequency domain complex cross-correlation

$$G \star H_n^* \stackrel{\text{def}}{=} \frac{1}{N\Delta_t} \sum_{u=0}^{N-1} G_{u+n} H_u^* \qquad (102b)$$

is

$$\frac{1}{N\Delta_t} \sum_{n=0}^{N-1} G \star H_n^* e^{i2\pi nt/N} = g_t h_t^*, \qquad (102c)$$

so $\{g_t h_t^*\} \longleftrightarrow \{G \star H_n^*\}$. Letting $\{H_n\} = \{G_n\}$ yields an autocorrelation, for which $\{|g_t|^2\} \longleftrightarrow \{G \star G_n^*\}$.

(e) Equivalent widths and autocorrelation widths: In the time and frequency domains we have

$$\text{width}_e \{g_t\} \stackrel{\text{def}}{=} \frac{\Delta_t \sum_{t=0}^{N-1} g_t}{g_0} = \frac{G_0}{\frac{1}{N\Delta_t} \sum_{n=0}^{N-1} G_n} = \frac{1}{\text{width}_e \{G_n\}}. \qquad (102d)$$

In the time domain we have

$$\text{width}_a \{g_t\} \stackrel{\text{def}}{=} \frac{\Delta_t \sum_{t=0}^{N-1} g \star g_t^*}{g \star g_0^*} = \frac{\Delta_t \left|\sum_{t=0}^{N-1} g_t\right|^2}{\sum_{t=0}^{N-1} |g_t|^2}. \qquad (102e)$$

In the frequency domain we have

$$\text{width}_a \{G_n\} \stackrel{\text{def}}{=} \frac{\frac{1}{N\Delta_t} \sum_{n=0}^{N-1} G \star G_n^*}{G \star G_0^*} = \frac{\left|\sum_{n=0}^{N-1} G_n\right|^2}{N\Delta_t \sum_{n=0}^{N-1} |G_n|^2}. \qquad (102f)$$

3.13 Exercises

[3.1] (a) Express a_n and b_n in terms of the G_n defined in Equation (48c).

(b) Assume now that the a_n and b_n in Equation (48a) are real-valued. Show that this equation can be written in the following form:

$$\tilde{g}_p(t) = \frac{c_0}{2} + \sum_{n=1}^{\infty} c_n \cos\left(\frac{2\pi nt}{T} + \phi_n\right),$$

where $\{c_n\}$ and $\{\phi_n\}$ are sequences of real-valued constants. Express c_n and ϕ_n in terms of the G_n of Equation (48c) and vice versa.

[3.2] Use parts (b) and (a) of Exercise [1.2] to show that Equation (50c) can be rewritten as

$$g_{p,m}(t) = 1 + 2\Re\left(\frac{\phi e^{it} - \phi^{m+1} e^{i(m+1)t}}{1 - \phi e^{it}}\right),$$

where, for a complex number $z = x + iy$, $\Re(z) \stackrel{\text{def}}{=} x$. Conclude that

$$g_{p,m}(t) = \frac{1 - \phi^2 + f_m}{1 + \phi^2 - 2\phi \cos(t)}$$

with an appropriate definition for f_m. Use this to show that, for all t,

$$\lim_{m \to \infty} g_{p,m}(t) = g_p(t),$$

where $g_p(\cdot)$ is given by Equation (50b) and $|\phi| < 1$. (This result shows that we can in fact replace the $\stackrel{\text{ms}}{=}$ in the Fourier series representation of $g_p(\cdot)$ by $=$; i.e., the Fourier series representation of $g_p(t)$ is in fact equal to $g_p(t)$ itself for all t.)

[3.3] Consider a periodic function $g_p(\cdot)$ with a period of $T = 2$ such that

$$g_p(t) = e^{a|t|}, \quad -1 < t \leq 1,$$

where a is a nonzero real-valued constant.
  (a) What are the Fourier coefficients for this function? You might find the following indefinite integral useful:
  $$\int e^{ax} \cos(px) \, dx = e^{ax} \frac{a \cos(px) + p \sin(px)}{a^2 + p^2}.$$
  (b) What is the discrete power spectrum for $g_p(\cdot)$?
  (c) Determine the mth-order Fourier series approximation $g_{p,m}(\cdot)$ to $g_p(\cdot)$.
  (d) Create plots (similar to those in Figure 51a) showing how well $g_{p,m}(\cdot)$ approximates $g_p(\cdot)$ for $m = 2, 4, 8$ and 16 when $a = -1$.

[3.4] Repeat Exercise [3.3], but now use the following periodic function with period $T = 2$:

$$g_p(t) = 3t^2, \quad -1 < t \leq 1.$$

For part (a), you might find the following indefinite integral useful:

$$\int x^2 \cos(x) \, dx = 2x \cos(x) + (x^2 - 2) \sin(x).$$

For part (d), use $m = 4, 8, 16$ and 32 instead of $m = 2, 4, 8$ and 16.

[3.5] Repeat Exercise [3.3], but now use the following periodic function with period $T = 2$:

$$g_p(t) = \pi t, \quad -1 < t \leq 1.$$

For part (a), you might find the following indefinite integral useful:

$$\int x \sin(x) \, dx = \sin(x) - x \cos(x).$$

For part (d), use $m = 4, 8, 16$ and 32 instead of $m = 2, 4, 8$ and 16.

[3.6] Consider the function $g_p(t) = 4\cos^6(\pi t) + \sin^2(10\pi t)$.
  (a) Show that $g_p(\cdot)$ is a periodic function, and find its Fourier representation. Hint: make use of Exercise [1.2b] and of the fact that the Fourier representation for a function must be unique.
  (b) What is the discrete power spectrum for $g_p(\cdot)$?
  (c) What is the mth-order Fourier series approximation $g_{p,m}(\cdot)$ for this function?
  (d) Create plots (similar to those in Figure 51a) showing how well $g_{p,m}(\cdot)$ approximates $g_p(\cdot)$ for $m = 1, 2$ and 3. How well does $g_{p,m}(\cdot)$ approximate $g_p(\cdot)$ when $m \geq 4$?

[3.7] Exercise [49b] invites the reader to prove the "one-function" Parseval's theorem of Equation (49d). As complements,

(a) prove the "two-functions" Parseval's theorem stated in Equation (97e), and

(b) prove the "two-sequences" Parseval's theorem appropriate for the discrete time/continuous frequency case, namely, Equation (99a).

[3.8] Let $g(\cdot)$ be an $L^2(\mathbb{R})$ function with a Fourier transform $G(\cdot)$ that is band-limited to $[-W, W]$. For complex z, define

$$\tilde{g}(z) = \int_{-W}^{W} G(f) e^{i2\pi f z} \, df$$

(see Section 3.3). Use the Schwarz inequality in Equation (55d) to prove that the kth derivative of $\tilde{g}(z)$

$$\frac{d^k \tilde{g}(z)}{dz^k} = \int_{-W}^{W} G(f) (i2\pi f)^k \, e^{i2\pi f z} \, df$$

satisfies, for all complex numbers z,

$$\left| \frac{d^k \tilde{g}(z)}{dz^k} \right|^2 < (2\pi W)^{2k} \frac{e^{4\pi W |y|}}{4\pi |y|} \int_{-\infty}^{\infty} |g(t)|^2 \, dt,$$

where y is the imaginary part of z.

[3.9] Prove that, if $g(\cdot) \longleftrightarrow G(\cdot)$, then, for all $a \neq 0$,

$$|a|^{1/2} g(at) \longleftrightarrow \frac{1}{|a|^{1/2}} G(f/a)$$

(this is Equation (58a)).

[3.10] (a) Generalize Heisenberg's uncertainty principle in Section 3.4 to the case where the function $g(\cdot)$ does not necessarily have unit energy; i.e.,

$$\int_{-\infty}^{\infty} |g(t)|^2 \, dt = C < \infty.$$

To simplify matters, assume that

$$\mu_g = \int_{-\infty}^{\infty} t \, |g(t)|^2 \, dt = 0 \text{ and } \mu_G = \int_{-\infty}^{\infty} f \, |G(f)|^2 \, df = 0.$$

(b) Generalize to the case where μ_g and μ_G are not necessarily zero.

[3.11] Verify that the three convolutions depicted in the right-hand column of Figure 69 are given by, from top to bottom,

$$g * h(t) = \begin{cases} 1 - |t|, & |t| < 1; \\ 0, & |t| \geq 1; \end{cases} \qquad g * h(t) = \begin{cases} \frac{3}{4} - t^2, & |t| \leq \frac{1}{2}; \\ \frac{(3-2|t|)^2}{8}, & \frac{1}{2} < |t| < \frac{3}{2}; \\ 0, & |t| \geq \frac{3}{2}; \end{cases}$$

and

$$g * h(t) = \begin{cases} \frac{2}{3} - t^2 + \frac{|t|^3}{2}, & |t| \leq 1; \\ \frac{(2-|t|)^3}{6}, & 1 < |t| < 2; \\ 0, & |t| \geq 2. \end{cases} \qquad (104)$$

[3.12] Let $g(\cdot)$ and $h(\cdot)$ be two square integrable functions defined on the real axis, and let $G(\cdot)$ and $H(\cdot)$ denote their Fourier transforms. Exercise [67] calls for manipulating the order in which integrations are carried out to show that the Fourier transform of the convolution $g * h(\cdot)$ is the function defined by $G(f)H(f)$ (see the solution in the Appendix). Show this same result by first verifying that, for a fixed t,

$$h^*(t - \cdot) \longleftrightarrow H^*(\cdot) e^{-i2\pi f t}$$

3.13 Exercises

and then manipulating the "two-functions" Parseval's theorem of Equation (97e).

[3.13] (a) Show that the relationships in Equations (72b) and (73a) involving complex cross-correlation and complex autocorrelation are true.

(b) Because of the duality of the time and frequency domains, convolution in the *frequency* domain is equivalent to multiplication in the *time* domain. Prove this result; i.e., $g(\cdot)h(\cdot) \longleftrightarrow G * H(\cdot)$, as stated by Equations (98a) and (98b).

[3.14] Figure 59b indicates that width$_e \{g(\cdot)\}$ for the function defined by $g(t) = \exp(-|t|)$ is 2. For comparison, compute the width measures, width$_v \{g(\cdot)\}$ of Equation (60b) and width$_a \{g(\cdot)\}$ of Equation (73b).

[3.15] (a) Prove the version of the convolution theorem indicated by Equations (99b) and (99c) without recourse to use of δ functions.

(b) Since the Fourier transforms $G_p(\cdot)$ and $H_p(\cdot)$ are periodic functions with a period of $2f_\mathcal{N}$, the appropriate definition for a convolution $G_p * H_p(\cdot)$ in the frequency domain is given by Equation (99d). Verify Equation (99e), which states that the inverse Fourier transform of $G_p * H_p(\cdot)$ is $\{g_t h_t\}$.

[3.16] Use the following three facts to verify Equation (85c): first, $\{g_t\}$ has the Fourier integral representation

$$g_t = \int_{-W}^{W} G_p(f) e^{i2\pi ft} \, df$$

(assuming $\Delta_t = 1$); second, $|g_t|^2 = g_t g_t^*$; and, finally, the summation $\sum_{t=0}^{N-1} e^{i2\pi ft}$ can be reduced using Exercise [1.2c].

[3.17] For real-valued g_t, the concentration of energy $\beta^2(W)$ in the frequency range $|f| \leq W < 1/2$ (defined in Equation (85b)) takes the form

$$\beta^2(W) = \sum_{t'=0}^{N-1} \sum_{t=0}^{N-1} g_t \frac{\sin\left(2\pi W[t'-t]\right)}{\pi(t'-t)} g_{t'} \Bigg/ \sum_{t=0}^{N-1} g_t^2$$

(cf. Equation (88a)). Rewrite the above as the Rayleigh quotient $\beta^2(W) = \boldsymbol{g}^T \boldsymbol{A} \boldsymbol{g}/\boldsymbol{g}^T \boldsymbol{g}$ (here $\boldsymbol{g}$ is an N-dimensional vector, and $\boldsymbol{A}$ is an $N \times N$ matrix), and differentiate both sides with respect to $\boldsymbol{g}$ to show that the sequence that maximizes $\beta^2(W)$ must satisfy Equation (88b).

[3.18] As stated in the text following Equation (87b), the kth-order Slepian sequence is given by

$$v_{k,t}(N,W) = \frac{(-1)^k}{\epsilon_k \lambda_k(N,W)} \int_{-W}^{W} U_k(f; N, W) e^{i2\pi f[t-(N-1)/2]} \, df$$

for $t \in \mathbb{Z}$. Use this to show that

$$U_k(f; N, W) = (-1)^k \epsilon_k \sum_{t=0}^{N-1} v_{k,t}(N, W) e^{-i2\pi f[t-(N-1)/2]}.$$

Show that $U_k(\cdot; N, W)$ is symmetric for even k, i.e., $U_k(-f; N, W) = U_k(f; N, W)$, and skew-symmetric for odd k, i.e., $U_k(-f; N, W) = -U_k(f; N, W)$, so that we can write the above more compactly as

$$U_k(f; N, W) = \epsilon_k \sum_{t=0}^{N-1} v_{k,t}(N, W) e^{i2\pi f[t-(N-1)/2]}$$

(Slepian, 1978).

[3.19] The aim of this exercise is to derive some properties of the DPSWF $U_k(\cdot; N, W)$ and the Slepian sequence $\{v_{k,t}(N, W)\}$ stated in Section 3.10. Let $V_k(f; N, W) \stackrel{\text{def}}{=} I(f; W) U_k(f; N, W)$ on $[-1/2, 1/2]$, where $I(\cdot; W)$ is as defined in part (a) of the previous exercise. (Note that all functions here are periodic with period 1.) Then the integral equation shown in Equation (86b) with $U(\cdot) = U_k(\cdot; N, W)$ and $\lambda = \lambda_k(N, W)$ can be written

$$N\mathcal{D}_N * V_k(f; N, W) = \lambda_k(N, W)U_k(f; N, W).$$

The sequences $\{v_{k,t}(N, W)\}$ and $\{u_{k,t}(N, W)\}$ are defined by Fourier transform relationships for the discrete time/continuous frequency:

$$\{(-1)^k \epsilon_k \lambda_k(N, W) v_{k,t}(N, W)\} \longleftrightarrow e^{-i\pi(N-1)f} V_k(f; N, W)$$

$$\{(-1)^k \epsilon_k u_{k,t}(N, W)\} \longleftrightarrow e^{-i\pi(N-1)f} U_k(f; N, W).$$

(a) By expanding the Dirichlet kernel in the convolution above, show that

$$u_{k,t}(N, W) = \begin{cases} v_{k,t}(N, W), & t = 0, 1, \ldots, N-1; \\ 0, & \text{otherwise.} \end{cases}$$

(b) The properties of the solutions to integral equation shown in Equation (86b) imply that the $U_k(\cdot; N, W)$ functions are orthogonal on $[-W, W]$. They have been normalized to have unit energy on $[-1/2, 1/2]$. Use the fact that

$$\sum_{t=-\infty}^{\infty} u_{j,t} u_{k,t}^* = \sum_{t=-\infty}^{\infty} v_{j,t} u_{k,t}^*$$

and the "two-sequences" version of Parseval's theorem (Equation (99a)) to show that the $U_k(\cdot; N, W)$ functions are orthonormal on $[-1/2, 1/2]$ and that their energy on $[-W, W]$ is $\lambda_k(N, W)$.

(c) Show that the $\{v_{k,t}(N, W)\}$ sequences are orthonormal over $t = 0, 1, \ldots, N-1$.
(d) Show that the $\{v_{k,t}(N, W)\}$ sequences are orthogonal over $t \in \mathbb{Z}$ with energy $1/\lambda_k(N, W)$.
(e) Show that the $\{v_{k,t}(N, W)\}$ sequences satisfy the $N \times N$ linear system shown in Equation (87c).

[3.20] (a) Suppose that $g_0, g_1, \ldots, g_{N-1}$ and $h_0, h_1, \ldots, h_{N-1}$ are two finite sequences of length N. Prove the *cyclic convolution theorem* of Equation (101e), and describe how $g * h_0, g * h_1, \ldots, g * h_{N-1}$ can be computed using two DFTs and one inverse DFT. (As an aside, we note that this result has some important practical implications. If N is a sample size acceptable to an FFT algorithm, computation of $\{g * h_t\}$ directly from Equation (101d) would require on the order of N^2 floating point operations on a digital computer, whereas use of three FFTs would typically require on the order of $N \log_2(N)$ such operations. It is thus often more efficient (particularly for large N) to compute cyclic convolutions via FFTs rather than via the defining Equation (101d).)

(b) Suppose now that $g_0, g_1, \ldots, g_{N_g-1}$ and $h_0, h_1, \ldots, h_{N_h-1}$ are two finite sequences of lengths N_g and N_h with $N_g < N_h$. Suppose we are interested in computing the following sequence of length $N_h - N_g + 1$ (a portion of a noncyclic convolution):

$$c_n = \sum_{t=0}^{N_g-1} g_t h_{n-t} \text{ for } n = N_g - 1, N_g, \ldots, N_h - 1.$$

Show that, if N is large enough, the c_n can be computed via cyclic convolution of two sequences of length N, namely,

$$g'_t = \begin{cases} g_t, & 0 \leq t \leq N_g - 1; \\ 0, & N_g \leq t \leq N - 1; \end{cases} \text{ and } h'_t = \begin{cases} h_t, & 0 \leq t \leq N_h - 1; \\ 0, & N_h \leq t \leq N - 1. \end{cases}$$

How large is "large enough?" (This is another example of the use of zero padding.)

[3.21] (a) For $a > 0$, define $I(t; a) = 1$ for $|t| \leq a$ and zero otherwise. Show that

$$I(t; a) \longleftrightarrow \frac{\sin(2\pi f a)}{\pi f}.$$

(b) For $N > 0$, define the sequence $\{i_{N,t}\}$ by $i_{N,t} = 1$ for $t = 0, 1, \ldots, N-1$ and zero otherwise. Show that

$$\{i_{N,t}\} \longleftrightarrow e^{-i(N-1)\pi f} \frac{\sin(N\pi f)}{\sin(\pi f)}$$

(hint: consider Exercise [1.2c]).

4

Foundations for Stochastic Spectral Analysis

4.0 Introduction

In the previous chapter we produced representations for various deterministic functions and sequences in terms of linear combinations of sinusoids with different frequencies (for mathematical convenience we actually used complex exponentials instead of sinusoids directly). These representations allow us to easily define various energy and power spectra and to attach a physical meaning to them. For example, subject to square integrability conditions, we found that periodic functions are representable (at least in the mean square sense) by sums of sinusoids over a discrete set of frequency components, while nonperiodic functions are representable (also in the mean square sense) by an integral of sinusoids over a continuous range of frequencies. For periodic functions, the energy from $-\infty$ to ∞ is infinite, so we can define their spectral properties in terms of distributions of power over a discrete set of frequencies. For nonperiodic functions, the energy from $-\infty$ to ∞ is finite, so we can define their properties in terms of an energy distribution over a continuous range of frequencies.

We now want to find some way of representing a stationary process in terms of a "sum" of sinusoids so that we can meaningfully define an appropriate spectrum for it; i.e., we want to be able to directly relate our representation for a stationary process to its spectrum in much the same way we did for deterministic functions. Now a stationary process has associated with it an ensemble of realizations that describe the possible outcomes of a random experiment. Each realization is a fixed function or sequence of time, so we could try to apply the theory for deterministic functions and sequences on a realization by realization basis and then extend it somehow to the entire ensemble. However, the functions and sequences we can handle easily either have finite energy or are periodic. Because the variance of a stationary process is constant over time, a typical realization has infinite energy; moreover, with the important exception of harmonic processes, a typical realization is also not periodic. This formally rules out merely applying the deterministic theory on a realization by realization basis (by use of a limiting argument, this approach can be carried out to a certain extent for some stationary processes – see Section 4.2 for details). There is a different approach that works for stationary processes and is discussed in the next section. It is based upon the "important exception" of the harmonic processes and leads to a spectral representation theorem for stationary processes.

4.1 Spectral Representation of Stationary Processes

In this section we motivate and state (but do not prove) the spectral representation theorem for stationary processes due to Cramér (1942). A rigorous proof is rather involved – the interested reader is referred to Priestley (1981, section 4.11) for details. This theorem is fundamental in the spectral analysis of stationary processes since it allows us to relate the spectrum of such a process directly to a representation for the process itself. Indeed, in the words of Koopmans (1974, p. 36), "One of the essential reasons for the central position held by stationary stochastic processes in time series analysis is the existence of a spectral representation for the process from which [the spectrum] can be directly computed."

We motivate the spectral representation theorem for discrete parameter stationary processes by considering the special case of a real-valued discrete time harmonic process that was formulated in Exercise [37] and involves fixed amplitudes and random phases:

$$X_t = \sum_{l=1}^{L} D_l \cos(2\pi f_l t + \phi_l), \quad t \in \mathbb{Z}, \tag{108a}$$

where $L \geq 1$; D_l and f_l are real-valued constants; the phases $\{\phi_l\}$ are independent random variables (RVs), each having a rectangular distribution on the interval $[-\pi, \pi]$ (this is Equation (35d) with the process mean μ assumed to be zero); and $\mathbb{Z}$ is the set of all integers. We assume that the frequencies f_l are ordered such that $f_l < f_{l+1}$ and that $0 < f_l < 1/2$ for all l. Except for excluding frequencies 0 and $1/2$ (which simplifies our discussion somewhat), this latter stipulation is not really a restriction due to the aliasing phenomenon (see Sections 3.9 and 4.5) – here the Nyquist frequency is $1/2$ because we assume that Δ_t is 1. Since

$$D_l \cos(2\pi f_l t + \phi_l) = \frac{D_l}{2} \left(e^{i\phi_l} e^{i 2\pi f_l t} + e^{-i\phi_l} e^{-i 2\pi f_l t} \right),$$

we can rewrite Equation (108a) as

$$X_t = \sum_{l=-L}^{L} C_l e^{i 2\pi f_l t}, \tag{108b}$$

where

$$C_l \stackrel{\text{def}}{=} D_l e^{i\phi_l}/2 \text{ and } C_{-l} \stackrel{\text{def}}{=} D_l e^{-i\phi_l}/2, \quad l = 1, \ldots, L;$$

$C_0 \stackrel{\text{def}}{=} 0$; $f_0 \stackrel{\text{def}}{=} 0$; and $f_{-l} \stackrel{\text{def}}{=} -f_l$. Since the ϕ_l RVs are assumed to be independent, the RVs $C_1, C_2, \ldots, C_L$ are also independent. Since $C_{-l} = C_l^*$, the RVs C_{-l} and C_l are certainly not independent, but – rather surprisingly – they are uncorrelated (see Exercise [4.1]). Thus the $2L+1$ RVs $C_{-L}, C_{-L+1}, \ldots, C_L$ are all mutually uncorrelated with means and variances given by

$$E\{C_l\} = 0 \text{ and } \text{var}\{C_l\} = E\{|C_l|^2\} = D_l^2/4$$

for all l if we define $D_0 = 0$ and $D_{-l} = D_l$ (these results follow from Section 2.6). Thus

$$\text{var}\{X_t\} = \sum_{l=-L}^{L} \text{var}\{C_l e^{i 2\pi f_l t}\} = \sum_{l=-L}^{L} E\{|C_l|^2\} = \sum_{l=-L}^{L} D_l^2/4; \tag{108c}$$

i.e., the variance of the stationary process $\{X_t\}$ can be decomposed into a sum of components $E\{|C_l|^2\}$, each of which is the expected square amplitude of the complex exponential of frequency f_l in Equation (108b). We can thus define a useful variance spectrum by

$$S^{(V)}(f) = \begin{cases} D_l^2/4, & f = f_l, l = 0, \pm 1, \ldots, \pm L; \\ 0, & \text{otherwise.} \end{cases}$$

4.1 Spectral Representation of Stationary Processes

Let us now define the complex-valued stochastic process

$$Z(f) = \sum_{j=0}^{l} C_j, \quad f_l < f \leq f_{l+1} \text{ with } l = 0, \ldots, L, \qquad (109a)$$

where $f_{L+1} \stackrel{\text{def}}{=} 1/2$. For completeness, we define $Z(0) = 0$. With this definition, $\{Z(f)\}$ is a "jump" process that is defined on the interval $[0, 1/2]$ and has a random complex-valued jump at each f_l. Thus

$$Z(f) = \begin{cases} 0, & 0 \leq f \leq f_1; \\ C_1, & f_1 < f \leq f_2; \\ C_1 + C_2, & f_2 < f \leq f_3; \\ C_1 + C_2 + C_3, & f_3 < f \leq f_4, \end{cases}$$

and so forth. Note that the "time" variable for this process is actually frequency.

Let us define

$$dZ(f) = \begin{cases} Z(f + df) - Z(f), & 0 \leq f < 1/2; \\ 0, & f = 1/2; \\ dZ^*(-f), & -1/2 \leq f < 0, \end{cases} \qquad (109b)$$

where df is a small positive increment such that $0 < f + df < 1/2$ when $0 < f < 1/2$. For $l \geq 0$ we thus have

$$dZ(f_l) = Z(f_l + df) - Z(f_l) = \sum_{j=0}^{l} C_j - \sum_{j=0}^{l-1} C_j = C_l,$$

and, for any $f \neq f_l$ for some l, $dZ(f) = 0$ for df sufficiently small. Since $E\{C_l\} = 0$, it follows that $E\{dZ(f)\} = 0$ for all f. Exercise [4.2] says that, if f, f', df and df' are such that the intervals $[f, f+df]$ and $[f', f'+df']$ are nonintersecting subintervals of $[-1/2, 1/2]$, then the RVs $dZ(f)$ and $dZ(f')$ are uncorrelated; i.e.,

$$\text{cov}\{dZ(f'), dZ(f)\} = E\{dZ(f')\,dZ^*(f)\} = 0.$$

Because of this property, the process $\{Z(f)\}$ is said to have *orthogonal increments*, and the process itself is called an *orthogonal process*. Note that $\{Z(f)\}$ gives us each C_l in $\{X_t\}$ and that the expected square magnitude of the jump in $\{Z(f)\}$ at f_l is just $E\{|C_l|^2\}$; i.e.,

$$E\{|dZ(f_l)|^2\} = E\{|C_l|^2\} = D_l^2/4.$$

Now let $g(\cdot)$ be a function that is continuous over the interval $[-1/2, 1/2]$, and let $H(\cdot)$ be a step function defined over that same interval with jumps at

$$-1/2 < f_1' < f_2' < \cdots < f_N' < 1/2$$

of finite sizes $a_1, a_2, \ldots, a_N$. From the definition of the Riemann–Stieltjes integral it follows that

$$\int_{-1/2}^{1/2} g(f)\,dH(f) = \sum_{k=1}^{N} a_k g(f_k')$$

(see the Comments and Extensions [C&Es] to Section 2.3). If we match $g(f)$ and $H(f)$ up with, respectively, $\exp(i2\pi ft)$ and $Z(f)$, we can rewrite Equation (108b) as follows:

$$X_t = \int_{-1/2}^{1/2} e^{i2\pi ft} \, dZ(f). \tag{110a}$$

This is a stochastic version of the Riemann–Stieltjes integral, which is defined in a way analogous to the usual version (see Equation (25b) and Exercise [4.3]). Equation (110a) is called the *spectral representation* for the stationary process shown in Equation (108b). It is useful mainly due to the properties of the orthogonal increments process. As an example, let us show how we can derive Equation (108c) using Equation (110a). Since $\{X_t\}$ is a real-valued process with $E\{X_t\} = 0$,

$$\begin{aligned} \operatorname{var}\{X_t\} &= E\{X_t^2\} = E\{X_t X_t^*\} \\ &= E\left\{\int_{-1/2}^{1/2} e^{i2\pi ft} \, dZ(f) \int_{-1/2}^{1/2} e^{-i2\pi f't} \, dZ^*(f')\right\} \\ &= \int_{-1/2}^{1/2} \int_{-1/2}^{1/2} e^{i2\pi (f-f')t} E\{dZ(f) \, dZ^*(f')\}. \end{aligned} \tag{110b}$$

Because of the properties of $\{Z(f)\}$, the only time the expectation within the double integral is nonzero occurs when $f = f' = f_l$ for $l = -L, \ldots, L$, in which case we have

$$E\{dZ(f_l) \, dZ^*(f_l)\} = E\{|dZ(f_l)|^2\} = E\{|C_l|^2\}.$$

Since $\exp[-i2\pi(f_l - f_l)t] = 1$, the Riemann–Stieltjes integral reduces to Equation (108c). (The above manipulations are purely formal; see Exercise [4.3] for some justification.)

Rather surprisingly, we can develop a spectral representation for *any* discrete parameter stationary process by considering what happens to Equation (108b) in the limit as $L \to \infty$ in such a way that the maximum difference between adjacent frequencies f_{l-1} and f_l goes to 0 and the process variance shown in Equation (108c) converges to a finite number. We are now in a position to state (without proof) the *spectral representation theorem for discrete parameter stationary processes* (for more details, see Priestley, 1981, p. 251). Let $\{X_t\}$ be a real-valued discrete parameter stationary process with zero mean. There exists an orthogonal process, $\{Z(f)\}$, defined on the interval $[-1/2, 1/2]$, such that

$$X_t = \int_{-1/2}^{1/2} e^{i2\pi ft} \, dZ(f) \tag{110c}$$

for all integers t, where the above equality is in the mean square sense (in fact, $\{Z(f)\}$ is unique – see Theorem 4.2, Chonavel, 2002). The process $\{Z(f)\}$ has the following properties:

[1] $E\{dZ(f)\} = 0$ for all $|f| \leq 1/2$;
[2] $dS^{(I)}(f) \stackrel{\text{def}}{=} E\{|dZ(f)|^2\}$, say, for all $|f| \leq 1/2$, where the bounded nondecreasing function $S^{(I)}(\cdot)$ is called the *integrated spectrum* of $\{X_t\}$; and
[3] for any two distinct frequencies f and f' contained in the interval $[-1/2, 1/2]$,

$$\operatorname{cov}\{dZ(f), dZ(f')\} = E\{dZ(f) \, dZ^*(f')\} = 0.$$

Equation (110c) is called the *spectral representation* of the process $\{X_t\}$. It says that we can represent *any* discrete parameter stationary process as an infinite sum of complex exponentials

4.1 Spectral Representation of Stationary Processes

(i.e., sine and cosine functions) at frequencies f with associated random amplitudes $|\mathrm{d}Z(f)|$ and random phases $\arg(\mathrm{d}Z(f))$, where $\arg(z)$ is the phase angle of the complex number z. Moreover, the expected value of the square modulus of $\mathrm{d}Z(f)$ *defines* an integrated spectrum $S^{(\mathrm{I})}(\cdot)$ for $\{X_t\}$, and the spectrum for $\{X_t\}$ is unique.

The property of $\{Z(f)\}$ that makes the representation of Equation (110c) helpful in the proof of many results in spectral analysis is that its increments at different frequencies are uncorrelated. As an example of the use of this property, we now show a fundamental relationship between the autocovariance sequence (ACVS) $\{s_\tau\}$ and the integrated spectrum $S^{(\mathrm{I})}(\cdot)$ for a stationary process $\{X_t\}$. Since

$$X_{t+\tau}X_t^* = \int_{-1/2}^{1/2} e^{i2\pi f(t+\tau)} \, \mathrm{d}Z(f) \int_{-1/2}^{1/2} e^{-i2\pi f't} \, \mathrm{d}Z^*(f')$$

$$= \int_{-1/2}^{1/2}\int_{-1/2}^{1/2} e^{i2\pi f(t+\tau)} e^{-i2\pi f't} \, \mathrm{d}Z(f) \, \mathrm{d}Z^*(f'),$$

it follows that the ACVS can be written as

$$s_\tau = E\{X_{t+\tau}X_t\} = E\{X_{t+\tau}X_t^*\} = \int_{-1/2}^{1/2}\int_{-1/2}^{1/2} e^{i2\pi(f-f')t} e^{i2\pi f\tau} E\{\mathrm{d}Z(f)\,\mathrm{d}Z^*(f')\}.$$

Because of the orthogonality property of $\{Z(f)\}$, the only contribution to the double integral occurs when $f = f'$, so we have

$$s_\tau = \int_{-1/2}^{1/2} e^{i2\pi f\tau} E\{|\mathrm{d}Z(f)|^2\} = \int_{-1/2}^{1/2} e^{i2\pi f\tau} \, \mathrm{d}S^{(\mathrm{I})}(f), \tag{111a}$$

which shows that the integrated spectrum determines the ACVS.

If in fact $S^{(\mathrm{I})}(\cdot)$ is differentiable everywhere with a derivative denoted by $S(\cdot)$, we have

$$E\{|\mathrm{d}Z(f)|^2\} = \mathrm{d}S^{(\mathrm{I})}(f) = S(f)\,\mathrm{d}f. \tag{111b}$$

The function $S(\cdot)$ is called the *spectral density function* (SDF). We can now rewrite Equation (111a) in terms of this SDF as

$$s_\tau = \int_{-1/2}^{1/2} S(f) e^{i2\pi f\tau} \, \mathrm{d}f. \tag{111c}$$

From Section 3.8, we know that a square summable deterministic sequence $\{g_t\}$ has the Fourier representation

$$g_t = \int_{-1/2}^{1/2} G_p(f) e^{i2\pi ft} \, \mathrm{d}f, \quad \text{where } G_p(f) \stackrel{\mathrm{def}}{=} \sum_{t=-\infty}^{\infty} g_t e^{-i2\pi ft} \tag{111d}$$

(these are Equations (75a) and (74a) with $\Delta_t = 1$). Since $\{s_\tau\}$ is just a deterministic sequence, Equation (111c) indicates that it is the inverse Fourier transform of $S(\cdot)$. If we assume that $S(\cdot)$ is square integrable (and thus, from Parseval's theorem, that the ACVS is square summable), we have the important fact that $S(\cdot)$ is the Fourier transform of $\{s_\tau\}$:

$$S(f) = \sum_{\tau=-\infty}^{\infty} s_\tau e^{-i2\pi f\tau}, \quad f \in \mathbb{R}, \tag{111e}$$

where $S(\cdot)$ is an even periodic function with a period of unity (strictly speaking, the above equality is in the mean square sense, but it can be regarded as a pointwise equality in almost all practical applications). Hence we have

$$\{s_\tau\} \longleftrightarrow S(\cdot). \tag{112a}$$

Note that we can regard $S(\cdot)$ as an amplitude spectrum for $\{s_\tau\}$.

The stationary process $\{X_t\}$ and the orthogonal process $\{Z(f)\}$ have a Fourier relationship, and so do the ACVS $\{s_\tau\}$ and the SDF $S(\cdot)$. Let us compare these relationships with the spectral representation for a square summable sequence $\{g_t\}$ developed in Section 3.8 and repeated in Equation (111d). For this discussion we will let $\{g_t\}$ play the role of $\{X_t\}$ even though the former is deterministic with finite energy and the latter is stochastic with infinite expected energy. If we define

$$G_p^{(\mathrm{I})}(f) = \int_{-1/2}^{f} G_p(f')\,\mathrm{d}f' \text{ so that } \frac{\mathrm{d}G_p^{(\mathrm{I})}(f)}{\mathrm{d}f} = G_p(f),$$

then we can represent g_t using an ordinary Riemann–Stieltjes integral involving increments of $G_p^{(\mathrm{I})}(\cdot)$:

$$g_t = \int_{-1/2}^{1/2} e^{\mathrm{i}2\pi f t}\,\mathrm{d}G_p^{(\mathrm{I})}(f).$$

This can be compared with the spectral representation for $\{X_t\}$ in Equation (110c). The reason we cannot reduce Equation (110c) to just a stochastic Riemann integral is that, in contrast to $G_p^{(\mathrm{I})}(\cdot)$, the stochastic process $\{Z(f)\}$ does *not* possess a derivative in any well-defined sense. In particular, this prevents us from establishing a relationship for $\{Z(f)\}$ in terms of $\{X_t\}$; in contrast, the derivative of $G_p^{(\mathrm{I})}(\cdot)$ – namely, $G_p(\cdot)$ – is the Fourier transform of $\{g_t\}$.

Let us now examine the autocorrelation of $\{g_t\}$, which is the sequence $\{g \star g_\tau^*\}$ defined by

$$g \star g_\tau^* = \sum_{u=-\infty}^{\infty} g_u g_{u+\tau}^*$$

(note that we have called $g \star g_\tau^*$ an auto*correlation* rather than an auto*covariance*, in agreement with engineering practice for deterministic sequences – since $|g \star g_\tau^*| > 1$ is not precluded, this terminology conflicts with that of the statistical community). As noted just following Equation (100a), the autocorrelation of $\{g_t\}$ is the inverse Fourier transform of $|G_p(\cdot)|^2$. Hence

$$g \star g_\tau^* = \int_{-1/2}^{1/2} |G_p(f)|^2 e^{\mathrm{i}2\pi f \tau}\,\mathrm{d}f. \tag{112b}$$

If we define

$$H_p^{(\mathrm{I})}(f) = \int_{-1/2}^{f} |G_p(f')|^2\,\mathrm{d}f' \text{ so that } \frac{\mathrm{d}H_p^{(\mathrm{I})}(f)}{\mathrm{d}f} = |G_p(f)|^2,$$

we can rewrite the autocorrelation as an ordinary Riemann–Stieltjes integral:

$$g \star g_\tau^* = \int_{-1/2}^{1/2} e^{\mathrm{i}2\pi f \tau}\,\mathrm{d}H_p^{(\mathrm{I})}(f).$$

If we compare the above to Equation (111a), we see that $E\{|dZ(f)|^2\} = dS^{(I)}(f)$ in the stochastic model plays the role of $dH_p^{(I)}(f)$ in the deterministic case. The reason we cannot in general express Equation (111a) as an ordinary Riemann integral analogous to Equation (112b) is that the integrated spectrum $S^{(I)}(\cdot)$ does not always possess a derivative; when its derivative $S(\cdot)$ – the SDF – does exist, we then have Equation (111c), which is analogous to Equation (112b). (If we allow the use of the Dirac delta function, we can define a derivative for $S^{(I)}(\cdot)$ – see the remarks at the end of Section 4.4.)

As a final comparison, in the deterministic case, we have claimed that $|G_p(\cdot)|^2$ is an energy spectral density function so that $|G_p(f)|^2\,df$ gives the contribution to the energy in $\{g_t\}$ due to frequencies in a small interval about f. Since it follows from Equations (111a) and (111c) that

$$\operatorname{var}\{X_t\} = s_0 = \int_{-1/2}^{1/2} dS^{(I)}(f) = \int_{-1/2}^{1/2} S(f)\,df, \qquad (113\text{a})$$

clearly $S(f)\,df$ is the contribution to the variance in the stationary process due to frequencies in a small interval about f. In the next section we note that the variance of a process is in fact closely related to the concept of power so that $S(\cdot)$ is often called a *power spectral density function* (PSDF).

Comments and Extensions to Section 4.1

[1] Why does the integral in Equation (110c) only include the range of frequencies from $-1/2$ to $1/2$? The answer is aliasing (see Sections 3.9 and 4.5): the values of $\exp(i2\pi ft)$ and $\exp(i2\pi[f \pm k]t)$ are identical for all integers t and k, and so it is only necessary to have complex exponentials with frequencies in the range $[-1/2, 1/2]$ in the representation of Equation (110c).

[2] In the more general case when $E\{X_t\} = \mu \neq 0$ and the sampling interval Δ_t is not unity, we can write

$$X_t = \mu + \int_{-f_\mathcal{N}}^{f_\mathcal{N}} e^{i2\pi ft\Delta_t}\,dZ(f) \quad \text{for } t \in \mathbb{Z}, \text{ where, as usual, } f_\mathcal{N} \stackrel{\text{def}}{=} \frac{1}{2\Delta_t}, \qquad (113\text{b})$$

and $\{Z(f)\}$ is the orthogonal process corresponding to the zero mean process $\{X_t - \mu\}$. Equations (111a) and (111c) are replaced by

$$s_\tau = \int_{-f_\mathcal{N}}^{f_\mathcal{N}} e^{i2\pi f\tau\Delta_t}\,dS^{(I)}(f) = \int_{-f_\mathcal{N}}^{f_\mathcal{N}} S(f) e^{i2\pi f\tau\Delta_t}\,df \qquad (113\text{c})$$

for $\tau \in \mathbb{Z}$, while Equation (111e) becomes

$$S(f) = \Delta_t \sum_{\tau=-\infty}^{\infty} s_\tau e^{-i2\pi f\tau\Delta_t}, \quad f \in \mathbb{R}, \qquad (113\text{d})$$

where $S(\cdot)$ is an even periodic function with a period of $2f_\mathcal{N}$ (the SDFs for $\{X_t\}$ and $\{X_t - \mu\}$ are necessarily the same because their ACVSs are identical due to the fact that $\operatorname{cov}\{X_{t+\tau}, X_t\}$ is equal to $\operatorname{cov}\{X_{t+\tau} - \mu, X_t - \mu\}$). Note also that Equation (113a) becomes

$$\operatorname{var}\{X_t\} = s_0 = \int_{-f_\mathcal{N}}^{f_\mathcal{N}} dS^{(I)}(f) = \int_{-f_\mathcal{N}}^{f_\mathcal{N}} S(f)\,df. \qquad (113\text{e})$$

[3] The *spectral representation theorem for continuous parameter stationary processes* is quite similar to the discrete parameter theorem. We merely state it and refer the reader to Priestley (1981, p. 246) for details. Let $\{X(t)\}$ be a real-valued, *stochastically continuous*, continuous parameter stationary process with zero mean. Then there exists an orthogonal process, $\{Z(f)\}$, such that, for all $t \in \mathbb{R}$, $X(t)$ can be expressed in the form

$$X(t) = \int_{-\infty}^{\infty} e^{i2\pi ft} \, dZ(f).$$

The process $\{Z(f)\}$ has the following properties:

(a) $E\{dZ(f)\} = 0$ for all f;
(b) $dS^{(I)}(f) \stackrel{\text{def}}{=} E\{|dZ(f)|^2\}$, say, for all f, where the bounded nondecreasing function $S^{(I)}(\cdot)$ is called the *integrated spectrum* of $\{X(t)\}$; and
(c) $\text{cov}\{dZ(f), dZ(f')\} = E\{dZ(f) \, dZ^*(f')\} = 0$ for any two frequencies $f \neq f'$.

Note that this theorem only holds for a subclass of continuous parameter stationary processes, namely, those that are stochastically continuous. Sine we do not use this concept elsewhere in this book, it suffices to say that stochastic continuity is a mild regularity condition that rules out some pathological processes of little interest in practical applications. It holds if and only if the autocovariance function (ACVF) $s(\cdot)$ for the process is continuous at the origin (see Priestley, 1981, p. 151, for details).

For the continuous parameter case, Equation (111a) becomes

$$s(\tau) = \int_{-\infty}^{\infty} e^{i2\pi f\tau} \, dS^{(I)}(f).$$

If $S^{(I)}(\cdot)$ is differentiable and $s(\cdot)$ is square integrable, Equations (111c) and (111e) become

$$s(\tau) = \int_{-\infty}^{\infty} S(f) e^{i2\pi f\tau} \, df \quad \text{and} \quad S(f) = \int_{-\infty}^{\infty} s(\tau) e^{-i2\pi f\tau} \, d\tau. \tag{114}$$

[4] With a few changes, the approach we have outlined above can also be used to motivate a spectral representation theorem for complex-valued stationary processes. In particular, we let the C_l RVs for $l \neq 0$ in Equation (108b) be *any* collection of $2L$ uncorrelated complex-valued RVs with zero means (i.e., we drop the restriction that $C_{-l} = C_l^*$), and in Equation (109a) we use the C_l RVs with $l < 0$ to construct $Z(f)$ for $f < 0$ in an obvious way – this in turn is used to define $dZ(f)$ for $f < 0$ rather than stipulating that $dZ(f) = dZ^*(-f)$ for $f < 0$ as is done in Equation (109b). The statement of the spectral representation theorem itself is the same except that "real-valued" is replaced by "complex-valued."

Because we no longer have the constraint $dZ(f) = dZ^*(-f)$ for a complex-valued process, it follows that the SDF (when it exists) need not be an even function; i.e., in general, we no longer have $S(-f) = S(f)$. An example of this is the so-called *analytic series* for a real-valued stationary process $\{X_t\}$, which, by definition, is constructed by taking the Hilbert transform $\mathcal{HT}\{X_t\}$ of the process and then forming the complex-valued process given by $X_t + i\mathcal{HT}\{X_t\}$. The SDF of this complex-valued process is proportional to the SDF of $\{X_t\}$ for $f \geq 0$, but is identically zero for $f < 0$. For details (including the definition of the Hilbert transform), see Section 10.13 of this book or section 9–3 of Papoulis and Pillai (2002).

4.2 Alternative Definitions for the Spectral Density Function

In the previous section we defined the integrated spectrum and the SDF (when it exists) in terms of the spectral representation of a stationary process. This approach allows us to relate the integrated spectrum and SDF directly to the representation of the process itself. It is sometimes stated that by *definition* the SDF is the Fourier transform of the ACVS, i.e., that Equation (111e) for discrete parameter stationary processes (or Equation (114) for continuous parameter processes) *defines* $S(\cdot)$ (see, for example, section 9–3 of Papoulis and Pillai, 2002). It is difficult to attach much meaning to $S(\cdot)$ from this definition alone. The usual approach is to appeal to the theory of linear filters in order to establish a physical meaning for the SDF (see Section 5.6).

4.2 Alternative Definitions for the Spectral Density Function

Another way of defining the SDF makes use of the Fourier theory for deterministic sequences with finite energy (Section 3.8). Here, instead of just drawing comparisons between square summable sequences and stationary processes as we did in the previous section, we treat portions of realizations of a stationary process as a square summable sequence and apply some limiting arguments. We present this definition here because it is particularly informative (see sections 4.7 and 4.8 of Priestley, 1981). Let $\{x_t\}$ be any realization of the discrete parameter stationary process $\{X_t\}$ with zero mean. Then we should have

$$\frac{1}{N}\sum_{t=0}^{N-1} x_t^2 \approx \sigma^2 \stackrel{\text{def}}{=} E\{X_t^2\}$$

for large N. It is thus unlikely that $\{x_t\}$ has finite energy since we should have

$$\lim_{N\to\infty} \sum_{t=0}^{N-1} x_t^2 = \infty,$$

but it is plausible that it has finite power since we should have

$$\lim_{N\to\infty} \frac{1}{N} \sum_{t=0}^{N-1} x_t^2 = \sigma^2$$

(this argument says that the variance of a stationary process is in fact just its power). Let us define a new sequence $\{x_{N,t}\}$ by

$$x_{N,t} = \begin{cases} x_t, & 0 \leq t \leq N-1; \\ 0, & \text{otherwise.} \end{cases}$$

Since $\{x_{N,t}\}$ is square summable (the number of nonzero terms is finite), the theory of Section 3.8 says that it has the following Fourier integral representation:

$$x_{N,t} = \int_{-1/2}^{1/2} G_N(f) e^{i2\pi f t}\, df,$$

where

$$G_N(f) \stackrel{\text{def}}{=} \sum_{t=-\infty}^{\infty} x_{N,t} e^{-i2\pi f t} = \sum_{t=0}^{N-1} x_t e^{-i2\pi f t}. \tag{115}$$

Since Parseval's theorem says that

$$\sum_{t=0}^{N-1} x_{N,t}^2 = \int_{-1/2}^{1/2} |G_N(f)|^2\, df,$$

it follows that $|G_N(f)|^2\, df$ is equal to the contribution to the energy of $\{x_{N,t}\}$ from components with frequencies in a small interval about f. This suggests that, even if

$$\lim_{N\to\infty} |G_N(f)|^2$$

were to define a reasonable function of f, it would not be finitely integrable since the energy of $\{x_t\}$ is infinite. Previously, in cases where energy was infinite, we have found power to be well defined. This suggests that, under suitable conditions, the function of f defined via a proper interpretation of

$$\lim_{N\to\infty} \frac{|G_N(f)|^2}{N} \tag{116a}$$

might be well behaved enough to be finitely integrable. If this is true, an integral of a proper interpretation of the above quantity over a small interval of length df that is centered about f will equal the contribution to the power (variance) of $\{x_t\}$ from components in that frequency range.

Now the quantity in Equation (116a) obviously depends on $\{x_t\}$, the particular realization of the stationary process we have chosen. To construct a corresponding quantity that reflects the power per frequency properties of the entire stochastic process (and not just one realization), it is natural to average the values of $|G_N(f)|^2/N$ over all realizations. If we now redefine $G_N(f)$ to be an RV defined for each realization of $\{X_t\}$ by Equation (115), we are led to the quantity

$$S(f) \stackrel{\text{def}}{=} \lim_{N\to\infty} E\left\{\frac{|G_N(f)|^2}{N}\right\} = \lim_{N\to\infty} E\left\{\frac{1}{N}\left|\sum_{t=0}^{N-1} X_t e^{-i2\pi ft}\right|^2\right\}. \tag{116b}$$

When it exists, $S(\cdot)$ has the following interpretation: $S(f)\,df$ is the average contribution (over all realizations) to the power from components with frequencies in a small interval about f. The power (variance) is just the integral of $S(\cdot)$ over $f \in [-1/2, 1/2]$. The function $S(\cdot)$ is called the *power spectral density function* of the process $\{X_t\}$. This definition for $S(\cdot)$ requires that $\{X_t\}$ satisfies certain conditions to validate the manipulations that lead to Equation (116b). When these conditions do hold, the function $S(\cdot)$ is identical to the SDF defined via the spectral representation theorem. (In Chapter 6 we will learn that the quantity inside the expectation operation in Equation (116b) is known as the *periodogram*; hence this definition of the SDF is just the limit – as the sample size N gets large – of the expected value of the periodogram.)

4.3 Basic Properties of the Spectrum

In this section we go over some of the basic properties of the integrated spectrum $S^{(I)}(\cdot)$ and the SDF $S(\cdot)$. When the latter exists, the integrated spectrum is related to it by

$$S^{(I)}(f) = \int_{-1/2}^{f} S(f')\,df'.$$

In this case, $S^{(I)}(\cdot)$ and $S(\cdot)$ are seen to have the following properties.

[1] $S^{(I)}(-1/2) = 0$ because $S^{(I)}(-1/2) = \int_{-1/2}^{-1/2} S(f')\,df' = 0$.

[2] $S^{(I)}(1/2) = s_0$ because $S^{(I)}(1/2) = \int_{-1/2}^{1/2} S(f')\,df' = s_0$ (from Equation (111c) with τ set to zero).

[3] $S(f) \geq 0$ because Equation (111b) says that $S(f)\,df = E\{|dZ(f)|^2\}$, and we must have $df > 0$ and $E\{|dZ(f)|^2\} \geq 0$.

[4] $f < f'$ implies $S^{(I)}(f) \leq S^{(I)}(f')$ because $S^{(I)}(f') - S^{(I)}(f)$ is equal to the integral of a nonnegative function $S(\cdot)$ over the interval $[f, f']$ (this also follows because the spectral representation theorem states that $S^{(I)}(\cdot)$ is a nondecreasing function of f).

Except for the scaling factor s_0 (the variance of $\{X_t\}$) and (possibly) the relatively unimportant convention of right continuity, properties [1], [2] and [4] say that $S^{(I)}(\cdot)$ resembles a probability distribution function over the interval $[-1/2, 1/2]$ and, for that reason, is sometimes called a *spectral distribution function*. Likewise, except for the same scaling factor, $S(\cdot)$ resembles a probability density function because of properties [2] and [3]. The differentiability condition needed for Equation (111b) to hold suggests that $S(\cdot)$ need not exist for all stationary processes. Similarly, there are certain probability distribution functions (such as the binomial distribution) that do not have a corresponding probability density function. However, all RVs do possess a probability distribution function, so this function is considered more fundamental than a probability density function. The relationship is similar between $S^{(I)}(\cdot)$ and $S(\cdot)$: the former exists for all discrete parameter stationary processes, whereas the latter exists only in certain cases.

The process mean and the integrated spectrum encapsulate the first- and second-order properties of a stationary process. Equation (111a) tells us that the ACVS follows from knowledge of the integrated spectrum. The connection between the integrated spectrum and the ACVS is formalized in *Wold's theorem* (Priestley, 1981, p. 222). This theorem states that a necessary and sufficient condition for the sequence $\{s_\tau\}$ to be the ACVS for some discrete parameter stationary process with variance s_0 is that there exists a function $S^{(I)}(\cdot)$ – the integrated spectrum for the process – defined on $[-1/2, 1/2]$ such that $S^{(I)}(-1/2) = 0$; $S^{(I)}(1/2) = s_0$; $S^{(I)}(\cdot)$ is nondecreasing on $[-1/2, 1/2]$; and for all integers τ

$$s_\tau = \int_{-1/2}^{1/2} e^{i2\pi f\tau} \, dS^{(I)}(f).$$

Let's use Wold's theorem to determine the integrated spectrum for a white noise process with variance σ^2. Such a process is stationary and has an ACVS defined by $s_0 = \sigma^2$ and $s_\tau = 0$ for $|\tau| \neq 0$ (see Section 2.6). Because the function defined by

$$S^{(I)}(f) = (f + 1/2)\sigma^2$$

satisfies the requirements of Wold's theorem including that

$$s_\tau = \int_{-1/2}^{1/2} e^{i2\pi f\tau} \, dS^{(I)}(f) = \int_{-1/2}^{1/2} \sigma^2 e^{i2\pi f\tau} \, df$$

for all τ, we can conclude that $S^{(I)}(\cdot)$ is in fact the integrated spectrum for a white noise process. Since it is differentiable with derivative

$$S(f) = \frac{dS^{(I)}(f)}{df} = \sigma^2, \tag{117}$$

an SDF exists for this process and is constant over f (a constant SDF is analogous to the electromagnetic spectrum of white light, which has equal contributions of radiation from all visible frequencies of light).

As a second application of Wold's theorem, consider the harmonic process of Equation (35c). Starting with the expression for its ACVS in Equation (36a), we have

$$s_\tau = \sum_{l=1}^{L} \sigma_l^2 \cos(2\pi f_l \tau) = \sum_{l=1}^{L} \sigma_l^2 \left(e^{i2\pi f_l \tau} + e^{-i2\pi f_l \tau}\right)/2 = \sum_{l=-L}^{L} \sigma_l^2 e^{i2\pi f_l \tau}/2,$$

where $\sigma_{-l}^2 \stackrel{\text{def}}{=} \sigma_l^2$, $\sigma_0^2 \stackrel{\text{def}}{=} 0$ and $f_{-l} \stackrel{\text{def}}{=} -f_l$. If we define $S^{(I)}(\cdot)$ to be a step function with jumps at $\pm f_l$ of size $\sigma_l^2/2$, we have

$$s_\tau = \int_{-1/2}^{1/2} e^{i2\pi f\tau} \, dS^{(I)}(f),$$

so Wold's theorem tells us that $S^{(\mathrm{I})}(\cdot)$ must be the integrated spectrum for a harmonic process. An SDF does not exist in this case since $S^{(\mathrm{I})}(\cdot)$ is not differentiable everywhere.

As a final application, let $\tilde{S}(\cdot)$ be any real-valued nonnegative function of f defined for $f \in [-1/2, 1/2]$ such that $\tilde{S}(-f) = \tilde{S}(f)$ and the integral $\int_{-1/2}^{1/2} \tilde{S}(f)\,\mathrm{d}f$ exists and is finite. The last stipulation implies that

$$\tilde{s}_\tau \stackrel{\mathrm{def}}{=} \int_{-1/2}^{1/2} \tilde{S}(f)\mathrm{e}^{\mathrm{i}2\pi f \tau}\,\mathrm{d}f$$

exists and is finite for all $\tau \in \mathbb{Z}$ ($\tilde{s}_\tau$ is real-valued because $\tilde{S}(\cdot)$ is an even function). Define

$$\tilde{S}^{(\mathrm{I})}(f) = \int_{-1/2}^{f} \tilde{S}(f')\,\mathrm{d}f'.$$

Then $\tilde{S}^{(\mathrm{I})}(\cdot)$ satisfies the properties of an integrated spectrum for a stationary process, and Wold's theorem says that there exists some stationary process $\{X_t\}$ that has $\{\tilde{s}_\tau\}$ as its ACVS; moreover, because the derivative of $\tilde{S}^{(\mathrm{I})}(\cdot)$ exists and is given by $\tilde{S}(\cdot)$, this process has $\tilde{S}(\cdot)$ as its SDF. Thus, if we are given any real-valued symmetric function defined over $f \in [-1/2, 1/2]$, we need only check that it is nonnegative and integrates to a finite value for it to qualify as the SDF for some stationary process. By contrast checking that a given symmetric sequence is the ACVS for some stationary process is a more difficult task. Without the frequency domain characterization provided by Wold's theorem, we would need to verify the time domain characterization given by Equation (28b), namely, that the sequence is positive semidefinite. Direct verification of positive semidefiniteness is not an easy task in general. The relative ease of specifying SDFs as compared to specifying ACVSs is a hint that the frequency domain is more useful than the time domain when it comes to characterizing stationary processes, a theme that we pick up again in Section 4.6.

Comments and Extensions to Section 4.3

[1] All of the results in this section hold, by and large, for continuous parameter stationary processes $\{X(t)\}$ if we replace each $1/2$ by ∞ throughout. The main difference is in Wold's theorem, which is known as the *Wiener–Khintchine theorem* in the continuous parameter case (Priestley, 1981, p. 219): a necessary and sufficient condition for $s(\cdot)$ to be the ACVF of some *stochastically continuous*, continuous parameter stationary process with variance σ^2 is that there exists a function $S^{(\mathrm{I})}(\cdot)$ – the integrated spectrum – such that $S^{(\mathrm{I})}(-\infty) = 0$; $S^{(\mathrm{I})}(\infty) = \sigma^2$; $S^{(\mathrm{I})}(\cdot)$ is nondecreasing; and, for all τ,

$$s(\tau) = \int_{-\infty}^{\infty} \mathrm{e}^{\mathrm{i}2\pi f \tau}\,\mathrm{d}S^{(\mathrm{I})}(f).$$

[2] The corresponding theory for complex-valued stationary processes closely parallels the real-valued case. The only major difference is that $S(-f) \neq S(f)$ in general; i.e., the SDF for a complex-valued stationary process is not necessarily symmetric about the origin. The stated Wold and Wiener–Khintchine theorems in fact encompass both the real-valued and complex-valued stationary processes.

[3] Throughout this book all SDFs $S(\cdot)$ for real-valued processes are *two-sided* (or *double-sided*) in the sense that they are symmetric about the origin; i.e., $S(-f) = S(f)$. In some communities *one-sided* (or *single-sided*) SDFs are the convention. In terms of our SDFs, these are defined as

$$S_{\mathrm{single}}(f) = \begin{cases} 2S(f), & f \geq 0; \\ 0, & f < 0, \end{cases}$$

so that we have, for example,

$$s_0 = \int_0^{1/2} S_{\text{single}}(f)\,df \quad \text{in comparison to} \quad s_0 = \int_{-1/2}^{1/2} S(f)\,df.$$

Because SDFs are frequently plotted on a decibel scale, and since a factor of two translates approximately into a 3 dB shift, failure to pay attention to the distinction between one-sided and two-sided SDFs can lead to mysterious discrepancies of 3 dB!

[4] Discrepancies between SDFs can also be due to expressing frequencies in standardized units rather than physically meaningful units. When the sampling interval is Δ_t, the relationship between the process variance and the SDF is given by

$$s_0 = \int_{-f_\mathcal{N}}^{f_\mathcal{N}} S(f)\,df, \quad \text{where} \quad f_\mathcal{N} = \frac{1}{2\Delta_t}$$

(this is Equation (113c) with $\tau = 0$). If Δ_t has units of, say, seconds, then the SDF $S(\cdot)$ is a function of a variable whose units are cycles per second. In some textbooks and software packages, the frequency variable is standardized to be either $\nu \stackrel{\text{def}}{=} f\Delta_t$ or $\omega \stackrel{\text{def}}{=} 2\pi f \Delta_t$, where ν and ω are expressed in, respectively, cycles per unit time and radians per unit time. With these change of variables, we have

$$s_0 = \frac{1}{\Delta_t}\int_{-1/2}^{1/2} S\left(\frac{\nu}{\Delta_t}\right) d\nu = \int_{-1/2}^{1/2} S_\nu(\nu)\,d\nu, \quad \text{where} \quad S_\nu(\nu) \stackrel{\text{def}}{=} \frac{1}{\Delta_t} S\left(\frac{\nu}{\Delta_t}\right),$$

and

$$s_0 = \frac{1}{2\pi \Delta_t}\int_{-\pi}^{\pi} S\left(\frac{\omega}{2\pi \Delta_t}\right) d\omega = \int_{-\pi}^{\pi} S_\omega(\omega)\,d\omega, \quad \text{where} \quad S_\omega(\omega) \stackrel{\text{def}}{=} \frac{1}{2\pi \Delta_t} S\left(\frac{\omega}{2\pi \Delta_t}\right).$$

The functions $S(\cdot)$, $S_\nu(\cdot)$ and $S_\omega(\cdot)$ are all valid SDFs because integration over the appropriate range of frequencies yields s_0 in each case; however, depending upon what Δ_t is, the values of $S_\nu(\nu)$ at $\nu = f\Delta_t$ and $S_\omega(\omega)$ at $\omega = 2\pi f \Delta_t$ can differ from $S(f)$. Care must be taken to take into account the factors $1/\Delta_t$ and $1/(2\pi \Delta_t)$ when comparing these SDFs. (Yet another – but less common – source of confusion is a convention in which the SDF always integrates to unity rather than s_0. This amounts to standardizing the process variance s_0 to be unity and has the appeal of making SDFs have exactly the same properties as probability density functions. Conversion between this convention and the one we have adopted involves a multiplication or division by s_0, which can lead to discrepancies if accidentally forgotten.)

For future reference, we note that

$$S\left(\frac{\nu}{\Delta_t}\right) = S_\nu(\nu)\,\Delta_t \quad \text{implies that} \quad S(f) = S_\nu(f\Delta_t)\,\Delta_t.$$

For example, if $S_\nu(\cdot)$ is a periodic function with a period of unity such that $S_\nu(\nu) = \nu^2$ for $-1/2 \leq \nu \leq 1/2$, then $S(f) = f^2 \Delta_t^3$ is an SDF with a period of $2f_\mathcal{N}$ such that

$$\int_{-f_\mathcal{N}}^{f_\mathcal{N}} S(f)\,df = \frac{1}{12} \tag{119}$$

for all choices of the sampling interval Δ_t.

4.4 Classification of Spectra

Since $S^{(I)}(\cdot)$ is quite similar to a probability distribution function, we have the following theorem, which is an analog to the *Lebesgue decomposition theorem* for distribution functions (Chung, 1974, theorem 1.3.2). Any integrated spectrum $S^{(I)}(\cdot)$ can be written as the unique sum of up to three canonical forms $S_1^{(I)}(\cdot)$, $S_2^{(I)}(\cdot)$ and $S_3^{(I)}(\cdot)$, each of which is an integrated spectrum in its own right. These canonical forms have the following properties.

[1] $S_1^{(I)}(\cdot)$ is "absolutely continuous" – this means that its derivative exists for almost all f and is equal almost everywhere to an SDF $S(\cdot)$ such that

$$S_1^{(I)}(f) = \int_{-1/2}^{f} S(f')\,\mathrm{d}f'.$$

[2] $S_2^{(I)}(\cdot)$ is a step function with jumps of size $\{p_l : l = 1, 2, \ldots\}$ at the frequencies $\{f_l : l = 1, 2, \ldots\}$.

[3] $S_3^{(I)}(\cdot)$ is a continuous singular function – although, by definition, a singular function has a derivative of zero almost everywhere, such a function can in fact be continuous and, rather surprisingly, strictly increasing (Chung, 1974, section 1.3).

Now $S_3^{(I)}(\cdot)$ is pathological and of no practical use in spectral analysis, so we won't consider it any further. We thus consider all integrated spectra $S^{(I)}(\cdot)$ of practical importance to consist of either (i) $S_1^{(I)}(\cdot)$ by itself, (ii) $S_2^{(I)}(\cdot)$ by itself or (iii) a combination of these two "pure" forms, i.e., $S^{(I)}(f) = S_1^{(I)}(f) + S_2^{(I)}(f)$. The two pure forms are described as follows.

[1] When $S^{(I)}(f) = S_1^{(I)}(f)$, the process $\{X_t\}$ is said to have a *purely continuous spectrum*. The SDF $S(\cdot)$ exists in this case for all f and is fully equivalent to the integrated spectrum. A standard result from Fourier analysis (the Riemann–Lebesgue lemma, Titchmarsh, 1939, p. 403) says that the ACVS $\{s_\tau\}$ of the process decays to zero as $|\tau| \to \infty$. Most of the standard models for stationary processes belong to this class. These include white noise and ARMA processes.

[2] When $S^{(I)}(f) = S_2^{(I)}(f)$, the integrated spectrum consists entirely of a step function, and the stationary process is said to have a *purely discrete spectrum*, a *line spectrum* or a *spectrum with line components*. The harmonic process has this type of spectrum, which, in contrast to a purely continuous spectrum, has an ACVS that does not damp down to zero as $|\tau| \to \infty$.

A process for which $S^{(I)}(f) = S_1^{(I)}(f) + S_2^{(I)}(f)$ is said to have a *discrete spectrum* (rather than purely discrete) if the SDF $S(\cdot)$ corresponding to $S_1^{(I)}(\cdot)$ is that of a white noise process. We say it has a *mixed spectrum* if $S(\cdot)$ is not the SDF for a white noise process. An example of a process with a discrete spectrum can be constructed by adding a harmonic process and an uncorrelated white noise process together; one with a mixed spectrum can be made by replacing the white noise process with any purely continuous stationary process having a colored (i.e., nonwhite) SDF (see Exercise [2.7a]). In both cases, due to the $S_2^{(I)}(\cdot)$ component, the ACVS never damps down to zero.

For all processes other than those with purely continuous spectra, the integrated spectrum $S^{(I)}(\cdot)$ is not differentiable due to the presence of the step function $S_2^{(I)}(\cdot)$. However, it is a useful fiction to say that $S^{(I)}(\cdot)$ is differentiable by using a Dirac delta function to define an "SDF" for $S^{(I)}(\cdot)$ by, say,

$$S_0(f) = S(f) + \sum_l p_l \delta(f - f_l), \tag{120}$$

4.4 Classification of Spectra

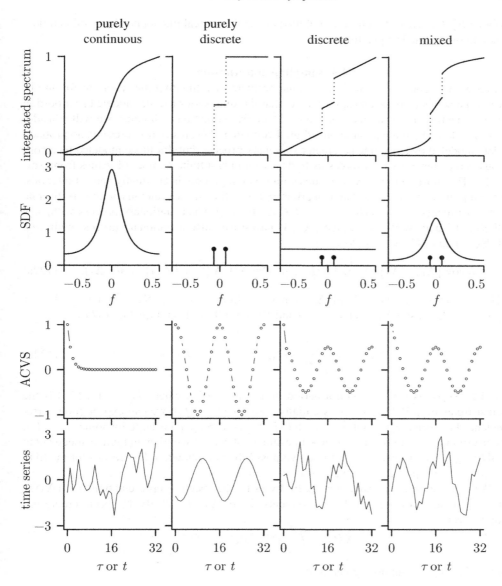

Figure 121 Examples of four integrated spectra $S^{(I)}(\cdot)$ (top row of plots), corresponding spectral density functions $S(\cdot)$ (second row), autocovariance sequences $\{s_\tau\}$ from lag 0 to 32 (third row) and one simulated time series of length 33 from each process (bottom row).

where $S(\cdot)$ is the SDF associated with the purely continuous portion of the integrated spectrum, and $\{p_l\}$ and $\{f_l\}$ are, respectively, the size of the steps and the frequencies at which the steps occur in $S_2^{(I)}(\cdot)$. If we ignore concerns about the right and left continuity of $S^{(I)}(\cdot)$ at the frequencies f_l, the definition of the delta function yields, as desired,

$$\int_{-1/2}^{f} S_0(f') \, df' = S^{(I)}(f).$$

The four spectral classes we deal with from now on are illustrated in Figure 121. In the second row, we have made use of delta functions to depict an SDF for the three classes involving $S_2^{(I)}(\cdot)$. These functions are plotted as a solid vertical line topped by a solid circle – the height of the line is equal to p_l.

4.5 Sampling and Aliasing

Starting with Chapter 6, we will discuss the problem of estimating the spectrum for an observed time series (this assumes, of course, that the time series can be modeled as a portion of one realization of a stationary process). The only estimation techniques we will consider are ones that can be implemented on a digital computer. This restriction presents no problem for data modeled by a discrete parameter stationary process, but, for those modeled by a continuous parameter process, it forces us to deal with only a finite number of values from, say, $\{X(t)\}$. The most common way to subsample such a process is at equally spaced intervals. For a sampling interval $\Delta_t > 0$ and an arbitrary time offset t_0, we can thus define the discrete parameter process $X_t = X(t_0 + t\Delta_t)$, $t \in \mathbb{Z}$. If $\{X(t)\}$ is a stationary process with, say, SDF $S_{X(t)}(\cdot)$ and ACVF $s(\tau)$, then $\{X_t\}$ is necessarily also a stationary process with, say, SDF $S_{X_t}(\cdot)$ and ACVS $\{s_\tau\}$. Now,

$$s_\tau = \text{cov}\{X_{t+\tau}, X_t\} = \text{cov}\{X(t_0 + [t+\tau]\Delta_t), X(t_0 + t\Delta_t)\} = s(\tau\Delta_t), \quad (122a)$$

so $\{s_\tau\}$ is sampled at intervals of Δ_t from $s(\cdot)$. Since $s(\cdot) \longleftrightarrow S_{X(t)}(\cdot)$ and $\{s_\tau\} \longleftrightarrow S_{X_t}(\cdot)$, the relationship between $S_{X(t)}(\cdot)$ and $S_{X_t}(\cdot)$ follows from Equation (82a):

$$S_{X_t}(f) = \sum_{k=-\infty}^{\infty} S_{X(t)}(f + \tfrac{k}{\Delta_t}), \quad f \in \mathbb{R}. \quad (122b)$$

Note that $S_{X_t}(\cdot)$ is periodic with a period of $1/\Delta_t = 2f_\mathcal{N}$, where $f_\mathcal{N} = 1/(2\Delta_t)$ is the Nyquist frequency. If $S_{X(t)}(\cdot)$ is essentially zero for $|f| > f_\mathcal{N}$, we can expect good correspondence between $S_{X_t}(\cdot)$ and $S_{X(t)}(\cdot)$ for $|f| \leq f_\mathcal{N}$; if $S_{X(t)}(\cdot)$ is large for some $|f| > f_\mathcal{N}$, the correspondence can be quite poor – an estimate of $S_{X_t}(\cdot)$ will then not tell us much about the SDF $S_{X(t)}(\cdot)$. An example of these two possibilities is shown in Figure 123 for the SDF $S_{X(t)}(f) = 1.9\exp(-2f^2) + \exp(-6[|f| - 1.25]^2)$.

We can also arrive at Equation (122b) by using the spectral representations for $\{X(t)\}$ and $\{X_t\}$ in the following way. For convenience, let $E\{X(t)\} = 0$. By the spectral representation theorem

$$X(t) = \int_{-\infty}^{\infty} e^{i2\pi ft}\, dZ_{X(t)}(f),$$

where $\{Z_{X(t)}(f)\}$ is an orthogonal process.

▷ **Exercise [122]** Show that

$$X_t = \int_{-f_\mathcal{N}}^{f_\mathcal{N}} e^{i2\pi ft\Delta_t}\, dZ(f), \quad \text{where } dZ(f) \stackrel{\text{def}}{=} \sum_{k=-\infty}^{\infty} e^{i2\pi(f+\tfrac{k}{\Delta_t})t_0}\, dZ_{X(t)}(f + \tfrac{k}{\Delta_t}). \quad ◁$$

Note that, by the properties of $\{Z_{X(t)}(f)\}$, we have $E\{dZ(f)\} = 0$ for all $|f| \leq f_\mathcal{N}$. Suppose now that f' and f satisfy $-f_\mathcal{N} \leq f' < f \leq f_\mathcal{N}$. Then

$$\text{cov}\{dZ(f'), dZ(f)\} = \sum_{k=-\infty}^{\infty}\sum_{l=-\infty}^{\infty} e^{i2\pi(f'+\tfrac{k}{\Delta_t})t_0} e^{-i2\pi(f+\tfrac{l}{\Delta_t})t_0}$$
$$\times E\{dZ_{X(t)}(f' + \tfrac{k}{\Delta_t})\, dZ_{X(t)}^*(f + \tfrac{l}{\Delta_t})\}$$
$$= 0.$$

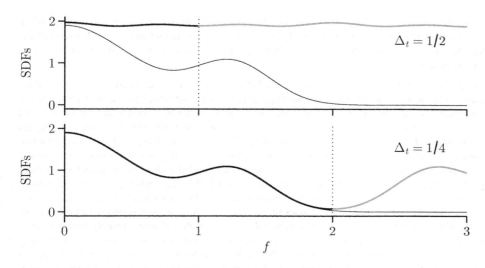

Figure 123 Spectral density functions for a continuous parameter process $\{X(t)\}$ (thin curves on the two plots) and for two discrete parameter processes sampled from $\{X(t)\}$ (thick curves, partly black and partly gray). The sampling interval Δ_t associated with the top plot is $1/2$ (yielding a Nyquist frequency $f_\mathcal{N} = 1$, indicated by the dotted vertical line), while Δ_t is $1/4$ for the bottom plot (now $f_\mathcal{N} = 2$). Because there is substantial power in the SDF of $\{X(t)\}$ for frequencies between 1 and 2 but little beyond $f = 2$, the SDF for the $\Delta_t = 1/2$ sampled process does not agree well with the SDF for the continuous parameter process for $f \in [0, f_\mathcal{N}]$, whereas there is good agreement when $\Delta_t = 1/4$ – in fact, the agreement is so good for $f \in [0, f_\mathcal{N}]$ that the thin curve for the SDF of $\{X(t)\}$ is only visible starting a bit before $f_\mathcal{N} = 2$. Because the SDFs for the discrete parameter processes are even functions with a periodicity of $2f_\mathcal{N}$, knowledge of these SDFs over $f \in [0, f_\mathcal{N}]$ fully determines them for all $f \in \mathbb{R}$. In the figure, black thick curves show these SDFs over $f \in [0, f_\mathcal{N}]$, while gray ones shown them over $f > f_\mathcal{N}$.

We can thus take $\{dZ(f)\}$ to be the increments of the orthogonal process $\{Z(f)\}$ in the spectral representation for $\{X_t\}$ in Equation (110c). The integrated spectrum $S^{(I)}_{X_t}(\cdot)$ for $\{X_t\}$ is specified via

$$dS^{(I)}_{X_t}(f) = E\{|dZ(f)|^2\} = \sum_{k=-\infty}^{\infty} E\{|dZ_{X(t)}(f + \tfrac{k}{\Delta_t})|^2\} = \sum_{k=-\infty}^{\infty} dS^{(I)}_{X(t)}(f + \tfrac{k}{\Delta_t}),$$

$|f| \leq f_\mathcal{N}$, where $S^{(I)}_{X(t)}(\cdot)$ is the integrated spectrum for $\{X(t)\}$. This expression reduces to Equation (122b) when $S^{(I)}_{X(t)}(\cdot)$ is differentiable.

Comments and Extensions to Section 4.5

[1] Although a sampling interval of $\Delta_t = 1/4$ seems natural for the example considered in Figure 123, there are valid reasons for using both a larger and smaller value. For example, use of $\Delta_t = 1/2$ yields a discrete parameter process $\{X_t\}$ that – as can be be seen from the thick curve in the top plot – is quite close to white noise. Sampling at this rate thus yields approximately uncorrelated variates, which might be helpful in applying certain statistical procedures. On the other hand, Kay (1981a) shows that, to approximately minimize the mean square error of an estimator of the ACVF based on discretely sampled values, we should oversample at *twice* the natural Nyquist rate, which for the example of Figure 123 would yield $\Delta_t = 1/8$ approximately. Another rationale for oversampling is to eliminate potential spectral estimation problems near the Nyquist frequency – see Hardin (1986) for details. For a discussion of the potential problems that undersampling can cause in interpreting SDFs, see Wunsch and Gunn (2003), where the focus is on climatological time series formed by sampling measurements from a core extracted from the bottom of the ocean.

[2] Given a discrete parameter stationary process $\{X_t\}$ possessing an SDF $S_X(\cdot)$, it is sometimes of interest to subsample this process to create, say, $Y_t = X_{nt}$, $t \in \mathbb{Z}$, for some integer $n \geq 2$. Exercise [4.14] considers the case $n = 2$ and calls for showing that $\{Y_t\}$ is a stationary process with an SDF that is an aliased version of $S_X(\cdot)$.

[3] Taking aliasing into account is sometimes needed to properly interpret an SDF. This need arises particularly in the study of proxy-based climatological time series, for which the sampling interval between observations is dictated by the nature of the proxy and need not be a good match for the physical process of interest. Pisias and Mix (1988) consider how aliasing affects the SDF for a Miocene oxygen isotope time series. If the series is presumed to be sampled at an interval of $\Delta_t = 25,000$ years, fluctuations associated with orbital precession with a shorter period of $23,000$ years (commonly found in late Quaternary paleoclimatic time series) can be significantly confounded with a low-frequency component usually attributed to eccentricity. Due to the manner in which the time series is constructed, a more realistic assumption is that the sampling interval is not fixed but rather is subject to fluctuations ("jitter") about a nominal value of $25,000$ years. Under this assumption there are potentially significant distortions in the SDF that unfortunately depend on the unknown strength and nature (purely random versus autocorrelated) of the fluctuations in sampling intervals. These distortions make it difficult to physically interpret various features in the SDF. Wunsch (2000) examines a sharp low-frequency peak that appears in an estimated SDF for an interpolated time series derived from ice and ocean cores. The period associated with the sharp peak is approximately 1500 years, and physical mechanisms that would explain the peak are lacking. In creating the interpolated series, the sampling interval Δ_t was set to the convenient choice of 365 days. This choice induces a subtle aliasing effect because Δ_t differs slightly from the lengths of physically significant years, namely, the tropical year (365.2422 days) and the anomalistic year (365.2596 days). The aliases associated with these two physically significant years nicely trap the sharp low-frequency peak, hence offering a sensible explanation for the peak not involving fluctuations at a physically difficult-to-explain low frequency.

Section 10.4 has another example of how aliasing impacts the interpretation of an SDF.

4.6 Comparison of SDFs and ACVSs as Characterizations

One of the reasons spectral analysis is so widely used in the physical sciences is that the SDF for a purely continuous stationary process gives us an easy way of summarizing – and visualizing – the important second-order properties of the process. The spectral representation theorem provides the necessary theoretical basis for this statement. It is sometimes argued that, since the SDF and ACVS contain the same amount of "information" (in the sense that, if we know one of them, we can calculate the other), the two characterizations are equally informative. This is certainly true in some cases. A white noise process is equally well characterized by either its SDF or ACVS. In other cases, however, the SDF is clearly the preferred characterization, as we demonstrate with the following simple example.

The top row of Figure 125 shows two plots – the SDF for the AR(4) process defined in Equation (35a) and its corresponding ACVS. If we examine the SDF, we know from the spectral representation theorem that realizations from a process with this SDF should show a tendency to be a combination of oscillations close to two different frequencies, namely, 0.11 and 0.14 cycles per unit time (see the right-hand column of plots in Figure 34). The power levels of these oscillations are within 3 dB of each other (this corresponds approximately to a factor of 2). Note that it is very difficult to draw these same conclusions from an examination of the ACVS plot alone.

The bottom row of Figure 125 shows two similar plots for a second stationary process. This has essentially the same characteristics as the one in the top row, with the exception of a small additional tendency to oscillate at frequencies close to 0.35. The magnitude of this component is about 21 dB less than the dominant component centered around $f = 0.11$. Note, however, that the ACVS for this process is virtually identical to that in the top row – the effect of the additional component does not show up in any appreciable way in the ACVS! (For

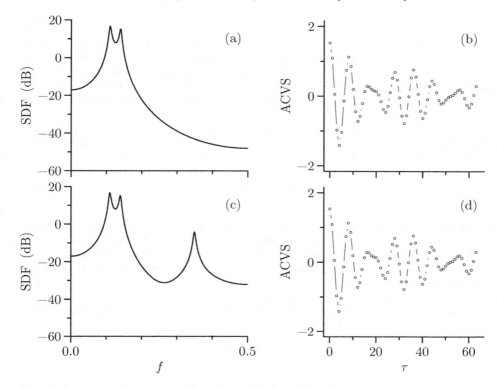

Figure 125 Two spectral density functions (left-hand column) and their corresponding autocovariance sequences (right-hand column).

the record, the second process is the sum of two stationary processes that are uncorrelated with each other – Exercise [2.7a] says that the resulting process is indeed stationary. The first process is the AR(4) process of Equation (35a), and the second is an AR(2) process dictated by the model

$$X_t = -1.15 X_{t-1} - 0.96 X_{t-2} + \epsilon_t, \tag{125}$$

where $\{\epsilon_t\}$ is a Gaussian white noise process with mean zero and variance 0.0004.)

The SDF usually proves to be the more sensitive and interpretable diagnostic or exploratory tool. Also, since physical time series are usually recorded using mechanical or electronic devices, one must know the way in which the measuring equipment affects the recorded data. This knowledge is vital for understanding the limitations of the recorded data. The easiest way to express these limitations is in the frequency domain by means of the transfer (or frequency response) function for the measuring instrument (see Section 5.1). Thus the simplest way to characterize recorded data with significant measurement noise is usually in the frequency domain.

4.7 Summary of Foundations for Stochastic Spectral Analysis

The spectral representation theorem expresses a discrete parameter stationary process $\{X_t : t \in \mathbb{Z}\}$ as a "weighted sum" of complex exponentials in a manner that permits the definition of an integrated spectrum for the process. Assuming for convenience that $E\{X_t\} = 0$ and that the sampling interval Δ_t between X_t and X_{t+1} is unity, the theorem says that, in the mean square sense,

$$X_t = \int_{-1/2}^{1/2} e^{i2\pi f t} \, dZ(f),$$

where $\{dZ(f)\}$ (the "weights") corresponds to the increments of an orthogonal process and has three key properties. First, $E\{dZ(f)\} = 0$ for $f \in [-1/2, 1/2]$; second, $dS^{(I)}(f) \stackrel{\text{def}}{=} E\{|dZ(f)|^2\}$, where the function $S^{(I)}(\cdot)$ is the integrated spectrum for $\{X_t\}$; and third, for any two distinct frequencies f and f' contained in the interval $[-1/2, 1/2]$, the covariance between $dZ(f)$ and $dZ(f')$ is zero. The integrated spectrum is bounded and nondecreasing over the interval $[-1/2, 1/2]$, with $S^{(I)}(-1/2) = 0$ and $S^{(I)}(1/2) = \text{var}\{X_t\}$. For real-valued processes, $S^{(I)}(\cdot)$ is symmetric about zero in the sense that, if $0 \leq f' < f \leq 1/2$, then $S^{(I)}(f) - S^{(I)}(f') = S^{(I)}(-f') - S^{(I)}(-f)$; in addition, $S^{(I)}(f) - S^{(I)}(f')$ represents the contribution to $\text{var}\{X_t\}$ due to frequencies in the interval $(f', f]$ (there is an equal contribution due to the corresponding negative frequencies in the interval $[-f, -f')$). The integrated spectrum can be used to obtain the autocovariance sequence (ACVS) $\{s_\tau\}$ for $\{X_t\}$ because

$$s_\tau = \text{cov}\{X_{t+\tau}, X_t\} = \int_{-1/2}^{1/2} e^{i2\pi f\tau} E\{|dZ(f)|^2\} = \int_{-1/2}^{1/2} e^{i2\pi f\tau} dS^{(I)}(f). \tag{126}$$

Whereas the spectral representation theorem gives a representation of the process itself, Wold's theorem gives a representation for an ACVS. It provides a formal connection between the integrated spectrum and the ACVS, namely, that a necessary and sufficient condition for $\{s_\tau\}$ to be the ACVS for some discrete parameter stationary process is that there exists a function $S^{(I)}(\cdot)$ – the integrated spectrum – defined on $[-1/2, 1/2]$ such that $S^{(I)}(-1/2) = 0$; $S^{(I)}(1/2) = s_0$; $S^{(I)}(\cdot)$ is nondecreasing on $[-1/2, 1/2]$; and s_τ is related to $S^{(I)}(\cdot)$ as per Equation (126) for all integers τ.

Let us now specialize to the important case where the integrated spectrum is differentiable everywhere in the interval $[-1/2, 1/2]$. Its derivative, namely,

$$S(f) \stackrel{\text{def}}{=} \frac{dS^{(I)}(f)}{df}$$

is known as the spectral density function (SDF). By letting $dS^{(I)}(f) = S(f)\,df$ in Equation (126), we see that the ACVS is related to the SDF via

$$s_\tau = \int_{-1/2}^{1/2} S(f) e^{i2\pi f\tau}\, df.$$

If we set $\tau = 0$ in the above, we obtain the fundamental relationship

$$\text{var}\{X_t\} = s_0 = \int_{-1/2}^{1/2} dS^{(I)}(f) = \int_{-1/2}^{1/2} S(f)\,df,$$

which tells us that $S(f)\,df$ can be interpreted as the contribution to the process variance due to frequencies in a small interval about f. If $\{s_\tau\}$ is square summable (a sufficient, but not necessary, condition), we can recover the SDF from the ACVS via

$$S(f) = \sum_{\tau=-\infty}^{\infty} s_\tau e^{-i2\pi f\tau} \quad \text{and hence } \{s_\tau\} \longleftrightarrow S(\cdot).$$

The SDF is a periodic function of f, with a period of unity. For real-valued processes, the SDF is an even function; i.e., $S(-f) = S(f)$.

For practical purposes, all second-order stationary processes can be considered to have one of four types of integrated spectra, namely, either a purely continuous spectrum, a purely discrete spectrum (sometimes called a line spectrum), a discrete spectrum or a mixed spectrum. A process with a purely continuous spectrum has an integrated spectrum that is fully equivalent to an SDF, while the integrated spectrum for a process with a purely discrete spectrum is a step function. Processes with either a discrete or mixed spectrum can be regarded as the sum of two uncorrelated stationary processes, one of which has a purely continuous spectrum, and the other, a purely discrete spectrum. In the case of a process with a discrete spectrum, the component with the purely continuous spectrum is white noise, while this component is colored for a process with a mixed spectrum.

4.8 Exercises

[4.1] Let ϕ be an RV with a uniform (i.e., rectangular) distribution over the interval $[-\pi, \pi]$. Consider the complex-valued RV $C \stackrel{\text{def}}{=} De^{i\phi}/2$, where D is a real-valued constant. Show that C and C^* are uncorrelated, where $C^* = De^{-i\phi}/2$ is the complex conjugate of C. (The result of this exercise can be used to support the claim that the $2L + 1$ RVs in the summation in Equation (108b) are uncorrelated.)

[4.2] Show that the stochastic process defined by Equation (109a) has orthogonal increments.

[4.3] Formulate a definition for the stochastic Riemann–Stieltjes integral based upon the definition in Equation (25b) for the ordinary Riemann–Stieltjes integral. Use this definition in Equation (110b) to show that it does in fact reduce to Equation (108c). (Hint: interchange the operations of $E\{\cdot\}$ and lim in the definition of the integral.)

[4.4] Define the complex-valued stochastic process

$$W_t = \sum_{l=-L}^{L} C_l e^{i2\pi f_l t},$$

where C_l is a complex-valued RV such that

$$E\{C_l\} = 0 \text{ and } E\{|C_l|^2\} < \infty, \quad |l| \leq L < \infty,$$

and f_l is a fixed real-valued constant such that $f_{-l} = -f_l$. Show that, whereas $\{W_t\}$ must be a stationary process if the C_l RVs are uncorrelated, $\{W_t\}$ need not be so if these RVs are correlated.

[4.5] Let $\{X_t\}$ be a discrete time stationary process with zero mean and integrated spectrum $S^{(I)}(\cdot)$. Define

$$J(f) = \frac{1}{\sqrt{N}} \sum_{t=0}^{N-1} X_t e^{-i2\pi f t}.$$

(a) Show that

$$J(f) = \sqrt{N} \int_{-1/2}^{1/2} e^{i(N-1)\pi(f'-f)} \mathcal{D}_N(f'-f) \, dZ(f'),$$

where $\{Z(f)\}$ is an orthogonal process and $\mathcal{D}_N(\cdot)$ is Dirichlet's kernel (see Exercise [1.2c]).

(b) Based upon part (a), show that, if $S^{(I)}(\cdot)$ has derivative $S(\cdot)$, then

$$E\{|J(f)|^2\} = N \int_{-1/2}^{1/2} \mathcal{D}_N^2(f-f') S(f') \, df'.$$

[4.6] Consider the following "rectangular" sequence:

$$s_\tau = \begin{cases} 1, & |\tau| \leq K; \\ 0, & |\tau| > K, \end{cases}$$

where $K \geq 1$ is an integer. Is $\{s_\tau\}$ the ACVS for some discrete parameter stationary process $\{X_t\}$ with SDF $S(\cdot)$?

[4.7] Consider the following "triangular" sequence:

$$s_\tau = \begin{cases} 1 - \frac{|\tau|}{K}, & \text{if } |\tau| \leq K; \\ 0, & \text{if } |\tau| > K, \end{cases}$$

where $K \geq 1$ is an integer. Is $\{s_\tau\}$ the ACVS for some discrete parameter stationary process $\{X_t\}$ with SDF $S(\cdot)$?

[4.8] (a) Consider the following "rectangular" SDF:

$$S_R(f) = \begin{cases} 2, & |f| \leq 1/4 \text{ and} \\ 0, & 1/4 < |f| \leq 1/2, \end{cases}$$

where the Nyquist frequency $f_\mathcal{N}$ is here $1/2$ (the SDF is defined outside of $f \in [-\frac{1}{2}, \frac{1}{2}]$ based on the fact that it is periodic with a period of $2f_\mathcal{N} = 1$). A process with this just-discussed SDF is an example of what is known as band-limited white noise. Show that the ACVS for this process is given by

$$s_{R,\tau} = \begin{cases} 1, & \tau = 0; \\ 2/|\pi\tau|, & |\tau| = 1, 5, 9, \ldots; \\ -2/|\pi\tau|, & |\tau| = 3, 7, 11, \ldots; \text{ and} \\ 0, & |\tau| = 2, 4, 6 \ldots. \end{cases}$$

(b) Now consider the following "triangular" SDF:

$$S_T(f) = 2 - 4|f|, \quad |f| \leq 1/2.$$

Show that the ACVS for a process with this SDF is given by

$$s_{T,\tau} = \begin{cases} 1, & \tau = 0; \\ 4/(\pi\tau)^2, & |\tau| = 1, 3, 5, \ldots; \text{ and} \\ 0 & \text{otherwise;} \end{cases}$$

i.e., $s_{T,\tau} = s_{R,\tau}^2$. (Hint: if both $G_p(\cdot)$ and $H_p(\cdot)$ in Equation (100b) are set equal to $S_R(\cdot)$, then $G_p \star H_p^*(\cdot)$ is the same as $S_T(\cdot)$.)

(c) Suppose $\{R_{1,t}\}$ and $\{R_{2,t}\}$ are zero mean stationary processes, both of which have $S_R(\cdot)$ of part (a) as their SDF. If we assume that the RVs R_{1,t_0} and R_{2,t_1} are independent of each other for all $t_0, t_1 \in \mathbb{Z}$ and that $R_{1,t_0} R_{1,t_1}$ and $R_{2,t_2} R_{2,t_3}$ are likewise independent for all $t_0, t_1, t_2, t_3 \in \mathbb{Z}$, how can we use $\{R_{1,t}\}$ and $\{R_{2,t}\}$ to define a new process $\{T_t\}$ whose SDF is given by $S_T(\cdot)$ of part (b)?

[4.9] Suppose that $\{X_t\}$ is a stationary process with zero mean, ACVS $\{s_{X,\tau}\}$, SDF $S_X(\cdot)$ and sampling interval $\Delta_t = 1$. Let C be an RV with mean zero and finite nonzero variance σ_C^2. Suppose that C is uncorrelated with X_t for all t. Show that $Y_t \stackrel{\text{def}}{=} X_t + C$ is a stationary process, and determine its ACVS and its integrated spectrum. (This is an example of a *nonergodic* stationary process – see C&E [2] for Section 6.1.)

[4.10] Figures 129a and 129b show, respectively, four SDFs and four realizations of time series, each of which is generated from a zero mean stationary process having one of these SDFs. The order in which the SDFs are presented ((a) to (d)) does not necessarily match up with the ordering of the time series. Determine which time series goes with which SDF, and state the rationale for the chosen pairings. (The SDFs in Figure 129a correspond to the following stationary processes: (a) an AR(1) process $X_t = 0.8X_{t-1} + \epsilon_t$ with $\sigma_\epsilon^2 = 0.36$; (b) an AR(2) process $X_t = -0.8X_{t-1} - 0.6X_{t-2} + \epsilon_t$ with $\sigma_\epsilon^2 = 0.48$; (c) a white noise process with variance 3; and (d) a white noise process with variance 8.)

[4.11] Equation (32b) says that the ACVS of a qth-order moving average process is

$$s_\tau = \begin{cases} \sigma_\epsilon^2 \sum_{j=0}^{q-|\tau|} \theta_{q,j+|\tau|} \theta_{q,j}, & |\tau| \leq q; \\ 0, & |\tau| > q. \end{cases}$$

(a) Show that, if we define $\theta_{q,j} = 0$ for $j < 0$ and $j > q$, we can rewrite the above as

$$s_\tau = \sigma_\epsilon^2 \sum_{j=-\infty}^{\infty} \theta_{q,j+\tau} \theta_{q,j}, \quad \tau \in \mathbb{Z}.$$

4.8 Exercises

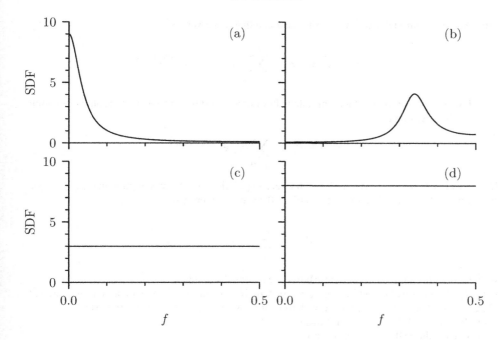

Figure 129a Plots of $S(f)$ versus f for four theoretical SDFs (see Exercise [4.10]).

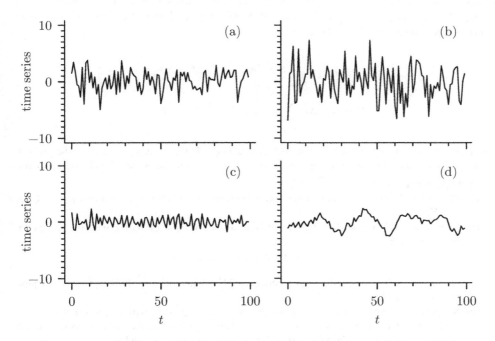

Figure 129b Plots of four time series, each of which is a realization of a portion $X_0, X_1, \ldots, X_{99}$ of a different zero mean stationary process (see Exercise [4.10]).

(b) Argue that an SDF exists for this process and that it is given by

$$S(f) = \sigma_\epsilon^2 \, \Delta_t \sum_{\tau=-q}^{q} \sum_{j=-\infty}^{\infty} \theta_{q,j+\tau} \theta_{q,j} e^{-i2\pi f \tau \Delta_t}.$$

(c) Use the autocorrelation theorem stated after Equation (100a) to show that the above equation can be rewritten as

$$S(f) = \sigma_\epsilon^2 \, \Delta_t \left| \sum_{j=0}^{q} \theta_{q,j} e^{-i2\pi f j \Delta_t} \right|^2.$$

[4.12] Let $\{X_{1,t}\}, \{X_{2,t}\}, \ldots, \{X_{m,t}\}$ be as in Exercise [2.7a]. Under the assumption that $\{s_{j,\tau}\} \longleftrightarrow S_j(\cdot)$ for $j = 1, \ldots, m$, determine the SDF for the stationary process

$$X_t \stackrel{\text{def}}{=} \sum_{j=1}^{m} c_j X_{j,t},$$

where, as before, $c_1, c_2, \ldots, c_m$ are arbitrary real-valued variables.

[4.13] Let $\{X_t\}$ be a real-valued white noise process with finite variance σ^2 and with a sampling interval of $\Delta_t = 2$ sec between adjacent observations (implying that $f_\mathcal{N} = 1/(2\,\Delta_t) = 0.25$ Hz).
  (a) What is the ACVS $\{s_{X,\tau}\}$ for this process?
  (b) What is the SDF $S_X(\cdot)$ for this process?
  (c) Let $Y_t = X_{2t}$, so that $\{Y_t\}$ has a sampling interval of $2\Delta_t = 4$ sec. What is the ACVS $\{s_{Y,\tau}\}$ for $\{Y_t\}$?
  (d) What is the SDF $S_Y(\cdot)$ for $\{Y_t\}$?

[4.14] Suppose that $\{X_t\}$ is a real-valued stationary process with sampling interval Δ_t and corresponding Nyquist frequency $f_\mathcal{N} = \frac{1}{2\Delta_t}$. Assume that its ACVS $\{s_{X,\tau}\}$ and SDF $S_X(\cdot)$ are related as dictated by Equations (113c) and (113d):

$$s_{X,\tau} = \int_{-f_\mathcal{N}}^{f_\mathcal{N}} S_X(f) e^{i2\pi f \tau \Delta_t} \, df, \quad \tau \in \mathbb{Z}$$

$$S_X(f) = \Delta_t \sum_{\tau=-\infty}^{\infty} s_{X,\tau} e^{-i2\pi f \tau \Delta_t}, \quad f \in \mathbb{R}$$

(recall that $S_X(\cdot)$ is a periodic function with period $2f_\mathcal{N}$). Let $Y_t = X_{2t}, t \in \mathbb{Z}$; i.e., the process $\{Y_t\}$ is formed by subsampling every other random variable from the process $\{X_t\}$, and hence the sampling interval for $\{Y_t\}$ is $\Delta_t' \stackrel{\text{def}}{=} 2\Delta_t$.
  (a) Show that $\{Y_t\}$ is a stationary process, and determine its ACVS $\{s_{Y,\tau}\}$ in terms of the ACVS of $\{X_t\}$.
  (b) Show that $\{Y_t\}$ has an SDF $S_Y(\cdot)$ that is an aliased version of $S_X(\cdot)$, with the two SDFs being related by

$$S_Y(f) = S_X(f) + S_X(f + f_\mathcal{N}).$$

  Argue that $S_Y(\cdot)$ is periodic with a period of $2f_\mathcal{N}'$, where $f_\mathcal{N}' \stackrel{\text{def}}{=} 1/(2\,\Delta_t')$ is the Nyquist frequency for $\{Y_t\}$.
  (c) Suppose now that $\{X_t\}$ is a first-order moving average process with mean zero and coefficient θ; i.e., we can write $X_t = \epsilon_t - \theta \epsilon_{t-1}$, where $\{\epsilon_t\}$ is a white noise process with mean zero and variance σ_ϵ^2. The SDF and ACVS for this process are given by

$$S_X(f) = \sigma_\epsilon^2 \, \Delta_t \left| 1 - \theta e^{-i2\pi f \Delta_t} \right|^2 \quad \text{and} \quad s_{X,\tau} = \begin{cases} \sigma_\epsilon^2 (1 + \theta^2), & \tau = 0; \\ -\sigma_\epsilon^2 \theta, & \tau = \pm 1; \\ 0, & |\tau| \geq 2; \end{cases}$$

(see Exercise [4.11] with $q = 1$, $\theta_{1,0} = -1$ and $\theta_{1,1} = \theta$). Use part (a) of this exercise to show that $\{Y_t\}$ has an ACVS of a white noise process. Use part (b) to show that $\{Y_t\}$ has an SDF of a white noise process.

[4.15] Consider a *continuous* parameter complex-valued stationary process $\{X(t)\}$ with the following "Gaussian-shaped" ACVF:

$$s(\tau) = Ae^{-B\tau^2 + i2\pi f_0 \tau}, \quad |\tau| < \infty, \tag{131a}$$

where $A > 0$, $B > 0$ and f_0 are all real-valued constants. (For appropriate A, B and f_0, this process has been found to be an adequate model for certain unwanted echoes in radar – "clutter" noise – and for Doppler measurements of blood flow; see, for example, Jacovitti and Scarano, 1987.) Show that the corresponding SDF for this process is given by

$$S(f) = A\left(\frac{\pi}{B}\right)^{1/2} e^{-\pi^2(f-f_0)^2/B}, \quad |f| < \infty.$$

(This SDF is a rescaled version of the Gaussian (or normal) probability density function with mean f_0 and is thus sometimes called a *Gaussian SDF*. Note that the term "Gaussian" here refers to the shape of the SDF and *not* to the probability distribution of the RV $X(t)$, which need not have a Gaussian distribution.)

[4.16] Consider a *continuous* parameter real-valued stationary process $\{X(t)\}$ with a *Lorenzian* SDF:

$$S(f) = \frac{2L\sigma^2}{1 + (2\pi f L)^2} \quad \text{for } f \in \mathbb{R}, \tag{131b}$$

where $\sigma^2 > 0$ is the process variance, and $L > 0$ is a parameter known as the correlation length or the integral time scale. Show that the ACVF for $\{X(t)\}$ is given by

$$s(\tau) = \sigma^2 e^{-|\tau|/L} \quad \text{for } \tau \in \mathbb{R}. \tag{131c}$$

Hint: make use of the definite integral

$$\int_0^\infty \frac{\cos(mx)}{1 + x^2} \, dx = \frac{\pi}{2} e^{-|m|},$$

where m is any real-valued variable.

[4.17] Consider a *continuous* parameter real-valued stationary process $\{X(t)\}$ with an SDF given by

$$S(f) = \begin{cases} C, & |f| \leq 2; \\ 0, & |f| > 2, \end{cases}$$

where $C > 0$ is a real-valued constant (a process with the above SDF is known as band-limited white noise). For a given sampling interval $\Delta_t > 0$, define the associated discrete parameter process $X_t = X(t\Delta_t)$, $t \in \mathbb{Z}$. Determine the SDF for $\{X_t\}$ when (a) $\Delta_t = \frac{1}{2}$, (b) $\Delta_t = \frac{1}{3}$, (c) $\Delta_t = \frac{1}{4}$ and (d) $\Delta_t = \frac{1}{5}$. Verify that the integral for each of these four SDFs over $[-f_\mathcal{N}, f_\mathcal{N}]$ is the same as the integral of $S(\cdot)$ over $f \in \mathbb{R}$, where, as usual, $f_\mathcal{N} = \frac{1}{2\Delta_t}$ is the Nyquist frequency. Comment upon the forms that these four SDFs take with regard to which frequencies (if any) are the dominant contributors to the variance of the sampled process.

[4.18] Consider a discrete parameter stochastic process $\{X_t\}$ whose ensemble of realizations consists of just two infinite sequences, namely, $x_{1,t} = (-1)^t$ and $x_{2,t} = (-1)^{t+1}$, $t \in \mathbb{Z}$.
(a) If both realizations are equally likely to be chosen, show that $\{X_t\}$ is a stationary process, and determine its ACVS.
(b) Determine the integrated spectrum for $\{X_t\}$.

[4.19] Let $\{X_t\}$ be a real-valued stationary process with variance σ^2, integrated spectrum $S^{(I)}(\cdot)$, SDF $S(\cdot)$ and a sampling interval of Δ_t. Show that, for any $0 \leq f_0 \leq f_\mathcal{N}$, we must have $S^{(I)}(f_0) + S^{(I)}(-f_0) = \sigma^2$.

5

Linear Time-Invariant Filters

5.0 Introduction

In Section 3.6 we discussed the convolution of two functions $g(\cdot)$ and $h(\cdot)$ defined over the entire real axis $\mathbb{R} = \{t : -\infty < t < \infty\}$ and found that the Fourier transform of their convolution,

$$g * h(t) \stackrel{\text{def}}{=} \int_{-\infty}^{\infty} g(u)h(t-u)\,du, \tag{132}$$

was simply the product of $G(\cdot)$ and $H(\cdot)$, our notation for the Fourier transforms of $g(\cdot)$ and $h(\cdot)$. In Section 3.2 we argued that the quantity $|H(f)|^2$ defines an energy spectral density function (SDF) for $h(\cdot)$. Let us now regard the convolution in Equation (132) as a manipulation of a function $h(\cdot)$ that produces a new function $g * h(\cdot)$. The original function $h(\cdot)$ has a distribution of energy with respect to frequency given by its energy SDF, and the new function has a distribution of energy given by

$$|G(f)H(f)|^2 = |G(f)|^2 |H(f)|^2.$$

Thus the energy SDF of the new function $g * h(\cdot)$ is multiplicatively related to the energy SDF of the original function $h(\cdot)$ via the Fourier transform of the manipulator $g(\cdot)$.

In this chapter we investigate and extend these ideas through the theory of linear time-invariant (LTI) filters. Our goal is to formalize ways of relating the spectra associated with inputs to an LTI filter to the spectra associated with outputs from the filter. The feature that makes LTI filters so easy to work with is our ability to represent various functions and stationary processes as linear combinations of complex exponentials.

LTI filters play a key role in the spectral analysis of stationary processes. As we shall see, an LTI filter is nothing more than a linear time-invariant transformation of some function of time. If the function of time is a realization of a stationary process, an LTI filter transforms the process into a new process that, under very mild conditions, is also stationary. An important feature of LTI filters is that, given the integrated spectrum of the original process, there is an easy way to determine the integrated spectrum of the new process. The theory of LTI filters thus gives us a powerful means for determining the SDFs of a wide class of stationary processes. We will also make extensive use of LTI filters in Chapter 7 to smooth out the inherent variability in certain estimates of the SDF.

5.1 Basic Theory of LTI Analog Filters

Let us define a continuous parameter *filter* L as a mapping, or association, between an input function $x(\cdot)$ and an output function $y(\cdot)$. Symbolically we write

$$L\{x(\cdot)\} = y(\cdot). \tag{133a}$$

Since we regard both $x(\cdot)$ and $y(\cdot)$ as functions of time $t \in \mathbb{R}$, the qualifier "continuous parameter" is appropriate – in the engineering literature a continuous parameter filter is often called an *analog* filter. In mathematics L is known as a *transformation* or *operator*. It is important to realize that a filter is *not* just an ordinary function: for example, a real-valued function that is defined over $\mathbb{R}$ associates a *point* in $\mathbb{R}$ with another point in $\mathbb{R}$, whereas a filter associates a *function* from some – so far unidentified – abstract space of functions with another function in that same space.

For the remainder of this section we need the following special notation. If α is a real or complex-valued scalar and $x(\cdot)$ is a function, the notation $\alpha x(\cdot)$ refers to the function defined by $\alpha x(t)$ for $t \in \mathbb{R}$. If $x_1(\cdot)$ and $x_2(\cdot)$ are two functions, then $x_1(\cdot) + x_2(\cdot)$ denotes the function defined by $x_1(t) + x_2(t)$. Finally, if τ is a real-valued scalar and $x(\cdot)$ is a function, then $x(\cdot; \tau)$ denotes the function whose value at time t is given by $x(t + \tau)$; i.e.,

$$x(t; \tau) = x(t + \tau), \quad t \in \mathbb{R}.$$

Thus the filter defined by

$$L\{x(\cdot)\} = x(\cdot; \tau)$$

is a shift filter, which takes as input a certain function $x(\cdot)$ and returns a function that is defined by shifting the original function by τ units to the left. For example, if the input to the shift filter is the function defined by $x(t) = \sin(t)$ and $\tau = \pi/2$, the output is the function defined by $x(t; \pi/2) = \cos(t)$.

An analog filter L is called a *linear time-invariant* (LTI) analog filter if it has the following three properties:

[1] Scale preservation:

$$L\{\alpha x(\cdot)\} = \alpha L\{x(\cdot)\};$$

i.e., multiplication of the input by the factor α results in multiplication of the output by α also.

[2] Superposition:

$$L\{x_1(\cdot) + x_2(\cdot)\} = L\{x_1(\cdot)\} + L\{x_2(\cdot)\};$$

i.e., if we define a new function by adding $x_1(\cdot)$ and $x_2(\cdot)$ and if we use it as input to the LTI filter L, the output from L is simply that function defined by adding together the outputs of L when $x_1(\cdot)$ and $x_2(\cdot)$ are separately used as inputs to L.

[3] Time invariance:

$$\text{if } L\{x(\cdot)\} = y(\cdot), \text{ then } L\{x(\cdot; \tau)\} = y(\cdot; \tau); \tag{133b}$$

i.e., if two inputs to the LTI filter are the same except for a shift in time, the outputs will also be the same except for the same shift in time.

Properties [1] and [2] together express the linearity of L:

$$L\{\alpha x_1(\cdot) + \beta x_2(\cdot)\} = \alpha L\{x_1(\cdot)\} + \beta L\{x_2(\cdot)\}.$$

By induction, it follows that

$$L\left\{\sum_{j=1}^{N}\alpha_j x_j(\cdot)\right\} = \sum_{j=1}^{N}\alpha_j L\{x_j(\cdot)\}. \tag{134}$$

With suitable conditions the above holds "in the limit" so that Equation (134) is valid when the finite summation is replaced by an infinite summation.

Now suppose we take an $L^2(\mathbb{R})$ deterministic function (i.e., one that is square integrable over the entire real axis) or a realization of a stationary process and use it as input to some LTI filter. As we shall see, under mild conditions, the output from this LTI filter is also, respectively, an $L^2(\mathbb{R})$ function or a realization of a different stationary process (defined on a realization by realization basis by the LTI filter). It is interesting that Equations (133b) and (134) are all we need to derive the relationship between the spectra of the input and output to the LTI filter. What follows are the details (this material is based on Koopmans, 1974).

Let the input into the LTI filter be the complex exponential

$$\mathcal{E}_f(t) = e^{i2\pi f t}, \quad t \in \mathbb{R},$$

where f is some fixed frequency, and let $y_f(\cdot)$ denote the output function; i.e.,

$$y_f(\cdot) = L\{\mathcal{E}_f(\cdot)\}.$$

The rationale for this approach is simple. If we regard L as a "black box" that accepts an input and transforms it somehow, and if we want to learn something about L, we might feed it simple test functions such as complex exponentials to learn how it reacts. The complex exponentials are of particular interest because all the representations we have examined for deterministic functions and stationary processes involve linear combinations of one kind or another of complex exponentials. Now, by the properties [1] and [3] of an LTI filter, we have for all τ

$$y_f(\cdot;\tau) \stackrel{[3]}{=} L\{\mathcal{E}_f(\cdot;\tau)\} = L\{e^{i2\pi f\tau}\mathcal{E}_f(\cdot)\} \stackrel{[1]}{=} e^{i2\pi f\tau}L\{\mathcal{E}_f(\cdot)\} = e^{i2\pi f\tau}y_f(\cdot).$$

This implies that

$$y_f(t;\tau) = y_f(t+\tau) = e^{i2\pi f\tau}y_f(t) \text{ for all } t \text{ and } \tau.$$

In particular, for $t = 0$ we obtain

$$y_f(\tau) = e^{i2\pi f\tau}y_f(0).$$

Since τ can assume any real value, the above implies that

$$y_f(t) = e^{i2\pi f t}y_f(0), \text{ for all } t; \text{ i.e., } y_f(\cdot) = y_f(0)\mathcal{E}_f(\cdot).$$

Thus, when the function $\mathcal{E}_f(\cdot)$ is used as input to the LTI filter L, the output is the same function multiplied by some constant, $y_f(0)$, which is independent of time but will depend in general on the frequency f. To keep track of this frequency dependence, define

$$G(f) = y_f(0).$$

5.1 Basic Theory of LTI Analog Filters

We thus have shown that

$$L\{\mathcal{E}_f(\cdot)\} = G(f)\mathcal{E}_f(\cdot). \tag{135a}$$

In mathematical terms, the complex exponentials (regarded as functions of t with f fixed) would be called the *eigenfunctions* for the LTI filter L, and each $G(f)$ would be called an associated *eigenvalue*. The relationship expressed by Equation (135a) is of fundamental importance: if the input to an LTI filter is a complex exponential, the output is also a complex exponential with the exact same frequency multiplied by $G(f)$. Why is this important? Suppose that $x(\cdot)$ can be represented by

$$x(t) = \sum_f \alpha_f e^{i2\pi ft}; \text{ i.e., } x(\cdot) = \sum_f \alpha_f \mathcal{E}_f(\cdot). \tag{135b}$$

Then Equations (134) and (135a) tell us that the output from the LTI filter is just

$$y(\cdot) = L\{x(\cdot)\} = \sum_f \alpha_f G(f)\mathcal{E}_f(\cdot);$$

i.e., $y(\cdot)$ can be represented by

$$y(t) = \sum_f \alpha_f G(f) e^{i2\pi ft}.$$

In particular, the spectral representation theorem for continuous parameter stationary processes says that, if $x(\cdot)$ is a realization of a stationary process $\{X(t)\}$ with zero mean, then

$$x(t) = \int_{-\infty}^{\infty} e^{i2\pi ft} \, dZ_x(f), \quad t \in \mathbb{R},$$

where $Z_x(\cdot)$ is the corresponding realization of the orthogonal process $\{Z_X(f)\}$. As a function of t, we can regard the above equation as a special case of Equation (135b) if we equate α_f with the increments $dZ_x(f)$ (this step requires some justification because we have passed from a finite to an infinite summation). Hence we have

$$y(t) = \int_{-\infty}^{\infty} e^{i2\pi ft} G(f) \, dZ_x(f) \text{ and } Y(t) = \int_{-\infty}^{\infty} e^{i2\pi ft} G(f) \, dZ_X(f).$$

It follows that, under a mild "matching" condition, $\{Y(t)\}$ is a stationary process and hence has a spectral representation given by

$$Y(t) = \int_{-\infty}^{\infty} e^{i2\pi ft} \, dZ_Y(f).$$

The matching condition essentially says that $\text{var}\{Y(t)\}$ must be finite. This condition is needed because it is possible to construct an LTI filter that maps a stationary process to a process having infinite variance, which is a problem because our definition of stationarity requires such a process to have finite variance. If we denote the integrated spectra for $\{X(t)\}$ and $\{Y(t)\}$ by $S_X^{(I)}(\cdot)$ and $S_Y^{(I)}(\cdot)$, respectively, the two expressions above for $Y(t)$ – combined with the fact that a stationary process has a unique integrated spectrum – tell us that

$$dS_Y^{(I)}(f) = E\{|dZ_Y(f)|^2\} = |G(f)|^2 E\{|dZ_X(f)|^2\} = |G(f)|^2 \, dS_X^{(I)}(f). \tag{135c}$$

When SDFs exist for all f,

$$\mathrm{d}S_Y^{(\mathrm{I})}(f) = S_Y(f)\,\mathrm{d}f \text{ and } \mathrm{d}S_X^{(\mathrm{I})}(f) = S_X(f)\,\mathrm{d}f,$$

so Equation (135c) reduces to

$$S_Y(f) = |G(f)|^2 S_X(f).$$

The function $G(\cdot)$ is called the *transfer function* (or *frequency response function*) of the LTI filter L. The transfer function relates the integrated spectra of the input to – and the output from – an LTI filter in a very simple fashion. In particular, the relationship is independent of time, and power is not transferred from one frequency to another. These facts show the importance of LTI filters within spectral analysis, and, conversely, the usefulness of spectral analysis when LTI transformations are applied to a time series.

In general, $G(\cdot)$ is a complex-valued function, so we can write

$$G(f) = |G(f)|e^{\mathrm{i}\theta(f)}, \tag{136}$$

where $|G(\cdot)|$ and $\theta(\cdot)$ are called, respectively, the *gain function* and the *phase function* of the LTI filter. Note that $\theta(f) = \arg(G(f))$. The quantities

$$-\frac{1}{2\pi} \cdot \frac{\mathrm{d}\theta(f)}{\mathrm{d}f} \text{ and } -\theta(f)$$

define the *group delay* and the *phase shift function*.

Equation (135a) gives us a simple rule for computing the transfer function of an LTI filter: if we apply the function $\mathcal{E}_f(\cdot)$ as input to an LTI filter, the coefficient of $\exp(\mathrm{i}2\pi ft)$ in the resulting output from the filter *defines* $G(f)$, the transfer function at frequency f.

The results of this section that pertain to stationary processes can be summarized by the following theorem, which we call the *LTI analog filtering theorem*: if $\{X(t)\}$ is a continuous parameter stationary process with zero mean and integrated spectrum $S_X^{(\mathrm{I})}(\cdot)$ and if L is an LTI analog filter with transfer function $G(\cdot)$ such that the matching condition

$$\int_{-\infty}^{\infty} |G(f)|^2 \,\mathrm{d}S_X^{(\mathrm{I})}(f) < \infty$$

holds, then $\{Y(t)\} \stackrel{\mathrm{def}}{=} L\{\{X(t)\}\}$ is a continuous parameter stationary process with zero mean and integrated spectrum $S_Y^{(\mathrm{I})}(\cdot)$ such that

$$\mathrm{d}S_Y^{(\mathrm{I})}(f) = |G(f)|^2\,\mathrm{d}S_X^{(\mathrm{I})}(f).$$

The proof of this theorem is Exercise [5.1].

In the remainder of this book we prefer to use the informal notation

$$L\{x(t)\} = y(t) \text{ instead of } L\{x(\cdot)\} = y(\cdot)$$

to define an LTI analog filter. This is done merely for notational convenience – it allows us to define functions implicitly on a point by point basis without having to come up with an explicit notation for them (as we did for $\mathcal{E}_f(\cdot)$). In all cases our informal notation means that the LTI filter L maps the function defined on a point by point basis by $x(t)$ to the function defined on a point by point basis by $y(t)$. For example, Equation (135a) in this informal notation is

$$L\{e^{\mathrm{i}2\pi ft}\} = G(f)e^{\mathrm{i}2\pi ft}.$$

5.1 Basic Theory of LTI Analog Filters

Comments and Extensions to Section 5.1

[1] A general class of linear transformations of $x(\cdot)$ is given by

$$y(t) = \int_{-\infty}^{\infty} K(t,t')x(t')\,\mathrm{d}t'$$

for some $K(\cdot,\cdot)$ (Bracewell, 2000, p. 210; Champeney, 1987). Under the assumption of time invariance (see Equation (133b)), we have

$$y(t-\tau) = \int_{-\infty}^{\infty} K(t,t')x(t'-\tau)\,\mathrm{d}t'.$$

If we make the change of variable $t'' = t' - \tau$, we get

$$y(t-\tau) = \int_{-\infty}^{\infty} K(t,t''+\tau)x(t'')\,\mathrm{d}t'',$$

and if we replace $t - \tau$ by t and relabel t'' as t', we have

$$y(t) = \int_{-\infty}^{\infty} K(t+\tau, t'+\tau)x(t')\,\mathrm{d}t'.$$

A comparison of the two expressions for $y(t)$ shows that time invariance implies $K(t,t') = K(t+\tau, t'+\tau)$ for all t and τ. If we let $\tau = -t'$, we have $K(t,t') = K(t-t',0)$ for all t and t' so that we can write, say, $K(t,t') = g(t-t')$, a function purely of $t - t'$. Hence $y(\cdot)$ can be expressed as a convolution:

$$y(t) = \int_{-\infty}^{\infty} g(t-t')x(t')\,\mathrm{d}t'.$$

Linearity plus time invariance thus *implies* the convolution relationship. However, to be able to include the trivial case $y(t) = x(t)$, we must allow $g(\cdot) = \delta(\cdot)$, the Dirac delta function. If we want to exclude generalized functions like the delta function from the convolution expression, then "linearity plus time invariance" gives a set of filters that are larger than those that can be expressed as a convolution.

[2] The theory of LTI filters also justifies the form in which we have chosen to represent functions and stationary processes, namely, as linear combinations of complex exponentials. For example, suppose $g_p(\cdot)$ is a member of the $L^2(-T/2, T/2)$ class of functions, i.e., the class of all complex-valued continuous time functions such that

$$\int_{-T/2}^{T/2} |g_p(t)|^2\,\mathrm{d}t < \infty.$$

From the results of Section 3.1, we can represent $g_p(\cdot)$ over the interval $[-T/2, T/2]$ as

$$g_p(t) = \sum_{n=-\infty}^{\infty} G_n e^{i2\pi f_n t}, \quad \text{where } f_n \stackrel{\text{def}}{=} \frac{n}{T} \text{ and } G_n \stackrel{\text{def}}{=} \frac{1}{T}\int_{-T/2}^{T/2} g_p(t)e^{-i2\pi f_n t}\,\mathrm{d}t.$$

Since we can easily extend $g_p(\cdot)$ to the whole real axis as a periodic function with period T, Parseval's theorem

$$\int_{-T/2}^{T/2} |g_p(t)|^2\,\mathrm{d}t = T\sum_{n=-\infty}^{\infty} |G_n|^2$$

allows us to use $|G_n|^2$ to define a discrete power spectrum.

Let us now define

$$\phi_n(t) = e^{i2\pi f_n t}/\sqrt{T}, \quad n \in \mathbb{Z}. \tag{137}$$

We say the collection of functions $\phi_n(\cdot)$ forms an *orthonormal basis* for $L^2(-T/2, T/2)$ because, first,

$$\int_{-T/2}^{T/2} \phi_m(t)\phi_n^*(t)\,dt = \begin{cases} 0, & m \neq n; \\ 1, & m = n, \end{cases}$$

and, second, any function $g_p(\cdot)$ in $L^2(-T/2, T/2)$ can be written as

$$g_p(t) = \sum_{n=-\infty}^{\infty} \mathcal{G}_n \phi_n(t), \quad \text{where } \mathcal{G}_n \stackrel{\text{def}}{=} \int_{-T/2}^{T/2} g_p(t)\phi_n^*(t)\,dt. \tag{138a}$$

In this new notation Parseval's theorem becomes

$$\int_{-T/2}^{T/2} |g_p(t)|^2\,dt = \sum_{n=-\infty}^{\infty} |\mathcal{G}_n|^2, \tag{138b}$$

which allows us to use $|\mathcal{G}_n|^2$ (divided by T) to define a discrete power spectrum.

The orthonormal basis for $L^2(-T/2, T/2)$ that we have just defined is not unique. There are many other orthonormal bases that we could define such that both Equations (138a) and (138b) would still hold and $|\mathcal{G}_n|^2$ would thus define a discrete power spectrum with respect to this new basis. The theory of LTI filters tells us there is something special about the orthonormal basis defined by Equation (137). Thus, if we define an LTI filter L as

$$L\{g_p(t)\} = g_p * h_p(t) \stackrel{\text{def}}{=} \frac{1}{T} \int_{-T/2}^{T/2} g_p(u) h_p(t-u)\,du,$$

we know that the discrete power spectrum for the output $g_p * h_p(\cdot)$ is

$$\frac{1}{T^2} |\mathcal{G}_n|^2 |\mathcal{H}_n|^2, \quad \text{where } \mathcal{H}_n \stackrel{\text{def}}{=} \int_{-T/2}^{T/2} h_p(t)\phi_n^*(t)\,dt$$

(assuming that $h_p(\cdot)$ is a periodic $L^2(-T/2, T/2)$ function). Such a simple relationship is unique to the basis defined by Equation (137); for any other nontrivially different basis, LTI filtering results in a more complicated relationship between the discrete power spectra of the input and the output.

For example, a nonsinusoidal orthonormal basis that has proved useful in practical applications can be formed from the *Walsh functions* (Beauchamp, 1984). When $T = 1$, these functions assume only the values ± 1. The first nine of one version of these functions are shown in Figure 139. They can be labeled from top to bottom as $W_n(\cdot)$ for $n = 0, 1, \ldots, 8$ – see Beauchamp (1984) for details on how to generate these and higher-order Walsh functions. For Walsh functions, the concept corresponding to frequency is *sequency*, which is defined as half the number of zero crossings over one period (taken to be unity in Figure 139). This is just the number of transitions from ± 1 to ∓ 1 with the convention that the endpoints count as one transition if

$$\lim_{\substack{\delta \to 0 \\ \delta > 0}} W_n\left(-\tfrac{1}{2} + \delta\right) \neq \lim_{\substack{\delta \to 0 \\ \delta > 0}} W_n\left(\tfrac{1}{2} - \delta\right).$$

With this definition the sequency of $W_n(\cdot)$ is $\lfloor (n+1)/2 \rfloor$, so each nonzero sequency is associated with exactly two Walsh functions. Figure 139 shows all the Walsh functions for sequencies 0 to 4. By way of comparison, note that each nonzero frequency f_m is associated with exactly two orthogonal sinusoids, namely, $\cos(2\pi f_m t)$ and $\sin(2\pi f_m t)$, and that both sinusoids have sequency m.

If we now redefine

$$\phi_n(t) = \begin{cases} W_{2n}(t)/\sqrt{T}, & n \geq 0; \\ W_{2|n|-1}(t)/\sqrt{T}, & n < 0, \end{cases}$$

it can be shown that Equations (138a) and (138b) still hold. We can thus define a Walsh discrete power spectrum via $|\mathcal{G}_n|^2$. Since $\phi_n(t)$ and $\phi_{-n}(t)$ have sequency n, the $\pm n$ components of this spectrum tell us the contribution to the power from Walsh functions of sequency $|n|$.

5.1 Basic Theory of LTI Analog Filters

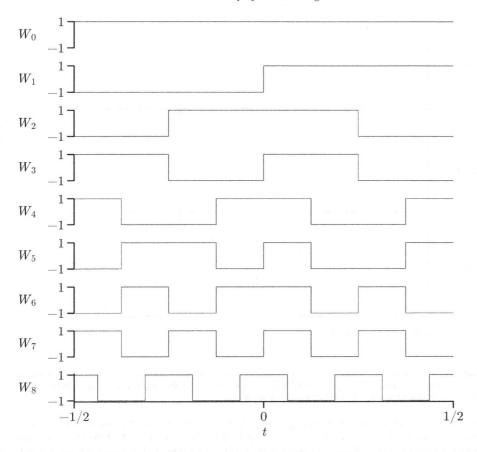

Figure 139 Walsh functions $W_n(\cdot)$ from $n = 0$ (top) to $n = 8$ (bottom). This figure was adapted from Figures 1.4 and 1.7 of Beauchamp (1984).

Now suppose that $g_p(\cdot) = W_2(\cdot)$. Since our periodic function is just the second Walsh function, its Walsh discrete power spectrum is concentrated entirely at sequency 1. If we pass this function through the LTI filter defined by, say,

$$L\{g_p(t)\} = \int_{-1/2}^{1/2} g_p(u) h_p(t-u) \, \mathrm{d}u \stackrel{\mathrm{def}}{=} g_p * h_p(t)$$

(see Section 5.3 for details on this type of filter), where $h_p(\cdot)$ has a period of unity and is defined over $[-1/2, 1/2]$ by

$$h_p(t) = \begin{cases} 4, & |t| \leq 1/8; \\ 0, & 1/8 < |t| \leq 1/2 \end{cases}$$

(see Figure 140), it is obvious that the Walsh discrete power spectrum for the output from this filter is no longer concentrated just at sequency 1; i.e., it cannot be expressed as just a linear combination of $W_1(\cdot)$ and $W_2(\cdot)$ (see Figure 139). In fact it can be shown that the power is now distributed over an infinite range of sequencies. By contrast, if we pass a band-limited function through an LTI filter, then the output must also be band-limited – the (Fourier) discrete power spectrum for the output from any LTI filter can have nonzero contributions only at those frequencies that are nonzero in the discrete power spectrum of the input (see Figure 141).

Walsh spectral analysis provides some interesting contrasts to the usual Fourier spectral analysis and has proven useful for time series whose values shift suddenly from one level to another. For example,

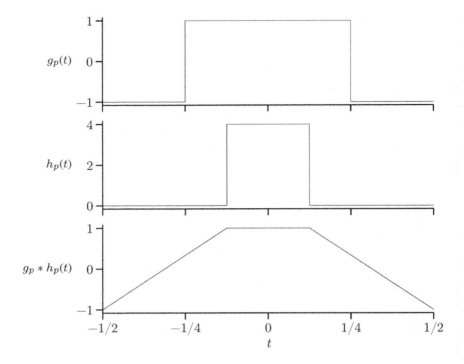

Figure 140 Filtering of $W_2(\cdot)$. When $g_p(\cdot) = W_2(\cdot)$ (top plot) – which has sequency 1 – is convolved with a periodic rectangular smoothing kernel $h_p(\cdot)$ (middle plot), the result is a periodic function $g_p * h_p(\cdot)$ (bottom plot) that cannot be expressed in terms of Walsh functions of sequency 1.

a Fourier power spectrum is invariant with respect to a shift in the time origin of a function, but a Walsh power spectrum need not be; although it is not possible for a function to be both band-limited and time-limited, a function can be both sequency-limited and time-limited; and, if a function has a band-limited Fourier representation, it cannot have a sequency-limited Walsh representation and vice versa. For further details, the reader is referred to Morettin (1981), Beauchamp (1984) and Stoffer (1991); for interesting applications of Walsh spectral analysis, see Stoffer et al. (1988), Lanning and Johnson (1983) and Kowalski et al. (2000).

5.2 Basic Theory of LTI Digital Filters

In the previous section we defined an analog (or continuous parameter) filter as a transformation that maps a function of time to another such function. A parallel theory exists for a transformation that associates a sequence with another sequence – such a transformation is referred to as a discrete parameter filter or *digital* filter. The theory of linear time-invariant digital filters closely parallels that of LTI analog filters, so we only sketch the key points for sequences in this section.

A digital filter L that transforms an input sequence $\{x_t\}$ into an output sequence $\{y_t\}$ is called a linear time-invariant digital filter if it has the following three properties:

[1] Scale preservation:
$$L\{\{\alpha x_t\}\} = \alpha L\{\{x_t\}\}.$$

[2] Superposition:
$$L\{\{x_{1,t} + x_{2,t}\}\} = L\{\{x_{1,t}\}\} + L\{\{x_{2,t}\}\}.$$

[3] Time invariance:
$$\text{if } L\{\{x_t\}\} = \{y_t\}, \text{ then } L\{\{x_{t+\tau}\}\} = \{y_{t+\tau}\},$$

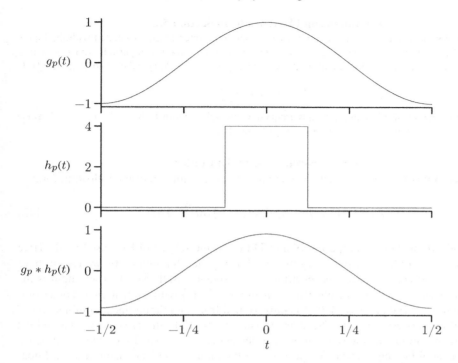

Figure 141 Filtering of a sinusoid with unit frequency. When $g_p(t) = \cos(2\pi t)$ (top plot) is convolved with the same rectangular smoothing kernel $h_p(\cdot)$ as in Figure 140 (middle plot), the result is $g_p * h_p(t) = \frac{2\sqrt{2}}{\pi}\cos(2\pi t) \doteq 0.9\cos(2\pi t)$ (bottom plot), so the output also involves a sinusoid with unit frequency. This figure and Figure 140 together illustrate that, while filtering can widen the sequency range of an input function, it can at most reduce its frequency range.

where τ is integer-valued and the notation $\{x_{t+\tau}\}$ refers to the sequence whose tth element is $x_{t+\tau}$.

By using $\{\exp(\mathrm{i}2\pi ft)\}$ – a sequence with tth element $\exp(\mathrm{i}2\pi ft)$ – as the input to an LTI digital filter, it follows from the same arguments as before that

$$L\{\{e^{\mathrm{i}2\pi ft}\}\} = G(f)\{e^{\mathrm{i}2\pi ft}\},$$

where $G(\cdot)$ is the transfer function (cf. Equation (135a)). Note that $G(\cdot)$ is periodic with a unit period since $\exp(\mathrm{i}2\pi[f+k]t) = \exp(\mathrm{i}2\pi ft)$ for all integers k. The result corresponding to the LTI analog filtering theorem is the following *LTI digital filtering theorem*: if $\{X_t\}$ is a discrete parameter stationary process with zero mean and integrated spectrum $S_X^{(\mathrm{I})}(\cdot)$ and if L is an LTI digital filter with transfer function $G(\cdot)$ such that the matching condition

$$\int_{-1/2}^{1/2} |G(f)|^2 \, \mathrm{d}S_X^{(\mathrm{I})}(f) < \infty$$

holds, then $\{Y_t\} \stackrel{\mathrm{def}}{=} L\{\{X_t\}\}$ is a discrete parameter stationary process with zero mean and integrated spectrum $S_Y^{(\mathrm{I})}(\cdot)$ such that

$$\mathrm{d}S_Y^{(\mathrm{I})}(f) = |G(f)|^2 \, \mathrm{d}S_X^{(\mathrm{I})}(f).$$

As before, we shall find it convenient to use the succinct – but formally incorrect – notation

$$L\{x_t\} = y_t \text{ as shorthand for } L\{\{x_t\}\} = \{y_t\}.$$

Comments and Extensions to Section 5.2

[1] The development in this section assumes that the sampling interval Δ_t associated with the inputs $\{x_t\}$ and $\{X_t\}$ is unity. For general Δ_t, the transfer function $G(\cdot)$ for L is a periodic function with a period of $1/\Delta_t$ and is obtained by inputing the sequence $\{\exp(\mathrm{i}2\pi ft\Delta_t)\}$ rather than $\{\exp(\mathrm{i}2\pi ft)\}$; i.e.,

$$L\{e^{\mathrm{i}2\pi ft\Delta_t}\} = G(f)e^{\mathrm{i}2\pi ft\Delta_t}.$$

Once we know the transfer function over any frequency interval of width $1/\Delta_t$, e.g., $|f| \leq 1/(2\Delta_t) = f_\mathcal{N}$ (the Nyquist frequency), periodicity give us $G(f)$ for all $f \in \mathbb{R}$.

5.3 Convolution as an LTI Filter

We consider in this section some details about an LTI analog filter L of the following form:

$$L\{X(t)\} = \int_{-\infty}^{\infty} g(u)X(t-u)\,\mathrm{d}u \stackrel{\mathrm{def}}{=} Y(t) \tag{142}$$

(that this indeed satisfies the properties of an LTI filter is the subject of Exercise [5.2a]). Here the input to the LTI filter is a stationary process $\{X(t)\}$ that, for simplicity, we take to have zero mean and a purely continuous spectrum with associated SDF $S_X(\cdot)$. The output is the stochastic process $\{Y(t)\}$ that results from convolving $\{X(t)\}$ with the real-valued deterministic function $g(\cdot)$. The process $\{Y(t)\}$ is thus formed from an infinite linear combination of members of the process $\{X(t)\}$. The characteristics of the LTI filter are entirely determined by $g(\cdot)$, which – in the analog (continuous parameter) case – is called the *impulse response function* for the following reason. Suppose we let the input to the LTI analog filter in Equation (142) be $\delta(\cdot)$, the Dirac delta function with an infinite spike at the origin. By the properties of that function, we have

$$L\{\delta(t)\} = \int_{-\infty}^{\infty} g(u)\delta(t-u)\,\mathrm{d}u = g(t).$$

Hence, the input of an "impulse" (the delta function) into the LTI filter yields an output that defines the impulse response function.

To find the transfer function for L in Equation (142), we apply the function of t defined by $\exp(\mathrm{i}2\pi ft)$ as input to the LTI filter to get

$$L\{e^{\mathrm{i}2\pi ft}\} = \int_{-\infty}^{\infty} g(u)e^{\mathrm{i}2\pi f(t-u)}\,\mathrm{d}u = e^{\mathrm{i}2\pi ft}\int_{-\infty}^{\infty} g(u)e^{-\mathrm{i}2\pi fu}\,\mathrm{d}u.$$

Hence the transfer function

$$G(f) \stackrel{\mathrm{def}}{=} \int_{-\infty}^{\infty} g(u)e^{-\mathrm{i}2\pi fu}\,\mathrm{d}u$$

is just the Fourier transform of the impulse response function $g(\cdot)$. Since $S_Y(\cdot)$ and $S_X(\cdot)$ are related by $S_Y(f) = |G(f)|^2 S_X(f)$, the matching condition says that we must have

$$\int_{-\infty}^{\infty} |G(f)|^2 S_X(f)\,\mathrm{d}f < \infty$$

for $\{Y_t\}$ to have a finite variance as required for a stationary process. If $S_X(\cdot)$ happens to be a bounded function, the matching condition becomes

$$\int_{-\infty}^{\infty} |G(f)|^2 \,\mathrm{d}f < \infty.$$

5.3 Convolution as an LTI Filter

In this case, Parseval's relationship for deterministic nonperiodic functions with finite energy says that, because $g(\cdot)$ is real-valued, the matching condition is equivalent to

$$\int_{-\infty}^{\infty} g^2(u)\,\mathrm{d}u < \infty$$

(this is also a sufficient condition for the existence of $G(\cdot)$).

The corresponding results for the discrete parameter case are quite similar. Recalling the definition for the convolution of two sequences with sampling interval Δ_t (Equation (99b)), we can write

$$L\{X_t\} = \Delta_t \sum_{u=-\infty}^{\infty} g_u X_{t-u} = Y_t, \tag{143a}$$

say, where $\{g_u\}$ is a real-valued deterministic sequence called the *impulse response sequence* (note that X_t and Y_t have different units). The right-hand side of the above is a Riemann sum approximation to the integral in Equation (142) with a grid size of Δ_t. If $\{X_t\}$ is a stationary process with a bounded SDF $S_X(\cdot)$ and sampling interval Δ_t and if $\{g_u\}$ is square summable, it follows from Exercise [5.3] that $\{Y_t\}$ is a stationary process with an SDF given by

$$S_Y(f) = |G(f)|^2 S_X(f), \tag{143b}$$

where the transfer function is now obtained by using the sequence $\{\exp(\mathrm{i}2\pi ft\Delta_t)\}$ as input to the LTI digital filter:

$$L\{\mathrm{e}^{\mathrm{i}2\pi ft\Delta_t}\} = \Delta_t \sum_{u=-\infty}^{\infty} g_u \mathrm{e}^{\mathrm{i}2\pi f(t-u)\Delta_t} = \mathrm{e}^{\mathrm{i}2\pi ft\Delta_t} G(f),$$

with

$$G(f) = \Delta_t \sum_{u=-\infty}^{\infty} g_u \mathrm{e}^{-\mathrm{i}2\pi fu\Delta_t} \tag{143c}$$

being the Fourier transform of $\{g_u\}$ as defined by Equation (74a). If the SDF for $\{X_t\}$ is not bounded, then the matching condition

$$\int_{-f_\mathcal{N}}^{f_\mathcal{N}} |G(f)|^2 S_X(f)\,\mathrm{d}f < \infty$$

is required for $\{Y_t\}$ to be a stationary process.

If, as is commonly done is practice, we choose to define the digital filter as

$$L\{X_t\} = \sum_{u=-\infty}^{\infty} g_u X_{t-u} = Y_t, \tag{143d}$$

i.e., we drop the Δ_t so that that X_t and Y_t have the same units, then the transfer function becomes

$$G(f) = \sum_{u=-\infty}^{\infty} g_u \mathrm{e}^{-\mathrm{i}2\pi fu\Delta_t} \tag{143e}$$

and is now merely proportional to the Fourier transform of $\{g_u\}$ unless Δ_t happens to be unity (see Exercise [5.3]). The relationship $S_Y(f) = |G(f)|^2 S_X(f)$ continues to hold.

5.4 Determination of SDFs by LTI Digital Filtering

Let us now determine the SDFs of some discrete parameter stationary processes using the theory described at the end of the previous section (we assume $\Delta_t = 1$ for convenience).

Moving Average Process

An MA(q) process with zero mean by definition satisfies the equation

$$X_t = \epsilon_t - \sum_{j=1}^{q} \theta_{q,j} \epsilon_{t-j}, \tag{144a}$$

where each $\theta_{q,j}$ is a constant ($\theta_{q,q} \neq 0$) and $\{\epsilon_t\}$ is a white noise process with zero mean and variance σ_ϵ^2 (this is Equation (32a) with $\mu = 0$ and $\theta_{q,0}$ replaced by its defining value of -1). Define the filter L for a sequence $\{y_t\}$ by

$$L\{y_t\} = y_t - \sum_{j=1}^{q} \theta_{q,j} y_{t-j},$$

and note that $X_t = L\{\epsilon_t\}$. It is easy to argue that L is an LTI digital filter. To determine its transfer function, we input the sequence defined by $\exp(\mathrm{i}2\pi f t)$:

$$L\{\mathrm{e}^{\mathrm{i}2\pi ft}\} = \mathrm{e}^{\mathrm{i}2\pi ft} - \sum_{j=1}^{q} \theta_{q,j} \mathrm{e}^{\mathrm{i}2\pi f(t-j)} = \mathrm{e}^{\mathrm{i}2\pi ft}\left(1 - \sum_{j=1}^{q}\theta_{q,j}\mathrm{e}^{-\mathrm{i}2\pi fj}\right),$$

so the transfer function is given by

$$G(f) = 1 - \sum_{j=1}^{q} \theta_{q,j} \mathrm{e}^{-\mathrm{i}2\pi fj}.$$

From the relationships $S_X(f) = |G(f)|^2 S_\epsilon(f)$ and $S_\epsilon(f) = \sigma_\epsilon^2$, we have, for an MA($q$) process,

$$S_X(f) = \sigma_\epsilon^2 \left|1 - \sum_{j=1}^{q} \theta_{q,j} \mathrm{e}^{-\mathrm{i}2\pi fj}\right|^2. \tag{144b}$$

Autoregressive Process

A stationary AR(p) process with zero mean satisfies the equation

$$X_t = \sum_{j=1}^{p} \phi_{p,j} X_{t-j} + \epsilon_t, \tag{144c}$$

where each $\phi_{p,j}$ is a constant ($\phi_{p,p} \neq 0$) and $\{\epsilon_t\}$ is as in the previous example (see Section 2.6). Define the filter L for a sequence $\{y_t\}$ by

$$L\{y_t\} = y_t - \sum_{j=1}^{p} \phi_{p,j} y_{t-j},$$

and note that $L\{X_t\} = \epsilon_t$. The transfer function of the filter is

$$G(f) = 1 - \sum_{j=1}^{p} \phi_{p,j} \mathrm{e}^{-\mathrm{i}2\pi fj}. \tag{144d}$$

From the relationship $S_\epsilon(f) = |G(f)|^2 S_X(f)$, we conclude that

$$S_X(f) = \frac{\sigma_\epsilon^2}{\left|1 - \sum_{j=1}^p \phi_{p,j} e^{-i2\pi f j}\right|^2}, \qquad (145a)$$

as long as $|G(f)|^2 \neq 0$. In Section 2.6 we mentioned that the $\phi_{p,j}$ coefficients must satisfy certain conditions for $\{X_t\}$ to be stationary. It can be shown (Grenander and Rosenblatt, 1984, pp. 36–8) that a necessary and sufficient condition for the existence of a stationary solution to $\{X_t\}$ of Equation (144c) is that $G(\cdot)$ never vanish; i.e., $G(f) \neq 0$ for any f (for additional discussion, see Section 9.2). In this case, the stationary solution to Equation (144c) is unique and is given by

$$X_t = \int_{-1/2}^{1/2} \frac{e^{i2\pi f t}}{G(f)} \, dZ_\epsilon(f), \qquad (145b)$$

where $\{Z_\epsilon(f)\}$ is the orthogonal process in the spectral representation for $\{\epsilon_t\}$.

Autoregressive Moving Average Process
An ARMA(p, q) process with zero mean by definition satisfies the difference equation

$$X_t = \sum_{j=1}^p \phi_{p,j} X_{t-j} + \epsilon_t - \sum_{j=1}^q \theta_{q,j} \epsilon_{t-j},$$

where $\phi_{p,j}$ and $\theta_{q,j}$ are constants (with $\phi_{p,p} \neq 0$ and $\theta_{q,q} \neq 0$) and $\{\epsilon_t\}$ is as in the previous two examples. Exercise [5.8] is to show that, when this process is stationary, its SDF is given by

$$S_X(f) = \sigma_\epsilon^2 \frac{\left|1 - \sum_{j=1}^q \theta_{q,j} e^{-i2\pi f j}\right|^2}{\left|1 - \sum_{j=1}^p \phi_{p,j} e^{-i2\pi f j}\right|^2}. \qquad (145c)$$

5.5 Some Filter Terminology

So far our discussion of filters has been rather theoretical and of limited practical use. The remainder of this chapter concentrates on defining some terminology used in filtering theory, developing an interpretation of the spectrum via filters and considering some applications. In particular we shall see that the Slepian sequences introduced in Chapter 3 arise naturally in filter design. The literature on filtering is vast, and hence the discussion given here is not intended to do much more than introduce some ideas of importance to spectral analysis (useful references for filtering theory are Hamming, 1983, and Rabiner and Gold, 1975).

Let us begin by defining the concepts of cascaded filters, low-pass filters, high-pass filters and band-pass filters. A *cascaded filter* is an arrangement of a set of n filters such that the output from the first filter is the input to the second filter and so forth. This notion applies to both analog and digital filters and is important because it is often possible to regard a seemingly complicated filter as a cascaded filter involving n components with easily determined transfer functions. If all n filters are LTI filters and if the input to the system is considered a realization of a stationary process, it is easy to describe the spectral relationship between the input to the first filter and the output from the nth filter. Let $S_j^{(\mathrm{I})}(\cdot)$ be the integrated spectrum of the input to the jth filter, and let $S_{j+1}^{(\mathrm{I})}(\cdot)$ be the integrated spectrum of the output from the jth filter. (We assume that all these spectra exist.) Let $G_j(\cdot)$ be the transfer function of the jth filter. From Equation (135c) we have

$$dS_{j+1}^{(\mathrm{I})}(f) = |G_j(f)|^2 \, dS_j^{(\mathrm{I})}(f) \text{ for } j = 1, 2, \ldots, n.$$

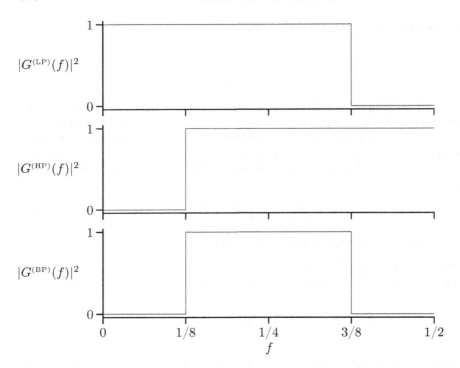

Figure 146 Squared gain functions for an ideal low-pass filter with a cutoff frequency of $f_0 = 3/8$ (upper plot); an ideal high-pass filter that eliminates frequencies whose magnitudes are less than $f_0 = 1/8$ (middle); and an ideal band-pass filter that is formed by cascading the low-pass and high-pass filters together (bottom plot) – note that $|G^{(\text{BP})}(f)|^2 = |G^{(\text{LP})}(f)|^2 |G^{(\text{HP})}(f)|^2$.

The relationship between the integrated spectra of the input and the output to the cascaded filter is thus given by

$$dS_{n+1}^{(\text{I})}(f) = |G_n(f)|^2 \cdot |G_{n-1}(f)|^2 \cdots |G_1(f)|^2 \, dS_1^{(\text{I})}(f).$$

Note that the integrated spectrum of the output does not depend on the order in which the filters occur in the cascade (as long as the integrated spectra exist at each step in the cascade).

If the squared gain function of an LTI analog filter has the form

$$|G^{(\text{LP})}(f)|^2 = \begin{cases} 1, & |f| \leq f_0; \\ 0, & \text{otherwise,} \end{cases}$$

where $f_0 > 0$, we call the filter an *ideal low-pass filter*, since it passes all components with frequencies lower than f_0, but completely suppresses all components with higher frequencies. Conversely, if the squared gain function has the form

$$|G^{(\text{HP})}(f)|^2 = \begin{cases} 1, & |f| \geq f_0; \\ 0, & \text{otherwise,} \end{cases}$$

we call the filter an *ideal high-pass filter*. Finally, if the squared gain function is of the form

$$|G^{(\text{BP})}(f)|^2 = \begin{cases} 1, & 0 < f_0 \leq |f| \leq f_1; \\ 0, & \text{otherwise,} \end{cases}$$

where $f_1 > f_0$, we call the associated filter an *ideal band-pass filter*. The effect of this filter is to allow all frequencies whose magnitudes are in the band $[f_0, f_1]$ to pass through unattenuated, but all other frequency components are completely eliminated. As is illustrated in Figure 146, we can create a band-pass filter by cascading appropriate low-pass and high-pass filters together.

Assuming $0 < f_0 < f_1 < 1/2$, the corresponding definitions for LTI digital filters are

$$|G^{(\text{LP})}(f)|^2 = \begin{cases} 1, & |f| \leq f_0; \\ 0, & f_0 < |f| \leq 1/2; \end{cases} \tag{147a}$$

$$|G^{(\text{HP})}(f)|^2 = \begin{cases} 0, & |f| \leq f_0; \\ 1, & f_0 < |f| \leq 1/2; \end{cases} \tag{147b}$$

and

$$|G^{(\text{BP})}(f)|^2 = \begin{cases} 0, & |f| < f_0 \text{ or } f_1 < |f| \leq 1/2; \\ 1, & f_0 \leq |f| \leq f_1. \end{cases}$$

For all three filters the squared gain functions are periodic with a period of unity and hence can be determined over $\mathbb{R}$ based upon their definitions over $[-1/2, 1/2]$. In addition to illustrating ideal analog filters, Figure 146 can be taken as an illustration in the digital case as well.

Let us focus now on an LTI digital filter L with impulse response sequence $\{g_u\}$:

$$L\{X_t\} = \sum_{u=-\infty}^{\infty} g_u X_{t-u}.$$

This filter is said to be *causal* if $g_u = 0$ for all $u < 0$ and *acausal* otherwise. If $g_u = 0$ outside a finite range for u, it is called a *finite impulse response* (FIR) filter; otherwise, it is called an *infinite impulse response* (IIR) filter. Ideal low-, high- and band-pass filters are not realizable as FIR filters (see Section 5.8).

5.6 Interpretation of Spectrum via Band-Pass Filtering

We can use ideal band-pass filters to motivate a physical interpretation for the integrated spectrum. Consider the real-valued continuous parameter stationary process $\{X(t)\}$ with integrated spectrum $S_X^{(\text{I})}(\cdot)$. Its variance (a measure of the power for the process) is given by

$$\sigma_X^2 \stackrel{\text{def}}{=} \int_{-\infty}^{\infty} \mathrm{d}S_X^{(\text{I})}(f) = \int_{-\infty}^{\infty} S_X(f) \, \mathrm{d}f,$$

where the second equality holds in the case where $S_X^{(\text{I})}(\cdot)$ is differentiable with derivative $S_X(\cdot)$ – the SDF of $\{X(t)\}$. Suppose we use this process as input to an ideal band-pass filter with a transfer function that satisfies

$$|G(f)|^2 = \begin{cases} 1, & f' \leq |f| \leq f' + \Delta_{\mathrm{f}}; \\ 0, & \text{otherwise,} \end{cases}$$

where $f' > 0$ and Δ_{f} is a small positive increment in frequency. Let $\{Y(t)\}$ represent the output from this filter, and let $S_Y^{(\text{I})}(f)$ and $S_Y(f)$ be, respectively, its integrated spectrum and SDF (when the latter exists). Its variance is given by

$$\sigma_Y^2 \stackrel{\text{def}}{=} \int_{-\infty}^{\infty} \mathrm{d}S_Y^{(\text{I})}(f) = \int_{-\infty}^{\infty} S_Y(f) \, \mathrm{d}f.$$

By the LTI analog filtering theorem, the relationship between the spectrum of the output $S_Y^{(\text{I})}(\cdot)$ and the spectrum of the input $S_X^{(\text{I})}(\cdot)$ is

$$\mathrm{d}S_Y^{(\text{I})}(f) = |G(f)|^2 \, \mathrm{d}S_X^{(\text{I})}(f),$$

or, in the case where the SDFs exist,

$$S_Y(f) = |G(f)|^2 S_X(f).$$

▷ **Exercise [148]** Show that

$$\sigma_Y^2 = 2\left[S_X^{(\mathrm{I})}(f' + \Delta_\mathrm{f}) - S_X^{(\mathrm{I})}(f')\right]. \tag{148a}$$ ◁

Hence the incremental difference in the integrated spectrum at the frequency f' is simply the variance associated with the output of a narrow band-pass filter bordering that frequency (the factor of 2 is needed due to the two-sided nature of the integrated spectrum). When an SDF exists and does not vary much over $[f', f' + \Delta_\mathrm{f}]$, Equation (148a) yields

$$\sigma_Y^2 \approx 2 S_X(f')\Delta_\mathrm{f}.$$

Since the variance of the input to the filter is just $\int_{-\infty}^{\infty} S_X(f')\,\mathrm{d}f'$, the above tells us that $2S_X(f')\Delta_\mathrm{f}$ is approximately the contribution to the variance from frequencies $|f| \in [f', f' + \Delta_\mathrm{f}]$.

5.7 An Example of LTI Digital Filtering

As a simple example of an LTI digital filter, let us consider the following impulse response sequence:

$$g_u^{(1)} = \begin{cases} 1/2, & u = 0; \\ 1/4, & u = \pm 1; \\ 0, & \text{otherwise.} \end{cases} \tag{148b}$$

This sequence defines an acausal FIR filter with a transfer function given by

$$G^{(1)}(f) = \cos^2(\pi f)$$

for $|f| \leq 1/2$ (see Exercise [5.14b]). The squared modulus of this function is shown by the thick curve in the upper plot of Figure 149. Its shape resembles the squared gain function for an ideal low-pass filter in that high-frequency components are attenuated – i.e., reduced in magnitude – in comparison to low-frequency components. The results of applying this filter to some data related to the rotation of the earth are shown in the lower plot. Here the pluses represent the original unfiltered data x_t, and the solid curve represents the filtered data, namely,

$$y_t^{(1)} \stackrel{\mathrm{def}}{=} g_{-1}^{(1)} x_{t+1} + g_0^{(1)} x_t + g_1^{(1)} x_{t-1}.$$

Note that the filtered data have the same shape as the "backbone" of the original data (the low-frequency components) and have less "local variability" (the high-frequency components) than the original data. This is consistent with what we would expect from the shape of the transfer function for the filter.

We now examine the difference between the original data and the filtered data, say,

$$r_t^{(1)} \stackrel{\mathrm{def}}{=} x_t - y_t^{(1)}.$$

This residual series (the dots in the lower plot of Figure 149) can also be expressed as a filtering of our original data because

$$r_t^{(1)} = -g_{-1}^{(1)} x_{t+1} + (1 - g_0^{(1)}) x_t - g_1^{(1)} x_{t-1} = -(x_{t+1} - 2x_t + x_{t-1})/4.$$

5.7 An Example of LTI Digital Filtering

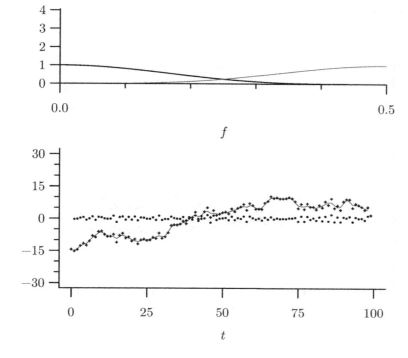

Figure 149 Example of LTI digital filtering. The top plot shows $|G^{(1)}(f)|^2$ versus f (thick curve) and the corresponding quantity for the "residual filter" (thin curve), while the bottom plot shows the unfiltered data (the pluses), the filtered data (solid curve) and the residuals (the dots).

The transfer function for the filter defined by the impulse response sequence $h_0^{(1)} = 1 - g_0^{(1)}$ and $h_u^{(1)} = -g_u^{(1)}$ for $u \neq 0$ is just $\sin^2(\pi f)$ for $|f| \leq 1/2$, and its squared modulus is shown by the thin curve in the upper plot of Figure 149. The shape of this transfer function is that of a nonideal high-pass filter: the filter attenuates the low-frequency components and leaves the high-frequency components relatively unaltered.

In the just-discussed example we were able to construct a useful high-pass filter by subtracting the output of a low-pass filter from its input. This is a useful procedure because it allows us to decompose a time series into two parts, one containing primarily its low-frequency components, and the other its high-frequency components. This decomposition is also possible with other low-pass filters if certain criteria are satisfied. One important property concerns the normalization used for the impulse response sequence of the filter. Note that the *shape* of the squared modulus of a transfer function is unchanged if we multiply each member of the impulse response sequence by a constant. For example, the impulse response sequence defined by $g_u^{(2)} = 2g_u^{(1)}$ has a transfer function given by $2G^{(1)}(\cdot)$. Its squared modulus is shown by the thick curve in the upper plot of Figure 150. The corresponding filtered and residual series are defined, respectively, by

$$y_t^{(2)} = g_{-1}^{(2)} x_{t+1} + g_0^{(2)} x_t + g_1^{(2)} x_{t-1} \text{ and } r_t^{(2)} = x_t - y_t^{(2)}.$$

They are shown by the solid curve and the dots in the lower plot of Figure 150. The filter associated with the residual series, namely,

$$r_t^{(2)} \stackrel{\text{def}}{=} -g_{-1}^{(2)} x_{t+1} + (1 - g_0^{(2)}) x_t - g_1^{(2)} x_{t-1} = -(x_{t+1} + x_{t-1})/2,$$

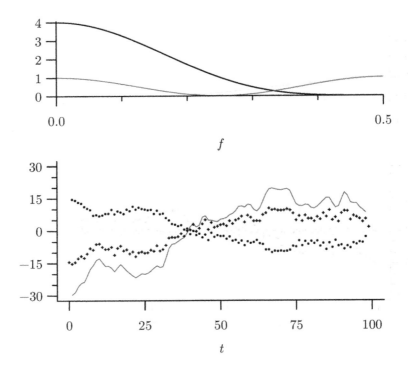

Figure 150 Second example of digital filtering.

has a transfer function whose squared modulus is given by the thin curve in the upper plot. Note that its frequency response characteristics are not at all like those of a high-pass filter and that the resulting residual series is not composed of just high-frequency components.

If we want a low-pass filter whose output traces the backbone of a given time series rather than a rescaled version of the series, the examples of Figures 149 and 150 show that normalization of the impulse response sequence is important. The proper normalization can be specified by insisting that, if the time series is locally smooth around x_t, the output from the filter should be the value x_t itself. We define locally smooth as meaning locally linear. Thus, if
$$x_t = \alpha + \beta t$$
for constants α and β, and if the impulse response sequence $\{g_u\}$ is nonzero only for $|u| \leq K$, we want
$$\sum_{u=-K}^{K} g_u x_{t-u} = x_t; \text{ i.e., } \sum_{u=-K}^{K} g_u[\alpha + \beta(t-u)] = \alpha + \beta t. \tag{150}$$

▷ **Exercise [150]** Show that, if the impulse response sequence is symmetric about $u = 0$, i.e., $g_{-u} = g_u$, Equation (150) is satisfied for all α, β and t if
$$\sum_{u=-K}^{K} g_u = 1.$$
◁

Note that in our examples this normalization holds for the impulse response sequence $\{g_u^{(1)}\}$ but not for $\{g_u^{(2)}\}$.

If an impulse response sequence is symmetric about $u = 0$, its associated transfer function must be real-valued and is in fact given by

$$G(f) = g_0 + 2 \sum_{u=1}^{K} g_u \cos(2\pi f u).$$

If in addition the impulse response sequence sums to unity, then

$$G(0) = g_0 + 2 \sum_{u=1}^{K} g_u = 1;$$

hence $|G(0)|^2 = 1$ and $10 \log_{10}(|G(0)|^2) = 0$; i.e., the response is 0 dB down at $f = 0$ – another way of stating the requirement $\sum g_u = 1$. If $\{g_u\}$ defines a low-pass filter, $G(\cdot)$ should be of the following form: first, it should be close to 1 for all f less than, say, f_L in magnitude (f_L is called the [nominal] *cutoff frequency* of the filter); and, second, it should be close to 0 for f between, say, f_H and $1/2$ in magnitude (where $f_H > f_L$ but the two are as close as possible). If we now form the residual series

$$r_t = x_t - \sum_{u=-K}^{K} g_u x_{t-u} = \sum_{u=-K}^{K} h_u x_{t-u},$$

where $h_0 = 1 - g_0$ and $h_u = -g_u$ for $|u| \geq 1$, the filter associated with $\{h_u\}$ has a transfer function given by

$$H(f) = h_0 + 2 \sum_{u=1}^{K} h_u \cos(2\pi f u) = 1 - g_0 - 2 \sum_{u=1}^{K} g_u \cos(2\pi f u) = 1 - G(f).$$

Thus, under the assumptions we have made about $\{g_u\}$, the residual filter $\{h_u\}$ should resemble a high-pass filter: first, it should be close to zero for all f less than f_L in magnitude; and, second, it should be close to one for f between f_H and $1/2$ in magnitude.

To summarize our discussion, two conditions on a low-pass FIR filter are sufficient for its associated residual filter to be a reasonable high-pass filter: first, the impulse response sequence of the filter sums to one; and, second, the sequence is symmetric about $u = 0$. The first condition is essential (as the filter $\{g_u^{(2)}\}$ shows); the second condition was needed for our mathematical development, but it can be circumvented – it is possible to construct examples of asymmetric low-pass filters whose associated residual filter is a reasonable high-pass filter.

Comments and Extensions to Section 5.7

[1] Our example can also be used to point out the strong resemblance between detrending a time series and subjecting it to a high-pass filter. *Trend* can be loosely defined as a "long term change in the mean level" (Chatfield, 2004, p. 12; Granger, 1966). We might argue that the solid curve in the lower plot of Figure 149 represents the trend of the time series, while the dots represent the detrended series. Alternatively, we might fit a least squares line to the series and argue that the fitted regression line is the trend, while the residuals from the regression fit are the detrended data. (As indicated by the results of Exercise [5.18], fitting a regression line is *not* an example of LTI filtering, so unfortunately we cannot use the notion of a transfer function to help us assess the effect of this type of detrending.) This and other forms of detrending act like high-pass filters – detrending effectively removes low-frequency components in a time series (an interesting example of detrending a time series of radar returns from a moving aircraft is given by Alavi and Jenkins, 1965).

5.8 Least Squares Filter Design

In this and the next section we consider some simple effective approaches to approximating an ideal low-pass digital filter (with obvious modifications, these same methods can be used to approximate ideal high-pass and band-pass filters also). The transfer function for an ideal low-pass filter of bandwidth $2W < 1$ is given by

$$G_I(f) \stackrel{\text{def}}{=} \begin{cases} 1, & |f| \leq W; \\ 0, & W < |f| \leq 1/2. \end{cases} \tag{152a}$$

Since $G_I(\cdot)$ is square integrable over the region $[-1/2, 1/2]$, we can periodically extend it outside that region and then appeal to the theory of Section 3.8 to consider it as the Fourier transform of a sequence defined by

$$g_{I,u} = \int_{-1/2}^{1/2} G_I(f) e^{i2\pi fu} \, df = \int_{-W}^{W} e^{i2\pi fu} \, df = \begin{cases} 2W, & u = 0; \\ \dfrac{\sin(2\pi Wu)}{\pi u}, & u \neq 0. \end{cases} \tag{152b}$$

From this expression we see that the ideal low-pass digital filter defined by Equation (152a) is a symmetric (in the sense that $g_{I,-u} = g_{I,u}$) acausal IIR filter – this limits its usefulness in practical applications and motivates us to look at various approximations. For simplicity, we only consider approximations from within the class of symmetric acausal FIR filters.

Our first approach is called *least squares filter design* and is based upon the following result: if K is a nonnegative integer, then among all functions of the form

$$H_K(f) \stackrel{\text{def}}{=} \sum_{u=-K}^{K} h_u e^{-i2\pi fu},$$

the one that minimizes

$$\int_{-1/2}^{1/2} |G_I(f) - H_K(f)|^2 \, df$$

is obtained by letting $h_u = g_{I,u}$ for $|u| \leq K$ (see the discussion surrounding Equation (52)). Thus the acausal FIR filter defined by

$$g_{K,u} = \begin{cases} g_{I,u}, & |u| \leq K; \\ 0, & \text{otherwise}, \end{cases}$$

has a transfer function

$$G_K(f) \stackrel{\text{def}}{=} \sum_{u=-K}^{K} g_{K,u} e^{-i2\pi fu}$$

that is the best Kth-order approximation – in the least squares sense – to the transfer function of an ideal low-pass digital filter.

As an example, Figure 153a shows $|G_K(f)|^2$ versus f for $W = 0.1$ and $K = 32$. This figure points out a potential problem with least squares filters, namely, ripples in the region where $G_I(\cdot)$ has a discontinuity – in this example, this occurs at $W = 0.1$ (indicated by the dotted vertical line). These ripples are due to the Gibbs phenomenon and constitute a form of leakage in the region where $G_I(f) = 0$. We introduced these ideas in Section 3.8 and discussed there the use of convergence factors $\{c_u\}$ to reduce the ripples. In the present context these factors would define a new impulse response sequence given by, say,

$$g_{K,u}^{(C)} \stackrel{\text{def}}{=} c_u g_{K,u} = \begin{cases} c_u g_{I,u}, & |u| \leq K; \\ 0, & \text{otherwise}, \end{cases}$$

5.8 Least Squares Filter Design

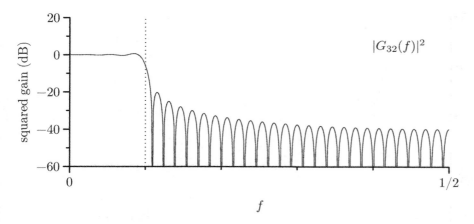

Figure 153a Squared modulus of $G_{32}(\cdot)$, the squared gain function for the least squares approximation of order $K = 32$ to an ideal low-pass filter with pass-band $[-0.1, 0.1]$. The dotted vertical line marks the cutoff frequency $W = 0.1$.

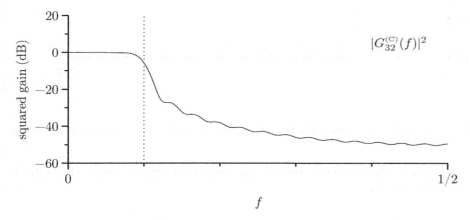

Figure 153b Squared modulus of $G_{32}^{(C)}(\cdot)$ with triangular convergence factors – compare this with $G_{32}(\cdot)$ in Figure 153a.

with a corresponding transfer function given by

$$G_K^{(C)}(f) \stackrel{\text{def}}{=} \sum_{u=-K}^{K} g_{K,u}^{(C)} e^{-i2\pi f u}.$$

For example, the use of Cesàro sums yields triangular convergence factors:

$$c_u = \begin{cases} \gamma\left(1 - \dfrac{|u|}{K+1}\right), & |u| \leq K; \\ 0, & \text{otherwise}; \end{cases}$$

here γ is a gain factor that allows us to, say, force the normalization $\sum g_{K,u}^{(C)} = G_K^{(C)}(0) = 1$ (see the discussion in the previous section). Figure 153b shows $|G_K^{(C)}(f)|^2$ versus f for the same values of W and K as in Figure 153a. Note that the ripples are substantially reduced in $|G_K^{(C)}(\cdot)|^2$ but also that the discontinuity at $W = 0.1$ is now less accurately rendered than in $|G_K(\cdot)|^2$.

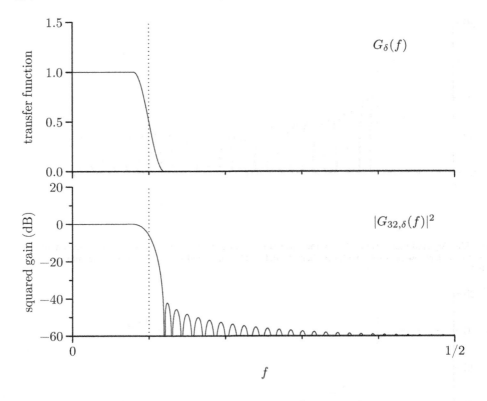

Figure 154 $G_\delta(\cdot)$ with $W = 0.1$ (indicated by the dotted vertical lines) and $\delta = 0.02$ (top plot) and the squared modulus of its approximation $G_{32,\delta}(\cdot)$ (bottom plot).

Let us note the following interesting interpretation of convergence factors (Bloomfield, 2000). Since we can write

$$G_K^{(C)}(f) = \int_{-1/2}^{1/2} G_I(f') C_K(f - f') \, df' \text{ for } C_K(f) \stackrel{\text{def}}{=} \sum_{u=-K}^{K} c_u e^{-i2\pi f u},$$

$G_K^{(C)}(\cdot)$ is a "smoothed" version of $G_I(\cdot)$ (see Section 3.6). Because $G_I(\cdot)$ is already smooth everywhere except at the discontinuities at $\pm W$, the main effect of convolving it with the smoothing kernel $C_K(\cdot)$ is to eliminate these discontinuities, but we could just as well do this directly. For example, instead of seeking least squares approximations to $G_I(\cdot)$ itself, we could look at such approximations to the following smoothed version of $G_I(\cdot)$:

$$G_\delta(f) \stackrel{\text{def}}{=} \begin{cases} 1, & |f| \leq W - \delta; \\ \frac{1}{2}\left[1 + \cos\left(\pi \frac{|f| - W + \delta}{2\delta}\right)\right], & W - \delta < |f| \leq W + \delta; \\ 0, & W + \delta < |f| \leq 1/2 \end{cases} \quad (154)$$

under the constraints $0 < \delta < W$ and $W + \delta < 1/2$ (see the top plot of Figure 154 for the case $W = 0.1$ and $\delta = 0.02$). This yields

$$G_{K,\delta}(f) \stackrel{\text{def}}{=} \sum_{u=-K}^{K} g_{\delta,u} e^{-i2\pi f u} \text{ with } g_{\delta,u} \stackrel{\text{def}}{=} \int_{-1/2}^{1/2} G_\delta(f) e^{i2\pi f u} \, df$$

as an approximation to $G_\delta(\cdot)$ (calculation of $g_{\delta,u}$ is the subject of Exercise [5.20]). The squared modulus of the order $K = 32$ approximation for the case depicted in the top plot of Figure 154 is shown in the lower plot there – this should be compared with Figures 153a and 153b. Note in particular that the sidelobes of $G_{K,\delta}(\cdot)$ are considerably lower than those of the other two examples.

5.9 Use of Slepian Sequences in Low-Pass Filter Design

Here we consider a symmetric acausal FIR filter of width $2K+1$ that approximates an ideal low-pass filter as well as possible in the following sense: among all filters $\{g_u\}$ such that $g_u = 0$ for $|u| > K$ and normalized such that $\sum g_u = 1$, we want the filter whose transfer function

$$G_K(f) \stackrel{\text{def}}{=} \sum_{u=-K}^{K} g_u e^{-i2\pi f u}$$

is as concentrated as possible in the range $|f| \leq W$. This is very close to the index-limited concentration problem we considered in Section 3.10. We again use the measure of concentration

$$\beta^2(W) \stackrel{\text{def}}{=} \int_{-W}^{W} |G_K(f)|^2 \, df \bigg/ \int_{-1/2}^{1/2} |G_K(f)|^2 \, df$$

$$= \sum_{u'=-K}^{K} \sum_{u=-K}^{K} g_u^* \frac{\sin\left(2\pi W(u' - u)\right)}{\pi(u' - u)} g_{u'} \bigg/ \sum_{u=-K}^{K} |g_u|^2$$

(see Equation (88a)). Following the approach of Section 3.10, the solution to this concentration problem is any eigenvector associated with the largest eigenvalue of the following set of matrix eigenvalue equations:

$$\sum_{u'=-K}^{K} \frac{\sin\left(2\pi W(u' - u)\right)}{\pi(u' - u)} g_{u'} = \lambda g_u, \quad u = -K, \ldots, K. \tag{155}$$

Let $\lambda_0(W)$ denote this eigenvalue, and let

$$\tilde{\boldsymbol{v}}_0(W) \stackrel{\text{def}}{=} [\tilde{v}_{0,-K}(W), \ldots, \tilde{v}_{0,0}(W), \ldots, \tilde{v}_{0,K}(W)]^T$$

be the corresponding eigenvector normalized so that its elements sum to unity. The elements of this eigenvector are renormalized and reindexed portions of the zeroth-order *Slepian sequence*, which we denoted as $\{v_{0,t}(N,W)\}$ in Section 3.10 (Slepian sequences are also known as discrete prolate spheroidal sequences). If we let $\boldsymbol{A}$ be the $(2K+1) \times (2K+1)$ matrix whose (u', u)th element is $\sin[2\pi W(u' - u)]/[\pi(u' - u)]$ (defined to be $2W$ when $u' = u$), then the set of $2K+1$ equations in Equation (155) can be written as $\boldsymbol{A}\tilde{\boldsymbol{v}}_0(W) = \lambda_0(W)\tilde{\boldsymbol{v}}_0(W)$, and the concentration measure is

$$\beta^2(W) = \frac{\tilde{\boldsymbol{v}}_0^T(W) \boldsymbol{A} \tilde{\boldsymbol{v}}_0(W)}{\tilde{\boldsymbol{v}}_0^T(W) \tilde{\boldsymbol{v}}_0(W)} = \frac{\lambda_0(W) \tilde{\boldsymbol{v}}_0^T(W) \tilde{\boldsymbol{v}}_0(W)}{\tilde{\boldsymbol{v}}_0^T(W) \tilde{\boldsymbol{v}}_0(W)} = \lambda_0(W).$$

The solid curves in Figure 156a show the squared magnitude of the resulting transfer function $G_K(\cdot)$ for $W = 0.1$ and the cases $K = 8$ (thin curve) and 32 (thick). The dotted vertical line indicates the location of the cutoff frequency $W = 0.1$. This plot shows that each $G_K(\cdot)$ captures one important aspect of the ideal low-pass digital filter – a small amount of energy in the frequency range $W < |f| \leq 1/2$. However, both fail miserably in another

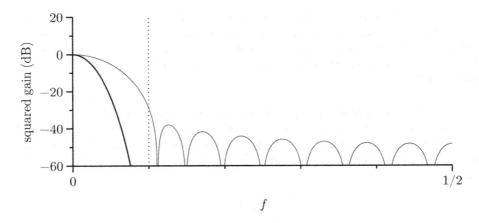

Figure 156a Squared gain functions corresponding to $\tilde{v}_0(W)$ with $W = 0.1$ and $K = 8$ (thin curve) and 32 (thick curve). The dotted vertical line marks the frequency W.

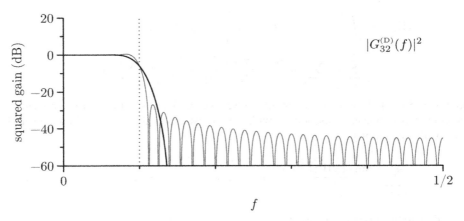

Figure 156b Squared modulus of $G_{32}^{(\mathrm{D})}(\cdot)$ using Slepian sequences as convergence factors for the cases $\delta = 0.04$ (thick curve) and $\delta = 0.01$ (thin). The dotted vertical line marks the position of the cutoff frequency $W = 0.1$.

aspect: the squared modulus of the transfer function of the ideal filter is not anywhere near constant for $0 \leq |f| \leq W$.

The $(2K+1) \times (2K+1)$ matrix $\boldsymbol{A}$ from which the eigenvectors are derived is a symmetric Toeplitz matrix (see Section 2.4). Makhoul (1981a) has shown that the transfer function for an eigenvector of a symmetric Toeplitz matrix will typically be zero at $2K$ different frequencies f such that $-1/2 < f \leq 1/2$ if the eigenvector corresponds to either the maximum or minimum eigenvalue. Thus we expect the transfer function of our FIR filter to have zero response (a notch) at certain frequencies. For example, 8 notches are apparent in the thin solid curve in Figure 156a for $K = 8$, and by symmetry there must be 8 additional notches in the frequency range $[-1/2, 0]$ for a total of $2K = 16$ notches.

A more reasonable use for Slepian sequences in low-pass filter design is as convergence factors (Rabiner and Gold, 1975, sections 3.8 and 3.11). Let us define

$$d_u = \begin{cases} \gamma \tilde{v}_{0,u}(\delta), & |u| \leq K; \\ 0, & \text{otherwise,} \end{cases}$$

where, as before, γ is a gain factor; however, note that the cutoff frequency for the Slepian sequence is now δ instead of W. We can now define a FIR filter with impulse response sequence

$$g^{(D)}_{K,u} = d_u g_{I,u},$$

where, as in the previous section, $\{g_{I,u}\}$ is the impulse response sequence of an ideal low-pass filter of bandwidth $2W$. The transfer function corresponding to $\{g^{(D)}_{K,u}\}$ is given by

$$G^{(D)}_K(f) = \int_{-1/2}^{1/2} G_I(f') D_K(f-f') \, df' \text{ for } D_K(f) \stackrel{\text{def}}{=} \sum_{u=-K}^{K} d_u e^{-i2\pi f u},$$

where $G_I(\cdot)$ is the transfer function given in Equation (152a) for the ideal low-pass digital filter. Because $G_I(\cdot)$ is convolved with $D_K(\cdot)$ to form $G^{(D)}_K(\cdot)$, the discontinuity in $G_I(\cdot)$ is smeared out into a transition band. The width of this transition band is proportional to the width of the central lobe of $D_K(\cdot)$, while the sidelobes of the latter contribute to leakage outside the pass-band. For a given filter width $2K+1$, by adjusting the bandwidth parameter δ for $\{\tilde{v}_{0,u}(\delta)\}$, we can trade off between a small central lobe (its width decreases as δ decreases) and small sidelobes (their heights decrease as δ increases). The Slepian sequence is a natural choice for the convergence factors because the squared modulus of its transfer function is a good approximation to a Dirac delta function.

As an example, Figure 156b shows the squared modulus of $G^{(D)}_K(\cdot)$ for $K=32$, $W=0.1$ and the cases $\delta = 0.04$ (thick solid curve) and $\delta = 0.01$ (thin). In each case the gain factor γ was set so that $G^{(D)}_K(0) = 1$. As expected, the transfer function for $\delta = 0.04$ has a wider transition region about $W = 0.1$ than does the one for $\delta = 0.01$, whereas the sidelobe level is much higher in the latter case than in the former. Of course, if we allow ourselves the liberty of increasing K beyond 32, the trade-off between transition region width and sidelobe level becomes easier to manage.

5.10 Exercises

[5.1] Let $\{X(t)\}$ be a stochastically continuous, continuous parameter stationary process with zero mean and spectral representation

$$X(t) = \int_{-\infty}^{\infty} e^{i2\pi f t} \, dZ_X(f),$$

where $\{Z_X(f)\}$ is an orthogonal process such that $E\{|dZ_X(f)|^2\} = dS^{(I)}_X(f)$. If $G(\cdot)$ is a complex-valued function such that

$$\int_{-\infty}^{\infty} |G(f)|^2 \, dS^{(I)}_X(f) < \infty, \qquad (157)$$

show that

$$Y(t) \stackrel{\text{def}}{=} \int_{-\infty}^{\infty} e^{i2\pi f t} G(f) \, dZ_X(f)$$

is a stationary process. Why is the condition shown in Equation (157) necessary?

[5.2] (a) Show that Equation (142) defines an LTI filter.
(b) Show that

$$L\{X(t)\} \stackrel{\text{def}}{=} \frac{d^p X(t)}{dt^p}$$

is an LTI filter, and determine its transfer function. This is an example of an LTI analog filter that must make use of Dirac delta functions if it is to be expressed as a convolution.

[5.3] Consider a zero mean stationary process $\{X_t\}$ with a sampling interval of Δ_t, an associated Nyquist frequency $f_\mathcal{N} = 1/(2\Delta_t)$, a spectral representation given by

$$X_t = \int_{-f_\mathcal{N}}^{f_\mathcal{N}} e^{i2\pi ft\Delta_t} \, dZ_X(f)$$

and an integrated spectrum $S_X^{(I)}(\cdot)$ specified by $E\{|dZ_X(f)|^2\} = dS_X^{(I)}(f)$. Define

$$Y_t = C \sum_{u=-\infty}^{\infty} g_u X_{t-u},$$

where $C \neq 0$ is a real-valued constant, and $\{g_u\}$ is a real-valued sequence satisfying $\sum g_u^2 < \infty$. Show that $\{Y_t\}$ is a stationary process whose integrated spectrum $S_Y^{(I)}(\cdot)$ is specified by

$$dS_Y^{(I)}(f) = |G(f)|^2 \, dS_X^{(I)}(f),$$

where

$$G(f) \stackrel{\text{def}}{=} C \sum_{u=-\infty}^{\infty} g_u e^{-i2\pi fu\Delta_t}.$$

(Note that, if $C = \Delta_t$, then $G(\cdot)$ is the Fourier transform of $\{g_u\}$ if we regard that sequence as having a sampling interval of Δ_t, whereas it is just proportional to the transform if C is anything else. Note also that, if $S_X^{(I)}(\cdot)$ is differentiable with derivative $S_X(\cdot)$, then $S_Y^{(I)}(\cdot)$ is also differentiable, and its derivative $S_Y(\cdot)$ is given by $S_Y(f) = |G(f)|^2 S_X(f)$.)

[5.4] Let $\{X_t\}$ be a discrete parameter stationary process with SDF given by $S_X(\cdot)$ defined over the interval $[-1/2, 1/2]$.

(a) Let the first difference process $\{Y_t\}$ be given by $Y_t \stackrel{\text{def}}{=} X_t - X_{t-1}$. Show that the SDF of $\{Y_t\}$ is given by $S_Y(f) = 4\sin^2(\pi f) S_X(f)$. Does a first difference filter resemble a low-pass or high-pass filter?

(b) If a two coefficient filter is now applied to $\{Y_t\}$ to yield $\{Z_t\}$ according to

$$Z_t = aY_t + bY_{t-1}, \text{ where } a = \frac{-1+\sqrt{3}}{4} \text{ and } b = \frac{1+\sqrt{3}}{4},$$

show that the SDF of $\{Z_t\}$ is given by

$$S_Z(f) = \sin^2(\pi f) \left[1 + 2\cos^2(\pi f)\right] S_X(f).$$

[5.5] Let $\{X_t\}$ be a discrete parameter stationary process with SDF given by $S_X(\cdot)$ defined over the interval $[-1/2, 1/2]$. With p taken to be a positive integer, show that the SDF for the process $\{Y_t\}$ defined by $Y_t = X_t - \alpha X_{t-p}$ is

$$S_Y(f) = 4\alpha^2 \sin^2(\pi f p) S_X(f). \tag{158}$$

[5.6] Determine the SDF for the process $\{X_t\}$ defined by

$$X_t = \alpha X_{t-d} + \epsilon_t,$$

where $0 < \alpha < 1$; $d \geq 2$ is an integer; and $\{\epsilon_t\}$ is a white noise process with mean zero and variance σ_ϵ^2. At what frequency (or frequencies) does this SDF have its maximum value? What does this suggest about the oscillatory behavior of $\{X_t\}$? (The first part of exercise 4.16, Brockwell and Davis, 1991, is a special case of this exercise.)

5.10 Exercises

[5.7] Consider the MA(2) process

$$X_t = \epsilon_t - \theta_{2,1}\epsilon_{t-1} - \theta_{2,2}\epsilon_{t-2},$$

where, as usual, $\{\epsilon_t\}$ is a white noise process with mean zero and variance σ_ϵ^2.
(a) Show that the SDF for $\{X_t\}$ is given by

$$S_X(f) = \sigma_\epsilon^2 \left[1 + \theta_{2,1}^2 + \theta_{2,2}^2 - 2\theta_{2,1}\left(1 - \theta_{2,2}\right)\cos\left(2\pi f\right) - 2\theta_{2,2}\cos\left(4\pi f\right)\right].$$

(b) The autocovariance sequence (ACVS) $\{s_{X,\tau}\}$ for an MA process and its SDF form a Fourier transform pair. Use this relationship to show that the variance of $\{X_t\}$ is given by

$$\sigma_X^2 = \left(1 + \theta_{2,1}^2 + \theta_{2,2}^2\right)\sigma_\epsilon^2$$

and that the lag $\tau = 1$ and 2 members of its ACVS are

$$s_{X,1} = -\theta_{2,1}\left(1 - \theta_{2,2}\right)\sigma_\epsilon^2 \text{ and } s_{X,2} = -\theta_{2,2}\sigma_\epsilon^2.$$

(c) Let $\rho_{X,\tau} = s_{X,\tau}/s_{X,0}$ represent the ACS for $\{X_t\}$ at lag τ. Compute $\rho_{X,\tau}$ at lags $\tau = 1$ and $\tau = 2$ for the cases (i) $\theta_{2,1} = 5/6$ and $\theta_{2,2} = -1/6$ and (ii) $\theta_{2,1} = 5$ and $\theta_{2,2} = -6$. Comment upon the result.

[5.8] Let $\{X_t\}$ be a discrete parameter stationary ARMA(p,q) process, which implies that $\{X_t\}$ satisfies the equation

$$X_t = \sum_{j=1}^{p}\phi_{p,j}X_{t-j} + \epsilon_t - \sum_{j=1}^{q}\theta_{q,j}\epsilon_{t-j},$$

where $\{\epsilon_t\}$ is a white noise process with zero mean and variance σ_ϵ^2. Show that the SDF of this process is given by Equation (145c).

[5.9] (a) Let $X_t = \phi X_{t-1} + \epsilon_t$ be a stationary first-order autoregressive process, where $|\phi| < 1$, and $\{\epsilon_t\}$ is a zero mean white noise process with variance σ_ϵ^2. Let $\{W_t\}$ be a white noise process with zero mean and variance σ_W^2. Suppose that X_t and $W_{t'}$ are uncorrelated for all $t, t' \in \mathbb{Z}$. Let $Y_t \stackrel{\text{def}}{=} X_t + W_t$. Determine the SDF for $\{Y_t\}$.
(b) Determine the SDF for the stationary ARMA(1,1) process $\{U_t\}$ defined by $U_t = \phi U_{t-1} + \eta_t - \theta\eta_{t-1}$, where ϕ is the same as in part (a), and $\{\eta_t\}$ is a zero mean white noise process with variance σ_η^2.
(c) Let $\sigma_\epsilon^2 = \sigma_W^2 = 1$ and $\phi = 1/2$. By equating the SDFs in parts (a) and (b), show that $\{Y_t\}$ in (a) can be written as $\{U_t\}$ in (b), provided $\theta = 1/(2\sigma_\eta^2)$, where σ_η^2 is the solution to the equation

$$\sigma_\eta^2\left(1 + \frac{1}{4\sigma_\eta^4}\right) = \frac{9}{4}.$$

[5.10] Let $\{X_t\}$ be a discrete parameter stationary process with an SDF $S_X(\cdot)$ defined over the interval $[-1/2, 1/2]$. Let

$$Y_t \stackrel{\text{def}}{=} X_t - \frac{1}{2K+1}\sum_{j=-K}^{K}X_{t+j},$$

where $K > 0$ is an integer. Find the SDF for $\{Y_t\}$. Plot this SDF for the cases $K = 4$ and 16 when $\{X_t\}$ is a white noise process.

[5.11] Let $\{X_t\}$ be a discrete parameter stationary process with an SDF $S_X(\cdot)$ defined over the interval $[-1/2, 1/2]$.
(a) Let

$$\overline{X}_{K,t} \stackrel{\text{def}}{=} \frac{1}{K}\sum_{k=0}^{K-1}X_{t-k} \text{ and } Y_{K,t} \stackrel{\text{def}}{=} \overline{X}_{K,t} - \overline{X}_{K,t-K}.$$

Show that $\{Y_{K,t}\}$ is a zero mean stationary process with SDF

$$S_{Y,K}(f) = |G_K(f)|^2 S_X(f) \text{ with } |G_K(f)|^2 = \frac{4\sin^4(K\pi f)}{K^2 \sin^2(\pi f)}.$$

(b) Plot $|G_K(\cdot)|^2$ for $K = 1, 2, 4, 8$ and 16 to see that $|G_K(\cdot)|^2$ is an approximation to the squared magnitude of a transfer function for a band-pass filter with a pass-band defined by $|f| \in [1/(4K), 1/(2K)]$. (This sequence of filters is the basis for a crude method of estimating the SDF called *pilot analysis*. This method was useful as a simple way of computing a rough estimate of the SDF in days before modern computers and the popularization of the fast Fourier transform. For some details, see section 7.3.2 of Jenkins and Watts, 1968.)

[5.12] Let $\{\epsilon_t\}$ be a white noise process with mean zero and variance σ_ϵ^2. Define the stationary processes $\{X_t\}$ and $\{Y_t\}$ by

$$X_t = \tfrac{2}{9}\epsilon_t + \tfrac{5}{9}\epsilon_{t-1} + \tfrac{2}{9}\epsilon_{t-2} \text{ and } Y_t = \tfrac{4}{9}\epsilon_t + \tfrac{4}{9}\epsilon_{t-1} + \tfrac{1}{9}\epsilon_{t-2}.$$

(a) Let $L_X\{\cdot\}$ and $L_Y\{\cdot\}$ be LTI digital filters such that $L_X\{\epsilon_t\} = X_t$ and $L_Y\{\epsilon_t\} = Y_t$. Show that the gain function $|G_X(\cdot)|$ corresponding to $L_X\{\cdot\}$ is given by

$$|G_X(f)| = \tfrac{5}{9} + \tfrac{4}{9}\cos(2\pi f),$$

and find the gain function $|G_Y(\cdot)|$ corresponding to $L_Y\{\cdot\}$. In what respects do the associated transfer functions $G_X(\cdot)$ and $G_Y(\cdot)$ differ?

(b) Compare the SDF of $\{X_t\}$ with the SDF of $\{Y_t\}$.

(c) How do $\{X_t\}$ and $\{Y_t\}$ differ from a moving average process as defined by Equation (32a)? Are there moving average processes that are equivalent to $\{X_t\}$ and $\{Y_t\}$ (i.e., have the same SDFs)?

[5.13] (a) Let $\{X_t\}$ be a real-valued zero mean stationary process with ACVS $\{s_{X,\tau}\}$ and SDF $S_X(\cdot)$ such that $\{s_{X,\tau}\}$ and $S_X(\cdot)$ are a Fourier transform pair (assume $\Delta_t = 1$ for convenience). Define the complex-valued process $\{Z_t\}$ by

$$Z_t = X_t e^{-i2\pi f_0 t},$$

where f_0 is any fixed frequency such that $0 < |f_0| \leq 1/2$; i.e., the tth element of the sequence $\{X_t\}$ is multiplied by $\exp(-i2\pi f_0 t)$ to create the tth element of $\{Z_t\}$. Show that $\{Z_t\}$ is a zero mean stationary process with ACVS and SDF given, respectively, by

$$s_{Z,\tau} = s_{X,\tau} e^{-i2\pi f_0 \tau} \text{ and } S_Z(f) = S_X(f_0 + f).$$

Does the transformation from $\{X_t\}$ to $\{Z_t\}$ constitute an LTI filter? (Creation of $\{Z_t\}$ is an example of a technique called *complex demodulation*; see Tukey, 1961, Hasan, 1983, Bloomfield, 2000 and Stoica and Moses, 2005.)

(b) Suppose $f_0 = 1/2$ so that $\exp(-i2\pi f_0 t) = (-1)^t$. Show that we can effectively turn a low-pass filter into a high-pass filter by the following three operations: (i) form $Z_t = (-1)^t X_t$; (ii) pass $\{Z_t\}$ through a low-pass filter to produce, say, $\{Z'_t\}$; and (iii) form $X'_t \stackrel{\text{def}}{=} (-1)^t Z'_t$, where $\{X'_t\}$ can be regarded as a high-pass filtered version of $\{X_t\}$.

(c) Let $\{X_t\}$ and f_0 be as defined in part (a). Define the real-valued process $\{Y_t\}$ by $Y_t = X_t \cos(2\pi f_0 t)$. Is $\{Y_t\}$ a stationary process?

[5.14] Part (b) of the previous exercise demonstrated that, by demodulating the input to a low-pass filter at $f_0 = 1/2$ and then demodulating the output at $f_0 = 1/2$ also, we can effectively turn a low-pass filter into a high-pass filter. This complementary exercise shows that we can achieve the same result by demodulating the filter itself at $f_0 = 1/2$.

(a) Consider two filters with real-valued impulse response sequences $\{g_u\}$ and $\{h_u\}$ related via

$$h_u = (-1)^u g_u. \qquad (161a)$$

Show that the corresponding transfer functions $G(\cdot)$ and $H(\cdot)$ are related by

$$H(f) = G(f - \tfrac{1}{2}). \qquad (161b)$$

Use the above to argue that, if $\{g_u\}$ resembles a low-pass filter, then $\{h_u\}$ must resemble a high-pass filter.

(b) As a concrete example, consider the impulse response sequence $\{g_u^{(1)}\}$ of Equation (148b). Show that its transfer function is given by

$$G^{(1)}(f) = \cos^2(\pi f).$$

Define $\{h_u^{(1)}\}$ in terms of $\{g_u^{(1)}\}$ as per Equation (161a), and let $H^{(1)}(\cdot)$ denote its transfer function. Determine the transfer function $H^{(1)}(\cdot)$ for $\{h_u^{(1)}\}$ using the fact that $\{h_u^{(1)}\} \longleftrightarrow H^{(1)}(\cdot)$, and demonstrate that $H^{(1)}(\cdot)$ is indeed related to $G^{(1)}(\cdot)$ as per Equation (161b).

(c) Suppose now that the filters $\{g_u\}$ and $\{h_u\}$ are related by $h_u = (-1)^u g_{-u}$ rather than by Equation (161a). What is the relationship between the transfer functions for $\{g_u\}$ and $\{h_u\}$? If $\{g_u\}$ resembles a low-pass filter, does $\{h_u\}$ still resemble a high-pass filter?

(d) Finally consider the causal FIR filter $\{g_u\}$, where $g_u = 0$ for $u < 0$ and $u \geq L$ for some positive integer L. Define $h_u = (-1)^u g_{L-1-u}$. What is the relationship between the transfer functions for $\{g_u\}$ and $\{h_u\}$? If $\{g_u\}$ resembles a low-pass filter, does $\{h_u\}$ still resemble a high-pass filter?

[5.15] Let $\{X(t)\}$ be a continuous parameter stationary process with SDF $S_X(\cdot)$ defined over $\mathbb{R}$. For $\Delta_t > 0$ let

$$\overline{X}(t) \stackrel{\text{def}}{=} \frac{1}{\Delta_t} \int_{t-\Delta_t}^{t} X(u)\, du$$

represent the average value of the process over the interval $[t - \Delta_t, t]$.

(a) Show that $\{\overline{X}(t)\}$ is a stationary process, and find its SDF $S_{\overline{X}}(\cdot)$.

(b) Let $\overline{X}_{\Delta_t',t} \stackrel{\text{def}}{=} \overline{X}(t\Delta_t')$, $t \in \mathbb{Z}$, be a discrete parameter process formed by taking samples Δ_t' time units apart from $\{\overline{X}(t)\}$. Determine the SDF $S_{\overline{X}_{\Delta_t'}}(\cdot)$ for $\{\overline{X}_{\Delta_t',t}\}$ in terms of $S_X(\cdot)$.

(c) Suppose now that $S_X(f) = C$ for $f \in \mathbb{R}$, where C is a positive constant. Since $S_X(\cdot)$ is flat, we can regard it as the SDF for a continuous parameter version of white noise. As noted in Section 2.7, such a process technically does not exist but is sometimes a convenient fiction. This fiction makes sense in part (a) because $S_{\overline{X}}(\cdot)$ is a well-defined SDF when $S_X(f) = C$ (i.e., as opposed to $S_X(\cdot)$, the integral of $S_{\overline{X}}(\cdot)$ over $\mathbb{R}$ is finite). Show that, if $\Delta_t' = \Delta_t$ in part (b), then the SDF $S_{\overline{X}_{\Delta_t}}(\cdot)$ is flat, i.e., corresponds to the SDF for white noise. Why might we have anticipated this result? Hint: use the fact that

$$\sum_{k=-\infty}^{\infty} (a+k)^{-2} = \pi^2 \mathrm{cosec}^2(\pi a).$$

(d) As a follow-on to part (c), show that, if $\Delta_t' = \Delta_t/2$, then the SDF is given by

$$S_{\overline{X}_{\Delta_t'}}(f) = C\cos^2(\pi f \Delta_t/2), \quad |f| \leq f_{\mathcal{N}} \text{ where here } f_{\mathcal{N}} = 1/\Delta_t.$$

Show that this SDF is that of a first-order moving average process whose unit lag autocorrelation is $1/2$. Why might we have anticipated this result?

(e) In part (b), let $\Delta'_t = \Delta_t$, and define the discrete parameter process $\overline{Y}_t = \overline{X}_{\Delta_t,t} - \overline{X}_{\Delta_t,t-1}$; i.e., $\{\overline{Y}_t\}$ is the first difference process corresponding to $\{\overline{X}_{\Delta_t,t}\}$ (see Exercise [5.4]). Find the SDF $S_{\overline{Y}_t}(\cdot)$ for $\{\overline{Y}_t\}$.

(f) Let $\overline{Y}(t) = \overline{X}(t) - \overline{X}(t - \Delta_t)$, $t \in \mathbb{R}$, define a continuous parameter process representing the difference between average values of $\{X(t)\}$ that are Δ_t time units apart. Let $\overline{Y}'_t = \overline{Y}(t\Delta_t)$, $t \in \mathbb{Z}$, define a discrete parameter process formed by taking samples Δ_t time units apart from $\{\overline{Y}(t)\}$. Show that the SDF for $\{\overline{Y}'_t\}$ is the same as that of $\{\overline{Y}_t\}$ of part (e); i.e., we can interchange the order in which we carry out differencing and sampling. (This way of processing $\{X(t)\}$ arises in the study of the timekeeping properties of atomic clocks; see Barnes et al., 1971.)

[5.16] If the action of L on complex exponentials is described by Equation (135a), and if $G(-f) = G^*(f)$ for all f, what does L do to sines and cosines?

[5.17] Suppose that $g_p(\cdot)$ and $h_p(\cdot)$ are two periodic functions with period T such that

$$\int_{-T/2}^{T/2} |g_p(t)|^2 \, dt < \infty \quad \text{and} \quad \int_{-T/2}^{T/2} |h_p(t)|^2 \, dt < \infty.$$

Let $\{G_n\}$ and $\{H_n\}$ be their Fourier coefficients as defined by Equation (49b). If we regard $h_p(\cdot)$ as a function of interest and $g_p(\cdot)/T$ as a filter, their convolution

$$g_p * h_p(t) \stackrel{\text{def}}{=} \frac{1}{T} \int_{-T/2}^{T/2} g_p(u) h_p(t-u) \, du$$

can be regarded as a filtered version of $h_p(\cdot)$ (the above is Equation (96b)). What plays the role of the transfer function here?

[5.18] For the filter

$$L\{X(t)\} \stackrel{\text{def}}{=} \alpha + \beta X(t),$$

where α and β are arbitrary nonzero constants, consider all three conditions needed for it to be an LTI filter. Which conditions (if any) fail to hold?

[5.19] Suppose that $g(\cdot)$ is an $L^2(\mathbb{R})$ function whose Fourier transform $G(\cdot)$ is also such. The mapping from $g(\cdot)$ to $G(\cdot)$ is an example of an analog filter, but is it an LTI filter?

[5.20] Show that the impulse response sequence $\{g_{\delta,u}\}$ corresponding to $G_\delta(\cdot)$ of Equation (154) is given by

$$g_{\delta,u} = \begin{cases} 2W, & \text{when } u = 0; \\ \delta \sin(\pi W/2\delta), & |u| = 1/4\delta; \\ \dfrac{\sin(2\pi W u)\cos(2\pi \delta u)}{\pi(u - 16u^3\delta^2)}, & \text{otherwise.} \end{cases}$$

Hint: consider the inverse Fourier transform of the following periodic function with a period of unity:

$$D(f) \stackrel{\text{def}}{=} G_\delta(f) - G_I(f) = \begin{cases} 0, & |f| \leq W - \delta; \\ \dfrac{1}{2}\left[\cos\left(\pi \dfrac{|f| - W + \delta}{2\delta}\right) - 1\right], & W - \delta < |f| \leq W; \\ \dfrac{1}{2}\left[\cos\left(\pi \dfrac{|f| - W + \delta}{2\delta}\right) + 1\right], & W < |f| \leq W + \delta; \\ 0, & W + \delta < |f| \leq 1/2, \end{cases}$$

where $G_\delta(\cdot)$ and $G_I(\cdot)$ are defined by Equations (154) and (152a).

6

Periodogram and Other Direct Spectral Estimators

6.0 Introduction

Here we begin our study of nonparametric estimators of the spectral density function (SDF) for stationary processes with purely continuous spectra. The spectral properties for these processes are more readily seen from their SDFs than from their integrated spectra. Of course in this case the SDF and the integrated spectrum contain equivalent information, but the former is usually easier to interpret – just as probability density functions (rather than cumulative probability distribution functions) give a better visual indication of the distribution of probabilities for a random variable (RV). Our discussion will thus concentrate on estimation of the SDF.

We shall base our SDF estimates on the observed values of a time series. We consider a time series of sample size N as one realization of a contiguous portion $X_0, X_1, \ldots, X_{N-1}$ of a discrete parameter stationary process $\{X_t\}$ with SDF $S(\cdot)$ (see Section 4.5 for a discussion on sampling from a continuous parameter process). The motivation for nonparametric SDF estimators is the relationship between the SDF and the autocovariance sequence (ACVS) $\{s_\tau\}$ given by Equation (113d), namely,

$$S(f) = \Delta_t \sum_{\tau=-\infty}^{\infty} s_\tau e^{-i2\pi f \tau \Delta_t}, \tag{163}$$

where, as before, Δ_t is the sampling interval (the time that passes between when X_t and X_{t+1} are recorded). The key idea is to estimate $\{s_\tau\}$ in some fashion and then apply Equation (163) to estimate the SDF. For a real-valued stationary process,

$$s_\tau = \operatorname{cov}\{X_{t+\tau}, X_t\} = E\{(X_{t+\tau} - \mu)(X_t - \mu)\},$$

where $\mu = E\{X_t\}$ is the mean of the process. In Section 6.1, we consider an obvious estimator of μ (the sample mean), and then we look at two estimators of $\{s_\tau\}$ in Section 6.2. Plugging one of these ACVS estimators into the right-hand side of Equation (163) leads to an SDF estimator known as the *periodogram*, which is the subject of Section 6.3. While easy to compute, the periodogram can have two statistical deficiencies. First, its expected value is not guaranteed to be in reasonable agreement with the true SDF (when this is true, the periodogram is said to be a *biased estimator* of $S(\cdot)$). Second, it has a degree of variability that is unacceptable in many applications (statistically speaking, the periodogram is an *inconsistent estimator*). Section 6.4 looks at one scheme (tapering) for alleviating bias in the periodogram, leading to *direct spectral estimators* of $S(\cdot)$ (the periodogram is a special case of a direct spectral estimator). Section 6.5 considers a second bias-reduction scheme (prewhitening). The statistical properties of direct spectral estimators are the subject of Section 6.6. The remaining sections of this chapter give details about how to compute direct spectral estimates (Section 6.7), examples of direct spectral estimates for actual time series (6.8), comments on direct spectral estimators for complex-valued time series (6.9), a summary (6.10) and some exercises (6.11). (None of the estimators we consider in this chapter improves upon the inconsistency of the periodogram, but they form building blocks for nonparametric estimators that do address this statistical deficiency, as discussed in subsequent chapters.)

6.1 Estimation of the Mean

Let $\{X_t\}$ be a discrete parameter (second-order) stationary process with mean μ and ACVS $\{s_\tau\}$. We want to estimate μ and $\{s_\tau\}$ from a sample of N observations of our process, which we consider as observed values of the RVs $X_0, X_1, \ldots, X_{N-1}$.

A natural estimator of μ is just the sample mean of $X_0, X_1, \ldots, X_{N-1}$, given by the RV

$$\overline{X} \stackrel{\text{def}}{=} \frac{1}{N} \sum_{t=0}^{N-1} X_t.$$

Since

$$E\{\overline{X}\} = \frac{1}{N} \sum_{t=0}^{N-1} E\{X_t\} = \mu,$$

this estimator is unbiased. Moreover, under certain mild conditions (see section 6.1 of Fuller, 1996), $\overline{X}$ is a consistent estimator of μ, by which we mean that for every $\delta > 0$,

$$\lim_{N \to \infty} \mathbf{P}\left[|\overline{X} - \mu| > \delta\right] = 0, \tag{164a}$$

where $\mathbf{P}[\cdot]$ is the probability measure associated with the stationary process. Consistency follows directly from Chebyshev's inequality

$$\mathbf{P}\left[|\overline{X} - \mu| > \delta\right] \leq \frac{E\{(\overline{X} - \mu)^2\}}{\delta^2} = \frac{\text{var}\{\overline{X}\}}{\delta^2},$$

as soon as we can show that var $\{\overline{X}\} \to 0$ as $N \to \infty$. For the important special case of an absolutely summable ACVS,

$$\text{var}\{\overline{X}\} \approx \frac{S(0)}{N \Delta_t} \tag{164b}$$

for large N, where $S(0)$ is a finite number (if $S(0) = 0$, this approximation is not particularly useful). The above implies that the variance of the sample mean decreases to 0 as N goes to ∞. A proof of this fact is outlined in the first of the Comments and Extensions (C&Es) for this section. The effectiveness of the approximation for selected first-order moving average and autoregressive processes is explored in Exercises [6.1] and [9.2c].

Comments and Extensions to Section 6.1

[1] We wish to show that approximation (164b) is valid for large N for a stationary process with an ACVS that is absolutely summable; i.e., $\sum_{\tau=-\infty}^{\infty} |s_\tau| < \infty$.

▷ **Exercise [165]** Show that, for any stationary process (whether or not its ACVS is absolutely summable),

$$\text{var}\{\overline{X}\} = \frac{1}{N} \sum_{\tau=-(N-1)}^{N-1} \left(1 - \frac{|\tau|}{N}\right) s_\tau. \quad (165a) \triangleleft$$

As $N \to \infty$, we can reduce this expression to a compact form by appealing to the Cesàro summability theorem stated in Equation (79). In the present context, this theorem says that, if $\sum_{\tau=-(N-1)}^{N-1} s_\tau$ converges to a limit as $N \to \infty$ (this must be true since the ACVS is assumed to be absolutely summable), then $\sum_{\tau=-(N-1)}^{N-1} \left(1 - |\tau|/N\right) s_\tau$ converges to the same limit. We can thus conclude that

$$\lim_{N\to\infty} N \,\text{var}\{\overline{X}\} = \lim_{N\to\infty} \sum_{\tau=-(N-1)}^{N-1} \left(1 - \frac{|\tau|}{N}\right) s_\tau = \lim_{N\to\infty} \sum_{\tau=-(N-1)}^{N-1} s_\tau = \sum_{\tau=-\infty}^{\infty} s_\tau. \quad (165b)$$

The assumption of absolute summability implies that $\{X_t\}$ has a purely continuous spectrum with SDF given by

$$S(f) = \Delta_t \sum_{\tau=-\infty}^{\infty} s_\tau e^{-i2\pi f \tau \Delta_t} \quad \text{and hence} \quad S(0) = \Delta_t \sum_{\tau=-\infty}^{\infty} s_\tau$$

(this result follows from theorems 15.10 and 15.2, Champeney, 1987). Equation (165b) now becomes

$$\lim_{N\to\infty} N \,\text{var}\{\overline{X}\} = \frac{S(0)}{\Delta_t}; \quad \text{i.e.,} \quad \text{var}\{\overline{X}\} \approx \frac{S(0)}{N\Delta_t} \quad (165c)$$

for large N, which completes the proof (theorem 18.2.1 of Ibragimov and Linnik, 1971, has a proof that assumes only the existence of an SDF continuous at $f = 0$).

[2] There are stationary processes for which the sample mean is *not* a consistent estimator of the true mean of the process (see Exercise [4.9] for a simple example). Theorems that state conditions under which the sample mean converges in some sense to the true mean fall under the category of *ergodic theorems* (see section 16 of Yaglom, 1987a, for a nice discussion). We do not pay much attention to ergodicity in this book – our experience is that it is of limited use in practical applications (this point is expanded upon in section 1.1.6, Chilès and Delfiner, 2012).

[3] If we denote the covariance matrix of $\boldsymbol{X} \stackrel{\text{def}}{=} [X_0, X_1, \ldots, X_{N-1}]^T$ by $\boldsymbol{\Sigma}$ (see Section 2.4), and if we assume that $\boldsymbol{\Sigma}^{-1}$ exists, the best (in the sense of having minimum variance) linear unbiased estimator of μ is given by

$$\hat{\mu} = \frac{\boldsymbol{O}^T \boldsymbol{\Sigma}^{-1} \boldsymbol{X}}{\boldsymbol{O}^T \boldsymbol{\Sigma}^{-1} \boldsymbol{O}}, \quad \text{where} \quad \boldsymbol{O} \stackrel{\text{def}}{=} [\underbrace{1, 1, \ldots, 1}_{N \text{ of these}}]^T.$$

Note that this estimator requires knowledge of the ACVS out to lag $N-1$, which is rarely the case in situations where estimating the SDF is of interest. One way to quantify the gain in using $\hat{\mu}$ rather than $\overline{X}$ is via the ratio

$$e_N(\overline{X}, \hat{\mu}) \stackrel{\text{def}}{=} \frac{\text{var}\{\hat{\mu}\}}{\text{var}\{\overline{X}\}},$$

which is called the *relative efficiency* of $\overline{X}$ with respect to $\hat{\mu}$. Necessarily $e_N(\overline{X}, \hat{\mu}) \leq 1$, but typically $e_N(\overline{X}, \hat{\mu}) < 1$ (with white noise being a notable exception). A related concept is the *asymptotic relative efficiency*:

$$e(\overline{X}, \hat{\mu}) \stackrel{\text{def}}{=} \lim_{N\to\infty} e_N(\overline{X}, \hat{\mu}).$$

It can be argued that, if the SDF $S(\cdot)$ is piecewise continuous with no discontinuity at $f = 0$ and $0 < S(f) < \infty$ for all f, then $e(\overline{X}, \hat{\mu}) = 1$ (see chapter 7 of Grenander and Rosenblatt, 1984, and theorem 6.1.3 of Fuller, 1996). Because $e(\overline{X}, \hat{\mu})$ is unity, we say that $\overline{X}$ is an *asymptotically efficient* estimator of μ. Samarov and Taqqu (1988) discuss some interesting cases in which $\overline{X}$ is not asymptotically efficient; i.e., $e(\overline{X}, \hat{\mu}) < 1$. They consider the class of *fractionally differenced processes*, for which the SDF is given by

$$S(f) = C \left|\sin(\pi f \Delta_t)\right|^\alpha, \text{ where } C > 0 \text{ and } \alpha > -1.$$

Note that, for small $|f|$, we have $S(f) \propto |f|^\alpha$ approximately, so $S(f) \to 0$ as $f \to 0$ if $\alpha > 0$ while $S(f) \to \infty$ as $f \to 0$ if $-1 < \alpha < 0$. Samarov and Taqqu show that $0.98 < e(\overline{X}, \hat{\mu}) < 1$ when $-1 < \alpha < 0$; $e(\overline{X}, \hat{\mu})$ decreases from 1 to 0 as α increases from 0 to 1; and $e(\overline{X}, \hat{\mu}) = 0$ for all $\alpha \geq 1$. Hence the sample mean can be a poor estimator of μ when $S(\cdot)$ decreases to 0 rapidly as $f \to 0$.

6.2 Estimation of the Autocovariance Sequence

Because

$$s_\tau \stackrel{\text{def}}{=} E\left\{(X_{t+\tau} - \mu)(X_t - \mu)\right\},$$

a natural estimator for the ACVS is

$$\hat{s}_\tau^{(U)} \stackrel{\text{def}}{=} \frac{1}{N - |\tau|} \sum_{t=0}^{N-|\tau|-1} (X_{t+|\tau|} - \overline{X})(X_t - \overline{X}), \tag{166a}$$

$\tau = -(N-1), \ldots, -1, 0, 1, \ldots, N-1$; i.e., $|\tau| < N$. Note that $\hat{s}_{-\tau}^{(U)} = \hat{s}_\tau^{(U)}$, so this estimator mimics the symmetry property of $\{s_\tau\}$. For a sample of N observations there are no observations more than $N - 1$ lag units apart, so we cannot estimate s_τ for $|\tau| \geq N$ by an estimator with a form dictated by the right-hand side of Equation (166a). If we momentarily replace $\overline{X}$ by μ in Equation (166a), we see that

$$E\{\hat{s}_\tau^{(U)}\} = \frac{1}{N - |\tau|} \sum_{t=0}^{N-|\tau|-1} E\{(X_{t+|\tau|} - \mu)(X_t - \mu)\} = \frac{1}{N - |\tau|} \sum_{t=0}^{N-|\tau|-1} s_\tau = s_\tau$$

for all $|\tau| < N$. Thus $\hat{s}_\tau^{(U)}$ is an unbiased estimator of s_τ when μ is known. If μ is estimated by $\overline{X}$, it is a straightforward (but messy) exercise to derive $E\{\hat{s}_\tau^{(U)}\}$ exactly (Anderson, 1971, pp. 448–9). Such calculations can be used to argue that, when the process mean must be estimated from the sample mean, $\hat{s}_\tau^{(U)}$ is typically a biased estimator of s_τ (see Exercise [6.2]).

There is a second estimator of s_τ that is usually preferred to $\hat{s}_\tau^{(U)}$, namely,

$$\hat{s}_\tau^{(P)} \stackrel{\text{def}}{=} \frac{1}{N} \sum_{t=0}^{N-|\tau|-1} (X_{t+|\tau|} - \overline{X})(X_t - \overline{X}) \tag{166b}$$

(the rationale for the superscript on $\hat{s}_\tau^{(P)}$ is given in the next section). The only difference between the two estimators is the multiplicative factor in front of the summation, i.e., $1/(N-|\tau|)$ for $\hat{s}_\tau^{(U)}$, and $1/N$ for $\hat{s}_\tau^{(P)}$. If we again momentarily replace $\overline{X}$ by μ, it follows that

$$E\{\hat{s}_\tau^{(P)}\} = \frac{1}{N} \sum_{t=0}^{N-|\tau|-1} s_\tau = \left(1 - \frac{|\tau|}{N}\right) s_\tau, \tag{166c}$$

so that $\hat{s}_\tau^{(P)}$ is a biased estimator, and the magnitude of its bias increases as $|\tau|$ increases. In the literature $\hat{s}_\tau^{(U)}$ and $\hat{s}_\tau^{(P)}$ are frequently called, respectively, the "unbiased" and "biased"

6.2 Estimation of the Autocovariance Sequence

estimators of s_τ (even though, when $\overline{X}$ is used, both are typically biased, and, in fact, the magnitude of the bias in $\hat{s}_\tau^{(U)}$ can be *greater* than that in $\hat{s}_\tau^{(P)}$ – see Exercise [6.2] and the example shown in the left-hand column of Figure 168).

Why should we prefer the biased estimator $\hat{s}_\tau^{(P)}$ to the unbiased estimator $\hat{s}_\tau^{(U)}$? Here are some thoughts:

[1] For many stationary processes of practical interest, the mean square error (MSE) of $\hat{s}_\tau^{(P)}$ is smaller than that of $\hat{s}_\tau^{(U)}$; i.e.,

$$\text{MSE}\{\hat{s}_\tau^{(P)}\} \stackrel{\text{def}}{=} E\{(\hat{s}_\tau^{(P)} - s_\tau)^2\} < E\{(\hat{s}_\tau^{(U)} - s_\tau)^2\} = \text{MSE}\{\hat{s}_\tau^{(U)}\} \qquad (167a)$$

(Exercise [6.3] concerns a special case for which the above holds). Here is the basis for this statement. Since

$$\text{MSE} = \text{variance} + (\text{bias})^2,$$

where the bias for an estimator $\hat{\theta}$ of the parameter θ is defined as the quantity $E\{\hat{\theta}\} - \theta$, inequality (167a) evidently says that the variability in $\hat{s}_\tau^{(U)}$ is more harmful than the bias in $\hat{s}_\tau^{(P)}$. This effect is particularly marked for $|\tau|$ close to N, where the variability of $\hat{s}_\tau^{(U)}$ is inherently large. Consider, for example, $\hat{s}_{N-1}^{(U)}$ and $\hat{s}_{N-1}^{(P)}$ – these two quantities are, respectively,

$$(X_{N-1} - \overline{X})(X_0 - \overline{X}) \quad \text{and} \quad \frac{1}{N}(X_{N-1} - \overline{X})(X_0 - \overline{X}), \qquad (167b)$$

from which we can see that the variance of $\hat{s}_{N-1}^{(U)}$ is N^2 times the variance of $\hat{s}_{N-1}^{(P)}$. As concrete examples, Figure 168 shows the true ACVS for the second-order autoregressive (AR(2)) process defined by Equation (34) (upper left-hand plot) and for the AR(4) process defined by (35a) (upper right-hand plot) for lags 0 to 63 (see Exercise [9.6] for how to compute these ACVSs). Under the assumptions that these processes are Gaussian, that the sample mean is used to estimate the process mean and that the sample size N is 64, we can compute exactly the square bias, variance and MSE at various lags τ for both $\{\hat{s}_\tau^{(U)}\}$ and $\{\hat{s}_\tau^{(P)}\}$. These three quantities are shown beneath the ACVS plots for the two AR processes (light curves for $\{\hat{s}_\tau^{(U)}\}$, dark for $\{\hat{s}_\tau^{(P)}\}$). Except for the biased estimator in the AR(4) case at large lags, the curves for the variance and the MSE are practically the same – this supports the assertion that the variance is the dominant contributor to the MSE. For both cases, the MSE of $\hat{s}_\tau^{(P)}$ is strictly smaller than that of $\hat{s}_\tau^{(U)}$ for all nonzero τ (note also that, except at a few small lags, the square bias of $\{\hat{s}_\tau^{(P)}\}$ is smaller than that of $\{\hat{s}_\tau^{(U)}\}$ in the AR(2) case).

[2] If $\{X_t\}$ has a purely continuous spectrum, it follows from the Riemann–Lebesgue lemma (Titchmarsh, 1939, p. 403) that

$$s_\tau \to 0 \text{ as } |\tau| \to \infty$$

(the condition $\sum_{\tau=-\infty}^{\infty} |s_\tau| < \infty$ also implies the above, as does $\sum_{\tau=-\infty}^{\infty} s_\tau^2 < \infty$). It makes some sense – but not very much – to prefer an estimator of the sequence $\{s_\tau\}$ that decreases to 0 nicely as $|\tau|$ approaches $N-1$. Consider, again, Equation (167b), which shows that $\hat{s}_{N-1}^{(P)}$ is much closer to 0 than $\hat{s}_{N-1}^{(U)}$ is.

[3] We have noted that the ACVS of a stationary process must be positive semidefinite. Whereas the sequence $\{\hat{s}_\tau^{(P)}\}$ is always positive semidefinite, the sequence $\{\hat{s}_\tau^{(U)}\}$ need not be so (here we assume that both $\hat{s}_\tau^{(P)}$ and $\hat{s}_\tau^{(U)}$ are defined to be zero for all $|\tau| \geq N$).

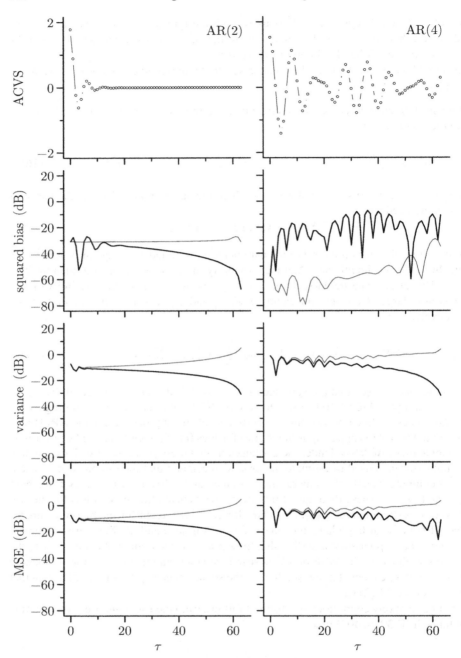

Figure 168. Comparison of square bias, variance and MSE of the unbiased and biased estimators of the ACVS for the AR(2) process of Equation (34) (left column) and the AR(4) process of Equation (35a) (right). The true ACVSs are shown (to lag 63) in the first row of plots. In the other three rows, the light and dark curves refer to, respectively, the unbiased and biased estimators. The sample size is 64; the process means are assumed unknown; both processes are Gaussian; and all curves are based on exact computations.

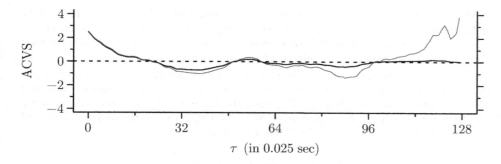

Figure 169. Comparison of unbiased and biased estimates of ACVS for wind speed time series shown in Figure 2(a). The light and dark curves show, respectively, the unbiased and biased estimates.

The following simple proof of the first claim is given in McLeod and Jiménez (1984, 1985). (The second claim is the subject of Exercise [6.6].) Let

$$d_k \stackrel{\text{def}}{=} x_k - \bar{x},$$

where the right-hand side is any observed value of $X_k - \overline{X}$. Let $\{\epsilon_t\}$ be a zero mean white noise process with variance $\sigma_\epsilon^2 = 1/N$, and construct the following Nth-order moving average process:

$$Y_t = \sum_{k=0}^{N-1} d_k \epsilon_{t-k}.$$

Then the *theoretical* (and hence necessarily positive semidefinite) ACVS of $\{Y_t\}$,

$$s_{Y,\tau} = \sigma_\epsilon^2 \sum_{t=0}^{N-|\tau|-1} d_{t+|\tau|} d_t = \frac{1}{N} \sum_{t=0}^{N-|\tau|-1} (x_{t+|\tau|} - \bar{x})(x_t - \bar{x}),$$

equals the *estimated* ACVS of $\{X_t\}$ using $\{\hat{s}_\tau^{(P)}\}$ (cf. Equation (32b)). Because this result holds for all realizations, we can conclude that the estimated ACVS $\{\hat{s}_\tau^{(P)}\}$ is positive semidefinite. One practical consequence of the possible failure of $\{\hat{s}_\tau^{(U)}\}$ to be positive semidefinite is "curious behavior of the estimates of the spectrum" (Jenkins and Watts, 1968, p. 183); for example, use of $\{\hat{s}_\tau^{(U)}\}$ can lead to estimates of the SDF – a nonnegative quantity, recall – that are negative for some frequencies.

Figure 169 illustrates the just-discussed points [2] and [3]. The light and dark curves show, respectively, $\hat{s}_\tau^{(U)}$ and $\hat{s}_\tau^{(P)}$ versus the lag τ. Note, first, that $\{\hat{s}_\tau^{(P)}\}$ damps to zero gracefully while $\{\hat{s}_\tau^{(U)}\}$ does not and, second, that $\hat{s}_\tau^{(U)} > \hat{s}_0^{(U)}$ for some large values of τ – this cannot happen for a valid theoretical ACVS.

For these reasons, statisticians prefer $\hat{s}_\tau^{(P)}$ over $\hat{s}_\tau^{(U)}$ as the estimator of s_τ. Hereinafter we shall use it as our standard "raw" estimator of the ACVS.

Comments and Extensions to Section 6.2

[1] A comparison of Equations (166b) and (101h) shows that $\hat{s}_\tau^{(P)}$ might correspond to an autocorrelation of the finite sequence $X_0 - \overline{X}, X_1 - \overline{X}, \ldots, X_{N-1} - \overline{X}$, but with modifications needed for the beginning and end of the sequence. To within a multiplicative constant, this correspondence can be made exact by considering the autocorrelation of the zero padded sequence

$$X_0 - \overline{X}, X_1 - \overline{X}, \ldots, X_{N-1} - \overline{X}, \underbrace{0, \ldots, 0}_{N-1 \text{ zeros}}.$$

The fact that the discrete Fourier transform (DFT) of an autocorrelation is the square magnitude of the DFT of the sequence involved in the autocorrelation can be exploited to good use: we can compute $\hat{s}_\tau^{(P)}$ for $\tau = 0, \ldots, N-1$ using substantially fewer numerical operations than would be needed if we were to use (166b) directly. The details are the subject of Exercise [6.7d].

[2] We have shown that the sequence $\{\hat{s}_\tau^{(P)}\}$ is necessarily positive semidefinite. Since the $\hat{s}_\tau^{(P)}$ terms are a sequence of RVs, this statement means that each realization of these RVs (a sequence of numbers) is positive semidefinite. Newton (1988, p. 165) gives a stronger result: a realization of the sequence $\{\hat{s}_\tau^{(P)}\}$ as defined by Equation (166b) is positive definite if and only if the corresponding realizations of $X_0, \ldots, X_{N-1}$ are not all identical (see also McLeod and Jiménez, 1985).

6.3 A Naive Spectral Estimator – the Periodogram

Suppose the discrete parameter real-valued stationary process $\{X_t\}$ with zero mean has a purely continuous spectrum with SDF $S(\cdot)$. For the remainder of this chapter, we assume that the relationship

$$S(f) = \Delta_t \sum_{\tau=-\infty}^{\infty} s_\tau e^{-i2\pi f \tau \Delta_t} \qquad (170a)$$

holds and that $S(\cdot)$ is in fact continuous for all f. If $\sum |s_\tau| < \infty$, theorem 15.10 of Champeney (1987) says that $S(\cdot)$ must be continuous everywhere, so the assumption of continuity is restrictive but not overly so.

Given a time series that can be regarded as a realization of $X_0, X_1, \ldots, X_{N-1}$, our problem is to estimate $S(\cdot)$. Now it is possible to estimate s_τ for $|\tau| = 0, 1, \ldots, N-1$, but *not* for $|\tau| \geq N$, by

$$\hat{s}_\tau^{(P)} = \frac{1}{N} \sum_{t=0}^{N-|\tau|-1} X_{t+|\tau|} X_t \qquad (170b)$$

(this uses the assumption that the process mean is known to be 0). It thus seems natural to replace s_τ in Equation (170a) by $\hat{s}_\tau^{(P)}$ for $|\tau| \leq N-1$ and to truncate the summation over τ at the points $\pm(N-1)$ – this amounts to *defining* $\hat{s}_\tau^{(P)} = 0$ for $|\tau| \geq N$. These substitutions lead to an estimator known as the *periodogram* (even though it is a function of frequency and not period):

$$\hat{S}^{(P)}(f) \stackrel{\text{def}}{=} \Delta_t \sum_{\tau=-(N-1)}^{N-1} \hat{s}_\tau^{(P)} e^{-i2\pi f \tau \Delta_t}. \qquad (170c)$$

Note that the periodogram mimics an SDF in that it is an even function and is periodic with a period of $2f_\mathcal{N}$, where $f_\mathcal{N} \stackrel{\text{def}}{=} 1/(2\Delta_t)$ is the Nyquist frequency (i.e., $\hat{S}^{(P)}(-f) = \hat{S}^{(P)}(f)$ and $\hat{S}^{(P)}(f + 2nf_\mathcal{N}) = \hat{S}^{(P)}(f)$ for all f and all $n \in \mathbb{Z}$).

The following exercise points out a second way we can obtain the periodogram.

▷ **Exercise [170]** Show that

$$\hat{S}^{(P)}(f) = \frac{\Delta_t}{N} \left| \sum_{t=0}^{N-1} X_t e^{-i2\pi f t \Delta_t} \right|^2. \qquad (170d) \triangleleft$$

This equation makes it clear that, as is true for theoretical SDFs, the periodogram is nonnegative. If we compare the summation above to Equation (74a), we can regard it as being proportional to the Fourier transform of the infinite sequence $\ldots, 0, X_0, X_1, \ldots, X_{N-1}, 0, \ldots$, i.e., one for which X_t is defined to be zero for $t < 0$ and $t \geq N$. Thus the periodogram can be taken to be either the Fourier transform of the biased estimator of the ACVS (with $\hat{s}_\tau^{(P)}$

6.3 A Naive Spectral Estimator – the Periodogram

defined to be zero for $|\tau| \geq N$) or, to within a multiplicative factor, the square modulus of the Fourier transform of $X_0, \ldots, X_{N-1}$ padded before and after with an infinite number of zeros.

Equation (170d) can be linked to the DFT of the finite sequence $X_0, X_1, \ldots, X_{N-1}$, thus providing a convenient means of computing the periodogram. To see this connection, note that this DFT is given by

$$\mathcal{X}_k = \Delta_t \sum_{t=0}^{N-1} X_t e^{-i2\pi kt/N}, \quad k = 0, 1, \ldots, N-1 \tag{171a}$$

(see Equation (91b)), noting that $\mathcal{X}_k$ is associated with Fourier frequency $f_k \stackrel{\text{def}}{=} k/(N\Delta_t)$. Evaluation of (170d) at $f = f_k$ yields

$$\hat{S}^{(P)}(f_k) = \frac{\Delta_t}{N} \left| \sum_{t=0}^{N-1} X_t e^{-i2\pi f_k t \Delta_t} \right|^2 = \frac{\Delta_t}{N} \left| \sum_{t=0}^{N-1} X_t e^{-i2\pi kt/N} \right|^2 = \frac{|\mathcal{X}_k|^2}{N\Delta_t}, \tag{171b}$$

which gives us the periodogram over the grid of Fourier frequencies. Because we are typically interested in plotting the periodogram over the frequency interval $[0, f_{\mathcal{N}}]$, we need only consider f_k such that $0 \leq f_k \leq f_{\mathcal{N}}$. These are the Fourier frequencies indexed by $k = 0, 1, \ldots, \lfloor N/2 \rfloor$, where $\lfloor N/2 \rfloor$ is the greatest integer less than or equal to $N/2$. (C&E [1] discusses how to compute the periodogram over a grid of frequencies finer than what is given by the Fourier frequencies.)

Just as the ACVS and the SDF form a Fourier transform pair, so do $\{\hat{s}_\tau^{(P)}\}$ and $\hat{S}^{(P)}(\cdot)$:

$$\{\hat{s}_\tau^{(P)}\} \longleftrightarrow \hat{S}^{(P)}(\cdot)$$

(the above is the rationale for the superscript on $\hat{s}_\tau^{(P)}$). The inverse Fourier transform corresponding to Equation (170c) says that

$$\hat{s}_\tau^{(P)} = \int_{-f_{\mathcal{N}}}^{f_{\mathcal{N}}} \hat{S}^{(P)}(f) e^{i2\pi f \tau \Delta_t} \, df, \quad \tau \in \mathbb{Z} \tag{171c}$$

(see Equation (75a), recalling that $f_{\mathcal{N}} = 1/(2\Delta_t)$). In particular we have

$$\hat{s}_0^{(P)} = \int_{-f_{\mathcal{N}}}^{f_{\mathcal{N}}} \hat{S}^{(P)}(f) \, df, \quad \text{which mimics the relationship} \quad s_0 = \int_{-f_{\mathcal{N}}}^{f_{\mathcal{N}}} S(f) \, df. \tag{171d}$$

Note that the integral in Equation (171c) must evaluate to zero for all $|\tau| \geq N$ since $\hat{s}_\tau^{(P)}$ is by definition zero at those indices.

The relationship $\{\hat{s}_\tau^{(P)}\} \longleftrightarrow \hat{S}^{(P)}(\cdot)$ assumes that $\{\hat{s}_\tau^{(P)}\}$ is an infinite sequence with $\hat{s}_\tau^{(P)} \stackrel{\text{def}}{=} 0$ for $|\tau| \geq N$. If we consider the finite sequence $\{\hat{s}_\tau^{(P)} : \tau = -(N-1), \ldots, N-1\}$ of $2N-1$ variables, we can argue from Equations (100g) and (101b) that

$$\{\hat{s}_\tau^{(P)} : \tau = -(N-1), \ldots, N-1\} \longleftrightarrow \{\hat{S}^{(P)}(f_k') : k = -(N-1), \ldots, N-1\} \tag{171e}$$

with $f_k' = k/[(2N-1)\Delta_t]$. Hence, in addition to (171c), which makes use of continuous frequencies, the inverse DFT of Equation (101a) says that

$$\hat{s}_\tau^{(P)} = \frac{1}{(2N-1)\Delta_t} \sum_{k=-(N-1)}^{N-1} \hat{S}^{(P)}(f_k') e^{i2\pi f_k' \tau \Delta_t}, \quad \tau = -(N-1), \ldots, N-1, \tag{171f}$$

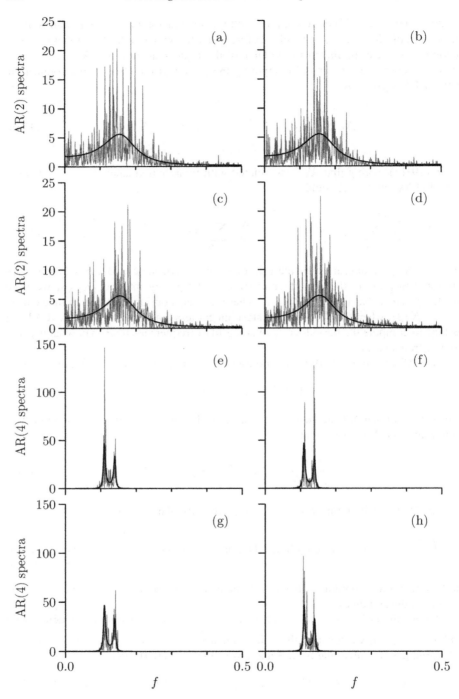

Figure 172. Periodograms (rough-looking light curves) for eight autoregressive time series of length $N = 1024$ shown in Figure 34, along with true SDFs (smooth thick curves). The periodograms are computed over the grid of Fourier frequencies $f_k = k/N$ using Equation (171b) with $\Delta_t = 1$ and are plotted on a linear/linear scale.

6.3 A Naive Spectral Estimator – the Periodogram

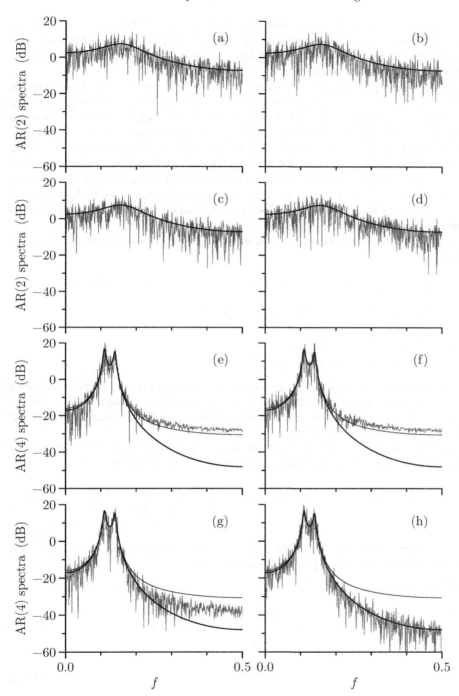

Figure 173. As in Figure 172, but now plotted on a decibel/linear scale. The additional smooth thin curves in plots (e) to (h) show the expected value of the periodogram.

which uses discrete frequencies. Note that the right-hand side of Equation (171f) can be regarded as a Riemann sum approximation to the integral in Equation (171c) based upon an evenly spaced grid of frequencies with a spacing of $1/[(2N-1)\Delta_t]$; however, in this case the "approximation" is in fact exact!

Eight examples of periodograms are shown in Figure 172, one for each of the eight AR time series of length $N = 1024$ shown in Figure 34. The periodograms (choppy-looking light curves) are evaluated over the grid of Fourier frequencies (as per Equation (171b)) and are compared with the true SDFs for the AR processes (smooth dark curves, as per Equations (145a), (33), (34) and (35a)). Figure 173 shows the same periodograms and true SDFs, but now on a decibel scale; i.e., we plot $10 \log_{10}(\hat{S}^{(\text{P})}(f))$ rather than $\hat{S}^{(\text{P})}(f)$ versus f. We will use these examples to illustrate certain undesirable properties of the periodogram, which led Broersen (2002) to refer to it as a quick and dirty estimator of the SDF, and caused Tukey (1980) to comment that "more lives have been lost looking at the raw periodogram than by any other action involving time series!"

Let us now investigate the sampling properties of $\hat{S}^{(\text{P})}(\cdot)$. We will be chiefly concerned at first with

$$E\{\hat{S}^{(\text{P})}(f)\}, \quad \text{var}\{\hat{S}^{(\text{P})}(f)\} \quad \text{and} \quad \text{cov}\{\hat{S}^{(\text{P})}(f'), \hat{S}^{(\text{P})}(f)\}, \quad f' \neq f.$$

If $\hat{S}^{(\text{P})}(\cdot)$ were an ideal estimator of $S(\cdot)$, we would have

[1] $E\{\hat{S}^{(\text{P})}(f)\} \approx S(f)$ for all f,
[2] $\text{var}\{\hat{S}^{(\text{P})}(f)\} \to 0$ as $N \to \infty$ and
[3] $\text{cov}\{\hat{S}^{(\text{P})}(f'), \hat{S}^{(\text{P})}(f)\} \approx 0$ for $f' \neq f$.

We shall find, however, that, while [1] is a good approximation for some processes and some values of f, it can be very poor in other cases; [2] is blatantly false when $S(f) > 0$; and [3] holds if f' and f are certain distinct frequencies, namely, the Fourier frequencies f_k, and if the bias in the periodogram is negligible.

We start by examining $E\{\hat{S}^{(\text{P})}(f)\}$ and determining how close it is to $S(f)$, the quantity it is supposed to estimate. From Equations (170d) and (166c) it follows that

$$E\{\hat{S}^{(\text{P})}(f)\} = \Delta_t \sum_{\tau=-(N-1)}^{N-1} E\{\hat{s}_\tau^{(\text{P})}\} e^{-i2\pi f \tau \Delta_t} = \Delta_t \sum_{\tau=-(N-1)}^{N-1} \left(1 - \frac{|\tau|}{N}\right) s_\tau e^{-i2\pi f \tau \Delta_t}. \tag{174a}$$

This is a convenient formula for computing $E\{\hat{S}^{(\text{P})}(f)\}$ for various stationary processes with known ACVS, but the following expression gives more insight.

▷ **Exercise [174]** Show that

$$E\{\hat{S}^{(\text{P})}(f)\} = \int_{-f_\mathcal{N}}^{f_\mathcal{N}} \mathcal{F}(f - f') S(f') \, df', \tag{174b}$$

where

$$\mathcal{F}(f) \stackrel{\text{def}}{=} \Delta_t N \mathcal{D}_N^2(f \Delta_t); \tag{174c}$$

here $\mathcal{D}_N(\cdot)$ is Dirichlet's kernel, which is defined in Equation (17c) and is discussed in Section 3.8. ◁

The function $\mathcal{F}(\cdot)$ is called *Fejér's kernel*. We note that, for $0 < |f| \leq f_\mathcal{N}$,

$$\mathcal{F}(f) = \frac{\Delta_t \sin^2(N\pi f \Delta_t)}{N \sin^2(\pi f \Delta_t)}, \tag{174d}$$

6.3 A Naive Spectral Estimator – the Periodogram

which follows directly from Equations (174c) and (17c). The true SDF is thus convolved with Fejér's kernel to give $E\{\hat{S}^{(\mathrm{P})}(\cdot)\}$. Because it dictates how well we can view $S(\cdot)$, Fejér's kernel is referred to as the *spectral window* for $\hat{S}^{(\mathrm{P})}(\cdot)$.

To understand the implications of Equation (174b), we need to investigate the properties of Fejér's kernel.

[1] $\mathcal{F}(\cdot)$ is even, periodic with a period of $2f_\mathcal{N}$ and continuous (these follow directly from (174c) and the properties of Dirichlet's kernel);

[2] it follows from Exercise [1.2c] that

$$N^2 \mathcal{D}_N^2(f\,\Delta_\mathrm{t}) = \left|\sum_{t=0}^{N-1} \mathrm{e}^{\mathrm{i}2\pi f t\,\Delta_\mathrm{t}}\right|^2, \text{ so we have } \mathcal{F}(f) = \frac{\Delta_\mathrm{t}}{N} \sum_{t=0}^{N-1}\sum_{u=0}^{N-1} \mathrm{e}^{\mathrm{i}2\pi f(t-u)\,\Delta_\mathrm{t}}; \quad (175\mathrm{a})$$

a term-by-term integration of the right-hand side shows that

$$\int_{-f_\mathcal{N}}^{f_\mathcal{N}} \mathcal{F}(f)\,\mathrm{d}f = 1; \quad (175\mathrm{b})$$

[3] for all sample sizes N, $\mathcal{F}(0) = N\,\Delta_\mathrm{t}$ (this follows readily from Equations (174c) and (17c));

[4] for $N > 1$, $f \in [-f_\mathcal{N}, f_\mathcal{N}]$ and $f \neq 0$, $\mathcal{F}(f) < \mathcal{F}(0)$ (a calculus problem);

[5] for any fixed $f \in [-f_\mathcal{N}, f_\mathcal{N}]$ except for $f = 0$, $\mathcal{F}(f) \to 0$ as $N \to \infty$ (the N in the denominator of (174d) dominates); and

[6] for any integer $k \neq 0$ such that $f_k = k/(N\,\Delta_\mathrm{t})$ satisfies $|f_k| \leq f_\mathcal{N}$, we have $\mathcal{F}(f_k) = 0$ (because $\sin(k\pi) = 0$).

The quantities $10\log_{10}(\mathcal{F}(f))$ and $\mathcal{F}(f)$ are plotted versus f in, respectively, the left- and right-hand columns of Figure 176 for $\Delta_\mathrm{t} = 1$ and $N = 4$, 16 and 64. We see that $\mathcal{F}(\cdot)$ is indeed an even function and consists of a broad central peak (also called the *central lobe*) and $N-2$ sidelobes, which decrease in magnitude as $|f|$ increases. The sidelobes just before and after the central lobe are only about 13 dB below that of the central lobe. This pattern holds for all N, but note that the location of these sidelobes gets closer and closer to zero as N increases (this is consistent with property [5]).

From properties [2], [3] and [5], it follows that, as $N \to \infty$, Fejér's kernel acts as a Dirac delta function with an infinite spike at $f = 0$. Since $S(\cdot)$ is assumed to be continuous, we have from Equation (174b)

$$\lim_{N \to \infty} E\{\hat{S}^{(\mathrm{P})}(f)\} = S(f); \quad (175\mathrm{c})$$

i.e., for all f, $\hat{S}^{(\mathrm{P})}(f)$ is an asymptotically unbiased estimator of $S(f)$. In addition, the following theorem gives us a rate of decrease for the bias under a certain regularity condition (Brillinger, 1981a, p. 123): if $\{X_t\}$ is a stationary process with ACVS $\{s_\tau\}$ such that

$$\sum_{\tau=-\infty}^{\infty} |\tau s_\tau| < \infty, \text{ then } E\{\hat{S}^{(\mathrm{P})}(f)\} = S(f) + O\left(\frac{1}{N}\right).$$

The regularity condition essentially implies that the ACVS dies down to zero quickly; alternatively, it implies that $S(\cdot)$ is a smooth function (for example, the condition holds if $S(\cdot)$ has a continuous first derivative). When this is true, the bias in the periodogram decreases at the rate of $1/N$ – but note carefully that nothing is said about the *absolute* magnitude of the bias.

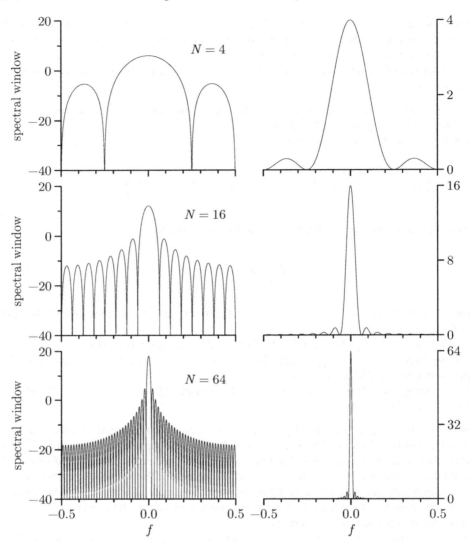

Figure 176. Fejér's kernel for sample sizes $N = 4$, 16 and 64 on a decibel scale (left-hand plots) and a linear scale (right-hand). This kernel is the spectral window for the periodogram.

With a stronger regularity condition (say, $\sum_{\tau=-\infty}^{\infty} |\tau^2 s_\tau| < \infty$), the rate of decrease is even faster (see Brillinger, 1981b, for details).

While there are some stationary processes for which the bias in the periodogram is negligible for small N (see, for example, the discussion surrounding Figure 177), unfortunately asymptotic unbiasedness does not imply that, for an arbitrary process, the bias in $\hat{S}^{(P)}(f)$ is necessarily small for any particular N. A quote from Thomson (1982) is appropriate here:

> ... for processes with spectra typical of those encountered in engineering, the sample size must be extraordinarily large for the periodogram to be reasonably unbiased. While it is not clear what sample size, if any, gives reasonably valid results, in my experience periodogram estimates computed from 1.2 million data points on the WT4 waveguide project were too badly biased to be useful. The best that could be said for them is that they were so obviously incorrect as not to be dangerously misleading. In other applications

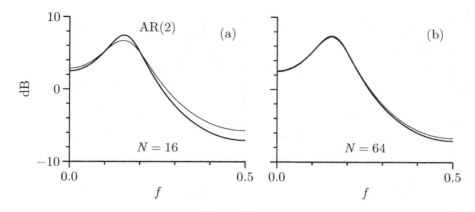

Figure 177. Bias properties of periodogram for a process with low dynamic range. The thick curve in each plot is the true SDF $S(\cdot)$ for the AR(2) process described by Equation (34). The thin curves are $E\{\hat{S}^{(P)}(\cdot)\}$ for sample sizes N = 16 and 64 plotted versus frequency. The vertical axis is in decibels – either $10 \log_{10}(S(f))$ or $10 \log_{10}(\hat{S}^{(P)}(f))$.

where less is known about the process, such errors may not be so obvious.

As examples of processes for which bias in the periodogram is – and is not – of concern for sample sizes in the range of N = 16 to 1024, let us consider three specific stationary processes (each with $\Delta_t = 1$ so that $f_\mathcal{N} = 1/2$). The first process is white noise with a variance of 1; the second is the AR(2) process defined by Equation (34); and the third, the AR(4) process defined by Equation (35a). The SDFs for the AR(2) and AR(4) processes are shown by the thick curves in, respectively, Figures 177 and 178. A useful crude characterization of these three processes is in terms of their *dynamic range*, which we define by the ratio

$$10 \log_{10}\left(\frac{\max_f S(f)}{\min_f S(f)}\right). \tag{177a}$$

The dynamic range of a white noise process is 0 dB; the AR(2) and AR(4) processes have dynamic ranges of about 14 dB and 65 dB, respectively (the latter is typical of the spectra for many geophysical processes).

The SDF for the white noise process is just 1 for all $f \in [-1/2, 1/2]$. It follows from Equations (174b), (175b) and the fact that $\mathcal{F}(\cdot)$ is periodic with a period of unity in this case that

$$E\{\hat{S}^{(P)}(f)\} = \int_{-1/2}^{1/2} \mathcal{F}(f - f')\,df' = 1, \tag{177b}$$

so the periodogram is unbiased in the case of white noise.

Figure 177 shows $E\{\hat{S}^{(P)}(f)\}$ as a function of f for N = 16 and 64 for the AR(2) process. For both plots in the figure the thin curve shows $E\{\hat{S}^{(P)}(f)\}$, while the thick curve is the true SDF. For N = 16 we see that $E\{\hat{S}^{(P)}(f)\}$ is within 2 dB of $S(f)$ for all f, while the two functions nearly coincide for N = 64. Thus the periodogram is already largely unbiased by N = 64 for this process with a rather small dynamic range. This unbiasedness is evident in the top two rows of Figures 172 and 173: although highly erratic, the periodograms for the four AR(2) time series of length N = 1024 generally follow the true SDF.

The situation is quite different for the AR(4) process, which has the largest dynamic range of the three. For N = 64 (Figure 178(a)) there are biases of more than 30 dB (i.e., three orders of magnitude) for f close to 1/2. Even for N = 1024 (Figure 178(d)), $E\{\hat{S}^{(P)}(f)\}$

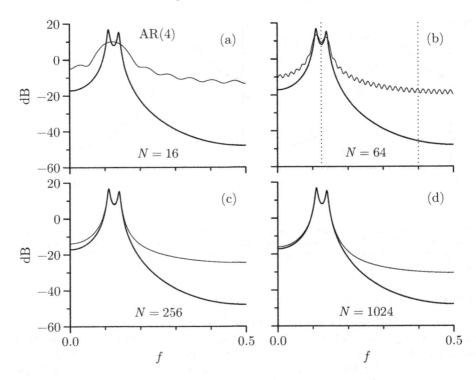

Figure 178. Bias properties of periodogram for a process with high dynamic range.

and $S(f)$ differ by almost 20 dB (two orders of magnitude) for values of f for which $S(f)$ is small. This bias is apparent in the bottom two rows of Figure 173, where we see that the actual degree of bias can vary considerably from one AR(4) time series to the next. While $E\{\hat{S}^{(\mathrm{P})}(f)\}$ should reflect an average over many realizations, it need not be a good match for any particular realization.

The bias in the periodogram for processes with high dynamic range can be attributed to the sidelobes of Fejér's kernel. Equation (174b) says that $E\{\hat{S}^{(\mathrm{P})}(f)\}$ is the integral of the product of that kernel centered at f with the true SDF. If Fejér's kernel consisted of just a central lobe, the product would involve just frequencies close to $S(f)$; because of the sidelobes, the product also involves distant frequencies. For the AR(4) process, $E\{\hat{S}^{(\mathrm{P})}(f)\}$ is badly biased at those frequencies for which $S(f)$ is small compared with the rest of the SDF. This transfer of power from one region of the SDF to another via the kernel of the convolution is known as *leakage*.

As an illustration of how leakage arises in the AR(4) example, let's consider the case $N = 64$ at the frequencies $f = 1/8$ and $f = 2/5$ (marked by dotted lines in Figure 178(b)). The first row of Figure 180 shows both $S(f)$ (thick curves) and $\mathcal{F}(\frac{1}{8} - f)$ (thin) versus $f \in [-1/2, 1/2]$ with the frequency $f = 1/8$ indicated by dotted lines (plots (a) and (b) depict the same content, but the vertical scale in (a) is expressed in decibels, whereas it is linear in (b) – the remaining rows of this figure have a similar layout). The second row shows the product $S(f)\mathcal{F}(\frac{1}{8} - f)$, which is concentrated about $f = 1/8$. Equation (174b) says that the integral of this product is equal to $E\{\hat{S}^{(\mathrm{P})}(1/8)\}$, and Figure 178(b) shows that $E\{\hat{S}^{(\mathrm{P})}(1/8)\}$ agrees fairly well with $S(1/8)$ (note that the agreement becomes even better as we increase the sample size to $N = 256$ or 1024). This good agreement is due the fact that the frequency $f = 1/8$ falls in the high-power "twin-peaks" portion of the SDF. By contrast, the bottom

6.3 A Naive Spectral Estimator – the Periodogram

two rows consider instead the frequency $f = 2/5$, which falls in the low-power portion of the AR(4) SDF. The alignment of $\mathcal{F}(\frac{2}{5} - f)$ with $S(f)$ is such that their product (shown in plots (g) and (h) of Figure 180) is not concentrated about $f = 2/5$, but rather is dominated by the interaction of the sidelobes of Fejér's kernel with the twin-peaks part of the SDF. The integral of $S(f)\mathcal{F}(\frac{2}{5} - f)$ over $f \in [-1/2, 1/2]$ is thus not a good approximation to $S(2/5)$, as is indicated by Figure 178(b).

The thin curve in Figure 178(d) showing $10 \log_{10}(E\{\hat{S}^{(P)}(f)\})$ versus f is replicated in plots (e) to (h) of Figure 173 so that we can compare the periodograms for the four AR(4) time series with their expected values. We see that, over the portions of the SDF subject to leakage (i.e., where $E\{\hat{S}^{(P)}(f)\}$ differs substantially from $S(f)$), the actual periodogram can be systematically above or below its expected value (in fact the periodogram in Figure 173(h) is in better agreement with the true SDF than with its expected value). Thus different realizations of the AR(4) process can exhibit a degree of leakage that differs from what $E\{\hat{S}^{(P)}(\cdot)\}$ predicts. C&E [2] offers an explanation as to why the four AR(4) time series have varying amounts of leakage based upon how well the beginning and end of each series match up.

There are two common techniques for lessening the bias in the periodogram – tapering and prewhitening. The former modifies the kernel in the convolution in Equation (174b); the latter preprocesses a time series to reduce the dynamic range of the SDF to be estimated. These techniques are discussed in the next two sections.

Comments and Extensions to Section 6.3

[1] Equation (171b) says that we can compute the periodogram over the grid of Fourier frequencies by suitably scaling the square magnitude of the DFT of the time series. The periodogram, however, is defined for all $f \in \mathbb{R}$, and, as we shall see, there are good reasons for wanting to evaluate the periodogram over finer grids of frequencies. The key to doing so is zero padding, as the following exercise indicates.

▷ **Exercise [179]** For any integer $N' > N$, show that the DFT of the zero padded series

$$X_0, X_1, \ldots, X_{N-1}, \underbrace{0, \ldots, 0}_{N'-N \text{ zeros}}$$

can be used to give the periodogram over a grid of frequencies dictated by $f'_k \stackrel{\text{def}}{=} k/(N' \Delta_t)$, which has a finer spacing than that given by the Fourier frequencies f_k. Show additionally that the placement of the zeros doesn't matter in the sense that, if N_1 and N_2 are any nonnegative integers such that $N_1 + N_2 = N' - N$, then the DFT of

$$\underbrace{0, \ldots, 0}_{N_1 \text{ zeros}}, X_0, X_1, \ldots, X_{N-1}, \underbrace{0, \ldots, 0}_{N_2 \text{ zeros}}$$

can again be used to evaluate the periodogram over the frequencies f'_k. ◁

Thus, by padding a time series with enough zeros, we can use the DFT to compute its periodogram over an arbitrarily dense grid of frequencies.

Equation (171e) hints at one reason for wanting to evaluate the periodogram over a grid of frequencies other than the Fourier frequencies: if we know the periodogram over the grid defined by $f'_k = k/[(2N-1)\Delta_t]$, then we can use an inverse DFT to compute the biased estimator of the ACVS (see Section 6.7 for more details). We can readily compute the periodogram over f'_k by zero padding the time series with $N - 1$ zeros; however, since the advent of the fast Fourier transform (FFT) algorithm, sample sizes of digitally recorded time series are often chosen to be a power of two, say, $N = 2^J$, where J is a positive integer. If we pad such a series with $N - 1$ zeros, the length $N' = 2^{J+1} - 1$ of the padded series is not a power of two, and we cannot take advantage of some simple implementations of

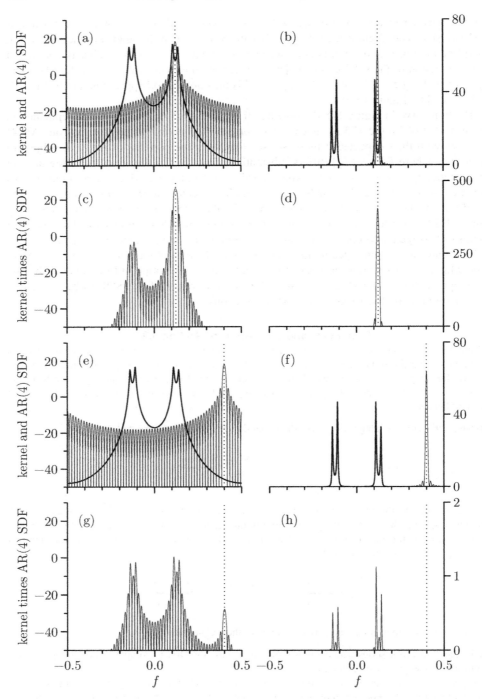

Figure 180. Illustration of how leakage arises in the periodogram (left-hand plots have vertical scales in decibels; right-hand plots show the same content, but on linear scales). The first and third rows show the AR(4) SDF (thick curves) along with Fejér's kernel (thin) with its central lobe located at either $f = 1/8$ (first row) or $f = 2/5$ (third). The second and fourth rows show the product of the Fejér's kernel and the SDF in the rows just above.

6.3 A Naive Spectral Estimator – the Periodogram

the FFT algorithm. For this reason, it is convenient to add another zero so that N' becomes 2^{J+1}, again a power of two. In effect we are evaluating the periodogram over the grid defined by $\tilde{f}_k = k/(2N\,\Delta_t)$, which has a spacing twice as fine as the one given by the Fourier frequencies $f_k = k/(N\,\Delta_t)$. Using an argument similar to the one leading to Equation (171e), we find that

$$\{\hat{s}_\tau^{(P)} : \tau = -(N-1),\ldots,N\} \longleftrightarrow \{\hat{S}^{(P)}(\tilde{f}_k) : k = -(N-1),\ldots,N\}, \tag{181}$$

where we recall that $\hat{s}_N^{(P)} = 0$ by definition. Hence grids defined by either $k/[(2N-1)\,\Delta_t]$ or $k/(2N\,\Delta_t)$ allow computation of $\{\hat{s}_\tau^{(P)}\}$ via an inverse DFT. Exercise [6.7b] notes that if we use a grid at least as finely spaced as $1/[(2N-1)\,\Delta_t]$, then we can recover $\{\hat{s}_\tau^{(P)}\}$ from $\{\hat{S}^{(P)}(f'_k)\}$; on the other hand, we cannot fully recover $\{\hat{s}_\tau^{(P)}\}$ if we use a grid with any coarser spacing – this includes the one defined by the Fourier frequencies f_k. (We note in C&E [4] for Section 7.1 that the common practice of evaluating the periodogram over just the Fourier frequencies can lead to a curiosity that disappears if the $\tilde{f}_k$ grid is used instead.)

For comparison, Figure 182 shows the same eight periodograms as depicted in 173, but now evaluated over the "twice-as-fine" grid $\tilde{f}_k$ rather than the Fourier frequencies. While most of the periodograms are qualitatively similar, the ones in plots (e), (g) and (h) look different in the high-frequency portion of the AR(4) SDF, which is subject to leakage. The amount of leakage is greater in 182(g) and (h) than in 173(g) and (h), but appears to be less in 182(e) than in 173(e). Thus examining the periodogram at just the Fourier frequencies might not give a good picture of how it varies over $f \in [0, f_\mathcal{N}]$.

[2] Leakage in the periodogram has sometimes been attributed to a mismatch between the beginning and end of a time series (see, for example, Fougere, 1985b). This argument is based on the interpretation that the DFT treats a time series of length N as it were a periodic sequence with a period of N (see Section 3.11). Credence for this interpretation is supplied by Exercise [6.12], which notes that the periodograms for a time series and all of its $N-1$ circular shifts are identical at the Fourier frequencies (Equation (45a) displays the circular shifts for a time series of length $N = 4$). The reciprocity relationships between the time and frequency domains suggest that a sharp feature in one domain manifests itself as a diffuse feature in the other. Because of the implied periodicity, a mismatch between the beginning and end of a time series is a sharp feature that manifests itself in the periodogram as leakage (a diffuse feature).

To explore this interpretation of leakage, the circles in Figure 183a show the ends X_{1021}, X_{1022}, X_{1023} and beginnings X_0, X_1, X_2 of the four AR(4) time series plotted in Figure 34. We can use the definition for the AR(4) process in Equation (35a) to forecast where we would expect X_{1024} to be for each series, given the observed values $X_{1020},\ldots,X_{1023}$ (this is done by letting $t = 1024$ and setting ϵ_{1024} to its expected value of zero). These forecasts are indicated by pluses in the figure. Given the presumed periodicity, the magnitude of the difference between the forecast of X_{1024} and the observed X_0 is a rough measure of the mismatch between the beginning and end of each series. The mismatch is largest for series (e) and smallest for series (h). A comparison with Figures 173 and 182 shows that leakage in the four periodograms does seem to increase as the mismatch increases. Figure 183b looks at this relationship further. Here we generated 100 additional AR(4) time series of length $N = 1024$. For each series we computed its periodogram over the $\tilde{f}_k = k/(2N)$ grid of frequencies and then formed the average of $\hat{S}^{(P)}(\tilde{f}_k)$ over $0.4 < \tilde{f}_k \leq 0.5$ (for this range of frequencies, the true SDF is relatively flat, and the periodogram is prone to leakage). The mean levels for the true SDF and the expected value of the periodogram over $f \in [0.4, 0.5]$ are -47.2 dB and -30.3 dB (indicated by the dotted and dashed lines in the figure). The circles show the 100 averages from the periodograms versus the absolute values of the difference between X_0 and the prediction of X_{1024}. The curve is a smoothing of the circles using a locally weighted regression (Cleveland, 1979). We see that, while the degree of leakage does tend to increase as the absolute prediction error increases, a small error does not necessarily imply the absence of leakage.

As we note in the next section, tapering essentially forces both the beginning and the end of a time series toward zero and hence can be thought of as one way of compensating for a potential end-point mismatch. Two other methods that achieve a similar effect are based on end-point matching and use of reflecting boundary conditions. End point matching consists of subtracting off a line prior to computing

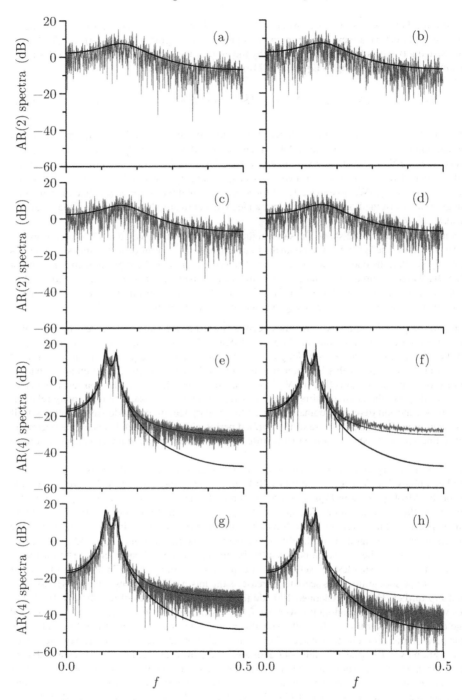

Figure 182. As in Figure 173, but now with the periodograms evaluated over a grid of frequencies twice as fine as the Fourier frequencies, i.e., $\tilde{f}_k = k/(2N)$ rather than $f_k = k/N$.

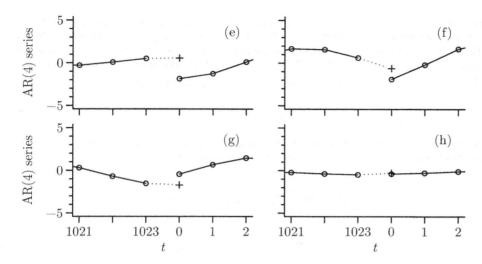

Figure 183a. Detailed look at the beginning and end of each of the four AR(4) time series shown in the bottom of Figure 34 (circles), along with a forecast of X_{1024} (plotted as a plus at $t = 0$). The distance between the plus and circle at $t = 0$ is a measure of the mismatch between the extremes of the time series. This measure matches up well with the degree of leakage evident in the corresponding periodogram (see Figures 173 and 182).

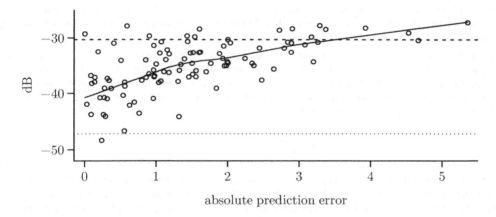

Figure 183b. Demonstration that degree of observed leakage in AR(4) periodograms is influenced by the mismatch between the beginning and end of the time series.

the periodogram so that the ends of the adjusted series $\{\widetilde{X}_t\}$ are both zero:

$$\widetilde{X}_t = X_t - X_0 - \frac{X_{N-1} - X_0}{N-1} t, \quad t = 0, \ldots, N-1. \tag{183a}$$

While this method adequately corrects the bias in the periodogram for some time series (Fougere, 1985b), computer experiments indicate that it fails to do so for others, including rather spectacularly in our AR(4) example – see Exercise [6.14] (for these series, tapering accomplishes more than just endpoint matching because it forces a smooth transition between the beginning and end of the series). The idea behind reflecting boundary conditions is to use either the series of length $2N$

$$X_0, X_1, X_2, \ldots, X_{N-2}, X_{N-1}, X_{N-1}, X_{N-2}, \ldots, X_2, X_1, X_0 \tag{183b}$$

or the series of length $2N - 2$

$$X_0, X_1, X_2, \ldots, X_{N-2}, X_{N-1}, X_{N-2}, \ldots, X_2, X_1 \tag{183c}$$

in place of the original time series. By construction, both these extended series have end points that match up well and hence are a better match for the periodic treatment imposed on a time series by the DFT. The use of reflecting boundary conditions is related to a method for estimating the SDF based upon the discrete cosine transform of type II (DCT–II) – see C&E [7] for Section 6.6 for further discussion.

[3] If $E\{X_t\} = \mu \neq 0$ as we have assumed so far, we can replace X_t by either

$$X_t - \mu \text{ or } X_t - \overline{X} \tag{184}$$

accordingly as μ is either known or unknown. Three interesting points about this centering of $\{X_t\}$ are the subject of Exercise [6.15]. First, for *any* value C independent of t,

$$\tilde{S}^{(\mathrm{P})}(f_k) \stackrel{\text{def}}{=} \frac{\Delta_t}{N} \left| \sum_{t=0}^{N-1} (X_t - C) e^{-i 2\pi f_k t \Delta_t} \right|^2$$

$$= \frac{\Delta_t}{N} \left| \sum_{t=0}^{N-1} X_t e^{-i 2\pi f_k t \Delta_t} \right|^2 = \hat{S}^{(\mathrm{P})}(f_k),$$

where $f_k = k/(N \Delta_t)$ for integers k such that $0 < f_k \leq f_{\mathcal{N}}$; i.e., the frequency f_k is one of the nonzero Fourier frequencies. Second, $\tilde{S}^{(\mathrm{P})}(0) = 0$ when $C = \overline{X}$, the sample mean. Since the periodogram is a continuous function of frequency, this latter fact implies that the periodogram of a centered time series can be a badly biased estimate of $S(f)$ for f close to 0 (say, $|f| < f_1 = 1/(N \Delta_t)$). Third, failure to center a time series that has a nonzero mean can lead to serious bias in the periodogram at non-Fourier frequencies, with the bias being dictated by Fejér's kernel.

[4] In older works on spectral analysis, the spectrum at frequency f of a segment $X_0, X_1, \ldots, X_{N-1}$ of a stationary process is sometimes *defined* – at least implicitly – to be $E\{\hat{S}^{(\mathrm{P})}(f)\}$. This matches our definition in the case of white noise, but the plots in Figure 178 shows that the two definitions can be quite different in general. In fact, if we translate the definition (116b) for the SDF discussed in Section 4.2 into the notation of this chapter, we have

$$S(f) = \lim_{N \to \infty} E\{\hat{S}^{(\mathrm{P})}(f)\}.$$

Thus an operational difficulty with using $E\{\hat{S}^{(\mathrm{P})}(f)\}$ as the definition for the spectrum is that the spectrum can depend critically on the sample size. This is undesirable, if for no other reason than the difficulty it introduces in meaningfully comparing the spectra from data collected in different experiments concerning the same physical phenomenon.

[5] We have chosen to label our sample of size N as $X_0, X_1, \ldots, X_{N-1}$; a second common convention is $X_1, X_2, \ldots, X_N$. The periodogram is the same for both conventions: if we let $X'_t \stackrel{\text{def}}{=} X_{t-1}$ for $t = 1, \ldots, N$, then

$$\left| \sum_{t=1}^{N} X'_t e^{-i 2\pi f t \Delta_t} \right|^2 = \left| \sum_{t=0}^{N-1} X_t e^{-i 2\pi f t \Delta_t} \right|^2$$

even though in general

$$\sum_{t=1}^{N} X'_t e^{-i 2\pi f t \Delta_t} \neq \sum_{t=0}^{N-1} X_t e^{-i 2\pi f t \Delta_t}$$

due to differences in phase.

[6] Suppose we let Σ denote the covariance matrix of $\left[X_0, X_1, \ldots, X_{N-1} \right]^T$, i.e., a vector whose elements are a segment of size N from a stationary process with SDF $S(\cdot)$ (see Section 2.4). The dynamic range of $S(\cdot)$ has an interesting relationship to an accepted measure of the ill-conditioning of

6.4 Bias Reduction – Tapering

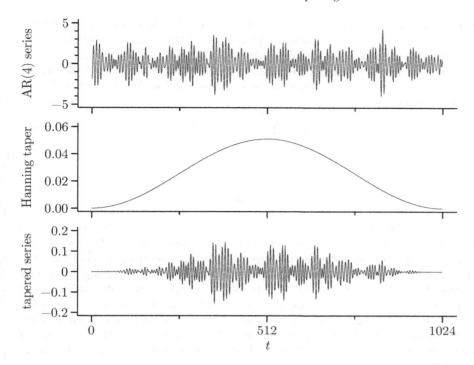

Figure 185. AR(4) time series $\{X_t\}$ from Figure 34(e), a Hanning data taper $\{h_t\}$ and their product of $\{h_t X_t\}$ (the Hanning taper is given by Equation (189b)).

the matrix Σ (this is important to know if we wish to compute its inverse). Let λ_N and ν_N represent the largest and smallest eigenvalues of Σ. This measure – called the condition number – is

$$d_N \stackrel{\text{def}}{=} \frac{\lambda_N}{\nu_N}.$$

Grenander and Szegő (1984) show that

$$\lim_{N \to \infty} d_N = \frac{\max_f S(f)}{\min_f S(f)}.$$

Hence the dynamic range of $S(\cdot)$ is approximately equal to $10 \log_{10}(d_N)$, and, conversely, an approximation to the condition number d_N can be obtained from the dynamic range.

[7] In Section 6.2 we used an argument due to McLeod and Jiménez (1984, 1985) to show that any realization of $\{\hat{s}_\tau^{(\mathrm{P})}\}$ must be positive semidefinite. In this section we have shown that $\{\hat{s}_\tau^{(\mathrm{P})}\} \longleftrightarrow \hat{S}^{(\mathrm{P})}(\cdot)$. Given that any realization of $\hat{S}^{(\mathrm{P})}(\cdot)$ satisfies the properties of an SDF (a real-valued nonnegative function of frequency that is even and integrates to a finite value), it follows from Wold's theorem that the corresponding realization of $\{\hat{s}_\tau^{(\mathrm{P})}\}$ is the ACVS for some stationary process and hence must be positive semidefinite (see the discussion at the end of Section 4.3).

6.4 Bias Reduction – Tapering

As noted in the previous section, much of the bias in the periodogram can be attributed to the sidelobes of Fejér's kernel $\mathcal{F}(\cdot)$. If these sidelobes were substantially smaller, there would be considerably less bias in $\hat{S}^{(\mathrm{P})}(\cdot)$ due to the leakage of power from one portion of the SDF to another. Tapering is a technique that effectively reduces the sidelobes associated with Fejér's kernel (with some trade-offs). It was introduced in the context of spectral analysis by Blackman and Tukey (1958, p. 93).

Suppose that we have a time series that can be regarded as a realization of a portion $X_0, X_1, \ldots, X_{N-1}$ of a zero mean stationary process with SDF $S(\cdot)$. As illustrated in Figure 185, tapering a time series consists of forming the product $h_t X_t$ for $t = 0, \ldots, N-1$, where $\{h_t\}$ is a suitable sequence of real-valued constants called a *data taper* (other names in the literature are *data window*, *temporal window*, *linear window*, *linear taper*, *fader* and *shading sequence*). Let

$$J(f) \stackrel{\text{def}}{=} \Delta_t^{1/2} \sum_{t=0}^{N-1} h_t X_t e^{-i2\pi ft\Delta_t}, \tag{186a}$$

and let

$$\hat{S}^{(\text{D})}(f) \stackrel{\text{def}}{=} |J(f)|^2 = \Delta_t \left| \sum_{t=0}^{N-1} h_t X_t e^{-i2\pi ft\Delta_t} \right|^2. \tag{186b}$$

Spectral estimators of the above form are referred to as *direct spectral estimators*, hence the superscript on $\hat{S}^{(\text{D})}(\cdot)$; they are also called *modified periodograms*. Figure 187 shows eight examples of direct spectral estimates that correspond to the eight periodograms in Figure 182 (each of these estimates makes use of the data taper shown in Figure 185). In comparison to the periodograms, the estimates for the AR(2) series still show highly erratic fluctuations about the true SDF, with variations ranging approximately 20 dB (i.e., two orders of magnitude); however, the estimates for the four AR(4) series are in better agreement with their true SDF.

To understand why the direct spectral estimator seems to offer some improvement over the periodogram in the AR(4) case, we need to consider its first-moment properties, for which we need the following result.

▷ **Exercise [186]** Using the spectral representation for $\{X_t\}$ given by Equation (113b) with $\mu = 0$, show that

$$J(f) = \frac{1}{\Delta_t^{1/2}} \int_{-f_N}^{f_N} H(f - f') \, dZ(f'), \tag{186c}$$

where $\{h_t\} \longleftrightarrow H(\cdot)$ under the assumption that $\{h_t\}$ is an infinite sequence with $h_t = 0$ for $t < 0$ and $t \geq N$; i.e.,

$$H(f) = \Delta_t \sum_{t=0}^{N-1} h_t e^{-i2\pi ft\Delta_t}. \tag{186d}$$ ◁

Because the process $\{Z(f)\}$ has orthogonal increments, we have (by an argument identical to that used in Exercise [4.5b])

$$E\{\hat{S}^{(\text{D})}(f)\} = \int_{-f_N}^{f_N} \mathcal{H}(f - f') S(f') \, df', \tag{186e}$$

where

$$\mathcal{H}(f) \stackrel{\text{def}}{=} \frac{1}{\Delta_t} |H(f)|^2 = \Delta_t \left| \sum_{t=0}^{N-1} h_t e^{-i2\pi ft\Delta_t} \right|^2. \tag{186f}$$

For computational purposes, we note that

$$E\{\hat{S}^{(\text{D})}(f)\} = \Delta_t \sum_{\tau=-(N-1)}^{N-1} \left(\sum_{t=0}^{N-|\tau|-1} h_{t+|\tau|} h_t \right) s_\tau e^{-i2\pi f\tau\Delta_t}, \tag{186g}$$

6.4 Bias Reduction – Tapering

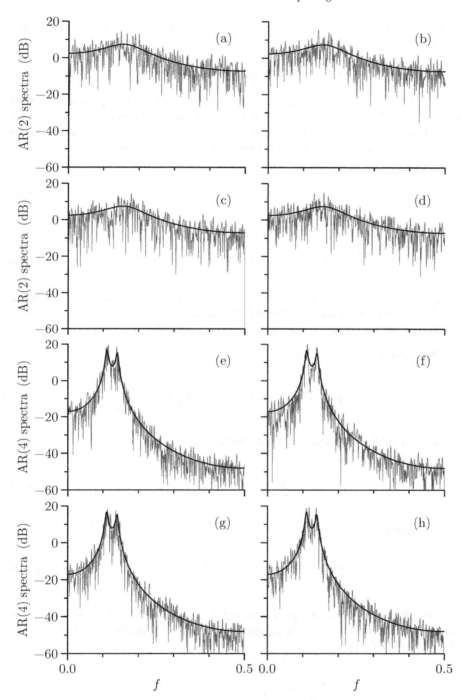

Figure 187. As in Figure 182, but now showing direct spectral estimates rather than periodograms (the direct spectral estimates use the Hanning data taper shown in Figure 185 and defined in Equation (189b)).

(see Exercise [6.16]). If $h_t = 1/\sqrt{N}$ for $0 \le t \le N-1$ (the so-called rectangular or default taper), we have

$$\hat{S}^{(D)}(f) = \hat{S}^{(P)}(f) \text{ and } \mathcal{H}(f) = \mathcal{F}(f).$$

That is, $\hat{S}^{(D)}(\cdot)$ reverts to the periodogram; $\mathcal{H}(\cdot)$ becomes Fejér's kernel; and Equation (186g) reduces to Equation (174a) because the inner sum in the former reduces to $1 - |\tau|/N$ when $h_t = 1/\sqrt{N}$. Just as we referred to $\mathcal{F}(\cdot)$ as the spectral window for the periodogram, so do we call $\mathcal{H}(\cdot)$ the spectral window for the direct spectral estimator. We note that

$$\hat{S}^{(D)}(f) = \Delta_t \sum_{\tau=-(N-1)}^{N-1} \hat{s}_\tau^{(D)} e^{-i2\pi f \tau \Delta_t}, \tag{188a}$$

where

$$\hat{s}_\tau^{(D)} \stackrel{\text{def}}{=} \begin{cases} \sum_{t=0}^{N-|\tau|-1} h_{t+|\tau|} X_{t+|\tau|} h_t X_t, & |\tau| \le N-1; \\ 0, & |\tau| \ge N \end{cases} \tag{188b}$$

(see Exercise [6.17]). Since $\{\hat{s}_\tau^{(D)}\} \longleftrightarrow \hat{S}^{(D)}(\cdot)$, we see that the sequence $\{\hat{s}_\tau^{(D)}\}$ is the estimator of the ACVS corresponding to the SDF estimator $\hat{S}^{(D)}(\cdot)$.

The following exercise points to the need to properly normalize a data taper.

▷ **Exercise [188]** Show that

$$\int_{-f_\mathcal{N}}^{f_\mathcal{N}} E\{\hat{S}^{(D)}(f)\} \, df = s_0 \sum_{t=0}^{N-1} h_t^2. \qquad \triangleleft$$

Thus, if we normalize $\{h_t\}$ such that it has unit energy, i.e.,

$$\sum_{t=0}^{N-1} h_t^2 = 1, \tag{188c}$$

we have the desirable property

$$\int_{-f_\mathcal{N}}^{f_\mathcal{N}} E\{\hat{S}^{(D)}(f)\} \, df = s_0 = \int_{-f_\mathcal{N}}^{f_\mathcal{N}} S(f) \, df \tag{188d}$$

(note that (188c) holds for the rectangular data taper $h_t = 1/\sqrt{N}$). Because $\{h_t\} \longleftrightarrow H(\cdot)$ and since $\mathcal{H}(\cdot)$ is proportional to the square modulus of $H(\cdot)$, it follows from Parseval's theorem (Equation (75b)) that

$$\sum_{t=0}^{N-1} h_t^2 = \int_{-f_\mathcal{N}}^{f_\mathcal{N}} \mathcal{H}(f) \, df.$$

The normalization (188c) is thus equivalent to

$$\int_{-f_\mathcal{N}}^{f_\mathcal{N}} \mathcal{H}(f) \, df = 1, \tag{188e}$$

which becomes Equation (175b) in the case of a rectangular data taper. For a white noise process with variance σ^2 and SDF $S(f) = \sigma^2 \Delta_t$, it follows from (186e) that

$$E\{\hat{S}^{(D)}(f)\} = \sigma^2 \Delta_t \int_{-f_\mathcal{N}}^{f_\mathcal{N}} \mathcal{H}(f - f') \, df' = \sigma^2 \Delta_t \int_{-f_\mathcal{N}}^{f_\mathcal{N}} \mathcal{H}(f) \, df = \sigma^2 \Delta_t$$

(the second integral above is the same as the first since $\mathcal{H}(\cdot)$ is periodic with period $2f_\mathcal{N}$). Thus direct spectral estimators that satisfy (188c) are unbiased in the case of white noise – a fact that we have already noted for the special case of a rectangular data taper.

The key idea behind tapering is to select $\{h_t\}$ so that $\mathcal{H}(\cdot)$ has much smaller sidelobes than $\mathcal{F}(\cdot)$. A discontinuity in a sequence of numbers means that there will be ripples in its Fourier transform (see Section 3.8). Since $h_t = 0$ for $t < 0$ and $t \geq N$, we can have small sidelobes in $\mathcal{H}(\cdot)$ by making h_t decay to 0 in a smoother fashion than the abrupt change in the rectangular data taper. That this indeed can be done is the subject of Figure 190, which shows eight data tapers for sample size $N = 64$, and in Figure 191, which shows their associated spectral windows. For the sake of comparison, the first taper (Figure 190(a)) is just the rectangular data taper, so its spectral window is Fejér's kernel. The taper below it is the so-called 20% cosine taper. The $p \times 100\%$ cosine taper is defined by

$$h_t = \begin{cases} C\left[1 - \cos\left(\frac{\pi(t+1)}{\lfloor pN/2 \rfloor + 1}\right)\right], & 0 \leq t \leq \lfloor \frac{pN}{2} \rfloor - 1; \\ 2C, & \lfloor \frac{pN}{2} \rfloor \leq t \leq N - \lfloor \frac{pN}{2} \rfloor - 1; \\ h_{N-t-1}, & N - \lfloor \frac{pN}{2} \rfloor \leq t \leq N - 1, \end{cases} \quad (189a)$$

where C is a normalizing constant that forces $\sum_{t=0}^{N-1} h_t^2 = 1$ (when $p = 0$, the above reduces to $h_t = 1/\sqrt{N}$, i.e., the rectangular data taper). Thus the 20% cosine taper is like the rectangular taper, except the first 10% and last 10% of the points have been replaced by portions of a raised cosine. Figures 190(c) and (d) show the cases $p = 0.5$ and $p = 1$, i.e., a 50% cosine taper and a 100% cosine taper, the latter being one formulation of the *Hanning data taper*:

$$h_t = \begin{cases} C\left[1 - \cos\left(\frac{\pi(t+1)}{\lfloor N/2 \rfloor + 1}\right)\right], & 0 \leq t \leq \lfloor \frac{N+1}{2} \rfloor - 1; \\ h_{N-t-1}, & \lfloor \frac{N+1}{2} \rfloor \leq t \leq N - 1, \end{cases} \text{ and } C = \begin{cases} \frac{\sqrt{2}}{\sqrt{(3N-2)}}, & N \text{ even}; \\ \frac{\sqrt{2}}{\sqrt{(3N+3)}}, & N \text{ odd} \end{cases}$$
(189b)

(see Exercise [6.18a] for an alternative definition). The effect of the 20%, 50% and 100% cosine tapers is to "force" the tapered series $\{h_t X_t\}$ smoothly toward zero (cf. Figure 185). The left-hand column of Figure 191 shows the spectral windows for these tapers. All have significantly smaller sidelobes than those of Fejér's kernel (upper left-hand plot); however, the widths of the central lobes are all larger than that of Fejér's kernel. We have suppressed the sidelobes at the expense of wider central lobes.

Another important family of data tapers is based upon the zeroth-order Slepian sequences (also called discrete prolate spheroidal sequences). These sequences were introduced in Section 3.10 (see the discussion following Equation (88b)) and also play a prominent role in Chapter 8. Recall that the zeroth-order Slepian sequence with parameters N and W is the sequence of length N that has the largest possible concentration of its energy in the frequency interval $[-W, W]$ as measured by the ratio

$$\beta^2(W) \stackrel{\text{def}}{=} \int_{-W}^{W} |H(f)|^2 \, df \bigg/ \int_{-1/2}^{1/2} |H(f)|^2 \, df$$

(cf. Equation (85b)). When $\Delta_t = 1$ (as is assumed in the above equation), the spectral window $\mathcal{H}(\cdot)$ is equal to $|H(\cdot)|^2$ and hence is highly concentrated over $f \in [-W, W]$, which contains

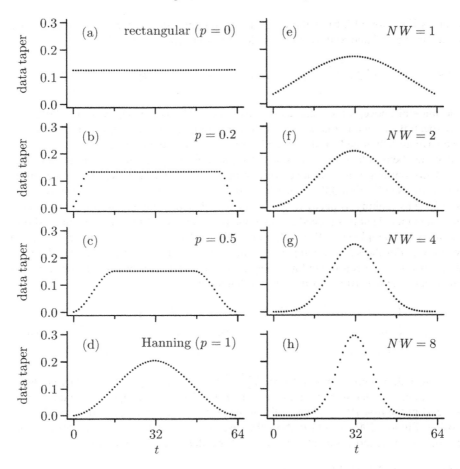

Figure 190. Eight data tapers for use with a time series of length $N = 64$. The left-hand column shows four $p \times 100\%$ cosine tapers ($p = 0$, which is the same as the default or rectangular taper, $p = 0.2$, 0.5 and 1.0, which is the Hanning taper). The right-hand column has approximations to four zeroth-order Slepian data tapers ($NW = 1, 2, 4$ and 8).

the central lobe. The concentration of energy in the region of the central lobe implies that the sidelobes of $\mathcal{H}(\cdot)$ over $|f| \in [W, 1/2]$ should be small. There is a natural trade-off between the width $2W$ of the region containing the central lobe and the energy in the sidelobes – as the former increases, the latter decreases. This point is illustrated in Figure 191. The right-hand column shows spectral windows for the four data tapers plotted in the corresponding panels of Figure 190. These are *approximations* to the zeroth-order Slepian data tapers for $N = 64$ and $NW = 1, 2, 4$ and 8 (Equation (196b) gives the approximation, which is discussed in C&E [5]). Note that the energy in the sidelobes of the spectral windows decreases markedly as W (and hence the width of the central lobe) increases. It is particularly interesting to compare the spectral window for the $NW = 1$ Slepian data taper with that for the periodogram (Figure 191(a)). The central lobe widths of these two windows are nearly the same, yet the sidelobes for the window associated with the Slepian data taper are about 10 dB below those of Fejér's kernel.

Figure 193 illustrates how the trade-off between the central lobe width and the magnitude of the sidelobes affects the first moment of direct spectral estimators. The thin curve in each plot shows $E\{\hat{S}^{(D)}(f)\}$ as a function of f for the AR(4) process with $N = 64$. The thick curves are the true SDF. The estimators in the eight plots use the data tapers depicted

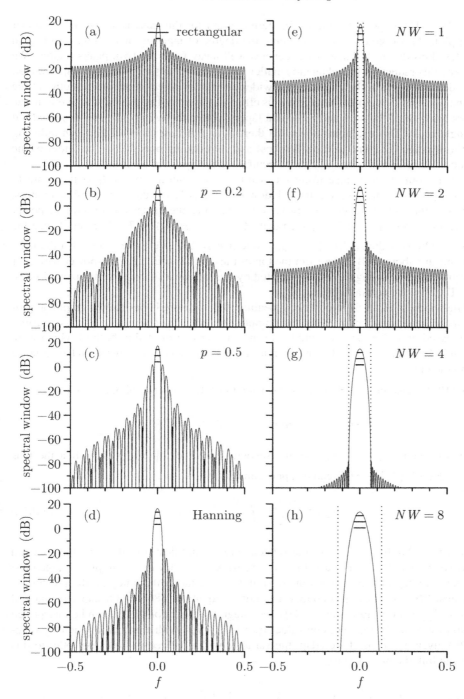

Figure 191. Spectral windows $\mathcal{H}(f)$ on a decibel scale (i.e., $10 \log_{10}(\mathcal{H}(f))$) versus f in cycles per unit time for the eight data tapers shown on Figure 190. The dotted lines in the right-hand column indicate the locations of $f = \pm W$ for the Slepian tapers with W set via $NW = 1, 2, 4$ and 8. The three horizontal solid lines just below the central lobes are bandwidth measures based upon, from the top on down, the half-power, variance and autocorrelation widths.

in Figure 190. The expected value of the periodogram for this case is shown in the upper left-hand plot. As we go down each column, the degree of tapering increasing, resulting in less and less bias due to leakage at frequencies well away from the twin-peaks region. We can attribute this reduction in leakage to smaller sidelobes in the spectral windows; however, smaller sidelobes come at the expense of a wider central lobe, which acts to smooth out the SDF in the twin-peaks region. The deleterious effect of this smoothing is most apparent in the estimators using the $NW = 4$ and 8 Slepian tapers, where the twin peaks are smoothed out into a single peak. This example shows that there is an inherent trade-off between reducing bias due to leakage (i.e., due to the sidelobes) and introducing bias due to the width of the central lobe. We refer to the bias due to the central lobe as *spectral window bias*.

We need some way of measuring the width of the central lobe of $\mathcal{H}(\cdot)$ if we are to compare the potential effect of spectral window bias for different direct spectral estimators. A standard engineering practice is to measure the width as the distance between half-power points, i.e., those points f_0 and $-f_0$ such that $\mathcal{H}(-f_0) = \mathcal{H}(f_0) = \mathcal{H}(0)/2$, and hence the half-power width is $\text{width}_{\text{hp}}\{\mathcal{H}(\cdot)\} \stackrel{\text{def}}{=} 2f_0$. Note that $10 \log_{10}(1/2) \approx -3$ so that the half-power points are about 3 dB down from the top of the central lobe. The half-power width for each spectral window in Figure 191 is depicted by the top-most of the three horizontal lines, drawn 3 dB below the central lobe.

While $\text{width}_{\text{hp}}\{\mathcal{H}(\cdot)\}$ is intuitively appealing, this measure is basically a graphical procedure that in general is inconvenient to compute. As potential alternatives, let us consider measures based upon the variance width and the autocorrelation width, both of which were introduced in Chapter 3. The following exercise points out a connection between the variance width and the bias in $\hat{S}^{(\text{D})}(f)$.

▷ **Exercise [192]** Assume that $S(\cdot)$ can be expanded in a Taylor series about f such that

$$S(f + \phi) \approx S(f) + \phi S'(f) + \frac{\phi^2}{2} S''(f),$$

where $S'(\cdot)$ and $S''(\cdot)$ are the first and second derivatives of $S(\cdot)$. Show that the bias $b(f) \stackrel{\text{def}}{=} E\{\hat{S}^{(\text{D})}(f)\} - S(f)$ can be approximated by

$$b(f) \approx \frac{S''(f)}{2} \int_{-f_\mathcal{N}}^{f_\mathcal{N}} \phi^2 \mathcal{H}(\phi) \, d\phi. \tag{192a}$$ ◁

The bias in $\hat{S}^{(\text{D})}(\cdot)$ at frequency f is thus related to two quantities: first, the curvature (second derivative) of $S(\cdot)$ at f – a large curvature due to a peak or a valley in $S(\cdot)$ implies a large bias; and second, an integral that depends on $\mathcal{H}(\cdot)$ alone – the larger this integral, the greater the bias. Because $\mathcal{H}(\cdot)$ is necessarily nonnegative due to Equation (186f) and because its integral over $[-f_\mathcal{N}, f_\mathcal{N}]$ is unity (Equation (188e)), the integral in Equation (192a) can be interpreted as the variance of an RV that takes on values over $[-f_\mathcal{N}, f_\mathcal{N}]$ with a probability density function (PDF) given by $\mathcal{H}(\cdot)$. If we adapt the variance width defined in Equation (60c) to this PDF, we can link the bias $b(f)$ to

$$\text{width}_{\text{v}}\{\mathcal{H}(\cdot)\} = \left(12 \int_{-f_\mathcal{N}}^{f_\mathcal{N}} \phi^2 \mathcal{H}(\phi) \, d\phi\right)^{1/2} = \frac{1}{\Delta_{\text{t}}} \left(1 + \frac{12}{\pi^2 \Delta_{\text{t}}} \sum_{\tau=1}^{N-1} \frac{(-1)^\tau}{\tau^2} h \star h_\tau\right)^{1/2}, \tag{192b}$$

where

$$h \star h_\tau = \Delta_{\text{t}} \sum_{t=0}^{N-|\tau|-1} h_{t+|\tau|} h_t \tag{192c}$$

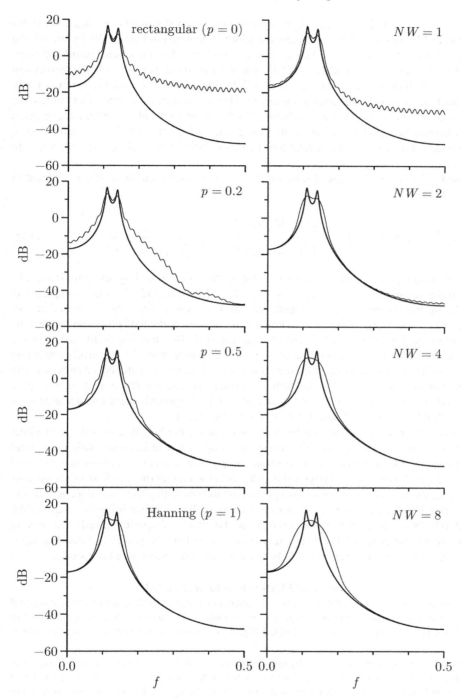

Figure 193. Comparison of true SDF $S(\cdot)$ and $E\{\hat{S}^{(D)}(\cdot)\}$. The thick curves are the true SDF of the AR(4) process of Equation (35a). The thin curves show $E\{\hat{S}^{(D)}(\cdot)\}$ for eight direct spectral estimators using the tapers of length $N = 64$ in Figure 190. Both $S(\cdot)$ and $E\{\hat{S}^{(D)}(\cdot)\}$ are displayed in decibels (i.e., $10\log_{10}(S(f))$ versus f, where f is in cycles/Δ_t with $\Delta_t = 1$).

is the autocorrelation of $\{h_t\}$ as per Equation (99f) (the second equality in Equation (192b) is the subject of Exercise [6.19]). The variance width is depicted in Figure 191 by the middle of the three horizontal lines drawn below each central lobe. For the cases considered here, width$_v\{\mathcal{H}(\cdot)\}$ is larger than width$_{hp}\{\mathcal{H}(\cdot)\}$, but the two measures are in better agreement as the degree of tapering increases, i.e., as p or NW increases. The measures are markedly different for Fejér's kernel (the rectangular taper), whose prominent sidelobes act to inflate the variance measure above what would be dictated by the central lobe alone. With the exception of Fejér's kernel, the approximation to the bias given by Equation (192a) is a reflection of spectral window bias (due to the central lobe) and has little to do with leakage (bias due to sidelobes).

A third measure of the central lobe of $\mathcal{H}(\cdot)$ is its autocorrelation width, as defined in Equation (100f):

$$B_{\mathcal{H}} \stackrel{\text{def}}{=} \text{width}_a\{\mathcal{H}(\cdot)\} = \frac{\left(\int_{-f_\mathcal{N}}^{f_\mathcal{N}} \mathcal{H}(f)\,df\right)^2}{\int_{-f_\mathcal{N}}^{f_\mathcal{N}} \mathcal{H}^2(f)\,df} = \frac{\Delta_t}{\sum_{\tau=-(N-1)}^{N-1}(h \star h_\tau)^2}, \tag{194}$$

where the second equality holds because $\mathcal{H}(\cdot)$ integrates to unity and by applying Parseval's theorem to the relationship $\{h \star h_\tau\} \longleftrightarrow \mathcal{H}(\cdot)\Delta_t$. This measure of width is depicted in Figure 191 by the bottom-most horizontal lines and is seen to fall between the other two measures. The autocorrelation width is in better agreement with the half-power width than the variance width is, particularly in the case of Fejér's kernel. Because $B_\mathcal{H}$ is the easiest of the three measures to compute and because of its good agreement with the intuitively appealing half-power width, we henceforth adopt it as our standard measure of the *effective bandwidth of the spectral estimator* $\hat{S}^{(D)}(\cdot)$. (An additional interpretation for $B_\mathcal{H}$ is that, if $0 < f < f' < f_\mathcal{N}$ and $|f - f'| \geq B_\mathcal{H}$, then $\hat{S}^{(D)}(f)$ and $\hat{S}^{(D)}(f')$ are approximately uncorrelated – for details, see Walden et al., 1995, and C&E [4] for Section 6.6.)

In summary, tapering is a useful method for reducing the bias due to leakage in direct spectral estimators. This is particularly important for spectra with large dynamic range (and of less importance for those with small dynamic range). When spectral analysis is being used as a tool for an exploratory data analysis (i.e., little is known about the true SDF of the process generating the data), it is important to use spectral estimators with good bias characteristics to make sure that it is the data and *not* the spectral window that is determining the shape of the estimated SDF. Note that so far we have considered the effects of tapering as reflected only by the first-moment properties of direct spectral estimators. Other advantages and disadvantages of tapering are presented in Section 6.6, where we discuss second-moment properties.

Comments and Extensions to Section 6.4

[1] As we shall see in Chapter 8, tapering plays a fundamental role in modern nonparametric spectral estimation. Some researchers, however, regard it as an undesirable (and suspicious) operation. For example, Yuen (1979) states the following (edited slightly to follow the notation and nomenclature of this chapter):

> ... the idea [of tapering] is statistically quite unsound. ... the most serious criticism is this: [tapering] effectively "throws data away" because some of the values of X_t are multiplied into small weighting factors. ... Why is this statistically bad? In the theory of estimation it is the basic idiom of the subject that the weight given to any individual value should decrease with its variance. That is, the more uncertainty there is in the data point, the less weight it should be given. Now, X_t is a time-invariance [sic] random process. Its variance is constant. Thus all its values should receive equal weight.

In advocating maximum entropy spectral analysis (see Chapter 9) in preference to nonparametric spectral analysis, Fougere (1985a) stated

6.4 Bias Reduction – Tapering

Also there is never any tapering – which really ought to be called tampering – of the data with resulting loss of information!

Evidently these researchers regard tapering as an operation that effectively modifies a time series. From our viewpoint, tapering is an operation that replaces Fejér's kernel in Equation (174b) with one having better sidelobe properties. Its actions should thus be judged in the frequency domain rather than the time domain. There are some caveats and trade-offs in using data tapers, but, if these are observed, tapering can be a useful way to control the bias due to leakage in direct spectral estimators for an SDF with a large dynamic range. Moreover, the multitaper approach discussed in Chapter 8 largely overcomes Yuen's objection (see also Brillinger, 1981b, for a good overall discussion of tapering in the context of a response to Yuen, 1979). Also, we note in Chapter 9 that, in one of its formulations (the Yule–Walker method), maximum entropy spectral analysis can actually be improved when used in combination with data tapering.

It should be clear from our AR(2) example that there are certain processes for which tapering with a single data taper accomplishes nothing useful (the same statement does not apply to the multitaper approach, which can be used to obtain estimates of the variability of spectral estimators even in situations where bias is not an issue – see Chapter 8). The most extreme case is a white noise process. Sloane (1969) shows that tapering such a process does little more than reduce the effective sample size; i.e., it amounts to throwing a certain portion of the data away.

[2] We have used the normalization $\sum_{t=0}^{N-1} h_t^2 = 1$ (Equation (188c)) to ensure that $\hat{S}^{(D)}(\cdot)$ mimics the true SDF in that the expected value of the integral of $\hat{S}^{(D)}(\cdot)$ equals the variance of the process. An alternative normalization is to require that the sum of the squares of $h_t X_t$ be equal to that of $X_t/\sqrt{N}$, i.e., to require that h_t be such that

$$\sum_{t=0}^{N-1} (h_t X_t)^2 = \frac{1}{N} \sum_{t=0}^{N-1} X_t^2 = \hat{s}_0^{(P)}. \tag{195a}$$

This makes the normalization of h_t data dependent. By taking the expected value of both sides above, we find that

$$\sum_{t=0}^{N-1} E\{h_t^2 X_t^2\} = s_0.$$

Hence the expected sum of squares of $h_t X_t$ is an unbiased estimator of the variance s_0 of the process (this assumes that X_t is known to have a mean of zero). If $\hat{s}_0^{(P)}$ is a consistent estimator of s_0, this variance normalization (sometimes called *restoration of power* in the literature) is equivalent to that of (188c) for large N. Note that any (reasonable) normalization of h_t affects only the level of $\hat{S}^{(D)}(\cdot)$ and not its shape.

[3] If $E\{X_t\} = \mu \neq 0$, we need to modify $\hat{S}^{(D)}(\cdot)$ somehow. The obvious changes are

$$\Delta_t \left| \sum_{t=0}^{N-1} h_t(X_t - \mu) e^{-i2\pi f t \Delta_t} \right|^2 \quad \text{or} \quad \Delta_t \left| \sum_{t=0}^{N-1} h_t(X_t - \overline{X}) e^{-i2\pi f t \Delta_t} \right|^2$$

accordingly as μ is either known or unknown. If both centering and tapering of a time series are needed, centering must be done before tapering (see Exercise [6.20]). There is, however, an alternative estimator to $\overline{X}$ for μ that is sometimes used in conjunction with tapering. Exercise [6.15b] shows that the periodogram $\hat{S}^{(P)}(\cdot)$ of a centered time series is zero at zero frequency. This is arguably appropriate – since zero frequency corresponds to a constant term, the power of a centered series at zero frequency should be zero. If we now taper a centered series with a nonrectangular data taper, the resulting direct estimator $\hat{S}^{(D)}(\cdot)$ need not be zero at zero frequency. Suppose, however, that we consider the following estimator of μ:

$$\tilde{\mu} \stackrel{\text{def}}{=} \sum_{t=0}^{N-1} h_t X_t \bigg/ \sum_{t=0}^{N-1} h_t \tag{195b}$$

(this assumes that $\sum h_t \neq 0$, which is true for all the specific tapers we have considered so far, but is not true in general). Note that $\tilde{\mu}$ is an unbiased estimator of μ and that it reduces to $\overline{X}$ when $\{h_t\}$ is the rectangular data taper. If we center our time series with $\tilde{\mu}$ instead of $\overline{X}$ and then taper it, our direct spectral estimator becomes

$$\Delta_t \left| \sum_{t=0}^{N-1} h_t (X_t - \tilde{\mu}) e^{-i2\pi f t \Delta_t} \right|^2. \tag{196a}$$

When $f = 0$ the quantity between the absolute value signs becomes

$$\sum_{t=0}^{N-1} h_t (X_t - \tilde{\mu}) = \sum_{t=0}^{N-1} h_t X_t - \sum_{t=0}^{N-1} h_t \left(\frac{\sum_{t=0}^{N-1} h_t X_t}{\sum_{t=0}^{N-1} h_t} \right) = 0,$$

as a result of which the direct spectral estimator is again identically zero at zero frequency. (We reconsider this alternative centering scheme in Section 10.10.)

[4] Since the formulae we have developed in this section show that the effect of tapering depends upon both the data taper and the true SDF, it might seem that, without knowledge of the SDF, there is no way of determining whether a data taper is required in practical applications. For fine details this is correct, but the need for tapering for a particular time series can usually be established by comparing its periodogram with a direct spectral estimate that uses a data taper with good sidelobe characteristics. Consider the bottom two rows of Figures 182 and 187, which show, respectively, periodograms and direct spectral estimates using a Hanning data taper for the four simulated AR(4) time series. If we compare the two estimates for each series, the main indicator that leakage is a concern is the difference in their levels in the high frequency region; moreover, as we would expect in the presence of substantial bias, the level of the periodogram is the higher of the two.

A secondary potential indication of the presence of leakage in a periodogram is a marked decrease in the amount of local variability in a low power region – compared with the high power region – when plotted on a logarithmic scale. This possibility is demonstrated in Figure 182(f). As we shall see in Section 6.4, nonparametric spectral estimators should have the same local variability at all frequencies when plotted on a logarithmic scale. The decrease in local variability in the high frequency region of Figure 182(f) is an indication that something is not right here – evidently this region is dominated by leakage of power from other portions of the SDF. However, it is important to note several things about this simple example. First, a decrease in local variability can also be due to other causes, a common one being the presence of outliers (discordant values) in a time series – see Martin and Thomson (1982) and Chave et al. (1987). Second, a *lack* of decrease of local variability in regions of low power does *not* automatically imply that tapering is unnecessary – consider, for example, Figure 182(g). Third, the need for tapering sometimes manifests itself as a ringing (oscillations) in a low power portion of an SDF (we illustrate this later on in Figures 226 and 227). Finally, leakage is not limited to the high frequency region of an SDF – it can potentially occur in any portion with relatively low power. The best advice we can offer is to try different degrees of tapering and to carefully check the corresponding SDF estimates for evidence of leakage – or lack thereof.

[5] Walden (1989) shows that, for a given N and W, the zeroth-order Slepian data taper can be approximated by

$$h_t = C I_0 \left(\pi W (N-1) \left(1 - \frac{(2t+1-N)^2}{N^2} \right)^{1/2} \right), \tag{196b}$$

where C is a normalizing constant forcing $\sum h_t^2 = 1$, and $I_0(\cdot)$ is the modified Bessel function of the first kind and zeroth order (this is an improvement on an approximation originally due to Kaiser, 1966, 1974). The advantage of Equation (196b) is that functions exist for calculating $I_0(\cdot)$ in many high-level languages.

[6] The two families of data tapers we have focused on ($p \times 100\%$ cosine and zeroth-order Slepian) are by no means the only ones that have been used extensively in spectral analysis – in fact, a bewildering number of tapers has been proposed and discussed in the literature. Harris (1978) compares 23 different classes of tapers (see also Nuttall, 1981, and Geçkinli and Yavuz, 1978 and 1983). Some of these tapers arise as solutions to different optimality criteria. For example, Adams (1991) proposes an "optimal" taper that provides a quantified "best" trade-off between peak sidelobe level and total sidelobe energy. Researchers have also developed tapers with particular time series in mind. Hurvich and Chen (2000) provide an example in their development of a taper that is efficient for potentially overdifferenced time series arising from a fractionally differenced or related process. Despite this plethora of choices, we have yet to encounter a practical application that provides a compelling case for using a taper outside of the cosine or Slepian families.

6.5 Bias Reduction – Prewhitening

In addition to tapering, another common method of controlling the bias in direct spectral estimators is *prewhitening* (Press and Tukey, 1956). This technique is based upon linear filtering theory. Suppose that we filter a portion $X_0, \ldots, X_{N-1}$ of a stationary process $\{X_t\}$ with SDF $S_X(\cdot)$ to create

$$Y_t \stackrel{\text{def}}{=} \sum_{l=0}^{L-1} g_l X_{t+L-1-l}, \quad 0 \leq t \leq M-1, \tag{197a}$$

where $L \geq 2$ is the width of the filter $\{g_l\}$, and $M \stackrel{\text{def}}{=} N - L + 1$. This yields a portion of length M of a stationary process $\{Y_t\}$ with SDF $S_Y(\cdot)$. The two SDFs are related by

$$S_Y(f) = \left| \sum_{l=0}^{L-1} g_l e^{-i2\pi f l \Delta_t} \right|^2 S_X(f), \tag{197b}$$

where the multiplier of $S_X(f)$ defines the square gain function for the filter.

Now, if $S_X(\cdot)$ has a wide dynamic range, it might be possible to choose $\{g_l\}$ such that the dynamic range of $S_Y(\cdot)$ is much less. This means we might be able to construct a direct spectral estimator of $S_X(\cdot)$ with low bias by using the following procedure. First, we use the filtered data to produce

$$\hat{S}_Y^{(\text{D})}(f) \stackrel{\text{def}}{=} \Delta_t \left| \sum_{t=0}^{M-1} h_t Y_t e^{-i2\pi f t \Delta_t} \right|^2,$$

where $\{h_t\}$ is a data taper with a fairly narrow central lobe (such as the rectangular taper or the Slepian data taper with $NW = 1$). If $S_Y(\cdot)$ has a small dynamic range, we should have $E\{\hat{S}_Y^{(\text{D})}(f)\} \approx S_Y(f)$ to within a few decibels. This implies that

$$E\left\{ \frac{\hat{S}_Y^{(\text{D})}(f)}{\left| \sum_{l=0}^{L-1} g_l e^{-i2\pi f l \Delta_t} \right|^2} \right\} \approx \frac{S_Y(f)}{\left| \sum_{l=0}^{L-1} g_l e^{-i2\pi f l \Delta_t} \right|^2} = S_X(f)$$

also to within a few decibels. Since we have effectively reduced the dynamic range of the SDF to be estimated and hence can now use a data taper with a narrow central lobe, this procedure could prevent the introduction of spectral window bias caused by data tapers with wide central lobes (see Figure 193). In the ideal case, the filter $\{g_l\}$ would reduce $\{X_t\}$ to white noise. It is thus called a *prewhitening filter*.

There are, however, some potential trade-offs and problems with prewhitening. There is an inevitable decrease in the sample size of the time series used in the direct spectral estimator (from N to $M = N - L + 1$). This decrease can adversely affect the first-moment and other statistical properties of the spectral estimator if L is large relative to N. This can happen in practice – an SDF with a large amount of structure can require a prewhitening filter of a large length to effectively reduce its dynamic range. Just such a situation occurs in exploration seismology where the power spectrum of seismic time series decays very rapidly with decreasing frequency because of the effect of analog recording filters.

There is also a "cart and horse" problem with prewhitening. Design of a prewhitening filter requires at least some knowledge of the shape of the underlying SDF, the very thing we are trying to estimate! Sometimes it is possible to make a reasonable guess about the general shape of the SDF from previous experiments or physical theory. This is obviously not a viable general solution, particularly since spectral analysis is often used as an exploratory data analysis tool. An alternative way of obtaining a prewhitening filter is to *estimate* it from the time series itself. One popular method for doing so is to fit an autoregressive model to the data and then use the fitted model as a prewhitening filter. The details of this approach are given in Section 9.10.

As examples, let's consider four prewhitening filters for the AR(4) time series of Figure 34(f), for which $\Delta_t = 1$. The periodogram for this particular series has pronounced leakage (see Figures 173(f) and 182(f)). Letting $\phi_{4,1}, \ldots, \phi_{4,4}$ represent the coefficients of $X_{t-1}, \ldots, X_{t-4}$ in Equation (35a), we can write

$$Y_t = X_{t+4} - \sum_{l=1}^{4} \phi_{4,l} X_{t+4-l} = \epsilon_{t+4},$$

where $\{\epsilon_t\}$ is a white noise process. Hence perfect prewhitening is provided by $\{1, -\phi_{4,1}, \ldots, -\phi_{4,4}\}$, which is a *prediction error filter* of length $L = 5$: we can regard $\sum_l \phi_{4,l} X_{t+4-l}$ as a predictor of X_{t+4} based upon $X_{t+3}, \ldots, X_t$, and the difference between X_{t+4} and its prediction is the prediction error. The top row of plots in Figure 199 shows the periodogram $\hat{S}_Y^{(P)}(\cdot)$ for $Y_0, Y_1, \ldots, Y_{1019}$ (left-hand column) along with the theoretical SDF for $\{Y_t\}$, which is flat because it corresponds to a white noise process. Note here, with $N = 1024$, we have $M = N - L + 1 = 1020$. The right-hand plot shows the "postcolored" estimate of the SDF for $\{X_t\}$, namely,

$$\hat{S}_X^{(PC)}(f) = \frac{\hat{S}_Y^{(P)}(f)}{\left|1 - \sum_{l=1}^{4} \phi_{4,l} e^{-i2\pi fl}\right|^2}, \qquad (198)$$

along with the true AR(4) SDF $S_X(\cdot)$. Leakage has been eliminated at the expense of trivially shortening the time series by four values. (The fact that such a short filter prewhitens perfectly is obviously due to the AR(4) assumption and is an exception – rather than the rule – in practical applications.)

The bottom three rows of Figure 199 illustrate what can happen if we use a prewhitening filter that is less than perfect. The first of these is

$$Y_t = X_{t+1} - 0.99 X_t,$$

which is a simple omnibus prewhitening filter that has proven to be effective in a number of applications (Tukey, 1967). This filter is close to the first difference filter $\{g_0 = 1, g_1 = -1\}$ described in Exercise [5.4a], which is problematic as a prewhitening filter because postcoloring involves division by $4\sin^2(\pi f)$, and this is zero when $f = 0$. Redefining g_1 to be -0.99 (or other values close to -1 such as -0.98) arguably improves the postcolored SDF estimate at frequencies close to zero. Figure 199(c) shows the true SDF $S_Y(\cdot)$ for the prewhitened process (smooth curve) along with the periodogram based upon $Y_0, Y_1, \ldots, Y_{1022}$ (jagged curve).

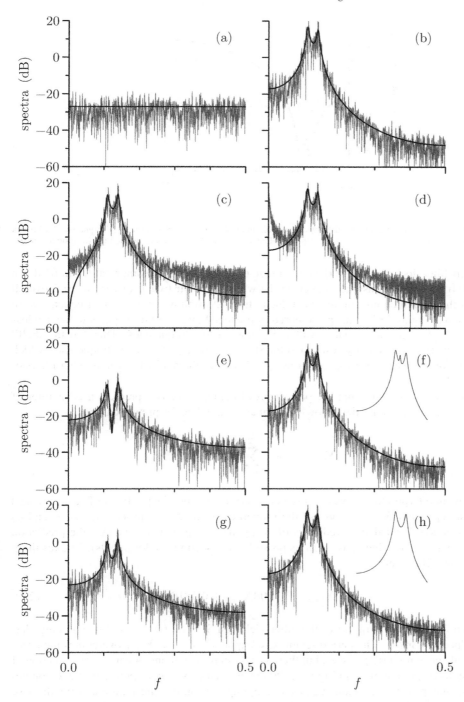

Figure 199. Periodograms computed after prewhitening the AR(4) series of Figure 34(f) along with true SDFs for the prewhitened AR(4) process (left-hand column), and corresponding postcolored periodograms and SDF for the AR(4) process (right-hand). Each row uses a different prewhitening filter (see main text for details). The extra curves in (f) and (h) are the expected values of the postcolored periodograms at low frequencies (plotted versus $f + 0.25$).

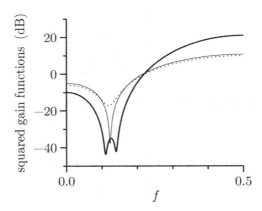

Figure 200. Square gain functions for the perfect prewhitening filter for the AR(4) process of Equation (35a) (thick solid curve), the filter used in Equation (200) (thin solid curve) and the filter used in Equation (201a) (dotted curve).

While this filter does reduce the dynamic range of the AR(4) SDF somewhat (from 65 down to 56 dB) if we consider just the frequencies between where the twin peaks are and 0.5, overall it pushes the dynamic range up to 71 dB due to the distortion it causes at low frequencies. In comparison to Figure 182(f), the postcolored estimate in Figure 199(d) displays somewhat less leakage at high frequencies, but considerably more at low frequencies. This "off-the-shelf" prewhitening filter is evidently best suited for SDFs with high power at low frequencies, which is not true for our AR(4) process. (Exercise [6.22] invites the reader to explore other filters of the form $\{1, g_1\}$.)

Motivated by the fact that the perfect prewhitening filter is a prediction error filter of width $L = 5$, let's consider the following form as a candidate for prewhitening:

$$Y_t = X_{t+2} - \sum_{l=1}^{2} g_l X_{t+2-l}.$$

We can again interpret the above as a prediction error filter, but the predictor of X_{t+2} is based upon just X_{t+1} and X_t. We can use the properties of the AR(4) process to set g_1 and g_2 such that, amongst all linear predictors of X_{t+2} that use just X_{t+1} and X_t, the prediction errors have a variance that is as small as possible (see Section 9.4 for details). The resulting prewhitening filter yields

$$Y_t \doteq X_{t+2} - 1.4198 X_{t+1} + 0.9816 X_t. \tag{200}$$

As shown in Figure 199(e), use of this filter yields an SDF $S_Y(\cdot)$ whose dynamic range has been reduced to 36 dB. The periodogram based upon $Y_0, Y_1, \ldots, Y_{1021}$ overcomes the problem of leakage at high frequencies, but there is a fly in the ointment. A careful examination of the postcolored periodogram in Figure 199(f) reveals an elevation of power in the region between the twin peaks. This is not due just to sampling variability – the additional smooth curve in this plot shows the expected value of the postcolored periodogram at low frequencies f, but plotted versus $f + 0.25$. The new peak that emerges is caused by the fact that the prewhitening filter has created a region of low power between the twin peaks, and this region is prone to leakage. Thus, while we have successfully eliminated leakage at the high frequencies, we have *induced* leakage elsewhere! (If we use a direct spectral estimator with an $NW = 1$ Slepian taper rather than the periodogram, the new peak disappears almost completely.)

An important clue as to why the prewhitening filter of Equation (200) gives mixed results is offered by a plot of its square gain function in Figure 200 (thin solid curve). The corresponding square gain function for the perfect prewhitening filter is shown by the thick solid curve. We see that the square gain function for the filter of Equation (200) has a pronounced dip in the midst of the range of frequencies trapping the twin peaks. As Figure 199(e) shows, this dip produces a deep valley in the prewhitened SDF $S_Y(\cdot)$, which is susceptible to leakage. A filter that has roughly the same square gain function but with a less pronounced dip is given by

$$Y_t = X_{t+2} - 1.3 X_{t+1} + 0.8 X_t. \tag{201a}$$

The square gain function for this prewhitening filter is shown as the dotted curve in Figure 200. The bottom row of Figure 199 shows that this filter works much better overall – there is no visual difference between the expected value of the postcolored periodogram and the true AR(4) SDF. This improvement comes about even though the dynamic range for $S_Y(\cdot)$ has increased to 40 dB over the value of 36 dB obtained using the prewhitening filter of Equation (200).

These examples demonstrate the efficacy of prewhitening as a way of compensating for bias in the periodogram due to leakage, but they also show that care must be taken in choosing the prewhitening filter – an improper choice can actually increase the bias! On the other hand, a less than perfect prewhitening filter can work just as well as a perfect filter in eliminating leakage. The key requirement is that the filter needs to compensate for the general shape of the SDF for the time series under study. (Further examples and discussion of prewhitening are given in Section 9.10.)

Comments and Extensions to Section 6.5

[1] There are some time series $\{X_t\}$ for which we are interested in estimating the SDF $S_X(\cdot)$ only over a selected range of frequencies, say, $[f_L, f_H]$. In this case we can subject $\{X_t\}$ to a band-pass filter that attenuates all spectral components with frequencies $|f| \notin [f_L, f_H]$ – this is sometimes known as *rejection filtration* (Blackman and Tukey, 1958, pp. 42–3). If $S_Y(\cdot)$ is the SDF of the filtered time series and if the gain of the band-pass filter is approximately unity over the pass-band, then we should have $S_Y(f) \approx S_X(f)$ for $f_L \leq |f| \leq f_H$. Because of the filtering operation, estimation of $S_Y(f)$ over $|f| \in [f_L, f_H]$ should be free of leakage from frequencies outside that range. This can be particularly valuable if there is a large contribution to the variance of $\{X_t\}$ due to frequencies $|f| \notin [f_L, f_H]$. Alavi and Jenkins (1965) and Tsakiroglou and Walden (2002) give examples of rejection filtration, the former in the context of measurements of the error of a radar tracking a moving aircraft, and the latter in the context of the AR(4) process.

As with prewhitening, filtering of a time series to suppress certain frequencies decreases the number of data points available to estimate $S_Y(\cdot)$. Since we might have to make the filter width large to get a decent approximation to a band-pass filter, this procedure might not be worthwhile if leakage from frequencies $|f| \notin [f_L, f_H]$ is small or can be controlled effectively with a data taper having a fairly narrow central lobe.

6.6 Statistical Properties of Direct Spectral Estimators

We investigated the first-moment (bias) properties of direct spectral estimators in Section 6.4. We now want to investigate some of their other statistical properties. We begin by considering the very special – but important – case of a Gaussian white noise process $\{G_t\}$ with zero mean and variance σ^2 (the letter G is used here to emphasize the Gaussian assumption). Let

$$A(f) = \Delta_t^{1/2} \sum_{t=0}^{N-1} h_t G_t \cos(2\pi f t \Delta_t) \text{ and } B(f) = \Delta_t^{1/2} \sum_{t=0}^{N-1} h_t G_t \sin(2\pi f t \Delta_t) \tag{201b}$$

be the real and imaginary parts of the complex conjugate of

$$J(f) \stackrel{\text{def}}{=} \Delta_t^{1/2} \sum_{t=0}^{N-1} h_t G_t e^{-i2\pi f t \Delta_t},$$

where $\{h_t\}$ is a data taper. Since $\{G_t\}$ has zero mean, it follows that

$$E\{A(f)\} = E\{B(f)\} = 0 \text{ for all } f.$$

From Exercise [2.1e] and the properties of a white noise process, we have

$$\text{cov}\{A(f), A(f')\} = \sigma^2 \Delta_t \sum_{t=0}^{N-1} h_t^2 \cos(2\pi f t \Delta_t) \cos(2\pi f' t \Delta_t),$$

$$\text{cov}\{B(f), B(f')\} = \sigma^2 \Delta_t \sum_{t=0}^{N-1} h_t^2 \sin(2\pi f t \Delta_t) \sin(2\pi f' t \Delta_t);$$

and

$$\text{cov}\{A(f), B(f')\} = \sigma^2 \Delta_t \sum_{t=0}^{N-1} h_t^2 \cos(2\pi f t \Delta_t) \sin(2\pi f' t \Delta_t).$$

Letting $f' = f$ in the first two equations above yields

$$\text{var}\{A(f)\} = \sigma^2 \Delta_t \sum_{t=0}^{N-1} h_t^2 \cos^2(2\pi f t \Delta_t), \quad \text{var}\{B(f)\} = \sigma^2 \Delta_t \sum_{t=0}^{N-1} h_t^2 \sin^2(2\pi f t \Delta_t).$$

The above equations simplify drastically if $h_t = 1/\sqrt{N}$ for all t (i.e., the direct spectral estimator is in fact just the periodogram) and if f and f' are Fourier (or standard) frequencies $f_k = k/(N \Delta_t)$ satisfying $0 \leq f_k \leq f_\mathcal{N}$, where k is an integer. In this case we have

$$\text{cov}\{A(f_j), A(f_k)\} = \text{cov}\{B(f_j), B(f_k)\} = 0 \text{ for all } f_j \neq f_k; \quad (202a)$$

$$\text{cov}\{A(f_j), B(f_k)\} = 0 \text{ for all } f_j \text{ and } f_k; \quad (202b)$$

$$\text{var}\{A(f_k)\} = \begin{cases} \sigma^2 \Delta_t/2, & f_k \neq 0 \text{ or } f_\mathcal{N}; \\ \sigma^2 \Delta_t, & f_k = 0 \text{ or } f_\mathcal{N}; \end{cases} \quad (202c)$$

and

$$\text{var}\{B(f_k)\} = \begin{cases} \sigma^2 \Delta_t/2, & f_k \neq 0 \text{ or } f_\mathcal{N}; \\ 0, & f_k = 0 \text{ or } f_\mathcal{N}, \end{cases} \quad (202d)$$

all of which follow from the orthogonality relationships of Exercise [1.3].

Since $A(f_k)$ and $B(f_k)$ are linear combinations of Gaussian RVs, they in turn are Gaussian RVs. Because uncorrelatedness implies independence for Gaussian RVs, the $A(f_k)$ and $B(f_k)$ terms are zero mean independent Gaussian RVs with variances given by Equations (202c) and (202d). Recall that, if $Y_0, Y_1, \ldots, Y_{\nu-1}$ are independent Gaussian RVs with zero means and unit variances, then the RV

$$\chi_\nu^2 = Y_0^2 + Y_1^2 + \cdots + Y_{\nu-1}^2$$

has a chi-square distribution with ν degrees of freedom. When $\nu = 2$, this distribution is the same as an exponential distribution with a mean of 2 (see C&E [1] at the end of this section). Since

$$A^2(f_k) + B^2(f_k) = \hat{S}_G^{(\text{P})}(f_k),$$

6.6 Statistical Properties of Direct Spectral Estimators

i.e., the periodogram, and since for $f_k \neq 0$ or $f_{\mathcal{N}}$

$$\left(\frac{2}{\sigma^2 \Delta_\text{t}}\right)^{1/2} A(f_k) \text{ and } \left(\frac{2}{\sigma^2 \Delta_\text{t}}\right)^{1/2} B(f_k)$$

are independent Gaussian RVs with zero mean and unit variance, it follows that

$$\frac{2}{\sigma^2 \Delta_\text{t}} \hat{S}_G^{(\text{P})}(f_k) \stackrel{\text{d}}{=} \chi_2^2; \text{ i.e., } \hat{S}_G^{(\text{P})}(f_k) \stackrel{\text{d}}{=} \frac{\sigma^2 \Delta_\text{t}}{2} \chi_2^2 \text{ for } f_k \neq 0 \text{ or } f_{\mathcal{N}}. \quad (203\text{a})$$

Here $\stackrel{\text{d}}{=}$ means "equal in distribution," so the statement $Y \stackrel{\text{d}}{=} c\chi_\nu^2$ means that the RV Y has the same distribution as a chi-square RV (with ν degrees of freedom) after multiplication by a constant c. The corresponding result for $f_k = 0$ or $f_{\mathcal{N}}$ is

$$\hat{S}_G^{(\text{P})}(f_k) \stackrel{\text{d}}{=} \sigma^2 \Delta_\text{t} \chi_1^2.$$

Since $S_G(f) = \sigma^2 \Delta_\text{t}$ is the SDF for $\{G_t\}$, we can rewrite the above as

$$\hat{S}_G^{(\text{P})}(f_k) \stackrel{\text{d}}{=} \begin{cases} S_G(f_k)\chi_2^2/2, & f_k \neq 0 \text{ or } f_{\mathcal{N}}; \\ S_G(f_k)\chi_1^2, & f_k = 0 \text{ or } f_{\mathcal{N}}. \end{cases} \quad (203\text{b})$$

Because $E\{\chi_\nu^2\} = \nu$ and $\text{var}\{\chi_\nu^2\} = 2\nu$, it follows from the above that

$$E\{\hat{S}_G^{(\text{P})}(f_k)\} = \sigma^2 \Delta_\text{t} = S_G(f_k) \text{ for all } f_k \quad (203\text{c})$$

and that

$$\text{var}\{\hat{S}_G^{(\text{P})}(f_k)\} = \begin{cases} \sigma^4 \Delta_\text{t}^2 = S_G^2(f_k), & f_k \neq 0 \text{ or } f_{\mathcal{N}}; \\ 2\sigma^4 \Delta_\text{t}^2 = 2S_G^2(f_k), & f_k = 0 \text{ or } f_{\mathcal{N}} \end{cases} \quad (203\text{d})$$

(see Exercise [6.23] for an extension of the above to $0 < f < f_{\mathcal{N}}$). Moreover, all the RVs in the set $\{\hat{S}_G^{(\text{P})}(f_k) : 0 \leq k \leq \lfloor N/2 \rfloor\}$ are independent. Thus, the sampling properties of $\hat{S}_G^{(\text{P})}(f_k)$ on the grid of Fourier frequencies are completely known and have a simple form for a zero mean Gaussian white noise process $\{G_t\}$.

For a stationary process $\{X_t\}$ that is not necessarily Gaussian with an SDF $S(\cdot)$ that is not necessarily constant, it can be shown – subject to certain technical assumptions – that

$$\hat{S}^{(\text{P})}(f) \stackrel{\text{d}}{=} \begin{cases} S(f)\chi_2^2/2, & 0 < |f| < f_{\mathcal{N}}; \\ S(f)\chi_1^2, & |f| = 0 \text{ or } f_{\mathcal{N}}, \end{cases} \quad (203\text{e})$$

asymptotically as $N \to \infty$ (see, for example, Brillinger, 1981a, section 5.2; Brockwell and Davis, 1991, section 10.3; Fuller, 1996, section 7.1; and Shumway and Stoffer, 2017, appendix C.2). Furthermore, for $0 \leq |f'| < |f| \leq f_{\mathcal{N}}$, $\hat{S}^{(\text{P})}(f)$ and $\hat{S}^{(\text{P})}(f')$ are asymptotically independent. We thus have, in addition to $E\{\hat{S}^{(\text{P})}(f)\} = S(f)$, that

$$\text{var}\{\hat{S}^{(\text{P})}(f)\} = \begin{cases} S^2(f), & 0 < |f| < f_{\mathcal{N}}; \\ 2S^2(f), & f = 0 \text{ or } |f| = f_{\mathcal{N}}; \end{cases} \quad (203\text{f})$$

$$\text{cov}\{\hat{S}^{(\text{P})}(f'), \hat{S}^{(\text{P})}(f)\} = 0, \ 0 \leq |f'| < |f| \leq f_{\mathcal{N}}, \quad (203\text{g})$$

asymptotically as $N \to \infty$ (for finite sample sizes N, these asymptotic results are useful approximations if certain restrictions are observed – see C&E [3] for this section and Exercise [6.27f]). If we restrict ourselves to the nonnegative Fourier frequencies f_k and place certain constraints on $\{X_t\}$ (see C&E [5]), we also find that the $\lfloor N/2 \rfloor + 1$ RVs

$$\hat{S}^{(\text{P})}(f_0), \hat{S}^{(\text{P})}(f_1), \ldots, \hat{S}^{(\text{P})}(f_{\lfloor N/2 \rfloor})$$

are all approximately pairwise uncorrelated for N large enough; i.e.,

$$\text{cov}\{\hat{S}^{(\text{P})}(f_j), \hat{S}^{(\text{P})}(f_k)\} \approx 0, \quad j \neq k \text{ and } 0 \leq j, k \leq \lfloor N/2 \rfloor. \tag{204a}$$

As we shall see, this last result plays an important role in the derivation of the sampling properties of spectral estimators based upon the periodogram (even though it is a large-sample result, it is often regarded as a reasonable approximation for finite N – see, for example, Bolt and Brillinger, 1979, and Brillinger, 1987).

Let us now turn our attention to the large sample theory for direct spectral estimators $\hat{S}^{(\text{D})}(\cdot)$ other than the periodogram (for details, see C&E [2]). For a particular sample size N, the variance of $\hat{S}^{(\text{D})}(f)$ can actually be larger or smaller than that of $\hat{S}^{(\text{P})}(f)$, depending upon the particular data taper used and the characteristics of the true underlying SDF; however, for processes whose ACVS satisfies a summability condition and for a taper $\{h_t\}$ that is reasonable in form, Brillinger (1981a, p. 127) concludes that the large-sample univariate distributional properties of $\hat{S}^{(\text{D})}(f)$ are the *same* as those of $\hat{S}^{(\text{P})}(f)$ – this implies that

$$\hat{S}^{(\text{D})}(f) \stackrel{\text{d}}{=} \begin{cases} S(f)\chi_2^2/2, & 0 < |f| < f_\mathcal{N}; \\ S(f)\chi_1^2, & |f| = 0 \text{ or } f_\mathcal{N}, \end{cases} \tag{204b}$$

and

$$\text{var}\{\hat{S}^{(\text{D})}(f)\} = \begin{cases} S^2(f), & 0 < |f| < f_\mathcal{N}; \\ 2S^2(f), & |f| = 0 \text{ or } f_\mathcal{N} \end{cases} \tag{204c}$$

asymptotically as $N \to \infty$ (see C&E [3] for caveats on applying these asymptotic results).

Equation (204b) allows us to construct an asymptotically correct confidence interval (CI) for $S(f)$ at a fixed f. For $0 < |f| < f_\mathcal{N}$, the large-sample distribution of $\hat{S}^{(\text{D})}(f)$ is related to the distribution of a chi-square RV χ_2^2 with two degrees of freedom, which has a PDF given by

$$f(u) = \begin{cases} \frac{1}{2}e^{-u/2}, & u \geq 0; \\ 0, & u < 0. \end{cases} \tag{204d}$$

Let $Q_2(p)$ represent the $p \times 100\%$ percentage point of this distribution; i.e.,

$$\mathbf{P}\left[\chi_2^2 \leq Q_2(p)\right] = \int_0^{Q_2(p)} f(u)\,du = 1 - e^{-Q_2(p)/2} = p \text{ so that } Q_2(p) = -2\log(1-p),$$

where $\mathbf{P}[A]$ is the probability of the event A occurring. We thus have, for $0 \leq p \leq 1/2$,

$$\mathbf{P}\left[Q_2(p) \leq \chi_2^2 \leq Q_2(1-p)\right] = 1 - 2p.$$

When $S(f) > 0$, Equation (204b) says that $2\hat{S}^{(\text{D})}(f)/S(f)$ is a χ_2^2 RV. Hence

$$\mathbf{P}\left[Q_2(p) \leq \frac{2\hat{S}^{(\text{D})}(f)}{S(f)} \leq Q_2(1-p)\right] = 1 - 2p.$$

The event in the brackets above occurs if and only if the events $Q_2(p) \leq 2\hat{S}^{(\text{D})}(f)/S(f)$ and $2\hat{S}^{(\text{D})}(f)/S(f) \leq Q_2(1-p)$ both occur. These events in turn are equivalent to the single event in the brackets in the next equation, implying that

$$\mathbf{P}\left[\frac{2\hat{S}^{(\text{D})}(f)}{Q_2(1-p)} \leq S(f) \leq \frac{2\hat{S}^{(\text{D})}(f)}{Q_2(p)}\right] = 1 - 2p.$$

6.6 Statistical Properties of Direct Spectral Estimators

Hence

$$\left[\frac{2\hat{S}^{(\mathrm{D})}(f)}{Q_2(1-p)}, \frac{2\hat{S}^{(\mathrm{D})}(f)}{Q_2(p)}\right] = \left[-\frac{\hat{S}^{(\mathrm{D})}(f)}{\log{(p)}}, -\frac{\hat{S}^{(\mathrm{D})}(f)}{\log{(1-p)}}\right]$$

is a $100(1-2p)\%$ CI for $S(f)$. Since $0 < x < y$ implies that $10\log_{10}(x) < 10\log_{10}(y)$, we also have

$$\mathbf{P}\left[10\log_{10}\left(\frac{2\hat{S}^{(\mathrm{D})}(f)}{Q_2(1-p)}\right) \leq 10\log_{10}(S(f)) \leq 10\log_{10}\left(\frac{2\hat{S}^{(\mathrm{D})}(f)}{Q_2(p)}\right)\right] = 1 - 2p,$$

and hence

$$\left[10\log_{10}(\hat{S}^{(\mathrm{D})}(f)) + 10\log_{10}\left(\frac{2}{Q_2(1-p)}\right), 10\log_{10}(\hat{S}^{(\mathrm{D})}(f)) + 10\log_{10}\left(\frac{2}{Q_2(p)}\right)\right]$$

is a $100(1-2p)\%$ CI for $10\log_{10}(S(f))$. The length of this CI is

$$10\log_{10}\left(\frac{Q_2(1-p)}{Q_2(p)}\right) = 10\log_{10}\left(\frac{\log{(p)}}{\log{(1-p)}}\right), \tag{205a}$$

which is independent of the value of $\hat{S}^{(\mathrm{D})}(f)$, a fact we will put to good use shortly. For future reference we note that, if we set $p = 0.025$ to obtain a 95% CI, the interval becomes

$$\left[10\log_{10}(\hat{S}^{(\mathrm{D})}(f)) - 5.67, 10\log_{10}(\hat{S}^{(\mathrm{D})}(f)) + 15.97\right]. \tag{205b}$$

Its width is 21.63 dB, a result that holds for all direct spectral estimators, the periodogram included. (Note that the CIs presented here apply only to the SDF or its log at one particular frequency, and do not provide a simultaneous confidence band across all $|f| \in (0, f_\mathcal{N})$ for either $S(\cdot)$ or its log.)

While $\hat{S}^{(\mathrm{P})}(f)$ and $\hat{S}^{(\mathrm{D})}(f)$ have the same large-sample univariate distribution, the relationship between $\hat{S}^{(\mathrm{D})}(f)$ and $\hat{S}^{(\mathrm{D})}(f')$ at distinct frequencies f and f' is in general different from that between $\hat{S}^{(\mathrm{P})}(f)$ and $\hat{S}^{(\mathrm{P})}(f')$. In particular, the grid of frequencies over which $\mathrm{cov}\{\hat{S}^{(\mathrm{D})}(f_j), \hat{S}^{(\mathrm{D})}(f_k)\} \approx 0$ is coarser than the one defined by the Fourier frequencies; thus, at adjacent Fourier frequencies f_k and f_{k+1}, the RVs $\hat{S}^{(\mathrm{D})}(f_k)$ and $\hat{S}^{(\mathrm{D})}(f_{k+1})$ can be significantly correlated. As a concrete example, the burden of Exercise [6.18b] is to show that $\mathrm{corr}\{\hat{S}_G^{(\mathrm{D})}(f_k), \hat{S}_G^{(\mathrm{D})}(f_{k+1})\} = 4/9$ for Gaussian white noise with $\hat{S}_G^{(\mathrm{D})}(\cdot)$ based on a Hanning-like data taper (this result presumes that f_k is not too close to either zero or $f_\mathcal{N}$). More generally, in the case of a Gaussian process with SDF $S_G(\cdot)$, it follows from Equation (211c) in the C&Es along with an assumption that $S_G(\cdot)$ is continuous on the interval $[-f_\mathcal{N}, f_\mathcal{N}]$ that

$$\mathrm{cov}\{\hat{S}_G^{(\mathrm{D})}(f), \hat{S}_G^{(\mathrm{D})}(f')\} \approx \frac{1}{\Delta_t^2}\left|\int_{-f_\mathcal{N}}^{f_\mathcal{N}} H^*(f-u)H(f'-u)S_G(u)\,\mathrm{d}u\right|^2$$

$$\leq \frac{S_{\max}^2}{\Delta_t^2}\left|\int_{-f_\mathcal{N}}^{f_\mathcal{N}} |H(f-u)| \cdot |H(f'-u)|\,\mathrm{d}u\right|^2$$

$$= \frac{S_{\max}^2}{\Delta_t^2}\left|\int_{-f_\mathcal{N}}^{f_\mathcal{N}} |H(f-f'+v)| \cdot |H(v)|\,\mathrm{d}v\right|^2, \tag{205c}$$

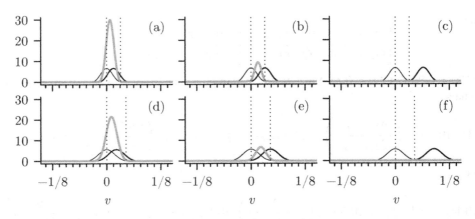

Figure 206. Examples of the integrand (thick gray curves) dictating the bound in Equation (205c) for the covariance between $\hat{S}_G^{(D)}(f)$ and $\hat{S}_G^{(D)}(f')$. The integrand is the product of $|H(\cdot)|$ and $|H(f - f' + \cdot)|$ (the thin and thick dark curves), where $H(\cdot)$ is the Fourier transform of a Slepian data taper for $N = 64$ with $W = 2/N$ (top row) and $4/N$ (bottom). The distance between the dotted lines indicates the standard bandwidth measure $B_{\mathcal{H}}$ of Equation (194) ($B_{\mathcal{H}} \doteq 0.0323$ in the top row and 0.0447 in the bottom). In plots (a), (b) and (c) (and similarly in (d), (e) and (f)), $f' - f$ is set to, respectively, $B_{\mathcal{H}}/2$, $B_{\mathcal{H}}$ and $2B_{\mathcal{H}}$. The spacing between the tick marks on the v axis is $1/64$ (the spacing between the Fourier frequencies $f_j = j/N$).

where $v = f' - u$ and $S_{\max} \stackrel{\text{def}}{=} \max_f S_G(f)$. Suppose, for example, that $H(\cdot)$ corresponds to a zeroth-order Slepian data taper. The above integrand will then be small whenever f and f' are far enough apart so that the central lobes of $|H(f - f' + \cdot)|$ and $|H(\cdot)|$ do not overlap significantly. Figure 206 illustrates the case $N = 64$ with $\Delta_t = 1$ and W set so that $NW = 2$ (top row of plots) and $NW = 4$ (bottom). The thin dark curve in the plots shows $|H(v)|$ versus v while the thick dark curves show $|H(f - f' + v)|$ versus v with $f' - f$ set to, from left to right in each row, $B_{\mathcal{H}}/2$, $B_{\mathcal{H}}$ and $2B_{\mathcal{H}}$, where $B_{\mathcal{H}}$ is our standard bandwidth measure for a direct spectral estimator (see Equation (194)). The distance between the vertical dotted lines in each plot is $B_{\mathcal{H}}$, while the minor tick marks depict the grid of Fourier frequencies. The thick gray curves show the integrands $|H(f - f' + v)| \cdot |H(v)|$. As $|f - f'|$ increases, the area under the gray curve decreases, implying a lower upper bound for the covariance between $\hat{S}_G^{(D)}(f)$ and $\hat{S}_G^{(D)}(f')$ and hence a decrease in their correlation. If we specialize to white noise, we can interpret the square of the area under the gray curve as an approximation to $\text{corr}\{\hat{S}_G^{(D)}(f), \hat{S}_G^{(D)}(f')\}$. For the six cases shown in Figure 206, these approximations are (a) 0.43, (b) 0.03 and (c) 0.00 for the top row and, for the bottom, (d) 0.44, (e) 0.04 and (f) 0.00. Hence the correlation is quite small when $|f - f'| \geq B_{\mathcal{H}}$.

As discussed in C&Es [2] and [4], for a direct spectral estimator with an associated spectral window $\mathcal{H}(\cdot)$, the bandwidth measure $B_{\mathcal{H}}$ quantifies the spacing $|f - f'|$ required for $\hat{S}^{(D)}(f)$ and $\hat{S}^{(D)}(f')$ to be approximately uncorrelated. In particular, if $0 \leq f < f' \leq f_{\mathcal{N}}$ and $|f - f'| \geq B_{\mathcal{H}}$, then $\text{corr}\{\hat{S}^{(D)}(f), \hat{S}^{(D)}(f')\} \approx 0$, in agreement with the conclusion we drew from Figure 206. A grid of frequencies over which the direct spectral estimator $\hat{S}^{(D)}(\cdot)$ is approximately pairwise uncorrelated is thus given by $0, B_{\mathcal{H}}, 2B_{\mathcal{H}}, \ldots, KB_{\mathcal{H}}$, where K is the largest integer such that $KB_{\mathcal{H}} \leq f_{\mathcal{N}}$. Alternatively, this grid can be expressed by $kc\Delta_f = cf_k$, $k = 0, \ldots, K$, where $\Delta_f \stackrel{\text{def}}{=} 1/(N\Delta_t)$ is the distance between adjacent Fourier frequencies, and $c = B_{\mathcal{H}}/\Delta_f$. For the Hanning, $NW = 2$ Slepian and $NW = 4$ Slepian tapers, approximate values for c are, respectively, 2, 2 and 3 – these yield a grid size twice or three times as large as the Fourier frequencies (see the top row of Table 214, which is discussed in C&E [4]). As Brillinger (1981b) points out, although the grid size is now larger,

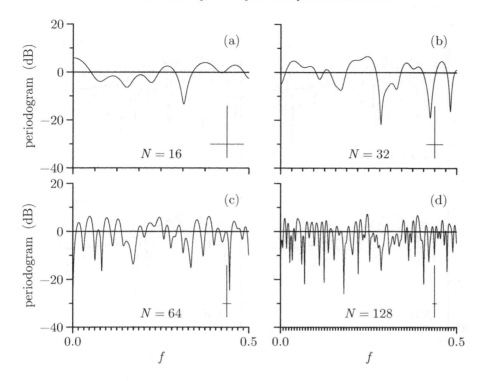

Figure 207. Inconsistency of the periodogram. The plots show the periodogram (on a decibel scale) versus frequency for samples of Gaussian white noise with $N = 16$, 32, 64 and 128. We take Δ_t to be unity so the Nyquist frequency is $f_\mathcal{N} = 1/2$. Each periodogram is evaluated over the grid of frequencies $k/2048$, $k = 0, 1, \ldots, 1024$. Tick marks on the horizontal axes indicate the locations of the Fourier frequencies. The horizontal lines at 0 dB show the true SDF. The widths of the crisscrosses depict the bandwidth measure $B_\mathcal{H}$ of Equation (194), while their heights are the lengths of a 95% CI for $10 \log_{10}(S(f))$ based upon $10 \log_{10}(\hat{S}^{(\mathrm{P})}(f))$.

the pairwise covariances of the $K + 1$ RVs

$$\hat{S}^{(\mathrm{D})}(0), \hat{S}^{(\mathrm{D})}(c/(N \Delta_t)), \ldots, \hat{S}^{(\mathrm{D})}(Kc/(N \Delta_t)) \tag{207}$$

are zero to a better approximation for finite N than the corresponding large-sample result stated in (204a) for the periodogram. This better approximation is useful for constructing certain statistical tests and is another benefit of tapering.

Since $S(f) > 0$ typically, Equations (203f) and (204c) show that the variances of $\hat{S}^{(\mathrm{P})}(f)$ and $\hat{S}^{(\mathrm{D})}(f)$ do *not* decrease to 0 as $N \to \infty$. Thus the probability that $\hat{S}^{(\mathrm{P})}(f)$ or $\hat{S}^{(\mathrm{D})}(f)$ becomes arbitrarily close to its asymptotic expected value of $S(f)$ is zero – in statistical terminology, neither $\hat{S}^{(\mathrm{P})}(f)$ nor $\hat{S}^{(\mathrm{D})}(f)$ is a consistent estimator of $S(f)$. Figure 207 illustrates the inconsistency of the periodogram for a Gaussian white noise process with unit variance and unit sampling interval Δ_t: as the sample size increases from 16 to 128, the periodogram shows no sign of converging to the SDF for the process, namely, $S(f) = 1$ (indicated by the horizontal lines at 0 dB). As N increases, there is an increase in the visual roughness of the periodogram, which is due to the decrease in covariance between $\hat{S}^{(\mathrm{P})}(f)$ and $\hat{S}^{(\mathrm{P})}(f')$ separated by a fixed distance $|f - f'|$. Inconsistency and the decrease in covariance are also reflected by the crisscrosses in the lower right-hand corners. In keeping with Equation (205b), the height of each crisscross at $f = \frac{7}{16}$ would have depicted a 95% CI for $S(\frac{7}{16})$ if $10 \log_{10}(\hat{S}^{(\mathrm{P})}(\frac{7}{16}))$ had been equal to -30 dB (never the case for the four periodograms in the figure). Mentally moving the center of a crisscross to an actual value for $10 \log_{10}(\hat{S}^{(\mathrm{P})}(f))$ with $0 < f < 0.5$

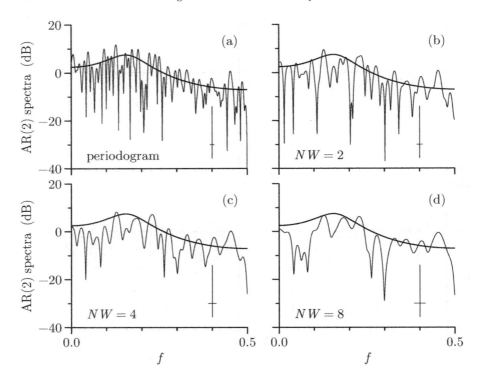

Figure 208. Variability of direct spectral estimators. The bumpy curve in plot (a) is the periodogram for a realization of size $N = 128$ from the AR(2) process of Equation (34) (the realization itself is the first 128 points in Figure 34(a)). The corresponding curves in the remaining plots show direct spectral estimates based upon a Slepian data taper with NW set to (b) 2, (c) 4 and (d) 8. The thick curve in each plot is the true SDF (cf. Figure 177). All spectra are plotted on a decibel scale versus frequencies $k/2048$, $k = 0, 1, \ldots, 1024$. The crisscrosses depict the bandwidth measure $B_\mathcal{H}$ (widths) and the length of a 95% CI for $10 \log_{10}(S(f))$ (heights).

gives a 95% CI for $S(f)$ because, as indicated by Equation (205b), the length of the CI does not depend on $10 \log_{10}(\hat{S}^{(\mathrm{P})}(f))$ itself. The fact that the height of the crisscross remains the same as N increases is a manifestation of the inconsistency of the periodogram. On the other hand, the width of each crisscross is equal to the bandwidth measure $B_\mathcal{H}$ of Equation (194). As N increases, this measure decreases, which indicates that the separation $|f - f'|$ needed for $\hat{S}^{(\mathrm{P})}(f)$ and $\hat{S}^{(\mathrm{P})}(f')$ to be approximately uncorrelated is also decreasing.

Figure 208 shows four direct spectral estimates for the AR(2) process $\{X_t\}$ defined in Equation (34). These estimates are based on the first $N = 128$ points of the time series shown in Figure 34(a). The bumpy curve in plot (a) is the periodogram. Similar curves in the remaining plots are direct spectral estimates based upon zeroth-order Slepian data tapers with (b) $NW = 2$, (c) $NW = 4$ and (d) $NW = 8$. The true SDF for $\{X_t\}$ is shown by the smooth curve in each plot. Because the dynamic range of this SDF is small, the bias in the periodogram is already negligible for a sample size as small as 64 (see Figure 177). There is thus no need for a data taper here. The smoother appearance of the direct spectral estimates with the Slepian data tapers is due to the increased distance between uncorrelated spectral estimates. As before, we can quantify this distance using $B_\mathcal{H}$, which is shown by the width of the crisscross in each plot. A complementary interpretation is that tapering in this example has effectively reduced the size of our time series without any gain in bias reduction (since there really is no bias to be reduced!). Based upon the grid size of approximately uncorrelated spectral estimates, we can argue that the statistical properties of the Slepian-based direct

6.6 Statistical Properties of Direct Spectral Estimators

spectral estimators are approximately equivalent to those of a periodogram with $128/(NW)$ data points, yielding 64, 32 and 16 values for cases (b), (c) and (d). The smoother appearance as NW increases can be attributed to this effective decrease in sample size. This effect is also evident in Figure 207 – as the sample size decreases, the periodogram has a smoother appearance. Indeed the degrees of smoothness in $\hat{S}^{(P)}(\cdot)$ in plots (a) to (d) of Figure 207 are a fairly good visual match to the $\hat{S}^{(D)}(\cdot)$ in plots (d) to (a) of Figure 208, indicating that the covariance structure across frequencies in the direct spectral estimates is controlled by the effective sample size $128/(NW)$. Finally we note that the local variability of all four direct spectral estimators in Figure 208 is about the same. This supports the claim that all direct spectral estimators of $S(f)$ have approximately the same variance (see Equations (203f) and (204c)) and reflects the fact that, as indicated by the height of the crisscrosses and by Equation (205b), the lengths of the 95% CIs for $10\log_{10}(S(f))$ are the same for all four spectral estimates.

Comments and Extensions to Section 6.6

[1] Equation (204b) says that, except at $f = 0$ and $f_{\mathcal{N}}$, a direct spectral estimator $\hat{S}^{(D)}(f)$ approximately has the distribution of a scaled chi-square RV with two degrees of freedom, which is a special case of an exponential RV. Letting $U \stackrel{\text{def}}{=} \hat{S}^{(D)}(f)$, the PDF for this distribution is given by

$$f_U(u; S(f)) = \begin{cases} \frac{1}{S(f)} e^{-u/S(f)}, & u \geq 0; \\ 0, & u < 0. \end{cases} \quad (209a)$$

This function is shown in Figure 210a(a) for the case $S(f) = 2$, for which U becomes an ordinary (i.e., unscaled) chi-square RV (see Equation (204d)). Since this PDF has an upper tail that decays at a slower rate than that of a Gaussian PDF, a random sample of variates with this distribution will typically contain "upshoots," i.e., a few values that appear to be unusually large compared with the bulk of the variates (see Figure 210b(a); cf. Figure 172); however, when direct spectral estimates are plotted on a decibel (or logarithmic) scale, the visual appearance is quite different – we typically see prominent "downshoots" instead (see Figure 210b(b); cf. Figures 173, 182 and 187). The explanation for this "paradox" is that we are actually examining variates with the PDF of $V \stackrel{\text{def}}{=} 10\log_{10}(U)$. This PDF is given by

$$f_V(v; S(f)) = \frac{\log(10)}{10\,S(f)} 10^{v/10} e^{-(10^{v/10})/S(f)}, \quad -\infty < v < \infty, \quad (209b)$$

and is shown Figure 210a(b), again for the case $S(f) = 2$. Note that, in contrast to $f_U(\cdot)$, this PDF has a prominent *lower* tail. Note also that

$$f_V(v; S(f)) = f_V(v - 10\log_{10}(S(f)/2); 2).$$

Thus if $S(f)$ is something other than 2, the PDF depicted in 210a(b) merely shifts along the v-axis. This result implies that the variance (and other central moments) of $10\log_{10}(\hat{S}^{(D)}(f))$ does not depend on $S(f)$. Another way to establish this result is to note that the assumption $\hat{S}^{(D)}(f) \stackrel{d}{=} S(f)\chi_2^2/2$ implies that

$$10\log_{10}(\hat{S}^{(D)}(f)) \stackrel{d}{=} 10\log_{10}(S(f)/2) + 10\log_{10}(\chi_2^2)$$

and hence that, whereas $\text{var}\{\hat{S}^{(D)}(f)\} = S^2(f)$,

$$\text{var}\{10\log_{10}(\hat{S}^{(D)}(f))\} = \text{var}\{10\log_{10}(\chi_2^2)\}.$$

Taking the log of $\hat{S}^{(D)}(f)$ acts as a variance-stabilizing transform in that the dependence of the variance on $S(f)$ is removed.

[2] For use in Section 6.3, we develop here a more accurate expression for the covariance of direct spectral estimators than given by the qualitative "grid size" argument surrounding Equation (207) (Jones,

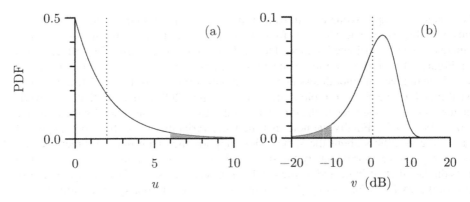

Figure 210a. Probability density functions for (a) a chi-square RV with two degrees of freedom and (b) 10 $\log_{10}$ of the same RV. The shaded areas depict the upper 5% tail area in (a) and the lower 5% tail area in (b). The vertical dotted lines indicate the expected values of RVs with these PDFs, i.e., 2 in (a) and 10 $\log_{10}(2/e^\gamma)$ in (b), where $\gamma \doteq 0.5772$ is Euler's constant (this can be obtained from the digamma function $\psi(\cdot)$ since $\psi(1) = -\gamma$).

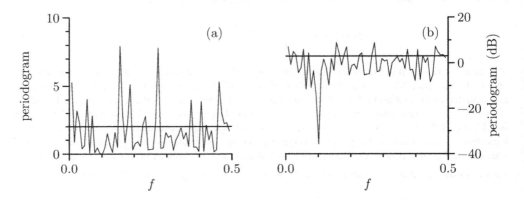

Figure 210b. Periodogram $\hat{S}_G^{(P)}(\cdot)$ at Fourier frequencies $f_k = k/128$, $k = 1, 2, \ldots, 63$, for a time series of length $N = 128$ drawn from a Gaussian white noise process with mean $\mu = 0$ and variance $\sigma^2 = 2$ and plotted on (a) a linear scale and (b) a decibel scale. The variates $\hat{S}_G^{(P)}(f_k)$ in (a) can be regarded as a random sample of 63 independent and identically distributed RVs from a chi-square distribution with two degrees of freedom. The PDF for this distribution is shown in Figure 210a(a). The variates $10 \log_{10}(\hat{S}_G^{(P)}(f_k))$ in (b) constitute a random sample from a distribution whose PDF is shown in Figure 210a(b). The horizontal lines in the plots depict the SDF $S_G(f) = 2$ for the white noise process in (a) and $10 \log_{10}(S_G(f)) = 10 \log_{10}(2) \doteq 3.01$ in (b).

1971; Koopmans, 1974, Chapter 8; and Walden, 1990a). Suppose that $\{G_t\}$ is a stationary Gaussian process – not necessarily white noise – with zero mean and SDF $S_G(\cdot)$ and that, as before,

$$J(f) \stackrel{\text{def}}{=} \Delta_t^{1/2} \sum_{t=0}^{N-1} h_t G_t e^{-i2\pi f t \Delta_t} \quad \text{with } \{h_t\} \longleftrightarrow H(\cdot).$$

Now,

$$\text{cov}\{|J(f')|^2, |J(f)|^2\} = \text{cov}\{J(f')J^*(f'), J(f)J^*(f)\}.$$

Since $\{G_t\}$ is a Gaussian process, the distribution of $J(f)$ is complex-valued Gaussian for all f (for non-Gaussian processes, the distribution of $J(f)$ can still be reasonably approximated as such, but this depends somewhat on the effect of the taper). We can now use the Isserlis theorem (Equation (30)) to

6.6 Statistical Properties of Direct Spectral Estimators

show that

$$\begin{aligned}\operatorname{cov}\{|J(f')|^2, |J(f)|^2\} &= \operatorname{cov}\{J(f'), J(f)\}\operatorname{cov}\{J^*(f'), J^*(f)\} \\ &\quad + \operatorname{cov}\{J(f'), J^*(f)\}\operatorname{cov}\{J^*(f'), J(f)\} \\ &= E\{J(f')J^*(f)\}E\{J^*(f')J(f)\} \\ &\quad + E\{J(f')J(f)\}E\{J^*(f')J^*(f)\} \\ &= \left|E\{J(f')J^*(f)\}\right|^2 + \left|E\{J(f')J(f)\}\right|^2. \end{aligned} \quad (211a)$$

From the fact that

$$J(f) = \frac{1}{\Delta_t^{1/2}} \int_{-f_\mathcal{N}}^{f_\mathcal{N}} H(f-u)\,dZ(u)$$

(this is Equation (186c)) and the properties of $dZ(\cdot)$, we have

$$E\{J(f')J^*(f)\} = \frac{1}{\Delta_t}\int_{-f_\mathcal{N}}^{f_\mathcal{N}} H(f'-u)H^*(f-u)S_G(u)\,du.$$

Since $dZ(-u) = dZ^*(u)$ (see Equation (109b)), it is also true that

$$J(f) = -\frac{1}{\Delta_t^{1/2}} \int_{-f_\mathcal{N}}^{f_\mathcal{N}} H(f+u)\,dZ^*(u),$$

and hence

$$E\{J(f')J(f)\} = -\frac{1}{\Delta_t}\int_{-f_\mathcal{N}}^{f_\mathcal{N}} H(f'+u)H(f-u)S_G(u)\,du.$$

With these expressions and by substituting $f + \eta$ for f', we can now write

$$\begin{aligned}\operatorname{cov}\{\hat{S}_G^{(D)}(f+\eta), \hat{S}_G^{(D)}(f)\} &= \operatorname{cov}\{|J(f+\eta)|^2, |J(f)|^2\} \\ &= \frac{1}{\Delta_t^2}\left|\int_{-f_\mathcal{N}}^{f_\mathcal{N}} H^*(f+\eta-u)H(f-u)S_G(u)\,du\right|^2 \\ &\quad + \frac{1}{\Delta_t^2}\left|\int_{-f_\mathcal{N}}^{f_\mathcal{N}} H(f+\eta+u)H(f-u)S_G(u)\,du\right|^2. \end{aligned} \quad (211b)$$

Although this result is exact for Gaussian processes (and, as is shown in Exercise [6.28a], can be computed using the inverse Fourier transforms of $H(\cdot)$ and $S_G(\cdot)$), it is unwieldy in practice, so we seek an approximation to it. Thomson (1977) points out that the second integral in Equation (211b) is large only for f near 0 and $f_\mathcal{N}$ – we exclude these cases here – so that the first integral is usually dominant. Suppose the central lobe of $H(\cdot)$ goes from $-W$ to W approximately. If the frequency separation η is less than $2W$ so that the central lobes of the shifted versions of $H(\cdot)$ in the first term overlap (cf. Figure 206(a) and (b)), then the covariance of Equation (211b) depends primarily on the SDF in the domain $[f + \eta - W, f + W]$. If we take the SDF to be locally constant about f, we obtain

$$\begin{aligned}\operatorname{cov}\{\hat{S}_G^{(D)}(f+\eta), \hat{S}_G^{(D)}(f)\} &\approx \frac{1}{\Delta_t^2}\left|\int_{-f_\mathcal{N}}^{f_\mathcal{N}} H^*(f+\eta-u)H(f-u)S_G(u)\,du\right|^2 \quad (211c)\\ &\approx \frac{S_G^2(f)}{\Delta_t^2}\left|\int_{-f_\mathcal{N}}^{f_\mathcal{N}} H(u)H(\eta-u)\,du\right|^2 = \frac{S_G^2(f)}{\Delta_t^2}|H * H(\eta)|^2,\end{aligned}$$

where $*$ denotes convolution as defined by Equation (99b) – here we have made use of the facts that $H(\cdot)$ is periodic with period $2f_\mathcal{N} = 1/\Delta_t$ and that $H^*(-u) = H(u)$.

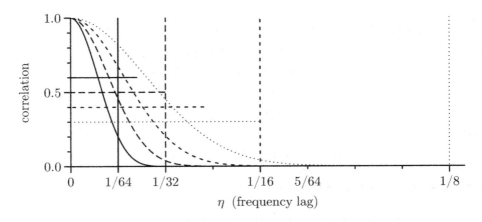

Figure 212. Correlation of direct spectral estimators. The frequency domain correlation $R(\eta)$ is plotted versus the frequency lag η for the four Slepian data tapers shown in the right-hand column of Figure 190 (recall that $N = 64$ and $\Delta_t = 1$). The cases $NW = 1, 2, 4$ and 8 are shown, respectively, by the solid, long dashed, short dashed and dotted curves; the corresponding vertical lines indicate the location of W for these four cases (1/64, 1/32, 1/16 and 1/8 cycles per unit time); and the horizontal lines depict the bandwidth measure width$_e \{R(\cdot)\}$ defined by Equation (214a).

▷ **Exercise [212]** Show that the above covariance can be written in terms of the (real-valued) data taper as

$$\text{cov}\{\hat{S}_G^{(D)}(f+\eta), \hat{S}_G^{(D)}(f)\} \approx S_G^2(f) \left|\sum_{t=0}^{N-1} h_t^2 e^{-i2\pi\eta t \Delta_t}\right|^2 \stackrel{\text{def}}{=} R(\eta, f).$$

◁

If we let $\eta = 0$, we obtain $R(0, f) = S_G^2(f)$ because of the normalization $\sum_{t=0}^{N-1} h_t^2 = 1$. This implies $\text{var}\{\hat{S}_G^{(D)}(f)\} \approx R(0, f) = S_G^2(f)$, in agreement with our earlier quoted result for $|f| \neq 0$ or $\pm f_\mathcal{N}$ (see Equation (204c)). Under our working assumption that $S_G(f+\eta) \approx S_G(f)$, we have $\text{var}\{\hat{S}_G^{(D)}(f+\eta)\} \approx \text{var}\{\hat{S}_G^{(D)}(f)\} \approx R(0, f)$. It follows that the correlation between $\hat{S}_G^{(D)}(f+\eta)$ and $\hat{S}_G^{(D)}(f)$ is given approximately by

$$R(\eta) \stackrel{\text{def}}{=} \frac{R(\eta, f)}{R(0, f)} = \left|\sum_{t=0}^{N-1} h_t^2 e^{-i2\pi\eta t \Delta_t}\right|^2, \quad (212a)$$

which depends only on the data taper and not upon either f or $S_G(\cdot)$ (Thomson, 1982).

The value $R(\eta)$ thus gives us an approximation to the correlation between $\hat{S}_G^{(D)}(f+\eta)$ and $\hat{S}_G^{(D)}(f)$ under the assumption that $S_G(\cdot)$ is constant in the interval $[f+\eta-W, f+W]$, where W is the half-width of the central lobe of the spectral window $\mathcal{H}(\cdot)$ associated with $\hat{S}_G^{(D)}(\cdot)$. Figure 212 shows $R(\cdot)$ for the four Slepian data tapers shown in the right-hand column of Figure 190 (here $N = 64$ and $\Delta_t = 1$). The grid of the Fourier frequencies $f_k = k/N$ for this example has a spacing (or grid size) of $1/N = 1/64$. The tick marks on the η-axis show the grid over $k = 0, \ldots, 8$. For a white noise process – a case where there is no bias in the periodogram so tapering is not needed – values of the periodogram at Fourier frequencies $f_k = k/N$ and $f_{k+1} = (k+1)/N$ are uncorrelated, whereas the corresponding values for a direct spectral estimator using a Slepian data taper with $NW = K$ are positively correlated, with the correlation increasing as K increases (this assumes f_k and f_{k+1} are not too close to either 0 or $f_\mathcal{N} = 1/2$).

Since $R(\cdot)$ is a $2f_\mathcal{N}$ periodic function (see Equation (212a)), its inverse Fourier transform is a sequence, say, $\{r_\tau\}$:

$$\{r_\tau\} \longleftrightarrow R(\cdot). \quad (212b)$$

To obtain an expression for r_τ in terms of $\{h_t\}$, recall that $\{\hat{s}_\tau^{(D)}\} \longleftrightarrow \hat{S}^{(D)}(\cdot)$ and note that, if we replace X_t by $h_t/\Delta_t^{1/2}$ in Equation (186b) for $\hat{S}^{(D)}(\cdot)$, the latter reduces to $R(\cdot)$. With this substitution,

Equation (188b) tells us that

$$r_\tau = \begin{cases} (1/\Delta_t) \sum_{t=0}^{N-|\tau|-1} h_{t+|\tau|}^2 h_t^2, & |\tau| \leq N-1; \\ 0, & |\tau| \geq N. \end{cases} \quad (213a)$$

We note three facts: (1) the analog of Equation (188a) yields

$$R(\eta) = \Delta_t \sum_{\tau=-(N-1)}^{N-1} r_\tau e^{-i2\pi\eta\tau\Delta_t}; \quad (213b)$$

(2) since $\{r_\tau\}$ is a "rescaled ACVS" for $\{h_t^2\}$, it can be calculated efficiently using techniques discussed in Section 6.7; and (3) since Equation (212a) tells us that $R(0) = 1$, we have

$$\Delta_t \sum_{\tau=-(N-1)}^{N-1} r_\tau = 1.$$

In addition, under our working assumption that $S_G(\cdot)$ is locally constant, $\hat{S}_G^{(D)}(f)$ and $\hat{S}_G^{(D)}(f)/S_G(f)$ differ only by a constant in the interval $[f + \eta - W, f + W]$. Since

$$R(\eta) \approx \text{cov}\,\{\hat{S}_G^{(D)}(f+\eta)/S_G(f), \hat{S}_G^{(D)}(f)/S_G(f)\}$$

(i.e., the direct spectral estimator is locally stationary), it follows that the "spectrum" of the spectral estimator $\hat{S}_G^{(D)}(\cdot)$ is given by $\{S_G^2(f)r_\tau\}$, where τ is a pseudo-lag variable. Setting $\eta = 0$ in Equation (213b) yields

$$R(0) = \Delta_t \sum_{\tau=-(N-1)}^{N-1} r_\tau \quad \text{and hence} \quad S_G^2(f)R(0) = \Delta_t \sum_{\tau=-(N-1)}^{N-1} S_G^2(f)r_\tau,$$

which is a relationship analogous to the usual one between an ACVS and its SDF at zero frequency (see Equation (113d)).

[3] For the practitioner, the effect of the restriction in the preceding discussion to frequencies "not too close to 0 or Nyquist" is that – for *finite* sample sizes N from a process with zero mean – the χ_2^2-based distribution for the periodogram in Equation (203e) is only appropriate for $1/(N\Delta_t) \leq |f| \leq f_\mathcal{N} - 1/(N\Delta_t)$. Likewise, the χ_2^2 result in Equation (204b) for general direct spectral estimators $\hat{S}^{(D)}(f)$ is reasonable for $B_\mathcal{H}/2 \leq |f| \leq f_\mathcal{N} - B_\mathcal{H}/2$ or, more conservatively, $B_\mathcal{H} \leq |f| \leq f_\mathcal{N} - B_\mathcal{H}$, where $B_\mathcal{H}$ is the effective bandwidth of $\hat{S}^{(D)}(\cdot)$ (see Exercise [6.30] for a specific example).

Another warning about Equations (203e) and (204b) is that the χ_1^2 result for $f = 0$ is no longer valid if we center the time series by subtracting the sample mean from each observation prior to forming the periodogram or other direct spectral estimators, as would be called for in practical applications where typically we do not know the process mean (again, see Exercise [6.30] for an example). In particular, as we noted in C&E [3] for Section 6.3, centering a time series forces the value of the periodogram at $f = 0$ to be identically equal to zero, indicating that the result $\hat{S}^{(P)}(0) \stackrel{d}{=} S(0)\chi_1^2$ cannot hold (with a possible exception if $S(0) = 0$).

[4] The bandwidth of an SDF estimator is sometimes regarded as a measure of the minimal separation in frequency between approximately uncorrelated SDF estimates (Priestley, 1981, p. 519; Walden et al., 1995). Since, in the Gaussian case, the correlation between $\hat{S}^{(D)}(f+\eta)$ and $\hat{S}^{(D)}(f)$ is approximated by $R(\eta)$ of Equation (212a), we need to determine the point at which $R(\cdot)$ can be regarded as being close to zero (see Figure 212). A convenient measure of this point is $\text{width}_e\,\{R(\cdot)\}$, the equivalent width of $R(\cdot)$. This measure is best suited for a function that is real-valued, positive everywhere, peaked about 0 and continuous at 0. The function $R(\cdot)$ satisfies these conditions. From Equation (100d), we have

$$\text{width}_e\,\{R(\cdot)\} = \int_{-f_\mathcal{N}}^{f_\mathcal{N}} R(\eta)\,d\eta \Big/ R(0) = \int_{-f_\mathcal{N}}^{f_\mathcal{N}} R(\eta)\,d\eta$$

	\multicolumn{4}{c	}{$p \times 100\%$ Cosine Taper}	\multicolumn{4}{c}{Slepian Taper}
$p =$	0.0	0.2	0.5	1.0	$NW =$ 1	2	4	8
$B_\mathcal{H}/\Delta_f$	1.50	1.56	1.72	2.06	1.59	2.07	2.86	4.01
width$_e\{R(\cdot)\}/\Delta_f$	1.00	1.11	1.35	1.93	1.40	1.99	2.81	3.97
$B_\mathcal{H}$/width$_e\{R(\cdot)\}$	1.50	1.41	1.27	1.06	1.13	1.04	1.02	1.01

Table 214. Comparison of two bandwidth measures for direct spectral estimators $\hat{S}^{(D)}(\cdot)$. The first measure $B_\mathcal{H}$ is given by Equation (194) and is the autocorrelation width of the spectral window associated with $\hat{S}^{(D)}(\cdot)$. The second measure width$_e\{R(\cdot)\}$ is given by Equation (214a) and is essentially the minimal distance η such that $\hat{S}^{(D)}(f+\eta)$ and $\hat{S}^{(D)}(f)$ are approximately uncorrelated (assuming that the process is Gaussian and that f is not too close to either zero or Nyquist frequency). The tabulated values show the ratio of these measures to either Δ_f (first two rows) or one another (third row) for the eight data tapers of length $N = 64$ shown in Figure 190, where $\Delta_f = 1/(N\,\Delta_t)$ is the spacing between adjacent Fourier frequencies $f_k = k/(N\,\Delta_t)$.

because $R(0) = 1$. The right-hand side is related to the notion of *correlation time*, which, in its time-domain formulation, measures the time needed for any correlation in a stationary process to die out (Yaglom, 1987a, p. 113; other names for this notion are *decorrelation time* and *integral time scale*). Since $\{r_\tau\} \longleftrightarrow R(\cdot)$ implies that

$$r_\tau = \int_{-f_\mathcal{N}}^{f_\mathcal{N}} R(\eta) e^{i2\pi\eta\tau\,\Delta_t}\,d\eta \quad \text{and hence} \quad r_0 = \int_{-f_\mathcal{N}}^{f_\mathcal{N}} R(\eta)\,d\eta,$$

we find that, in combination with Equation (213a),

$$\text{width}_e\{R(\cdot)\} = r_0 = \frac{1}{\Delta_t}\sum_{t=0}^{N-1} h_t^4, \qquad (214a)$$

which is very easy to calculate. The horizontal lines on Figure 212 show width$_e\{R(\cdot)\}$ for four Slepian data tapers.

It is of interest to compare width$_e\{R(\cdot)\}$ with our standard measure of the effective bandwidth of a direct spectral estimator, namely, $B_\mathcal{H}$ of Equation (194), where $\mathcal{H}(\cdot)$ is the spectral window associated with $\hat{S}^{(D)}(\cdot)$. The first line of Table 214 shows the ratio of $B_\mathcal{H}$ to Δ_f for the eight data tapers of length $N = 64$ shown in Figure 190, where $\Delta_f = 1/(N\,\Delta_t)$ is the distance between adjacent Fourier frequencies. The second and third lines show the ratios width$_e\{R(\cdot)\}/\Delta_f$ and $B_\mathcal{H}$/width$_e\{R(\cdot)\}$. In all cases $B_\mathcal{H}$ is greater than width$_e\{R(\cdot)\}$, implying our standard measure is such that, if $0 < f < f' < f_\mathcal{N}$ and $|f - f'| \geq B_\mathcal{H}$, then corr$\{\hat{S}^{(D)}(f), \hat{S}^{(D)}(f')\} \approx 0$. The difference between the two measures is greatest for the default (rectangular) data taper and decreases as the degree of tapering increases, i.e., as either p or NW increases. (The values in the table remain virtually the same if we redo the computations using $N = 1024$ rather than $N = 64$.)

[5] The result stated in Equation (204a), namely, that the periodogram is asymptotically uncorrelated over the grid of Fourier frequencies, needs some clarification. A firm basis for this result is Corollary 7.2.1 of Fuller (1996), from which we can deduce that, if $\{X_t\}$ can be expressed as

$$X_t = \sum_{j=0}^{\infty} \theta_j \epsilon_{t-j} \quad \text{with} \quad \sum_{j=1}^{\infty} j|\theta_j| < \infty, \qquad (214b)$$

where $\{\epsilon_t\}$ is a Gaussian white noise process with zero mean and finite variance, then

$$\left|\text{cov}\{\hat{S}^{(P)}(f_j), \hat{S}^{(P)}(f_k)\}\right| \leq \frac{C}{N^2},$$

where C is a finite constant independent of f_j or f_k. The conditions for this result to hold are restrictive, but reasonable to assume in certain practical applications. For example, a Gaussian AR(1) process $X_t = \phi X_{t-1} + \epsilon_t$ with $|\phi| < 1$ can be represented as above with $\theta_j = \phi^j$ (see Equation (44b)). This stationary process is commonly used as a simple model for climate time series, in part due to its interpretation as a discretized first-order differential equation (von Storch and Zwiers, 1999).

▷ **Exercise [215]** Show that the summation condition on $\{\theta_j\}$ holds here since

$$\sum_{j=1}^{\infty} j|\theta_j| = \frac{|\phi|}{(1-|\phi|)^2}.$$

◁

If some of the stipulations surrounding Equation (214b) cannot reasonably be assumed to hold, subtle issues arise. For example, if $\{X_t\}$ can be expressed as

$$X_t = \sum_{j=0}^{\infty} \theta_j \epsilon_{t-j} \quad \text{with} \quad \sum_{j=1}^{\infty} |\theta_j| < \infty,$$

where $\{\epsilon_t\}$ are now independent and identically distributed RVs with zero mean and finite variance, and if the SDF for $\{X_t\}$ is positive at all frequencies, then Theorem 7.1.2 of Fuller (1996) implies that, for a *finite* set of *fixed* frequencies, say $0 < f_{(1)} < f_{(2)} < \cdots < f_{(J)} < f_\mathcal{N}$, the RVs $\hat{S}^{(\mathrm{P})}(f_{(1)})$, $\hat{S}^{(\mathrm{P})}(f_{(2)}), \ldots, \hat{S}^{(\mathrm{P})}(f_{(J)})$, are asymptotically independent with distributions dictated by $S(f_{(j)})\chi_2^2/2$, $j = 1, \ldots, J$. This same result follows from Theorem 5.2.6 of Brillinger (1981a), but now under the assumption that $\{X_t\}$ is a completely stationary process with all of its moments existing and with its cumulants of all orders being absolutely summable (cumulants can be constructed directly from moments). Unfortunately this result is sometimes misinterpreted as being valid over the grid of Fourier frequencies (see the discussion by Anderson of Diggle and al Wasel, 1997).

For additional discussion on the large-sample distribution of the periodogram and other direct spectral estimators under various assumptions, see Fay and Soulier (2001), Kokoszka and Mikosch (2000) and Lahiri (2003a).

[6] It is sometimes of interest to test the hypothesis that a particular time series can be regarded as a segment $X_0, \ldots, X_{N-1}$ of a realization of a white noise process (see Section 10.15). One of the most popular tests for white noise is known as the *cumulative periodogram test*. This test is based upon the periodogram over the set of Fourier frequencies $f_k = k/(N \Delta_t)$ such that $0 < f_k < f_\mathcal{N}$, i.e., $f_1, \ldots, f_M$ with $M = \lfloor (N-1)/2 \rfloor$. As noted in the discussion surrounding Equation (203b), under the null hypothesis that $\{X_t\}$ is a Gaussian white noise process, $\hat{S}^{(\mathrm{P})}(f_1), \ldots, \hat{S}^{(\mathrm{P})}(f_M)$ constitute a set of independent and identically distributed RVs having a scaled chi-square distribution with two degrees of freedom, i.e., an exponential distribution. If we form the normalized cumulative periodogram

$$\mathcal{P}_k \stackrel{\mathrm{def}}{=} \frac{\sum_{j=1}^{k} \hat{S}^{(\mathrm{P})}(f_j)}{\sum_{j=1}^{M} \hat{S}^{(\mathrm{P})}(f_j)}, \qquad k = 1, \ldots, M-1, \tag{215a}$$

the $\mathcal{P}_k$ terms have the same distribution as an ordered random sample of $M-1$ RVs from the uniform distribution over the interval $[0, 1]$ (Bartlett, 1955). We can thus base our test for white noise on the well-known *Kolmogorov goodness-of-fit test* for a completely specified distribution (Conover, 1999). If we let

$$D^+ \stackrel{\mathrm{def}}{=} \max_{1 \le k \le M-1} \left(\frac{k}{M-1} - \mathcal{P}_k \right) \quad \text{and} \quad D^- \stackrel{\mathrm{def}}{=} \max_{1 \le k \le M-1} \left(\mathcal{P}_k - \frac{k-1}{M-1} \right),$$

we can reject the null hypothesis of white noise at the α level of significance if the Kolmogorov test statistic $D \stackrel{\mathrm{def}}{=} \max(D^+, D^-)$ exceeds $D(\alpha)$, the upper $\alpha \times 100\%$ percentage point for the distribution of D under the null hypothesis. A simple approximation to $D(\alpha)$ is given by

$$\tilde{D}(\alpha) \stackrel{\mathrm{def}}{=} \frac{C(\alpha)}{(M-1)^{1/2} + 0.12 + 0.11/(M-1)^{1/2}}, \tag{215b}$$

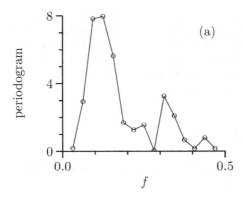 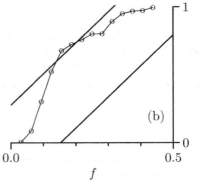

Figure 216. Periodogram (a) and associated normalized cumulative periodogram (b) for the first 32 values of the AR(2) series in Figure 34(a). The periodogram is plotted on a linear/linear scale.

where $C(0.10) = 1.224$, $C(0.05) = 1.358$ and $C(0.01) = 1.628$ (Stephens, 1974). For all $M \geq 7$, this approximation is good to within 1% for the three stated values of α (i.e., $0.99 < \tilde{D}(\alpha)/D(\alpha) < 1.01$ for $\alpha = 0.10, 0.05$ and 0.01). Values for $D(\alpha)$ can be found in Table A13 of Conover (1999).

We can obtain a useful graphical equivalent of this test by plotting $\mathcal{P}_k$ versus f_k for $k = 1, \ldots, M-1$ along with the lines defined by

$$L_u(f) \stackrel{\text{def}}{=} \frac{fN\Delta_{\text{t}} - 1}{M - 1} + \tilde{D}(\alpha) \text{ and } L_l(f) \stackrel{\text{def}}{=} \frac{fN\Delta_{\text{t}}}{M - 1} - \tilde{D}(\alpha). \tag{216}$$

If any of the points given by $(f_k, \mathcal{P}_k)$ falls outside the region between these two lines, then we reject the null hypothesis at a level of significance of α; conversely, if all of the points fall inside this region, then we fail to reject the null hypothesis at the stated level.

As a concrete example, let us consider a short time series consisting of the first $N = 32$ values of the AR(2) series of Figure 34(a). Figure 216(a) shows $\hat{S}^{(\text{P})}(f_j)$ for this series versus $f_j = j/32$ for $j = 1, \ldots, M = 15$ (recall that $\Delta_{\text{t}} = 1$). The jagged thin curve in 216(b) depicts the corresponding normalized cumulative periodogram $\mathcal{P}_k$ versus f_k for $k = 1, \ldots, M - 1 = 14$. The two thick lines are $L_u(\cdot)$ and $L_l(\cdot)$ corresponding to a significance level of $\alpha = 0.05$; specifically, since here $\tilde{D}(0.05) \doteq 0.349$ (for the record, $D(0.05) \doteq 0.349$ also), we have

$$L_u(f) \doteq \frac{32}{14}f - \frac{1}{14} + 0.349 \text{ and } L_l(f) \doteq \frac{32}{14}f - 0.349.$$

Since the normalized cumulative periodogram is not entirely contained in the region between these two lines, we reject the null hypothesis of white noise at the 0.05 level of significance. Equivalently, the values of D^+ and D^- are, respectively, 0.066 and 0.392, so the value of D is 0.392, which – since it exceeds $\tilde{D}(0.05) \doteq 0.349$ – tells us to reject the null hypothesis. We note that D^- is the value associated with $f_5 = 5/32$; i.e., $D^- = \mathcal{P}_5 - 4/14$ with $\mathcal{P}_5 \doteq 0.6773$. (For other usages of this test, see the discussions surrounding Figure 231, C&E [1] for Section 10.9, Figure 574 and Exercise [10.13].)

Finally, we note that the cumulative periodogram test requires computation of the periodogram over the Fourier frequencies. When, for some sample sizes N, we cannot use an FFT algorithm directly to compute the required DFT of $X_0, \ldots, X_{N-1}$, we can use the chirp transform algorithm to obtain the required DFT efficiently (see the discussion in C&E [1] for Section 3.11).

[7] We noted in C&E [2] for Section 6.3 that leakage in the periodogram can be attributed to a mismatch between the extremes of a time series. While tapering is one approach for eliminating this mismatch, another is to extend the series $X_0, X_1, \ldots, X_{N-1}$ via reflection to obtain the series of length $2N$ depicted in Equation (183b) and to compute the periodogram of the reflection-extended series (throughout this discussion, we assume $\Delta_{\text{t}} = 1$ for convenience).

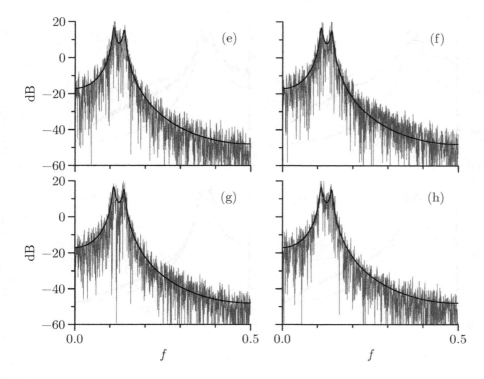

Figure 217. DCT-based periodograms (rough-looking light curves) for the four AR(4) time series of length $N = 1024$ shown in Figure 34, along with the true SDF for the AR(4) process (smooth thick curves). The DCT-periodograms are computed over the grid of frequencies $\tilde{f}_k = k/(2N)$. The usual periodograms are shown in Figure 182 over this same grid and in Figure 173 over the grid of Fourier frequencies $f_k = k/N$.

▷ **Exercise [217]** Show that the periodogram of the reflection-extended series over its Fourier frequencies $\tilde{f}_k = k/(2N)$ is given by

$$\tilde{S}^{(\text{DCT})}(\tilde{f}_k) = \frac{2}{N}\left(\sum_{t=0}^{N-1} X_t \cos\left(\pi\tilde{f}_k[2t+1]\right)\right)^2 = \frac{2}{N}\left(\sum_{t=0}^{N-1} X_t \cos\left(\frac{\pi k[2t+1]}{2N}\right)\right)^2.$$

Show also that this estimator is useless at $\tilde{f}_k = 1/2$ since $\tilde{S}^{(\text{DCT})}(1/2) = 0$ always.   ◁

The rationale for the superscript on $\tilde{S}^{(\text{DCT})}(\cdot)$ is that this estimator can related to the so-called *type-II discrete cosine transform* (DCT–II; see Ahmed et al., 1974, and Strang, 1999, for background). One definition for the DCT–II of $X_0, X_1, \ldots, X_{N-1}$ is

$$\mathcal{C}_k = \left(\frac{2-\delta_k}{N}\right)^{1/2} \sum_{t=0}^{N-1} X_t \cos\left(\frac{\pi k[2t+1]}{2N}\right), \quad k = 0, 1, \ldots, N-1$$

(Gonzalez and Woods, 2007), where δ_k is Kronecker's delta function (i.e., $\delta_k = 1$ if $k = 0$ and $= 0$ otherwise). Hence $\tilde{S}^{(\text{DCT})}(\tilde{f}_k) = \mathcal{C}_k^2$ for $k = 1, 2, \ldots, N-1$, while $\tilde{S}^{(\text{DCT})}(\tilde{f}_0) = 2\mathcal{C}_0^2$. Because of this connection, we define the *DCT-based periodogram* to be

$$\hat{S}^{(\text{DCT})}(\tilde{f}_k) = \mathcal{C}_k^2, \quad k = 0, 1, \ldots, N-1,$$

which differs from $\tilde{S}^{(\text{DCT})}(\tilde{f}_k)$ only at $k = 0$ (Chan et al., 1995, use the term *semiperiodogram* for a related concept).

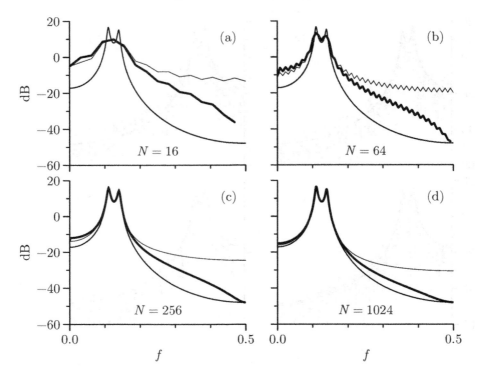

Figure 218. Expected value of periodogram $\hat{S}^{(\mathrm{P})}(\tilde{f}_k)$ (thinnest curve) and of DCT-based periodogram $\hat{S}^{(\mathrm{DCT})}(\tilde{f}_k)$ (thickest) versus $\tilde{f}_k = k/(2N)$ for AR(4) process of Equation (35a) with sample sizes $N = 16, 64, 256$ and 1024. In each plot the true SDF is the smooth curve of medium thickness.

Figure 217 shows $\hat{S}^{(\mathrm{DCT})}(\cdot)$ for the four AR(4) series displayed in Figure 34, along with the true SDF for the AR(4) process. If we compare these estimates to the corresponding periodograms depicted in Figures 173 and 182, the DCT-based periodograms appear to have less bias overall; however, they are also noisier looking, particularly in the twin-peaks area of the SDF.

To explore the bias and variance properties of $\hat{S}^{(\mathrm{DCT})}(\cdot)$ in more depth, suppose that $\{X_t\}$ is a Gaussian stationary process with zero mean and ACVS $\{s_\tau\}$. Then, for all k, $\mathcal{C}_k$ also obeys a Gaussian distribution with mean zero and variance, say, σ_k^2.

▷ **Exercise [218]** Show that

$$\sigma_0^2 = \sum_{\tau=-(N-1)}^{N-1} \left(1 - \frac{|\tau|}{N}\right) s_\tau,$$

while, for $k = 1, \ldots, N-1$,

$$\sigma_k^2 = E\{\hat{S}^{(\mathrm{P})}(\tilde{f}_k)\} - \frac{2}{N \sin(\pi k/N)} \sum_{\tau=1}^{N-1} s_\tau \sin(\pi \tau k/N). \qquad \triangleleft$$

Under the assumption that the ACVS is absolutely summable, the results of this exercise can be used to argue that, as was true for the usual periodogram, the DCT-based periodogram is asymptotically unbiased for $0 \leq f < 1/2$:

$$\lim_{N \to \infty} E\{\hat{S}^{(\mathrm{DCT})}(f)\} \to S(f); \qquad (218)$$

however, in contrast to the periodogram, $\hat{S}^{(\mathrm{DCT})}(1/2)$ does not have a useful definition.

Figure 218 compares the expected value of the usual periodogram $\hat{S}^{(\mathrm{P})}(\cdot)$ (thinnest curve in each plot) with that of the DCT-based periodogram $\hat{S}^{(\mathrm{DCT})}(\cdot)$ (thickest) over the frequencies $\tilde{f}_k$ for the AR(4) process of Equation (35a), along with the true SDF. Each plot depicts a different sample size, namely, $N = 16, 64, 256$ and 1024 (cf. Figure 178). At high frequencies $\hat{S}^{(\mathrm{DCT})}(\cdot)$ is considerably less biased than $\hat{S}^{(\mathrm{P})}(\cdot)$, but its bias is somewhat larger at low frequencies. The two estimators have very similar first moments in the high-power twin-peaks portion of the SDF. In keeping with Equation (218), the bias in $\hat{S}^{(\mathrm{DCT})}(\cdot)$ decreases as N increases.

The theory above implies that

$$\left(\frac{c_k}{\sigma_k}\right)^2 \stackrel{\mathrm{d}}{=} \chi_1^2, \quad \text{i.e.,} \quad \hat{S}^{(\mathrm{DCT})}(\tilde{f}_k) \stackrel{\mathrm{d}}{=} \sigma_k^2 \chi_1^2, \tag{219a}$$

and hence $E\{\hat{S}^{(\mathrm{DCT})}(\tilde{f}_k)\} = \sigma_k^2$ and $\mathrm{var}\,\{\hat{S}^{(\mathrm{DCT})}(\tilde{f}_k)\} = 2\sigma_k^4$ for all k. Specializing to the case of Gaussian white noise with variance $\sigma^2 = S(f)$, we find

$$E\{\hat{S}^{(\mathrm{DCT})}(\tilde{f}_k)\} = S(\tilde{f}_k) \quad \text{and} \quad \mathrm{var}\,\{\hat{S}^{(\mathrm{DCT})}(\tilde{f}_k)\} = 2S^2(\tilde{f}_k)$$

for $0 \leq \tilde{f}_k < 1/2$; by contrast, we have

$$E\{\hat{S}^{(\mathrm{P})}(f_k)\} = S(f_k) \quad \text{and} \quad \mathrm{var}\,\{\hat{S}^{(\mathrm{P})}(f_k)\} = S^2(f_k)$$

for $0 < f_k < 1/2$. The DCT-based periodogram thus has twice the variance of the usual periodogram. This result holds approximately for colored Gaussian processes, which explains why the DCT-based periodograms in Figure 217 are markedly choppier looking than their $\hat{S}^{(\mathrm{P})}(\cdot)$ counterparts in Figures 173 and 182.

We leave it as an exercise for the reader to show that, for Gaussian white noise with mean zero and variance σ^2, the RVs $\hat{S}^{(\mathrm{DCT})}(\tilde{f}_k)$, $k = 0, 1, \ldots, N-1$, are independent and identically distributed with a distribution dictated by $\sigma^2 \chi_1^2$. The collective number of degrees of freedom is thus N. By contrast, the RVs $\hat{S}^{(\mathrm{P})}(f_k)$, $k = 0, 1, \ldots, \lfloor N/2 \rfloor$, are independent, but not identically distributed: the distribution for $k = 1, \ldots, \lfloor (N-1)/2 \rfloor$ is given by $\sigma^2 \chi_2^2/2$, whereas that for $k = 0$ and $k = N/2$ (when N is an even sample size) is dictated by $\sigma^2 \chi_1^2$. For both even and odd sample sizes, the collective number of degrees of freedom is still N. Ignoring $f = 0$ and $f = 1/2$, while the RVs in the DCT-based periodogram have twice the variability as those of the usual periodogram, this deficiency is compensated for by the fact that there are (approximately) twice as many independent RVs in the DCT-based periodogram.

For additional discussion on the use of the DCT–II for SDF estimation, see Davies (2001) and Narasimhan and Harish (2006). Also the celebrated physicist Albert Einstein wrote a short article on spectral analysis (Einstein, 1914), in which he proposed smoothing a basic SDF estimator that is essentially based on a discrete cosine transform; moreover, his use of this transform evidently arose from the (then) common practice of using reflection-extended time series in conjunction with Fourier analysis (Yaglom, 1987b).

Exercise [6.31] invites the reader to redo the key results above for the alternative way of constructing a reflected series displayed in Equation (183c).

6.7 Computational Details

Here we give some details about computing the periodogram and other direct spectral estimators. We shall strive for clarity rather than get into issues of computational efficiency. All computations are done in terms of DFTs and inverse DFTs, so we first review the following pertinent results.

Recall that the DFT for a finite sequence $g_0, \ldots, g_{M-1}$ with a sampling interval of unity is the sequence $G_0, \ldots, G_{M-1}$, where, with $\tilde{f}'_k = k/M$,

$$G_k = \sum_{t=0}^{M-1} g_t e^{-\mathrm{i}2\pi kt/M} = \sum_{t=0}^{M-1} g_t e^{-\mathrm{i}2\pi \tilde{f}'_k t} \tag{219b}$$

and
$$g_t = \frac{1}{M}\sum_{k=0}^{M-1} G_k e^{i2\pi kt/M} = \frac{1}{M}\sum_{k=0}^{M-1} G_k e^{i2\pi \tilde{f}'_k t} \qquad (220a)$$

(see Equations (91b) and (92a)). This relationship is summarized by the notation $\{g_t\} \longleftrightarrow \{G_k\}$. Both of these sequences are defined outside the range 0 to $M-1$ by cyclic (periodic) extension; i.e., if $t < 0$ or $t \geq M$, then $g_t \stackrel{\text{def}}{=} g_{t \bmod M}$ so that, for example, $g_{-1} = g_{M-1}$ and $g_M = g_0$. We define the cyclic autocorrelation of $\{g_t\}$ by

$$g \star g_t^* = \sum_{s=0}^{M-1} g_{s+t} g_s^*, \qquad t = 0, \ldots, M-1 \qquad (220b)$$

(cf. Equation (101h)). If $\{h_t\}$ is another sequence of length M with DFT $\{H_k\}$, i.e., $\{h_t\} \longleftrightarrow \{H_k\}$, we define the cyclic convolution of $\{g_t\}$ and $\{h_t\}$ by

$$g * h_t = \sum_{s=0}^{M-1} g_s h_{t-s}, \qquad t = 0, \ldots, M-1 \qquad (220c)$$

(see Equation (101d)). We define the frequency domain cyclic convolution of $\{G_k\}$ and $\{H_k\}$ by

$$G * H_k = \frac{1}{M} \sum_{l=0}^{M-1} G_l H_{k-l}, \qquad k = 0, \ldots, M-1 \qquad (220d)$$

(see Equation (101f)). We note that

$$\{g \star g_t^*\} \longleftrightarrow \{|G_k|^2\}, \quad \{g * h_t\} \longleftrightarrow \{G_k H_k\} \text{ and } \{g_t h_t\} \longleftrightarrow \{G * H_k\}$$

(see Equations (102a), (101e) and (101g)). Exercise [3.20a] outlines an efficient method for computing cyclic autocorrelations and convolutions.

We start with a time series of length N that can be regarded as a realization of $X_0, X_1, \ldots, X_{N-1}$, a segment of length N of the stationary process $\{X_t\}$ with sampling interval Δ_t and unknown mean μ, SDF $S(\cdot)$ and ACVS $\{s_\tau\}$. Let M be any integer satisfying $M \geq 2N-1$ (the choice $M = 2N$ is often used, particularly when N is a power of two). We let

$$\tilde{X}_t \stackrel{\text{def}}{=} \begin{cases} X_t - \overline{X}, & 0 \leq t \leq N-1; \\ 0, & N \leq t \leq M-1; \end{cases} \text{ and } \tilde{h}_t \stackrel{\text{def}}{=} \begin{cases} h_t, & 0 \leq t \leq N-1; \\ 0, & N \leq t \leq M-1, \end{cases} \qquad (220e)$$

where $\overline{X} = \sum_{t=0}^{N-1} X_t/N$ is the sample mean, and $\{h_t\}$ is a data taper of length N with normalization $\sum_{t=0}^{N-1} h_t^2 = 1$. The sequences $\{\tilde{X}_t\}$ and $\{\tilde{h}_t\}$ are each of length M and are implicitly defined outside the range $t = 0$ to $M - 1$ by cyclic extension. For convenience, we perform our computations assuming that the sampling interval for these sequences – and all others referred to in Figure 221 – is unity so that the DFT formulae of Equations (219b) and (220a) can be used (after completion of the computations, we indicate the adjustments needed to handle $\Delta_t \neq 1$). We create these two zero padded sequences so that we can compute various noncyclic convolutions using cyclic convolution (see Exercise [3.20b]). We denote the DFTs of $\{\tilde{X}_t\}$ and $\{\tilde{h}_t\}$ by, respectively, the sequences $\{\tilde{\mathcal{X}}_k\}$ and $\{\tilde{H}(\tilde{f}'_k)\}$, both of which are of length M and indexed by $k = 0, \ldots, M-1$ (note that $\tilde{\mathcal{X}}_k$ is different from $\mathcal{X}_k$ of

6.7 Computational Details

$$\{\tilde{X}_t\} \xleftrightarrow{(1)} \{\tilde{\mathcal{X}}_k\}$$

$$\downarrow \text{mult} \qquad \qquad \downarrow \text{conv}$$

$$\{\tilde{h}_t \tilde{X}_t\} \xleftrightarrow{(2)} \{\tilde{H} * \tilde{\mathcal{X}}_k\}$$

$$\downarrow \text{auto} \qquad \qquad \downarrow \text{mod sq}$$

$$\{\tilde{s}_\tau^{(\mathrm{D})}\} = \{\tilde{h}\tilde{X} \star \tilde{h}\tilde{X}_\tau\} \xleftrightarrow{(3)} \{|\tilde{H} * \tilde{\mathcal{X}}_k|^2\} = \{\tilde{S}^{(\mathrm{D})}(\tilde{f}'_k)\}$$

Figure 221. Pathways for computing $\tilde{S}^{(\mathrm{D})}(\cdot)$, from which, upon multiplication by Δ_t, we can obtain the direct spectral estimate $\hat{S}^{(\mathrm{D})}(f'_k) = \Delta_t \tilde{S}^{(\mathrm{D})}(\tilde{f}'_k)$, where $f'_k = k/(M \Delta_t)$ and $\tilde{f}'_k = k/M$ (adapted from figure 1 of Van Schooneveld and Frijling, 1981).

Equation (171a) because we have forced $\{\tilde{X}_t\}$ to have a sampling interval of unity). As usual, these Fourier relationships are summarized by

$$\{\tilde{X}_t\} \longleftrightarrow \{\tilde{\mathcal{X}}_k\} \text{ and } \{\tilde{h}_t\} \longleftrightarrow \{\tilde{H}(\tilde{f}'_k)\}.$$

Our goal is to compute either the periodogram $\hat{S}^{(\mathrm{P})}(\cdot)$ or some other direct spectral estimate $\hat{S}^{(\mathrm{D})}(\cdot)$ at frequencies $f'_k = k/(M \Delta_t)$ (note the difference between f'_k and $\tilde{f}'_k = k/M$: the former has physically meaningful units, whereas we take the latter to be unitless). Since the periodogram is a special case of $\hat{S}^{(\mathrm{D})}(\cdot)$, we first consider computation of direct spectral estimates in general and then comment afterwards on shortcuts that can be made if only the periodogram is of interest. Figure 221 depicts pathways to compute $\hat{S}^{(\mathrm{D})}(\cdot)$. These involve three Fourier transform pairs, the first of which we have already noted. Here are comments regarding the remaining two.

(2) The left-hand side is obtained from a term by term multiplication of the sequences $\{\tilde{h}_t\}$ and $\{\tilde{X}_t\}$, while its DFT (the right-hand side) can be obtained either directly or by convolving the sequences $\{\tilde{H}(\tilde{f}'_k)\}$ and $\{\tilde{\mathcal{X}}_k\}$ using Equation (220d).

(3) The left-hand side is the autocorrelation of $\{\tilde{h}_t \tilde{X}_t\}$ and can be obtained using Equation (220b). Its DFT can be obtained either directly or by computing the square modulus of $\{\tilde{H} * \tilde{\mathcal{X}}_k\}$ term by term. If we let $\tilde{s}_\tau^{(\mathrm{D})}$ be the τth element of the autocorrelation of $\{\tilde{h}_t \tilde{X}_t\}$, then we have

$$\hat{s}_\tau^{(\mathrm{D})} = \begin{cases} \tilde{s}_{|\tau|}^{(\mathrm{D})}, & |\tau| \leq N - 1; \\ 0, & \text{otherwise.} \end{cases} \qquad (221)$$

On the right-hand side, if we let $\tilde{S}^{(\mathrm{D})}(\tilde{f}'_k) \stackrel{\text{def}}{=} |\tilde{H} * \tilde{\mathcal{X}}_k|^2$, then we have

$$\hat{S}^{(\mathrm{D})}(f'_k) = \Delta_t \tilde{S}^{(\mathrm{D})}(\tilde{f}'_k), \qquad k = 0, \ldots, \lfloor M/2 \rfloor.$$

As shown in Figure 221, there are four computational pathways to get $\{\tilde{S}^{(\mathrm{D})}(\tilde{f}'_k)\}$ (and hence $\{\hat{S}^{(\mathrm{D})}(f'_k)\}$) from $\{X_t\}$. Here are two of them:

$$\{\tilde{X}_t\} \qquad\qquad\qquad \{\tilde{X}_t\} \xrightarrow{(1)} \{\tilde{\mathcal{X}}_k\}$$

$$\downarrow \text{mult} \qquad\qquad\qquad\qquad\qquad\qquad \downarrow \text{conv}$$

$$\{\tilde{h}_t \tilde{X}_t\} \xrightarrow{(2)} \{\tilde{H} * \tilde{\mathcal{X}}_k\} \qquad\qquad \{\tilde{H} * \tilde{\mathcal{X}}_k\}$$

$$\qquad\qquad\qquad \downarrow \text{mod sq} \qquad\qquad\qquad \downarrow \text{mod sq}$$

$$\qquad\qquad\qquad \tilde{S}^{(\mathrm{D})}(\tilde{f}'_k)\} \qquad\qquad\qquad \{\tilde{S}^{(\mathrm{D})}(\tilde{f}'_k)\}$$

The one on the left avoids convolution and autocorrelation (these operations can be time consuming on a digital computer if implemented via their definitions in Equations (220b) and (220d)). The right-hand pathway makes use of a convolution, which, from a computational point of view, makes it less favorable; however, Exercise [6.18a] notes the theoretically interesting fact that one version of the Hanning data taper has a DFT with just a few nonzero terms, and hence the convolution can be done relatively quickly. If we are also interested in the ACVS estimate $\{\hat{s}^{(\mathrm{D})}\}$, we can augment either pathway to obtain it by taking the inverse Fourier transform of $\{|\tilde{H} * \tilde{\mathcal{X}}_k|^2\}$. In the final two pathways, we need to create $\{\hat{s}^{(\mathrm{D})}\}$ to get to $\{\tilde{S}^{(\mathrm{D})}(\tilde{f}'_k)\}$:

$$\begin{array}{ccc}
\{\tilde{X}_t\} & & \{\tilde{X}_t\} \xrightarrow{(1)} \{\tilde{\mathcal{X}}_k\} \\
\downarrow \text{mult} & & \downarrow \text{conv} \\
\{\tilde{h}_t \tilde{X}_t\} & & \{\tilde{h}_t \tilde{X}_t\} \xleftarrow{(2)} \{\tilde{H} * \tilde{\mathcal{X}}_k\} \\
\downarrow \text{auto} & & \downarrow \text{auto} \\
\{\tilde{s}^{(\mathrm{D})}_\tau\} \xrightarrow{(3)} \tilde{S}^{(\mathrm{D})}(\tilde{f}'_k) & & \{\tilde{s}^{(\mathrm{D})}_\tau\} \xrightarrow{(3)} \tilde{S}^{(\mathrm{D})}(\tilde{f}'_k)
\end{array}$$

Neither pathway is attractive computationally, but both are of pedagogical interest.

As an example of these manipulations, let us consider a time series related to the rotation of the earth. The entire series has 100 values. These are shown, after centering by subtracting off 21.573, by the pluses in the bottoms of Figures 149 and 150. We make use of the first $N = 20$ values prior to centering. For the record, these are 7.1, 6.3, 7.0, 8.8, 9.9, 9.0, 11.0, 13.5, 12.8, 15.4, 15.6, 14.1, 13.1, 13.2, 14.1, 10.4, 13.6, 14.6, 12.4 and 12.9. The sampling interval is $\Delta_t = 1/4$ year. Figure 223 shows four Fourier transform pairs of interest, one pair per row of plots: the sequence in each right-hand plot is the DFT of the left-hand sequence. The three Fourier transform pairs in Figure 221 are shown in rows 1, 3 and 4 of Figure 223. To use a conventional "power of two" FFT, we let $M = 64$ (the smallest power of two greater than $2N - 1 = 39$). The left-hand plot of the first row thus shows $\{\tilde{X}_t\}$, a sequence of length 64, the first 20 values of which constitute the series centered using its sample mean (11.74), and the last 44 values are all zeros. The corresponding right-hand plot is its DFT – a sequence of 64 complex-valued numbers – the real parts of which are shown by the dark connected dots, and the imaginary parts by the gray ones. The second row of plots shows $\{\tilde{h}_t\}$ (here a Hanning data taper of length 20 padded with 44 zeros) and its DFT (another complex-valued sequence). The left-hand plot on the third row is $\{\tilde{h}_t \tilde{X}_t\}$. The left-hand plot on the bottom row is $\{\tilde{s}^{(\mathrm{D})}_\tau\}$, from which we can pick out $\{\hat{s}^{(\mathrm{D})}_\tau\}$; the right-hand plot is $\{\tilde{S}^{(\mathrm{D})}(\tilde{f}'_k)\}$, which is real-valued. For $k = 0$ to 32, multiplication of $\tilde{S}^{(\mathrm{D})}(\tilde{f}'_k)$ by $\Delta_t = 1/4$ yields $\hat{S}^{(\mathrm{D})}(f'_k)$, a direct spectral estimator of $S(\cdot)$ at frequency $f'_k = k/16$ cycles per year (note that $k = 32$ corresponds to the Nyquist frequency of 2 cycles per year). One comment: a data taper is of questionable utility for this short series and is employed here primarily for illustrative purposes.

Since the periodogram is a direct spectral estimator that makes use of the default (rectangular) data taper $h_t = 1/\sqrt{N}$, we could compute it as described by Figure 221. In this case $\{\tilde{H} * \tilde{\mathcal{X}}_k\}$ ends up being just the Fourier transform of $\{\tilde{X}_t/\sqrt{N}\}$, and the periodogram is gotten by multiplying $\{|\tilde{H} * \tilde{\mathcal{X}}_k|\}$ by Δ_t. We can simplify Figure 221 by dropping the second row since it really does nothing more than provide a division by $\sqrt{N}$. This leads to the pathways shown in Figure 224a, which yields $\{N \cdot \tilde{S}^{(\mathrm{P})}(\tilde{f}'_k)\}$, whereas the pathways in Figure 221 give us $\{\tilde{S}^{(\mathrm{D})}(\tilde{f}'_k)\} = \{\tilde{S}^{(\mathrm{P})}(\tilde{f}'_k)\}$. Dropping the division by $\sqrt{N}$ thus turns into a need to eventually divide by N. When $\Delta_t \neq 1$, we can combine the division by N with the multiplication by Δ_t needed to get the actual periodogram $\{\hat{S}^{(\mathrm{P})}(f'_k)\} = \{\Delta_t \tilde{S}^{(\mathrm{P})}(\tilde{f}'_k)\}$; i.e., a single multiplication by Δ_t/N gets us from $\{N \cdot \tilde{S}^{(\mathrm{P})}(\tilde{f}'_k)\}$ to the desired $\{\hat{S}^{(\mathrm{P})}(f'_k)\}$.

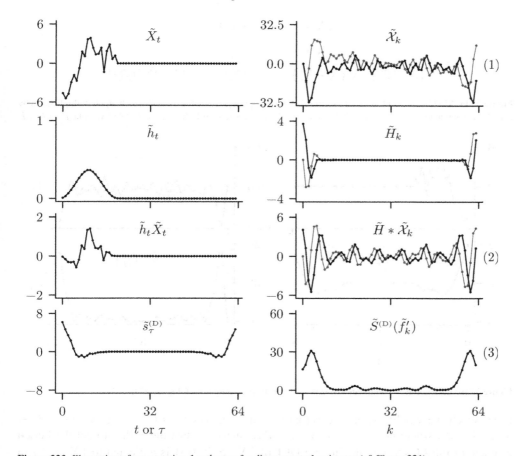

Figure 223. Illustration of computational pathways for direct spectral estimates (cf. Figure 221).

While there are two pathways from $\{\tilde{X}_t\}$ to $\{N \cdot \tilde{S}^{(P)}(\tilde{f}'_k)\}$ embedded in Figure 224a, the one of main computation interest is

$$\{\tilde{X}_t\} \xrightarrow{(1)} \{\tilde{\mathcal{X}}_k\}$$
$$\downarrow \text{mod sq}$$
$$\{N \cdot \tilde{S}^{(P)}(\tilde{f}'_k)\}.$$

This pathway avoids the autocorrelation embedded in the other pathway, but does not give us the biased ACVS estimate directly. We can obtain this estimate via the inverse DFT of $\{N \cdot \tilde{S}^{(P)}(\tilde{f}'_k)\}$, which yields $\{N \cdot \tilde{s}^{(P)}_\tau\}$, from which we can extract $\{\hat{s}^{(P)}_\tau\}$ through an analog of Equation (221). If we are only interested in computing the periodogram over the grid of Fourier frequencies $f_k = k/(N\Delta_t)$, then we could start by defining $\widetilde{\widetilde{X}}_t = X_t - \overline{X}$, $0 \leq t \leq N-1$; i.e., we eliminate the zero padding in $\{\tilde{X}_t\}$. Letting $\{\widetilde{\widetilde{\mathcal{X}}}_k\}$ denote the DFT of $\{\widetilde{\widetilde{X}}_t\}$, we would then entertain the pathway

$$\{\widetilde{\widetilde{X}}_t\} \xleftrightarrow{(1)} \{\widetilde{\widetilde{\mathcal{X}}}_k\}$$
$$\downarrow \text{mod sq}$$
$$\{|\widetilde{\widetilde{\mathcal{X}}}_k|^2\} = \{N \cdot \widetilde{\widetilde{S}}^{(P)}(\tilde{f}_k)\},$$

$$\{\tilde{X}_t\} \xleftrightarrow{(1)} \{\tilde{\mathcal{X}}_k\}$$

$$\downarrow \text{auto} \qquad \qquad \downarrow \text{mod sq}$$

$$\{N \cdot \tilde{s}_\tau^{(\text{P})}\} = \{\tilde{X} \star \tilde{X}_\tau\} \xleftrightarrow{(3)} \{|\tilde{\mathcal{X}}_k|^2\} = \{N \cdot \tilde{S}^{(\text{P})}(\tilde{f}'_k)\}$$

Figure 224a. Modification of Figure 221 appropriate for computing $N \cdot \tilde{S}^{(\text{P})}(\cdot)$, from which, upon multiplication by Δ_t/N, we obtain the periodogram $\hat{S}^{(\text{P})}(f'_k) = \Delta_t \tilde{S}^{(\text{P})}(\tilde{f}'_k)$, where, as before, $f'_k = k/(M\,\Delta_t)$ and $\tilde{f}'_k = k/M$.

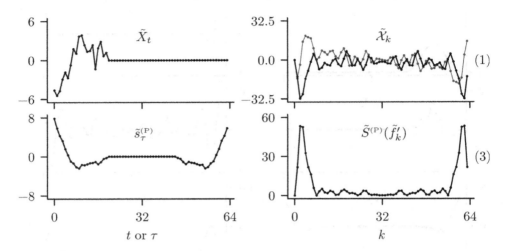

Figure 224b. Illustration of computational pathways for the periodogram (cf. Figure 224a).

from which we would extract the actual periodogram via $\hat{S}^{(\text{P})}(f_k) = \Delta_t \tilde{S}^{(\text{P})}(\tilde{f}_k)$, $k = 0, \ldots, \lfloor N/2 \rfloor$, where here $\tilde{f}_k = k/N$. We have, however, lost the ability to get the biased estimate of the ACVS; i.e., the inverse DFT of $\{N \cdot \tilde{S}^{(\text{P})}(\tilde{f}_k)\}$ is *not* equal to $\{N \cdot \tilde{s}_\tau^{(\text{P})}\}$ in general. We must stick with zero padding if we want to get the biased ACVS estimate.

Figure 224b illustrates the pathways in Figure 224a using the same 20-point time series as before. The top row is the same as that in Figure 223. The left-hand plot on the bottom row is $\{\tilde{s}_\tau^{(\text{P})}\}$, from which we can get the values for $\{\hat{s}_\tau^{(\text{P})}\}$ via an analog of Equation (221); the right-hand plot is the real-valued sequence $\{\tilde{S}^{(\text{P})}(\tilde{f}'_k)\}$ (we show $\{\tilde{s}_\tau^{(\text{P})}\}$ and $\{\tilde{S}^{(\text{P})}(\tilde{f}'_k)\}$ rather than $\{N \cdot \tilde{s}_\tau^{(\text{P})}\}$ and $\{N \cdot \tilde{S}^{(\text{P})}(\tilde{f}'_k)\}$ to facilitate comparison with Figure 223). For $k = 0$ to 32, multiplying $\tilde{S}^{(\text{P})}(\tilde{f}'_k)$ by $\Delta_t = 1/4$ yields $\hat{S}^{(\text{P})}(f'_k)$, the periodogram at frequency $f'_k = k/16$ cycles per year.

6.8 Examples of Periodogram and Other Direct Spectral Estimators

Ocean Wave Data

Figure 225 shows a plot of a time series recorded in the Pacific Ocean by a wave-follower. As the wave-follower moved up and down with the waves in the ocean, it measured the surface displacement (i.e., sea level) as a function of time. The frequency response of the wave-follower was such that – mainly due to its inertia – frequencies higher than 1 Hz could not be reliably measured. The data were originally recorded using an analog device. They were low-pass filtered in analog form using an antialiasing filter with a cutoff of approximately 1 Hz and then sampled every $1/4$ sec, yielding a Nyquist frequency of $f_\mathcal{N} = 2\,\text{Hz}$. The plotted series consists of a 256 sec portion of this data, so there are $N = 1024$ data points in all.

Figure 226 shows four direct spectral estimates for this ocean wave data, each evaluated on the grid of 512 positive Fourier frequencies. The top plot is the periodogram $\hat{S}^{(\text{P})}(\cdot)$,

6.8 Examples of Periodogram and Other Direct Spectral Estimators

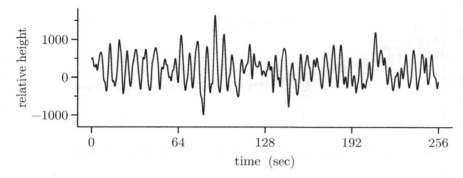

Figure 225. Ocean wave data (courtesy of A. Jessup, Applied Physics Laboratory, University of Washington). There are $N = 1024$ samples taken $\Delta_t = 1/4$ sec apart.

while the bottom three plots are direct spectral estimates $\hat{S}^{(D)}(\cdot)$ using a Slepian data taper with NW parameters of $1/\Delta_t$, $2/\Delta_t$ and $4/\Delta_t$. We centered the time series by subtracting off the sample mean $\overline{X} \doteq 209.1$ prior to computing all four spectral estimates. We used Equation (196b) to form the Slepian tapers $\{h_t\}$ and normalized them such that $\sum h_t^2 = 1$.

All four spectral estimates show a broad, low frequency peak at 0.160 Hz, corresponding to a period of 6.2 sec (the location of this frequency is marked by a thin vertical line in the upper left-hand corner of the top plot). While the dominant features of the time series can be attributed to this broad peak and other features in the frequency range 0 to 0.2 Hz, the data were actually collected to investigate whether the rate at which the SDF decreases over the range 0.2 to 1.0 Hz is in fact consistent with calculations based upon a physical model. The range from 1 to 2 Hz is of little physical interest because it is dominated by instrumentation and preprocessing (i.e., inertia in the wave-follower and the effect of the antialiasing filter); nonetheless, it is of operational interest to examine this portion of the SDF to check that the spectral levels and shape are in accord with the frequency response claimed by the manufacturer of the wave-follower, the transfer function of the antialiasing filter, and a rough guess at the SDF for ocean waves from 1 to 2 Hz.

An examination of the high frequency portions of the SDFs in Figure 226 shows evidence of bias due to leakage in both the periodogram and the direct spectral estimate with the $NW = 1/\Delta_t$ data taper. The evidence is twofold. First, the levels of these two spectral estimates at high frequencies are considerably higher than those for the $NW = 2/\Delta_t$ and $NW = 4/\Delta_t$ direct spectral estimates (by about 25 dB for the periodogram and 10 dB for the $NW = 1/\Delta_t$ spectral estimate). Second, the local variability in the periodogram is markedly less in the high frequencies as compared to the low frequencies, an indication of leakage we have seen previously (see Figures 173(e) and (f)). The same is true to a lesser extent for the $NW = 1/\Delta_t$ spectral estimate.

As an interesting aside, in Figure 227 we have replotted the periodogram but now over a grid of frequencies twice as fine as the Fourier frequencies – recall that the plots of Figure 226 involve just the Fourier frequencies. This new plot shows an *increase* in the local variability of the periodogram in the high frequencies as compared to the low frequencies. If we were to expand the frequency scale on the plot, we would see that the increased variability is in the form of a ringing; i.e., the SDF alternates between high and low values. Had we seen this plot first, this ringing would have told us that leakage might be a problem. The key point here is that, when plotted on a logarithmic scale (such as decibels), the periodogram and other direct spectral estimators should exhibit approximately the same local variability across all frequencies – if this is not so, it is important to understand why not. Leakage causes the

226 Periodogram and Other Direct Spectral Estimators

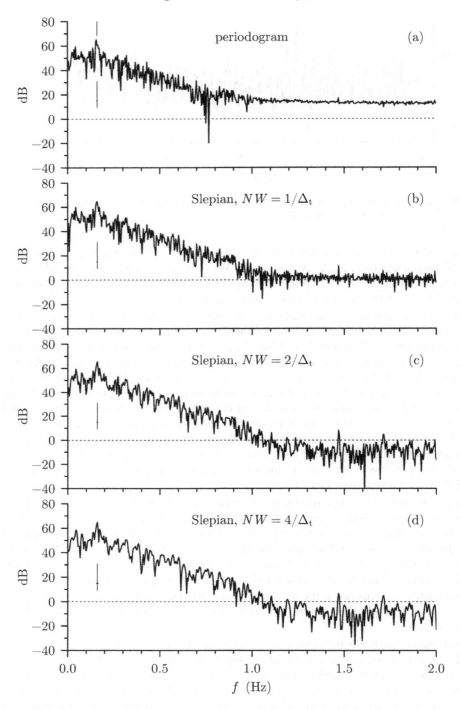

Figure 226. Direct spectral estimates for ocean wave data. The (barely visible) width of the crisscross below each spectral estimate gives the bandwidth measure $B_{\mathcal{H}}$ as per Equation (194), while its height gives the length of a 95% CI for $10\,\log_{10}(S(f))$ as per Equation (205b).

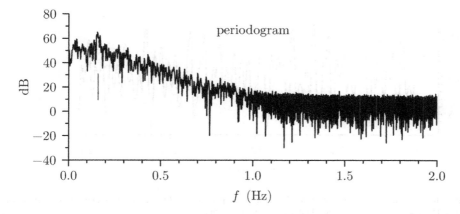

Figure 227. Periodogram for ocean wave data. Here the periodogram is plotted over a grid of frequencies twice as fine as the Fourier frequencies, whereas in the top plot of Figure 226 it is plotted over just the Fourier frequencies. The strange appearance of this plot is due to the fact that, because of leakage, the periodogram has a nearly constant value of approximately 12 dB at all high Fourier frequencies, whereas its values are considerably below 12 dB over most of the grid of high frequencies halfway between the Fourier frequencies.

ringing here, but other potential causes are outliers (see, for example, Figure 21 of Martin and Thomson, 1982) and echoes (Bogert et al., 1963).

As in Figures 207 and 208, we have added crisscrosses to the left-hand portions of all the plots in Figures 226 and 227. These are a useful guide in assessing the sampling variability in the depicted spectral estimates because they provide, after being mentally centered on a particular value for $10\log_{10}(\hat{S}^{(P)}(f))$ or $10\log_{10}(\hat{S}^{(D)}(f))$, a 95% CI for $10\log_{10}(S(f))$. The horizontal width of each crisscross shows $B_{\mathcal{H}}$ of Equation (194), which is our standard measure of the effective bandwidth of a direct spectral estimator. Recall that, if $0 < f < f' < f_{\mathcal{N}}$ and $|f - f'| \geq B_{\mathcal{H}}$, we can regard $\hat{S}^{(D)}(f)$ and $\hat{S}^{(D)}(f')$ as being approximately uncorrelated. The widths (just barely visible!) of the crisscrosses in the four plots of Figure 226 are, from top to bottom, 0.0059 Hz, 0.0062 Hz, 0.0081 Hz and 0.0113 Hz. Under the assumption that $S(\cdot)$ is slowly varying, the crisscross gives us a rough idea of the local variability we can expect to see in $\hat{S}^{(D)}(\cdot)$ across frequencies if all of the assumptions behind spectral analysis hold. If f and f' are closer together than the width of the crisscross, we can expect $\hat{S}^{(D)}(f)$ and $\hat{S}^{(D)}(f')$ to be similar to some degree because they are positively correlated; on the other hand, if f and f' are farther apart than this width, $\hat{S}^{(D)}(f)$ and $\hat{S}^{(D)}(f')$ should exhibit a variation consistent with the height of the crisscross. For example, the high frequency portion of the periodogram in the upper plot of Figure 226 shows much less variability from frequency to frequency than we would expect from the height of the crisscross, while, on the finer scale of Figure 227, the local variability is both too large and on too fine a scale (i.e., rapid variations occur on a scale less than the effective bandwidth $B_{\mathcal{H}}$). In contrast, the low frequency portion of the periodogram does not exhibit any gross departure from reasonable variability.

If we now concentrate on the bottom two plots of Figure 226, we see that the $NW = 2/\Delta_t$ and $NW = 4/\Delta_t$ spectral estimates look quite similar overall. Increasing the degree of tapering beyond that given by the $NW = 2/\Delta_t$ data taper thus does not appear to gain us anything. We can conclude that a Slepian data taper with $NW = 2/\Delta_t$ is sufficient to control leakage over all frequencies for the ocean wave data (if, however, we only want to study the SDF for frequencies less than 1 Hz, the $NW = 1/\Delta_t$ spectral estimate also has acceptable bias properties).

A careful examination of the $NW = 2/\Delta_t$ direct spectral estimate shows one interesting feature barely evident in the periodogram, namely, a small peak at $f = 1.469$ Hz about 15 dB

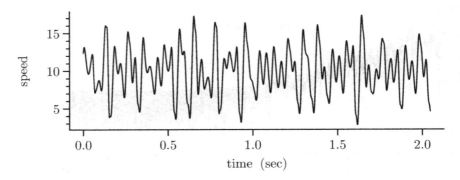

Figure 228. Chaotic beam data (courtesy of W. Constantine, Department of Mechanical Engineering, University of Washington). There are $N = 2048$ samples taken $\Delta_t = 0.001$ sec apart.

above the local background. This peak evidently corresponds to a 0.7 sec resonance in the wave-follower. That this peak is not attributable to just statistical fluctuations can be assessed by the F-test for periodicity discussed in Section 10.10 and applied to the ocean wave data in Section 10.15; however, a rough indication that it is significant can be seen by mentally moving the crisscross on the plot to this peak and noting that the lower end of a 95% CI for $S(1.469)$ is about 10 dB above a reasonable guess at the local level of the SDF. Other than this peak (and a smaller, less significant one at 1.715 Hz), the spectral estimate is relatively flat between 1.0 and 2.0 Hz, a conclusion that does not disagree with factors known to influence this portion of the SDF.

To conclude, this example shows that bias in the periodogram can seriously impact its usefulness as an SDF estimator. Tapering helps reduce the bias, but the sampling variability in the resulting direct spectral estimates is no better than that of the periodogram. Getting a better visual representation of the rate at which the power in the SDF decreases over the frequency range 0.2 to 1.0 Hz requires using SDF estimators with reduced variability. After a discussion of such estimators in the next chapter, we continue our analysis of the ocean wave data in Section 7.12.

Chaotic Beam Data

Our next example is a time series from an experiment designed to capture chaotic motion (Constantine, 1999). The data (Figure 228) are speeds of the bottom of a thin steel beam recorded over time. The beam was suspended vertically by clamping its top to an electromechanical shaker, which subjected the beam to a horizontal excitation through rapid sinusoidal variations, thus causing the bottom of the beam to vibrate. The vertical mounting allowed the beam's bottom to be suspended between two magnets. These magnets provided nonlinear buckling forces as the beam was being excited at a frequency close to its first natural mode of vibration (approximately 10 Hz). The amplitude of the sinusoidal variations was adjusted to yield seemingly chaotic motion. The bottom speeds were recorded at $\Delta_t = 0.001$ sec intervals for a little over 2 sec ($N = 2048$ measurements overall). For more details about this time series, see Constantine (1999); for discussions on chaotic dynamics in general, see, e.g., Moon (1992), Acheson (1997) or Chan and Tong (2001).

We start by examining the periodogram $\{\hat{S}^{(\text{P})}(\cdot)\}$ over the positive Fourier frequencies $f_k = k/(N \Delta_t)$ (Figure 229(a)). This periodogram is somewhat reminiscent of the one for the ocean wave data (Figure 226(a)): both have fluctuations that are markedly smaller at high frequencies than at low frequencies. As was the case for the ocean wave data, this pattern leads us to suspect that the periodogram for the chaotic beam data suffers from bias due to leakage. By examining direct spectral estimates based upon Slepian data tapers with $NW = 1/\Delta_t$

6.8 Examples of Periodogram and Other Direct Spectral Estimators

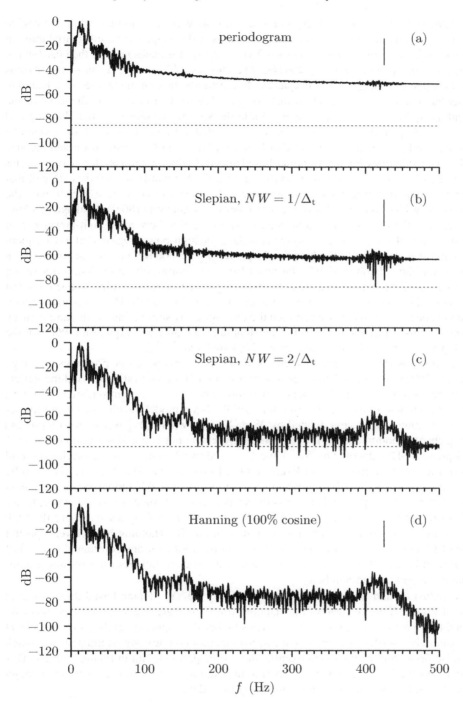

Figure 229. Direct spectral estimates for chaotic beam data. The (barely visible) width of the crisscrosses gives the bandwidth measure $B_{\mathcal{H}}$ (Equation (194)), while its height gives the length of a 95% CI for $10\log_{10}(S(f))$ (Equation (205b)). The horizontal dashed lines aid in comparing the four estimates (see the text for the rationale for placing them at -86 dB).

(Figure 229(b)), $2/\Delta_t$ (229(c)), $3/\Delta_t$ and $4/\Delta_t$ (not shown), we find that, while the choice $NW = 1/\Delta_t$ results in some decrease in bias, there is still leakage over a substantial interval of frequencies; on the other hand, the choice $NW = 2/\Delta_t$ eliminates most of the bias, but not at frequencies above approximately 450 Hz. The choices $3/\Delta_t$ and $4/\Delta_t$ show no obvious signs of bias due to leakage at any frequency. Interestingly enough, the choice of a Hanning data taper leads to a direct spectral estimate (Figure 229(d)) that is virtually the same as the $3/\Delta_t$ and $4/\Delta_t$ Slepian-based estimates. While the spectral windows for the Hanning and $NW = 2/\Delta_t$ Slepian data tapers have central lobes that are almost identical, their sidelobe patterns are markedly different, a fact that Figures 191(d) and (f) illustrate for sample size $N = 64$. The sidelobes for the Hanning-based spectral window are smaller at frequencies well separated from the central lobe than they are for the Slepian-based window. Recall that the expected value of a direct spectral estimator is the convolution of the true SDF with the spectral window $\mathcal{H}(\cdot)$ associated with the data taper (see Equation (186e)). Leakage in a low-power region of the true SDF arises because of an interaction between a high-power region and the sidelobes of $\mathcal{H}(\cdot)$. For the chaotic beam data, the high-power region is at low frequencies. For frequencies close to the Nyquist frequency $f_\mathcal{N} = 500$ Hz, the interaction between this region and the sidelobes of $\mathcal{H}(\cdot)$ far away from the central lobe is evidently not strong enough to produce noticeable leakage in the Hanning-based estimate at high frequencies, but the opposite holds for the Slepian-based estimate. (Prior to forming the Hanning- and Slepian-based direct spectral estimates, we centered the time series by subtracting off its sample mean $\overline{X} \doteq 10.0$. We used Equation (189b) for the Hanning data taper and Equation (196b) for the Slepian tapers. We normalized all data tapers $\{h_t\}$ such that $\sum_{t=0}^{N-1} h_t^2 = 1$.)

While it is possible to determine theoretical SDFs for certain chaotic processes (see, e.g., Lopes et al., 1997), the goal of this experiment was not to compare estimated and theoretical SDFs, but rather to investigate the effect of measurement noise on certain statistics of interest in assessing nonlinear dynamics (Constantine, 1999). If we model the chaotic beam time series as a realization of a process defined by $X_t = Y_t + \epsilon_t$, where $\{Y_t\}$ is a stationary process exhibiting chaotic motion and $\{\epsilon_t\}$ represents measurement noise, and if we assume that $\{Y_t\}$ and $\{\epsilon_t\}$ possess SDFs denoted by $S_Y(\cdot)$ and $S_\epsilon(\cdot)$ and that Y_t and ϵ_u are uncorrelated for all choices of t and u, then the result of Exercise [4.12] says that the SDF for $\{X_t\}$ is given by $S_X(f) = S_Y(f) + S_\epsilon(f)$. If $\{\epsilon_t\}$ is white noise (often a reasonable model for measurement noise) so that $S_\epsilon(f) = \sigma_\epsilon^2 \Delta_t$, where σ_ϵ^2 is the variance of $\{\epsilon_t\}$, then we would expect to see $S_X(\cdot)$ bottom out close to a value of $\sigma_\epsilon^2 \Delta_t$ (the noise floor) at frequencies where the SDF for measurement noise dominates that for chaotic motion. The Hanning-based direct spectral estimate of Figure 229(d) shows no evidence of a noise floor over any substantial interval of frequencies, indicating that the effect of measurement noise on statistics for assessing nonlinear dynamics might be minimal here.

In conclusion, had we looked at, say, the $NW = 2/\Delta_t$ Slepian-based direct spectral estimate of Figure 229(c) instead of the Hanning-based estimate, we might have erroneously concluded that a noise floor exists at frequencies higher than approximately 450 Hz at a level of approximately -86 dB (indicated by the horizontal dashed line). We would deduce a much higher noise floor from an uncritical examination of the periodogram in Figure 229(a)). This example shows the key role played by tapering in obtaining SDF estimates with low enough bias to allow proper assessment of the underlying true SDF.

Comments and Extensions to Section 6.8

[1] In C&E [6] for Section 6.6, we discussed a test for white noise based upon the normalized cumulative periodogram (see Equation (215a)). Here we apply this test to the ocean noise time series (Figure 2(d)), which, for pedagogical purposes, we regarded in Chapter 1 as a realization of a white noise process. Figure 231(a) shows the periodogram for this time series over the grid of Fourier frequencies. The

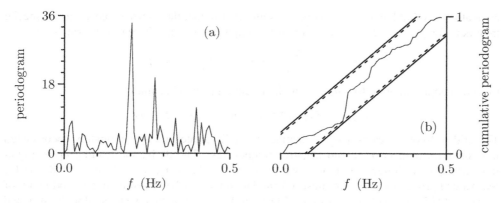

Figure 231. Periodogram (a) and associated normalized cumulative periodogram (b) for the ocean noise series in Figure 2(d). The periodogram is plotted on a linear/linear scale.

jagged curve in Figure 231(b) is the cumulative periodogram, for which the corresponding test statistic D is 0.1536. This statistic is less than the upper $\alpha = 0.05$ percentage point $\tilde{D}(0.05) \doteq 0.1696$ for the distribution of D under the null hypothesis of white noise, but greater than the $\alpha = 0.1$ point $\tilde{D}(0.1) \doteq 0.1528$. We thus cannot reject the null hypothesis at a level of significance of $\alpha = 0.05$, but we can at the higher level of 0.1. The upper and lower solid lines in Figure 231(b) give a visualization of the cumulative periodogram test at the $\alpha = 0.05$ level of significance. The fact that the cumulative periodogram never crosses either line corresponds to our not being able to reject the null hypothesis at that level of significance. On the other hand, the dashed lines correspond to the $\alpha = 0.1$ test, and the cumulative periodogram *does* dip below the lower line, but just barely at two frequencies ($\frac{23}{128} \doteq 0.180$ and $\frac{24}{128} \doteq 0.188$ Hz). The observed level of significance is quite close to 0.1, which provides cause for caution in regarding the ocean noise as a realization of white noise. (We return to this example in Section 10.15, where we apply a second – arguably more powerful – test for white noise, which casts serious doubt on the white noise hypothesis.)

6.9 Comments on Complex-Valued Time Series

All of the previous sections in this chapter assumed that we are dealing with a real-valued time series. Suppose now that our time series is complex-valued and can be regarded as a realization of a portion $Z_0, Z_1, \ldots, Z_{N-1}$ of a complex-valued discrete parameter stationary process $\{Z_t\}$ with zero mean and SDF $S(\cdot)$. If we can make the simplifying assumption that the real and imaginary components of $\{Z_t\}$ are uncorrelated processes with equal variances, we need only make a surprisingly few minor modifications to the material in the previous sections in order to estimate $S(\cdot)$. For example, the biased estimator of the ACVS in Equation (170b) now becomes

$$\hat{s}^{(\mathrm{P})}_\tau = \begin{cases} \sum_{t=0}^{N-\tau-1} Z_{t+\tau} Z_t^* / N, & 0 \leq \tau \leq N-1; \\ \left(\hat{s}^{(\mathrm{P})}_{-\tau}\right)^*, & -(N-1) \leq \tau < 0; \\ 0, & \text{otherwise.} \end{cases} \quad (231)$$

The main result of Equation (170d) still holds, namely,

$$\Delta_t \sum_{\tau=-(N-1)}^{N-1} \hat{s}^{(\mathrm{P})}_\tau e^{-i2\pi f\tau \Delta_t} = \frac{\Delta_t}{N} \left| \sum_{t=0}^{N-1} Z_t e^{-i2\pi ft \Delta_t} \right|^2 = \hat{S}^{(\mathrm{P})}(f),$$

but a slightly different proof is required. All of the stated results concerning tapering and prewhitening in Sections 6.4 and 6.5 still hold, with obvious changes to Equations (188b) and (195a). The main results of Section 6.6 concerning the statistical properties of direct spectral estimators also hold and, in fact, simplify somewhat in that the annoying necessity of special treatment for $f = 0$ and $f_{\mathcal{N}}$ disappears; for example, Equation (204b) now becomes

$$\hat{S}^{(\mathrm{D})}(f) \stackrel{\mathrm{d}}{=} \frac{S(f)}{2}\chi_2^2 \text{ for } -f_{\mathcal{N}} \leq f \leq f_{\mathcal{N}}$$

to a good approximation for large N, while Equation (204c) becomes

$$\mathrm{var}\,\{\hat{S}^{(\mathrm{D})}(f)\} \approx S^2(f) \text{ for } -f_{\mathcal{N}} \leq f \leq f_{\mathcal{N}}.$$

The grid of frequencies over which distinct $\hat{S}^{(\mathrm{D})}(f)$ variates are approximately uncorrelated now includes both positive and negative frequencies. The derivation leading up to Equation (211b) is no longer valid since we made use of the fact that $\mathrm{d}Z(-u) = \mathrm{d}Z^*(u)$ for real-valued processes; also, we must reformulate the normalized cumulative periodogram of Equation (215a) in an obvious two-sided fashion. Finally, note that we must plot all estimated SDFs over the set of frequencies ranging from $-f_{\mathcal{N}}$ to $f_{\mathcal{N}}$ instead of just from 0 to $f_{\mathcal{N}}$ because the SDF is no longer an even function as in the case of real-valued processes.

The assumption that the real and imaginary components of $\{Z_t\}$ are uncorrelated with equal variances is key to our ability to make straightforward adjustments to the real-valued theory to handle the complex-valued case. This assumption is often reasonable when, e.g., we create $\{Z_t\}$ via complex demodulation and low-pass filtering of a real-valued process (for details, see Bloomfield, 2000, Hasan, 1983, or Stoica and Moses, 2005); however, examples abound of complex-valued processes for which this assumption is *not* reasonable, rendering the corresponding complex-valued theory markedly different from the real-valued theory (Picinbono and Bondon, 1997, and Schreier and Scharf, 2003). One example is a process whose real and imaginary components are geometrically related to motion taking place in a plane. Here the physics governing the motion can dictate a nonzero correlation between the components. For details about these types of complex-valued time series, see Schreier and Scharf (2010, chapter 10) and Walden (2013).

6.10 Summary of Periodogram and Other Direct Spectral Estimators

Given a time series that we can regard as a portion $X_0, X_1, \ldots, X_{N-1}$ of a real-valued stationary process $\{X_t\}$ with a sampling interval of Δ_t and with an autocovariance sequence (ACVS) $\{s_{X,\tau}\}$ and spectral density function (SDF) $S_X(\cdot)$ related by $\{s_{X,\tau}\} \longleftrightarrow S_X(\cdot)$, we can use the fact that

$$S_X(f) = \Delta_t \sum_{\tau=-\infty}^{\infty} s_{X,\tau} e^{-\mathrm{i}2\pi f\tau\Delta_t} \quad (232)$$

to formulate an estimator for $S_X(\cdot)$ based upon an estimator of $\{s_{X,\tau}\}$ (see Equation (163)). Under the simplifying assumption that $\{X_t\}$ has zero mean, we can estimate $s_{X,\tau}$ for $|\tau| < N$ using the so-called biased ACVS estimator

$$\hat{s}_{X,\tau}^{(\mathrm{P})} = \frac{1}{N} \sum_{t=0}^{N-|\tau|-1} X_{t+|\tau|} X_t \quad \text{(see (170b))}$$

(when, as is usually the case, we cannot assume a priori that $\{X_t\}$ has zero mean, we must center the time series by subtracting off its sample mean $\overline{X} = \frac{1}{N}\sum_{t=0}^{N-1} X_t$ so that the summand $X_{t+|\tau|}X_t$ becomes $(X_{t+|\tau|} - \overline{X})(X_t - \overline{X})$). If we define $\hat{s}_{X,\tau}^{(\mathrm{P})} = 0$ for $|\tau| \geq N$, the resulting estimator $\{\hat{s}_{X,\tau}^{(\mathrm{P})}\}$ is always a positive semidefinite sequence, which means that it is a valid theoretical ACVS; i.e., it is the ACVS for a well-defined stationary process. Substituting $\{\hat{s}_{X,\tau}^{(\mathrm{P})}\}$ for $\{s_{X,\tau}\}$ in Equation (232) yields the SDF estimator known as the periodogram:

$$\hat{S}_X^{(\mathrm{P})}(f) \stackrel{\mathrm{def}}{=} \Delta_t \sum_{\tau=-(N-1)}^{N-1} \hat{s}_{X,\tau}^{(\mathrm{P})} e^{-\mathrm{i}2\pi f\tau\Delta_t}. \quad \text{(see (170c))}$$

6.10 Summary of Periodogram and Other Direct Spectral Estimators

Another formulation for the periodogram that is fully equivalent to the above is

$$\hat{S}_X^{(P)}(f) = \frac{\Delta_t}{N} \left| \sum_{t=0}^{N-1} X_t e^{-i2\pi f t \Delta_t} \right|^2. \qquad \text{(see (170d))}$$

Just as the theoretical ACVS and SDF satisfy $\{s_{X,\tau}\} \longleftrightarrow S_X(\cdot)$, we have $\{\hat{s}_{X,\tau}^{(P)}\} \longleftrightarrow \hat{S}_X^{(P)}(\cdot)$. In this Fourier transform pair, the inverse relationship tells us that

$$\hat{s}_{X,0}^{(P)} = \int_{-f_\mathcal{N}}^{f_\mathcal{N}} \hat{S}_X^{(P)}(f)\,\mathrm{d}f, \text{ which mimics } s_{X,0} = \int_{-f_\mathcal{N}}^{f_\mathcal{N}} S_X(f)\,\mathrm{d}f. \qquad \text{(see (171d))}$$

Thus, just as $S_X(\cdot)$ is an analysis of the process variance $s_{X,0}$, so is $\hat{S}_X^{(P)}(\cdot)$ an analysis of the sample variance $\hat{s}_{X,0}^{(P)}$ across a continuum of frequencies. We also have

$$\{\hat{s}_{X,\tau}^{(P)} : \tau = -(N-1), \ldots, N-1\} \longleftrightarrow \{\hat{S}_X^{(P)}(f_k') : k = -(N-1), \ldots, N-1\}, \qquad \text{(see (171e))}$$

where $f_k' \stackrel{\text{def}}{=} k/[(2N-1)\Delta_t]$. The inverse Fourier transform (Equation (171f)) leads to

$$\hat{s}_{X,0}^{(P)} = \frac{1}{(2N-1)\Delta_t} \sum_{k=-(N-1)}^{N-1} \hat{S}_X^{(P)}(f_k'),$$

and hence $\{\hat{S}_X^{(P)}(f_k')\}$ is an analysis of $\hat{s}_{X,0}^{(P)}$ across a discrete set of frequencies. More generally, for any $N' \geq N$ and with f_k' redefined to be $k/(N'\Delta_t)$, we have

$$\hat{s}_{X,0}^{(P)} = \frac{1}{N'\Delta_t} \sum_{k=0}^{N'-1} \hat{S}_X^{(P)}(f_k'). \qquad \text{(see (237b))}$$

The statistical properties of the periodogram are unfortunately such as to limit its value as an estimator of the SDF. Turning first to its first-moment properties, we have

$$E\{\hat{S}_X^{(P)}(f)\} = \Delta_t \sum_{\tau=-(N-1)}^{N-1} \left(1 - \frac{|\tau|}{N}\right) s_{X,\tau} = \int_{-f_\mathcal{N}}^{f_\mathcal{N}} \mathcal{F}(f-f') S_X(f')\,\mathrm{d}f', \qquad (233)$$

where

$$\mathcal{F}(f) \stackrel{\text{def}}{=} \Delta_t N \mathcal{D}_N^2(f\Delta_t)$$

is Fejér's kernel, which is defined in terms of Dirichlet's kernel $\mathcal{D}_N(\cdot)$ (see Equations (174a), (174b) and (174c)) As $N \to \infty$, this kernel acts as a Dirac delta function, and hence, for N large enough, $E\{\hat{S}_X^{(P)}(f)\} \approx S_X(f)$; however, for certain processes, this approximation is not accurate even for samples sizes that superficially seem large. The periodogram is badly biased for finite N when there is a deleterious interaction in the integrand of Equation (233) between a highly structured $S_X(\cdot)$ and the sidelobes of Fejér's kernel (these sidelobes are in evidence in Figure 176). In other words, $\mathcal{F}(\cdot)$ fails to act as a decent approximation to a delta function in Equation (233) because it is not sufficiently concentrated around $f - f' = 0$. This form of bias goes by the name of leakage.

Two approaches for dealing with leakage are tapering and prewhitening. Tapering consists of multiplying the values of a time series by a data taper $\{h_t : t = 0, 1, \ldots, N-1\}$, a suitably chosen sequence of real-valued constants usually normalized such that $\sum_t h_t^2 = 1$ (when needed, we define h_t to be zero for $t < 0$ and $t \geq N$). Tapering leads to the family of so-called direct spectral estimators:

$$\hat{S}_X^{(D)}(f) \stackrel{\text{def}}{=} \Delta_t \left| \sum_{t=0}^{N-1} h_t X_t e^{-i2\pi f t \Delta_t} \right|^2 \qquad \text{(see (186b))}$$

(note that use of the rectangular or default taper $h_t = 1/\sqrt{N}$ turns $\hat{S}_X^{(\mathrm{D})}(\cdot)$ back into the periodogram). In contrast to the periodogram, we now have

$$E\{\hat{S}_X^{(\mathrm{D})}(f)\} = \Delta_t \sum_{\tau=-(N-1)}^{N-1} \left(\sum_{t=0}^{N-|\tau|-1} h_{t+|\tau|} h_t \right) s_{X,\tau} e^{-i2\pi f \tau \Delta_t}$$

$$= \int_{-f_{\mathcal{N}}}^{f_{\mathcal{N}}} \mathcal{H}(f - f') S_X(f') \, \mathrm{d}f', \tag{234}$$

where

$$\mathcal{H}(f) \stackrel{\mathrm{def}}{=} \Delta_t \left| \sum_{t=0}^{N-1} h_t e^{-i2\pi f t \Delta_t} \right|^2$$

is the spectral window associated with $\{h_t\}$ (see Equations (186g), (186e) and (186f)). For certain processes and sample sizes for which the periodogram suffers from unacceptable leakage, we can have $E\{\hat{S}_X^{(\mathrm{D})}(f)\} \approx S_X(f)$ to a good approximation for a suitably chosen data taper. Examples of suitable tapers include the Hanning data taper

$$h_t = \begin{cases} C\left[1 - \cos\left(\frac{\pi(t+1)}{\lfloor N/2 \rfloor + 1}\right)\right], & 0 \le t \le \lfloor \frac{N+1}{2} \rfloor - 1; \\ h_{N-t-1}, & \lfloor \frac{N+1}{2} \rfloor \le t \le N - 1, \end{cases} \tag{see (189b)}$$

and tapers based on an approximation to zeroth-order Slepian sequences, namely,

$$h_t = C I_0 \left(\pi W(N-1) \left(1 - \frac{(2t+1-N)^2}{N^2} \right)^{1/2} \right) \tag{see (196b)}$$

(in the above equations, C is a generic constant enforcing the normalization $\sum_t h_t^2 = 1$; $\lfloor x \rfloor$ is the largest integer less than or equal to x; $I_0(\cdot)$ is the modified Bessel function of the first kind and zeroth order; and W defines the interval of frequencies $[-W, W]$ over which the spectral window $\mathcal{H}(\cdot)$ is maximally concentrated – this is usually specified indirectly by setting NW, for which typical values are 1, 2, 4 and 8).

Tapering attempts to compensate for leakage in the periodogram by replacing Fejér's kernel in Equation (233) with a spectral window $\mathcal{H}(\cdot)$ that is effectively a better approximation to a delta function (cf. Equations (233) and (234)). Prewhitening takes the complementary approach of replacing a highly structured SDF with one having less structure. The idea is to apply a prewhitening filter $\{g_l : l = 0, 1, \ldots, L - 1\}$ to the time series, yielding a new series

$$Y_t \stackrel{\mathrm{def}}{=} \sum_{l=0}^{L-1} g_l X_{t+L-1-l}, \quad 0 \le t \le N - L. \tag{see (197a)}$$

The new series is a realization of a stationary process with an SDF given by

$$S_Y(f) = \left| \sum_{l=0}^{L-1} g_l e^{-i2\pi f l \Delta_t} \right|^2 S_X(f). \tag{see (197b)}$$

If $\{g_l\}$ is chosen such that $S_Y(\cdot)$ is less highly structured than $S_X(\cdot)$, the periodogram $\hat{S}_Y^{(\mathrm{P})}(\cdot)$ for the Y_t's might have acceptable bias properties, in which case we can entertain a "postcolored" estimate of $S_X(\cdot)$, namely,

$$\hat{S}_X^{(\mathrm{PC})}(f) = \frac{\hat{S}_Y^{(\mathrm{P})}(f)}{\left| \sum_{l=0}^{L-1} g_l e^{-i2\pi f l \Delta_t} \right|^2}. \tag{cf. (198)}$$

Turning now to statistical properties other than the first moment, we can state that, subject to certain technical assumptions,

$$\hat{S}_X^{(D)}(f) \stackrel{d}{=} \begin{cases} S_X(f)\chi_2^2/2, & 0 < |f| < f_\mathcal{N}; \\ S_X(f)\chi_1^2, & |f| = 0 \text{ or } f_\mathcal{N}, \end{cases} \qquad \text{(see (204b))}$$

asymptotically as $N \to \infty$ for all direct spectral estimators (including the periodogram). Here $\stackrel{d}{=}$ means "equal in distribution," and χ_ν^2 is a chi-square random variable (RV) with ν degrees of freedom (thus the statement $U \stackrel{d}{=} c\chi_\nu^2$ means that the RV U has the same distribution as a chi-square RV after multiplication by a constant c). While the univariate properties of direct spectral estimators (including the periodogram) are the same, their bivariate properties are taper-dependent. Focusing on the periodogram first and considering the subset of Fourier frequencies $f_k \stackrel{\text{def}}{=} k/(N \Delta_t)$ satisfying $0 \le f_k \le f_\mathcal{N}$, we can claim that, under suitable conditions, the $\lfloor N/2 \rfloor + 1$ RVs $\hat{S}_X^{(P)}(f_0), \hat{S}_X^{(P)}(f_1), \ldots, \hat{S}_X^{(P)}(f_{\lfloor N/2 \rfloor})$ are all approximately pairwise uncorrelated for N large enough. A complementary claim is that, if $0 \le f < f' \le f_\mathcal{N}$, then

$$\text{cov}\{\hat{S}_X^{(P)}(f), \hat{S}_X^{(P)}(f')\} \approx 0 \text{ as along as } |f - f'| \ge \frac{1}{N \Delta_t},$$

i.e., a spacing greater than between adjacent Fourier frequencies. For other direct spectral estimators, the corresponding statement is that

$$\text{cov}\{\hat{S}_X^{(D)}(f), \hat{S}_X^{(D)}(f')\} \approx 0 \text{ as along as } |f - f'| \ge B_\mathcal{H},$$

where $B_\mathcal{H}$ is our standard measure of the effective bandwidth of $\hat{S}_X^{(D)}(\cdot)$:

$$B_\mathcal{H} \stackrel{\text{def}}{=} \frac{\left(\int_{-f_\mathcal{N}}^{f_\mathcal{N}} \mathcal{H}(f)\,df\right)^2}{\int_{-f_\mathcal{N}}^{f_\mathcal{N}} \mathcal{H}^2(f)\,df} = \frac{\Delta_t}{\sum_{\tau=-(N-1)}^{N-1}(h \star h_\tau)^2} \qquad \text{(see (194))}$$

(here $\{h \star h_\tau\}$ is the autocorrelation of $\{h_t\}$ defined as per Equation (99f)).

An important aspect of the large-sample distribution for direct spectral estimators is its lack of dependence on the sample size N. Since the variance of χ_ν^2 is 2ν, the distributional properties of $\hat{S}_X^{(D)}(\cdot)$ imply that

$$\text{var}\{\hat{S}_X^{(D)}(f)\} = \begin{cases} S_X^2(f), & 0 < |f| < f_\mathcal{N}; \\ 2S_X^2(f), & |f| = 0 \text{ or } f_\mathcal{N}. \end{cases} \qquad \text{(see (204c))}$$

Ignoring the pathological case $S_X(f) = 0$, neither the periodogram nor any other direct spectral estimator is such that $\text{var}\{\hat{S}_X^{(D)}(f)\} \to 0$ as $N \to \infty$. In statistical parlance, these SDF estimators are *inconsistent* because they do not get closer and closer to the true SDF with high probability as the sample size gets arbitrarily large. While direct spectral estimators are fundamentally flawed in this aspect, they are the basic building blocks for consistent SDF estimators to be discussed in subsequent chapters.

6.11 Exercises

[6.1] Suppose $\{\epsilon_t\}$ is a white noise process with mean zero and variance $0 < \sigma_\epsilon^2 < \infty$, and consider the following three stationary processes constructed from it:

(a) $X_t = \mu + \epsilon_0$;
(b) $X_t = \mu + \epsilon_t + \epsilon_{t-1}$; and
(c) $X_t = \mu + \epsilon_t - \epsilon_{t-1}$.

Let $\{s_\tau\}$ be the ACVS for any one of the three processes, and let $\overline{X} = \frac{1}{N}\sum_{t=0}^{N-1} X_t$ be the sample mean based upon a portion of one of the processes. For all three processes, express $\overline{X}$ as a linear combination of RVs from $\{\epsilon_t\}$, and determine $\text{var}\{\overline{X}\}$ directly with this expression. In each case, verify that the equation for var $\{\overline{X}\}$ is in agreement with Equation (165a). Equations (165b) and (165c) suggest the approximation

$$\operatorname{var}\{\overline{X}\} \approx \frac{1}{N} \sum_{\tau=-\infty}^{\infty} s_\tau.$$

Comment on how well the above works in each case.

[6.2] Suppose that $X_0, \ldots, X_{N-1}$ is a sample of size N from a white noise process with unknown mean μ and variance σ^2. With $\hat{s}_\tau^{(U)}$ and $\hat{s}_\tau^{(P)}$ defined as in Equations (166a) and (166b), show that, for $0 < |\tau| \le N-1$,

$$E\{\hat{s}_\tau^{(U)}\} = -\frac{\sigma^2}{N} \quad \text{and} \quad E\{\hat{s}_\tau^{(P)}\} = -\left(1 - \frac{|\tau|}{N}\right)\frac{\sigma^2}{N}.$$

Use this to argue (a) that the magnitude of the bias of the biased estimator $\hat{s}_\tau^{(P)}$ can be less than that of the "unbiased" estimator $\hat{s}_\tau^{(U)}$ when μ is estimated by $\overline{X}$ and (b) that, for this example, the mean square error of $\hat{s}_\tau^{(P)}$ is less than that of $\hat{s}_\tau^{(U)}$ for $0 < |\tau| \le N-1$.

[6.3] Suppose that the RVs X_0 and X_1 are part of a real-valued Gaussian stationary process $\{X_t\}$ with unknown mean μ and unknown ACVS $\{s_\tau\}$, which is arbitrary except for the mild restriction $s_1 < s_0$. Based upon just these two RVs, consider the so-called unbiased and biased estimators of s_1 dictated by, respectively, Equations (166a) and (166b):

$$\hat{s}_1^{(U)} = (X_0 - \overline{X})(X_1 - \overline{X}) \quad \text{and} \quad \hat{s}_1^{(P)} = \frac{1}{2}(X_0 - \overline{X})(X_1 - \overline{X}), \quad \text{where } \overline{X} = \frac{X_0 + X_1}{2}.$$

(a) For this special case of $N = 2$, determine $E\{\hat{s}_1^{(U)}\}$ and $E\{\hat{s}_1^{(P)}\}$, and use these to derive the biases of $\hat{s}_1^{(U)}$ and $\hat{s}_1^{(P)}$ as estimators of s_1. How does the squared bias of $\hat{s}_1^{(U)}$ compare with that of $\hat{s}_1^{(P)}$? Under what conditions (if any) are $\hat{s}_1^{(U)}$ and $\hat{s}_1^{(P)}$ unbiased?

(b) Determine the variances of $\hat{s}_1^{(U)}$ and $\hat{s}_1^{(P)}$. How does var$\{\hat{s}_1^{(U)}\}$ compare with var$\{\hat{s}_1^{(P)}\}$? Hint: recall the Isserlis theorem (Equation (30)).

(c) Using the results of parts (a) and (b), verify that, for this special case of $N = 2$, the inequality of Equation (167a) always holds:

$$\operatorname{MSE}\{\hat{s}_1^{(P)}\} = E\{(\hat{s}_1^{(P)} - s_1)^2\} < E\{(\hat{s}_1^{(U)} - s_1)^2\} = \operatorname{MSE}\{\hat{s}_1^{(U)}\}.$$

Investigate the relative contribution of var$\{\hat{s}_1^{(U)}\}$ to MSE$\{\hat{s}_1^{(U)}\}$, and do the same for $\hat{s}_1^{(P)}$.

(d) Calculate $\mathbf{P}\left[\hat{s}_1^{(U)} < -1\right]$ when var$\{X_0\} = 4/5$ and $\rho_1 \stackrel{\text{def}}{=} s_1/s_0 = 1/2$, and calculate $\mathbf{P}\left[\hat{s}_1^{(U)} > 1\right]$ also. Hint: if Z is a Gaussian RV with zero mean and unit variance, then Z^2 obeys a chi-square distribution with one degree of freedom.

(e) With $\hat{s}_\tau^{(U)}$ and $\hat{s}_\tau^{(P)}$ being dictated by Equations (166a) and (166b) for $\tau = 0$ in addition to $\tau = 1$, verify that, for the special case of $N = 2$,

$$\hat{\rho}_1^{(U)} \stackrel{\text{def}}{=} \frac{\hat{s}_1^{(U)}}{\hat{s}_0^{(U)}} = -1 \quad \text{and} \quad \hat{\rho}_1^{(P)} \stackrel{\text{def}}{=} \frac{\hat{s}_1^{(P)}}{\hat{s}_0^{(P)}} = -\frac{1}{2}$$

no matter what values the RVs X_0 and X_1 assume (we can ignore the special case $X_0 = X_1 = 0$ since it has probability zero). Thus, while part (d) indicates that realizations of $\hat{s}_1^{(U)}$ can take on different values, those of $\hat{\rho}_1^{(U)}$ cannot.

[6.4] Equations (174d) and (175a) show two ways of expressing Fejér's kernel. Verify the following third expression:

$$\mathcal{F}(f) = \Delta_t \sum_{\tau=-(N-1)}^{N-1} \left(1 - \frac{|\tau|}{N}\right) e^{-i2\pi f \tau \Delta_t}. \tag{236}$$

[6.5] Let $X_0, \ldots, X_{N-1}$ be a sample of size N from a stationary process with an unknown mean and an SDF $S(\cdot)$ defined over $[-1/2, 1/2]$ (i.e., the sampling interval Δ_t is taken to be 1). At lag $\tau = 0$ both the unbiased and biased estimators of the ACVS reduce to

$$\hat{s}_0 = \frac{1}{N} \sum_{t=0}^{N-1} (X_t - \overline{X})^2.$$

(a) Show that $E\{\hat{s}_0\} = s_0 - \text{var}\{\overline{X}\} \leq s_0$.

(b) Show that

$$E\{\hat{s}_0\} = \int_{-1/2}^{1/2} \left(1 - \frac{1}{N}\mathcal{F}(f)\right) S(f)\,df,$$

where $\mathcal{F}(\cdot)$ is Fejér's kernel (see Equation (174c)). Hint: consider Equations (165a), (174a) and (174b).

(c) Plot $(1 - \mathcal{F}(f)/N)$ versus f for, say, $N = 64$. Based upon this plot, for what kind of $S(\cdot)$ would there be a large discrepancy between $E\{\hat{s}_0\}$ and s_0?

[6.6] Construct an example of a time series of one specific length N for which the unbiased estimate of the ACVS

$$\hat{s}^{(U)}_{-(N-1)}, \hat{s}^{(U)}_{-(N-2)}, \ldots, \hat{s}^{(U)}_{-1}, \hat{s}^{(U)}_0, \hat{s}^{(U)}_1, \ldots, \hat{s}^{(U)}_{N-2}, \hat{s}^{(U)}_{N-1}$$

given by Equation (166a) is *not* a valid ACVS for some stationary process (here $\hat{s}^{(U)}_\tau \stackrel{\text{def}}{=} 0$ for $|\tau| \geq N$). Note: N can be chosen in whatever manner so desired.

[6.7] Let $X_0, \ldots, X_{N-1}$ be a portion of a stationary process with sampling interval Δ_t, and let $\hat{s}^{(P)}_\tau$ for $\tau = 0, \pm 1, \ldots, \pm(N-1)$ be the biased estimator of the ACVS defined by either Equation (170b) – if $E\{X_t\}$ is known to be zero – or Equation (166b) – if $E\{X_t\}$ is unknown and hence estimated by the sample mean $\overline{X}$. For $|\tau| \geq N$, let $\hat{s}^{(P)}_\tau \stackrel{\text{def}}{=} 0$. Let $f'_k = k/(N' \Delta_t)$, where N' is any integer greater than or equal to N.

(a) Show that

$$\hat{S}^{(P)}(f'_k) = \Delta_t \sum_{\tau=0}^{N'-1} \left(\hat{s}^{(P)}_\tau + \hat{s}^{(P)}_{N'-\tau}\right) e^{-i2\pi k\tau/N'}$$

and hence that

$$\hat{s}^{(P)}_\tau + \hat{s}^{(P)}_{N'-\tau} = \frac{1}{N' \Delta_t} \sum_{k=0}^{N'-1} \hat{S}^{(P)}(f'_k) e^{i2\pi k\tau/N'} \qquad (237a)$$

(Bloomfield, 2000, section 8.4).

(b) Show that, when $N' = 2N - 1$, part (a) can be manipulated to yield Equation (171f). Argue that, if $N' \geq 2N - 2$, we can still recover $\{\hat{s}^{(P)}_\tau\}$ from the inverse DFT of $\{\hat{S}^{(P)}(f'_k)\}$, but that, if $N' < 2N - 2$, we cannot do so in general.

(c) Use part (a) to note that, for any $N' \geq N$,

$$\hat{s}^{(P)}_0 = \frac{1}{N' \Delta_t} \sum_{k=0}^{N'-1} \hat{S}^{(P)}(f'_k). \qquad (237b)$$

(d) A DFT or inverse DFT of a sequence with N' terms often can be computed efficiently using an FFT algorithm (see Section 3.11). Such an algorithm typically requires a number of arithmetic operations (additions, subtractions, multiplications and divisions) proportional to $N' \log_2(N')$. Compare the computational procedure for evaluating $\hat{s}^{(P)}_\tau$ for $\tau = 0, \ldots, N-1$ via Equation (171f) with direct use of Equation (166b) in terms of the number of arithmetic operations required.

[6.8] From Equation (170d) we know that, given $\hat{s}_0^{(P)}, \ldots, \hat{s}_{N-1}^{(P)}$, we can compute the periodogram $\hat{S}^{(P)}(\cdot)$ at any given frequency. Suppose that we are given the periodogram on the grid of Fourier frequencies, i.e., $\hat{S}^{(P)}(f_k)$ for $k = 0, 1, \ldots, \lfloor N/2 \rfloor$, where, as usual, $f_k = k/(N \Delta_t)$ and $\lfloor N/2 \rfloor$ is the largest integer less than or equal to $N/2$. Using these $\lfloor N/2 \rfloor + 1$ values, is it possible in general to recover $\hat{s}_0^{(P)}, \ldots, \hat{s}_{N-1}^{(P)}$ in some manner (i.e., not necessarily via an inverse DFT of $\{\hat{S}^{(P)}(f_k)\}$)?

[6.9] Consider the sequence $\{\hat{s}_\tau^{(P)} : \tau = -(N-1), \ldots, N-1\}$, which contains the biased estimator of the ACVS. Note that, because $\hat{s}_{-\tau}^{(P)} = \hat{s}_\tau^{(P)}$, there are only N distinct variables in this sequence of length $2N - 1$. Equation (171e) says that the DFT of this sequence yields the periodogram over the grid of frequencies $f_k' = k/[(2N-1)\Delta_t]$, $k = -(N-1), \ldots, N-1$. Let $\{\tilde{S}_k : k = 0, \ldots, N-1\}$ be the DFT of $\{\hat{s}_\tau^{(P)} : \tau = 0, \ldots, N-1\}$. Show that we can form the periodogram over the grid of Fourier frequencies $f_k = k/(N\Delta_t)$ directly in terms of the $\tilde{S}_k$ variables. Use this formulation to argue that, given the periodogram at just the Fourier frequencies, it is not in general possible to recover $\{\hat{s}_\tau^{(P)} : \tau = 0, \ldots, N-1\}$ (you may assume the validity of Equation (237b)).

[6.10] The intent of this exercise is to verify some key results about the periodogram using a simple example. Let a be any real-valued nonzero constant, and suppose that $\{a, 0, -a\}$ is a realization of length $N = 3$ of a portion X_0, X_1, X_2 of a stationary process with sampling interval $\Delta_t = 1$, unknown mean μ, SDF $S(\cdot)$ and ACVS $\{s_\tau\}$.
 (a) Determine the biased estimator $\{\hat{s}_\tau^{(P)}\}$ of the ACVS for $\{a, 0, -a\}$. For comparison, also determine the unbiased estimator $\{\hat{s}_\tau^{(U)}\}$ of the ACVS. Is $\{\hat{s}_\tau^{(U)}\}$ necessarily a valid ACVS for some stationary process? Explain why or why not.
 (b) Determine the periodogram $\hat{S}^{(P)}(\cdot)$ for $\{a, 0, -a\}$ using its definition in Equation (170c) as the Fourier transform of $\{\hat{s}_\tau^{(P)}\}$.
 (c) An equivalent way of obtaining the periodogram is via Equation (170d), which is based on the DFT of $\{a, 0, -a\}$. Verify that computing the periodogram in this alternative manner gives the same result as in part (b).
 (d) As noted in Equations (171c) and (171f), the biased estimator of the ACVS can be recovered from the periodogram via either an integral of $\hat{S}^{(P)}(\cdot)$ over $[-1/2, 1/2]$ or a summation involving $\hat{S}^{(P)}(f_k')$, where $f_k' = k/(2N - 1)$. Verify that these two equations hold for the example considered here when $\tau = 0$ and $\tau = 2$.
 (e) Based upon Equation (181), show that the biased estimator of the ACVS can also be recovered from a summation involving $\hat{S}^{(P)}(\tilde{f}_k)$, where $\tilde{f}_k = k/(2N)$. Verify that the equation expressing $\hat{s}_\tau^{(P)}$ in terms of the $\hat{S}^{(P)}(\tilde{f}_k)$'s holds for the example considered here when $\tau = 0$ and $\tau = 2$.

[6.11] Let $x_0, x_1, \ldots, x_{N-1}$ represent a particular time series of interest. Let $\hat{S}_x^{(P)}(\cdot)$ be the periodogram for $\{x_t\}$. Define two new time series $\{v_t\}$ and $\{w_t\}$ via $v_t = -10x_t$ and $w_t = cx_t$, where c is a real-valued constant such that $w_t \neq x_t$ and $w_t \neq v_t$. Let $\hat{S}_v^{(P)}(\cdot)$ and $\hat{S}_w^{(P)}(\cdot)$ denote the periodograms for $\{v_t\}$ and $\{w_t\}$. The periodograms for $\{x_t\}$, $\{v_t\}$ and $\{w_t\}$ are each shown in one of the four plots making up Figure 239 (as usual, the periodograms are displayed on a decibel scale). Which plot corresponds to which periodogram, and what value(s) can the constant c assume? Briefly explain the rationale for your choices.

[6.12] Suppose that $x_0, x_1, \ldots, x_{N-1}$ represents one particular time series and that we form its periodogram using

$$\hat{S}_x^{(P)}(f) = \frac{1}{N}\left|\sum_{t=0}^{N-1} x_t e^{-i2\pi f t}\right|^2.$$

For any integer $0 < m \leq N - 1$, define $y_t = x_{t+m \bmod N}$; i.e., $y_0, y_1, \ldots, y_{N-1}$ is a circularly shifted version of $x_0, x_1, \ldots, x_{N-1}$ (see Exercise [2.19] for a discussion of the "mod" operator). Let

$$\hat{S}_y^{(P)}(f) = \frac{1}{N}\left|\sum_{t=0}^{N-1} y_t e^{-i2\pi f t}\right|^2$$

be the periodogram for y_t.

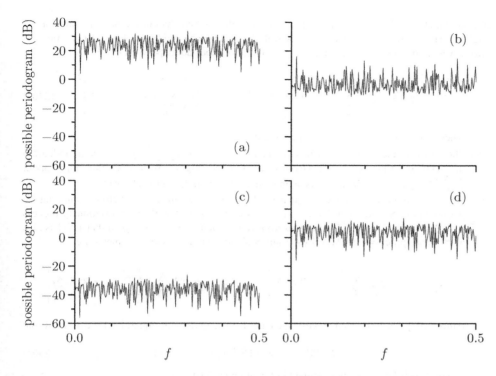

Figure 239. Plots of periodograms for a time series $\{x_t\}$ and for two rescaled versions given by $v_t = -10x_t$ and $w_t = cx_t$, along with a spurious plot.

    (a) Show that $\hat{S}_y^{(\mathrm{P})}(f_k) = \hat{S}_x^{(\mathrm{P})}(f_k)$, where $f_k = k/N$ is a Fourier frequency.
    (b) Prove or disprove the claim that $\hat{S}_y^{(\mathrm{P})}(f) = \hat{S}_x^{(\mathrm{P})}(f)$ at all frequencies (i.e., f need not be a Fourier frequency).

[6.13] Let $X_0, X_1, \ldots, X_{N-1}$ be a portion of a zero mean stationary process with SDF $S(\cdot)$ and sampling interval Δ_t, from which we form a periodogram $\hat{S}^{(\mathrm{P})}(f)$. Let

$$b(f) \stackrel{\mathrm{def}}{=} E\{\hat{S}^{(\mathrm{P})}(f)\} - S(f)$$

represent the bias in the periodogram at frequency f. Recall that, if $g(\cdot)$ is a function defined over the interval $[a, b]$, then, by definition,

$$\frac{1}{b-a} \int_a^b g(x)\, \mathrm{d}x \tag{239}$$

is the average value of $g(\cdot)$ over $[a, b]$.
    (a) Show that the average value of the bias in the periodogram over $f \in [-f_\mathcal{N}, f_\mathcal{N}]$ is zero.
    (b) Show that the average value of the bias in the periodogram over $f \in [0, f_\mathcal{N}]$ is also zero.

[6.14] For each of the four AR(4) time series of length $N = 1024$ displayed in Figures 34(e), (f), (g) and (h), compute the periodogram after subjecting the series to the end-point matching dictated by Equation (183a), followed by centering the end-point matched series by subtracting off its sample mean (the four time series are downloadable from the "Data" part of the website for the book). Compute each periodogram in two ways, namely, across just the nonzero Fourier frequencies $f_k = k/N$ and across a grid of nonzero frequencies $\tilde{f}_k = k/(2N)$ whose spacing is twice as fine as the Fourier frequencies. Plot your results along with the true AR(4) SDF on a decibel scale for comparison with Figures 173(e), (f), (g) and (h) and Figures 182(e), (f), (g) and (h). The goal of

240 *Periodogram and Other Direct Spectral Estimators*

end-point matching is to create an SDF estimator with less bias than in the periodogram for the original unaltered series. Comment on how well the goal is achieved for the four AR(4) series.

[6.15] (a) Given a time series $Y_0, \ldots, Y_{N-1}$, show that, if C does not depend on t,

$$\tilde{S}^{(P)}(f_k) \stackrel{\text{def}}{=} \frac{\Delta_t}{N} \left| \sum_{t=0}^{N-1} (Y_t - C) e^{-i2\pi f_k t \Delta_t} \right|^2 = \frac{\Delta_t}{N} \left| \sum_{t=0}^{N-1} Y_t e^{-i2\pi f_k t \Delta_t} \right|^2, \quad (240a)$$

where k is an integer, $f_k = k/(N\Delta_t)$ and $0 < f_k \leq f_\mathcal{N}$.

(b) Show that, in the above, $\tilde{S}^{(P)}(0) = 0$ when $C = \overline{Y} = \frac{1}{N}\sum_{t=0}^{N-1} Y_t$; i.e., the periodogram of a centered time series is necessarily zero at the origin. What does this imply about $\sum_{\tau=-(N-1)}^{N-1} \hat{s}_\tau^{(P)}$, where $\hat{s}_\tau^{(P)}$ is defined in keeping with Equation (166b)?

(c) Here we examine one of the consequences of *not* centering a time series before computing the periodogram. Suppose that $Y_0, \ldots, Y_{N-1}$ is a segment of length N of a stationary process with ACVS $\{s_\tau\}$, SDF $S(\cdot)$ and nonzero mean μ. Define $X_t = Y_t - \mu$ so that $\{X_t\}$ is also a stationary process with ACVS $\{s_\tau\}$ and SDF $S(\cdot)$, but now with zero mean. Let

$$\tilde{S}_Y^{(P)}(f) \stackrel{\text{def}}{=} \frac{\Delta_t}{N} \left| \sum_{t=0}^{N-1} Y_t e^{-i2\pi f t \Delta_t} \right|^2 \quad \text{and} \quad \hat{S}_X^{(P)}(f) \stackrel{\text{def}}{=} \frac{\Delta_t}{N} \left| \sum_{t=0}^{N-1} X_t e^{-i2\pi f t \Delta_t} \right|^2$$

(since $\{X_t\}$ is a zero mean process, the latter is in keeping with Equation (170d)). Show that

$$E\{\tilde{S}_Y^{(P)}(f)\} = E\{\hat{S}_X^{(P)}(f)\} + \mu^2 \mathcal{F}(f), \quad (240b)$$

where $\mathcal{F}(\cdot)$ is Fejér's kernel (defined in Equation (175a)).

(d) As an example of the practical impact of (240a) and (240b), consider the Willamette River time series, which is downloadable from the "Data" part of the website for the book (Figures 2(c) and 568a show, respectively, the first 128 values and all $N = 395$ values of this series – the entire series is to be considered for this exercise). Here the sampling interval and Nyquist frequency are $\Delta_t = 1/12$ year and $f_\mathcal{N} = 6$ cycles per year. Figure 568b shows the periodogram over the grid of frequencies $k/(1024\Delta_t)$, $k = 1, \ldots, 512$, after centering the series by subtracting off its sample mean of 9.825 (see Exercise [179] for details about how to compute the periodogram over this grid of frequencies). Compute and plot the periodogram for the time series *without* centering it (1) over the nonzero Fourier frequencies $k/(395\Delta_t)$, $k = 1, \ldots, 197$, and (2) over the same grid of frequencies used in Figure 568b. Taking into consideration Equations (240a) and (240b), comment on how these two plots compare with Figure 568b.

[6.16] Verify Equation (186g). Hint: consider replacing $X_t/\sqrt{N}$ in the solution to Exercise [170] with $h_t X_t$.

[6.17] Show that Equation (188a) is true.

[6.18] (a) Define $h_t = C\left[1 - \cos(2\pi t/N)\right]$, $t = 0, 1, \ldots, N-1$, where C is a constant chosen such that $\sum_{t=0}^{N-1} h_t^2 = 1$. This is essentially the definition for the Hanning data taper given by Priestley (1981, p. 562), but it is slightly different from the one in Equation (189b); in particular, here $h_0 = 0$, and the taper is asymmetric in the sense that $h_t \neq h_{N-t-1}$ in general, whereas in (189b) $h_t > 0$ and $h_t = h_{N-t-1}$ for $0 \leq t \leq N-1$. Let

$$I(f_k) \stackrel{\text{def}}{=} \left(\frac{\Delta_t}{N}\right)^{1/2} \sum_{t=0}^{N-1} X_t e^{-i2\pi f_k t \Delta_t} \quad \text{and} \quad J(f_k) \stackrel{\text{def}}{=} \Delta_t^{1/2} \sum_{t=0}^{N-1} h_t X_t e^{-i2\pi f_k t \Delta_t},$$

where, as usual, $f_k = k/(N\Delta_t)$. Show that

$$J(f_k) = CN^{1/2}\left(-\frac{1}{2}I(f_{k-1}) + I(f_k) - \frac{1}{2}I(f_{k+1})\right).$$

(b) Noting that $\hat{S}_G^{(P)}(f_k) = |I(f_k)|^2$ and $\hat{S}_G^{(D)}(f_k) = |J(f_k)|^2$, show that, when $\{X_t\}$ is Gaussian white noise with mean zero and variance σ^2,

$$\text{corr}\{\hat{S}_G^{(D)}(f_{k+\tau}), \hat{S}_G^{(D)}(f_k)\} = \begin{cases} 4/9, & \tau = 1; \\ 1/36, & \tau = 2; \\ 0, & \tau = 3, \end{cases} \quad (241)$$

for $2 \leq k < \lfloor N/2 \rfloor - 3$, whereas $\text{corr}\{\hat{S}_G^{(P)}(f_{k+\tau}), \hat{S}_G^{(P)}(f_k)\} = 0$ for $\tau = 1, 2$ and 3. Hint: make use of the Isserlis theorem (Equation (30)) in a manner similar to how it is used to derive Equation (211a).

[6.19] Verify the computational formula in Equation (192b).

[6.20] Consider a time series $X_0, \ldots, X_{N-1}$ that is a portion of a stationary process with mean μ. Let $\{h_t\}$ be a data taper whose associated spectral window is $\mathcal{H}(\cdot)$. Consider

$$\hat{S}^{(D)}(f) = \Delta_t \left| \sum_{t=0}^{N-1} h_t(X_t - \mu)e^{-i2\pi ft\Delta_t} \right|^2 \quad \text{and} \quad \tilde{S}^{(D)}(f) \stackrel{\text{def}}{=} \Delta_t \left| \sum_{t=0}^{N-1} (h_t X_t - \mu)e^{-i2\pi ft\Delta_t} \right|^2,$$

i.e., a direct spectral estimator and and a related estimator in which the process mean is subtracted *after* the time series is tapered.

(a) Develop an equation that relates $E\{\hat{S}^{(D)}(f)\}$ and $E\{\tilde{S}^{(D)}(f)\}$ for $0 \leq f \leq f_\mathcal{N}$, where, as usual, $f_\mathcal{N} = 1/(2\Delta_t)$ is the Nyquist frequency.

(b) Using the equation called for in part (a), show that, at Fourier frequencies $f_k = k/(N\Delta_t)$ satisfying $0 < f_k \leq f_\mathcal{N}$, we have $E\{\tilde{S}^{(D)}(f_k)\} = E\{\hat{S}^{(D)}(f_k)\} + \mu^2 \mathcal{H}(f_k)$.

(c) Briefly discuss the implication of part (a) for SDF estimation when the operations of centering and tapering a white noise process with nonzero mean are inadvertently interchanged. Briefly discuss the implications of part (b) for both white and colored processes $\{X_t\}$ when $\{h_t\}$ is the default (rectangular) data taper.

[6.21] Recall the definition for the average value of a function given in Equation (239). Let $\hat{S}^{(P)}(\cdot)$ and $\hat{S}^{(D)}(\cdot)$ represent, respectively, the periodogram and a direct spectral estimator based upon the Hanning data taper for a time series with Nyquist frequency $f_\mathcal{N}$. Can we potentially determine the need for tapering by comparing the average values of $\hat{S}^{(P)}(\cdot)$ and $\hat{S}^{(D)}(\cdot)$ over the interval $[-f_\mathcal{N}, f_\mathcal{N}]$? If we replace $\hat{S}^{(P)}(\cdot)$ and $\hat{S}^{(D)}(\cdot)$ with $\log(\hat{S}^{(P)}(\cdot))$ and $\log(\hat{S}^{(D)}(\cdot))$, can we potentially determine the need for tapering by comparing their average values? Illustrate your answers by considering the periodograms (Figure 173) and Hanning-based direct spectral estimates (Figure 187) for the four AR(4) time series displayed in Figure 34 (these are available on the website for the book).

[6.22] Here we explore the use of short prewhitening filters to compensate for bias in the periodogram of the AR(4) time series $\{X_t\}$ displayed in Figure 34(f) (this series is available on the website for the book). Use the prewhitening filter $\{g_0 = 1, g_1 = -0.9\}$ to obtain a prewhitened series $\{Y_t\}$. Compute the periodogram for $\{Y_t\}$ over the grid of frequencies $f'_k = k/2048$, $k = 0, \ldots, 1024$, and then postcolor the periodogram to create an estimate for the SDF of $\{X_t\}$. Plot this estimate in a manner similar to Figure 199(d). Repeat using the prewhitening filter $\{g_0 = 1, g_1 = -1\}$ in place of $\{g_0 = 1, g_1 = -0.9\}$. Finally, repeat using the filter $\{g_0 = 1, g_1 = -0.7165\}$. How well do these three estimates of the SDF for $\{X_t\}$ compare to the estimate shown in Figure 199(d)?

[6.23] (a) Show that, if we let $0 < f < f_\mathcal{N}$ rather than restricting f to be a Fourier frequency, then Equation (203c) becomes $E\{\hat{S}_G^{(P)}(f)\} = \sigma^2 \Delta_t = S_G(f)$ while Equation (203d) is then

$$\text{var}\{\hat{S}_G^{(P)}(f)\} = \sigma^4 \Delta_t^2 \left(1 + \mathcal{D}_N^2(2f\Delta_t)\right) = S_G^2(f)\left(1 + \mathcal{D}_N^2(2f\Delta_t)\right),$$

where $\mathcal{D}_N(\cdot)$ is Dirichlet's kernel as defined by Equation (17c) (Proakis and Manolakis, 2007, Problem 14.3(c)).

(b) For what values of f does $\text{var}\{\hat{S}_G^{(P)}(f)\}$ achieve its minimum value of $\sigma^4 \Delta_t^2$?

(c) Argue that, as $N \to \infty$, $\text{var}\{\hat{S}_G^{(P)}(f)\} \to \sigma^4 \Delta_t^2$ for all $0 < f < f_\mathcal{N}$.

(d) Plot $\text{var}\{\hat{S}_G^{(P)}(f)\}$ versus f for the cases $N = 16$ and 64 assuming that $\sigma^2 = 1$ and $\Delta_t = 1$.

[6.24] (a) Generate a realization of a portion $X_0, X_1, \ldots, X_{63}$ of the zero mean AR(2) process of Equation (34) (see Exercise [597] for a description of how to generate realizations from this process). Assuming for convenience that $\Delta_t = 1$, compute the periodogram for $\{X_t\}$ at three adjacent Fourier frequencies, namely, $f_6 = 6/64$, $f_7 = 7/64$ and $f_8 = 8/64$, and call these values $\hat{S}_0^{(P)}(f_6)$, $\hat{S}_0^{(P)}(f_7)$ and $\hat{S}_0^{(P)}(f_8)$. Repeat what we just described a large number N_R of times (using a different realization of $\{X_t\}$ each time) to obtain the sequences $\{\hat{S}_j^{(P)}(f_6) : j = 0, \ldots, N_R - 1\}$, $\{\hat{S}_j^{(P)}(f_7) : j = 0, \ldots, N_R - 1\}$ and $\{\hat{S}_j^{(P)}(f_8) : j = 0, \ldots, N_R - 1\}$ (take "large" to mean something from 1,000 up to 100,000). Compute the sample mean and sample variance for the three sequences, and compute the sample correlation coefficient $\hat{\rho}$ between $\{\hat{S}_j^{(P)}(f_6)\}$ and $\{\hat{S}_j^{(P)}(f_7)\}$, between $\{\hat{S}_j^{(P)}(f_6)\}$ and $\{\hat{S}_j^{(P)}(f_8)\}$ and between $\{\hat{S}_j^{(P)}(f_7)\}$ and $\{\hat{S}_j^{(P)}(f_8)\}$ (see Equation (3a)). Compare these sample values with theoretical values suggested by Equations (174a), (175c), (145a), (203f) and (203g) (Equation (174a) calls for the ACVS for the AR(2) process – Exercise [9.6] states how to form it). Form histograms for $\{\hat{S}_j^{(P)}(f_6)\}$, $\{\hat{S}_j^{(P)}(f_7)\}$ and $\{\hat{S}_j^{(P)}(f_8)\}$, and compare them to the PDFs suggested by Equations (203e) and (209a). Form histograms also for $\{10 \log_{10}(\hat{S}_j^{(P)}(f_6))\}$, $\{10 \log_{10}(\hat{S}_j^{(P)}(f_7))\}$ and $\{10 \log_{10}(\hat{S}_j^{(P)}(f_8))\}$, and compare them to the PDFs suggested by Equation (209b).

(b) Repeat part (a), but now use the Fourier frequencies $f_{29} = 29/64$, $f_{30} = 30/64$ and $f_{31} = 31/64$ in place of f_6, f_7 and f_8.

[6.25] Repeat Exercise [6.24], but with the following modifications:

(a) use a direct spectral estimate with a Hanning data taper (Equation (189b)) instead of the periodogram and,

(b) in comparing the sample values to theory, use Equation (186g) rather than (174a); use Equation (204c) rather than (203f); and Equation (207) with $c = 2$ rather than (204a).

[6.26] Repeat Exercises [6.24] and [6.25], but with the following modifications. In place of the AR(2) process, generate realizations of a portion $X_0, X_1, \ldots, X_{255}$ of the zero mean AR(4) process of Equation (35a) (Exercise [11.1] describes how to generate these realizations). For part (a) of Exercise [6.24] take the three adjacent Fourier frequencies to be $27/256$, $28/256$ and $29/256$. Exercise [9.6] states how to form the ACVS for the AR(4) process, which is needed for part (a). For part (b) take the three frequencies to be $125/256$, $126/256$ and $127/256$.

[6.27] Suppose that $\{X_t\}$ is a real-valued stationary process with zero mean, sampling interval Δ_t, ACVS $\{s_\tau\}$ and SDF $S(\cdot)$. Let $a_0, \ldots, a_{N-1}$ and $b_0, \ldots, b_{N-1}$ be any two sequences of complex-valued numbers of finite length N. Define

$$A(f) = \Delta_t \sum_{t=0}^{N-1} a_t e^{-i2\pi f t \Delta_t} \text{ and } B(f) = \Delta_t \sum_{t=0}^{N-1} b_t e^{-i2\pi f t \Delta_t};$$

i.e., with $a_t \stackrel{\text{def}}{=} 0$ and $b_t \stackrel{\text{def}}{=} 0$ for all $t < 0$ and $t \geq N$, then $\{a_t\} \longleftrightarrow A(\cdot)$ and $\{b_t\} \longleftrightarrow B(\cdot)$.

(a) Use the spectral representation theorem for $\{X_t\}$ to show that

$$\text{cov}\left\{\sum_{t=0}^{N-1} a_t X_t, \sum_{t=0}^{N-1} b_t X_t\right\} = \frac{1}{\Delta_t^2} \int_{-f_\mathcal{N}}^{f_\mathcal{N}} A(f) B^*(f) S(f) \, df. \qquad (242a)$$

(b) Show that we also have

$$\text{cov}\left\{\sum_{t=0}^{N-1} a_t X_t, \sum_{t=0}^{N-1} b_t X_t\right\} = \sum_{t=0}^{N-1} \sum_{u=0}^{N-1} a_t b_u^* s_{t-u}. \qquad (242b)$$

(c) Now suppose that $\{X_t\}$ is a Gaussian process. Show that

$$\text{cov}\left\{\left|\sum_{t=0}^{N-1} a_t X_t\right|^2, \left|\sum_{t=0}^{N-1} b_t X_t\right|^2\right\}$$

$$= \frac{1}{\Delta_t^4}\left|\int_{-f_\mathcal{N}}^{f_\mathcal{N}} A(f)B^*(f)S(f)\,\mathrm{d}f\right|^2 + \frac{1}{\Delta_t^4}\left|\int_{-f_\mathcal{N}}^{f_\mathcal{N}} A(f)B(-f)S(f)\,\mathrm{d}f\right|^2 \quad (243a)$$

(hint: generalize the argument leading to Equation (211a)).

(d) Show that Equation (243a) can be rewritten as

$$\text{cov}\left\{\left|\sum_{t=0}^{N-1} a_t X_t\right|^2, \left|\sum_{t=0}^{N-1} b_t X_t\right|^2\right\} = \left|\sum_{t=0}^{N-1}\sum_{u=0}^{N-1} a_t b_u^* s_{t-u}\right|^2 + \left|\sum_{t=0}^{N-1}\sum_{u=0}^{N-1} a_t b_u s_{t-u}\right|^2. \quad (243b)$$

(e) Use the results of part (c) to verify Equation (211b) (hint: let a_t and b_t be, respectively, $\Delta_t^{1/2} h_t e^{-i2\pi(f+\eta)t\,\Delta_t}$ and $\Delta_t^{1/2} h_t e^{-i2\pi f t\,\Delta_t}$).

(f) Consider a time series $X_0, X_1, \ldots, X_{N-1}$ that is a portion of a Gaussian AR(1) process $X_t = \phi X_{t-1} + \epsilon_t$ with an ACVS given by $s_\tau = \phi^{|\tau|}/(1-\phi^2)$, where $\phi = 0.8$, and $\{\epsilon_t\}$ is a zero mean Gaussian white noise process with variance $\sigma_\epsilon^2 = 1$ (see Exercise [2.17a]). Given this series (and assuming $\Delta_t = 1$ for convenience), consider the periodogram $\hat{S}^{(\text{P})}(\tilde{f}_k)$ of Equation (170d) over the grid of frequencies $\tilde{f}_k = k/(2N)$, $k = 0, 1, \ldots, N$, with N set initially to 16, then to 64 and finally to 256. Use part (d) of this exercise to compute the variance of $\hat{S}^{(\text{P})}(\tilde{f}_k)$ for all $N+1$ settings for $\tilde{f}_k$ and for all three choices of sample size N. An approximation to this variance is obtainable from Equation (203f) in conjunction with (145a) (in the latter, p should be set to 1, and $\phi_{1,1}$ to ϕ). Plot the ratio of the approximate to the exact variance versus $\tilde{f}_k$ for each of the three sample sizes. Repeat what we just described, but now with ϕ set to 0.9, 0.99 and finally -0.99. Comment on how well the approximation works.

[6.28] (a) If $\{h_t\} \longleftrightarrow H(\cdot)$ and $\{s_{G,\tau}\} \longleftrightarrow S_G(\cdot)$, show that Equation (211b) can be rewritten as

$$\text{cov}\{\hat{S}_G^{(\text{D})}(f+\eta), \hat{S}_G^{(\text{D})}(f)\}$$

$$= \Delta_t^2\left|\sum_{j=0}^{N-1} h_j C_j^*(f+\eta)e^{-i2\pi f j\,\Delta_t}\right|^2 + \Delta_t^2\left|\sum_{j=0}^{N-1} h_j C_j(f+\eta)e^{-i2\pi f j\,\Delta_t}\right|^2,$$

where

$$C_j(f) \stackrel{\text{def}}{=} \sum_{k=0}^{N-1} h_k s_{G,j-k} e^{-i2\pi f k\,\Delta_t}.$$

(b) As per Equation (212a), let

$$R(\eta) \stackrel{\text{def}}{=} \left|\sum_{t=0}^{N-1} h_t^2 e^{-i2\pi\eta t\,\Delta_t}\right|^2, \quad \text{and, as per usual, assume } \sum_{t=0}^{N-1} h_t^2 = 1.$$

Show that, for Gaussian white noise with variance σ^2,

$$\text{cov}\{\hat{S}_G^{(\text{D})}(f+\eta), \hat{S}_G^{(\text{D})}(f)\} = \sigma^4\left[R(\eta) + R(2f+\eta)\right]\Delta_t^2$$

and hence that

$$\text{corr}\{\hat{S}_G^{(\text{D})}(f+\eta), \hat{S}_G^{(\text{D})}(f)\} = \frac{R(\eta) + R(2f+\eta)}{\sqrt{[\{1+R(2f)\}\{1+R(2f+2\eta)\}]}}.$$

(c) Using the Hanning data taper, create a plot of corr $\{\hat{S}_G^{(\text{D})}(f+\eta), \hat{S}_G^{(\text{D})}(f)\}$ versus $\eta \in [0, 1/4]$ for the case $N = 64$ and $f = 1/4$ with Δ_t assumed to be unity so that $f_{\mathcal{N}} = 1/2$.

[6.29] Suppose that $\{h_t\}$ is the rectangular data taper. What does $R(\eta)$ in Equation (212a) reduce to? For what values of η is $R(\eta) = 0$?

[6.30] Generate a realization of a portion $X_0, X_1, \ldots, X_{63}$ of the zero mean AR(2) process of Equation (34) (see Exercise [597] for a description of how to generate realizations from this process). Assuming $\Delta_t = 1$ and using the Hanning data taper $\{h_t\}$, compute a direct spectral estimate for $\{X_t\}$ at the frequencies $f'_k = k/1024$, $k = 0, \ldots, 48$, and denote the kth of these 49 values by $\hat{S}_0^{(\text{D})}(f'_k)$ (hint: to compute $\hat{S}_0^{(\text{D})}(\cdot)$ over the required grid of frequencies, consider the effect of zero padding $\{h_t X_t\}$ with $1024 - 64 = 960$ zeros.) Repeat what we just described a large number N_{R} of times to obtain $\{\hat{S}_j^{(\text{D})}(f'_k) : j = 0, \ldots, N_{\text{R}} - 1\}$, $k = 0, \ldots, 48$ (take "large" to mean something from 1,000 up to 100,000). Compute the sample mean and sample variance for each of these 49 sequences. Compare these sample values with theoretical values suggested by Equation (204b). Repeat again, but now center each simulated series by subtracting off its sample mean prior to applying the Hanning data taper. Compute the effective bandwidth $B_{\mathcal{H}}$ for $\hat{S}_j^{(\text{D})}(\cdot)$ as per Equations (194) and (192c), and comment upon the claim that the χ_2^2-based result stated in Equation (204b) is reasonable at low frequencies but with the restriction that $f \geq B_{\mathcal{H}}/2$ or, more conservatively, $f \geq B_{\mathcal{H}}$ (see C&E [3] for Section 6.6). Comment also on the applicability of the χ_1^2-based result stated in Equation (204b) when the series are centered.

[6.31] Here we consider some properties of an SDF estimator based on the periodogram of the reflected series shown in Equation (183c) (C&E [7] for Section 6.6 discusses use of the alternative scheme displayed in Equation (183b), which leads to the estimator $\tilde{S}^{(\text{DCT})}(\cdot)$ defined in Exercise [217]).

(a) Show that the periodogram of the reflected series over $\check{f}_k \stackrel{\text{def}}{=} k/(2N-2)$ takes the form

$$\hat{S}(\check{f}_k) = \frac{1}{2N-2}\left(X_0 + 2\sum_{t=1}^{N-2} X_t \cos(2\pi \check{f}_k t) + (-1)^k X_{N-1}\right)^2.$$

(b) Create a plot similar to Figure 217 that shows $\hat{S}(\check{f}_k)$ versus $\check{f}_k$ for the four AR(4) time series displayed in Figure 34 (these series are available on the website for the book). How does $\hat{S}(\cdot)$ compare to the DCT-based periodograms $\tilde{S}^{(\text{DCT})}(\cdot)$ shown in Figure 217?

(c) Assuming $\{X_t\}$ is a stationary process with mean zero and ACVS $\{s_\tau\}$, show that

$$E\{\hat{S}(\check{f}_k)\} = \frac{1}{N-1}\left(C_k + D_k + 2\sum_{\tau=1}^{N-2}\left(s_\tau + (-1)^k s_{N-1-\tau}\right)\cos(2\pi \check{f}_k \tau)\right)$$

for $k = 0, 1, \ldots, N-1$, where $C_k \stackrel{\text{def}}{=} s_0 + (-1)^k s_{N-1}$ and

$$D_0 \stackrel{\text{def}}{=} 2\sum_{\tau=-(N-3)}^{N-3}(N - |\tau| - 2)s_\tau,$$

$$D_k \stackrel{\text{def}}{=} (N-3)s_0 + 2\sum_{\tau=1}^{N-3}(N - \tau - 2)s_\tau \cos(2\pi \check{f}_k \tau)$$

$$- \frac{2}{\sin(2\pi \check{f}_k)}\sum_{\tau=1}^{N-3} s_\tau \sin(2\pi \check{f}_k[\tau+1]), \quad k = 1, \ldots, N-2,$$

$$D_{N-1} \stackrel{\text{def}}{=} 2\sum_{\tau=-(N-3)}^{N-3}(N - |\tau| - 2)(-1)^\tau s_\tau.$$

(d) Create a plot similar to Figure 218 that shows $E\{\hat{S}(\check{f}_k)\}$ versus $\check{f}_k$ for sample sizes $N = 16, 64, 256$ and 1024 for the AR(4) process of Equation (35a). How does $E\{\hat{S}(\cdot)\}$ compare to $E\{\tilde{S}^{(\text{DCT})}(\cdot)\}$ shown in Figure 218? Note: Exercise [9.6] tells how to compute the ACVS for the AR(4) process.

7

Lag Window Spectral Estimators

7.0 Introduction

In the previous chapter, we considered the periodogram $\hat{S}^{(\text{P})}(\cdot)$ as a seemingly "natural" estimator of a spectral density function (SDF) based upon a time series we regard as a realization of a portion $X_0, X_1, \ldots, X_{N-1}$ of a stationary process. The periodogram unfortunately is of limited use, both because it can be badly biased and because of its unduly large variance, which does not decrease as the sample size N increases. With an appropriate data taper, use of a direct spectral estimator $\hat{S}^{(\text{D})}(\cdot)$ can reduce the bias considerably; however, the variance of $\hat{S}^{(\text{D})}(f)$ has the same properties as that of the periodogram. This is problematic from at least two viewpoints. First, we need to be able to see the basic structure in spectral estimates when plotted versus frequency – the large variability in $\hat{S}^{(\text{D})}(\cdot)$ can mask potentially important features. Second, spectral estimators with better variance properties than $\hat{S}^{(\text{D})}(\cdot)$ can lead to more powerful statistical tests of certain hypotheses of interest.

In this chapter and the next, we look at schemes that use direct spectral estimators as building blocks to form estimators with better variance properties. This chapter is devoted to *lag window spectral estimators*, which reduce variance by smoothing $\hat{S}^{(\text{D})}(\cdot)$ across frequencies (smoothing is actually done by applying a so-called lag window to the estimator $\{\hat{s}_\tau^{(\text{D})}\}$ of the autocovariance sequence (ACVS) corresponding to $\hat{S}^{(\text{D})}(\cdot)$). In Section 7.1 we discuss the rationale for smoothing $\hat{S}^{(\text{D})}(\cdot)$, which leads to the lag window formulation. This estimator is associated with a *smoothing window* (a filter related to the lag window, but operating directly on $\hat{S}^{(\text{D})}(\cdot)$). We discuss the connection between lag and smoothing windows, consider some of their basic properties and introduce the notion of *smoothing window bandwidth*, which characterizes the degree of smoothing to which $\hat{S}^{(\text{D})}(\cdot)$ is subjected. We devote Sections 7.2 and 7.3 to studying the first- and second-moment properties of lag window estimators, after which we consider their large-sample distributions (Section 7.4). Section 7.5 has six examples of lag windows and their associated smoothing windows. We discuss how to choose amongst them in Section 7.6. The common theme of Sections 7.7, 7.8 and 7.9 is setting of the smoothing window bandwidth, which controls the overall quality of the SDF estimator for a particular time series. Section 7.11 discusses computational details. Section 7.12 has five examples of lag window spectral estimation based on actual time series. We close the chapter with a summary (Section 7.13) and exercises for the reader (Section 7.14).

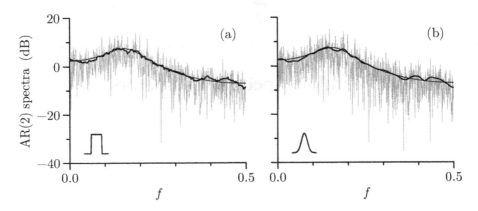

Figure 246 Examples of discretely smoothed direct spectral estimates $\hat{S}^{(\text{DS})}(\cdot)$. The rough-looking light curve in (a) shows the periodogram evaluated over the Fourier frequencies $f_k = k/(N\,\Delta_t)$ for the AR(2) time series shown in Figure 34(a), for which $\Delta_t = 1$. The thick dark curve is $\hat{S}^{(\text{DS})}(\cdot) = \overline{S}(\cdot)$, which is based on a "running average" filter with a span of $2M + 1 = 31$. The impulse response sequence for this filter is depicted in the lower left-hand corner on a linear/linear scale (its vertical axis is not depicted). The thin dark curve is the true SDF for the AR(2) process. Plot (b) shows a second example of $\hat{S}^{(\text{DS})}(\cdot)$. This estimate is based on smoothing the periodogram (now evaluated over the frequencies $f'_k = k/(2N\,\Delta_t)$) using a filter whose impulse response sequence is proportional to samples from a Gaussian probability density function (PDF). Both these examples of $\hat{S}^{(\text{DS})}(\cdot)$ are clearly better overall estimates of the true SDF than the periodogram.

7.1 Smoothing Direct Spectral Estimators

As mentioned in the introduction, lag window estimators are based on smoothing a direct spectral estimator $\hat{S}^{(\text{D})}(\cdot)$ across frequencies. The justification for this approach is the following. Suppose N is large enough so that the periodogram $\hat{S}^{(\text{P})}(\cdot)$ is essentially an unbiased estimator of $S(\cdot)$ and is pairwise uncorrelated at the Fourier frequencies f_k (i.e., the approximations stated by Equation (204a) are valid). If $S(\cdot)$ is slowly varying in the neighborhood of, say, f_k, then

$$S(f_{k-M}) \approx \cdots \approx S(f_k) \approx \cdots \approx S(f_{k+M})$$

for some integer $M > 0$. Thus $\hat{S}^{(\text{P})}(f_{k-M}), \ldots, \hat{S}^{(\text{P})}(f_k), \ldots, \hat{S}^{(\text{P})}(f_{k+M})$ are a set of $2M + 1$ approximately unbiased and uncorrelated estimators of the same quantity, namely, $S(f_k)$. Suppose we average them to form the estimator

$$\overline{S}(f_k) \stackrel{\text{def}}{=} \frac{1}{2M+1} \sum_{j=-M}^{M} \hat{S}^{(\text{P})}(f_{k-j}). \tag{246a}$$

(Figure 246(a) shows an example of $\overline{S}(\cdot)$). Under our assumptions we have

$$E\{\overline{S}(f_k)\} \approx S(f_k) \quad \text{and} \quad \text{var}\{\overline{S}(f_k)\} \approx \frac{S^2(f_k)}{2M+1} \approx \frac{\text{var}\{\hat{S}^{(\text{P})}(f_k)\}}{2M+1},$$

where we have made use of Equation (203f) under the simplifying assumption that $f_{k-M} > 0$ and $f_{k+M} < f_{\mathcal{N}}$. If we now consider increasing both the sample size N and the index k in such a way that $k/(N\,\Delta_t) = f_k$ is held constant, we can then let M get large also and claim that $\text{var}\{\overline{S}(f_k)\}$ can be made arbitrarily small so that $\overline{S}(f_k)$ is a consistent estimator of $S(f_k)$.

The estimator $\overline{S}(f_k)$ is a special case of a more general spectral estimator of the form

$$\hat{S}^{(\text{DS})}(f'_k) \stackrel{\text{def}}{=} \sum_{j=-M}^{M} g_j \hat{S}^{(\text{D})}(f'_{k-j}) \quad \text{with} \quad f'_k = \frac{k}{N'\,\Delta_t}, \tag{246b}$$

7.1 Smoothing Direct Spectral Estimators

where $\{g_j\}$ is a sequence of $2M+1$ smoothing coefficients with $g_{-j}=g_j$; $\hat{S}^{(\mathrm{D})}(\cdot)$ is a direct spectral estimator (Equation (186b)); and N' is a positive integer that controls the spacing of the frequencies over which the smoothing occurs. Typically we choose $N' \geq N$ (the sample size) so that the frequencies f'_k are at least as closely spaced as the Fourier frequencies $f_k = k/(N\,\Delta_{\mathrm{t}})$. We call $\hat{S}^{(\mathrm{DS})}(f'_k)$ a *discretely smoothed direct spectral estimator* (see Figure 246 for two examples). Note that we can regard $\{g_j\}$ as the coefficients of an LTI digital filter.

Now the estimator $\hat{S}^{(\mathrm{DS})}(f'_k)$ is formed by smoothing the direct spectral estimator $\hat{S}^{(\mathrm{D})}(\cdot)$ with a *discrete* convolution over a *discrete* set of frequencies. Because $\hat{S}^{(\mathrm{D})}(\cdot)$ is defined for all $f \in [-f_\mathcal{N}, f_\mathcal{N}]$, we can also smooth it using a *continuous* convolution over a *continuous* set of frequencies. We thus consider an estimator of the form

$$\hat{S}_m^{(\mathrm{LW})}(f) = \int_{-f_\mathcal{N}}^{f_\mathcal{N}} V_m(f-\phi)\hat{S}^{(\mathrm{D})}(\phi)\,\mathrm{d}\phi, \tag{247a}$$

where the *design window* $V_m(\cdot)$ is a symmetric real-valued $2f_\mathcal{N}$ periodic function that is square integrable over $[-f_\mathcal{N}, f_\mathcal{N}]$ and whose smoothing properties can be controlled by a parameter m (the rationale for putting LW on $\hat{S}_m^{(\mathrm{LW})}(\cdot)$ is given following Equation (248b)).

▷ **Exercise [247]** Recalling from Equations (188a) and (188b) that $\{\hat{s}_\tau^{(\mathrm{D})}\} \longleftrightarrow \hat{S}^{(\mathrm{D})}(\cdot)$, show that we can rewrite Equation (247a) as

$$\hat{S}_m^{(\mathrm{LW})}(f) = \Delta_{\mathrm{t}} \sum_{\tau=-(N-1)}^{N-1} v_{m,\tau}\hat{s}_\tau^{(\mathrm{D})}\mathrm{e}^{-\mathrm{i}2\pi f \tau\,\Delta_{\mathrm{t}}}, \tag{247b}$$

where

$$v_{m,\tau} \stackrel{\mathrm{def}}{=} \int_{-f_\mathcal{N}}^{f_\mathcal{N}} V_m(\phi)\mathrm{e}^{\mathrm{i}2\pi\phi\tau\,\Delta_{\mathrm{t}}}\,\mathrm{d}\phi. \tag{247c}$$ ◁

We note that $\{v_{m,\tau}\}$ and the design window $V_m(\cdot)$ are a Fourier transform pair:

$$\{v_{m,\tau}\} \longleftrightarrow V_m(\cdot).$$

Equation (247b) tells us that we can compute $\hat{S}_m^{(\mathrm{LW})}(f)$ using a finite number of numerical operations even though it is defined in Equation (247a) via a continuous convolution over a continuous set of frequencies. We note that Equation (247b) does not involve $v_{m,\tau}$ for $|\tau| \geq N$ but that $v_{m,\tau}$ need *not* be zero for these values of τ. For theoretical purposes, we will find it convenient to set

$$w_{m,\tau} \stackrel{\mathrm{def}}{=} \begin{cases} v_{m,\tau}, & |\tau| < N; \\ 0, & |\tau| \geq N, \end{cases} \tag{247d}$$

and

$$W_m(f) \stackrel{\mathrm{def}}{=} \Delta_{\mathrm{t}} \sum_{\tau=-(N-1)}^{N-1} w_{m,\tau}\mathrm{e}^{-\mathrm{i}2\pi f\tau\,\Delta_{\mathrm{t}}} \tag{247e}$$

so that

$$\{w_{m,\tau}\} \longleftrightarrow W_m(\cdot).$$

Note that $W_m(\cdot)$ is identical to the design window $V_m(\cdot)$ if $v_{m,\tau} = 0$ for $|\tau| \geq N$; however, even if this is not true, $V_m(\cdot)$ and $W_m(\cdot)$ are equivalent as far as Equation (247a) is concerned because we have

$$\int_{-f_\mathcal{N}}^{f_\mathcal{N}} W_m(f-\phi)\hat{S}^{(\mathrm{D})}(\phi)\,\mathrm{d}\phi = \int_{-f_\mathcal{N}}^{f_\mathcal{N}} V_m(f-\phi)\hat{S}^{(\mathrm{D})}(\phi)\,\mathrm{d}\phi = \hat{S}_m^{(\mathrm{LW})}(f) \tag{247f}$$

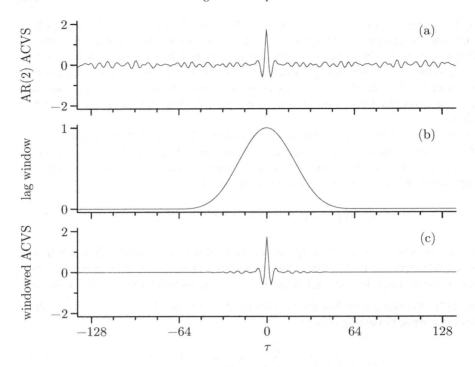

Figure 248 Portion of the biased estimate $\{\hat{s}_\tau^{(\mathrm{P})}\}$ of the ACVS for AR(2) time series shown in Figure 34(a), a Parzen lag window $\{w_{m,\tau}\}$ and their product of $\{w_{m,\tau}\hat{s}_\tau^{(\mathrm{P})}\}$ (the Parzen lag window is given by Equation (275a), and here $m = 64$). The corresponding spectral estimate $\hat{S}_m^{(\mathrm{LW})}(\cdot)$ is shown in Figure 249.

for all f; i.e., the estimator $\hat{S}_m^{(\mathrm{LW})}(\cdot)$ will be *identically* the same whether we use $V_m(\cdot)$ or $W_m(\cdot)$ (see Exercise [7.1]). The essential distinction between $V_m(\cdot)$ and $W_m(\cdot)$ is that $w_{m,\tau}$ is guaranteed to be 0 for $|\tau| \geq N$ whereas $v_{m,\tau}$ need not be so.

We can thus assume that the estimator $\hat{S}_m^{(\mathrm{LW})}(\cdot)$ can be written as

$$\hat{S}_m^{(\mathrm{LW})}(f) \stackrel{\text{def}}{=} \int_{-f_\mathcal{N}}^{f_\mathcal{N}} W_m(f - \phi)\hat{S}^{(\mathrm{D})}(\phi)\,\mathrm{d}\phi = \Delta_\mathrm{t} \sum_{\tau=-(N-1)}^{N-1} w_{m,\tau}\hat{s}_\tau^{(\mathrm{D})} \mathrm{e}^{-\mathrm{i}2\pi f\tau \Delta_\mathrm{t}}, \quad (248\mathrm{a})$$

where $\{w_{m,\tau}\} \longleftrightarrow W_m(\cdot)$ with $w_{m,\tau} = 0$ for $|\tau| \geq N$. If we set

$$\hat{s}_{m,\tau}^{(\mathrm{LW})} \stackrel{\text{def}}{=} \begin{cases} w_{m,\tau}\hat{s}_\tau^{(\mathrm{D})}, & |\tau| \leq N-1; \\ 0, & |\tau| \geq N, \end{cases} \quad (248\mathrm{b})$$

we have $\{\hat{s}_{m,\tau}^{(\mathrm{LW})}\} \longleftrightarrow \hat{S}_m^{(\mathrm{LW})}(\cdot)$; i.e., the sequence $\{\hat{s}_{m,\tau}^{(\mathrm{LW})}\}$ is the estimator of the ACVS corresponding to $\hat{S}_m^{(\mathrm{LW})}(\cdot)$.

We call the function $W_m(\cdot)$ defined in Equation (247e) a *smoothing window* (although most other authors would call it a spectral window, a term we reserve for another concept); its inverse Fourier transform $\{w_{m,\tau}\}$ is called a *lag window* (other names for it are *quadratic window* and *quadratic taper*). We call $\hat{S}_m^{(\mathrm{LW})}(f)$ a *lag window spectral estimator* of $S(f)$ (see item [1] in the Comments and Extensions [C&Es]). In practice, we specify a lag window spectral estimator via either a design window $V_m(\cdot)$ or a lag window $\{w_{m,\tau}\}$ – the smoothing window $W_m(\cdot)$ follows once one of these is given. Figure 248 shows examples of an ACVS estimate, a lag window and their product, while Figure 249 shows the corresponding lag window spectral estimate.

How are the spectral estimators $\{\hat{S}^{(\mathrm{DS})}(f_k')\}$ and $\hat{S}_m^{(\mathrm{LW})}(\cdot)$ related?

7.1 Smoothing Direct Spectral Estimators

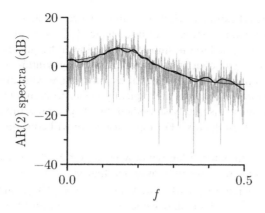

Figure 249 Example of lag window estimate $\hat{S}_m^{(\text{LW})}(\cdot)$ (thick dark curve). This estimate is based on smoothing the periodogram $\hat{S}^{(\text{P})}(\cdot)$ for the AR(2) time series of Figure 34(a) based upon a Parzen lag window $\{w_{m,\tau}\}$ with $m = 64$. The periodogram, evaluated over frequencies $\tilde{f}_k = k/(2N)$, is the rough-looking light curve, while the true SDF for the AR(2) process is the thin dark curve. Note that this lag window estimate is virtually identical to the discretely smoothed direct spectral estimate shown in Figure 246(b). A portion of the ACVS estimate corresponding to $\hat{S}_m^{(\text{LW})}(\cdot)$ is shown in the bottom panel of Figure 248.

▷ **Exercise [249a]** Show that

$$\hat{S}^{(\text{DS})}(f'_k) = \Delta_t \sum_{\tau=-(N-1)}^{N-1} v_{g,\tau} \hat{s}_\tau^{(\text{D})} e^{-i2\pi f'_k \tau \Delta_t}, \tag{249a}$$

where

$$v_{g,\tau} \stackrel{\text{def}}{=} \sum_{j=-M}^{M} g_j e^{i2\pi f'_j \tau \Delta_t} = \sum_{j=-M}^{M} g_j e^{i2\pi j\tau/N'}. \tag{249b} \triangleleft$$

Hence any discretely smoothed direct spectral estimator can be expressed as a lag window spectral estimator with an implicitly defined lag window given by

$$w_{g,\tau} \stackrel{\text{def}}{=} \begin{cases} v_{g,\tau}, & |\tau| \leq N-1; \\ 0, & |\tau| \geq N, \end{cases} \tag{249c}$$

and a corresponding smoothing window given by Equation (247e) with $w_{m,\tau}$ replaced by $w_{g,\tau}$.

Now consider any lag window spectral estimator $\hat{S}_m^{(\text{LW})}(\cdot)$ based upon $\{w_{m,\tau}\} \longleftrightarrow W_m(\cdot)$. Although this estimator is defined for all $f \in \mathbb{R}$, it is completely determined if we know $\{\hat{s}_{m,\tau}^{(\text{LW})}\}$ because $\{\hat{s}_{m,\tau}^{(\text{LW})}\} \longleftrightarrow \hat{S}_m^{(\text{LW})}(\cdot)$. By definition $\hat{s}_{m,\tau}^{(\text{LW})} = 0$ when $|\tau| \geq N$. For $|\tau| < N$, we find, using the same argument that led to Equation (171e), that

$$\{\hat{s}_{m,\tau}^{(\text{LW})} : \tau = -(N-1), \ldots, N-1\} \longleftrightarrow \{\hat{S}_m^{(\text{LW})}(f'_k) : k = -(N-1), \ldots, N-1\},$$

where, as before, $f'_k = k/[(2N-1)\Delta_t]$ defines a grid of frequencies almost twice as fine as the Fourier frequencies $f_k \stackrel{\text{def}}{=} k/(N\Delta_t)$. Hence $\hat{S}_m^{(\text{LW})}(\cdot)$ is completely determined once we know it over the finite set of frequencies f'_k.

▷ **Exercise [249b]** Show that

$$\hat{S}_m^{(\text{LW})}(f'_k) = \sum_{j=-(N-1)}^{N-1} g_j \hat{S}^{(\text{D})}(f'_{k-j}), \text{ where } g_j \stackrel{\text{def}}{=} \frac{W_m(f'_j)}{(2N-1)\Delta_t}. \tag{249d} \triangleleft$$

Hence any lag window spectral estimator $\hat{S}_m^{(\text{LW})}(\cdot)$ can be expressed at $f = f'_k$ as a discretely smoothed direct spectral estimator over the set of frequencies $\{f'_j : j = -(N-1), \ldots, N-1\}$ with weights g_j that are proportional to the smoothing window at frequency f'_j.

From these arguments we can conclude that the class of lag window spectral estimators and the class of discretely smoothed direct spectral estimators are equivalent. From a practical point of view, lag window estimators are typically easier to compute because they are based on multiplications whereas discretely smoothed direct spectral estimators are based on convolutions (see Section 7.11 for details). For this reason, we will concentrate mainly on lag window spectral estimators.

Given a direct spectral estimator, a lag window estimator $\hat{S}_m^{(\text{LW})}(\cdot)$ is fully specified by its lag window $\{w_{m,\tau}\}$. So far we have only assumed that $w_{m,\tau} = 0$ for $|\tau| \geq N$, but this condition by itself is not sufficient to get what we want out of $\hat{S}_m^{(\text{LW})}(\cdot)$, namely, an estimator with smaller variance than $\hat{S}^{(\text{D})}(\cdot)$. Indeed, if we let $w_{m,\tau} = 1$ for all $|\tau| \leq N - 1$ in Equation (248a), we find that $\hat{S}_m^{(\text{LW})}(\cdot)$ reduces to $\hat{S}^{(\text{D})}(\cdot)$ (compare Equations (188a) and (248a)). Not surprisingly, we must impose some conditions on the lag window $\{w_{m,\tau}\}$ or, equivalently, the smoothing window $W_m(\cdot)$ to ensure that $\hat{S}_m^{(\text{LW})}(\cdot)$ is a reasonable SDF estimator with better statistical properties than $\hat{S}^{(\text{D})}(\cdot)$. We thus make the following three assumptions.

[1] Since $\hat{s}_{-\tau}^{(\text{D})} = \hat{s}_\tau^{(\text{D})}$, we require that $w_{m,-\tau} = w_{m,\tau}$ for all τ and all choices of m. This implies that $W_m(\cdot)$ is an even real-valued function that is periodic with a period of $2f_\mathcal{N}$.

[2] We require the normalization

$$\int_{-f_\mathcal{N}}^{f_\mathcal{N}} W_m(f)\,df = 1, \text{ or, equivalently, } w_{m,0} = 1, \text{ for all } m \qquad (250\text{a})$$

(the equivalence follows from $\{w_{m,\tau}\} \longleftrightarrow W_m(\cdot)$). This stipulation is similar to one we made in Section 5.7 to ensure that a low-pass digital filter passes unaltered a locally linear portion of a sequence. In terms of the g_j weights of Equation (249d), the condition $w_{m,0} = 1$ is equivalent to

$$\sum_{j=-(N-1)}^{N-1} g_j = 1. \qquad (250\text{b})$$

(Exercise [7.2] is to prove this statement).

[3] We require that $W_m(\cdot)$ acts more and more like a Dirac delta function as $m \to \infty$. In particular, for any $\delta > 0$ and for all $|f| \in [\delta, f_\mathcal{N}]$, we require

$$W_m(f) \to 0 \text{ as } m \to \infty. \qquad (250\text{c})$$

For a more technical statement of this and other requirements needed to prove various theorems about lag window estimators, see Priestley (1981, pp. 450–1).

In addition to these assumptions, note that, if

$$W_m(f) \geq 0 \text{ for all } m \text{ and } f, \qquad (250\text{d})$$

then it follows from Equation (248a) that, since $\hat{S}^{(\text{D})}(f) \geq 0$ necessarily (consider Equation (186b)), we must have $\hat{S}_m^{(\text{LW})}(f) \geq 0$. Interestingly, the condition of Equation (250d) is sufficient, but not necessary, to ensure the nonnegativity of $\hat{S}_m^{(\text{LW})}(f)$; see the discussion on the Daniell smoothing window in Section 7.5. As we shall see in the discussion concerning Figure 282, there are valid reasons for considering smoothing windows for which $W_m(f) < 0$ for some values of f. Thus, whereas we always require the just-listed items [1] to [3] to hold in addition to $w_{m,\tau} = 0$ for $|\tau| \geq N$, the condition of Equation (250d) is desirable but not required. Note that, if this condition were to hold, the assumption of Equation (250a) implies that $W_m(\cdot)$ could then be regarded as a PDF for an RV distributed over the interval $[-f_{\mathcal{N}}, f_{\mathcal{N}}]$.

We also need to define a bandwidth for the smoothing window $W_m(\cdot)$ in order to have some idea about the range of frequencies that influences the value of $\hat{S}_m^{(\text{LW})}(f)$. When the condition of Equation (250d) holds, we can regard $W_m(\cdot)$ as a PDF for an RV that assumes values restricted to the interval $[-f_{\mathcal{N}}, f_{\mathcal{N}}]$. A convenient measure of width is then given by Equation (60c):

$$\beta_W \stackrel{\text{def}}{=} \left(12 \int_{-f_{\mathcal{N}}}^{f_{\mathcal{N}}} f^2 W_m(f) \, df \right)^{1/2}. \tag{251a}$$

This definition of window bandwidth is essentially due to Grenander (1951, p. 525) although he omits the factor of 12. As noted in Section 3.4, inclusion of this factor makes the bandwidth of a rectangular smoothing window equal to its natural width (this is related to the Daniell smoothing window – see Section 7.5). A convenient computational formula for β_W is

$$\beta_W = \frac{1}{\Delta_t} \left(1 + \frac{12}{\pi^2} \sum_{\tau=1}^{N-1} \frac{(-1)^\tau}{\tau^2} w_{m,\tau} \right)^{1/2} \tag{251b}$$

(cf. Equation (192b); the extra Δ_t in the right-hand side of Equation (192b) can be explained by the fact that $\{w_{m,\tau}\} \longleftrightarrow W_m(\cdot)$, whereas $\{h \star h_\tau\} \longleftrightarrow \Delta_t \mathcal{H}(\cdot)$).

There is, however, a potential danger in using β_W to define the smoothing window bandwidth. If the nonnegativity condition of Equation (250d) does not hold, the integral in Equation (251a) can actually be negative, in which case β_W is imaginary (C&E [3] for Section 7.5 has an example of this happening)! Because of this potential problem, we prefer the following measure for the smoothing window bandwidth due to Jenkins (1961):

$$B_W \stackrel{\text{def}}{=} \frac{1}{\int_{-f_{\mathcal{N}}}^{f_{\mathcal{N}}} W_m^2(f) \, df}. \tag{251c}$$

Since $W_m(\cdot)$ is real-valued, so must be B_W. This bandwidth measure is just the autocorrelation width of $W_m(\cdot)$ as given by Equation (100f) (the integral in the numerator of Equation (100f) is unity due to Equation (250a)). Since $\{w_{m,\tau}\}$ and $W_m(\cdot)$ are a Fourier transform pair, the Parseval's relationship of Equation (75b) states that

$$\int_{-f_{\mathcal{N}}}^{f_{\mathcal{N}}} W_m^2(f) \, df = \Delta_t \sum_{\tau=-(N-1)}^{N-1} w_{m,\tau}^2, \tag{251d}$$

so we also have

$$B_W = \frac{1}{\Delta_t \sum_{\tau=-(N-1)}^{N-1} w_{m,\tau}^2}. \tag{251e}$$

(See C&E [2] for an additional way of viewing B_W.)

Several other measures of window bandwidth are used widely in the literature (Priestley, 1981, p. 520). For example, Parzen (1957) proposed using the equivalent width of $W_m(\cdot)$:

$$B_{W,P} \stackrel{\text{def}}{=} \text{width}_e \{W_m(\cdot)\} = \frac{1}{W_m(0)} = \frac{1}{\Delta_t \sum_{\tau=-(N-1)}^{N-1} w_{m,\tau}} \tag{251f}$$

(the above follows from Equation (100d) after recalling that $W_m(\cdot)$ integrates to unity and that $w_{m,0} = 1$). Figure 252 shows an example of a smoothing window along with depictions of its bandwidth as measured by β_W, B_W and $B_{W,P}$.

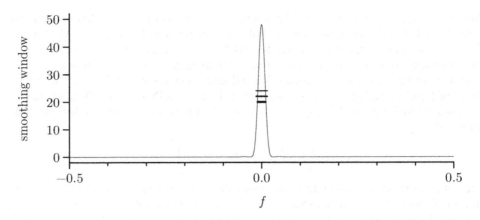

Figure 252 Example of a nonnegative smoothing window $W_m(\cdot)$ and its bandwidth as measured by β_W, B_W and $B_{W,P}$ (given by, from top to bottom, the widths of the three horizontal lines). Here $W_m(\cdot)$ is the Parzen smoothing window of Equation (275b) with $m = 64$. Noting that $W_{64}(0) = 48$, we can visually deduce the half-power width of $W_m(\cdot)$ (another measure of its bandwidth) from the top-most line since it is plotted vertically at 24.

Comments and Extensions to Section 7.1

[1] The lag window spectral estimator $\hat{S}_m^{(\mathrm{LW})}(\cdot)$ is known by many other names in the literature: quadratic window estimator, weighted covariance estimator, indirect nonparametric estimator, windowed periodogram estimator, smoothed periodogram estimator, Blackman–Tukey estimator, spectrograph estimator, spectral estimator of the Grenander–Rosenblatt type and correlogram method power spectral density estimator – and this list is by no means exhaustive! To compound the confusion, some of these names are also used in the literature to refer to estimators that are slightly different from our definition of $\hat{S}_m^{(\mathrm{LW})}(\cdot)$. For example, our definition (Equation (248a)) is essentially a weighting applied to the ACVS estimator $\{\hat{s}^{(\mathrm{D})}\}$ corresponding to *any* direct spectral estimator; other authors define it – at least implicitly – for just the special case of $\{\hat{s}^{(\mathrm{P})}\}$, the ACVS estimator corresponding to the periodogram.

[2] It can be argued from the material in Sections 7.3 and 7.5 that Jenkins's smoothing window bandwidth B_W is the width of the Daniell smoothing window that yields an estimator with the same large-sample variance as one employing $W_m(\cdot)$. Both Priestley (1981) and Bloomfield (2000) argue *against* use of B_W because of this link with variance – they suggest that window bandwidth should be tied instead to the bias introduced by the smoothing window. This property is enjoyed by β_W (see Equation (256d)), but, as noted prior to Equation (251c), we prefer B_W mainly because of its superior practical utility. In most applications, however, as long as the nonnegativity condition of Equation (250d) holds, the measures β_W, B_W and $B_{W,P}$ are interchangeable (see Figure 252).

[3] Considering the defining equality in Equation (248a), we can interpret a lag window estimator $\hat{S}_m^{(\mathrm{LW})}(\cdot)$ as just applying a linear filter to $\hat{S}^{(\mathrm{D})}(\cdot)$, with the intent of extracting a function (the true SDF) buried in rather substantial noise (dictated by the properties of a χ_2^2 distribution). The linear filter in question is specified by the smoothing window $W_m(\cdot)$ and typically acts as a low-pass filter. A complementary interpretation of $\hat{S}_m^{(\mathrm{LW})}(\cdot)$ comes from comparing the Fourier representation for a periodic function $g_p(\cdot)$ with the representation for $\hat{S}^{(\mathrm{D})}(\cdot)$ in terms of $\{\hat{s}_\tau^{(\mathrm{D})}\}$:

$$g_p(t) = \sum_{n=-\infty}^{\infty} G_n e^{i2\pi f_n t} \quad \text{and} \quad \hat{S}^{(\mathrm{D})}(f) = \sum_{\tau=-\infty}^{\infty} \hat{s}_\tau^{(\mathrm{D})} e^{-i2\pi f \tau},$$

where we have set $\Delta_t = 1$ to help with the comparison (see Equations (49c) and (188a), recalling that $\hat{s}_\tau^{(\mathrm{D})} = 0$ when $|\tau| \geq N$). We can attribute any visual choppiness in $g_p(\cdot)$ as being due to the influence of the G_n coefficients with large $|n|$, i.e., those associated with high frequencies f_n. We can alleviate this choppiness by damping down the offending G_n, with the degree of damping ideally increasing as $|n|$ increases (cf. Equation (69b) and the discussion surrounding Figure 70). Similarly, we can attribute the inherently noisy appearance of $\hat{S}^{(\mathrm{D})}(\cdot)$ as being due to the $\hat{s}_\tau^{(\mathrm{D})}$ terms with large $|\tau|$. Hence we can smooth out $\hat{S}^{(\mathrm{D})}(\cdot)$ by damping down the offending $\hat{s}_\tau^{(\mathrm{D})}$. In view of

$$\sum_{\tau=-\infty}^{\infty} w_{m,\tau}\hat{s}_\tau^{(\mathrm{D})}\mathrm{e}^{-\mathrm{i}2\pi f\tau} = \hat{S}_m^{(\mathrm{LW})}(f)$$

we see that a lag window $\{w_{m,\tau}\}$ does exactly that, assuming that it is similar to the one depicted in the middle panel of Figure 248 (the above is essentially the same as Equation (248a)).

[4] We have noted that any discretely smoothed direct spectral estimator $\hat{S}^{(\mathrm{DS})}(\cdot)$ can be reexpressed as a lag window estimator, with the connection between the smoothing coefficients $\{g_j\}$ and the implicit lag window $\{w_{g,\tau}\}$ given by Equations (249b) and (249c). In view of this, let us reconsider the $\hat{S}^{(\mathrm{DS})}(\cdot)$ shown in Figure 246(a), which is based on rectangular coefficients $\{g_j = 1/(2M+1) : j = -M, \ldots, M\}$ applied over the grid of Fourier frequencies $f_k = k/N$ (here $M = 15$). The implicit lag window is given by

$$w_{g,\tau} = \frac{1}{2M+1}\sum_{j=-M}^{M}\mathrm{e}^{\mathrm{i}2\pi j\tau/N} = \mathcal{D}_{2M+1}\left(\tau/N\right), \quad |\tau| \leq N-1,$$

where $\mathcal{D}_{2M+1}(\cdot)$ is Dirichlet's kernel (the above follows from Exercise [1.2e]). The function $\mathcal{D}_{2M+1}(\cdot)$ is an even periodic function with a period of unity. Periodicity implies that $w_{g,N-\tau} = w_{g,\tau}$ for $1 \leq \tau \leq N-1$, as is illustrated by the thick wiggly curve in Figure 254(a). In view of the previous C&E, the fact that this lag window does not damp down the $\hat{s}_\tau^{(\mathrm{P})}$ at the most extreme lags is cause for concern (however, this is mitigated by the fact that $\hat{s}_\tau^{(\mathrm{P})} \to 0$ as $|\tau| \to N-1$). This questionable property is not inherent to rectangular coefficients, but rather holds for all $\{w_{g,\tau}\}$ arising from use of $\{g_j\}$ over the grid of Fourier frequencies: with N' set to N in Equation (249b), we have

$$w_{g,N-\tau} = \sum_{j=-M}^{M} g_j \mathrm{e}^{\mathrm{i}2\pi j(N-\tau)/N} = \sum_{j=-M}^{M} g_j \mathrm{e}^{\mathrm{i}2\pi j}\mathrm{e}^{-\mathrm{i}2\pi j\tau/N} = \sum_{j=-M}^{M} g_j \mathrm{e}^{\mathrm{i}2\pi j\tau/N} = w_{g,\tau},$$

since $g_{-j} = g_j$. As a second example, the thin smooth curve in Figure 254(a) shows $\{w_{g,\tau}\}$ for Gaussian coefficients g_j that are proportional to $\exp(-j^2/150)$ for $j = -35, \ldots, 35$.

A more reasonable implicit lag window arises if we produce $\hat{S}^{(\mathrm{DS})}(\cdot)$ by smoothing the direct SDF estimator over a grid of frequencies that is denser than what is afforded by the Fourier frequencies f_k. In particular, if we use $\tilde{f}_k = k/(2N)$, the implicit lag window corresponding to rectangular coefficients is

$$w_{g,\tau} = \frac{1}{2M+1}\sum_{j=-M}^{M}\mathrm{e}^{\mathrm{i}2\pi j\tau/(2N)} = \mathcal{D}_{2M+1}\left(\tau/(2N)\right), \quad |\tau| \leq N-1.$$

The thick wiggly curve in Figure 254(b) shows this lag window when $M = 30$, while the thin smooth curve is for Gaussian coefficients $g_j \propto \exp(-j^2/600)$, $j = -70, \ldots, 70$. In both cases the aberrant increase in $w_{g,\tau}$ as $|\tau| \to N-1$ is gone. The Gaussian coefficients produced the $\hat{S}^{(\mathrm{DS})}(\cdot)$ shown in Figure 246(b), which is markedly smoother than the one in Figure 246(a), despite the fact that both estimates have comparable bandwidths (see the next C&E). While the primary reason for the improved smoothness is the fact that the implicit lag window for Figure 246(b) no longer has ripples, a secondary reason is that the lag window does not ramp up toward unity for $|\tau|$ near $N-1$, as must be the case if we choose to smooth a direct spectral estimator over the Fourier frequencies f_k rather than over $\tilde{f}_k$. Hence it is better to formulate discretely smoothed direct spectral estimators over a grid of frequencies that is finer than the Fourier frequencies.

[5] The measures of smoothing window bandwidth for a lag window estimator we have considered so far (β_W, B_W and $B_{W,P}$) can all be computed once we know the lag window $\{w_{m,\tau}\}$. One idea for defining corresponding bandwidth measures for a discretely smoothed direct spectral estimator $\hat{S}^{(\mathrm{DS})}(\cdot)$ would be to take its defining smoothing coefficients $\{g_j\}$, form the implicit lag window $\{w_{g,\tau}\}$ via

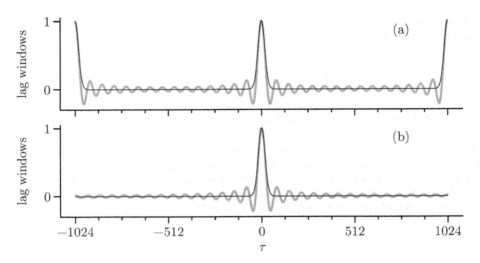

Figure 254 Implicit lag windows $\{w_{g,\tau}\}$ for four discretely smoothed direct spectral estimators $\hat{S}^{(\mathrm{DS})}(\cdot)$. The windows in the top plot correspond to rectangular smoothing coefficients $g_j = 1/31, j = -15, \ldots, 15$, (thick wiggly curve) and Gaussian coefficients $g_j \propto \exp(-j^2/150), j = -35, \ldots, 35$, (thin smooth curve), where the constant of proportionality is set such that $\sum_j g_j = 1$. In both cases, the grid of frequencies over which the direct spectral estimator is smoothed is taken to be the Fourier frequencies $f_k = k/(N\,\Delta_{\mathrm{t}})$, where $N = 1024$. The corresponding curves in the bottom plot are for $g_j = 1/61, j = -30, \ldots, 30$, and $g_j \propto \exp(-j^2/600), j = -70, \ldots, 70$, but now the grid of frequencies is taken to be $\tilde{f}_k = k/(2N\,\Delta_{\mathrm{t}})$ and hence is twice as fine as the grid dictated by the Fourier frequencies.

Equations (249c) and (249b) and then use this as the basis for computing either β_W, B_W or $B_{W,P}$. This approach works in some instances, but can fail when $\{g_j\}$ is used over the grid of Fourier frequencies. As an example, let us reconsider the $\hat{S}^{(\mathrm{DS})}(\cdot)$ shown in Figure 246(a), which is based on the rectangular smoothing coefficients $\{g_j = 1/(2M+1) : j = -M, \ldots, M\}$ with $M = 15$. Since the spacing of the Fourier frequencies is $1/(N\,\Delta_{\mathrm{t}})$, the natural bandwidth is $(2M+1)/(N\,\Delta_{\mathrm{t}})$, which is equal to 0.03027 when $N = 1024$ and $\Delta_{\mathrm{t}} = 1$. Using the implicit lag window (shown by the thick wiggly curve in Figure 254(a)), we obtain the bandwidths listed in the first row of Table 255. While β_W agrees well with the natural bandwidth, the other two measures are off by a factor of two. This disparity can be attributed to how $w_{g,\tau}$ is treated at large $|\tau|$ in Equations (251b), (251e) and (251f). The second row of the table shows that a discrepancy between β_W and the other two measures arises again for the $\hat{S}^{(\mathrm{DS})}(\cdot)$ shown in Figure 246(b), which is based on Gaussian smoothing coefficients $\{g_j\}$ (the implicit lag window is the thin smooth curve in Figure 254(a)). On the other hand, if we use the grid of frequencies $\tilde{f}_k$ with spacing $1/(2N\,\Delta_{\mathrm{t}})$, the natural bandwidth is $(2M+1)/(2N\,\Delta_{\mathrm{t}}) \doteq 0.02979$ for the rectangular smoothing coefficients (M is now set to 30). The third row shows that all three measures now agree well with the natural bandwidth (the corresponding implicit lag window is the thick wiggly curve in Figure 254(b)). The fourth row indicates good agreement between β_W and B_W for the Gaussian coefficients when used with $\tilde{f}_k$, but $B_{W,P}$ is noticeably smaller (the implicit lag window is the thin smooth curve of Figure 254(b)).

Since attaching a bandwidth measure to $\hat{S}^{(\mathrm{DS})}(\cdot)$ via its implicit lag window is problematic, let us consider measures that are based directly on $\{g_j\}$. Let $\Delta_{\mathrm{f}} = 1/(N'\,\Delta_{\mathrm{t}})$ denote the spacing between the frequencies f'_k over which the estimators $\hat{S}^{(\mathrm{D})}(\cdot)$ and $\hat{S}^{(\mathrm{DS})}(\cdot)$ are evaluated in Equation (246b). As suggested by Equation (249d), set $W_m(f'_j) = g_j/\Delta_{\mathrm{f}}$. Use of a Riemann sum to approximate the integrals in Equations (251a) and (251c) yields the following analogs to β_W and B_W:

$$\beta_g \stackrel{\mathrm{def}}{=} \Delta_{\mathrm{f}} \left(24 \sum_{j=1}^{M} j^2 g_j \right)^{1/2} \quad \text{and} \quad B_g \stackrel{\mathrm{def}}{=} \frac{\Delta_{\mathrm{f}}}{\sum_{j=-M}^{M} g_j^2}. \tag{254}$$

		Bandwidth Measure						
g_j	Grid	β_W	B_W	$B_{W,P}$	β_g	B_g	$B_{g,P}$	Natural
Rectangular	$1/N$	0.03025	0.01537	0.01537	0.03026	0.03027	0.03027	0.03027
Gaussian	$1/N$	0.02928	0.01522	0.01071	0.02929	0.02998	0.02120	—
Rectangular	$1/(2N)$	0.02978	0.02979	0.02980	0.02978	0.02979	0.02979	0.02979
Gaussian	$1/(2N)$	0.02928	0.02998	0.02120	0.02928	0.02998	0.02120	—

Table 255 Comparison of window bandwidth measures for discretely smoothed direct spectral estimators $\hat{S}^{(DS)}(\cdot)$. The measures β_W, B_W and $B_{W,P}$ use the implicit lag window associated with $\hat{S}^{(DS)}(\cdot)$, while β_g, B_g and $B_{g,P}$ are based directly on the smoothing coefficients $\{g_j\}$. The smoothing coefficients are either rectangular $g_j = 1/(2M+1)$, $j = -M, \ldots, M$ (with $M = 15$ and 30, respectively, in the first and third rows), or Gaussian $g_j \propto \exp(-j^2/(2\sigma^2))$, $j = -M, \ldots, M$ (with $M = 35$ and $\sigma^2 = 75$ in the second row and $M = 70$ and $\sigma^2 = 300$ in the fourth row). Assuming $\Delta_t = 1$, the "natural" bandwidth for rectangular smoothing coefficients is $(2M+1)/N$, where $N = 1024$. The grid of frequencies over which $\hat{S}^{(D)}(\cdot)$ is smoothed to produce $\hat{S}^{(DS)}(\cdot)$ corresponds to either the Fourier frequencies $f_k = k/N$ with spacing $1/N$ or $\tilde{f}_k = k/(2N)$ with spacing $1/(2N)$.

The measure B_g is an example of the autocorrelation width for a sequence (see Equation (100e)). The analog of the Parzen measure of bandwidth of Equation (251f) is $B_{g,P} = \Delta_f/g_0$, which is related to the equivalent width for a sequence (see Equation (100d)). For the rectangular smoothing coefficients $g_j = 1/(2M+1)$, an easy exercise shows that both B_g and $B_{g,P}$ are equal to the natural bandwidth $(2M+1)\Delta_f$, whereas β_g is so to a good approximation since

$$\beta_g = 2[M(M+1)]^{1/2}\Delta_f \approx 2(M+\tfrac{1}{2})\Delta_f = (2M+1)\Delta_f.$$

Table 255 shows that β_g, B_g and $B_{g,P}$ closely agree with, respectively, β_W, B_W and $B_{W,P}$ when the $1/(2N)$ grid is used, but that B_g and $B_{g,P}$ are to be preferred over B_W and $B_{W,P}$ when the $1/N$ grid is used.

7.2 First-Moment Properties of Lag Window Estimators

We now consider the first-moment properties of $\hat{S}_m^{(LW)}(\cdot)$. It follows from Equation (248a) that

$$E\{\hat{S}_m^{(LW)}(f)\} = \int_{-f_\mathcal{N}}^{f_\mathcal{N}} W_m(f-\phi) E\{\hat{S}^{(D)}(\phi)\} \, d\phi. \tag{255a}$$

▷ **Exercise [255]** Use Equation (186e) to show that

$$E\{\hat{S}_m^{(LW)}(f)\} = \int_{-f_\mathcal{N}}^{f_\mathcal{N}} \mathcal{U}_m(f-f') S(f') \, df', \tag{255b}$$

where

$$\mathcal{U}_m(f) \stackrel{\text{def}}{=} \int_{-f_\mathcal{N}}^{f_\mathcal{N}} W_m(f-f'') \mathcal{H}(f'') \, df''. \tag{255c}$$ ◁

We call $\mathcal{U}_m(\cdot)$ the *spectral window* for the lag window spectral estimator $\hat{S}_m^{(LW)}(\cdot)$ (just as we called $\mathcal{H}(\cdot)$ the spectral window of $\hat{S}^{(D)}(\cdot)$ – compare Equations (186e) and (255b)). Exercise [7.3] is to show that

$$\mathcal{U}_m(f) = \sum_{\tau=-(N-1)}^{N-1} w_{m,\tau}\, h \star h_\tau\, e^{-i2\pi f \tau \Delta_t}, \tag{255d}$$

where $\{h \star h_\tau\}$ is the autocorrelation of $\{h_t\}$ (see Equation (192c)). The above is useful for calculating $\mathcal{U}_m(\cdot)$ in practice.

The estimator $\hat{S}_m^{(\text{LW})}(\cdot)$ has several properties that parallel those for $\hat{S}^{(\text{D})}(\cdot)$ described in Section 6.4 – see Exercise [7.4]. In addition, because Equations (186e) and (255b) indicate that $E\{\hat{S}^{(\text{D})}(f)\}$ and $E\{\hat{S}_m^{(\text{LW})}(f)\}$ are both related to $S(f)$ via convolution of a spectral window and the true SDF, we can extend our standard measure of the effective bandwidth of $\hat{S}^{(\text{D})}(\cdot)$ to work with $\hat{S}_m^{(\text{LW})}(\cdot)$ by replacing $\mathcal{H}(f)$ with $\mathcal{U}_m(f)$ in Equation (194).

▷ **Exercise [256]** Taking Equation (255d) as a given, show that

$$B_\mathcal{U} \stackrel{\text{def}}{=} \text{width}_a\{\mathcal{U}_m(\cdot)\} = \frac{\left(\int_{-f_\mathcal{N}}^{f_\mathcal{N}} \mathcal{U}_m(f)\,df\right)^2}{\int_{-f_\mathcal{N}}^{f_\mathcal{N}} \mathcal{U}_m^2(f)\,df} = \frac{\Delta_t}{\sum_{\tau=-(N-1)}^{N-1} w_{m,\tau}^2 (h \star h_\tau)^2}. \quad (256a) \triangleleft$$

Let us now assume that $\hat{S}^{(\text{D})}(f)$ is an approximately unbiased estimator of $S(f)$, which is a reasonable assumption as long as we have made effective use of tapering (or prewhitening). This allows us to examine the effect of the smoothing window separately on the first-moment properties of $\hat{S}_m^{(\text{LW})}(\cdot)$. From Equation (255a) we then have

$$E\{\hat{S}_m^{(\text{LW})}(f)\} \approx \int_{-f_\mathcal{N}}^{f_\mathcal{N}} W_m(f-f')S(f')\,df'. \quad (256b)$$

The stated requirement involving Equation (250c) says that $W_m(\cdot)$ must act like a Dirac delta function as m gets large. As we shall see in Section 7.5, we cannot make m arbitrarily large if the sample size N is fixed; however, if we let $N \to \infty$, we can also let $m \to \infty$, so we can claim that $\hat{S}_m^{(\text{LW})}(f)$ is an asymptotically unbiased estimator of $S(f)$:

$$E\{\hat{S}_m^{(\text{LW})}(f)\} \to S(f) \text{ as } m, N \to \infty. \quad (256c)$$

For finite sample sizes N and finite values of the smoothing parameter m, however, even if $\hat{S}^{(\text{D})}(\cdot)$ is an approximately unbiased estimator, the smoothing window $W_m(\cdot)$ can introduce significant bias in $\hat{S}_m^{(\text{LW})}(\cdot)$ by inadvertently smoothing together adjacent features in $\hat{S}^{(\text{D})}(\cdot)$ – this happens when the true SDF is not slowly varying. This bias obviously depends upon both the true SDF and the smoothing window. It is informative to derive an expression to quantify this bias (Priestley, 1981, p. 458). Let us take the bias due to the smoothing window alone to be

$$b_W(f) \stackrel{\text{def}}{=} \int_{-f_\mathcal{N}}^{f_\mathcal{N}} W_m(f-f')S(f')\,df' - S(f),$$

a reasonable definition in view of Equation (256b). An argument that exactly parallels the one leading to Equation (192a) yields

$$b_W(f) \approx \frac{S''(f)}{2} \int_{-f_\mathcal{N}}^{f_\mathcal{N}} \phi^2 W_m(\phi)\,d\phi = \frac{S''(f)}{24}\beta_W^2, \quad (256d)$$

where β_W is the measure of smoothing window bandwidth given in Equation (251a). The bias due to the smoothing window in $\hat{S}_m^{(\text{LW})}(f)$ is thus large if $S(\cdot)$ is varying rapidly at f as measured by its second derivative, and the bias increases as the smoothing window bandwidth increases.

7.2 First-Moment Properties of Lag Window Estimators

Comments and Extensions to Section 7.2

[1] The smoothing window bandwidth B_W measures the width of the central lobe of a smoothing window $W_m(\cdot)$. The measure $B_{\mathcal{U}}$ does the same for a spectral window $\mathcal{U}_m(\cdot)$, which is the convolution of $W_m(\cdot)$ and a spectral window $\mathcal{H}(\cdot)$. Since convolution is a smoothing operation, one interpretation for $\mathcal{U}_m(\cdot)$ is as a smoothed-out version of $W_m(\cdot)$. The central lobe of $\mathcal{U}_m(\cdot)$ should thus be wider than that of $W_m(\cdot)$, so it is reasonable to expect $B_{\mathcal{U}}$ to be bigger than B_W. The following exercise shows this must be true.

▷ **Exercise [257]** Show that $B_{\mathcal{U}} \geq B_W$ (assuming the usual normalization $\sum_t h_t^2 = 1$). Hint: make use of the Cauchy inequality, namely,

$$\left| \sum_{t=0}^{N-1} a_t b_t \right|^2 \leq \sum_{t=0}^{N-1} |a_t|^2 \sum_{t=0}^{N-1} |b_t|^2, \qquad (257a)$$

where the sequences $\{a_t\}$ and $\{b_t\}$ are either real- or complex-valued. (A proof of this inequality can be formulated along the same lines as the one for the Schwarz inequality (Equation (55d)). The proof allows us to note that equality holds if and only if the sequence $\{a_t\}$ is proportional to $\{b_t\}$.) ◁

[2] We have introduced data tapers as a means of compensating for the bias (first moment) of $\hat{S}^{(\mathrm{P})}(\cdot)$ and lag windows as a means of decreasing the variance (second central moment) of direct spectral estimators $\hat{S}^{(\mathrm{D})}(\cdot)$ (including the periodogram). It is sometimes claimed in the literature that a lag window can be used to control both the bias and the variance in $\hat{S}^{(\mathrm{P})}(\cdot)$ (see, for example, Grenander and Rosenblatt, 1984, section 4.2); i.e., we can avoid the use of a nonrectangular data taper. The basis for this claim can be seen by considering the spectral window $\mathcal{U}_m(\cdot)$ in Equation (255d). Equation (255b) says that this spectral window describes the bias properties of a lag window spectral estimator consisting of a lag window $\{w_{m,\tau}\}$ used in combination with a data taper $\{h_t\}$. The following argument shows we can produce the same spectral window by using a different lag window in combination with just the rectangular data taper. First, we note that, in the case of a rectangular data taper, i.e., $h_t = 1/\sqrt{N}$, the spectral window $\mathcal{U}_m(\cdot)$ reduces to

$$\Delta_{\mathrm{t}} \sum_{\tau=-(N-1)}^{N-1} w_{m,\tau} \left(1 - |\tau|/N\right) e^{-i 2\pi f \tau \Delta_{\mathrm{t}}}.$$

For any data taper $\{h_t\}$, we can write, using Equation (255d),

$$\mathcal{U}_m(f) = \Delta_{\mathrm{t}} \sum_{\tau=-(N-1)}^{N-1} w_{m,\tau} \frac{\sum_{t=0}^{N-|\tau|-1} h_{t+|\tau|} h_t}{1 - |\tau|/N} \left(1 - |\tau|/N\right) e^{-i 2\pi f \tau \Delta_{\mathrm{t}}}$$

$$= \Delta_{\mathrm{t}} \sum_{\tau=-(N-1)}^{N-1} w'_{m,\tau} \left(1 - |\tau|/N\right) e^{-i 2\pi f \tau \Delta_{\mathrm{t}}},$$

where

$$w'_{m,\tau} \stackrel{\mathrm{def}}{=} w_{m,\tau} \frac{\sum_{t=0}^{N-|\tau|-1} h_{t+|\tau|} h_t}{1 - |\tau|/N}. \qquad (257b)$$

Thus use of the lag window $\{w'_{m,\tau}\}$ with the rectangular data taper produces the same spectral window $\mathcal{U}_m(\cdot)$ as use of the lag window $\{w_{m,\tau}\}$ with the nonrectangular data taper $\{h_t\}$. Nuttall and Carter (1982) refer to $\{w'_{m,\tau}\}$ as a *reshaped lag window*. As we show by example in C&E [2] for Section 7.5, a potential problem with this scheme is that the smoothing window associated with $\{w'_{m,\tau}\}$ (i.e., its Fourier transform) can have some undesirable properties, including prominent *negative* sidelobes. This means that, for some time series, the associated $\hat{S}_m^{(\mathrm{LW})}(\cdot)$ can be negative at some frequencies.

7.3 Second-Moment Properties of Lag Window Estimators

We sketch here a derivation of one approximation to the large-sample variance of $\hat{S}_m^{(\text{LW})}(f)$ – a second, more accurate approximation is considered in C&E [1]. We assume that the direct spectral estimator $\hat{S}^{(\text{D})}(\cdot)$ is approximately uncorrelated on a grid of frequencies defined by $f'_k = k/(N' \Delta_t)$, where here $N' \leq N$, and k is an integer such that $0 < f'_k < f_\mathcal{N}$ – see the discussion concerning Equation (207). From Equation (248a) and an obvious change of variable, we have

$$\hat{S}_m^{(\text{LW})}(f) = \int_{-f_\mathcal{N}-f}^{f_\mathcal{N}-f} W_m(-\phi) \hat{S}^{(\text{D})}(\phi + f) \, d\phi = \int_{-f_\mathcal{N}}^{f_\mathcal{N}} W_m(\phi) \hat{S}^{(\text{D})}(\phi + f) \, d\phi;$$

here we use the facts that $W_m(\cdot)$ is an even function and that both it and $\hat{S}^{(\text{D})}(\cdot)$ are $2f_\mathcal{N}$ periodic. From Equation (250c) we can assume that, for large N, $W_m(\phi) \approx 0$ for all $|\phi| > J/(N' \Delta_t) = f'_J$ for some positive integer J. Hence

$$\hat{S}_m^{(\text{LW})}(f'_k) \approx \int_{-f'_J}^{f'_J} W_m(\phi) \hat{S}^{(\text{D})}(\phi + f'_k) \, d\phi \approx \sum_{j=-J}^{J} W_m(f'_j) \hat{S}^{(\text{D})}(f'_j + f'_k) \frac{1}{N' \Delta_t},$$

where we have approximated the Riemann integral by a Riemann sum (see Figure 259). If we note that $f'_j + f'_k = f'_{j+k}$, we now have

$$\begin{aligned}
\text{var}\{\hat{S}_m^{(\text{LW})}(f'_k)\} &\approx \frac{1}{(N' \Delta_t)^2} \text{var} \left\{ \sum_{j=-J}^{J} W_m(f'_j) \hat{S}^{(\text{D})}(f'_{j+k}) \right\} \\
&\stackrel{(1)}{\approx} \frac{1}{(N' \Delta_t)^2} \sum_{j=-J}^{J} W_m^2(f'_j) \, \text{var}\left\{ \hat{S}^{(\text{D})}(f'_{j+k}) \right\} \\
&\stackrel{(2)}{\approx} \frac{1}{(N' \Delta_t)^2} \sum_{j=-J}^{J} W_m^2(f'_j) S^2(f'_{j+k}) \\
&\stackrel{(3)}{\approx} \frac{S^2(f'_k)}{(N' \Delta_t)^2} \sum_{j=-J}^{J} W_m^2(f'_j) \\
&\stackrel{(4)}{\approx} \frac{S^2(f'_k)}{(N' \Delta_t)^2} \sum_{j=-\lfloor N'/2 \rfloor}^{\lfloor N'/2 \rfloor} W_m^2(f'_j) \\
&\stackrel{(5)}{\approx} \frac{S^2(f'_k)}{N' \Delta_t} \int_{-f_\mathcal{N}}^{f_\mathcal{N}} W_m^2(\phi) \, d\phi,
\end{aligned} \tag{258}$$

where we have made use of five approximations:

(1) pairwise uncorrelatedness of the components of $\hat{S}^{(\text{D})}(\cdot)$ defined by the grid of frequencies f'_k;
(2) the large-sample variance of $\hat{S}^{(\text{D})}(\cdot)$ as given by Equation (204c), for which we need $0 < f'_{j+k} < f_\mathcal{N}$ for all j;
(3) a smoothness assumption on $S(\cdot)$ asserting that $S(f'_{j+k}) \approx S(f'_k)$ locally;
(4) the previous assumption that $W_m(\phi) \approx 0$ for $|\phi| > f'_J$; and
(5) an approximation of the summation by a Riemann integral.

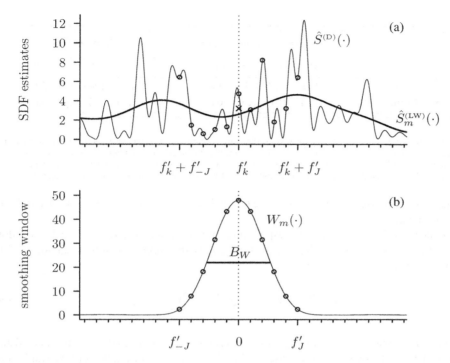

Figure 259 Approximating a lag window estimator via a Riemann sum. The thin curve in plot (a) is a direct spectral estimate $\hat{S}^{(\mathrm{D})}(\cdot)$. The thick curve is a lag window estimate $\hat{S}_m^{(\mathrm{LW})}(\cdot)$, which is the result of convolving $\hat{S}^{(\mathrm{D})}(\cdot)$ with the smoothing window $W_m(\cdot)$ shown in plot (b). The dotted vertical line in (a) indicates the location of the frequency f'_k, with the corresponding estimate $\hat{S}^{(\mathrm{D})}(f'_k)$ marked by a circle. The estimate $\hat{S}_m^{(\mathrm{LW})}(f'_k)$ is formed by multiplying the thin curves in (a) and (b) together and then integrating the product from $-f_\mathcal{N}$ to $f_\mathcal{N}$. The grid of frequencies over which $\hat{S}^{(\mathrm{D})}(\cdot)$ is approximately both uncorrelated and dictated by a scaled χ_2^2 distribution is given by $f'_j + f'_k$ for integers j such that $0 < f'_j + f'_k < f_\mathcal{N}$. The tick marks on the upper horizontal axis show this grid. The smoothing window is essentially zero outside of the interval $[f'_{-J}, f'_J]$ when we set $J = 5$. The eleven values from $\hat{S}^{(\mathrm{D})}(\cdot)$ and $W_m(\cdot)$ used to approximate $\hat{S}_m^{(\mathrm{LW})}(f'_k)$ by a Riemann sum are indicated by the circles. The approximation itself is indicated by the x in plot (a). The width of the thick horizontal line in (b) depicts the smoothing window bandwidth B_W as calculated by Equation (251e). (Here the direct spectral estimator is a periodogram based on the first $N = 256$ values of the AR(2) series of Figure 34(a); the grid is defined by the Fourier frequencies so $N' = N$; $f'_k = k/N' = 38/256$; and $W_m(\cdot)$ is the Parzen smoothing window of Equation (275b) with $m = 64$.)

Finally, we assume that Equation (258) holds for all frequencies (and not just the ones defined by f'_k) and that N' can be related to N by $N' \approx N/C_h$, where C_h is greater than or equal to unity and depends only upon the data taper for the direct spectral estimator (Table 260 gives the value of C_h for some common data tapers, and Equation (262) is a computational formula for it). With these modifications we now have

$$\mathrm{var}\{\hat{S}_m^{(\mathrm{LW})}(f)\} \approx \frac{C_h S^2(f)}{N\,\Delta_\mathrm{t}} \int_{-f_\mathcal{N}}^{f_\mathcal{N}} W_m^2(\phi)\,\mathrm{d}\phi. \tag{259a}$$

Using the definition for the smoothing window bandwidth B_W in Equation (251c), we rewrite the above as

$$\mathrm{var}\{\hat{S}_m^{(\mathrm{LW})}(f)\} \approx \frac{C_h S^2(f)}{B_W N\,\Delta_\mathrm{t}}. \tag{259b}$$

Thus, as one would expect, increasing (decreasing) the smoothing window bandwidth causes

	$p \times 100\%$ Cosine Taper					Slepian Taper			
	$p =$	0.0	0.2	0.5	1.0	$NW =$ 1	2	4	8
C_h		1.00	1.12	1.35	1.94	1.34	1.96	2.80	3.98
See Figure 190		(a)	(b)	(c)	(d)	(e)	(f)	(g)	(h)

Table 260 Variance inflation factor C_h for various data tapers. Note that C_h increases as the width of the central lobe of the associated spectral window increases. The values listed here are approximations – valid for large N – to the right-hand side of Equation (262). When $p = 0$, the $p \times 100\%$ cosine taper is the same as the rectangular data taper, which is the taper used implicitly by the periodogram. The choice $p = 1$ yields the Hanning data taper.

the variance to decrease (increase). Because $C_h \geq 1$, we can interpret it as a variance inflation factor due to tapering.

It is also useful to derive a corresponding approximation in terms of the lag window. Because of the Parseval relationship in Equation (251d), the approximation of Equation (259a) can be rewritten as

$$\text{var}\{\hat{S}_m^{(\text{LW})}(f)\} \approx \frac{C_h S^2(f)}{N} \sum_{\tau=-(N-1)}^{N-1} w_{m,\tau}^2. \qquad (260)$$

Since under our assumptions $\hat{S}_m^{(\text{LW})}(f)$ is asymptotically unbiased, it is also a consistent estimator of $S(f)$ if its variance decreases to 0 as $N \to \infty$. From Equations (259a) and (260), the large-sample variance of $\hat{S}_m^{(\text{LW})}(f)$ approaches zero if we make one of the following additional – but equivalent – assumptions:

$$\lim_{N \to \infty} \frac{1}{N \Delta_t} \int_{-f_\mathcal{N}}^{f_\mathcal{N}} W_m^2(\phi) \, d\phi = 0 \quad \text{or} \quad \lim_{N \to \infty} \frac{1}{N} \sum_{\tau=-(N-1)}^{N-1} w_{m,\tau}^2 = 0.$$

We can use Equation (251c) to reexpress the above in terms of B_W as

$$\lim_{N \to \infty} \frac{B_W}{\Delta_f} = \infty, \quad \text{where } \Delta_f \stackrel{\text{def}}{=} \frac{1}{N \Delta_t}.$$

In other words, the smoothing window bandwidth B_W must grow to cover many different Fourier frequency bins. Thus one of these three equivalent assumptions is enough to ensure that $\hat{S}_m^{(\text{LW})}(f)$ is a consistent estimator of $S(f)$ (see also the discussion in Section 7.4).

We now consider the covariance between $\hat{S}_m^{(\text{LW})}(f'_k)$ and $\hat{S}_m^{(\text{LW})}(f'_{k'})$. By the asymptotic unbiasedness of $\hat{S}_m^{(\text{LW})}(f)$, it follows that

$$\text{cov}\{\hat{S}_m^{(\text{LW})}(f'_k), \hat{S}_m^{(\text{LW})}(f'_{k'})\} \approx E\{\hat{S}_m^{(\text{LW})}(f'_k)\hat{S}_m^{(\text{LW})}(f'_{k'})\} - S(f'_k)S(f'_{k'}).$$

We can proceed as before to argue that

$$\hat{S}_m^{(\text{LW})}(f'_k) \approx \frac{1}{N' \Delta_t} \sum_{j=-J}^{J} W_m(f'_j) \hat{S}^{(\text{D})}(f'_{j+k}).$$

Since $\hat{S}_m^{(\text{LW})}(f'_{k'})$ has a similar expression, it follows that

$$E\{\hat{S}_m^{(\text{LW})}(f'_k)\hat{S}_m^{(\text{LW})}(f'_{k'})\} \approx \frac{1}{(N' \Delta_t)^2} \sum_{j,j'=-J}^{J} W_m(f'_j) W_m(f'_{j'}) E\{\hat{S}^{(\text{D})}(f'_{j+k}) \hat{S}^{(\text{D})}(f'_{j'+k'})\}.$$

If f'_k and $f'_{k'}$ are far enough apart so that $|k-k'| > 2J$ (recall that J defines the point at which $W_m(f) \approx 0$ so that $2J$ is proportional to the "width" of $W_m(\cdot)$), we can use the asymptotic uncorrelatedness and unbiasedness of $\hat{S}^{(D)}(\cdot)$ on the grid frequencies defined by f'_k to argue that

$$E\{\hat{S}_m^{(\text{LW})}(f'_k)\hat{S}_m^{(\text{LW})}(f'_{k'})\} \approx \frac{1}{(N'\Delta_t)^2} \sum_{j,j'=-J}^{J} W_m(f'_j)W_m(f'_{j'})S(f'_{j+k})S(f'_{j'+k'}).$$

By the same assumption on the smoothness of $S(\cdot)$ as before,

$$E\{\hat{S}_m^{(\text{LW})}(f'_k)\hat{S}_m^{(\text{LW})}(f'_{k'})\} \approx \frac{1}{(N'\Delta_t)^2} S(f'_k)S(f'_{k'})\left(\sum_{j=-J}^{J} W_m(f'_j)\right)^2$$

$$\approx S(f'_k)S(f'_{k'})\left(\int_{-f_\mathcal{N}}^{f_\mathcal{N}} W_m(\phi)\,\mathrm{d}\phi\right)^2$$

$$= S(f'_k)S(f'_{k'}).$$

Here we have used the Riemann integral to approximate the summation and the assumption of Equation (250a) to show that the integral is identically 1. It now follows that

$$\text{cov}\{\hat{S}_m^{(\text{LW})}(f'_k), \hat{S}_m^{(\text{LW})}(f'_{k'})\} \approx 0, \quad f'_k \neq f'_{k'}, \tag{261a}$$

for N large enough. By an extension of the above arguments, one can show that the same result holds for any two fixed frequencies $f' \neq f''$ for N large enough. Note carefully that uncorrelatedness need not be approximately true when f' and f'' are separated by a distance less than the "width" of $W_m(\cdot)$, which we can conveniently assume to be measured by the smoothing window bandwidth B_W.

Comments and Extensions to Section 7.3

[1] We here derive Equation (260) by a route that makes more explicit the form of the factor C_h. Let $f'_k = k/[(2N-1)\Delta_t]$ be a set of frequencies on a grid almost twice as finely spaced as the Fourier frequencies. From Equation (249d) we have the exact result

$$\hat{S}_m^{(\text{LW})}(f'_k) = \frac{1}{(2N-1)\Delta_t} \sum_{j=-(N-1)}^{N-1} W_m(f'_j)\hat{S}^{(D)}(f'_{k-j}). \tag{261b}$$

This is simply the convolution of $\{W_m(f'_j)\}$ with the "frequency" series $\{\hat{S}^{(D)}(f'_j)\}$. If, for large N, we assume that $W_m(f'_j) \approx 0$ for all $f'_j > J'/[2(N-1)\Delta_t] = f'_{J'}$ for some integer J' and that in the interval $[f'_{k-J'}, f'_{k+J'}]$ the true SDF $S(\cdot)$ is locally constant, then the "frequency" series $\{\hat{S}^{(D)}(f'_j)\}$ is locally stationary. The spectrum of the resultant series $\{\hat{S}_m^{(\text{LW})}(f'_j)\}$ – locally to f'_k – is thus the product of the square of the inverse Fourier transform of $\{W_m(f'_j)\}$ – i.e., $\{w_{m,\tau}^2 : \tau = -(N-1), \ldots, N-1\}$ – with the spectrum of $\hat{S}^{(D)}(f'_j)$, which, under the Gaussian assumption, is $\{S^2(f'_k)r_\tau\}$ (see the discussion following Equation (212b)). Hence the spectrum of $\hat{S}_m^{(\text{LW})}(f'_j)$ – locally to f'_k – is given by $S^2(f'_k)w_{m,\tau}^2 r_\tau$, $\tau = -(N-1), \ldots, N-1$. Since a spectrum decomposes a variance, the variance of $\hat{S}_m^{(\text{LW})}(f'_k)$ is given by

$$\text{var}\{\hat{S}_m^{(\text{LW})}(f'_k)\} \approx \Delta_t\, S^2(f'_k) \sum_{\tau=-(N-1)}^{N-1} w_{m,\tau}^2 r_\tau = S^2(f'_k) \sum_{\tau=-(N-1)}^{N-1} w_{m,\tau}^2 \sum_{t=0}^{N-|\tau|-1} h_{t+|\tau|}^2 h_t^2, \tag{261c}$$

where we have made use of Equation (213a); see Walden (1990a) for details. For the rectangular taper, $h_t = 1/\sqrt{N}$ for $0 \leq t \leq N-1$, and hence we have

$$\sum_{t=0}^{N-|\tau|-1} h_{t+|\tau|}^2 h_t^2 = \sum_{t=0}^{N-|\tau|-1} \frac{1}{N^2} = \left(1 - \frac{|\tau|}{N}\right),$$

so that

$$\text{var}\{\hat{S}_m^{(\text{LW})}(f_k')\} \approx \frac{S^2(f_k')}{N} \sum_{\tau=-(N-1)}^{N-1} w_{m,\tau}^2 \left(1 - \frac{|\tau|}{N}\right),$$

a result of some utility (see, for example, Walden and White, 1990).

If we now simplify Equation (261c) by setting $r_\tau = r_0$ for all τ, we obtain

$$\text{var}\{\hat{S}_m^{(\text{LW})}(f_k')\} \approx S^2(f_k') \sum_{t=0}^{N-1} h_t^4 \sum_{\tau=-(N-1)}^{N-1} w_{m,\tau}^2 \text{ since } r_0 = \frac{1}{\Delta_t} \sum_{t=0}^{N-1} h_t^4$$

from Equation (214a). As is discussed in Walden (1990a), this approximation is really only good for relatively "modest" data tapers, i.e., those for which the Fourier transform of $\{r_\tau\}$ – namely, $R(\cdot)$ of Equation (212a) – damps down to zero rapidly (see Figure 212). Comparison of the above with Equation (260) yields

$$C_h = N \sum_{t=0}^{N-1} h_t^4 \tag{262}$$

(see also Hannan, 1970, pp. 271–2, and Brillinger, 1981a, pp. 150–1). The Cauchy inequality of Equation (257a) with $a_t = h_t^2$ and $b_t = 1$ tells us that $C_h \geq 1$ (since we always assume the normalization $\sum_{t=0}^{N-1} h_t^2 = 1$). We also note that $C_h = 1$ only for the rectangular data taper $h_t = 1/\sqrt{N}$. For large N, we can compute C_h approximately for various data tapers by approximating the above summation with an integral. The results for several common data tapers are shown in Table 260.

[2] We give here more details concerning – and an alternative expression for – the approximation to the variance of a lag window spectral estimator given in Equation (261c). First, we note a desirable property that the approximation given by Equation (261c) has, but the less accurate approximation of Equation (260) does not. Now we can regard a direct spectral estimator as a special case of a lag window estimator if we take the lag window to be $w_{m,\tau} = 1$ for all τ (to see this, just compare Equations (188a) and (248a)). With this choice for $w_{m,\tau}$, Amrein and Künsch (2011) point out that the less accurate approximation to $\text{var}\{\hat{S}_m^{(\text{LW})}(f)\}$ does not in general match up with the usual approximation to $\text{var}\{\hat{S}^{(D)}(f)\}$ given by Equation (204c): making use of Equation (262), the former (Equation (260)) gives

$$\text{var}\{\hat{S}_m^{(\text{LW})}(f)\} \approx \frac{C_h S^2(f)}{N} \sum_{\tau=-(N-1)}^{N-1} w_{m,\tau}^2 = S^2(f)(2N-1) \sum_{t=0}^{N-1} h_t^4,$$

whereas the latter states that $\text{var}\{\hat{S}^{(D)}(f)\} \approx S^2(f)$. In particular, if we consider the rectangular taper $h_t = 1/\sqrt{N}$, we obtain $\text{var}\{\hat{S}_m^{(\text{LW})}(f)\} \approx 2S^2(f)$ in comparison to the usual $\text{var}\{\hat{S}^{(D)}(f)\} \approx S^2(f)$. By contrast, the more accurate approximation does not exhibit this mismatch because, when $w_{m,\tau} = 1$ (and letting $f = f_k'$), Equation (261c) becomes

$$\text{var}\{\hat{S}_m^{(\text{LW})}(f)\} \approx S^2(f) \sum_{\tau=-(N-1)}^{N-1} \sum_{t=0}^{N-|\tau|-1} h_{t+|\tau|}^2 h_t^2 = S^2(f)$$

in light of the following result.

▷ **Exercise [263]** Show that, if $\{h_t\}$ is a data taper with the usual normalization $\sum_t h_t^2 = 1$, then

$$\sum_{\tau=-(N-1)}^{N-1} \sum_{t=0}^{N-|\tau|-1} h_{t+|\tau|}^2 h_t^2 = 1.$$

Hint: consider Equations (212a), (213a) and (213b). ◁

Amrein and Künsch (2011) note that an implication of this mismatch is to call into question the use of the approximation given by Equation (260) when the degree of smoothing induced by a lag window is quite small.

Second, we give an alternative expression for the more accurate approximation of Equation (261c). This approximation is based upon Equation (261b), from which it directly follows from Exercise [2.1e] that

$$\operatorname{var}\{\hat{S}_m^{(\mathrm{LW})}(f_k')\} = \frac{1}{(2N-1)^2 \Delta_t^2} \sum_{j,j'=-(N-1)}^{N-1} W_m(f_j') W_m(f_{j'}') \operatorname{cov}\{\hat{S}^{(\mathrm{D})}(f_{k-j}'), \hat{S}^{(\mathrm{D})}(f_{k-j'}')\}.$$

From Equation (250c) we can assume that $W_m(f_j') \approx 0$ for all $|j| > J$. In practice we can determine J by setting $f_J' \approx cB_W$, where B_W is the smoothing window bandwidth of Equation (251e), and c is, say, 1 or 2 (J is the integer closest to $(2N-1)B_W \Delta_t$ when $c = 1$). We thus have

$$\operatorname{var}\{\hat{S}_m^{(\mathrm{LW})}(f_k')\} \approx \frac{1}{(2N-1)^2 \Delta_t^2} \sum_{j,j'=-J}^{J} W_m(f_j') W_m(f_{j'}') \operatorname{cov}\{\hat{S}^{(\mathrm{D})}(f_{k-j}'), \hat{S}^{(\mathrm{D})}(f_{k-j'}')\}.$$

Under the assumptions that we are dealing with a Gaussian stationary process, that $S(\cdot)$ is locally flat from f_{k-J}' to f_{k+J}' and that $0 < f_{k-J}'$ and $f_{k+J}' < f_\mathcal{N}$ (with f_{k-J}' not being too near to 0 and f_{k+J}' not being too near to $f_\mathcal{N}$), we can use the result of Exercise [212] to simplify the above to

$$\operatorname{var}\{\hat{S}_m^{(\mathrm{LW})}(f_k')\} \approx \frac{S^2(f_k')}{(2N-1)^2 \Delta_t^2} \sum_{j=-J}^{J} W_m(f_j') \sum_{j'=-J}^{J} W_m(f_{j'}') R(f_{j-j'}'),$$

where $R(\cdot)$ is defined in Equation (212a) and depends only on the data taper $\{h_t\}$. By a change of variables similar to that used in the solution to Exercise [170], we can rewrite the above as

$$\operatorname{var}\{\hat{S}_m^{(\mathrm{LW})}(f_k')\} \approx \frac{S^2(f_k')}{(2N-1)^2 \Delta_t^2} \sum_{l=-2J}^{2J} R(f_l') \sum_{l'=0}^{2J-|l|} W_m(f_{J-l'-|l|}') W_m(f_{J-l'}').$$

The inner summation is essentially an autocorrelation, so it can be efficiently calculated for all lags l using discrete Fourier transforms (see Walden, 1990a, for details); however, as is demonstrated in Figure 212, $R(\cdot)$ typically decreases effectively to 0 rather quickly so that we can shorten the outer summation to, say, its $2L + 1$ innermost terms:

$$\operatorname{var}\{\hat{S}_m^{(\mathrm{LW})}(f_k')\} \approx \frac{S^2(f_k')}{(2N-1)^2 \Delta_t^2} \sum_{l=-L}^{L} R(f_l') \sum_{l'=0}^{2J-|l|} W_m(f_{J-l'-|l|}') W_m(f_{J-l'}').$$

For example, if we use a Slepian data taper with $NW \leq 8$, $N = 64$ and $\Delta_t = 1$, we can conveniently let $L = \min(10, 2J)$, where $l = 10$ corresponds to $\eta = l/[(2N-1)\Delta_t] = 10/127 \approx 5/64$ in Figure 212.

Finally, we note that, with a trivial adjustment to the limits of the summations, the above expression is a special case of a more general formula given by Amrein and Künsch (2011), which is valid for arbitrary discretely smoothed direct spectral estimators (see their Equation (5)). Their paper gives a rigorous theoretical justification for the more accurate approximation, along with examples demonstrating concretely the improvement it affords.

7.4 Asymptotic Distribution of Lag Window Estimators

We consider here the asymptotic (large-sample) distribution of $\hat{S}_m^{(\text{LW})}(f)$. We can write, using Equations (248a) and (249d),

$$\hat{S}_m^{(\text{LW})}(f) = \int_{-f_\mathcal{N}}^{f_\mathcal{N}} W_m(f - \phi) \hat{S}^{(\text{D})}(\phi) \, d\phi$$

$$\approx \frac{1}{(2N-1)\Delta_\text{t}} \sum_{j=-(N-1)}^{N-1} W_m(f'_j) \hat{S}^{(\text{D})}(f - f'_{-j}),$$

with the approximation becoming an equality when $f = f'_k = k/[(2N-1)\Delta_\text{t}]$ for some integer k. As noted in Section 6.6, under mild conditions and for large N, the $\hat{S}^{(\text{D})}(f - \tilde{f}_{-j})$ terms can be regarded as a set of χ^2 RVs scaled by appropriate multiplicative constants. It follows that $\hat{S}_m^{(\text{LW})}(f)$ is asymptotically a linear combination of χ^2 RVs with weights that depend on the smoothing window $W_m(\cdot)$. The exact form of such a distribution is hard to determine, but there is a well-known approximation we can use (Welch, 1936; Fairfield Smith, 1936; Welch, 1938; Satterthwaite, 1941; Satterthwaite, 1946; Box, 1954; Blackman and Tukey, 1958; and Jenkins, 1961). Assume

$$\hat{S}_m^{(\text{LW})}(f) \stackrel{\text{d}}{=} a\chi_\nu^2;$$

i.e., $\hat{S}_m^{(\text{LW})}(f)$ has the distribution of a chi-square RV scaled by a constant a, where the degrees of freedom ν and a are both unknown. By the properties of the χ_ν^2 distribution, we have

$$E\{\hat{S}_m^{(\text{LW})}(f)\} = E\{a\chi_\nu^2\} = a\nu \quad \text{and} \quad \text{var}\{\hat{S}_m^{(\text{LW})}(f)\} = \text{var}\{a\chi_\nu^2\} = 2a^2\nu.$$

We can use these two expressions to derive equations for ν and a in terms of the expected value and variance of $\hat{S}_m^{(\text{LW})}(f)$:

$$\nu = \frac{2\left(E\{\hat{S}_m^{(\text{LW})}(f)\}\right)^2}{\text{var}\{\hat{S}_m^{(\text{LW})}(f)\}} \quad \text{and} \quad a = \frac{E\{\hat{S}_m^{(\text{LW})}(f)\}}{\nu}. \tag{264a}$$

Under the assumptions of the previous two sections, the large-sample expected value and variance for $\hat{S}_m^{(\text{LW})}(f)$ are, respectively, $S(f)$ and

$$\frac{C_h S^2(f)}{N \Delta_\text{t}} \int_{-f_\mathcal{N}}^{f_\mathcal{N}} W_m^2(\phi) \, d\phi = \frac{C_h S^2(f)}{N} \sum_{\tau=-(N-1)}^{N-1} w_{m,\tau}^2$$

(see Equations (256c), (259a) and (260)). If we substitute these values into Equation (264a), we find that, for large samples,

$$\nu = \frac{2N\Delta_\text{t}}{C_h \int_{-f_\mathcal{N}}^{f_\mathcal{N}} W_m^2(\phi) \, d\phi} = \frac{2N}{C_h \sum_{\tau=-(N-1)}^{N-1} w_{m,\tau}^2} \quad \text{and} \quad a = \frac{S(f)}{\nu}. \tag{264b}$$

Hence we assume

$$\hat{S}_m^{(\text{LW})}(f) \stackrel{\text{d}}{=} S(f) \chi_\nu^2 / \nu. \tag{264c}$$

The quantity ν is often called the *equivalent degrees of freedom* (EDOFs) for the estimator $\hat{S}_m^{(\text{LW})}(f)$. From Equation (251e) for the smoothing window bandwidth B_W, we obtain

$$\nu = \frac{2N B_W \Delta_\text{t}}{C_h}, \tag{264d}$$

so that an increase in B_W (i.e., a greater degree of smoothing) yields an increase in ν. Using var $\{a\chi_\nu^2\} = 2a^2\nu$, we have (for large samples)

$$\text{var}\{\hat{S}_m^{(\text{LW})}(f)\} \approx \frac{2S^2(f)}{\nu}.$$

As ν increases, this variance decreases, but, for a fixed sample size N, a decrease in the variance of $\hat{S}_m^{(\text{LW})}(f)$ comes at the potential expense of an increase in its bias. Here we must be careful. The above expression for var $\{\hat{S}_m^{(\text{LW})}(f)\}$ is based on the assumption that $\hat{S}_m^{(\text{LW})}(f)$ is approximately distributed as $a\chi_\nu^2$ with a and ν given above. If, by increasing the bandwidth B_W with the idea of making ν large and hence the variance small, we inadvertently introduce significant bias in $\hat{S}_m^{(\text{LW})}(f)$, then $\hat{S}_m^{(\text{LW})}(f)$ will deviate substantially from its assumed distribution, and the above expression for var $\{\hat{S}_m^{(\text{LW})}(f)\}$ can be misleading.

The $a\chi_\nu^2$ approximation to the distribution of $\hat{S}_m^{(\text{LW})}(f)$ allows us to construct an approximate confidence interval (CI) for $S(f)$ at a fixed f. Let $Q_\nu(p)$ represent the $p \times 100\%$ percentage point of the χ_ν^2 distribution; i.e., $\mathbf{P}\left[\chi_\nu^2 \leq Q_\nu(p)\right] = p$. We thus have, for $0 \leq p \leq 1/2$,

$$\mathbf{P}\left[Q_\nu(p) \leq \chi_\nu^2 \leq Q_\nu(1-p)\right] = 1 - 2p.$$

Since the distribution of $\hat{S}_m^{(\text{LW})}(f)$ is approximately that of the RV $a\chi_\nu^2$ with $a = S(f)/\nu$, it follows that $\nu\hat{S}_m^{(\text{LW})}(f)/S(f)$ is approximately distributed as χ_ν^2. Hence

$$\mathbf{P}\left[Q_\nu(p) \leq \frac{\nu\hat{S}_m^{(\text{LW})}(f)}{S(f)} \leq Q_\nu(1-p)\right] = \mathbf{P}\left[\frac{\nu\hat{S}_m^{(\text{LW})}(f)}{Q_\nu(1-p)} \leq S(f) \leq \frac{\nu\hat{S}_m^{(\text{LW})}(f)}{Q_\nu(p)}\right] = 1 - 2p,$$

from which it follows that

$$\left[\frac{\nu\hat{S}_m^{(\text{LW})}(f)}{Q_\nu(1-p)}, \frac{\nu\hat{S}_m^{(\text{LW})}(f)}{Q_\nu(p)}\right] \tag{265}$$

is a $100(1 - 2p)\%$ CI for $S(f)$. Note that this CI applies only to $S(f)$ at one particular value of f. We cannot get a confidence band for the entire function $S(\cdot)$ by this method.

Since $\hat{S}_m^{(\text{LW})}(f)$ is approximately distributed as an $a\chi_\nu^2$ RV, we can employ – when ν is large (say greater than 30) – the usual scheme of approximating a χ_ν^2 distribution by a Gaussian distribution with the same mean and variance; i.e., the RV

$$\frac{\hat{S}_m^{(\text{LW})}(f) - S(f)}{S(f)(2/\nu)^{1/2}} \stackrel{\text{d}}{=} \frac{a\chi_\nu^2 - a\nu}{(2a^2\nu)^{1/2}}$$

is approximately Gaussian distributed with zero mean and unit variance. Under these circumstances, an approximate $100(1 - 2p)\%$ CI for $S(f)$ has the form

$$\left[\frac{\hat{S}_m^{(\text{LW})}(f)}{1 + \Phi^{-1}(1-p)(2/\nu)^{1/2}}, \frac{\hat{S}_m^{(\text{LW})}(f)}{1 + \Phi^{-1}(p)(2/\nu)^{1/2}}\right],$$

where $\Phi^{-1}(p)$ is the $p \times 100\%$ percentage point of the standard Gaussian distribution.

The CI for $S(f)$ given in Equation (265) has a width of

$$\hat{S}_m^{(\text{LW})}(f)\left[\frac{\nu}{Q_\nu(p)} - \frac{\nu}{Q_\nu(1-p)}\right],$$

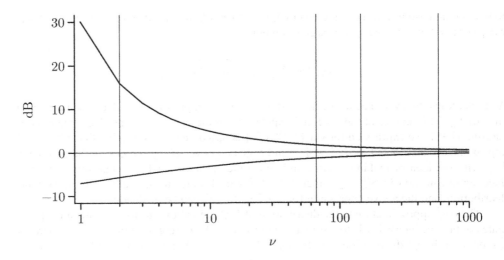

Figure 266 Upper and lower additive factors needed to create a 95% confidence interval (CI) for $10\log_{10}(S(f))$ based upon a spectral estimator with ν equivalent degrees of freedom (EDOFs). Here ν ranges from 1 up to 1000 in increments of 1. The thin vertical lines indicate $\nu = 2, 66, 146$ and 581. The latter three values are the smallest integer-valued EDOFs needed to achieve a 95% CI for $10\log_{10}(S(f))$ whose width is less than, respectively, 3 dB, 2 dB and 1 dB. The case $\nu = 2$ is associated with direct SDF estimators $\hat{S}^{(\text{D})}(\cdot)$ at frequencies $0 < f < f_{\mathcal{N}}$ (see Equations (204b) and (205b)), while $\nu = 1$ is associated with $\hat{S}^{(\text{D})}(0)$, $\hat{S}^{(\text{D})}(f_{\mathcal{N}})$ and DCT-based periodograms (see Equations (204b) and (219a)).

which unfortunately depends on $\hat{S}_m^{(\text{LW})}(f)$. This fact makes plots of $\hat{S}_m^{(\text{LW})}(f)$ versus f difficult to interpret: the width of a CI for $S(f)$ is proportional to $\hat{S}_m^{(\text{LW})}(f)$ and thus varies from frequency to frequency. If we assume that $S(f) > 0$ and $\hat{S}_m^{(\text{LW})}(f) > 0$, we can write

$$\mathbf{P}\left[\frac{\nu \hat{S}_m^{(\text{LW})}(f)}{Q_\nu(1-p)} \leq S(f) \leq \frac{\nu \hat{S}_m^{(\text{LW})}(f)}{Q_\nu(p)}\right]$$
$$= \mathbf{P}\left[10\log_{10}\left(\frac{\nu \hat{S}_m^{(\text{LW})}(f)}{Q_\nu(1-p)}\right) \leq 10\log_{10}(S(f)) \leq 10\log_{10}\left(\frac{\nu \hat{S}_m^{(\text{LW})}(f)}{Q_\nu(p)}\right)\right].$$

It follows that

$$\left[10\log_{10}(\hat{S}_m^{(\text{LW})}(f)) + 10\log_{10}\left(\frac{\nu}{Q_\nu(1-p)}\right), 10\log_{10}(\hat{S}_m^{(\text{LW})}(f)) + 10\log_{10}\left(\frac{\nu}{Q_\nu(p)}\right)\right] \tag{266a}$$

is a $100(1-2p)\%$ CI for $10\log_{10}(S(f))$. Note that the width of this CI, namely

$$10\log_{10}\left(\frac{\nu}{Q_\nu(p)}\right) - 10\log_{10}\left(\frac{\nu}{Q_\nu(1-p)}\right) = 10\log_{10}\left(\frac{Q_\nu(1-p)}{Q_\nu(p)}\right), \tag{266b}$$

is now independent of $10\log_{10}(\hat{S}_m^{(\text{LW})}(f))$. This is a rationale for plotting SDF estimates on a decibel scale. Figure 266 shows $10\log_{10}(\nu/Q_\nu(p))$ and $10\log_{10}(\nu/Q_\nu(1-p))$ versus ν (upper and lower thick curves, respectively) for $p = 0.025$ and integer-valued ν from 1 to 1000. (Note that, when $\nu = 2$, the CI width given by Equation (266b) reduces to what is given in Equation (205a), which is appropriate for a CI based upon a direct spectral estimator.)

As a concrete example, let us use the Parzen lag window estimate $\hat{S}_m^{(\text{LW})}(\cdot)$ shown in Figure 249 to create 95% CIs for both $S(f)$ and $10\log_{10}(S(f))$ (in contrast to what happens

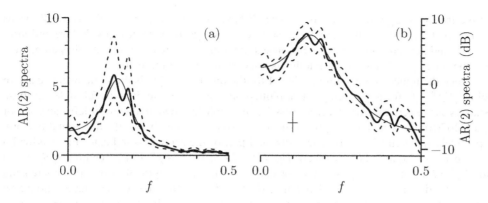

Figure 267 95% confidence intervals (CIs) for an SDF based upon a lag window estimate. The thick solid curves in both plots are the Parzen lag window estimate $\hat{S}_m^{(\mathrm{LW})}(\cdot)$ shown in Figure 249 (this estimate makes use of the AR(2) time series of Figure 34(a)). The dashed curves show the upper and lower limits of 95% CIs for the true AR(2) SDF based upon $\hat{S}_m^{(\mathrm{LW})}(\cdot)$. The true AR(2) SDF is shown as the thin solid curve. The two plots are the same except for the vertical axes: in (a), this axis is on a linear scale, while it is on a decibel scale in (b). The true SDF is trapped by the CIs at most frequencies, but it does fall barely below them before and after $f = 0.4$ and above them around the Nyquist frequency $f_{\mathcal{N}} = 0.5$ (while readily apparent in plot (b), these patterns are almost impossible to see in (a)). The height and width of the crisscross in the lower left-hand portion of plot (b) are equal to, respectively, the width of each 95% CI and the bandwidth measure $B_{\mathcal{U}}$ of Equation (256a).

in practical applications, here we know $S(f)$ because $\hat{S}_m^{(\mathrm{LW})}(\cdot)$ is based upon the AR(2) time series of Figure 34(a), with an SDF dictated by Equations (34), (33) and (145a)). We first need to determine the EDOFs ν associated with the lag window estimate, which, according to Equation (264b), depends upon the sample size N, the variance inflation factor C_h and the lag window $\{w_{m,\tau}\}$. Here $N = 1024$, $C_h = 1$ because $\hat{S}_m^{(\mathrm{LW})}(\cdot)$ is based upon the periodogram (see Table 260), and $w_{m,\tau}$ is given by the forthcoming Equation (275a) with m set to 64, yielding

$$\sum_{\tau=-(N-1)}^{N-1} w_{m,\tau}^2 \doteq 34.51.$$

Hence $\nu \doteq 59.34$. From Equation (265) we obtain

$$\left[\frac{\nu \hat{S}_m^{(\mathrm{LW})}(f)}{Q_\nu(0.975)}, \frac{\nu \hat{S}_m^{(\mathrm{LW})}(f)}{Q_\nu(0.025)}\right] \doteq \left[0.72\,\hat{S}_m^{(\mathrm{LW})}(f), 1.49\,\hat{S}_m^{(\mathrm{LW})}(f)\right]$$

as a 95% CI for $S(f)$, where $Q_\nu(0.025) \doteq 39.94$ and $Q_\nu(0.975) \doteq 82.52$ are the lower and upper 2.5% percentage points of a chi-square distribution with $\nu \doteq 59.34$ degrees of freedom. Figure 267(a) shows $\hat{S}_m^{(\mathrm{LW})}(f)$ versus f (thick solid curve) along with the corresponding CIs (upper and lower dashed curves) and the true SDF (thin solid curve). Note that the width of the CI at f is proportional to $\hat{S}_m^{(\mathrm{LW})}(f)$. Simple adjustments to Equation (266a) give the lower and upper parts of the CI for $10\,\log_{10}(S(f))$:

$$10\,\log_{10}\left(\frac{\nu}{Q_\nu(0.975)}\right) + 10\,\log_{10}(\hat{S}_m^{(\mathrm{LW})}(f)) \doteq -1.43 + 10\,\log_{10}(\hat{S}_m^{(\mathrm{LW})}(f)),$$

$$10\,\log_{10}\left(\frac{\nu}{Q_\nu(0.025)}\right) + 10\,\log_{10}(\hat{S}_m^{(\mathrm{LW})}(f)) \doteq 1.72 + 10\,\log_{10}(\hat{S}_m^{(\mathrm{LW})}(f)).$$

The width of each CI (3.15 dB) no longer depends upon the particular value of $\hat{S}_m^{(\mathrm{LW})}(f)$. Figure 267(b) is similar to Figure 267(a), but now shows $10\,\log_{10}(\hat{S}_m^{(\mathrm{LW})}(f))$, its related CI and the true $10\log_{10}(S(f))$ versus f. The CI based on $10\log_{10}(\hat{S}_m^{(\text{LW})}(f))$ generally traps the true $10\log_{10}(S(f))$, but fails to do so at certain high frequencies (just before and after $f = 0.4$, and also around $f = 0.5$). Because a log transform preserves order, this statement is also true for $\hat{S}_m^{(\text{LW})}(f)$ and $S(f)$, but this is not easy to see in Figure 267(a).

Since the upper (lower) limit of the CI for the unknown $S(f)$ is the same distance above (below) $\hat{S}_m^{(\text{LW})}(f)$ for any given f when plotted on a decibel scale, it is useful to depict these distances as the vertical part of a crisscross. An example is shown in the lower left-hand part of Figure 267(b), for which the lengths of the crisscross above and below the horizontal line are, respectively, 1.72 and 1.43 dB. The horizontal part of the crisscross indicates the bandwidth measure $B_{\mathcal{U}}$ of Equation (256a) (here $B_{\mathcal{U}} \doteq 0.0296$). If $|f - f'| < B_{\mathcal{U}}$, estimates $\hat{S}_m^{(\text{LW})}(f)$ and $\hat{S}_m^{(\text{LW})}(f')$ tend to be positively correlated, with the strength of correlation increasing as $|f - f'|$ decreases; on the other hand, if $|f - f'| \geq B_{\mathcal{U}}$, the two estimates should be approximately uncorrelated. We cannot expect $\hat{S}_m^{(\text{LW})}(\cdot)$ to be able to resolve features in $S(\cdot)$ whose width is smaller than $B_{\mathcal{U}}$; on the other hand, we can expect to see bumps and valleys in $\hat{S}_m^{(\text{LW})}(\cdot)$ whose widths are characterized by $B_{\mathcal{U}}$, as the example in Figure 267(b) illustrates. We shall show similar crisscrosses on most forthcoming linear/decibel plots of $\hat{S}_m^{(\text{LW})}(\cdot)$ and other nonparametric SDF estimates.

7.5 Examples of Lag Windows

We consider here six examples of lag and smoothing windows that can be used in practice to form a specific lag window spectral estimator $\hat{S}_m^{(\text{LW})}(\cdot)$. Five of our examples (the Bartlett, Daniell, Bartlett–Priestley, Parzen and Papoulis lag windows) have been used extensively in practical applications. As we document by example in Section 7.12, the remaining example (the Gaussian lag window) has some attractive properties that can recommend its use over the five standard windows even though it has not seen much use in practice. Two other lag windows – mainly of historical interest – are the subjects of Exercises [7.7] and [7.8], and we discuss some other nonstandard windows in C&E [5].

For each of the specific lag windows, we show a figure with four plots (Figures 270, 272b, 274b, 276, 278 and 279). Plot (a) in each figure shows an example of the lag window $w_{m,\tau}$ versus lag τ for a sample size $N = 64$ and a particular setting of the window parameter m. Plot (b) shows the corresponding smoothing window $W_m(\cdot)$ on a decibel scale versus frequency (the sampling interval Δ_t is taken to be 1, so that the Nyquist frequency $f_{\mathcal{N}}$ is 1/2). Because $w_{m,-\tau} = w_{m,\tau}$ and $W_m(-f) = W_m(f)$ by assumption, we need only plot $w_{m,\tau}$ for $\tau \geq 0$ and $W_m(f)$ for $f \geq 0$. We indicate the smoothing window bandwidths β_W (Equation (251b)) and B_W (Equation (251e)) by horizontal lines plotted, respectively, 3 and 6 dB down from $W_m(0)$ (the widths of the displayed lines are equal to $\beta_W/2$ and $B_W/2$). Plots (c) and (d) show two different spectral windows $\mathcal{U}_m(\cdot)$ for the same m and N. From Equation (255c) we see that $\mathcal{U}_m(\cdot)$ depends upon both the smoothing window $W_m(\cdot)$ and the data taper $\{h_t\}$ through its associated spectral window $\mathcal{H}(\cdot)$. Plots (c) and (d) show $\mathcal{U}_m(\cdot)$ corresponding to, respectively, the rectangular data taper (see Figures 190(a) and 191(a)) and a Slepian data taper with $NW = 4$ (Figures 190(g) and 191(g)). We have included plots (c) and (d) to emphasize the point that the expected value of a lag window spectral estimator $\hat{S}_m^{(\text{LW})}(\cdot)$ depends upon both the smoothing window $W_m(\cdot)$ and the data taper used in the corresponding direct spectral estimator $\hat{S}^{(\text{D})}(\cdot)$. Plots (c) and (d) also depict our standard measure of the effective bandwidth of $\hat{S}_m^{(\text{LW})}(\cdot)$, namely, $B_{\mathcal{U}}$ (Equation (256a)). This measure is shown as a horizontal line plotted 3 dB down from $\mathcal{U}_m(0)$ – the width of the line is $B_{\mathcal{U}}/2$.

In all six examples, the smoothing window $W_m(\cdot)$ consists of a central lobe (i.e., the one centered about zero frequency) and several sidelobes. As we note in the next section, these sidelobes ideally should be as small as possible (to minimize what is defined there as smoothing window leakage). The magnitude of the peak in the first sidelobe relative to the magnitude of the peak in the central lobe is thus of interest, as is the rate of decay of the envelope formed by the peaks of the sidelobes. This envelope typically decays approximately as f^α for some $\alpha < 0$. For example, if $\alpha = -1$, a doubling of frequency corresponds to a doubling in the decay of the magnitude of the peaks in the sidelobes of $W_m(\cdot)$. On a decibel (i.e., $10 \log_{10}$) scale such a doubling corresponds to a $10 \log_{10}(2) \approx 3$ dB drop in a frequency octave (doubling of frequency). The magnitude of the envelope thus decays at 3 dB per octave. We quote this decay rate for each of the lag windows because it is a good guide to how fast the sidelobes decay with increasing frequency (see, however, the discussions about the Daniell, Bartlett–Priestley and Gaussian windows shown in Figures 272b(b), 274b(b) and 278(b)).

Table 279 at the end of this section lists approximations for the asymptotic variance of $\hat{S}_m^{(\mathrm{LW})}(f)$, the EDOFs ν, the smoothing window bandwidth B_W and the alternative bandwidth measure β_W for the six lag windows. This table shows a number of interesting relationships. For example, for fixed m, we have $\mathrm{var}\{\hat{S}_m^{(\mathrm{LW})}(f)\} \to 0$ as $N \to \infty$ for lag window estimators based on these six windows. Also we can deduce that $\mathrm{var}\{\hat{S}_m^{(\mathrm{LW})}(f)\}$ is inversely proportional either to β_W (for the Daniell, Bartlett–Priestly, Parzen, Gaussian and Papoulis windows) or to β_W^2 (for the Bartlett window). Since the bias due to the smoothing window alone is proportional to β_W^2 (see Equation (256d)), there is a trade-off between the bias and variance in lag window spectral estimators, or, to state it more picturesquely, "bias and variance are antagonistic" (Priestley, 1981, p. 528).

Bartlett Window (Figure 270)

The lag window for Bartlett's spectral estimator (Bartlett, 1950) is given by

$$w_{m,\tau} = \begin{cases} 1 - |\tau|/m, & |\tau| < m; \\ 0, & |\tau| \geq m, \end{cases} \tag{269a}$$

where m is an integer-valued window parameter that can assume values from 1 up to N. Bartlett's window applies linearly decreasing weights to the sample ACVS up to lag m and zero weights thereafter. The parameter m can be interpreted as a truncation point beyond which the ACVS is considered to be zero. The corresponding smoothing window is

$$W_m(f) = \Delta_\mathrm{t} \sum_{\tau=-m}^{m} \left(1 - \frac{|\tau|}{m}\right) e^{-i2\pi f \tau \Delta_\mathrm{t}} = \mathcal{F}(f), \tag{269b}$$

where the latter quantity is Fejér's kernel (see Equation (174c)). Note that the smoothing window is of the same form as the two-sided Cesàro partial sum we met in Equation (79). From the plots of $\mathcal{F}(\cdot)$ in Figure 176, we see that the width of the central lobe decreases as m increases, so the amount of smoothing decreases as m increases (a convention for m we maintain for the other lag windows we discuss). Since Fejér's kernel is nonnegative everywhere, it follows that Bartlett's estimator of an SDF is also nonnegative everywhere. The magnitude of the peak of the first sidelobe of Fejér's kernel is down about 13 dB from its value at $f = 0$. The envelope of Bartlett's smoothing window decays as approximately f^{-2}, which corresponds to a decay of 6 dB per frequency octave.

Note from plots (c) and (d) in Figure 270 that the two spectral windows $\mathcal{U}_m(\cdot)$ corresponding to the default data taper and the Slepian data taper both have a decay rate similar to that of the smoothing window shown in plot (b). This is due to the fact that $\mathcal{U}_m(\cdot)$ is the convolution of the smoothing window $W_m(\cdot)$ and the spectral window $\mathcal{H}(\cdot)$ associated with the data taper. In the case of the default data taper (plot (c)), both $W_m(\cdot)$ and $\mathcal{H}(\cdot)$ are in fact Fejér's kernel, so the decay rate of their convolution must be the same. In the case of the Slepian data taper, a glance at its spectral window $\mathcal{H}(\cdot)$ in Figure 191(g) shows that $\mathcal{U}_m(\cdot)$ is

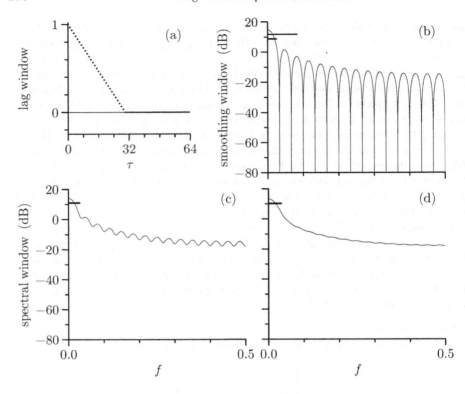

Figure 270 Bartlett lag, smoothing and spectral windows for $m = 30$ and $N = 64$ (the smoothing window bandwidth is $B_W \doteq 0.05$). For a description of the contents of each of the four plots, see the second paragraph from the beginning of this section (Section 7.5).

essentially a smoothed version of Fejér's kernel and hence has the same decay rate as Fejér's kernel. Now let us make a practical interpretation. The user who selected a Slepian data taper with $NW = 4$ in order to achieve low bias in a direct spectral estimate $\hat{S}^{(D)}(\cdot)$ of an SDF with a suspected large dynamic range might be shocked to find that smoothing $\hat{S}^{(D)}(\cdot)$ with Bartlett's window yields an *overall* spectral window $\mathcal{U}_m(\cdot)$ that decays at the same rate as the spectral window for the periodogram!

The rationale Bartlett (1950) used to come up with the lag window of Equation (269a) has an interesting connection to Welch's overlapped segment averaging spectral estimator (see Section 8.8). Bartlett did not apply any tapering (other than the default rectangular taper) so that the direct spectral estimator of Equation (188a) to be smoothed was just the periodogram. Bartlett argued that one could reduce the sampling fluctuations of the periodogram by

[1] splitting the original sample of N observations into N/m contiguous blocks, each block containing m observations (we assume that N/m is an integer);

[2] forming the periodogram for each block; and

[3] averaging the N/m periodograms together.

Let $\hat{S}^{(P)}_{k,m}(\cdot)$ be the periodogram for the kth block. Bartlett reasoned that $\hat{S}^{(P)}_{j,m}(f)$ and $\hat{S}^{(P)}_{k,m}(f)$ for $j \neq k$ should be approximately uncorrelated (with the approximation improving as m gets larger – Exercise [7.9] investigates this claim). The reduction in variance should thus be inversely proportional to the number of blocks N/m. Bartlett worked with the unbiased

ACVS estimator

$$\hat{s}_\tau^{(U)} = \frac{N}{N - |\tau|} \hat{s}_\tau^{(P)},$$

in terms of which we can write

$$\hat{S}_{k,m}^{(P)}(f) = \Delta_t \sum_{\tau=-m}^{m} \left(1 - \frac{|\tau|}{m}\right) \hat{s}_{k,\tau}^{(U)} e^{-i2\pi f \tau \Delta_t},$$

where $\{\hat{s}_{k,\tau}^{(U)}\}$ denotes the unbiased ACVS estimator based upon just the data in the kth block. If we average the N/m periodogram estimates together, we get

$$\frac{1}{N/m} \sum_{k=1}^{N/m} \hat{S}_{k,m}^{(P)}(f) = \Delta_t \sum_{\tau=-m}^{m} \left(1 - \frac{|\tau|}{m}\right) \left[\frac{1}{N/m} \sum_{k=1}^{N/m} \hat{s}_{k,\tau}^{(U)}\right] e^{-i2\pi f \tau \Delta_t}$$

$$= \Delta_t \sum_{\tau=-m}^{m} \left(1 - \frac{|\tau|}{m}\right) \bar{s}_\tau^{(U)} e^{-i2\pi f \tau \Delta_t},$$

where $\bar{s}_\tau^{(U)}$ is the average of the $\hat{s}_{k,\tau}^{(U)}$ terms. Bartlett then argued that $\bar{s}_\tau^{(U)}$ ignores information concerning s_τ that is contained in pairs of data values in adjacent blocks. He suggested replacing $\bar{s}_\tau^{(U)}$ with $\hat{s}_\tau^{(U)}$, the unbiased estimator of s_τ obtained by using all N observations. This substitution yields the estimator of $S(f)$ given by

$$\Delta_t \sum_{\tau=-m}^{m} \left(1 - \frac{|\tau|}{m}\right) \hat{s}_\tau^{(U)} e^{-i2\pi f \tau \Delta_t} = \Delta_t \sum_{\tau=-m}^{m} \frac{1 - |\tau|/m}{1 - |\tau|/N} \hat{s}_\tau^{(P)} e^{-i2\pi f \tau \Delta_t}.$$

The common form for Bartlett's lag window spectral estimator (in the case of a default rectangular taper), namely,

$$\hat{S}_m^{(LW)}(f) = \Delta_t \sum_{\tau=-m}^{m} \left(1 - \frac{|\tau|}{m}\right) \hat{s}_\tau^{(P)} e^{-i2\pi f \tau \Delta_t},$$

can be regarded as an approximation to the above under the assumption that N is much larger than m since then

$$\frac{1 - |\tau|/m}{1 - |\tau|/N} = \left(1 - \frac{|\tau|}{m}\right)\left(1 + \frac{|\tau|}{N} + \left[\frac{|\tau|}{N}\right]^2 + \cdots\right) \approx 1 - \frac{|\tau|}{m}.$$

Bartlett noted that the choice of m is a compromise between reducing the variance of $\hat{S}_m^{(LW)}(f)$ and reducing its resolution (i.e., smearing out local features).

Daniell (or Rectangular) Window (Figures 272a and 272b)
As we noted after Equation (71), the quantity

$$\frac{1}{b-a} \int_a^b f(x)\,dx$$

is often called the average value of the function $f(\cdot)$ in the interval $[a, b]$. Daniell's lag window spectral estimator (Daniell, 1946) is simply the average value of the direct spectral estimator $\hat{S}^{(D)}(\cdot)$ in an interval of length $1/(m\,\Delta_t)$ around each value of f:

$$\hat{S}_m^{(LW)}(f) = m\,\Delta_t \int_{f - 1/(2m\,\Delta_t)}^{f + 1/(2m\,\Delta_t)} \hat{S}^{(D)}(f)\,df,$$

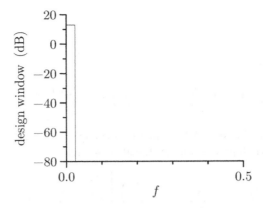

Figure 272a Daniell design window for $m = 20$.

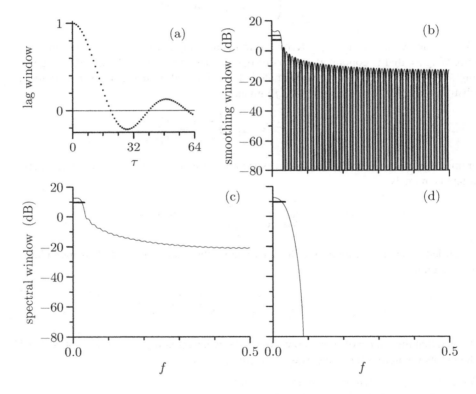

Figure 272b Daniell windows for $m = 20$ and $N = 64$ ($B_W \doteq 0.05$).

where, for f near $\pm f_{\mathcal{N}}$, we use the fact that $\hat{S}^{(D)}(\cdot)$ is $2f_{\mathcal{N}}$ periodic. Here the parameter $m \geq 1$ does not correspond to a truncation point – and need not be an integer – as in the case of the Bartlett lag window; however, it does control the degree of averaging to which $\hat{S}^{(D)}(\cdot)$ is subjected: the smaller m is, the greater the amount of smoothing. Comparison with Equation (247a) shows that the design window for Daniell's estimator is

$$V_m(f) = \begin{cases} m\,\Delta_{\mathrm{t}}, & |f| \leq 1/(2m\,\Delta_{\mathrm{t}}); \\ 0, & 1/(2m\,\Delta_{\mathrm{t}}) < |f| \leq f_{\mathcal{N}} \end{cases} \quad (272)$$

7.5 Examples of Lag Windows

(the above also defines the design window for $|f| > f_\mathcal{N}$ since it is periodic with a period of $2f_\mathcal{N}$). The corresponding lag window is given by Equation (247d) in conjunction with Equation (247c):

$$w_{m,\tau} = \begin{cases} m\,\Delta_t \int_{-1/(2m\,\Delta_t)}^{1/(2m\,\Delta_t)} e^{i2\pi f\tau\,\Delta_t}\, df = \dfrac{\sin(\pi\tau/m)}{\pi\tau/m}, & |\tau| < N; \\ 0, & |\tau| \geq N \end{cases} \quad (273a)$$

(we interpret the ratio above as unity when $\tau = 0$). The smoothing window for Daniell's spectral estimator is thus

$$W_m(f) = \Delta_t \sum_{\tau=-(N-1)}^{N-1} \frac{\sin(\pi\tau/m)}{\pi\tau/m} e^{-i2\pi f\tau\,\Delta_t}. \quad (273b)$$

Figure 272a shows the Daniell design window $V_m(\cdot)$ for $m = 20$. Because the sidelobes of the corresponding smoothing window $W_m(\cdot)$ alternate between positive and negative values, we have plotted the quantity $|W_m(\cdot)|$ in Figure 272b(b) rather than $W_m(\cdot)$. The negative sidelobes are shaded – unfortunately, there are so many sidelobes that this shading is barely visible. The magnitude of the peak of the first sidelobe in $|W_m(\cdot)|$ is down about 11 dB from its value at $f = 0$. The envelope of the sidelobes decays as approximately f^{-1}, which corresponds to a decay of 3 dB per frequency octave. However, since Equation (247f) implies that we can just as well regard $V_m(\cdot)$ in Equation (272) as Daniell's smoothing window, we can also argue that the latter has effectively *no* sidelobes! This example shows that the smoothing window $W_m(\cdot)$ has to be interpreted with some care. Because $\hat{S}_m^{(\text{LW})}(\cdot)$ is the cyclic convolution of two periodic functions, one of them (namely, $\hat{S}^{(\text{D})}(\cdot)$) with Fourier coefficients $\{\hat{s}_\tau^{(\text{D})}\}$ that are identically 0 for lags $|\tau| \geq N$, there is an inherent ambiguity in $W_m(\cdot)$, which we have sidestepped by arbitrarily setting the Fourier coefficients of $W_m(\cdot)$ (namely, $\{w_{m,\tau}\}$) to 0 for lags $|\tau| \geq N$. Thus, the fact that a smoothing window has negative sidelobes does *not* automatically imply that the corresponding lag window spectral estimator can sometimes take on negative values!

Note that, in contrast to what happened with Bartlett's window (Figure 270), the two spectral windows $\mathcal{U}_m(\cdot)$ in Figure 272b now have a markedly different appearance. The first of these is again dominated by Fejér's kernel (the spectral window for the rectangular data taper shown in Figure 191(a)), but the second reflects the convolution of $V_m(\cdot)$ in Equation (272) and the spectral window for the Slepian data taper (shown in Figure 191(g)). We can make another practical interpretation. The user who chose, say, the Daniell smoothing window because of its lack of sidelobes and used the default rectangular taper might be surprised to find that because of the effect of Fejér's kernel – basically reflecting the finiteness of the data – the *overall* spectral window $\mathcal{U}_m(\cdot)$ again decays as just f^{-2}, i.e., the same rate as the sidelobes of the periodogram!

Bartlett–Priestley (Quadratic or Epanechnikov) Window (Figures 274a and 274b)
The Bartlett–Priestley window is specified via the following design window:

$$V_m(f) = \begin{cases} 3m\,\Delta_t \left[1 - (2fm\,\Delta_t)^2\right]/2, & |f| \leq 1/(2m\,\Delta_t); \\ 0, & 1/(2m\,\Delta_t) < |f| \leq f_\mathcal{N}, \end{cases} \quad (273c)$$

where the parameter m can assume any value greater than or equal to 1 (the window is defined for $|f| > f_\mathcal{N}$ by periodic extension). The burden of Exercise [7.10a] is to show that the

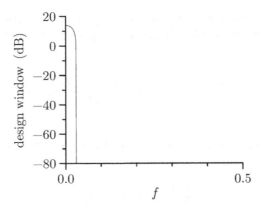

Figure 274a Bartlett–Priestley design window for $m = 16.67$.

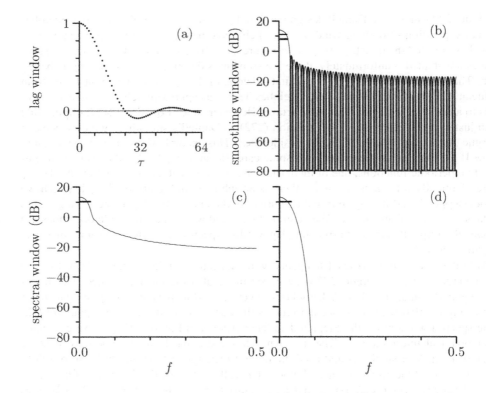

Figure 274b Bartlett–Priestley windows for $m = 16.67$ and $N = 64$ ($B_W \doteq 0.05$).

corresponding lag window is given by

$$w_{m,\tau} = \begin{cases} 1, & \tau = 0; \\ \dfrac{3m^2}{\pi^2 \tau^2}\left[\dfrac{\sin(\pi\tau/m)}{\pi\tau/m} - \cos(\pi\tau/m)\right], & 1 \leq |\tau| < N; \\ 0, & |\tau| \geq N. \end{cases} \quad (274)$$

Figure 274a shows the Bartlett–Priestley design window for $m = 16.67$, while Figure 274b depicts the corresponding lag, smoothing and spectral windows for this same value of m and for $N = 64$. The design window visually resembles the Daniell design window (Figure 272a), but, while the latter transitions to zero abruptly, the Bartlett–Priestley window does so in a continuous manner. The Daniell and Bartlett–Priestley smoothing windows are also similar (plot (b) of Figures 272b and 274b), but the central lobe of the latter is smoother than the former. As is the case for the Daniell smoothing window, the Bartlett–Priestley smoothing window has negative sidelobes. The magnitudes of the sidelobes in the Daniell case are approximately 5 dB higher than in the Bartlett–Priestley case, but the decay rates are essentially the same; however, in view of Equation (247f), we can argue that, like the Daniell window, the Bartlett–Priestley smoothing window is equivalent to one with no sidelobes. The Daniell and Bartlett–Priestley spectral windows that are associated with the rectangular data taper (plot (c) of Figures 272b and 274b) are quite similar, with the exception of small ripples in the Daniell window not present in the Bartlett–Priestley. The two windows that are associated with the Slepian taper (plot (d) of Figures 272b and 274b) are virtually identical. As demonstrated in Figure 319 in Section 7.12, the discontinuous nature of the Daniell design window can result in a lag window estimate that is choppier looking than the comparable Bartlett–Priestley estimate, even though overall the two estimates track each other well because the first-moment properties of the Daniell and Bartlett–Priestley windows are nearly identical.

As discussed in Priestley (1962), the Bartlett–Priestley window arises as the solution to an optimality problem that involves minimizing an approximation to the normalized mean square error $E\{[\hat{S}_m^{(\text{LW})}(f) - S(f)]^2/S^2(f)\}$ at one fixed frequency f with respect to a restricted class of design windows proposed in Parzen (1957). Priestley (1981, pp. 444–5) notes that Bartlett (1963) independently introduced this window but that Epanechnikov (1969) established its optimality properties in a more general context.

Parzen Window (Figure 276)
Parzen (1961) suggested the following lag window:

$$w_{m,\tau} = \begin{cases} 1 - 6\,(\tau/m)^2 + 6\,(|\tau|/m)^3, & |\tau| \leq m/2; \\ 2\,(1 - |\tau|/m)^3, & m/2 < |\tau| < m; \\ 0, & |\tau| \geq m, \end{cases} \quad (275a)$$

where the window parameter m is an integer that can range from 1 up to N. The corresponding smoothing window is given by

$$W_m(f) = \frac{A_m(f)\,\Delta_t}{m^3 \sin^4(\pi f\,\Delta_t)}, \quad (275b)$$

where, for m even,

$$A_m(f) \stackrel{\text{def}}{=} 4\left[3 - 2\sin^2(\pi f\,\Delta_t)\right] \sin^4(m\pi f\,\Delta_t/2)$$

(Priestley, 1981, p. 444), while, for m odd,

$$A_m(f) \stackrel{\text{def}}{=} [3 - 2\sin^2(\pi f\,\Delta_t)][2 - \sin^2(m\pi f\,\Delta_t)] \\ - \cos(\pi f\,\Delta_t)\cos(m\pi f\,\Delta_t)[6 - \sin^2(\pi f\,\Delta_t)]$$

(Exercise [7.11] gives equations for $W_m(0)$, which needs special care since the above equations are indeterminate at $f = 0$). The Parzen lag window can be derived by taking the Bartlett lag window (treated as a continuous function of τ) with parameter $m/2$, convolving it with itself, and then rescaling and sampling the resulting function (see the lower right-hand plot

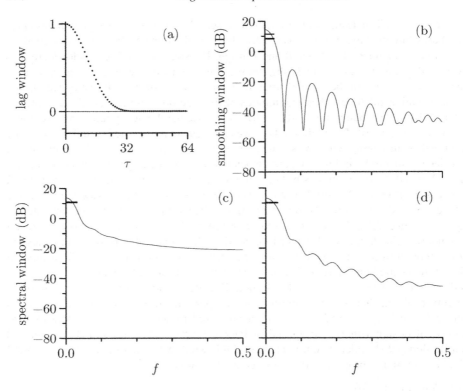

Figure 276 Parzen windows for $m = 37$ and $N = 64$ ($B_W \doteq 0.05$).

of Figure 69 and Equation (104)). In fact, for continuous τ, the Bartlett lag window and the Parzen lag window are related to the PDFs of the sum of, respectively, two and four uniformly distributed RVs (the functions shown in the right-hand column of Figure 69 are proportional to the PDFs for the sum of two, three and four such RVs). Because of the central limit theorem, we can regard the Parzen lag window – and hence its smoothing window – as having approximately the shape of a Gaussian PDF (cf. the lower right-hand plot of Figure 69).

Parzen SDF estimates, like the Bartlett, Daniell and Bartlett–Priestley estimates, are always nonnegative. As f increases, the envelope of Parzen's smoothing window decreases as approximately f^{-4}, i.e., 12 dB per octave. The magnitude of the peak of the first sidelobe is down about 28 dB from the magnitude of the central lobe. Thus the sidelobes decay much more rapidly than those of Bartlett's smoothing window (6 dB per octave), and the first sidelobe is also much smaller (28 dB down as compared to 13 dB).

Note that the two spectral windows $\mathcal{U}_m(\cdot)$ in Figure 276 are quite different. As for the Bartlett, Daniell and Bartlett–Priestley windows, the first of these is again dominated by Fejér's kernel, but the second is the convolution of Parzen's $W_m(\cdot)$ with the spectral window for the Slepian data taper and has sidelobes about 25 dB below that of the first at $f = 1/2$.

Gaussian Window (Figure 278)
The fact that the Parzen lag window was designed to have approximately the shape of a Gaussian PDF leads us to consider the following *Gaussian lag window*:

$$w_{m,\tau} = \begin{cases} e^{-\tau^2/m^2}, & |\tau| < N; \\ 0, & |\tau| \geq N, \end{cases} \quad (276)$$

where the parameter m can assume *any* positive value (the smoothing window bandwidth

B_W decreases as m increases, as is true for all the other lag windows we have considered so far). Figure 278(a) shows an example of $\{w_{m,\tau}\}$ for the case $N = 64$ and $m = 16$. As expected, this window is visually quite similar to the Parzen lag window of Figure 276(a). The smoothing window for $\{w_{m,\tau}\}$ is

$$W_m(f) = \Delta_t \sum_{\tau=-(N-1)}^{N-1} e^{-\tau^2/m^2} e^{-i2\pi f \tau \Delta_t}. \qquad (277)$$

As is also true for the Daniell and Bartlett–Priestley smoothing windows, $W_m(\cdot)$ has some negative sidelobes. This fact is demonstrated for our example in Figure 278(b), which plots $|W_m(f)|$ versus f with shading used to indicate negative sidelobes. The presence of negative sidelobes opens up the possibility that a lag window estimator based on $\{w_{m,\tau}\}$ might potentially be negative at some frequencies; however, as is also true in the Daniell and Bartlett–Priestley cases, this cannot happen, as the following exercise indicates.

▷ **Exercise [277]** Given any direct spectral estimator $\hat{S}^{(D)}(\cdot)$ based upon a time series of length N, show that, for all f,

$$\hat{S}_m^{(LW)}(f) = \int_{-f_\mathcal{N}}^{f_\mathcal{N}} W_m(f-\phi) \hat{S}^{(D)}(\phi) \, d\phi \geq 0,$$

where $W_m(\cdot)$ is given by Equation (277). ◁

Note that the magnitudes of the sidelobes of $W_m(\cdot)$ in Figure 278(b) are substantially smaller than those shown in Figure 276(b) for the Parzen smoothing window. The magnitude of the peak of the first sidelobe in $|W_m(\cdot)|$ is down about 79 dB from its value at $f = 0$, as compared to 28 dB for the Parzen window. As a result, the spectral window shown in Figure 278(d) more closely resembles the ones associated with the sidelobe-free Daniell and and Bartlett–Priestley windows (plot (d) of Figures 272b and 274b) than that associated with the Parzen window (Figure 276(d)). The envelope of the sidelobes for the Gaussian smoothing window decays as approximately $f^{-0.7}$, which corresponds to a decay of 2 dB per frequency octave. This poor decay rate (the slowest we have seen so far) is not particular important because the sidelobes are so small in magnitude.

Although, as we shall see in Section 7.12, the Gaussian lag window is competitive with the Daniell, Bartlett–Priestley and Parzen windows, it has not seen much use in practical applications. Neave (1972) and Harris (1978) include a Gaussian window $\exp(-\tau^2/m^2)$, $\tau \in \mathbb{R}$, in their investigation of windows in an abstract context (i.e., not necessarily for use as lag windows). One disadvantage they point out is the discontinuity that arises when this window is truncated to a finite interval ($[-(N-1), N-1]$ in our case). Depending upon the settings for m and N, this discontinuity can be large enough to cause a prominent ringing in the Fourier transform of either the truncated window or samples thereof. This ringing might adversely impact the performance of the Gaussian window in comparison to windows exhibiting continuity (an example being the Parzen and window). This discontinuity is, however, a moot point for a Gaussian *lag* window due to the nature of $\{\hat{s}_\tau^{(D)}\}$. By construction, $\hat{s}_\tau^{(D)} = 0$ for all $|\tau| \geq N$, which means that eliminating the discontinuity in Equation (276) by redefining $w_{m,\tau}$ to be $e^{-\tau^2/m^2}$ for all τ yields the same lag window estimator $\hat{S}_m^{(LW)}(\cdot)$ as with the existing definition. Thus, while the Gaussian window might not be appealing when viewed abstractly, it is well-suited for use as a lag window (the forthcoming Figures 319 and 320 document an example where the Gaussian lag window outperforms the Daniell, Bartlett–Priestley and Parzen lag windows on an actual time series).

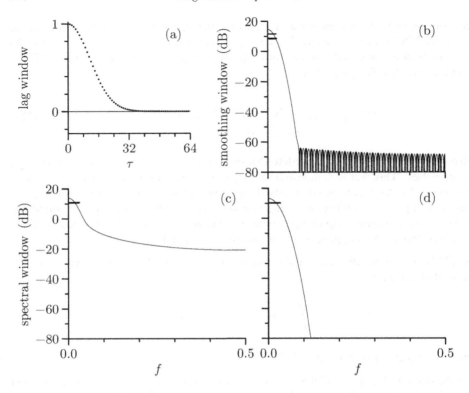

Figure 278 Gaussian windows for $m = 16$ and $N = 64$ ($B_W \doteq 0.05$).

Papoulis Window (Figure 279)
Papoulis (1973) found that the continuous τ analog of the following lag window produces a window with a certain minimum bias property:

$$w_{m,\tau} = \begin{cases} \dfrac{1}{\pi} |\sin(\pi\tau/m)| + (1 - |\tau|/m)\cos(\pi\tau/m), & |\tau| < m; \\ 0, & |\tau| \geq m, \end{cases} \quad (278)$$

where m is an integer ranging from 1 up to N (Bohman, 1961, derived this window earlier in the context of characteristic functions). The rationale behind this lag window is as follows. Equation (256d) tells us that the bias in a lag window spectral estimator due to the smoothing window alone is proportional to β_W^2, where β_W is defined in Equation (251a). The Papoulis window is the solution to the continuous τ analog of the following problem: for fixed $m > 0$, amongst all lag windows $\{w_{m,\tau}\}$ with $w_{m,\tau} = 0$ for $|\tau| \geq m$ and with a corresponding smoothing window $W_m(\cdot)$ such that $W_m(f) \geq 0$ for all f, find the window such that $\int_{-f_N}^{f_N} f^2 W_m(f) \, df \propto \beta_W^2$ is minimized.

The derivation of the Papoulis smoothing window is left as an exercise. Comparison of Figures 276 and 279 shows that the Parzen and the Papoulis windows have quite similar characteristics.

Comments and Extensions to Section 7.5

[1] We motivated our discussion of lag window estimators by first considering discretely smoothed direct spectral estimators $\{\hat{S}^{(\text{DS})}(f_k')\}$. As shown by Equation (246b), the latter are based on a set of smoothing coefficients $\{g_j\}$, which implicitly define a lag window $\{w_{g,\tau}\}$ (see Equations (249a),

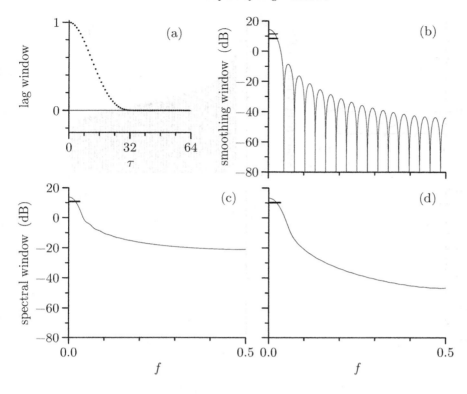

Figure 279 Papoulis windows for $m = 34$ and $N = 64$ ($B_W \doteq 0.05$).

Estimator	Asymptotic Variance	ν	B_W	β_W
Bartlett	$\dfrac{0.67mC_hS^2(f)}{N}$	$\dfrac{3N}{mC_h}$	$\dfrac{1.5}{m\,\Delta_t}$	$\dfrac{0.92}{\sqrt{m}\,\Delta_t}$
Daniell	$\dfrac{mC_hS^2(f)}{N}$	$\dfrac{2N}{mC_h}$	$\dfrac{1}{m\,\Delta_t}$	$\dfrac{1}{m\,\Delta_t}$
Bartlett–Priestley	$\dfrac{1.2mC_hS^2(f)}{N}$	$\dfrac{1.67N}{mC_h}$	$\dfrac{0.83}{m\,\Delta_t}$	$\dfrac{0.77}{m\,\Delta_t}$
Parzen	$\dfrac{0.54mC_hS^2(f)}{N}$	$\dfrac{3.71N}{mC_h}$	$\dfrac{1.85}{m\,\Delta_t}$	$\dfrac{1.91}{m\,\Delta_t}$
Gaussian	$\dfrac{1.25mC_hS^2(f)}{N}$	$\dfrac{1.60N}{mC_h}$	$\dfrac{0.80}{m\,\Delta_t}$	$\dfrac{0.78}{m\,\Delta_t}$
Papoulis	$\dfrac{0.59mC_hS^2(f)}{N}$	$\dfrac{3.41N}{mC_h}$	$\dfrac{1.70}{m\,\Delta_t}$	$\dfrac{1.73}{m\,\Delta_t}$

Table 279 Asymptotic variance, equivalent degrees of freedom ν, smoothing window bandwidth B_W and alternative bandwidth measure β_W (used in Equation (256d) for the bias due to the smoothing window alone) for six lag window spectral density estimators. The tabulated quantities are approximations to the formulae given in Equations (260), (264b), (251e) and (251b). The quantity C_h – a variance inflation factor – depends upon the data taper used in the corresponding direct spectral estimator $\hat{S}^{(D)}(\cdot)$ (see Table 260 and Equation (262)).

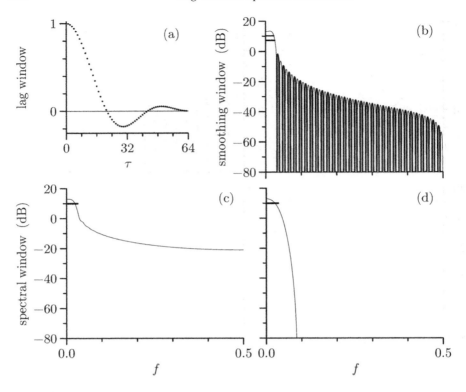

Figure 280 Modified Daniell windows for $M = 3$ and $N = 64$ ($B_W \doteq 0.05$).

(249b) and (249c)). The relationship between $\{w_{g,\tau}\}$ and $\{g_j\}$ is similar to that between a lag window and its smoothing window, suggesting that we can sample a smoothing window to obtain smoothing coefficients.

As an example of obtaining $\{g_j\}$ in this manner, let us consider smoothing coefficients of the form

$$g_j = \begin{cases} 1/(2M), & |j| < M; \\ 1/(4M), & |j| = M; \\ 0, & \text{otherwise.} \end{cases}$$

We can consider these weights as being generated by sampling from a Daniell smoothing window $V_m(\cdot)$ (see Equation (272)) with an "end point" adjustment. Bloomfield (2000) calls the resulting estimator $\{\hat{S}^{(\text{DS})}(f'_k)\}$ a *modified Daniell* spectral estimator. Note that $\{g_j\}$ can be regarded as a low-pass LTI filter. Since $\sum_{j=-M}^{M} g_j = 1$, this filter has the normalization that we argued in Section 5.7 is appropriate for a smoother. Assume the frequencies involved in $\{\hat{S}^{(\text{DS})}(f'_k)\}$ are given by $f'_k = \tilde{f}_k = k/(2N\,\Delta_t)$. We can use Equations (249b) and (249c) to reexpress $\hat{S}^{(\text{DS})}(\cdot)$ as a lag window spectral estimator $\hat{S}_m^{(\text{LW})}(\cdot)$. The corresponding lag and smoothing windows are shown, respectively, in Figures 280(a) and (b) for the case $M = 3$ and $N = 64$. As for the usual Daniell window in Figure 272b, the smoothing window has negative sidelobes (shaded in plot (b)), but, because the estimate is necessarily nonnegative, these sidelobes are again an artifact caused by truncation. The spectral windows for the modified Daniell window (plots (c) and (d)) closely resemble those for the usual Daniell window ((c) and (d) of Figure 272b).

The device of sampling from a smoothing window to produce the coefficients for a discretely smoothed direct spectral estimator can obviously be applied to other smoothing windows besides the Daniell window. A variation on this idea is to generate the coefficients from, say, the Parzen smoothing window (with a renormalization to ensure that the weights sum to unity). This form of spectral estimator is discussed in Cleveland and Parzen (1975) and Walden (1990a).

7.5 Examples of Lag Windows

[2] We introduced the idea of a reshaped lag window in C&E [2] for Section 7.2 (see Equation (257b)). Figure 282 gives an example for such a window formed from the Parzen lag window with $m = 37$ and the Slepian data taper with $NW = 4$ (a combination appearing in Figure 276). When used in conjunction with the rectangular data taper, the lag window in Figure 282(a) thus yields exactly the same spectral window as shown in Figure 276(d). The corresponding smoothing window is shown in plot (b), where now we plot $|W_m(\cdot)|$ rather than $W_m(\cdot)$. It has a prominent single *negative* sidelobe (indicated by the shaded area). That this lag window can lead to negative spectral estimates is demonstrated in plot (c). This plot shows two lag window spectral estimates for the first 64 values of the realization of the AR(4) process $\{X_t\}$ depicted in Figure 34(e). The thin curve is the lag window estimate formed using the Parzen lag window in combination with the Slepian data taper; the thick curve uses the reshaped lag window with the rectangular data taper. The shaded areas under the latter curve indicate the frequencies where $\hat{S}_m^{(\text{LW})}(f)$ is negative so that $|\hat{S}_m^{(\text{LW})}(f)|$ is plotted rather than $\hat{S}_m^{(\text{LW})}(f)$. The true SDF for the AR(4) process is shown as the thick curve in plot (d); the thin curve there is $E\{\hat{S}_m^{(\text{LW})}(\cdot)\}$, which is the result of convolving the spectral window in Figure 276(d) with the true SDF and hence is the same for both spectral estimates depicted in 282(c). (In view of Equation (255a), there are two other ways of constructing $E\{\hat{S}_m^{(\text{LW})}(\cdot)\}$. First, it is the convolution of $E\{\hat{S}^{(\text{D})}(\cdot)\}$ given by the thin curve in the $NW = 4$ plot of Figure 193 with the smoothing window of Figure 276(b). Second, it is the convolution of $E\{\hat{S}^{(\text{P})}(\cdot)\}$ shown by the thin curve in Figure 178(b) with the smoothing window whose absolute value is shown in Figure 282(b). Note in the latter case that, even though the smoothing window has a prominent negative sidelobe, the resulting $E\{\hat{S}_m^{(\text{LW})}(\cdot)\}$ is still totally positive.)

While it is possible to create a lag window estimator using a rectangular data taper whose *expected value* is the same as another lag window estimator using a non-rectangular taper, this example shows that, for a particular time series, the *actual* SDF estimates can be qualitatively quite different: the reshaped lag window estimator can assume undesirable negative values, whereas, for the Bartlett, Daniell, Bartlett–Priestley, Parzen, Gaussian and Papoulis lag windows in combination with an arbitrary data taper, the lag window estimator is guaranteed to be nonnegative, thus matching a basic property of true SDFs.

[3] We noted following our description of Grenander's measure β_W of smoothing window bandwidth (Equation (251a)) that β_W could assume an imaginary value if the smoothing window is negative at some frequencies. The smoothing window shown in Figure 282(b) is negative for $f \in [0.05, 0.5]$. A computation indicates that $\beta_W \doteq \sqrt{(-0.0087)} \doteq 0.093i$, hence illustrating this potential problem. By contrast, Jenkin's measure B_W of smoothing window bandwidth (Equation (251c)) is 0.057 and, as depicted by the short horizontal line in Figure 282(b), is intuitively sensible.

[4] As pointed out in our discussion of the Daniell window, smoothing a direct spectral estimator using its rectangular design window (see Figure 272a) is entirely equivalent to smoothing it using its corresponding smoothing window (see Figure 272b(b)), even though the latter has sidelobes that rapidly oscillate between positive and negative values. Because of this equivalence we can regard the Daniell smoothing window as having no sidelobes. Taking this point of view, there is an interesting pattern in plots (c) and (d) in the figures for the Bartlett, Daniell, Bartlett–Priestley, Parzen, Gaussian and Papoulis lag windows (Figures 270, 272b, 274b, 276, 278 and 279). These plots show the spectral window $\mathcal{U}_m(\cdot)$, which, from Equation (255c), is the convolution of the smoothing window $W_m(\cdot)$ and the spectral window $\mathcal{H}(\cdot)$ determined by the data taper of the underlying direct spectral estimator. The rolloff in each plot (c) reflects the slower of the rolloffs of $W_m(\cdot)$ and $\mathcal{H}(\cdot)$ for the rectangular data taper, i.e., Fejér's kernel $\mathcal{F}(\cdot)$. In all six cases, Fejér's kernel dominates the rolloff, so all of the (c) plots look similar. The rolloff in each plot (d) reflects the slower of the rolloffs of $W_m(\cdot)$ and $\mathcal{H}(\cdot)$ for a Slepian data taper. Since $\mathcal{H}(\cdot)$ now damps down so fast, the rolloff is dominated by the smoothing window $W_m(\cdot)$ for the Bartlett, Parzen and Papoulis windows, but *not* for the Daniell and Bartlett–Priestley windows. In these two cases, the rolloff is dominated by $\mathcal{H}(\cdot)$ since $W_m(\cdot)$ can be regarded as either a rectangular or a quadratic window with an "infinitely fast" rolloff. The Gaussian case requires special attention. A comparison of Figures 191(g) and 278(b) shows that $\mathcal{H}(\cdot)$ and $W_m(\cdot)$ are similar in that each has a prominent central lobe surrounded by small sidelobes. Their convolution $\mathcal{U}(\cdot)$ has a central lobe that is necessarily wider than that of $W_m(\cdot)$, with an accompanying suppression of its sidelobes.

The important point to keep in mind is that the central lobe and the decay rate of the sidelobes of the spectral window $\mathcal{U}_m(\cdot)$ depend on both the spectral window corresponding to the data taper and the smoothing window. Thus, for example, use of the Parzen smoothing window with a sidelobe decay rate

282 *Lag Window Spectral Estimators*

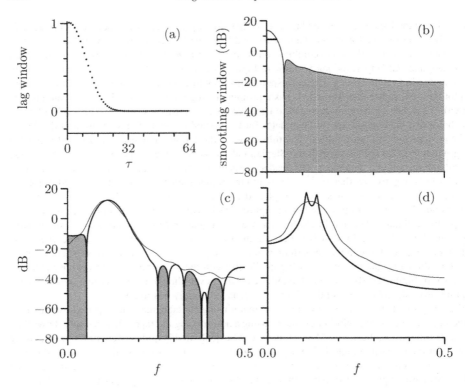

Figure 282 Reshaped lag window formed from a Parzen lag window with $m = 37$ and a Slepian data taper with $NW = 4$ for sample size $N = 64$ (cf. Figure 276). Plot (a) shows the reshaped lag window, while (b) shows the absolute value of the corresponding smoothing window $W_m(\cdot)$, with shaded portions indicating where $W_m(f) < 0$ (the width of the short horizontal line is equal to half of $B_W \doteq 0.057$). The thin curve in plot (c) shows an ordinary SDF estimate based upon a Parzen lag window with $m = 37$ and a Slepian data taper with $NW = 4$ for the first 64 values of the AR(4) time series shown in Figure 34(e); the thick curve shows the absolute value of the corresponding reshaped lag window estimate, with shaded portions indicating where this estimate is negative. The thin curve in (d) shows the theoretical expected value corresponding to the two lag window estimates in (c), while the thick curve is the true AR(4) SDF.

of 12 dB per octave does not imply that the sidelobes of $\mathcal{U}_m(\cdot)$ also have this decay rate *unless* the data taper for the direct spectral estimator is suitably chosen.

[5] We have presented data tapers as a means of reducing the bias in direct spectral estimators, and lag windows as a way of decreasing their variance. Despite differences in their intended purposes, a glance at Figures 185 and 248 shows that tapering and lag windowing have strong similarities in how they are implemented. In both cases, there is a sequence (either a time series $\{X_t : t = 0, 1, \ldots, N-1\}$ or an ACVS estimator $\{\hat{s}_\tau^{(D)} : \tau = -(N-1), -(N-2), \ldots, N-1\}$) whose beginning and end we would like to damp down to zero. We accomplish this by a point-by-point multiplication of the sequence by another sequence (the data taper or the lag window). There are, of course, obvious differences between data tapers and lag windows. First, the indices t for $\{h_t\}$ and τ for $\{w_{m,\tau}\}$ have different ranges. Second, the usual normalization for a data taper is $\sum_t h_t^2 = 1$, whereas the requirement $w_{m,0} = 1$ sets the normalization for a lag window. If we ignore these two differences, certain data tapers and lag windows are remarkably similar. Figure 283 shows curves for a Slepian data taper and a Parzen lag window that give the appearance of being almost identical.

The similarity in appearance between certain data tapers and lag windows suggests that, with adjustments in indexing and normalization, a good data taper can be transformed into a good lag window and *vice versa*. To some extent, this suggestion is correct; however, because data tapers and lag windows are designed with different goals in mind (lessening leakage and smoothing bumpy direct spectral estimators, respectively), there are subtle differences between what constitutes a good data taper and a

7.5 Examples of Lag Windows 283

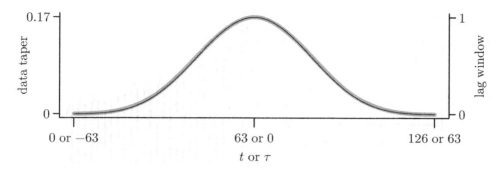

Figure 283 Comparison of Parzen lag window $\{w_{m,\tau}\}$ with $m = 64$ (thick gray curve) and Slepian data taper $\{h_t\}$ for $N = 127$ and $NW = 3.5$ (thin black curve). When regarded as lag windows, both displayed curves have associated smoothing window bandwidth $B_W \doteq 0.029$.

good lag window. For example, a lag window whose smoothing window is nonnegative would seem to be advantageous because this constraint rules out the possibility of a negative SDF estimate (however, as demonstrated by the Daniell, Bartlett–Priestley and Gaussian smoothing windows shown in parts (b) of Figures 272b, 274b and 278, this constraint is sufficient – but not necessary – to guarantee $\hat{S}_m^{(\text{LW})}(f) \geq 0$). Since the smoothing window is the DFT of the lag window, this constraint takes the form of requiring the DFT to be nonnegative. The DFTs for all eight data tapers shown in Figure 190 have sidelobes that are negative. Transforming a data taper into a lag window can thus potentially lead to a negative SDF estimate at some frequencies. On the other hand, we typically do not judge the quality of a data taper based directly on its DFT, but rather on its spectral window (the squared magnitude of its DFT).

Let us first consider converting data tapers into lag windows, starting with the $p \times 100\%$ cosine tapers of Equation (189a). Adaptation of the defining equation yields the lag window

$$w_{m,\tau} = \begin{cases} 1, & 0 \leq |\tau| \leq m - \lfloor pM/2 \rfloor - 1; \\ \frac{1}{2}\left[1 - \cos\left(\frac{\pi(m - |\tau|)}{\lfloor pM/2 \rfloor + 1}\right)\right], & m - \lfloor pM/2 \rfloor \leq |\tau| \leq m - 1; \\ 0, & \tau \geq m, \end{cases} \quad (283)$$

where m can be any positive integer, and $M \stackrel{\text{def}}{=} 2m - 1$. Figure 284 shows (a) the lag window and (b) the smoothing window for the case $p = 1$ and $m = 27$ with $N = 64$, in addition to the spectral windows corresponding to using (c) the rectangular data taper and (d) the $NW = 4$ Slepian data taper. The smoothing window has some negative sidelobes, so Figure 284(b) is a plot of $|W_m(f)|$ in decibels versus f, with shading indicating the sidelobes for which $W_m(f) < 0$. To ascertain the effect of these negative sidelobes on $\hat{S}_m^{(\text{LW})}(\cdot)$, we generated 100,000 independent realizations of length $N = 64$ from the AR(4) process of Equation (35a). Figure 193 indicates the Hanning data taper yields a direct spectral estimator $\hat{S}^{(\text{D})}(\cdot)$ with overall bias properties that are good in comparison to the other seven tapers considered in that figure, so we computed this $\hat{S}^{(\text{D})}(\cdot)$ for each realization and then formed $\hat{S}_m^{(\text{LW})}(\cdot)$ using the 100% cosine lag window shown in Figure 284(a). We found 99,859 of the lag window SDF estimates to be negative at some frequencies (Figure 291 shows an additional case). Thus, in contrast to the negative sidelobes of the Daniell, Bartlett–Priestley and Gaussian smoothing windows, those of the 100% cosine smoothing window really can lead to negative SDF estimates. (Similar results are found for $p \times 100\%$ cosine lag windows with $0 \leq p < 1$.)

The fact that the spectral window for a data taper is necessarily nonnegative suggests another approach for generating a lag window from a data taper, with the advantage that the resulting smoothing window is nonnegative. Recall that, if $\{h_t\} \longleftrightarrow H(\cdot)$, then $\{h \star h_t\} \longleftrightarrow |H(\cdot)|^2 = \mathcal{H}(\cdot)$ (the spectral window). Additionally, if $\{h_t : t = 0, 1, \ldots, N - 1\}$ has the usual normalization $\sum_t h_t^2 = 1$, then its autocorrelation $\{h \star h_\tau : \tau = -(N - 1), -(N - 2), \ldots, N - 1\}$ has the proper length, indexing and normalization (since $h \star h_0 = \sum_t h_t^2$) to be a lag window for a time series of length N. The

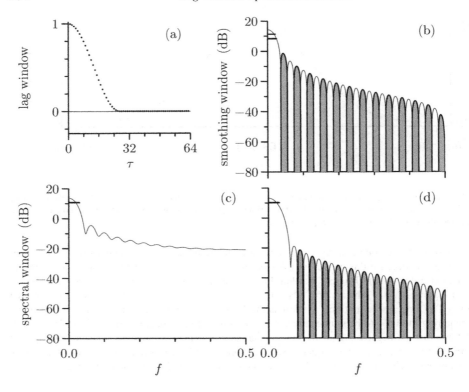

Figure 284 100% cosine windows for $N = 64$ and $m = 27$ ($B_W \doteq 0.05$).

smoothing window for the lag window defined by $w_{m,\tau} = h \star h_\tau$ is the same as the spectral window $\mathcal{H}(\cdot)$ and hence is nonnegative. If we use the $p \times 100\%$ cosine tapers with this scheme, we can create a different lag window for each choice of $p \in [0, 1]$ for a given N; however, the smoothing window bandwidths B_W for these windows are limited in range. For example, when $N = 64$, we find that $0.023 < B_W < 0.033$, so we cannot, for example, generate a lag window whose smoothing window bandwidth is approximately 0.05 (the nominal value used in our examples of other lag windows).

A variation on the autocorrelation scheme that yields a wider range for B_W is to generate a taper of length $m \leq N$ for a given p and then pad this taper with $N - m$ zeros. This padded taper of length N is then autocorrelated to create the desired lag window. With $N = 64$ and $p = 1$, we now find $0.032 < B_W \leq 1$ as we vary m from 1 to 64, while the choice $p = 0.2$ yields $0.024 < B_W \leq 1$ (for comparison, note that the spacing between Fourier frequencies is $1/N \doteq 0.016$). The parameter m is a truncation point since $w_{m,m} = 0$ whereas $w_{m,\tau} \neq 0$ for $|\tau| < m$. Figure 285 shows an example of a lag window created by autocorrelating a 100% cosine data taper of length $m = 41$ (zero padded to length $N = 64$). This lag window compares quite favorably with Parzen and Papoulis lag windows with similar smoothing window bandwidths of $B_W = 0.05$ (see Figures 276 and 279).

The greater flexibility of the Slepian tapers facilitates the creation of suitable lag windows either directly or via an autocorrelation. Figure 191 makes it clear that, by increasing NW, we can increase the width of the central lobe of the Fourier transform for the Slepian tapers well beyond what we can achieve by varying p within the $p \times 100\%$ cosine tapers. We can thus use a Slepian taper of length $2N - 1$ to directly create a lag window appropriate for a time series of length N, with the smoothing window bandwidth B_W being controlled by adjusting $m = (2N - 1)W$ (note that B_W increases as m increases, which is the opposite of what happens for the lag windows listed in Table 279). Figure 286 shows an example of a lag window created in this manner with $m = 10.2$. The Slepian lag window is quite competitive with other lag windows we have examined. Its smoothing window does have some negative sidelobes, but, over these sidelobes, $|W(f)|$ is always less than $-110\,\text{dB}$ and hence are not

7.5 Examples of Lag Windows 285

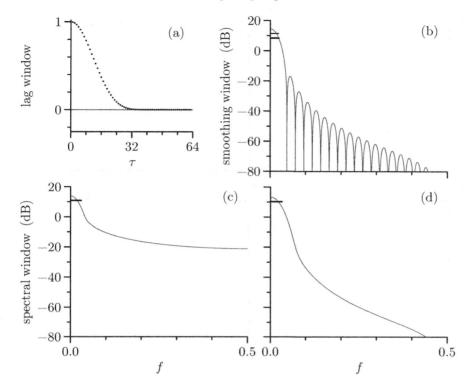

Figure 285 Autocorrelated 100% cosine windows for $N = 64$ and $m = 41$ ($B_W \doteq 0.05$).

visible in Figure 286(b). There is thus the possibility of negative SDF estimates, but a repetition of our experiment involving 100,000 realizations of the AR(4) process of Equation (35a) failed to find *any* cases for which $\hat{S}_m^{(LW)}(f) < 0$. (Exercise [7.13] invites the reader to form a Slepian-based lag window with a nonnegative smoothing window by autocorrelating a Slepian taper of length N.)

Two comments are in order. First, it is desirable in principle for the upper limit of the smoothing window bandwidth B_W to be twice the Nyquist frequency (this width would be appropriate for smoothing a periodogram of a white noise process since it would recover $\hat{S}_m^{(LW)}(f) = \Delta_t \hat{s}_0^{(P)}$ as the SDF estimator). Thus, when $\Delta_t = 1$, we would ideally have B_W range all the way up to unity. By making $m = (2N-1)W$ sufficiently large for a Slepian lag window, we can get B_W close to unity, but we cannot get $B_W = 1$ exactly, as can be done with, for example, the Parzen and Papoulis lag windows by setting $m = 1$. Second, while the scheme of creating a lag window by reindexing and renormalizing a data taper works well with the Slepian taper, it is not so successful with other tapers. Suppose we set m equal to N in Equation (283) to create a lag window based upon a $p \times 100\%$ cosine taper. Our only option for controlling the smoothing window bandwidth is to vary p. In our $N = 64$ example, B_W can only vary from 0.008 to 0.021, which is too small for this scheme to be practical.

Now let's consider converting lag windows into data tapers. A lag window always has an odd length, whereas a data taper needs to be of the same length as the time series under study. Unfortunately time series with even lengths are more often than not the case these days because measurement systems preferentially collect time series with lengths set to a power of two. When N is even, an *ad hoc* solution to this mismatch is to base the data taper on a lag window of odd length $N - 1$ after padding the latter with a single zero. This approach unfortunately has the effect of shortening the time series by one value, but we note that a commonly used alternative to our definition of the Hanning data taper does this also (see Exercise [6.18a]).

Another approach for extracting a data taper is possible for the Bartlett, Parzen and Papoulis lag windows, for which the lag window parameter m represents a truncation point. We can regard these

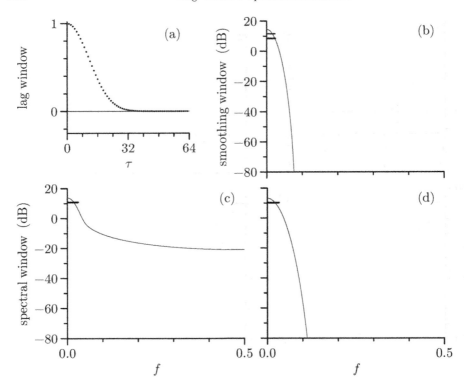

Figure 286 Slepian windows for $N = 64$ and $m = 10.2$ ($B_W \doteq 0.05$).

windows as arising from samples of a continuous function, say $h(\cdot)$, whose support is over the finite interval $(-1, 1)$; i.e., $h(t) = 0$ when $|t| \geq 1$. We can create a data taper appropriate for any sample size by setting

$$\tilde{h}_t = h\left(\frac{2(t+1)}{N+1} - 1\right), \quad t = 0, 1, \ldots, N-1, \quad (286)$$

and then defining $h_t = \tilde{h}_t / \sqrt{\sum_t \tilde{h}_t^2}$ to obtain a properly normalized taper. We can also regard the Daniell, Bartlett–Priestley and Gaussian lag windows as arising from samples of continuous functions, but not ones with finite support; however, here we can use the parameter m to define a suitable $h(\cdot)$ so that we can obtain $\tilde{h}_t$ as per Equation (286).

Figures 287(a) to (e) show examples of data tapers based upon the (a) Daniell, (b) Bartlett–Priestley, (c) Parzen, (d) Gaussian and (e) Papoulis lag windows, with corresponding plots in Figure 288 showing their spectral windows. For comparison, Figures 287(f) and 288(f) show a conventional Slepian taper with $NW = 3.14$. The Parzen and Papoulis tapers have a similar autocorrelation width of $B_{\mathcal{H}} \doteq 0.04$, so we set the smoothing window parameters m in the Daniell, Bartlett–Priestley and Gaussian cases (and NW in the Slepian case) to achieve this same spectral window bandwidth. We can judge the relative merits of these tapers by studying the sidelobes in their spectral windows. Since smaller sidelobes imply better protection against leakage, the Daniell and Bartlett–Priestley tapers are clearly inferior choices. The spectral window for the Slepian taper has sidelobes uniformly smaller than those for the Gaussian taper, but the same cannot be said for the Parzen and Papoulis cases. While the Slepian taper should yield a $\hat{S}^{(D)}(\cdot)$ with less bias because the sidelobes of its spectral window closest to the central lobe are quite a bit smaller than in the Parzen and Papoulis cases, conceivably there are SDFs for which use of the latter two tapers would result in less bias because their more distant sidelobes are smaller.

In conclusion, with sufficient care, it is possible to convert data tapers into decent lag windows, and lag windows into acceptable data tapers. C&E [1] in the next section discusses the relative merits

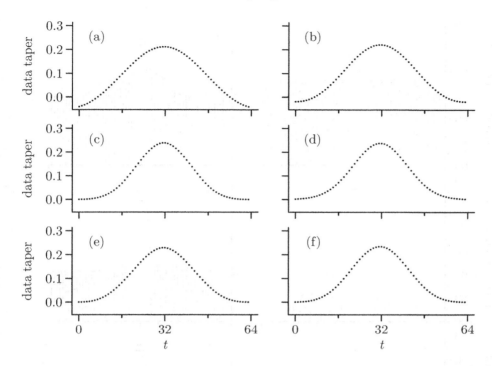

Figure 287 Six data tapers for use with a time series of length $N = 64$ (cf. Figure 190). The first five tapers are extracted from the (a) Daniell, (b) Bartlett–Priestley, (c) Parzen, (d) Gaussian and (e) Papoulis lag windows, while the final taper (f) is a zeroth-order Slepian data taper with $NW = 3.14$. The lag window parameter m is set to, respectively, 0.754, 0.528 and 0.435 for the Daniell, Bartlett–Priestley and Gaussian windows.

of the specific lag windows we generated from data tapers, pointing out that the Slepian lag window and the autocorrelated 100% cosine lag window are quite competitive with conventional lag windows. The Parzen and Papoulis data tapers we generated from lag windows might also have some appeal over conventional data tapers.

7.6 Choice of Lag Window

In the previous section we considered the Bartlett, Daniell, Bartlett–Priestley, Parzen, Gaussian and Papoulis lag window spectral estimators, each of which has a window parameter m controlling its lag window and smoothing window. For a particular SDF estimation problem, which estimator should we use, and how should we set m? We address the first of these questions in this section, and the second in the next section.

Several different criteria have been proposed in the literature for evaluating different lag window spectral estimators. One of the more useful is based on the concept of *smoothing window leakage*. Since

$$\hat{S}_m^{(\mathrm{LW})}(f) = \int_{-f_{\mathcal{N}}}^{f_{\mathcal{N}}} W_m(f - \phi)\hat{S}^{(\mathrm{D})}(\phi)\,d\phi$$

(Equation (248a)), a lag window estimator of $S(f)$ is the result of integrating the product of the direct spectral estimator $\hat{S}^{(\mathrm{D})}(\cdot)$ and the smoothing window $W_m(\cdot)$ after the latter has been shifted so that its central lobe is centered at frequency f. Under our operational assumption that $S(\cdot)$ is slowly varying, we want $\hat{S}_m^{(\mathrm{LW})}(f)$ to be influenced mainly by values in $\hat{S}^{(\mathrm{D})}(\cdot)$ with frequencies "close" to f. We define "close" here to mean those frequencies lying within

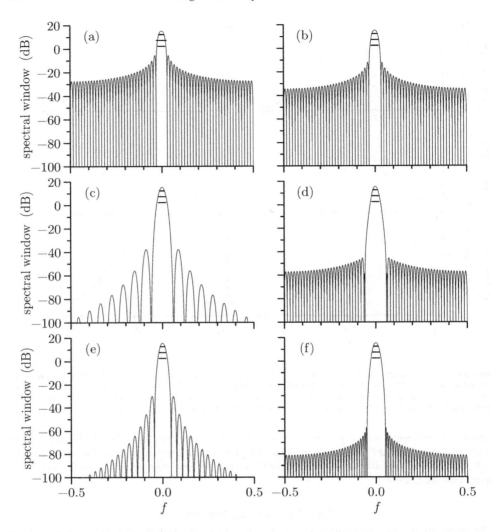

Figure 288 Spectral windows for the six data tapers shown on Figure 287 (cf. Figure 191). The three horizontal solid lines below the central lobes are bandwidth measures based upon, from the top on down, the half-power, variance and autocorrelation widths. The autocorrelation width is $B_{\mathcal{H}} \doteq 0.04$ for all six spectral windows.

the central lobe of the shifted smoothing window $W_m(f - \cdot)$. If this smoothing window has significant sidelobes and if the dynamic range of $\hat{S}^{(D)}(\cdot)$ is large, $\hat{S}_m^{(LW)}(f)$ can be unduly influenced by values in $\hat{S}^{(D)}(\cdot)$ lying under one or more of the sidelobes of the smoothing window. If this in fact happens, we say that the estimate $\hat{S}_m^{(LW)}(f)$ suffers from smoothing window leakage.

One criterion for window selection is thus to insist that the smoothing window leakage be small. If we have two different lag window estimators whose smoothing windows have the same bandwidth B_W (a measure of the width of the central lobe of $W_m(\cdot)$), this criterion would dictate picking the window whose sidelobes are in some sense smaller. Considering the six smoothing windows discussed in the previous section, the first sidelobe for the Bartlett window is 13 dB down from its central lobe, and the decay rate for its sidelobes is 6 dB per octave; the first sidelobes of the Parzen and Papoulis windows are 28 and 23 dB down, and both have decay rates of 12 dB per octave; for the Gaussian window, the comparable measures are

7.6 Choice of Lag Window

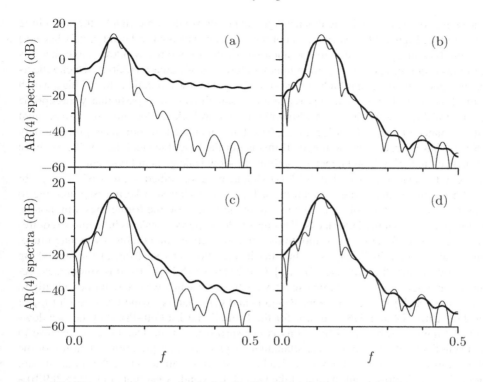

Figure 289 Direct spectral estimate for the first 64 values of the AR(4) time series of Figure 34(e) (thin curves in each plot), along with lag window estimates based upon (a) Bartlett, (b) Daniell, (c) Parzen and (d) Gaussian lag windows (thick curves). The direct spectral estimate is based upon the Hanning data taper (Equation (189a) with $p = 1$ and $N = 64$). The Bartlett lag window is defined by Equation (269a) with $m = 30$ (Figure 270(a) shows this window); the Daniell, by Equation (273a) with $m = 20$ and $N = 64$ (Figure 272b(a)); the Parzen, by Equation (275a) with $m = 37$ (Figure 276(a)); and the Gaussian, by Equation (276) with $m = 16$ (Figure 278(a)). The lag window parameter m is set in each case so that the smoothing window bandwidth is the same ($B_W \doteq 0.05$).

79 dB down and 2 dB per octave; and arguably the Daniell and Bartlett–Priestley smoothing windows have no sidelobes, so our first choice would be one of these. Based upon how far down their first sidelobes are from their central lobes, our choices amongst the four remaining windows would be, from first to last, the Gaussian, Parzen, Papoulis and Bartlett smoothing windows.

Figure 289 shows two examples of lag window estimates where smoothing window leakage is evident, and two for which it is not. The thin curve in each plot is a direct spectral estimate $\hat{S}^{(D)}(\cdot)$ based upon the Hanning data taper for the first 64 values of the AR(4) time series displayed in Figure 34(e). The thick curves show corresponding lag window estimates based upon (a) Bartlett, (b) Daniell, (c) Parzen and (d) Gaussian lag windows. The windows themselves are exactly the ones shown in, respectively, Figures 270(a), 272b(a), 276(a) and 278(a), all of which have a smoothing window bandwidth of $B_W \doteq 0.05$. The lag window estimates $\hat{S}_m^{(LW)}(\cdot)$ should ideally be a smoothed version of $\hat{S}^{(D)}(\cdot)$, but $\hat{S}_m^{(LW)}(f)$ is systematically above $\hat{S}^{(D)}(f)$ at high frequencies for the Bartlett and Parzen lag windows. This deviation is due to smoothing window leakage and is more prominent in the Bartlett estimate than in the Parzen estimate, a pattern consistent with the sidelobes of their smoothing windows. By contrast, the Daniell and Gaussian estimates track $\hat{S}^{(D)}(\cdot)$ nicely at high frequencies, with the Gaussian estimate being a little smoother in appearance. (Figure 319(b) later in this chapter has an additional illustration of smoothing window leakage.)

There are valid reasons for *not* making smoothing window leakage our only criterion.

First, the degree of distortion due to this leakage can be controlled somewhat by the smoothing parameter m (see Figure 292). Second, smoothing window leakage is relevant largely because it affects the bias in $\hat{S}_m^{(\text{LW})}(f)$ due to the sidelobes of the smoothing window. Note that this source of bias is *not* the same as the smoothing window bias of Equation (256d), which reflects the bias introduced by the central lobe of the smoothing window (so-called "local" bias, which results in a loss of resolution). The very rectangular Daniell window is particularly susceptible to local bias (Walden and White, 1984). Smoothing window leakage does not take into account either the variance of $\hat{S}_m^{(\text{LW})}(f)$ or local bias. Third, this leakage is a significant problem only when $\hat{S}^{(\text{D})}(\cdot)$ has a large dynamic range. If this is not the case, the sidelobes of $W_m(\cdot)$ have little influence on $\hat{S}_m^{(\text{LW})}(\cdot)$, and hence smoothing window leakage is not relevant.

A second consideration is that $\hat{S}_m^{(\text{LW})}(\cdot)$ should be a smoothed version of $\hat{S}^{(\text{D})}(\cdot)$. In Section 3.6 we noted that, when smoothing a function, a Gaussian kernel is preferable to a rectangular one from the point of view of smoothness because the former has a monotone attenuation property that the latter lacks. This desirable property translates here into a requirement that the transfer function for a smoothing window decrease monotonically in magnitude. Now the "transfer function" here is in fact the inverse Fourier transform of the smoothing window, which is just the sequence $\{w_{m,\tau}\}$ (see Exercise [5.17]). By this smoothness requirement, we would prefer smoothing windows whose lag windows decay monotonically to 0. This is true for the Bartlett, Parzen, Papoulis and Gaussian lag windows (see plot (a) of Figures 270, 276, 279 and 278), but *not* for the Daniell and Bartlett–Priestley lag windows (plot (a) of Figures 272b and 274b) – this is not surprising in the Daniell case because in fact the Daniell smoothing window is rectangular. A practical interpretation is that use of the Daniell or Bartlett–Priestley windows can lead to undesirable ripples in $\hat{S}_m^{(\text{LW})}(\cdot)$ that are not present in $\hat{S}^{(\text{D})}(\cdot)$. A close examination of the Daniell lag window estimate in Figure 289(b) – aided by a comparison with the ripple-free Gaussian estimate below it – reveals a hint of these ripples (the forthcoming Figure 319 has a better example).

If smoothing window leakage does not come into play for the direct spectral estimate to be smoothed, our preference is for either the Parzen or Papoulis smoothing window from amongst the five standard windows we have discussed; if leakage proves significant for these windows, our choice from amongst these five is then either the Daniell or the Bartlett–Priestley smoothing window, but the Gaussian window can outperform both of these windows by providing a ripple-free estimate with comparable protection against leakage. However, except in cases where the dynamic range of $\hat{S}^{(\text{D})}(\cdot)$ is large enough to make smoothing window leakage an issue, the following quote from Jenkins and Watts (1968, p. 273) is germane: "... the important question in empirical spectral analysis is the choice of [smoothing window] bandwidth and *not* the choice of [smoothing] window."

Comments and Extensions to Section 7.6

[1] In addition to the five standard lag windows and the Gaussian window, we have also considered three additional lag windows in previous C&Es, namely, the 100% cosine, autocorrelated 100% cosine and Slepian lag windows (see Figures 284, 285 and 286). When applied to the Hanning-based direct spectral estimate shown in Figure 289, autocorrelated 100% cosine and Slepian lag window estimates with parameters chosen such that $B_W \doteq 0.05$ are visually identical to the Gaussian estimate shown in Figure 289(d). Figure 291 shows the corresponding 100% cosine lag window estimate, which suffers from smoothing window leakage and also is negative over certain frequency intervals, rendering this lag window the least attractive of the three additional windows. (Exercise [7.14] invites the reader to compute the expected value of the five lag window estimators shown in Figures 289 and 291, with the take-home message being that the smoothing window leakage evident in these figures is quite likely to be repeated for other AR(4) realizations.)

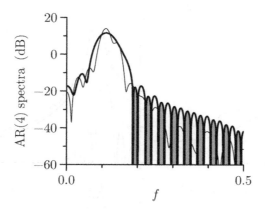

Figure 291 As in Figure 289, but now showing a lag window estimate based upon the 100% cosine lag window as defined by Equation (283) with $p = 1$, $m = 27$ and $M = 53$, yielding a smoothing window bandwidth of $B_W \doteq 0.05$ (Figure 284(a) shows this window). This estimate is negative at some frequency intervals, over which $10 \log_{10}(|\hat{S}_m^{(\mathrm{LW})}(f)|)$ is shown with shading underneath.

7.7 Choice of Lag Window Parameter

In the previous section, we argued that any one of several specific lag windows would be a reasonable choice. Unfortunately, the choice of the window parameter m is *not* so easy: a sensible choice of m depends in part upon the shape of the unknown SDF we want to estimate.

To illustrate the importance in selecting m properly, Figure 292 plots

$$\int_{-f_{\mathcal{N}}}^{f_{\mathcal{N}}} W_m(f - f')S(f')\,\mathrm{d}f'$$

versus f for the case where $S(\cdot)$ is the SDF of the AR(4) process $\{X_t\}$ of Equation (35a); $W_m(\cdot)$ is Parzen's smoothing window; and $m = 32, 64, 128$ and 256, with corresponding smoothing window bandwidths $B_W \doteq 0.058, 0.029, 0.014$ and 0.007. The true SDF for this AR(4) process (indicated by the thick curves in Figure 292) has two sharp peaks close to each other. For both $m = 32$ and 64, the amount of smoothing done by Parzen's lag window spectral estimator is such that the expected value of the estimator (given by the thin curves) does not accurately represent the true SDF in the region of the two peaks. The amount of smoothing is related to the width of the central lobe of Parzen's smoothing window (shown – centered about $f = 1/10$ – by the thin curves in the lower left-hand part of each plot). As we increase m to 128 or 256, the width of this central lobe decreases enough so that the expected value of the estimator more accurately represents the true SDF – in fact, the agreement is so good in the $m = 256$ case that the thick and thin curves are indistinguishable in the plot. (Note that this example considers the bias due *only* to the smoothing window. As we have seen in Figure 193, the bias in the direct spectral estimator itself can be a significant factor for small N even if we are careful to use a good data taper.)

To make a choice of m that yields an approximately unbiased estimator, we must have some idea of the "size" of the important features in $S(\cdot)$ – the very SDF we are trying to estimate! In many situations this is simply not known a priori. For example, spectral analysis is often used as an exploratory data analysis tool, in which case little is known about $S(\cdot)$ in advance. There are cases, however, where it is possible to make an intelligent guess about the typical shape of $S(\cdot)$ either from physical arguments or from analyses of previous data of a similar nature. As an example, let us assume that the typical SDF to be expected is the same as the SDF for the AR(2) process of Equation (34). This SDF is shown in all four plots of

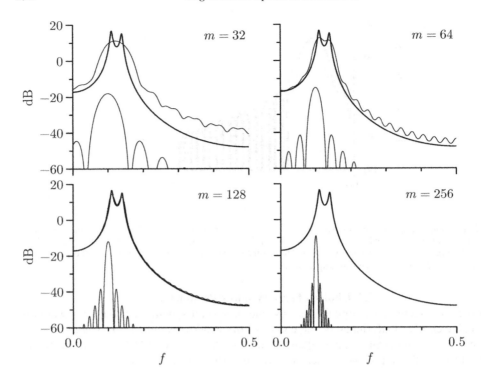

Figure 292 Effect of window parameter m on the expected value of an unbiased direct spectral estimator that has been smoothed using Parzen's lag window. The thick curve in each plot is the SDF for the AR(4) process of Equation (35a); the thin curves that follow the thick curves to varying degrees show the component of $E\{\hat{S}_m^{(\text{LW})}(\cdot)\}$ solely due to Parzen's smoothing window for m = 32, 64, 128 and 256; and the thin curves in the lower left-hand corners show the shape of the smoothing window itself.

the forthcoming Figure 294 as a thin dark curve. We can define the *spectral bandwidth* B_S as roughly the width of the *smallest* feature of interest in the projected SDF $S(\cdot)$. Here we take this feature to be the broad peak, the maximum for which is at $f_0 \doteq 0.155$. We can measure the peak's width in terms of half-power points; i.e., those points $f_1 < f_0$ and $f_2 > f_0$ such that

$$S(f_1) = S(f_2) = \frac{S(f_0)}{2} \quad \text{and hence} \quad B_S = f_2 - f_1.$$

Here $S(f_0) \doteq 7.5$ dB, and, since the frequencies $f_1 \doteq 0.085$ and $f_2 \doteq 0.208$ yield the required conditions, we have $B_S \doteq 0.123$. As a second example, suppose that the expected SDF is the AR(4) SDF shown in Figure 292, which has two prominent peaks of comparable width. If we take the one centered about $f_0 \doteq 0.110$ to be the smallest feature of interest, we find $f_1 \doteq 0.107$ and $f_2 \doteq 0.114$, yielding $B_S \doteq 0.007$. For either example, we can then use either Table 279 or Equation (251e) to find a value of m such that the smoothing window bandwidth B_W is acceptable (note that the bandwidths B_S and B_W have the same units as f, namely, cycles per unit time). Priestley (1981) recommends setting m such that $B_W = B_S/2$ so that the smoothing window does not smooth out any important features in the SDF. Assuming use of the Parzen window for our two examples, Table 279 says that

$$m = \frac{1.85}{B_W} = \frac{3.6}{B_S} \doteq 30 \quad \text{for the AR(2) case, and} \quad m = 565 \quad \text{for the AR(4) case.}$$

This advice, however, must be used with caution, since the amount of data N that is available can be such that the variance of $\hat{S}_m^{(\text{LW})}(\cdot)$ is unacceptably large if we were to use the value of m chosen by this bandwidth matching criterion. (Priestley has an extensive discussion on selecting m such that a trade-off is made between the bias – due to smoothing window leakage – and the variance of a lag window spectral estimator.)

In the more common case where there is little prior knowledge of the true SDF, we can entertain choosing the window parameter m using the objective criteria discussed in Sections 7.8, 7.9 and 7.10. The following three subjective methods also merit discussion (Priestley, 1981, pp. 539–42).

The first method is known as *window closing*. The idea is to compute a sequence of different SDF estimates for the same time series using different smoothing window bandwidths that range from large to small. For large values of B_W, the estimates will look smooth, but, as B_W decreases, the estimates will progressively exhibit more detail until a point is reached where the estimates are more "erratic" in form. Based upon a subjective evaluation of these estimates, we can hopefully pick a value of m that is appropriate in the sense that the resulting estimate is neither too smooth nor too erratic. Since $\hat{S}_m^{(\text{LW})}(\cdot)$ is supposed to be a smoothed version of $\hat{S}^{(\text{D})}(\cdot)$, our reference point in these comparisons should be how well the former captures – and points out – the important features of the latter. These visual comparisons can reject some estimates as obviously being "too smooth" or "too erratic."

As an example, let us consider four Parzen lag window estimates for the AR(2) time series shown in Figure 34(a). There is no need for tapering since the SDF is simple enough that the periodogram is essentially bias free when $N = 1024$ (cf. Figure 177). Figure 294 shows $\hat{S}_m^{(\text{LW})}(\cdot)$ with $m = 4, 16, 29$ and 64 (thick dark curves) along with the periodogram (rough-looking light curves) and the true SDF (thin dark curves). The width of the crisscross in each plot depicts the bandwidth measure $B_{\mathcal{U}}$, while its height shows the length of a 95% CI for $10\log_{10}(S(f))$ based upon $10\log_{10}(\hat{S}_m^{(\text{LW})}(f))$. A comparison of the four $\hat{S}_m^{(\text{LW})}(\cdot)$ with the true SDF shows the importance of properly choosing the smoothing parameter m: whereas the $m = 4$ estimate is oversmoothed and the $m = 64$ estimate is undersmoothed, the $m = 16$ and $m = 29$ estimates agree with the true SDF much better. Note that a 95% CI for $10\log_{10}(S(f))$ based upon the oversmoothed $m = 4$ estimate would in fact *fail* to include the true value of the SDF over the majority of the frequencies between 0 and $1/2$, whereas the opposite is true for CIs based upon the other three estimates. We can quantify how well a particular $\hat{S}_m^{(\text{LW})}(\cdot)$ agrees with the true SDF $S(\cdot)$ by computing the mean square error

$$\text{MSE}(m) = \frac{1}{N+1}\sum_{k=0}^{N}\left(\hat{S}_m^{(\text{LW})}(\tilde{f}_k) - S(\tilde{f}_k)\right)^2, \quad \text{where} \quad \tilde{f}_k = k/(2N). \tag{293}$$

The above gives 1.309, 0.105, 0.049 and 0.110 for $m = 4, 16, 29$ and 64. In fact, the choice with the smallest MSE(m) amongst all possible m for this particular AR(2) time series is the $m = 29$ lag window estimate, which fortuitously is quite close to the setting $m = 30$ dictated by the $B_W = B_S/2$ criterion. Here the subjective criterion of window closing would likely rule out the $m = 4$ choice because $\hat{S}_m^{(\text{LW})}(f)$ is systematically above $\hat{S}^{(\text{D})}(f)$ at high frequencies; however, it would be difficult to choose visually between the $m = 16$ and $m = 29$ estimates even though MSE(16) is approximately twice as large as MSE(29). We might also rule out the $m = 64$ choice as being too erratic even though the visually more pleasing $m = 16$ choice has approximately the same MSE(m). On the other hand, the maximum difference between any two of the $m = 16, 29$ and 64 estimates is less than 2.2 dB in magnitude (a factor of 1.7), so these estimates all agree to well within the same order of magnitude.

As a second example, let us consider four Parzen lag window estimates for the AR(4) time series of length $N = 1024$ shown in Figure 34(e). Figures 182(e) and 187(e) show the periodogram and a direct spectral estimate based upon the Hanning data taper for this series. It is evident that the periodogram suffers from leakage, so we form lag window estimates

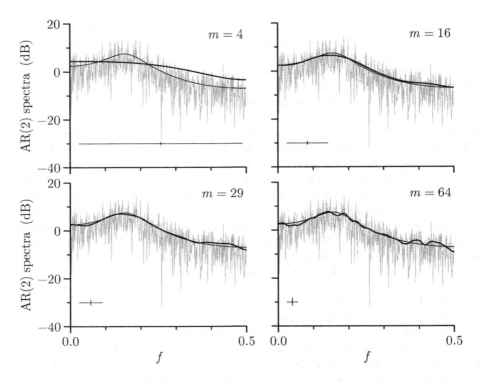

Figure 294 Lag window spectral estimates $\hat{S}_m^{(\text{LW})}(\cdot)$ for AR(2) series of Figure 34(a) using a Parzen smoothing window with m = 4, 16, 29 and 64 (thick dark curves). The thin dark curve on each plot is the true AR(2) SDF, while the rough-looking light curve is the periodogram at the Fourier frequencies. The width of the crisscross emanating from the lower left-hand corner of each plot gives the bandwidth measure $B_{\mathcal{U}}$ as per Equation (256a), while its height gives the length of a 95% CI for $10\log_{10}(S(f))$ as per Equation (266a).

using the Hanning-based $\hat{S}^{(\text{D})}(\cdot)$. Four such estimates with m set to 64, 128, 179 and 565 are shown in Figure 295, along with the underlying direct spectral estimate and the true SDF. Here we can rule out the $m = 64$ estimate because it oversmooths the twin peaks and is systematically above the direct spectral estimate at high frequencies, which is indicative of smoothing window leakage (the ripples in $\hat{S}_m^{(\text{LW})}(\cdot)$ at those frequencies are another indication – cf. the $m = 64$ case in Figure 292). The $m = 565$ estimate is much more erratic looking than the remaining two estimates, of which we might prefer the $m = 128$ estimate over $m = 179$ because the latter is a bit bumpier looking. Calculations indicate that

$$\text{MSE}(m) = \begin{cases} 10.241, & m = 64; \\ 4.886, & m = 128; \\ 4.228, & m = 179; \\ 10.329, & m = 565. \end{cases}$$

Visibly $m = 128$ might seem to be a better choice than $m = 179$, but it actually isn't. The subjectively chosen $m = 128$ estimate is, however, not too far off the mark since MSE(128) is only about 15% larger than MSE(179), which is the minimizer of MSE(m) over all possible m for this particular AR(4) time series. (It is of interest to note that the $m = 64$ and $m = 565$ estimates have approximately the same MSE, with the former being bias dominated, and the latter, variance dominated. The setting $m = 565$ comes from the $B_W = B_S/2$ criterion, which is geared toward eliminating bias in $\hat{S}_m^{(\text{LW})}(\cdot)$ and does not explicitly take into account its variance.)

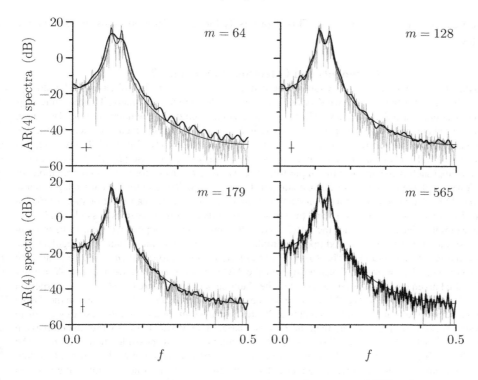

Figure 295 As in Figure 294, but now showing lag window spectral estimates $\hat{S}_m^{(\mathrm{LW})}(\cdot)$ for AR(4) series of Figure 34(e) using a Parzen smoothing window with m = 64, 128, 179 and 565 (thick dark curves). The thin dark curve in each plot is now the true AR(4) SDF, while the rough-looking light curve is the Hanning-based direct spectral estimate.

The second subjective method is based upon the sample ACVS. If our spectral estimator is of the truncation type (such as the Bartlett, Parzen or Papoulis estimators), we can argue that we should choose m so that $s_\tau \approx 0$ for all $|\tau| \geq m$. Thus we could plot $\hat{s}_\tau^{(\mathrm{P})}$ versus τ and pick m as that point starting at which the sample ACVS shows only small fluctuations around 0. Although crude, this method has evidently been quite effective in a number of problems and is arguably preferable to choosing a completely arbitrary value for m. There are two dangers with this approach. First, sampling fluctuations can cause the sample ACVS to deviate substantially from the true ACVS. Second, if $S(\cdot)$ contains a large peak with a wide bandwidth and a small peak with a narrow bandwidth, the effect of the small peak will not be apparent at all in the sample ACVS (see Figure 125 supports this assertion). The value of m selected by this method will reflect only the bandwidth of the large peak. We can thus miss some important features in the SDF by using this method.

As an example, let us reconsider Figure 248(a), which shows $\{\hat{s}_\tau^{(\mathrm{P})}\}$ for the AR(2) time series behind the four lag window estimates shown in Figure 248. Here $\hat{s}_\tau^{(\mathrm{P})}$ over the range $\tau = 5, \ldots, 128$ fluctuates about zero with no particularly dominant value. We might be tempted to pick $m = 5$ by the subjective sample ACVS method. An $m = 5$ Parzen lag window estimate is only a marginal improvement over the $m = 4$ estimate shown in Figure 248. This subjective method does not serve us as well as window closing for this particular example.

The final subjective method is simply to let m be some fixed proportion of N, say 20% or 30%, or to let $m = \sqrt{N}$. Such recommendations are fundamentally unsound since they do not take into account the underlying process that generated the time series. Although we might have no prior knowledge of the spectral bandwidth, it is desirable to choose m based upon some properties of the process. This is precisely the aim of the previous two subjective techniques.

Comments and Extensions to Section 7.7

[1] While B_W is a measure of the width of the central lobe of $W_m(\cdot)$, the effective bandwidth $B_\mathcal{U}$ measures the same for the spectral window associated with a lag window estimator. Hence, while B_W takes into account just the lag window, $B_\mathcal{U}$ takes into account both the lag window and the data taper. In situations where we know the spectral bandwidth B_S, a possible refinement to selecting the lag window parameter m via the relationship $B_W = B_S/2$ is to use $B_\mathcal{U} = B_S/2$ instead. If anything, this refinement will call for *less* smoothing since $B_\mathcal{U} \geq B_W$ always (see Exercise [257]). A disadvantage of this new approach is the absence of simple formulae that link $B_\mathcal{U}$ to m for various combinations of lag windows and data tapers, i.e., formulae analogous to $B_W = 1.85/(m \, \Delta_t)$ for the Parzen lag window (see Table 279). Nonetheless, for a given data taper and lag window, we can use Equation (256a) to compute $B_\mathcal{U}$ for various choices of m and thus to set m such that $B_\mathcal{U}$ is as close as possible to $B_S/2$.

As examples, let us reconsider the AR(2) and AR(4) processes. Suppose we want to form a lag window estimate for the former based upon a sample of size $N = 1024$, the default (rectangular) data taper and the Parzen lag window. The $B_\mathcal{U} = B_S/2$ criterion leads us to select $m = 30$, i.e., the same value we got using $B_W = B_S/2$. Considering now the AR(4) process, suppose we again set $N = 1024$ and select the Parzen lag window, but now we use the Hanning data taper. Here the $B_\mathcal{U} = B_S/2$ criterion sets $m = 714$, whereas $B_W = B_S/2$ leads to $m = 565$. When the choice $m = 714$ is used on the AR(4) time series behind the four Parzen lag window estimates displayed in Figure 295, the resulting estimate (not shown) is even bumpier looking than the $m = 565$ case. Here MSE(714) $\doteq$ 13.156, in comparison to MSE(565) $\doteq$ 10.329 and MSE(179) $\doteq$ 4.228 (the minimum over all possible choices of m). The criteria $B_W = B_S/2$ and $B_\mathcal{U} = B_S/2$ are both geared toward eliminating bias, but do not take into consideration variance. Hence, if we use either criteria for setting m, there is no guarantee of getting an estimate with a sensible trade-off between bias and variance.

[2] The MSE of Equation (293) is only one of many possibilities for quantifying how well a particular SDF estimate matches up with the true SDF. Two variations on the MSE appearing in the literature are the normalized mean square error

$$\text{NMSE}(m) = \frac{1}{N+1} \sum_{k=0}^{N} \left(\frac{\hat{S}_m^{(\text{LW})}(\tilde{f}_k) - S(\tilde{f}_k)}{S(\tilde{f}_k)} \right)^2 \tag{296a}$$

and the mean square log error

$$\text{MSLE}(m) = \frac{1}{N+1} \sum_{k=0}^{N} \left(\log\left(\hat{S}_m^{(\text{LW})}(\tilde{f}_k)\right) - \log\left(S(\tilde{f}_k)\right) \right)^2 \tag{296b}$$

(see, e.g., Lee, 1997, who points out that both are problematic when $S(\tilde{f}_k) = 0$). If we momentarily assume that $\hat{S}_m^{(\text{LW})}(\tilde{f}_k) \stackrel{d}{=} S(\tilde{f}_k)\chi_\nu^2/\nu$ for all k (a dicey assumption for $\tilde{f}_k$ close to 0 or $f_\mathcal{N}$), then

$$E\{\text{NMSE}(m)\} = \frac{2}{\nu} \quad \text{and} \quad E\{\text{MSLE}(m)\} = \left[\psi\left(\frac{\nu}{2}\right) - \log\left(\frac{\nu}{2}\right)\right]^2 + \psi'\left(\frac{\nu}{2}\right), \tag{296c}$$

where $\psi(\cdot)$ and $\psi'(\cdot)$ are the digamma and trigamma functions. Neither of these expected values involves the true SDF, whereas

$$E\{\text{MSE}(m)\} = \frac{2}{(N+1)\nu} \sum_{k=0}^{N} S^2(\tilde{f}_k) \tag{296d}$$

does (Exercise [7.15] is to verify Equations (296c) and (296d)). While NMSE and MSLE do not depend preferentially on large values of the SDF, MSE does. Returning to the AR(2) example, for which $m = 29$

minimized MSE(m), we find that NMSE(m) and MSLE(m) are both minimized when $m = 22$ (the estimate $\hat{S}_{22}^{(\mathrm{LW})}(\cdot)$ is shown in the right-hand plot of the forthcoming Figure 310 and is a compromise between $\hat{S}_{16}^{(\mathrm{LW})}(\cdot)$ and $\hat{S}_{29}^{(\mathrm{LW})}(\cdot)$ shown in Figure 294; its maximum deviations from these are, respectively, 0.6 dB and 0.5 dB). For the AR(4) example, the minimum values are MSE(179), NMSE(154) and MSLE(130) (the estimate $\hat{S}_{154}^{(\mathrm{LW})}(\cdot)$ is in appearance halfway between $\hat{S}_{128}^{(\mathrm{LW})}(\cdot)$ and $\hat{S}_{179}^{(\mathrm{LW})}(\cdot)$ in Figure 295, with maximum deviations less than 1.7 dB and 1.2 dB; the $\hat{S}_{130}^{(\mathrm{LW})}(\cdot)$ estimate is always within 0.4 dB of $\hat{S}_{128}^{(\mathrm{LW})}(\cdot)$, which visual inspection of Figure 295 arguably suggests to be the most appropriate amongst the four estimates displayed there).

The *Kullback–Leibler (KL) discrepancy* between two PDFs motivates another measure of interest (see, e.g., Ombao et al., 2001). If $p(\cdot)$ and $q(\cdot)$ are the PDFs for two nonnegative RVs, the KL discrepancy takes the form

$$d(p(\cdot), q(\cdot)) \stackrel{\mathrm{def}}{=} \int_0^\infty p(x) \log\left(\frac{p(x)}{q(x)}\right) \, \mathrm{d}x \tag{297a}$$

(Kullback and Leibler, 1951). The above is minimized when $p(\cdot)$ and $q(\cdot)$ are the same. We note that $d(p(\cdot), q(\cdot)) \neq d(q(\cdot), p(\cdot))$ in general, so the discrepancy depends on the ordering. The PDF corresponding to the assumption $\hat{S}_m^{(\mathrm{LW})}(\tilde{f}_k) \stackrel{\mathrm{d}}{=} S(f)\chi_\nu^2/\nu$ is

$$f(x) = \left(\frac{\nu}{2S(\tilde{f}_k)}\right)^{\nu/2} \frac{x^{\nu/2-1} e^{-\nu x/(2S(\tilde{f}_k))}}{\Gamma(\nu/2)}, \quad 0 \leq x < \infty, \tag{297b}$$

which is the same as the PDF for a gamma RV with shape parameter $\nu/2$ and rate parameter $\nu/(2S(\tilde{f}_k))$. Setting $p(\cdot)$ and $q(\cdot)$ in Equation (297a) to two gamma PDFs – with ν taken to be the same but allowing the SDFs to be distinct – inspires the following measure:

$$\mathrm{KL}(m) = \frac{1}{N+1} \sum_{k=0}^N \left[\frac{S(\tilde{f}_k)}{\hat{S}_m^{(\mathrm{LW})}(\tilde{f}_k)} - \log\left(\frac{S(\tilde{f}_k)}{\hat{S}_m^{(\mathrm{LW})}(\tilde{f}_k)}\right) - 1\right] \tag{297c}$$

(see Exercise [7.16a] for details). The fact that $x - \log(x) - 1 \geq 0$ for all $x > 0$ tells us that the above must be nonnegative (here we ignore the special cases $S(\tilde{f}_k) = 0$ and $\hat{S}_m^{(\mathrm{LW})}(\tilde{f}_k) = 0$); in addition, it attains its minimum value of zero when each $\hat{S}_m^{(\mathrm{LW})}(\tilde{f}_k)$ is identical to $S(f_k)$. Exercise [7.16b] is to show that

$$E\{\mathrm{KL}(m)\} = \frac{\nu}{\nu - 2} - \log\left(\frac{\nu}{2}\right) + \psi\left(\frac{\nu}{2}\right) - 1 \tag{297d}$$

under the previous assumption $\hat{S}_m^{(\mathrm{LW})}(\tilde{f}_k) \stackrel{\mathrm{d}}{=} S(\tilde{f}_k)\chi_\nu^2/\nu$, but now with the additional constraint that $\nu > 2$. The above is similar to $E\{\mathrm{NMSE}(m)\}$ and $E\{\mathrm{MSLE}(m)\}$ in its lack of dependence on the true SDF. For the AR(2) example, KL(m) is minimized when $m = 22$, in agreement with the minimizers for both NMSE(m) and MSLE(m). For the AR(4) example, the minimizer is $m = 129$, which is in close agreement with $m = 130$, the minimizer for MSLE(m) (the $\hat{S}_{129}^{(\mathrm{LW})}(\cdot)$ estimate is always within 0.2 dB of $\hat{S}_{128}^{(\mathrm{LW})}(\cdot)$ displayed in Figure 295).

7.8 Estimation of Spectral Bandwidth

We have looked at the bandwidth B_W of a smoothing window and how to compute it using Equation (251e). To compute the spectral bandwidth B_S, we must have good knowledge of the SDF we are trying to estimate. Can we estimate the spectral bandwidth directly from a time series? If this were possible, it would clearly be beneficial since we could use the estimated spectral bandwidth to guide the setting of B_W. The following argument shows that, if we assume the time series is drawn from a Gaussian stationary process with a dominantly unimodal SDF, it is possible to produce a satisfactory estimator for the spectral bandwidth (for details, see Walden and White, 1990).

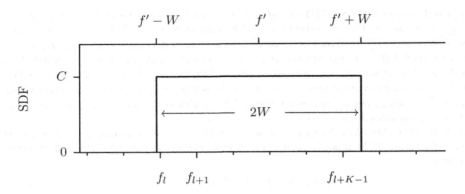

Figure 298 SDF for ideal band-pass process with bandwidth of $2W$ centered about frequency f'. The tick marks on the bottom horizontal axis show the locations of the Fourier frequencies, three of which are labelled.

The standard estimator of the variance s_0 of a time series drawn from a stationary process with unknown mean is

$$\hat{s}_0^{(P)} = \frac{1}{N} \sum_{t=0}^{N-1} \left(X_t - \overline{X}\right)^2.$$

Assume for convenience that the process mean is known to be 0 so that we can take

$$\hat{s}_0^{(P)} = \frac{1}{N} \sum_{t=0}^{N-1} X_t^2$$

(this allows us to simplify the calculations in the rest of this section considerably – the resulting estimator of the spectral bandwidth is still useful even if the process mean must be estimated). The expected value of $\hat{s}_0^{(P)}$ is

$$E\{\hat{s}_0^{(P)}\} = \frac{1}{N} \sum_{t=0}^{N-1} E\{X_t^2\} = s_0.$$

▷ **Exercise [298]** By evoking the Isserlis theorem (Equation (30)), show that

$$\text{var}\,\{\hat{s}_0^{(P)}\} = \frac{2}{N} \sum_{\tau=-(N-1)}^{N-1} \left(1 - \frac{|\tau|}{N}\right) s_\tau^2. \tag{298a}$$ ◁

In Section 7.4 the distribution of the spectral estimator $\hat{S}_m^{(LW)}(f)$ was approximated by that of a χ_ν^2 RV times a constant a. If we make the same approximation here, namely,

$$\hat{s}_0^{(P)} \stackrel{d}{=} b\chi_\eta^2 \tag{298b}$$

(Rice, 1945), we obtain

$$\eta = \frac{2\left(E\{\hat{s}_0^{(P)}\}\right)^2}{\text{var}\,\{\hat{s}_0^{(P)}\}} = \frac{Ns_0^2}{\sum_{\tau=-(N-1)}^{N-1} (1 - |\tau|/N)\, s_\tau^2} \tag{298c}$$

as an expression for the *degrees of freedom in a time series* of length N (analogous to the degrees of freedom in a spectral estimator – see Equation (264a)). Note that both concepts involve quadratic forms (see Chapter 8), and some researchers make this point explicit by calling η the number of degrees of freedom of order 2 (Kikkawa and Ishida, 1988). Exercise [9.2e] uses a first-order autoregressive process to explore how η depends on $\{s_\tau\}$.

We now need to relate η somehow to the width of the dominant hump of a unimodal SDF. Suppose for the moment that $\{X_t\}$ is an ideal band-pass process with an SDF defined over $f \in [-f_\mathcal{N}, f_\mathcal{N}]$ by

$$S(f) = \begin{cases} C, & f' - W \leq |f| \leq f' + W; \\ 0, & \text{otherwise,} \end{cases} \qquad (299)$$

where $C > 0$; $f' > 0$ is the center frequency of the pass-band; and $2W > 0$ is the corresponding bandwidth (see Figure 298). We assume that $f' - W > 0$ and $f' + W < f_\mathcal{N}$. For this simple example we can relate the bandwidth $2W$ to the degrees of freedom in a sample of length N from $\{X_t\}$ by the following argument. From a special case of Equation (237b) and from the fact that $\hat{S}^{(\mathrm{P})}(\cdot)$ is $2f_\mathcal{N}$ periodic, we can write, with $f_k = k/(N\Delta_t)$ (i.e., the Fourier frequencies),

$$\hat{s}_0^{(\mathrm{P})} = \frac{1}{N\Delta_t} \sum_{k=0}^{N-1} \hat{S}^{(\mathrm{P})}(f_k)$$

$$= \frac{1}{N\Delta_t} \sum_{k=-\lfloor (N-1)/2 \rfloor}^{\lfloor N/2 \rfloor} \hat{S}^{(\mathrm{P})}(f_k) \approx \frac{1}{N\Delta_t} \sum_{f'-W \leq |f_k| \leq f'+W} \hat{S}^{(\mathrm{P})}(f_k)$$

since $\hat{S}^{(\mathrm{P})}(f_k)$ should be small for those f_k such that $S(f_k) = 0$. Because $\hat{S}^{(\mathrm{P})}(\cdot)$ is symmetric about zero, we also have

$$\hat{s}_0^{(\mathrm{P})} \approx \frac{2}{N\Delta_t} \sum_{f'-W \leq f_k \leq f'+W} \hat{S}^{(\mathrm{P})}(f_k) = \frac{2}{N\Delta_t} \sum_{k=0}^{K-1} \hat{S}^{(\mathrm{P})}(f_{l+k}),$$

where f_l and K are such that $f_{l-1} < f' - W$ and $f_{l+K} > f' + W$ but $f' - W \leq f_l < \cdots < f_{l+K-1} \leq f' + W$. Since the spacing between adjacent Fourier frequencies is $\Delta_f = 1/(N\Delta_t)$, we have $K \approx 2W/\Delta_f = 2WN\Delta_t$. From Section 6.6 we know that, for those f_{l+k} in the pass-band,

$$\hat{S}^{(\mathrm{P})}(f_{l+k}) \stackrel{\mathrm{d}}{=} \frac{S(f_{l+k})}{2}\chi_2^2 = \frac{C}{2}\chi_2^2$$

to a good approximation and that the $\hat{S}^{(\mathrm{P})}(f_{l+k})$ RVs at distinct Fourier frequencies are approximately independent of each other. Since the sum of K independent χ_2^2 RVs has a χ_{2K}^2 distribution, we can conclude that, to a good approximation,

$$\hat{s}_0^{(\mathrm{P})} \stackrel{\mathrm{d}}{=} \frac{C}{N\Delta_t}\chi_{2K}^2.$$

Matching the above with Equation (298b) yields $\eta = 2K \approx 4WN\Delta_t$, so the width of the pass-band $2W$ is approximately $\eta/(2N\Delta_t)$. For an ideal band-pass process, we can thus take the spectral bandwidth B_S to be $\eta/(2N\Delta_t)$.

Suppose now that $\{X_t\}$ is either an ideal low-pass or high-pass process; i.e., its SDF is given by Equation (299) with either $f' = 0$ (low-pass case) or $f' = f_\mathcal{N}$ (high-pass), and we assume that $W < f_\mathcal{N}$. The width of the passband for both processes is $2W$, the same as for an ideal band-pass process (to support this claim, recall that the SDF is an even periodic function with a period of $2f_\mathcal{N}$); however, the burden of Exercise [7.19] is to show that $2W$ is now approximately $\eta/(N\Delta_t)$. For ideal low-pass or high-pass processes, we can thus take the spectral bandwidth B_S to be $\eta/(N\Delta_t)$.

For a time series of length N drawn from a process with an SDF whose shape is dominantly unimodal and resembles that of a band-pass, low-pass or high-pass process, we can use the arguments we have just made to *define* the bandwidth of the series as

$$B_T = \frac{\eta}{(1+\delta_{\text{BP}})N\,\Delta_{\text{t}}} = \frac{s_0^2}{(1+\delta_{\text{BP}})\,\Delta_{\text{t}}\sum_{\tau=-(N-1)}^{N-1}(1-|\tau|/N)\,s_\tau^2}, \quad (300\text{a})$$

where $\delta_{\text{BP}} = 1$ for the band-pass case and $= 0$ otherwise. An obvious estimator of B_T is

$$\hat{B}_T \stackrel{\text{def}}{=} \frac{(\hat{s}_0^{(\text{P})})^2}{(1+\delta_{\text{BP}})\,\Delta_{\text{t}}\sum_{\tau=-(N-1)}^{N-1}(1-|\tau|/N)(\hat{s}_\tau^{(\text{P})})^2}. \quad (300\text{b})$$

Walden and White (1990) find that, for the band-pass case, $\hat{B}_T$ is a biased estimator of B_T, but that $5\hat{B}_T/3 - 1/(N\,\Delta_{\text{t}})$ is approximately unbiased.

▷ **Exercise [300]** Using their result, show that

$$\tilde{B}_T \stackrel{\text{def}}{=} \frac{5}{6}\left(\frac{(\hat{s}_0^{(\text{P})})^2}{\Delta_{\text{t}}\sum_{\tau=-(N-1)}^{N-1}(1-|\tau|/N)(\hat{s}_\tau^{(\text{P})})^2}\right) - \frac{1}{N\,\Delta_{\text{t}}} \quad (300\text{c})$$

is an approximately unbiased estimator of B_T in the band-pass, low-pass and high-pass cases. ◁

Thus the unbiased estimator of B_T does not depend on the nature of the unimodality.

We hope it is clear that, just as the smoothing window should be unimodal for its bandwidth measure B_W to be useful, so should the SDF for B_T to be useful. Kikkawa and Ishida (1988) say similarly that second-order parameters (such as η) "... are good measures for both low-pass and band-pass processes." Distinct bimodality in a SDF will lead to B_S and B_T being very different since B_S will be a measure of the width of the smaller mode, while B_T will be an average over the entire SDF (see Exercise [7.20]). If B_T is meaningful and if we replace B_S by $\tilde{B}_T$ and adopt the recommendation $B_W = \tilde{B}_T/2$ as before, the practical results are generally quite satisfactory; however, as we emphasized in the previous section, the $B_W = B_S/2$ criterion for setting the smoothing window bandwidth is geared toward eliminating bias, but does not pay attention to the resulting variance of the lag window estimator – the same comment holds for the $B_W = \tilde{B}_T/2$ criterion.

As a first example, consider the AR(2) process of Equation (34), which has an SDF that is dominantly unimodal and has a shape reminiscent of a band-pass process (this SDF is shown as a thin dark curve in each of the plots in Figure 294). Using the AR(2) series of Figure 34(a), we obtain $\tilde{B}_T \doteq 0.248$. Setting $B_W = 1.85/m = \tilde{B}_T/2$ yields $m = 15$ for the Parzen lag window. This value produces a lag window estimate that is almost identical to the $m = 16$ case shown in Figure 294 and hence is quite acceptable. The three AR(2) series in Figures 34(b), (c) and (d) lead to comparable results, namely, $\tilde{B}_T \doteq 0.247, 0.267$ and 0.265, resulting in $m = 15, 14$ and 14. For comparison, Exercise [7.20] says that $\tilde{B}_T \to 0.265$ as $N \to \infty$, yielding $m = 14$, while the $B_W = B_S/2$ criterion recommends $m = 30$ based on $B_S \doteq 0.123$.

For a second example, let us consider the AR(4) process of Equation (35a). The thin dark curve in Figure 301 shows that its SDF is *not* dominantly unimodal because of the prominent twin peaks, but let's consider using $\tilde{B}_T$ anyway. For the AR(4) series of Figure 34(e), we get $\tilde{B}_T \doteq 0.036$, so equating $B_W = 1.85/m$ to $\tilde{B}_T/2$ yields $m = 102$. The corresponding Parzen spectral estimate is shown in Figure 301 and – except for the fact that the twin peaks

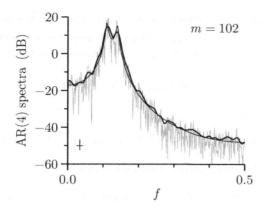

Figure 301 As in Figure 295, but with lag window spectral estimate for AR(4) series of Figure 34(e) using a Parzen smoothing window with $m = 102$. This choice comes from setting the smoothing window bandwidth B_W to be half of the estimated time series bandwidth measure $\tilde{B}_T$.

are slightly blurred out in comparison to the $m = 128$ and $m = 179$ estimates shown in Figure 295 – is quite acceptable. The AR(4) series in Figures 34(f), (g) and (h) give $\tilde{B}_T \doteq 0.030$, 0.043 and 0.034, resulting in $m = 121$, 85 and 107. Exercise [7.20] says that here $B_T \to 0.033$ as $N \to \infty$, which corresponds to $m = 111$. Recall that the $B_W = B_S/2$ criterion (based on the higher of the twin peaks) yields $m = 565$ from $B_S \doteq 0.007$, which is more than four times smaller than the large sample B_T. We can attribute this discrepancy to lack of unimodality in this example.

7.9 Automatic Smoothing of Log Spectral Estimators

In the previous sections we have considered smoothing a direct spectral estimator by convolving it with a smoothing window. We present here an alternative approach in which we smooth $\log(\hat{S}^{(D)}(\cdot))$ instead of $\hat{S}^{(D)}(\cdot)$. This approach leads to an objective procedure for determining the amount of smoothing that is optimal in a certain mean square sense. For further details, see Wahba (1980), who originally proposed this technique for smoothing the log of the periodogram, and Sjoholm (1989), who extended it to the case of all direct spectral estimators.

For a given SDF $S(\cdot)$, let us define the *log spectral density function* as

$$C(f) = \lambda \log(S(f)), \tag{301a}$$

where λ is a constant facilitating use of different logarithmic scales (for example, if we set $\lambda = 10 \log_{10}(e)$, then $C(\cdot)$ is expressed in decibels). We assume that $C(\cdot)$ has a Fourier series representation

$$C(f) = \Delta_t \sum_{\tau=-\infty}^{\infty} c_\tau e^{-i2\pi f\tau \Delta_t}, \quad \text{where } c_\tau \stackrel{\text{def}}{=} \int_{-f_\mathcal{N}}^{f_\mathcal{N}} C(f) e^{i2\pi f\tau \Delta_t}\, df,$$

so that $\{c_\tau\} \longleftrightarrow C(\cdot)$. If $\lambda = 1$, the sequence $\{c_\tau\}$ is sometimes referred to as the *cepstrum* (Bogert et al., 1963) and is in many ways analogous to the ACVS.

Given observed values of $X_0, \ldots, X_{N-1}$ from a stationary process with SDF $S(\cdot)$, a natural estimator of $C(f)$ is

$$\hat{C}^{(D)}(f) \stackrel{\text{def}}{=} \lambda \log(\hat{S}^{(D)}(f)), \tag{301b}$$

where $\hat{S}^{(\mathrm{D})}(\cdot)$ is a direct spectral estimator. If sufficient care is taken in selecting a data taper when needed, then $\hat{S}^{(\mathrm{D})}(f) \stackrel{\mathrm{d}}{=} S(f)\chi_\nu^2/\nu$ to a good approximation, where $\nu = 2$ when $0 < f < f_\mathcal{N}$, while $\nu = 1$ when $f = 0$ or $f = f_\mathcal{N}$ (see Equation (204b)). It follows that

$$\lambda \log\left(\hat{S}^{(\mathrm{D})}(f)\right) \stackrel{\mathrm{d}}{=} \lambda \log\left(S(f)\chi_\nu^2/\nu\right) \text{ and hence } \hat{C}^{(\mathrm{D})}(f) \stackrel{\mathrm{d}}{=} C(f) - \lambda \log(\nu) + \lambda \log(\chi_\nu^2)$$

approximately, from which we get

$$E\{\hat{C}^{(\mathrm{D})}(f)\} \approx C(f) - \lambda \log(\nu) + \lambda E\{\log(\chi_\nu^2)\} \text{ and } \mathrm{var}\{\hat{C}^{(\mathrm{D})}(f)\} \approx \lambda^2 \mathrm{var}\{\log(\chi_\nu^2)\}.$$

For positive integers ν, Bartlett and Kendall (1946) show that

$$E\{\log(\chi_\nu^2)\} = \log(2) + \psi(\nu/2) \text{ and } \mathrm{var}\{\log(\chi_\nu^2)\} = \psi'(\nu/2), \qquad (302\mathrm{a})$$

where $\psi(\cdot)$ and $\psi'(\cdot)$ are the digamma and trigamma functions. In particular, we have

$$E\{\log(\chi_\nu^2)\} = \begin{cases} -\log(2) - \gamma, & \nu = 1; \\ \log(2) - \gamma, & \nu = 2; \end{cases} \text{ and } \mathrm{var}\{\log(\chi_\nu^2)\} = \begin{cases} \pi^2/2, & \nu = 1; \\ \pi^2/6, & \nu = 2, \end{cases} \quad (302\mathrm{b})$$

where $\gamma \doteq 0.5772$ is Euler's constant. These results yield

$$E\{\hat{C}^{(\mathrm{D})}(f)\} \approx \begin{cases} C(f) - \lambda \log(2) - \lambda\gamma, & f = 0 \text{ or } f_\mathcal{N}; \\ C(f) - \lambda\gamma, & 0 < f < f_\mathcal{N}; \end{cases} \qquad (302\mathrm{c})$$

and

$$\mathrm{var}\{\hat{C}^{(\mathrm{D})}(f)\} \approx \begin{cases} \lambda^2 \pi^2/2, & f = 0 \text{ or } f_\mathcal{N}; \\ \lambda^2 \pi^2/6, & 0 < f < f_\mathcal{N}. \end{cases} \qquad (302\mathrm{d})$$

This variance involves *known* constants – in particular it does *not* depend on $C(f)$.

As noted in the discussion surrounding Equation (207), there exists a grid of frequencies defined by $f_k' = k/(N' \Delta_t)$ with $N' \leq N$ such that the $\hat{S}^{(\mathrm{D})}(f_k')$ RVs are approximately pairwise uncorrelated for $k = 0, \ldots, \lfloor N'/2 \rfloor$. With the additional assumption that $\{X_t\}$ is a Gaussian process, we can argue that the $\hat{C}^{(\mathrm{D})}(f_k')$ RVs are also approximately pairwise uncorrelated. Let us now define the following estimator of c_τ:

$$\hat{c}_\tau'^{(\mathrm{D})} = \frac{1}{N' \Delta_t} \sum_{k=-(N_L-1)}^{N_U-1} \hat{C}^{(\mathrm{D})}(f_k') e^{\mathrm{i}2\pi f_k' \tau \Delta_t}, \quad \tau = -(N_L - 1), \ldots, N_U - 1, \qquad (302\mathrm{e})$$

where $N_L = \lfloor (N'-1)/2 \rfloor + 1$ and $N_U = \lfloor N'/2 \rfloor + 1$ so that $N_U + N_L - 1 = N'$ (recall that $\lfloor x \rfloor$ is the integer part of x). Note that, if N' is odd so that $N_L = N_U$, then Equation (302e) mimics the expression for $\hat{s}_\tau^{(\mathrm{P})}$ in Equation (171f). In view of Equations (100g), (101a) and (101b), we can write

$$\hat{C}^{(\mathrm{D})}(f_k') = \Delta_t \sum_{\tau=-(N_L-1)}^{N_U-1} \hat{c}_\tau'^{(\mathrm{D})} e^{-\mathrm{i}2\pi f_k' \tau \Delta_t}, \qquad k = -(N_L - 1), \ldots, N_U - 1,$$

and we have $\{\hat{c}_\tau'^{(\mathrm{D})}\} \longleftrightarrow \{\hat{C}^{(\mathrm{D})}(f_k')\}$. In view of Equations (74a) and (75a), by defining $\hat{c}_\tau'^{(\mathrm{D})} = 0$ for $\tau \geq N_U$ and $\tau \leq -N_L$, we also have $\{\hat{c}_\tau'^{(\mathrm{D})}\} \longleftrightarrow \hat{C}^{(\mathrm{D})}(\cdot)$, and hence

$$\hat{C}^{(\mathrm{D})}(f) = \Delta_t \sum_{\tau=-(N_L-1)}^{N_U-1} \hat{c}_\tau'^{(\mathrm{D})} e^{-\mathrm{i}2\pi f \tau \Delta_t}.$$

7.9 Automatic Smoothing of Log Spectral Estimators

For a given lag window $\{w_{m,\tau}\}$, let us now define the following analog to the lag window spectral estimator $\hat{S}_m^{(\text{LW})}(f)$:

$$\hat{C}_m^{(\text{LW})}(f) = \Delta_t \sum_{\tau=-(N_L-1)}^{N_U-1} w_{m,\tau} \hat{c}_\tau^{\prime(\text{D})} e^{-i2\pi f \tau \Delta_t}. \tag{303a}$$

Just as $\hat{S}_m^{(\text{LW})}(\cdot)$ is a smoothed version of $\hat{S}^{(\text{D})}(\cdot)$, so is $\hat{C}_m^{(\text{LW})}(\cdot)$ a smoothed version of $\hat{C}^{(\text{D})}(\cdot)$. As a measure of how well $\hat{C}_m^{(\text{LW})}(\cdot)$ estimates $C(\cdot)$, let us consider the mean integrated square error (MISE):

$$E\left\{\int_{-f_\mathcal{N}}^{f_\mathcal{N}} \left(\hat{C}_m^{(\text{LW})}(f) - C(f)\right)^2 \, df\right\} = I_m + \Delta_t \sum_{\substack{\tau \geq N_U \\ \tau \leq -N_L}} c_\tau^2, \tag{303b}$$

where

$$I_m \stackrel{\text{def}}{=} \Delta_t \sum_{\tau=-(N_L-1)}^{N_U-1} E\left\{(w_{m,\tau}\hat{c}_\tau^{\prime(\text{D})} - c_\tau)^2\right\}$$

(this is an application of Parseval's theorem as stated by Equation (75b); see also the discussion surrounding Equation (52)). For a given lag window, the value of I_m depends on the window parameter m. The idea is to estimate I_m by, say, $\hat{I}_m$ and then to pick m such that $\hat{I}_m$ is minimized. Note that, if the right-most term in Equation (303b) is small relative to I_m, then I_m is a good approximation to the MISE. In any case, I_m is the only part of this error depending on the lag window and hence under our control. Now

$$I_m = \Delta_t \sum_{\tau=-(N_L-1)}^{N_U-1} \left[w_{m,\tau}^2 E\{(\hat{c}_\tau^{\prime(\text{D})})^2\} - 2w_{m,\tau}c_\tau E\{\hat{c}_\tau^{\prime(\text{D})}\} + c_\tau^2\right]. \tag{303c}$$

The burden of Exercise [7.21] is to show that, for large N',

$$E\{\hat{c}_\tau^{\prime(\text{D})}\} \approx c_\tau - \frac{\lambda\gamma\delta_\tau}{\Delta_t} \quad \text{and} \quad E\{(\hat{c}_\tau^{\prime(\text{D})})^2\} \approx \left(c_\tau - \frac{\lambda\gamma\delta_\tau}{\Delta_t}\right)^2 + \frac{\lambda^2\pi^2(1 + \delta_\tau + \delta_{\tau-\frac{N'}{2}})}{6N'\Delta_t^2}, \tag{303d}$$

where δ_τ is Kronecker's delta function (thus $\delta_\tau = 1$ if $\tau = 0$ and $= 0$ otherwise).

▷ **Exercise [303]** Show that Equations (303c) and (303d) lead to the approximation

$$I_m \approx \frac{\lambda^2\gamma^2}{\Delta_t} + \Delta_t \sum_{\tau=-(N_L-1)}^{N_U-1} c_\tau^2(1 - w_{m,\tau})^2 + \frac{\lambda^2\pi^2}{6N'\Delta_t} \sum_{\tau=-(N_L-1)}^{N_U-1} w_{m,\tau}^2\left(1 + \delta_\tau + \delta_{\tau-\frac{N'}{2}}\right).$$

◁

Since $w_{m,0} = 1$ always, c_0^2 does not influence the above approximation; in view of Equation (303d), an approximately unbiased estimator of c_τ^2 for $0 < |\tau| < N'/2$ is given by $(\hat{c}_\tau^{\prime(\text{D})})^2 - (\lambda^2\pi^2/(6N'\Delta_t^2))$. This suggests the following estimator for I_m:

$$\hat{I}_m \stackrel{\text{def}}{=} \frac{\lambda^2\gamma^2}{\Delta_t} + \Delta_t \sum_{\tau=-(N_L-1)}^{N_U-1} \left((\hat{c}_\tau^{\prime(\text{D})})^2 - \frac{\lambda^2\pi^2}{6N'\Delta_t^2}\right)(1 - w_{m,\tau})^2$$

$$+ \frac{\lambda^2\pi^2}{6N'\Delta_t} \sum_{\tau=-(N_L-1)}^{N_U-1} w_{m,\tau}^2\left(1 + \delta_\tau + \delta_{\tau-\frac{N'}{2}}\right). \tag{303e}$$

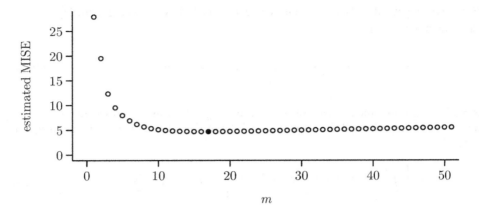

Figure 304 $\hat{I}_m$ of Equation (303e) versus m for the AR(2) time series of Figure 34(a) based upon the log periodogram and the Parzen lag window. The minimum value is at $m = 17$ (solid circle).

We can then objectively select m by picking the value for which $\hat{I}_m$ is smallest.

Sjoholm (1989) points out that an estimate of $S(\cdot)$ based on smoothing $\log(\hat{S}^{(D)}(\cdot))$ might have poor quantitative properties in that power is not necessarily preserved: since smoothing is done in log space, a high-power region can be disproportionately modified if it is adjacent to a low-power region. If $\hat{C}_m^{(LW)}(\cdot)$ is to be used for other than qualitative purposes, the just-described procedure for selecting m can be regarded merely as a way of establishing the appropriate smoothing window bandwidth, so that, once m is known, we can use it in computing $\hat{S}_m^{(LW)}(\cdot)$. To see the justification for using m as determined by the $\hat{I}_m$ criterion to form $\hat{S}_m^{(LW)}(\cdot)$, consider, for example, the Parzen lag window for which $w_{m,\tau} = 0$ for $|\tau| \geq m$ as per Equation (275a). Assuming $m \leq N_L$ (always the case in practice) and recalling that $\{\hat{c}_\tau^{\prime(D)}\} \longleftrightarrow \hat{C}^{(D)}(\cdot)$ and $\{\hat{s}_\tau^{(D)}\} \longleftrightarrow \hat{S}^{(D)}(\cdot)$, a study of Equations (303a) and (248a) indicates that

$$\hat{C}_m^{(LW)}(f) = \Delta_t \sum_{\tau=-(m-1)}^{m-1} w_{m,\tau} \hat{c}_\tau^{\prime(D)} e^{-i2\pi f \tau \Delta_t} = \int_{-f_\mathcal{N}}^{f_\mathcal{N}} W_m(f-\phi) \hat{C}^{(D)}(\phi) \, d\phi$$

and

$$\hat{S}_m^{(LW)}(f) = \Delta_t \sum_{\tau=-(m-1)}^{m-1} w_{m,\tau} \hat{s}_\tau^{(D)} e^{-i2\pi f \tau \Delta_t} = \int_{-f_\mathcal{N}}^{f_\mathcal{N}} W_m(f-\phi) \hat{S}^{(D)}(\phi) \, d\phi.$$

Use of the same smoothing window $W_m(\cdot)$ to create both $\hat{C}_m^{(LW)}(\cdot)$ and $\hat{S}_m^{(LW)}(\cdot)$ makes sense since our choice of the smoothing window bandwidth should take into consideration both the bandwidths of the important features in $C(\cdot)$ and $S(\cdot)$ and how smooth we want the estimators to be. The bandwidths for $C(\cdot)$ and $S(\cdot)$ are arguably comparable since $C(\cdot)$ is just $S(\cdot)$ after a log transformation, and the degrees of smoothness are also relatively the same for $\hat{C}_m^{(LW)}(\cdot)$ and $\hat{S}_m^{(LW)}(\cdot)$ when the latter is not subject to smoothing window leakage.

As a first example, let us consider the AR(2) time series of length $N = 1024$ shown in Figure 34(a). Figure 304 shows a plot of $\hat{I}_m$ versus m based upon the periodogram at the Fourier frequencies k/N and the Parzen lag window (here we can let $N' = N$ since a coarser grid is not needed because tapering is not needed). The minimizing value is $m = 17$ (the other three AR(2) time series shown in Figure 34 give comparable values of 16, 17 and

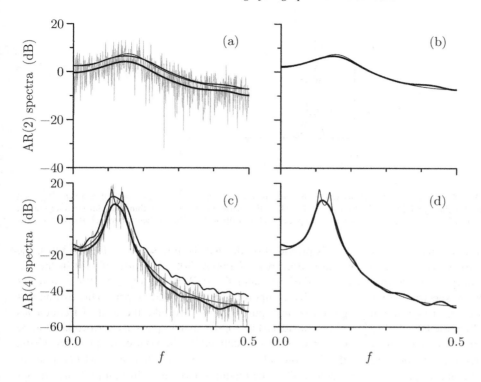

Figure 305 Comparison of estimates $\hat{C}_m^{(\text{LW})}(\cdot)$ of log SDFs $C(\cdot) = 10 \log_{10}(S(\cdot))$ with estimates $\hat{S}_m^{(\text{LW})}(\cdot)$ of SDFs $S(\cdot)$. The rough-looking light curve in plot (a) is the periodogram $\hat{S}^{(\text{P})}(\cdot)$ for the AR(2) series of Figure 34(a), upon which thin, medium and thick dark curves are superimposed. The thin and medium curves track each other closely and depict, respectively, the true AR(2) SDF and an $m = 17$ Parzen lag window SDF estimate $\hat{S}_m^{(\text{LW})}(\cdot)$ (this setting for m is dictated by the $\hat{I}_m$ criterion – see Figure 304). The thick dark curve below these two curves shows the log SDF estimate $\hat{C}_m^{(\text{LW})}(\cdot)$, which is also based on the $m = 17$ Parzen lag window. Plot (b) compares the true AR(2) SDF with $\hat{C}_m^{(\text{LW})}(\cdot)$ after correcting the latter for bias by adding $10 \log_{10}(e)\gamma$. The bottom row of plots has the same structure as the top row, but now involve the AR(4) series of Figure 34(e). The rough-looking light curve in (c) is a direct spectral estimate $\hat{S}^{(\text{D})}(\cdot)$ based on the Hanning data taper rather than the periodogram, and the estimates $\hat{S}_m^{(\text{LW})}(\cdot)$ and $\hat{C}_m^{(\text{LW})}(\cdot)$ are based on an $m = 45$ Parzen lag window (again the $\hat{I}_m$ criterion dictates this choice).

18). The erratic-looking light curve in Figure 305(a) shows the periodogram $\hat{S}^{(\text{P})}(\cdot)$ for this AR(2) time series on a decibel scale, which is the same as displaying – on a linear scale – the estimate $\hat{C}^{(\text{D})}(\cdot)$ of the log SDF defined by $\hat{C}^{(\text{D})}(f) = 10 \log_{10}(\hat{S}^{(\text{P})}(f))$. The true AR(2) SDF $S(\cdot)$ is the thin black curve, which, because we are displaying it on a decibel scale, is the same as the true log SDF $C(\cdot)$. This SDF is intertwined with a medium black curve showing an $m = 17$ Parzen lag window estimate $\hat{S}_m^{(\text{LW})}(\cdot)$, i.e., $\hat{S}^{(\text{P})}(\cdot)$ convolved with a Parzen smoothing window. The thick black curve below the thin and medium curves is the smoothed log periodogram $\hat{C}_m^{(\text{LW})}(\cdot)$, i.e., $\hat{C}^{(\text{D})}(\cdot)$ convolved with an $m = 17$ Parzen smoothing window. The estimate $\hat{C}_m^{(\text{LW})}(\cdot)$ is systematically below the true log SDF. This bias arises because $\hat{C}_m^{(\text{LW})}(\cdot)$ is a smoothed version of $\hat{C}^{(\text{D})}(\cdot)$, which is a biased estimator of $C(\cdot)$; i.e., $E\{\hat{C}^{(\text{D})}(f)\} \approx C(f) - \lambda\gamma$ for $0 < f < f_\mathcal{N}$, where here $\lambda = 10 \log_{10}(e)$. Figure 305(b) shows the true AR(2) log SDF (again as a thin black curve) with $\hat{C}_m^{(\text{LW})}(\cdot)$ after correcting it for bias by the addition of $10 \log_{10}(e)\gamma$. The agreement between this bias-corrected $\hat{C}_m^{(\text{LW})}(\cdot)$ and the true log SDF $C(\cdot)$ is as good as the agreement between $\hat{S}_m^{(\text{LW})}(\cdot)$ and the true SDF $S(\cdot)$.

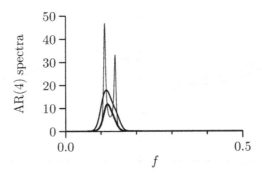

Figure 306 Comparison of true AR(4) SDF with a lag window estimate $\hat{S}_m^{(LW)}(\cdot)$ and a log SDF estimate $\hat{C}_m^{(LW)}(\cdot)$ (corrected for bias by addition of $10 \log_{10}(e)\,\gamma$) for AR(4) time series of Figure 34(e) (thin, medium and thick curves, respectively). Figures 305(c) and (d) show these spectra on linear/dB scales, whereas here we use a linear/linear scale.

A more challenging example is provided by the AR(4) time series of Figure 34(e), the results for which are shown in the bottom row of Figure 305. Because of bias in the periodogram, we define the basic log SDF estimator $\hat{C}^{(D)}(f) = 10 \log_{10}(\hat{S}^{(D)}(f))$ in terms of a direct spectral estimator $\hat{S}^{(D)}(\cdot)$ based upon the Hanning data taper. This estimate is shown as the rough-looking light curve in Figure 305(c). We take the grid of frequencies over which the RVs $\hat{S}^{(D)}(f'_k)$ can be considered as approximately pairwise uncorrelated to be $f'_k = 2k/N$, i.e., a grid twice as coarse as the one defined by the Fourier frequencies. Using the Parzen lag window again, the $\hat{I}_m$ now selects $m = 45$ (the other three AR(4) time series in Figure 34 give $m = 86$, 46 and 50). The medium dark curve in Figure 305(c) shows the $m = 45$ Parzen lag window estimate $\hat{S}_m^{(LW)}(\cdot)$, which, when compared to the true AR(4) SDF (thin dark curve), is seen to suffer from smoothing window leakage at high frequencies and to smear out the high-power twin-peaks portion of the SDF. In comparison, the $m = 45$ Parzen estimate $\hat{C}_m^{(LW)}(\cdot)$ (thick dark curve) does not suffer from smoothing window leakage, but still blurs the twin peaks. In addition, $\hat{C}_m^{(LW)}(\cdot)$ is systematically below the true log SDF, which again can be attributed to the bias in $\hat{C}^{(D)}(\cdot)$. Figure 305(d) compares $\hat{C}_m^{(LW)}(\cdot)$ with the true SDF after correcting for bias by addition of $10 \log_{10}(e)\,\gamma$. The agreement between the bias-corrected $\hat{C}_m^{(LW)}(\cdot)$ and the log SDF $C(\cdot)$ is now better than that between $\hat{S}_m^{(LW)}(\cdot)$ and $S(\cdot)$ except in the twin peaks region, where the bias-corrected $\hat{C}_m^{(LW)}(\cdot)$ undershoots $C(\cdot)$. To focus on this discrepancy in more detail, Figure 306 shows plots of $S(f)$, $\hat{S}_m^{(LW)}(f)$ and $10^{(\hat{C}_m^{(LW)}(f)+10\log_{10}(e)\gamma)/10}$ versus f; i.e., all spectra are now shown on a linear scale rather on the decibel scales of Figures 305(c) and (d). While both estimates smear out the twin peaks badly, the integral of $\hat{S}_m^{(LW)}(\cdot)$ is very close to the sample variance of the AR(4) time series (which in turn is reasonably close to the AR(4) process variance, i.e., the integral of the true SDF), whereas the integral of the estimate based on $\hat{C}_m^{(LW)}(\cdot)$ is about 40% smaller than the sample variance. This illustrates that smoothing $\log(\hat{S}^{(D)}(\cdot))$ can lead to an SDF estimate that does *not* constitute an analysis of the sample variance of a time series, which is one of the key goals of a proper spectral analysis.

Comments and Extensions to Section 7.9

[1] The starting point in estimating the log spectrum is the computation of $\hat{C}^{(D)}(f'_k) = \lambda \log(\hat{S}^{(D)}(f'_k))$, which is problematic if $\hat{S}^{(D)}(f'_k) = 0$. Although generally $\hat{S}^{(D)}(f) > 0$ for $f \neq 0$, we can have $\hat{S}^{(D)}(0) = 0$ due to centering the data (see Exercise [6.15b] and the discussion concerning Equation (195b)). This would force $\hat{C}^{(D)}(0)$ to be $-\infty$, in which case we need to redefine $\hat{C}^{(D)}(0)$ somehow. A simple solution is to set $\hat{C}^{(D)}(0) = \hat{C}^{(D)}(f'_1)$, but the merits of this approach are currently unknown.

[2] In addition to Wahba (1980), the log periodogram is used in a number of papers as the starting point

for SDF estimation. Pawitan and O'Sullivan (1994) use Wahba's smoothed log-periodogram estimator as the first iteration in a scheme leading to a penalized quasi-likelihood estimator. This estimator is defined as the maximizer of a roughness penalty combined with the Whittle likelihood function (Whittle, 1953; Dzhaparidze and Yaglom, 1983). This likelihood function is based on the assumption that the periodogram over the Fourier frequencies $f_k = k/(N\Delta_t)$, $k = 1, 2, \ldots, \lfloor (N-1)/2 \rfloor$, consists of independent $S(f)\chi_2^2/2$ RVs (C&E [5] for Section 6.6 has a discussion related to this assumption). Kooperberg et al. (1995) also make use of the Whittle likelihood, but now in combination both with cubic splines to express the log of the SDF and with indicator functions to handle the line components in processes with mixed spectra (see Section 4.4). Other papers that make the same assumption about the periodogram are Stoica and Sandgren (2006), who advocate smoothing the log periodogram by thresholding the corresponding estimated cepstrum, and Moulin (1994) and Gao (1993, 1997), who develop wavelet-based techniques for smoothing the log periodogram (see also section 10.6 of Percival and Walden, 2000). A common problem with these and other periodogram-based estimators is that, as noted in Section 6.3, the periodogram can be severely biased even for sample sizes that might seem to be large, and smoothing its log does nothing to compensate for this bias.

7.10 Bandwidth Selection for Periodogram Smoothing

In Section 7.1 we looked briefly at the periodogram to motivate discretely smoothed direct spectral estimators $\hat{S}^{(\text{DS})}(\cdot)$ (see Equations (246a) and (246b)). Here we consider the periodogram as the basis for the SDF estimator

$$\hat{S}_m^{(\text{DSP})}(f_k) \stackrel{\text{def}}{=} \sum_{j=-\lfloor N/2 \rfloor}^{N-\lfloor N/2 \rfloor - 1} g_{m,j} \hat{S}^{(\text{P})}(f_{k-j}), \qquad (307)$$

where $f_k = k/(N\Delta_t)$ is a Fourier frequency, and $\{g_{m,j}\}$ is a set of N nonnegative smoothing coefficients such that $g_{m,-j} = g_{m,j}$ when the index j above is positive, $g_{m,-N/2} = 0$ when N is even, and $\sum_j g_{m,j} = 1$. The parameter m controls the degree of smoothing, and, once m is set, the smoothing coefficients are fully determined.

▷ **Exercise [307]** Show that $\hat{S}_m^{(\text{DSP})}(\cdot)$ is a special case of $\hat{S}^{(\text{DS})}(\cdot)$ of Equation (246b). ◁

We refer to $\hat{S}_m^{(\text{DSP})}(f_k)$ as a *discretely smoothed periodogram*. Note that Equation (307) makes use of the periodogram over all possible Fourier frequencies f_k while recognizing that the periodogram is an even periodic function of frequency with a period of $2f_\mathcal{N}$ (see Equation (170c)). The particular form this equation takes is computationally appealing (this is the subject of Exercise [7.22a]). The smoothing coefficients are usually based on samples from a symmetric PDF. For example, a Gaussian PDF yields coefficients proportional to $\exp(-(mj)^2)$, where, in keeping with earlier use of m in this chapter (see, e.g., Table 279), the bandwidth associated with $\hat{S}_m^{(\text{DSP})}(\cdot)$ decreases as m increases. Exercises [7.22b] to [7.22d] consider the EDOFs, the standard measure of effective bandwidth and the smoothing window bandwidth for $\hat{S}_m^{(\text{DSP})}(\cdot)$. (In addition to the first part of Section 7.1, the following items in C&Es for previous sections deal with material of relevance to $\hat{S}_m^{(\text{DSP})}(\cdot)$: items [4] and [5] of Section 7.1; [1] and [2] of Section 7.3; and [1] of Section 7.5.)

Our task is to objectively select an estimator from amongst the set of estimators $\hat{S}_m^{(\text{DSP})}(\cdot)$ indexed by the smoothing parameter m. A number of papers have considered this problem under the rubric of bandwidth or span selection. These include Hurvich (1985), Beltrão and Bloomfield (1987), Hurvich and Beltrão (1990), Lee (1997, 2001), Fan and Kreutzberger (1998), Ombao et al. (2001), Hannig and Lee (2004) and Mombeni et al. (2017). All of these define estimators that are essentially special cases of $\hat{S}_m^{(\text{DSP})}(\cdot)$. Here we present the method of Ombao et al. (2001). As is commonly done in the literature advocating smoothing of the periodogram, they assume that the periodogram $\hat{S}^{(\text{P})}(f_k)$ over the $\lfloor (N-1)/2 \rfloor$ Fourier frequencies satisfying $0 < f_k < f_{\mathcal{N}}$ constitute a set of independent $S(f_k)\chi_2^2/2$ RVs; in addition $\hat{S}^{(\mathrm{P})}(0)$ and $\hat{S}^{(\mathrm{P})}(f_{N/2})$ are assumed to be $S(f_k)\chi_1^2$ RVs that are independent both of one another and with the RVs associated with $0 < f_k < f_{\mathcal{N}}$ (note that $\hat{S}^{(\mathrm{P})}(f_{N/2})$ disappears when N is odd). The periodogram for Gaussian white noise agrees with this assumption (see the discussion surrounding Equation (203b)); for other processes satisfying certain stringent conditions, it is asymptotically reasonable (see C&E [5] for Section 6.6); but it can be dicey for time series encountered in practical applications (in particular, the assumption implies that the periodogram is unbiased, and there is ample evidence of its bias at certain frequencies for the ocean wave and chaotic beam data of Figures 225 and 228).

To select m, Ombao et al. (2001) propose minimizing a generalized cross-validated criterion that arises from the following considerations. McCullagh and Nelder (1989) discuss *deviance* as a measure of the goodness of fit between observations and a model for the observations (see their sections 2.3 and 8.3). Deviance is defined as the log of the ratio of two likelihood functions, one associated with a fully saturated model (i.e., the number of parameters in the model is equal to the number of observations), and the other, with a non-saturated model. In the current context, the observations consist of the periodogram at the Fourier frequencies, and the parameters in the fully saturated model are the values of the SDF at these frequencies. The non-saturated model is associated with the discretely smoothed periodogram. For $0 < f_k < f_{\mathcal{N}}$, the distribution of $\hat{S}^{(\mathrm{P})}(f_k)$ is taken to be that of the RV $S(f_k)\chi_2^2/2$, which has a PDF given by $f(u; S(f_k)) = \exp(-u/S(f_k))/S(f_k)$ (cf. Equation (209a)). The log likelihood function for a single observation $\hat{S}^{(\mathrm{P})}(f_k)$ is

$$\ell(S(f_k) \mid \hat{S}^{(\mathrm{P})}(f_k)) \stackrel{\mathrm{def}}{=} \log\left(f(\hat{S}^{(\mathrm{P})}(f_k); S(f_k))\right) = -\frac{\hat{S}^{(\mathrm{P})}(f_k)}{S(f_k)} - \log(S(f_k)).$$

The maximum likelihood estimator of $S(f_k)$, i.e., the value $S(f_k)$ that maximizes the log likelihood, is just $\hat{S}^{(\mathrm{P})}(f_k)$. The corresponding maximized log likelihood is

$$\ell(\hat{S}^{(\mathrm{P})}(f_k) \mid \hat{S}^{(\mathrm{P})}(f_k)) = -1 - \log(\hat{S}^{(\mathrm{P})}(f_k)).$$

Deviance is defined to be the difference between the above and the same log likelihood, but now evaluated at $\hat{S}_m^{(\mathrm{DSP})}(f_k)$:

$$D(\hat{S}^{(\mathrm{P})}(f_k), \hat{S}_m^{(\mathrm{DSP})}(f_k)) \stackrel{\mathrm{def}}{=} \ell(\hat{S}^{(\mathrm{P})}(f_k) \mid \hat{S}^{(\mathrm{P})}(f_k)) - \ell(\hat{S}_m^{(\mathrm{DSP})}(f_k) \mid \hat{S}^{(\mathrm{P})}(f_k))$$
$$= \frac{\hat{S}^{(\mathrm{P})}(f_k)}{\hat{S}_m^{(\mathrm{DSP})}(f_k)} - \log\left(\frac{\hat{S}^{(\mathrm{P})}(f_k)}{\hat{S}_m^{(\mathrm{DSP})}(f_k)}\right) - 1.$$

Given the assumption that the periodogram across the Fourier frequencies constitutes a set of independent RVs, Ombao et al. (2001) argue that an appropriate overall measure of deviance is

$$\frac{1}{M} \sum_{k=0}^{M-1} q_k D(\hat{S}^{(\mathrm{P})}(f_k), \hat{S}_m^{(\mathrm{DSP})}(f_k)), \tag{308}$$

where $M = \lfloor (N+1)/2 \rfloor$, and $q_k = 1$ with two exceptions, namely, $q_0 = 1/2$ and, if N is even, $q_{M-1} = 1/2$ also (these exceptions handle the periodogram at zero and Nyquist frequencies, where its distribution is assumed to be that of a $S(f_k)\chi_1^2$ RV rather than a $S(f_k)\chi_2^2/2$ RV; however, if the time series has been centered by subtracting off its sample mean (usually done in practical applications), then Exercise [6.15b] says that $\hat{S}^{(\mathrm{P})}(0) = 0$, in which case

7.10 Bandwidth Selection for Periodogram Smoothing

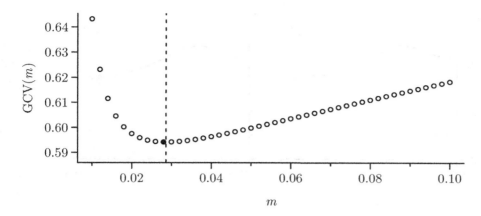

Figure 309 GCV(m) of Equation (309a) (with $\hat{S}_m^{(\mathrm{DSP})}(f_k)$ formed using weights $g_{m,j} \propto \exp(-(mj)^2)$) versus $m = 0.01, 0.012, \ldots, 0.1$ for the AR(2) time series of Figure 34(a). The minimum value over the displayed grid is at $m = 0.028$ (solid circle). The vertical dashed line indicates the minimizer of GCV(m) picked out by the function optimize in R, namely, $m \doteq 0.0286$.

$k = 0$ in Equation (308) must be changed to $k = 1$). Their generalized cross-validated deviance method for selecting m is to minimize Equation (308) after division by $(1 - g_{m,0})^2$ (the so-called model degrees of freedom), i.e., to minimize

$$\mathrm{GCV}(m) \stackrel{\mathrm{def}}{=} \frac{1}{M(1-g_{m,0})^2} \sum_{k=0}^{M-1} q_k \left[\frac{\hat{S}^{(\mathrm{P})}(f_k)}{\hat{S}_m^{(\mathrm{DSP})}(f_k)} - \log\left(\frac{\hat{S}^{(\mathrm{P})}(f_k)}{\hat{S}_m^{(\mathrm{DSP})}(f_k)} \right) - 1 \right] \quad (309\mathrm{a})$$

(if, for example, $g_{m,j}$ is proportional to $\exp(-(mj)^2)$, then $g_{m,0} = 1/\sum_j \exp(-(mj)^2)$; if $k = 0$ in Equation (308) is changed to $k = 1$, then the same must be done in Equation (309a)). The role of $(1 - g_{m,0})^2$ is vital: without it, the fact that the minimum value of $x - \log(x) - 1$ over all positive x is zero and occurs at $x = 1$ says that the criterion would always set $\hat{S}_m^{(\mathrm{DSP})}(\cdot)$ to $\hat{S}^{(\mathrm{P})}(\cdot)$, i.e., no smoothing at all. If $\hat{S}_m^{(\mathrm{DSP})}(\cdot)$ is the same as $\hat{S}^{(\mathrm{P})}(\cdot)$, we must have $g_{m,0} = 1$, and $1/M(1 - g_{m,0})^2$ is infinite. As $\hat{S}_m^{(\mathrm{DSP})}(\cdot)$ moves away from the periodogram, the weight $g_{m,0}$ decreases from unity, and $1/M(1 - g_{m,0})^2$ gets smaller. By contrast the summation in Equation (309a) gets larger, so the two components of that equation play off against each other. In effect $1/(1 - g_{m,0})^2$ manages a trade-off between bias and variance in the discretely smoothed periodogram: as $\hat{S}_m^{(\mathrm{DSP})}(\cdot)$ gets smoother, the magnitude of its bias in $\hat{S}_m^{(\mathrm{DSP})}(\cdot)$ increases, while its variance decreases. Ombao et al. (2001) recommend section 6.9 of Hastie and Tibshirani (1990) for a discussion that illuminates the connection between GCV(m) and cross validation.

As a first example, let us consider again the AR(2) time series shown in Figure 34(a). The circles in Figure 309 show GCV(m) versus $m = 0.01, 0.012, \ldots, 0.1$ for a discretely smoothed periodogram using weights g_j proportional to $\exp(-(mj)^2)$. The value $m = 0.028$ minimizes GCV(m) over this grid (shown by the solid circle). The best value for m need not be on this grid, and indeed a nonlinear optimization routine picks the slightly higher off-grid value of $m \doteq 0.0286$ (indicated in the figure by the vertical dashed line). The left-hand plot of Figure 310 shows $\hat{S}_m^{(\mathrm{DSP})}(\cdot)$ for this setting of m (thick dark curve), along with the true AR(2) SDF (thin dark curve) and the periodogram (rough-looking light curve). The mean square error for $\hat{S}_m^{(\mathrm{DSP})}(\cdot)$, namely,

$$\mathrm{MSE}(m) = \frac{1}{N/2 + 1} \sum_{k=0}^{N/2} \left(\hat{S}_m^{(\mathrm{DSP})}(f_k) - S(f_k) \right)^2, \quad (309\mathrm{b})$$

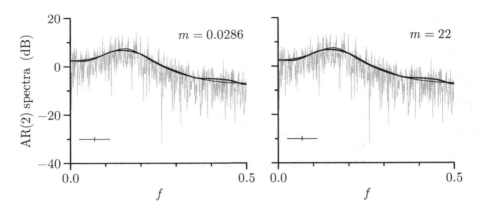

Figure 310 Discretely smoothed periodogram $\hat{S}_m^{(\text{DSP})}(\cdot)$ using weights $g_{m,j} \propto \exp{(-(mj)^2)}$ with $m \doteq 0.0286$ for AR(2) series of Figure 34(a) (thick dark curve in left-hand plot). The thin dark curve (barely distinguishable from the thick curve) is the true AR(2) SDF, while the rough-looking light curve is the periodogram at the Fourier frequencies. The right-hand plot is similar, but now shows an $m = 22$ Parzen lag window estimate $\hat{S}_m^{(\text{LW})}(\cdot)$. The widths of the crisscrosses indicate the bandwidth measure $B_{\mathcal{U}}$ (via Equation (347b) for $\hat{S}_m^{(\text{DSP})}(\cdot)$ and via (256a) for $\hat{S}_m^{(\text{LW})}(\cdot)$), while their heights display a 95% CI for $10 \log_{10}(S(f))$ along the lines of Equation (266a) (with ν set by Equation (347a) for $\hat{S}_m^{(\text{DSP})}(\cdot)$ and by Equation (264b) for $\hat{S}_m^{(\text{LW})}(\cdot)$).

is 0.0562 for the m picked out by the GCV criterion (recall that $N = 1024$ for the AR(2) time series). The minimizer of MSE(m) is $m \doteq 0.0381$, for which MSE(m) $\doteq 0.0453$. The ratio of these two MSEs is 1.240, so $\hat{S}_m^{(\text{DSP})}(\cdot)$ with m chosen by the GCV criterion has an MSE that is 25% larger than the best possible m (the three AR(2) time series in Figures 34(b), (c) and (d) lead to ratios of 2.385, 1.321 and 1.004).

The discretely smoothed periodogram selected by GCV(m) for the AR(2) time series has associated EDOFs $\nu \doteq 175.2$ and bandwidth measure $B_{\mathcal{U}} \doteq 0.0862$ (see Equations (347a) and (347b)). These quantities dictate the height and width of the crisscross in the left-hand plot of Figure 310. The right-hand plot illustrates the close relationship discussed in Section 7.1 between discretely smoothed periodograms and lag windows estimates. The thick dark curve shows an $m = 22$ Parzen lag window estimate $\hat{S}_m^{(\text{LW})}(\cdot)$ based on the periodogram (the two other curves are the same as in the left-hand plot). The setting $m = 22$ yields a bandwidth measure of $B_{\mathcal{U}} \doteq 0.0849$, which is as close as we can get to the measure associated with the discretely smoothed periodogram $\hat{S}_m^{(\text{DSP})}(\cdot)$. The maximum difference between the two estimates is minuscule (0.07 dB). The crisscrosses in the two plots are also virtually identical (the EDOFs for the lag window estimate are $\nu \doteq 172.6$).

As a second example, we turn again to the AR(4) time series of Figure 34(e). Figure 173(e) shows its periodogram, which is badly biased over a substantial interval of frequencies. Since smoothing this periodogram cannot lead to an acceptable SDF estimate, let us instead consider a Hanning-based direct spectral estimate $\hat{S}^{(\text{D})}(\cdot)$. Figure 187(e) indicates the bias in this estimate is substantially less than the periodogram's, and hence $\hat{S}^{(\text{D})}(\cdot)$ is a more suitable candidate for smoothing. To do so, we need only replace $\hat{S}^{(\text{P})}(f_k)$ with $\hat{S}^{(\text{D})}(f_k)$ in Equation (307) and then use the resulting estimator along with $\hat{S}^{(\text{D})}(f_k)$ in Equation (309a); however, minimizing this redefined GCV(m) to set m lacks theoretical justification because the $\hat{S}^{(\text{D})}(f_k)$ RVs at the Fourier frequencies are not independent. Ignoring this fact and applying the GCV criterion anyway, we obtain the estimate shown in the left-hand side of Figure 311, for which $m \doteq 0.2313$ (the minimizer of GCV(m)). As determined by Equation (309b) after replacing $\hat{S}_m^{(\text{DSP})}(\cdot)$ with the discretely smoothed direct spectral estimate, the corresponding MSE is 4.258. The best possible MSE is 4.256, which corresponds

7.10 Bandwidth Selection for Periodogram Smoothing

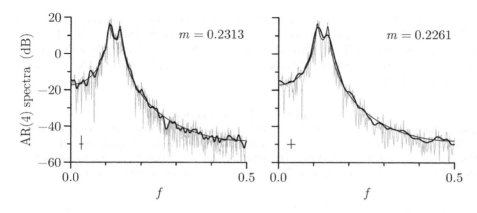

Figure 311 Discretely smoothed Hanning-based direct spectral estimate using weights $g_{m,j} \propto \exp(-(mj)^2)$ with $m \doteq 0.2313$ for AR(4) series of Figure 34(e) (thick dark curve in left-hand plot). The thin dark curve is the true AR(4) SDF, while the rough-looking light curve is $\hat{S}^{(D)}(\cdot)$ at the Fourier frequencies f_k, $k = 0, 1, \ldots, N/2$. The right-hand plot is similar, but now the light curve is $\hat{S}^{(D)}(\cdot)$ at f_{2k}, $k = 0, 1, \ldots, N/4$ (i.e., a grid twice as coarse as the Fourier frequencies), while the thick dark curve is $\hat{S}^{(D)}(\cdot)$ after smoothing over this coarser grid. The widths of the crisscrosses indicate the bandwidth measure $B_{\mathcal{U}}$, and their heights display a 95% CI for $10\log_{10}(S(f))$ along the lines of Equation (266a) with ν set appropriately. Exercise [7.23] addresses creation of the left-hand crisscross; for the right-hand plot, $B_{\mathcal{U}}$ and ν are determined using Equations (347b) and (347a) with N replaced by $N/2$.

to setting m to 0.2348. The ratio between these MSEs is 1.0003 (the three AR(4) time series in Figures 34(f), (g) and (h) lead to ratios of 2.810, 1.001 and 1.013).

Despite lacking firm theoretical backing, the GCV criterion appears to select a reasonable bandwidth for a Hanning-based direct spectral estimator $\hat{S}^{(D)}(\cdot)$, at least for the AR(4) process of Equation (35a). Nonetheless it is of interest to consider the following modification to the GCV scheme because it has some theoretical appeal. When a Hanning data taper is used, it is reasonable to take $\hat{S}^{(D)}(f_{2k})$, $k = 0, 1, \ldots, \lfloor N/4 \rfloor$, to be approximately independent RVs obeying rescaled chi-square distributions with either one or two degrees of freedom; i.e., we entertain the same assumption that Ombao et al. (2001) make about the periodogram, but now over a grid of frequencies twice as coarse as the one defined by the Fourier frequencies f_k. If we regard the direct spectral estimates $\hat{S}^{(D)}(f_{2k})$ as surrogates for a periodogram for a time series of length $\lfloor N/2 \rfloor$ and if we plug them into straightforward modifications of Equations (307) and (309a), we are led to the discretely smoothed direct spectral estimate displayed in the right-hand plot of Figure 311. The minimizer of GCV(m) is $m \doteq 0.2261$, and the associated MSE over the grid of frequencies f_{2k} is 7.916. The minimizer of MSE(m) is $m \doteq 0.4148$. Its associate MSE is 5.752, so the ratio of these MSEs is 1.376 (the ratios are 1.560, 1.418 and 2.836 for the AR(4) series in Figures 34(f), (g) and (h)). Superficially it seems that the modified GCV scheme is not working as well as the one that just ignores theoretical considerations! (Exercise [7.24] asks the reader to verify parts of Figures 310 and 311 and then to generate other AR(2) and AR(4) time series and analyze these for comparison.)

A critique of the modified GCV scheme is that it smooths over only half as many values of $\hat{S}^{(D)}(\cdot)$ as the scheme that just ignores correlations between $\hat{S}^{(D)}(f_k)$ and $\hat{S}^{(D)}(f_{k\pm 1})$. A possible refinement is to use the modified scheme to pick m and then smooth $\hat{S}^{(D)}(\cdot)$ over the Fourier frequencies with a newly determined m set such that the resulting bandwidth $B_{\mathcal{U}}$ matches the one associated with the modified scheme. The bandwidth associated with the estimate shown in the right-hand plot of Figure 311 is $B_{\mathcal{U}} \doteq 0.02291$, which is gotten from Equation (347b) by using $N/2 = 512$ in place of N and by using $m \doteq 0.2261$ to set the $N/2$ weights $g_{m,j}$. Regarding the right-hand side of Equation (347b) as a function of m with N

set to 1024, we can find the value of m yielding N weights such that the corresponding $B_{\mathcal{U}}$ is also 0.02291. This leads to using $m \doteq 0.1099$ and to an associated MSE of 7.567. The ratio of this to be best possible MSE is 1.316, which is worse than the value of close to unity for the first scheme we considered (the ratios are 1.464, 1.341 and 2.943 for the AR(4) series in Figures 34(f), (g) and (h)).

Comments and Extensions to Section 7.10

[1] The terms between the brackets in Equation (309a) defining GCV(m) are quite similar to the ones between the brackets in Equation (297c), which defines a discrepancy measure inspired by the Kullback–Leibler (KL) discrepancy between two PDFs. Hannig and Lee (2004) propose a method for selecting m that is based directly on the KL discrepancy. Their method selects m as the minimizer of

$$\frac{1}{M-1}\sum_{k=1}^{M-1}\left[\sum_{j=-J}^{J} g_{m,j} + \frac{\hat{S}_m^{(\mathrm{DSP})}(f_k) - \sum_{j=-J}^{J} g_{m,j}\hat{S}^{(\mathrm{P})}(f_{k-j})}{\frac{1}{2J+1}\sum_{j=-J}^{J}\hat{S}^{(\mathrm{P})}(f_{k-j})} - \log\left(\frac{\hat{S}_m^{(\mathrm{DSP})}(f_k)}{\hat{S}^{(\mathrm{P})}(f_k)}\right)\right],$$

where J is a small integer that controls local averaging of the periodogram (both with weights $g_{m,j}$ and with weights $1/(2J+1)$). If we set $J = 0$ so that local averaging is eliminated, the above is proportional to

$$\sum_{k=1}^{M-1}\left[\frac{\hat{S}_m^{(\mathrm{DSP})}(f_k)}{\hat{S}^{(\mathrm{P})}(f_k)} - \log\left(\frac{\hat{S}_m^{(\mathrm{DSP})}(f_k)}{\hat{S}^{(\mathrm{P})}(f_k)}\right)\right]. \tag{312a}$$

If we eliminate the terms corresponding to zero and Nyquist frequency in GCV(m), the result is proportional to

$$\frac{1}{(1-g_{m,0})^2}\sum_{k=1}^{M-1}\left[\frac{\hat{S}^{(\mathrm{P})}(f_k)}{\hat{S}_m^{(\mathrm{DSP})}(f_k)} - \log\left(\frac{\hat{S}^{(\mathrm{P})}(f_k)}{\hat{S}_m^{(\mathrm{DSP})}(f_k)}\right)\right]. \tag{312b}$$

The two methods swap the roles of $\hat{S}^{(\mathrm{P})}(f_k)$ and $\hat{S}_m^{(\mathrm{DSP})}(f_k)$; moreover, since $x - \log(x)$ is minimized for positive x when $x = 1$, it follows that Equation (312a) is minimized when $\hat{S}_m^{(\mathrm{DSP})}(f_k)$ is set to $\hat{S}^{(\mathrm{P})}(f_k)$, i.e., the discretely smoothed periodogram degenerates to the periodogram. Local averaging evidently plays a role comparable to that governed by $g_{m,0}$ in Equation (312b) (without $1/(1-g_{m,0})^2$, Equation (312b) is also minimized when $\hat{S}_m^{(\mathrm{DSP})}(f_k)$ is set to $\hat{S}^{(\mathrm{P})}(f_k)$). Hannig and Lee (2004) note that J manages the trade-off between bias and variance in $\hat{S}_m^{(\mathrm{DSP})}(f_k)$, but its use introduces an element of subjectivity to their procedure that does not exist in the Ombao et al. (2001) procedure. The Hannig and Lee (2004) procedure has some asymptotic appeal since the authors were able to show that their discrepancy estimator is consistent.

[2] The fact that all lag window estimators have a corresponding well-defined ACVS is put to good use in Chapter 11 when we discuss generating realizations of time series with an SDF specified by a lag window estimator (an important topic if, e.g., bootstrapping is of interest). It would be nice to do the same with discretely smoothed periodograms, but defining an ACVS for these is not straightforward. To understand why this is so, let us first review some facts about the periodogram.

Since $\{\hat{s}_\tau^{(\mathrm{P})}\} \longleftrightarrow \hat{S}^{(\mathrm{P})}(\cdot)$, the ACVS estimator corresponding to the periodogram is the biased estimator of the ACVS. Given the periodogram, we can retrieve its corresponding ACVS estimator using the inverse Fourier transform:

$$\hat{s}_\tau^{(\mathrm{P})} = \int_{-f_{\mathcal{N}}}^{f_{\mathcal{N}}} \hat{S}^{(\mathrm{P})}(f) e^{i2\pi f\tau \Delta_t}\, df, \quad \tau \in \mathbb{Z}$$

(this is Equation (171c)). The above might suggest that we need to know the periodogram for all $f \in [-f_{\mathcal{N}}, f_{\mathcal{N}}]$ to get $\{\hat{s}_\tau^{(\mathrm{P})}\}$; however, by definition $\hat{s}_\tau^{(\mathrm{P})} = 0$ for $|\tau| \geq N$, and we also have

$$\{\hat{s}_\tau^{(\mathrm{P})} : \tau = -N, \ldots, N-1\} \longleftrightarrow \{\hat{S}^{(\mathrm{P})}(\tilde{f}_k) : k = -N, \ldots, N-1\} \tag{312c}$$

7.10 Bandwidth Selection for Periodogram Smoothing

where $\tilde{f}_k = k/(2N\,\Delta_t)$ (the above is a slight variation on Equation (171e) – it follows from an argument analogous to the one leading to that equation). We can thus retrieve the nonzero portion of $\{\hat{s}_\tau^{(P)}\}$ using the inverse DFT:

$$\hat{s}_\tau^{(P)} = \frac{1}{2N\,\Delta_t} \sum_{k=-N}^{N-1} \hat{S}^{(P)}(\tilde{f}_k) e^{i2\pi \tilde{f}_k \tau \Delta_t} \tag{313a}$$

(cf. Equation (171f)). Thus, given the periodogram over the grid of frequencies $\tilde{f}_k$ twice as fine as that of the Fourier frequencies f_k, we can get $\{\hat{s}_\tau^{(P)} : \tau \in \mathbb{Z}\}$. Exercise [6.12] demonstrates that it is not possible to recover $\{\hat{s}_\tau^{(P)}\}$ given the periodogram over just the Fourier frequencies.

Defining an ACVS estimator, say $\{\hat{s}_\tau^{(DSP)}\}$, to go along with a discretely smoothed periodogram is complicated by the fact that $\hat{S}_m^{(DSP)}(\cdot)$ is specified only at the Fourier frequencies. Just as having access to $\hat{S}^{(P)}(\cdot)$ only at these frequencies is not enough to uniquely specify an ACVS estimator, the same is true for $\hat{S}_m^{(DSP)}(\cdot)$. If, however, we were to define $\hat{S}_m^{(DSP)}(\tilde{f}_k)$ for odd indices k, we could then obtain $\{\hat{s}_\tau^{(DSP)} : \tau = 0, \ldots, N\}$ by replacing $\hat{S}^{(P)}(\tilde{f}_k)$ with $\hat{S}_m^{(DSP)}(\tilde{f}_k)$ on the right-hand side of Equation (312c) (we need only worry about odd k because $\hat{S}_m^{(DSP)}(\tilde{f}_k) = \hat{S}_m^{(DSP)}(f_{k/2})$ for even k). Linear interpolation is one option: given the estimates at the Fourier frequencies provided by Equation (307), define estimates between these frequencies (i.e., at frequencies $\tilde{f}_k$ indexed by odd k) using

$$\hat{S}_m^{(DSP)}(\tilde{f}_k) = \frac{\hat{S}_m^{(DSP)}(\tilde{f}_{k-1}) + \hat{S}_m^{(DSP)}(\tilde{f}_{k+1})}{2} = \frac{\hat{S}_m^{(DSP)}(f_{(k-1)/2}) + \hat{S}_m^{(DSP)}(f_{(k+1)/2})}{2}, \tag{313b}$$

and, if N is odd, additionally define $\hat{S}_m^{(DSP)}(\tilde{f}_{-N}) = \hat{S}_m^{(DSP)}(f_{-(N-1)/2})$ (note that $\tilde{f}_{-N} = -f_\mathcal{N}$). A second option is a modification of Equation (307) that makes use of the periodogram over frequencies $\tilde{f}_{k-j}$ rather than just the Fourier frequencies:

$$\hat{S}_m^{(DSP)}(\tilde{f}_k) = \sum_{j=-N}^{N-1} \tilde{g}_{m,j} \hat{S}^{(P)}(\tilde{f}_{k-j}), \quad \text{where } \tilde{g}_{m,j} = \begin{cases} g_{m,j/2}, & j \text{ even}; \\ 0, & j \text{ odd}. \end{cases} \tag{313c}$$

When k is even so that $\tilde{f}_k$ is a Fourier frequency, the above reduces to Equation (307); when k is odd, the above defines a discretely smoothed periodogram at the non-Fourier frequencies. This definition combines values of the periodogram at non-Fourier frequencies using the same weighting scheme employed to create $\hat{S}_m^{(DSP)}(f_k)$ at the Fourier frequencies. Exercise [7.25] explores these and other options for defining $\hat{s}_\tau^{(DSP)}$.

[3] A lag window estimator $\hat{S}_m^{(LW)}(\cdot)$ that is based on the periodogram is an alternative to a discretely smoothed periodogram $\hat{S}_m^{(DSP)}(\cdot)$; however, even when confined to just the Fourier frequencies, $\hat{S}_m^{(LW)}(\cdot)$ is not a special case of $\hat{S}_m^{(DSP)}(\cdot)$. We thus cannot use the GCV procedure as described by Equation (309a) to set the parameter m for the lag window $\{w_{m,\tau}\}$, but the procedure can be adapted to do so, as follows. We start by reworking Exercise [249b] so that Equation (249d) uses the periodogram over the grid of frequencies $\tilde{f}_k = k/(2N\,\Delta_t)$ (the Fourier frequencies $f_k = k/(N\,\Delta_t)$ are a subset of these):

$$\hat{S}_m^{(LW)}(\tilde{f}_k) = \sum_{j=-(N-1)}^{N-1} g_j \hat{S}^{(P)}(\tilde{f}_{k-j}), \quad \text{where } g_j = \frac{1}{2N-1} \sum_{\tau=-(N-1)}^{N-1} w_{m,\tau} e^{-i2\pi \tilde{f}_j \tau \Delta_t}$$

(the expression for g_j above involves an appeal to Equation (247e)). We then make two adjustments to the GCV criterion. On the right-hand side of Equation (309a), we replace $\hat{S}_m^{(DSP)}(f_k)$ with $\hat{S}_m^{(LW)}(f_k)$. We also change $g_{m,0}$ to $2g_0$, where g_0 is dictated by the above, and the factor of 2 is introduced to correct for the $\tilde{f}_k$ grid having twice as many frequencies as the f_k grid. These adjustments yield the adaptation

$$\text{GCV}(m) = \frac{1}{M} \sum_{k=0}^{M-1} \frac{q_k}{(1-2g_0)^2} \left[\frac{\hat{S}^{(P)}(f_k)}{\hat{S}_m^{(LW)}(f_k)} - \log\left(\frac{\hat{S}^{(P)}(f_k)}{\hat{S}_m^{(LW)}(f_k)}\right) - 1 \right]. \tag{313d}$$

Application of this criterion to the AR(2) time series of Figure 34(a) picks out an $m = 22$ Parzen lag window estimate, which is shown in the right-hand plot of Figure 310 and is virtually identical to the discretely smoothed periodogram picked out by the GCV criterion of Equation (309a) (left-hand plot). (Exercise [7.26] addresses the extent to which other realizations of the AR(2) process yields similar results.)

[4] We noted in C&E [2] for Section 7.7 that the MSE is but one way to measure the discrepancy between estimated and true SDFs. The normalized mean square error NMSE(m) of Equation (296a) is a viable alternative if the true SDF is positive at all frequencies, as is the case for the AR(2) and AR(4) SDFs we've used as examples. For the estimate $\hat{S}^{(\text{DSP})}(\cdot)$ of the AR(2) SDF shown in the left-hand plot of Figure 310, the theoretical minimizer of NMSE(m) is $m \doteq 0.0283$ for which NMSE(m) $\doteq 0.014939$. The NMSE is 0.014944 for the m picked out by the GCV criterion (0.0286). The ratio of these two NMSEs is 1.0003, and the corresponding ratios for the other three AR(2) time series are 1.012, 1.133 and 1.002; by contrast, the MSE-based ratios for the four series are 1.240, 2.385, 1.321 and 1.004. The NMSE measure thus views the GCV bandwidth selection procedure more favorably than does the MSE. Turning now to the AR(4) SDF estimate in the left-hand plot of Figure 311, the theoretical minimizer of NMSE(m) is $m \doteq 0.1858$ for which NMSE(m) $\doteq 0.1433$. The NMSE is 0.1537 for the m picked out by the GCV criterion (0.2313). The ratio of these NMSEs is 1.072, and the ratios for the other AR(4) series are 1.647, 1.364 and 1.323. The MSE-based ratios are 1.0003, 2.810, 1.001 and 1.013. With the exception of the second AR(4) series, the NMSE measure views the GCV bandwidth selection procedure less favorably than does the MSE – the opposite of what happened in the AR(2) case! Exercise [7.27] invites the reader to look at many other realizations of the AR(2) and AR(4) processes in part to explore more thoroughly the evaluation of the GCV procedure by different discrepancy measures.

7.11 Computational Details

Here we give details about computation of a lag window spectral estimate $\hat{S}_m^{(\text{LW})}(\cdot)$ given a time series we regard as a realization of a portion $X_0, X_1, \ldots, X_{N-1}$ of a stationary process with sampling interval Δ_t and unknown mean μ, SDF $S(\cdot)$ and ACVS $\{s_\tau\}$. Pathways for computing $\hat{S}_m^{(\text{LW})}(\cdot)$ from $\{X_t\}$ are shown in Figure 315 and involve four Fourier transform pairs. Because a lag window estimator is built upon a direct spectral estimator $\hat{S}^{(\text{D})}(\cdot)$, the first three pairs are the same as the ones shown in Figure 221 for computing $\hat{S}^{(\text{D})}(\cdot)$. The left-hand sides of these three pairs are

(1) $\{\tilde{X}_t : t = 0, 1, \ldots, M-1\}$, the time series after centering by subtracting off its sample mean $\overline{X}$ and then padding with at least $N-1$ zeros to form a sequence of length $M \geq 2N-1$ (see Equation (220e));
(2) $\{\tilde{h}_t \tilde{X}_t\}$, the product of $\tilde{X}_t$ and the zero-padded data taper $\tilde{h}_t$ (Equation (220e)); and
(3) $\{\tilde{h}\tilde{X} \star \tilde{h}\tilde{X}_\tau\} \stackrel{\text{def}}{=} \{\tilde{s}_\tau^{(\text{D})}\}$, the cyclic autocorrelation of $\{\tilde{h}_t \tilde{X}_t\}$ as per Equation (220b).

The right-hand sides are

(1) $\{\tilde{\mathcal{X}}_k : k = 0, 1, \ldots, M-1\}$, the DFT of $\{\tilde{X}_t\}$;
(2) $\{\tilde{H} \ast \tilde{\mathcal{X}}_k\}$, the frequency domain cyclic convolution of the DFT $\{\tilde{H}_k\}$ of $\{\tilde{h}_t\}$ and the DFT $\{\tilde{\mathcal{X}}_k\}$ as per Equation (220d); and
(3) $\{|\tilde{H} \ast \tilde{\mathcal{X}}_k|^2\} \stackrel{\text{def}}{=} \{\tilde{S}^{(\text{D})}(\tilde{f}_k')\}$, the square modulus of $\{\tilde{H} \ast \tilde{\mathcal{X}}_k\}$, where $\tilde{f}_k' = k/M$.

The lag window $\{w_{m,\tau}\}$ enters the picture as a contributor to the fourth Fourier transform pair in Figure 315. Using this window, we define the sequence $\{\tilde{w}_{m,\tau}\}$ of length M by

$$\tilde{w}_{m,\tau} = \begin{cases} w_{m,\tau}, & 0 \leq \tau \leq N-1; \\ 0, & N \leq \tau \leq M-N; \\ w_{m,M-\tau}, & M-N+1 \leq \tau \leq M-1. \end{cases} \tag{314}$$

We denote its DFT by $\{\tilde{W}(\tilde{f}_k')\}$. The left-hand side of (4) is obtained by multiplying the sequences $\{\tilde{w}_{m,\tau}\}$ and $\{\tilde{s}_\tau^{(\text{D})}\}$ term by term to produce $\{\tilde{w}_{m,\tau}\tilde{s}_\tau^{(\text{D})}\} \stackrel{\text{def}}{=} \{\tilde{s}_{m,\tau}^{(\text{LW})}\}$. Its DFT –

7.11 Computational Details

$$\begin{array}{ccc}
\{\tilde{X}_t\} & \xleftrightarrow{(1)} & \{\tilde{\mathcal{X}}_k\} \\
\downarrow \text{mult} & & \downarrow \text{conv} \\
\{\tilde{h}_t \tilde{X}_t\} & \xleftrightarrow{(2)} & \{\tilde{H} * \tilde{\mathcal{X}}_k\} \\
\downarrow \text{auto} & & \downarrow \text{mod sq} \\
\{\tilde{s}_\tau^{(\text{D})}\} = \{\tilde{h}\tilde{X} \star \tilde{h}\tilde{X}_\tau\} & \xleftrightarrow{(3)} & \{|\tilde{H} * \tilde{\mathcal{X}}_k|^2\} = \{\tilde{S}^{(\text{D})}(\tilde{f}_k')\} \\
\downarrow \text{mult} & & \downarrow \text{conv} \\
\{\tilde{s}_{m,\tau}^{(\text{LW})}\} = \{\tilde{w}_{m,\tau}\tilde{s}_\tau^{(\text{D})}\} & \xleftrightarrow{(4)} & \{\tilde{W} * \tilde{S}^{(\text{D})}(\tilde{f}_k')\} = \{\tilde{S}_m^{(\text{LW})}(\tilde{f}_k')\}
\end{array}$$

Figure 315 Pathways for computing $\tilde{S}_m^{(\text{LW})}(\cdot)$, from which, upon multiplication by Δ_t, we can obtain the lag window estimate $\hat{S}_m^{(\text{LW})}(f_k') = \Delta_t \tilde{S}_m^{(\text{LW})}(\tilde{f}_k')$, where $f_k' = k/(M\,\Delta_t)$ and $\tilde{f}_k' = k/M$ (adapted from figure 1 of Van Schooneveld and Frijling, 1981). This figure incorporates Figure 221, which shows computational pathways for direct spectral estimates.

denoted as $\{\tilde{W} * \tilde{S}^{(\text{D})}(\tilde{f}_k')\} = \{\tilde{S}_m^{(\text{LW})}(\tilde{f}_k')\}$ – can be obtained directly or indirectly via the frequency domain cyclic convolution of $\{\tilde{W}(\tilde{f}_k')\}$ and $\{\tilde{S}^{(\text{D})}(\tilde{f}_k')\}$ using Equation (220d). From the left-hand side of (4), we obtain the lag window-based ACVS estimate:

$$\hat{s}_{m,\tau}^{(\text{LW})} = \begin{cases} \tilde{s}_{|\tau|}^{(\text{LW})}, & |\tau| \leq N-1; \\ 0, & \text{otherwise.} \end{cases}$$

From the right-hand side, we obtain the desired lag window spectral estimate:

$$\hat{S}_m^{(\text{LW})}(f_k') = \Delta_t\, \tilde{S}_m^{(\text{LW})}(\tilde{f}_k'), \qquad k = 0,\ldots,\lfloor M/2 \rfloor.$$

As shown in Figure 315, we can use any one of eight computational pathways to obtain $\{\hat{S}_m^{(\text{LW})}(f_k')\}$ via $\{\tilde{S}_m^{(\text{LW})}(\tilde{f}_k')\}$ from $\{X_t\}$, each of which we might prefer under certain conditions. For example, the pathway

$$\begin{array}{c}
\{\tilde{X}_t\} \\
\downarrow \text{mult} \\
\{\tilde{h}_t \tilde{X}_t\} \xrightarrow{(2)} \{\tilde{H} * \tilde{\mathcal{X}}_k\} \\
\hspace{4cm} \downarrow \text{mod sq} \\
\{\tilde{h}\tilde{X} \star \tilde{h}\tilde{X}_\tau\} \xleftarrow{(3)} \{|\tilde{H} * \tilde{\mathcal{X}}_k|^2\} \\
\downarrow \text{mult} \\
\{\tilde{w}_{m,\tau}\tilde{s}_\tau^{(\text{D})}\} \xrightarrow{(4)} \{\tilde{W} * \tilde{S}^{(\text{D})}(\tilde{f}_k')\} = \{\tilde{S}_m^{(\text{LW})}(\tilde{f}_k')\}
\end{array}$$

requires two DFTs and one inverse DFT and allows us to easily obtain $\{\hat{S}^{(\text{D})}(f_k')\}$, $\{\hat{s}_\tau^{(\text{D})}\}$, $\{\hat{s}_{m,\tau}^{(\text{LW})}\}$ and $\{\hat{S}_m^{(\text{LW})}(f_k')\}$ (in that order). If we are not interested in examining the ACVS estimates $\{\hat{s}_\tau^{(\mathrm{D})}\}$ and $\{\hat{s}_{m,\tau}^{(\mathrm{LW})}\}$, then we could use the pathway

$$\{\tilde{X}_t\} \\ \downarrow \text{mult} \\ \{\tilde{h}_t \tilde{X}_t\} \xrightarrow{(2)} \{\tilde{H} * \tilde{\mathcal{X}}_k\} \\ \downarrow \text{mod sq} \\ \{|\tilde{H} * \tilde{\mathcal{X}}_k|^2\} \\ \downarrow \text{conv} \\ \{\tilde{W} * \tilde{S}^{(\mathrm{D})}(\tilde{f}'_k)\}\} = \{\tilde{S}_m^{(\mathrm{LW})}(\tilde{f}'_k)\},$$

which involves one DFT and one convolution. This second pathway might require fewer numerical operations than the first if the sequence $\{\tilde{W}(\tilde{f}'_k)\}$ is sufficiently short so that the convolution can be computed efficiently with Equation (220d) (in fact, we now only need $M \geq N$ instead of $M \geq 2N - 1$ because padding with $N - 1$ or more zeros is only required to correctly compute $\{\hat{s}_\tau^{(\mathrm{D})}\}$ and $\{\hat{s}_{m,\tau}^{(\mathrm{LW})}\}$). Note also that this pathway is exactly how in practice we would compute the discretely smoothed direct spectral estimator $\hat{S}^{(\mathrm{DS})}(\cdot)$ of Equation (246b).

Figure 317 gives an example of computing a lag window spectral estimate by extending the example of computing a direct spectral estimate shown in Figure 223. The four top rows in both figures are the same and are described in Section 6.7, to which we refer the reader for a description. Turning to the final two rows of Figure 317, the left-hand plot on the fifth row is $\{\tilde{w}_{m,\tau}\}$, formed here using a Parzen lag window with $m = 10$; the right-hand plot is the real-valued sequence $\{\tilde{W}(\tilde{f}'_k)\}$, the DFT of $\{\tilde{w}_{m,\tau}\}$. The left-hand plot on the last row is $\{\tilde{s}_{m,\tau}^{(\mathrm{LW})}\}$, from which we can form $\{\hat{s}_{m,\tau}^{(\mathrm{LW})}\}$; the right-hand plot is $\{\tilde{S}_m^{(\mathrm{LW})}(\tilde{f}'_k)\}$, from which we can compute $\{\hat{S}_m^{(\mathrm{LW})}(f'_k)\}$, a smoothed version of $\{\hat{S}^{(\mathrm{D})}(f'_k)\}$.

7.12 Examples of Lag Window Spectral Estimators

Ocean Wave Data

Here we continue the analysis of the ocean wave time series that we started in Section 6.8. Figure 225 shows the time series, while Figure 226 displays four direct spectral estimates for this series, namely, the periodogram and three estimates based upon Slepian tapers with NW set to $1/\Delta_t$, $2/\Delta_t$ and $4/\Delta_t$. There is evidence of bias due to leakage in the periodogram and – to a lesser extent – in the estimate based on the $NW = 1/\Delta_t$ Slepian taper. As discussed in Section 6.8, part of the rationale for collecting this time series was to investigate the rate at which the SDF is decreasing over the frequency range 0.2 to 1.0 Hz. While the periodogram exhibits leakage in this frequency range, none of the Slepian-based direct spectral estimates do; however, to illustrate a point concerning smoothing window leakage (see the discussion surrounding Figure 319(a)), we will concentrate on the $NW = 2/\Delta_t$ direct spectral estimate (Figure 226(c)) even though the $NW = 1/\Delta_t$ estimate (Figure 226(b)) is adequate for this range of frequencies. We want to smooth this direct spectral estimate to get a better visual representation of the rate at which the power in the SDF decreases over this frequency range. We do so using a lag window approach.

Figure 318 shows a Parzen lag window estimate with parameter $m = 150$ (dark curve). This SDF estimate is a smoothed version of the $NW = 2/\Delta_t$ direct spectral estimate (the rough-looking light curve). We set $m = 150$ to achieve a smoothing window bandwidth B_W comparable to a visual determination of the width of the broad peak in $\hat{S}^{(\mathrm{D})}(\cdot)$ at $f =$

7.12 *Examples of Lag Window Spectral Estimators* 317

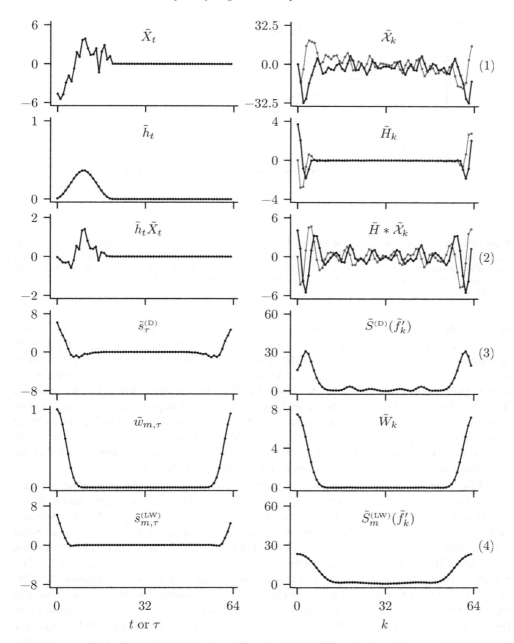

Figure 317 Illustration of computational pathways for lag window spectral estimates (cf. Figure 315). The upper four rows are identical to Figure 223, which illustrates computational pathways for direct spectral estimates.

0.160 Hz. This choice produces a smoothed version of $\hat{S}^{(D)}(\cdot)$ that smears out this peak somewhat (and the one at $f = 1.469$ Hz). To help assess this lag window estimate, we display a crisscross with an interpretation similar to the crisscrosses for the direct spectral estimates in Figure 226; however, the horizontal width of the crisscross here now depicts $B_{\mathcal{U}} \doteq 0.050$ Hz, the measure of effective bandwidth appropriate for a lag window estimator (Equation (256a)). This bandwidth can again be interpreted as a rough measure of the distance in frequency

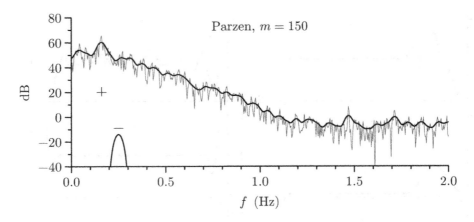

Figure 318 Parzen lag window spectral estimate $\hat{S}_m^{(\mathrm{LW})}(\cdot)$ with $m = 150$ for ocean wave data (dark curve). The light curve shows the underlying $\hat{S}^{(\mathrm{D})}(\cdot)$ (also shown in Figure 226(c)). The width and height of the crisscross gives the bandwidth measure $B_\mathcal{U}$ (Equation (256a)) and the length of a 95% CI for $10\log_{10} S(f))$ (Equation (266a)). The central lobe for the smoothing window $W_m(\cdot)$ is shown in the lower left-hand corner with a horizontal line depicting its bandwidth B_W.

between adjacent uncorrelated spectral estimates. The vertical height of this crisscross represents the length of a 95% CI for $10\log_{10}(S(f))$ based upon $10\log_{10}(\hat{S}_m^{(\mathrm{LW})}(f))$. This length is computed as described at the end of Section 7.4 (the associated EDOFs for this estimate are $\nu \doteq 12.6$). The central lobe of the corresponding smoothing window $W_m(\cdot)$ (shifted so that its peak value is plotted at $1/4$ Hz and -14 dB) is shown in the lower left-hand corner. There is a horizontal line above this smoothing window, which depicts the smoothing window bandwidth B_W. This bandwidth can be computed either via its definition in Equation (251e) or via the simple approximating formula $1.85/(m\,\Delta_\mathrm{t})$ stated in Table 279, both of which yield 0.049 Hz (note that this is slightly less than $B_\mathcal{U} \doteq 0.050$ Hz, as is reasonable given that the latter takes into account the data taper in addition to the lag window).

While the $m = 150$ Parzen estimate tracks the $NW = 2/\Delta_\mathrm{t}$ direct spectral estimate nicely and is considerably smoother, it is arguably too bumpy to adequately represent the rolloff pattern over the range of frequencies of main interest ($0.2\,\mathrm{Hz} \leq f \leq 1.0\,\mathrm{Hz}$). We thus consider a second Parzen estimate, but now one that is associated with a larger smoothing window bandwidth B_W. If we let $m = 55$, we obtain the estimate shown in Figure 319(a) (this figure has a layout similar to that of Figure 318). For this estimate we have $B_W \doteq 0.135$ Hz, which is almost three times larger than the one associated with the $m = 150$ estimate. The $m = 55$ Parzen estimate is much smoother looking (a comparison of the crisscrosses in Figures 318 and 319(a) shows the inherent trade-off between variability – as measured by the length of a 95% CI – and the effective bandwidth as measured by $B_\mathcal{U}$). While this new estimate arguably offers a better representation of the rolloff pattern for $0.2\,\mathrm{Hz} \leq f \leq 1.0\,\mathrm{Hz}$, it does not do well at other frequencies. The peaks at 0.160 and 1.469 Hz have been smeared out almost to the point of being unrecognizable. There is also evidence of smoothing window leakage between frequencies 1.0 Hz and 2.0 Hz: the dark curve in Figure 319(a) should be a smoothed version of the $NW = 2$ direct spectral estimate (the light curve), but note that the dark curve is consistently *higher* than the light curve (by contrast, this is not the case in the high-power portion of the spectral estimate, which in this example corresponds to low-frequency values).

To confirm that the $m = 55$ Parzen estimate does indeed suffer from smoothing window leakage at high frequencies, we can compare it to Daniell and Bartlett–Priestley lag window

7.12 Examples of Lag Window Spectral Estimators

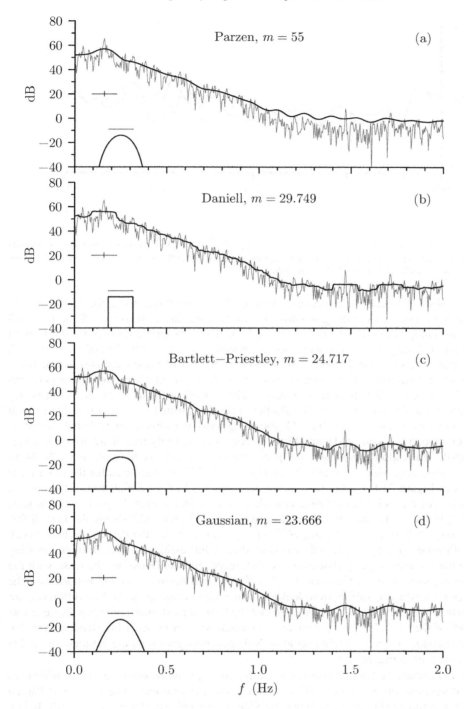

Figure 319 Four lag window spectral estimates $\hat{S}_m^{(\mathrm{LW})}(\cdot)$ for ocean wave data (dark curves). The layout is the same as for Figure 318, with the exception that the central lobe for the design window $V_m(\cdot)$ – rather than the smoothing window $W_m(\cdot)$ – is shown in the lower left-hand corners for the Daniell and Bartlett–Priestley estimates (plots (b) and (c)).

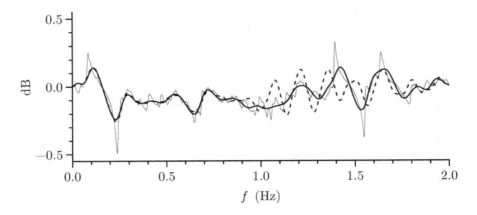

Figure 320 First difference $10\log_{10}(\hat{S}_m^{(\text{LW})}(\tilde{f}_k)) - 10\log_{10}(\hat{S}_m^{(\text{LW})}(\tilde{f}_{k-1}))$ of lag window SDF estimates versus $\tilde{f}_k$. The dashed, light jagged and dark smooth curves are for, respectively, the Parzen, Bartlett–Priestley and Gaussian estimates shown in parts (a), (c) and (d) of Figure 319.

estimates, both of which are necessarily leakage free. For these two estimates we set m such that the resulting smoothing window bandwidths are the same as that of the $m = 55$ Parzen smoothing window (i.e., $B_W \doteq 0.135\,\text{Hz}$). The appropriate choices for the Daniell and Bartlett–Priestley smoothing windows are, respectively, $m = 29.749$ and $m = 24.717$. The estimates are shown in Figures 319(b) and (c). Neither of these estimates is systematically above the underlying direct spectral estimate in the 1.0 to 2.0 Hz region, which confirms that the Parzen estimate suffers from leakage. The Daniell estimate, however, is not nearly as smooth as the Parzen estimate. The Bartlett–Priestley estimate is superior to the Daniell estimate in terms of smoothness, but a Gaussian lag window estimate offers further improvement even though its associated smoothing window is not entirely free of sidelobes (see Figure 278(b)). This estimate is shown in Figure 319(d), for which we have set $m = 23.666$ to achieve a smoothing window bandwidth of $B_W \doteq 0.135\,\text{Hz}$ (the same as for the other three estimates in Figure 319; for the record, we note that the EDOFs for all four estimates are $\nu \doteq 34.4$). All the estimates we have considered are evaluated over the grid of frequencies $\tilde{f}_k = k/(2N\Delta_\text{t})$. To qualitatively assess smoothness, Figure 320 shows the first difference $10\log_{10}(\hat{S}_m^{(\text{LW})}(\tilde{f}_k)) - 10\log_{10}(\hat{S}_m^{(\text{LW})}(\tilde{f}_{k-1}))$ versus $\tilde{f}_k$ for the Parzen (dashed curve), Bartlett–Priestley (light jagged) and Gaussian (dark solid) estimates. The Bartlett–Priestley estimate has the most rapidly changing first difference, and the Gaussian, the least, with the Parzen being similar to the Gaussian for $f < 0.9\,\text{Hz}$ and having a ringing pattern over the frequencies for which it suffers from leakage. Since rapid changes in the first difference are an indication of lack of smoothness, we see that the Gaussian estimate combines the desirable properties of the $m = 55$ Parzen estimate (smoothness) with those of the Bartlett–Priestley estimate (freedom from smoothing window leakage), thus giving us the estimate of choice amongst the four in Figure 319.

To summarize, we have considered five different lag window estimates, all of which are smoothed versions of an erratic $NW = 2$ direct spectral estimate. The $m = 150$ Parzen lag window estimate (Figure 318) shows the SDF rolling off over $0.2\,\text{Hz} \leq f \leq 1.0\,\text{Hz}$ in a complicated non-monotonic manner, which does not make it easy to compare with physical theories. The $m = 55$ Parzen estimate (Figure 319(a)) is a more useful estimate over the frequency range of interest, but a comparison of this estimate with its corresponding $\hat{S}^{(\text{D})}(\cdot)$ indicates that it suffers from smoothing window leakage over $1.0\,\text{Hz} \leq f \leq 2.0\,\text{Hz}$ (an important point to note here is that we should routinely compare any $\hat{S}_m^{(\text{LW})}(\cdot)$ we are entertaining with its corresponding $\hat{S}^{(\mathrm{D})}(\cdot)$ to make sure that the former is a reasonably smoothed version of the latter). The Daniell, Bartlett–Priestley and Gaussian estimates (Figures 319(b), (c) and (d)) show no evidence of leakage, with the Gaussian being the estimate of choice because of its superior smoothness properties. Note that the crisscrosses in all four plots in Figures 319 are virtually identical, which says that the four estimates shown there have the same bandwidths and EDOFs. Thus, while all four estimates share some statistical properties, there are important differences amongst them, and these differences cannot be deduced based on bandwidths and EDOFs alone. (We return to an analysis of these data in Sections 8.9, 9.12, 10.15 and 11.6)

Ice Profile Data
Let us now consider two arctic sea-ice profiles (this analysis expands upon an unpublished 1991 report by W. Fox, Applied Physics Laboratory, University of Washington). Ice morphology was measured by a profiling device moving at a constant speed in a straight line roughly parallel to the surface of the ice. The distance between the profiler and the ice surface was recorded at a constant rate in time to obtain measurements spaced every $\Delta_t = 1.7712\,\mathrm{m}$ apart. We can regard measurements along this transect as a "time" series related to variations in the surface height of the ice over distance, i.e., a sea-ice profile. The two top plots of Figure 322 employ similar vertical scales to show two profiles. The first profile (upper plot) is fairly rough and has $N_R = 1121$ data points, while the second (middle) is relatively smooth and has $N_S = 288$ points. The bottom plot gives an expanded view of the smooth profile. In what follows, we center the profiles by subtracting off their sample means – these are $\overline{X}_R \doteq -19.92$ for the rough profile and $\overline{X}_S \doteq -8.93$ for the smooth.

One reason for collecting these profiles was to develop a way of generating representative artificial profiles for use in simulations. It is clear from the top two plots of Figure 322 that it is unrealistic to regard both profiles as portions of two realizations from a single stationary process – the mean values and the variances of the two profiles are quite different. Superficially, however, the profiles in the bottom and top plots appear to have similar "spikiness," so one simple model worth investigating is that the spectral content of the two profiles is the same except for a scaling factor; i.e., if we let $S_R(\cdot)$ and $S_S(\cdot)$ represent the SDFs for the rough and smooth profiles, we have $S_R(f) = cS_S(f)$ over all f for some constant $c > 0$. Here we informally test this hypothesis by comparing estimated spectra for the two series.

The light choppy curve in Figure 323(a) – replicated in plots (c) and (e) – is the periodogram $\hat{S}_R^{(\mathrm{P})}(\cdot)$ for the rough profile; similarly, the periodogram $\hat{S}_S^{(\mathrm{P})}(\cdot)$ for the smooth profile appears in Figures 323(b), (d) and (f). An examination of direct spectral estimates using Slepian data tapers with $NW = 1/\Delta_t, 2/\Delta_t$ and $4/\Delta_t$ shows that, for both profiles, these estimates do not differ in any substantial way from the periodograms. Thus, in contrast to the ocean wave series, the periodograms $\hat{S}_R^{(\mathrm{P})}(\cdot)$ and $\hat{S}_S^{(\mathrm{P})}(\cdot)$ appear to be bias free. Figure 323 displays both periodograms over positive Fourier frequencies less than or equal to the Nyquist frequency $f_{\mathcal{N}} \doteq 0.2823$. The spacing between these frequencies is $1/(N_R\,\Delta_t) \doteq 0.0005$ and $1/(N_S\,\Delta_t) \doteq 0.0020$ cycles/meter for the rough and smooth profiles. The fact that $\hat{S}_S^{(\mathrm{P})}(\cdot)$ looks smoother than $\hat{S}_R^{(\mathrm{P})}(\cdot)$ is merely because the smooth profile has a smaller sample size (cf. Figure 207). Note that, since the units for the profiles are in feet and since the "time" variable is measured in meters, the SDFs has units of squared feet per cycle per meter.

We next smooth the two periodograms to better compare them. Since the dynamic ranges of the periodograms are not large (less than about 40 dB), smoothing window leakage should not be an issue here, so we can safely use the Parzen lag window. The two periodograms suggest that the underlying SDFs are dominantly unimodal with the mode at zero frequency, so we can use the estimator $\hat{B}_T$ of Equation (300c) to compute the bandwidth for the two profiles. For the rough profile we obtain $\tilde{B}_T = 0.0239$, and for the smooth profile, $\tilde{B}_T = $

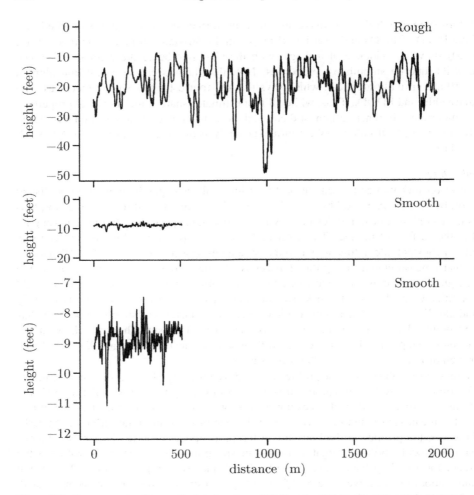

Figure 322 Two segments of ice profile data (courtesy of W. Fox, Applied Physics Laboratory, University of Washington). The top plot shows a relatively rough profile, while the middle, a shorter but relatively smooth one (also shown on the bottom plot with an expanded scale). The "time" values are really distances measured in meters, while the time series values are in feet. The distance between observations is $\Delta_t = 1.7712$ m. The rough profile has 1121 data points, while the smooth one has 288 points.

0.1069. Following the advice of Section 7.8, we select the lag window parameter m for each profile such that the resulting smoothing window bandwidth B_W is equal to $\tilde{B}_T/2$. Using the approximation $B_W = 1.85/(m\,\Delta_t)$ from Table 279, we obtain

$$m = \left[\frac{3.7}{\tilde{B}_T\,\Delta_t}\right] = \begin{cases} 88, & \text{for the rough profile;} \\ 20, & \text{for the smooth profile,} \end{cases} \quad (322)$$

where $[x]$ here denotes x rounded to the nearest integer. The resulting lag window estimators $\hat{S}^{(LW)}_{R,88}(\cdot)$ and $\hat{S}^{(LW)}_{S,20}(\cdot)$ are shown as the dark smoother curves in Figures 323(a) and (b).

Rather than objectively smoothing the periodograms through estimation of the spectral bandwidth, we can also entertain the objective methods described in Sections 7.9 (automatic smoothing after a log transformation) and 7.10 (bandwidth selection via generalized cross-validation). Using the first of these two methods with the Parzen lag window, we set m to the

7.12 Examples of Lag Window Spectral Estimators

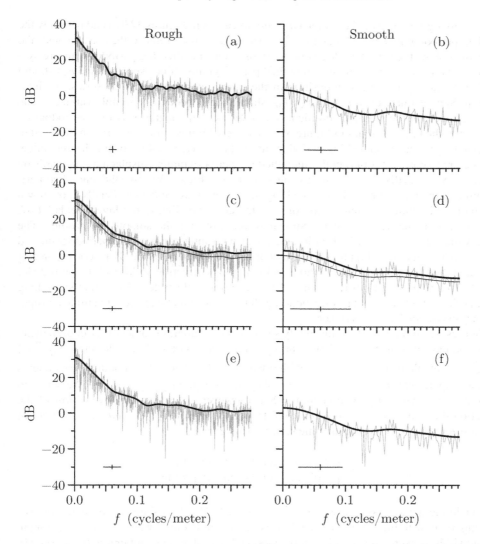

Figure 323 Periodogram for rough ice profile data (light choppy curves replicated in three left-hand plots) and for smooth profile (right-hand plots). The darkest curve in each plot is a smoothed version of the corresponding periodogram. Plots (a) and (b) show Parzen lag window estimates with the lag window parameter m set by equating the smoothing window bandwidth B_W to $\tilde{B}_T/2$, where $\tilde{B}_T$ is the estimator of the bandwidth of the time series given by Equation (300c); the darkest curves in plots (c) and (d) are Parzen lag window estimates with m set to the minimizer of $\hat{I}_m$ of Equation (303e), while the smooth curves just below these are smoothed versions of the log periodogram, again based on the Parzen lag window; and, finally, plots (e) and (f) are discretely smoothed periodograms with the smoothing parameters m set to the minimizer of $\mathrm{GCV}(m)$ of Equation (309a). The crisscross in the lower left-hand corner of each plot shows the bandwidth measure $B_{\mathcal{U}}$ and the length of 95% CIs.

minimizer of $\hat{I}_m$ of Equation (303e), yielding

$$m = \begin{cases} 35, & \text{for the rough profile;} \\ 11, & \text{for the smooth profile} \end{cases} \tag{323}$$

(due to centering of the time series, the periodograms at $f = 0$ are zero, so their logs are problematic – C&E [1] for Section 7.9 discusses how to handle this difficulty). We can use the selected m in conjunction with the Parzen lag window to smooth either the periodogram or the log periodogram. The thicker of the two smooth curves in Figures 323(c) and (d) show the smoothed periodograms $\hat{S}_{R,35}^{(\text{LW})}(\cdot)$ and $\hat{S}_{S,11}^{(\text{LW})}(\cdot)$ (the crisscrosses refer to these estimates); the thinner ones show the smoothed log periodograms. The smoothed log periodograms are systematically below the corresponding smoothed periodograms. Since spectral analysis should provide a frequency-based decomposition of the sample variance of a time series, the integral of an SDF estimate over $f \in [-f_\mathcal{N}, f_\mathcal{N}]$ should ideally equal the sample variance. The sample variance for the rough ice profile is 44.95. Numerical integrations of the periodogram and the smoothed periodogram in Figure 323(c) agree with this value; integration of the SDF corresponding to the smoothed log periodogram yields 19.41, i.e., 43% of the desired value. If we attempt to correct for bias in the smoothed log periodogram by displacing it upwards by $10\log_{10}(e)\gamma \doteq 2.5\,\text{dB}$, (see the discussion surrounding Figure 305(b)), the smoothed periodogram and the bias-corrected smoothed log periodogram agree better visually (not shown in the plots), but numerical integration of the SDF corresponding to the latter yields 34.57, which is 77% of the desired value. Similar results hold for the smooth ice profile. The sample variance is 0.2345, which is in agreement with integration of the periodogram and smoothed periodogram in Figure 323(d); by contrast, integration of the SDFs corresponding to the smoothed log periodogram without and with bias correction yield 0.1150 and 0.2048, i.e., 49% and 87% of the desired value. These examples illustrate the fact that smoothing the log periodogram can yield a corresponding SDF estimate that does not constitute a proper frequency-based analysis of variance.

Figures 323(e) and (f) show discretely smoothed periodograms objectively chosen using the method described in Section 7.10. For a given m, the smoothed periodograms $\hat{S}_{R,m}^{(\text{DSP})}(\cdot)$ and $\hat{S}_{S,m}^{(\text{DSP})}(\cdot)$ are defined in accordance with Equation (307), where the smoothing coefficients $g_{m,j}$ are proportional to $\exp(-(mj)^2)$ and obey the constraint $\sum_j g_{m,j} = 1$. We set m to the minimizer of $\text{GCV}(m)$ of Equation (309a) (however, we adjust the lower limit of the summation there to be $k = 1$ so as to avoid taking the log of $\hat{S}_R^{(\text{P})}(0)$ and $\hat{S}_S^{(\text{P})}(0)$, both of which are zero due to centering the time series). This gives

$$m \doteq \begin{cases} 0.0459, & \text{for the rough profile;} \\ 0.0712, & \text{for the smooth profile,} \end{cases}$$

which yield the dark smooth curves $\hat{S}_{R,0.0459}^{(\text{DSP})}(\cdot)$ and $\hat{S}_{S,0.0712}^{(\text{DSP})}(\cdot)$ shown in plots (e) and (f). For comparison with the Parzen lag window estimates shown in plots (a) to (d), we compute the bandwidth measures for the discretely smoothed periodograms, obtaining $B_\mathcal{U} \doteq 0.0278$ for the rough profile and $B_\mathcal{U} \doteq 0.0703$ for the smooth (see Equation (347b)). We then determine Parzen lag window estimates $\hat{S}_{R,m}^{(\text{LW})}(\cdot)$ and $\hat{S}_{S,m}^{(\text{LW})}(\cdot)$ with m chosen so that the associated bandwidth measures for these estimates are as close as possible to the measures for the discretely smoothed periodograms. This gives

$$m = \begin{cases} 38, & \text{for the rough profile;} \\ 15, & \text{for the smooth profile,} \end{cases} \qquad (324)$$

which are comparable to the selections stated in Equation (323) that yield the Parzen lag window estimates shown in plots (c) and (d). The estimates $\hat{S}_{R,0.0459}^{(\text{DSP})}(\cdot)$ and $\hat{S}_{R,38}^{(\text{LW})}(\cdot)$ are almost the same, differing by less than 0.24 dB across the Fourier frequencies; likewise, $\hat{S}_{S,0.0712}^{(\text{DSP})}(\cdot)$ and $\hat{S}_{S,15}^{(\text{LW})}(\cdot)$ differ overall by less than 0.13 dB.

The crisscrosses in the lower left-hand corners of Figures 323(a) to (f) show, as usual, the bandwidth measure $B_\mathcal{U}$ (Equation (256a) for plots (a) to (d) and Equation (347b) for (e) and (f)) and the length in decibels of 95% CIs for the true SDFs at a given frequency (these

7.12 Examples of Lag Window Spectral Estimators

	Rough Profile			Smooth Profile		
	(a)	(c)	(e)	(b)	(d)	(f)
$B_{\mathcal{U}}$ (cycles/meter)	0.0122	0.0302	0.0278	0.0536	0.0964	0.0703
95% CI Length (dB)	3.54	2.22	2.31	3.33	2.46	2.89

Table 325 Lengths of horizontal and vertical components of crisscrosses in Figure 323.

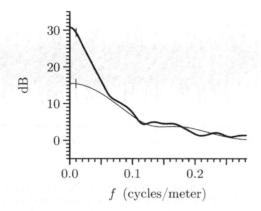

Figure 325 Parzen lag window estimates $\hat{S}_{R,35}^{(\text{LW})}(\cdot)$ (dark curve, which is the same as the one in Figure 323(c)) and $\hat{S}_{S,11}^{(\text{LW})}(\cdot)$ (light curve, which reproduces the one in Figure 323(d), but with an upward displacement of 13 dB).

length are dictated by Equation (266b) with EDOFs η computed using Equation (264b) with $C_h = 1$ for plots (a) to (d) and using Equation (347a) for (e) and (f)). Table 325 lists $B_{\mathcal{U}}$ and the 95% CI lengths for the six smoothed periodograms. The CI lengths range from 2.22 to 3.54 dB, which are nearly two orders of magnitude smaller than the length associated with an unsmoothed periodogram (21.6 dB). The smoothed periodograms for the smooth series look similar, but, for the rough series, the smoothed periodogram in Figure 323(a) looks noticeably bumpier than the ones in (c) and (e). All six SDF smoothed periodograms capture the germane patterns in the corresponding raw periodograms.

Figure 325 compares the Parzen lag window estimates $\hat{S}_{R,35}^{(\text{LW})}(\cdot)$ and $\hat{S}_{S,11}^{(\text{LW})}(\cdot)$ of Figures 323(c) and (d). The thick solid curve shows $\hat{S}_{R,35}^{(\text{LW})}(\cdot)$ while the thin curve is $\hat{S}_{S,11}^{(\text{LW})}(\cdot)$ after it has been moved up by 13 dB – this corresponds to multiplying $\hat{S}_{S,11}^{(\text{LW})}(\cdot)$ by a factor of 20. With this adjustment, there is reasonably good agreement between the two SDF estimates from about 0.05 cycles/meter up to the Nyquist frequency; however, for frequencies less than 0.05 cycles/meter, $\hat{S}_{R,35}^{(\text{LW})}(\cdot)$ and $20\,\hat{S}_{S,11}^{(\text{LW})}(\cdot)$ diverge. Vertical lines emanating from the lag window estimates at $f = 0.01$ cycles/meter indicate what 95% CIs would be when based upon the unadjusted and adjusted lag window estimates. These lines show that the difference between the two spectral estimates is much larger than can be reasonably explained by their inherent variability. Based upon these qualitative results, we can conclude that the simple model $S_R(f) = cS_S(f)$ is not viable.

Atomic Clock Data

Our third example involves the differences in times as kept by two atomic clocks. Figure 326(a) shows these time differences after adjustment for some isolated so-called phase

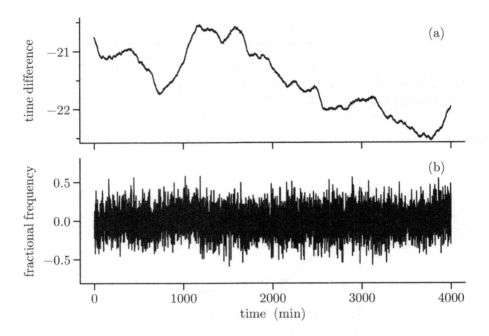

Figure 326 Atomic clock data (courtesy of L. Schmidt and D. Matsakis, US Naval Observatory). Plot (a) shows the time differences $\{X_t\}$ between two atomic clocks after removing a small number of phase glitches, while plot (b) shows the corresponding fractional frequency deviates $\{Y_t\}$ after multiplication by 10^{12}. The time differences are measured in nanoseconds, while the fractional frequency deviates are unitless.

glitches, which are small anomalous steps in time that are easy to detect and that do not reflect the inherent timekeeping ability of these clocks. The time differences are measured in nanoseconds (i.e., 10^{-9} of a second) at a sampling interval of $\Delta_t = 1\,\text{min}$. Since there are $N = 4000$ measurements in all, the total span of the data is close to 2.8 days. An atomic clock keeps time by counting the number of oscillations from an embedded frequency standard, which is a hydrogen maser for both clocks under study. This standard oscillates at a nominal frequency. If the standard were to generate oscillations that never deviated from its nominal frequency, the clock could keep time perfectly. The joint imperfection in the frequency standards embedded in the two atomic clocks is reflected by their fractional frequency deviates. If we consider the time differences to be a portion of a realization of a stochastic process $\{X_t\}$, the fractional frequency deviates are a realization of the first difference of this process divided by Δ_t, but only after we reexpress the latter in the same units as X_t (i.e., $\Delta_t = 60 \times 10^9$ nanoseconds). These deviates are thus given by

$$Y_t \stackrel{\text{def}}{=} \frac{X_t - X_{t-1}}{\Delta_t}$$

and are shown in Figure 326(b) after multiplication by 10^{12} to simplify labeling the vertical axis. We can regard $\{Y_t\}$ as a filtered version of $\{X_t\}$.

▷ **Exercise [326]** Appeal to Equations (143d), (143e) and (143b) to claim that, if $\{X_t\}$ possesses an SDF $S_X(\cdot)$, then $\{Y_t\}$ has an SDF given by

$$S_Y(f) = \frac{4\sin^2(\pi f \Delta_t)}{\Delta_t^2} S_X(f).$$

◁

7.12 Examples of Lag Window Spectral Estimators

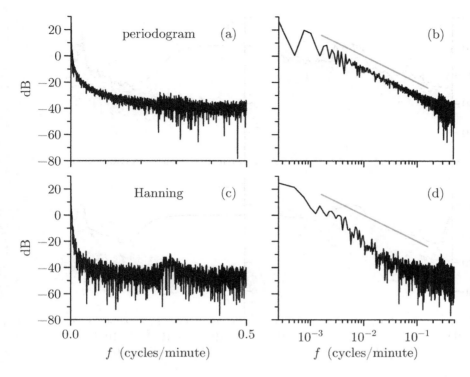

Figure 327 Periodogram (top row of plots) and direct spectral estimate based upon a Hanning data taper (bottom row) for atomic clock time differences. The left-hand column shows the SDF estimates in decibels versus a linear frequency scale, whereas the right-hand column show the same estimates, but versus a logarithmic scale. The gray lines in the right-hand column indicate the rolloff of an SDF proportional to f^{-2}.

Let us first consider estimation of the SDF $S_X(\cdot)$ for the time differences $\{X_t\}$. The upper row of plots in Figure 327 shows the periodogram $\hat{S}_X^{(P)}(\cdot)$ versus the positive Fourier frequencies on both a decibel versus linear scale (plot (a)) and a decibel versus log scale (b). The bottom row shows corresponding plots for a direct SDF estimate $\hat{S}_X^{(D)}(\cdot)$ based upon the Hanning data taper $\{h_t\}$ (Equation (189a) with $p = 1$; experimentation with other tapers indicates that the Hanning taper is a good choice here). To form this estimate, we centered $\{X_t\}$ by subtracting off its sample mean $\overline{X} \doteq -21.5$ and normalized $\{h_t\}$ such that $\sum h_t^2 = 1$. A comparison of $\hat{S}_X^{(P)}(\cdot)$ with $\hat{S}_X^{(D)}(\cdot)$ shows that there is substantial leakage in the periodogram. The leakage here is potentially misleading. A common practice amongst analysts of atomic clocks is to model the SDF $S_X(\cdot)$ for $\{X_t\}$ over selected ranges of frequencies in terms of canonical power-law processes. A power-law process dictates that $S_X(f) = C|f|^\alpha$. Restricting our attention to $f > 0$, this model implies that $\log_{10}(S_X(f)) = \log_{10}(C) + \alpha \log_{10}(f)$ and hence that, when $\log_{10}(S_X(f))$ is plotted versus $\log_{10}(f)$, we should see a line with a slope dictated by the power-law exponent α. One canonical choice for the power-law exponent is $\alpha = -2$, which corresponds to so-called random-walk phase noise (other canonical choices are 0, -1, -3 and -4; see Barnes et al., 1971). Since a decibel scale is merely a relabeling of a log scale, the format of Figure 327(b) allows placement of a gray line to depict the pattern for the SDF of a power-law process with $\alpha = -2$. The rolloff rate corresponding to this process is in good agreement with that exhibited by the periodogram over low frequencies ($f \leq 0.1$ cycles/minute); however, as Figure 327(d) shows, this agreement vanishes when the periodogram is replaced by a leakage-reduced direct SDF estimate. The fact that the

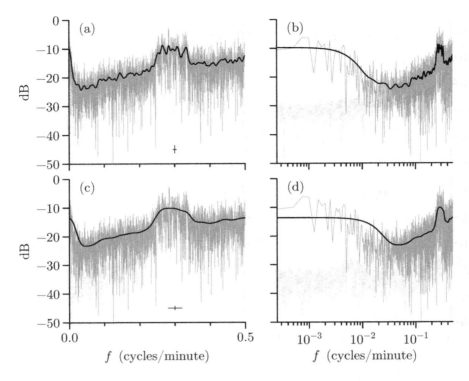

Figure 328 Periodogram for atomic clock fractional frequency deviates (light choppy curves) along with corresponding Gaussian lag window estimates (dark curves) with $m = 80$ (top row) and 20 (bottom). The horizontal (frequency) axes are linear in the left-hand column and logarithmic in the right. The crisscrosses in plots (a) and (c) show the bandwidth measure $B_\mathcal{U}$ (Equation (256a)) and the length of 95% CIs based upon the lag window estimates.

periodogram falsely suggests that a power-law process with $\alpha = -2$ might be an appropriate model for $\{X_t\}$ over low frequencies is attributable to its spectral window, which is Fejér's kernel. This window has sidelobes whose envelope decays at a rate of approximately f^{-2} (see the discussion surrounding Figure 270, where Fejér's kernel turns up as the smoothing window associated with the Bartlett lag window). Figure 327(d) suggests that, at frequencies lower than about 0.05 cycles/minute, the true SDF for $\{X_t\}$ decays at a faster rate than f^{-2}. Hence the overall pattern exhibited by the periodogram is really due to Fejér's kernel rather than the true SDF.

Let us now look at the estimation of the SDF $S_Y(\cdot)$ for the fractional frequency deviates $\{Y_t\}$. Figure 328 shows the periodogram $\hat{S}_Y^{(\mathrm{P})}(\cdot)$ in decibels versus the positive Fourier frequencies (light choppy curves), both on linear (left-hand column) and log (right-hand scales. In contrast to $\hat{S}_X^{(\mathrm{P})}(\cdot)$, here the periodogram shows no evidence of leakage, an assessment based on comparing it to a Hanning-based direct SDF estimate (not shown). Because the periodogram is so erratic, it is helpful to compute associated Gaussian lag window estimates (Equation (276)) to better see the overall pattern of the SDF. Two such estimates are shown in Figure 328 as dark curves, the one in the top row with m set to 80, and in the bottom row, to 20. In contrast to Figure 319, here Parzen lag window estimates (not shown) are virtually identical to the displayed Gaussian lag window estimates once the lag window parameter has been set to achieve similar smoothing window bandwidths B_W. There are two prominent peaks evident in all the estimates, a narrow one centered at $f = 0$ cycles/minute, and a broad one at $f = 0.28$ cycles/minute. The bandwidth of the lag window estimate in the top row is narrow enough to preserve the narrow low-frequency peak, but at the cost of introducing

7.12 Examples of Lag Window Spectral Estimators

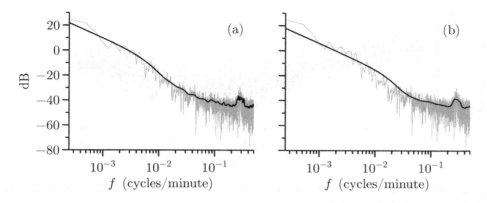

Figure 329 Hanning-based direct spectral estimate $\hat{S}_X^{(\mathrm{D})}(\cdot)$ for atomic clock time differences (light choppy curves, a reproduction of the dark choppy curve in Figure 327(d)), along with postcolored estimates $\hat{S}_X^{(\mathrm{PC})}(\cdot)$ based upon Gaussian lag window estimates $\hat{S}_{Y,m}^{(\mathrm{LW})}(\cdot)$ (thick curves). For plots (a) and (b), the underlying lag window estimates are the dark curves shown in, respectively, Figures 328(b) and (d).

unappealing bumpiness elsewhere; by contrast, the wider bandwidth in use in the bottom row yields a much smoother looking SDF estimate overall, but blurs out the low-frequency peak. This example illustrates that, because its bandwidth is constant across frequencies, a single lag window estimator cannot be easily adjusted to uniformly handle an SDF with peaks having quite different bandwidths.

Since the fractional frequency deviates $\{Y_t\}$ are the result of filtering the time differences $\{X_t\}$ with a rescaled first difference filter, and since the dynamic range of $S_Y(\cdot)$ is evidently markedly smaller than that of $S_X(\cdot)$, we can regard the first difference filter as a prewhitening filter. This suggests that we can estimate $S_X(\cdot)$ based on a lag window estimator $\hat{S}_{Y,m}^{(\mathrm{LW})}(\cdot)$ using the postcoloring procedure described in Section 6.5, thus obtaining the estimator

$$\hat{S}_X^{(\mathrm{PC})}(f) \stackrel{\text{def}}{=} \frac{\Delta_t^2}{4\sin^2(\pi f \Delta_t)} \hat{S}_{Y,m}^{(\mathrm{LW})}(f). \tag{329}$$

Figures 329(a) and (b) show postcolored estimates (dark curves) based upon, respectively, the $m = 80$ and 20 Gaussian lag window estimates shown in Figures 328(b) and (d). Both postcolored estimates are overlaid on the Hanning-based direct SDF estimate $\hat{S}_X^{(\mathrm{D})}(\cdot)$ (light choppy curves) shown in Figure 328(d). The postcolored estimates appear to be smoother versions of $\hat{S}_X^{(\mathrm{D})}(\cdot)$, with the differences between the $\hat{S}_X^{(\mathrm{PC})}(\cdot)$'s being due to the bandwidth properties of the underlying lag window estimates. The close agreement illustrates the fact that tapering and prewhitening are alternative ways of creating SDF estimates that do not suffer from severe bias as does the periodogram. (C&E [3] discusses a subtle issue that arises in interpreting the postcolored estimator $\tilde{S}_X^{(\mathrm{PC})}(\cdot)$, which calls into question the assumption that $\{X_t\}$ is a stationary process.)

As a concluding comment, we note that Barnes et al. (1971) and numerous subsequent papers about the analysis of clock data advocate the use of power-law processes to model both time differences and fractional frequency deviates. Such processes are reasonable models if SDF estimates, when plotted on a log/log scale, appear to obey a piecewise linear pattern, with each line spanning at least a decade of frequencies; i.e., letting $[f_\mathrm{L}, f_\mathrm{H}]$ denote an interval over which linearity hold, we desire $f_\mathrm{H}/f_\mathrm{L} \geq 10$. While there is compelling evidence that atomic clocks prevalent in the 1970s tended to have SDFs with this pattern, Figures 327(d), 328(b), 328(d) and 329 call into question piecewise linearity as a useful characterization for atomic

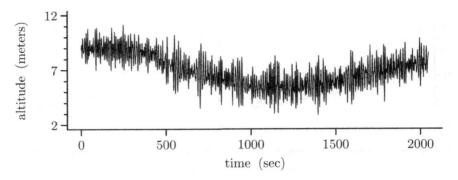

Figure 330 Ship altitude data (courtesy of J. Mercer, Applied Physics Laboratory, University of Washington). There are $N = 2048$ samples taken $\Delta_t = 1$ sec apart.

clocks of recent vintage. This example shows the need to carefully assess through spectral analysis the appropriateness of power-law processes as models for particular time series.

Ship Altitude Data
As our fifth example, we consider a time series collected on an oceanographic cruise in open waters of the Pacific Ocean (Mercer et al., 2009). To carry out certain experiments, the ship was maneuvered to stay at a fixed latitude and longitude (lat/long), but moderate winds and seas hampered this goal. The ship's lat/long position was monitored using the Global Positioning System, which also recorded the ship's altitude as a function of time. Figure 330 shows $N = 2048$ measurements of altitude collected at $\Delta_t = 1$ sec intervals (the series thus spans a little over 34 min). Prominent low- and high-frequency variations are visible – the former are due in part to tides and currents, and the latter, to swells. Understanding the nature of this time series is of interest to help interpret various oceanographic measurements collected aboard the ship.

We begin our analysis by considering the periodogram. Figure 331(a) shows $\hat{S}^{(\text{P})}(\cdot)$ on a decibel scale over the grid of positive Fourier frequencies. In contrast to the periodograms shown in Figures 226(a), 327(a) and 229(a), here we do not see obvious indications of inhomogeneous variability across frequencies, which is a tip-off that the periodogram might suffer from leakage. Figure 331(b) shows a corresponding direct spectral estimate $\hat{S}^{(\text{D})}(\cdot)$ based upon the Hanning data taper $\{h_t\}$ (Equation (189a), with $p = 1$ and C set such that $\sum_t h_t^2 = 1$). As usual, we centered the time series by subtracting off its sample mean prior to applying the Hanning taper. Here $\hat{S}^{(\text{P})}(\cdot)$ and $\hat{S}^{(\text{D})}(\cdot)$ have the same general pattern across frequencies, but there are some subtle differences that become more apparent once we reduce the inherent variability by applying a lag window. The light and dark curves in Figure 331(c) show lag window estimates based upon, respectively, $\hat{S}^{(\text{P})}(\cdot)$ and $\hat{S}^{(\text{D})}(\cdot)$. For both estimates, we used the Gaussian lag window of Equation (276) with the parameter m set to 80 (we chose this setting subjectively using the window-closing method described in Section 7.7). There are two intervals of frequencies over which we see systematic differences between the two lag window estimates: one at low frequencies, and the other at high. In both intervals, the estimate based upon the periodogram is higher than the one employing the Hanning-based direct spectral estimate. Both intervals trap low-power regions that are adjacent to high-power regions, so we can chalk up the discrepancy between the two estimates as being due to leakage in the periodogram. Comparison of the Hanning-based direct spectral estimate with ones using Slepian data tapers with $NW = 2, 3, 4$ and 5 (not shown) indicates that the Hanning-based estimate adequately ameliorates bias in the periodogram. The bias in the low-frequency region is close to $10\,\text{dB}$ in places (i.e., a potentially serious discrepancy of an order of magnitude), whereas

7.12 Examples of Lag Window Spectral Estimators

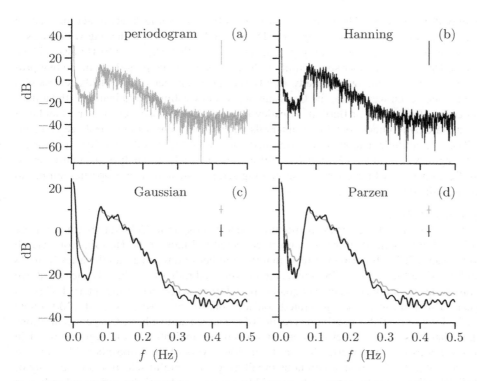

Figure 331 Periodogram (a) and a direct spectral estimate based upon a Hanning data taper (b) for the ship altitude data. Plot (c) shows two lag window estimates using a Gaussian lag window with parameter $m = 80$. The light and dark curves are smoothed versions of, respectively, the periodogram and the Hanning-based direct spectral estimate. Plot (d) is similar to (c), but now uses a Parzen lag window with parameter $m = 185$. The widths of the crisscrosses show either the bandwidth measure $B_\mathcal{H}$ of Equation (194) (plots (a) and (b)) or $B_\mathcal{U}$ of Equation (256a) ((c) and (d)); their heights show the length of a 95% CI for $10\log_{10}(S(f))$ (Equation (266a), with ν set to 2 in plots (a) and (b) and with ν set via Equations (264b) and (262) in (c) and (d)). In plots (c) and (d), the top and bottom crisscrosses are associated with, respectively, the periodogram-based and Hanning-based lag window estimates.

it is smaller in the high-frequency region (around 5 dB overall). This example of leakage contrasts in two ways with what we found for the ocean wave, atomic clock and chaotic beam data. First, leakage here is not confined to a single interval whose upper limit is the Nyquist frequency, but rather two distinct intervals. An SDF can theoretically have low power regions anywhere within the frequency range $[0, f_\mathcal{N}]$, opening up the possibility of leakage in disjoint intervals. Second, the variability in the periodogram for the ship altitude data appears to be consistent with what statistical theory dictates, whereas this was not the case with the other three examples. Apparent homogeneity of variance in the periodogram (when displayed on a log-based scale) is thus not a reliable indicator that the periodogram does not suffer from significant leakage.

Finally let us comment upon our choice of a Gaussian lag window. Figure 331(d) is similar to Figure 331(c) but is based on the Parzen lag window (Equation (275a)). Here we set the smoothing parameter m to 185, which forces the Parzen smoothing window bandwidth B_W to be the same as that of the Gaussian window (0.01 Hz – see Table 279). The Parzen lag window estimates are in good agreement with the Gaussian estimates over frequencies higher than about 0.05 Hz, but there are noticeable differences over the low-frequency interval $[0.01, 0.05]$. In particular, there is a pronounced ringing in the Parzen estimate associated with the Hanning data taper that is absent in the corresponding Gaussian estimate. This ringing is attributable to smoothing window leakage, which is caused by the sidelobes of the

Parzen smoothing window. The Gaussian estimate is free of such leakage, a fact that can be confirmed by comparing it with leakage-free Daniell and Bartlett–Priestley lag window estimates (not shown) with the smoothing parameter set such that $B_W = 0.01$ Hz also. Thus, whereas the Hanning data taper successfully reduces leakage in the periodogram, an inappropriate choice of a lag window can inadvertently introduce another form of leakage. As was also true for the ocean wave data, the Gaussian lag window estimate amalgamates the attractive properties of a Daniell or Bartlett–Priestley lag window estimate (leakage-free) and of a Parzen estimate (smoothness outside of intervals where smoothing window leakage is a concern). (The ship altitude data we have examined consists of the first 2048 values in a longer series of length $N = 14{,}697$ – Exercise [7.29] invites the reader to analyze the entire series. See also Section 11.6, which considers simulating series with statistical properties similar to those of the ship altitude series).

Atmospheric CO_2 Data

Our final example is a time series of monthly mean atmospheric CO_2 values (in parts per million by volume (ppmv)) from air samples collected at Mauna Loa, Hawaii, from March 1958 to February 2017. Ignoring the facts that the number of days in a month varies from 28 to 31 and that data for 7 of the 708 months are interpolated rather than directly measured, we regard this series as being collected at equally spaced times with a sampling interval of $\Delta_t = 1/12$ year. This time series has generated considerable interest. Keeling et al. (2009) note that it is the longest available continuous record of measured atmospheric CO_2 concentrations and argue that it is "... a precise record and a reliable indicator of the regional trend in the concentrations of atmospheric CO_2 in the middle layers of the troposphere" because of "... the favorable site location, continuous monitoring, and careful selection and scrutiny of the data." The light wiggly curve in Figure 333(a) shows a plot of the series and indicates an increasing trend upon which are superimposed seasonal variations. The dramatic upward trend is part of the scientific evidence that the concentration in atmospheric CO_2 is on the increase over the 59-year period depicted.

Because of the importance of this time series, it is of interest to consider statistical models for it as an aid in understanding its nature. Here we explore how far we can get with a simple model in which we assume the series to be a realization of a nonstationary process $\{X_t\}$ that is the sum of three components, namely, a low-order polynomial trend, seasonal variations and a stochastic noise process:

$$X_t = T_t + S_t + Y_t,$$

where $\{T_t\}$ is the trend, $\{S_t\}$ is a deterministic seasonal component (i.e., $S_{t+12} = S_t$ for all t) and $\{Y_t\}$ is the noise, which we assume to be stationary with zero mean and with SDF $S_Y(\cdot)$. The dark smooth curve in Figure 333(a) shows the result of fitting a quadratic trend, i.e.,

$$T_t = \alpha + \beta t + \gamma t^2,$$

to $\{X_t\}$, with the three parameters estimated by least squares. The estimated trend $\widehat{T}_t$ agrees reasonably well with the backbone of the data. With the trend defined in this manner, we could proceed to investigate the residuals $R_t \stackrel{\text{def}}{=} X_t - \widehat{T}_t$, a pathway we return to in Section 10.15. Here we take an alternative approach in which we employ linear time-invariant (LTI) filters to eliminate both the trend (assumed to be well-approximated by a quadratic polynomial) and the seasonal component. This approach is appealing because we can easily keep track of the effect of LTI filtering operations on the noise process, thus allowing qualitative evaluation of simple hypotheses about its SDF. Because subtracting off a linear least squares fit is not an LTI filter operation, it is harder to deduce the spectral properties of $\{Y_t\}$ from $\{R_t\}$ (in addition the residuals do not encapsulate just the noise process, but the seasonal component as well).

7.12 Examples of Lag Window Spectral Estimators 333

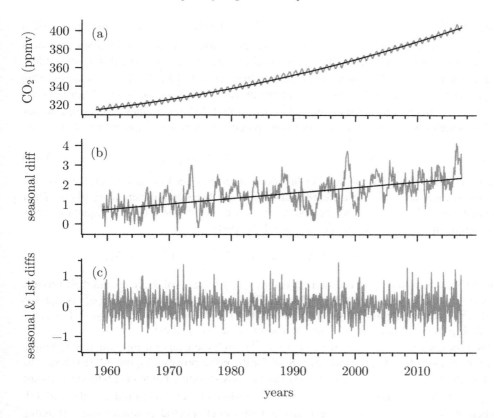

Figure 333 Atmospheric CO_2 data (P. Tans, Earth System Research Laboratory, National Oceanic & Atmospheric Administration, and R. Keeling, Scripps Institution of Oceanography – see the "Data" part of the website for the book for details on how to obtain these data). The light wiggly line in plot (a) depicts monthly values $\{X_t\}$ of atmospheric CO_2 concentration at Mauna Loa, Hawaii, from March 1958 to February 2017, while the dark smooth curve is a quadratic polynomial fit using least squares. The similarly displayed parts of plot (b) are the result $\{X_t^{(1)}\}$ of applying a lag $p = 12$ seasonal differencing filter to $\{X_t\}$ and a line fit via least squares. Plot (c) shows the result $\{X_t^{(2)}\}$ of taking $\{X_t^{(1)}\}$ and subjecting it to a first difference filter.

The task at hand is thus to find LTI filters that eliminate a quadratic trend and the seasonal component. As discussed in Section 2.8, we can remove seasonality in data having a period of $p = 12$ using a lag p seasonal differencing operation (Box et al., 2015; Brockwell and Davis, 2016). This procedure yields

$$X_t^{(1)} \stackrel{\text{def}}{=} X_t - X_{t-p} = T_t^{(1)} + Y_t^{(1)},$$

where $T_t^{(1)} \stackrel{\text{def}}{=} T_t - T_{t-p}$ and $Y_t^{(1)} \stackrel{\text{def}}{=} Y_t - Y_{t-p}$. For a quadratic trend we have

$$T_t^{(1)} = \alpha + \beta t + \gamma t^2 - \left(\alpha + \beta(t-p) + \gamma(t-p)^2\right) = (\beta - \gamma p)p + 2\gamma p t \stackrel{\text{def}}{=} \alpha' + \beta' t;$$

i.e., a lag p seasonal differencing filter reduces a quadratic trend to a linear trend. As demonstrated by Equations (39) and (40), we can reduce a linear trend to a constant by applying a first difference filter, leading us to consider

$$X_t^{(2)} \stackrel{\text{def}}{=} X_t^{(1)} - X_{t-1}^{(1)} = T_t^{(2)} + Y_t^{(2)} = \beta' + Y_t^{(2)},$$

where $T_t^{(2)} \stackrel{\text{def}}{=} T_t^{(1)} - T_{t-1}^{(1)}$ and $Y_t^{(2)} \stackrel{\text{def}}{=} Y_t^{(1)} - Y_{t-1}^{(1)}$. The light rough curve in Figure 333(b) shows $\{X_t^{(1)}\}$. Under our assumed model, this series should be the sum of a linear trend and a realization of a stationary process that has been subjected to a lag p seasonal differencing filter. The dark line in Figure 333(b) is what we get by fitting a line to $\{X_t^{(1)}\}$ via least squares. This line appears to be a reasonable summary of the upward pattern in the series. Figure 333(c) shows $\{X_t^{(2)}\}$, the result of subjecting $\{X_t^{(1)}\}$ to a first difference filter. According to our model, this series is a realization of a stationary process with mean β' and with an SDF given by

$$S_{X^{(2)}}(f) = |G_1(f)|^2 \cdot |G_{12}(f)|^2 S_Y(f), \qquad (334a)$$

where $G_1(\cdot)$ and $G_{12}(\cdot)$ are the transfer functions defined as per Equation (143e) and associated with, respectively, the first difference and lag $p = 12$ seasonal differencing filters. Their associated squared gain functions are both special cases of the one arising in Equation (158), which, upon adjusting for the fact that $\Delta_t \neq 1$ here, gives

$$|G_1(f)|^2 = 4\sin^2(\pi f \Delta_t) \text{ and } |G_{12}(f)|^2 = 4\sin^2(12\pi f \Delta_t).$$

Figure 335(a) shows $|G_1(f)|^2$ (dashed curve), $|G_{12}(f)|^2$ (dark solid curve) and their product (light curve) versus f on a linear/decibel scale. The squared gain function $|G_1(\cdot)|^2$ is equal to zero at zero frequency, whereas $|G_{12}(\cdot)|^2$ is zero for $f = k$ cycles per year, $k = 0, 1, \ldots, 6$ – hence their product is also equal to zero at integer-valued frequencies.

We now consider a spectral analysis of $\{X_t^{(2)}\}$, with the idea of using Equation (334a) to deduce the nature of $S_Y(\cdot)$, the unknown SDF for the noise process. Figure 335(b) shows the periodogram $\hat{S}_{X^{(2)}}^{(P)}(f)$ versus f (light curve) and a corresponding Parzen lag window estimate $\hat{S}_{X^{(2)},m}^{(LW)}(f)$ with lag window parameter $m = 185$ (dark curve), which reduces the variability inherent in the periodogram (comparisons with direct spectral estimates using Hanning and Slepian data tapers indicate that the periodogram suffices for our purposes here). Equation (334a) suggests estimating $S_Y(f)$ via

$$\hat{S}_Y(f) \stackrel{\text{def}}{=} \frac{\hat{S}_{X^{(2)},m}^{(LW)}(f)}{|G_1(f)|^2 \cdot |G_{12}(f)|^2};$$

however, this estimate is problematic because it is infinite at integer-valued frequencies. Instead we take an approach whereby we hypothesize an SDF for $\{Y_t\}$, use Equation (334a) to determine the corresponding SDF for $\{X_t^{(2)}\}$ and then compare this theoretical SDF with the estimate $\hat{S}_{X^{(2)},m}^{(LW)}(\cdot)$. If the comparison is favorable, we have informal evidence in favor of the hypothesized SDF.

Let us start with a very simple hypothesis, namely, that $\{Y_t\}$ is a white noise process with unknown variance σ_Y^2, which says that its SDF is given by $S_Y(f) = \sigma_Y^2 \Delta_t$. Under this hypothesis, the variances of $\{X_t^{(2)}\}$ and $\{Y_t\}$ are related by $\sigma_{X^{(2)}}^2 = 4\sigma_Y^2$ (an easy exercise, but this result also follows from Equation (42) within Exercise [2.8]). We can use the sample variance $\hat{\sigma}_{X^{(2)}}^2$ for $\{X_t^{(2)}\}$ to estimate the unknown variance σ_Y^2 via $\hat{\sigma}_Y^2 = \hat{\sigma}_{X^{(2)}}^2/4$, from which we get the SDF estimate $\hat{S}_Y(f) = \hat{\sigma}_Y^2 \Delta_t$. Plugging this estimate into Equation (334a) yields

$$\hat{S}_{X^{(2)}}(f) \stackrel{\text{def}}{=} |G_1(f)|^2 \cdot |G_{12}(f)|^2 \hat{S}_Y(f). \qquad (334b)$$

This estimate is shown as the light curve in Figure 335(c) along with the Parzen lag window estimate (dark curve). If we use the crisscross to assess the sampling variability in $\hat{S}_{X^{(2)},m}^{(LW)}(\cdot)$, we see significant mismatches between this estimate and $\hat{S}_{X^{(2)}}(\cdot)$ over a considerable range of

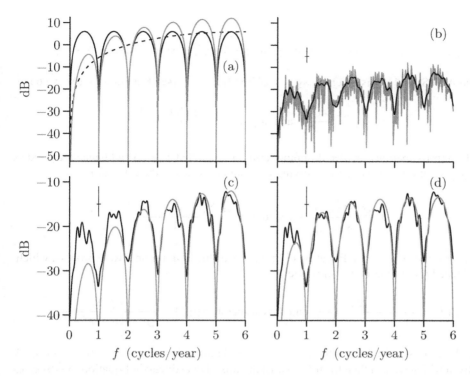

Figure 335 Squared gain functions, spectral estimates and hypothesized SDFs for CO_2 data. Plot (a) shows the squared gain functions for the lag $p = 12$ seasonal differencing filter (dark solid curve) and the first difference filter (dashed curve), along with their product (light curve). Plot (b) shows the periodogram (light bumpy curve) for $\{X_t^{(2)}\}$, i.e., the CO_2 data after subjection to both seasonal and first differencing. Superimposed upon the periodogram is a corresponding Parzen lag window estimate (dark curve) with smoothing window parameter $m = 185$. The width of the crisscross shows the bandwidth measure $B_{\mathcal{U}}$ (Equation (256a)) for the Parzen estimate, while its height is the length of a 95% CI for $10\log_{10}(S_{X^{(2)}}(f))$ (Equation (266a), with $\nu \doteq 14$). The curve for the Parzen estimate and the crisscross are reproduced in plots (c) and (d). The light curve in (c) shows a theoretical SDF for $\{X_t^{(2)}\}$ under the hypothesis that the underlying noise process $\{Y_t\}$ for the CO_2 data is white noise, while the one in (d) assumes noise dictated by a first-order autoregressive process.

low frequencies. This informal test suggests that the hypothesis that $\{Y_t\}$ is white noise is not tenable – we need to consider some form of colored noise. (There are also mismatches at the integer-valued frequencies, but these are of less concern: the bandwidth of the Parzen estimate – indicated by the width of the crisscross in Figure 335(c) – is wider than the narrow dips in $\hat{S}_{X^{(2)}}(\cdot)$, so we cannot expect the Parzen estimate to faithfully capture these narrow-band features.)

Accordingly, let us now entertain the hypothesis that $\{Y_t\}$ is a first-order autoregressive (AR(1)) process, which is a popular model for colored noise in climate research (von Storch and Zwiers, 1999). In keeping with Equation (144c), we presume $Y_t = \phi Y_{t-1} + \epsilon_t$, where $|\phi| < 1$ is an unknown parameter, and $\{\epsilon_t\}$ is a white noise process with mean zero and unknown variance σ_ϵ^2. We need to estimate ϕ and σ_ϵ^2 somehow based upon the observed filtered series $\{X_t^{(2)}\}$. A simple approach is to obtain expressions for ϕ and σ_ϵ^2 in terms of the lag $\tau = 0$ and $\tau = 1$ elements of the ACVS $\{s_{X^{(2)},\tau}\}$ for $\{X_t^{(2)}\}$. We can then use these expressions in conjunction with the standard estimators $\hat{s}_{X^{(2)},0}^{(P)}$ and $\hat{s}_{X^{(2)},1}^{(P)}$ to estimate ϕ and σ_ϵ^2. Two applications of Equation (42) yield

$$s_{X^{(2)},0} = 4s_{Y,0} - 4s_{Y,1} + 2s_{Y,11} - 4s_{Y,12} + 2s_{Y,13}$$

and

$$s_{X^{(2)},1} = -2s_{Y,0} + 4s_{Y,1} - 2s_{Y,2} + s_{Y,10} - 2s_{Y,11} + s_{Y,12} - 2s_{Y,13} + s_{Y,14}.$$

Since the ACVS for $\{Y_t\}$ is given by $s_{Y,\tau} = \phi^{|\tau|}\sigma_\epsilon^2/(1-\phi^2)$ (see Equation (44c)), we have

$$s_{X^{(2)},0} \approx \frac{4\sigma_\epsilon^2}{1+\phi} \quad \text{and} \quad s_{X^{(2)},1} \approx -\frac{2\sigma_\epsilon^2(1-\phi)}{1+\phi},$$

where the approximations involve dropping terms ϕ^τ such that $\tau \geq 10$ and hence are good as long as $|\phi|$ is not too close to unity. Solving for ϕ and σ_ϵ^2 leads to the estimators

$$\hat{\phi} \stackrel{\text{def}}{=} \frac{2\hat{s}_{X^{(2)},1}^{(P)}}{\hat{s}_{X^{(2)},0}^{(P)}} + 1 \quad \text{and} \quad \hat{\sigma}_\epsilon^2 \stackrel{\text{def}}{=} \frac{\hat{s}_{X^{(2)},0}^{(P)}(1+\hat{\phi})}{4}.$$

Setting $p = 1$ and $\phi_{1,1} = \phi$ in Equation (145a) gives us the SDF for an AR(1) process, which, when adjusted for Δ_t not being unity, yields

$$S_Y(f) = \frac{\sigma_\epsilon^2 \Delta_t}{|1 - \phi e^{-i2\pi f \Delta_t}|^2} \quad \text{and the estimator} \quad \hat{S}_Y(f) = \frac{\hat{\sigma}_\epsilon^2 \Delta_t}{|1 - \hat{\phi} e^{-i2\pi f \Delta_t}|^2}.$$

Applying this procedure to the CO_2 data, we obtain the estimates $\hat{\phi} \doteq 0.36$ and $\hat{\sigma}_\epsilon^2 \doteq 0.064$, from which we can then form $\hat{S}_Y(\cdot)$. Using this SDF estimate in Equation (334b) gives us the imputed SDF $\hat{S}_{X^{(2)}}(\cdot)$ for the filtered process $\{X_t^{(2)}\}$ under the hypothesis that the noise process $\{Y_t\}$ is an AR(1) process. The light curve in Figure 335(d) shows this imputed SDF, which we need to compare with the lag window estimate $\hat{S}_{X^{(2)},m}^{(\text{LW})}(\cdot)$ (dark curve). While there is markedly better agreement between $\hat{S}_{X^{(2)},m}^{(\text{LW})}(\cdot)$ and the imputed SDF here than under the white noise hypothesis (Figure 335(c)), there are still significant mismatches at low frequencies (i.e., less than about 0.5 cycles/year) as assessed by the crisscross. We can conclude that a potentially important part of the covariance structure of the noise process is not being captured by an AR(1) model.

To conclude, this example demonstrates that spectral analysis is useful in assessing the effect of LTI filtering operations on a time series. If our operating model for the series has additive trend, seasonal and noise components, removal of the first two by operations other than filtering (e.g., least squares) can make it difficult to determine the SDF for the noise component. For the CO_2 data, we found an AR(1) model to be adequate except at capturing certain low-frequency fluctuations, where the model underestimates the contribution due to these frequencies. To compensate for this underestimation, our spectral analysis suggests that an alternative model worth exploring is one exhibiting a power law over low frequencies. We introduced the notion of power-law processes in our discussion of the atomic clock data. They come into play here because processes obeying a power law at low frequencies can attach more importance to those frequencies than an AR(1) process is capable of doing.

Comments and Extensions to Section 7.12

[1] The usual way we have plotted SDFs and their estimates is on a decibel scale versus frequency on a linear scale, i.e., $10 \log_{10}(S(f))$ versus f (as examples, see Figures 319, 323 and 327(c)). Occasionally we have utilized a linear/linear scale (Figures 172 and 267(a)) and a decibel/log scale (Figures 327(b) and (d) and Figure 329). Figure 338a illustrates some strengths and weaknesses of these three ways of displaying an SDF and, for completeness, shows a fourth possibility we have not considered so far,

7.12 Examples of Lag Window Spectral Estimators

namely, a linear/log scale (plot (b)). All four plots in this figure depict the same SDF $S(\cdot)$, which is the sum of the SDF shown in Figure 125(c) and of the SDF for an AR(1) process dictated by the model $X_t = 0.95 X_{t-1} + \epsilon_t$, where $\{\epsilon_t\}$ is a white noise process with zero mean and variance 0.25 (for convenience, we take the sampling interval Δ_t to be unity so that the Nyquist frequency is $f_\mathcal{N} = 0.5$). Plot (a) is on a linear/linear scale and clearly shows that the SDF has a prominent low-frequency component in addition to twin peaks located near 0.11 and 0.14 cycles per unit time. Since the integral of $S(\cdot)$ over $[-f_\mathcal{N}, f_\mathcal{N}]$ is equal to the variance of the underlying stationary process, the area under the curve in plot (a) is equal to half the process variance. This plot clearly shows that low-frequency fluctuations contribute more to the overall variance than do fluctuations attributable to either one of the twin peaks individually. By contrast, the usual decibel/linear plot (c) shows an additional peak near $f = 0.35$, which is insignificant on the linear/linear plot. The ability to showcase interesting features in an SDF that are overshadowed by more dominant components is a strength of a decibel/linear plot, but a downside is that we must exercise discipline when using this plot to interpret the relative importance of various portions of the SDF: visually it appears that the peak at $f = 0.35$ is of greater importance as a contributor to the overall variance than actually is the case. (When plotting a nonparametric SDF estimate rather than a theoretical SDF, an additional advantage to a decibel/linear scale is the ease with which we can display pointwise confidence intervals for the true unknown SDF using a crisscross, as is demonstrated in Figure 267.) The decibel/log scale of plot (d) allows us to study how the SDF varies as f decreases to zero while still pulling out the small peak at $f = 0.35$ in addition to the twin peaks. This particular SDF flattens out with decreasing frequency, an aspect of $S(\cdot)$ not readily apparent in plots (a) and (c). In addition, we noted in our discussion of the atomic clock data that a linear feature in an SDF displayed on decibel/log scale indicates that a power-law process might be a useful characterization for the SDF over a certain range of frequencies. In plot (d) there is a linear feature at low frequencies. Its slope is zero, thus indicating that $S(\cdot)$ varies as a power-law process with exponent zero (i.e., white noise) as f decreases to zero. Depending on what we wish to emphasize, some combination of plots (a), (c) and (d) might be appropriate to use.

Figure 338a(b) shows the SDF on a linear/log scale, which we have not found occasion to use previously. Such a plot might be of interest if there were fine structure in the SDF at low frequencies that is jammed together on a linear/linear plot. This is not the case in Figure 338a(b), so there is no clear rationale for advocating this plot over (a), (c) or (d). In particular, if we want to point out which frequencies are the dominant contributors to the overall variance, it would be better to use Figure 338a(a) rather than Figure 338a(b) because the latter is open to the misinterpretation that low-frequency fluctuations are much more important than those coming from the twin peaks. There is, however, an interesting variation on a linear/log plot called a *variance-preserving SDF* that seeks to prevent such a misinterpretation and that has found use in practical applications (see, for example, Emery and Thomson, 2001). For a stationary process $\{X_t\}$ possessing an SDF $S(\cdot)$ and a sampling interval of unity, the fundamental relationship

$$\operatorname{var}\{X_t\} = \int_{-1/2}^{1/2} S(f)\, df = 2 \int_0^{1/2} S(f)\, df$$

motivates plotting $S(f)$ versus $f \in [0, 1/2]$ with the interpretation that $S(f)\, df$ is the contribution to the process variance due to frequencies in a small interval about f. Letting λ denote an arbitrary nonzero constant, use of

$$\frac{d(\lambda \log_e(f))}{df} = \frac{\lambda}{f}$$

allows us to reexpress the fundamental relationship as

$$\operatorname{var}\{X_t\} = \frac{2}{\lambda} \int_0^{1/2} f S(f)\, d(\lambda \log_e(f)).$$

The above motivates plotting the so-called variance-preserving SDF $fS(f)/\lambda$ versus $\lambda \log_e(f)$ with the interpretation that $\frac{fS(f)}{\lambda} d(\lambda \log_e(f))$ also represents the contribution to the process variance due to frequencies in a small interval about f. Figure 338b shows the same SDF depicted in Figure 338a, but now in a variance-preserving manner (here we set $\lambda = \log_{10}(e)$, leading to $\lambda \log_e(f) = \log_{10}(f)$). Since

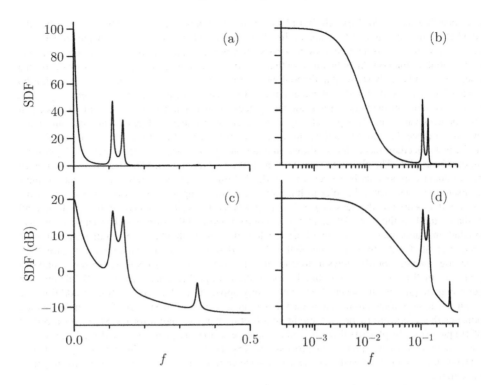

Figure 338a Four ways of plotting the same spectral density function $S(f)$. Plot (a) shows the SDF on a linear scale versus frequency, also on a linear scale; (b) uses a linear/log scale; (c), a decibel/linear scale; and (d), a decibel/log scale. The displayed SDF is the sum of three autoregressive SDFs, namely, ones corresponding to the AR(4) process of Equation (35a), the AR(2) process of Equation (125) and the AR(1) process defined by $X_t = 0.95 X_{t-1} + \epsilon_t$, where $\{\epsilon_t\}$ is zero mean white noise process with variance 0.25.

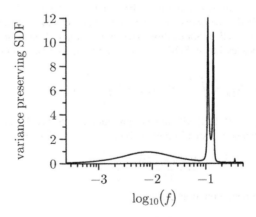

Figure 338b As in Figure 338a, but with the spectral density function displayed as a variance-preserving spectrum; i.e., $fS(f)/\log_{10}(e)$ is plotted versus $\log_{10}(f)$.

7.12 Examples of Lag Window Spectral Estimators

$fS(f)$ is virtually zero for values of $\log_{10}(f)$ smaller than what are shown in Figure 338b, the area under the curve depicted there is a tiny amount less than the area under the curve in Figure 338a(a), i.e., half of the process variance. Advocates of this type of plot might argue that it is to be preferred over the plot shown in Figure 338a(b) because, while both offer depictions against frequency on a log scale, the variance-preserving SDF allows us to see the relative contributions to the overall variance more clearly; however, the example we have put together illustrates an important caveat, namely, that distortions can arise, leading to an improper interpretation of the properties of the underlying SDF. Specifically, the variance-preserving SDF in Figure 338b distorts the low-frequency content of the SDF such that it appears as a broad peak centered near $\log_{10}(f) = -2.1$ (i.e., $f \doteq 0.008$). A practitioner might be tempted to regard this peak as evidence for a weak tendency in the process to have oscillations centered about this frequency, an interpretation that is not consistent with the actual SDF for the process. Because of the danger of similar misinterpretations (which, unfortunately, have occurred in the literature), we do not advocate use of the variance-preserving SDF when unaccompanied by one or more of the standard plots illustrated in Figure 338a (for example, the false impression created by Figure 338b could be corrected by considering it alongside Figure 338a(b), which shows that the SDF does not have an actual peak around $f \doteq 0.008$).

[2] In C&E [3] for Section 7.10 we noted an adaptation of the GCV criterion for use in determining the smoothing parameter m for a lag window estimator. Using this adaptation with the Parzen lag window and with the rough and smooth sea-ice profiles yields $m = 39$ for the former and 15 for the latter. These choices are almost identical to the choices $m = 38$ and 15 displayed in Equation (324) – these yield the Parzen lag window estimates that are the closest in bandwidth to those of the discretely smoothed periodograms of Figures 323(e) and (f). For these two time series the GCV criteria for discretely smoothed periodograms and for lag window estimators yield smoothed periodograms that are almost identical.

[3] Here we reconsider the postcolored SDF estimator $\hat{S}_X^{(\mathrm{PC})}(\cdot)$ for the atomic clock time differences $\{X_t\}$ (the thick curves in Figure 329 show two such estimates). As indicated by its definition in Equation (329), this estimator is the product of a lag window estimator and the squared gain function for a rescaled first difference filter. We can write the lag window estimator as

$$\hat{S}_{Y,m}^{(\mathrm{LW})}(f) = \Delta_t \sum_{\tau=-(N-1)}^{N-1} w_{m,\tau} \hat{s}_\tau^{(\mathrm{D})} e^{-i 2\pi f \tau \Delta_t} = \Delta_t \left(\hat{s}_0^{(\mathrm{D})} + 2 \sum_{\tau=1}^{N-1} w_{m,\tau} \hat{s}_\tau^{(\mathrm{D})} \cos(2\pi f \tau \Delta_t) \right)$$

(see Equation (248a)). Since the cosine function is bounded and continuous with derivatives of all orders, this estimator must also be such since it is just a linear combination of cosines with different frequencies. In addition, its derivative at $f = 0$ must be zero (an easy exercise), which implies that we can approximate $\hat{S}_{Y,m}^{(\mathrm{LW})}(f)$ by a constant, say C, over a small interval of frequencies centered about zero. We can use the small angle approximation $\sin(x) \approx x$ to approximate the squared gain function over this same interval, yielding

$$\hat{S}_X^{(\mathrm{PC})}(f) \approx \frac{C}{4\pi^2 f^2}$$

for f close to zero. Thus $\hat{S}_X^{(\mathrm{PC})}(f) \to \infty$ as $|f| \to 0$ in a manner such that the integral of $\hat{S}_X^{(\mathrm{PC})}(f)$ over $[-f_\mathcal{N}, f_\mathcal{N}]$ necessarily diverges to infinity. A fundamental property of an SDF is that it integrates to the variance of the associated stationary process, which must be finite. Evidently the SDF estimator $\hat{S}_X^{(\mathrm{PC})}(\cdot)$ does not correspond to the SDF for a stationary process, which suggests that $\{X_t\}$ is not a stationary process. Note that we would *not* reach this same conclusion based upon the Hanning-based direct spectral estimate $\hat{S}_X^{(\mathrm{D})}(\cdot)$ (the light choppy curves in Figure 329), which necessarily integrates to a finite value. Even though Figure 329 shows $\hat{S}_X^{(\mathrm{PC})}(\cdot)$ and $\hat{S}_X^{(\mathrm{D})}(\cdot)$ to be similar over the displayed frequencies, there is a key difference between them: whereas $\hat{S}_X^{(\mathrm{PC})}(f)$ explodes as $|f| \to 0$, $\hat{S}_X^{(\mathrm{D})}(f)$ converges to a finite value (this follows from an argument similar to the one we used for $\hat{S}_{Y,m}^{(\mathrm{LW})}(\cdot)$, which says that $\hat{S}_X^{(\mathrm{D})}(f)$ is well approximated by a constant over a sufficiently small interval about zero).

Because $\hat{S}_X^{(\mathrm{PC})}(\cdot)$ does not integrate to a finite value, it is not a valid SDF; on the other hand, the lag window estimate $\hat{S}_{Y,m}^{(\mathrm{LW})}(\cdot)$ for its rescaled first difference does integrate finitely and is a valid

SDF. This suggests that, even if we cannot regard $\{X_t\}$ as a stationary process, we can take its first difference (or a rescaling thereof) to be such. Yaglom (1958) expands the definition of an SDF to include stochastic processes $\{X_t\}$ that are nonstationary, but whose first- or higher-order differences are stationary. Suppose that $\{X_t\}$ is a stochastic process whose dth-order difference

$$X_t^{(d)} \stackrel{\text{def}}{=} \sum_{k=0}^{d} \binom{d}{k} (-1)^k X_{t-k}$$

is a stationary process with SDF $S_{X^{(d)}}(\cdot)$, where d is a nonnegative integer (for convenience, we take the sampling interval Δ_t to be unity for $\{X_t\}$ and hence also for $\{X_t^{(d)}\}$). Thus $X^{(0)} = X_t$, $X^{(1)} = X_t - X_{t-1}$, $X^{(2)} = X_t - 2X_{t-1} + X_{t-2}$ and so forth. Note that $X^{(d)} = X_t^{(d-1)} - X_{t-1}^{(d-1)}$ for all $d \geq 1$. Now, if $\{X_t\}$ were a stationary process with SDF $S_X(\cdot)$, then Exercise [5.4a] and the notion of a cascaded filter (see Section 5.5) tell us that the SDFs for $\{X_t^{(d)}\}$ and $S_X(\cdot)$ would be related by

$$S_{X^{(d)}}(f) = 4^d \sin^{2d}(\pi f) S_X(f).$$

If $\{X_t\}$ is instead nonstationary, Yaglom (1958) uses the above relationship to *define* an SDF for it:

$$S_X(f) \stackrel{\text{def}}{=} \frac{S_{X^{(d)}}(f)}{4^d \sin^{2d}(\pi f)}.$$

Here the nonstationary process $\{X_t\}$ is said to have *stationary differences of order d*. In general a process with dth-order stationary differences can be either stationary or nonstationary. A related concept is an *intrinsically stationary process of order d*: $\{X_t\}$ is said to be such if $\{X_t\}$, $\{X_t^{(1)}\}$, ..., $\{X_t^{(d-1)}\}$ are all nonstationary, but $\{X_t^{(d)}\}$ is stationary (see Chilès and Delfiner, 2012, or Stein, 1999). While $S_X(\cdot)$ need not possess all of the properties of a proper SDF (in particular, integration to a finite value), it can – with care – be dealt with in the same manner as the SDF for a stationary process. This theory allows us to interpret the postcolored SDF estimator of Equation (329) as an SDF for a first-order intrinsically stationary process.

7.13 Summary of Lag Window Spectral Estimators

Suppose we have a time series that is a portion $X_0, X_1, \ldots, X_{N-1}$ of a real-valued zero mean stationary process $\{X_t\}$ with a sampling interval of Δ_t (and hence a Nyquist frequency of $f_\mathcal{N} = 1/(2\Delta_t)$) and with an autocovariance sequence (ACVS) $\{s_\tau\}$ and spectral density function (SDF) $S(\cdot)$ related by $\{s_\tau\} \longleftrightarrow S(\cdot)$. (If $\{X_t\}$ has an unknown mean, we must replace X_t with $X_t' = X_t - \overline{X}$ in all computational formulae, where $\overline{X} = \sum_{t=0}^{N-1} X_t/N$ is the sample mean.) Using a data taper $\{h_t\} \longleftrightarrow H(\cdot)$ with normalization $\sum_t h_t^2 = 1$, we start by forming a direct spectral estimator for $S(\cdot)$:

$$\hat{S}^{(\text{D})}(f) \stackrel{\text{def}}{=} \Delta_t \left| \sum_{t=0}^{N-1} h_t X_t e^{-i2\pi f t \Delta_t} \right|^2 = \Delta_t \sum_{\tau=-(N-1)}^{N-1} \hat{s}_\tau^{(\text{D})} e^{-i2\pi f \tau \Delta_t}, \quad \text{(see (186b) and (188a))}$$

where the ACVS estimator $\{\hat{s}_\tau^{(\text{D})}\}$ is given by

$$\hat{s}_\tau^{(\text{D})} \stackrel{\text{def}}{=} \begin{cases} \sum_{t=0}^{N-|\tau|-1} h_{t+|\tau|} X_{t+|\tau|} h_t X_t, & |\tau| \leq N-1; \\ 0, & |\tau| \geq N \end{cases} \quad \text{(see (188b))}$$

(note that $\{\hat{s}_\tau^{(\text{D})}\} \longleftrightarrow \hat{S}^{(\text{D})}(\cdot)$). A lag window estimator $\hat{S}_m^{(\text{LW})}(\cdot)$ is the result of smoothing this direct spectral estimator across frequencies:

$$\hat{S}_m^{(\text{LW})}(f) \stackrel{\text{def}}{=} \int_{-f_\mathcal{N}}^{f_\mathcal{N}} W_m(f-\phi) \hat{S}^{(\text{D})}(\phi) \, d\phi, \quad \text{(see (248a))}$$

7.13 Summary of Lag Window Spectral Estimators

where $W_m(\cdot)$ is a smoothing window with a parameter m that controls the degree of smoothing to be imposed upon $\hat{S}^{(D)}(\cdot)$. The rationale for smoothing $\hat{S}^{(D)}(\cdot)$ is to obtain an estimator with reduced variance, which is desirable because direct spectral estimators are highly variable (see, e.g., the choppy curves in Figure 187). While the above equation emphasizes that $\hat{S}_m^{(LW)}(\cdot)$ is a filtered version of $\hat{S}^{(D)}(\phi)$, actual computation of a lag window estimator makes use of the inverse Fourier transform of $W_m(\cdot)$, namely, $\{w_{m,\tau}\} \longleftrightarrow W_m(\cdot)$:

$$\hat{S}_m^{(LW)}(f) = \Delta_t \sum_{\tau=-(N-1)}^{N-1} w_{m,\tau} \hat{s}_\tau^{(D)} e^{-i2\pi f \tau \Delta_t} \qquad \text{(see (248a))}$$

(note that $\{w_{m,\tau} \hat{s}_\tau^{(D)}\} \longleftrightarrow \hat{S}_m^{(LW)}(\cdot)$). The sequence $\{w_{m,\tau}\}$ is known as a lag window, from which $\hat{S}_m^{(LW)}(\cdot)$ gets its name. We assume that $w_{m,0} = 1$, that $w_{m,\tau} = 0$ for $|\tau| \geq N$ and that $w_{m,-\tau} = w_{m,\tau}$ for all τ. These conditions imply that the smoothing window $W_m(\cdot)$ is an even real-valued periodic function with a period of $2f_\mathcal{N}$ whose integral over $[-f_\mathcal{N}, f_\mathcal{N}]$ is unity. We can characterize the degree of smoothing that $W_m(\cdot)$ imparts to $\hat{S}^{(D)}(\cdot)$ using the smoothing window bandwidth measure given by

$$B_W \stackrel{\text{def}}{=} \frac{1}{\int_{-f_\mathcal{N}}^{f_\mathcal{N}} W_m^2(f)\,df} = \frac{1}{\Delta_t \sum_{\tau=-(N-1)}^{N-1} w_{m,\tau}^2}. \qquad \text{(see (251c) and (251e))}$$

A typical smoothing window resembles a bell-shaped curve (see Figure 259(b)), and B_W is a measure of the width of the curve.

Turning now to an examination of the statistical properties of $\hat{S}_m^{(LW)}(\cdot)$, we have

$$E\{\hat{S}_m^{(LW)}(f)\} = \int_{-f_\mathcal{N}}^{f_\mathcal{N}} \mathcal{U}_m(f - f') S(f')\,df', \qquad \text{(see (255b))}$$

where

$$\mathcal{U}_m(f) \stackrel{\text{def}}{=} \int_{-f_\mathcal{N}}^{f_\mathcal{N}} W_m(f - f'') \mathcal{H}(f'')\,df'' = \sum_{\tau=-(N-1)}^{N-1} w_{m,\tau}\, h \star h_\tau\, e^{-i2\pi f\tau \Delta_t},$$
$$\text{(see (255c) and (255d))}$$

is the spectral window associated with $\hat{S}_m^{(LW)}(\cdot)$; in the above, $\mathcal{H}(\cdot)$ is the spectral window for the underlying direct spectral estimator $\hat{S}^{(D)}(\cdot)$ (by definition, this window is the square modulus of the Fourier transform $H(\cdot)$ of the data taper $\{h_t\}$ used to create $\hat{S}^{(D)}(\cdot)$), and $\{h \star h_\tau\}$ is the autocorrelation of $\{h_t\}$ (see Equation (192c)). Under appropriate conditions, by letting the smoothing window bandwidth B_W decrease towards zero as the sample size N increases to infinity, the spectral window $\mathcal{U}_m(\cdot)$ acts as a Dirac delta function, allowing us to claim that $\hat{S}_m^{(LW)}(f)$ is asymptotically an unbiased estimator of $S(f)$. With proper care in selecting the data taper and lag window, it is often possible for $E\{\hat{S}_m^{(LW)}(f)\}$ to be approximately equal to $S(f)$ for finite N, but the quality of the approximation for a particular N depends highly upon the complexity of the true SDF. We take the effective bandwidth $B_\mathcal{U}$ of $\hat{S}_m^{(LW)}(\cdot)$ to be the autocorrelation width of its spectral window:

$$B_\mathcal{U} \stackrel{\text{def}}{=} \text{width}_a\{\mathcal{U}_m(\cdot)\} = \frac{\left(\int_{-f_\mathcal{N}}^{f_\mathcal{N}} \mathcal{U}_m(f)\,df\right)^2}{\int_{-f_\mathcal{N}}^{f_\mathcal{N}} \mathcal{U}_m^2(f)\,df} = \frac{\Delta_t}{\sum_{\tau=-(N-1)}^{N-1} w_{m,\tau}^2 (h \star h_\tau)^2}. \qquad \text{(see (256a))}$$

Note that $B_\mathcal{U}$ depends upon both the lag window and the data taper.

As to the second-moment properties of $\hat{S}_m^{(LW)}(\cdot)$, under suitable conditions,

$$\text{var}\{\hat{S}_m^{(LW)}(f)\} \approx \frac{C_h S^2(f)}{B_W N \Delta_t}, \qquad \text{(see (259b))}$$

where $C_h \geq 1$ is a variance inflation factor due solely to tapering:

$$C_h = N \sum_{t=0}^{N-1} h_t^4. \qquad \text{(see (262))}$$

This factor is unity if and only if $h_t = 1/\sqrt{N}$ (the default data taper, which turns $\hat{S}^{(D)}(\cdot)$ into the periodogram). Table 260 gives large sample approximations to C_h for some commonly used data tapers. If $B_W \to 0$ and $N \to \infty$ in such a way that $B_W N \to \infty$, then $\text{var}\{\hat{S}_m^{(LW)}(f)\}$ decreases to 0, which, when coupled with asymptotic unbiasedness, says that $\hat{S}_m^{(LW)}(f)$ is a consistent estimator of $S(f)$. Also, for $f'_j = j/(N'\Delta_t)$, $\text{cov}\{\hat{S}_m^{(LW)}(f'_j), \hat{S}_m^{(LW)}(f'_k)\} \approx 0$, where j and k are distinct integers such that $0 \leq j, k \leq \lfloor N'/2 \rfloor$; here $N' = N/c$ where $c \geq 1$ is a factor that depends primarily upon the bandwidth B_W of the smoothing window $W_m(\cdot)$ (see Section 7.3).

A large-sample approximation to the distribution of $\hat{S}_m^{(LW)}(\cdot)$ is that of a scaled chi-square random variable:

$$\hat{S}_m^{(LW)}(f) \stackrel{d}{=} \frac{S(f)}{\nu} \chi_\nu^2,$$

where ν is the equivalent degrees of freedom for $\hat{S}_m^{(LW)}(\cdot)$ given by

$$\nu = \frac{2 N B_W \Delta_t}{C_h} \qquad \text{(see (264d))}$$

(as ν gets large, $\hat{S}_m^{(LW)}(f)$ becomes asymptotically Gaussian distributed with mean $S(f)$ and variance $2S^2(f)/\nu$. The above leads to approximate $100(1 - 2p)\%$ pointwise confidence intervals (CIs) of the form

$$\left[\frac{\nu \hat{S}_m^{(LW)}(f)}{Q_\nu(1-p)}, \frac{\nu \hat{S}_m^{(LW)}(f)}{Q_\nu(p)}\right] \qquad \text{(see (265))}$$

for $S(f)$ and of the form

$$\left[10 \log_{10}(\hat{S}_m^{(LW)}(f)) + 10 \log_{10}\left(\frac{\nu}{Q_\nu(1-p)}\right), 10 \log_{10}(\hat{S}_m^{(LW)}(f)) + 10 \log_{10}\left(\frac{\nu}{Q_\nu(p)}\right)\right]$$
$$\text{(see (266a))}$$

for $10 \log_{10}(S(f))$ (the width of the latter CI is independent of $\hat{S}_m^{(LW)}(f)$).

Specific lag windows we've considered in detail are the Bartlett (Equation (269a)), Daniell (Equation (273a)), Bartlett–Priestley (Equation (274)), Parzen (Equation (275a)), Gaussian (Equation (276)) and Papoulis (Equation (278)) windows, with the Gaussian having particularly attractive properties. The smoothing window bandwidth B_W for each of these windows is controlled by a smoothing window parameter m whose setting is critical for obtaining a $\hat{S}_m^{(LW)}(\cdot)$ that is appropriate for a particular time series. A subjective method for setting m is window closing, which consists of visually comparing lag window estimates with different choices for m to the underlying $\hat{S}^{(D)}(\cdot)$ and picking m such that the resulting $\hat{S}_m^{(LW)}(\cdot)$ best captures the salient features in $\hat{S}^{(D)}(\cdot)$. One objective method for setting m is to estimate the bandwidth B_S of the unknown SDF and to pick m such that $B_W = B_S/2$. The theory behind this method assumes that the unknown SDF is dominantly unimodal, which is often – but certainly not always – the case in practical applications. A second method is based upon smoothing an estimator of the log of the SDF rather than SDF itself, which allows for m to be set by minimizing an appealing mean integrated square error criterion. While this method yields acceptable results for certain SDFs, it is problematic in other cases. A third method uses generalized cross-validation to objectively pick the appropriate degree of smoothing for a discretely smoothed periodogram.

7.14 Exercises

[7.1] Show that Equation (247f) is true.

[7.2] Show that Equation (250b) is equivalent to $w_{m,0} = 1$.

[7.3] Show that Equation (255d) is true. Does this equation still hold if we drop our assumption that $w_{m,\tau} = 0$ for $|\tau| \geq N$? If we replace $W_m(\cdot)$ with $V_m(\cdot)$ in Equation (255c), why does this not change the definition of $\mathcal{U}_m(\cdot)$? (Section 7.1 describes the distinction between $W_m(\cdot)$ and $V_m(\cdot)$.)

[7.4] (a) As an extension to Equation (186g), show that

$$E\{\hat{S}_m^{(\mathrm{LW})}(f)\} = \Delta_{\mathrm{t}} \sum_{\tau=-(N-1)}^{N-1} \left(w_{m,\tau} \sum_{t=0}^{N-|\tau|-1} h_{t+|\tau|} h_t \right) s_\tau \mathrm{e}^{-\mathrm{i}2\pi f \tau \Delta_{\mathrm{t}}}. \qquad (343\mathrm{a})$$

(b) As extensions to Equations (188d) and (188e), show that

$$\int_{-f_{\mathcal{N}}}^{f_{\mathcal{N}}} E\{\hat{S}_m^{(\mathrm{LW})}(f)\}\,\mathrm{d}f = s_0 \quad \text{and} \quad \int_{-f_{\mathcal{N}}}^{f_{\mathcal{N}}} \mathcal{U}_m(f)\,\mathrm{d}f = 1$$

under the usual assumptions $\sum_t h_t^2 = 1$ and $w_{m,0} = 1$.

[7.5] Use the approach of Section 7.4 to show that the equivalent number of degrees of freedom for the spectral estimator defined by Equation (246b) is

$$\nu = \frac{2N}{C_h \sum_{\tau=-(N-1)}^{N-1} \left| \sum_{j=-M}^{M} g_j \mathrm{e}^{\mathrm{i}2\pi j\tau/N'} \right|^2}.$$

[7.6] Suppose $\hat{S}(f)$ is a nonparametric SDF estimator with a distribution that is well approximated by that of the RV $a\chi_\nu^2$, where, as usual, χ_ν^2 is a chi-square RV with ν degrees of freedom, and a is a constant. If $\hat{S}(f)$ is an unbiased estimator of the true SDF $S(f)$ at frequency f, what must a be equal to? Using this value for a, provide a brief argument that, if $S(f)$ is finite, then $\mathrm{var}\,\{\hat{S}(f)\} \to 0$ as $\nu \to \infty$.

[7.7] An early attempt to produce a spectral estimator with better variance properties than the periodogram was to truncate the summation in Equation (170d) at $m < N$ to produce

$$\hat{S}^{(\mathrm{TP})}(f) \stackrel{\mathrm{def}}{=} \Delta_{\mathrm{t}} \sum_{\tau=-m}^{m} \hat{s}_\tau^{(\mathrm{P})} \mathrm{e}^{-\mathrm{i}2\pi f \tau \Delta_{\mathrm{t}}}. \qquad (343\mathrm{b})$$

The idea is to discard estimates of the ACVS for which there are relatively little data. This estimator is known as the *truncated periodogram*. Show that $\hat{S}^{(\mathrm{TP})}(\cdot)$ can be regarded as a special case of a lag window spectral estimator, and determine its lag window and its smoothing window. Argue that the truncated periodogram need *not* be nonnegative at all frequencies.

[7.8] Blackman and Tukey (1958, section B.5) discuss the following class of lag windows:

$$w_{m,\tau} = \begin{cases} 1 - 2a + 2a\cos\left(\pi\tau/m\right), & |\tau| \leq m; \\ 0, & |\tau| > m, \end{cases} \qquad (343\mathrm{c})$$

where $m < N$ and a is a parameter in the range $0 < a \leq 1/4$. If we let $a = 0.23$, then, among all smoothing windows corresponding to lag windows of the above form, we obtain the smoothing window that minimizes the magnitude in the first sidelobe relative to the magnitude of the peak in the central lobe. This particular case is called the *Hamming lag window*. A slight variation is to let $a = 1/4$. The resulting lag window is called the *Hanning lag window*.

(a) Determine the smoothing window for the lag window of Equation (343c).

(b) Show that, if we use the rectangular data taper, then $\hat{S}_m^{(\mathrm{LW})}(f)$ can be expressed as a linear combination of the truncated periodogram of Equation (343b) at three frequencies.

[7.9] This exercise examines an assumption of approximate uncorrelatedness made in Bartlett (1950) as part of formulating the Bartlett lag window of Equation (269a) (some aspects of Exercise [6.27f] parallel this exercise). Let N_S denote a block size, and consider a time series $X_0, X_1, \ldots, X_{N-1}$ of sample size $N = 2N_S$ from which we construct two periodograms, one making use of the first block of N_S values, and the other, the second block (we assume $\Delta_t = 1$ for convenience):

$$\hat{S}_0^{(\text{P})}(f) \stackrel{\text{def}}{=} \frac{1}{N_S} \left| \sum_{t=0}^{N_S-1} X_t e^{-i2\pi ft} \right|^2 \quad \text{and} \quad \hat{S}_{N_S}^{(\text{P})}(f) \stackrel{\text{def}}{=} \frac{1}{N_S} \left| \sum_{t=N_S}^{2N_S-1} X_t e^{-i2\pi ft} \right|^2.$$

Suppose the series is a portion of a Gaussian AR(1) process $X_t = \phi X_{t-1} + \epsilon_t$ with an ACVS given by $s_\tau = \phi^{|\tau|}/(1-\phi^2)$, where $\phi = 0.8$, and $\{\epsilon_t\}$ is a zero mean Gaussian white noise process with variance $\sigma_\epsilon^2 = 1$ (see Exercise [2.17a]). With N_S set to 16, use Equation (243b) to obtain the covariances and variances needed to compute the correlation ρ_k between $\hat{S}_0^{(\text{P})}(\tilde{f}_k)$ and $\hat{S}_{N_S}^{(\text{P})}(\tilde{f}_k)$ at frequencies $\tilde{f}_k = k/(2N_S)$, $k = 0, 1, \ldots, N_S$ (i.e., a grid twice as fine as that of the Fourier frequencies). Plot ρ_k versus $\tilde{f}_k$. Create similar plots with N_S set to 64 and then to 256. Finally repeat this exercise with ϕ set to 0.9, then 0.99 and finally -0.99. Comment on how well $\hat{S}_0^{(\text{P})}(\tilde{f}_k)$ and $\hat{S}_{N_S}^{(\text{P})}(\tilde{f}_k)$ are approximately uncorrelated for the twelve pairs of ϕ and N_S under consideration.

[7.10] (a) Verify that the lag window $\{w_{m,\tau}\}$ of Equation (274) corresponds to the Bartlett–Priestley design window of Equation (273c).

(b) Using $V_m(\cdot)$ as a proxy for $W_m(\cdot)$ in Equation (251c), show that B_W is approximately $5/(6m\,\Delta_t) \doteq 0.83/(m\,\Delta_t)$ for the Bartlett–Priestley window (this approximation appears in Table 279). Explore the accuracy of this approximation by making use of Equation (251e).

(c) Using $V_m(\cdot)$ as a proxy for $W_m(\cdot)$ in Equation (251a), show that β_W is approximately $\sqrt{3/5}/(m\,\Delta_t) \doteq 0.77/(m\,\Delta_t)$ for the Bartlett–Priestley window (this approximation appears in Table 279). Explore the accuracy of this approximation by making use of Equation (251b).

[7.11] Show that, for the Parzen smoothing window of Equation (275b), $W_m(0) = 3m\,\Delta_t/4$ for even m and $W_m(0) = (1 + 3m^4)\,\Delta_t/(4m^3)$ for odd m.

[7.12] The Parzen lag window $\{w_{m,\tau}\}$ of Equation (275a) is defined such that $w_{m,\tau} = 0$ for $|\tau| \geq m$, where m is a positive integer we restricted to be such that $m \leq N$. With this restriction, the corresponding smoothing window $W_m(\cdot)$ of Equation (275b) does not depend on N. To emphasize this fact, denote this window henceforth as $W_{m,N\text{-free}}(\cdot)$; i.e., assuming $m \leq N$ and $w_{m,\tau} = 0$ for all $|\tau| \geq m$, we have $\{w_{m,\tau} : \tau \in \mathbb{Z}\} \longleftrightarrow W_{m,N\text{-free}}(\cdot)$.

Now consider the case $m > N$. With the smoothing window defined as per Equation (247e), show that

$$W_m(f) = \Delta_t(2N-1) \int_{-f_\mathcal{N}}^{f_\mathcal{N}} \mathcal{D}_{2N-1}(f'\Delta_t) W_{m,N\text{-free}}(f - f')\, df', \quad (344)$$

where $\mathcal{D}_{2N-1}(\cdot)$ is Dirichlet's kernel (Equation (17c)). Argue that the corresponding lag window estimator $\hat{S}_m^{(\text{LW})}(f)$ can be expressed not only in the usual way per Equation (247f) as

$$\int_{-f_\mathcal{N}}^{f_\mathcal{N}} W_m(f - f') \hat{S}^{(\text{D})}(f')\, df', \quad \text{but also as} \quad \int_{-f_\mathcal{N}}^{f_\mathcal{N}} W_{m,N\text{-free}}(f - f') \hat{S}^{(\text{D})}(f')\, df'.$$

Create plots similar to those in Figure 276 with the only difference being that m is set to 100 rather than 37 (recall that $N = 64$, and hence we are in the $m > N$ case). For comparison with $W_m(\cdot)$, also plot $W_{m,N\text{-free}}(\cdot)$ in a manner similar to Figure 276(b). Comment upon your findings. (While we have focused on the Parzen lag window here, the key results of this exercise also hold for other windows defined such that $w_{m,\tau} = 0$ for all $|\tau| \geq m$. Two examples are the Bartlett and Papoulis windows (Equations (269a) and (278)).)

[7.13] Create lag windows appropriate for a time series of length $N = 64$ by autocorrelating a Slepian data taper with NW set to each of the following 161 values: 1, 1.05, ..., 8.95, 9 (C&E [5] for

7.14 Exercises

Section 7.5 has a discussion on formulating a lag window by autocorrelating a data taper). For each of the resulting 161 lag windows, compute the smoothing window bandwidth B_W using Equation (251e). Using the setting for NW with B_W closest to 0.05, create plots similar to those comprising Figure 286. Comment upon your findings.

[7.14] (a) Compute and plot $E\{\hat{S}_m^{(\text{LW})}(f)\}$ versus f for the five lag window estimators involved in Figures 289 and 291. For comparison, display also in each plot a reproduction of the thin curve in the lower left-hand plot of Figure 193, which depicts the expected value of the underlying direct spectral estimator $\hat{S}^{(\text{D})}(\cdot)$ that involves the Hanning data taper and the AR(4) process of Equation (35a). Hints: use Equations (343a) and (186g) along with the procedure given in Exercise [6.31d] for forming the ACVS for the AR(4) process.

(b) Repeat part (a) for the Papoulis, autocorrelated 100% cosine and Slepian lag windows.

[7.15] Verify Equations (296c) and (296d). Hint: make use of Equation (302a).

[7.16] (a) Show that, if we take the PDFs $p(\cdot)$ and $q(\cdot)$ in the Kullback–Leibler (KL) discrepancy of Equation (297a) to correspond to those of the RVs $S_p(f)\chi^2_{\nu_p}/\nu_p$ and $S_q(f)\chi^2_{\nu_q}/\nu_q$, where $S_p(f) > 0$, $S_q(f) > 0$, $\nu_p > 0$ and $\nu_q > 0$, then

$$d(p(\cdot), q(\cdot)) = \frac{\nu_q}{2}\left[\frac{S_p(f)}{S_q(f)} - \log\left(\frac{S_p(f)}{S_q(f)}\right)\right] + \eta(\nu_p, \nu_q), \quad (345a)$$

where $\eta(\nu_p, \nu_q)$ is to be determined as part of the exercise (the forms for $p(\cdot)$ and $q(\cdot)$ can be deduced from Equation (297b)). Show that, if ν_p and ν_q are both equal to, say, ν, then the above becomes

$$d(p(\cdot), q(\cdot)) = \frac{\nu}{2}\left(\left[\frac{S_p(f)}{S_q(f)} - \log\left(\frac{S_p(f)}{S_q(f)}\right)\right] - 1\right).$$

Hint: Gradshteyn and Ryzhik (1980) state the following definite integrals:

$$\int_0^\infty x^a e^{-bx}\, dx = \frac{\Gamma(a+1)}{b^{a+1}}, \quad \text{when } a > -1 \text{ and } b > 0; \text{ and}$$

$$\int_0^\infty x^{c-1} e^{-dx} \log(x)\, dx = \frac{\Gamma(c)}{d^c}\left[\psi(c) - \log(d)\right], \quad \text{when } c > 0 \text{ and } d > 0,$$

where $\psi(\cdot)$ is the digamma function. (Note that, conditional on $\nu_p = \nu_q = \nu$, the KL discrepancy is the same to within a constant of proportionality $\nu/2$. The choice $\nu = 2$ conveniently sets this constant to be unity and leads to the measure given by Equation (297c).)

(b) Under the assumptions that $\hat{S}_m^{(\text{LW})}(\tilde{f}_k) \stackrel{d}{=} S(\tilde{f}_k)\chi^2_\nu/\nu$ for all k and that $\nu > 2$, verify Equation (297d). Hint: make use of Equation (302a).

(c) In addition to $d(p(\cdot), q(\cdot))$, Kullback and Leibler (1951) explore $d(q(\cdot), p(\cdot))$ as a measure of discrepancy. In our context, the measure inspires

$$\text{LK}(m) \stackrel{\text{def}}{=} \frac{1}{N+1}\sum_{k=0}^{N}\left[\frac{\hat{S}_m^{(\text{LW})}(\tilde{f}_k)}{S(\tilde{f}_k)} - \log\left(\frac{\hat{S}_m^{(\text{LW})}(\tilde{f}_k)}{S(\tilde{f}_k)}\right) - 1\right] \quad (345b)$$

as an alternative to KL(m); i.e., we swap $S(\tilde{f}_k)$ and $\hat{S}_m^{(\text{LW})}(\tilde{f}_k)$ in Equation (297c). Assuming that $\hat{S}_m^{(\text{LW})}(\tilde{f}_k) \stackrel{d}{=} S(\tilde{f}_k)\chi^2_\nu/\nu$ for all $\tilde{f}_k$, determine $E\{\text{LK}(m)\}$, and comment briefly on how it compares to $E\{\text{KL}(m)\}$ (Equation (297d)).

(d) Kullback and Leibler (1951) also explore $d(p(\cdot), q(\cdot)) + d(q(\cdot), p(\cdot))$ as a symmetric measure of discrepancy that does not depend on the ordering of $p(\cdot)$ and $q(\cdot)$. This measure inspires consideration of

$$\text{KLLK}(m) \stackrel{\text{def}}{=} \text{KL}(m) + \text{LK}(m). \quad (345c)$$

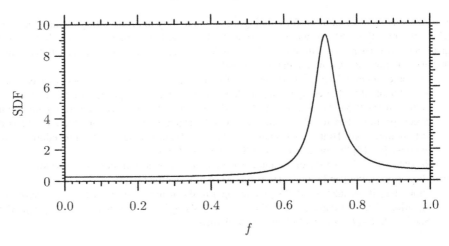

Figure 346 Hypothesized SDF. Note that $S(f)$ versus f is plotted on a *linear/linear* scale, not on the usual linear/decibel scale.

Formulate a succinct expression for $\text{KL}(m) + \text{LK}(m)$, and determine its expected value by appealing to previous parts of this exercise.

[7.17] (a) For each of the four AR(2) time series displayed in plots (a) to (d) of Figure 34 (downloadable from the "Data" part of the website for the book), compute $\text{MSE}(m)$, $\text{NMSE}(m)$, $\text{MSLE}(m)$, $\text{KL}(m)$, $\text{LK}(m)$ and $\text{KLLK}(m)$ as per Equations (293), (296a), (296b), (297c), (345b) and (345c) for Parzen lag window estimates with $m = 16$, 32 and finally 64. Compare these measures to their corresponding expected values (see Equations (296d), (296c) and (297d) and parts (c) and (d) of Exercise [7.16]). Comment briefly on your findings.

(b) Generate a large number, say N_R, of simulated time series of length $N = 1024$ from the AR(2) process of Equation (34) (take "large" to be at least 1000 – see Exercise [597] for a description of how to generate realizations from this process). For each simulated series, compute $\text{MSE}(m)$ for Parzen lag window estimates with $m = 16$, 32 and finally 64. Do the same for $\text{NMSE}(m)$, $\text{MSLE}(m)$, $\text{KL}(m)$, $\text{LK}(m)$ and $\text{KLLK}(m)$. Create a scatterplot of the N_R values for $\text{NMSE}(16)$ versus those for $\text{MSE}(16)$ and then also for $m = 32$ and $m = 64$. Do the same for the remaining 14 pairings of the six measures. For a given m, compute the sample correlation coefficients corresponding to the 15 scatterplots. The sample correlation coefficients for one of the 15 pairings should be almost indistinguishable from unity for all three settings of m – offer an explanation as to why this happens. For each m, average all N_R values of $\text{MSE}(m)$. Compare this average with its expected value. Do the same for the remaining five measures. Comment briefly on your findings.

[7.18] Suppose an investigator plans to collect a time series that can be modeled by a stationary process with SDF $S(\cdot)$. The sampling interval Δ_t is set at $1/2$ sec so the Nyquist frequency $f_\mathcal{N}$ is 1 Hz. Suppose that the rough shape of $S(\cdot)$ is known from either physical theory or previous experiments and is given by Figure 346. The investigator plans to use the Parzen lag window spectral estimator.

(a) What data taper would you recommend if the sample size N of the collected time series is 128? Would you change your recommendation if the sample size were increased to 1024? State the reasons for your recommendations.

(b) Would prewhitening be useful here? State the reasons for your answer.

(c) Determine the spectral bandwidth of $S(\cdot)$ by examining Figure 346. What does this imply about the size of the window bandwidth of the Parzen smoothing window? For the data taper(s) you selected in part (a) for sample sizes of $N = 128$ and 1024, determine what values of m (if any) achieve the desired window bandwidth. Determine the corresponding EDOFs ν.

[7.19] Assuming that $f' = 0$ and $W < f_\mathcal{N}$ in Equation (299) so that the SDF is that of an ideal low-pass rather than an ideal band-pass process, suitably modify the argument following that equation to conclude that $2W$ is now approximately equal to $\eta/(N\,\Delta_{\mathrm{t}})$ rather than the value derived for the band-pass case, namely, $\eta/(2N\,\Delta_{\mathrm{t}})$. In doing so, make the assumption that $\hat{S}^{(\mathrm{P})}(0)$ is negligible (as would be the case for a centered time series – see Exercise [6.15b]). Repeat this exercise, but now assuming that $f' = f_{\mathcal{N}}$ so that the SDF is that of an ideal high-pass process (assume that $\hat{S}^{(\mathrm{P})}(f_{\mathcal{N}})$ is negligible when N is even).

[7.20] Show that, as $N \to \infty$,
$$B_T \to \frac{s_0^2}{2\int_{-f_{\mathcal{N}}}^{f_{\mathcal{N}}} S^2(f)\,\mathrm{d}f} \stackrel{\mathrm{def}}{=} B_T^{(\infty)},$$
where B_T is the time series bandwidth measure of Equation (300a) with δ_{BP} set to unity. What is $B_T^{(\infty)}$ when $S(\cdot)$ is the SDF for the ideal band-pass process of Equation (299)? Show via computations that $B_T^{(\infty)} \doteq 0.265$ for the AR(2) process of Equation (34) and $\doteq 0.033$ for the AR(4) process of Equation (35a). (The forthcoming Equations (508a) and (508b) state the ACVSs for these two processes.)

[7.21] Verify the two approximations stated in Equation (303d) under the assumption that N' is large.

[7.22] This exercise considers computational details and three measures associated with the discretely smoothed periodogram $\hat{S}_m^{(\mathrm{DSP})}(\cdot)$ of Equation (307).

(a) Sections 6.7 and 7.11 give details on how to efficiently compute direct spectral estimates (including the periodogram) and lag window estimates using discrete Fourier transforms. Provide similar details for computing $\hat{S}_m^{(\mathrm{DSP})}(\cdot)$ over the grid of Fourier frequencies.

(b) Starting with the obvious analog of Equation (264a), use an approach similar to what was used in Section 7.4 to argue that, by making certain assumptions (to be stated as carefully as possible as part of the exercise), we can take the EDOFs for $\hat{S}_m^{(\mathrm{DSP})}(f_k)$ to be
$$\nu = \frac{2}{\sum_{j=-\lfloor N/2 \rfloor}^{N-\lfloor N/2 \rfloor -1} g_{m,j}^2}. \tag{347a}$$

How reasonable are the assumptions behind this expression for ν?

(c) Use an approach similar to the one leading to Equation (256a) to show that the standard measure for the effective bandwidth of $\hat{S}_m^{(\mathrm{DSP})}(\cdot)$ is
$$B_{\mathcal{U}} = \frac{1}{\Delta_{\mathrm{t}} \sum_{\tau=-(N-1)}^{N-1} \left| \sum_{j=0}^{N-1} g_{m,j-\lfloor N/2\rfloor}\, \mathrm{e}^{\mathrm{i}2\pi f_j \tau \Delta_{\mathrm{t}}} \right|^2 \left(1 - \frac{|\tau|}{N}\right)^2}. \tag{347b}$$

Comment briefly upon how the above compares to Equation (256a).

(d) Use Equations (251c) and (258) to motivate the following as a definition for the smoothing window bandwidth associated with $\{g_{m,j}\}$:
$$\frac{1}{N\,\Delta_{\mathrm{t}} \sum_{j=-\lfloor N/2 \rfloor}^{N-\lfloor N/2 \rfloor -1} g_{m,j}^2}.$$

[7.23] Adapt Equations (347a) and (347b) to handle discrete smoothing of a Hanning-based direct SDF estimator $\hat{S}^{(\mathrm{D})}(\cdot)$ across the Fourier frequencies, i.e., an estimator formed as per the right-hand side of Equation (307), but with $\hat{S}^{(\mathrm{P})}(f_{k-j})$ replaced by $\hat{S}^{(\mathrm{D})}(f_{k-j})$. In adapting Equation (347a), assume that the correlation between $\hat{S}^{(\mathrm{D})}(f_k)$ and $\hat{S}^{(\mathrm{D})}(f_{k\pm l})$ is $1/2$ when $l = 1$ and is 0 for all other positive integers l of interest (Equation (241) suggests this approximation if we simplify 4/9 and 1/36 to 1/2 and 0). Using weights $g_{m,j} \propto \exp(-(mj)^2)$ with $m = 0.2313$, verify the left-hand plot of Figure 311, including the crisscross (the required AR(4) time series is available via the website for the book). The crisscross depends in part on the value of ν from your adaptation of Equation (347a). By appealing to Table 279, use this value to determine a setting m for a Parzen lag window estimator such that the resulting EDOF agrees with ν as closely as possible. Using this setting for m, compute a Parzen lag window estimate for the AR(4) time series that also utilizes the Hanning-based $\hat{S}^{(D)}(\cdot)$. How well does this estimate match up with the estimate shown in the left-hand plot of Figure 311?

[7.24] Verify the contents of the left-hand plot of Figure 310 and both plots in Figure 311 (the required AR(2) and AR(4) time series are available on the book's website). Verify that the GCV-based MSEs associated with the SDFs estimates depicted in these three plots are correctly stated in the discussion surrounding these plots (namely, 0.0562 for the left-hand plot of Figure 310, and 4.258 and 7.916 for the left- and right-hand plots in Figure 311). Verify that the ratios of the GCV-based MSEs to the associated best possible MSEs are also stated correctly (1.240, 1.0003 and 1.376, respectively). Generate a large number N_R of additional realizations of length $N = 1024$ of the AR(2) and AR(4) processes of Equations (34) and (35a), and compute the ratio of the GCV-based MSE to the best possible MSE for each of these realizations. Are the ratios associated with Figures 310 and 311 typical of what you obtained from the additional realizations? (For the above, take N_R to be at least 1000, and see Exercises [597] and [11.1] for descriptions of how to generate realizations of the AR(2) and AR(4) processes.)

[7.25] Suppose that, given a time series of length N, we have calculated its periodogram $\hat{S}^{(P)}(f_k)$ over the N Fourier frequencies $f_k = k/(N\Delta_t)$ satisfying $-f_\mathcal{N} \leq f_k < f_\mathcal{N}$. We then use these periodogram values to form a discretely smoothed periodogram $\hat{S}_m^{(\text{DSP})}(f_k)$ over these same frequencies using a set of weights $\{g_{m,j}\}$ (see Equation (307)). Let $\tilde{f}_k \stackrel{\text{def}}{=} k/(2N\Delta_t)$, and note that $\tilde{f}_{2k} = f_k$. As discussed in C&E [2] for Section 7.10, we can take

$$\hat{s}_\tau^{(\text{DSP})} \stackrel{\text{def}}{=} \frac{1}{2N\Delta_t} \sum_{k=-N}^{N-1} \hat{S}_m^{(\text{DSP})}(\tilde{f}_k) e^{i2\pi \tilde{f}_k \tau \Delta_t} \qquad (348)$$

to be an ACVS estimator corresponding to the discretely smoothed periodogram once we define $\hat{S}_m^{(\text{DSP})}(\tilde{f}_k)$ when k is odd (we already know $\hat{S}_m^{(\text{DSP})}(\tilde{f}_k)$ for even k since $\hat{S}_m^{(\text{DSP})}(\tilde{f}_k) = \hat{S}_m^{(\text{DSP})}(f_{k/2})$). Here we explore options for doing so.

(a) Show that, if we set $\hat{S}_m^{(\text{DSP})}(\tilde{f}_k)$ for positive odd indices k to *any* nonnegative values so desired and if we set $\hat{S}_m^{(\text{DSP})}(\tilde{f}_{-k}) = \hat{S}_m^{(\text{DSP})}(\tilde{f}_k)$, then $\{\hat{s}_\tau^{(\text{DSP})} : \tau \in \mathbb{Z}\}$ corresponds to the ACVS for a harmonic process.

(b) Compute ACVS estimates $\{\hat{s}_\tau^{(\text{DSP})}(\cdot) : \tau = 0, 1, \ldots, 1024\}$ corresponding to the discretely smoothed periodogram of Figure 310(a) by setting $\hat{S}_m^{(\text{DSP})}(\tilde{f}_k)$ at odd indices k in three different ways: using (i) zeros, (ii) the linear interpolation scheme of Equation (313b) and (iii) the convolution scheme of Equation (313c). Comment briefly on how these three estimates compare. Given that each of these ACVS estimates is associated with a harmonic process, how would you go about estimating s_τ for $\tau > 1024$? (Figure 34(a) shows the AR(2) series used to form the discretely smoothed periodogram of interest here – this series is downloadable from the book's website).

(c) A harmonic process does not possess a proper SDF, so the ACVS given in part (a) is disconcerting since the intent of a discretely smoothed periodogram is to serve as an SDF estimator. Construct a proper SDF, say $S_{\text{PC}}(\cdot)$, that is piecewise constant over intervals of the form $(\tilde{f}_k - 1/(4N), \tilde{f}_k + 1/(4N))$ such that $S_{\text{PC}}(\tilde{f}_k) = \hat{S}_m^{(\text{DSP})}(\tilde{f}_k)$ for all k, and show that its ACVS is given by

$$s_{\text{PC},\tau} = \frac{\sin(\pi\tau/(2N))}{\pi\tau/(2N)} \hat{s}_\tau^{(\text{DSP})}, \qquad \tau \in \mathbb{Z}$$

(when $\tau = 0$, interpret the above ratio to be unity).

(d) For each of the three ACVS estimates computed in part (b), compute corresponding $s_{\text{PC},\tau}$ sequences for $\tau = 0, 1, \ldots, 4096$. Comment briefly on how $s_{\text{PC},\tau}$ compares to $\hat{s}_\tau^{(\text{DSP})}$ in each case and how the three $s_{\text{PC},\tau}$ sequences compare to one another. Comment on how these $s_{\text{PC},\tau}$ sequences compare to the usual biased estimator $\hat{s}_\tau^{(P)}$ at lags $\tau \geq 1024$.

[7.26] Figure 310 shows the periodogram for the AR(2) time series of Figure 34(a) along with two smoothed versions thereof. The left-hand plot depicts a discretely smoothed periodogram $\hat{S}_m^{(\text{DSP})}(\cdot)$ with the smoothing parameter m chosen by the GCV criterion of Equation (309a); the right-hand,

7.14 Exercises 349

a Parzen lag window estimate $\hat{S}_m^{(\mathrm{LW})}(\cdot)$ with m chosen via Equation (313d). Create similar figures for the AR(2) time series shown in Figures 34(b), (c) and (d) (downloadable from the book's website). For a given series, how well do $\hat{S}_m^{(\mathrm{DSP})}(\cdot)$ and $\hat{S}_m^{(\mathrm{LW})}(\cdot)$ agree?

[7.27] We have discussed three objective methods for determining the amount of smoothing to be applied to a direct spectral estimator: estimation of the spectral bandwidth (Section 7.8), estimation of a mean integrated square error in log space (Section 7.9) and a generalized cross-validation criterion (Section 7.10). This exercise seeks to evaluate these three methods using different measures of how well a particular SDF estimate matches up with the true SDF.

  (a) Generate a large number N_{R} of realizations of length $N = 1024$ from the AR(2) process of Equation (34) (take N_{R} to be at least 1000, and see Exercise [597] for a description of how to generate the realizations). For each realization, form periodogram-based Parzen lag window estimates $\hat{S}_m^{(\mathrm{LW})}(f_k)$ over the Fourier frequencies $f_k = k/N$, $k = 0, 1, \ldots, N/2$, with m set, first, to $[3.7/\tilde{B}_T]$, where $\tilde{B}_T$ is the estimator of the time series bandwidth given by Equation (300c), and $[x]$ is x rounded to the nearest integer (cf. the discussion surrounding Equation (322)); and, second, to the minimizer of $\hat{I}_m$ of Equation (303e). For each realization, also form a discretely smoothed periodogram $\hat{S}_m^{(\mathrm{DSP})}(f_k)$ as per Equation (307), with weights $g_{m,j} \propto \exp(-(mj)^2)$ such that $\sum_j g_{m,j} = 1$, and with m set to be the minimizer of Equation (309a). Evaluate the quality of the two lag window estimates and the discretely smoothed periodogram using at least two of the following seven measures $M_1, \ldots, M_7$:

$$M_1 = \frac{1}{N/2 + 1} \sum_{k=0}^{N/2} \left(\hat{S}_m(f_k) - S(f_k) \right)^2$$

$$M_2 = \frac{1}{N/2 + 1} \sum_{k=0}^{N/2} \left(\log\left(\hat{S}_m(f_k)\right) - \log\left(S(f_k)\right) \right)^2$$

$$M_3 = \frac{1}{N/2 + 1} \sum_{k=0}^{N/2} \left(\frac{\hat{S}_m(f_k) - S(f_k)}{S(f_k)} \right)^2$$

$$M_4 = \frac{1}{N/2 + 1} \sum_{k=0}^{N/2} \left[\frac{S(f_k)}{\hat{S}_m(f_k)} - \log\left(\frac{S(f_k)}{\hat{S}_m(f_k)}\right) - 1 \right]$$

$$M_5 = \frac{1}{N/2 + 1} \sum_{k=0}^{N/2} \left[\frac{\hat{S}_m(f_k)}{S(f_k)} - \log\left(\frac{\hat{S}_m(f_k)}{S(f_k)}\right) - 1 \right]$$

$$M_6 = M_4 + M_5$$

$$M_7 = \max_k |10 \log_{10}\left(\hat{S}_m(f_k)/S(f_k)\right)|,$$

where $\hat{S}_m(f_k)$ is either $\hat{S}_m^{(\mathrm{LW})}(f_k)$ or $\hat{S}_m^{(\mathrm{DSP})}(f_k)$, and $S(\cdot)$ is the true AR(2) SDF (measures M_1 to M_6 are adaptations of MSE of Equation (293), MSLE of Equation (296b), NMSE of Equation (296a), KL of Equation (297c), LK of Equation (345b) and KLLK of Equation (345c)). Average each of your chosen measures over all N_{R} realizations. Comment on your findings.

  (b) Repeat part (a), but now using the AR(4) process of Equation (35a) (Exercise [11.1] describes generation of the required realizations). Replace the periodogram with a Hanning-based direct SDF estimator throughout, blindly using exactly the same methods for selecting m and the same measures for evaluating the resulting SDF estimates as before. Comment on your findings.

[7.28] Figure 319 shows four lag window estimates $\hat{S}_m^{(\mathrm{LW})}(\cdot)$ for the ocean wave time series, all based upon a direct spectral estimate $\hat{S}^{(\mathrm{D})}(\cdot)$ using a $NW = 2/\Delta_t$ Slepian data taper. Conceptually each $\hat{S}_m^{(\mathrm{LW})}(\cdot)$ is the convolution of $\hat{S}^{(\mathrm{D})}(\cdot)$ with a smoothing window $W_m(\cdot)$, i.e., the Fourier transform of the chosen lag window $\{w_{m,\tau}\}$. Suppose we subject $\hat{S}_m^{(\mathrm{LW})}(\cdot)$ to an additional smoothing operation identical to the first; i.e., we form $\hat{S}^{(\text{RLW})}(\cdot)$ by convolving $\hat{S}_m^{(\text{LW})}(\cdot)$ with $W_m(\cdot)$, where "RLW" stands for "repeated lag window."

(a) Show that the repeated lag window estimator is an ordinary lag window estimator (i.e., one based directly on $\hat{S}^{(D)}(\cdot)$) by a suitable definition of a new lag window. In terms of $W_m(\cdot)$, what is the smoothing window that is used to produce $\hat{S}^{(\text{RLW})}(\cdot)$ from $\hat{S}^{(D)}(\cdot)$?

(b) For the lag windows used in Figure 319, show that the smoothing window bandwidth for $\hat{S}^{(\text{RLW})}(\cdot)$ is greater than that for $\hat{S}_m^{(\text{LW})}(\cdot)$. What conditions do we need to impose on other lag windows for this relationship to hold?

(c) Recreate the contents of Figure 319 but using the appropriate repeated lag windows in place of the original lag windows and showing the central lobes of the smoothing windows for all four lag windows (rather than the lobes of the design windows for the Daniell and Bartlett–Priestley lag windows). Compare the smoothing window bandwidths (Equation (251e)), the standard measure of effective bandwidths (Equation (256a)) and EDOFs (Equation (264b)) associated with the original and repeated lag window estimates. Comment upon your findings. (The book's website has the required data.)

[7.29] Figure 330 shows the ship altitude data we used in the analysis presented in Section 7.12. This series has 2048 values, but it is the first part of a longer series of length $N = 14{,}697$ that is available on the book's website. Repeat the analysis of the ship altitude data, but now using this longer series. Do any of the conclusions we made based on the shorter series need to be changed?

8

Combining Direct Spectral Estimators

8.0 Introduction

In Chapter 6 we introduced the notion of tapering a time series $\{X_t\}$ to create a direct spectral estimator $\hat{S}^{(D)}(\cdot)$ for a spectral density function (SDF). Such an estimator is an alternative to the most basic SDF estimator, namely, the periodogram $\hat{S}^{(P)}(\cdot)$, which can be badly biased. By carefully choosing a data taper, we can often form a direct spectral estimator whose overall bias properties improve upon those of the periodogram. Both $\hat{S}^{(P)}(\cdot)$ and $\hat{S}^{(D)}(\cdot)$ are inherently noisy and inconsistent SDF estimators. The lag window estimators of Chapter 7 reduce noise and achieve consistency by smoothing either $\hat{S}^{(P)}(\cdot)$ or $\hat{S}^{(D)}(\cdot)$ across frequencies. In this chapter we consider two attractive alternatives to lag window estimators, namely, multitaper estimators and Welch's overlapped segment averaging (WOSA) estimator, which were introduced in seminal papers by Thomson (1982) and Welch (1967), respectively. Both are based on creating multiple direct spectral estimators from a single time series and then combining them together. Both offer potential improvements over lag window estimators, particularly those based on a direct spectral estimator involving substantive tapering (as offered by the Hanning data taper, for example). While tapering does reduce bias due to leakage, the sample size of $\{X_t\}$ is also effectively reduced. The price of these reductions appears in lag window estimators as an increase in variance, which is quantified by the taper-dependent factor C_h (see Equation (260), and note that Table 260 says C_h is close to 2 for the Hanning taper). This inflated variance is acceptable in some practical applications, but not in other cases. Like lag window estimators, the multitaper and WOSA spectral estimators achieve a reduction in variance from what a direct spectral estimator offers, but without paying a notable price due to tapering.

Section 8.1 motivates the multitaper approach and presents its basic properties. Section 8.2 gives an initial look at multitapering using the Slepian tapers as applied to the AR(4) time series of Figure 34(e). Section 8.3 investigates the statistical properties of the multitaper spectral estimator for the simple case of white noise. Section 8.4 formally justifies the multitaper approach by showing its connection to quadratic spectral estimators (Section 8.5 presents a second justification based upon projections and the concept of regularization). Section 8.6 discusses a second approach to multitapering that uses so-called sinusoidal tapers. Section 8.7 illustrates how two periodogram-based estimation procedures can be improved upon by using multitaper estimators as the foundation rather than periodograms. Section 8.8 discusses WOSA, after which the chapter concludes with examples in Section 8.9, a summary in 8.10 and exercises in 8.11.

8.1 Multitaper Spectral Estimators – Overview

In this section, we begin our discussion of multitaper spectral estimators, which were introduced in a seminal paper by Thomson (1982). Our starting point is a computational formula for the estimator, after which we look into some of its basic statistical properties.

Suppose we have a time series that is a realization of a portion $X_0, X_1, \ldots, X_{N-1}$ of the stationary process $\{X_t\}$ with zero mean and SDF $S(\cdot)$. As usual, we denote the sampling interval between observations as Δ_t so that the Nyquist frequency is $f_{\mathcal{N}} \stackrel{\text{def}}{=} 1/(2\Delta_t)$. As its name implies, the multitaper spectral estimator utilizes more than just a single data taper. The *basic multitaper estimator* is the average of K direct spectral estimators and hence takes the form

$$\hat{S}^{(\mathrm{MT})}(f) \stackrel{\text{def}}{=} \frac{1}{K} \sum_{k=0}^{K-1} \hat{S}_k^{(\mathrm{MT})}(f) \text{ with } \hat{S}_k^{(\mathrm{MT})}(f) \stackrel{\text{def}}{=} \Delta_t \left| \sum_{t=0}^{N-1} h_{k,t} X_t \mathrm{e}^{-\mathrm{i}2\pi f t \Delta_t} \right|^2, \quad (352\mathrm{a})$$

where $\{h_{k,t}\}$ is the data taper for the kth direct spectral estimator $\hat{S}_k^{(\mathrm{MT})}(\cdot)$ (as before, we assume the unit-energy normalization $\sum_{t=0}^{N-1} h_{k,t}^2 = 1$; see the discussion surrounding Equation (188c)). A generalization that will be of interest later on is the *weighted multitaper estimator*

$$\hat{S}^{(\mathrm{WMT})}(f) \stackrel{\text{def}}{=} \sum_{k=0}^{K-1} d_k \hat{S}_k^{(\mathrm{MT})}(f), \quad (352\mathrm{b})$$

where the weights d_k are nonnegative and are assumed to sum to unity (the rationale for this assumption is the subject of Exercise [8.1]; criteria for setting the weights are discussed in Sections 8.4, 8.5 and 8.6). Note that the basic multitaper estimator is a special case of the weighted estimator, since $\hat{S}^{(\mathrm{WMT})}(f)$ reduces to $\hat{S}^{(\mathrm{MT})}(f)$ if we set $d_k = 1/K$ for all k. In Thomson (1982) the estimator $\hat{S}_k^{(\mathrm{MT})}(\cdot)$ is called the kth *eigenspectrum*, but note that it is just a special case of the familiar direct spectral estimator $\hat{S}^{(\mathrm{D})}(\cdot)$ of Equation (186b). Let $H_k(\cdot)$ denote the Fourier transform of $\{h_{k,t}\}$ under the usual assumption that $h_{k,t} = 0$ for $t < 0$ and $t \geq N$:

$$H_k(f) \stackrel{\text{def}}{=} \Delta_t \sum_{t=0}^{N-1} h_{k,t} \mathrm{e}^{-\mathrm{i}2\pi f t \Delta_t}.$$

For each data taper, we can define an associated spectral window

$$\mathcal{H}_k(f) \stackrel{\text{def}}{=} \frac{1}{\Delta_t} |H_k(f)|^2 = \Delta_t \left| \sum_{t=0}^{N-1} h_{k,t} \mathrm{e}^{-\mathrm{i}2\pi f t \Delta_t} \right|^2. \quad (352\mathrm{c})$$

Equation (186e) says that the first moment of $\hat{S}_k^{(\mathrm{MT})}(\cdot)$ is given by

$$E\{\hat{S}_k^{(\mathrm{MT})}(f)\} = \int_{-f_{\mathcal{N}}}^{f_{\mathcal{N}}} \mathcal{H}_k(f - f') S(f') \, \mathrm{d}f'.$$

Hence

$$E\{\hat{S}^{(\mathrm{WMT})}(f)\} = \sum_{k=0}^{K-1} d_k E\{\hat{S}_k^{(\mathrm{MT})}(f)\} = \sum_{k=0}^{K-1} d_k \int_{-f_{\mathcal{N}}}^{f_{\mathcal{N}}} \mathcal{H}_k(f - f') S(f') \, \mathrm{d}f'$$

$$= \int_{-f_{\mathcal{N}}}^{f_{\mathcal{N}}} \widetilde{\mathcal{H}}(f - f') S(f') \, \mathrm{d}f', \quad (352\mathrm{d})$$

where

$$\widetilde{\mathcal{H}}(f) \stackrel{\text{def}}{=} \sum_{k=0}^{K-1} d_k \mathcal{H}_k(f). \tag{353a}$$

For the basic multitaper estimator, the above becomes

$$E\{\hat{S}^{(\text{MT})}(f)\} = \int_{-f_\mathcal{N}}^{f_\mathcal{N}} \overline{\mathcal{H}}(f-f')S(f')\,\mathrm{d}f' \text{ with } \overline{\mathcal{H}}(f) \stackrel{\text{def}}{=} \frac{1}{K}\sum_{k=0}^{K-1} \mathcal{H}_k(f). \tag{353b}$$

The function $\overline{\mathcal{H}}(\cdot)$ is the spectral window for $\hat{S}^{(\text{MT})}(\cdot)$, as $\widetilde{\mathcal{H}}(\cdot)$ is for $\hat{S}^{(\text{WMT})}(\cdot)$. Because the sidelobe level of $\mathcal{H}_k(\cdot)$ dictates whether or not $\hat{S}_k^{(\text{MT})}(\cdot)$ is approximately free of leakage, it follows that, if each $\mathcal{H}_k(\cdot)$ provides good protection against leakage, then the basic and weighted multitaper estimators will also be largely free of leakage.

As dictated by Equation (194), an appropriate measure of the bandwidth for the kth eigenspectrum is

$$B_{\mathcal{H}_k} \stackrel{\text{def}}{=} \text{width}_\text{a}\{\mathcal{H}_k(\cdot)\} = \frac{\left(\int_{-f_\mathcal{N}}^{f_\mathcal{N}} \mathcal{H}_k(f)\,\mathrm{d}f\right)^2}{\int_{-f_\mathcal{N}}^{f_\mathcal{N}} \mathcal{H}_k^2(f)\,\mathrm{d}f} = \frac{\Delta_t}{\sum_{\tau=-(N-1)}^{N-1}(h_k \star h_{k,\tau})^2} \tag{353c}$$

(Walden et al., 1995). For the weighted multitaper estimator, this measure becomes

$$B_{\widetilde{\mathcal{H}}} \stackrel{\text{def}}{=} \text{width}_\text{a}\{\widetilde{\mathcal{H}}(\cdot)\} = \frac{\left(\int_{-f_\mathcal{N}}^{f_\mathcal{N}} \widetilde{\mathcal{H}}(f)\,\mathrm{d}f\right)^2}{\int_{-f_\mathcal{N}}^{f_\mathcal{N}} \widetilde{\mathcal{H}}^2(f)\,\mathrm{d}f} = \frac{\Delta_t}{\sum_{\tau=-(N-1)}^{N-1}\left(\sum_{k=0}^{K-1} d_k h_k \star h_{k,\tau}\right)^2} \tag{353d}$$

(see Exercise [8.2]). By setting $d_k = 1/K$ in the above, we obtain the corresponding measure for the basic multitaper estimator:

$$B_{\overline{\mathcal{H}}} \stackrel{\text{def}}{=} \text{width}_\text{a}\{\overline{\mathcal{H}}(\cdot)\} = \frac{\Delta_t}{\sum_{\tau=-(N-1)}^{N-1}\left(\frac{1}{K}\sum_{k=0}^{K-1} h_k \star h_{k,\tau}\right)^2}. \tag{353e}$$

If each of the K eigenspectra is approximately unbiased, then we can argue that the basic multitaper estimator $\hat{S}^{(\text{MT})}(\cdot)$ is also approximately unbiased (see Exercise [8.1]). Thus a basic multitaper estimator does not have any obvious advantage over any of its constituent eigenspectra in terms of its first-moment properties. The rationale for averaging the K different eigenspectra is to produce an estimator of $S(f)$ with smaller variance than that of any individual $\hat{S}_k^{(\text{MT})}(f)$. To find a condition on the tapers that achieves this reduction in variance, let us now consider the variance of $\hat{S}^{(\text{MT})}(f)$ under the assumption that each constituent eigenspectrum $\hat{S}_k^{(\text{MT})}(f)$ is approximately unbiased. Now

$$\text{var}\{\hat{S}^{(\text{MT})}(f)\} = \frac{1}{K^2}\sum_{k=0}^{K-1}\text{var}\{\hat{S}_k^{(\text{MT})}(f)\} + \frac{2}{K^2}\sum_{j<k}\text{cov}\{\hat{S}_j^{(\text{MT})}(f), \hat{S}_k^{(\text{MT})}(f)\} \tag{353f}$$

(see Exercise [2.2]). Since each eigenspectrum is a direct spectral estimator, an application of Equation (204c) says that $\text{var}\{\hat{S}_k^{(\text{MT})}(f)\} \approx S^2(f)$ for $0 < f < f_\mathcal{N}$, with the validity of the approximation increasing as $N \to \infty$. Under the additional assumptions that $\{X_t\}$ is a

Gaussian process, that $S(\cdot)$ is locally constant about f and that f is not too close to zero or the Nyquist frequency, the result of Exercise [8.3] says that

$$\mathrm{cov}\{\hat{S}_j^{(\mathrm{MT})}(f), \hat{S}_k^{(\mathrm{MT})}(f)\} \approx S^2(f) \left| \sum_{t=0}^{N-1} h_{j,t} h_{k,t} \right|^2. \quad (354\mathrm{a})$$

Approximate uncorrelatedness of $\hat{S}_j^{(\mathrm{MT})}(f)$ and $\hat{S}_k^{(\mathrm{MT})}(f)$ when $j \neq k$ follows if the data tapers are orthogonal in the sense that

$$\sum_{t=0}^{N-1} h_{j,t} h_{k,t} = 0. \quad (354\mathrm{b})$$

If the above holds for all $j \neq k$, we obtain

$$\mathrm{var}\{\hat{S}^{(\mathrm{MT})}(f)\} \approx \frac{S^2(f)}{K}; \quad (354\mathrm{c})$$

i.e., the variance of $\hat{S}^{(\mathrm{MT})}(f)$ is smaller than that of each $\hat{S}_k^{(\mathrm{MT})}(f)$ by a factor of $1/K$. Exercise [8.4] is to show that the corresponding approximation for the weighted multitaper estimator is

$$\mathrm{var}\{\hat{S}^{(\mathrm{WMT})}(f)\} \approx S^2(f) \sum_{k=0}^{K-1} d_k^2, \quad (354\mathrm{d})$$

which reduces to Equation (354c) if we set $d_k = 1/K$ for all k.

Turning now to the distribution of $\hat{S}^{(\mathrm{MT})}(f)$, since each eigenspectrum $\hat{S}_k^{(\mathrm{MT})}(\cdot)$ is a direct spectral estimator, we can appeal to Equation (204b) to argue that, to a good approximation,

$$\hat{S}_k^{(\mathrm{MT})}(f) \stackrel{\mathrm{d}}{=} \begin{cases} S(f)\chi_2^2/2, & 0 < |f| < f_{\mathcal{N}}; \\ S(f)\chi_1^2, & |f| = 0 \text{ or } f_{\mathcal{N}} \end{cases} \quad (354\mathrm{e})$$

asymptotically as $N \to \infty$ (here $\stackrel{\mathrm{d}}{=}$ means "equal in distribution"). Recall that, if $\chi_{\nu_1}^2$ and $\chi_{\nu_2}^2$ are two independent chi-square random variables (RVs) with, respectively, ν_1 and ν_2 degrees of freedom, then their sum $\chi_{\nu_1}^2 + \chi_{\nu_2}^2$ is a chi-square RV with $\nu_1 + \nu_2$ degrees of freedom. To the degree that the eigenspectra can be taken to be mutually independent, it follows that

$$\hat{S}^{(\mathrm{MT})}(f) \stackrel{\mathrm{d}}{=} \begin{cases} S(f)\chi_{2K}^2/2K, & 0 < |f| < f_{\mathcal{N}}; \\ S(f)\chi_K^2/K, & |f| = 0 \text{ or } f_{\mathcal{N}} \end{cases} \quad (354\mathrm{f})$$

asymptotically as $N \to \infty$. Thus, as long as f is not too close to zero frequency or $\pm f_{\mathcal{N}}$, the equivalent degrees of freedom (EDOFs) ν associated with $\hat{S}^{(\mathrm{MT})}(f)$ is $2K$ (we discuss what constitutes "not too close" in item [2] of the Comments and Extensions [C&Es]). Just as for the lag window spectral estimates in Section 7.4, we can use the chi-square approximations to the distribution of $\hat{S}^{(\mathrm{MT})}(f)$ to compute, say, a 95% confidence interval (CI) for $S(f)$ based on $\hat{S}^{(\mathrm{MT})}(f)$, or, equivalently, a CI for $10 \log_{10}(S(f))$ based on $10 \log_{10}(\hat{S}^{(\mathrm{MT})}(f))$. The forms of the CIs for $S(f)$ and $10 \log_{10}(S(f))$ are the same as that shown in Equations (265) and (266a), but with $\hat{S}^{(\mathrm{MT})}(f)$ substituted for $\hat{S}_m^{(\mathrm{LW})}(f)$, and ν set equal to $2K$ when f is not too close to zero frequency or $\pm f_{\mathcal{N}}$. Exercise [8.5] says that, for the weighted multitaper estimator $\hat{S}^{(\mathrm{WMT})}(f)$,

$$\nu = \frac{2}{\sum_{k=0}^{K-1} d_k^2} \quad (354\mathrm{g})$$

when f is sufficiently away from zero frequency or $\pm f_\mathcal{N}$, which reduces to $2K$ when $d_k = 1/K$.

The key to a multitaper spectral estimator with good bias and variance properties is thus a set of K orthogonal data tapers with each taper providing good protection against leakage. Two prominent sets of such tapers have been studied extensively in the literature. The seminal paper on multitaper spectral estimation (Thomson, 1982) advocated use of *Slepian multitapers*. We begin our study of Slepian multitaper estimators in the next section, with a more detailed discussion of their rationale being given in Sections 8.3 to 8.5. The second set – known as the *sinusoidal multitapers* – was introduced in Riedel and Sidorenko (1995); these are discussed in detail in Section 8.6. Both sets lead to examples of *constructed* multitaper estimators, in that we construct $\hat{S}^{(\text{MT})}(\cdot)$ starting from K predefined orthogonal tapers. In Section 8.4, we describe another way in which multitapers estimators arise. In this approach we start with a certain predefined nonparametric estimator (a lag window estimator being an example), from which we can extract a set of orthogonal tapers. These tapers lead to a *deconstructed* weighted multitaper estimator that is fully equivalent to the nonparametric estimator used to derive the orthogonal tapers. The statistical theory is the same for both constructed and deconstructed estimators. The multitaper formulation thus offers a unified means of comparing seemingly different nonparametric estimators.

As we note in Sections 8.2 and 8.6, constructed multitaper estimators have distinct advantages over other nonparametric estimators for a time series that is a realization of a portion $X_0, X_1, \ldots, X_{N-1}$ of a stationary process. In addition, there are natural extensions to multitapering for both spectral estimation of time series with missing or irregularly sampled observations and the problem of estimating univariate spectra that depend upon multiple variables, e.g., spatial variables or multiple times (Bronez, 1985, 1986, 1988; Liu and Van Veen, 1992; Hanssen, 1997; Fodor and Stark, 2000).

Comments and Extensions to Section 8.1

[1] In our discussion of direct spectral estimators $\hat{S}^{(\text{D})}(\cdot)$, we noted in C&E [4] for Section 6.6 that, under a Gaussian assumption, we can base a measure of the bandwidth of $\hat{S}^{(\text{D})}(\cdot)$ on the equivalent width of an approximation $R(\eta)$ to the correlation between $\hat{S}^{(\text{D})}(f)$ and $\hat{S}^{(\text{D})}(f + \eta)$ (see Equation (214a)). Greenhall (2006) extends this idea to work with multitaper estimators by deriving an approximation $R^{(\text{MT})}(\eta)$ to the correlation between $\hat{S}^{(\text{MT})}(f)$ and $\hat{S}^{(\text{MT})}(f + \eta)$. The resulting bandwidth measure is

$$\text{width}_e\{R^{(\text{MT})}(\cdot)\} = \frac{1}{K\,\Delta_t} \sum_{t=0}^{N-1} \left(\sum_{k=0}^{K-1} h_{k,t}^2 \right)^2, \tag{355a}$$

where the width measure integrates out the frequency domain lag η. Note that, when $K = 1$, the above is in agreement with Equation (214a). This bandwidth measure is easier to compute than the standard bandwidth measure $B_{\overline{\mathcal{H}}}$ of Equation (353e). We will compare the performance of $\text{width}_e\{R^{(\text{MT})}(\cdot)\}$ with that of $B_{\overline{\mathcal{H}}}$ in Sections 8.2 and 8.6 for, respectively, the Slepian and sinusoidal multitaper estimators. For weighted multitaper estimators, the corresponding result is

$$\text{width}_e\left\{R^{(\text{WMT})}(\cdot)\right\} = \frac{\sum_{t=0}^{N-1} \left(\sum_{k=0}^{K-1} d_k h_{k,t}^2\right)^2}{\Delta_t \sum_{k=0}^{K-1} d_k^2}, \tag{355b}$$

(see Exercise [8.6b]); in the above, $R^{(\text{WMT})}(\eta)$ is approximately the correlation between $\hat{S}^{(\text{WMT})}(f)$ and $\hat{S}^{(\text{WMT})}(f + \eta)$.

[2] We noted in C&E [3] for Section 6.6 that, for finite sample sizes N, care must be exercised when using the asymptotic result that, for $0 < |f| < f_\mathcal{N}$, the distribution of a direct spectral estimator

$\hat{S}^{(\mathrm{D})}(f)$ is the same as a $S(f)\chi_2^2/2$ RV. Meaningful application of this result must take into account the effective bandwidth $B_{\mathcal{H}}$ of $\hat{S}^{(\mathrm{D})}(\cdot)$, which decreases as the sample size increases. For finite N the asymptotic result is a decent approximation when $B_{\mathcal{H}}/2 \leq |f| \leq f_{\mathcal{N}} - B_{\mathcal{H}}/2$ (or, more conservatively, when $B_{\mathcal{H}} \leq |f| \leq f_{\mathcal{N}} - B_{\mathcal{H}}$). An analogous qualification holds for the multitaper SDF estimator $\hat{S}^{(\mathrm{MT})}(f)$, which asymptotically has the distribution of a $S(f)\chi_{2K}^2/2K$ RV when $0 < |f| < f_{\mathcal{N}}$. The bandwidth measure for this estimator is $B_{\overline{\mathcal{H}}}$ (Equation (353e)), and we can regard the asymptotic result as being valid for finite N when $B_{\overline{\mathcal{H}}}/2 \leq |f| \leq f_{\mathcal{N}} - B_{\overline{\mathcal{H}}}/2$ (or, more conservatively, when $B_{\overline{\mathcal{H}}} \leq |f| \leq f_{\mathcal{N}} - B_{\overline{\mathcal{H}}}$).

[3] Here we explore a way to assess the variability in multitaper SDF estimators that does not make use of the chi-square distribution. Recall that the basic multitaper estimator $\hat{S}^{(\mathrm{MT})}(f)$ of Equation (352a) is the average of the eigenspectra $\hat{S}_k^{(\mathrm{MT})}(f)$, $k = 0, \ldots, K-1$. These K RVs are approximately independent and identically distributed, and, when $0 < |f| < f_{\mathcal{N}}$, their common distribution is related to that of a χ_2^2 RV as per Equation (354e). These approximations allow us in turn to approximate the distribution of $\hat{S}^{(\mathrm{MT})}(f)$ in terms of a χ_{2K}^2 RV as per Equation (354f) and to create, e.g., an approximate 95% CI for the unknown true SDF $S(\cdot)$ based upon its estimator $\hat{S}^{(\mathrm{MT})}(f)$. Thomson and Chave (1991) and Thomson (2007) note that one of the appealing aspects of multitaper SDF estimation is that it offers another approach for creating CIs other than via the χ_{2K}^2 pathway. This alternative approach – *jackknifing* – essentially assesses the variability in the basic multitaper estimator by manipulating its underlying eigenspectra. Since we might question the assumptions behind the χ_{2K}^2-based CIs (e.g., the assumption that the distributions of the eigenspectra are adequately modeled in terms of a χ_2^2 distribution), jackknifing offers a potentially valuable check on the reasonableness of CIs based on the χ_{2K}^2 distribution. Here we outline the key ideas behind this alternative approach (in addition to Thomson and Chave, 1991, and Thomson, 2007, see, e.g., Miller, 1974b, or Efron and Gong, 1983, for discussions on jackknifing).

The jackknifing approach starts by considering

$$\log(\hat{S}^{(\mathrm{MT})}(f)) = \log\left\{\frac{1}{K}\sum_{k=0}^{K-1} \hat{S}_k^{(\mathrm{MT})}(f)\right\}$$

as an estimator for $\log(S(f))$. Under the assumptions that the eigenspectra are close to being independent and identically distributed, the jackknife procedure looks at delete-one estimators

$$\log(\hat{S}_{-j}^{(\mathrm{MT})}(f)) \stackrel{\mathrm{def}}{=} \log\left\{\frac{1}{K-1}\sum_{\substack{k=0\\k\neq j}}^{K-1} \hat{S}_k^{(\mathrm{MT})}(f)\right\}, \quad j = 0, \ldots, K-1$$

and their average

$$\frac{1}{K}\sum_{j=0}^{K-1}\log(\hat{S}_{-j}^{(\mathrm{MT})}(f)) \stackrel{\mathrm{def}}{=} \log(\hat{S}_-^{(\mathrm{MT})}(f)).$$

The jackknife estimate of standard error in estimating the logarithmic spectrum is given by

$$\hat{\sigma} = \left\{\frac{K-1}{K}\sum_{j=0}^{K-1}\left[\log(\hat{S}_{-j}^{(\mathrm{MT})}(f)) - \log(\hat{S}_-^{(\mathrm{MT})}(f))\right]^2\right\}^{1/2}. \tag{356}$$

Let t_{K-1} denote an RV with a Student's t-distribution with $K-1$ degrees of freedom, and let $T_{K-1}(p)$ represent its $p \times 100\%$ percentage point; i.e., $\mathbf{P}\left[t_{K-1} \leq T_{K-1}(p)\right] = p$. Thomson and Chave (1991) claim that $[\log(\hat{S}^{(\mathrm{MT})}(f)) - \log(S(f))]/\hat{\sigma}$ has approximately the same distribution as t_{K-1}. They note that the validity of this claim is partially due to their use of the log transform, which, when applied to chi-square RVs, has the effect of driving the distribution closer to normality (cf. the left- and right-hand plots in Figure 210a). Since, for $0 \leq p \leq 1/2$,

$$\mathbf{P}\left[T_{K-1}(p) \leq t_{K-1} \leq T_{K-1}(1-p)\right] = 1 - 2p,$$

we have

$$\mathbf{P}\left[T_{K-1}(p) \leq \frac{\log(\hat{S}^{(\mathrm{MT})}(f)) - \log(S(f))}{\hat{\sigma}} \leq T_{K-1}(1-p)\right]$$

$$= \mathbf{P}\left[e^{-\hat{\sigma} T_{K-1}(1-p)} \hat{S}^{(\mathrm{MT})}(f) \leq S(f) \leq e^{-\hat{\sigma} T_{K-1}(p)} \hat{S}^{(\mathrm{MT})}(f)\right] = 1 - 2p,$$

from which it follows that

$$\left[e^{-\hat{\sigma} T_{K-1}(1-p)} \hat{S}^{(\mathrm{MT})}(f),\ e^{-\hat{\sigma} T_{K-1}(p)} \hat{S}^{(\mathrm{MT})}(f)\right] \qquad (357a)$$

is an approximate $100(1-2p)\%$ CI for $S(f)$. C&E [4] for Section 8.2 compares the jackknifing and χ^2_{2K}-based approaches to forming CIs using an example (see also Exercise [8.8]).

8.2 Slepian Multitaper Estimators

As noted in the previous section, the key to multitaper spectral estimation is a set of K orthogonal data tapers with good leakage properties. Thomson (1982) advocates use of portions of *Slepian sequences* (also known as *discrete prolate spheroidal sequences*). These sequences are characterized by a user-chosen parameter W, with $2W$ defining the bandwidth for the concentration problems posed in Section 3.10, for which the zeroth-order Slepian sequence is the optimum solution (the reader should note that, in that section, W is expressed in standardized units so that $0 < W < 1/2$, whereas here W has physically meaningful units so that now $0 < W < 1/(2\Delta_t) = f_\mathcal{N}$). For $k \geq 1$, the kth-order Slepian sequence is the optimum solution under the constraint of being orthogonal to all Slepian sequences of orders zero to $k-1$. In the Slepian multitaper scheme, we hence define the kth-order taper $\{h_{k,t}\}$ to be

$$h_{k,t} = \begin{cases} v_{k,t}(N, W), & t = 0, 1, \ldots, N-1; \\ 0, & \text{otherwise,} \end{cases} \qquad (357b)$$

where $\{v_{k,t}(N,W)\}$ is the notation we used in Section 3.10 for the kth-order Slepian sequence (as usual, we assume the unit-energy normalization $\sum_t h_{k,t}^2 = 1$). We refer to $\{h_{k,t}\}$ as the *kth-order Slepian data taper*. Slepian tapers of orders $k=0$ to $K-1$ generally provide good protection against leakage if K is chosen to be less than the Shannon number $2NW\Delta_t$. Here is a recipe for computing the Slepian multitaper estimator $\hat{S}^{(\mathrm{MT})}(\cdot)$.

[1] We first select the *regularization bandwidth* $2W$ (see Equation (377a) for an interpretation of this quantity). For a sample size N, the *fundamental Fourier frequency* is defined to be $1/(N\Delta_t)$ (note that this is just the smallest nonzero Fourier frequency). Typically W is taken to be a small multiple $j > 1$ of the fundamental frequency so that $W = j/(N\Delta_t)$ for, say, $j = 2, 3$ or 4. We note, however, that noninteger multiples are sometimes used (for example, Thomson, 1982, p. 1086, uses 3.5) and that larger values than 4 are sometimes of interest. The value of W is usually expressed indirectly via the duration × half bandwidth product $NW = j/\Delta_t$ (or just $NW = j$ with Δ_t standardized to unity). In the example to follow, we picked $NW = 4/\Delta_t$, but a reasonable choice of W must deal with the following trade-off. If we make W large, the number of tapers with good leakage properties increases – hence we can make K large and decrease the variance of $\hat{S}^{(\mathrm{MT})}(f)$. On the other hand, we shall see that the spectral bandwidth of $\hat{S}^{(\mathrm{MT})}(\cdot)$ increases (i.e., the resolution decreases) as K increases – if we use too many tapers, we can inadvertently smear out fine features in the SDF.

[2] With W so specified, we compute the eigenspectra $\hat{S}_k^{(\mathrm{MT})}(\cdot)$ for $k = 0, \ldots, K_{\max} - 1$ with $K_{\max} \stackrel{\text{def}}{=} \lfloor 2NW\Delta_t \rfloor - 1$. Unless we are dealing with a process that is very close

to white noise, these will be all of the eigenspectra with potentially good first-moment properties. (For interest in the example that follows, we also compute the eigenspectrum for $k = K_{\max}$, but this need not be done in general.)

[3] Finally we average together $K \leq K_{\max}$ of these eigenspectra, where K is determined by examining the individual eigenspectra for evidence of leakage.

As an example, let us construct Slepian multitaper spectral estimates for the AR(4) time series shown in Figure 34(e). Here $N = 1024$ and $\Delta_t = 1$, and we have set W by letting $NW = 4$; i.e., $W = 4/N = 1/256$. Since $2NW = 8$, we need to compute the eigenspectra of orders $k = 0$ to 6 because these are the ones with potentially good bias properties. Additionally, we compute the eigenspectrum for $k = 7$. The left-hand plots of Figures 360 and 362 show the Slepian data tapers $\{h_{k,t}\}$ of orders 0 to 7; the corresponding right-hand plots show the product $\{h_{k,t}X_t\}$ of these tapers with the AR(4) time series. Note carefully what happens to the values at the beginning and end of this time series as k varies from 0 to 7: for $k = 0$, these extremes are severely attenuated, but, as k increases, the attenuation becomes less and less until the extremes are actually *accentuated* for $k = 7$. One interpretation of the Slepian multitaper scheme is that the higher order tapers pick up "information" that is lost by just using the zeroth-order Slepian taper alone (this point is discussed in more detail in C&E [1]).

The left-hand plots of Figures 361 and 363 show the low frequency portion of the spectral windows $\mathcal{H}_k(\cdot)$ for $k = 0$ to 7. The solid vertical line in each plot indicates the location of the regularization half-bandwidth $W = 1/256 \approx 0.004$; on the other hand, the dashed vertical line shows half of the standard bandwidth measure, i.e., $B_{\mathcal{H}_k}/2$, which depends upon k, but is always less than W. Note that, as k increases, the level of the sidelobes of $\mathcal{H}_k(\cdot)$ also increases until at $k = 7$ the main sidelobe level is just barely below the lowest lobe in $[-W, W]$. In fact, the spectral windows for Slepian data tapers of higher order than 7 resemble the squared modulus of a band-pass filter with a center frequency outside the interval $[-W, W]$ – see C&E [2]. Use of these tapers would yield direct spectral estimates $\hat{S}_k^{(\mathrm{MT})}(f)$ whose expectation is controlled by values of the true SDF *outside* $[f - W, f + W]$!

The jagged curves in the right-hand plots of Figures 361 and 363 show the eigenspectra $\hat{S}_k^{(\mathrm{MT})}(\cdot)$. The thick smooth curve in each of these plots shows the true SDF for the AR(4) process. The crisscross in each plot (the horizontal portion of which is barely visible) has the usual interpretation: its width shows the effective bandwidth $B_{\mathcal{H}_k}$ of $\hat{S}_k^{(\mathrm{MT})}(\cdot)$, and its height, the length of a 95% CI for $10 \log_{10}(S(f))$ based upon $10 \log_{10}(\hat{S}_k^{(\mathrm{MT})}(f))$. Note that, whereas the eigenspectra of orders $k = 0$ to 3 (Figure 361) show no indication of leakage, those of order 4 and higher (Figure 363) show increasing evidence of leakage at high frequencies. For example, the $k = 4$ estimate (top right-hand plot of Figure 363) is about 5 dB above the true SDF for frequencies in the range 0.3 to 0.5, whereas the $k = 7$ estimate is elevated there by about 20 dB (had we not known the true SDF, we could have discovered this leakage by comparison with the eigenspectra of orders 0 to 3). The lack of homogeneity of variance across frequencies in the $k = 6$ and 7 eigenspectra is another indication of leakage in the high frequencies (see the discussion in C&E [4] for Section 6.4).

Let us now consider the multitaper spectral estimates $\hat{S}^{(\mathrm{MT})}(\cdot)$ and corresponding spectral windows $\overline{\mathcal{H}}(\cdot)$ for different choices of K, the number of eigenspectra to be averaged together. Figures 364 and 365 show these two functions for $K = 1$ up to 8 (the case $K = 1$ corresponds to using just a single data taper, so the two plots in the first row of Figure 364 are identical to the ones in the first row of Figure 361). The thick curves in the left-hand plots are the low frequency portions of the spectral windows, while the solid and dashed vertical lines mark $W = 1/256$ and $B_{\overline{\mathcal{H}}}/2$, respectively. Note that, as K increases, the bandwidth measure $B_{\overline{\mathcal{H}}}$ increases steadily, becoming almost equal to the regularization bandwidth $2W$ for $K = 7$ and exceeding it when $K = 8$. The jagged curves in the right-hand plots are the corresponding multitaper spectral estimates, and the thick smooth curve in each of these plots is the true SDF. As K increases, the level of the sidelobes in $\overline{\mathcal{H}}(\cdot)$ increases (as discussed in Sections 8.4 and 8.5, this is related to an increase in what we refer to there as broad-band bias); note also that $\overline{\mathcal{H}}(\cdot)$ becomes noticeably closer to being constant over frequencies less than W in magnitude (this is related to a decrease in the – yet to be defined – local bias).

Let us now study how well the $\hat{S}^{(\text{MT})}(\cdot)$ in the right-hand plots do in terms of bias and variance. For $K = 1$ up to 5, we see that there is little evidence of any bias in $\hat{S}^{(\text{MT})}(\cdot)$ at any frequency; for $K = 6$ there is evidence of a small bias of about 3 dB for frequencies between 0.4 and 0.5; and, for $K = 7$ and 8, there is significant bias in the high frequencies. On the other hand, as K increases, we see that the variability in $\hat{S}^{(\text{MT})}(\cdot)$ steadily decreases, as Equation (354c) suggests it should. The crisscrosses demonstrate that, as K increases, the length of a 95% CI for $10\log_{10}(S(f))$ decreases, while the bandwidth measure $B_{\overline{\mathcal{H}}}$ increases (recall that this measure is the approximate distance in frequency between uncorrelated spectral estimators). Note that, as K increases, this length decreases and that, except in the high frequency regions of $\hat{S}^{(\text{MT})}(\cdot)$ for $K = 7$ and 8 where bias is dominant, the amount of variability in $\hat{S}^{(\text{MT})}(\cdot)$ is roughly consistent with these heights.

For this particular example, the Slepian multitaper spectral estimator with $K = 5$ data tapers visually gives us the best compromise between good protection against leakage and a marked reduction in variance over that offered by any individual eigenspectrum. In addition to leakage protection and variance reduction, a third consideration is the resolution of the estimator. Note that the variations in the true AR(4) SDF are on a scale roughly comparable to our selected regularization bandwidth $2W$ (compare the width of the crisscross in Figure 365 for the $K = 7$ case with the true SDF, recalling that here $B_{\overline{\mathcal{H}}}$ is almost equal to $2W$). Had we increased the regularization bandwidth by a factor of, say, 2, we would then have more than 5 data tapers with acceptable leakage properties, resulting in a leakage-suppressed estimator with a reduction in variance beyond what we achieved here; however, care would be needed to ensure that this additional reduction in variance does not come at the expense of a loss of resolution due to the increase in $B_{\overline{\mathcal{H}}}$ that comes about with increasing K.

In addition to its ability to produce SDF estimates with good leakage, variance and resolution properties, Slepian multitaper spectral estimation has a number of additional points in its favor including the following.

[1] In contrast to prewhitening, which typically requires careful design of a prewhitening filter, the Slepian multitaper scheme can be used in a fairly automatic fashion. Hence it is useful in situations where thousands – or millions – of individual time series must be processed so that the pure volume of data precludes a careful analysis of individual series (this occurs routinely in exploration geophysics).

[2] As discussed in Sections 8.4 and 8.5, the bias of Slepian multitaper spectral estimators can be broken down into two quantifiable components: the local bias (due to frequency components within a user-selectable pass-band $[f - W, f + W]$) and the broad-band bias (due to components outside this pass-band).

[3] As we shall discuss in Section 10.10, there is an appealing statistical test based upon the Slepian multitaper spectral estimator for the presence of sinusoidal (line) components in a time series. Slepian multitapering thus offers a unified approach to estimating both mixed spectra and SDFs (we defined processes with mixed spectra in Section 4.4 and will discuss them in detail in Chapter 10).

The seminal references for Slepian multitapering are Thomson (1982) and Bronez (1985, 1988). Complementary introductions include Park et al. (1987b), Walden (1990b), Mullis and Scharf (1991), Hogan and Lakey (2012) and Babadi and Brown (2014). Slepian multitapering has been used on a number of interesting time series, including ones concerning terrestrial

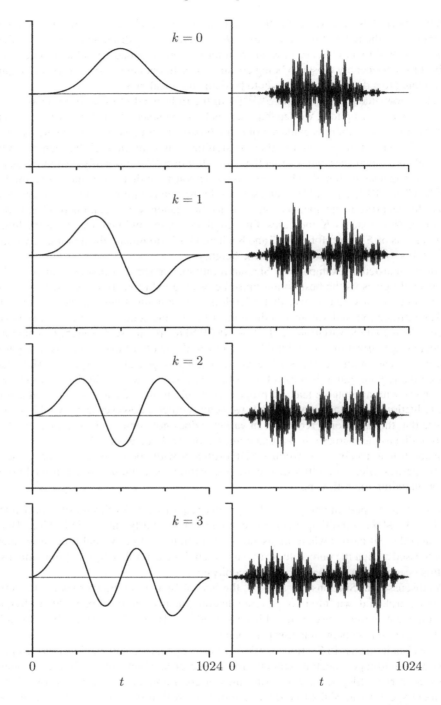

Figure 360 Slepian data tapers $\{h_{k,t}\}$ (left-hand plots) and products $\{h_{k,t}X_t\}$ of these tapers and AR(4) time series $\{X_t\}$ shown in Figure 34(e) (right-hand plots), part 1. Here $NW = 4$ with $N = 1024$. The orders k for the Slepian data tapers are 0, 1, 2 and 3 (top to bottom).

8.2 Slepian Multitaper Estimators

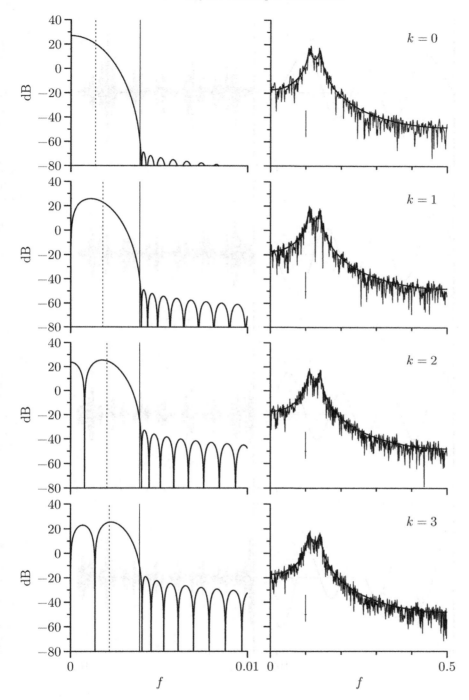

Figure 361 Spectral windows $\mathcal{H}_k(\cdot)$ for Slepian multitapers $\{h_{k,t}\}$ (left-hand plots) and eigenspectra $\hat{S}_k^{(\mathrm{MT})}(\cdot)$ for AR(4) time series (right-hand, thin curves), part 1. The orders k are the same as in Figure 360. The true AR(4) SDF is the thick curve in each right-hand plot. See the text for details about the vertical lines (crisscrosses) in the left-hand (right-hand) plots.

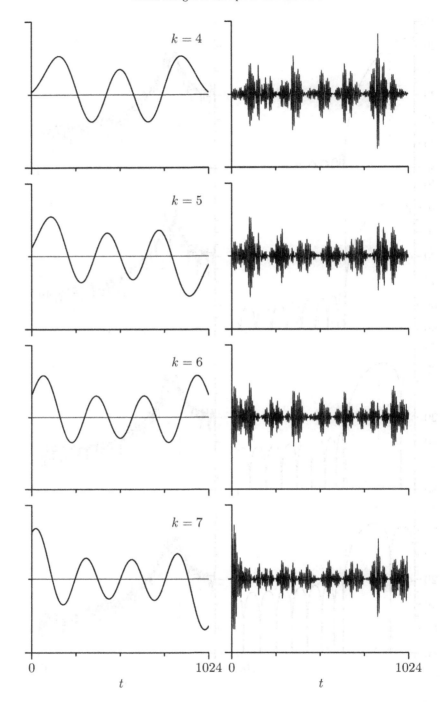

Figure 362 Slepian data tapers and products of these tapers and a time series, part 2 (cf. Figure 360). Here the orders k for the tapers are 4, 5, 6 and 7 (top to bottom). Note that, for $k = 0$ in the top row of Figure 360, the extremes of the time series are severely attenuated, but that, as k increases, portions of the extremes are accentuated.

8.2 Slepian Multitaper Estimators

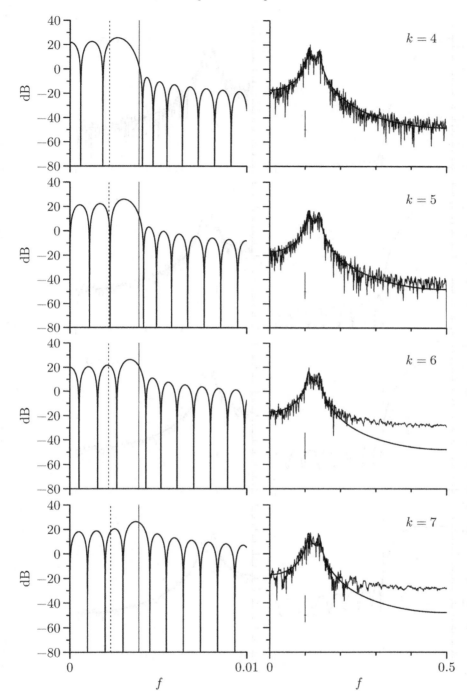

Figure 363 Spectral windows and eigenspectra for AR(4) time series, part 2 (cf. Figure 361). The orders k are the same as in Figure 362. Note the marked increase in bias in the eigenspectra as k increases.

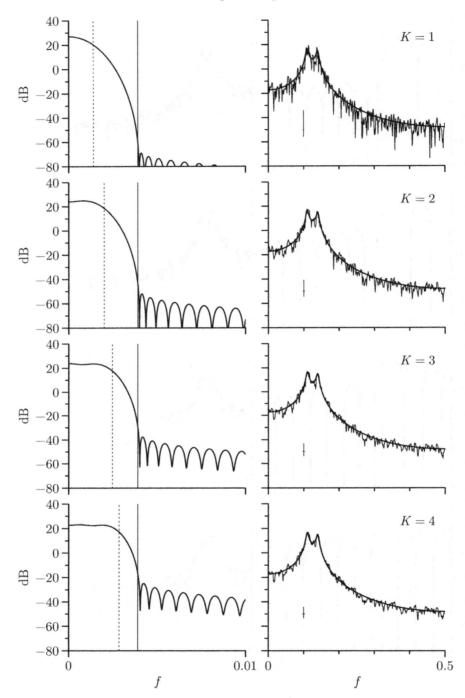

Figure 364 Slepian multitaper spectral estimates $\hat{S}^{(\mathrm{MT})}(\cdot)$ formed by averaging K eigenspectra (right-hand plots) and corresponding spectral windows $\overline{\mathcal{H}}(\cdot)$ (left-hand plots) for $K = 1, 2, 3$ and 4. See the text for details about the vertical lines (crisscrosses) in the left-hand (right-hand) plots.

8.2 Slepian Multitaper Estimators

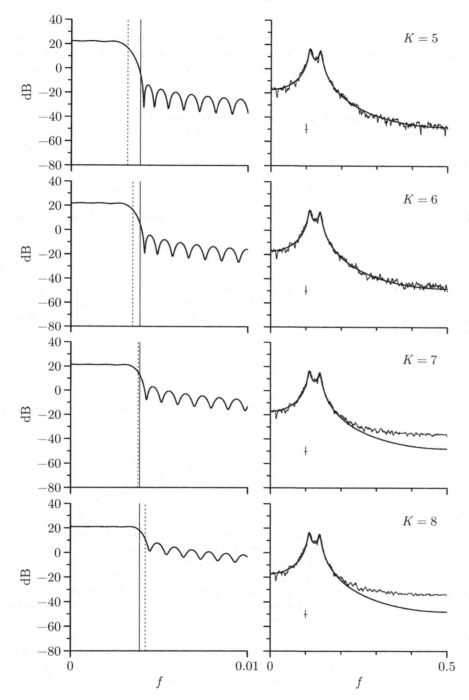

Figure 365 Slepian multitaper spectral estimates $\hat{S}^{(\mathrm{MT})}(\cdot)$ formed by averaging K eigenspectra (right-hand plots) and corresponding spectral windows $\overline{\mathcal{H}}(\cdot)$ (left-hand plots) for $K = 5, 6, 7$ and 8. Note the increasing discrepancy at high frequencies between $\hat{S}^{(\mathrm{MT})}(\cdot)$ and the true SDF as K gets large.

free oscillations (Park et al., 1987a, and Lindberg and Park, 1987), the relationship between atmospheric carbon dioxide and global temperature (Kuo et al. 1990), oxygen isotope ratios from deep-sea cores (Thomson, 1990b) and the rotation of the earth (King, 1990). Van Veen and Scharf (1990) discuss the relationship between structured covariance matrices and multi-tapering.

Comments and Extensions to Section 8.2

[1] We present here support for the statement that Slepian multitapering is a scheme for recovering information usually lost when but a single data taper is used (Walden, 1990b). In Section 3.10 we found that the vector

$$\boldsymbol{v}_k(N,W) = [v_{k,0}(N,W), v_{k,1}(N,W), \ldots, v_{k,N-1}(N,W)]^T = [h_{k,0}, h_{k,1}, \ldots, h_{k,N-1}]^T$$

is the kth-order eigenvector for the $N \times N$ real symmetric matrix $\boldsymbol{A}$ with (t', t)th element given by $\sin[2\pi W(t'-t)]/[\pi(t'-t)]$. Let

$$\boldsymbol{\Sigma} \stackrel{\text{def}}{=} [\boldsymbol{v}_0(N,W), \boldsymbol{v}_1(N,W), \ldots, \boldsymbol{v}_{N-1}(N,W)]$$

be a matrix whose columns are the eigenvectors. Since each eigenvector is assumed to be normalized to have a unit sum of squares, we have $\boldsymbol{\Sigma}^T\boldsymbol{\Sigma} = \boldsymbol{I}_N$, where $\boldsymbol{I}_N$ is the $N \times N$ identity matrix; i.e., $\boldsymbol{\Sigma}$ is an orthonormal matrix. Because the transpose of $\boldsymbol{\Sigma}$ is its inverse, it follows that $\boldsymbol{\Sigma}\boldsymbol{\Sigma}^T = \boldsymbol{I}_N$ also. This yields the interesting result that

$$\sum_{k=0}^{N-1} h_{k,t}^2 = 1 \text{ for } t = 0, 1, \ldots, N-1;$$

i.e., the sum over all orders k of the squares of the tth element of each taper is unity. The energy from all N tapered series $\{h_{k,t}X_t\}$ is thus

$$\sum_{k=0}^{N-1}\sum_{t=0}^{N-1}(h_{k,t}X_t)^2 = \sum_{t=0}^{N-1}X_t^2\sum_{k=0}^{N-1}h_{k,t}^2 = \sum_{t=0}^{N-1}X_t^2;$$

i.e., the total energy using all N possible tapers equals the energy in the time series. Figure 367 shows the build-up of $\sum_{k=0}^{K-1}h_{k,t}^2$ for $t = 0, 1, \ldots, N-1$ as K increases from 1 to 8 (as before, $N = 1024$, and W is set such that $NW = 4$). Loosely speaking, the plot for a particular K shows the relative influence of X_t – as t varies – on the multitaper spectral estimator $\hat{S}^{(\text{MT})}(\cdot)$ formed using K eigenspectra. The $K = 1$ estimator (i.e., a single Slepian taper) is not influenced much by values of $\{X_t\}$ near $t = 0$ or 1023. As K increases to 7, however, the amount of data in the extremes that has little influence decreases steadily, supporting the claim that multitapering is picking up information from the extremes of a time series that is lost with use of a single data taper. Increasing K further to 8 shows a breakdown in which certain values near the extremes start to have *greater* influence than the central portion of the time series (see the case $K = 9$ in the bottom plot of Figure 368).

[2] The previous figures in this chapter refer to the Slepian data tapers $\{h_{k,t}\}$ of orders $k = 0$ to 7 for $N = 1024$ and $NW = 4$. For this particular choice of W, these are all the tapers whose associated spectral windows $\mathcal{H}_k(\cdot)$ have the majority of their energy concentrated in the frequency interval $[-W, W]$; i.e., we have

$$\frac{\int_{-W}^{W}\mathcal{H}_k(f)\,df}{\int_{-1/2}^{1/2}\mathcal{H}_k(f)\,df} = \lambda_k(N,W) > 1/2 \text{ for } k = 0, \ldots, 7,$$

where $\lambda_k(N,W)$ is the kth-order eigenvalue defined in Section 3.10 (the usual normalization $\sum_t h_{k,t}^2 = 1$ implies that $\int_{-1/2}^{1/2}\mathcal{H}_k(f)\,df = 1$, so the denominator in the ratio above is just unity). The eigenvalue

8.2 Slepian Multitaper Estimators

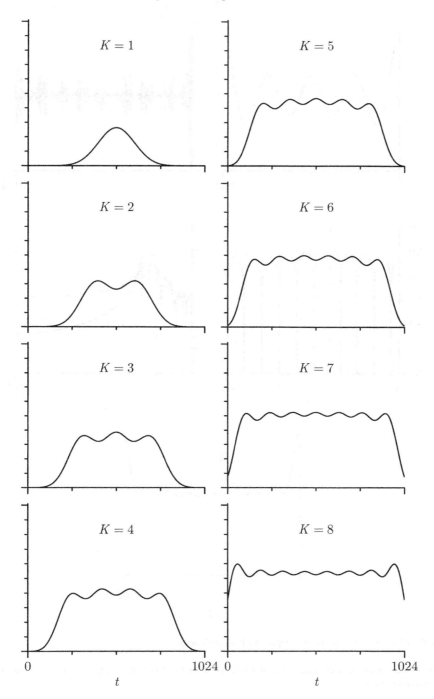

Figure 367 Decomposition of Slepian taper energy across t as K (the number of tapers used in the multitaper scheme) increases.

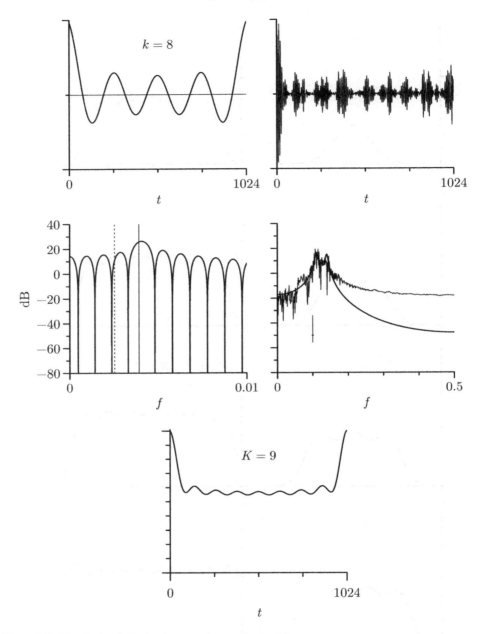

Figure 368 Plots for $k = 8$ Slepian data taper (see text for details).

$\lambda_k(N, W)$ thus measures the concentration of $\mathcal{H}_k(\cdot)$ over the frequency range $[-W, W]$. For our particular example, we have

$$\lambda_0(N, W) = 0.999\,999\,999\,7 \qquad \lambda_3(N, W) = 0.999\,97$$
$$\lambda_1(N, W) = 0.999\,999\,97 \qquad \lambda_4(N, W) = 0.999\,4 \qquad \lambda_6(N, W) = 0.94$$
$$\lambda_2(N, W) = 0.999\,998\,8 \qquad \lambda_5(N, W) = 0.993 \qquad \lambda_7(N, W) = 0.70.$$

For the sake of comparison, Figure 368 shows corresponding plots for the eighth-order Slepian data taper, for which $\lambda_8(N, W) = 0.30$ (for $k > 8$, the concentration measures $\lambda_k(N, W)$ are all close

NW	\multicolumn{16}{c}{K}
	1	2	3	4	5	6	7	8	9	10	11	12	13	14	15	16
1	1.16	1.24														
2	1.04	1.09	1.11	1.14												
4	1.02	1.06	1.07	1.07	1.07	1.06	1.07	1.08								
8	1.01	1.06	1.07	1.07	1.07	1.07	1.07	1.07	1.06	1.06	1.05	1.05	1.04	1.04	1.04	1.04

Table 369 Comparison of two bandwidth measures for Slepian multitaper spectral estimators $\hat{S}^{(\mathrm{MT})}(\cdot)$. The first measure $B_{\overline{\mathcal{H}}}$ is given by Equation (353e) and is the autocorrelation width of the spectral window associated with $\hat{S}^{(\mathrm{MT})}(\cdot)$. The second measure $\mathrm{width}_e\{R^{(\mathrm{MT})}(\cdot)\}$ is given by Equation (355a) and can be interpreted as the minimal distance η such that $\hat{S}^{(\mathrm{MT})}(f+\eta)$ and $\hat{S}^{(\mathrm{MT})}(f)$ are approximately uncorrelated (assuming that the process is Gaussian and that f is not too close to either zero or Nyquist frequency). The tabulated values show the ratio $B_{\overline{\mathcal{H}}}/\mathrm{width}_e\{R^{(\mathrm{MT})}(\cdot)\}$ for sample size $N = 1024$ and for various setting of the duration $\times$ half bandwidth product NW and of the number of tapers K used to form $\hat{S}^{(\mathrm{MT})}(\cdot)$.

to zero – see Figure 91). The two plots in the top row show the taper itself and the product of the taper and the AR(4) series. Note, in particular, the accentuation of the extremes of the series. The left-hand plot in the middle row shows that the peak value of the spectral window $\mathcal{H}_8(\cdot)$ occurs just *outside* the frequency interval $[-W, W]$; the right-hand plot shows the poor leakage properties of $\hat{S}^{(\mathrm{MT})}_8(\cdot)$. Finally, as noted previously, the plot on the last row shows that the distribution of energy over t is skewed to accentuate the extremes of $\{X_t\}$ when $K = 9$ tapers are used to form $\hat{S}^{(\mathrm{MT})}(\cdot)$.

[3] In C&E [1] for Section 8.1 we noted an alternative to our standard bandwidth measure $B_{\overline{\mathcal{H}}}$ for a basic multitaper SDF estimator, namely, $\mathrm{width}_e\{R^{(\mathrm{MT})}(\cdot)\}$ of Equation (355a), which is the equivalent width of an approximation $R^{(\mathrm{MT})}(\eta)$ to the correlation between $\hat{S}^{(\mathrm{MT})}(f)$ and $\hat{S}^{(\mathrm{MT})}(f+\eta)$ (Greenhall, 2006). Table 369 compares these two measures by listing their ratio $B_{\overline{\mathcal{H}}}/\mathrm{width}_e\{R^{(\mathrm{MT})}(\cdot)\}$ for Slepian multitapers with $NW = 1, 2, 4$ and 8 and with K (the number of eigenspectra averaged together) ranging from 1 to $2NW$. The measure $B_{\overline{\mathcal{H}}}$ is consistently larger than $\mathrm{width}_e\{R^{(\mathrm{MT})}(\cdot)\}$, but no more than 24% and typically less than 10%. Thus, for most practical applications, we can regard the two bandwidth measures as being interchangeable. (The values in the table are for sample size $N = 1024$. Similar calculations for sample sizes of powers of two ranging from $2^5 = 32$ up to $2^{15} = 32{,}768$ yield tables identical to Table 369 except for a few selected differences of ± 0.01.)

[4] In C&E [3] for Section 8.1 we described jackknifing as an alternative to the usual χ^2-based procedure for formulating a CI for $S(f)$ based upon a multitaper estimator $\hat{S}^{(\mathrm{MT})}(f)$. Figure 370 offers a comparison of these two ways of producing CIs using the $K = 5$ multitaper estimate shown in the upper right-hand plot of Figure 365 as an example (its underlying eigenspectra $\hat{S}^{(\mathrm{MT})}_k(\cdot)$, $k = 0, 1, \ldots, 4$, are shown in the right-hand columns of Figures 361 and 363.) The smooth curve in both plots of Figure 370 is the SDF for the AR(4) process of Equation (35a), which is what $\hat{S}^{(\mathrm{MT})}(\cdot)$ is attempting to estimate. The jagged curves in the upper plot show the upper and lower limits of 95% CIs on a decibel scale for $S(f)$ versus f. These CIs are based upon a χ^2_{10} distribution as per Equation (266a), with ν set to 10 and with $\hat{S}^{(\mathrm{LW})}_m(f)$ replaced by $\hat{S}^{(\mathrm{MT})}(f)$. Because of the decibel scale, the difference between the upper and lower limits is the same at all frequencies f. Equation (266a) says this difference is $10 \log_{10}(10/Q_{10}(0.025)) - 10 \log_{10}(10/Q_{10}(0.975)) \doteq 8.0\,\mathrm{dB}$, which is depicted by the height of the crisscross on the plot. The two dashed vertical lines mark the frequencies $B_{\overline{\mathcal{H}}}/2$ and $f_{\mathcal{N}} - B_{\overline{\mathcal{H}}}/2 = 0.5 - B_{\overline{\mathcal{H}}}/2$, where, as usual, $B_{\overline{\mathcal{H}}}$ is the bandwidth associated with $\hat{S}^{(\mathrm{MT})}(\cdot)$ and is equal to the width of the crisscross. As noted in C&E [2] for Section 8.1, it is inappropriate to construct a CI based upon a χ^2_ν distribution with $\nu = 2K = 10$ for $f \in [0, B_{\overline{\mathcal{H}}}/2]$ or $f \in [f_{\mathcal{N}} - B_{\overline{\mathcal{H}}}/2, f_{\mathcal{N}}]$ (pending further research, appropriate settings for ν would arguably sweep from $K = 5$ to $2K = 10$ as f sweeps from 0 to $B_{\overline{\mathcal{H}}}/2$ or from $f_{\mathcal{N}}$ to $f_{\mathcal{N}} - B_{\overline{\mathcal{H}}}/2$). The CIs for $f \in [B_{\overline{\mathcal{H}}}/2, f_{\mathcal{N}} - B_{\overline{\mathcal{H}}}/2]$ are pointwise; i.e., they apply only to the SDF at one particular frequency, and do not provide a simultaneous confidence band. If we consider the Fourier frequencies $f_k = k/1024$ in this interval, we find that 96.0% of the corresponding CIs trap the true SDF. This percentage lends credence to the χ^2-based procedure

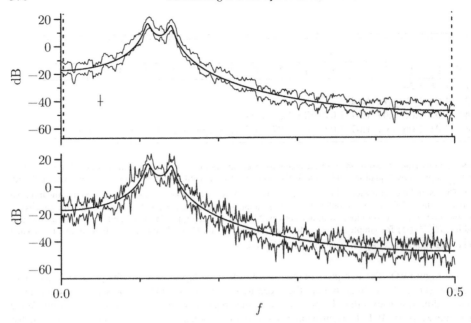

Figure 370 95% CIs on a decibel scale (jagged curves) for an SDF versus frequency f generated using either a χ^2_{10} distribution (upper plot) or jackknifing (lower) – see text for details.

since it is not inconsistent with a pointwise coverage of 95%.

The lower plot in Figure 370 shows upper and lower limits for a 95% CI for $S(f)$ versus f using jackknifing as per Equation (357a). Although the CIs are again shown on a decibel scale, their widths are no longer constant, but rather depend upon f. Focusing on the Fourier frequencies f_k, $k = 0, \ldots, 512$, the corresponding CIs have widths ranging from 0.8 dB to 36.7 dB, with a mean of 10.9 dB and a median of 11.9 dB. These CIs include the true SDF at 94.1% of the Fourier frequencies, which lends credence to the jackknifing procedure.

[5] Lii and Rosenblatt (2008) establish rigorous conditions under which the mean square error of a Slepian-based basic multitaper estimator converges to zero as $N \to \infty$. The conditions are intuitively reasonable. In particular, the regularization bandwidth $2W$ must decrease to zero but in such a manner that $NW\Delta_t \to \infty$ also; and the number of tapers K must always be less than or equal to $2NW\Delta_t$, but we also need $K \to \infty$.

8.3 Multitapering of Gaussian White Noise

The statistical properties of the basic multitaper estimator $\hat{S}^{(\text{MT})}(\cdot)$ are easy to derive in the case of Gaussian white noise and serve to illustrate the decrease in variability that multitapering affords. Accordingly, for this section only, let us assume that the time series we are dealing with is a realization of a portion $X_0, X_1, \ldots, X_{N-1}$ of a Gaussian white noise process with zero mean and unknown variance s_0. The SDF is thus just $S(f) = s_0 \Delta_t$. Standard statistical theory suggests that, in several senses, the best estimator of s_0 for this model is just the sample variance

$$\frac{1}{N} \sum_{t=0}^{N-1} X_t^2 = \hat{s}_0^{(\text{P})}.$$

Thus the best estimator for the SDF is just $\hat{s}_0^{(\text{P})} \Delta_t$. Equation (298a) tells us that

$$\text{var}\{\hat{s}_0^{(\text{P})} \Delta_t\} = \frac{2s_0^2 \Delta_t^2}{N}. \tag{370}$$

8.3 Multitapering of Gaussian White Noise

As we discussed in Section 6.3, the periodogram $\hat{S}^{(\mathrm{P})}(f)$ is an unbiased estimator of $S(f)$ in the case of white noise, and hence a data taper is certainly not needed to control bias. Moreover, because $S(\cdot)$ is flat, we can smooth $\hat{S}^{(\mathrm{P})}(\cdot)$ with a rectangular smoothing window of width $2f_{\mathcal{N}}$ and height $1/(2f_{\mathcal{N}}) = \Delta_{\mathrm{t}}$ to obtain

$$\int_{-f_{\mathcal{N}}}^{f_{\mathcal{N}}} \hat{S}^{(\mathrm{P})}(f)\, \Delta_{\mathrm{t}}\, \mathrm{d}f = \hat{s}_0^{(\mathrm{P})} \Delta_{\mathrm{t}}$$

(this follows from Equation (171d)). We can thus easily recover the best spectral estimator $\hat{s}_0^{(\mathrm{P})} \Delta_{\mathrm{t}}$ by averaging $\hat{S}^{(\mathrm{P})}(\cdot)$ over $[-f_{\mathcal{N}}, f_{\mathcal{N}}]$. On the other hand, suppose that we use any nonrectangular taper $\{\tilde{h}_{0,t}\}$ – with the usual normalization $\sum_{t=0}^{N-1} \tilde{h}_{0,t}^2 = 1$ – to produce the direct spectral estimator $\hat{S}^{(\mathrm{D})}(\cdot)$. If we smooth this estimator, we obtain

$$\int_{-f_{\mathcal{N}}}^{f_{\mathcal{N}}} \hat{S}^{(\mathrm{D})}(f)\, \Delta_{\mathrm{t}}\, \mathrm{d}f = \sum_{t=0}^{N-1} \tilde{h}_{0,t}^2 X_t^2\, \Delta_{\mathrm{t}} = \hat{s}_0^{(\mathrm{D})} \Delta_{\mathrm{t}}.$$

Now

$$\mathrm{var}\{\hat{s}_0^{(\mathrm{D})} \Delta_{\mathrm{t}}\} = \Delta_{\mathrm{t}}^2 \sum_{t=0}^{N-1} \mathrm{var}\{\tilde{h}_{0,t}^2 X_t^2\} = 2 s_0^2 \Delta_{\mathrm{t}}^2 \sum_{t=0}^{N-1} \tilde{h}_{0,t}^4$$

since we have $\mathrm{var}\{X_t^2\} = 2 s_0^2$ from the Isserlis theorem (Equation (30)). The Cauchy inequality, namely,

$$\left| \sum_{t=0}^{N-1} a_t b_t \right|^2 \leq \sum_{t=0}^{N-1} |a_t|^2 \sum_{t=0}^{N-1} |b_t|^2,$$

with $a_t = \tilde{h}_{0,t}^2$ and $b_t = 1$, tells us that $\sum_{t=0}^{N-1} \tilde{h}_{0,t}^4 \geq 1/N$ with equality holding if and only if $\tilde{h}_{0,t} = \pm 1/\sqrt{N}$ (i.e., $\{\tilde{h}_{0,t}\}$ must be – disregarding an innocuous change of sign – the rectangular data taper). We can conclude that $\mathrm{var}\{\hat{s}_0^{(\mathrm{P})} \Delta_{\mathrm{t}}\} < \mathrm{var}\{\hat{s}_0^{(\mathrm{D})} \Delta_{\mathrm{t}}\}$ for any nonrectangular data taper. Use of a single data taper on white noise yields a smoothed spectral estimator with increased variance and thus amounts to throwing away a certain portion of the data (Sloane, 1969).

We now show that we can compensate for this loss through multitapering. Accordingly, let $\{\tilde{h}_{0,t}\}, \{\tilde{h}_{1,t}\}, \ldots, \{\tilde{h}_{N-1,t}\}$ be *any* set of orthonormal data tapers, each of length N. If we let $\widetilde{\Sigma}$ be the $N \times N$ matrix whose columns are formed from these tapers, i.e.,

$$\widetilde{\Sigma} \stackrel{\mathrm{def}}{=} \begin{bmatrix} \tilde{h}_{0,0} & \tilde{h}_{1,0} & \cdots & \tilde{h}_{N-1,0} \\ \tilde{h}_{0,1} & \tilde{h}_{1,1} & \cdots & \tilde{h}_{N-1,1} \\ \vdots & \vdots & \ddots & \vdots \\ \tilde{h}_{0,N-1} & \tilde{h}_{1,N-1} & \cdots & \tilde{h}_{N-1,N-1} \end{bmatrix},$$

orthonormality says that $\widetilde{\Sigma}^T \widetilde{\Sigma} = I_N$, where I_N is the $N \times N$ identity matrix. Since the transpose of $\widetilde{\Sigma}$ is thus its inverse, we also have $\widetilde{\Sigma} \widetilde{\Sigma}^T = I_N$; i.e.,

$$\sum_{k=0}^{N-1} \tilde{h}_{k,t} \tilde{h}_{k,u} = \begin{cases} 1, & t = u; \\ 0, & \text{otherwise} \end{cases} \qquad (371)$$

(see C&E [1] for the previous section). Let

$$\tilde{S}_k^{(\mathrm{MT})}(f) = \Delta_{\mathrm{t}} \left| \sum_{t=0}^{N-1} \tilde{h}_{k,t} X_t e^{-\mathrm{i}2\pi f t \Delta_{\mathrm{t}}} \right|^2 \quad (372\mathrm{a})$$

be the kth direct spectral estimator, and consider the multitaper estimator formed by averaging all N of the $\tilde{S}_k^{(\mathrm{MT})}(f)$:

$$\tilde{S}^{(\mathrm{MT})}(f) = \frac{1}{N} \sum_{k=0}^{N-1} \tilde{S}_k^{(\mathrm{MT})}(f).$$

We can write

$$\tilde{S}^{(\mathrm{MT})}(f) = \frac{\Delta_{\mathrm{t}}}{N} \sum_{k=0}^{N-1} \left(\sum_{t=0}^{N-1} \tilde{h}_{k,t} X_t e^{-\mathrm{i}2\pi f t \Delta_{\mathrm{t}}} \right) \left(\sum_{u=0}^{N-1} \tilde{h}_{k,u} X_u e^{\mathrm{i}2\pi f u \Delta_{\mathrm{t}}} \right)$$

$$= \frac{\Delta_{\mathrm{t}}}{N} \sum_{t=0}^{N-1} \sum_{u=0}^{N-1} X_t X_u \left(\sum_{k=0}^{N-1} \tilde{h}_{k,t} \tilde{h}_{k,u} \right) e^{-\mathrm{i}2\pi f(t-u) \Delta_{\mathrm{t}}}.$$

From Equation (371), the term in the parentheses is unity if $t = u$ and is zero otherwise, so we obtain

$$\tilde{S}^{(\mathrm{MT})}(f) = \frac{\Delta_{\mathrm{t}}}{N} \sum_{t=0}^{N-1} X_t^2 = \hat{s}_0^{(\mathrm{P})} \Delta_{\mathrm{t}}. \quad (372\mathrm{b})$$

A multitaper spectral estimator using N orthogonal data tapers is thus identical to the best spectral estimator for a white noise process, so multitapering effectively restores the information normally lost when but a single taper is used.

We emphasize that the argument leading to Equation (372b) is valid for *any* set of orthogonal data tapers. This cavalier approach works only in the very special case of white noise because then any data taper yields an unbiased direct spectral estimator. For colored processes, this is not the case – we must then insist upon a set of orthogonal tapers, at least some of which provide decent leakage protection. With this restriction, we are led to the Slepian tapers (Section 8.2) and the sinusoidal tapers (Section 8.6) as reasonable sets to choose.

Finally, let us study the rate at which the variance of the multitaper estimator decreases as the number of eigenspectra K that are averaged together increases. To do so, we now assume that we use the Slepian data tapers $\{h_{k,t}\}$ to form the eigenspectra. Equation (353f) gives an expression for the variance of $\hat{S}^{(\mathrm{MT})}(f)$. Exercise [8.9b] indicates how to compute the variance and covariance terms in this equation for a Gaussian white noise process. The exercise also shows that var $\{\hat{S}^{(\mathrm{MT})}(f)\}$ depends on f. As a specific example, let us set $f = 1/4$. The thick curve in Figure 373a shows var $\{\hat{S}^{(\mathrm{MT})}(1/4)\}$ versus the number of eigenspectra K for $N = 64$, $NW = 4$, $\Delta_{\mathrm{t}} = 1$ and $s_0 = 1$. Because $S(1/4) = s_0 \Delta_{\mathrm{t}} = 1$, the result of Exercise [8.9b] says that var $\{\hat{S}^{(\mathrm{MT})}(1/4)\} = 1 = 0 \,\mathrm{dB}$ for $K = 1$, while Equations (372b) and (370) tell us that var $\{\hat{S}^{(\mathrm{MT})}(1/4)\} = 2/64 \doteq 0.03 \doteq -15 \,\mathrm{dB}$ for $K = N = 64$. The extremes of the thick curve agree with these values. The thin curve shows var $\{\hat{S}^{(\mathrm{MT})}(1/4)\}$ versus K under the incorrect assumption that cov $\{\hat{S}_j^{(\mathrm{MT})}(1/4), \hat{S}_k^{(\mathrm{MT})}(1/4)\}$ is zero for all $j \neq k$. The thin vertical line marks the Shannon number $2NW = 8$. We see that, for $K \leq 8$, the variance decreases at a rate consistent with cov $\{\hat{S}_j^{(\mathrm{MT})}(1/4), \hat{S}_k^{(\mathrm{MT})}(1/4)\} = 0$ to a good approximation for all $j \neq k$ such that $0 \leq j,k < 8$. As K increases beyond the Shannon number, the rate of decrease in variance becomes slower than that implied by uncorrelatedness of the $\hat{S}_k^{(\mathrm{MT})}(1/4)$ terms. For colored processes, this result suggests that the modest decrease

8.3 Multitapering of Gaussian White Noise

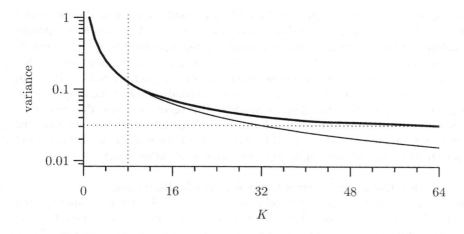

Figure 373a Decrease in var $\{\hat{S}^{(\mathrm{MT})}(f)\}$ for $f = 1/4$ as the number K of eigenspectra used to form the Slepian multitaper spectral estimator increases from 1 up to the sample size $N = 64$ (thick curve). Here a Gaussian white noise process is assumed with unit variance and a sampling interval $\Delta_t = 1$. The data tapers are Slepian tapers with $NW = 4$. The thin curve indicates the decrease in variance to be expected if the eigenspectra were all uncorrelated. The vertical dotted line marks the Shannon number $2NW = 8$, and the horizontal, $2/64$, which is equal to var $\{\hat{S}^{(\mathrm{MT})}(1/4)\}$ for $K = N = 64$.

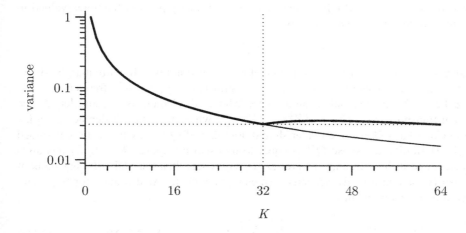

Figure 373b As in Figure 373a, but now for Slepian tapers with $NW = 16$. As before, the vertical dotted line marks the Shannon number, which is now $2NW = 32$.

in the variance of $\hat{S}^{(\mathrm{MT})}(1/4)$ afforded by including high-order eigenspectra will be readily offset by the poor bias properties of these eigenspectra.

Figure 373b shows a second example, which differs from the first only in that we now set $NW = 16$. The two examples are similar in some aspects, but differ in the following ways. First, in the $NW = 4$ case, var $\{\hat{S}^{(\mathrm{MT})}(1/4)\}$ achieves its minimum value $(2/64)$ only when $K = 64$, whereas, for $NW = 16$, this same minimum is achieved at both $K = 32$ and $K = 64$. Thus, for even N and for certain sets of orthonormal tapers, the best estimator of a white noise SDF can be recovered using $N/2$ rather than N tapers. To see why this makes sense, recall that the best estimator $\hat{s}_0^{(\mathrm{P})} \Delta_t$ is based on the sum of squares of N independent Gaussian RVs, while each eigenspectrum is based on the sum of two squared Gaussian RVs (these come from the real and imaginary parts of the DFT of the underlying tapered time series). Use of $K = N/2$ eigenspectra would thus involve the sum of N Gaussian RVs, opening up the possibility of the resulting multitaper estimator being equivalent to $\hat{S}_0^{(\text{P})}$. Second, in the $NW = 4$ case, values of $\text{var}\{\hat{S}^{(\text{MT})}(1/4)\}$ for K greater than the Shannon number $2NW = 8$ are *smaller* than the one at the Shannon number; by contrast, in the $NW = 16$ case for which the Shannon number is now 32, the corresponding values are *larger* except at $K = 64$, where, as already noted, there is equality. This second example illustrates that, for correlated RVs with the same expected value, averaging more RVs can actually cause the variance of their mean to increase! (C&E [4] for Section 8.6 along with Figure 403 considers what happens if we replace Slepian multitapers with sinusoidal multitapers.)

8.4 Quadratic Spectral Estimators and Multitapering

We now present two complementary approaches to justifying the Slepian multitaper spectral estimator. The first is due to Bronez (1985, 1988) and is based upon a study of the first and second moments of quadratic spectral estimators (defined in the upcoming Equation (374a)); the second is closely related to the integral equation approach in Thomson's 1982 paper and is discussed in the next section.

Suppose that $X_0, X_1, \ldots, X_{N-1}$ is a segment of length N of a real-valued stationary process with zero mean and SDF $S(\cdot)$. For a fixed frequency f, let us define the complex-valued process

$$Z_t = X_t e^{i 2\pi f t \Delta_t}.$$

The results of Exercise [5.13a] tell us that $\{Z_t\}$ is a stationary process (complex-valued in general) with zero mean and SDF given by

$$S_Z(f') = S(f - f')$$

(recall that $S(\cdot)$ is an even function, but note that $S_Z(\cdot)$ need not be). In particular, $S_Z(0) = S(f)$, so we can obtain an estimate of $S(f)$ via an estimate of $S_Z(\cdot)$ at zero frequency.

Let Z be an N-dimensional column vector whose tth element is Z_t, and let Z^H be its Hermitian transpose; i.e., $Z^H = (Z_0^*, \ldots, Z_{N-1}^*)$, where the asterisk denotes complex conjugation (for what follows, the reader should note that, if Q is a matrix with real-valued elements, its Hermitian transpose Q^H is equal to its usual transpose Q^T). If X_t has units of, say, meters and if the sampling interval Δ_t is, say, 1 sec, then the SDF $S(\cdot)$ has units of (meters)^2 per Hz or, equivalently, seconds $\times$ (meters)^2 per cycle. An estimator for $S(f)$ with the correct dimensionality is thus

$$\hat{S}^{(\text{Q})}(f) \stackrel{\text{def}}{=} \hat{S}_Z^{(\text{Q})}(0) = \Delta_t \sum_{s=0}^{N-1} \sum_{t=0}^{N-1} Z_s^* Q_{s,t} Z_t = \Delta_t \, Z^H Q Z, \qquad (374a)$$

where $Q_{s,t}$ is the (s, t)th element of the $N \times N$ matrix Q of weights. We assume that Q is real-valued and symmetric; with these provisos, $\hat{S}^{(\text{Q})}(f)$ is necessarily real-valued, in agreement with the fact that $S(f)$ is real-valued (see Exercise [8.10]). We also assume that Q does not depend on $\{Z_t\}$. An estimator of the form of Equation (374a) is known in the literature as a *quadratic estimator*. Note that, if the matrix Q is positive semidefinite (sometimes called nonnegative definite), we have the desirable property $\hat{S}^{(\text{Q})}(f) \geq 0$.

Let us now show that we have already met an important example of a quadratic estimator in Chapter 7, namely, the lag window spectral estimator

$$\hat{S}_m^{(\text{LW})}(f) \stackrel{\text{def}}{=} \Delta_t \sum_{\tau=-(N-1)}^{N-1} w_{m,\tau} \hat{s}_\tau^{(\text{D})} e^{-i 2\pi f \tau \Delta_t}, \qquad (374b)$$

8.4 Quadratic Spectral Estimators and Multitapering

where $\{w_{m,\tau}\}$ is a lag window;

$$\hat{S}_\tau^{(D)} \stackrel{\text{def}}{=} \begin{cases} \sum_{t=0}^{N-|\tau|-1} h_{t+|\tau|} X_{t+|\tau|} h_t X_t, & \tau = -(N-1),\ldots, N-1; \\ 0, & \text{otherwise}; \end{cases}$$

and $\{h_t\}$ is a data taper.

▷ **Exercise [375a]** Show that we can rewrite Equation (374b) as

$$\hat{S}_m^{(\text{LW})}(f) = \Delta_t \sum_{s=0}^{N-1} \sum_{t=0}^{N-1} w_{m,s-t} h_s X_s h_t X_t e^{-i2\pi f(s-t)\Delta_t}.$$

◁

Using the above, it follows that

$$\hat{S}_m^{(\text{LW})}(f) = \Delta_t \sum_{s=0}^{N-1} \sum_{t=0}^{N-1} Z_s^* h_s w_{m,s-t} h_t Z_t, \qquad (375\text{a})$$

which says that $\hat{S}_m^{(\text{LW})}(f)$ is a quadratic estimator with $Q_{s,t} = h_s w_{m,s-t} h_t$. Since a lag window spectral estimator can sometimes be negative (consider the reshaped lag window estimate of Figure 282(c)), the matrix Q for such an estimator need not be positive semidefinite. By letting $w_{m,\tau} = 1$ for all τ, a lag window spectral estimator reduces to a direct spectral estimator

$$\hat{S}^{(D)}(f) = \Delta_t \left| \sum_{t=0}^{N-1} h_t X_t e^{-i2\pi f t \Delta_t} \right|^2, \qquad (375\text{b})$$

for which $Q_{s,t} = h_s h_t$. Since a direct spectral estimator can never be negative, the matrix Q for such an estimator must be positive semidefinite. (The burden of Exercise [8.11] is to show that the weighted multitaper estimator $\hat{S}^{(\text{WMT})}(\cdot)$ of Equation (352b) is another example of a quadratic estimator.)

We want to find a quadratic estimator $\hat{S}^{(Q)}(f)$ for $S(f)$ with good bias and variance properties. To ensure the condition $\hat{S}(f) \geq 0$, we insist from now on that Q in Equation (374a) be positive semidefinite. If Q were positive definite, its rank would be N; since we are only assuming it to be positive semidefinite, its rank – call it K – can be taken to satisfy $1 \leq K \leq N$ (the case $K = 0$ is uninteresting). Since Q is a $N \times N$ symmetric matrix of real-valued numbers, there exists an $N \times N$ real-valued orthonormal matrix H_N such that $H_N^T Q H_N = D_N$, where D_N is an $N \times N$ diagonal matrix with diagonal elements $d_0, \ldots, d_{N-1}$ (see, for example, theorem 1.8.8 of Graybill, 1983, or theorem 8.1.1 of Golub and Van Loan, 2013). Each d_k is an eigenvalue of Q and is necessarily real-valued. We assume them to be ordered such that $d_0 \geq d_1 \geq \cdots \geq d_{N-2} \geq d_{N-1}$. Because Q is positive semidefinite with rank K, the eigenvalues of Q are all nonnegative, and the number of positive eigenvalues is equal to K so that $d_0 \geq d_1 \geq \cdots \geq d_{K-1} > 0$ whereas $d_k = 0$ for $K \leq k \leq N-1$. Because H_N is an orthonormal matrix, we have $H_N^T H_N = H_N H_N^T = I_N$, where I_N is the $N \times N$ identity matrix. This implies that we can write $Q = H_N D_N H_N^T$, which is known as the spectral decomposition of Q. Let h_k be an N-dimensional column vector representing the kth column of H_N, and let $h_{k,t}$ be its tth element, $t = 0, \ldots, N-1$.

▷ **Exercise [375b]** Show that we can also write

$$Q = H D_K H^T = \sum_{k=0}^{K-1} d_k h_k h_k^T, \qquad (375\text{c})$$

where H is an $N \times K$ matrix consisting of the first K columns $h_0, \ldots, h_{K-1}$ of H_N, and D_K is a diagonal matrix with diagonal entries $d_0, d_1, \ldots, d_{K-1}$.

◁

For future reference we note that $H^T H = I_K$ (the $K \times K$ identity matrix).

▷ **Exercise [376]** With $HD_K H^T$ substituted for Q in Equation (374a), show that

$$\hat{S}^{(\mathrm{Q})}(f) = \Delta_{\mathrm{t}} Z^H HD_K H^T Z = \Delta_{\mathrm{t}} \sum_{k=0}^{K-1} d_k \left| \sum_{t=0}^{N-1} h_{k,t} X_t \mathrm{e}^{-\mathrm{i}2\pi f t \Delta_{\mathrm{t}}} \right|^2. \tag{376a}$$ ◁

Recalling the definition

$$\hat{S}_k^{(\mathrm{MT})}(f) = \Delta_{\mathrm{t}} \left| \sum_{t=0}^{N-1} h_{k,t} X_t \mathrm{e}^{-\mathrm{i}2\pi f t \Delta_{\mathrm{t}}} \right|^2$$

from Equation (352a), it follows that we can write

$$\hat{S}^{(\mathrm{Q})}(f) = \sum_{k=0}^{K-1} d_k \hat{S}_k^{(\mathrm{MT})}(f) = \hat{S}^{(\mathrm{WMT})}(f), \tag{376b}$$

where $\hat{S}^{(\mathrm{WMT})}(f)$ is the weighted multitaper estimator of Equation (352b). We thus have the important result that all quadratic estimators with a real-valued, symmetric, positive semidefinite weight matrix Q can be written as a weighted sum of K direct spectral estimators $\hat{S}_k^{(\mathrm{MT})}(\cdot)$ (with K being the rank of Q), the kth one of which uses the kth column $h_{k,0}, h_{k,1}, \ldots, h_{k,N-1}$ of H as its data taper; moreover, because the h_k vectors are mutually orthonormal, it follows that the data tapers $\{h_{j,t}\}$ and $\{h_{k,t}\}$ are orthogonal for $j \neq k$. The right-hand side of Equation (376a) is an example of a *deconstructed* multitaper estimator because the tapers arise from a decomposition of the Q matrix; by contrast, the Slepian scheme of Section 8.2 forms a *constructed* multitaper estimator because it starts with a definition for its tapers (see C&E [4] for Section 8.8 for explicit examples of tapers associated with deconstructed multitaper estimators).

We need expressions for the expected value of $\hat{S}^{(\mathrm{Q})}(f)$. It follows from Equations (376b), (352d), (353a) and (352c) that

$$E\{\hat{S}^{(\mathrm{Q})}(f)\} = \int_{-f_\mathcal{N}}^{f_\mathcal{N}} \widetilde{\mathcal{H}}(f - f') S(f') \, \mathrm{d}f', \tag{376c}$$

where

$$\widetilde{\mathcal{H}}(f) = \sum_{k=0}^{K-1} d_k \mathcal{H}_k(f) \text{ and } \mathcal{H}_k(f) = \Delta_{\mathrm{t}} \left| \sum_{t=0}^{N-1} h_{k,t} \mathrm{e}^{-\mathrm{i}2\pi f t \Delta_{\mathrm{t}}} \right|^2.$$

It is also convenient to have a "covariance domain" expression for $E\{\hat{S}^{(\mathrm{Q})}(f)\}$. If we let $\{s_\tau\}$ represent the ACVS for $\{X_t\}$, Exercise [5.13a] says that the ACVS for $\{Z_t\}$ is given by $s_{Z,\tau} = s_\tau \mathrm{e}^{\mathrm{i}2\pi f \tau \Delta_{\mathrm{t}}}$. If we let Σ_Z be the $N \times N$ covariance matrix for $Z_0, \ldots, Z_{N-1}$ (i.e., its (j,k)th element is $s_{Z,j-k}$), it follows from Exercise [8.12] that

$$E\{\hat{S}^{(\mathrm{Q})}(f)\} = \Delta_{\mathrm{t}} \, \mathrm{tr}\{Q \Sigma_Z\} = \Delta_{\mathrm{t}} \, \mathrm{tr}\{HD_K H^T \Sigma_Z\} = \Delta_{\mathrm{t}} \, \mathrm{tr}\{D_K H^T \Sigma_Z H\}, \tag{376d}$$

where $\mathrm{tr}\{\cdot\}$ is the matrix trace operator (for the last equality we have used the fact that, if A and B are matrices with dimensions $M \times N$ and $N \times M$, then $\mathrm{tr}\{AB\} = \mathrm{tr}\{BA\}$). Equating the above to Equation (376c) gives

$$\int_{-f_\mathcal{N}}^{f_\mathcal{N}} \widetilde{\mathcal{H}}(f - f') S(f') \, \mathrm{d}f' = \Delta_{\mathrm{t}} \, \mathrm{tr}\{D_K H^T \Sigma_Z H\}. \tag{376e}$$

8.4 Quadratic Spectral Estimators and Multitapering

We can make substantial progress in obtaining a quadratic estimator with quantifiably good first-moment properties for a wide class of SDFs if we define "good" in a manner different from what we used in Chapter 6, which would lead to $E\{\hat{S}^{(Q)}(f)\} \approx S(f)$. Here we adopt a criterion that incorporates the important notion of *regularization*: for a selected $W > 0$, we want

$$E\{\hat{S}^{(Q)}(f)\} \approx \frac{1}{2W} \int_{f-W}^{f+W} S(f')\,df' \stackrel{\text{def}}{=} \overline{S}(f), \qquad (377a)$$

where $2W > 0$ defines the regularization bandwidth. Note that $\overline{S}(f)$ is the average value of $S(\cdot)$ over the interval from $f - W$ to $f + W$ (see Equation (71)). The requirement that Equation (377a) holds rather than just $E\{\hat{S}^{(Q)}(f)\} \approx S(f)$ can be justified two ways. First, we noted in Section 5.6 that one interpretation of $S(f)$ is that $2S(f)\,df$ is approximately the variance of the process created by passing $\{X_t\}$ through a sufficiently narrow band-pass filter. If we were to actually use such a filter with bandwidth $2W$ and center frequency f, the spectral estimate we would obtain via the sample variance of the output from the filter would be more closely related to $\overline{S}(f)$ than to $S(f)$. The requirement $E\{\hat{S}^{(Q)}(f)\} \approx \overline{S}(f)$ thus incorporates the limitations imposed by the filter bandwidth. Second, because we must base our estimate of the function $S(\cdot)$ on the finite sample $X_0, X_1, \ldots, X_{N-1}$, we face an inherently ill-posed problem in the sense that, with a finite number of observations, we cannot hope to estimate $S(\cdot)$ well over an infinite number of frequencies unless $S(\cdot)$ is sufficiently smooth. Since $\overline{S}(\cdot)$ is a smoothed version of $S(\cdot)$, it should be easier to find an approximately unbiased estimator for $\overline{S}(f)$ than for $S(f)$; moreover, if $S(\cdot)$ is itself sufficiently smooth, then $\overline{S}(\cdot)$ and $S(\cdot)$ should be approximately equal for W small enough. In effect, we are redefining the function to be estimated from one that need not be smooth, namely, $S(\cdot)$, to one that is guaranteed to be smooth to a certain degree, namely, $\overline{S}(\cdot)$.

With these comments in mind, let us see what constraints we need to impose on the matrix $\mathbf{Q}$ in Equation (374a) so that $\hat{S}^{(Q)}(f)$ is an approximately unbiased estimator of $\overline{S}(f)$, i.e., that $E\{\hat{S}^{(Q)}(f)\} \approx \overline{S}(f)$. Can we insist upon having

$$E\{\hat{S}^{(Q)}(f)\} = \overline{S}(f) = \frac{1}{2W} \int_{f-W}^{f+W} S(f')\,df'$$

for all possible $\overline{S}(\cdot)$? From Equation (376c) we would then need

$$\widetilde{\mathcal{H}}(f') = \begin{cases} 1/(2W), & |f'| \leq W; \\ 0, & W < |f'| \leq f_{\mathcal{N}}. \end{cases} \qquad (377b)$$

Since $\widetilde{\mathcal{H}}(\cdot)$ is a sum of the functions $d_k \mathcal{H}_k(\cdot)$, each one of which is proportional to the modulus squared of the Fourier transform of a finite sequence of values $\{h_{k,t}\}$, we cannot attain the simple form stated in Equation (377b) (see the discussion at the beginning of Section 5.8). We must therefore be satisfied with unbiasedness to some level of approximation. The approach that we take is, first, to insist that $\hat{S}^{(Q)}(f)$ be an unbiased estimator of $\overline{S}(f)$ for white noise processes and, second, to develop bounds for two components of the bias for colored noise.

If $\{X_t\}$ is a white noise process with variance s_0, then $S(f) = s_0 \Delta_t$ for all f, so

$$\overline{S}(f) = \frac{1}{2W} \int_{f-W}^{f+W} s_0 \Delta_t\,df' = s_0 \Delta_t.$$

If $\hat{S}^{(Q)}(f)$ is to be an unbiased estimator of $\overline{S}(f)$ for white noise, Equation (376c) tells us that

$$E\{\hat{S}^{(Q)}(f)\} = s_0 \Delta_t \int_{-f_{\mathcal{N}}}^{f_{\mathcal{N}}} \widetilde{\mathcal{H}}(f - f')\,df' = s_0 \Delta_t,$$

i.e., that
$$\int_{-f_N}^{f_N} \widetilde{\mathcal{H}}(f - f') \, df' = \int_{-f_N}^{f_N} \widetilde{\mathcal{H}}(f') \, df = 1$$

(see Equation (188e)). Since Σ_Z for a white noise process is a diagonal matrix, all of whose elements are equal to s_0, it follows from Equation (376e) that the above requirement is equivalent to $\operatorname{tr}\{D_K H^T H\} = \operatorname{tr}\{D_K\} = 1$; i.e., we need

$$\sum_{k=0}^{K-1} d_k = 1 \qquad (378a)$$

(see also Exercise [8.1], which justified the assumption that the weights d_k for a weighted multitaper estimator $\hat{S}^{(\text{WMT})}(f)$ sum to unity).

For a colored noise process let us define the bias in the estimator $\hat{S}^{(Q)}(f)$ as

$$b(f) = E\{\hat{S}^{(Q)}(f)\} - \overline{S}(f) = \int_{-f_N}^{f_N} \widetilde{\mathcal{H}}(f - f') S(f') \, df' - \frac{1}{2W} \int_{f-W}^{f+W} S(f') \, df'.$$

It is impossible to give any sort of reasonable bound for $b(f)$ since the second integral above does not depend upon the matrix Q. It is useful, however, to split the bias into two components, one due to frequencies from $f - W$ to $f + W$ – the *local bias* – and the other, to frequencies outside this interval – the *broad-band bias*. Accordingly, we write

$$b(f) = b^{(\text{L})}(f) + b^{(\text{B})}(f),$$

where

$$b^{(\text{L})}(f) \stackrel{\text{def}}{=} \int_{f-W}^{f+W} \widetilde{\mathcal{H}}(f - f') S(f') \, df' - \frac{1}{2W} \int_{f-W}^{f+W} S(f') \, df'$$
$$= \int_{f-W}^{f+W} \left(\widetilde{\mathcal{H}}(f - f') - \frac{1}{2W} \right) S(f') \, df'$$

and

$$b^{(\text{B})}(f) \stackrel{\text{def}}{=} \int_{-f_N}^{f-W} \widetilde{\mathcal{H}}(f - f') S(f') \, df' + \int_{f+W}^{f_N} \widetilde{\mathcal{H}}(f - f') S(f') \, df'.$$

To obtain useful upper bounds on the local and broad-band bias, let us assume that $S(\cdot)$ is bounded by $S_{\max}$, i.e., that $S(f) \leq S_{\max} < \infty$ for all f in the interval $[-f_N, f_N]$. For the local bias we then have

$$|b^{(\text{L})}(f)| \leq \int_{f-W}^{f+W} \left| \widetilde{\mathcal{H}}(f - f') - \frac{1}{2W} \right| S(f') \, df'$$
$$\leq S_{\max} \int_{-W}^{W} \left| \widetilde{\mathcal{H}}(-f'') - \frac{1}{2W} \right| df'',$$

where $f'' = f' - f$. We thus can take the quantity

$$\beta^{(\text{L})}\{\hat{S}^{(Q)}(\cdot)\} \stackrel{\text{def}}{=} \int_{-W}^{W} \left| \widetilde{\mathcal{H}}(f'') - \frac{1}{2W} \right| df'' \qquad (378b)$$

8.4 Quadratic Spectral Estimators and Multitapering

as a useful indicator of the magnitude of the local bias in $\hat{S}^{(Q)}(f)$. Note that we can control the local bias by approximating the unattainable ideal $\widetilde{\mathcal{H}}(\cdot)$ of Equation (377b) as closely as possible over the regularization band.

Let us now obtain a bound for the broad-band bias (this is related to leakage and is usually a more important concern than the local bias). With $f'' = f' - f$ as before, we have

$$b^{(B)}(f) \leq S_{\max} \left(\int_{-f_{\mathcal{N}}}^{f-W} \widetilde{\mathcal{H}}(f - f') \, df' + \int_{f+W}^{f_{\mathcal{N}}} \widetilde{\mathcal{H}}(f - f') \, df' \right)$$

$$= S_{\max} \left(\int_{-f_{\mathcal{N}}}^{f_{\mathcal{N}}} \widetilde{\mathcal{H}}(f - f') \, df' - \int_{f-W}^{f+W} \widetilde{\mathcal{H}}(f - f') \, df' \right)$$

$$= S_{\max} \left(\int_{-f_{\mathcal{N}}}^{f_{\mathcal{N}}} \widetilde{\mathcal{H}}(f - f') \, df' - \int_{-W}^{W} \widetilde{\mathcal{H}}(-f'') \, df'' \right).$$

By considering Equation (376e) for a white noise process with unit variance (i.e., $S(f') = \Delta_{\mathrm{t}}$), we can rewrite the first integral above as

$$\int_{-f_{\mathcal{N}}}^{f_{\mathcal{N}}} \widetilde{\mathcal{H}}(f - f') \, df' = \mathrm{tr}\{\boldsymbol{D}_K \boldsymbol{H}^T \boldsymbol{H}\} = \mathrm{tr}\{\boldsymbol{D}_K\} = \sum_{k=0}^{K-1} d_k$$

(this follows because $\boldsymbol{\Sigma}_Z$ then becomes the identity matrix). The second integral can be rewritten using Equation (376e) again, but this time with an SDF $S^{(\mathrm{BL})}(\cdot)$ for band-limited white noise; i.e.,

$$S^{(\mathrm{BL})}(f) = \begin{cases} \Delta_{\mathrm{t}}, & |f| \leq W; \\ 0, & W < |f| \leq f_{\mathcal{N}}. \end{cases} \quad (379\mathrm{a})$$

The corresponding ACVS $\{s_\tau^{(\mathrm{BL})}\}$ at lag τ is given by Equation (113c):

$$s_\tau^{(\mathrm{BL})} = \int_{-f_{\mathcal{N}}}^{f_{\mathcal{N}}} e^{i2\pi f' \tau \Delta_{\mathrm{t}}} S^{(\mathrm{BL})}(f') \, df' = \int_{-W}^{W} e^{i2\pi f' \tau \Delta_{\mathrm{t}}} \Delta_{\mathrm{t}} \, df' = \frac{\sin(2\pi W \tau \Delta_{\mathrm{t}})}{\pi \tau}, \quad (379\mathrm{b})$$

where, as usual, this ratio is defined to be $2W\Delta_{\mathrm{t}}$ when $\tau = 0$. If we let $\boldsymbol{\Sigma}^{(\mathrm{BL})}$ be the $N \times N$ matrix whose (j,k)th element is $s_{j-k}^{(\mathrm{BL})}$, we obtain from Equation (376e) with $f = 0$

$$\int_{-W}^{W} \widetilde{\mathcal{H}}(-f'') \, df'' = \mathrm{tr}\{\boldsymbol{D}_K \boldsymbol{H}^T \boldsymbol{\Sigma}^{(\mathrm{BL})} \boldsymbol{H}\}.$$

We thus have that

$$b^{(B)}(f) \leq S_{\max} \left(\sum_{k=0}^{K-1} d_k - \mathrm{tr}\{\boldsymbol{D}_K \boldsymbol{H}^T \boldsymbol{\Sigma}^{(\mathrm{BL})} \boldsymbol{H}\} \right),$$

so we can take the quantity

$$\beta^{(B)}\{\hat{S}^{(Q)}(\cdot)\} \stackrel{\mathrm{def}}{=} \sum_{k=0}^{K-1} d_k - \mathrm{tr}\{\boldsymbol{D}_K \boldsymbol{H}^T \boldsymbol{\Sigma}^{(\mathrm{BL})} \boldsymbol{H}\} \quad (379\mathrm{c})$$

to be a useful indicator of the broad-band bias in $\hat{S}^{(Q)}(f)$.

Suppose for the moment that our only criterion for choosing the matrix Q is that our broad-band bias indicator be made as small as possible, subject to the constraint that the resulting estimator $\hat{S}^{(Q)}(f)$ be unbiased for white noise processes. This means that we want to

$$\text{maximize} \ \ \text{tr}\,\{D_K H^T \Sigma^{(\text{BL})} H\} \ \text{subject to} \ \sum_{k=0}^{K-1} d_k = 1. \tag{380a}$$

Exercise [8.13] indicates that, if $\Sigma^{(\text{BL})}$ were positive definite with a distinct largest eigenvalue, then a solution to this maximization problem would be to set $K = 1$ and to set h_0 equal to a normalized eigenvector associated with the largest eigenvalue of the matrix $\Sigma^{(\text{BL})}$. In fact, $\Sigma^{(\text{BL})}$ satisfies the stated condition: because its (j,k)th element is given by Equation (379b) with τ set to $j - k$, its eigenvalues and eigenvectors are the solutions to the problem posed previously by Equation (88b) in Section 3.10. As before, we denote these eigenvalues as $\lambda_k(N, W)$, but note carefully the difference between these and the d_k's: both are eigenvalues, but the former is associated with $\Sigma^{(\text{BL})}$, and the latter, with Q. From the discussion surrounding Equation (88b), we know that all of the eigenvalues $\lambda_k(N, W)$ are positive (and hence $\Sigma^{(\text{BL})}$ is positive definite) and that there is only one eigenvector corresponding to the largest eigenvalue $\lambda_0(N, W)$. Moreover, the elements of the normalized eigenvector corresponding to $\lambda_0(N, W)$ can be taken to be a finite subsequence of the zeroth-order Slepian data taper, namely, $v_{0,0}(N, W), v_{0,1}(N, W), \ldots, v_{0,N-1}(N, W)$. Because of the definition for the zeroth-order Slepian data taper $\{h_{0,t}\}$ in Equation (357b), the solution to the maximization problem is to construct Q by setting $K = 1$ and $d_0 = 1$ to obtain

$$Q = H D_K H^T = h_0 h_0^T, \tag{380b}$$

where h_0 contains the nonzero portion of $\{h_{0,t}\}$ as defined in Equation (357b).

▷ **Exercise [380]** Show that corresponding value of the broad-band bias indicator of Equation (379c) is

$$\beta^{(\text{B})}\{\hat{S}^{(Q)}(\cdot)\} = 1 - \lambda_0(N, W). \qquad \triangleleft$$

In summary, under the constraint $\sum_{k=0}^{K-1} d_k = 1$, we can minimize the broad-band bias indicator of Equation (379c) by setting Q equal to the outer product of the N-dimensional vector h_0 with itself, where the elements of h_0 are the nonzero portion of a zeroth-order Slepian data taper (normalized, as usual, to have unit energy). The (s,t)th element of the rank $K = 1$ weight matrix $Q = h_0 h_0^T$ is $h_{0,s} h_{0,t}$, and the estimator is just a direct spectral estimator of the form of Equation (375b). Note that this method of specifying a quadratic estimator obviously does *not* take into consideration either the variance of the estimator or its local bias, both of which we now examine in more detail.

Suppose now that we wish to obtain a quadratic estimator (with a positive semidefinite weight matrix Q) that has good variance properties, subject to the mild condition that the estimator be unbiased in the case of white noise, i.e., that the eigenvalues of Q sum to unity as per Equation (378a). From Equation (376b) we know that a rank K quadratic estimator $\hat{S}^{(Q)}(f)$ can be written as the weighted sum of K direct spectral estimators $\hat{S}_k^{(\text{MT})}(f)$ employing a set of K orthonormal data tapers. In Section 8.3 we found that, for Gaussian white noise and for uniform weights $d_k = 1/K$, the variance of such an estimator is minimized by using *any* set of $K = N$ orthonormal data tapers. This result is in direct conflict with the recommendation we came up with for minimizing our broad-band bias indicator, namely, that we use just a single zeroth-order Slepian data taper.

An obvious compromise estimator that attempts to balance this conflict between variance and broad-band bias is to use as many members K of the set of orthonormal Slepian data

8.4 Quadratic Spectral Estimators and Multitapering

tapers as possible. With this choice and with d_k set equal to $1/K$ for $k = 0, \ldots, K - 1$, our quadratic estimator $\hat{S}^{(\mathrm{Q})}(f)$ becomes the basic multitaper estimator $\hat{S}^{(\mathrm{MT})}(f)$ defined by Equations (352a) and (357b). Let us now discuss how this compromise affects the broad-band bias indicator, the variance and the local bias indicator for the resulting quadratic estimator.

[1] If we use K orthonormal Slepian data tapers, the broad-band bias indicator of Equation (379c) becomes

$$\beta^{(\mathrm{B})}\{\hat{S}^{(\mathrm{Q})}(\cdot)\} = 1 - \frac{1}{K} \sum_{k=0}^{K-1} \lambda_k(N, W) \tag{381}$$

(this is Exercise [8.14]). Because the $\lambda_k(N, W)$ terms are all close to unity as long as we set K to an integer less than the Shannon number $2NW\,\Delta_\mathrm{t}$, the broad-band bias indicator must be close to zero if we impose the restriction $K < 2NW\,\Delta_\mathrm{t}$; however, this indicator necessarily increases as K increases.

[2] For Gaussian white noise, we saw in Figures 373a and 373b that, for $K \leq 2NW\,\Delta_\mathrm{t}$, the variance of the average $\hat{S}^{(\mathrm{MT})}(f)$ of K direct spectral estimators $\hat{S}_k^{(\mathrm{MT})}(f)$ is consistent with the approximation that all K of the $\hat{S}_k^{(\mathrm{MT})}(f)$ RVs are pairwise uncorrelated. Since $\hat{S}_k^{(\mathrm{MT})}(f)$ is approximately distributed as $S(f)\chi_2^2/2$ for $k = 0, \ldots, K - 1$ (as long as f is not too close to 0 or $f_\mathcal{N}$), these figures suggests that, with K so chosen, the estimator $\hat{S}^{(\mathrm{MT})}(f)$ should be approximately distributed as $S(f)\chi_{2K}^2/2K$. The variance of $\hat{S}^{(\mathrm{MT})}(f)$ is hence approximately $S^2(f)/K$. For colored Gaussian processes, this approximation is still good as long as $S(f)$ does not vary too rapidly over the interval $[f - W, f + W]$ (see the next section or Thomson, 1982, section IV).

[3] The local bias indicator $\beta^{(\mathrm{L})}\{\hat{S}^{(\mathrm{Q})}(\cdot)\}$ of Equation (378b) is small if the spectral window for $\hat{S}^{(\mathrm{MT})}(\cdot)$ is as close to $1/(2W)$ as possible over the regularization band. This window is denoted as $\overline{\mathcal{H}}(\cdot)$ in Equation (353b) and is shown in the left-hand plots of Figures 364 and 365 (where $\Delta_\mathrm{t} = 1$) for $K = 1$ to 8, $N = 1024$ and $W = 4/N$. Since $10 \log_{10}(1/(2W)) = 21$ dB here, these plots indicate that, as K increases, the spectral window $\overline{\mathcal{H}}(\cdot)$ becomes closer to $1/(2W)$ over the regularization band and hence that the local bias indicator decreases as K increases.

In summary, we have shown that all quadratic estimators with a symmetric real-valued positive semidefinite weight matrix $\mathbf{Q}$ of rank K can be written as a weighted sum of K direct spectral estimators with orthonormal data tapers, i.e., as a weighted multitaper spectral estimator $\hat{S}^{(\mathrm{WMT})}(f)$ (Equation (352b)). The requirement that the quadratic estimator be unbiased for white noise is satisfied if the sum of the eigenvalues d_k for $\mathbf{Q}$ is unity. We then redefine (or "regularize") our spectral estimation problem so that the quantity to be estimated is $S(f)$ averaged over an interval of selectable width $2W$ (i.e., $\overline{S}(f)$ of Equation (377a)) rather than just $S(f)$ itself. This redefinition allows us to profitably split the bias of a quadratic estimator into two parts, denoted as the local bias and the broad-band bias. Minimization of the broad-band bias indicator $\beta^{(\mathrm{B})}\{\hat{S}^{(\mathrm{Q})}(\cdot)\}$ dictates that we set $K = 1$, with the single data taper being a zeroth-order Slepian data taper; on the other hand, the variance of a quadratic estimator is minimized in the special case of Gaussian white noise by choosing K to be N, but any set of N orthonormal data tapers yields the same minimum value. The obvious compromise for colored noise is thus to use as many of the low-order members of the family of Slepian data tapers as possible, yielding (with $d_k = 1/K$) the basic multitaper spectral estimator $\hat{S}^{(\mathrm{MT})}(f)$. The broad-band bias indicator suggests restricting K to be less than the Shannon number $2NW\,\Delta_\mathrm{t}$. With K thus restricted, the estimator $\hat{S}^{(\mathrm{MT})}(f)$ follows approximately a rescaled χ_{2K}^2 distribution (provided that f is not too close to either 0 or $f_\mathcal{N}$ and that $S(f)$ does not vary too rapidly over the interval $[f - W, f + W]$). The variance of $\hat{S}^{(\mathrm{MT})}(f)$ is inversely proportional to K – increasing K thus decreases the variance. The local bias indicator $\beta^{(\mathrm{L})}\{\hat{S}^{(\mathrm{Q})}(\cdot)\}$ also decreases as K increases. An increase in K above unity thus improves both the variance and local bias of $\hat{S}^{(\mathrm{MT})}(f)$ at the cost of increasing its broad-band bias. (Bronez 1985, 1988, extends the development in this section to the case of continuous parameter complex-valued stationary processes sampled at irregular time intervals.)

Comments and Extensions to Section 8.4

[1] As noted above, the kth-order Slepian taper $\{h_{k,t}\}$ is a normalized version of the eigenvector $\boldsymbol{h}_k$ associated with the kth largest eigenvalue $\lambda_k(N, W)$ of the $N \times N$ covariance matrix $\boldsymbol{\Sigma}^{(\mathrm{BL})}$. The (j, k)th element of this matrix is $s_{j-k}^{(\mathrm{BL})}$, where $\{s_\tau^{(\mathrm{BL})}\}$ is the ACVS for band-limited white noise (see Equation (379b)). The SDF for this process is flat and nonzero over $|f| \leq W$ (see Equation (379a)). Hansson and Salomonsson (1997) propose an alternative to Slepian multitapering in which the underlying SDF is still band-limited over $|f| \leq W$, but is proportional to $\exp(-C|f|)$ over the pass-band rather than being flat (the constant $C \geq 0$ controls the degree of non-flatness). Let $\boldsymbol{\Sigma}^{(\mathrm{C})}$ denote the corresponding $N \times N$ covariance matrix, and let $\lambda_k(N, W, C)$ be its k-th largest eigenvalue. Because eigenvectors are orthogonal, the solutions $\boldsymbol{h}_k$ to the matrix equations $\boldsymbol{\Sigma}^{(\mathrm{C})}\boldsymbol{h}_k = \lambda_k(N, W, C)\boldsymbol{h}_k$, $k = 0, \ldots, K - 1$, define a multitapering scheme. The intended goal of this scheme – called *peak-matched multitapering* – is to reduce the bias and variance in estimating the locations of the peaks in an SDF that has one or more peaks (examples being the AR(4) "twin peaks" SDF displayed in Figure 361 and elsewhere, and the fractionally differenced SDFs shown in the forthcoming Figure 408). Similar to Slepian multitapering (to which peak-matched multitapering reduces when $C = 0$), the spectral windows for the associated tapers are highly concentrated over the pass-band, but, to better manage how the concentration is manifested inside and outside of the pass-band, Hansson and Salomonsson (1997) introduce a second SDF defined to be $D \geq 1$ for $|f| \leq W$ and to be 1 for $W < |f| \leq f_\mathcal{N}$. Let $\boldsymbol{\Sigma}^{(\mathrm{D})}$ denote the corresponding $N \times N$ covariance matrix. The k-th order multitaper of interest is proportional to the eigenvector $\boldsymbol{h}_k$ associated with the k-th largest eigenvalue $\lambda_k(N, W, C, D)$ of the generalized eigenvalue problem $\boldsymbol{\Sigma}^{(\mathrm{C})}\boldsymbol{h}_k = \lambda_k(N, W, C, D)\boldsymbol{\Sigma}^{(\mathrm{D})}\boldsymbol{h}_k$. Simulation studies show that peak-matched multitapering is successful in achieving its intended goal; however, in addition to setting the regularization bandwidth $2W$ and the number of tapers K to be used, this scheme also requires setting C and D. Hansson and Salomonsson (1997) simplify matters by setting C to just one specific value, but this still leaves one more parameter to contend with than what Slepian multitapering involves. In a subsequent paper, Hansson (1999) forms a weighted multitaper estimator (Equation (352b)) using peak-matched multitapers with the weights d_k chosen to optimize normalized bias, variance or mean square error in an interval of frequencies surrounding a predetermined peaked SDF.

8.5 Regularization and Multitapering

In this section we present a rationale for the Slepian multitaper spectral estimator that closely follows Thomson's original 1982 approach. It makes extensive use of the spectral representation theorem and presents the spectral estimation problem as a search for a calculable approximation to a desirable quantity that unfortunately cannot be calculated from observable data alone. We first review some key concepts.

A way of representing any stationary process $\{X_t\}$ with zero mean is given by the spectral representation theorem introduced in Chapter 4:

$$X_t = \int_{-1/2}^{1/2} e^{i2\pi ft} \, \mathrm{d}Z(f)$$

(this is Equation (110c); for convenience, we assume that $\Delta_t = 1$ in this section, yielding a Nyquist frequency of $f_\mathcal{N} = 1/2$). The increments of the orthogonal process $\{Z(f)\}$ define the SDF of $\{X_t\}$ as

$$S(f) \, \mathrm{d}f = E\{|\mathrm{d}Z(f)|^2\}. \tag{382}$$

8.5 Regularization and Multitapering

Thomson (1982) suggests – purely for convenience – changing the definition of $\mathrm{d}Z(\cdot)$ by a phase factor, which of course does not affect Equation (382):

$$X_t = \int_{-1/2}^{1/2} e^{i2\pi f[t-(N-1)/2]} \,\mathrm{d}Z(f). \tag{383a}$$

Given $X_0, X_1, \ldots, X_{N-1}$, we want to relate the Fourier transform of these N values to the spectral representation for $\{X_t\}$. For convenience, we shall work with the following phase-shifted Fourier transform:

$$\Upsilon(f) \stackrel{\mathrm{def}}{=} \sum_{t=0}^{N-1} X_t e^{-i2\pi f[t-(N-1)/2]}. \tag{383b}$$

Thus, from Equations (383a) and (383b), we have

$$\Upsilon(f) = \int_{-1/2}^{1/2} \left(\sum_{t=0}^{N-1} e^{-i2\pi(f-f')[t-(N-1)/2]} \right) \mathrm{d}Z(f'). \tag{383c}$$

The summation in the parentheses is proportional to Dirichlet's kernel (see Exercise [1.2c]). Hence we arrive at a Fredholm integral equation of the first kind:

$$\Upsilon(f) = \int_{-1/2}^{1/2} \frac{\sin[N\pi(f-f')]}{\sin[\pi(f-f')]} \,\mathrm{d}Z(f')$$

(Thomson, 1982). Since $\mathrm{d}Z(\cdot)$ is by no stretch of the imagination a smooth function, it is not possible to solve for $\mathrm{d}Z(\cdot)$ using standard inverse theory. We adopt an approach that emphasizes regularization of the spectral estimation problem and uses the same building blocks as Thomson. This leads us to consider projecting the Fourier transform $\Upsilon(\cdot)$ onto an appropriate set of basis functions, to which we now turn.

We introduced the discrete prolate spheroidal wave functions (DPSWFs) in Section 3.10, where we denoted them as $U_k(\cdot; N, W)$, but here we simplify the notation to just $U_k(\cdot)$. As indicated in Exercise [3.18],

$$U_k(f) = (-1)^k \epsilon_k \sum_{t=0}^{N-1} v_{k,t}(N,W) e^{-i2\pi f[t-(N-1)/2]}, \quad k = 0, 1, \ldots, N-1,$$

$$= (-1)^k \epsilon_k \sum_{t=0}^{N-1} h_{k,t} e^{-i2\pi f[t-(N-1)/2]}, \tag{383d}$$

where

$$\epsilon_k = \begin{cases} 1, & k \text{ even}; \\ i, & k \text{ odd}, \end{cases}$$

and we define the kth-order Slepian data taper $\{h_{k,t}\}$ as in Equation (357b). If we recall that the $U_k(\cdot)$ functions are orthogonal over $[-W, W]$, i.e.,

$$\int_{-W}^{W} U_j(f) U_k(f) \,\mathrm{d}f = \begin{cases} \lambda_k, & j = k; \\ 0, & \text{otherwise,} \end{cases} \tag{383e}$$

then the rescaled functions $U_k(\cdot)/\sqrt{\lambda_k}$ are orthonormal:

$$\int_{-W}^{W} \frac{U_j(f)}{\sqrt{\lambda_j}} \frac{U_k(f)}{\sqrt{\lambda_k}} \, df = \begin{cases} 1, & j = k; \\ 0, & \text{otherwise} \end{cases}$$

(here λ_k is shorthand for $\lambda_k(N,W)$ as defined in Section 3.10). Slepian (1978, section 4.1) shows that the finite-dimensional space of functions of the form of Equation (383b) is spanned over the interval $[-W, W]$ by the rescaled DPSWFs $U_k(\cdot)/\sqrt{\lambda_k}$. Hence we can write $\Upsilon(\cdot)$ as

$$\Upsilon(f) = \sum_{k=0}^{N-1} \Upsilon_k \frac{U_k(f)}{\sqrt{\lambda_k}} \quad \text{for } \Upsilon_k \stackrel{\text{def}}{=} \int_{-W}^{W} \Upsilon(f) \frac{U_k(f)}{\sqrt{\lambda_k}} \, df$$

(cf. the discussion surrounding Equation (138a)). If, for a given frequency f', we now define a new function $\Upsilon_{f'}(\cdot)$ via $\Upsilon_{f'}(f) = \Upsilon(f + f')$, we can express it as

$$\Upsilon_{f'}(f) = \sum_{k=0}^{N-1} \Upsilon_k(f') \frac{U_k(f)}{\sqrt{\lambda_k}} \quad \text{for } \Upsilon_k(f') \stackrel{\text{def}}{=} \int_{-W}^{W} \Upsilon_{f'}(f) \frac{U_k(f)}{\sqrt{\lambda_k}} \, df;$$

i.e., we have

$$\Upsilon(f + f') = \sum_{k=0}^{N-1} \Upsilon_k(f') \frac{U_k(f)}{\sqrt{\lambda_k}} \quad \text{for } \Upsilon_k(f') = \int_{-W}^{W} \Upsilon(f + f') \frac{U_k(f)}{\sqrt{\lambda_k}} \, df. \tag{384a}$$

If we substitute the expression for $\Upsilon(\cdot)$ given in Equation (383b) into the above (after switching the roles of the variables f and f'), we obtain

$$\Upsilon_k(f) = \int_{-W}^{W} \left(\sum_{t=0}^{N-1} X_t e^{-i2\pi(f'+f)[t-(N-1)/2]} \right) \frac{U_k(f')}{\sqrt{\lambda_k}} \, df'$$

$$= \sum_{t=0}^{N-1} X_t e^{-i2\pi f[t-(N-1)/2]} \int_{-W}^{W} e^{-i2\pi f'[t-(N-1)/2]} \frac{U_k(f')}{\sqrt{\lambda_k}} \, df'.$$

From Exercise [3.18] and recalling that $h_{k,t} = v_{k,t}(N,W)$ for $t = 0, 1, \ldots, N-1$ (see Equation (357b)), we have the following relationship between the kth-order Slepian data taper $\{h_{k,t}\}$ and the kth-order DPSWF $U_k(\cdot)$:

$$h_{k,t} = \frac{(-1)^k}{\epsilon_k \sqrt{\lambda_k}} \int_{-W}^{W} e^{i2\pi f'[t-(N-1)/2]} \frac{U_k(f')}{\sqrt{\lambda_k}} \, df', \quad t = 0, 1, \ldots, N-1.$$

Because both $h_{k,t}$ and $U_k(\cdot)$ are real-valued and because $(-1)^k/\epsilon_k^* = 1/\epsilon_k$, we also have

$$h_{k,t} = \frac{1}{\epsilon_k \sqrt{\lambda_k}} \int_{-W}^{W} e^{-i2\pi f'[t-(N-1)/2]} \frac{U_k(f')}{\sqrt{\lambda_k}} \, df'. \tag{384b}$$

This yields

$$\Upsilon_k(f) = \epsilon_k \sqrt{\lambda_k} \sum_{t=0}^{N-1} h_{k,t} X_t e^{-i2\pi f[t-(N-1)/2]}. \tag{384c}$$

8.5 Regularization and Multitapering

Thus the projection $\Upsilon_k(\cdot)$ of the phase-shifted Fourier transform $\Upsilon(\cdot)$ onto the kth basis function $U_k(\cdot)/\sqrt{\lambda_k}$ is just the (phase-shifted and rescaled) Fourier transform of $X_0, X_1, \ldots, X_{N-1}$ multiplied by the kth-order Slepian data taper.

Exercise [8.15] says that another representation for $\Upsilon_k(f)$ is

$$\Upsilon_k(f) = \sqrt{\lambda_k} \int_{-1/2}^{1/2} U_k(f') \, dZ(f+f'). \tag{385a}$$

In Equation (384a) the phase-shifted Fourier transform $\Upsilon_k(\cdot)$ is "seen" through the DPSWF with smoothing carried out only over the interval $[-W, W]$, while, in the above, $dZ(\cdot)$ is seen through the DPSWF with smoothing carried out over the whole Nyquist interval $[-1/2, 1/2]$.

Now, from Equation (352a) with $\Delta_t = 1$ (as we assume in this section), we have

$$\hat{S}_k^{(\mathrm{MT})}(f) = \left| \sum_{t=0}^{N-1} h_{k,t} X_t e^{-i2\pi f t} \right|^2 = \left| \epsilon_k \sum_{t=0}^{N-1} h_{k,t} X_t e^{-i2\pi f [t-(N-1)/2]} \right|^2.$$

It follows from Equation (384c) that

$$\hat{S}_k^{(\mathrm{MT})}(f) = \left| \Upsilon_k(f)/\sqrt{\lambda_k} \right|^2. \tag{385b}$$

As noted in Section 8.1, the estimator $\hat{S}_k^{(\mathrm{MT})}(\cdot)$ is called the *kth eigenspectrum*. The terminology eigenspectrum is motivated by the fact that the taper $\{h_{k,t}\}$ is an eigenvector for Equation (88b). As defined per Equation (357b), the Slepian taper $\{h_{k,t}\}$ has the usual normalization $\sum_{t=0}^{N-1} h_{k,t}^2 = 1$ (this follows from Parseval's theorem of Equation (75b) and from Equation (87a), which says that $U_k(\cdot)$ has unit energy over the interval $[-1/2, 1/2]$). When $\{X_t\}$ is white noise, we noted in Section 8.3 that $\hat{S}_k^{(\mathrm{MT})}(f)$ is an unbiased estimator of $S(f)$; i.e., $E\{\hat{S}_k^{(\mathrm{MT})}(f)\} = S(f)$ for all f and k.

We are now in a position to consider a method of "regularizing" the spectral estimation problem. Our approach is first to introduce a set of desirable (but unobservable) projections and then to outline how these can best be approximated by the observable $\Upsilon_k(\cdot)$ – recall that these were formed in Equation (384a) by projecting the rescaled Fourier transform $\Upsilon(\cdot)$ onto the kth basis function $U_k(\cdot)/\sqrt{\lambda_k}$. In Equation (385a) the projection $\Upsilon_k(\cdot)$ was then expressed in terms of $dZ(\cdot)$, which determines the SDF via Equation (382). Now consider the equivalent *direct* projection for $dZ(\cdot)$ onto the same kth basis function:

$$Z_k(f) \stackrel{\mathrm{def}}{=} \int_{-W}^{W} \frac{U_k(f')}{\sqrt{\lambda_k}} \, dZ(f+f'). \tag{385c}$$

Comparison of Equations (385a) and (385c) shows that – apart from different uses of the scaling factor $\sqrt{\lambda_k}$ – the difference between $\Upsilon_k(\cdot)$ and $Z_k(\cdot)$ lies in the integration limits. Of course, there is a large practical difference – $\Upsilon_k(\cdot)$ is calculable from $X_0, X_1, \ldots, X_{N-1}$ via Equation (384c), while $Z_k(\cdot)$ is not! Also $dZ(\cdot)$ cannot be written as a finite linear combination of the DPSWFs: unlike $\Upsilon(\cdot)$, $dZ(\cdot)$ does not fall in an N-dimensional space due to its highly discontinuous nature. Nevertheless, we are still at liberty to project $dZ(\cdot)$ onto each of the N DPSWFs to obtain $Z_k(\cdot)$ in Equation (385c).

Note that Equation (385c) can be rewritten in terms of a convolution:

$$Z_k(f) = (-1)^k \int_{f-W}^{f+W} \frac{U_k(f-f')}{\sqrt{\lambda_k}} \, dZ(f'), \tag{385d}$$

so that clearly $Z_k(\cdot)$ represents smoothing $dZ(\cdot)$ by a rescaled kth-order DPSWF over an interval of width $2W$ (here we have used the fact that $U_k(\cdot)$ is symmetric for even k and skew-symmetric for odd k – see Exercise [3.18]). For a chosen W, $Z_k(\cdot)$ represents the best – in the sense that the DPSWFs maximize the concentration measure (see Section 3.10) – glimpse at dZ in each orthogonal direction. It is straightforward to show that

$$E\{|Z_k(f)|^2\} = \int_{f-W}^{f+W} \left[\frac{U_k(f-f')}{\sqrt{\lambda_k}}\right]^2 S(f')\,df'. \tag{386}$$

For a general spectral shape this expected value can be interpreted as a smoothed version of the spectrum, with the smoothing by the scaled DPSWF (squared) over the interval of width $2W$. When the true SDF is that of a white noise process, we see, using Equation (383e), that

$$E\{|Z_k(f)|^2\} = S(f);$$

i.e., $|Z_k(f)|^2$ is an unbiased (but incalculable) estimator of $S(f)$ for white noise.

Thomson's 1982 approach puts emphasis on finding a calculable expression close to $Z_k(f)$. Why is $|Z_k(f)|^2$ so appealing when its expected value is a smoothed version of the SDF? The answer illuminates a key difficulty with spectral estimation. Estimation of $S(\cdot)$ from a time series $X_0, X_1, \ldots, X_{N-1}$ is an ill-posed problem because, given a finite number N of observations, we cannot uniquely determine $S(\cdot)$ over the infinite number of points for which it is defined. As we indicated in the previous section, the problem can be "regularized" by instead calculating an average of the function over a small interval, i.e., the integral convolution of the unknown quantity with a good smoother. The smoother in Equation (386) is good because it smooths only over the main lobe of width $2W$ and thus avoids sidelobe leakage.

Let us now establish a connection between the projections $Z_k(\cdot)$ and $\Upsilon_k(\cdot)$. As has already been pointed out, $Z_k(\cdot)$ cannot be computed from the data whereas $\Upsilon_k(\cdot)$ can. Following the approach of Thomson (1982), we introduce the weight $b_k(f)$ and look at the difference

$$Z_k(f) - b_k(f)\Upsilon_k(f).$$

What weight should we use to make the right-hand side – incorporating the calculable but defective quantity $\Upsilon_k(f)$ – most like the incalculable but desirable (leakage-free) $Z_k(f)$? From Equations (385c) and (385a) we can write this difference as

$$\begin{aligned}
Z_k(f) - b_k(f)\Upsilon_k(f) &= \int_{-W}^{W} \frac{U_k(f')}{\sqrt{\lambda_k}}\,dZ(f+f') - b_k(f)\sqrt{\lambda_k}\int_{-1/2}^{1/2} U_k(f')\,dZ(f+f') \\
&= \left(\frac{1}{\sqrt{\lambda_k}} - b_k(f)\sqrt{\lambda_k}\right)\int_{-W}^{W} U_k(f')\,dZ(f+f') \\
&\quad - b_k(f)\sqrt{\lambda_k}\int_{f'\notin[-W,W]} U_k(f')\,dZ(f+f'),
\end{aligned}$$

where the second integral is over all frequencies in the disjoint intervals $[-1/2, -W]$ and $[W, 1/2]$. Since both these integrals are with respect to $dZ(f)$, they are uncorrelated over these disjoint domains of integration (because for $f \neq f'$ we know $E\{dZ(f)\,dZ^*(f')\} = 0$). Thus the mean square error (MSE) between $Z_k(f)$ and $b_k(f)\Upsilon_k(f)$, namely,

$$\mathrm{MSE}_k(f) \stackrel{\mathrm{def}}{=} E\{|Z_k(f) - b_k(f)\Upsilon_k(f)|^2\},$$

is given by

$$\left(\frac{1}{\sqrt{\lambda_k}} - b_k(f)\sqrt{\lambda_k}\right)^2 E\left\{\left|\int_{-W}^{W} U_k(f')\,\mathrm{d}Z(f+f')\right|^2\right\}$$
$$+ b_k^2(f)\lambda_k E\left\{\left|\int_{f\notin[-W,W]} U_k(f')\,\mathrm{d}Z(f+f')\right|^2\right\}. \tag{387a}$$

Considering the first expectation above and recalling that $|z|^2 = zz^*$ for any complex-valued variable z, we have

$$E\left\{\left|\int_{-W}^{W} U_k(f')\,\mathrm{d}Z(f+f')\right|^2\right\}$$
$$= \int_{-W}^{W}\int_{-W}^{W} U_k(f')U_k(f'')E\{\mathrm{d}Z(f+f')\,\mathrm{d}Z^*(f+f'')\}$$
$$= \int_{-W}^{W} U_k^2(f')S(f+f')\,\mathrm{d}f' \approx S(f)\int_{-W}^{W} U_k^2(f')\,\mathrm{d}f' = \lambda_k S(f),$$

provided the SDF $S(\cdot)$ is slowly varying in $[f-W, f+W]$ (the approximation becomes an equality for a white noise SDF).

The second expectation is the expected value of the part of the smoothed $\mathrm{d}Z(\cdot)$ outside the primary smoothing band $[-W, W]$. As in the previous section, this can be described as the broad-band bias of the kth eigenspectrum. This bias depends on details of the SDF outside $[f-W, f+W]$, but, as a useful approximation, let us consider its average value – in the sense of Equation (71) – over the frequency interval $[-1/2, 1/2]$ of unit length:

$$\int_{-1/2}^{1/2} E\left\{\left|\int_{f'\notin[-W,W]} U_k(f')\,\mathrm{d}Z(f+f')\right|^2\right\}\,\mathrm{d}f$$
$$= \int_{f'\notin[-W,W]} U_k^2(f')\int_{-1/2}^{1/2} S(f+f')\,\mathrm{d}f\,\mathrm{d}f' = s_0 \int_{f'\notin[-W,W]} U_k^2(f')\,\mathrm{d}f',$$

where we use the fundamental property that $S(\cdot)$ is periodic with unit periodicity, and hence its integral over any interval of unit length is equal to $s_0 = \mathrm{var}\{X_t\}$. Because the $U_k(\cdot)$ functions are orthonormal over $[-1/2, 1/2]$ and orthogonal over $[-W, W]$, we have

$$\int_{f'\notin[-W,W]} U_k^2(f')\,\mathrm{d}f' = \int_{-1/2}^{1/2} U_k^2(f')\,\mathrm{d}f' - \int_{-W}^{W} U_k^2(f')\,\mathrm{d}f' = 1 - \lambda_k.$$

Hence the average value of the broad-band bias for general spectra is given by $(1-\lambda_k)s_0$; for white noise the second expectation in Equation (387a) need not be approximated by averaging over $[-1/2, 1/2]$, but can be evaluated directly, giving $(1-\lambda_k)s_0$ again, so that this expression is exact for the broad-band bias for white noise. For general spectra a useful approximation to the MSE is thus

$$\mathrm{MSE}_k(f) \approx \left(\frac{1}{\sqrt{\lambda_k}} - b_k(f)\sqrt{\lambda_k}\right)^2 \lambda_k S(f) + b_k^2(f)\lambda_k(1-\lambda_k)s_0, \tag{387b}$$

but note that this approximation is an equality for a white noise SDF. To find the value of $b_k(f)$ that minimizes the approximate MSE, we differentiate with respect to $b_k(f)$ and set the result to zero to obtain

$$b_k(f) = \frac{S(f)}{\lambda_k S(f) + (1 - \lambda_k)s_0}. \tag{388a}$$

With this optimum value for $b_k(f)$, we can substitute it into the approximation to the MSE to obtain

$$\text{MSE}_k(f) \approx \frac{S(f)(1 - \lambda_k)s_0}{\lambda_k S(f) + (1 - \lambda_k)s_0}.$$

Let us now consider the special case in which $\{X_t\}$ is white noise; i.e., $S(f) = s_0$. From Equation (388a) we see that $b_k(f)$ is identically unity for all frequencies. Hence, for white noise,

$$\Upsilon_k(f) \text{ estimates } Z_k(f)$$

in a minimum MSE sense. Since the approximation in Equation (387b) is an equality in the case of white noise, we have

$$\text{MSE}_k(f) = \left(\frac{1}{\sqrt{\lambda_k}} - \sqrt{\lambda_k}\right)^2 \lambda_k s_0 + \lambda_k(1 - \lambda_k)s_0 = (1 - \lambda_k)s_0;$$

i.e., the MSE is identical to the broad-band bias. If k is less than $2NW - 1$ so that $\lambda_k \approx 1$, then the MSE is negligible.

What is the best way to combine the individual eigenspectra in the special case of white noise? From Equation (385b) we have

$$\lambda_k \hat{S}_k^{(\text{MT})}(f) = |\Upsilon_k(f)|^2. \tag{388b}$$

Since $\Upsilon_k(f)$ estimates $Z_k(f)$, it follows that

$$\frac{1}{K} \sum_{k=0}^{K-1} \lambda_k \hat{S}_k^{(\text{MT})}(f) \text{ estimates } \frac{1}{K} \sum_{k=0}^{K-1} |Z_k(f)|^2. \tag{388c}$$

What is $E\{\hat{S}_k^{(\text{MT})}(f)\}$? Equation (388b) says that we can get this by rescaling $E\{|\Upsilon_k(f)|^2\}$. It follows from Equation (385a) that

$$\frac{E\{|\Upsilon_k(f)|^2\}}{\lambda_k} = \int_{-1/2}^{1/2} \int_{-1/2}^{1/2} U_k(f')U_k(f'')E\{dZ(f + f')\,dZ^*(f + f'')\}$$
$$= \int_{-1/2}^{1/2} U_k^2(f')S(f + f')\,df' = \int_{-1/2}^{1/2} U_k^2(f - f')S(f')\,df'. \tag{388d}$$

For a white noise process, because each $U_k(\cdot)$ has unit energy over $[-1/2, 1/2]$, it follows that

$$E\{\hat{S}_k^{(\text{MT})}(f)\} = S(f).$$

Hence, in view of Equation (388c), a natural weighted multitaper spectral estimator when the SDF is white – based on the first few eigenspectra (i.e., those with least sidelobe leakage) – is given by

$$\bar{S}^{(\text{WMT})}(f) \stackrel{\text{def}}{=} \frac{\sum_{k=0}^{K-1} \lambda_k \hat{S}_k^{(\text{MT})}(f)}{\sum_{k'=0}^{K-1} \lambda_{k'}} = \sum_{k=0}^{K-1} d_k \hat{S}_k^{(\text{MT})}(f), \tag{388e}$$

8.5 Regularization and Multitapering

where here
$$d_k = \frac{\lambda_k}{\sum_{k'=0}^{K-1} \lambda_{k'}}.$$

Division of the eigenvalues by their sum forces the sum of the weights d_k to be unity, which is required for $\bar{S}^{(\text{WMT})}(f)$ to be an unbiased spectral estimator (cf. Exercise [8.1]). Note also that $K \approx \sum_{k=0}^{K-1} \lambda_k$ provided K is chosen to be less than or equal to $2NW - 1$ so that each λ_k for $k = 0, 1, \ldots, K-1$ is close to unity.

The estimator in Equation (388e) is considered here to be the initial weighted spectral estimator resulting from the theory. It is intuitively attractive since, as the order k of the Slepian tapers increases, the corresponding eigenvalues will decrease, and the eigenspectra will become more contaminated with leakage; i.e., $\text{MSE}_k(f) = (1 - \lambda_k)s_0$ will become larger. The eigenvalue-based weights in Equation (388e) will help to lessen the contribution of the higher leakage eigenspectra. In practice this effect will be negligible provided K is chosen no larger than $2NW - 1$.

The estimator in Equation (388e) can be refined to take into account a colored SDF, as follows. If we combine Equations (388b), (388d) and (388e), we obtain

$$E\{\bar{S}^{(\text{WMT})}(f)\} = \int_{-1/2}^{1/2} \left(\frac{1}{\sum_{k'=0}^{K-1} \lambda_{k'}} \sum_{k=0}^{K-1} \lambda_k U_k^2(f - f') \right) S(f') \, df'. \tag{389a}$$

Note that this expectation integrates over the full band $[-1/2, 1/2]$ and not just $[-W, W]$. If $S(\cdot)$ has a large dynamic range, notable spectral leakage could occur in this smoothing expression due to sidelobes of the DPSWF outside $[-W, W]$ for larger values of k, e.g., those values of k approaching $2NW - 1$. Hence Equation (388e) might not be a satisfactory estimator for colored SDFs. By contrast, the integration range for the expectation of $|Z_k(f)|^2$ in Equation (386) is only $[-W, W]$, so that such sidelobe problems are avoided for $|Z_k(f)|^2$. Now

$$b_k^2(f)|\Upsilon_k(f)|^2 \text{ estimates } |Z_k(f)|^2,$$

where $b_k(f)$ will take the general form of Equation (388a), involving the true SDF $S(f)$. Hence

$$\frac{1}{K} \sum_{k=0}^{K-1} b_k^2(f) \lambda_k \hat{S}_k^{(\text{MT})}(f) \text{ estimates } \frac{1}{K} \sum_{k=0}^{K-1} |Z_k(f)|^2$$

for a colored SDF. By noting that $K \approx \sum_{k=0}^{K-1} b_k^2(f) \lambda_k$ provided K does not exceed $2NW - 1$, we arrive at the *adaptive multitaper spectral estimator*

$$\hat{S}^{(\text{AMT})}(f) \stackrel{\text{def}}{=} \frac{\sum_{k=0}^{K-1} b_k^2(f) \lambda_k \hat{S}_k^{(\text{MT})}(f)}{\sum_{k'=0}^{K-1} b_{k'}^2(f) \lambda_{k'}} = \sum_{k=0}^{K-1} d_k(f) \hat{S}_k^{(\text{MT})}(f), \tag{389b}$$

which has the same form as the estimator of Equation (388e), but now with frequency-dependent weights

$$d_k(f) = \frac{b_k^2(f) \lambda_k}{\sum_{k'=0}^{K-1} b_{k'}^2(f) \lambda_{k'}} \tag{389c}$$

(Equation (389b) reduces to Equation (388e) upon setting the $b_k^2(f)$ weights to unity). We cannot claim that $E\{\hat{S}^{(\text{AMT})}(f)\}$ is exactly equal to $S(f)$ since Equation (388d) implies that in general $E\{\hat{S}_k^{(\text{MT})}(f)\}$ is not identical to $S(f)$ for a colored SDF; however, $E\{\hat{S}_k^{(\text{MT})}(f)\}$

and $S(f)$ will generally be close provided the latter does not vary rapidly over the interval $[f - W, f + W]$. The arguments leading to Equation (389b) justify the same result in Thomson (1982, equation (5.3)), but the definition of the weights $b_k(f)$ used here is more appealing since these weights are unity for a white noise SDF.

Now the $b_k(f)$ weight formula of Equation (388a) involves the true unknown spectrum and variance. Spectral estimation via Equation (389b) must thus be carried out in an iterative fashion such as the following. To proceed, we assume $NW \geq 2$. We start with a spectral estimate of the form of Equation (388e) with K set equal to 1 or 2. This initial estimate involves only the one or two tapers with lowest sidelobe leakage and hence will preserve rapid spectral decays. This spectral estimate is then substituted – along with the estimated variance – into Equation (388a) to obtain the weights for orders $k = 0, 1, \ldots, K - 1$ with K typically $2NW - 1$. These weights are then substituted into Equation (389b) to obtain the new spectral estimate with K again $2NW - 1$. The spectral estimate given by Equation (389b) is next substituted back into Equation (388a) to get new weights and so forth. Usually two executions of Equation (389b) are sufficient, but more might be called for (and the number of iterations needed might be different for different frequencies). Thomson (1982) describes a method for estimating the broad-band bias (rather than its average over $[-1/2, 1/2]$) within the iterative stage, but this additional complication often leads to very marginal changes and suffers from additional estimation problems.

An EDOFs argument says that $\hat{S}^{(\text{AMT})}(f)$ is approximately equal in distribution to the RV $S(f)\chi_\nu^2/\nu$, where the chi-square RV has degrees of freedom

$$\nu \approx \frac{2 \left(\sum_{k=0}^{K-1} b_k^2(f) \lambda_k \right)^2}{\sum_{k=0}^{K-1} b_k^4(f) \lambda_k^2} \tag{390}$$

when $0 < f < f_\mathcal{N}$ (this is an application of Equation (354g) with the weights d_k set as per Equation (389c)).

Comments and Extensions to Section 8.5

[1] The estimators of Equations (388e) and (389b) differ significantly from each other in two important properties. First, the weights in the spectral estimator of Equation (388e) do not depend on frequency – as a result, Parseval's theorem is satisfied in expected value (this is also true for the simple average of Equation (352a)). To see this, use Equation (389a) to write

$$\int_{-1/2}^{1/2} E\{\bar{S}^{(\text{WMT})}(f)\} \, df = \sum_{k=0}^{K-1} \lambda_k \int_{-1/2}^{1/2} \frac{1}{\sum_{k'=0}^{K-1} \lambda_{k'}} \left(\int_{-1/2}^{1/2} \mathcal{U}_k^2(f - f') \, df \right) S(f') \, df'.$$

The integral in the parentheses is unity, so we obtain

$$\int_{-1/2}^{1/2} E\{\bar{S}^{(\text{WMT})}(f)\} \, df = \int_{-1/2}^{1/2} S(f') \, df' = s_0,$$

as required. In contrast, Exercise [8.16] shows that Parseval's theorem is in general *not* satisfied exactly in expected value for the adaptive multitaper spectral estimator of Equation (389b).

Second, the weighted multitaper estimator of Equation (388e) (and also Equation (352b)) has approximately constant variance across all the Fourier frequencies (excluding 0 and Nyquist), whereas the adaptive weighting scheme of Equation (389b) can give rise to appreciable variations of the variance throughout the spectrum, making interpretation of the plot of the spectral estimate more difficult (see the example involving Figure 426 in Section 8.9).

8.6 Sinusoidal Multitaper Estimators

As we noted in Section 8.2, the original multitaper SDF estimator introduced by Thomson (1982) employs the Slepian data tapers. Subsequently Riedel and Sidorenko (1995) proposed a second family of tapers – known as the *sinusoidal tapers* – to form a multitaper SDF estimator. The Slepian tapers arise as the solution to a certain optimization problem, and the sinusoidal tapers do also, but as approximate solutions to an entirely different problem.

To formulate the optimization problem that leads to the sinusoidal tapers, let us reconsider Exercise [192], which, given a direct spectral estimator $\hat{S}^{(\mathrm{D})}(\cdot)$ based upon a data taper whose spectral window is $\mathcal{H}(\cdot)$, suggests the following approximation to the bias in the estimator:

$$E\{\hat{S}^{(\mathrm{D})}(f)\} - S(f) \approx \frac{S''(f)}{24}\beta_{\mathcal{H}}^2, \quad \text{where } \beta_{\mathcal{H}} = \left(12\int_{-f_\mathcal{N}}^{f_\mathcal{N}} \phi^2 \mathcal{H}(\phi)\,\mathrm{d}\phi\right)^{1/2} \quad (391\mathrm{a})$$

(as discussed following Exercise [192], $\beta_{\mathcal{H}}$ is the variance width of $\mathcal{H}(\cdot)$ and is one of several measures of the bandwidth of $\hat{S}^{(\mathrm{D})}(\cdot)$). We can thus take $\beta_{\mathcal{H}}^2$ to be an indicator of the bias in the estimator $\hat{S}^{(\mathrm{D})}(\cdot)$, with larger values indicating more bias. Consider a weighted multitaper estimator $\hat{S}^{(\mathrm{WMT})}(\cdot)$ defined as per Equation (352b) and based on K direct spectral estimators $\hat{S}_k^{(\mathrm{MT})}(\cdot)$ that are associated with data tapers $\{h_{k,t}\}$, spectral windows $\mathcal{H}_k(\cdot)$ and weights d_k. A derivation paralleling the one used in Exercise [192] leads to

$$E\{\hat{S}^{(\mathrm{WMT})}(f)\} - S(f) \approx \frac{S''(f)}{24}\beta_{\widetilde{\mathcal{H}}}^2, \quad \text{where } \beta_{\widetilde{\mathcal{H}}} = \left(12\int_{-f_\mathcal{N}}^{f_\mathcal{N}} \phi^2 \widetilde{\mathcal{H}}(\phi)\,\mathrm{d}\phi\right)^{1/2}, \quad (391\mathrm{b})$$

and $\widetilde{\mathcal{H}}(f) = \sum_{k=0}^{K-1} d_k \mathcal{H}_k(f)$ defines the spectral window associated with $\hat{S}^{(\mathrm{WMT})}(\cdot)$. Thus $\beta_{\widetilde{\mathcal{H}}}^2$ is an indicator of the bias in the estimator $\hat{S}^{(\mathrm{WMT})}(\cdot)$. Exercise [8.21] says that

$$\beta_{\widetilde{\mathcal{H}}}^2 = \frac{12}{\Delta_\mathrm{t}^2}\,\mathrm{tr}\,\{\boldsymbol{D}_K \boldsymbol{H}^T \boldsymbol{\Sigma}^{(\mathrm{PL})} \boldsymbol{H}\}, \quad (391\mathrm{c})$$

where $\boldsymbol{D}_K$ is an $K \times K$ diagonal matrix whose diagonal entries are the weights d_k; $\boldsymbol{H}$ is an $N \times K$ matrix whose columns $\boldsymbol{h}_k$ contain the data tapers $\{h_{k,t}\}$, $k = 0, 1, \ldots, K-1$; and $\boldsymbol{\Sigma}^{(\mathrm{PL})}$ is a covariance matrix whose (j,k)th element is $s_{j-k}^{(\mathrm{PL})}$, where

$$s_\tau^{(\mathrm{PL})} \stackrel{\mathrm{def}}{=} \Delta_\mathrm{t}^3 \int_{-f_\mathcal{N}}^{f_\mathcal{N}} f^2 e^{i2\pi f\tau\Delta_\mathrm{t}}\,\mathrm{d}f = \begin{cases} 1/12, & \tau = 0; \\ (-1)^\tau/(2\pi^2\tau^2), & \text{otherwise.} \end{cases} \quad (391\mathrm{d})$$

The sequence $\{s_\tau^{(\mathrm{PL})}\}$ is the ACVS for a stationary process with a power-law SDF $S^{(\mathrm{PL})}(f) = f^2 \Delta_\mathrm{t}^3$ (the presence of Δ_t^3 here might strike the reader as a bit odd – if so, see the discussion surrounding Equation (119)).

As previously noted, a weighted multitaper spectral estimator $\hat{S}^{(\mathrm{WMT})}(\cdot)$ will be unbiased for a white noise process if the weights d_k sum to unity (see Equation (378a)). Subject to this constraint, in view of Equation (391c), we need to

$$\text{minimize}\ \mathrm{tr}\,\{\boldsymbol{D}_K \boldsymbol{H}^T \boldsymbol{\Sigma}^{(\mathrm{PL})} \boldsymbol{H}\}\ \text{subject to}\ \sum_{k=0}^{K-1} d_k = 1 \quad (391\mathrm{e})$$

if we want to minimize the bias indicator $\beta^2_{\mathcal{H}}$. The above is similar in spirit to the optimization problem that led to the Slepian multitaper scheme, namely,

$$\text{maximize }\operatorname{tr}\{\boldsymbol{D}_K \boldsymbol{H}^T \boldsymbol{\Sigma}^{(\mathrm{BL})} \boldsymbol{H}\} \text{ subject to } \sum_{k=0}^{K-1} d_k = 1 \qquad (392\mathrm{a})$$

(this is Equation (380a)). The solution to both problems is to set $K = 1$ and $d_0 = 1$ and to form $h_{0,0}, h_{0,1}, \ldots, h_{0,N-1}$ based upon a particular eigenvector $\boldsymbol{h}_0$ for the associated covariance matrix. In the case of Equation (392a), the eigenvector of interest corresponds to the *largest* eigenvalue λ_0 for the matrix $\boldsymbol{\Sigma}^{(\mathrm{BL})}$ and leads to the zeroth-order Slepian taper; in the case of Equation (391e), the eigenvector is associated with the *smallest* eigenvalue $\tilde{\lambda}_0$ of $\boldsymbol{\Sigma}^{(\mathrm{PL})}$ and leads to what Riedel and Sidorenko (1995) call the *zeroth-order minimum-bias taper*.

The Kth-order basic minimum-bias multitaper estimator $\hat{S}^{(\mathrm{MT})}(\cdot)$ makes use of tapers based on the eigenvectors $\boldsymbol{h}_k$ of $\boldsymbol{\Sigma}^{(\mathrm{PL})}$ that are associated with the K *smallest* eigenvalues $\tilde{\lambda}_0 < \tilde{\lambda}_1 < \cdots < \tilde{\lambda}_{K-1}$ (this is in contrast to the Slepian-based estimator, which makes use of the eigenvectors of $\boldsymbol{\Sigma}^{(\mathrm{BL})}$ associated with the K *largest* eigenvalues $\lambda_0(N, W) > \lambda_1(N, W) > \cdots > \lambda_{K-1}(N, W)$). We refer to $\{h_{k,t}\}$ as the *kth-order minimum-bias taper*. We assume the usual normalization $\boldsymbol{h}_k^T \boldsymbol{h}_k = \sum_t h_{k,t}^2 = 1$. Because the $\boldsymbol{h}_k$'s are eigenvectors, we have $\boldsymbol{h}_j^T \boldsymbol{h}_k = \sum_t h_{j,t} h_{k,t} = 0$ for $j \neq k$, which is the orthogonality condition of Equation (354b). Given minimum-bias tapers of orders $k = 0, 1, \ldots, K-1$, we form K direct spectral estimators (the eigenspectra) and average them together to form the basic minimum-bias multitaper estimator $\hat{S}^{(\mathrm{MT})}(\cdot)$, which takes the form of Equation (352a).

▷ **Exercise [392]** Show that the bias indicator for the kth-order minimum-bias eigenspectrum is

$$\beta^2_{\mathcal{H}_k} = \frac{12\tilde{\lambda}_k}{\Delta_t^2}. \qquad (392\mathrm{b}) \triangleleft$$

The above implies that, because $\tilde{\lambda}_k$ increases as k increases, the bias indicator increases as the order of the eigenspectrum increases. Exercise [8.22] says that the bias indicator for the basic multitaper estimator $\hat{S}^{(\mathrm{WMT})}(\cdot) = \hat{S}^{(\mathrm{MT})}(\cdot)$ is

$$\beta^2_{\mathcal{H}} = \frac{12 \sum_{k=0}^{K-1} \tilde{\lambda}_k}{K \Delta_t^2}, \qquad (392\mathrm{c})$$

which increases as K increases since the eigenvalues $\tilde{\lambda}_k$ are all positive and increase as k increases. A form of the usual trade-off between bias and variance is thus in effect here: as K increases, the variance of the minimum-bias multitaper estimator $\hat{S}^{(\mathrm{MT})}(\cdot)$ decreases (as per Equation (354c)) but at the expense of an increase in its bias indicator.

The solid curves in Figure 393 show the minimum-bias tapers of orders $k = 0, 1, 2$ and 3 (top to bottom rows) for $N = 32$ (left-hand column) and $N = 99$ (right-hand). Riedel and Sidorenko (1995) note that the minimum-bias tapers can be approximated by the *sinusoidal tapers*, which can be generated using the following simple and easily computable formula:

$$h_{k,t} = \left(\frac{2}{N+1}\right)^{1/2} \sin\left[\frac{(k+1)\pi(t+1)}{N+1}\right], \quad t = 0, 1\ldots, N-1 \qquad (392\mathrm{d})$$

(this approximation comes from a discretization of a continuous t solution to the minimum-bias problem). The sinusoidal tapers are shown as dashed curves in Figure 393; however, these curves are difficult to see (particularly for $N = 99$) because they agree so well with

8.6 Sinusoidal Multitaper Estimators

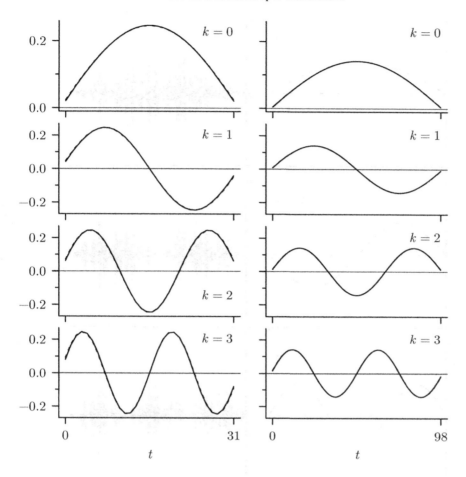

Figure 393 Minimum-bias (solid curves) and sinusoidal (dashed) tapers with $N = 32$ (left-hand column) and with $N = 99$ (right-hand) for orders $k = 0, 1, 2$ and 3 (top to bottom). The fact that the solid and dashed curves are almost indistinguishable (particularly for $N = 99$) shows that the sinusoidal tapers are accurate approximations to the minimum-bias tapers.

the solid curves, an indication that the sinusoidal tapers are in excellent agreement with the minimum-bias tapers!

Figures 394 through 399 show plots for the sinusoidal multitaper scheme analogous to those in Figures 360 to 365 for the Slepian scheme (these figures all use the AR(4) time series of Figure 34(e), for which $N = 1024$ and $\Delta_t = 1$). Let us first compare the plots of the sinusoidal tapers in the left-hand columns of Figures 394 and 396 to the ones for the Slepian tapers (Figures 360 and 362). Each plot has a thin horizontal line marking zero on the vertical axis. The kth-order sinusoidal and Slepian tapers are similar in having exactly k zero crossings, which partition the tapers into $k + 1$ segments. For a given $k \geq 2$, the partitionings differ in that, whereas the segments are of essentially the same length for the sinusoidal tapers, the two outermost segments are markedly longer than the $k - 1$ interior segment(s) for the Slepian tapers. In addition, the values of the sinusoidal tapers at $t = 0$ and $t = 1023$ are close to zero for all k, but those for the Slepian tapers noticeably deviate from zero starting with $k = 4$, with the deviations increasing for $k = 5$, 6 and 7 (Figure 362). As a result, all of the sinusoidally tapered AR(4) time series damp down at the extremes (right-hand columns of Figures 394 and 396), whereas the $k = 6$ and 7 Slepian tapers noticeably enhance

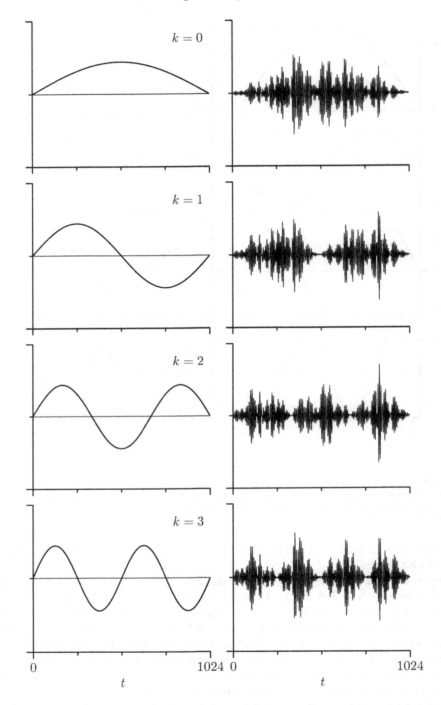

Figure 394 Sinusoidal data tapers $\{h_{k,t}\}$ (left-hand plots) and products $\{h_{k,t}X_t\}$ of these tapers and AR(4) time series $\{X_t\}$ shown in Figure 34(e) (right-hand plots), part 1. The orders k for the sinusoidal data tapers are 0, 1, 2 and 3 (top to bottom).

8.6 Sinusoidal Multitaper Estimators

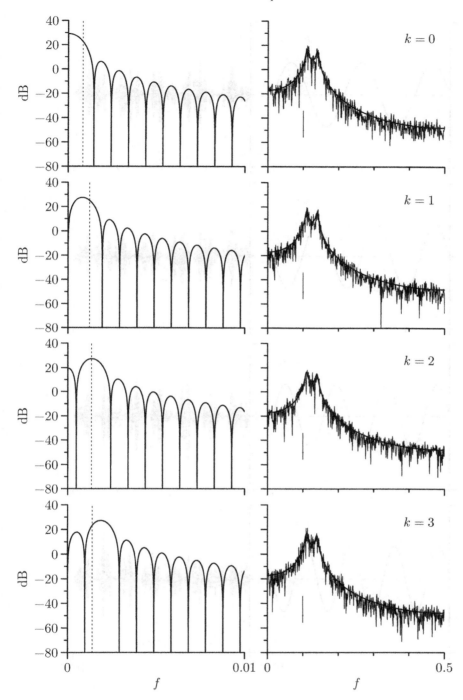

Figure 395 Spectral windows $\mathcal{H}_k(\cdot)$ corresponding to sinusoidal multitapers $\{h_{k,t}\}$ (left-hand plots) and eigenspectra $\hat{S}_k^{(\mathrm{MT})}(\cdot)$ for AR(4) time series (right-hand, thin curves), part 1. The taper orders k are 0, 1, 2 and 3. The dashed line in the left-hand plot marks $B_{\mathcal{H}_k}/2$ (half the standard bandwidth measure). In each right-hand plot, the thick curve is the true SDF for the AR(4) process, while the horizontal and vertical portions of the crisscross depict, respectively, $B_{\mathcal{H}_k}$ and the width of a 95% CI for the true SDF.

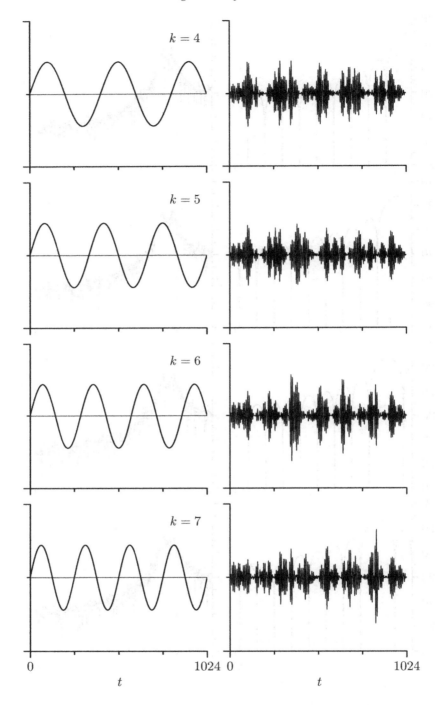

Figure 396 Sinusoidal data tapers and products of these tapers and a time series, part 2 (cf. Figure 394). Here the orders k for the tapers are 4, 5, 6 and 7 (top to bottom).

8.6 Sinusoidal Multitaper Estimators

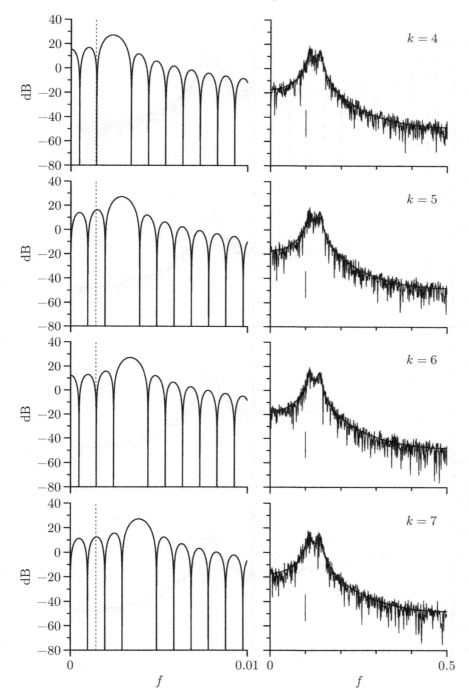

Figure 397 Spectral windows $\mathcal{H}_k(\cdot)$ corresponding to sinusoidal multitapers $\{h_{k,t}\}$ (left-hand plots) and eigenspectra $\hat{S}_k^{(\mathrm{MT})}(\cdot)$ for AR(4) time series (right-hand plots, thin curves), part 2 (cf. Figure 395). Here the taper orders k are 4, 5, 6 and 7.

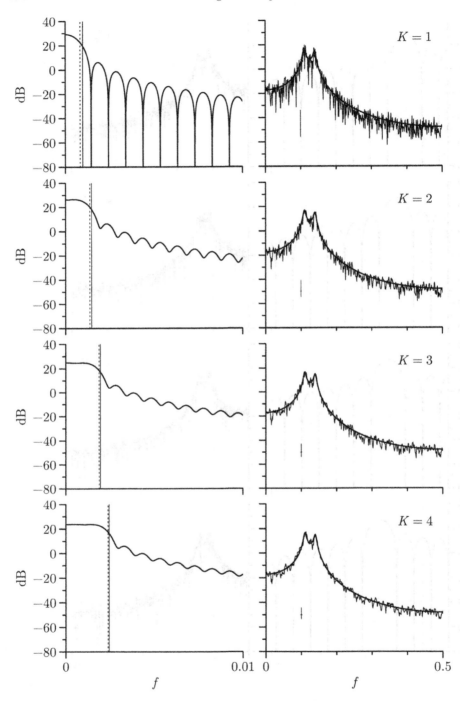

Figure 398 Sinusoidal multitaper spectral estimates $\hat{S}^{(\mathrm{MT})}(\cdot)$ formed by averaging K eigenspectra (right-hand plots) and corresponding spectral windows $\overline{\mathcal{H}}(\cdot)$ (left-hand plots) for $K = 1, 2, 3$ and 4. The dashed and solid vertical lines on each left-hand plot mark, respectively, $B_{\overline{\mathcal{H}}}/2$ (half of the standard bandwidth measure) and $(K + 1)/[2(N + 1)\Delta_{\mathrm{t}}]$ (a simple approximation to $B_{\overline{\mathcal{H}}}/2$, where here $\Delta_{\mathrm{t}} = 1$). In each right-hand plot, the thick curve is the true SDF for the AR(4) process, while the horizontal and vertical portions of the crisscross depict, respectively, $B_{\overline{\mathcal{H}}}$ and the width of a 95% CI for the true SDF.

8.6 Sinusoidal Multitaper Estimators

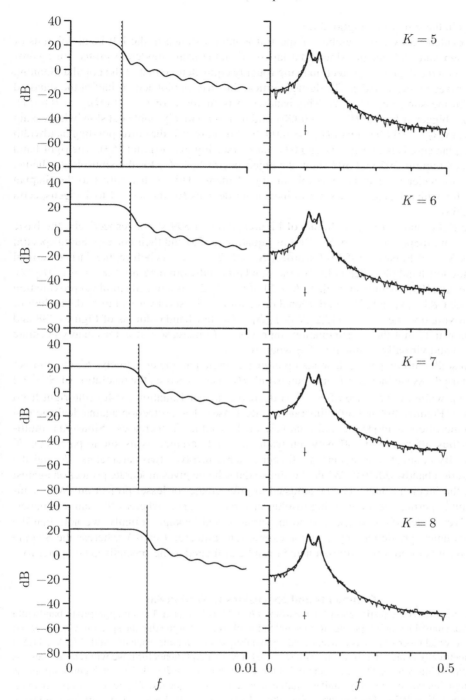

Figure 399 Sinusoidal multitaper spectral estimates $\hat{S}^{(\mathrm{MT})}(\cdot)$ formed by averaging K eigenspectra (right-hand plots) and corresponding spectral windows $\overline{\mathcal{H}}(\cdot)$ (left-hand plots) for $K = 5, 6, 7$ and 8 (cf. Figure 398).

the series in these regions (Figure 362).

As evidenced by the sidelobes of spectral windows shown in the left-hand columns of Figures 395 and 397, all the sinusoidal tapers offer the same moderate protection against leakage, with the degree of protection being roughly equivalent to that of the popular Hanning data taper (see Figure 191(d)). The dashed vertical line in each plot depicts half of the standard bandwidth measure, i.e., $B_{\mathcal{H}_k}/2$. This bandwidth is smallest for $k = 0$ ($B_{\mathcal{H}_0} \doteq 0.0017$), about 50% bigger at $k = 1$ ($B_{\mathcal{H}_1} \doteq 0.0025$), and then is relatively stable but slowly increasing (going from $B_{\mathcal{H}_2} \doteq 0.0027$ up to $B_{\mathcal{H}_7} \doteq 0.0029$). For a given k, the corresponding bandwidth in the Slepian case is between 48% and 67% larger (see Figures 361 and 363). The right-hand columns of Figures 395 and 397 show the eigenspectra $\hat{S}_k^{(\mathrm{MT})}(\cdot)$ for the sinusoidal scheme. There is no evidence of leakage visible in any of them. This is in contrast to the Slepian scheme, for which leakage is visible in increasing degrees for the $k = 4$ to 7 eigenspectra (Figure 363).

The right- and left-hand columns of Figures 398 and 399 show, respectively, the basic sinusoidal multitaper estimators $\hat{S}^{(\mathrm{MT})}(\cdot)$ of Equation (352a) and their corresponding spectral windows $\overline{\mathcal{H}}(\cdot)$ of Equation (353b) for orders $K = 1, 2, \ldots, 8$. As before, the dashed vertical lines in the left-hand plots depict half of the standard bandwidth measure (here this is $B_{\overline{\mathcal{H}}}/2$). Walden et al. (1995) demonstrate that $(K+1)/[(N+1)\Delta_t]$ is a useful simple approximation to $B_{\overline{\mathcal{H}}}$ (see C&Es [1] and [2] for additional discussion). The solid vertical lines depict half of this approximation, i.e., $(K+1)/[2(N+1)\Delta_t]$. The left-hand columns of Figures 398 and 399 demonstrate that the approximation improves as K increases: it is reasonably accurate for $K \leq 3$ and virtually identical to $B_{\overline{\mathcal{H}}}$ when $K \geq 4$.

Sinusoidal multitapering depends upon just a single parameter, namely, K, the number of tapers used. As we increase K, the bandwidth $B_{\overline{\mathcal{H}}}$ increases (i.e., the resolution of $\hat{S}^{(\mathrm{MT})}(\cdot)$ decreases), while the variance of $\hat{S}^{(\mathrm{MT})}(\cdot)$ decreases as per Equation (354c) (the right-hand columns of Figures 398 and 399 illustrate this decrease). The protection against leakage that this scheme offers is moderate and relatively unchanged as K increases. Sinusoidal multitapering thus offers a trade-off between resolution and variance as its single parameter K changes. By contrast, the Slepian multitaper scheme involves two parameters: K and the regularization bandwidth $2W$. While the sinusoidal scheme gives moderate protection against leakage, the Slepian scheme can be adjusted to offer greater or lesser protection. With sinusoidal multitapering, we are juggling resolution versus variance, whereas Slepian multitapering involves trade-offs amongst resolution, variance and leakage. Finally, we note that the sinusoidal tapers are defined by a simple expression (Equation (392d)), whereas the Slepian tapers have no closed form expression and must be evaluated using specialized computer routines.

Comments and Extensions to Section 8.6

[1] Here we offer some justification for regarding $(K+1)/[(N+1)\Delta_t]$ as an appropriate bandwidth measure for sinusoidal multitapering. Riedel and Sidorenko (1995) argue that the spectral window $\mathcal{H}_k(\cdot)$ for the sinusoidal taper $\{h_{k,t}\}$ is concentrated on two frequency intervals, each of width $1/[(N+1)\Delta_t]$ and symmetrically centered at frequencies $\pm(k+1)/[2(N+1)\Delta_t]$ (henceforth we will refer to these as "concentration intervals"). This argument holds for $k \geq 1$, but not for $k = 0$, for which concentration is over a single interval of arguably similar width centered at zero frequency. Figure 401 offers a demonstration when $N = 1024$. The top three plots show $\mathcal{H}_k(\cdot)$ for $k = 0, 1$ and 7 (these are also shown in the left-hand columns of Figures 395 and 397). The thick horizontal lines toward the tops of these plots indicate the concentration intervals – one interval for $k = 0$, and two each for $k = 1$ and 7. For comparison, the thin horizontal line approximately 3 dB below each dominant lobe depicts width$_{\mathrm{hp}}\{\mathcal{H}_k(\cdot)\}$, i.e., its half-power width (see the discussion surrounding Figure 191). The half-power widths roughly agree with $1/[(N+1)\Delta_t]$, which lends credence to the latter as a bandwidth measure (width$_{\mathrm{hp}}\{\mathcal{H}_k(\cdot)\}$ is approximately 12% smaller than $1/[(N+1)\Delta_t]$ for $k \geq 1$ and 19% bigger for $k = 0$). The horizontal

8.6 Sinusoidal Multitaper Estimators

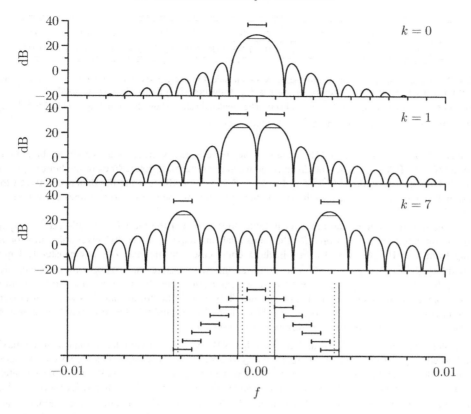

Figure 401 Spectral windows $\mathcal{H}_k(f)$ on a decibel scale versus $f \in [-0.01, 0.01]$ for sinusoidal tapers of length $N = 1024$ and of orders $k = 0, 1$ and 7 (bumpy curves in top three plots), along with horizontal lines indicating either the frequency intervals over which the spectral windows are mainly concentrated (thick lines above the bumpy curves) or the half-power widths of the dominant lobes (thin lines approximately 3 dB down from the peak values). The horizontal lines in the bottom plot show the concentration intervals for $k = 0$ (there is one such interval) and for $k = 1, \ldots, 7$ (two intervals for each k, for a total of 15 intervals). The distance between the outermost (innermost) two solid vertical lines is equal to the simple bandwidth measure $(K+1)/[(N+1)\Delta_t]$ when $K = 8$ ($K = 1$). The distances between the corresponding pairs of dotted vertical lines show the alternative bandwidth measure of Equation (401).

lines in the bottom plot of Figure 401 indicate the concentration intervals for $k = 0, 1, \ldots, 7$. The distance between the two outermost vertical solid lines is equal to the simple measure $(K+1)/[(N+1)\Delta_t]$ when $K = 8$ (see the discussion surrounding Equation (401) for an explanation of the dotted lines). The interval given by the union of all 15 displayed concentration intervals has a width *exactly* equal to the simple measure for $K = 8$. An analogous equality holds for all $K \geq 2$, but not for $K = 1$. The distance between the two innermost solid lines shows the simple measure when $K = 1$. Here the measure is equal to *twice* the width of the single concentration interval associated with $k = 0$, so we might be tempted to use half the simple measure when $K = 1$. If, however, we regard the simple measure as an approximation to $B_{\overline{\mathcal{H}}}$, it is clear from the upper left-hand plot of Figure 398 what the effect would be of halving the simple measure: the dashed and solid vertical lines are dictated by, respectively, the standard and simple bandwidth measures, and halving the latter would result in a less accurate approximation.

[2] In C&E [1] for Section 8.1 we discussed a bandwidth measure $\text{width}_e\{R^{(\text{MT})}(\cdot)\}$ (Equation (355a)) for multitaper estimators that is an alternative to the standard bandwidth measure $B_{\overline{\mathcal{H}}}$ of Equation (353e). For sinusoidal multitaper estimators using K tapers satisfying the mild constraint $2K \leq N$, this alternative measure has a simple form, namely,

$$\text{width}_e\{R^{(\text{MT})}(\cdot)\} = \frac{K + \frac{1}{2}}{(N+1)\Delta_t} \tag{401}$$

$K=$	1	2	3	4	5	6	7	8	9	10	11	12	13	14	15	16
	1.14	1.12	1.10	1.08	1.07	1.07	1.06	1.06	1.05	1.05	1.05	1.04	1.04	1.04	1.04	1.04

Table 402 Comparison of two bandwidth measures for sinusoidal multitaper spectral estimators $\hat{S}^{(\mathrm{MT})}(\cdot)$ (cf. Table 369 for Slepian multitapers). The first measure is $B_{\overline{\mathcal{H}}}$ (Equation (353e)), and the second, width$_e\{R^{(\mathrm{MT})}(\cdot)\}$ (Equation (355a), which reduces to Equation (401)). The tabulated values show the ratio $B_{\overline{\mathcal{H}}}/\mathrm{width}_e\{R^{(\mathrm{MT})}(\cdot)\}$ as the number of tapers K used to form $\hat{S}^{(\mathrm{MT})}(\cdot)$ ranges from 1 to 16.

(Greenhall, 2006; see Exercise [8.23]). Note that the above is nearly the same as our simple bandwidth measure $(K+1)/[(N+1)\Delta_t]$. The distance between the outermost dotted lines in the bottom plot of Figure 401 is equal to width$_e\{R^{(\mathrm{MT})}(\cdot)\}$ for $K=8$ and $N=1024$, while the distance between the corresponding solid lines is equal to the simple measure (the two sets of innermost lines correspond to the case $K=1$).

As we did for Slepian multitapers (see Table 369), consider the ratio $B_{\overline{\mathcal{H}}}/\mathrm{width}_e\{R^{(\mathrm{MT})}(\cdot)\}$ as a way to compare the two measures for the sinusoidal multitapers. Table 402 gives this ratio for K ranging from 1 to 16. As was also true in the Slepian case, the measure $B_{\overline{\mathcal{H}}}$ is consistently larger than width$_e\{R^{(\mathrm{MT})}(\cdot)\}$, with the difference between the two measures decreasing with increasing K; however, the ratio is never greater than 14%, so we can again argue that the two bandwidth measures are interchangeable for most practical applications. (Table 402 holds for all sample sizes N of powers of two ranging from $2^7 = 128$ up to $2^{15} = 32768$. For $N=32$ and 64, the corresponding tables are identical to Table 402 except for some differences of ± 0.01.)

[3] For Slepian tapers, given N, W and Δ_t such that $NW\Delta_t$ is a positive integer, the use of the maximum theoretically sensible number of tapers, $K_{\max} = 2NW\Delta_t - 1$, ensures that the spectral window $\overline{\mathcal{H}}(\cdot)$ resembles a rectangle in being nonzero and approximately flat over $[-W, W]$ and approximately zero outside that interval. If $\overline{\mathcal{H}}(\cdot)$ were exactly a rectangle over $[-W,W]$, its bandwidth measure $B_{\overline{\mathcal{H}}}$ would be equal to the regularization bandwidth $2W$. The fact that $\overline{\mathcal{H}}(\cdot)$ is almost rectangular means that the regularization bandwidth $2W$ and $B_{\overline{\mathcal{H}}}$ are almost identical. The left-hand plot in the $K=7$ row of Figure 365 illustrates this fact. Here $K_{\max} = 7$, and the solid and dashed vertical lines indicating, respectively, W and $B_{\overline{\mathcal{H}}}/2$ are quite close to each other. If, however, we had chosen a smaller number of tapers $K < K_{\max}$ so that $\overline{\mathcal{H}}(\cdot)$ does not fill out the interval $[-W, W]$, then the bandwidth measure $B_{\overline{\mathcal{H}}}$ for a Slepian-based $\hat{S}^{(\mathrm{MT})}(\cdot)$ can be markedly less than its regularization bandwidth $2W$. The left-hand plots in the rows corresponding to $K = 1, \ldots, 6$ in Figures 364 and 365 illustrate this fact: the solid and dashed vertical lines are no longer in good agreement, with the disagreement getting worse as K decreases. Contrariwise, for sinusoidal tapers, $B_{\overline{\mathcal{H}}}$ and the simple bandwidth measure $(K+1)/[(N+1)\Delta_t]$ are approximately equal no matter what K is (see the vertical dashed and solid lines in the left-hand columns of Figures 398 and 399). As K increases, the width of a sinusoidal-based spectral window $\overline{\mathcal{H}}(\cdot)$ increases commensurate with the simple measure. The width of a Slepian-based $\overline{\mathcal{H}}(\cdot)$ is only in keeping with the regularization bandwidth $2W$ when $K = K_{\max}$. The fact that the regularization bandwidth $2W$ is often not an adequate measure of bandwidth for Slepian-based $\hat{S}^{(\mathrm{MT})}(\cdot)$ – coupled with the desire to have a common way of measuring bandwidth that works across a wide range of nonparametric SDF estimators – motivates our advocacy for measures based on the autocorrelation width of spectral windows (see Walden et al., 1995, for details).

[4] In Section 8.3 we discussed multitapering of white noise using an arbitrary set of orthonormal data tapers. We then specialized to the case of Slepian data tapers and considered the rate at which the variance of the multitaper estimator $\hat{S}^{(\mathrm{MT})}(f)$ decreases as a function of K (the number of eigenspectra averaged) for two specific examples. These examples differ only in their settings for NW, namely, $NW = 4$ in Figure 373a and $NW = 16$ in Figure 373b. Figure 403 shows what happens if we use sinusoidal multitapers in place of Slepian tapers, leaving the rest of the setup the same as before (i.e., $f = 1/4$, $N = 64$, $\Delta_t = 1$ and $s_0 = 1$). The thick curves in all three figures show var$\{\hat{S}^{(\mathrm{MT})}(1/4)\}$ versus K. If we compare the ones in Figures 373b (Slepian with $NW = 16$) and 403 (sinusoidal), we find the striking result that, once we take numerical precision into account, they are identical!

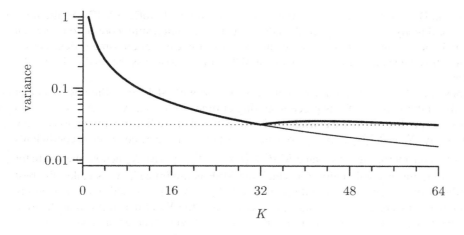

Figure 403 As in Figures 373a and 373b, but now using sinusoidal tapers in place of Slepian tapers. The thick curves in this and the other two figures show var $\{\hat{S}^{(\mathrm{MT})}(1/4)\}$ versus the number of eigenspectra averaged to form $\hat{S}^{(\mathrm{MT})}(1/4)$. The thick curve here is numerically identical to the one in Figure 373b for the $NW = 16$ Slepian tapers.

8.7 Improving Periodogram-Based Methodology via Multitapering

As described in Section 6.3, the periodogram for a time series $X_0, X_1, \ldots, X_{N-1}$ is a primitive estimator of its underlying SDF. The inherent variability in the periodogram and its deleterious bias properties render it unattractive as a general purpose SDF estimator; however, in time series analysis, while its use as an SDF estimator per se is limited, the periodogram often plays a role in estimating certain functions of the SDF. If the function in question involves combining values of the SDF across frequencies, the problem with the variability in the periodogram at one particular frequency might be ameliorated when combined across frequencies. If so, potential bias in the periodogram is then its chief distraction, but Equation (175c) argues that the periodogram is asymptotically unbiased. For practitioners with a time series of finite length N, relying on this asymptotic argument to dismiss potential bias is questionable since the periodogram can still be badly biased even if N seems superficially large (see the discussion surrounding Figure 178).

In this section we consider two functions of an SDF (the innovation variance and the exponent of a power-law process) for which periodogram-based estimators have been advocated. We demonstrate that replacing the periodogram with a multitaper estimator leads to estimators of these functions that perform better for time series with sample sizes N of interest to practitioners. Why should this be the case? The periodogram has high variability, which is reflected in its association with a χ_2^2 distribution at almost all frequencies; by contrast, a basic multitaper estimator using $K \geq 2$ tapers is associated with a χ_{2K}^2 distribution. Figure 266 shows that, even for $2 \leq K \leq 5$, there is noticeable decrease in variability as quantified by the width of a 95% CI for an SDF at a fixed frequency. If bias is not a concern, a periodogram-based estimator that works well due to combining the periodogram across frequencies suggests that a multitaper-based estimator should work just as well even if K is small. If bias is a concern, then the multitaper-based estimator is potentially superior. In essence, we are replacing the highly variable and potentially biased periodogram with an estimator having reduced variability (modestly so if K is just 2) and reduced bias (potentially highly so).

Estimation of the Innovation Variance

A prime goal of time series analysis is forecasting a yet-to-be-observed value in a series given previous observations. Forecasting is an intricate subject and is covered thoroughly in, e.g.,

Brockwell and Davis (2016), Priestley (1981) and Shumway and Stoffer (2017). While forecasting is not directly related to spectral analysis, there is an interesting connection between the SDF and one measure of how well we can expect to forecast certain time series. As we now discuss, this connection allows us to use an SDF estimate to assess the predictability of a time series.

Suppose that $\{X_t\}$ is a zero mean stationary process with SDF $S(\cdot)$. Suppose also that we know the "infinite past" of this process starting at time $t-1$, i.e., $X_{t-1}, X_{t-2}, \ldots$, and that we want to predict X_t. If we consider only predictors that are linear combinations of the infinite past, say, $\widetilde{X}_t = \sum_{j=1}^{\infty} \varphi_j X_{t-j}$, we can assess how well a particular set of coefficients $\{\varphi_j\}$ does by considering its associated MSE $E\{(X_t - \widetilde{X}_t)^2\}$. Let $\widehat{X}_t$ denote the predictor formed from the linear combination for which the MSE is minimized; $\widehat{X}_t$ is called the best linear predictor of X_t, given the infinite past. If $E\{(X_t - \widehat{X}_t)^2\} > 0$ and if $\{X_t\}$ is purely nondeterministic (as are all ARMA(p,q) processes), then the Wold decomposition theorem states that we can express the process as an infinite-order moving average process:

$$X_t = -\sum_{j=0}^{\infty} \theta_j \epsilon_{t-j},$$

where $\{\epsilon_t\}$ is a white noise process with zero mean and finite variance $\sigma_\epsilon^2 > 0$ (see, e.g., Brockwell and Davis, 1991, for details on purely nondeterministic processes and Wold's decomposition theorem). We then have $\widehat{X}_t = -\sum_{j=1}^{\infty} \theta_j \epsilon_{t-j}$, and the *innovation variance* is

$$E\{(X_t - \widehat{X}_t)^2\} = E\{\epsilon_t^2\} = \sigma_\epsilon^2. \tag{404a}$$

A theorem due to Szegő and Kolmogorov (see, e.g., Priestley, 1981, p. 741, or Brockwell and Davis, 1991, p. 191) states that

$$\sigma_\epsilon^2 = \frac{1}{\Delta_t} \exp\left(\Delta_t \int_{-f_\mathcal{N}}^{f_\mathcal{N}} \log(S(f))\,df\right), \tag{404b}$$

so that σ_ϵ^2 depends only on the SDF of the process. If we consider predicting X_t given a finite amount of prior data, the innovation variance is of interest because it can give us insight into how the quality of our prediction is affected by data limitations.

Based upon the connection between σ_ϵ^2 and the SDF and given the time series $X_0, X_1, \ldots, X_{N-1}$, Davis and Jones (1968) propose estimating the innovation variance using the periodogram $\widehat{S}^{(P)}(\cdot)$. Their periodogram-based estimator of σ_ϵ^2 is

$$\widehat{\sigma}_{(DJ)}^2 \stackrel{\text{def}}{=} \frac{1}{\Delta_t} \exp\left\{\frac{\Delta_t}{M} \sum_{k=1}^{M} \left[\log\left(\widehat{S}^{(P)}(f_k)\right) + \gamma\right]\right\}, \tag{404c}$$

where $M = \lfloor (N-1)/2 \rfloor$; $f_k = k/(N\,\Delta_t)$ is a Fourier frequency; and $\gamma \doteq 0.5772$ is Euler's constant. Since

$$E\{\log(\widehat{S}^{(P)}(f))\} \approx \log(S(f)) - \gamma \quad \text{for } 0 < f < f_\mathcal{N}$$

(see Equations (301a), (301b) and (302c)), we can regard the summation in Equation (404c) as being proportional to a bias-corrected Riemann sum approximation to the integral in Equation (404b). Jones (1976a) and Hannan and Nicholls (1977) argue for using a smoothed version of the periodogram in forming the innovation variance estimator. In particular the latter consider

$$\widehat{\sigma}_{(HN)}^2 \stackrel{\text{def}}{=} \frac{L}{\Delta_t} \exp\left\{\frac{\Delta_t}{M} \sum_{k=1}^{M} \left[\log\left(\frac{1}{L}\sum_{l=1}^{L} \widehat{S}^{(P)}(f_{(k-1)L+l})\right) - \psi(L)\right]\right\}, \tag{404d}$$

where now $M = \lfloor (N-1)/2L \rfloor$, and $\psi(\cdot)$ is the digamma function (the derivative of the log of the gamma function), which enters as a bias correction. Note that the smoothed periodogram is a special case of the discretely smoothed direct spectral estimator of Equation (246b). If $L = 1$, the estimator $\hat{\sigma}^2_{(\text{HN})}$ collapses to $\hat{\sigma}^2_{(\text{DJ})}$ since $\psi(1) = -\gamma$.

Noting the potential for significant bias in the periodogram, Pukkila and Nyquist (1985) advocate estimating σ^2_ϵ based on a direct spectral estimator $\hat{S}^{(\text{D})}(\cdot)$ employing a nonrectangular data taper. Their estimator takes the form

$$\hat{\sigma}^2_{(\text{PN})} \stackrel{\text{def}}{=} \frac{1}{\Delta_t} \exp\left\{\frac{\Delta_t}{M} \sum_{k=1}^{M} \left[\log\left(\hat{S}^{(\text{D})}(f_k)\right) + \gamma - \frac{\pi^2}{12M}\right]\right\}, \qquad (405a)$$

where here $M = \lfloor (N-1)/2 \rfloor$. The above is similar to $\hat{\sigma}^2_{(\text{DJ})}$ with $\hat{S}^{(\text{P})}(f_k)$ replaced by $\hat{S}^{(\text{D})}(f_k)$, but the new term $\pi^2/12M$ compensates for an additional source of bias discussed by Pukkila and Nyquist. They recommend using a trapezoidal data taper to form $\hat{S}^{(\text{D})}(\cdot)$. This taper is quite similar to the $p \times 100\%$ cosine taper of Equation (189b) if we set p to $\min\{1, 12.5/N\}$ to mimic the parameter setting for their taper – we use the cosine taper to formulate $\hat{\sigma}^2_{(\text{PN})}$ in what follows.

In keeping with our theme that multitapering is an improvement over direct spectral estimators (particularly the periodogram), let us consider a multitaper-based innovation variance estimator that is essentially a generalization of the single-taper Pukkila–Nyquist estimator:

$$\hat{\sigma}^2_{(\text{MT})} \stackrel{\text{def}}{=} \frac{K}{\Delta_t} \exp\left\{\frac{\Delta_t}{M} \sum_{k=1}^{M} \left[\log\left(\hat{S}^{(\text{MT})}(f_k)\right) - \psi(K_k) - \frac{\psi'(K)}{2M}\right]\right\}, \qquad (405b)$$

where K is the number of multitapers employed; $M = \lfloor N/2 \rfloor$; $\hat{S}^{(\text{MT})}(\cdot)$ is a sinusoidal multitaper estimator; $K_k = K$ for all k except when N is even and $k = M$, in which case $K_M = K/2$; and $\psi'(\cdot)$ is the trigamma function. Note that, in contrast to $\hat{\sigma}^2_{(\text{DJ})}$, $\hat{\sigma}^2_{(\text{HN})}$ and $\hat{\sigma}^2_{(\text{PN})}$, the multitaper-based estimator makes use of SDF estimates at the Nyquist frequency when N is even (simulation studies indicate this to be a valuable refinement when N is small; Walden, 1995, considers a Slepian-based version of $\hat{\sigma}^2_{(\text{MT})}$, but without the Nyquist frequency refinement). If we were to revert to using $M = \lfloor (N-1)/2 \rfloor$ and if we set $K = 1$, the estimator $\hat{\sigma}^2_{(\text{MT})}$ collapses to $\hat{\sigma}^2_{(\text{PN})}$ since $\psi(1) = -\gamma$ and $\psi'(1) = \pi^2/6$.

As examples, let us consider the AR(2) and AR(4) processes defined by Equations (34) and (35a). For an AR(p) process

$$X_t = \sum_{j=1}^{p} \phi_{p,j} X_{t-j} + \epsilon_t,$$

where $\{\epsilon_t\}$ is a white noise process with finite variance $\sigma^2_p > 0$, the best linear predictor of X_t given the infinite past – and its associated innovation variance – are

$$\widehat{X}_t = \sum_{j=1}^{p} \phi_{p,j} X_{t-j} \quad \text{and} \quad E\{(X_t - \widehat{X}_t)^2\} = E\{\epsilon_t^2\} = \sigma^2_p.$$

After generating 10,000 independent realizations of sample size $N = 1024$ from one of the two AR processes of interest using the procedures described in Exercises [597] and [11.1], we can compute 10,000 estimates of the innovation variance using the periodogram-based estimator $\hat{\sigma}^2_{(\text{DJ})}$ of Equation (404c) – call the nth such estimate $\hat{\sigma}^2_{(\text{DJ}),n}$. We can quantify how

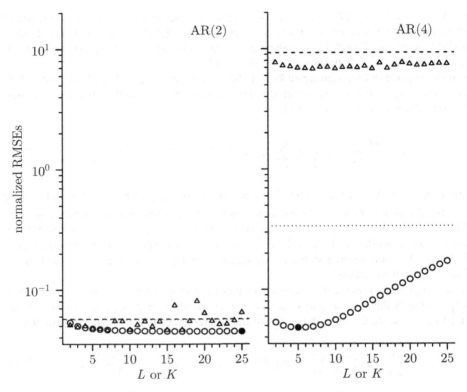

Figure 406 NRMSEs for estimators of innovation variance (see text for details).

well these estimates collectively agree with the true innovation variance by computing the normalized root mean-square error (NRMSE), namely,

$$\text{NRMSE} \stackrel{\text{def}}{=} \left[\frac{1}{10000} \sum_{n=1}^{10000} \left(\frac{\hat{\sigma}^2_{(\text{DJ}),n} - \sigma_p^2}{\sigma_p^2} \right)^2 \right]^{1/2}. \tag{406}$$

Similarly we can also the NRMSE criterion to evaluate estimates of the innovation variance based on the estimators $\hat{\sigma}^2_{(\text{HN})}$ with $L = 2, 3, \ldots, 64$, $\hat{\sigma}^2_{(\text{PN})}$ and, finally, $\hat{\sigma}^2_{(\text{MT})}$ with $K = 2, 3, \ldots, 64$.

Figure 406 summarizes the NRMSE evaluations for the AR(2) process (left-hand plot) and the AR(4) process (right-hand). Horizontal dashed lines show the NRMSEs for $\hat{\sigma}^2_{(\text{DJ})}$, and dotted lines, those for $\hat{\sigma}^2_{(\text{PN})}$ (in the AR(2) case, these NRMSEs are, respectively, 0.0572 and 0.0567, which are so close that the lines overlap). Triangles show the NRMSEs for $\hat{\sigma}^2_{(\text{HN})}$, $L = 2, 3, \ldots, 25$, and circles, for $\hat{\sigma}^2_{(\text{MT})}$, $K = 2, 3, \ldots, 25$ (the corresponding values for $26, 27, \ldots, 64$ are not shown – the NRMSEs for $\hat{\sigma}^2_{(\text{MT})}$ increase monotonically, while those for $\hat{\sigma}^2_{(\text{HN})}$ have an erratic pattern that is generally increasing and always above the corresponding values for $\hat{\sigma}^2_{(\text{MT})}$). For both processes, the smallest NRMSE is associated with a multitaper-based estimator $\hat{\sigma}^2_{(\text{MT})}$ ($K = 25$ in the AR(2) case and $K = 5$ in the AR(4) – these are indicated by filled-in circles).

For the AR(2) process, the periodogram is essentially bias free as an estimator of the SDF for a sample size of 64 and is even more so for the size ($N = 1024$) considered here (Figure 177(b) shows that the expected value of the periodogram is in close agreement with

8.7 Improving Periodogram-Based Methodology via Multitapering

the true SDF when $N = 64$). Nonetheless, as the left-hand plot of Figure 406 shows, while use of a single taper does not give a significant improvement over the periodogram, the estimator $\hat{\sigma}^2_{(\text{HN})}$ based on the smoothed periodogram does for carefully selected L, but multitapering leads to slightly better NRMSEs (and less care is needed in selecting K). In particular, the NRMSEs for $\hat{\sigma}^2_{(\text{DJ})}$ and $\hat{\sigma}^2_{(\text{PN})}$ (indicated by the co-located horizontal lines) are larger than the best NRMSEs for $\hat{\sigma}^2_{(\text{HN})}$ and $\hat{\sigma}^2_{(\text{MT})}$ (the former occurs when $L = 7$ and is equal to 0.046, which is quite close to the normalized NRMSE of 0.045 for $\hat{\sigma}^2_{(\text{MT})}$ with $K = 25$). The NRMSEs for $\hat{\sigma}^2_{(\text{MT})}$ when $K = 11, 12, \ldots, 24$ do not differ much from the $K = 25$ case (none of them are greater than by a factor of 1.01).

We might expect the AR(4) process to be problematic for periodogram-based estimators of the innovation variance, given that, in contrast to the AR(2) process, the expected value of the periodogram has substantial bias at high frequencies when $N = 1024$ (see Figure 178(d)). Indeed the right-hand plot of Figure 406 shows that the periodogram-based estimators $\hat{\sigma}^2_{(\text{DJ})}$ and $\hat{\sigma}^2_{(\text{HN})}$ have NRMSEs that are about two orders of magnitude larger than the one for the best multitaper estimate at $K = 5$. This NRMSE is also significantly less than the one associated with the single-taper based $\hat{\sigma}^2_{(\text{PN})}$ (indicated by the dotted line). Thus, for the two AR processes considered here, the multitaper estimator either compares favorably to periodogram-based estimators or offers a substantial improvement (Exercise [8.18] invites the reader to look at two additional AR(2) processes).

Estimation of the Exponent of a Power-Law Process

Time series in fields as diverse as metrology, astronomy, geophysics and econometrics often appear to have SDFs $S(\cdot)$ well approximated by Cf^α over some interval of positive frequencies, where $C > 0$ and $\alpha < 0$ are finite constants; i.e., $S(f)$ increases as f decreases to zero. The pattern of increase is highly structured: a plot of $\log(S(f))$ versus $\log(f)$ is linear with a slope of α. Such SDFs are said to obey a power law with exponent α. If linearity persists as $f \to 0$ and if $-1 < \alpha < 0$, the associated processes are said to have long-range dependence or long memory (Beran, 1994). Examples of time series for which power-law processes have been entertained as models include fractional frequency deviates in atomic clocks (see Figure 326(b) and the discussion surrounding it), X-ray variability of galaxies (McHardy and Czerny, 1987), fluctuations in the earth's rate of rotation (Munk and MacDonald, 1975), impedances of rock layers derived from boreholes (Walden and Hosken, 1985; Todoeschuck and Jensen, 1988; Kerner and Harris, 1994), and stock volatility (Ray and Tsay, 2000; Wright, 2002).

Consider a time series $X_0, X_1, \ldots, X_{N-1}$ that is a portion of a process with an SDF that, to a good approximation, obeys a power law Cf^α over $f \in (0, f_C]$, where $f_C \leq f_N$ is a high frequency cutoff. The task at hand is to estimate α given the time series. A stationary process that conveniently fits into this framework is a fractionally differenced (FD) process (Granger and Joyeux, 1980; Hosking, 1981). The SDF for such a process takes the form

$$S(f) = \frac{\sigma_\epsilon^2 \Delta_t}{[4\sin^2(\pi f \Delta_t)]^\delta}, \tag{407}$$

where $0 < \sigma_\epsilon^2 < \infty$. The process has long-range dependence when $0 < \delta < 1/2$ (Exercise [5.4a] says that the squared gain function associated with differencing a stationary process is given by $4\sin^2(\pi f \Delta_t)$, which motivates using "fractional differencing" to describe the above SDF). The small angle approximation $\sin(x) \approx x$ suggests that, at low frequencies,

$$S(f) \approx \frac{\sigma_\epsilon^2 \Delta_t}{[2\pi f \Delta_t]^{2\delta}} = C|f|^{-2\delta}, \quad \text{where } C \stackrel{\text{def}}{=} \frac{\sigma_\epsilon^2 \Delta_t}{[2\pi \Delta_t]^{2\delta}};$$

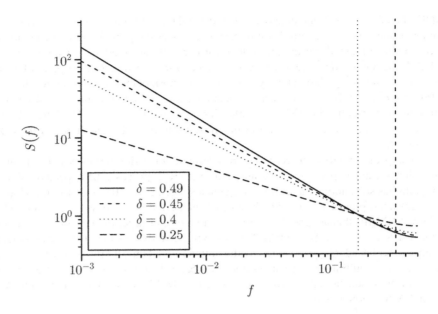

Figure 408 SDFs for fractionally differenced processes with $\delta = 0.25, 0.4, 0.45$ and 0.49 versus $f \in [1/10^3, 1/2]$ (here we assume $\Delta_t = 1$ so that $f_\mathcal{N} = 1/2$). The vertical dotted and dashed lines mark the frequencies $f = 1/6$ and $f_\mathcal{C} = 1/3$.

i.e., $S(\cdot)$ approximately obeys a power law with an exponent dictated by $\alpha = -2\delta$ (note that $-1 < \alpha < 0$, as required for long-range dependence). Figure 408 shows $S(f)$ versus f on a log/log scale for four FD processes with different settings for δ but with a common setting of unity for both Δ_t and σ_ϵ^2. The vertical dotted line marks $f = 1/6$, at which the SDFs are equal to unity no matter what δ is set to (this follows from the fact that $\sin(\pi/6) = 1/2$). This figure suggests that $\log(S(f))$ versus $\log(f)$ varies linearly to a good approximation for $f \in (0, 1/6]$ and – to a somewhat lesser extent – for $f \in (0, 1/3]$ (the vertical dashed line marks $f = 1/3$).

Initial efforts on estimating the power-law exponent $\alpha = -2\delta$ in setups similar to the above focused on using the log of the periodogram as the response in a linear regression model (the following is based on material in Mohr, 1981; Geweke and Porter-Hudak, 1983; Kasyap and Eom, 1988; Yajima, 1989; Percival, 1991; McCoy et al., 1998; and Weisberg, 2014). The key assumption behind this approach is that the periodogram $\hat{S}^{(\mathrm{P})}(f_k)$ has a distribution well approximated by that of $S(f_k)\chi_2^2/2$, where $f_k = k/(N\Delta_t)$ is a Fourier frequency satisfying $0 < f_k \leq f_\mathcal{C}$, and χ_2^2 is a chi-square RV with two degrees of freedom (see Equation (203e)). Using $S(f) \approx Cf^\alpha$, we have

$$\log\left(\hat{S}^{(\mathrm{P})}(f_k)\right) \stackrel{\mathrm{d}}{=} \log(C/2) + \alpha \log(f_k) + \log(\chi_2^2) \tag{408a}$$

approximately. Equation (302b) tells us that $E\{\log(\chi_2^2)\} = \log(2) - \gamma$, where $\gamma \doteq 0.5772$ is Euler's constant. Letting $x_k \stackrel{\text{def}}{=} \log(f_k)$, we can entertain – as a good approximation – a simple linear regression model, namely,

$$Y_k^{(\mathrm{P})} \stackrel{\text{def}}{=} \log\left(\hat{S}^{(\mathrm{P})}(f_k)\right) \approx \beta^{(\mathrm{P})} + \alpha x_k + \varepsilon_k^{(\mathrm{P})}, \quad k = 1, 2, \ldots, M, \tag{408b}$$

where $\beta^{(\mathrm{P})} \stackrel{\text{def}}{=} \log(C) - \gamma$; the error term is such that $\varepsilon_k^{(\mathrm{P})} \stackrel{\mathrm{d}}{=} \log(\chi_2^2) - \log(2) + \gamma$ and hence $E\{\varepsilon_k^{(\mathrm{P})}\} = 0$; and M is the largest integer such that $f_M \leq f_\mathcal{C}$ (motivated by Figure 408, we set

8.7 Improving Periodogram-Based Methodology via Multitapering

the cutoff $f_{\mathcal{C}}$ to be $1/(3\,\Delta_{\mathrm{t}})$). The periodogram-based ordinary least squares (OLS) estimator of the power-law exponent α is

$$\hat{\alpha}^{(\mathrm{P})} = \frac{\sum_{k=1}^{M} x_k^{(\mathrm{P})} Y_k^{(\mathrm{P})}}{\sum_{k=1}^{M} (x_k^{(\mathrm{P})})^2}, \quad \text{where } x_k^{(\mathrm{P})} \stackrel{\text{def}}{=} x_k - \frac{1}{M}\sum_{k=1}^{M} x_k. \quad (409\mathrm{a})$$

Another look at Equation (302b) says that $\mathrm{var}\,\{\log{(\chi_2^2)}\} = \pi^2/6$ and hence that $\mathrm{var}\,\{\varepsilon_k^{(\mathrm{P})}\} = \pi^2/6$ also. Under the additional assumption that distinct errors $\varepsilon_j^{(\mathrm{P})}$ and $\varepsilon_k^{(\mathrm{P})}$ are uncorrelated, standard OLS theory says

$$\mathrm{var}\,\{\hat{\alpha}^{(\mathrm{P})}\} = \frac{\mathrm{var}\,\{\varepsilon_k^{(\mathrm{P})}\}}{\sum_{k=1}^{M} (x_k^{(\mathrm{P})})^2} = \frac{\pi^2}{6\sum_{k=1}^{M} (x_k^{(\mathrm{P})})^2} \quad (409\mathrm{b})$$

(the assumption of uncorrelatedness is similar in spirit to the approximate uncorrelatedness of the periodogram over the grid of Fourier frequencies stated in Equation (204a), but ideally uncorrelatedness would fall out theoretically from the FD process itself under suitable conditions; the lack of supporting theory calls into question the validity of the above expression for $\mathrm{var}\,\{\hat{\alpha}^{(\mathrm{P})}\}$, but the simulation study presented below suggests it is reasonably accurate). Remarkably there is no need to estimate $\mathrm{var}\,\{\varepsilon_k^{(\mathrm{P})}\}$ using residuals as is usually the case in a simple linear regression setting.

Let us now consider a second way to estimate α that is similar in spirit to the above, but for which we replace the periodogram $\hat{S}^{(\mathrm{P})}(f_k)$ with a multitaper estimator $\hat{S}^{(\mathrm{MT})}(f_k)$ using K sinusoidal tapers (McCoy et al., 1998). Equation (354f) says that the distribution of $\hat{S}^{(\mathrm{MT})}(f_k)$ is approximately that of $S(f_k)\chi_{2K}^2/2K$. The analog of Equation (408a) is

$$\log\left(\hat{S}^{(\mathrm{MT})}(f_k)\right) \stackrel{\mathrm{d}}{=} \log{(C/2K)} + \alpha\log{(f_k)} + \log{(\chi_{2K}^2)}. \quad (409\mathrm{c})$$

Equation (302a) says that $E\{\log{(\chi_{2K}^2)}\} = \log{(2)} + \psi(K)$, where $\psi(\cdot)$ is the digamma function. The multitaper equivalent of the linear regression of Equation (408b) is

$$Y_k^{(\mathrm{MT})} \stackrel{\text{def}}{=} \log\left(\hat{S}^{(\mathrm{MT})}(f_k)\right) \approx \beta^{(\mathrm{MT})} + \alpha x_k + \varepsilon_k^{(\mathrm{MT})}, \quad k = M_0, M_0+1, \ldots, M, \quad (409\mathrm{d})$$

where $\beta^{(\mathrm{MT})} \stackrel{\text{def}}{=} \log{(C/K)} + \psi(K)$; $\varepsilon_k^{(\mathrm{MT})} \stackrel{\mathrm{d}}{=} \log{(\chi_{2K}^2)} - \log{(2)} - \psi(K)$, which results in the errors $\varepsilon_k^{(\mathrm{MT})}$ having zero mean; $M_0 = \lceil(K+1)/2\rceil$, where $\lceil x \rceil$ denotes the smallest integer greater than or equal to x; and M is the same as in Equation (408b). Note that $M_0 \geq 2$ when $K \geq 2$, which means that some terms associated with the lowest nonzero Fourier frequencies are excluded in the above model whereas they are included in the periodogram-based model. The trimming of these terms is needed because Equation (409c) is a reasonable approximation only if f_k is sufficiently far away from zero frequency. As noted in C&E [2] for Section 8.1 we can regard any frequency satisfying the inequality $f_k \geq B_{\overline{\mathcal{H}}}/2$ as being far enough away, where $B_{\overline{\mathcal{H}}}$ is the bandwidth measure of Equation (353e). A simple approximation to $B_{\overline{\mathcal{H}}}$ for the sinusoidal multitapers is $(K+1)/[(N+1)\,\Delta_{\mathrm{t}}]$ (see C&E [1] for Section 8.6). Use of this approximation leads to the formula stated for M_0. The multitaper-based OLS estimator of α parallels the periodogram-based estimator of Equation (409a):

$$\hat{\alpha}^{(\mathrm{MT})} = \frac{\sum_{k=M_0}^{M} x_k^{(\mathrm{MT})} Y_k^{(\mathrm{MT})}}{\sum_{k=M_0}^{M} (x_k^{(\mathrm{MT})})^2}, \quad \text{where } x_k^{(\mathrm{MT})} \stackrel{\text{def}}{=} x_k - \frac{1}{M - M_0 + 1}\sum_{k=M_0}^{M} x_k. \quad (409\mathrm{e})$$

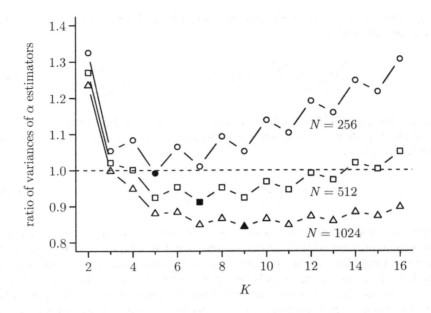

Figure 410 Variance ratio var $\{\hat{\alpha}^{(\mathrm{MT})}\}/$ var $\{\hat{\alpha}^{(\mathrm{P})}\}$ versus number of tapers K used to form the multitaper-based estimator $\hat{\alpha}^{(\mathrm{MT})}$ of the power-law exponent α (see text for details).

To get an expression for var $\{\hat{\alpha}^{(\mathrm{MT})}\}$, we first note that var $\{\varepsilon_k^{(\mathrm{MT})}\}$ is the same as var $\{\log(\chi_{2K}^2)\}$, which Equation (302a) says is equal to $\psi'(K)$, where $\psi'(\cdot)$ is the trigamma function. In contrast to the periodogram, we cannot claim that the errors in the multitaper case are approximately pairwise uncorrelated across the Fourier frequencies. Walden et al. (1998) argue that, for sinusoidal tapers, a convenient and reasonably accurate model for the covariance structure of the errors is

$$\operatorname{cov}\{\varepsilon_j^{(\mathrm{MT})}, \varepsilon_k^{(\mathrm{MT})}\} = \psi'(K)\Delta_{j-k}, \quad \text{where } \Delta_\tau \stackrel{\text{def}}{=} \begin{cases} \left(1 - \frac{|\tau|}{K+1}\right), & |\tau| \leq K; \\ 0, & \text{otherwise.} \end{cases} \quad (410a)$$

Exercise [8.19] uses the above along with standard theory from linear regression analysis to obtain, as an approximation,

$$\operatorname{var}\{\hat{\alpha}^{(\mathrm{MT})}\} = \psi'(K) \frac{\sum_{j=0}^{M-M_0} x_{M_0+j}^{(\mathrm{MT})} \sum_{k=\max\{-j,-K\}}^{\min\{K, M-M_0-j\}} x_{k+M_0+j}^{(\mathrm{MT})} \Delta_k}{\left[\sum_{j=M_0}^{M} \left(x_j^{(\mathrm{MT})}\right)^2\right]^2}. \quad (410b)$$

Figure 410 plots the ratio var $\{\hat{\alpha}^{(\mathrm{MT})}\}/$ var $\{\hat{\alpha}^{(\mathrm{P})}\}$ versus number of tapers $K = 2, 3, \ldots, 16$ used to form the multitaper-based estimator $\hat{\alpha}^{(\mathrm{MT})}$. This and the periodogram-based estimator $\hat{\alpha}^{(\mathrm{P})}$ are both based upon a time series $X_0, X_1, \ldots, X_{N-1}$ possessing an SDF that is assumed to be well approximated by Cf^α for $f \in (0, 1/(3\Delta_t)]$. We consider three settings for N: 256 (circles), 512 (squares) and 1024 (triangles). A filled-in symbol indicates the setting of K such that the ratio is minimized for a given N. When the ratio falls below unity (indicated by the horizontal dashed line), the multitaper-based estimator $\hat{\alpha}^{(\mathrm{MT})}$ in theory outperforms $\hat{\alpha}^{(\mathrm{P})}$. For all three sample sizes, the minimum ratio is below unity (but just barely so for $N = 256$). The relative performance of $\hat{\alpha}^{(\mathrm{MT})}$ improves as the sample size N increases.

We must regard the variances for $\hat{\alpha}^{(\mathrm{P})}$ and $\hat{\alpha}^{(\mathrm{MT})}$ stated in Equations (409b) and (410b) as approximations since both equations result from certain simplifying assumptions. We can

8.7 Improving Periodogram-Based Methodology via Multitapering

		α			
Estimator	N	-0.5	-0.8	-0.9	-0.98
Periodogram	256	1.00	1.01	1.02	1.03
Periodogram	512	1.02	1.03	1.04	1.06
Periodogram	1024	1.01	1.03	1.04	1.05
Multitaper, $K = 5$	256	0.84	0.84	0.86	1.19
Multitaper, $K = 7$	512	0.89	0.89	0.90	1.15
Multitaper, $K = 9$	1024	0.88	0.87	0.88	1.07

Table 411a Ratio of $\widehat{\text{var}} \{\hat{\alpha}_n^{(\text{P})}\}$ to var $\{\hat{\alpha}^{(\text{P})}\}$ of Equation (409b) (top three rows) for 12 combinations of sample size N paired with power-law exponent α, followed by ratio of $\widehat{\text{var}} \{\hat{\alpha}_n^{(\text{MT})}\}$ to var $\{\hat{\alpha}^{(\text{MT})}\}$ of Equation (410b) (bottom three rows; see text for details).

	α			
N	-0.5	-0.8	-0.9	-0.98
256	0.84	0.82	0.84	1.14
512	0.81	0.79	0.80	1.01
1024	0.76	0.75	0.75	0.89

Table 411b Ratio of $\widehat{\text{var}} \{\hat{\alpha}_n^{(\text{MT})}\}$ to $\widehat{\text{var}} \{\hat{\alpha}_n^{(\text{P})}\}$ for three choices of sample size N paired with four choices for power-law exponent α. A value less than unity indicates that the multitaper-based estimator outperforms the one based on the periodogram (see text for details).

check their validity through a simulation study that starts by generating 10,000 independent realizations – each with the same sample size N – from an FD process with a particular setting of the long-range dependence parameter δ. We set the sample size to either $N = 256$, 512 or 1024 (the same as used in Figure 410). The settings for δ are the same as in Figure 408, namely, 0.25, 0.4, 0.45 and 0.49, which correspond to power-law exponents $\alpha = -0.5, -0.8, -0.9$ and -0.98. For a given combination of N and α, we compute 10,000 periodogram-based and multitaper-based estimates of α. The number of tapers K used for the multitaper estimates is in keeping with Figure 410 (5 for $N = 256$, 7 for $N = 512$ and 9 for $N = 1024$). Let $\hat{\alpha}_n^{(\text{P})}$ and $\hat{\alpha}_n^{(\text{MT})}$, $n = 1, 2, \ldots, 10000$, denote the estimates, and let $\widehat{\text{var}} \{\hat{\alpha}_n^{(\text{P})}\}$ and $\widehat{\text{var}} \{\hat{\alpha}_n^{(\text{MT})}\}$ denote their sample variances.

Tables 411a and 411b summarize the results of the simulation study. The first three rows of Table 411a show the ratio $\widehat{\text{var}} \{\hat{\alpha}_n^{(\text{P})}\} / \text{var} \{\hat{\alpha}^{(\text{P})}\}$ for the 12 combinations of N and α. All 12 displayed ratios are within 6% of unity, which indicates that the expression for var $\{\hat{\alpha}^{(\text{P})}\}$ given by Equation (409b) is a good approximation (it does tend to underestimate the actual variance somewhat). The bottom three rows show similar ratios $\widehat{\text{var}} \{\hat{\alpha}_n^{(\text{MT})}\} / \text{var} \{\hat{\alpha}^{(\text{MT})}\}$ for the multitaper case. The approximation of Equation (410b) overestimates the true variance (by 10% up to 16%) with the exception of the case $\alpha = -0.98$, for which the approximation is an underestimate that improves as the sample size increases. These results suggest that Equations (409b) and (410b) are reasonable approximations to the variances of the OLS estimators.

Table 411b shows the ratio $\widehat{\text{var}} \{\hat{\alpha}_n^{(\text{MT})}\} / \widehat{\text{var}} \{\hat{\alpha}_n^{(\text{P})}\}$ for the various combinations of sample size N and power-law exponent α. A value less than unity indicates that $\hat{\alpha}^{(\text{MT})}$ outperforms $\hat{\alpha}^{(\text{P})}$ as an estimator of α. With the exception of $\alpha = -0.98$, the multitaper-based estimator is the better estimator, and its performance improves noticeably with increasing N, offering a decrease in variance of about 16% over the periodogram estimator when $N = 256$ and 25% when $N = 1024$. For the exceptional case, the performance of $\hat{\alpha}^{(\text{MT})}$ still improves as N increases, but it is markedly worse than $\hat{\alpha}^{(\text{P})}$ for the smallest sample size. Thus, given enough data, we can improve upon methodology involving the periodogram by replacing it with an appropriate multitaper estimator. (Exercise [8.20] invites the reader to replicate this simulation study and to further compare the two estimators by considering their biases and MSEs rather than just their variances.)

Comments and Extensions to Section 8.7

[1] Given that the regression model of Equation (409d) for estimating the power-law exponent α involves correlated errors whose covariance structure is approximated by Equation (410a), we can in theory improve upon the OLS estimator $\hat{\alpha}^{(\text{MT})}$ by using a generalized least squares estimator (see, e.g., Priestley, 1981, equation (5.2.64)); however, in the simulation study reported here, this computationally more complex estimator performs only marginally better, in keeping with a similar finding in McCoy et al. (1998).

[2] Estimation of the log SDF is another example where multitapering improves upon methodology originally based on the periodogram. Moulin (1994) and Gao (1993, 1997) estimate $\log(S(\cdot))$ by applying a discrete wavelet transform (DWT) to the log periodogram $\log(\hat{S}^{(\text{P})}(\cdot))$, thresholding or shrinking certain of the DWT coefficients and then applying an inverse DWT to get the desired estimator. This requires an adaptation of standard Gaussian-based thresholding/shrinkage because, ignoring zero and Nyquist frequencies, the log periodogram is related to a $\log(\chi_2^2)$ distribution, which is markedly non-Gaussian; however, while this adaptation addresses non-Gaussianity, the resulting estimator offers no protection against significant bias in the periodogram. Walden et al. (1998) investigate replacing $\log(\hat{S}^{(\text{P})}(\cdot))$ with $\log(\hat{S}^{(\text{MT})}(\cdot))$, which has a distribution related to a $\log(\chi_{2K}^2)$ distribution (see also section 10.6 of Percival and Walden, 2000). If a modest number of tapers is used, say $K \geq 5$, the associated $\log(\chi_{2K}^2)$ distribution is well approximated by a Gaussian distribution, so standard Gaussian-based thresholding/shrinkage can be evoked; more importantly, the multitaper-based estimator of $\log(S(\cdot))$ inherits the favorable first-moment properties of $\hat{S}^{(\text{MT})}(\cdot)$. Simulation studies using the same stationary processes considered in Moulin (1994) and Gao (1993, 1997) indicate that multitaper-based estimators outperform periodogram-based estimators even on time series with sample sizes such that the periodogram does not suffer from significant bias.

Riedel and Sidorenko (1996) also consider multitaper-based estimation of the log SDF. Their proposed estimator uses a smoothing window to smooth $\log(\hat{S}^{(\text{MT})}(\cdot))$ across frequencies. Because of potential significant bias in the periodogram, they opt to compare their proposed estimator to one that smooths $\log(\hat{S}^{(\text{D})}(\cdot))$, where $\hat{S}^{(\text{D})}(\cdot)$ is a direct SDF estimator other than the periodogram (cf. the discussion surrounding Equation (303a)). They find their multitaper-based estimator to be superior. They also find their estimator to be superior to one in which $\hat{S}^{(\text{MT})}(\cdot)$ is first smoothed across frequencies and then subjected to a log transform. Their finding illustrates the value of replacing the periodogram and other direct spectral estimators not by an estimator with greatly reduced variance, but rather by one with a modest reduction in variance – this is what multitapering accomplishes with a modest number of tapers.

8.8 Welch's Overlapped Segment Averaging (WOSA)

In discussing the rationale behind the Bartlett lag window of Equation (269a), we introduced the idea of breaking up a time series into a number of contiguous nonoverlapping blocks, computing a periodogram based on the data in each block alone, and then averaging the individual periodograms together to form an overall spectral estimate. The resulting estimator does not have the form of a lag window spectral estimator (although, as we have seen, it does approximately correspond to a Bartlett lag window estimator), but, because it is the average of several periodograms based upon different blocks of data, its variance is smaller than that

8.8 Welch's Overlapped Segment Averaging (WOSA)

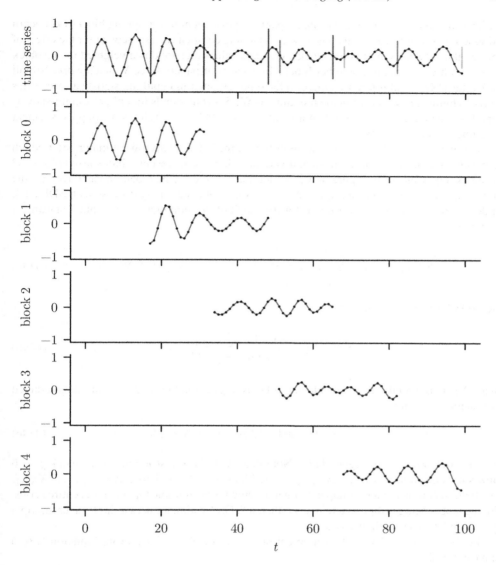

Figure 413 Breaking up a time series into overlapping blocks. The top plot shows a time series with 100 points (dots connected by lines). The ten vertical lines mark five blocks of $N_S = 32$ points each – lines of the same length and shading bracket one block. Each block is plotted separately in the bottom five plots. The blocks consist of points 0 to 31, 17 to 48, 34 to 65, 51 to 82 and 68 to 99. Overlapping blocks have 15 common points, yielding a $15/32 = 47\%$ overlap percentage.

of the individual periodograms. Hence block averaging is a viable alternative to lag windows as a way of controlling the inherent variability in a periodogram.

Welch (1967) further developed the idea of block averaging by introducing two important modifications. First, he advocated the use of a data taper on each block to reduce potential bias due to leakage in the periodogram; i.e., the spectral estimator for each block is now a direct spectral estimator as defined by Equation (186b). Second, he showed that allowing the blocks to overlap (see Figure 413) can produce a spectral estimator with better variance properties than one using contiguous nonoverlapping blocks. This reduction in variance occurs both because overlapping recovers some of the information concerning the autocovariance sequence

(ACVS) contained in pairs of data values spanning adjacent nonoverlapping blocks (a concern addressed by Bartlett, 1950), and because overlapping compensates somewhat for the effect of tapering the data in the individual blocks; i.e., data values that are downweighted in one block can have a higher weight in another block. In a series of papers since Welch's work, Nuttall and Carter (1982, and references therein) have investigated in detail the statistical properties of this technique, now known in the literature as WOSA (this stands for either "weighted overlapped segment averaging" – see Nuttall and Carter, 1982 – or "Welch's overlapped segment averaging" – see Carter, 1987).

Let us now formulate the precise definition of the WOSA spectral estimator, after which we develop expressions for its mean and variance. Let N_S represent a block size ($N_S = 32$ for the example shown in Figure 413), and let $h_0, \ldots, h_{N_S-1}$ be a data taper. Given a time series that is a realization of a portion $X_0, \ldots, X_{N-1}$ of a stationary process with SDF $S(\cdot)$, we define the direct spectral estimator for the block of N_S contiguous data values starting at index l as

$$\hat{S}_l^{(\text{D})}(f) = \Delta_t \left| \sum_{t=0}^{N_S-1} h_t X_{t+l} e^{-i2\pi f t \Delta_t} \right|^2, \qquad 0 \leq l \leq N - N_S. \tag{414a}$$

The WOSA spectral estimator is

$$\hat{S}^{(\text{WOSA})}(f) \stackrel{\text{def}}{=} \frac{1}{N_B} \sum_{j=0}^{N_B-1} \hat{S}_{jn}^{(\text{D})}(f), \tag{414b}$$

where N_B is the total number of blocks to be averaged together and n is an integer-valued shift factor satisfying

$$0 < n \leq N_S \quad \text{and} \quad n(N_B - 1) = N - N_S \tag{414c}$$

($n = 17$ and $N_B = 5$ in Figure 413). Note that, with these restrictions on n, block $j = 0$ utilizes the data values $X_0, \ldots, X_{N_S-1}$, while block $j = N_B - 1$ uses $X_{N-N_S}, \ldots, X_{N-1}$. Note that, in contrast to a multitaper estimator, which uses multiple tapers on the entire series, with the tapers being orthogonal to one another, a WOSA estimator uses a single taper multiple times on subseries of a time series.

Let us now consider the first-moment properties of $\hat{S}^{(\text{WOSA})}(\cdot)$. From Equation (186e) we have, for all j,

$$E\{\hat{S}_{jn}^{(\text{D})}(f)\} = \int_{-f_\mathcal{N}}^{f_\mathcal{N}} \mathcal{H}(f - f') S(f') \, df', \tag{414d}$$

where $\mathcal{H}(\cdot)$ is the spectral window corresponding to $h_0, \ldots, h_{N_S-1}$. It follows immediately that

$$E\{\hat{S}^{(\text{WOSA})}(f)\} = \int_{-f_\mathcal{N}}^{f_\mathcal{N}} \mathcal{H}(f - f') S(f') \, df' \tag{414e}$$

also. Note that the first-moment properties of $\hat{S}^{(\text{WOSA})}(\cdot)$ depend just on the block size N_S, the data taper $\{h_t\}$ and the true SDF $S(\cdot)$ – they do *not* depend on the total length of the time series N, the total number of blocks N_B or the shift factor n. This points out a potential pitfall in using WOSA, namely, that we must make sure that N_S is large enough so that $E\{\hat{S}^{(\text{WOSA})}(f)\} \approx S(f)$ for all f of interest.

In accordance with our standard practice of taking the effective bandwidth of a spectral estimator to be the autocorrelation width of its spectral window, it follows that the appropriate

8.8 Welch's Overlapped Segment Averaging (WOSA)

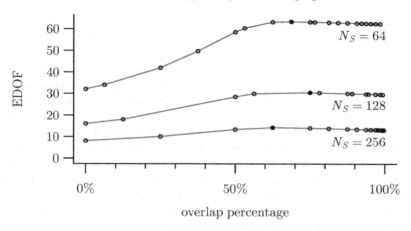

Figure 415 Equivalent degrees of freedom ν versus percentage of block overlap for sample size $N = 1024$, a Hanning data taper and block sizes $N_S = 64$, 128 and 256 (top, middle and bottom connected lines, respectively). The maximum value that ν achieves for each N_S is marked with a filled circle.

measure for $\hat{S}^{(\text{WOSA})}(\cdot)$ is the same as that for any of the constituent direct spectral estimators $\hat{S}^{(\text{D})}_{jn}(\cdot)$, namely, $B_{\mathcal{H}}$ of Equation (194).

Next, we consider the variance of $\hat{S}^{(\text{WOSA})}(f)$ under the assumption that this estimator is approximately unbiased. Since this estimator is the average of direct spectral estimators, we can obtain its variance based upon an application of Exercise [2.2]:

$$\text{var}\{\hat{S}^{(\text{WOSA})}(f)\} = \frac{1}{N_B^2} \sum_{j=0}^{N_B-1} \text{var}\{\hat{S}^{(\text{D})}_{jn}(f)\} + \frac{2}{N_B^2} \sum_{j<k} \text{cov}\{\hat{S}^{(\text{D})}_{jn}(f), \hat{S}^{(\text{D})}_{kn}(f)\}. \quad (415a)$$

For $0 < f < f_{\mathcal{N}}$, Equation (204c) tells us that, for all j, $\text{var}\{\hat{S}^{(\text{D})}_{jn}(f)\} \approx S^2(f)$. Under the assumption that $S(\cdot)$ is locally constant about f and that f is not too close to 0 or the Nyquist frequency, the result of Exercise [8.26] says that

$$\text{cov}\{\hat{S}^{(\text{D})}_{jn}(f), \hat{S}^{(\text{D})}_{kn}(f)\} \approx S^2(f) \left| \sum_{t=0}^{N_S-1} h_t h_{t+|j-k|n} \right|^2, \quad (415b)$$

where we define h_t to be zero for $t \geq N_S$ (see also Welch, 1967, or Thomson, 1977, section 3.3). Note that, if the jth and kth blocks have no data values in common, the summation above is identically zero, and hence so is the covariance. We thus have

$$\text{var}\{\hat{S}^{(\text{WOSA})}(f)\} \approx \frac{S^2(f)}{N_B} \left(1 + \frac{2}{N_B} \sum_{j<k} \left| \sum_{t=0}^{N_S-1} h_t h_{t+|j-k|n} \right|^2 \right)$$

$$= \frac{S^2(f)}{N_B} \left(1 + 2 \sum_{m=1}^{N_B-1} \left(1 - \frac{m}{N_B}\right) \left| \sum_{t=0}^{N_S-1} h_t h_{t+mn} \right|^2 \right). \quad (415c)$$

From an argument similar to that leading to Equation (264a), the EDOFs ν for $\hat{S}^{(\text{WOSA})}(f)$ are

$$\nu = \frac{2\left(E\{\hat{S}^{(\text{WOSA})}(f)\}\right)^2}{\text{var}\{\hat{S}^{(\text{WOSA})}(f)\}} \approx \frac{2N_B}{1 + 2 \sum_{m=1}^{N_B-1} \left(1 - \frac{m}{N_B}\right) \left| \sum_{t=0}^{N_S-1} h_t h_{t+mn} \right|^2}. \quad (415d)$$

As an example, suppose we have a time series with sample size $N = 1024$ that we want to break up into blocks of size N_S equal to either 64, 128 or 256 and that we plan to use the Hanning data taper on each block, i.e., h_t of Equation (189b) with N_S substituted for N. We can use Equation (415d) to compute approximately the EDOFs ν for $\hat{S}^{(\text{WOSA})}(f)$ for all acceptable shift factors n ranging from N_S (i.e., nonoverlapping blocks) down to 1 (i.e., maximally overlapped blocks, some of which have $N_S - 1$ data values in common). Figure 415 shows this approximation to ν versus the percentage of overlap $(1 - n/N_S) \times 100\%$ for block sizes $N_S = 64$ (top curve), 128 (middle) and 256 (bottom). For a fixed overlap percentage, ν is inversely proportional to N_S; however, in practical situations, we cannot just decrease the block size arbitrarily to increase ν because typically the bias in $\hat{S}^{(\text{WOSA})}(f)$ increases as N_S decreases. Note also that ν increases as the overlap percentage increases from 0% on up to about 70%, after which it decreases slightly (this decrease is counterintuitive and is possibly an artifact arising because Equation (415d) is based upon an approximation to $\text{var}\{\hat{S}^{(\text{WOSA})}(f)\}$). For all three choices of N_S, the value of ν at 50% overlap is within 10% of its maximum value. An operationally convenient recommendation in the engineering literature is to use a 50% overlap with the Hanning data taper, i.e., to set $n = \lfloor N_S/2 \rfloor$. With this substitution, Equation (415d) becomes

$$\nu \approx \frac{2N_B}{1 + 2(1 - 1/N_B)\left|\sum_{t=0}^{N_S/2-1} h_t h_{t+N_S/2}\right|^2} \approx \frac{36 N_B^2}{19 N_B - 1} \approx 1.89\left(\frac{2N}{N_S} - 1\right), \quad (416)$$

the verification of which is the subject of Exercise [8.27]. For a data taper more concentrated than the Hanning taper, an overlap percentage greater than 50% is needed to obtain ν to within 10% of its maximum value. For example, the appropriate overlap factor for the $NW = 4$ Slepian data taper is 65% (Thomson, 1977).

As a first example, let us look at WOSA spectral estimates for the AR(2) time series of Figure 34(a). Figure 417 shows four such estimates using a Hanning data taper and 50% overlap with block sizes of $N_S = 4, 16, 32$ and 64 (thick curves) along with the true SDF (thin curves). These estimates are remarkably similar to the four Parzen lag window estimates in Figure 294 with lag window parameters $m = 4, 16, 29$ and 64. Using $\nu \approx 1.89(\frac{2N}{N_S} - 1)$ from Equation (416) with $N = 1024$ yields EDOFs for the WOSA estimates of 965.8, 240.0, 119.1 and 58.6. With C_h taken to be unity (appropriate for the default rectangular taper used here), Table 279 says that $\nu \approx 3.71 N/m$ for the Parzen lag window estimates, which gives comparable EDOFs of 949.8, 237.4, 131.0 and 59.4. The bandwidth measures $B_\mathcal{H}$ for the four WOSA estimates are 0.433, 0.125, 0.064 and 0.032, which are also close to the bandwidth measures $B_\mathcal{U}$ for the lag windows estimates, namely, 0.463, 0.117, 0.065 and 0.030. The SDF estimates and crisscrosses in Figure 417 are thus quite similar to the ones in the corresponding plots of Figure 294. Thus, even though tapering is not really needed to deal with the AR(2) SDF, the use of tapering with overlapping blocks in the WOSA scheme yields estimators that are competitive with the lag window approach.

A qualitative difference between Figures 417 and 294 is the absence in the WOSA plots of the rough-looking light curves in the lag window plots. These curves show the direct spectral estimate $\hat{S}^{(\text{D})}(\cdot)$ that, after various degrees of smoothing, yields the four lag window estimates $\hat{S}_m^{(\text{LW})}(\cdot)$. A comparison of different $\hat{S}_m^{(\text{LW})}(\cdot)$ with $\hat{S}^{(\text{D})}(\cdot)$ can help determine appropriate values for the parameter m – this is the essence of the window closing technique. With the WOSA estimate, there is no natural equivalent to $\hat{S}^{(\text{D})}(\cdot)$ (although, of course, a WOSA estimate *is* the average of such estimates). The lack of a reference SDF estimate with which to evaluate different WOSA estimates makes it more difficult to select a reasonable value for N_S; however, side-by-side comparisons of the four WOSA estimates in Figure 417 would

8.8 Welch's Overlapped Segment Averaging (WOSA)

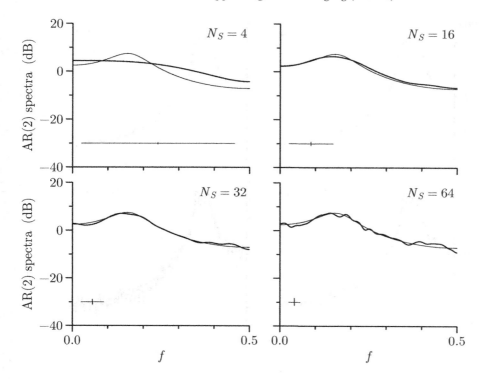

Figure 417 WOSA spectral estimates $\hat{S}^{(\text{WOSA})}(\cdot)$ for AR(2) time series of Figure 34(a) using a Hanning data taper (i.e., 100% cosine data taper) and 50% overlap with block sizes $N_S = 4$, 16, 32 and 64 (thick curves). The thin curve on each plot is the true AR(2) SDF. The width of the crisscross emanating from the lower left-hand corner on each plot shows the measure $B_\mathcal{H}$ of the effective bandwidth of $\hat{S}^{(\text{WOSA})}(\cdot)$ (this measure, given by Equation (194), can be interpreted as the approximate distance in frequency between uncorrelated spectral estimators). The height of the crisscross indicates the length of a 95% CI for 10 $\log_{10}(S(f))$ based upon 10 $\log_{10}(\hat{S}^{(\text{WOSA})}(f))$.

favor the $N_S = 16$ or 32 estimates because structure appears to be washed out of the $N_S = 4$ estimate, whereas the $N_S = 64$ estimate is too erratic looking.

As a second example, let us look at similar WOSA estimates for the AR(4) time series of Figure 34(e), but this time using block sizes of $N_S = 64$, 128, 256 and 512. Figure 418 shows the four estimates (thick curves) along with the true SDF (thin curves). Again using $\nu \approx 1.89(\frac{2N}{N_S} - 1)$ from Equation (416), we now obtain EDOFs of, respectively, 58.6, 28.3, 13.2 and 5.7. By comparison, these values are *greater* than the EDOFs for the four Parzen lag window estimates in Figure 295 computed using $\nu \approx 1.91N/m$, namely, 30.6, 15.3, 10.9 and 3.5. This approximation is based upon $\nu \approx 3.71N/(mC_h)$ from Table 279, where C_h depends upon the data taper used to construct the direct spectral estimate behind the Parzen lag window estimates. Here we have used the Hanning data taper, for which $C_h = 1.94$ (see Table 260, which states this as the value for the 100% cosine taper, another name for the Hanning taper). The bandwidth measures $B_\mathcal{H}$ for the four WOSA estimates are 0.032, 0.016, 0.008 and 0.004. These happen to be close to the bandwidth measures $B_\mathcal{U}$ for the lag windows estimates, namely, 0.029, 0.015, 0.011 and 0.004. Thus WOSA and Parzen lag window estimates with approximately the same bandwidths are such that the former have considerably larger values for the EDOFs and hence smaller variability. As we noted before, this reduction in variability is due to the fact that overlapping the blocks in the WOSA estimate compensates somewhat for the effect of tapering in lag window estimates

Let us now compare two of the WOSA estimates for the AR(4) time series with multi-

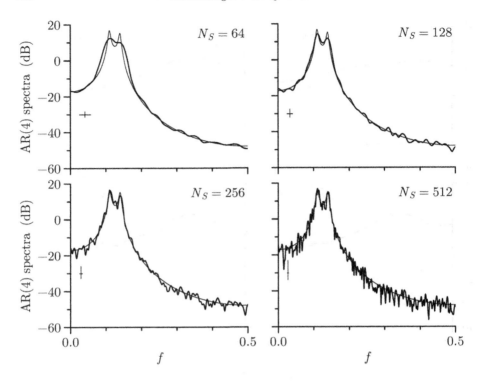

Figure 418 WOSA spectral estimates $\hat{S}^{(\text{WOSA})}(\cdot)$ for AR(4) time series of Figure 34(e) using a Hanning data taper and 50% overlap with block sizes N_S = 64, 128, 256 and 512 (thick bumpy curves). The thin smooth curve on each plot is the true AR(4) SDF. The crisscrosses below the curves have the same interpretation as in Figure 417.

taper estimates presented earlier. The $N_S = 512$ WOSA estimate shown in Figure 418 is approximately bias free, has 5.7 EDOFs, and hence is roughly comparable to the $K = 3$ Slepian and sinusoidal multitaper estimates in Figures 364 and 398, both of which are associated with $2K = 6$ EDOFs. The top row of Figure 419 reproduces these two multitaper estimates using a vertical scale that matches the one for the WOSA estimate. The two multitaper estimates look strikingly similar to the WOSA estimate. As noted before, the bandwidth measure $B_{\mathcal{H}}$ for the WOSA estimate is 0.004. The bandwidth measures $B_{\overline{\mathcal{H}}}$ for the Slepian and sinusoidal estimates are 0.005 and 0.004. All three estimates thus have comparable bias, variance and bandwidth properties. On the other hand, the approximately bias-free $N_S = 256$ WOSA estimate in Figure 418 has 13.2 EDOFs. It is natural to compare this estimate to $K = 7$ multitaper estimates since these would have $2K = 14$ EDOFs; however, while the $K = 7$ sinusoidal estimate in Figure 399 is approximately unbiased, the same cannot be said for the $K = 7$ Slepian estimate in Figure 365, which has significant leakage at high frequencies. The problem with this estimate is its usage of the $k = 6$ eigenspectrum, which has substantial bias at high frequencies (see Figure 363; the $k = 5$ eigenspectrum is also biased at high frequencies, but not nearly as severely). Since the choice $NW = 4$ does not give us $K = 7$ eigenspectra with acceptable bias properties, suppose we increase NW by increments of 0.5 until we do get the desired number of approximately leakage-free eigenspectra. The choice $NW = 5.5$ does the trick (the reader can confirm this by working Exercise [8.7]). The lower left-hand plot of Figure 419 shows the resulting $K = 7$ Slepian multitaper estimate. This estimate is similar in appearance both to the sinusoidal multitaper estimate plotted next to it and to the $N_S = 256$ WOSA estimate in Figure 418. The bandwidth measures for the WOSA, Slepian and sinusoidal estimates are, respectively, 0.008, 0.009 and 0.008. Again the three

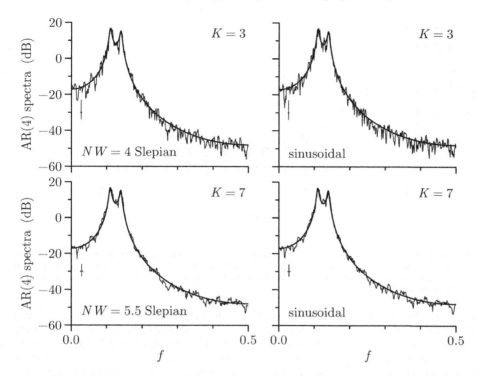

Figure 419 Slepian (left-hand column) and sinusoidal (right-hand) multitaper spectral estimates $\hat{S}^{(\mathrm{MT})}(\cdot)$ for AR(4) time series of Figure 34(e) using either $K = 3$ (top row) or $K = 7$ (bottom) tapers (bumpy curves). The smooth curve on each plot is the true AR(4) SDF. The crisscrosses below the curves have the same interpretation as in Figure 417.

estimates are comparable in terms of bias, variance and bandwidth. (Of course, rather than using the WOSA scheme or multitapering, the AR(4) time series could be handled just as well using prewhitening – in principle, all stationary and causal AR processes can be prewhitened *perfectly*, a nicety that never happens with real data!)

The WOSA spectral estimator is widely used because

[1] it can be implemented in a computationally efficient fashion (using DFTs of a fixed size based on an FFT algorithm);
[2] it can efficiently handle very long time series;
[3] there exist commercially available special purpose instruments – spectrum analyzers – that display spectral estimates based essentially upon WOSA; and
[4] a robust SDF estimator can be devised similar in spirit to WOSA except that the $\hat{S}_{jn}^{(\mathrm{D})}(f)$ terms in Equation (414b) are not just averaged together but rather are combined in such a fashion as to downweight individual estimators corresponding to blocks contaminated by outliers (for details, see Chave et al., 1987).

As mentioned above, the only real potential problem with WOSA is bias caused by an inappropriately small block size N_S (the $N_S = 64$ estimate of Figure 418 being an example). We can guard against this defect if it is possible to vary the block size.

Comments and Extensions to Section 8.8

[1] We have developed the theory for this section under the usual implicit assumption that the mean value of the stationary process $\{X_t\}$ is known to be zero. When the mean value is unknown (the usual case in practice), one appropriate modification is to substitute $X_l - \overline{X}, \ldots, X_{N_S+l-1} - \overline{X}$ for X_l,

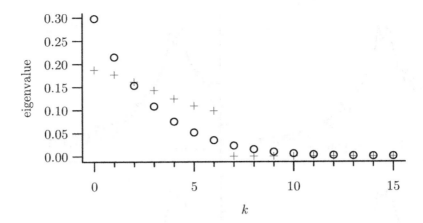

Figure 420 Eigenvalues d_k versus $k = 0, \ldots, 15$ for weight matrices Q associated with the $m = 179$ Parzen lag window estimate of Figure 295 (circles) and the $N_S = 256$ WOSA estimate of Figure 418 (pluses), both of which utilize the $N = 1024$ AR(4) time series of Figure 34(e). The rank K of the weight matrix is 1024 for the lag window estimate and 7 for the WOSA estimate.

$\ldots, X_{N_S+l-1}$ in Equation (414a), where (as usual) $\overline{X}$ is the sample mean of $X_0, \ldots, X_{N-1}$. With this modification, it is an easy exercise to show that $\hat{S}^{(\text{WOSA})}(0)$ is related to the variance of the sample means of the individual blocks of data, an interpretation of $S(0)$ that is in keeping with Equation (164b).

A second way of handling an unknown process mean is to use the sample means of each block of data instead of the sample mean $\overline{X}$ of all the data. Thus, to compute $\hat{S}_l^{(\text{D})}(f)$, we would substitute $X_l - \overline{X}_l, \ldots, X_{N_S+l-1} - \overline{X}_l$ for $X_l, \ldots, X_{N_S+l-1}$ in Equation (414a), where $\overline{X}_l \stackrel{\text{def}}{=} \sum_{t=0}^{N_S-1} X_{t+l}/N_S$. This procedure is particularly useful in situations where $\hat{S}_l^{(\text{D})}(f)$ is computed in near real-time (as is done in commercial spectrum analyzers). The relative merit of these two ways of handling the process mean is evidently an open question.

[2] The ACVS estimator $\{\hat{s}_\tau^{(\text{WOSA})}\}$ corresponding to $\hat{S}^{(\text{WOSA})}(\cdot)$ can be obtained readily since we have both

$$\{\hat{s}_\tau^{(\text{WOSA})} : \tau = -(N_S - 1), \ldots, N_S - 1\} \longleftrightarrow \{\hat{S}^{(\text{WOSA})}(f_k') : k = -(N_S - 1), \ldots, N_S - 1\},$$

where $f_k' = k/[(2N_S - 1)\Delta_\text{t}]$, and

$$\{\hat{s}_\tau^{(\text{WOSA})} : \tau = -(N_S - 1), \ldots, N_S\} \longleftrightarrow \{\hat{S}^{(\text{WOSA})}(\tilde{f}_k) : k = -(N_S - 1), \ldots, N_S\},$$

where $\tilde{f}_k = k/(2N_S \Delta_\text{t})$. If so desired, we can apply an appropriate lag window to $\{\hat{s}_\tau^{(\text{WOSA})}\}$ to smooth $\hat{S}^{(\text{WOSA})}(\cdot)$ (for details, see Nuttall and Carter, 1982).

[3] In forming the WOSA estimator we have insisted that the direct SDF estimator corresponding to $j = N_B - 1$ in Equation (414b) should involve a block of the time series ending with X_{N-1}. As a result, for a given block size N_S, there are only certain shift factors n satisfying the conditions of Equation (414c). This fact limits the percentage of overlaps that can be achieved, which explains the placement of the circles in Figure 415.

[4] WOSA and lag window estimators are both examples of quadratic spectral estimators and thus can be written as deconstructed multitaper estimators represented by Equation (376a). Here we explore these representations for two previously considered SDF estimates based on the AR(4) time series of Figure 34(e), namely, the $m = 179$ Parzen lag window estimate $\hat{S}_m^{(\text{LW})}(\cdot)$ of Figure 295 and the $N_S = 256$ WOSA estimate $\hat{S}^{(\text{WOSA})}(\cdot)$ of Figure 418. Visually these two estimates are similar, with $\hat{S}_m^{(\text{LW})}(\cdot)$ and $\hat{S}^{(\text{WOSA})}(\cdot)$ having EDOFs ν of, respectively, 10.9 and 13.2 and bandwidths $B_\mathcal{H}$ of 0.011 and 0.008.

Turning first to the Parzen lag window estimator, Equations (374a) and (375a) indicate that we can write $\hat{S}_m^{(\text{LW})}(f) = \Delta_\text{t} Z^H Q^{(\text{LW})} Z$, where the (s, t)th element of the $N \times N$ matrix $Q^{(\text{LW})}$ is

8.8 *Welch's Overlapped Segment Averaging (WOSA)* 421

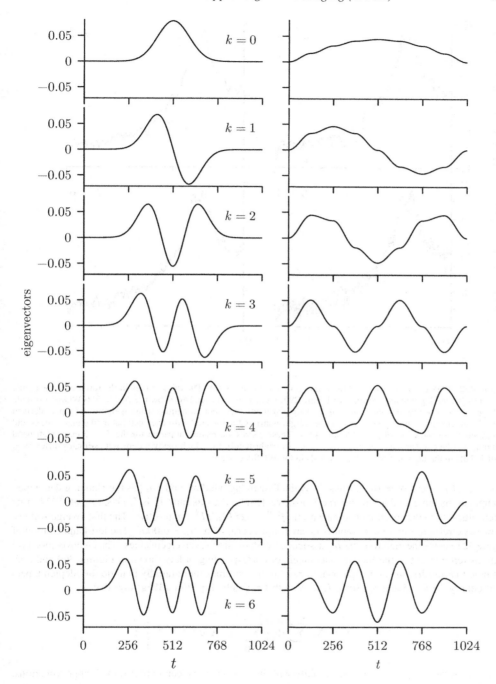

Figure 421 Eigenvectors h_k of orders $k = 0, \ldots, 6$ (top to bottom row) for weight matrices Q associated with the $m = 179$ Parzen lag window estimate of Figure 295 (left-hand column) and the $N_S = 256$ WOSA estimate of Figure 418 (right-hand).

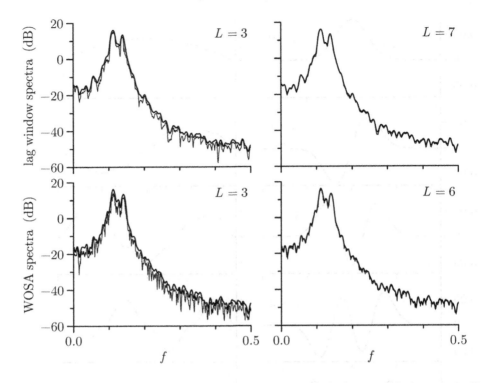

Figure 422 SDF estimates for the AR(4) time series of Figure 34(e). The thick curves in the top row replicate the $m = 179$ Parzen lag window estimate of Figure 295; those in the bottom row show the $N_S = 256$ WOSA estimate of Figure 418. The thin curves show rank L weighted multitaper approximations using weights and tapers derived from the eigenvalues and eigenvectors of the weight matrix $\boldsymbol{Q}$ associated with each estimator (the eigenvalues and eigenvectors are shown in Figures 420 and 421). These curves are readily apparent for the $L = 3$ cases (left-hand column), but less so for the $L = 7$ and $L = 6$ cases (right-hand), for which the two depicted estimates differ by at most 1.8 dB in the lag window case and 3.6 dB in the WOSA case.

$h_s w_{m,s-t} h_t$ – here $N = 1024$, $\{w_{m,\tau}\}$ is the Parzen lag window, and $\{h_t\}$ is the Hanning data taper. Computations indicate that this matrix is positive definite (i.e., its rank K is N) because all 1024 of its eigenvalues d_k in the spectral decomposition $\boldsymbol{Q}^{(\mathrm{LW})} = \boldsymbol{H} \boldsymbol{D}_K \boldsymbol{H}^T$ are positive. The first 16 eigenvalues, shown as circles in Figure 420, rapidly decay toward zero, with $d_{15} \doteq 0.0006$. The left-hand column of Figure 421 shows the eigenvectors $\boldsymbol{h}_k$ associated with the first seven eigenvalues. The eigenvectors have patterns reminiscent of the Slepian data tapers of corresponding orders shown in Figures 360 and 362, but note that the $\boldsymbol{h}_k$ tapers damp down to zero at the extremes more rapidly than do the Slepian tapers (particularly for $k = 5$ and 6). Since $K = N$ in this example, Equation (376a) says that

$$\hat{S}_m^{(\mathrm{LW})}(f) = \Delta_t \sum_{k=0}^{N-1} d_k \left| \sum_{t=0}^{N-1} h_{k,t} X_t e^{-i2\pi f t \Delta_t} \right|^2 ;$$

however, because d_k decays to zero rapidly with increasing k, we can expect rank L approximations, namely,

$$\tilde{S}_{m,L}^{(\mathrm{LW})}(f) \stackrel{\text{def}}{=} \Delta_t \sum_{k=0}^{L-1} d_k \left| \sum_{t=0}^{N-1} h_{k,t} X_t e^{-i2\pi f t \Delta_t} \right|^2 , \qquad (422)$$

to converge rapidly to $\hat{S}_m^{(\mathrm{LW})}(f)$ as we increase L (note that we must have $\tilde{S}_{m,L}^{(\mathrm{LW})}(f) \leq \hat{S}_m^{(\mathrm{LW})}(f)$ for all L and f). The top row of Figure 422 is consistent with this claim. The thick curves in both plots show $\hat{S}_m^{(\mathrm{LW})}(\cdot)$, while the thin curves show the approximations $\tilde{S}_{m,3}^{(\mathrm{LW})}(\cdot)$ (left-hand plot) and $\tilde{S}_{m,7}^{(\mathrm{LW})}(\cdot)$

8.8 Welch's Overlapped Segment Averaging (WOSA)

(right-hand). We see that $\tilde{S}_{m,3}^{(\text{LW})}(\cdot)$ is generally fairly close to $\hat{S}_m^{(\text{LW})}(\cdot)$; on the other hand, the agreement between $\tilde{S}_{m,7}^{(\text{LW})}(\cdot)$ and $\hat{S}_m^{(\text{LW})}(\cdot)$ is so good that the thin curve is only barely visible (the maximum error in the approximation is 1.8 dB). The fact that lag window estimators are sometimes well approximated by low-order weighted multitaper estimators is the basis for the statement "lag-windowing and multiple-data-windowing are roughly equivalent for smooth spectrum estimation," which is the title of McCloud et al. (1999), to which we refer the reader for a complementary discussion and further examples.

Let us now turn to the WOSA estimate $\hat{S}^{(\text{WOSA})}(\cdot)$, which uses a Hanning data taper appropriate for the block size $N_S = 256$ and an overlap between blocks of 50%, yielding a shift factor n of $N_S/2 = 128$ along with $N_B = 2(N/N_s - 1) + 1 = 7$ blocks. Equations (414b) and (414a) tell us that

$$\hat{S}^{(\text{WOSA})}(f) = \frac{\Delta_t}{N_B} \sum_{j=0}^{N_B-1} \left| \sum_{t=0}^{N_S-1} h_t X_{t+jN_S/2} e^{-i2\pi f t \Delta_t} \right|^2$$

$$= \frac{\Delta_t}{N_B} \sum_{j=0}^{N_B-1} \left| \sum_{t=0}^{N-1} h_{t-jN_S/2} X_t e^{-i2\pi f t \Delta_t} \right|^2 \stackrel{\text{def}}{=} \frac{1}{N_B} \sum_{j=0}^{N_B-1} \hat{S}_j^{(\text{WOSA})}(f),$$

for which we recall that $h_t = 0$ for $t < 0$ and $t \geq N_S$. We can interpret $\hat{S}_j^{(\text{WOSA})}(\cdot)$ as a direct spectral estimator that uses a length N data taper formed by taking a length N_S Hanning data taper and pre- and post-pending $jN_S/2$ and $N - N_S(1 + j/2)$ zeros respectively. Hence we can write

$$\hat{S}_j^{(\text{WOSA})}(f) = \Delta_t \, \boldsymbol{Z}^H \boldsymbol{Q}_j^{(\text{WOSA})} \boldsymbol{Z},$$

where the (s,t)th element of the $N \times N$ matrix $\boldsymbol{Q}_j^{(\text{WOSA})}$ is $h_{s-jN_S/2} h_{t-jN_S/2}$. Hence

$$\hat{S}^{(\text{WOSA})}(f) = \Delta_t \, \boldsymbol{Z}^H \boldsymbol{Q}^{(\text{WOSA})} \boldsymbol{Z}, \quad \text{where} \quad \boldsymbol{Q}^{(\text{WOSA})} \stackrel{\text{def}}{=} \frac{1}{N_B} \sum_{j=0}^{N_B-1} \boldsymbol{Q}_j^{(\text{WOSA})}. \qquad (423)$$

Each $\boldsymbol{Q}_j^{(\text{WOSA})}$ is the outer product of an N-dimensional vector with itself and thus has unit rank. The vectors forming the seven weight matrices $\boldsymbol{Q}_j^{(\text{WOSA})}$ are linearly independent, from which we can conclude that $\boldsymbol{Q}^{(\text{WOSA})}$ has rank $K = N_B = 7$. The pluses in Figure 420 show the first 16 eigenvalues in the spectral decomposition $\boldsymbol{Q}^{(\text{WOSA})} = \boldsymbol{H} \boldsymbol{D}_N \boldsymbol{H}^T$. Note that $d_k = 0$ for $k \geq 7$. The right-hand column of Figure 421 shows the eigenvectors $\boldsymbol{h}_k$ associated with the seven nonzero eigenvalues. The eigenvectors are visually quite different from those for the lag window estimate, although they are similar in certain aspects (e.g., the kth eigenvector for both estimates has k zero crossings). The rank L approximation $\tilde{S}_L^{(\text{WOSA})}(\cdot)$ to $\hat{S}^{(\text{WOSA})}(\cdot)$ takes a form analogous to Equation (422), but, because $\hat{S}^{(\text{WOSA})}(\cdot)$ has rank $K = 7$, we must have $\tilde{S}_L^{(\text{WOSA})}(\cdot) = \hat{S}^{(\text{WOSA})}(\cdot)$ for all $L \geq 7$. The thick curves in the bottom row of Figure 422 show $\hat{S}^{(\text{WOSA})}(\cdot)$, while the thin curves show the approximations $\tilde{S}_3^{(\text{WOSA})}(\cdot)$ (left-hand plot) and $\tilde{S}_6^{(\text{WOSA})}(\cdot)$ (right-hand). The $L = 3$ approximation is markedly poorer than the corresponding one for the lag window estimate, which can be explained by the fact that the sum of the first three eigenvalues is smaller in the WOSA case than in the lag window case (see Figure 420). In view of the fact that $\tilde{S}_7^{(\text{WOSA})}(\cdot) = \hat{S}^{(\text{WOSA})}(\cdot)$ the agreement between $\tilde{S}_6^{(\text{WOSA})}(\cdot)$ and $\hat{S}^{(\text{WOSA})}(\cdot)$ should be good. The bottom right-hand plot shows this to be true, with the maximum error in the approximation being 3.6 dB.

[5] As we've noted in this section and in Section 7.4, the EDOFs ν is an important concept for non-parametric SDF estimators because of the role it plays in forming CIs for the unknown true SDF. The expressions for ν given by Equations (264b) and (415d) for, respectively, lag window and WOSA estimators involve variables explicitly used in constructing these estimators. Because both estimators can be expressed as weighted multitaper estimators, Equation (354g) offers an alternative pathway for determining ν that depends only on the eigenvalues d_k – and not the eigenvectors – in the spectral decomposition of the weight matrix $\boldsymbol{Q}$.

As examples of the use of Equation (354g), let us reconsider the $m = 179$ Parzen lag window estimate $\hat{S}_m^{(\text{LW})}(\cdot)$ of Figure 295 and the $N_S = 256$ WOSA estimate $\hat{S}^{(\text{WOSA})}(\cdot)$ of Figure 418. If we use Equation (264b) to approximate ν for $\hat{S}_m^{(\text{LW})}(\cdot)$, we obtain $\nu \doteq 10.9$ (this calculation involves both the Parzen lag window $\{w_{m,\tau}\}$ and the variance inflation factor C_h, which depends on the Hanning data taper and is computed as per Equation (262)); on the other hand, using Equation (354g), we obtain $\nu \doteq 11.1$ (Figure 420 shows the first 16 of the required eigenvalues d_k as circles). The two approximations for the EDOFs are within 2% of each other. Turning now to the WOSA estimate, if we use the right-most approximation for ν in Equation (416) (an adaptation of Equation (415d) specific to WOSA estimates using the Hanning data taper with blocks overlapping by 50%), we get $\nu \doteq 13.2$; on the other hand, use of Equation (354g) with the seven nonzero eigenvalues d_k shown as pluses in Figure 420 yields $\nu \doteq 13.4$. The two approximations are within 1% of each other.

[6] We have presented prewhitening, multitapering and the WOSA scheme as three ways to create estimators with bias and variance properties that are potentially superior to those of direct spectral estimators and lag window estimators. While these approaches are often comparable in practical applications, we point the reader to two papers that present evidence favoring Slepian multitaper estimators over prewhitening and WOSA. Thomson (1990a, p. 614) argues that there are examples of processes with highly structured SDFs for which prewhitening is inferior to Slepian multitapering. Bronez (1992) shows, using bounds for bias, variance and resolution, that, if a Slepian multitaper estimator and a WOSA spectral estimator are fixed so that two of the three bounds are identical, the remaining bound favors the multitaper estimator.

[7] As mentioned above, it is counterintuitive that the EDOFs shown in Figure 415 do not monotonically increase as the overlap percentage between blocks increases, which implies that var $\{\hat{S}^{(\text{WOSA})}(f)\}$ is minimized at a percentage distinctly below 100%; moreover, the optimal percentage depends upon both the taper $\{h_t\}$ and the block size B_S chosen for use with WOSA. Barbé et al. (2010) investigate this paradox in detail and find that a simple modification to the WOSA scheme guarantees a monotonic decrease in variance as the overlap percentage increases. The modification is based on treating the time series under analysis as if it were circular, i.e., periodic with a period of N so that surrogates for the unobserved $X_N, X_{N+1}, \ldots$ are taken to be the observed $X_0, X_1, \ldots$. The usual WOSA estimator is then augmented by including a small number of blocks that splice the beginning and end of the time series together. This modification yields an estimator with slightly less variability than the usual WOSA estimator, but with an increase in bias that depends in part upon the potential mismatch between X_N and its surrogate X_0 (under reasonable conditions, the bias due to circularity is asymptotically negligible with increasing N). An untested conjecture is that replacing circularity with the reflecting boundary conditions of either Equation (183b) or (183c) would decrease this bias while maintaining monotonic decrease in variance.

[8] Hansson-Sandsten (2012) considers the problem of finding a WOSA estimator that approximates a given multitaper estimator. Whereas we can express all WOSA estimators as multitaper estimators, the converse is not true in general, but it is of interest to see how close we can come to expressing an arbitrary multitaper estimator as a WOSA estimator. Both estimators are examples of quadratic estimators. Equation (374a) thus says that both can be written as $\Delta_t \boldsymbol{Z}^H \boldsymbol{Q} \boldsymbol{Z}$ with an appropriate choice of the $N \times N$ weight matrix $\boldsymbol{Q}$. For a basic multitaper estimator $\hat{S}^{(\text{WMT})}(\cdot)$ employing tapers $\boldsymbol{h}_0, \ldots, \boldsymbol{h}_{K-1}$, the weight matrix takes the form

$$\boldsymbol{Q}^{(\text{MT})} = \frac{1}{K} \sum_{k=0}^{K-1} \boldsymbol{h}_k \boldsymbol{h}_k^T$$

(the above is in keeping with Equation (375c) since $d_k = 1/K$ in the basic multitaper scheme). For a WOSA estimator that is the average of N_B direct spectral estimators (each using the same WOSA taper $\{h_0, h_1, \ldots, h_{N_S-1}\}$), the weight matrix makes the form

$$\boldsymbol{Q}^{(\text{WOSA})} = \frac{1}{N_B} \sum_{j=0}^{N_B-1} \boldsymbol{q}_j \boldsymbol{q}_j^T,$$

where each N-dimensional vector $\boldsymbol{q}_j$ contains the WOSA taper along with $N - N_S$ zeros, with a certain number of zeros coming before h_0 and the rest after h_{N_S-1}, and with each $\boldsymbol{q}_j$ having a different allocation of the zeros (see the discussion surrounding Equation (423)). Letting $\boldsymbol{Q}_{j,k}$ denote the (j,k)th element of $\boldsymbol{Q}$, Hansson-Sandsten (2012) advocates picking the WOSA taper such that

$$\sum_{j=0}^{N-1}\sum_{k=0}^{N-1}\left(\boldsymbol{Q}_{j,k}^{(\mathrm{MT})} - \boldsymbol{Q}_{j,k}^{(\mathrm{WOSA})}\right)^2$$

is minimized over an appropriate space of possible tapers. Rather than letting the space be all possible properly normalized N_S-dimensional tapers, the focus is on linear combinations of the left eigenvectors from a singular value decomposition of a matrix of dimension $N_S \times KN_B$, whose columns are related to N_B snippets of length N_S from each of the K vectors $\boldsymbol{h}_0, \ldots, \boldsymbol{h}_{K-1}$ (the elements comprising the N_B snippets from a given $\boldsymbol{h}_k$ have indices in agreement with those for the nonzero elements in $\boldsymbol{q}_0, \ldots, \boldsymbol{q}_{N_B-1}$). This scheme is intended to pick a WOSA taper that collectively depends upon the K multitapers. Hansson-Sandsten (2012) finds that this scheme does not yield a decent WOSA approximation for certain $\hat{S}^{(\mathrm{WMT})}(\cdot)$, where the approximation is assessed by comparing the target $E\{\hat{S}^{(\mathrm{WMT})}(f)\}$ and var$\{\hat{S}^{(\mathrm{WMT})}(f)\}$ over a range of frequencies for a particular sample size N with the approximating $E\{\hat{S}^{(\mathrm{WOSA})}(f)\}$ and var$\{\hat{S}^{(\mathrm{WOSA})}(f)\}$. On the other hand, for certain stationary processes and sample sizes N, it is possible to find a decent approximation using a WOSA estimator with particular N_S and N_B, opening up the possibility of computing – in an efficient manner – approximate multitaper estimates for a time series as it is being collected.

8.9 Examples of Multitaper and WOSA Spectral Estimators

Ocean Wave Data

As examples of the multitaper and WOSA approaches to spectral analysis, we reconsider the ocean wave data discussed in Sections 6.8 and 7.12 (see Figure 225 for a plot of this time series and Figures 226, 227, 318 and 319 for plots of various nonparametric SDF estimates). As noted previously, we are mainly interested in the rate at which the SDF decreases over the range 0.2 to 1.0 Hz, but the SDF at higher frequencies (i.e., $1\,\mathrm{Hz} \leq f \leq 2\,\mathrm{Hz}$) is also of some interest. Amongst the SDF estimates looked at so far, the best is arguably the Gaussian lag window estimate (Figure 319(d)) since it nicely depicts the rolloff pattern of main interest.

We begin by considering multitaper estimation prior to which we center the time series by subtracting off its sample mean $\overline{X} \doteq 209.1$. Turning first to Slepian-based estimation, we can set the regularization bandwidth $2W$ to be fairly large due to the lack of sharp features in the SDF over the frequency range 0.2 to 1.0 Hz. We thus let $W = 4/(N\,\Delta_\mathrm{t})$ initially, yielding $2W \doteq 0.031$ Hz (recall that $N = 1024$ and $\Delta_\mathrm{t} = 1/4\,\mathrm{sec}$). Since the maximum number of potentially acceptable eigenspectra is $K_{\max} = \lfloor 2NW\,\Delta_\mathrm{t} \rfloor - 1 = 7$, we compute $\hat{S}_k^{(\mathrm{MT})}(\cdot)$ for orders $k = 0, 1, \ldots, 6$ over the grid of Fourier frequencies $f_j = j/(N\,\Delta_\mathrm{t})$, $j = 0, 1, \ldots, N/2$. A comparison of the high-order eigenspectra with those of orders $k = 0$ and 1 indicates that, while $\hat{S}_6^{(\mathrm{MT})}(\cdot)$ has unacceptable bias in the high frequency range, none of the eigenspectra of lower orders show evidence of bias. We thus form the basic multitaper estimate $\hat{S}^{(\mathrm{MT})}(\cdot)$ of Equation (352a) by averaging $K = 6$ eigenspectra together. Figure 426(a) shows this estimate along with a crisscross depicting the bandwidth measure $B_{\overline{\mathcal{H}}} \doteq 0.028$ Hz of Equation (353e) and the width (in decibels) of a 95% CI for $10\,\log_{10}(S(f_j))$ based on $10\,\log_{10}(\hat{S}^{(\mathrm{MT})}(f_j))$ with $2K = 12$ EDOFs. Note that $B_{\overline{\mathcal{H}}} \doteq 0.028$ Hz is less than the regularization bandwidth $2W \doteq 0.031$ Hz, as should be the case when we use fewer than $K_{\max}$ tapers (had we been able to use all 7 tapers, then $B_{\overline{\mathcal{H}}}$ would be 0.030 Hz, which is within 2% of $2W$).

If we increase the regularization half-bandwidth to $W = 7/(N\,\Delta_\mathrm{t})$, there are now $K = 12$ eigenspectra with acceptable bias properties. Averaging these together yields the multitaper estimate $\hat{S}^{(\mathrm{MT})}(\cdot)$ shown in Figure 426(b). This estimate has $2K = 24$ EDOFs and bandwidth

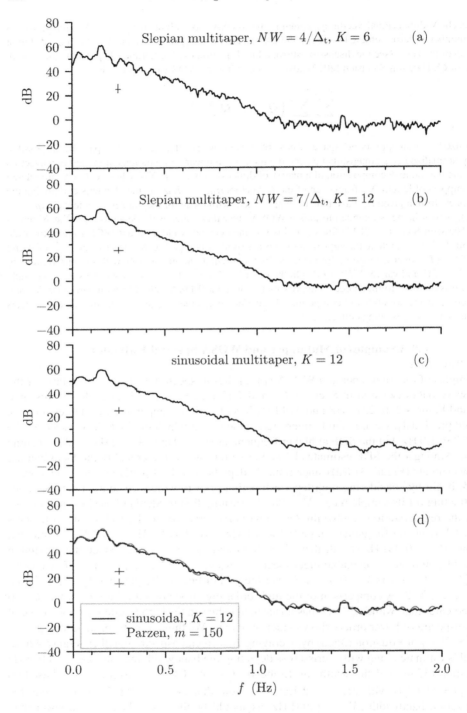

Figure 426 Multitaper spectral estimates for ocean wave data (all four plots) along with Parzen lag window estimate (bottom plot only).

8.9 Examples of Multitaper and WOSA Spectral Estimators 427

$B_{\overline{\mathcal{H}}} \doteq 0.052\,\mathrm{Hz}$, which is slightly less then the regularization bandwidth $2W = 0.055\,\mathrm{Hz}$ (use of $K_{\max} = 13$ tapers would yield $B_{\overline{\mathcal{H}}} \doteq 0.054\,\mathrm{Hz}$). A comparison of this estimate with the one for $W = 4/(N\,\Delta_{\mathrm{t}})$ indicates that the $W = 7/(N\,\Delta_{\mathrm{t}})$ estimate is smoother in appearance, as is to be expected.

Let us now consider a sinusoidal multitaper estimator that also uses $K = 12$ eigenspectra. Figure 426(c) shows this estimate, which has a bandwidth of the $B_{\overline{\mathcal{H}}} \doteq 0.051\,\mathrm{Hz}$ (the difference between $B_{\overline{\mathcal{H}}}$ and its approximation $(K+1)/[(N+1)\,\Delta_{\mathrm{t}}]$ is 0.00015). This estimate is quite similar to the Slepian-based estimate in Figure 426(b), whose bandwidth (0.052) is a nearly identical. It is also of interest to compare this estimate with the $m = 150$ Parzen lag window estimate shown by the thick curve in Figure 318. This lag window estimate has a comparable bandwidth of $B_{\mathcal{U}} \doteq 0.050\,\mathrm{Hz}$. We have replotted the sinusoidal multitaper and Parzen estimates in Figure 426(d) as thick and thin curves, but in places it is hard to see two distinct curves because the estimates are in such good agreement. There are two crisscrosses in Figure 426(d) – the upper one is for the multitaper estimate, and the lower, for the Parzen estimate. While the bandwidths are nearly identical, the width of a 95% CI for the multitaper estimate is somewhat smaller than that for the Parzen estimate (the former is based on EDOFs $\nu = 24$ yielding a width of $5.0\,\mathrm{dB}$, and the latter on $\nu = 12.6$ yielding $7.0\,\mathrm{dB}$). The estimate of choice would thus be the multitaper estimate, but, for this time series, lag window and multitaper spectral estimates are quite comparable.

Next we examine the performance of the Slepian-based adaptive multispectral estimator $\hat{S}^{(\mathrm{AMT})}(\cdot)$ of Equation (389b). As in Figure 426(a), we set $W = 4/(N\,\Delta_{\mathrm{t}})$, but now we entertain the maximum number $K_{\max} = 7$ of eigenspectra even though we previously excluded $\hat{S}^{(\mathrm{MT})}_6(\cdot)$ (recall that we did so because it suffers from leakage at high frequencies). Given the eigenspectra $\hat{S}^{(\mathrm{MT})}_k(\cdot)$ and their associated eigenvalues

$$\lambda_0 \doteq 0.999\,999\,999\,7, \quad \lambda_1 \doteq 0.999\,999\,97, \quad \ldots, \quad \lambda_5 \doteq 0.993, \quad \lambda_6 \doteq 0.94,$$

we use $K = 2$ tapers to compute the preliminary estimate $\bar{S}^{(\mathrm{WMT})}(f_j)$ of Equation (388e). We then use this estimate in place of $S(f_j)$ and $\hat{s}_0^{(\mathrm{P})} \doteq 143954$ in place of s_0 in Equation (388a) to obtain a first cut at the weights $b_k(f_j)$, $k = 0, 1, \ldots, K_{\max} - 1$. Use of these weights in Equation (389b) (with K henceforth set to $K_{\max}$) yields a corresponding first cut at the adaptive estimate – call it $\hat{S}^{(\mathrm{AMT})}_{(1)}(f_j)$. We now return to Equation (388a) and use $\hat{S}^{(\mathrm{AMT})}_{(1)}(f_j)$ to refine the weights, which we then use in Equation (389b) to obtain the second cut $\hat{S}^{(\mathrm{AMT})}_{(2)}(f_j)$. Letting $n = 2$, if

$$\frac{|\hat{S}^{(\mathrm{AMT})}_{(n)}(f_j) - \hat{S}^{(\mathrm{AMT})}_{(n-1)}(f_j)|}{\hat{S}^{(\mathrm{AMT})}_{(n-1)}(f_j)} < 0.05,$$

then we set the adaptive estimate $\hat{S}^{(\mathrm{AMT})}(f_j)$ equal to $\hat{S}^{(\mathrm{AMT})}_{(2)}(f_j)$, which happens at 463 of the 513 Fourier frequencies. For the 50 remaining f_j's, we need more iterations to satisfy the above stopping rule, but the most we need is $n = 7$. Figure 428(a) shows the resulting adaptive multitaper estimates $\hat{S}^{(\mathrm{AMT})}(\cdot)$ (if we set the tolerance to 0.001 rather than 0.05, we now need up to $n = 12$ iterations, but there is no visible difference between these estimates and the ones based on 0.05). A comparison of $\hat{S}^{(\mathrm{AMT})}(\cdot)$ with $\hat{S}^{(\mathrm{MT})}(\cdot)$ of Figure 426(a) indicates good agreement for frequencies between 0 and $1\,\mathrm{Hz}$. For higher frequencies, the estimate $\hat{S}^{(\mathrm{AMT})}(\cdot)$ shows more structure than $\hat{S}^{(\mathrm{MT})}(\cdot)$. What accounts for this difference? As indicated by Equation (390), the EDOFs ν for $\hat{S}^{(\mathrm{AMT})}(\cdot)$ are frequency-dependent – we have plotted ν as a function of frequency in Figure 428(c). Note that ν drops down to about 6 for f between 1 and $2\,\mathrm{Hz}$, which is half the degrees of freedom associated with $\hat{S}^{(\mathrm{MT})}(\cdot)$ in Figure 426(a). At least part of the additional structure in $\hat{S}^{(\mathrm{AMT})}(\cdot)$ can be attributed to this decrease in ν. To

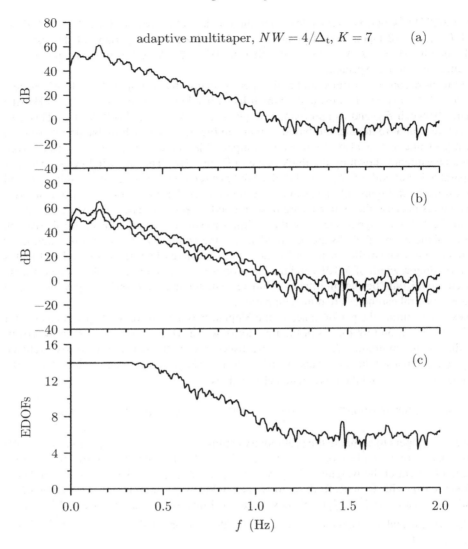

Figure 428 Adaptive multitaper spectral estimate for ocean wave data (top plot) along with 95% confidence intervals (middle) and EDOFs (bottom).

explore this point further, the upper and lower thin curves in Figure 428(b) show upper and lower 95% confidence limits for each $10\log_{10}(S(f_j))$ based upon $10\log_{10}(\hat{S}^{(\text{AMT})}(f_j))$ in plot (a). With the decrease of ν in the high frequency region, the widths of the CIs increase, indicating an increase of variability in $\hat{S}^{(\text{AMT})}(\cdot)$. The apparent additional spectral structure is in accordance with this increased variability.

Finally we note two additional differences between the basic estimator $\hat{S}^{(\text{MT})}(\cdot)$ and the adaptive estimator $\hat{S}^{(\text{AMT})}(\cdot)$. Once the regularization bandwidth $2W$ is set, the estimate $\hat{S}^{(\text{AMT})}(\cdot)$ follows automatically; i.e., we do not have to carefully examine the individual eigenspectra as we must to select a suitable K for $\hat{S}^{(\text{MT})}(\cdot)$. This illustrates the point that the adaptive multitaper estimator can be "turned loose" on SDFs with high dynamic ranges in an automatic fashion. Second, note that $\hat{S}^{(\text{AMT})}(f)$ has two more EDOFs than $\hat{S}^{(\text{MT})}(f)$ for f between 0 and 0.3 Hz ($\nu = 14$ as opposed to $\nu = 12$). This is because the adaptive estimator

8.9 Examples of Multitaper and WOSA Spectral Estimators 429

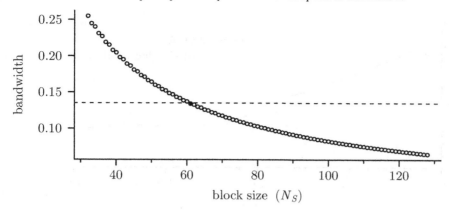

Figure 429 Effective bandwidth $B_{\mathcal{H}}$ of a WOSA SDF estimator of the ocean wave time series as a function of block size N_S. For a given N_S, we compute $B_{\mathcal{H}}$ as per Equation (194) using $\Delta_t = 0.25$ sec, N_S in place of N and the autocorrelation $\{h \star h_\tau\}$ (formed in keeping with Equation (99f)) of a Hanning data taper $\{h_t\}$ (Equation (189b), again with N_S replacing N). The horizontal dashed line indicates the effective bandwidth $B_{\mathcal{U}} \doteq 0.135$ Hz for the Gaussian lag window estimate shown in Figure 319(d). This bandwidth is computed using Equation (256a) using $N = 1024$, a Gaussian lag window $\{w_{m,\tau}\}$ (Equation (276) with m set to 23.67) and the autocorrelation $\{h \star h_\tau\}$ of our standard approximation to an $NW = 2/\Delta_t$ zeroth-order Slepian data taper $\{h_t\}$ (Equation (196b)). The setting $N_S = 61$ gives $B_{\mathcal{H}} \doteq 0.134$ Hz, which is the closest we can get to the desired $B_{\mathcal{U}}$.

makes use of $\hat{S}_6^{(\mathrm{MT})}(\cdot)$, which we rejected in forming $\hat{S}^{(\mathrm{MT})}(\cdot)$ only because of its poor bias properties at *high* frequencies. The inclusion of $\hat{S}_6^{(\mathrm{MT})}(\cdot)$ for low frequencies makes sense, a point clearly in favor of $\hat{S}^{(\mathrm{AMT})}(\cdot)$ (there is, however, very little *actual* difference between the two spectral estimates at low frequencies).

We now consider a Hanning-based WOSA approach. Our first task is to determine an appropriate block size N_S. For the sake of comparison, let us set N_S such that the effective bandwidth $B_{\mathcal{H}}$ for the resulting WOSA estimate matches the effective bandwidth $B_{\mathcal{U}} \doteq 0.135$ Hz for the Gaussian lag window estimate of Figure 319(d) as closely as possible. Figure 429 shows a plot of $B_{\mathcal{H}}$ versus block sizes N_S ranging from 32 to 128. The horizontal dashed line indicates $B_{\mathcal{U}}$. The closest we can get to $B_{\mathcal{U}}$ is by picking $N_S = 61$, for which $B_{\mathcal{H}} \doteq 0.134$ Hz. The usual recommendation for a Hanning-based WOSA estimator is to use blocks with a 50% overlap. We can achieve this approximately here by setting the shift factor n to 30. Since $N = 1024$, these settings for N_S and n give us $N_B = 33$ blocks in all. Let $X'_t = X_t - \overline{X}$ denote the centered time series. The first block consists of $X'_0, \ldots, X'_{60}$; the second, of $X'_{30}, \ldots, X'_{90}$; and, following this pattern, the next-to-last block would consist of $X'_{930}, \ldots, X'_{990}$, and the last, of $X'_{960}, \ldots, X'_{1020}$. With the blocks so defined, our WOSA estimate $\hat{S}^{(\mathrm{WOSA})}(\cdot)$ would make no use of the final three values of the time series. Although it makes no practical difference, let us correct this small – but annoying – deficiency by redefining the last block to be $X'_{963}, \ldots, X'_{1023}$ (this eliminates X'_{960}, X'_{961} and X'_{962}, but these are toward the middle of the next-to-last block).

Figure 430 shows the WOSA SDF estimate $\hat{S}^{(\mathrm{WOSA})}(\cdot)$ (thick curve) and the Gaussian lag window estimate $\hat{S}_m^{(\mathrm{LW})}(\cdot)$ of Figure 319(d) (thin). The upper/lower crisscrosses are for the WOSA and lag window estimates respectively. The width of the upper crisscross is equal to $B_{\mathcal{U}} \doteq 0.135$ Hz, whereas that for the lower one depicts $B_{\mathcal{H}} \doteq 0.134$ Hz (virtually the same by design). The two estimates are similar, with $\hat{S}_m^{(\mathrm{LW})}(\cdot)$ being slightly wobblier. One contributor to this extra wobbliness is the fact that the lag window estimate has associated EDOFs of $\nu \doteq 34.4$, whereas we have $\nu \doteq 62.6$ for WOSA, i.e., almost twice as large. Since the variance of an SDF estimator is inversely proportional to its EDOFs, we can expect the lag window estimate to be more variable. The extra variability is reflected in the heights of the

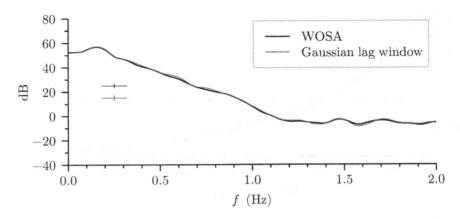

Figure 430 WOSA and Gaussian lag window spectral estimates for ocean wave data. The WOSA estimate is based on the Hanning data taper and $N_B = 33$ blocks (each with $N_S = 61$ data values) with 50% overlap between adjacent blocks. The lag window estimate is reproduced from Figure 319(d).

two crisscrosses, which indicate the widths of 95% CIs based upon one of the two SDF estimates. The width for the WOSA estimate is 3.1 dB as compared to 4.2 dB for the lag window estimate. The WOSA estimate is more stable because it has reclaimed information lost due to tapering in the direct spectral estimate underlying the lag window estimate. Nonetheless, the remarkable similarity of the two estimates lends credence to both.

Atomic Clock Data

Here we revisit the fractional frequency deviates $\{Y_t\}$ of Figure 326(b) and use them to illustrate multitaper-based methods for estimating the exponent of a power-law process and for estimating the innovation variance.

As discussed in Section 8.7, we can regard a portion of an SDF $S_Y(\cdot)$ as obeying a power law if there is an interval of frequencies f over which $\log(S_Y(f))$ versus $\log(f)$ varies linearly, with the slope of the line being equal to the exponent α of the power law. The light choppy curve in Figure 328(b) shows the periodogram for $\{Y_t\}$ on a decibel/log scale, which is equivalent to a log/log scale. The periodogram suggests a power law over $f \in (0, f_C]$ with a high frequency cutoff f_C somewhere between 0.01 and 0.1. A closer look at this periodogram and an examination of low-order sinusoidal multitaper estimates suggest setting f_C to the Fourier frequency $f_M = M/(N\Delta_t) \doteq 0.035$, where $M = 140$ (recall that $N = 3999$ and $\Delta_t = 1$ minute). With M thus determined, we only need to select the number K of sinusoidal tapers to fully specify the multitaper-based OLS estimator $\hat{\alpha}^{(\text{MT})}$ given by Equation (409e) (recall that $M_0 = \lceil (K+1)/2 \rceil$). We can objectively select K by considering the variance of $\hat{\alpha}^{(\text{MT})}$ given by Equation (410b). Evaluation of this variance for $K \geq 2$ indicates that it is minimized at $K = 7$, for which $\sqrt{\text{var}\{\hat{\alpha}^{(\text{MT})}\}} \doteq 0.11$. With this choice the estimator $\hat{\alpha}^{(\text{MT})}$ makes use of $\hat{S}_Y^{(\text{MT})}(\cdot)$ at Fourier frequencies ranging from $f_{M_0} = f_4$ to $f_M = f_{140}$.

Figure 431(a) shows the $K = 7$ sinusoidal multitaper SDF estimate $\hat{S}_Y^{(\text{MT})}(\cdot)$ of the fractional frequency deviates on a decibel/linear scale. Plot (b) also shows this estimate, but now on a decibel/log scale and with a focus on low frequencies. The estimate is plotted in three segments. The left-hand segment (light gray) shows $\hat{S}_Y^{(\text{MT})}(f_1)$ to $\hat{S}_Y^{(\text{MT})}(f_3)$; the middle (black jagged curve), $\hat{S}_Y^{(\text{MT})}(f_4)$ to $\hat{S}_Y^{(\text{MT})}(f_{140})$, for which the power law is in effect; and the right-hand (light gray), $\hat{S}_Y^{(\text{MT})}(\cdot)$ at frequencies just beyond the power-law interval. The black line through the middle segment is the line corresponding to the multitaper-based OLS fit. The slope of this line on a log/log scale is the desired estimate of α, namely, $\hat{\alpha}^{(\text{MT})} \doteq -1.21$ (the corresponding periodogram-based estimate is $\hat{\alpha}^{(\text{P})} \doteq -1.14$). If this power law were to

8.9 Examples of Multitaper and WOSA Spectral Estimators

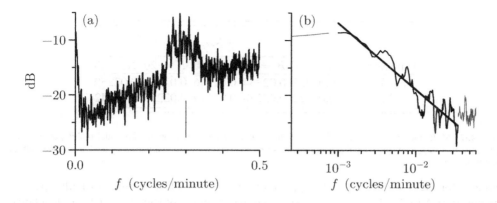

Figure 431 Sinusoidal multitaper SDF estimate $\hat{S}_Y^{(\mathrm{MT})}(\cdot)$ of order $K = 7$ for fractional frequency deviates $\{Y_t\}$ of Figure 326(b) (Figure 328 shows the periodogram for $\{Y_t\}$ and smoothed versions thereof). Plot (a) shows the estimate over $f \in (0, f_{\mathcal{N}}]$ on a decibel/linear scale, while (b) shows the estimate at low frequencies on a decibel/log scale. The portion of the estimate assumed to obey a power law is the black jagged curve in the center of (b), while the two thin curves show $\hat{S}_Y^{(\mathrm{MT})}(\cdot)$ before and after the power-law variation.

persist as $f \to 0$, the resulting SDF would not integrate to a finite value, which would imply an infinite variance, a condition in conflict with stationarity. We can, however, interpret an SDF of the form $C|f|^\alpha$ as $f \to 0$ with $-3 < \alpha \leq -1$ as being associated with a first-order intrinsically stationary process, i.e., a nonstationary process Y_t whose first difference $Y_t - Y_{t-1}$ is stationary (see C&E [3] for Section 7.12). In the context of fractional frequency deviates, intrinsically stationary processes associated with the canonical power laws $\alpha = -1$ and $\alpha = -2$ are referred to, respectively as flicker frequency noise and random walk frequency noise. Of these two, the estimate $\hat{\alpha}^{(\mathrm{MT})} \doteq -1.21$ is closer to the exponent for flicker frequency noise; however, if we take sampling variability into account, agreement with this type of noise is questionable since $\sqrt{\mathrm{var}\{\hat{\alpha}^{(\mathrm{MT})}\}} \doteq 0.11$ indicates that the estimate is almost two standard deviations away from $\alpha = -1$ (caution is in order here: the expression for $\mathrm{var}\{\hat{\alpha}^{(\mathrm{MT})}\}$ given by Equation (410b) is based on assumptions about multitaper SDF estimates that are reasonable for stationary processes but whose validity is a subject of conjecture for intrinsically stationary processes). We can conclude that interpreting the fractional frequencies deviates as flicker frequency noise is dicey (and using a stationary fractionally differenced process to model them over low frequencies is also problematic since stationarity would require $\alpha > -1$).

Let us now use the fractional frequency deviates $\{Y_t\}$ to illustrate multitaper-based estimation of the innovation variance σ_ϵ^2. This variance is one way to characterize frequency instability (Rutman, 1978). Thus, if we were to measure the fractional frequency deviates over one minute intervals for two clocks, we would declare the clock with the smaller innovation variance to be better at keeping time from one minute to the next (i.e., frequency instability is equated with unpredictability). The theory behind the estimator $\hat{\sigma}^2_{(\mathrm{MT})}$ of Equation (405b) presumes stationarity. Our analysis above of $\{Y_t\}$ as a low-frequency power-law process suggests that first-order intrinsic stationarity is a more appropriate model than stationarity. Percival (1983) shows that the innovation variance is invariant under differencing and proposes that the innovation variance for a first-order intrinsically stationary process be defined as the one connected with its first difference, which is necessarily stationary (this definition is in accordance with predictors for differenced ARMA(p,q) processes popularized in earlier editions of Box et al., 2015).

The first row of Table 432 shows multitaper-based estimates $\hat{\sigma}^2_{(\mathrm{MT})}$ of the innovation

	$\hat{\sigma}^2_{(\mathrm{MT})}$				
Time Series	$K=5$	$K=10$	$K=15$	$K=20$	$K=25$
$Y_t - Y_{t-1}$	0.02237	0.02244	0.02246	0.02251	0.02259
Y_t	0.02222	0.02228	0.02229	0.02235	0.02242

Table 432 Multitaper-based estimates $\hat{\sigma}^2_{(\mathrm{MT})}$ of innovation variance based on K sinusoidal multitapers for first difference $Y_t - Y_{t-1}$ of fractional frequency deviates (top row) and for fractional frequency deviates Y_t themselves (bottom).

variance for the first difference of $\{Y_t\}$ for various choices of the number of multitapers K ranging from 5 to 25. The differences between the various estimates are small, so the choice of K is not critical. The bottom row shows estimates based on $\{Y_t\}$ itself. There is little difference between these estimates and the ones based on the first differences, which indicates that the assessment of whether $\{Y_t\}$ is intrinsically stationary or stationary is also not critical in this example. (We reconsider these estimates of the innovation variance in Section 9.12, where we compare them to autoregressive procedures for forecasting the fractional frequency deviates.)

8.10 Summary of Combining Direct Spectral Estimators

We assume that the time series of interest is a realization of N contiguous random variables (RVs) $X_0, X_1, \ldots, X_{N-1}$ from a zero mean real-valued stationary process $\{X_t\}$ with sampling interval Δ_t (and hence Nyquist frequency $f_{\mathcal{N}} = 1/(2\Delta_t)$) and with a spectral density function (SDF) $S(\cdot)$ and an autocovariance sequence (ACVS) $\{s_\tau\}$ that form a Fourier transform pair: $\{s_\tau\} \longleftrightarrow S(\cdot)$. (If $\{X_t\}$ has an unknown mean, we replace X_t with $X'_t = X_t - \overline{X}$ in all computational formulae, where $\overline{X} = \sum_{t=0}^{N-1} X_t/N$ is the sample mean.) Based on this series, we create a collection of direct spectral estimators, each of which by itself is an estimator of $S(\cdot)$. The estimators in this collection are then combined together to form a new SDF estimator, with the intent that this new estimator will have statistical properties that improve upon those of any of the individual estimators in the collection. Two methods that follow this paradigm are multitapering and Welch's overlapped segment averaging (WOSA). In multitapering the direct spectral estimators in the collection are created by utilizing different data tapers over the entire time series (i.e., multiple tapers applied to one series); by contrast, in WOSA the estimators to be combined are created using the same taper, but applied to different subseries extracted from $X_0, X_1, \ldots, X_{N-1}$, each of the same length $N_S < N$ (i.e., one taper applied to multiple subseries). In both methods, the direct spectral estimators in the collection are usually combined using a simple average (weighted averages are sometimes used in multitapering). Individual summaries for multitapering and WOSA now follow. (We note in passing an interesting connection between the two methods, namely, that all WOSA estimators can be reexpressed as multitaper estimators (for details, see the discussion surrounding the right-hand column of Figure 421), but the converse is not true.)

Multitaper Spectral Analysis
Multitapering starts with a set of $K \geq 2$ orthonormal tapers $\{h_{k,t} : t = 0, 1, \ldots, N-1\}, k = 0, \ldots, K-1$; i.e.,

$$\sum_{t=0}^{N-1} h_{j,t} h_{k,t} = \begin{cases} 1 & j = k; \\ 0, & j \neq k. \end{cases} \qquad \text{(see (354b))}$$

These are used to create K direct spectral estimators $\hat{S}_k^{(\mathrm{MT})}(\cdot)$ – referred to as eigenspectra – that are then averaged together to form the basic multitaper estimator

$$\hat{S}^{(\mathrm{MT})}(f) \stackrel{\text{def}}{=} \frac{1}{K} \sum_{k=0}^{K-1} \hat{S}_k^{(\mathrm{MT})}(f) \text{ with } \hat{S}_k^{(\mathrm{MT})}(f) \stackrel{\text{def}}{=} \Delta_t \left| \sum_{t=0}^{N-1} h_{k,t} X_t e^{-i 2\pi f t \Delta_t} \right|^2. \quad \text{(see (352a))}$$

8.10 Summary of Combining Direct Spectral Estimators

The corresponding ACVS estimators are

$$\hat{s}_\tau^{(\mathrm{MT})} \stackrel{\mathrm{def}}{=} \frac{1}{K} \sum_{k=0}^{K-1} \hat{s}_{k,\tau}^{(\mathrm{MT})} \;\; \text{with} \;\; \hat{s}_{k,\tau}^{(\mathrm{MT})} \stackrel{\mathrm{def}}{=} \begin{cases} \sum_{t=0}^{N-|\tau|-1} h_{k,t+|\tau|} X_{t+|\tau|} h_{k,t} X_t, & |\tau| \leq N-1; \\ 0, & |\tau| \geq N, \end{cases}$$
(433)

and we have $\{\hat{s}_\tau^{(\mathrm{MT})}\} \longleftrightarrow \hat{S}^{(\mathrm{MT})}(\cdot)$ and $\{\hat{s}_{k,\tau}^{(\mathrm{MT})}\} \longleftrightarrow \hat{S}_k^{(\mathrm{MT})}(\cdot)$ (cf. Equation (188b)). The basic estimator is the most common form of multitapering, but a generalization of interest is a weighted multitaper estimator

$$\hat{S}^{(\mathrm{WMT})}(f) \stackrel{\mathrm{def}}{=} \sum_{k=0}^{K-1} d_k \hat{S}_k^{(\mathrm{MT})}(f), \qquad \text{(see (352b))}$$

where the weights d_k are nonnegative and are assumed to sum to unity (setting $d_k = 1/K$ yields the basic multitaper estimator).

Turning now to the statistical properties of the basic multitaper estimator, its first moment depends on the spectral windows associated with the tapers for the eigenspectra, namely,

$$\mathcal{H}_k(f) \stackrel{\mathrm{def}}{=} \Delta_t \left| \sum_{t=0}^{N-1} h_{k,t} e^{-i 2\pi f t \Delta_t} \right|^2. \qquad \text{(see (352c))}$$

In terms of these $\mathcal{H}_k(\cdot)$, we have

$$E\{\hat{S}^{(\mathrm{MT})}(f)\} = \int_{-f_\mathcal{N}}^{f_\mathcal{N}} \overline{\mathcal{H}}(f-f') S(f')\, df' \;\; \text{with} \;\; \overline{\mathcal{H}}(f) \stackrel{\mathrm{def}}{=} \frac{1}{K} \sum_{k=0}^{K-1} \mathcal{H}_k(f), \qquad \text{(see (353b))}$$

where $\overline{\mathcal{H}}(\cdot)$ is the spectral window for $\hat{S}^{(\mathrm{MT})}(\cdot)$. An appropriate measure of the bandwidth for the kth eigenspectrum is

$$B_{\mathcal{H}_k} \stackrel{\mathrm{def}}{=} \mathrm{width}_a\{\mathcal{H}_k(\cdot)\} = \frac{\left(\int_{-f_\mathcal{N}}^{f_\mathcal{N}} \mathcal{H}_k(f)\, df\right)^2}{\int_{-f_\mathcal{N}}^{f_\mathcal{N}} \mathcal{H}_k^2(f)\, df} = \frac{\Delta_t}{\sum_{\tau=-(N-1)}^{N-1} (h_k \star h_{k,\tau})^2}. \qquad \text{(see (353c))}$$

The corresponding measure for the basic multitaper estimator is

$$B_{\overline{\mathcal{H}}} \stackrel{\mathrm{def}}{=} \mathrm{width}_a\{\overline{\mathcal{H}}(\cdot)\} = \frac{\left(\int_{-f_\mathcal{N}}^{f_\mathcal{N}} \overline{\mathcal{H}}(f)\, df\right)^2}{\int_{-f_\mathcal{N}}^{f_\mathcal{N}} \overline{\mathcal{H}}^2(f)\, df} = \frac{\Delta_t}{\sum_{\tau=-(N-1)}^{N-1} \left(\frac{1}{K}\sum_{k=0}^{K-1} h_k \star h_{k,\tau}\right)^2}.$$
(see (353e))

Assuming that

[1] $\hat{S}^{(\mathrm{MT})}(f)$ is approximately unbiased,
[2] $\{X_t\}$ is a Gaussian process,
[3] $S(\cdot)$ is locally constant about f and
[4] f is not too close to zero or Nyquist frequency,

we have

$$\mathrm{var}\{\hat{S}^{(\mathrm{MT})}(f)\} \approx \frac{S^2(f)}{K}, \qquad \text{(see (354c))}$$

with the validity of the approximation increasing as $N \to \infty$. An appeal to Equation (204b) says that, asymptotically as $N \to \infty$,

$$\hat{S}_k^{(\mathrm{MT})}(f) \stackrel{\mathrm{d}}{=} \begin{cases} S(f)\chi_2^2/2, & 0 < |f| < f_\mathcal{N}; \\ S(f)\chi_1^2, & |f| = 0 \text{ or } f_\mathcal{N}. \end{cases} \qquad \text{(see (354e))}$$

To the degree that the eigenspectra can be taken to be mutually independent, the above implies that, as $N \to \infty$,

$$\hat{S}^{(\mathrm{MT})}(f) \stackrel{\mathrm{d}}{=} \begin{cases} S(f)\chi^2_{2K}/2K, & 0 < |f| < f_{\mathcal{N}}; \\ S(f)\chi^2_K/K, & |f| = 0 \text{ or } f_{\mathcal{N}}. \end{cases} \quad \text{(see (354f))}$$

Hence the equivalent degrees of freedom (EDOFs) ν associated with $\hat{S}^{(\mathrm{MT})}(f)$ is $2K$ for frequencies f not too close to either zero or Nyquist.

There are two sets of multitapers in wide use. The first is the Slepian multitapers, which Thomson (1982) advocated in his seminal paper on multitapering. Given a regularization bandwidth $2W$, the k-th taper $\{h_{k,t}\}$ is a normalized version of an eigenvector associated with k-th largest eigenvalue of the $N \times N$ covariance matrix $\boldsymbol{\Sigma}^{(\mathrm{BL})}$. The (i,j)th element of this matrix is $s^{(\mathrm{BL})}_{i-j}$, where $\{s^{(\mathrm{BL})}_\tau\}$ is the ACVS for band-limited white noise:

$$s^{(\mathrm{BL})}_\tau \begin{cases} 2W\,\Delta_t, & \tau = 0; \\ \sin(2\pi W \tau\,\Delta_t)/(\pi\tau), & \text{otherwise.} \end{cases} \quad \text{(see (379b))}$$

Typically W is set to a small multiple of the fundamental Fourier frequency $1/(N\,\Delta_t)$, i.e., $j/(N\,\Delta_t)$ with, e.g., $j = 2, 3, 4$ or 5, but sometimes non-integers such as 3.5 are used. The maximum number of usable multitapers is $\lfloor 2NW\,\Delta_t \rfloor - 1$; however, the actual number K of eigenspectra used to form $\hat{S}^{(\mathrm{MT})}(\cdot)$ is often set to less than the maximum if the eigenspectra $\hat{S}^{(\mathrm{MT})}_k(\cdot)$ for large k do not offer sufficient protection from leakage. Increasing W increases the number of leakage-free eigenspectra, thus allowing K to be increased; however, while increasing K decreases the variance of $\hat{S}^{(\mathrm{MT})}(\cdot)$, it also can result in a potential loss of resolution; i.e., $E\{\hat{S}^{(\mathrm{MT})}(\cdot)\}$ is an overly smoothed version of $S(\cdot)$. There is thus a bias/variance trade-off inherent in setting K. Unless K is close to the maximum number of usable multitapers, the bandwidth measure $B_{\overline{\mathcal{H}}}$ is markedly smaller than the regularization bandwidth $2W$.

The second set of multitapers in wide use is the sinusoidal multitapers, which were introduced by Riedel and Sidorenko (1995). In contrast to the Slepian tapers, there is a simple expression for the k-th order sinusoidal taper:

$$h_{k,t} = \left(\frac{2}{N+1}\right)^{1/2} \sin\left[\frac{(k+1)\pi(t+1)}{N+1}\right], \quad t = 0, 1\ldots, N-1; \quad \text{(see (392d))}$$

similar to the Slepian tapers, the sinusoidal tapers are related to the eigenvectors for an $N \times N$ covariance matrix, which we denote as $\boldsymbol{\Sigma}^{(\mathrm{PL})}$. The (i,j)th element of $\boldsymbol{\Sigma}^{(\mathrm{PL})}$ is $s^{(\mathrm{PL})}_{i-j}$, where

$$s^{(\mathrm{PL})}_\tau = \begin{cases} 1/12, & \tau = 0; \\ (-1)^\tau/(2\pi^2\tau^2), & \text{otherwise.} \end{cases} \quad \text{(see (391d))}$$

This ACVS corresponds to a power-law SDF $S(f) \propto f^2$. The eigenvectors for $\boldsymbol{\Sigma}^{(\mathrm{PL})}$ yield tapers that minimize a measure of bias in $\hat{S}^{(\mathrm{MT})}(\cdot)$ (see the discussion surrounding Equation (391b)). Riedel and Sidorenko (1995) refer to the multitapers arising from $\boldsymbol{\Sigma}^{(\mathrm{PL})}$ as minimum-bias multitapers. There is no simple expression for minimum-bias tapers, but the k-th order sinusoidal taper provides a remarkably good approximation with the simple expression displayed above. Use of the Slepian tapers requires specification of the regularization bandwidth $2W$, after which K is set to a value no greater than $\lfloor 2NW\,\Delta_t \rfloor - 1$. By contrast, the sinusoidal tapers only require specification of K. For any K, the sinusoidal tapers offer protection against leakage comparable to that offered by the popular Hanning data taper (by contrast, the Slepian tapers can offer substantially better protection, which is needed for certain SDFs, but not for many encountered in practice). With protection against leakage fixed, selection of K involves a bias/variance trade-off: increasing K decreases the variance of $\hat{S}^{(\mathrm{MT})}(\cdot)$, but can result in a potential loss of resolution. The bandwidth measure $B_{\overline{\mathcal{H}}}$ is well approximated by $(K+1)/[(N+1)\,\Delta_t]$.

WOSA Spectral Analysis
The WOSA spectral estimator starts with a single taper $\{h_t : t = 0, 1, \ldots, N_S - 1\}$ (a popular choice is the Hanning data taper of Equation (189b) – note that N in that equation must be replaced by N_S). The

8.10 Summary of Combining Direct Spectral Estimators

WOSA taper is used to create N_B direct spectral estimators that are then averaged together to form the WOSA estimator:

$$\hat{S}^{(\text{WOSA})}(f) \stackrel{\text{def}}{=} \frac{1}{N_B} \sum_{j=0}^{N_B-1} \hat{S}_{jn}^{(\text{D})}(f), \qquad \text{(see (414b))}$$

where $\hat{S}_{jn}^{(\text{D})}(\cdot)$ is the jth direct spectral estimator, which employs the single WOSA taper over a block of N_S contiguous time series values $X_{jn}, X_{jn+1}, \ldots, X_{jn+N_S-1}$:

$$\hat{S}_{jn}^{(\text{D})}(f) = \Delta_t \left| \sum_{t=0}^{N_S-1} h_t X_{t+jn} e^{-i2\pi f t \Delta_t} \right|^2 \qquad \text{(see (414a))}$$

(here $n > 0$ is a fixed integer-valued shift factor satisfying $0 < n \leq N_S$ and $n(N_B - 1) = N - N_S$). The corresponding ACVS estimator is

$$\hat{s}_\tau^{(\text{WOSA})} \stackrel{\text{def}}{=} \frac{1}{N_B} \sum_{j=0}^{N_B-1} \hat{s}_{jn,\tau}^{(\text{D})} \qquad (435a)$$

with

$$\hat{s}_{jn,\tau}^{(\text{D})} \stackrel{\text{def}}{=} \begin{cases} \sum_{t=0}^{N_S-|\tau|-1} h_{t+|\tau|} X_{t+jn+|\tau|} h_t X_{t+jn}, & |\tau| \leq N_S - 1; \\ 0, & |\tau| \geq N_S \end{cases} \qquad (435b)$$

and we have $\{\hat{s}_\tau^{(\text{WOSA})}\} \longleftrightarrow \hat{S}^{(\text{MT})}(\cdot)$ and $\{\hat{s}_{jn,\tau}^{(\text{D})}\} \longleftrightarrow \hat{S}_{jn}^{(\text{D})}(\cdot)$ (cf. Equation (188b)).

Turning now to the statistical properties of the WOSA estimator, its first moment depends on the spectral window associated with the WOSA taper, namely,

$$\mathcal{H}(f) = \Delta_t \left| \sum_{t=0}^{N_S-1} h_t e^{-i2\pi f t \Delta_t} \right|^2. \qquad \text{(cf. (186f))}$$

Since $E\{\hat{S}_{jn}^{(\text{D})}(f)\}$ is the same for all choices of j and n, it follows that

$$E\{\hat{S}^{(\text{WOSA})}(f)\} = E\{\hat{S}_{jn}^{(\text{D})}(f)\} = \int_{-f_\mathcal{N}}^{f_\mathcal{N}} \mathcal{H}(f - f') S(f') \, df'. \qquad \text{(cf. (414d) and (414e))}$$

An appropriate measure of the bandwidth for the WOSA estimator (and also for each of the individual direct spectral estimators used in creating this estimator) is

$$B_\mathcal{H} \stackrel{\text{def}}{=} \text{width}_a \{\mathcal{H}(\cdot)\} = \frac{\left(\int_{-f_\mathcal{N}}^{f_\mathcal{N}} \mathcal{H}(f) \, df \right)^2}{\int_{-f_\mathcal{N}}^{f_\mathcal{N}} \mathcal{H}^2(f) \, df} = \frac{\Delta_t}{\sum_{\tau=-(N_S-1)}^{N_S-1} (h \star h_\tau)^2}. \qquad \text{(cf. (194))}$$

Under assumptions [1] to [4] listed above for the multitaper estimator, we have

$$\text{var}\{\hat{S}^{(\text{WOSA})}(f)\} \approx \frac{S^2(f)}{N_B} \left(1 + 2 \sum_{m=1}^{N_B-1} \left(1 - \frac{m}{N_B} \right) \left| \sum_{t=0}^{N_S-1} h_t h_{t+mn} \right|^2 \right), \qquad \text{(see (415c))}$$

with the validity of the approximation increasing as $N \to \infty$. Using an argument similar to that leading to Equation (264a), the EDOFs ν for $\hat{S}^{(\text{WOSA})}(f)$ are

$$\nu \approx \frac{2N_B}{1 + 2 \sum_{m=1}^{N_B-1} \left(1 - \frac{m}{N_B} \right) \left| \sum_{t=0}^{N_S-1} h_t h_{t+mn} \right|^2}. \qquad \text{(see (415d))}$$

We can also state that, asymptotically as $N \to \infty$,

$$\hat{S}^{(\text{WOSA})}(f) \stackrel{\text{d}}{=} \begin{cases} S(f)\chi_\nu^2/\nu, & 0 < |f| < f_\mathcal{N}; \\ 2S(f)\chi_{\nu/2}^2/\nu, & |f| = 0 \text{ or } f_\mathcal{N}. \end{cases}$$

A popular way of specifying the WOSA estimator is to use the Hanning data taper with a 50% overlap between adjacent blocks; i.e., n is set to $\lfloor N_S/2 \rfloor$. In this case, we have

$$\nu \approx \frac{36 N_B^2}{19 N_B - 1} \approx 1.89 \left(\frac{2N}{N_S} - 1 \right). \qquad (\text{see } (416))$$

8.11 Exercises

[8.1] Show that, if each of the individual eigenspectra $\hat{S}_k^{(\text{MT})}(f)$ comprising the weighted multitaper estimator $\hat{S}^{(\text{WMT})}(f)$ of Equation (352b) is approximately unbiased, then $\hat{S}^{(\text{WMT})}(f)$ is also approximately unbiased if the weights d_k sum to unity. (Note that this result holds for the basic multitaper spectral estimator $\hat{S}^{(\text{MT})}(f)$ of Equation (352a): it is a special case of the weighted multitaper estimator, and its K weights $d_k = 1/K$ sum to unity.)

[8.2] Let the spectral window $\widetilde{\mathcal{H}}(\cdot)$ be defined as per Equation (353a) with $\sum_{k=0}^{K-1} d_k = 1$, and assume that the K data tapers $\{h_{0,t}\}, \ldots, \{h_{K-1,t}\}$ involved in the formulation of $\widetilde{\mathcal{H}}(\cdot)$ are orthonormal. Verify Equation (353d), which gives the bandwidth measure $B_{\widetilde{\mathcal{H}}}$ associated with $\widetilde{\mathcal{H}}(\cdot)$.

[8.3] Here we state a general result that leads to the approximation given in Equation (354a). We make use of an argument that closely parallels the one leading to Equation (211c) (the solution to the forthcoming Exercise [8.26] involves a similar argument). Assume that $\{X_t\}$ is a Gaussian stationary process with a spectral representation given by Equation (113b) with $\mu = 0$. Let

$$J_k(f) \stackrel{\text{def}}{=} \Delta_t^{1/2} \sum_{t=0}^{N-1} h_{k,t} X_t e^{-i2\pi f t \Delta_t} = \frac{1}{\Delta_t^{1/2}} \int_{-f_\mathcal{N}}^{f_\mathcal{N}} H_k(f - u)\, \mathrm{d}Z(u), \qquad (436a)$$

where the second equality follows from Equation (186c).

(a) Noting that $|J_k(f)|^2 = \hat{S}_k^{(\text{MT})}(f)$, show that

$$\operatorname{cov}\{\hat{S}_j^{(\text{MT})}(f'), \hat{S}_k^{(\text{MT})}(f)\} = \left| E\{J_j(f') J_k^*(f)\} \right|^2 + \left| E\{J_j(f') J_k(f)\} \right|^2,$$

where f and f' are any two fixed frequencies between 0 and $f_\mathcal{N}$.

(b) Show that the following two equalities hold:

$$E\{J_j(f') J_k^*(f)\} = \frac{1}{\Delta_t} \int_{-f_\mathcal{N}}^{f_\mathcal{N}} H_j(f' - u) H_k^*(f - u) S(u)\, \mathrm{d}u$$

$$E\{J_j(f') J_k(f)\} = \frac{1}{\Delta_t} \int_{-f_\mathcal{N}}^{f_\mathcal{N}} H_j(f' + u) H_k(f - u) S(u)\, \mathrm{d}u.$$

(c) Show that, if $S(\cdot)$ is locally constant about f'' and if f and f' are both close to f'' but with the stipulation that none of these frequencies are close to zero or $f_\mathcal{N}$, then

$$\operatorname{cov}\{\hat{S}_j^{(\text{MT})}(f'), \hat{S}_k^{(\text{MT})}(f)\} \approx S^2(f'') \left| \sum_{t=0}^{N-1} h_{j,t} h_{k,t} e^{-i2\pi(f'-f)t \Delta_t} \right|^2 \qquad (436b)$$

(note that setting $f'' = f' = f$ gives Equation (354a)).

[8.4] Using an argument analogous to the one leading to Equation (354c), show that Equation (354d) holds, which is an asymptotically valid approximation to the variance of the weighted multitaper estimator $\hat{S}^{(\text{WMT})}(f)$ of Equation (352b) for $0 < f < f_{\mathcal{N}}$. Argue that, if $K \geq 2$ and if the weights d_k are all positive, then this approximation to $\text{var}\{\hat{S}^{(\text{WMT})}(f)\}$ must be less than $S^2(f)$, i.e., the approximation to the variance of any of the individual eigenspectra used to form $\hat{S}^{(\text{WMT})}(f)$ (hint: recall the assumption that the weights sum to unity).

[8.5] Verify Equation (354g) under the assumptions that $E\{\hat{S}^{(\text{WMT})}(f)\} \approx S(f)$ and that the approximation for $\text{var}\{\hat{S}^{(\text{WMT})}(f)\}$ given by Equation (354d) holds.

[8.6] Here we derive the bandwidth measure of Equation (355b) (Greenhall, 2006). Assume that we have a time series that is a portion $X_0, X_1, \ldots, X_{N-1}$ of a zero mean Gaussian stationary process $\{X_t\}$ with sampling interval Δ_t and SDF $S(\cdot)$. Use this time series to form the weighted multitaper SDF estimator $\hat{S}^{(\text{WMT})}(\cdot)$ of Equation (352b), and assume that the tapers have unit energy and are pairwise orthogonal in the sense of Equation (354b).

  (a) Assuming that $S(\cdot)$, f and $f + \eta$ are such that we can use the approximation of Equation (436b), show that

$$\text{cov}\{\hat{S}^{(\text{WMT})}(f+\eta), \hat{S}^{(\text{WMT})}(f)\} \approx S^2(f) R^{(\text{WMT})}(\eta) \sum_{k=0}^{K-1} d_k^2,$$

where

$$R^{(\text{WMT})}(\eta) \stackrel{\text{def}}{=} \frac{\sum_{j=0}^{K-1} d_j \sum_{k=0}^{K-1} d_k \left|\sum_{t=0}^{N-1} h_{j,t} h_{k,t} e^{-i2\pi\eta t \Delta_t}\right|^2}{\sum_{k=0}^{K-1} d_k^2}.$$

Argue also that $R^{(\text{WMT})}(0) = 1$.

  (b) Using the definition of equivalent width given in Equation (100d), show that

$$\text{width}_e\{R^{(\text{WMT})}(\cdot)\} = \int_{-f_{\mathcal{N}}}^{f_{\mathcal{N}}} R^{(\text{WMT})}(\eta)\, d\eta = \frac{\sum_{t=0}^{N-1} \left(\sum_{k=0}^{K-1} d_k h_{k,t}^2\right)^2}{\Delta_t \sum_{k=0}^{K-1} d_k^2},$$

i.e., that Equation (355b) holds.

  (c) Consider now the special case $d_k = 1/K$, for which $\hat{S}^{(\text{WMT})}(\cdot)$ reduces to the basic multitaper SDF estimator $\hat{S}^{(\text{MT})}(\cdot)$ of Equation (352a). Show that the right-hand side of Equation (355b) reduces to the right-hand side of Equation (355a), and hence $\text{width}_e\{R^{(\text{WMT})}(\cdot)\}$ becomes $\text{width}_e\{R^{(\text{MT})}(\cdot)\}$.

[8.7] In Section 8.2 we presented an example of multitaper SDF estimation centered around the use of $NW = 4$ Slepian multitapers to estimate the SDF for the AR(4) time series shown in Figure 34(e). The focus of this exercise is to redo parts of this example, but now using $NW = 5.5$ Slepian multitapers, and to compare results for the two NW choices.

  (a) The right-hand columns of Figures 361 and 363 show eigenspectra of orders $k = 0, \ldots, 7$ based on $NW = 4$ multitapers. Compute and plot eigenspectra of the same orders but now using $NW = 5.5$ Slepian multitapers. Use these eigenspectra to compute and plot basic multitaper estimates of orders $K = 1, \ldots, 8$ for comparison with those shown in the right-hand columns of Figures 364 and 365 for the $NW = 4$ case. Comment on how the eigenspectra and basic multitaper estimates in the $NW = 4$ and $NW = 5.5$ cases compare.

  (b) The width of the horizontal part of the crisscross in each plot in the right-hand columns of Figures 364 and 365 shows the bandwidth measure $B_{\overline{\mathcal{H}}}$ for the corresponding basic $NW = 4$ Slepian multitaper estimate of orders $K = 1, \ldots, 8$. Recompute and tabulate these measures for comparison with similar measures based on $NW = 5.5$ Slepian multitapers. Comment on how the bandwidth measures in the $NW = 4$ and $NW = 5.5$ cases compare.

(Completion of this exercise requires access to the AR(4) time series, which is available from the "Data" part of the website for this book. The website also has the required $NW = 4$ and $NW = 5.5$ Slepian multitapers of length $N = 1024$ in case they are not otherwise readily available.)

[8.8] This exercise compares the χ^2-based approach and jackknifing for formulating a 95% CI of an unknown value $S(f)$ based upon multitaper SDF estimation. In what follows, set $f = 1/4$ throughout, and use $K = 5$ sinusoidal multitapers to create the basic multitaper estimator $\hat{S}^{(\mathrm{MT})}(f)$ and its associated eigenspectra $\hat{S}_k^{(\mathrm{MT})}(f)$.

(a) Figure 31(a) shows a realization of a Gaussian white noise process with zero mean, unit variance and SDF $S(f) = 1$ (this time series is accessible from the "Data" part of the website for this book). Using the first 64 values of this series, compute the eigenspectra $\hat{S}_k^{(\mathrm{MT})}(f)$, $k = 0, 1, \ldots, 4$, average them to form $\hat{S}^{(\mathrm{MT})}(f)$ and then create a 95% CI for $S(f)$ by using a modified version of Equation (265). Does this CI trap the true SDF for the white noise process at $f = 1/4$?

(b) Using the same $\hat{S}_k^{(\mathrm{MT})}(f)$ and $\hat{S}^{(\mathrm{MT})}(f)$ as in part (a), form a 95% CI for $S(f)$ via jackknifing as per Equation (357a). Does this second CI trap the true SDF?

(c) Generate a large number N_R (e.g., $N_\mathrm{R} = 10{,}000$) of realizations of Gaussian white noise with zero mean and unit variance, each of length $N = 64$. For each realization, repeat part (a). Determine the percentage of the N_R realizations for which the χ^2-based CI traps the true SDF at $f = 1/4$. Is this percentage in reasonable agreement with a coverage rate of 95%? Using the same N_R realizations, repeat part (b), and determine how often the jackknife CIs trap the true SDF. Is this percentage also in reasonable agreement with a 95% coverage rate? Repeat both coverage rate assessments using N_R realizations again, but now with lengths increased to $N = 256$ and $N = 1024$.

(d) Repeat part (c), but now using N_R realizations of the AR(2) process of Equation (34) (see Exercise [597] for a description of how to generate realizations from this process).

(e) Repeat part (c), but now for N_R realizations of the AR(4) process of Equation (35a) (see Exercise [11.1]).

[8.9] Suppose that $X_0, X_1, \ldots, X_{N-1}$ is a sample of a zero mean white noise process $\{X_t\}$ with variance s_0 and sampling interval Δ_t. Let $\{\tilde{h}_{j,t}\}$ and $\{\tilde{h}_{k,t}\}$ be any two orthonormal data tapers; i.e.,

$$\sum_{t=0}^{N-1} \tilde{h}_{j,t}\tilde{h}_{k,t} = 0 \quad \text{and} \quad \sum_{t=0}^{N-1} \tilde{h}_{j,t}^2 = \sum_{t=0}^{N-1} \tilde{h}_{k,t}^2 = 1.$$

For $l = j$ or k, define

$$\tilde{J}_l(f) = \Delta_t^{1/2} \sum_{t=0}^{N-1} \tilde{h}_{l,t} X_t e^{-i2\pi f t \Delta_t} \quad \text{so that} \quad \left|\tilde{J}_l(f)\right|^2 = \tilde{S}_l^{(\mathrm{MT})}(f)$$

(see Equation (372a)).

(a) Show that $\mathrm{cov}\{\tilde{J}_j(f), \tilde{J}_k(f)\} = 0$.

(b) Under the additional assumption that $\{X_t\}$ is a Gaussian process, use Equation (243b) to show that, for $l = j$ or k,

$$\mathrm{var}\{\tilde{S}_l^{(\mathrm{MT})}(f)\} = s_0^2 \Delta_t^2 \left(1 + \left|\sum_{t=0}^{N-1} \tilde{h}_{l,t}^2 e^{-i4\pi f t \Delta_t}\right|^2\right)$$

and that

$$\mathrm{cov}\{\tilde{S}_j^{(\mathrm{MT})}(f), \tilde{S}_k^{(\mathrm{MT})}(f)\} = s_0^2 \Delta_t^2 \left|\sum_{t=0}^{N-1} \tilde{h}_{j,t}\tilde{h}_{k,t} e^{-i4\pi f t \Delta_t}\right|^2.$$

What do the above equations reduce to if $f = 0$, $f_\mathcal{N}/2$ or $f_\mathcal{N}$?

[8.10] (a) Show that, if the real-valued matrix Q in Equation (374a) is symmetric, then $\hat{S}^{(Q)}(f)$ is real-valued.
(b) Suppose again that Q is real-valued, but now assume that $\hat{S}^{(Q)}(f)$ is real-valued rather than that Q is symmetric. Show that, without loss of generality, we can take Q to be symmetric.

[8.11] The quadratic estimator of Equation (374a) makes use of an $N \times N$ matrix Q of weights, the (s,t)th element of which is denoted by $Q_{s,t}$. The weighted multitaper estimator $\hat{S}^{(\text{WMT})}(\cdot)$ of Equation (352b), which is defined in terms of the eigenspectra $\hat{S}_k^{(\text{WMT})}(\cdot)$ of Equation (352a), is an example of a quadratic estimator. Determine $Q_{s,t}$ for $\hat{S}^{(\text{WMT})}(\cdot)$.

[8.12] Show that the first part of Equation (376d) holds, namely,

$$E\{\hat{S}^{(Q)}(f)\} = \Delta_t \operatorname{tr}\{Q \Sigma_Z\}.$$

[8.13] Suppose that Σ is an $N \times N$ real-valued symmetric positive definite matrix with eigenvalues $\lambda_0, \lambda_1, \ldots, \lambda_{N-1}$ (ordered from largest to smallest). Show that, if the largest eigenvalue λ_0 is distinct (i.e., $\lambda_0 > \lambda_1 \geq \cdots \geq \lambda_{N-1}$) and if A is any $N \times K$ real-valued matrix such that $\operatorname{tr}\{A^T A\} = 1$, then $\operatorname{tr}\{A^T \Sigma A\}$ is maximized when A is a normalized eigenvector associated with the eigenvalue λ_0. Hints:

(a) the eigenvalues of Σ must be positive; and
(b) if $v_0, v_1, \ldots, v_{N-1}$ are an orthonormal set of eigenvectors corresponding to the eigenvalues $\lambda_0, \lambda_1, \ldots, \lambda_{N-1}$, then each column of A can be expressed as a unique linear combination of the v_k terms.

[8.14] If A is an $N \times K$ matrix whose kth column is the rescaled Slepian data taper $\{h_{k,t}/\sqrt{K}\}$ of order k, verify that the broad-band bias indicator of Equation (379c) can be rewritten as

$$\operatorname{tr}\{A^T A\} - \operatorname{tr}\{A^T \Sigma^{(\text{BL})} A\} = 1 - \frac{1}{K} \sum_{k=0}^{K-1} \lambda_k(N, W)$$

(this result was stated in Equation (381)).

[8.15] Show that Equation (385a) holds. Hints: start with the expression for $\Upsilon_k(f')$ in Equation (384a), use Equation (383c), interchange the order of the integrals and the summation such that the innermost integral equals the integral on the right-hand side of Equation (384b), and then use the complex conjugate of Equation (383d) (recall that the DPSWF $U_k(\cdot)$ is a real-valued function). Finally, use the fact that both $dZ(\cdot)$ and $U_k(\cdot)$ are periodic functions with a period of unity.

[8.16] Show that, for the adaptive multitaper spectral estimator of Equation (389b),

$$\int_{-1/2}^{1/2} E\{\hat{S}^{(\text{AMT})}(f)\}\,df = \int_{-1/2}^{1/2} S(f') \left(\sum_{k=0}^{K-1} \lambda_k \int_{-1/2}^{1/2} \frac{b_k^2(f) U_k^2(f-f')}{\sum_{k=0}^{K-1} \lambda_k b_k^2(f)}\,df \right) df'.$$

Since the term in the parentheses above is not unity in general, it follows that, in contrast to both $\hat{S}^{(\text{MT})}(\cdot)$ and $\tilde{S}^{(\text{WMT})}(\cdot)$, Parseval's theorem is in general *not* satisfied exactly in expected value for $\hat{S}^{(\text{AMT})}(\cdot)$.

[8.17] Making use of the fact that the $U_k(\cdot)$ functions are orthonormal over $[-1/2, 1/2]$, we can replace the definition of the projections $\Upsilon_k(f')$ in Equation (384a) with

$$\tilde{\Upsilon}_k(f') = \int_{-1/2}^{1/2} \Upsilon(f+f') U_k(f)\,df.$$

(a) Show that $\tilde{\Upsilon}_k(f') = \Upsilon_k(f)/\sqrt{\lambda_k}$.
(b) We used the projection $\Upsilon_k(f')$ in Section 8.5 as partial motivation for the multitaper estimators of Equations (388e) and (389b). Explore what happens when $\tilde{\Upsilon}_k(f')$ is used instead.

[8.18] Figure 406 shows NRMSEs for various estimators of the innovation variance discussed in Section 8.7. The left-hand plot of that figure shows results based upon 10,000 realizations from the AR(2) process of Equation (34), and the right-hand plot, from the AR(4) process of Equation (35a). Consider now the AR(2) processes

$$X_t = 1.25 X_{t-1} - 0.75 X_{t-2} + \epsilon_t \text{ and } X_t = 1.8 X_{t-1} - 0.9 X_{t-2} + \epsilon_t,$$

where, in both cases, $\{\epsilon_t\}$ is a white noise process with unit variance (these are used as examples in Pukkila and Nyquist, 1985, and Walden, 1995).

  (a) Compute the SDFs for both processes, and plot them on a decibel scale versus frequency. Use these plots to determine the dynamic ranges of these SDFs as defined by Equation (177a) for comparison with those for the AR(2) process of Equation (34) (14 dB) and the AR(4) process of Equation (35a) (65 dB). (C&E [5] for Section 9.2 discusses computation of AR SDFs.)
  (b) For each of the two AR(2) process of interest here, create 10,000 realizations of sample size $N = 1024$ using the procedure described in Section 11.1. For each process, use these realizations to compute innovation variance estimates based upon the estimator $\hat{\sigma}^2_{(\text{DJ})}$ of Equation (404c); the estimator $\hat{\sigma}^2_{(\text{HN})}$ of Equation (404d) with L set to $2, 3, \ldots, 25$; the estimator $\hat{\sigma}^2_{(\text{PN})}$ of Equation (405a); and finally the estimator $\hat{\sigma}^2_{(\text{MT})}$ of Equation (405b) with K set to $2, 3, \ldots, 25$. For each of these estimates, compute NRMSEs analogous to that of Equation (406), and display the results in a manner similar to Figure 406. Comment upon your findings.

[8.19] Here we reformulate the multitaper-based simple linear regression model of Equation (409d) in standard vector/matrix notation (see, e.g., Weisberg, 2014), with the goal of verifying Equation (410b), which gives the variance of the OLS estimator $\hat{\alpha}^{(\text{MT})}$ of the power-law exponent α. Let Y be a column vector containing the responses $Y_k^{(\text{MT})}$, and let X be a matrix with two columns of predictors, the first column with elements all equal to unity, and the second, with elements equal to $x_k^{(\text{MT})}$ as defined in Equation (409e).

  (a) With β being a two-dimensional column vector of regression coefficients whose second element is α, argue that $Y = X\beta + \epsilon$ is equivalent to the model of Equation (409d), where ϵ is a column vector containing the error terms $\varepsilon_k^{(\text{MT})}$.
  (b) A standard result says that the OLS estimator of β is $\hat{\beta} = (X^T X)^{-1} X^T Y$ (see, e.g., Weisberg, 2014). Show that the second element of $\hat{\beta}$ is the multitaper-based OLS estimator $\hat{\alpha}^{(\text{MT})}$ of Equation (409e).
  (c) Taking the elements of the covariance matrix Σ for ϵ to be dictated by Equation (410a); noting that

$$\hat{\beta} = (X^T X)^{-1} X^T Y = (X^T X)^{-1} X^T (X\beta + \epsilon) = \beta + (X^T X)^{-1} X^T \epsilon;$$

and evoking a standard result from the theory of multivariate RVs, namely, that the covariance matrix for $M\epsilon$ is given by $M \Sigma M^T$, show that var$\{\hat{\alpha}^{(\text{MT})}\}$ is given by Equation (410b).

[8.20] Replicate the simulation study whose results are summarized in Tables 411a and 411b, which concern the variances of various estimators of the power-law exponent α associated with four different fractionally differenced (FD) processes. These tables make use of the sample variances of estimators of α based on 10,000 independent realizations of each FD process. In addition to variances, investigate the sample biases, squared biases and MSEs. To replicate this study, you need to generate simulated FD time series, which can be done using the circulant embedding method (Craigmile, 2003). This method is described in Section 11.2 and requires knowledge of the ACVS for an FD process, which can be computed recursively once the variance s_0 is known:

$$s_0 = \frac{\sigma_\epsilon^2 \Gamma(1 - 2\delta)}{\Gamma^2(1 - \delta)} \text{ and } s_\tau = s_{\tau-1} \frac{\tau + \delta - 1}{\tau - \delta}, \quad \tau = 1, 2, \ldots.$$

The ACVS involves the same two parameters σ_ϵ^2 and δ as does the FD SDF stated in Equation (407). The power-law exponent α and δ are related by $\alpha = -2\delta$. The statistical properties of

[8.21] Verify Equation (391c). Hint: argue that

$$\text{tr}\{\boldsymbol{D}_K \boldsymbol{H}^T \boldsymbol{\Sigma}^{(\text{PL})} \boldsymbol{H}\} = \Delta_t^2 \int_{-f_{\mathcal{N}}}^{f_{\mathcal{N}}} f^2 \widetilde{\mathcal{H}}(f)\, df.$$

You might also find the following indefinite integral to be helpful:

$$\int x^2 \cos(x)\, dx = 2x\cos(x) + (x^2 - 2)\sin(x).$$

[8.22] Show that, for a Kth-order basic minimum-bias estimator, the bias indicator of Equation (391c) takes the form of Equation (392c). Show also that the bias indicator increases as K increases.

[8.23] Given a time series of length N with sampling interval Δ_t, consider a multitaper SDF estimator based on K sinusoidal tapers $\{h_{0,t}\}, \{h_{1,t}\}, \ldots, \{h_{K-1,t}\}$, as defined by Equation (392d). Under the assumption that $2K \leq N$, show that the alternative bandwidth measure $\text{width}_e\{R^{(\text{MT})}(\cdot)\}$ of Equation (355a) reduces to $(K + \frac{1}{2})/[(N+1)\Delta_t]$ as stated in Equation (401) (Greenhall, 2006). Hint: define a periodic sequence $\{g_t\}$ of period $N+1$ such that, for $t = 0, 1, \ldots, N$,

$$g_t = \sum_{k=0}^{K-1} h_{k,t}^2;$$

make use of the following sequence of length $2K+1$:

$$G_k = \begin{cases} -\frac{1}{2}, & 1 \leq k \leq K; \\ K, & k = 0; \\ -\frac{1}{2}, & -K \leq k \leq -1; \end{cases}$$

and then consider Parseval's theorem (Equation (92b)).

[8.24] Let $\{X_{m,t}\}$, $m = 0, 1, \ldots, M-1$, be a set of $M \geq 2$ stationary processes with zero means, each of which has the same unknown ACVS $\{s_\tau\}$ and SDF $S(\cdot)$. Suppose that the RVs $X_{m,t}$ associated with distinct processes are independent of each other, which implies uncorrelatedness; i.e., $\text{cov}\{X_{m,t}, X_{m',t'}\} = 0$ for any t and t' when $m \neq m'$. Consider $\overline{X}_t \stackrel{\text{def}}{=} \frac{1}{\sqrt{M}} \sum_{m=0}^{M-1} X_{m,t}$.

(a) Show that $\{\overline{X}_t\}$ is also a zero mean stationary process, and determine its ACVS and SDF.

(b) Given a time series that is a realization of $X_{m,0}, X_{m,1}, \ldots, X_{m,N-1}$, we can form its periodogram, say $\hat{S}_m^{(\text{P})}(f)$. Doing this for $m = 0, 1, \ldots, M-1$ yields M periodograms, which we can average together to form

$$\bar{S}_X^{(\text{P})}(f) \stackrel{\text{def}}{=} \frac{1}{M} \sum_{m=0}^{M-1} \hat{S}_m^{(\text{P})}(f).$$

Given the M time series used to form the M periodograms, we can also construct a time series that is a realization of $\overline{X}_0, \overline{X}_1, \ldots, \overline{X}_{N-1}$ and form its periodogram, say $\hat{S}_{\overline{X}}^{(\text{P})}(f)$. Assuming that N is large enough and that f is sufficiently far enough from either zero or the Nyquist frequency such that all periodograms of interest are all approximately unbiased estimators with distributions dictated by a χ_2^2 RV times a multiplicative constant, compare the statistical properties of $\bar{S}_X^{(\text{P})}(f)$ (an average of periodograms from M independent time series) and $\hat{S}_{\overline{X}}^{(\text{P})}(f)$ (a periodogram formed by combining M independent time series).

[8.25] Suppose we have observed L time series, all of which can be regarded as independent realizations from the same stationary process with SDF $S(\cdot)$. Suppose that the lth such series has a unique sample size N_l and that the series are ordered such that $N_0 < N_1 < \cdots < N_{L-1}$. Suppose we use the lth series to form a direct spectral estimate $\hat{S}_l^{(D)}(f)$ at some frequency f such that $0 < f < f_\mathcal{N}$, where we assume that each $\hat{S}_l^{(D)}(f)$ is an unbiased estimator of $S(f)$. We then combine the L different direct spectral estimates together to form

$$\hat{S}(f) = \sum_{l=0}^{L-1} \alpha_l \hat{S}_l^{(D)}(f),$$

where $\alpha_l > 0$ for all l.

    (a) What condition do we need to impose on the weights α_l so that $\hat{S}(f)$ is also an unbiased estimator of $S(f)$?

    (b) Assuming that (i) the α_l's are chosen so that $\hat{S}(f)$ is unbiased and (ii) the usual approach to approximating the distribution of direct spectral estimates holds, determine the EDOFs ν for $\hat{S}(f)$.

    (c) For an unbiased estimator, how should the weights α_l be set so that ν is maximized? Hint: for real-valued constants a_l and b_l, the Cauchy inequality says that $|\sum_l a_l b_l|^2 \leq \sum_l a_l^2 \sum_l b_l^2$, with equality holding if and only if $a_l = c b_l$ for some constant c.

    (d) Does the fact that the time series all have different sample sizes N_l influence the EDOFs ν?

[8.26] Here we derive Equation (415b) using an argument closely paralleling the one leading to Equation (211c) (see also Exercise [8.3]). We assume that $\{X_t\}$ is a zero mean Gaussian stationary process. Let

$$J_l(f) \stackrel{\text{def}}{=} \Delta_t^{1/2} \sum_{t=0}^{N_S-1} h_t X_{t+l} e^{-i 2\pi f t \Delta_t} \quad \text{so} \quad |J_l(f)|^2 = \hat{S}_l^{(D)}(f).$$

    (a) Show that

$$\text{cov}\{\hat{S}_l^{(D)}(f), \hat{S}_m^{(D)}(f)\} = \left| E\{J_l(f) J_m^*(f)\} \right|^2 + \left| E\{J_l(f) J_m(f)\} \right|^2.$$

    (b) Show that

$$J_l(f) = \frac{1}{\Delta_t^{1/2}} \int_{-f_\mathcal{N}}^{f_\mathcal{N}} e^{i 2\pi f' l \Delta_t} H(f - f')\, dZ(f'). \qquad \text{(cf. (186c))}$$

    (c) Show that the following two equalities hold:

$$E\{J_l(f) J_m^*(f)\} = \frac{1}{\Delta_t} \int_{-f_\mathcal{N}}^{f_\mathcal{N}} e^{i 2\pi f'(l-m)\Delta_t} H(f-f') H^*(f-f') S(f')\, df'$$

$$E\{J_l(f) J_m(f)\} = \frac{1}{\Delta_t} \int_{-f_\mathcal{N}}^{f_\mathcal{N}} e^{i 2\pi f'(l-m)\Delta_t} H(f+f') H(f-f') S(f')\, df'.$$

    (d) Show that, if $S(\cdot)$ is locally constant about f and if f is not near 0 or $f_\mathcal{N}$,

$$\text{cov}\{\hat{S}_l^{(D)}(f), \hat{S}_m^{(D)}(f)\} \approx S^2(f) \left| \sum_{t=0}^{N_S-1} h_t h_{t+|l-m|} \right|^2,$$

where $h_t = 0$ for $t \geq N_S$.

[8.27] Verify the approximations for ν given in Equation (416). Hint: assume the block size N_S is large enough that any summations can be approximated by integrals.

[8.28] In C&E [5] for Section 8.8 we compared two ways of obtaining the EDOFs ν for a WOSA spectral estimator. The first way is via the right-hand approximation from Equation (416), and the second, via Equation (354g), which is based on the WOSA estimator being a special case of a quadratic

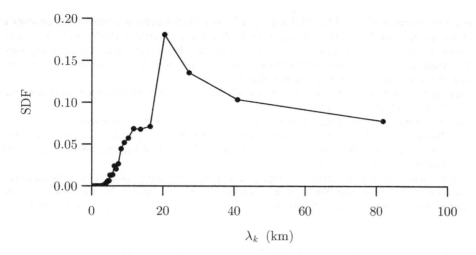

Figure 443 Estimated wavelength spectrum of ice sheet topography (Joughin, 1995, Figure 7.2(a)). Note that the plot of $\hat{S}_\lambda(\lambda_k)$ versus λ_k is on a linear/linear scale.

spectral estimator (Equation (374a)). Using the same setup as before (a sample size of $N = 1024$, a Hanning data taper and 50% overlap between blocks) but now considering seven different block sizes, namely, $N_S = 8, 16, 32, 64, 128, 256$ and 512, compute ν using each of the three approximations stated in Equation (416) and also via Equation (354g). How well do these four ways of determining ν match up?

[8.29] Consider the following variation on the WOSA SDF estimator for a time series of length $N = 2048$. We start with a standard WOSA estimator using $N_B = 15$ blocks extracted from the time series. Each block is of size $N_S = 256$; there is a 50% overlap between adjacent blocks; and we use the Hanning data taper. We evaluate the resulting SDF estimate over the grid of Fourier frequencies associated with the block size N_S. Let $\hat{S}^{(\text{WOSA})}(f_j)$ be the resulting WOSA estimator at $f_j = j/N_S$. We then smooth the the WOSA estimate across frequencies using a filter whose impulse response sequence is "triangular" and nonzero at five values:

$$\hat{S}(f_j) = \sum_{k=-2}^{2} \frac{3-|k|}{9} \hat{S}^{(\text{WOSA})}(f_{j-k}).$$

(a) Show that the weights used to form $\hat{S}(f_j)$ lead to an approximately unbiased estimator.
(b) Does a claim that the estimator $\hat{S}(f_j)$ has an EDOFs of 24 make sense?

[8.30] Figure 443 is taken from a study of ice sheet topography (Joughin, 1995) and shows an estimated spectrum $\hat{S}_\lambda(\lambda_k) \stackrel{\text{def}}{=} \hat{S}(1/\lambda_k)$ versus wavelength $\lambda_k \stackrel{\text{def}}{=} 1/f_k$ in kilometers (km), where $f_k = k/(N\Delta_t)$ is the kth Fourier frequency, $N = 512$, and $\Delta_t = 0.16$ km. The estimate $\hat{S}(f_k)$ is based upon averaging individual SDF estimates $\hat{S}_j^{(\text{D})}(f_k)$ for 11 "time" series $\{X_{j,t}\}$, $j = 1, \ldots, 11$, which can be regarded as pairwise uncorrelated; i.e.,

$$\hat{S}(f_k) = \frac{1}{11} \sum_{j=1}^{11} \hat{S}_j^{(\text{D})}(f_k).$$

Prior to computing each $\hat{S}_j^{(\text{D})}(\cdot)$, the time series was detrended by estimating a least squares quadratic fit and then subtracting this quadratic fit from the original series (i.e., $X_{j,t} - \hat{a}_j - \hat{b}_j t - \hat{c}_j t^2$ defines the jth detrended time series, where $\hat{a}_j$, $\hat{b}_j$ and $\hat{c}_j$ are the least squares estimates of the quadratic determined from $\{X_{j,t}\}$). Each $\hat{S}_j^{(D)}(f_k)$ is a direct spectral estimate computed using a Hanning data taper. The purpose of estimating the SDF was to determine which wavelengths make the most significant contributions to ice sheet topography. Two conclusions from the study were that "there are significant contributions to the topography at all wavelengths greater than 5 km" and that "there is a sharp peak in the spectrum near $\lambda_k = 20$ km."

    (a) At which wavenumbers is $\hat{S}_\lambda(\lambda_k)$ most likely to be affected by the detrending operation, and how would it be affected?

    (b) Is tapering useful for the purpose for which the spectrum is being estimated?

    (c) Based upon $\hat{S}_\lambda(\lambda_4) \doteq 0.18$, compute a 95% CI for the true spectrum at $\lambda_4 = N\Delta_t/4 = 20.48$ km. How would you use such a CI to assess the claim of a sharp peak in the spectrum near a wavelength of 20 km?

[8.31] Some commercial spectrum analyzers employ "exponential averaging" of direct spectral estimates based upon nonoverlapping blocks of data, with each block containing N_S data points. To be specific, the nth block of data, namely, $X_{nN_S}, X_{nN_S+1}, \ldots, X_{(n+1)N_S-1}$, is used to form

$$\hat{S}_n^{(D)}(f) = \Delta_t \left| \sum_{t=0}^{N_S-1} h_t X_{nN_S+t} e^{-i2\pi ft\Delta_t} \right|^2, \quad n = 0, 1, \ldots$$

(as usual, Δ_t is the sampling interval and $\{h_t\}$ is a properly normalized data taper). After the nth block of data is processed, the spectrum analyzer displays

$$\tilde{S}_n(f) \stackrel{\text{def}}{=} \alpha \tilde{S}_{n-1}(f) + (1-\alpha)\hat{S}_n^{(D)}(f)$$

for $n \geq 1$, where $0 < \alpha < 1$ and $\tilde{S}_0(f) = \hat{S}_0^{(D)}(f)$. Assume that $\{X_t\}$ is a zero mean stationary process with SDF $S(\cdot)$ and Nyquist frequency $f_\mathcal{N}$; that $\hat{S}_n^{(D)}(f) \stackrel{d}{=} S(f)\chi_2^2/2$ when $0 < f < f_\mathcal{N}$ (as per Equation (204b)); and that $\hat{S}_m^{(D)}(f)$ and $\hat{S}_n^{(D)}(f)$ are independent for $m \neq n$.

    (a) Find the mean, variance and EDOFs for $\tilde{S}_n(f)$ when $0 < f < f_\mathcal{N}$. What happens to these three quantities as $n \to \infty$?

    (b) Suppose we want to use the estimator $\tilde{S}_n(f)$ to construct a 95% CI for $S(f)$ such that the width of the interval is no greater than a certain fixed amount, which we take to be either 3 dB, 2 dB or 1 dB. The caption to Figure 266 indicates that we achieve these desired widths as soon as the EDOFs for $\tilde{S}_n(f)$ become, respectively, 66, 146 or 581. For each of these three cases, to what value should α be set such that we achieve the desired EDOFs as quickly as possible, i.e., when n is small as possible?

(This exercise is an expanded version of problem 11.5, Roberts and Mullis, 1987.)

9

Parametric Spectral Estimators

9.0 Introduction

In this chapter we discuss the basic theory behind parametric spectral density function (SDF) estimation. The main idea is simple. Suppose the discrete parameter stationary process $\{X_t\}$ has an SDF $S(\cdot)$ that is completely determined by K parameters $a_1, \ldots, a_K$:

$$S(f) = S(f; a_1, \ldots, a_K).$$

Using a time series that can be regarded as a realization of this process, suppose we can estimate the parameters of $S(\cdot)$ by, say, $\hat{a}_1, \ldots, \hat{a}_K$. If these parameter estimates are reasonable, then

$$\hat{S}(f) = S(f; \hat{a}_1, \ldots, \hat{a}_K)$$

should be a reasonable estimate of $S(f)$.

9.1 Notation

In what follows, it is important to keep in mind what basic assumptions are in effect. Accordingly, in this chapter we adopt these notational conventions for the following discrete parameter stationary processes, all of which are assumed to be real-valued and have zero mean:

[1] $\{X_t\}$ represents an arbitrary such process;
[2] $\{Y_t\}$, an *autoregressive* process of finite order p;
[3] $\{G_t\}$, a *Gaussian* process; and
[4] $\{H_t\}$, a *Gaussian autoregressive* process of finite order p.

For any of these four processes, we denote the autocovariance sequence (ACVS) by $\{s_\tau : \tau \in \mathbb{Z}\}$, and we assume there is a corresponding SDF $S(\cdot)$ with an associated Nyquist frequency $f_\mathcal{N} \stackrel{\text{def}}{=} 1/(2\Delta_t)$, where Δ_t is the sampling interval between values in the process (hence $S(\cdot)$ is periodic with a period of $2f_\mathcal{N}$).

9.2 The Autoregressive Model

The most widely used form of parametric SDF estimation involves an autoregressive model of order p, denoted as AR(p), as the underlying functional form for $S(\cdot)$. Recall that a stationary AR(p) process $\{Y_t\}$ with zero mean satisfies the equation

$$Y_t = \phi_{p,1} Y_{t-1} + \cdots + \phi_{p,p} Y_{t-p} + \epsilon_t = \sum_{j=1}^{p} \phi_{p,j} Y_{t-j} + \epsilon_t, \tag{446a}$$

where $\phi_{p,1}, \ldots, \phi_{p,p}$ are p fixed coefficients, and $\{\epsilon_t\}$ is a white noise process with zero mean and finite variance $\sigma_p^2 \stackrel{\text{def}}{=} \text{var}\{\epsilon_t\}$ (we also assume $\sigma_p^2 > 0$). The process $\{\epsilon_t\}$ is often called the *innovation process* that is associated with the AR(p) process, and σ_p^2 is called the *innovation variance*. Since we will have to refer to AR(p) models of different orders, we include the order p of the process in the notation for its parameters. The parameterization of the AR model given in Equation (446a) is the same as that used by, e.g., Box et al. (2015), but the reader should be aware that there is another common way of writing an AR(p) model, namely,

$$Y_t + \varphi_{p,1} Y_{t-1} + \cdots + \varphi_{p,p} Y_{t-p} = \epsilon_t,$$

where $\varphi_{p,j} = -\phi_{p,j}$. This convention is used, for example, by Priestley (1981) and is prevalent in the engineering literature. Equation (446a) emphasizes the analogy of an AR(p) model to a multiple linear regression model, but it is only an analogy and not a simple correspondence: if we regard Equation (446a) as a regression model, the predictors $Y_{t-1}, \ldots, Y_{t-p}$ are just lagged copies of the response Y_t. This setup does not fit into the usual way of thinking about regression models (e.g., the mean function in such models is the expected value of the response conditional on the predictors, and hence, when modeling a time series of length N, certain (unconditioned) responses would also need to serve as conditioned predictors).

The SDF for a stationary AR(p) process is given by

$$S(f) = \frac{\sigma_p^2 \, \Delta_t}{\left| 1 - \sum_{j=1}^{p} \phi_{p,j} e^{-i 2\pi f j \Delta_t} \right|^2} \tag{446b}$$

(cf. Equation (145a) with $\Delta_t = 1$; the above is a periodic function with a period of $2f_{\mathcal{N}}$). Here we have $p+1$ parameters, namely, the $\phi_{p,j}$ coefficients and σ_p^2, all of which we must estimate to produce an AR(p) SDF estimate. The coefficients cannot be chosen arbitrarily if $\{Y_t\}$ is to be a stationary process. As we noted in Section 5.4, a necessary and sufficient condition for the existence of a stationary solution to $\{Y_t\}$ of Equation (446a) is that

$$G(f) = 1 - \sum_{j=1}^{p} \phi_{p,j} e^{-i 2\pi f j} \neq 0 \text{ for any } f \in \mathbb{R} \tag{446c}$$

(see Equation (144d)). Another way to state this condition is that none of the solutions to the polynomial equation $1 - \sum_{j=1}^{p} \phi_{p,j} z^{-j} = 0$ lies *on* the unit circle in the complex plane (i.e., has an absolute value equal to unity). When $p = 1$, we can thus achieve stationarity as long as $\phi_{1,1} \neq \pm 1$. In addition to stationarity, an important property for $\{Y_t\}$ to possess is causality (see item [1] in the Comments and Extensions [C&Es] at the end of this section for a discussion of causality). The process $\{Y_t\}$ is causal if the roots of $1 - \sum_{j=1}^{p} \phi_{p,j} z^{-j}$ all lie *inside* the unit circle; if any are outside, it is acausal. Causality implies stationarity, but the converse is not true. An AR(1) process is causal (and hence stationary) if $|\phi_{1,1}| < 1$. For an

AR process of general order p, causality implies that cov $\{\epsilon_s, Y_t\} = 0$ when $s > t$, which is a key property we put to good use in the next section.

The rationale for this particular class of parametric SDFs is six-fold.

[1] Any continuous SDF $S(\cdot)$ can be approximated arbitrarily well by an AR(p) SDF if p is chosen large enough (Anderson, 1971, p. 411). Thus the class of AR processes is rich enough to approximate a wide range of processes. Unfortunately "large enough" can well mean an order p that is too large compared to the amount of available data.

[2] There exist efficient algorithms for fitting AR(p) models to time series. This might seem like a strange justification, but, since the early days of spectral analysis, recommended methodology has often been governed by what can in practice be calculated with commonly available computers.

[3] For a Gaussian process $\{G_t\}$ with autocovariances known up to lag p, the maximum entropy spectrum is identical to that of an AR(p) process. We discuss the principle of maximum entropy and comment upon its applicability to spectral analysis in Section 9.6.

[4] A side effect of fitting an AR(p) process to a time series for the purpose of SDF estimation is that we have simultaneously estimated a linear predictor and potentially identified a linear prewhitening filter for the series (the role of prewhitening in SDF estimation is discussed in Section 6.5). In Section 9.10 we describe an overall approach to SDF estimation that views parametric SDF estimation as a good method for determining appropriate prewhitening filters.

[5] For certain phenomena a physical argument can be made that an AR model is appropriate. A leading example is the acoustic properties of human speech (Rabiner and Schafer, 1978, chapter 3).

[6] Pure sinusoidal variations can be expressed as an AR model with $\sigma_p^2 = 0$. This fact – and its implications – are discussed in more detail in Section 10.12.

We have already encountered two examples of AR(p) processes in Chapter 2: the AR(2) process of Equation (34) and the AR(4) process of Equation (35a). The SDF for the AR(2) process is plotted as thick curves in the upper two rows of Figures 172 and 173; the SDF for the AR(4) process, in the lower two rows. Four realizations of each process are shown in Figure 34.

To form an AR(p) SDF estimate from a given set of data, we face two problems: first, determination of the order p that is most appropriate for the data and, second, estimation of the parameters $\phi_{p,1}, \ldots, \phi_{p,p}$ and σ_p^2. Typically we determine p by examining how well a range of AR models fits our data (as judged by some criterion), so it is necessary first to assume that p is known and to learn how to estimate the parameters for an AR(p) model.

Comments and Extensions to Section 9.2

[1] As noted in this section, as long as $|\phi_{1,1}| \neq 1$, there is a unique stationary solution to the equation $Y_t = \phi_{1,1} Y_{t-1} + \epsilon_t$, $t \in \mathbb{Z}$, where $\{\epsilon_t\}$ is a zero mean white noise process with finite variance $\sigma_p^2 > 0$. The burden of Exercise [9.1a] is to show that these solutions are

$$Y_t = \sum_{j=0}^{\infty} \phi_{1,1}^j \epsilon_{t-j} \text{ when } |\phi_{1,1}| < 1 \text{ and } Y_t = -\sum_{j=1}^{\infty} \phi_{1,1}^{-j} \epsilon_{t+j} \text{ when } |\phi_{1,1}| > 1 \qquad (447)$$

(note that Exercise [2.17a] eludes to the first solution). By definition the first solution is *causal* because Y_t only depends upon random variables (RVs) in the white noise process $\{\epsilon_s : s \in \mathbb{Z}\}$ such that $s \leq t$; by contrast, the second solution is *purely acausal* because Y_t depends solely upon RVs occurring *after* index t. For AR(p) processes in general, the condition that all of the roots of $1 - \sum_{j=1}^{p} \phi_{p,j} z^{-j}$ lie inside the unit circle guarantees that there is a unique stationary solution to Equation (446a) of the form

$$Y_t = \sum_{j=0}^{\infty} \psi_j \epsilon_{t-j},$$

where the ψ_j weights depend on $\phi_{p,1}, \ldots, \phi_{p,p}$ (in the AR(1) case, $\psi_j = \phi_{1,1}^j$). This causal solution implies that cov$\{\epsilon_s, Y_t\} = 0$ when $s > t$. When none of the roots of $1 - \sum_{j=1}^{p} \phi_{p,j} z^{-j}$ lies on the unit circle but one or more are outside of it, the unique stationary solution takes the form

$$Y_t = \sum_{j=-\infty}^{\infty} \psi_j \epsilon_{t-j},$$

where at least some of the ψ_j weights for $j < 0$ are nonzero (if all of the roots are outside of the unit circle, then $\psi_j = 0$ for all $j \geq 0$, and the solution is purely acausal). Acausality implies that we cannot guarantee cov$\{\epsilon_s, Y_t\}$ is zero when $s > t$.

When we have a time series that we regard as a realization of a portion $Y_0, Y_1, \ldots, Y_{N-1}$ of an AR(p) process, the question arises as to whether parameter estimates based upon this realization correspond to those for a causal (and hence stationary) AR process. Since causality implies stationarity (but not vice versa), causality is harder to satisfy than stationarity. Certain estimation procedures guarantee causality (e.g., the Yule–Walker estimator discussed in the next section – see Brockwell and Davis, 2016, section 5.1.1, which points to Brockwell and Davis, 1991, problem 8.3); others do not (e.g., least squares estimators – see Section 9.7). In cases where the parameter estimates correspond to an acausal stationary process, the estimated coefficients can still be substituted into Equation (446b) to produce an SDF estimator with all of the properties of a valid SDF (i.e., it is nonnegative everywhere, symmetric about the origin and integrates to a finite number); moreover, the argument that led us to deduce the form of an AR SDF assumes only stationarity and not causality (see the discussion following Equation (145a)). Additionally, for any SDF arising from an acausal stationary AR(p) process, there is a corresponding causal (and hence necessarily stationary) AR(p) process with *identically* the same SDF (see Exercise [9.1b] or Brockwell and Davis, 1991, section 4.4).

[2] We have stated the stationarity and causality conditions in terms of roots of the polynomial $1 - \sum_{j=1}^{p} \phi_{p,j} z^{-j}$, but an alternative formulation is to use $1 - \sum_{j=1}^{p} \phi_{p,j} z^{j}$ (see, e.g., Brockwell and Davis, 1991, section 4.4). In terms of this latter polynomial, the condition for causality is that all of its roots lie *outside* the unit circle; however, the stationarity condition is unchanged, namely, that none of the roots of the polynomial lies *on* the unit circle.

[3] Priestley (1981) relates the idea of causality to the concept of *asymptotic stationarity* (see his section 3.5.4). To understand the main idea behind this concept, consider an AR(1) process $Y_t = \phi_{1,1} Y_{t-1} + \epsilon_t$, for which the stationarity condition is $|\phi_{1,1}| \neq 1$. Given a realization of $\epsilon_0, \epsilon_1, \ldots$ of the white noise process $\{\epsilon_t\}$, suppose we define $\widetilde{Y}_0 = Z_0 + \epsilon_0$ and $\widetilde{Y}_t = \phi_{1,1} \widetilde{Y}_{t-1} + \epsilon_t$ for $t \geq 1$, where Z_0 is an RV with finite variance that is uncorrelated with $\{\epsilon_t\}$. The same argument leading to Equation (44a) says that

$$\widetilde{Y}_t = \phi_{1,1}^t Z_0 + \sum_{j=0}^{t-1} \phi_{1,1}^j \epsilon_{t-j}.$$

Since var$\{\widetilde{Y}_t\}$ depends on t, the process $\{\widetilde{Y}_t : t = 0, 1, \ldots\}$ is nonstationary; however, as t gets larger and larger, we can argue that, if $|\phi_{1,1}| < 1$ so that $\phi_{1,1}^t, \phi_{1,1}^{t+1}, \ldots$ become smaller and smaller, then $\widetilde{Y}_t$ resembles more and more

$$Y_t = \sum_{j=0}^{\infty} \phi_{1,1}^j \epsilon_{t-j},$$

which is the causal (and hence stationary) solution to $Y_t = \phi_{1,1} Y_{t-1} + \epsilon_t$; i.e., $\{\widetilde{Y}_t\}$ is asymptotically stationary. For an AR(1) process the condition for asymptotic stationarity is thus identical to the causality condition, namely, $|\phi_{1,1}| < 1$. This correspondence continues to hold for other AR(p) processes: the condition for asymptotic stationarity is the same as the condition for causality, namely, that the roots of $1 - \sum_{j=1}^{p} \phi_{p,j} z^{-j}$ all lie inside the unit circle. Asymptotic stationarity justifies setting $Y_{-p} = Y_{-p+1} = \cdots = Y_{-1} = 0$ when using Equation (446a) to simulate an AR time series – after a suitable burn-in period of, say, length M, the influence of the initial settings for Y_t when $t < 0$ should be small, and, to a certain degree of accuracy, we can regard $Y_M, Y_{M+1}, \ldots, Y_{N+M-1}$ as a realization of length N from a causal (and hence stationary) AR process (see Section 11.1 for a method for simulating AR processes that avoids a burn-in period).

After his discussion of asymptotic stationarity, Priestley (1981) notes the existence of acausal stationary solutions, but then states that henceforth "... when we refer to the 'condition for stationarity' for an AR model, we shall mean the condition under which a stationary solution exists with Y_t expressed in terms of present and past ϵ_t's only" (see his p. 135). This redefinition of stationarity is widely used; however, its usage is potentially confusing because stationarity by this new definition excludes certain AR processes deemed to be stationary under the technically correct definition. Throughout this chapter, whenever we refer to a stationary AR model, we do *not* restrict ourselves to a causal model; when we do need to deal with just causal models, we will use the qualifier "causal (and hence stationary)".

[4] In Section 8.7 we defined the innovation variance for a purely nondeterministic stationary process $\{X_t\}$ to be $E\{(X_t - \widehat{X}_t)^2\}$, where $\widehat{X}_t$ is the best linear predictor of X_t given the infinite past X_{t-1}, $X_{t-2}, \ldots$ (see Equation (404a)). A causal (and hence stationary) AR(p) process $\{Y_t\}$ is purely nondeterministic, and the best linear predictor of Y_t given the infinite past is $\widehat{Y}_t = \sum_{j=1}^{p} \phi_{p,j} Y_{t-j}$; i.e., the predictor depends just on the p most recent RVs and not on the distant past $Y_{t-p-1}, Y_{t-p-2}, \ldots$. Since $E\{(Y_t - \widehat{Y}_t)^2\} = E\{\epsilon_t^2\} = \sigma_p^2$, the use of "innovation variance" here to describe σ_p^2 is consistent with its definition in Equation (404a).

[5] Following ideas presented in C&E [1] for Section 3.11, we can efficiently compute the AR(p) SDF of Equation (446b) over a grid of frequencies by using zero padding in conjunction with an FFT algorithm. To do so, let $N' \geq p+1$ be any integer that the algorithm deems to be acceptable, and define $\phi_{p,0} = -1$ and $\phi_{p,j} = 0$ for $j = p+1, \ldots, N'-1$; i.e., we pad the sequence $\phi_{p,0}, \phi_{p,1}, \ldots, \phi_{p,p}$ with $N' - (p+1)$ zeros. Ignoring Δ_t in Equation (91b), the DFT of the zero padded sequence is

$$G_k = \sum_{j=0}^{N'-1} \phi_{p,j} e^{-i 2\pi k j/N'}, \quad k = 0, 1, \ldots, N'-1, \tag{449a}$$

and we have

$$\frac{\sigma_p^2 \Delta_t}{|G_k|^2} = \frac{\sigma_p^2 \Delta_t}{\left|1 - \sum_{j=1}^{p} \phi_{p,j} e^{-i 2\pi f'_k j \Delta_t}\right|^2} = S(f'_k), \tag{449b}$$

where $f'_k = k/(N' \Delta_t)$.

It should be noted that sharp peaks in the SDF might not be accurately represented if the grid of frequencies is not fine enough. C&E [3] for Section 10.12 discusses how to accurately locate such peaks, after which we can use Equation (446b) to check their accurate representation.

9.3 The Yule–Walker Equations

The oldest method of estimating the parameters for a causal (and hence stationary) AR(p) process $\{Y_t\}$ with zero mean and ACVS given by $s_\tau = E\{Y_{t+\tau} Y_t\}$ is based upon matching lagged moments, for which we need to express the parameters in terms of the ACVS. To do so, we first take Equation (446a) and multiply both sides of it by $Y_{t-\tau}$ to get

$$Y_{t-\tau} Y_t = \sum_{j=1}^{p} \phi_{p,j} Y_{t-\tau} Y_{t-j} + Y_{t-\tau} \epsilon_t. \tag{449c}$$

If we take the expectation of both sides, we have, for $\tau > 0$,

$$s_\tau = \sum_{j=1}^{p} \phi_{p,j} s_{\tau-j}, \qquad (450a)$$

where $E\{Y_{t-\tau}\epsilon_t\} = 0$ due to causality (as noted in the C&Es for the previous section, causality says that $Y_{t-\tau}$ can be written as an infinite linear combination of $\epsilon_{t-\tau}$, $\epsilon_{t-\tau-1}$, $\epsilon_{t-\tau-2}$, ..., but is uncorrelated with RVs in $\{\epsilon_t\}$ that occur after time $t-\tau$). Let $\tau = 1, 2, \ldots, p$ in Equation (450a) and recall that $s_{-j} = s_j$ to get the following p equations, known as the *Yule–Walker equations*:

$$\begin{aligned} s_1 &= \phi_{p,1} s_0 + \phi_{p,2} s_1 + \cdots + \phi_{p,p} s_{p-1} \\ s_2 &= \phi_{p,1} s_1 + \phi_{p,2} s_0 + \cdots + \phi_{p,p} s_{p-2} \\ &\vdots \quad\quad \vdots \quad\quad \vdots \quad\quad \ddots \quad\quad \vdots \\ s_p &= \phi_{p,1} s_{p-1} + \phi_{p,2} s_{p-2} + \cdots + \phi_{p,p} s_0 \end{aligned} \qquad (450b)$$

or, in matrix notation,

$$\boldsymbol{\gamma}_p = \boldsymbol{\Gamma}_p \boldsymbol{\Phi}_p,$$

where $\boldsymbol{\gamma}_p \stackrel{\text{def}}{=} [s_1, s_2, \ldots, s_p]^T$; $\boldsymbol{\Phi}_p \stackrel{\text{def}}{=} [\phi_{p,1}, \phi_{p,2}, \ldots, \phi_{p,p}]^T$; and

$$\boldsymbol{\Gamma}_p \stackrel{\text{def}}{=} \begin{bmatrix} s_0 & s_1 & \cdots & s_{p-1} \\ s_1 & s_0 & \cdots & s_{p-2} \\ \vdots & \vdots & \ddots & \vdots \\ s_{p-1} & s_{p-2} & \cdots & s_0 \end{bmatrix}. \qquad (450c)$$

If the covariance matrix $\boldsymbol{\Gamma}_p$ is positive definite (as is always the case in practical applications), we can now solve for $\phi_{p,1}, \phi_{p,2}, \ldots, \phi_{p,p}$ in terms of the lag 0 to lag p values of the ACVS:

$$\boldsymbol{\Phi}_p = \boldsymbol{\Gamma}_p^{-1} \boldsymbol{\gamma}_p. \qquad (450d)$$

Given a time series that is a realization of a portion $X_0, X_1, \ldots, X_{N-1}$ of *any* discrete parameter stationary process with zero mean and SDF $S(\cdot)$, one possible way to fit an AR(p) model to it is to replace s_τ in $\boldsymbol{\Gamma}_p$ and $\boldsymbol{\gamma}_p$ with the sample ACVS

$$\hat{s}_\tau^{(\text{P})} \stackrel{\text{def}}{=} \frac{1}{N} \sum_{t=0}^{N-|\tau|-1} X_{t+|\tau|} X_t$$

to produce estimates $\widetilde{\boldsymbol{\Gamma}}_p$ and $\widetilde{\boldsymbol{\gamma}}_p$. (How reasonable this procedure is for an arbitrary stationary process depends on how well it can be approximated by a stationary AR(p) process.) If the time series is not known to have zero mean (the usual case), we need to replace $\hat{s}_\tau^{(\text{P})}$ with

$$\frac{1}{N} \sum_{t=0}^{N-|\tau|-1} (X_{t+|\tau|} - \overline{X})(X_t - \overline{X}),$$

where $\overline{X}$ is the sample mean. We noted in C&E [2] for Section 6.2 that, with the above form for $\hat{s}_\tau^{(\text{P})}$ when $|\tau| \leq N-1$ and with $\hat{s}_\tau^{(\text{P})}$ set to zero when $|\tau| \geq N$, a realization of the sequence

$\{\hat{s}_\tau^{(\mathrm{P})}\}$ is positive definite if and only if the realizations of $X_0, \ldots, X_{N-1}$ are not all exactly the same (Newton, 1988, p. 165). This mild condition holds in all practical applications of interest. The positive definiteness of the sequence $\{\hat{s}_\tau^{(\mathrm{P})}\}$ in turn implies that the matrix $\widetilde{\boldsymbol{\Gamma}}_p$ is positive definite. Hence we can obtain estimates for AR(p) coefficients from

$$\widetilde{\boldsymbol{\Phi}}_p = \widetilde{\boldsymbol{\Gamma}}_p^{-1} \widetilde{\boldsymbol{\gamma}}_p. \tag{451a}$$

We are not quite done: we still need to estimate σ_p^2. To do so, let $\tau = 0$ in Equation (449c) and take expectations to get

$$s_0 = \sum_{j=1}^p \phi_{p,j} s_j + E\{Y_t \epsilon_t\}.$$

From the fact that $E\{Y_{t-j}\epsilon_t\} = 0$ for $j > 0$, it follows that

$$E\{Y_t \epsilon_t\} = E\left\{\left(\sum_{j=1}^p \phi_{p,j} Y_{t-j} + \epsilon_t\right)\epsilon_t\right\} = \sigma_p^2$$

and hence

$$\sigma_p^2 = s_0 - \sum_{j=1}^p \phi_{p,j} s_j. \tag{451b}$$

This equation suggests that we estimate σ_p^2 by

$$\tilde{\sigma}_p^2 \stackrel{\mathrm{def}}{=} \hat{s}_0^{(\mathrm{P})} - \sum_{j=1}^p \tilde{\phi}_{p,j} \hat{s}_j^{(\mathrm{P})}. \tag{451c}$$

We call the estimators $\widetilde{\boldsymbol{\Phi}}_p$ and $\tilde{\sigma}_p^2$ the *Yule–Walker estimators* of the AR(p) parameters. These estimators depend only on the sample ACVS up to lag p and can be used to estimate the SDF of $\{X_t\}$ by

$$\hat{S}^{(\mathrm{YW})}(f) \stackrel{\mathrm{def}}{=} \frac{\tilde{\sigma}_p^2 \Delta_t}{\left|1 - \sum_{j=1}^p \tilde{\phi}_{p,j} e^{-i2\pi f j \Delta_t}\right|^2}.$$

An important property of the Yule-Walker estimators is that the fitted AR(p) process has a theoretical ACVS that is *identical* to the sample ACVS up to lag p. This forced agreement implies that

$$\int_{-f_\mathcal{N}}^{f_\mathcal{N}} \hat{S}^{(\mathrm{YW})}(f) e^{i2\pi f \tau \Delta_t} \, df = \hat{s}_\tau^{(\mathrm{P})}, \quad \tau = 0, 1, \ldots, p,$$

which is of questionable value: as we have seen in Chapter 7, the Fourier transform of $\{\hat{s}_\tau^{(\mathrm{P})}\}$ is the periodogram, which can suffer from severe bias (Sakai et al., 1979, discuss connections between $\hat{S}^{(\mathrm{YW})}(\cdot)$ and the periodogram). For $\tau = 0$, however, the above says that

$$\int_{-f_\mathcal{N}}^{f_\mathcal{N}} \hat{S}^{(\mathrm{YW})}(f) \, df = \hat{s}_0^{(\mathrm{P})}, \tag{451d}$$

which is a useful property because it facilitates comparing the Yule–Walker SDF estimator with other estimators that also integrate to the sample variance. Under certain reasonable assumptions on a stationary process $\{X_t\}$, Berk (1974) has shown that $\hat{S}^{(\text{YW})}(f)$ is a consistent estimator of $S(f)$ for all f if the order p of the approximating AR process is allowed to increase as the sample size N increases. Unfortunately, this result is not particularly informative if one has a time series of fixed length N.

Equations (450b) and (451b) can be combined to produce the so-called *augmented Yule–Walker equations*:

$$\begin{bmatrix} s_0 & s_1 & \cdots & s_p \\ s_1 & s_0 & \cdots & s_{p-1} \\ \vdots & \vdots & \ddots & \vdots \\ s_p & s_{p-1} & \cdots & s_0 \end{bmatrix} \begin{bmatrix} 1 \\ -\phi_{p,1} \\ \vdots \\ -\phi_{p,p} \end{bmatrix} = \begin{bmatrix} \sigma_p^2 \\ 0 \\ \vdots \\ 0 \end{bmatrix}.$$

The first row above follows from Equation (451b), while the remaining p rows are just transposed versions of Equations (450b). This formulation is sometimes useful for expressing the estimation problem so that all the AR(p) parameters can be found simultaneously using a routine designed for solving a Toeplitz system of equations (see C&E [3] for Section 9.4).

Given the ACVS of $\{X_t\}$ or its estimator up to lag p, Equations (450d) and (451a) formally require matrix inversions to obtain the values of the actual or estimated AR(p) coefficients. As we shall show in the next section, there is an interesting alternative way to relate these coefficients to the true or estimated ACVS that avoids matrix inversion by brute force and clarifies the relationship between AR SDF estimation and the so-called maximum entropy method (see Section 9.6).

Comments and Extensions to Section 9.3

[1] There is no reason to insist upon substituting the usual biased estimator of the ACVS into Equation (450d) to produce estimated values for the AR coefficients. Any other reasonable estimates for the ACVS can be used – the only requirement is that the estimated sequence be positive definite so that the matrix inversion in Equation (450d) can be done. For example, if the process mean is assumed to be 0, the direct spectral estimator $\hat{S}^{(\text{D})}(\cdot)$ of Equation (186b) yields an estimate of the ACVS of the form

$$\hat{s}_\tau^{(\text{D})} = \sum_{t=0}^{N-|\tau|-1} h_{t+|\tau|} X_{t+|\tau|} h_t X_t$$

(see Equation (188b)), where $\{h_t\}$ is the data taper used with $\hat{S}^{(\text{D})}(\cdot)$. As is true for $\{\hat{s}_\tau^{(\text{P})}\}$, realizations of the sequence $\{\hat{s}_\tau^{(\text{D})}\}$ are always positive definite in practical applications. Since proper use of tapering ensures that the first-moment properties of $\hat{S}^{(\text{D})}(\cdot)$ are better overall than those of the periodogram, it makes some sense to use $\{\hat{s}_\tau^{(\text{D})}\}$ – the inverse Fourier transform of $\hat{S}^{(\text{D})}(\cdot)$ – rather than the usual biased estimator $\{\hat{s}_\tau^{(\text{P})}\}$ – the inverse Fourier transform of the periodogram. For an example, see C&E [2] for Section 9.4.

9.4 The Levinson–Durbin Recursions

The Levinson–Durbin recursions are an alternative way of solving for the AR coefficients in Equation (450b). Here we follow Papoulis (1985) and derive the equations that define the recursions by considering the following problem: given values of $X_{t-1}, \ldots, X_{t-k}$ of a stationary process $\{X_t\}$ with zero mean, how can we predict the value of X_t? A mathematically tractable solution is to find that *linear* function of $X_{t-1}, X_{t-2}, \ldots, X_{t-k}$, say

$$\vec{X}_t(k) \stackrel{\text{def}}{=} \sum_{j=1}^{k} \phi_{k,j} X_{t-j}, \tag{452}$$

9.4 The Levinson–Durbin Recursions

such that the *mean square linear prediction error*

$$\sigma_k^2 \stackrel{\text{def}}{=} E\left\{\left(X_t - \vec{X}_t(k)\right)^2\right\} = E\left\{\left(X_t - \sum_{j=1}^{k} \phi_{k,j} X_{t-j}\right)^2\right\} \qquad (453a)$$

is minimized. We call $\vec{X}_t(k)$ the *best linear predictor* of X_t, given $X_{t-1}, \ldots, X_{t-k}$. We note the following.

[1] For reasons that will become clear in the next few paragraphs, we are purposely using the same symbols for denoting the coefficients in Equation (452) and the coefficients of an AR(k) process, which $\{X_t\}$ need *not* be. For the time being, the reader should regard Equation (452) as a new definition of $\phi_{k,j}$ – we will show that this agrees with our old definition if in fact $\{X_t\}$ is an AR(k) process (see the discussion concerning Equation (454b) and its connection to Equation (450a)).

[2] The quantity $\vec{X}_t(k)$ is a scalar – it is not a vector as the arrow over the X might suggest to readers familiar with textbooks on physics. The arrow is meant to connote "forward prediction."

[3] When the stationary process in question is an AR(p) process, then, following our conventions, we should replace X with Y on the right-hand side of Equation (453a). If we also replace k with p and then appeal to Equation (446a), the right-hand side of Equation (453a) becomes var$\{\epsilon_t\}$, and, just below Equation (446a), we defined σ_p^2 to be equal to this variance. Hence this previous definition for σ_p^2 is in agreement with the definition in Equation (453a).

[4] If we don't have any values prior to X_t that we can use to predict it, we take its prediction to be just its expected value, namely, $E\{X_t\} = 0$. Hence we augment Equations (452) and (453a), which assume $k \geq 1$, by taking

$$\vec{X}_t(0) \stackrel{\text{def}}{=} 0 \text{ and } \sigma_0^2 \stackrel{\text{def}}{=} E\left\{\left(X_t - \vec{X}_t(0)\right)^2\right\} = E\{X_t^2\} = s_0. \qquad (453b)$$

If we denote the prediction error associated with the best linear predictor by

$$\vec{e}_t(k) \stackrel{\text{def}}{=} X_t - \vec{X}_t(k), \qquad (453c)$$

we can then derive the following result. For any set of real-valued numbers $\psi_1, \ldots, \psi_k$, define

$$P_k(\psi_1, \ldots, \psi_k) = E\left\{\left(X_t - \sum_{l=1}^{k} \psi_l X_{t-l}\right)^2\right\}.$$

Since $P_k(\cdot)$ is a quadratic function of the ψ_l terms and since the best linear predictor is defined as that set of ψ_l terms that minimizes the above, we can find the $\phi_{k,l}$ terms by differentiating the above equation and setting the derivatives to zero:

$$\frac{dP_k}{d\psi_j} = -2E\left\{\left(X_t - \sum_{l=1}^{k} \psi_l X_{t-l}\right) X_{t-j}\right\} = 0, \qquad 1 \leq j \leq k. \qquad (453d)$$

Since $\psi_j = \phi_{k,j}$ for all j at the solution and since

$$X_t - \sum_{l=1}^{k} \phi_{k,l} X_{t-l} = \vec{e}_t(k),$$

Equation (453d) reduces to the so-called *orthogonality principle*, namely,

$$E\{\vec{\epsilon}_t(k) X_{t-j}\} = \text{cov}\{\vec{\epsilon}_t(k), X_{t-j}\} = 0, \qquad 1 \leq j \leq k. \tag{454a}$$

In words, the orthogonality principle says that the prediction error is uncorrelated with all RVs utilized in the prediction. Note that, at the solution, $P_k(\phi_{k,1}, \ldots, \phi_{k,k}) = \sigma_k^2$.

If we rearrange Equations (453d), we are led to a series of equations that allows us to solve for $\phi_{k,l}$, namely,

$$E\{X_t X_{t-j}\} = \sum_{l=1}^{k} \phi_{k,l} E\{X_{t-l} X_{t-j}\}, \qquad 1 \leq j \leq k,$$

or, equivalently in terms of the ACVS,

$$s_j = \sum_{l=1}^{k} \phi_{k,l} s_{j-l}, \qquad 1 \leq j \leq k. \tag{454b}$$

Comparison of the above with Equation (450a), which generates the Yule–Walker equations, shows that they are identical! Thus, Equation (450d) here shows that, if the covariance matrix is positive definite, the $\phi_{k,l}$ terms are necessarily unique and hence $\vec{X}_t(k)$ is unique. This uniqueness leads to a useful corollary to the orthogonality principle, which is the subject of Exercise [9.5].

The reader should note that the $\phi_{k,l}$ terms of the best linear predictor are uniquely determined by the covariance properties of the process $\{X_t\}$ – we have not discussed so far the practical problem of estimating these coefficients for a process with an unknown covariance structure.

For the Yule–Walker equations for an AR(p) process, the innovation variance σ_p^2 can be related to values of the ACVS and the AR(p) coefficients (Equation (451b)). By an analogous argument, the mean square linear prediction error σ_k^2 can be expressed in terms of the ACVS and the $\phi_{k,j}$ terms:

$$\begin{aligned}\sigma_k^2 = E\{\vec{\epsilon}_t^2(k)\} &= E\left\{\vec{\epsilon}_t(k)\left(X_t - \vec{X}_t(k)\right)\right\} \\ &= E\{\vec{\epsilon}_t(k) X_t\} = E\left\{\left(X_t - \vec{X}_t(k)\right) X_t\right\} = s_0 - \sum_{j=1}^{k} \phi_{k,j} s_j,\end{aligned} \tag{454c}$$

where we have appealed to the orthogonality principle in order to go from the first line to the second. From the first expression on the second line, we note the important fact that, since both $\vec{\epsilon}_t(k)$ and X_t have zero mean,

$$\sigma_k^2 = E\{\vec{\epsilon}_t(k) X_t\} = \text{cov}\{\vec{\epsilon}_t(k), X_t\}; \tag{454d}$$

i.e., the mean square linear prediction error is just the covariance between the prediction error $\vec{\epsilon}_t(k)$ and X_t, the quantity being predicted.

To summarize our discussion to this point, the Yule–Walker equations arise in two related problems:

[1] For a stationary AR(p) process $\{Y_t\}$ with zero mean, ACVS $\{s_\tau\}$ and coefficients $\phi_{p,1}$, ..., $\phi_{p,p}$, the Yule–Walker equations relate the coefficients to the ACVS; the auxiliary

Equation (451b) gives the innovation variance in terms of $\{s_\tau\}$ and the AR(p) coefficients. Sampling versions of these equations allow us to use an AR(p) process to approximate a time series that can be regarded as a portion of an arbitrary stationary process $\{X_t\}$.

[2] For a stationary process $\{X_t\}$ with zero mean and ACVS $\{s_\tau\}$, the Yule–Walker equations relate the coefficients of the best linear predictor of X_t, given the k most recent prior values, to the ACVS; The auxiliary Equation (454c) gives the mean square linear prediction error in terms of $\{s_\tau\}$ and the coefficients of the best linear predictor.

The fact that the two problems are related implies that, for the AR(p) process $\{Y_t\}$, the mean square linear prediction error of its pth-order linear predictor is identical to its innovation variance, both of which are denoted by σ_p^2.

Before we get to the heart of our derivation of the Levinson–Durbin recursions, we note the following seemingly trivial fact: if $\{X_t\}$ is a stationary process, then so is $\{X_{-t}\}$, the process with time reversed. Since

$$E\{X_{-t+\tau}X_{-t}\} = E\{X_{t+\tau}X_t\} = s_\tau,$$

both $\{X_t\}$ and $\{X_{-t}\}$ have the same ACVS. It follows from Equation (454b) that $\overleftarrow{X}_t(k)$, the best linear "predictor" of X_t given the next k future values, can be written as

$$\overleftarrow{X}_t(k) \stackrel{\text{def}}{=} \sum_{j=1}^{k} \phi_{k,j} X_{t+j},$$

where $\phi_{k,j}$ is *exactly* the same coefficient occurring in the best linear predictor of X_t, given the k most recent prior values. The orthogonality principle applied to the reversed process tells us that

$$E\{\overleftarrow{\epsilon}_t(k) X_{t+j}\} = \text{cov}\{\overleftarrow{\epsilon}_t(k), X_{t+j}\} = 0, \quad 1 \leq j \leq k, \quad \text{where } \overleftarrow{\epsilon}_t(k) \stackrel{\text{def}}{=} X_t - \overleftarrow{X}_t(k).$$

From now on we refer to $\overrightarrow{X}_t(k)$ and $\overleftarrow{X}_t(k)$, respectively, as the *forward* and *backward* predictors of X_t of length k. We call $\overrightarrow{\epsilon}_t(k)$ and $\overleftarrow{\epsilon}_t(k)$ the corresponding *forward* and *backward prediction errors*. It follows from symmetry that

$$E\{\overleftarrow{\epsilon}_t^2(k)\} = E\{\overrightarrow{\epsilon}_t^2(k)\} = \sigma_k^2$$

and that the analog of Equation (454d) is

$$\sigma_k^2 = E\{\overleftarrow{\epsilon}_{t-k}(k) X_{t-k}\} = \text{cov}\{\overleftarrow{\epsilon}_{t-k}(k), X_{t-k}\}. \tag{455a}$$

The Levinson–Durbin algorithm follows from an examination of the following equation:

$$\overrightarrow{\epsilon}_t(k) = \overrightarrow{\epsilon}_t(k-1) - \theta_k \overleftarrow{\epsilon}_{t-k}(k-1), \tag{455b}$$

where θ_k is a constant to be determined. This equation is by no means obvious, but it is clearly plausible: $\overrightarrow{\epsilon}_t(k)$ depends upon one more variable than $\overrightarrow{\epsilon}_t(k-1)$, namely, X_{t-k}, upon which $\overleftarrow{\epsilon}_{t-k}(k-1)$ obviously depends. To prove Equation (455b), we first note that, by the orthogonality principle,

$$E\{\overrightarrow{\epsilon}_t(k-1) X_{t-j}\} = 0 \quad \text{and} \quad E\{\overleftarrow{\epsilon}_{t-k}(k-1) X_{t-j}\} = 0$$

for $j = 1, \ldots, k - 1$, and, hence, for *any* constant θ_k,

$$E\{[\vec{e}_t(k-1) - \theta_k \overleftarrow{e}_{t-k}(k-1)] X_{t-j}\} = 0, \qquad j = 1, \ldots, k-1. \tag{456a}$$

The equation above will also hold for $j = k$ if we let

$$\theta_k = \frac{E\{\vec{e}_t(k-1) X_{t-k}\}}{E\{\overleftarrow{e}_{t-k}(k-1) X_{t-k}\}} = \frac{E\{\vec{e}_t(k-1) X_{t-k}\}}{\sigma^2_{k-1}} \tag{456b}$$

(this follows from Equation (455a)). With this choice of θ_k and with

$$d_t(k) \stackrel{\text{def}}{=} \vec{e}_t(k-1) - \theta_k \overleftarrow{e}_{t-k}(k-1),$$

we have

$$E\{d_t(k) X_{t-j}\} = 0, \qquad 1 \leq j \leq k.$$

The corollary to the orthogonality principle stated in Exercise [9.5] now tells us that $d_t(k)$ is in fact the same as $\vec{e}_t(k)$, thus completing the proof of Equation (455b).

For later use, we note the time-reversed version of Equation (455b), namely,

$$\overleftarrow{e}_{t-k}(k) = \overleftarrow{e}_{t-k}(k-1) - \theta_k \vec{e}_t(k-1). \tag{456c}$$

We are now ready to extract the Levinson–Durbin recursions. It readily follows from Equation (455b) that

$$\vec{X}_t(k) = \vec{X}_t(k-1) + \theta_k \left(X_{t-k} - \overleftarrow{X}_{t-k}(k-1) \right);$$

the definitions of $\vec{X}_t(k)$, $\vec{X}_t(k-1)$ and $\overleftarrow{X}_{t-k}(k-1)$ further yield

$$\sum_{j=1}^{k} \phi_{k,j} X_{t-j} = \sum_{j=1}^{k-1} \phi_{k-1,j} X_{t-j} + \theta_k \left(X_{t-k} - \sum_{j=1}^{k-1} \phi_{k-1,j} X_{t-k+j} \right).$$

Equating coefficients of X_{t-j} (and appealing to the fundamental theorem of algebra) yields

$$\phi_{k,j} = \phi_{k-1,j} - \theta_k \phi_{k-1,k-j}, \qquad 1 \leq j \leq k-1, \text{ and } \phi_{k,k} = \theta_k.$$

Given $\phi_{k-1,1}, \ldots, \phi_{k-1,k-1}$ and $\phi_{k,k}$, we can therefore compute $\phi_{k,1}, \ldots, \phi_{k,k-1}$. Moreover, $\phi_{k,k}$ can be expressed using $\phi_{k-1,1}, \ldots, \phi_{k-1,k-1}$ and the ACVS up to lag k. To see this, reconsider Equation (456b) with θ_k replaced by $\phi_{k,k}$:

$$\phi_{k,k} = \frac{E\{\vec{e}_t(k-1) X_{t-k}\}}{\sigma^2_{k-1}} = \frac{E\left\{(X_t - \sum_{j=1}^{k-1} \phi_{k-1,j} X_{t-j}) X_{t-k}\right\}}{\sigma^2_{k-1}}$$

$$= \frac{s_k - \sum_{j=1}^{k-1} \phi_{k-1,j} s_{k-j}}{\sigma^2_{k-1}}.$$

We have now completed our derivation of what is known in the literature as the *Levinson–Durbin recursions* (sometimes called the *Levinson recursions* or the *Durbin–Levinson recursions*), which we can summarize as follows. Suppose $\phi_{k-1,1}, \ldots, \phi_{k-1,k-1}$ and σ^2_{k-1} are known. We can calculate $\phi_{k,1}, \ldots, \phi_{k,k}$ and σ^2_k using the following three equations:

$$\phi_{k,k} = \frac{s_k - \sum_{j=1}^{k-1} \phi_{k-1,j} s_{k-j}}{\sigma^2_{k-1}}; \tag{456d}$$

$$\phi_{k,j} = \phi_{k-1,j} - \phi_{k,k} \phi_{k-1,k-j}, \qquad 1 \leq j \leq k-1; \tag{456e}$$

$$\sigma^2_k = s_0 - \sum_{j=1}^{k} \phi_{k,j} s_j. \tag{456f}$$

9.4 The Levinson–Durbin Recursions

We can initiate the recursions by solving Equations (454b) and (454c) explicitly for the case $k = 1$:

$$\phi_{1,1} = s_1/s_0 \quad \text{and} \quad \sigma_1^2 = s_0 - \phi_{1,1} s_1. \tag{457a}$$

Since we defined σ_0^2 to be s_0 (see Equation (453b)), we can also think of $\phi_{1,1}$ as coming from Equation (456d) if we define the summation from $j = 1$ to 0 to be equal to zero.

There is an important variation on the Levinson–Durbin recursions (evidently due to Burg, 1975, p. 14). The difference between the two recursions is only in Equation (456f) for updating σ_k^2, but it is noteworthy for three reasons: first, it is a numerically better way of calculating σ_k^2 because it is not so sensitive to rounding errors; second, it requires fewer numerical operations and actually speeds up the recursions slightly; and third, it emphasizes the central role of $\phi_{k,k}$, the so-called *kth-order partial autocorrelation coefficient* (see C&E [4]). To derive the alternative to Equation (456f), we multiply both sides of Equation (455b) by X_t, recall that $\theta_k = \phi_{k,k}$, and take expectations to get

$$E\{\vec{\epsilon}_t(k) X_t\} = E\{\vec{\epsilon}_t(k-1) X_t\} - \phi_{k,k} E\{\overleftarrow{\epsilon}_{t-k}(k-1) X_t\}.$$

Using Equation (454d) we can rewrite the above as

$$\sigma_k^2 = \sigma_{k-1}^2 - \phi_{k,k} E\{\overleftarrow{\epsilon}_{t-k}(k-1) X_t\}.$$

Now

$$E\{\overleftarrow{\epsilon}_{t-k}(k-1) X_t\} = E\Big\{\Big(X_{t-k} - \sum_{j=1}^{k-1} \phi_{k-1,j} X_{t-k+j}\Big) X_t\Big\}$$

$$= s_k - \sum_{j=1}^{k-1} \phi_{k-1,j} s_{k-j} = \phi_{k,k} \sigma_{k-1}^2,$$

with the last equality following from Equation (456d). An alternative to Equation (456f) is

$$\sigma_k^2 = \sigma_{k-1}^2 (1 - \phi_{k,k}^2). \tag{457b}$$

The Levinson–Durbin recursions can thus be taken to be Equations (456d) and (456e) in combination with the above. Note that, since Equation (457a) says that $s_0 - \phi_{1,1} s_1 = s_0(1 - \phi_{1,1}^2)$, Equation (457b) for $k = 1$ is consistent with Equation (457a) when we make use of the definition $\sigma_0^2 = s_0$

Note that Equation (457b) tells us that, if $\sigma_{k-1}^2 > 0$, we must have $|\phi_{k,k}| \leq 1$ to ensure that σ_k^2 is nonnegative and, if in fact $|\phi_{k,k}| = 1$, the mean square linear prediction error σ_k^2 is 0. This implies that we could predict the process perfectly in the mean square sense with a kth-order linear predictor. Since a nontrivial linear combination of the RVs of the process $\{X_t\}$ thus has zero variance, the covariance matrix for $\{X_t\}$ is in fact positive semidefinite instead of positive definite (however, this cannot happen if $\{X_t\}$ is a purely continuous stationary process with nonzero variance; i.e., the derivative of its integrated spectrum (the SDF) exists – see Papoulis, 1985, for details).

We can now summarize explicitly the Levinson–Durbin recursive solution to the Yule–Walker equations for estimating the parameters of an AR(p) model from a sample ACVS. Although this method avoids the matrix inversion in Equation (451a), the two solutions are necessarily identical: the Levinson–Durbin recursions simply take advantage of the Toeplitz structure of $\widetilde{\Gamma}_p$ (see the discussion following Equation (29a)) to solve the problem more efficiently than brute force matrix inversion can. On a digital computer, however, the two solutions might be annoyingly different due to the vagaries of rounding error. We begin by solving Equations (451a) and (451c) explicitly for an AR(1) model to get

$$\tilde{\phi}_{1,1} = \hat{s}_1^{(\mathrm{P})}/\hat{s}_0^{(\mathrm{P})} \quad \text{and} \quad \tilde{\sigma}_1^2 = \hat{s}_0^{(\mathrm{P})} - \tilde{\phi}_{1,1} \hat{s}_1^{(\mathrm{P})} = \hat{s}_0^{(\mathrm{P})}(1 - \tilde{\phi}_{1,1}^2).$$

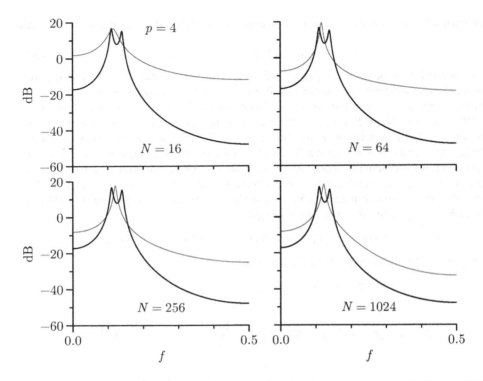

Figure 458 Yule–Walker AR(4) SDF estimates (thin curves) for portions of lengths 16, 64, 256 and 1024 of the realization of the AR(4) process shown in Figure 34(e) (the process is defined in Equation (35a)). The thick curve on each plot is the true SDF.

For $k = 2, \ldots, p$, we then recursively evaluate

$$\tilde{\phi}_{k,k} = \frac{\hat{s}_k^{(\text{P})} - \sum_{j=1}^{k-1} \tilde{\phi}_{k-1,j} \hat{s}_{k-j}^{(\text{P})}}{\tilde{\sigma}_{k-1}^2}; \tag{458a}$$

$$\tilde{\phi}_{k,j} = \tilde{\phi}_{k-1,j} - \tilde{\phi}_{k,k} \tilde{\phi}_{k-1,k-j}, \quad 1 \leq j \leq k-1; \tag{458b}$$

$$\tilde{\sigma}_k^2 = \tilde{\sigma}_{k-1}^2 (1 - \tilde{\phi}_{k,k}^2) \tag{458c}$$

to obtain finally $\tilde{\phi}_{p,1}, \ldots, \tilde{\phi}_{p,p}$ and $\tilde{\sigma}_p^2$.

As an example of autoregressive SDF estimation, we reconsider the time series shown in Figure 34(e). This series is a realization of the AR(4) process defined in Equation (35a) – this process has been cited in the literature as posing a difficult case for SDF estimation. Figure 458 shows the result of using the Yule–Walker method to fit an AR(4) model to the first 16 values in this time series, the first 64, the first 256 and finally the entire series (1024 values). In each case we calculated the biased estimator of the ACVS from the appropriate segment of data and used it as input to the Levinson–Durbin recursions. In each of the plots the thin curve is the SDF corresponding to the fitted AR(4) model, whereas the thick curve is the true AR(4) SDF for the process from which the time series was drawn. We see that the SDF estimates improve with increasing sample size N, but that there is still significant deviation from the true SDF even for $N = 1024$ – particularly in the region of the twin peaks, which collapse incorrectly to a single peak in the estimated SDFs.

Figure 459 shows the effect on the spectral estimates of increasing the order of the fitted model to 8. Although the process that generated this time series is an AR(4) process,

9.4 The Levinson–Durbin Recursions

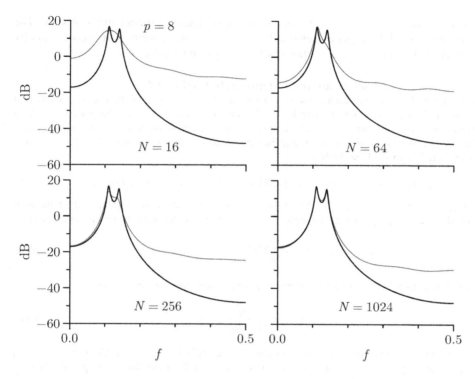

Figure 459 Yule–Walker AR(8) SDF estimates (see Figure 458).

	$j=1$	$j=2$	$j=3$	$j=4$	$j=5$	$j=6$	$j=7$	$j=8$
$\phi_{4,j}$	2.7607	−3.8106	2.6535	−0.9238	0	0	0	0
$\tilde{\phi}_{4,j}$	2.0218	−2.1202	1.0064	−0.2808	0	0	0	0
$\tilde{\phi}_{8,j}$	1.6144	−1.2381	−0.1504	0.1520	0.0224	−0.1086	−0.2018	0.0383
$\lvert\tilde{\phi}_{4,j} - \phi_{4,j}\rvert$	0.7389	1.6904	1.6471	0.6430	0	0	0	0
$\lvert\tilde{\phi}_{8,j} - \phi_{4,j}\rvert$	1.1463	2.5725	2.8039	1.0758	0.0224	0.1086	0.2018	0.0383

Table 459 Comparison of Yule–Walker AR(4) and AR(8) coefficient estimates based on realization of the AR(4) process in Figure 34(e) (the process is defined in Equation (35a)). Here all $N = 1024$ values of the time series are used. The top line shows the true AR coefficients; the next two lines, the AR(4) and AR(8) coefficient estimates; and the last two lines show the absolute differences between the estimates and the true coefficient values. The bottom right-hand plots of Figures 458 and 459 show the corresponding SDF estimates.

we generally get a much better fit to the true SDF by using a higher order model for this example – particularly in the low frequency portion of the SDF and around the twin peaks. Interestingly enough, the estimated coefficients for the AR(4) and AR(8) models superficially suggest otherwise. The top line of Table 459 shows the coefficients for the AR(4) process of Equation (35a) padded with four zeros to create an artificial AR(8) process. The next two lines show the Yule–Walker AR(4) and AR(8) coefficient estimates for the $N = 1024$ case (the corresponding SDF estimates are shown in the lower right-hand plots of Figures 458 and 459). The bottom two lines show the absolute differences between the estimated and true coefficients. These differences are uniformly *larger* for the fitted AR(8) coefficients than for the AR(4), even though the AR(8) SDF estimate is visually superior to the AR(4)! The improvement in the SDF estimate is evidently not due to better estimates on a coefficient by coefficient basis, but rather is tied to interactions amongst the coefficients.

Comments and Extensions to Section 9.4

[1] The Levinson–Durbin recursions allow us to build up the coefficients for the one-step-ahead best linear predictor of order k in terms of the order $k-1$ coefficients (combined with the order $k-1$ mean square linear prediction error and the ACVS up to lag k). It is also possible to go in the other direction: given the order k coefficients, we can determine the corresponding order $k-1$ quantities. To do so, note that we can write Equation (456e) both as

$$\phi_{k-1,j} = \phi_{k,j} + \phi_{k,k}\phi_{k-1,k-j} \quad \text{and} \quad \phi_{k-1,k-j} = \phi_{k,k-j} + \phi_{k,k}\phi_{k-1,j}$$

for $1 \leq j \leq k-1$. If, for $\phi_{k-1,k-j}$ in the left-hand equation, we substitute its value in the right-hand equation and solve for $\phi_{k-1,j}$, we get the order $k-1$ coefficients in terms of the order k coefficients:

$$\phi_{k-1,j} = \frac{\phi_{k,j} + \phi_{k,k}\phi_{k,k-j}}{1 - \phi_{k,k}^2}, \quad 1 \leq j \leq k-1. \tag{460a}$$

We can invert Equation (457b) to get the order $k-1$ mean square linear prediction error:

$$\sigma_{k-1}^2 = \frac{\sigma_k^2}{1 - \phi_{k,k}^2}. \tag{460b}$$

One use for the step-down procedure is to generate the ACVS $\{s_\tau\}$ for an AR(p) process given its $p+1$ parameters $\phi_{p,1}, \ldots, \phi_{p,p}$ and σ_p^2. To do so, we apply the procedure starting with the $\phi_{p,j}$ coefficients to obtain, after $p-1$ iterations,

$$\begin{aligned} \phi_{p-1,1}, \phi_{p-1,2}, \ldots, \phi_{p-1,p-1} \\ \vdots \\ \phi_{2,1}, \phi_{2,2} \\ \phi_{1,1}. \end{aligned} \tag{460c}$$

Next we take σ_p^2 and use Equation (460b) first with $\phi_{p,p}$ and then with the already obtained $\phi_{p-1,p-1}$, $\ldots, \phi_{2,2}, \phi_{1,1}$ to get $\sigma_{p-1}^2, \sigma_{p-2}^2, \ldots, \sigma_1^2, \sigma_0^2$. Noting that $s_0 = \sigma_0^2$, that $s_1 = \phi_{1,1}s_0$ (this comes from Equation (454b) upon setting $j = k = 1$) and that Equation (456d) can be rewritten as

$$s_k = \phi_{k,k}\sigma_{k-1}^2 + \sum_{j=1}^{k-1} \phi_{k-1,j}s_{k-j},$$

we have a recursive scheme for obtaining the ACVS:

$$\begin{aligned} s_0 &= \sigma_0^2 \\ s_1 &= \phi_{1,1}s_0 \\ s_2 &= \phi_{2,2}\sigma_1^2 + \phi_{1,1}s_1 \\ &\vdots \\ s_{p-1} &= \phi_{p-1,p-1}\sigma_{p-2}^2 + \sum_{j=1}^{p-2}\phi_{p-2,j}s_{p-1-j} \\ s_\tau &= \sum_{j=1}^{p}\phi_{p,j}s_{\tau-j}, \quad \tau = p, p+1, \ldots, \end{aligned} \tag{460d}$$

9.4 The Levinson–Durbin Recursions

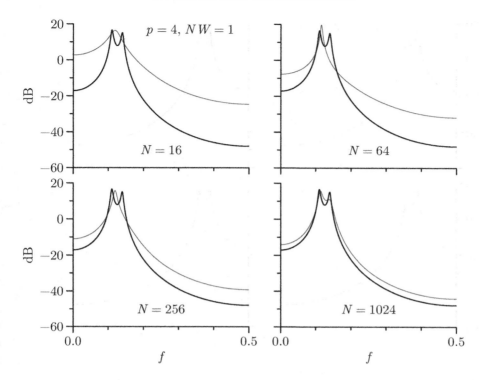

Figure 461 Yule–Walker AR(4) SDF estimates (thin curves) for portions of lengths 16, 64, 256 and 1024 of the realization of the AR(4) process shown in Figure 34(e). The thick curves show the true SDF. Instead of the usual biased ACVS estimator, the ACVS estimator $\{\hat{s}_\tau^{(D)}\}$ corresponding to a direct spectral estimator with an $NW = 1$ Slepian data taper was used.

where the final equation above is the same as Equation (450a). Exercise [9.6] is to use the above to determine the ACVSs for the AR(2) process of Equation (34) and the AR(4) process of Equation (34).

Two other uses for the step-down procedure are to formulate an efficient algorithm for simulating Gaussian ARMA(p, q) processes (see Section 11.1) and to determine if the coefficients $\phi_{p,1}, \ldots, \phi_{p,p}$ for an AR(p) process correspond to those for a causal (and hence stationary) process (see C&E [5]).

[2] The Yule–Walker estimator as usually defined uses the biased ACVS estimator $\{\hat{s}_\tau^{(P)}\}$; however, as noted in the C&Es for Section 9.3, there is no reason why we cannot use other positive definite estimates of the ACVS such as $\{\hat{s}_\tau^{(D)}\}$. Figures 461 and 462 illustrate the possible benefits of doing so (see also Zhang, 1992). These show SDF estimates for the same data used in Figures 458 and 459, but now the Yule–Walker estimator uses $\{\hat{s}_\tau^{(D)}\}$ corresponding to a direct spectral estimator with $NW = 1$ (in Figure 461) and $NW = 2$ (in Figure 462) Slepian data tapers. The improvements over the usual Yule–Walker estimates are dramatic. (If we increase the degree of tapering to, say, an $NW = 4$ Slepian taper, the estimates deteriorate slightly, but the maximum difference over all frequencies between these estimates and the corresponding $NW = 2$ estimates in Figure 462 is less than 3 dB. An estimate with excessive tapering is thus still much better than the usual Yule–Walker estimate.)

[3] Note that we did *not* need to assume Gaussianity in order to derive the Levinson–Durbin recursions. The recursions are also *not* tied explicitly to AR(p) processes, since they can also be used to find the coefficients of the best linear predictor of X_t, given the p prior values of the process, when the only assumption on $\{X_t\}$ is that it is a discrete parameter stationary process with zero mean. The recursions are simply an efficient method of solving a system of equations that possesses a Toeplitz structure. The $p \times p$ matrix associated with such a system (for example, $\boldsymbol{\Gamma}_p$ in Equation (450c)) consists of at most p distinct elements ($s_0, \ldots, s_{p-1}$ in this example). A computational analysis of the recursions shows that they require $O(p^2)$ or fewer operations to solve the equations instead of the $O(p^3)$ operations required

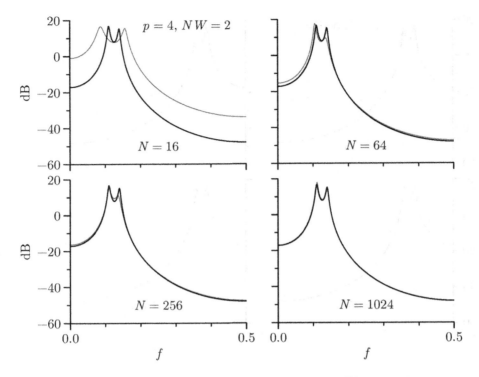

Figure 462 Yule–Walker AR(4) SDF estimates using the ACVS estimate $\{\hat{s}_\tau^{(D)}\}$ corresponding to a direct spectral estimator with an $NW = 2$ Slepian data taper.

by conventional methods using direct matrix inversion. The recursions also require less storage space on a computer (being proportional to p rather than p^2). Bunch (1985) gives an excellent discussion of Toeplitz matrices, their numerical properties and important refinements to the Levinson–Durbin recursions. Of particular interest, he discusses other algorithms for solving Toeplitz systems of equations that require just $O(p \log^2(p))$ operations. In practice, however, for p approximately less than 2000 (a rather high order for an AR model!), the Levinson–Durbin recursions are still the faster method because these alternatives require much more computer code. Morettin (1984) discusses the application of the recursions to other problems in time series analysis.

[4] The sequence $\{\phi_{k,k}\}$ indexed by k is known as the *partial autocorrelation sequence* (PACS) or the *reflection coefficient sequence*. The former terminology arises from the fact that $\phi_{k,k}$ is the correlation between X_t and X_{t-k} after they have been "adjusted" by the intervening $k - 1$ values of the process. The adjustment takes the form of subtracting off predictions based upon $X_{t-k+1}, \ldots, X_{t-1}$; i.e., the adjusted values are

$$X_t - \vec{X}_t(k-1) = \vec{\epsilon}_t(k-1) \text{ and } X_{t-k} - \overleftarrow{X}_{t-k}(k-1) = \overleftarrow{\epsilon}_{t-k}(k-1) \quad (462a)$$

(for $k = 1$ we define $\vec{X}_t(0) = 0$ and $\overleftarrow{X}_{t-1}(0) = 0$). The claim is thus that

$$\frac{\text{cov}\{\vec{\epsilon}_t(k-1), \overleftarrow{\epsilon}_{t-k}(k-1)\}}{\left(\text{var}\{\vec{\epsilon}_t(k-1)\} \text{var}\{\overleftarrow{\epsilon}_{t-k}(k-1)\}\right)^{1/2}} = \phi_{k,k} \quad (462b)$$

(Exercise [9.7] is to show that this is true).

Ramsey (1974) investigates properties of the PACS. He shows that, if $\phi_{p,p} \neq 0$ and $\phi_{k,k} = 0$ for all $k > p$ for a Gaussian stationary process, the process is necessarily an AR(p) process. Equation (457b) then tells us that

$$\sigma_{p-1}^2 > \sigma_p^2 = \sigma_{p+1}^2 = \cdots. \quad (462c)$$

9.4 The Levinson–Durbin Recursions

We use this fact later on to motivate a criterion for selecting the order of an AR model for a time series (see Equation (493a)).

[5] As noted in Section 9.2, the autoregressive process $\{Y_t\}$ of Equation (446a) is causal (and hence stationary) if the roots of $1 - \sum_{j=1}^{p} \phi_{p,j} z^{-j}$ all lie *inside* the unit circle (or, equivalently, if the roots of $1 - \sum_{j=1}^{p} \phi_{p,j} z^{j}$ all lie *outside* the unit circle). To assess causality for large p, we must compute the roots of a high-dimensional polynomial, which can be tricky due to the vagaries of finite-precision arithmetic. An alternative method for determining if an AR process is causal uses the step-down Levinson–Durbin recursions of Equation (460a). The idea is to use these numerically stable recursions to obtain what would constitute the partial autocorrelations $\phi_{p-1,p-1}, \phi_{p-2,p-2}, \ldots, \phi_{1,1}$ for a causal process. If $|\phi_{k,k}| < 1$ for $k = 1, \ldots, p$, then the AR(p) process is causal (Papoulis, 1985; Newton, 1988); on the other hand, if either $|\phi_{p,p}| \geq 1$ or if the recursions yield $|\phi_{k,k}| \geq 1$ for some $k < p$, we are dealing with either a nonstationary process or an acausal stationary process. As examples, Exercise [9.9] lists six different AR(4) processes whose causality – or lack thereof – can be determined using the step-down recursions.

[6] A more general way to predict X_t given the k most recent prior values of $\{X_t\}$ is to find that function of $X_{t-1}, X_{t-2}, \ldots, X_{t-k}$ – call it $g(X_{t-1}, \ldots, X_{t-k})$ – such that the *mean square prediction error*

$$M_k \stackrel{\text{def}}{=} E\{[X_t - g(X_{t-1}, \ldots, X_{t-k})]^2\}$$

is minimized. The resulting form of $g(\cdot)$ is appealing (for a derivation of this result, see Priestley, 1981, p. 76):

$$g(X_{t-1}, \ldots, X_{t-k}) = E\{X_t | X_{t-1}, \ldots, X_{t-k}\};$$

i.e., under the mean square error criterion, the predictor of X_t, given $X_{t-1}, \ldots, X_{t-k}$, is simply the conditional mean of X_t, given $X_{t-1}, \ldots, X_{t-k}$. We call $g(\cdot)$ the *best predictor* of X_t, given $X_{t-1}, \ldots, X_{t-k}$. As simple and appealing as $g(\cdot)$ is, it is unfortunately mathematically intractable in many cases of interest. An important exception is the case of a stationary Gaussian process $\{G_t\}$, for which the best predictor is identical to the best linear predictor:

$$E\{G_t | G_{t-1}, \ldots, G_{t-k}\} = \vec{G}_t(k) \stackrel{\text{def}}{=} \sum_{j=1}^{k} \phi_{k,j} G_{t-j}.$$

For a non-Gaussian process, the best predictor need not be linear. Moreover, for these processes, the symmetry between the forward and backward best *linear* predictors of X_t does not necessarily carry through to the forward and backward best predictors. Rosenblatt (1985, p. 52) gives an example of a rather extreme case of asymmetry, in which the best forward predictor corresponds to the best linear predictor and has a prediction error variance of 4, whereas the best backward predictor is nonlinear and has a prediction error variance of 0 (i.e., it predicts perfectly with probability 1)!

[7] It is also possible to derive the Levinson–Durbin recursions using a direct linear algebra argument. As before, let us write the Yule–Walker equations in matrix form

$$\boldsymbol{\Gamma}_p \boldsymbol{\Phi}_p = \boldsymbol{\gamma}_p, \tag{463}$$

where $\boldsymbol{\gamma}_p = [s_1, s_2, \ldots, s_p]^T$; $\boldsymbol{\Phi}_p = [\phi_{p,1}, \phi_{p,2}, \ldots, \phi_{p,p}]^T$; and $\boldsymbol{\Gamma}_p$ is as in Equation (450c). The trick is to solve for $\boldsymbol{\Phi}_{p+1}$ using the solution $\boldsymbol{\Phi}_p$. Suppose for the moment that it is possible to express $\phi_{p+1,j}$ for $j \leq p$ in the following manner:

$$\phi_{p+1,j} = \phi_{p,j} - \phi_{p+1,p+1} \theta_{p,j}.$$

Let $\boldsymbol{\Theta}_p = [\theta_{p,1}, \theta_{p,2}, \ldots, \theta_{p,p}]^T$ be a p-dimensional vector to be determined. With this recursion, Equation (463) becomes the following for order $p + 1$:

$$\begin{bmatrix} & & s_p \\ & \boldsymbol{\Gamma}_p & \vdots \\ & & s_1 \\ s_p & \cdots & s_1 & s_0 \end{bmatrix} \begin{bmatrix} \boldsymbol{\Phi}_p - \phi_{p+1,p+1} \boldsymbol{\Theta}_p \\ \phi_{p+1,p+1} \end{bmatrix} = \begin{bmatrix} \boldsymbol{\gamma}_p \\ s_{p+1} \end{bmatrix}.$$

If we separate the first p equations from the last one, we have

$$\boldsymbol{\Gamma}_p \boldsymbol{\Phi}_p - \phi_{p+1,p+1}\boldsymbol{\Gamma}_p\boldsymbol{\Theta}_p + \phi_{p+1,p+1}\begin{bmatrix} s_p \\ \vdots \\ s_1 \end{bmatrix} = \boldsymbol{\gamma}_p \qquad (464a)$$

and

$$[\,s_p \ \cdots \ s_1\,]\left(\boldsymbol{\Phi}_p - \phi_{p+1,p+1}\boldsymbol{\Theta}_p\right) + s_0\phi_{p+1,p+1} = s_{p+1}. \qquad (464b)$$

If we use the fact that $\boldsymbol{\Phi}_p$ satisfies Equation (463), we conclude that the following condition is sufficient for Equation (464a) to hold:

$$\boldsymbol{\Gamma}_p\boldsymbol{\Theta}_p = \begin{bmatrix} s_p \\ \vdots \\ s_1 \end{bmatrix}.$$

If we chose $\theta_{p,j} = \phi_{p,p+1-j}$, then the condition above is equivalent to

$$\begin{bmatrix} s_0 & s_1 & \cdots & s_{p-1} \\ s_1 & s_0 & \cdots & s_{p-2} \\ \vdots & \vdots & \ddots & \vdots \\ s_{p-1} & s_{p-2} & \cdots & s_0 \end{bmatrix} \begin{bmatrix} \phi_{p,p} \\ \phi_{p,p-1} \\ \vdots \\ \phi_{p,1} \end{bmatrix} = \begin{bmatrix} s_p \\ s_{p-1} \\ \vdots \\ s_1 \end{bmatrix}.$$

If we reverse the order of the rows in the equation above, we have that it is equivalent to

$$\begin{bmatrix} s_{p-1} & s_{p-2} & \cdots & s_0 \\ s_{p-2} & s_{p-3} & \cdots & s_1 \\ \vdots & \vdots & \ddots & \vdots \\ s_0 & s_1 & \cdots & s_{p-1} \end{bmatrix} \begin{bmatrix} \phi_{p,p} \\ \phi_{p,p-1} \\ \vdots \\ \phi_{p,1} \end{bmatrix} = \begin{bmatrix} s_1 \\ s_2 \\ \vdots \\ s_p \end{bmatrix}.$$

If we now reverse the order of the columns, we get Equation (463), which is assumed to hold. Hence our choice of $\boldsymbol{\Theta}_p$ satisfies Equation (464a). Equation (464b) can be satisfied by solving for $\phi_{p+1,p+1}$. In summary, if $\boldsymbol{\Phi}_p$ satisfies Equation (463),

$$\phi_{p+1,p+1} = \frac{s_{p+1} - \sum_{j=1}^{p}\phi_{p,j}s_{p+1-j}}{s_0 - \sum_{j=1}^{p}\phi_{p,j}s_j};$$

$$\phi_{p+1,j} = \phi_{p,j} - \phi_{p+1,p+1}\phi_{p,p+1-j}$$

for $j = 1, \ldots, p$, then $\boldsymbol{\Gamma}_{p+1}\boldsymbol{\Phi}_{p+1} = \boldsymbol{\gamma}_{p+1}$.

[8] The Levinson–Durbin recursions are closely related to the *modified Cholesky decomposition* of a positive definite covariance matrix $\boldsymbol{\Gamma}_N$ for a portion $\boldsymbol{X} = [X_0, X_1, \ldots, X_{N-1}]^T$ of a stationary process (Therrien, 1983; Newton, 1988, section A.1.1; Golub and Van Loan, 2013, theorem 4.1.3). Such a decomposition exists and is unique for any positive definite symmetric $N \times N$ matrix. In the case of $\boldsymbol{\Gamma}_N$ (defined as dictated by Equation (450c)), the decomposition states that we can write

$$\boldsymbol{\Gamma}_N = \boldsymbol{L}_N \boldsymbol{D}_N \boldsymbol{L}_N^T, \qquad (464c)$$

where $\boldsymbol{L}_N$ is an $N \times N$ lower triangular matrix with 1 as each of its diagonal elements, and $\boldsymbol{D}_N$ is an $N \times N$ diagonal matrix, each of whose diagonal elements are positive. This decomposition is equivalent to both

$$\boldsymbol{\Gamma}_N^{-1} = \boldsymbol{L}_N^{-T}\boldsymbol{D}_N^{-1}\boldsymbol{L}_N^{-1} \text{ and } \boldsymbol{L}_N^{-1}\boldsymbol{\Gamma}_N\boldsymbol{L}_N^{-T} = \boldsymbol{D}_N, \text{ where } \boldsymbol{L}_N^{-T} \stackrel{\text{def}}{=} (\boldsymbol{L}_N^{-1})^T = (\boldsymbol{L}_N^T)^{-1}. \qquad (464d)$$

9.4 The Levinson–Durbin Recursions

Because the inverse of a lower triangular matrix with diagonal elements equal to 1 is also lower triangular with 1's along its diagonal (as can be shown using a proof by induction), the first equivalence is the modified Cholesky decomposition for $\boldsymbol{\Gamma}_N^{-1}$. Due to the special structure of $\boldsymbol{\Gamma}_N$, the matrix $\boldsymbol{L}_N^{-1}$ takes the form

$$\boldsymbol{L}_N^{-1} = \begin{bmatrix} 1 & & & & & \\ -\phi_{1,1} & 1 & & & & \\ -\phi_{2,2} & -\phi_{2,1} & 1 & & & \\ \vdots & \vdots & \ddots & \ddots & & \\ -\phi_{k,k} & -\phi_{k,k-1} & \cdots & -\phi_{k,1} & 1 & \\ \vdots & \vdots & \ddots & \ddots & \ddots & \\ -\phi_{N-1,N-1} & -\phi_{N-1,N-2} & \cdots & & -\phi_{N-1,1} & 1 \end{bmatrix},$$

where the Levinson–Durbin recursions yield the $\phi_{k,j}$ coefficients. To see why this is the correct form for $\boldsymbol{L}_N^{-1}$, note first that, with $\vec{e}_0(0) \stackrel{\text{def}}{=} X_0$, we have

$$\boldsymbol{L}_N^{-1} \boldsymbol{X} = \left[\vec{e}_0(0), \vec{e}_1(1), \vec{e}_2(2), \ldots, \vec{e}_k(k), \ldots, \vec{e}_{N-1}(N-1)\right]^T; \qquad (465a)$$

i.e., the matrix $\boldsymbol{L}_N^{-1}$ transforms $\boldsymbol{X}$ into a vector of prediction errors. These prediction errors have two properties of immediate relevance. First, $\text{var}\{\vec{e}_k(k)\} = \sigma_k^2 > 0$, and these mean square prediction errors arise as part of the Levinson–Durbin recursions (Equation (453b) defines σ_0^2 to be s_0, the process variance). The second property is the subject of the following exercise.

▷ **Exercise [465]** Show that $\text{cov}\{\vec{e}_j(j), \vec{e}_k(k)\} = 0$ for $0 \leq j < k \leq N-1$. ◁

Since the prediction errors have variances dictated by σ_k^2 and are pairwise uncorrelated, it follows that the covariance matrix for $\boldsymbol{L}_N^{-1}\boldsymbol{X}$ is a diagonal matrix with $\sigma_0^2, \sigma_1^2, \ldots, \sigma_{N-1}^2$ as its diagonal elements – in fact this is the diagonal matrix $\boldsymbol{D}_N$ appearing in the modified Cholesky decomposition. A standard result from the theory of vectors of RVs says that, if $\boldsymbol{A}$ is a matrix with N columns, then the covariance matrix for $\boldsymbol{AX}$ is $\boldsymbol{A\Gamma}_N\boldsymbol{A}^T$ (see, e.g., Brockwell and Davis, 1991, proposition 1.6.1). Thus the covariance matrix of $\boldsymbol{L}_N^{-1}\boldsymbol{X}$ is $\boldsymbol{L}_N^{-1}\boldsymbol{\Gamma}_N\boldsymbol{L}_N^{-T}$, which, upon equating to $\boldsymbol{D}_N$, yields the second equivalent formulation of the modified Cholesky decomposition stated in Equation (464d).

For the AR(p) process $\{Y_t\}$ of Equation (446a), the matrix $\boldsymbol{L}_N^{-1}$ specializes to

$$\boldsymbol{L}_N^{-1} = \begin{bmatrix} 1 & & & & & & \\ -\phi_{1,1} & 1 & & & & & \\ -\phi_{2,2} & -\phi_{2,1} & 1 & & & & \\ \vdots & \vdots & \ddots & & & & \\ -\phi_{p,p} & \cdots & & -\phi_{p,1} & 1 & & \\ 0 & -\phi_{p,p} & \cdots & & -\phi_{p,1} & 1 & \\ \vdots & \ddots & \ddots & & \ddots & \ddots & \\ 0 & \cdots & 0 & -\phi_{p,p} & \cdots & -\phi_{p,1} & 1 \end{bmatrix}, \qquad (465b)$$

while the upper p diagonal elements of $\boldsymbol{D}_N$ are $\sigma_0^2, \sigma_1^2, \ldots, \sigma_{p-1}^2$, and the lower $N-p$ elements are all equal to σ_p^2. With $\boldsymbol{Y} = [Y_0, Y_1, \ldots, Y_{N-1}]^T$, we have

$$\boldsymbol{L}_N^{-1}\boldsymbol{Y} = \left[\vec{e}_0(0), \vec{e}_1(1), \vec{e}_2(2), \ldots, \vec{e}_{p-1}(p-1), \epsilon_p, \epsilon_{p+1}, \ldots, \epsilon_{N-1}\right]^T. \qquad (465c)$$

These facts will prove useful when we discuss maximum likelihood estimation in Section 9.8 and simulation of AR(p) processes in Section 11.1.

9.5 Burg's Algorithm

The Yule–Walker estimators of the AR(p) parameters are only one of many proposed and studied in the literature. Since the late 1960s, an estimation technique based upon *Burg's algorithm* has been in wide use (particularly in engineering and geophysics). The popularity of this method is due to a number of reasons.

[1] Burg's algorithm is a variation on the solution of the Yule–Walker equations via the Levinson–Durbin recursions and, like them, is computationally efficient and recursive. If we recall that the AR(p) process that is fit to a time series via the Yule–Walker method has a theoretical ACVS that agrees *identically* with that of the sample ACVS (i.e., the biased estimator of the ACVS), we can regard Burg's algorithm as providing an alternative estimator of the ACVS. Burg's modification is intuitively reasonable and relies heavily on the relationship of the recursions to the prediction error problem discussed in Section 9.4.

[2] Burg's algorithm arose from his work with the maximum entropy principle in connection with SDF estimation. We describe and critically examine this principle in the next section, but it is important to realize that, as shown in the remainder of this section, Burg's algorithm can be justified without appealing to entropy arguments.

[3] While in theory Burg's algorithm can yield an estimated AR(p) process that is nonstationary, in practice its estimates – like those of the Yule–Walker method – correspond to a causal (and hence stationary) AR(p) process. Although nonstationarity is a possibility, the coefficients produced by Burg's algorithm cannot correspond to an acausal stationary process (as C&E [1] shows, the Burg estimated PACS must be such that each PACS estimate is bounded by unity in magnitude, and C&E [5] for Section 9.4 notes that $|\phi_{k,k}| < 1$ for all k corresponds to a causal AR process).

[4] Monte Carlo studies and experience with actual data indicate that, particularly for short time series, Burg's algorithm produces more reasonable estimates than the Yule–Walker method (see, for example, Lysne and Tjøstheim, 1987). As we note in Section 10.13, it is not, unfortunately, free of problems of its own.

The key to Burg's algorithm is the central role of $\phi_{k,k}$ (the kth-order partial autocorrelation coefficient) in the Levinson–Durbin recursions: if we have $\phi_{k-1,1}, \ldots, \phi_{k-1,k-1}$ and σ_{k-1}^2, Equations (456e) and (456f) tell us we need only determine $\phi_{k,k}$ to calculate the remaining order k parameters (namely, $\phi_{k,1}, \ldots, \phi_{k,k-1}$ and σ_k^2). In the Yule–Walker scheme, we estimate $\phi_{k,k}$ using Equation (458a), which is based on Equation (456d) in the Levinson–Durbin recursions. Note that it utilizes the estimated ACVS up to lag p. Burg's algorithm takes a different approach. It estimates $\phi_{k,k}$ by minimizing a certain sum of squares of *observed* forward and backward prediction errors. Given we have a time series of length N taken to be a realization of a portion $X_0, X_1, \ldots, X_{N-1}$ of a discrete parameter stationary process with zero mean, and given Burg's estimators $\bar{\phi}_{k-1,1}, \ldots, \bar{\phi}_{k-1,k-1}$ of the coefficients for a model of order $k-1$, the errors we need to consider are

$$\vec{e}_t(k-1) \stackrel{\text{def}}{=} X_t - \sum_{j=1}^{k-1} \bar{\phi}_{k-1,j} X_{t-j}, \qquad k \leq t \leq N-1,$$

and

$$\overleftarrow{e}_{t-k}(k-1) \stackrel{\text{def}}{=} X_{t-k} - \sum_{j=1}^{k-1} \bar{\phi}_{k-1,j} X_{t-k+j}, \qquad k \leq t \leq N-1.$$

Suppose for the moment that $\bar{\phi}_{k,k}$ is *any* estimate of $\phi_{k,k}$ and that the remaining $\bar{\phi}_{k,j}$ terms are generated in a manner analogous to Equation (458b). We then have

$$\vec{e}_t(k) = X_t - \sum_{j=1}^{k} \bar{\phi}_{k,j} X_{t-j}$$

$$= X_t - \sum_{j=1}^{k-1} \left(\bar{\phi}_{k-1,j} - \bar{\phi}_{k,k} \bar{\phi}_{k-1,k-j} \right) X_{t-j} - \bar{\phi}_{k,k} X_{t-k}$$

$$= X_t - \sum_{j=1}^{k-1} \bar{\phi}_{k-1,j} X_{t-j} - \bar{\phi}_{k,k} \left(X_{t-k} - \sum_{j=1}^{k-1} \bar{\phi}_{k-1,k-j} X_{t-j} \right)$$

$$= \vec{e}_t(k-1) - \bar{\phi}_{k,k} \left(X_{t-k} - \sum_{j=1}^{k-1} \bar{\phi}_{k-1,j} X_{t-k+j} \right)$$

$$= \vec{e}_t(k-1) - \bar{\phi}_{k,k} \overleftarrow{e}_{t-k}(k-1), \quad k \leq t \leq N-1. \tag{467a}$$

We recognize this as a sampling version of Equation (455b). In a similar way, we can derive a sampling version of Equation (456c):

$$\overleftarrow{e}_{t-k}(k) = \overleftarrow{e}_{t-k}(k-1) - \bar{\phi}_{k,k} \vec{e}_t(k-1), \quad k \leq t \leq N-1. \tag{467b}$$

Equations (467a) and (467b) allow us to calculate the observed order k forward and backward prediction errors in terms of the order $k-1$ errors. These equations do not depend on *any* particular property of $\bar{\phi}_{k,k}$ and hence are valid no matter how we determine $\bar{\phi}_{k,k}$ as long as the remaining $\bar{\phi}_{k,j}$ terms are generated as dictated by Equation (458b). Burg's idea was to estimate $\phi_{k,k}$ such that the order k observed prediction errors are as small as possible by the appealing criterion that

$$\mathrm{SS}_k(\bar{\phi}_{k,k}) \stackrel{\text{def}}{=} \sum_{t=k}^{N-1} \{\vec{e}_t^{\,2}(k) + \overleftarrow{e}_{t-k}^{\,2}(k)\} = \sum_{t=k}^{N-1} \Big\{ \left[\vec{e}_t(k-1) - \bar{\phi}_{k,k} \overleftarrow{e}_{t-k}(k-1) \right]^2$$

$$+ \left[\overleftarrow{e}_{t-k}(k-1) - \bar{\phi}_{k,k} \vec{e}_t(k-1) \right]^2 \Big\}$$

$$= A_k - 2\bar{\phi}_{k,k} B_k + A_k \bar{\phi}_{k,k}^2 \tag{467c}$$

be as small as possible, where

$$A_k \stackrel{\text{def}}{=} \sum_{t=k}^{N-1} \{\vec{e}_t^{\,2}(k-1) + \overleftarrow{e}_{t-k}^{\,2}(k-1)\} \tag{467d}$$

and

$$B_k \stackrel{\text{def}}{=} 2 \sum_{t=k}^{N-1} \vec{e}_t(k-1) \overleftarrow{e}_{t-k}(k-1). \tag{467e}$$

Since $\mathrm{SS}_k(\cdot)$ is a quadratic function of $\bar{\phi}_{k,k}$, we can differentiate it and set the resulting expression to 0 to find the desired value of $\bar{\phi}_{k,k}$, namely,

$$\bar{\phi}_{k,k} = B_k / A_k. \tag{467f}$$

(Note that $\bar{\phi}_{k,k}$ has a natural interpretation as an estimator of the left-hand side of Equation (462b).) With $\bar{\phi}_{k,k}$ so determined, we can estimate the remaining $\bar{\phi}_{k,j}$ using an equation analogous to Equation (458b) – the second of the usual Levinson–Durbin recursions – and the corresponding observed prediction errors using Equations (467a) and (467b).

Given how $\bar{\phi}_{k,k}$ arises, it should be obvious that, with the proper initialization, we can apply Burg's algorithm recursively to work our way up to estimates for the coefficients of an AR(p) model. In the same spirit as the recursive step, we can initialize the algorithm by finding that value of $\bar{\phi}_{1,1}$ that minimizes

$$\text{SS}_1(\bar{\phi}_{1,1}) = \sum_{t=1}^{N-1} \vec{e}_t^{\,2}(1) + \overleftarrow{e}_{t-1}^{\,2}(1)$$

$$= \sum_{t=1}^{N-1} \left(X_t - \bar{\phi}_{1,1} X_{t-1}\right)^2 + \left(X_{t-1} - \bar{\phi}_{1,1} X_t\right)^2. \tag{468a}$$

This is equivalent to Equation (467c) with $k = 1$ if we adopt the conventions $\vec{e}_t(0) = X_t$ and $\overleftarrow{e}_{t-1}(0) = X_{t-1}$. We can interpret these as meaning that, with no prior (future) observations, we should predict (backward predict) X_t (X_{t-1}) by zero, the assumed known process mean.

Burg's algorithm also specifies a way of estimating the innovation variance σ_p^2. It is done recursively using an equation analogous to Equation (458c), namely,

$$\bar{\sigma}_k^2 = \bar{\sigma}_{k-1}^2 \left(1 - \bar{\phi}_{k,k}^2\right). \tag{468b}$$

This assumes that, at the kth step, $\bar{\sigma}_{k-1}^2$ is available. For $k = 1$ there is obviously a problem in that we have not defined what $\bar{\sigma}_0^2$ is. If we follow the logic of the previous paragraph and of Equation (453b) and consider the zeroth-order predictor of our time series to be zero (the known process mean), the obvious estimator of the zeroth-order mean square linear prediction error is just

$$\bar{\sigma}_0^2 \stackrel{\text{def}}{=} \frac{1}{N} \sum_{t=0}^{N-1} X_t^2 = \hat{s}_0^{(\text{P})},$$

the sample variance of the time series. If we let $\hat{S}^{(\text{BURG})}(\cdot)$ denote the resulting Burg SDF estimator, the above stipulation implies that in practice

$$\int_{-f_\mathcal{N}}^{f_\mathcal{N}} \hat{S}^{(\text{BURG})}(f) \, \mathrm{d}f = \hat{s}_0^{(\text{P})}. \tag{468c}$$

The above follows from an argument similar to the one used to show that the Yule–Walker estimator integrates to the sample variance (see Equation (451d)); however, whereas the Yule–Walker ACVS estimator agrees with $\hat{s}_0^{(\text{P})}, \hat{s}_1^{(\text{P})}, \ldots, \hat{s}_p^{(\text{P})}$, the same does *not* hold in general for the Burg estimator when $\tau \geq 1$.

As an example of the application of Burg's algorithm to autoregressive SDF estimation, Figure 469 shows the Burg AR(4) SDF estimates for the same series used to produce the Yule–Walker AR(4) and AR(8) SDF estimates of Figures 458 and 459. A comparison of these three figures shows that Burg's method is clearly superior to the Yule–Walker method (at least in the standard formulation of that method using the biased estimator of the ACVS and at least for this particular AR(4) time series).

9.5 Burg's Algorithm

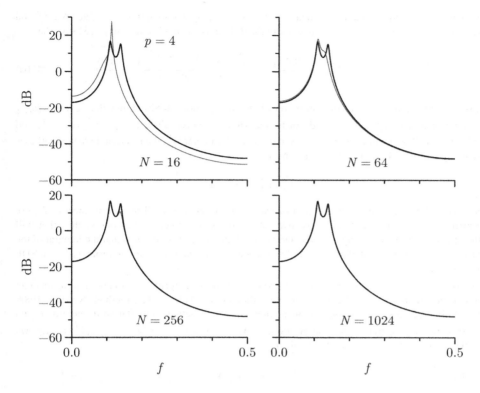

Figure 469 Burg AR(4) SDF estimates (thin curves) for portions of lengths 16, 64, 256 and 1024 of the realization of the AR(4) process shown in Figure 34(e). The thick curve in each plot is the true SDF. This figure should be compared with Figures 458 and 459.

Comments and Extensions to Section 9.5

[1] We have previously noted that the *theoretical* kth-order partial autocorrelation coefficient $\phi_{k,k}$ can be interpreted as a correlation coefficient and thus must lie in the interval $[-1, 1]$. One important property of Burg's algorithm is that the corresponding *estimated* value $\bar{\phi}_{k,k}$ is also necessarily in that range. To see this, note that

$$0 \leq \left(\vec{e}_t(k-1) \pm \overleftarrow{e}_{t-k}(k-1)\right)^2 = \vec{e}_t^{\,2}(k-1) \pm 2\vec{e}_t(k-1)\overleftarrow{e}_{t-k}(k-1) + \overleftarrow{e}_{t-k}^{\,2}(k-1)$$

implies that

$$|2\vec{e}_t(k-1)\overleftarrow{e}_{t-k}(k-1)| \leq \vec{e}_t^{\,2}(k-1) + \overleftarrow{e}_{t-k}^{\,2}(k-1)$$

and hence that

$$|B_k| \stackrel{\text{def}}{=} \left|2\sum_{t=k}^{N-1} \vec{e}_t(k-1)\overleftarrow{e}_{t-k}(k-1)\right| \leq \sum_{t=k}^{N-1} \left[\vec{e}_t^{\,2}(k-1) + \overleftarrow{e}_{t-k}^{\,2}(k-1)\right] \stackrel{\text{def}}{=} A_k.$$

Since $A_k \geq 0$ (with equality rarely occurring in practical cases) and $\bar{\phi}_{k,k} = B_k/A_k$, we have $|\bar{\phi}_{k,k}| \leq 1$ as claimed.

[2] We have described the estimator of σ_k^2 commonly associated with Burg's algorithm. There is a second obvious estimator. Since σ_k^2 is the kth-order mean square linear prediction error, this estimator is

$$\bar{\nu}_k^2 \stackrel{\text{def}}{=} \frac{\text{SS}_k(\bar{\phi}_{k,k})}{2(N-k)}.$$

In the numerator the sum of squares of Equation (467c) is evaluated at its minimum value. Are $\bar{\nu}_k^2$ and $\bar{\sigma}_k^2$ in fact different? The answer is yes, but usually the difference is small. Exercise [9.13] says that

$$\bar{\nu}_k^2 = \left(1 - \bar{\phi}_{k,k}^2\right) \left(\bar{\nu}_{k-1}^2 + \frac{2\bar{\nu}_{k-1}^2 - \vec{e}_{k-1}^{\,2}(k-1) - \overleftarrow{e}_{N-k}^{\,2}(k-1)}{2(N-k)}\right). \quad (470a)$$

Comparison of this equation with the definition of $\bar{\sigma}_k^2$ in Equation (468b) shows that, even if $\bar{\nu}_{k-1}^2$ and $\bar{\sigma}_{k-1}^2$ are identical, $\bar{\nu}_k^2$ and $\bar{\sigma}_k^2$ need not be the same; however, since $\bar{\nu}_{k-1}^2$, $\vec{e}_{k-1}^{\,2}(k-1)$ and $\overleftarrow{e}_{N-k}^{\,2}(k-1)$ can all be regarded as estimators of the same parameter, σ_k^2, the last term in the parentheses should usually be small. For large N, we thus have

$$\bar{\nu}_k^2 \approx \bar{\nu}_{k-1}^2 \left(1 - \bar{\phi}_{k,k}^2\right),$$

in agreement with the third equation of the Levinson–Durbin recursions. The relative merits of these two estimators are unknown, but, as far as SDF estimation is concerned, use of one or the other will only affect the level of the resulting estimate and not its shape. Use of $\bar{\sigma}_k^2$ ensures that the integral of the Burg-based SDF estimate is equal to the sample variance $\hat{s}_0^{(P)}$, which is a desirable property for an SDF estimator to have.

[3] There is a second – but equivalent – way of expressing Burg's algorithm that is both more succinct and also clarifies the relationship between this algorithm and the Yule–Walker method (Newton, 1988, section 3.4). Before presenting it, we establish some notation. Define a circular shift operator $\mathcal{L}$ and a subvector extraction operator $\mathcal{M}_{j,k}$ as follows: if $\mathbf{V}_0 = [v_0, v_1, \ldots, v_{M-2}, v_{M-1}]^T$ is any M-dimensional column vector of real numbers, then

$$\mathcal{L}\mathbf{V}_0 \stackrel{\text{def}}{=} [v_{M-1}, v_0, v_1, \ldots, v_{M-2}]^T \quad \text{and} \quad \mathcal{M}_{j,k}\mathbf{V}_0 \stackrel{\text{def}}{=} [v_j, v_{j+1}, \ldots, v_{k-1}, v_k]^T$$

(we assume that $0 \leq j < k \leq M - 1$). If $\mathbf{V}_1$ is any vector whose dimension is the same as that of $\mathbf{V}_0$, we define $\langle \mathbf{V}_0, \mathbf{V}_1 \rangle = \mathbf{V}_0^T \mathbf{V}_1$ to be their inner product. We also define the squared norm of $\mathbf{V}_0$ to be $\|\mathbf{V}_0\|^2 = \langle \mathbf{V}_0, \mathbf{V}_0 \rangle = \mathbf{V}_0^T \mathbf{V}_0$.

To fit an AR(p) model to $X_0, \ldots, X_{N-1}$ using Burg's algorithm, we first set up the following vectors of length $M = N + p$:

$$\vec{e}(0) \stackrel{\text{def}}{=} [X_0, X_1, \ldots, X_{N-1}, \underbrace{0, \ldots, 0}_{p \text{ of these}}]^T$$

and

$$\overleftarrow{e}(0) \stackrel{\text{def}}{=} \mathcal{L}\vec{e}(0) = [0, X_0, X_1, \ldots, X_{N-1}, \underbrace{0, \ldots, 0}_{p-1 \text{ of these}}]^T.$$

As before, we define $\bar{\sigma}_0^2 = \hat{s}_0^{(P)}$. For $k = 1, \ldots, p$, we then recursively compute

$$\bar{\phi}_{k,k} = \frac{2\langle \mathcal{M}_{k,N-1}\vec{e}(k-1), \mathcal{M}_{k,N-1}\overleftarrow{e}(k-1)\rangle}{\|\mathcal{M}_{k,N-1}\vec{e}(k-1)\|^2 + \|\mathcal{M}_{k,N-1}\overleftarrow{e}(k-1)\|^2} \quad (470b)$$

$$\bar{\sigma}_k^2 = \bar{\sigma}_{k-1}^2 \left(1 - \bar{\phi}_{k,k}^2\right)$$

$$\vec{e}(k) = \vec{e}(k-1) - \bar{\phi}_{k,k}\overleftarrow{e}(k-1)$$

$$\overleftarrow{e}(k) = \mathcal{L}(\overleftarrow{e}(k-1) - \bar{\phi}_{k,k}\vec{e}(k-1)).$$

This procedure yields the Burg estimators of $\phi_{k,k}$ and σ_k^2 for $k = 1, \ldots, p$ (if so desired, the remaining $\bar{\phi}_{k,j}$ can be generated using an equation similar to Equation (458b)). The $N - p$ forward prediction errors $\vec{e}_p(p), \ldots, \vec{e}_{N-1}(p)$ are the elements of $\mathcal{M}_{p,N-1}\vec{e}(p)$; the backward prediction errors $\overleftarrow{e}_0(p), \ldots, \overleftarrow{e}_{N-p-1}(p)$, those of $\mathcal{M}_{p+1,N}\overleftarrow{e}(p)$.

The Yule–Walker estimators $\tilde{\phi}_{k,k}$ and $\tilde{\sigma}_k^2$ can be generated by a scheme that is *identical* to the above, with one key modification: Equation (470b) becomes

$$\tilde{\phi}_{k,k} = \frac{2\langle \vec{\tilde{e}}(k-1), \cev{\tilde{e}}(k-1)\rangle}{\|\vec{\tilde{e}}(k-1)\|^2 + \|\cev{\tilde{e}}(k-1)\|^2} = \frac{\langle \vec{\tilde{e}}(k-1), \cev{\tilde{e}}(k-1)\rangle}{\|\vec{\tilde{e}}(k-1)\|^2} \qquad (471)$$

since $\|\cev{\tilde{e}}(k-1)\|^2 = \|\vec{\tilde{e}}(k-1)\|^2$ (all overbars in the three equations below Equation (470b) should also be changed to tildes). Note that, whereas Burg's algorithm uses an inner product tailored to involve just the actual data values, the inner product of the Yule–Walker estimator is influenced by the p zeros used to construct $\vec{\tilde{e}}(0)$ and $\cev{\tilde{e}}(0)$ (a similar interpretation of the Yule–Walker estimator appears in the context of least squares theory – see C&E [2] for Section 9.7). This finding supports Burg's contention (discussed in the next section) that the Yule–Walker method implicitly assumes $X_t = 0$ for $t < 0$ and $t \geq N$. Finally we note that the formulation involving Equation (471) allows us to compute the Yule–Walker estimators *directly* from a time series, thus bypassing the need to first compute $\{\hat{s}_\tau^{(\text{P})}\}$ as required by Equation (458a).

9.6 The Maximum Entropy Argument

Burg's algorithm is an outgrowth of his work on maximum entropy spectral analysis (MESA). Burg (1967) criticizes the use of lag window spectral estimators of the form

$$\Delta_t \sum_{\tau=-(N-1)}^{N-1} w_{m,\tau} \hat{s}_\tau^{(\text{P})} e^{-i2\pi f \tau \Delta_t}$$

because, first, we effectively force the sample ACVS to zero by multiplying it by a lag window $\{w_{m,\tau}\}$ and, second, we assume that $s_\tau = 0$ for $|\tau| \geq N$. To quote from Burg (1975),

> While window theory is interesting, it is actually a problem that has been artificially induced into the estimation problem by the assumption that $s_\tau = 0$ for $|\tau| \geq N$ and by the willingness to change perfectly good data by the weighting function. If one were not blinded by the mathematical elegance of the conventional approach, making unfounded assumptions as to the values of unmeasured data and changing the data values that one knows would be totally unacceptable from a common sense and, hopefully, from a scientific point of view. To overcome these problems, it is clear that a completely different philosophical approach to spectral analysis is required . . . .
>
> While readily understood, [the] conventional window function approach produces spectral estimates which are negative and/or spectral estimates which do not agree with their known autocorrelation values. These two affronts to common sense were the main reasons for the development of maximum entropy spectral analysis . . . .

Burg's "different philosophical approach" is to apply the principle of maximum entropy. In physics, entropy is a measure of disorder; for a stationary process, it can be usefully defined in terms of the predictability of the process. To be specific, suppose that, following our usual convention, $\{X_t\}$ is an arbitrary zero mean stationary process, but make the additional assumption that it is purely nondeterministic with SDF $S(\cdot)$ (as already mentioned in Section 8.7, purely nondeterministic processes play a role in the Wold decomposition theorem – see, e.g., Brockwell and Davis, 1991, for details). As in Section 8.7, consider the best linear predictor $\widehat{X}_t$ of X_t given the infinite past $X_{t-1}, X_{t-2}, \ldots$, and, in keeping with Equation (404a), denote the corresponding innovation variance by

$$E\{(X_t - \widehat{X}_t)^2\} \stackrel{\text{def}}{=} \sigma_\infty^2.$$

Equation (404b) states that

$$\sigma_\infty^2 = \frac{1}{\Delta_t} \exp\left(\Delta_t \int_{-f_\mathcal{N}}^{f_\mathcal{N}} \log(S(f))\, df\right).$$

In what follows, we take the entropy of the SDF $S(\cdot)$ to be

$$H\{S(\cdot)\} = \int_{-f_\mathcal{N}}^{f_\mathcal{N}} \log(S(f))\, df. \tag{472a}$$

Thus large entropy is equivalent to a large innovation variance, i.e., a large prediction error variance. (In fact, the concept of entropy is usually defined in such a way that the entropy measure above is only valid for the case of stationary *Gaussian* processes.)

We can now state Burg's maximum entropy argument. If we have *perfect* knowledge of the ACVS for a process just out to lag p, we know only that $S(\cdot)$ lies within a certain class of SDFs. One way to select a particular member of this class is to pick that SDF, call it $\tilde{S}(\cdot)$, that maximizes the entropy subject to the constraints

$$\int_{-f_\mathcal{N}}^{f_\mathcal{N}} \tilde{S}(f) e^{i2\pi f \tau \Delta_t}\, df = s_\tau, \qquad 0 \leq \tau \leq p. \tag{472b}$$

To quote again from Burg (1975),

> Maximum entropy spectral analysis is based on choosing the spectrum which corresponds to the most random or the most unpredictable time series whose autocorrelation function agrees with the known values.

The solution to this constrained maximization problem is

$$\tilde{S}(f) = \frac{\sigma_p^2 \Delta_t}{\left|1 - \sum_{k=1}^{p} \phi_{p,k} e^{-i2\pi f k \Delta_t}\right|^2},$$

where $\phi_{p,k}$ and σ_p^2 are the solutions to the augmented Yule–Walker equations of order p; i.e., the SDF is an AR(r) SDF with $r \leq p$ (r is not necessarily equal to p since $\phi_{p,r+1}, \ldots, \phi_{p,p}$ could possibly be equal to 0). To see that this is true, let us consider the AR(p) process $\{Y_t\}$ with SDF $S_Y(\cdot)$ and innovation variance σ_p^2. For such a process the best linear predictor of Y_t given the infinite past is just

$$\widehat{Y}_t = \sum_{k=1}^{p} \phi_{p,k} Y_{t-k},$$

and the corresponding prediction error variance is just the innovation variance σ_p^2. Let $\{X_t\}$ be a stationary process whose ACVS agrees with that of $\{Y_t\}$ up to lag p, and let $S(\cdot)$ denote its SDF. We know that the best linear predictor of order p for X_t is

$$\sum_{k=1}^{p} \phi_{p,k} X_{t-k},$$

where the $\phi_{p,k}$ coefficients depend only on the ACVS of $\{X_t\}$ up to lag p; moreover, the associated prediction error variance must be σ_p^2. Because it is based on the infinite past, the innovation variance for X_t cannot be larger than the prediction error variance for its pth-order predictor. Hence we have

$$\sigma_p^2 = \frac{1}{\Delta_t} \exp\left(\Delta_t \int_{-f_{\mathcal{N}}}^{f_{\mathcal{N}}} \log\left(S_Y(f)\right) \, df\right) \geq \frac{1}{\Delta_t} \exp\left(\Delta_t \int_{-f_{\mathcal{N}}}^{f_{\mathcal{N}}} \log\left(S(f)\right) \, df\right) = \sigma_\infty^2,$$

which implies that

$$H\{S_Y(\cdot)\} = \int_{-f_{\mathcal{N}}}^{f_{\mathcal{N}}} \log\left(S_Y(f)\right) \, df \geq \int_{-f_{\mathcal{N}}}^{f_{\mathcal{N}}} \log\left(S(f)\right) \, df = H\{S(\cdot)\};$$

i.e., the entropy for a class of stationary processes – all of whose members are known to have the same ACVS out to lag p – is maximized by an AR(p) process with the specified ACVS out to lag p (Brockwell and Davis, 1991, section 10.6).

Burg's original idea for MESA (1967) was to *assume* that the biased *estimator* of the ACVS up to some lag p was in fact equal to the *true* ACVS. The maximum entropy principle under these assumptions leads to a procedure that is identical to the Yule–Walker estimation method. Burg (1968) abandoned this approach. He argued that it was not justifiable because, from his viewpoint, the usual biased estimator for the ACVS makes the implicit assumption that $X_t = 0$ for all $t < 0$ and $t \geq N$ (for example, with this assumption, the limits in Equation (170b) defining $\hat{s}_\tau^{(P)}$ can be modified to range from $t = -\infty$ to $t = \infty$). Burg's algorithm was his attempt to overcome this defect by estimating the ACVS in a different manner. There is still a gap in this logic, because true MESA requires *exact* knowledge of the ACVS up to a certain lag, whereas Burg's algorithm is just another way of *estimating* the ACVS from the data. Burg's criticism that window estimators effectively change "known" low-order values of the ACVS is offset by the fact that his procedure estimates lags beyond some arbitrary order p by an extension based upon only the first p values of the estimated ACVS and hence completely ignores what our time series might be telling us about the ACVS beyond lag p. (Another claim that has been made is that Burg's algorithm implicitly extends the time series – through predictions – outside the range $t = 0$ to $N - 1$ so that it is the "most random" one consistent with the observed data, i.e., closest to white noise. There are formidable problems in making this statement rigorous. A good discussion of other claims and counterclaims concerning maximum entropy can be found in a 1986 article by Makhoul with the intriguing title "Maximum Confusion Spectral Analysis.")

MESA is equivalent to AR spectral analysis when (a) the constraints on the entropy maximization are in terms of low-order values of the ACVS and (b) Equation (472a) is used as the measure of entropy. However, as Jaynes (1982) points out, the maximum entropy principle is more general so they need not be equivalent. If either the constraints (or "given data") are in terms of quantities other than just the ACVS or the measure of entropy is different from the one appropriate for stationary Gaussian processes, the resulting maximum entropy spectrum can have a different analytical form than the AR model. For example, recall from Section 7.9 that the cepstrum for an SDF $\tilde{S}(\cdot)$ is the sequence $\{c_\tau\}$ defined by

$$c_\tau = \int_{-f_{\mathcal{N}}}^{f_{\mathcal{N}}} \log\left(\tilde{S}(f)\right) e^{i 2\pi f \tau \Delta_t} \, df \qquad (473)$$

(Bogert et al., 1963). If we maximize the entropy (472a) over $\tilde{S}(\cdot)$ subject to the $p + 1$ constraints of Equation (472b) and q additional constraints that Equation (473) holds for known $c_1, \ldots, c_q$, then – subject to an existence condition – the maximum entropy principle yields an

ARMA(p, q) SDF (Lagunas-Hernández et al., 1984; Makhoul, 1986, section 8). Ihara (1984) and Franke (1985) discuss other sets of equivalent constraints that yield an ARMA spectrum as the maximum entropy spectrum.

Is the maximum entropy principle a good criterion for SDF estimation? Figure 475 shows an example adapted from an article entitled "Spectral Estimation: An Impossibility?" (Nitzberg, 1979). Here we assume that the ACVS of a certain process is known to be

$$s_\tau = 1 - \frac{|\tau|}{8} \text{ for } |\tau| \leq 8 = p.$$

The left-hand column shows four possible extrapolations of the ACVS for $|\tau| > p$. The upper three plots assume that the ACVS is given by

$$s_\tau = \begin{cases} \alpha \left(1 - ||\tau| - 16|/8\right), & 8 < |\tau| \leq 24; \\ 0, & |\tau| > 24, \end{cases} \quad (474)$$

where, going down from the top, $\alpha = 0$, $1/2$ and $-1/2$, respectively. The bottom-most plot shows the extension (out to $\tau = 32$) dictated by the maximum entropy principle. The right-hand column shows the low frequency portions of the four corresponding SDFs, which are rather different from each other (here we assume that the sampling interval Δ_t is unity so that the Nyquist frequency is 0.5). This illustrates the point that many different SDFs can have the same ACVS up to a certain lag. Even if we grant Burg's claim that we know the ACVS perfectly up to lag p (we never do with real data!), it is not clear why the maximum entropy extension of the ACVS has more credibility than the other three extensions. In particular, setting $\alpha = 0$ in Equation (474) yields an SDF with a monotonic decay over $f \in [0, 0.1]$, whereas the maximum entropy extension has an SDF associated with a broad peak centered at $f \doteq 0.034$. Given the prevalence of SDFs with monotonic decay in physical applications, the maximum entropy SDF has arguably less intuitive appeal than the $\alpha = 0$ SDF.

Figure 476 explores this example from another perspective. Suppose that the true ACVS comes from setting $\alpha = 0$ in Equation (474). This ACVS corresponds to an MA(7) process given by

$$X_t = \sum_{j=0}^{7} \epsilon_{t-j},$$

where $\{\epsilon_t\}$ is white noise process with zero mean and variance $\sigma_\epsilon^2 = 1/8$ (cf. Equation (32a)). The top row of plots in Figure 476 shows this ACVS for $\tau = 0, \ldots, 32$, and the corresponding SDF for $f \in [0, 0.1]$. This MA process is noninvertible; i.e., it cannot be expressed as an AR process of infinite order (Exercise [2.15] has an example of an invertible MA process). Noninvertibility suggests that AR approximations of finite orders p cannot be particularly good. The bottom three rows show AR(p) approximations of orders $p = 8$, 9 and 16 – these are the maximum entropy extensions given the ACVS out to lags 8, 9 or 16. These extensions result in unappealing peaks in the maximum entropy SDF where none exist in the true SDF and on specifying nonzero correlations at lags beyond ones at which the process is known to have reached decorrelation. Nonetheless, there are many processes with spectra that can be well approximated by an AR(p) process, and the maximum entropy method can be expected to work well for these.

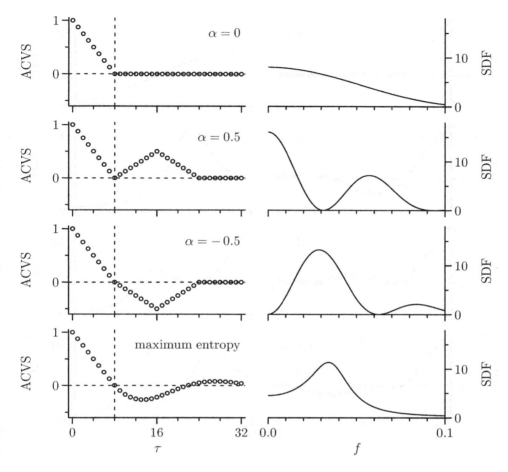

Figure 475 Example of four ACVSs (left-hand column) with identical values up to lag 8 and low frequency portions of corresponding SDFs (right-hand column – these are plotted on a linear/linear scale). The maximum entropy extension is shown on the bottom row.

9.7 Least Squares Estimators

In addition to the Yule–Walker estimator and Burg's algorithm, other ways for fitting an AR(p) model to a time series are least squares (LS) methods and the maximum likelihood method (discussed in the next section).

Suppose we have a time series of length N that can be regarded as a portion $Y_0, \ldots, Y_{N-1}$ of one realization of a stationary AR(p) process with zero mean. We can formulate an appropriate LS model in terms of our data as follows:

$$\boldsymbol{Y}_{\text{(FLS)}} = \mathcal{Y}_{\text{(FLS)}} \boldsymbol{\Phi} + \boldsymbol{\epsilon}_{\text{(FLS)}},$$

where $\boldsymbol{Y}_{\text{(FLS)}} = [Y_p, Y_{p+1}, \ldots, Y_{N-1}]^T$;

$$\mathcal{Y}_{\text{(FLS)}} = \begin{bmatrix} Y_{p-1} & Y_{p-2} & \cdots & Y_0 \\ Y_p & Y_{p-1} & \cdots & Y_1 \\ \vdots & \vdots & \ddots & \vdots \\ Y_{N-2} & Y_{N-3} & \cdots & Y_{N-p-1} \end{bmatrix};$$

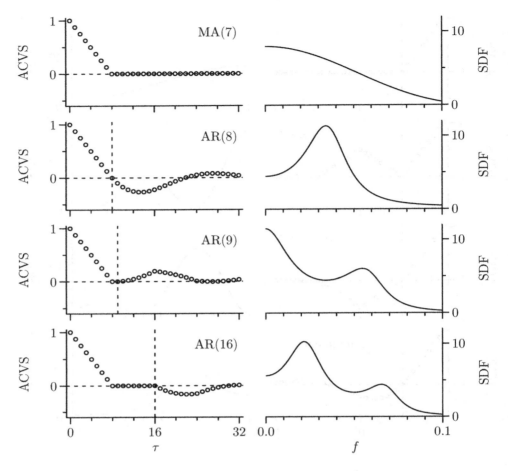

Figure 476 Four ACVSs (left-hand column) and low frequency portions of corresponding SDFs (right-hand column – these are plotted on a linear/linear scale). The top row of plots shows the ACVS for an MA(7) process at lags $\tau = 0, \ldots, 32$ and its corresponding SDF for $f \in [0, 0.1]$. The bottom three rows show corresponding maximum entropy extensions given knowledge of the MA(7) ACVS out to lags 8, 9 and 16 (marked by vertical dashed lines).

$\boldsymbol{\Phi} = [\phi_{p,1}, \phi_{p,2}, \ldots, \phi_{p,p}]^T$; and $\boldsymbol{\epsilon}_{\text{(FLS)}} = [\epsilon_p, \epsilon_{p+1}, \ldots, \epsilon_{N-1}]^T$ (the rationale for FLS rather than just LS in the subscripts will become apparent in a moment). We can thus estimate $\boldsymbol{\Phi}$ by finding the $\boldsymbol{\Phi}$ for which

$$\text{SS}_{\text{(FLS)}}(\boldsymbol{\Phi}) = \|\boldsymbol{Y}_{\text{(FLS)}} - \mathcal{Y}_{\text{(FLS)}}\boldsymbol{\Phi}\|^2 = \sum_{t=p}^{N-1} \left(Y_t - \sum_{k=1}^{p} \phi_{p,k} Y_{t-k} \right)^2 \qquad (476)$$

is minimized – as before, $\|\boldsymbol{V}\|^2 = \boldsymbol{V}^T \boldsymbol{V}$ for a vector $\boldsymbol{V}$. If we denote the vector that minimizes the above as $\widehat{\boldsymbol{\Phi}}_{\text{(FLS)}}$, standard LS theory tells us that it is given by

$$\widehat{\boldsymbol{\Phi}}_{\text{(FLS)}} = \left(\mathcal{Y}_{\text{(FLS)}}^T \mathcal{Y}_{\text{(FLS)}} \right)^{-1} \mathcal{Y}_{\text{(FLS)}}^T \boldsymbol{Y}_{\text{(FLS)}}$$

under the assumption that the matrix inverse exists (this assumption invariably holds in practice – if not, then $\widehat{\boldsymbol{\Phi}}_{\text{(FLS)}}$ is a solution to $\mathcal{Y}_{\text{(FLS)}}^T \mathcal{Y}_{\text{(FLS)}} \boldsymbol{\Phi} = \mathcal{Y}_{\text{(FLS)}}^T \boldsymbol{Y}_{\text{(FLS)}}$). We can estimate the innovation variance σ_p^2 by the usual (approximately unbiased) least squares estimator of the residual variation, namely,

$$\hat{\sigma}^2_{(\text{FLS})} \stackrel{\text{def}}{=} \frac{\text{SS}_{(\text{FLS})}(\widehat{\boldsymbol{\Phi}}_{(\text{FLS})})}{N - 2p}, \tag{477a}$$

where the divisor arises because there are effectively $N-p$ observations and p parameters to be estimated. Alternatively, if we condition on $Y_0, Y_1, \ldots, Y_{p-1}$, then we can interpret $\widehat{\boldsymbol{\Phi}}_{(\text{FLS})}$ as a conditional maximum likelihood estimator, for which the corresponding innovation variance estimator is

$$\tilde{\sigma}^2_{(\text{FLS})} \stackrel{\text{def}}{=} \frac{\text{SS}_{(\text{FLS})}(\widehat{\boldsymbol{\Phi}}_{(\text{FLS})})}{N - p} \tag{477b}$$

(Priestley, 1981, p. 352; McQuarrie and Tsai, 1998, p. 90).

The estimator $\widehat{\boldsymbol{\Phi}}_{(\text{FLS})}$ is known in the literature as the *forward least squares estimator* of $\boldsymbol{\Phi}$ to contrast it with two other closely related estimators. Motivated by the fact that the reversal of a stationary process is a stationary process with the same ACVS, we can reformulate the LS problem as

$$\boldsymbol{Y}_{(\text{BLS})} = \mathcal{Y}_{(\text{BLS})} \boldsymbol{\Phi} + \boldsymbol{\epsilon}_{(\text{BLS})},$$

where $\boldsymbol{Y}_{(\text{BLS})} = [Y_0, Y_1, \ldots, Y_{N-p-1}]^T$;

$$\mathcal{Y}_{(\text{BLS})} = \begin{bmatrix} Y_1 & Y_2 & \cdots & Y_p \\ Y_2 & Y_3 & \cdots & Y_{p+1} \\ \vdots & \vdots & \ddots & \vdots \\ Y_{N-p} & Y_{N-p+1} & \cdots & Y_{N-1} \end{bmatrix};$$

and $\boldsymbol{\epsilon}_{(\text{BLS})}$ is a vector of uncorrelated errors. The function of $\boldsymbol{\Phi}$ to be minimized is now

$$\text{SS}_{(\text{BLS})}(\boldsymbol{\Phi}) = \|\boldsymbol{Y}_{(\text{BLS})} - \mathcal{Y}_{(\text{BLS})} \boldsymbol{\Phi}\|^2 = \sum_{t=0}^{N-p-1} \left(Y_t - \sum_{k=1}^{p} \phi_{p,k} Y_{t+k}\right)^2. \tag{477c}$$

The *backward least squares estimator* of $\boldsymbol{\Phi}$ is thus given by

$$\widehat{\boldsymbol{\Phi}}_{(\text{BLS})} = \left(\mathcal{Y}_{(\text{BLS})}^T \mathcal{Y}_{(\text{BLS})}\right)^{-1} \mathcal{Y}_{(\text{BLS})}^T \boldsymbol{Y}_{(\text{BLS})},$$

and the appropriate estimate for the innovation variance is

$$\hat{\sigma}^2_{(\text{BLS})} = \frac{\text{SS}_{(\text{BLS})}(\widehat{\boldsymbol{\Phi}}_{(\text{BLS})})}{N - 2p}.$$

Finally, in the same spirit as Burg's algorithm, we can define a *forward/backward least squares estimator* $\widehat{\boldsymbol{\Phi}}_{(\text{FBLS})}$ of $\boldsymbol{\Phi}$ as that vector minimizing

$$\text{SS}_{(\text{FBLS})}(\boldsymbol{\Phi}) = \text{SS}_{(\text{FLS})}(\boldsymbol{\Phi}) + \text{SS}_{(\text{BLS})}(\boldsymbol{\Phi}). \tag{477d}$$

This estimator is associated with

$$\boldsymbol{Y}_{(\text{FBLS})} = \begin{bmatrix} \boldsymbol{Y}_{(\text{FLS})} \\ \boldsymbol{Y}_{(\text{BLS})} \end{bmatrix}, \quad \mathcal{Y}_{(\text{FBLS})} = \begin{bmatrix} \mathcal{Y}_{(\text{FLS})} \\ \mathcal{Y}_{(\text{BLS})} \end{bmatrix} \quad \text{and} \quad \hat{\sigma}^2_{(\text{FBLS})} = \frac{\text{SS}_{(\text{FBLS})}(\widehat{\boldsymbol{\Phi}}_{(\text{FBLS})})}{2N - 3p} \tag{477e}$$

(we use a divisor of $2N - 3p$ in creating $\hat{\sigma}^2_{(\text{FBLS})}$ because the length of $\boldsymbol{Y}_{(\text{FBLS})}$ is $2N - 2p$, and we have p coefficients to estimate). For a specified order p, we can compute the FBLS estimator as efficiently as the Burg estimator using a specialized algorithm due to Marple (1980);

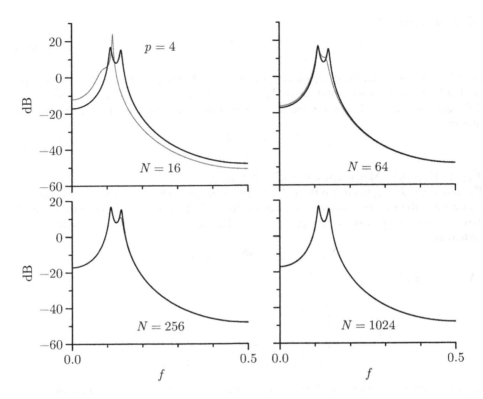

Figure 478 Forward/backward least squares AR(4) SDF estimates (thin curves) for portions of lengths 16, 64, 256 and 1024 of the realization of the AR(4) process shown in Figure 34(e). The thick curve in each plot is the true SDF. This figure should be compared with Figures 458 and 469.

however, in contrast to Burg's algorithm, the FBLS algorithm does *not*, as a side product, yield fitted models of orders $1, \ldots, p-1$. These additional fits become useful when dealing with criteria for order selection (see Section 9.11). In terms of spectral estimation, Monte Carlo studies indicate that the FBLS estimator generally performs better than the Yule–Walker, Burg, FLS and BLS estimators (see Marple, 1987, and Kay, 1988, and references therein).

As an example of FBLS SDF estimation, Figure 478 shows these estimates for the same time series used to produce the Yule–Walker and Burg AR(4) SDF estimates of Figures 458 and 469. A comparison of these three figures shows that, for this example, Burg's method and the FBLS estimator lead to SDF estimates that are comparable with, or superior to, the estimates produced by the Yule–Walker method. (Exercise [9.14] invites the reader to create similar figures for the FLS and BLS methods and for three other AR(4) time series.)

Comments and Extensions to Section 9.7

[1] A cautionary note is in order about SDF estimators formed from least squares estimators of the AR parameters. In contrast to the Yule–Walker and Burg estimators, these SDF estimators need *not* integrate to the sample variance $\hat{s}_0^{(P)}$ for the time series (integration to $\hat{s}_0^{(P)}$ is a desirable property shared by certain nonparametric SDF estimators, including the periodogram – see Equation (171d)). If the least squares estimator corresponds to a causal AR process, we can correct this deficiency by replacing the estimator of the innovation variance suggested by standard least squares theory with a different estimator. To formulate this alternative estimator, let $\widehat{\boldsymbol{\Phi}}_{(\text{LS})}$ and $\hat{\sigma}^2_{(\text{LS})}$ denote the least squares parameter estimators produced by any one of the three least squares schemes. Using $\widehat{\boldsymbol{\Phi}}_{(\text{LS})}$, we apply the step-down Levinson–Durbin recursions of Equation (460a) to obtain estimators, say $\hat{\phi}_{p-1,p-1}, \hat{\phi}_{p-2,p-2}, \ldots, \hat{\phi}_{1,1}$, of the

partial autocorrelations of all orders lower than p (the pth-order term is just the last element of $\widehat{\boldsymbol{\Phi}}_{\text{(LS)}}$ – we denote it as $\hat{\phi}_{p,p}$). The alternate to $\hat{\sigma}^2_{\text{(LS)}}$ is

$$\tilde{\sigma}^2_{\text{(LS)}} = \hat{s}^{\text{(P)}}_0 \prod_{j=1}^{p} \left(1 - \hat{\phi}^2_{j,j}\right) \tag{479}$$

(cf. Equation (509a), the subject of Exercise [9.8c]). The integral over $[-f_\mathcal{N}, f_\mathcal{N}]$ of the SDF estimator formed from $\widehat{\boldsymbol{\Phi}}_{\text{(LS)}}$ and $\tilde{\sigma}^2_{\text{(LS)}}$ is now guaranteed to be $\hat{s}^{\text{(P)}}_0$.

Unfortunately a least squares coefficient estimator $\widehat{\boldsymbol{\Phi}}_{\text{(LS)}}$ need *not* correspond to an estimated AR model that is causal (and hence stationary). Assuming that the estimated model is stationary (invariably the case in practical applications), the solution to Exercise [9.1b] points to a procedure for converting a stationary acausal AR model into a causal AR model, with the two models yielding exactly the same SDF estimators. In theory we could then take $\tilde{\sigma}^2_{\text{(LS)}}$ to be the innovation estimator for the causal model, which would ensure that the resulting SDF estimator integrates to $\hat{s}^{\text{(P)}}_0$. This corrective scheme is feasible for small p, but problems with numerical stability can arise as p increases because of the need to factor a polynomial of order p. (Exercise [9.15] in part concerns an FBLS estimate of a time series of length $N = 6$ and touches upon some of the issues raised here.)

[2] It should be noted that what we have defined as the Yule–Walker estimator in Section 9.3 is often called the least squares estimator in the statistical literature. To see the origin of this terminology, consider extending the time series $Y_0, Y_1, \ldots, Y_{N-1}$ by appending p dummy observations – each identically equal to zero – before and after it. The FLS model for this extended time series of length $N + 2p$ is

$$\boldsymbol{Y}_{\text{(YW)}} = \mathcal{Y}_{\text{(YW)}} \boldsymbol{\Phi} + \boldsymbol{\epsilon}_{\text{(YW)}},$$

where

$$\boldsymbol{Y}_{\text{(YW)}} \stackrel{\text{def}}{=} [Y_0, Y_1, \ldots, Y_{N-1}, \underbrace{0, \ldots, 0}_{p \text{ of these}}]^T;$$

$$\mathcal{Y}_{\text{(YW)}} \stackrel{\text{def}}{=} \begin{bmatrix} 0 & 0 & \cdots & 0 & 0 \\ Y_0 & 0 & \cdots & 0 & 0 \\ Y_1 & Y_0 & \cdots & 0 & 0 \\ \vdots & \vdots & \ddots & \vdots & \vdots \\ Y_{p-2} & Y_{p-3} & \cdots & Y_0 & 0 \\ Y_{p-1} & Y_{p-2} & \cdots & Y_1 & Y_0 \\ \vdots & \vdots & \ddots & \vdots & \vdots \\ Y_{N-1} & Y_{N-2} & \cdots & Y_{N-p+1} & Y_{N-p} \\ 0 & Y_{N-1} & \cdots & Y_{N-p+2} & Y_{N-p+1} \\ \vdots & \vdots & \ddots & \vdots & \vdots \\ 0 & 0 & \cdots & Y_{N-1} & Y_{N-2} \\ 0 & 0 & \cdots & 0 & Y_{N-1} \end{bmatrix};$$

and $\boldsymbol{\epsilon}_{\text{(YW)}} \stackrel{\text{def}}{=} [\epsilon'_0, \ldots, \epsilon'_{p-1}, \epsilon_p, \epsilon_{p+1}, \ldots, \epsilon_{N-1}, \epsilon'_N, \ldots, \epsilon'_{N+p-1}]^T$ – here we need the primes because the first and last p elements of this vector are *not* members of the innovation process $\{\epsilon_t\}$. Since

$$\frac{1}{N} \mathcal{Y}^T_{\text{(YW)}} \mathcal{Y}_{\text{(YW)}} = \begin{bmatrix} \hat{s}^{\text{(P)}}_0 & \hat{s}^{\text{(P)}}_1 & \cdots & \hat{s}^{\text{(P)}}_{p-1} \\ \hat{s}^{\text{(P)}}_1 & \hat{s}^{\text{(P)}}_0 & \cdots & \hat{s}^{\text{(P)}}_{p-2} \\ \vdots & \vdots & \ddots & \vdots \\ \hat{s}^{\text{(P)}}_{p-1} & \hat{s}^{\text{(P)}}_{p-2} & \cdots & \hat{s}^{\text{(P)}}_0 \end{bmatrix} = \widetilde{\boldsymbol{\Gamma}}_p$$

and $\mathcal{Y}^T_{\text{(YW)}} \boldsymbol{Y}_{\text{(YW)}}/N = [\hat{s}^{\text{(P)}}_1, \hat{s}^{\text{(P)}}_2, \ldots, \hat{s}^{\text{(P)}}_p]^T = \widetilde{\boldsymbol{\gamma}}_p$, it follows that the FLS estimator for this extended time series is

$$\left(\mathcal{Y}^T_{\text{(YW)}} \mathcal{Y}_{\text{(YW)}}\right)^{-1} \mathcal{Y}^T_{\text{(YW)}} \boldsymbol{Y}_{\text{(YW)}} = \widetilde{\boldsymbol{\Gamma}}^{-1}_p \widetilde{\boldsymbol{\gamma}}_p,$$

which is identical to the Yule–Walker estimator (cf. Equation (451a)).

[3] Some of the AR estimators that we have defined so far in this chapter are known in the engineering literature under a different name. For the record, we note the following correspondences (Kay and Marple, 1981):

$$\text{Yule–Walker} \iff \text{autocorrelation method}$$
$$\text{forward least squares} \iff \text{covariance method}$$
$$\text{forward/backward least squares} \iff \text{modified covariance method.}$$

9.8 Maximum Likelihood Estimators

We now consider maximum likelihood estimation of $\boldsymbol{\Phi}_p = [\phi_{p,1}, \phi_{p,2}, \ldots, \phi_{p,p}]^T$ and σ_p^2, the $p+1$ parameters in the AR(p) model. To do so, we assume that our observed time series is a realization of a portion $H_0, \ldots, H_{N-1}$ of a Gaussian AR(p) process with zero mean. Given these observations, the likelihood function for the unknown parameters is

$$L(\boldsymbol{\Phi}_p, \sigma_p^2 \mid \boldsymbol{H}) \stackrel{\text{def}}{=} \frac{1}{(2\pi)^{N/2}|\boldsymbol{\Gamma}_N|^{1/2}} \exp\left(-\boldsymbol{H}^T \boldsymbol{\Gamma}_N^{-1} \boldsymbol{H}/2\right), \tag{480a}$$

where $\boldsymbol{H} \stackrel{\text{def}}{=} [H_0, \ldots, H_{N-1}]^T$, $\boldsymbol{\Gamma}_N$ is the covariance matrix for $\boldsymbol{H}$ (i.e., its (j,k)th element is s_{j-k} – cf. Equation (450c)) and $|\boldsymbol{\Gamma}_N|$ is its determinant. For particular values of $\boldsymbol{\Phi}_p$ and σ_p^2, we can compute $L(\boldsymbol{\Phi}_p, \sigma_p^2 \mid \boldsymbol{H})$ by assuming that $\{H_t\}$ has these as its true parameter values. The maximum likelihood estimators (MLEs) are defined to be those values of the parameters that maximize $L(\boldsymbol{\Phi}_p, \sigma_p^2 \mid \boldsymbol{H})$ or, equivalently, that minimize the log likelihood function (after multiplication by -2 to eliminate a common factor):

$$l(\boldsymbol{\Phi}_p, \sigma_p^2 \mid \boldsymbol{H}) \stackrel{\text{def}}{=} -2\log\left(L(\boldsymbol{\Phi}_p, \sigma_p^2 \mid \boldsymbol{H})\right) = \log\left(|\boldsymbol{\Gamma}_N|\right) + \boldsymbol{H}^T \boldsymbol{\Gamma}_N^{-1} \boldsymbol{H} + N\log(2\pi). \tag{480b}$$

To obtain the MLEs, we must be able to compute $l(\boldsymbol{\Phi}_p, \sigma_p^2 \mid \boldsymbol{H})$ for particular choices of $\boldsymbol{\Phi}_p$ and σ_p^2, a task which at first glance appears to be formidable due to the necessity of finding the determinant for – and inverting – the $N \times N$-dimensional matrix $\boldsymbol{\Gamma}_N$. Fortunately, the Toeplitz structure of $\boldsymbol{\Gamma}_N$ allows us to simplify Equation (480b) considerably, as we now show (see, for example, Newton, 1988).

We first assume that $\boldsymbol{\Gamma}_N$ is positive definite so that we can make use of the modified Cholesky decomposition for it stated in Equation (464c), namely,

$$\boldsymbol{\Gamma}_N = \boldsymbol{L}_N \boldsymbol{D}_N \boldsymbol{L}_N^T,$$

where $\boldsymbol{L}_N$ is a lower triangular matrix, all of whose diagonal elements are 1, while $\boldsymbol{D}_N$ is a diagonal matrix. Because $\{H_t\}$ is assumed to be an AR(p) process, the first p diagonal elements of $\boldsymbol{D}_N$ are $\sigma_0^2, \sigma_1^2, \ldots, \sigma_{p-1}^2$, while the remaining $N-p$ elements are all equal to σ_p^2. Because the determinant of a product of square matrices is equal to the product of the individual determinants, and because the determinant of a triangular matrix is equal to the product of its diagonal elements, we now have

$$|\boldsymbol{\Gamma}_N| = |\boldsymbol{L}_N| \cdot |\boldsymbol{D}_N| \cdot |\boldsymbol{L}_N^T| = \sigma_p^{2(N-p)} \prod_{j=0}^{p-1} \sigma_j^2. \tag{480c}$$

The modified Cholesky decomposition immediately yields

$$\boldsymbol{\Gamma}_N^{-1} = \boldsymbol{L}_N^{-T} \boldsymbol{D}_N^{-1} \boldsymbol{L}_N^{-1},$$

9.8 Maximum Likelihood Estimators

where, for an AR(p) process, the matrix $\boldsymbol{L}_N^{-1}$ takes the form indicated by Equation (465b). Hence we have

$$\boldsymbol{H}^T \boldsymbol{\Gamma}_N^{-1} \boldsymbol{H} = \boldsymbol{H}^T \boldsymbol{L}_N^{-T} \boldsymbol{D}_N^{-1} \boldsymbol{L}_N^{-1} \boldsymbol{H} = (\boldsymbol{L}_N^{-1} \boldsymbol{H})^T \boldsymbol{D}_N^{-1} (\boldsymbol{L}_N^{-1} \boldsymbol{H}).$$

From Equation (465c) the first p elements of $\boldsymbol{L}_N^{-1} \boldsymbol{H}$ are simply $\vec{e}_0(0), \vec{e}_1(1), \ldots, \vec{e}_{p-1}(p-1)$, where, in accordance with earlier definitions,

$$\vec{e}_0(0) = H_0 \quad \text{and} \quad \vec{e}_t(t) = H_t - \sum_{j=1}^{t} \phi_{t,j} H_{t-j}, \qquad t = 1, \ldots, p-1.$$

The last $N - p$ elements follow from the assumed model for $\{H_t\}$ and are given by

$$\epsilon_t = H_t - \sum_{j=1}^{p} \phi_{p,j} H_{t-j}, \qquad t = p, \ldots, N-1.$$

Hence we can write

$$\boldsymbol{H}^T \boldsymbol{\Gamma}_N^{-1} \boldsymbol{H} = \sum_{t=0}^{p-1} \frac{\vec{e}_t^{\,2}(t)}{\sigma_t^2} + \sum_{t=p}^{N-1} \frac{\epsilon_t^2}{\sigma_p^2}. \tag{481a}$$

Combining this and Equation (480c) with Equation (480b) yields

$$l(\boldsymbol{\Phi}_p, \sigma_p^2 \mid \boldsymbol{H}) = \sum_{j=0}^{p-1} \log(\sigma_j^2) + (N-p) \log(\sigma_p^2) + \sum_{t=0}^{p-1} \frac{\vec{e}_t^{\,2}(t)}{\sigma_t^2} + \sum_{t=p}^{N-1} \frac{\epsilon_t^2}{\sigma_p^2} + N \log(2\pi).$$

If we note that we can write

$$\sigma_j^2 = \sigma_p^2 \lambda_j \quad \text{with} \quad \lambda_j \stackrel{\text{def}}{=} \left(\prod_{k=j+1}^{p} (1 - \phi_{k,k}^2) \right)^{-1}$$

for $j = 0, \ldots, p-1$ (this follows from Exercise [9.8c]), we now obtain a useful form for the log likelihood function:

$$l(\boldsymbol{\Phi}_p, \sigma_p^2 \mid \boldsymbol{H}) = \sum_{j=0}^{p-1} \log(\lambda_j) + N \log(\sigma_p^2) + \frac{\text{SS}_{(\text{ML})}(\boldsymbol{\Phi}_p)}{\sigma_p^2} + N \log(2\pi), \tag{481b}$$

where

$$\text{SS}_{(\text{ML})}(\boldsymbol{\Phi}_p) \stackrel{\text{def}}{=} \sum_{t=0}^{p-1} \frac{\vec{e}_t^{\,2}(t)}{\lambda_t} + \sum_{t=p}^{N-1} \epsilon_t^2. \tag{481c}$$

We can obtain an expression for the MLE $\hat{\sigma}^2_{(\text{ML})}$ of σ_p^2 by differentiating Equation (481b) with respect to σ_p^2 and setting it to zero. This shows that, no matter what the MLE $\widehat{\boldsymbol{\Phi}}_{(\text{ML})}$ of $\boldsymbol{\Phi}_p$ turns out to be, the estimator $\hat{\sigma}^2_{(\text{ML})}$ is given by

$$\hat{\sigma}^2_{(\text{ML})} \stackrel{\text{def}}{=} \text{SS}_{(\text{ML})}(\widehat{\boldsymbol{\Phi}}_{(\text{ML})})/N. \tag{481d}$$

The parameter σ_p^2 can thus be eliminated from Equation (481b), yielding what Brockwell and Davis (1991) refer to as the *reduced likelihood* (sometimes called the *profile likelihood*):

$$l(\boldsymbol{\Phi}_p \mid \boldsymbol{H}) \stackrel{\text{def}}{=} \sum_{j=0}^{p-1} \log(\lambda_j) + N \log(\text{SS}_{(\text{ML})}(\boldsymbol{\Phi}_p)/N) + N + N \log(2\pi). \quad (482\text{a})$$

We can determine $\widehat{\boldsymbol{\Phi}}_{(\text{ML})}$ by finding the value of $\boldsymbol{\Phi}_p$ that minimizes the reduced likelihood.

Let us now specialize to the case $p = 1$. To determine $\phi_{1,1}$, we must minimize

$$l(\phi_{1,1} \mid \boldsymbol{H}) = -\log(1 - \phi_{1,1}^2) + N \log(\text{SS}_{(\text{ML})}(\phi_{1,1})/N) + N + N \log(2\pi), \quad (482\text{b})$$

where

$$\text{SS}_{(\text{ML})}(\phi_{1,1}) = H_0^2(1 - \phi_{1,1}^2) + \sum_{t=1}^{N-1}(H_t - \phi_{1,1} H_{t-1})^2. \quad (482\text{c})$$

Differentiating Equation (482b) with respect to $\phi_{1,1}$ and setting the result equal to zero yields

$$\frac{\phi_{1,1} \text{SS}_{(\text{ML})}(\phi_{1,1})}{N} - (1 - \phi_{1,1}^2)\left(\sum_{t=1}^{N-1} H_{t-1} H_t - \phi_{1,1} \sum_{t=1}^{N-2} H_t^2\right) = 0, \quad (482\text{d})$$

where the second sum vanishes when $N = 2$. In general this is a cubic equation in $\phi_{1,1}$. The estimator $\hat{\phi}_{(\text{ML})}$ is thus equal to the root of this equation that minimizes $l(\phi_{1,1} \mid \boldsymbol{H})$.

For $p > 2$, we cannot obtain the MLEs so easily. We must resort to a nonlinear optimizer in order to numerically determine $\widehat{\boldsymbol{\Phi}}_{(\text{ML})}$. Jones (1980) describes in detail a scheme for computing $\widehat{\boldsymbol{\Phi}}_{(\text{ML})}$ that uses a transformation of variables to facilitate numerical optimization. An interesting feature of his scheme is the use of a Kalman filter to compute the reduced likelihood at each step in the numerical optimization. This approach has two important advantages: first, it can easily deal with time series for which some of the observations are missing; and, second, it can handle the case in which an AR(p) process is observed in the presence of additive noise (this is discussed in more detail in Section 10.12).

Maximum likelihood estimation has not seen much use in parametric spectral analysis for two reasons. First, in a pure parametric approach, we need to be able to routinely fit fairly high-order AR models, particularly in the initial stages of a data analysis (p in the range of 10 and higher is common). High-order models are often a necessity in order to adequately capture all of the important features of an SDF. The computational burden of the maximum likelihood method can become unbearable as p increases beyond even 4 or 5. Second, if we regard parametric spectral analysis as merely a way of designing low-order prewhitening filters (the approach we personally favor – see Section 9.10), then, even though the model order might now be small enough to allow use of the maximum likelihood method, other – more easily computed – estimators such as the Burg or FBLS estimators work perfectly well in practice. Any imperfections in the prewhitening filter can be compensated for in the subsequent nonparametric spectral analysis of the prewhitened series.

As an example, Figure 483 shows maximum likelihood SDF estimates for the same time series used to produce the Yule–Walker, Burg and FBLS AR(4) SDF estimates of Figures 458, 469 and 478. Visually the ML-based estimates correspond well to the Burg and FBLS AR(4) estimates. Table 483 tabulates the values of the reduced likelihood for the Yule-Walker, Burg, FBLS and ML estimates shown in Figures 458, 469, 478 and 483. As must be the case, the reduced likelihood is smallest for the ML estimates. For all sample sizes, the next smallest are for the FBLS estimates, with the Burg estimates being close behind (the Yule–Walker estimates are a distant fourth). These results are verification that the Burg and FBLS estimates are decent surrogates for the computationally more demanding ML estimates.

9.8 Maximum Likelihood Estimators

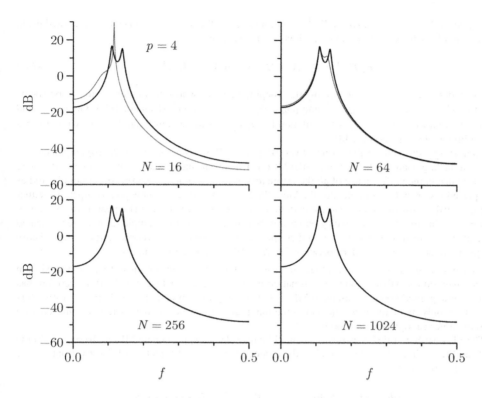

Figure 483 Maximum likelihood AR(4) SDF estimates (thin curves) for portions of lengths 16, 64, 256 and 1024 of the realization of the AR(4) process shown in Figure 34(e). The thick curve in each plot is the true SDF. This figure should be compared with Figures 458, 469 and 478.

	$l(\widehat{\boldsymbol{\Phi}}_p \mid \boldsymbol{H})$			
N	Yule–Walker	Burg	FBLS	ML
16	10.59193	−46.77426	−47.24593	−48.14880
64	−47.21592	−206.0418	−206.3352	−206.5077
256	−314.3198	−864.5200	−864.6390	−864.7399
1024	−1873.498	−3446.536	−3446.557	−3446.560

Table 483 Reduced (profile) likelihoods for estimated AR(4) coefficients based on Yule–Walker, Burg, FBLS and ML methods. The time series are of varying lengths N, and the corresponding SDF estimates are shown in Figures 458, 469, 478 and 483. As must be the case, the ML estimates have the lowest profile likelihood values, but those for the Burg and FBLS methods are comparable (the values for the Yule–Walker method are much higher, and the SDF estimates for this method are visually the poorest of the four methods).

Comments and Extensions to Section 9.8

[1] The large-sample distribution of the MLEs for the parameters of an AR(p) process has been worked out (see, for example, Newton, 1988, or Brockwell and Davis, 1991). For large N, the estimator $\widehat{\boldsymbol{\Phi}}_{(\text{ML})}$ is approximately distributed as a multivariate Gaussian RV with mean $\boldsymbol{\Phi}_p$ and covariance matrix $\sigma_p^2 \boldsymbol{\Gamma}_p^{-1}/N$, where $\boldsymbol{\Gamma}_p$ is the covariance matrix defined in Equation (450c); moreover, $\hat{\sigma}^2_{(\text{ML})}$ approximately follows a Gaussian distribution with mean σ_p^2 and variance $2\sigma_p^4/N$ and is approximately

independent of $\widehat{\boldsymbol{\Phi}}_{(\mathrm{ML})}$. The matrix $\sigma_p^2 \boldsymbol{\Gamma}_p^{-1}$ (sometimes called the *Schur matrix*) can be computed using Equation (464d), with N set to p; however, it is also the case that

$$\sigma_p^2 \boldsymbol{\Gamma}_p^{-1} = \boldsymbol{A}^T \boldsymbol{A} - \boldsymbol{B}^T \boldsymbol{B} = \boldsymbol{A}\boldsymbol{A}^T - \boldsymbol{B}\boldsymbol{B}^T, \qquad (484a)$$

where $\boldsymbol{A}$ and $\boldsymbol{B}$ are the $p \times p$ lower triangular Toeplitz matrices whose first columns are, respectively, the vectors $[1, -\phi_{p,1}, \ldots, -\phi_{p,p-1}]^T$ and $[\phi_{p,p}, \phi_{p,p-1}, \ldots, \phi_{p,1}]^T$ (Pagano, 1973; Godolphin and Unwin, 1983; Newton, 1988). This formulation for $\sigma_p^2 \boldsymbol{\Gamma}_p^{-1}$ is more convenient to work with at times than that implied by Equation (464d).

Interestingly the Yule–Walker, Burg, FLS, BLS and FBLS estimators of the AR(p) parameters *all* share the *same* large-sample distribution as the MLEs. This result suggests that, given enough data, there is no difference in the performance of all the estimators we have studied; however, small sample studies have repeatedly shown, for example, that the Yule–Walker estimator can be quite poor compared to either the Burg or FBLS estimators and that, for certain narrow-band processes, the Burg estimator is inferior to the FBLS estimator. This discrepancy points out the limitations of large-sample results (see Lysne and Tjøstheim, 1987, for further discussion on these points and also on a second-order large-sample theory that explains some of the differences between the Yule–Walker and LS estimators).

[2] As we have already noted, the fact that we must resort to a nonlinear optimizer to use the maximum likelihood method for AR(p) parameter estimation limits its use in practice. Much effort therefore has gone into finding good approximations to MLEs that are easy to compute. These alternative estimators are useful not only by themselves, but also as the initial guesses at the AR parameters that are required by certain numerical optimization routines.

The FLS, BLS and FBLS estimators can all be regarded as approximate MLEs. To see this in the case of the FLS estimator, we first simplify the reduced likelihood of Equation (482a) by dropping the $\sum \log(\lambda_j)$ term to obtain

$$l(\boldsymbol{\Phi}_p \mid \boldsymbol{H}) \approx N \log\left(\mathrm{SS}_{(\mathrm{ML})}(\boldsymbol{\Phi}_p)/N\right) + N + N \log(2\pi).$$

A justification for this simplification is that the deleted term becomes negligible as N gets large. The value of $\boldsymbol{\Phi}_p$ that minimizes the right-hand side above is identical to the value that minimizes the sum of squares $\mathrm{SS}_{(\mathrm{ML})}(\boldsymbol{\Phi}_p)$. If we in turn approximate $\mathrm{SS}_{(\mathrm{ML})}(\boldsymbol{\Phi}_p)$ by dropping the summation of the first p squared prediction errors in Equation (481c) (again negligible as N get large), we obtain an approximate MLE given by the value of $\boldsymbol{\Phi}_p$ that minimizes $\sum_{t=p}^{N-1} \epsilon_t^2$. However, this latter sum of squares is identical to the sum of squares $\mathrm{SS}_{(\mathrm{FLS})}(\cdot)$ of Equation (476), so our approximate estimator turns out to be just the FLS estimator.

Turning now to the BLS estimator, note that the log likelihood function in Equation (480b) for $\boldsymbol{H} = [H_0, \ldots, H_{N-1}]^T$ is *identical* to that for the "time reversed" series $\widetilde{\boldsymbol{H}} \stackrel{\text{def}}{=} [H_{N-1}, \ldots, H_0]^T$. This follows from the fact that

$$\widetilde{\boldsymbol{H}}^T \boldsymbol{\Gamma}_N^{-1} \widetilde{\boldsymbol{H}} = \boldsymbol{H}^T \boldsymbol{\Gamma}_N^{-1} \boldsymbol{H} \qquad (484b)$$

(Exercise [9.17] is to prove the above). Repeating the argument of the previous paragraph for this "time reversed" series leads to the BLS estimator as an approximate MLE. Finally, if we rewrite Equation (480b) as

$$l(\boldsymbol{\Phi}_p, \sigma_p^2 \mid \boldsymbol{H}) = \log\left(|\boldsymbol{\Gamma}_N|\right) + \frac{\boldsymbol{H}^T \boldsymbol{\Gamma}_N^{-1} \boldsymbol{H} + \widetilde{\boldsymbol{H}}^T \boldsymbol{\Gamma}_N^{-1} \widetilde{\boldsymbol{H}}}{2} + N \log(2\pi), \qquad (484c)$$

we can obtain the FBLS estimator as an approximate MLE (this is Exercise [9.18]). Due to the symmetric nature of the likelihood function, this particular estimator seems a more natural approximation for Gaussian processes than the other two estimators, both of which are direction dependent. (Here and elsewhere in this chapter, we have considered a time-reversed version $\{X_{-t}\}$ of a stationary process $\{X_t\}$, noting that both processes must have the same ACVS and that, in the Gaussian case, a time series and its reversal must have the same log likelihood function. Weiss, 1975, explores a technical concept called *time-reversibility* that involves invariance of the joint distribution of the RVs in a time series rather than of just their second-order properties. Lawrance, 1991, notes that "... the key contribution in Weiss (1975) is the proof that ARMA processes with an autoregressive component are reversible if and only if they are Gaussian." This contribution implies that a non-Gaussian time series and its time-reversed version need *not* have the same log likelihood function and that a time series showing evidence of directionality cannot be fully modeled by a Gaussian AR process.)

[3] Kay (1983) proposes a recursive maximum likelihood estimator (RMLE) for the AR(p) parameters that is similar in spirit to Burg's algorithm. One interpretation of Burg's algorithm is that it recursively fits one AR(1) model after another by minimizing the sum of squares of forward and backward prediction errors at each stage. The RMLE also recursively fits AR(1) models but by minimizing a reduced AR(1) likelihood analogous to that of Equation (482b). Fitting an AR(p) model by this method thus requires successively finding the roots of p different cubic equations, which, in contrast to the MLE method, is a computational improvement because it avoids the need for a nonlinear optimizer when $p \geq 2$. Kay (1983) reports on a Monte Carlo study that compares the RMLE method with the FBLS method and indicates that, while the former yields more accurate parameter estimates, the latter gives better estimates of the locations of peaks in an SDF (see the discussion of Table 459 for another example of better parameter estimates failing to translate into a better SDF estimate).

9.9 Confidence Intervals Using AR Spectral Estimators

For each of the nonparametric SDF estimators considered in Chapters 6, 7 and 8, we described how to construct a confidence interval (CI) for the unknown SDF based upon the estimator. Here we consider creating CIs based upon the statistical properties of a parametric AR SDF estimator. As Kaveh and Cooper (1976) point out, since "the AR spectral estimate is obtained through nonlinear operations..., analytical derivation of its statistical properties is, in general, formidable." Early work in this area include Kromer (1969), Berk (1974), Baggeroer (1976), Sakai (1979) and Reid (1979). Newton and Pagano (1984) use both inverse autocovariances and Scheffé projections to find simultaneous CIs, while Koslov and Jones (1985) and Burshtein and Weinstein (1987, 1988) develop similar approaches. We describe here an approach that is most similar to that of Burshtein and Weinstein, although differing in mathematical and statistical details. (The method described in this section for creating CIs is based on large-sample analytic theory. In Section 11.6 we present an entirely different approach using bootstrapping in our discussion of the ocean wave data.)

As we noted in C&E [1] for Section 9.8, all of the AR parameter estimators we have discussed in this chapter have the same large-sample distribution. Accordingly let $\widehat{\boldsymbol{\Phi}}_p = [\hat{\phi}_{p,1}, \hat{\phi}_{p,2}, \ldots, \hat{\phi}_{p,p}]^T$ and $\hat{\sigma}_p^2$ be any one of these estimators of $\boldsymbol{\Phi}_p = [\phi_{p,1}, \phi_{p,2}, \ldots, \phi_{p,p}]^T$ and σ_p^2 but with the stipulation that they are based upon a sample $H_0, H_1, \ldots, H_{N-1}$ from a Gaussian AR(p) process with zero mean. Recalling that $\stackrel{\mathrm{d}}{=}$ means "equal in distribution," we have, as $N \to \infty$,

$$\sqrt{N}(\widehat{\boldsymbol{\Phi}}_p - \boldsymbol{\Phi}_p) \stackrel{\mathrm{d}}{=} \mathcal{N}(\mathbf{0}, \sigma_p^2 \boldsymbol{\Gamma}_p^{-1}) \text{ and } \sqrt{N}(\hat{\sigma}_p^2 - \sigma_p^2) \stackrel{\mathrm{d}}{=} \mathcal{N}(0, 2\sigma_p^4), \tag{485a}$$

where $\mathcal{N}(\mathbf{0}, \sigma_p^2 \boldsymbol{\Gamma}_p^{-1})$ denotes a p-dimensional vector of RVs obeying a p-dimensional Gaussian distribution with zero mean vector and covariance matrix $\sigma_p^2 \boldsymbol{\Gamma}_p^{-1}$, and $\boldsymbol{\Gamma}_p$ is the $p \times p$ covariance matrix of Equation (450c); moreover, $\hat{\sigma}_p^2$ is independent of all the elements of $\widehat{\boldsymbol{\Phi}}_p$ (Newton, 1988; Brockwell and Davis, 1991). If we let $\hat{s}_\tau$ stand for the estimator of s_τ derived from $\widehat{\boldsymbol{\Phi}}_p$ and $\hat{\sigma}_p^2$, we can obtain a consistent estimator $\widehat{\boldsymbol{\Gamma}}_p$ of $\boldsymbol{\Gamma}_p$ by replacing s_τ in $\boldsymbol{\Gamma}_p$ with $\hat{s}_\tau$. These results are the key to the methods of calculating asymptotically correct CIs for AR spectra that we now present.

Let us rewrite the expression in Equation (446b) for the SDF of a stationary AR(p) process as

$$S(f) = \frac{\sigma_p^2 \Delta_{\mathrm{t}}}{|1 - e^H(f)\boldsymbol{\Phi}_p|^2}, \tag{485b}$$

where H denotes complex-conjugate (Hermitian) transpose and

$$e(f) \stackrel{\text{def}}{=} [e^{i2\pi f \Delta_t}, e^{i4\pi f \Delta_t}, \ldots, e^{i2p\pi f \Delta_t}]^T.$$

Let $e_\Re(f)$ and $e_\Im(f)$ be vectors containing, respectively, the real and imaginary parts of $e(f)$, and define the $p \times 2$ matrix E as

$$E^T = [e_\Re(f) \mid e_\Im(f)]^T = \begin{bmatrix} \cos(2\pi f \Delta_t) & \cdots & \cos(2p\pi f \Delta_t) \\ \sin(2\pi f \Delta_t) & \cdots & \sin(2p\pi f \Delta_t) \end{bmatrix}.$$

We assume for now that $0 < |f| < f_\mathcal{N}$. It follows from Exercise [2.2] and Equation (485a) that the 2×1 vector

$$B \stackrel{\text{def}}{=} E^T(\widehat{\boldsymbol{\Phi}}_p - \boldsymbol{\Phi}_p) \tag{486a}$$

has, for large N, a variance given by $\sigma_p^2 E^T(\Gamma_p^{-1}/N)E$ and hence asymptotically

$$B \stackrel{\text{d}}{=} \mathcal{N}(0, \sigma_p^2 \mathcal{Q}), \quad \text{where} \quad \mathcal{Q} \stackrel{\text{def}}{=} E^T(\Gamma_p^{-1}/N)E. \tag{486b}$$

Under the conditions specified above, we know from the Mann–Wald theorem (Mann and Wald, 1943; see also Bruce and Martin, 1989) that, asymptotically,

$$\frac{B^T \widehat{\mathcal{Q}}^{-1} B}{\hat{\sigma}_p^2} \stackrel{\text{d}}{=} \chi_2^2, \quad \text{where} \quad \widehat{\mathcal{Q}} \stackrel{\text{def}}{=} E^T(\widehat{\Gamma}_p^{-1}/N)E. \tag{486c}$$

From the definition of B in Equation (486a), we can write

$$\frac{B^T \widehat{\mathcal{Q}}^{-1} B}{\hat{\sigma}_p^2} = \frac{(E^T \widehat{\boldsymbol{\Phi}}_p - E^T \boldsymbol{\Phi}_p)^T \widehat{\mathcal{Q}}^{-1}(E^T \widehat{\boldsymbol{\Phi}}_p - E^T \boldsymbol{\Phi}_p)}{\hat{\sigma}_p^2} \stackrel{\text{d}}{=} \chi_2^2$$

when $0 < |f| < f_\mathcal{N}$. Hence,

$$\mathbf{P}\left[\frac{(E^T \widehat{\boldsymbol{\Phi}}_p - E^T \boldsymbol{\Phi}_p)^T \widehat{\mathcal{Q}}^{-1}(E^T \widehat{\boldsymbol{\Phi}}_p - E^T \boldsymbol{\Phi}_p)}{\hat{\sigma}_p^2} \leq Q_2(1-\alpha)\right] = 1 - \alpha, \tag{486d}$$

where $Q_\nu(\alpha)$ is the $\alpha \times 100\%$ percentage point of the χ_ν^2 distribution. Let $\mathcal{A}_0$ represent the event displayed between the brackets above. An equivalent description for $\mathcal{A}_0$ is that the true value of the 2×1 vector $E^T \boldsymbol{\Phi}_p$ lies inside the ellipsoid defined as the set of vectors, $E^T \widetilde{\boldsymbol{\Phi}}_p$ say, satisfying

$$(E^T \widehat{\boldsymbol{\Phi}}_p - E^T \widetilde{\boldsymbol{\Phi}}_p)^T \mathcal{M} (E^T \widehat{\boldsymbol{\Phi}}_p - E^T \widetilde{\boldsymbol{\Phi}}_p) \leq 1,$$

where $\mathcal{M} \stackrel{\text{def}}{=} \widehat{\mathcal{Q}}^{-1}/(\hat{\sigma}_p^2 Q_2(1-\alpha))$. Scheffé (1959, p. 407) shows that $E^T \widetilde{\boldsymbol{\Phi}}_p$ is in this ellipsoid if and only if

$$\left|a^T(E^T \widehat{\boldsymbol{\Phi}}_p - E^T \boldsymbol{\Phi}_p)\right|^2 \leq a^T \mathcal{M}^{-1} a \tag{486e}$$

for *all* two-dimensional vectors a, giving us another equivalent description for $\mathcal{A}_0$. Hence, specializing to two specific values for a, the occurrence of the event $\mathcal{A}_0$ implies the occurrence of the following two events:

$$\left|[1,0](E^T \widehat{\boldsymbol{\Phi}}_p - E^T \boldsymbol{\Phi}_p)\right|^2 = \left|e_\Re^T(f)(\widehat{\boldsymbol{\Phi}}_p - \boldsymbol{\Phi}_p)\right|^2 \leq [1,0]\mathcal{M}^{-1}[1,0]^T$$

9.9 Confidence Intervals Using AR Spectral Estimators

and

$$\left|[0,1](\boldsymbol{E}^T\widehat{\boldsymbol{\Phi}}_p - \boldsymbol{E}^T\boldsymbol{\Phi}_p)\right|^2 = \left|\boldsymbol{e}_\Im^T(f)(\widehat{\boldsymbol{\Phi}}_p - \boldsymbol{\Phi}_p)\right|^2 \leq [0,1]\mathcal{M}^{-1}[0,1]^T.$$

These two events in turn imply the occurrence of the event $\mathcal{A}_1$, defined by

$$\left|\boldsymbol{e}_\Re^T(f)(\widehat{\boldsymbol{\Phi}}_p - \boldsymbol{\Phi}_p)\right|^2 + \left|\boldsymbol{e}_\Im^T(f)(\widehat{\boldsymbol{\Phi}}_p - \boldsymbol{\Phi}_p)\right|^2 \leq [1,0]\mathcal{M}^{-1}[1,0]^T + [0,1]\mathcal{M}^{-1}[0,1]^T.$$

Let us rewrite both sides of this inequality. For the left-hand side, we can write

$$\left|\boldsymbol{e}_\Re^T(f)(\widehat{\boldsymbol{\Phi}}_p - \boldsymbol{\Phi}_p)\right|^2 + \left|\boldsymbol{e}_\Im^T(f)(\widehat{\boldsymbol{\Phi}}_p - \boldsymbol{\Phi}_p)\right|^2 = \left|\boldsymbol{e}_\Re^T(f)(\widehat{\boldsymbol{\Phi}}_p - \boldsymbol{\Phi}_p) - i\boldsymbol{e}_\Im^T(f)(\widehat{\boldsymbol{\Phi}}_p - \boldsymbol{\Phi}_p)\right|^2$$
$$= \left|\boldsymbol{e}^H(f)(\widehat{\boldsymbol{\Phi}}_p - \boldsymbol{\Phi}_p)\right|^2.$$

For the right-hand side, we let $\boldsymbol{d}^H = [1, -i]$ and use the fact that the 2×2 matrix $\mathcal{M}$ is symmetric to obtain

$$[1,0]\mathcal{M}^{-1}[1,0]^T + [0,1]\mathcal{M}^{-1}[0,1]^T = \boldsymbol{d}^H \mathcal{M}^{-1} \boldsymbol{d}.$$

Hence the event $\mathcal{A}_1$ is equivalent to

$$\left|\boldsymbol{e}^H(f)(\widehat{\boldsymbol{\Phi}}_p - \boldsymbol{\Phi}_p)\right|^2 \leq \boldsymbol{d}^H \mathcal{M}^{-1} \boldsymbol{d}, \quad \text{or,} \quad \frac{\left|\boldsymbol{e}^H(f)(\widehat{\boldsymbol{\Phi}}_p - \boldsymbol{\Phi}_p)\right|^2}{\boldsymbol{d}^H(\widehat{\mathcal{Q}}\hat{\sigma}_p^2)\boldsymbol{d}} \leq Q_2(1-\alpha).$$

Because the occurrence of the event $\mathcal{A}_0$ implies the occurrence of the event $\mathcal{A}_1$, it follows that $\mathcal{A}_0$ is a subset of $\mathcal{A}_1$. We thus must have $\mathbf{P}[\mathcal{A}_0] \leq \mathbf{P}[\mathcal{A}_1]$. Since Equation (486d) states that $\mathbf{P}[\mathcal{A}_0] = 1 - \alpha$, it follows that $\mathbf{P}[\mathcal{A}_1] \geq 1 - \alpha$, i.e., that

$$\mathbf{P}\left[\left|\boldsymbol{e}^H(f)(\widehat{\boldsymbol{\Phi}}_p - \boldsymbol{\Phi}_p)\right|^2 \leq Q_2(1-\alpha)\boldsymbol{d}^H(\widehat{\mathcal{Q}}\hat{\sigma}_p^2)\boldsymbol{d}\right] \geq 1 - \alpha. \tag{487}$$

Let us define

$$G(f) = 1 - \boldsymbol{e}^H(f)\boldsymbol{\Phi}_p \quad \text{so that} \quad |G(f)|^2 = \frac{\sigma_p^2 \Delta_t}{S(f)},$$

because of Equation (485b). With $\hat{G}(f) = 1 - \boldsymbol{e}^H(f)\widehat{\boldsymbol{\Phi}}_p$, we have

$$\left|\boldsymbol{e}^H(f)(\widehat{\boldsymbol{\Phi}}_p - \boldsymbol{\Phi}_p)\right|^2 = \left|\hat{G}(f) - G(f)\right|^2.$$

Recalling that $\boldsymbol{d}^H = [1, -i]$ and using the definition for $\widehat{\mathcal{Q}}$ in Equation (486c), we obtain

$$\boldsymbol{d}^H \widehat{\mathcal{Q}} \boldsymbol{d} = \boldsymbol{e}^H(f)(\widehat{\boldsymbol{\Gamma}}_p^{-1}/N)\boldsymbol{e}(f)$$

since $\boldsymbol{d}^H \boldsymbol{E}^T = \boldsymbol{e}^H(f)$. We can now rewrite Equation (487) as

$$\mathbf{P}\left[|\hat{G}(f) - G(f)|^2 \leq Q_2(1-\alpha)\hat{\sigma}_p^2 \boldsymbol{e}^H(f)(\widehat{\boldsymbol{\Gamma}}_p^{-1}/N)\boldsymbol{e}(f)\right] \geq 1 - \alpha,$$

from which we conclude that asymptotically the transfer function $G(f)$ of the true AR filter resides within a circle of radius

$$\hat{r}_2(f) \stackrel{\text{def}}{=} \left[Q_2(1-\alpha)\hat{\sigma}_p^2 \boldsymbol{e}^H(f)(\widehat{\boldsymbol{\Gamma}}_p^{-1}/N)\boldsymbol{e}(f)\right]^{1/2}$$

with probability at least $1 - \alpha$. Equivalently, as can readily be demonstrated geometrically,

$$\mathbf{P}\left[C_\text{L} \leq |G(f)|^{-2} \leq C_\text{U}\right] \geq 1 - \alpha, \qquad (488\text{a})$$

where

$$C_\text{L} \stackrel{\text{def}}{=} \left[|\hat{G}(f)| + \hat{r}_2(f)\right]^{-2} \quad \text{and} \quad C_\text{U} \stackrel{\text{def}}{=} \left[|\hat{G}(f)| - \hat{r}_2(f)\right]^{-2}.$$

Note that the quantity $\hat{\sigma}_p^2 \widehat{\boldsymbol{\Gamma}}_p^{-1}$ appearing in $\hat{r}_2(f)$ can be readily computed using the sampling version of Equation (484a) (other relevant computational details can be found in Burshtein and Weinstein, 1987).

Equation (488a) gives CIs for $|G(f)|^{-2} = S(f)/(\sigma_p^2 \Delta_\text{t})$, the normalized spectrum. These CIs come from an approach similar to that of Burshtein and Weinstein (1987), who note an adaptation to give asymptotically valid CIs for $S(f)$, i.e., without normalization. To see how, assume for the duration of this paragraph that $\hat{\sigma}_p^2$ is the MLE $\hat{\sigma}_{(\text{ML})}^2$ of Equation (481d). Brockwell and Davis (1991) suggest approximating the distribution of $N\hat{\sigma}_p^2/\sigma_p^2$ by that of a χ_{N-p}^2 RV, which leads to

$$\mathbf{P}\left[D_\text{L} \leq \sigma_p^2 \leq D_\text{U}\right] = 1 - \alpha, \quad \text{where} \quad D_\text{L} \stackrel{\text{def}}{=} \frac{N\hat{\sigma}_p^2}{Q_{N-p}\left(1 - \frac{\alpha}{2}\right)} \quad \text{and} \quad D_\text{U} \stackrel{\text{def}}{=} \frac{N\hat{\sigma}_p^2}{Q_{N-p}\left(\frac{\alpha}{2}\right)}.$$

Let $\mathcal{C}$ and $\mathcal{D}$ denote the events in the brackets in, respectively, Equation (488a) and the above equation. Because $\hat{\sigma}_p^2$ is independent of all the elements of $\widehat{\boldsymbol{\Phi}}_p$, the events $\mathcal{C}$ and $\mathcal{D}$ are independent and hence

$$\mathbf{P}\left[\mathcal{C} \cap \mathcal{D}\right] = \mathbf{P}\left[\mathcal{C}\right] \times \mathbf{P}\left[\mathcal{D}\right] \geq (1 - \alpha)^2.$$

The events $\mathcal{C}$ and $\mathcal{D}$ are equivalent to the events

$$\log\left(C_\text{L}\right) \leq \log\left(|G(f)|^{-2}\right) \leq \log\left(C_\text{U}\right) \quad \text{and} \quad \log\left(D_\text{L}\right) \leq \log\left(\sigma_p^2\right) \leq \log\left(D_\text{U}\right).$$

The occurrence of both $\mathcal{C}$ and $\mathcal{D}$ implies the occurrence of

$$\log\left(C_\text{L} D_\text{L} \Delta_\text{t}\right) \leq \log\left(\frac{\sigma_p^2 \Delta_\text{t}}{|G(f)|^2}\right) = \log\left(S(f)\right) \leq \log\left(C_\text{U} D_\text{U} \Delta_\text{t}\right),$$

which we denote as event $\mathcal{B}$ and which is equivalent to $C_\text{L} D_\text{L} \Delta_\text{t} \leq S(f) \leq C_\text{U} D_\text{U} \Delta_\text{t}$. Since $\mathcal{C} \cap \mathcal{D}$ is a subset of event $\mathcal{B}$, we have $\mathbf{P}\left[\mathcal{B}\right] \geq \mathbf{P}\left[\mathcal{C} \cap \mathcal{D}\right] \geq (1 - \alpha)^2$, which establishes

$$\mathbf{P}\left[\frac{\left[N\hat{\sigma}_p^2 \Delta_\text{t}/Q_{N-p}\left(1 - \frac{\alpha}{2}\right)\right]}{\left[|\hat{G}(f)| + \hat{r}_2(f)\right]^2} \leq S(f) \leq \frac{\left[N\hat{\sigma}_p^2 \Delta_\text{t}/Q_{N-p}\left(\frac{\alpha}{2}\right)\right]}{\left[|\hat{G}(f)| - \hat{r}_2(f)\right]^2}\right] \geq (1 - \alpha)^2. \qquad (488\text{b})$$

Note that the numerators in the event expressed above reflect the uncertainty due to estimation of the innovation variance, while the denominators reflect that of the AR coefficients. Setting $\alpha = 1 - \sqrt{0.95} \doteq 0.0253$ yields a CI for $S(f)$ with an associated probability of at least 0.95. (All the estimators of σ_p^2 we have discussed in this chapter are asymptotically equivalent to the MLE $\hat{\sigma}_{(\text{ML})}^2$, so Equation (488b) also holds when adapted to make use of them.)

Figures 489 and 490 show examples of pointwise CIs created using Equation (488b) with α set to $1 - \sqrt{0.95}$ to yield a probability of at least 0.95 (see the analysis of the ocean wave data in Section 9.12 for an additional example). The thin solid curve in each plot shows an MLE-based AR(p) SDF estimate using a time series generated by an AR(p) process; the solid

9.9 Confidence Intervals Using AR Spectral Estimators 489

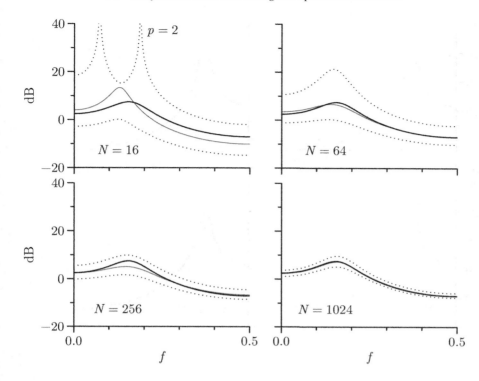

Figure 489 Maximum likelihood AR(2) SDF estimates (thin curves) for portions of lengths 16, 64, 256 and 1024 of the realization of the AR(2) process shown in Figure 34(a) (the process is defined in Equation (34)). The thick curve on each plot is the true SDF. The dotted curves show the lower and upper limits of pointwise confidence intervals with an associated probability of at least 0.95 based upon Equation (488b).

curve is the corresponding true AR(p) SDF; and the two dotted curves depict the lower and upper limits of the CIs. Figure 489 makes use of the AR(2) time series shown in Figure 34(a), with four AR(2) SDF estimates formed using the first $N = 16$, $N = 64$ and $N = 256$ values of the series and then all $N = 1024$ values. As is intuitively reasonable, the lower CI limits get progressively closer to the true AR(2) SDF as N increases. The upper limits do also *if* we ignore the $N = 16$ case. The latter has two peaks associated with two frequencies at which $|\hat{G}(f)| = \hat{r}_2(f)$, thus causing the upper limit in Equation (488b) to be infinite. While an infinite upper limit is not inconsistent with the inequality $\mathbf{P}[\mathcal{B}] \geq 0.95$ behind the CIs, this limit is not particularly informative and is an indication that the asymptotic theory behind Equation (488b) does not yield a decent approximation for such a small time series ($N = 16$). This deficiency in the upper limits for small sample sizes is also apparent in Figure 490, which makes use of the AR(4) time series of Figure 34(e). The $N = 16$ case is particularly unsettling: in addition to two frequencies at which the upper limits are infinite, there is an interval surrounding $f = 0.12$ over which the AR(4) SDF point estimates are more than an order of magnitude *outside* of the CIs! The $N = 64$ case also has point estimates outside of the CIs, but this strange pattern does not occur for the $N = 256$ and $N = 1024$ cases. The upper limits are all finite when the entire AR(4) time series is utilized ($N = 1024$), but these limits are infinite for the $N = 64$ case at two frequencies and for $N = 256$ at four frequencies (in the bizarre pattern of double twin peaks). Evidently we need a larger sample size in the AR(4) case than in the AR(2) for the asymptotic theory to offer a decent approximation, which is not surprising. For $N = 1024$ the widths of the CIs at frequencies just above 0 and just below 0.5 in the AR(2) case are, respectively, 2.6 dB and 1.9 dB; counterintuitively, they are somewhat

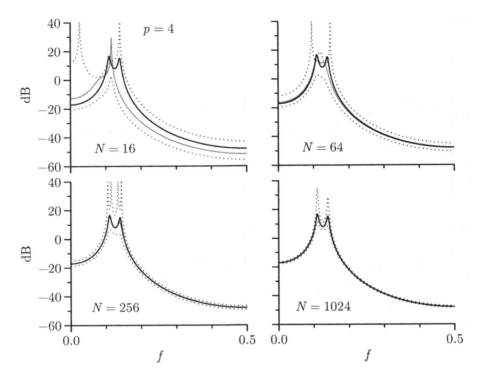

Figure 490 Maximum likelihood AR(4) SDF estimates (thin curves) for portions of lengths 16, 64, 256 and 1024 of the realization of the AR(4) process shown in Figure 34(e) (the process is defined in Equation (35a), and the estimates are also shown in Figure 483). The thick curve on each plot is the true SDF. The dotted curves show the lower and upper limits of pointwise confidence intervals with an associated probability of at least 0.95 based upon Equation (488b).

smaller in the more challenging AR(4) case (1.7 dB and 1.2 dB; for a possible explanation, see the second paragraph in C&E [2], noting that the spectral levels for the AR(4) process near $f = 0$ and $f = 0.5$ are considerably lower than those in the AR(2) case).

Comments and Extensions to Section 9.9

[1] In the theory presented in this section, we assumed that the frequency of interest f satisfies $0 < |f| < f_\mathcal{N}$. If in fact f is equal to zero or the Nyquist frequency, then $e_\Im(f) = \mathbf{0}$, and the rank of the quadratic form $\boldsymbol{B}^T \hat{\mathcal{Q}}^{-1} \boldsymbol{B}/\hat{\sigma}_p^2$ is one (rather than two as previously), so that now

$$\frac{\boldsymbol{B}^T \hat{\mathcal{Q}}^{-1} \boldsymbol{B}}{\hat{\sigma}_p^2} \stackrel{\mathrm{d}}{=} \chi_1^2, \qquad f = 0 \text{ or } \pm f_\mathcal{N}.$$

The only adjustment that is needed in our development is to replace $Q_2(1 - \alpha)$ with $Q_1(1 - \alpha)$, i.e., use the $(1 - \alpha) \times 100\%$ percentage point of the χ_1^2 distribution.

[2] For a Gaussian stationary process $\{G_t\}$ with an SDF $S_G(\cdot)$ that is strictly positive, bounded and sufficiently smooth, Kromer (1969) shows that, provided that $N \to \infty$ and $p \to \infty$ in such a manner that $N/p \to \infty$, then

$$\frac{\hat{S}_G^{(\mathrm{YW})}(f) - S_G(f)}{(2p/N)^{1/2} S_G(f)} \stackrel{\mathrm{d}}{=} \begin{cases} \mathcal{N}(0,1), & 0 < |f| < f_\mathcal{N}; \\ \mathcal{N}(0,2), & f = 0 \text{ or } \pm f_\mathcal{N}. \end{cases} \qquad (490)$$

Comparing the above with the asymptotic distribution of lag window spectral estimators (Section 7.4) reveals that, for $f \neq 0$ or $\pm f_\mathcal{N}$, the value N/p in AR spectral estimation plays the role of ν, the equivalent number of degrees of freedom, in lag window spectral estimation.

A comparison of Equation (488b) with Kromer's result of Equation (490) is not particularly easy. The reader is referred to Burshtein and Weinstein (1987, p. 508) for one possible approach, which leads to the conclusion that in both cases the width of the CI at f is proportional to the spectral level at the frequency; however, the constant of proportionality is somewhat different in the two cases.

9.10 Prewhitened Spectral Estimators

In the previous sections we have concentrated on estimating the parameters of an AR(p) process. We can in turn use these estimators to estimate the SDF for a time series by taking a *pure parametric approach*, in which we simply substitute the estimated AR parameters for the corresponding theoretical parameters in Equation (446b). Alternatively, we can employ a *prewhitening approach*, in which we use the estimates of the AR coefficients $\phi_{p,j}$ to form a prewhitening filter. We introduced the general idea behind prewhitening in Section 6.5. Here we give details on how to implement prewhitening using an estimated AR model and then comment on the relative merits of the pure parametric and prewhitening approaches.

Given a time series that is a realization of a portion $X_0, \ldots, X_{N-1}$ of a stationary process with zero mean and SDF $S_X(\cdot)$, the AR-based prewhitening approach consists of the following steps.

[1] We begin by fitting an AR(p) model to the time series using, say, Burg's algorithm to obtain $\bar{\phi}_{p,1}, \ldots, \bar{\phi}_{p,p}$ and $\bar{\sigma}_p^2$ (as noted by an example in Section 9.12, the particular estimator we use can be important; we have found the Burg method to work well in most practical situations). We set the order p to be just large enough to capture the general structure of the SDF $S_X(\cdot)$ (see Section 9.11 for a discussion of subjective and objective methods for selecting p). In any case, p should not be large compared to the sample size N.

[2] We then use the fitted AR(p) model as a prewhitening filter. The output from this filter is a time series of length $N - p$ given by

$$\vec{e}_t(p) = X_t - \bar{\phi}_{p,1} X_{t-1} - \cdots - \bar{\phi}_{p,p} X_{t-p}, \qquad t = p, \ldots, N-1,$$

where, as before, $\vec{e}_t(p)$ is the forward prediction error observed at time t. If we let $S_e(\cdot)$ denote the SDF for $\{\vec{e}_t(p)\}$, linear filtering theory tells us that the relationship between $S_e(\cdot)$ and $S_X(\cdot)$ is given by

$$S_e(f) = \left| 1 - \sum_{j=1}^p \bar{\phi}_{p,j} e^{-i2\pi f j \Delta_t} \right|^2 S_X(f). \qquad (491a)$$

[3] If the prewhitening filter has been appropriately chosen, the SDF $S_e(\cdot)$ for the observed forward prediction errors will have a smaller dynamic range than $S_X(\cdot)$. We can thus produce a direct spectral estimate $\hat{S}_e^{(D)}(\cdot)$ for $S_e(\cdot)$ with good bias properties using little or no tapering. If we let $\{\hat{s}_{e,\tau}^{(D)}\}$ represent the corresponding ACVS estimate, we can then produce an estimate of $S_e(\cdot)$ with decreased variability by applying a lag window $\{w_{m,\tau}\}$ to obtain

$$\hat{S}_{e,m}^{(LW)}(f) = \Delta_t \sum_{\tau=-(N-p-1)}^{N-p-1} w_{m,\tau} \hat{s}_{e,\tau}^{(D)} e^{-i2\pi f \tau \Delta_t}.$$

[4] Finally we estimate $S_X(\cdot)$ by "postcoloring" $\hat{S}_{e,m}^{(LW)}(\cdot)$ to obtain

$$\hat{S}_X^{(PC)}(f) = \frac{\hat{S}_{e,m}^{(LW)}(f)}{\left| 1 - \sum_{j=1}^p \bar{\phi}_{p,j} e^{-i2\pi f j \Delta_t} \right|^2}. \qquad (491b)$$

There are two obvious variations on this technique that have seen some use. First, we can make use of the observed backward prediction errors to produce a second estimator similar to $\hat{S}_{e,m}^{(\mathrm{LW})}(\cdot)$. These two estimators can be averaged together prior to postcoloring in step [4]. Second, we can obtain the first p values of the ACVS corresponding to $\hat{S}_X^{(\mathrm{PC})}(\cdot)$ and use them in the Yule–Walker method to produce a refined prewhitening filter for iterative use in step [2].

Here are five notes about this combined parametric/nonparametric approach to spectral estimation.

[1] The chief difference between this approach and a pure parametric approach is that we no longer regard the observed prediction errors $\{\overrightarrow{e}_t(p)\}$ as white noise, but rather we use a nonparametric approach to estimate their SDF.

[2] The problem of selecting p is lessened: any imperfections in the prewhitening filter can be compensated for in the nonparametric portion of this combined approach.

[3] Since all lag window spectral estimators correspond to an ACVS estimator that is identically zero after a finite lag q (see Equation (248b)), the numerator of $\hat{S}_X^{(\mathrm{PC})}(\cdot)$ has the form of an MA(q) SDF with $q \leq N - p - 1$. Hence the estimator $\hat{S}_X^{(\mathrm{PC})}(\cdot)$ is the SDF for some ARMA(p, q) model. Our combined approach is thus a way of implicitly fitting an ARMA model to a time series (see the end of Section 9.14 for an additional comment).

[4] The combined approach only makes sense in situations where tapering is normally required; i.e., the SDF $S_X(\cdot)$ has a high dynamic range (see the discussion in Section 6.5).

[5] Even if we use a pure parametric approach, it is useful to carefully examine the observed prediction errors $\{\overrightarrow{e}_t(p)\}$ because this is a valuable way to detect outliers in a time series (Martin and Thomson, 1982).

See Section 9.12 for an example of the prewhitening and pure parametric approaches to SDF estimation.

9.11 Order Selection for AR(p) Processes

In discussing the various estimators of the parameters for an AR(p) process, we have implicitly assumed that the model order p is known in advance, an assumption that is rarely valid in practice. To make a reasonable choice of p, various order selection criteria have been proposed and studied in the literature. We describe briefly here a few commonly used ones, but these are by no means the only ones that have been proposed or are in extensive use. The reader is referred to the articles by de Gooijer et al. (1985), Broersen (2000, 2002) and Stoica and Selén (2004) and to the books by Choi (1992), Stoica and Moses (2005) and Broersen (2006) for more comprehensive discussions.

In what follows, we first consider two subjective criteria (direct spectral estimate and partial autocorrelation) and then several objective criteria. Examples of their use are given in Section 9.12. In describing the criteria, we take $\hat{\phi}_{k,j}$ and $\hat{\sigma}_k^2$ to be any of the estimators of $\phi_{k,j}$ and σ_k^2 we have discussed so far in this chapter.

A comment is in order before we proceed. Any order selection method we use should be appropriate for what we intend to do with the fitted AR model. We usually fit AR models in the context of spectral analysis either to directly obtain an estimate of the SDF (a pure parametric approach) or to produce a prewhitening filter (a combined parametric/nonparametric approach). Some commonly used order selection criteria are geared toward selecting a low-order AR model that does well for one-step-ahead predictions. These criteria thus seem to be more appropriate for producing a prewhitening filter than for pure parametric spectral estimation. Unfortunately, selection of an appropriate order is vital for pure parametric spectral estimation: if p is too large, the resulting SDF estimate tends to exhibit spurious peaks; if p is too small, structure in the SDF can be smoothed over.

9.11 Order Selection for AR(p) Processes

Direct Spectral Estimate Criterion

If we are primarily interested in using a fitted AR model to produce a prewhitening filter, the following subjective criterion is viable. First, we compute a direct spectral estimate $\hat{S}^{(D)}(\cdot)$ using a data taper that provides good leakage protection (see Section 6.4). We then fit a sequence of relatively low-order AR models (starting with, say, $k = 2$ or 4) to our time series, compute the corresponding AR SDF estimates via Equation (446b), and compare these estimates with $\hat{S}^{(D)}(\cdot)$. We select our model order p as the smallest value k such that the corresponding AR SDF estimate generally captures the overall shape of $\hat{S}^{(D)}(\cdot)$.

Partial Autocorrelation Criterion

For an AR process of order p, the PACS $\{\phi_{k,k} : k = 1, 2, \ldots\}$ is nonzero for $k = p$ and zero for $k > p$. In other words, the PACS of a pth-order AR process has a cutoff after lag p. It is known that, for a Gaussian AR(p) process, the $\hat{\phi}_{k,k}$ terms for $k > p$ are approximately independently distributed with zero mean and a variance of approximately $1/N$ (see Kay and Makhoul, 1983, and references therein). Thus a rough procedure for testing $\phi_{k,k} = 0$ is to examine whether $\hat{\phi}_{k,k}$ lies between $\pm 2/\sqrt{N}$. By plotting $\hat{\phi}_{k,k}$ versus k, we can thus set p to a value beyond which $\phi_{k,k}$ can be regarded as being zero.

Final Prediction Error (FPE) Criterion

Equation (462c) notes that, for an AR(p) process, the mean square linear prediction errors of orders $p-1$ and higher obey the pattern

$$\sigma_{p-1}^2 > \sigma_p^2 = \sigma_{p+1}^2 = \sigma_{p+2}^2 = \cdots, \tag{493a}$$

where σ_k^2 for $k > p$ is defined via the augmented Yule–Walker equations of order k. This suggests a criterion for selecting p that consists of plotting $\hat{\sigma}_k^2$ versus k and setting p equal to that value of k such that

$$\hat{\sigma}_{k-1}^2 > \hat{\sigma}_k^2 \approx \hat{\sigma}_{k+1}^2 \approx \hat{\sigma}_{k+2}^2 \approx \cdots . \tag{493b}$$

Recall, however, that, for the Yule–Walker method and Burg's algorithm,

$$\hat{\sigma}_k^2 = \hat{\sigma}_{k-1}^2 \left(1 - \hat{\phi}_{k,k}^2\right),$$

showing that $\hat{\sigma}_k^2 < \hat{\sigma}_{k-1}^2$ except in the unlikely event that either $\hat{\phi}_{k,k}$ or $\hat{\sigma}_{k-1}^2$ happen to be identically zero. The sequence $\{\hat{\sigma}_k^2\}$ is thus nonincreasing, making it problematic at times to determine where the pattern of Equation (493b) occurs. The underlying problem is that $\hat{\sigma}_k^2$ tends to underestimate σ_k^2, i.e., to be biased toward zero. In the context of the Yule–Walker estimator, Akaike (1969, 1970) proposed the FPE estimator that in effect corrects for this bias. Theoretical justification for this estimator presumes the existence of two time series that are independent of each other, but drawn from the same AR process. The first series is used to estimate the AR coefficients, and these coefficients are in turn used to estimate the mean square linear prediction error – with this estimator being called the FPE – by forming one-step-ahead predictions through the second series. After appealing to the large-sample properties of the coefficient estimators, the FPE estimator for a kth-order AR model is taken to be

$$\text{FPE}(k) = \left(\frac{N+k+1}{N-k-1}\right) \hat{\sigma}_k^2 \tag{493c}$$

(Akaike, 1970, equations (4.7) and (7.3)). This form of the FPE assumes that the process mean is *unknown* (the usual case in practical applications) and that we have centered our time series by subtracting the sample mean prior to the estimation of the AR parameters (if the process mean is known, the number of unknown parameters decreases from $k+1$ to k, and the term in parentheses becomes $(N+k)/(N-k)$). Note that, as required, FPE(k) is an inflated version of $\hat{\sigma}_k^2$. The FPE order selection criterion is to set p equal to the value of k that minimizes FPE(k).

How well does the FPE criterion work in practice? Using simulated data, Landers and Lacoss (1977) found that the FPE – used with Burg's algorithm – selected a model order that was insufficient to resolve spectral details in an SDF with sharp peaks. By increasing the model order by a factor 3, they obtained adequate spectral estimates. Ulrych and Bishop (1975) and Jones (1976b) found that, when used with Burg's algorithm, the FPE tends to pick out spuriously high-order models. The same does not happen with the Yule–Walker estimator, illustrating that a particular order selection criterion need not perform equally well for different AR parameter estimators. Ulrych and Clayton (1976) found that the FPE does not work well with short time series, prompting Ulrych and Ooe (1983) to recommend that p be chosen between $N/3$ and $N/2$ for such series. Kay and Marple (1981) report that the results from using the FPE criteria have been mixed, particularly with actual time series rather than simulated AR processes.

Likelihood-Based Criteria
In the context of the maximum likelihood estimator, Akaike (1974) proposed another order selection criterion, known as *Akaike's information criterion* (AIC). For a kth-order AR process with an unknown process mean, the AIC is defined as

$$\text{AIC}(k) = -2 \log \{\text{maximized likelihood}\} + 2(k+1); \tag{494a}$$

if the process mean is known, the above becomes

$$\text{AIC}(k) = -2 \log \{\text{maximized likelihood}\} + 2k. \tag{494b}$$

The AIC, which is based on cross-entropy ideas, is very general and applicable in more than just a time series context. For a Gaussian AR(k) process $\{H_t\}$ with known mean, using the MLEs $\widehat{\boldsymbol{\Phi}}_{(\text{ML})}$ and $\hat{\sigma}^2_{(\text{ML})}$ in the right-hand side of Equation (481b) tells us – in conjunction with Equations (480b) and (481d) – that we can write

$$-2 \log \{\text{maximized likelihood}\} = \sum_{j=0}^{k-1} \log(\hat{\lambda}_j) + N \log(\hat{\sigma}^2_{(\text{ML})}) + N + N \log(2\pi),$$

where $\hat{\sigma}^2_{(\text{ML})}$ is the MLE of σ_k^2, while $\hat{\lambda}_j$ depends on the MLEs of the $\phi_{k,j}$ terms and not on $\hat{\sigma}^2_{(\text{ML})}$. The first term on the right-hand side becomes negligible for large N, while the last two terms are just constants. If we drop these three terms and allow the use of other estimators $\hat{\sigma}_k^2$ of σ_k^2 besides the MLE, we obtain the usual approximation to Equation (494b) that is quoted in the literature as the AIC for AR processes (see, for example, de Gooijer et al., 1985; Rosenblatt, 1985; or Kay, 1988):

$$\text{AIC}(k) = N \log(\hat{\sigma}_k^2) + 2k \tag{494c}$$

(alternatively, we can justify this formulation by using the conditional likelihood of H_0, ..., H_{N-1}, given $H_{-1} = H_{-2} = \cdots = H_{-k} = 0$). The AIC order selection criterion is to set p equal to the value of k that minimizes AIC(k). If the process mean is unknown, the appropriate equation is now

$$\text{AIC}(k) = N \log(\hat{\sigma}_k^2) + 2(k+1), \tag{494d}$$

which amounts to adding two to the right-hand side of Equation (494c) – hence the minimizers of Equations (494c) and (494d) are identical.

How well does the AIC criterion work in practice? Landers and Lacoss (1977), Ulrych and Clayton (1976) and Kay and Marple (1981) report results as mixed as those for the FPE criterion. Hurvich and Tsai (1989) note that the FPE and AIC both tend to select inappropriately large model orders. They argue that there is a bias in the AIC and that correcting for this bias yields a criterion that avoids the tendency to overfit in sample sizes of interest to practitioners. In the practical case of an unknown mean, correcting Equation (494d) for this bias yields the following criterion:

$$\text{AICC}(k) = N \log \left(\hat{\sigma}_k^2 \right) + 2 \frac{N}{N - (k+1) - 1} (k+1) \tag{495a}$$

(if the mean is known, replacing $k+1$ in the right-most term above with k yields the correction for Equation (494c)). As $N \to \infty$, the ratio in the above converges to unity, so asymptotically the AICC and the AIC of Equation (494d) are equivalent.

Comments and Extensions to Section 9.11

[1] When using MLEs for a Gaussian AR(k) process $\{H_t\}$ with an unknown mean μ, $N \log \left(\hat{\sigma}_k^2 \right)$ in Equations (494d) and (495a) for the AIC and the AICC is a stand-in for $-2 \log \{\text{maximized likelihood}\}$. The likelihood in question depends upon μ. Rather than actually finding the MLE for μ, a common practice in time series analysis is to use the sample mean to estimate μ. The time series is then centered by subtracting off its sample mean, and the centered time series is treated as if it were a realization of a process with a known mean of zero. After centering, a surrogate more accurate than $N \log \left(\hat{\sigma}_k^2 \right)$ would be $-2 \log (L(\widehat{\boldsymbol{\Phi}}_{(\text{ML})}, \hat{\sigma}_{(\text{ML})}^2 \mid \boldsymbol{H}))$, where $L(\widehat{\boldsymbol{\Phi}}_{(\text{ML})}, \hat{\sigma}_{(\text{ML})}^2 \mid \boldsymbol{H})$ is the maximum value of $L(\boldsymbol{\Phi}_k, \sigma_k^2 \mid \boldsymbol{H})$ – as defined by Equation (480a) – with $\boldsymbol{H} = [H_0, \ldots, H_{N-1}]^T$ taken to be the centered time series.

Now suppose we use the Yule–Walker estimators $\widetilde{\boldsymbol{\Phi}}_k$ and $\tilde{\sigma}_k^2$ with the centered series. Rather than using $N \log \left(\tilde{\sigma}_k^2 \right)$ in Equations (494d) and (495a) for the AIC and the AICC, we could use $-2 \log (L(\widetilde{\boldsymbol{\Phi}}_k, \tilde{\sigma}_k^2 \mid \boldsymbol{H}))$, which involves the likelihood function computed under the assumption that its true parameter values are the Yule–Walker estimators. We could also do the same for the Burg estimators $\overline{\boldsymbol{\Phi}}_k$ and $\bar{\sigma}_k^2$. Letting $\widehat{\boldsymbol{\Phi}}_k$ and $\hat{\sigma}_k^2$ now stand for either the MLEs, the Yule–Walker estimators or the Burg estimators, the more accurate AIC and the AICC for a centered time series are

$$\text{AIC}(k) = -2 \log (L(\widehat{\boldsymbol{\Phi}}_k, \hat{\sigma}_k^2 \mid \boldsymbol{H})) + 2(k+1) \tag{495b}$$

and

$$\text{AICC}(k) = -2 \log (L(\widehat{\boldsymbol{\Phi}}_k, \hat{\sigma}_k^2 \mid \boldsymbol{H})) + 2 \frac{N}{N - (k+1) - 1} (k+1). \tag{495c}$$

Since the MLEs are the minimizers of $-2 \log (L(\boldsymbol{\Phi}_k, \sigma_k^2 \mid \boldsymbol{H}))$, the smaller of the Yule–Walker-based $-2 \log (L(\widetilde{\boldsymbol{\Phi}}_k, \tilde{\sigma}_k^2 \mid \boldsymbol{H}))$ and of the Burg-based $-2 \log (L(\overline{\boldsymbol{\Phi}}_k, \bar{\sigma}_k^2 \mid \boldsymbol{H}))$ tells us which of these two estimators yields a likelihood closer to the MLE minimizer.

Can Equations (495b) and (495c) be used with the FLS, BLS and FBLS estimators? As noted in C&E [1] for Section 9.7, the FLS estimator of $\boldsymbol{\Phi}_k$ need not always correspond to a causal AR process (the same is true for BLS and FBLS estimators). The procedure that we have outlined in Section 9.8 for evaluating the likelihood function implicitly assumes causality and hence cannot be used when causality fails to hold. Evaluation of the likelihood function for acausal AR models is a topic in need of further exploration.

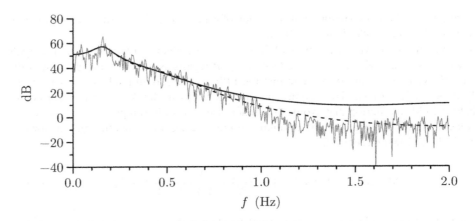

Figure 496 Determination of AR prewhitening filter and illustration of effect of different choices for parameter estimation. The jagged curve shows a Slepian-based direct spectral estimate of the ocean wave data (shown also in the plots of Figure 319). The dashed curve shows the estimated SDF corresponding to an AR(5) model fit to the data using Burg's algorithm. The solid smooth curve shows the estimated SDF for the same model, but now with parameters estimated using the Yule–Walker method.

9.12 Examples of Parametric Spectral Estimators

Ocean Wave Data

Here we return to the ocean wave data $\{X_t\}$ examined previously in Sections 6.8, 7.12 and 8.9. We consider both a prewhitening approach and a pure parametric approach (as in our previous analyses, we first center the time series by subtracting off its sample mean $\overline{X} \doteq 209.1$). For the prewhitening approach, we begin by finding a low-order AR model that can serve as a prewhitening filter. The jagged curve in Figure 496 shows a direct spectral estimate $\hat{S}_X^{(D)}(\cdot)$ for the ocean wave data using an $NW = 2/\Delta_t$ Slepian data taper (this curve is also shown in all four plots of Figure 319 – recall that, based upon a study of Figure 226, we argued that this estimate does not suffer unduly from leakage). Experimentation indicates that an AR(5) model with parameters estimated using Burg's algorithm captures the overall spectral structure indicated by $\hat{S}_X^{(D)}(\cdot)$. The SDF for this model is shown by the dashed curve in Figure 496. For comparison, the solid smooth curve in this plot shows the AR(5) SDF estimate based upon the Yule–Walker method. Note that, whereas the Burg estimate generally does a good job of tracking $\hat{S}_X^{(D)}(\cdot)$ across all frequencies, the same cannot be said for the Yule–Walker estimate – it overestimates the power in the high frequency region by as much as two orders of magnitude (20 dB).

Figure 497(a) shows the forward prediction errors $\{\vec{e}_t(5)\}$ obtained when we use the fitted AR(5) model to form a prewhitening filter. Since there were $N = 1024$ data points in the original ocean wave time series, there are $N - p = 1019$ points in the forward prediction error series. The jagged curve in Figure 497(b) shows the periodogram $\hat{S}_e^{(P)}(\cdot)$ for $\{\vec{e}_t(5)\}$ plotted versus the Fourier frequencies. A comparison of $\hat{S}_e^{(P)}(\cdot)$ with direct spectral estimates using a variety of Slepian data tapers indicates that the periodogram is evidently leakage-free, so we can take it to be an approximately unbiased estimate of the SDF $S_e(\cdot)$ associated with $\{\vec{e}_t(5)\}$ (Equation (491a) relates this SDF to the SDF $S_X(\cdot)$ for the ocean wave series). We now smooth $\hat{S}_e^{(P)}(\cdot)$ using a Parzen lag window $\{w_{m,\tau}\}$ with $m = 55$ to match what we did in Figure 319(a). The smooth curve in Figure 497(b) shows the Parzen lag window estimate $\hat{S}_{e,m}^{(LW)}(\cdot)$. The crisscross in this figure is analogous to the one in Figure 319(a). Its width and height here are based upon the statistical properties of a Parzen lag window with $m = 55$ applied to a time series of length 1019 with no tapering (see Table 279). This yields a lag window spectral estimate with $\nu \doteq 68.7$ equivalent degrees of freedom, whereas we found

9.12 Examples of Parametric Spectral Estimators

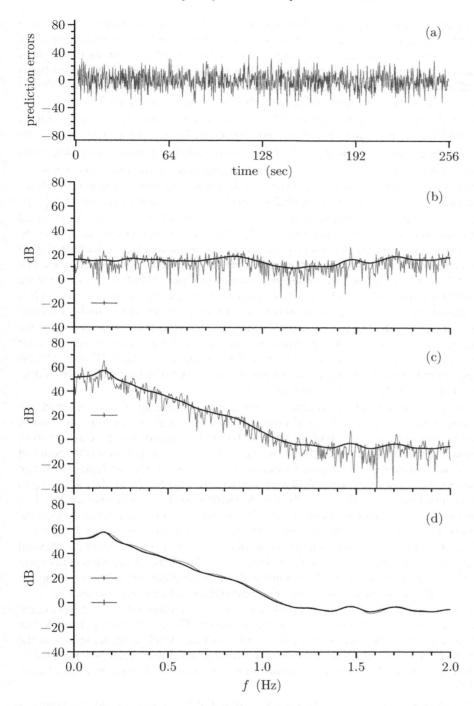

Figure 497 Illustration of prewhitening applied to ocean wave data (see main text for details).

$\nu \doteq 35.2$ for the estimate in Figure 319(a), which involved a Slepian data taper. The increase in the number of degrees of freedom occurs because prewhitening has obviated the need for tapering. This increase translates into a decrease in the length of a 95% CI, so that the height of the crisscross in Figure 497(b) is noticeably smaller than the one in Figure 319(a) (the heights are 2.9 dB and 4.1 dB, respectively; these crisscrosses also are replicated in Figure 497(d) – the top one is the same as in Figure 497(b), and the bottom one, as in Figure 319(a)).

The thick smooth curve in Figure 497(c) shows the postcolored lag window estimate $\hat{S}_X^{(\text{PC})}(\cdot)$, which is formed from $\hat{S}_{e,m}^{(\text{LW})}(\cdot)$ as per Equation (491b). The jagged curve is the same as in Figure 496 and shows that $\hat{S}_X^{(\text{PC})}(\cdot)$ can be regarded as a smoothed version of the leakage-free direct spectral estimate $\hat{S}_X^{(\text{D})}(\cdot)$ of the original ocean wave series. The amount of smoothing is appropriate given that our main objective is to determine the rate at which the SDF decreases over the frequency range 0.2 to 1.0 Hz. It is of interest to compare $\hat{S}_X^{(\text{PC})}(\cdot)$ with the $m = 55$ Parzen lag window estimate $\hat{S}_{X,m}^{(\text{LW})}(\cdot)$ shown in Figure 319(a). As we noted in our discussion of that figure, this estimate suffers from smoothing window leakage at the high frequencies. By contrast, there is no evidence of such leakage in $\hat{S}_X^{(\text{PC})}(\cdot)$, illustrating an additional advantage to the combined parametric/nonparametric approach. We also noted that the $m = 23.666$ Gaussian lag window estimate plotted in Figure 319(d) shows no evidence of smoothing window leakage. This estimate is replicated as the thin curve in Figure 497(d) for comparison with $\hat{S}_X^{(\text{PC})}(\cdot)$ (the thick curve). The two estimates are quite similar, differing by no more than 2.3 dB. The top crisscross is for $\hat{S}_X^{(\text{PC})}(\cdot)$, while the bottom one is for the Gaussian lag window estimate. The prewhitening approach yields an estimate with the same bandwidth as the Gaussian estimate, but its 95% CI is noticeably tighter. (Exercise [9.19] invites the reader to redo this analysis using a prewhitening filter dictated by the Yule–Walker method rather than the Burg-based filter used here.)

Let us now consider a pure parametric approach for estimating $S_X(\cdot)$, for which we need to determine an appropriate model order p. As a first step, we examine the sample PACS for the ocean wave data. With K taken to be the highest order of potential interest, the standard definition of the sample PACS is $\{\tilde{\phi}_{k,k} : k = 1, \ldots, K\}$, which is the result of recursively manipulating the usual biased estimator $\{\hat{s}_\tau^{(\text{P})}\}$ of the ACVS (see Equation (458a); these manipulations also yield Yule–Walker parameter estimates for the models AR(1) up to AR(K). As discussed in Section 9.11, the PACS criterion compares the estimated PACS to the approximate 95% confidence bounds $\pm 2/\sqrt{N}$ and sets p to the order beyond which the bulk of the estimates fall within these bounds. The top plot of Figure 499 shows $\{\tilde{\phi}_{k,k} : k = 1, \ldots, 40\}$. The horizontal dashed lines indicate the bounds $\pm 2/\sqrt{N} = \pm 0.0625$. Beyond $k = 6$ only two $\tilde{\phi}_{k,k}$ estimates exceed these limits ($k = 15$ and 26). Since we would expect roughly 5% of the deviates to exceed the limits just due to sampling variability, it is tempting to chalk two rather small exceedances out of 34 possibilities as being due to chance ($2/34 \doteq 6\%$). Indeed minimization of FPE(k), AIC(k) and AICC(k) in Equations (493c), (494a) and (495a) all pick out $k = 6$ as the appropriate model order. The corresponding Yule–Walker AR(6) SDF estimate differs by no more than 1 dB from the AR(5) estimate shown as the solid line in Figure 496. The rate at which this AR(6) SDF estimate decreases from 0.8 to 1.0 Hz differs markedly from the lag windows estimates shown in the bottom three plots of Figure 319.

Rather than using the Yule–Walker-based sample PACS, we can instead use estimates $\{\bar{\phi}_{k,k}\}$ associated with Burg's algorithm (Equation (467f)). The bottom plot of Figure 499 shows these estimates, which give an impression different from the standard estimates $\{\tilde{\phi}_{k,k}\}$. Here $\bar{\phi}_{k,k}$ exceeds the $\pm 2/\sqrt{N}$ bounds at lags 1–5, 7, 9–13, 16, 18, 20, 22 and 24–25. The Burg-based PACS criterion thus suggests picking $p = 25$. Minimization of Burg-based FPE(k), AIC(k) and AICC(k) picks orders of 27, 27 and 25. Since the AICC is the most

9.12 Examples of Parametric Spectral Estimators

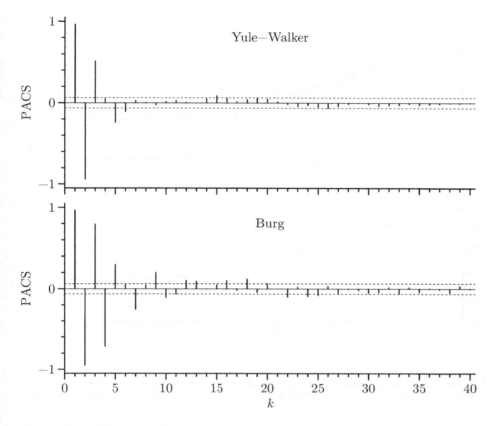

Figure 499 Partial autocorrelation sequences estimates at lags $k = 1, \ldots, 40$ for the ocean wave data. The upper plot shows estimates $\{\tilde{\phi}_{k,k}\}$ based on the Yule–Walker method (i.e., the standard sample PACS); the bottom shows estimates $\{\bar{\phi}_{k,k}\}$ based on Burg's algorithm.

sophisticated of these criteria, we select $p = 25$, in agreement with the PACS criterion. The AR(25) Burg SDF estimate is shown as the thick curves in Figure 500(a) and (b) at all Fourier frequencies satisfying $0 < f_j < f_{\mathcal{N}}$. The two thin curves above and below the thick curve in (a) represent no less than 95% CIs for a hypothesized true AR(25) SDF (see Equation (488b); the plot clips the upper CI, which goes up to 125 dB at $f_{41} \doteq 0.160\,\text{Hz}$). The thin curve in (b) is the Burg-based AR(5) postcolored Parzen lag window estimate $\hat{S}_X^{(\text{PC})}(\cdot)$ (also shown as the thick curves in Figures 497(c) and (d)). The two estimates agree well for $f \geq 0.4\,\text{Hz}$, over which they differ by no more than 1.5 dB. The AR(25) estimate portrays more structure over $f < 0.4\,\text{Hz}$ and, in particular, captures the peak centered at $f = 0.16\,\text{Hz}$ nicely, but the postcolored lag window estimate is better at portraying a hypothesized monotonic decay over 0.2 to $1.0\,\text{Hz}$, a primary interest of the investigators.

Atomic Clock Data

Here we revisit the atomic clock fractional frequency deviates $\{Y_t\}$, which are shown in Figure 326(b). In Section 8.9 we considered multitaper-based estimates of the innovation variance both for the deviates and for the first difference of the deviates $\{Y_t - Y_{t-1}\}$, the results of which are summarized in Table 432. Here we compare this nonparametric approach with an AR-based parametric approach. Using both the Yule–Walker method and Burg's algorithm, we fit AR models of orders $k = 1, 2, \ldots, 100$ to the deviates and also to their first differences (fitting was done after centering each series by subtracting off its sample mean). This gives us

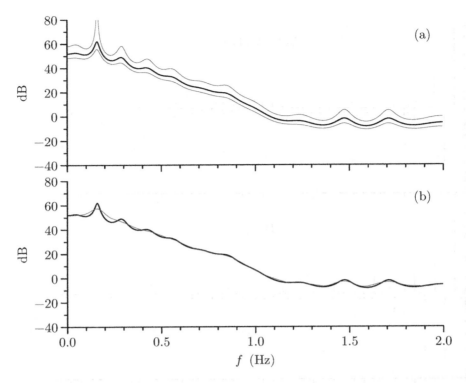

Figure 500 Burg AR(25) SDF estimate for ocean wave data (thick curves in both plots), along with – in upper plot – no less than 95% CIs (two thin curves above and below the thick curve) and – in lower plot – the Burg-based AR(5) postcolored Parzen lag window estimate.

	Yule–Walker				Burg			
Time Series	p	$\tilde{\sigma}_p^2$	sd $\{\tilde{\sigma}_p^2\}$	FPE(p)	p	$\bar{\sigma}_p^2$	sd $\{\bar{\sigma}_p^2\}$	FPE(p)
$Y_t - Y_{t-1}$	41	0.02259	0.00051	0.02307	41	0.02230	0.00050	0.02277
Y_t	42	0.02222	0.00050	0.02270	42	0.02216	0.00050	0.02264

Table 500 AR-based estimates of innovation variance for first difference $Y_t - Y_{t-1}$ of fractional frequency deviates (top row) and for fractional frequency deviates Y_t themselves (bottom).

100 different innovation variance estimates $\tilde{\sigma}_k^2$ for each of the two estimation methods. The multitaper estimates of the innovation variance depend on the parameter K, but not markedly so; on the other hand, the AR estimates depend critically on k, so we look at minimization of FPE(k), AIC(k) and AICC(k) in Equations (493c), (494a) and (495a) for guidance as to its choice. For both the Yule–Walker method and Burg's algorithm, all three criteria pick the rather large model orders of $p = 42$ for $\{Y_t\}$ and of $p = 41$ for $\{Y_t - Y_{t-1}\}$. The top and bottom rows of Table 500 show the Yule–Walker and Burg estimates of σ_p^2 for $\{Y_t\}$ and $\{Y_t - Y_{t-1}\}$, respectively. All four estimate are close to each other and to the ten multitaper-based estimates shown in Table 432 (the ratio of the largest to the smallest of the fourteen estimates is 1.02). Large-sample statistical theory suggests that the variance of of AR-based estimators of the innovation variance is approximately $2\sigma_p^4/N$ (see C&E [1] for Section 9.8). Substituting estimates for σ_p^4 and taking square roots give us an estimated standard deviation

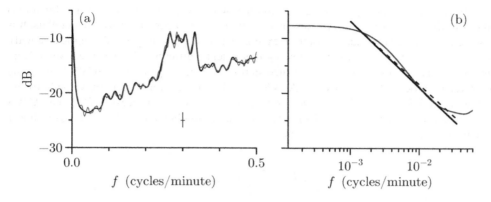

Figure 501 Burg AR(42) SDF estimate for fractional frequency deviates $\{Y_t\}$ (thick curve in both plots). Plot (a) shows the estimate over $f \in (0, f_{\mathcal{N}}]$ on a linear/dB scale while (b) shows the estimate at low frequencies on a log/dB scale. The thin curve in (a) is the same $m = 80$ Gaussian lag window estimate as is displayed in Figure 328(a) (the crisscross is copied from that plot). There are two lines in plot (b). The solid line is copied from Figure 431(b) and is a simple linear regression fit in log/dB space to a $K = 7$ sinusoidal multitaper SDF estimate over Fourier frequencies ranging from $f_4 = 4/(N\,\Delta_t) \doteq 0.010$ cycles/minute to $f_{140} = 140/(N\,\Delta_t) \doteq 0.035$ cycles/minute. The dashed line is a similar fit to the AR SDF estimate.

for the estimates – these are noted in Table 500 (the difference between the largest and smallest of the fourteen estimates is smaller than these standard deviations). Finally we note that the idea behind the FPE is to correct estimates of the innovation variance for a downward bias. This correction is noted in Table 500 (the standard deviations associated with the FPE estimates are only slightly bigger than the ones for the estimates of σ_p^2, increasing by no more than 0.00002).

Figure 501(a) shows the Burg AR(42) SDF estimate (thick curve; the Yule–Walker estimate is virtually the same, differing from the Burg estimate by no more than 0.2 dB). Of the estimates considered previously, the AR estimate most resembles the $m = 80$ Gaussian lag window estimate shown in Figure 328(a) and replicated here as the thin curve (the crisscross is also a replicate from that figure). While the two estimates differ in detail, they both give the same overall impression and have roughly the same bumpiness. Figure 501(b) zooms into the same low-frequency region as was considered in Figure 431(b) for the $K = 7$ sinusoidal multitaper estimate, for which we did a simple linear regression fit in log/dB space over a subset of Fourier frequencies in the low-frequency region – the resulting fitted line was shown in Figure 431(b) and is replicated in Figure 501(b) as the solid line. If we do a similar regression fit to the AR(42) estimate, we obtain the dashed line in Figure 501(b). The two lines are in reasonable agreement. The slopes of these lines can be translated into estimates of the exponent α for an SDF obeying a power law over the subsetted low-frequency region. The multitaper based estimate is $\hat{\alpha}^{(\text{MT})} \doteq -1.21$, while the Burg-based estimate is $\hat{\alpha}^{(\text{AR})} \doteq -1.15$. Given that we estimated $\sqrt{\text{var}\{\hat{\alpha}^{(\text{MT})}\}}$ to be 0.11, the two estimates of α are indeed in reasonable agreement (theory is lacking as to how to quantify the variance of $\hat{\alpha}^{(\text{AR})}$).

In conclusion, the AR-based parametric estimates of the innovation variance and of the SDF for the atomic clock data agree well with previously considered nonparametric estimates. Also, in contrast to the ocean wave example, here the Yule–Walker and Burg SDF estimates give comparable results.

9.13 Comments on Complex-Valued Time Series

We have assumed throughout this chapter that our time series of interest is real-valued. If we wish to fit an AR(p) model to a complex-valued series that is a portion of a realization of a complex-valued stationary process $\{Z_t\}$ with zero mean, we need to make some modifications in the definitions of our various AR parameter estimators (the form of the AR(p) model itself is the same as indicated by Equation (446a), except that $\phi_{p,j}$ is now in general complex-valued). Most of the changes are due to the following fundamental difference between real-valued and complex-valued processes. For a real-valued stationary process $\{X_t\}$ with ACVS $\{s_{X,\tau}\}$, the time reversed process $\{X_{-t}\}$ also has ACVS $\{s_{X,\tau}\}$, whereas, for a complex-valued process $\{Z_t\}$ with ACVS $\{s_{Z,\tau}\}$, the *complex conjugate* time reversed process $\{Z^*_{-t}\}$ has ACVS $\{s_{Z,\tau}\}$ (the proof is an easy exercise). One immediate implication is that, if the best linear predictor of Z_t, given $Z_{t-1}, \ldots, Z_{t-k}$, is

$$\overrightarrow{Z}_t(k) = \sum_{j=1}^{k} \phi_{k,j} Z_{t-j},$$

then the best linear "predictor" of Z_t, given $Z_{t+1}, \ldots, Z_{t+k}$, has the form

$$\overleftarrow{Z}_t(k) = \sum_{j=1}^{k} \phi^*_{k,j} Z_{t+j}. \tag{502}$$

Given these facts, a rederivation of the Levinson–Durbin recursions shows that we must make the following minor changes to obtain the proper Yule–Walker and Burg parameter estimators for complex-valued time series. For the Yule–Walker estimator, Equations (458b) and (458c) now become

$$\tilde{\phi}_{k,j} = \tilde{\phi}_{k-1,j} - \tilde{\phi}_{k,k} \tilde{\phi}^*_{k-1,k-j}, \quad 1 \le j \le k-1$$
$$\tilde{\sigma}_k^2 = \tilde{\sigma}_{k-1}^2 (1 - |\tilde{\phi}_{k,k}|^2)$$

(Equation (458a) remains the same). As before, we initialize the recursions using $\tilde{\phi}_{1,1} = \hat{s}_1^{(\text{P})}/\hat{s}_0^{(\text{P})}$, but now we set $\tilde{\sigma}_1^2 = \hat{s}_0^{(\text{P})}(1-|\tilde{\phi}_{1,1}|^2)$. In lieu of Equation (467f), the Burg estimator of the kth-order partial autocorrelation coefficient now takes the form

$$\bar{\phi}_{k,k} = \frac{2 \sum_{t=k}^{N-1} \overrightarrow{e}_t(k-1) \overleftarrow{e}^*_{t-k}(k-1)}{\sum_{t=k}^{N-1} |\overrightarrow{e}_t(k-1)|^2 + |\overleftarrow{e}_{t-k}(k-1)|^2}.$$

Equation (467a) still gives the recursive formula for the forward prediction errors, but the formula for the backward prediction errors is now given by

$$\overleftarrow{e}_{t-k}(k) = \overleftarrow{e}_{t-k}(k-1) - \bar{\phi}^*_{k,k} \overrightarrow{e}_t(k-1), \quad k \le t \le N-1,$$

instead of by Equation (467b).

Finally we note that

[1] the various least squares estimators are now obtained by minimizing an appropriate sum of squared moduli of prediction errors rather than sum of squares; and
[2] once we have obtained the AR parameter estimates, the parametric spectral estimate is obtained in the same manner as in the real-valued case, namely, by substituting the estimated parameters into Equation (446b).

For further details on parametric spectral estimation using complex-valued time series, see Kay (1988).

9.14 Use of Other Models for Parametric SDF Estimation

So far we have concentrated entirely on the use of AR(p) models for parametric SDF estimation. There are, however, other classes of models that could be used. We discuss briefly one such class here – namely, the class of moving average processes of order q (MA(q)) – because it provides an interesting contrast to AR(p) processes.

An MA(q) process $\{X_t\}$ with zero mean is defined by

$$X_t = \epsilon_t - \theta_{q,1}\epsilon_{t-1} - \theta_{q,2}\epsilon_{t-2} - \cdots - \theta_{q,q}\epsilon_{t-q},$$

where $\theta_{q,1}, \theta_{q,2}, \ldots, \theta_{q,q}$ are q fixed coefficients, and $\{\epsilon_t\}$ is a white noise process with zero mean and variance σ_ϵ^2 (see Equation (32a)). The above is stationary no matter what values the coefficients assume. Its SDF is given by

$$S(f) = \sigma_\epsilon^2 \, \Delta_t \left| 1 - \sum_{j=1}^{q} \theta_{q,j} e^{-i2\pi f j \, \Delta_t} \right|^2.$$

There are $q+1$ parameters (the q coefficients $\theta_{q,j}$ and the variance σ_ϵ^2) that we need to estimate to produce an MA(q) parametric SDF estimate.

It is interesting to reconsider the six-fold rationale for using AR(p) models discussed in Section 9.2 and to contrast it with the MA(q) case.

[1] As in the case of AR(p) processes, any continuous SDF $S(\cdot)$ can be approximated arbitrarily well by an MA(q) SDF if q is chosen large enough (Anderson, 1971, p. 411). In this sense the class of MA processes is as rich as that of AR processes, but in another sense it is *not* as rich (see the discussion surrounding Equation (504a)).

[2] Algorithms for explicitly fitting MA(q) models to time series tend to be computationally intensive compared with those needed to fit AR(p) models. However, if our interest is primarily in spectral estimation so that we do not require explicit estimates of $\theta_{q,j}$, there is an analog to the Yule–Walker approach that is computationally efficient but unfortunately has poor statistical properties (see the discussion surrounding Equation (504b)).

[3] From the discussion following Equation (473), we know that, for a Gaussian process with known variance s_0 and known cepstral values $c_1, \ldots, c_q$, the maximum entropy spectrum is identical to that of an MA(q) process (there is always a Gaussian process satisfying these $q+1$ constraints).

[4] Fitting an MA(q) model to a time series leads to an IIR filter, which – because of the problem of turn-on transients – is not as easy to work with as a FIR filter for prewhitening a time series of finite length.

[5] Physical arguments sometimes suggest that an MA model is appropriate for a particular time series. For example, we have already noted in Section 2.6 that Spencer-Smith and Todd (1941) argued thus for the thickness of textile slivers as a function of displacement along a sliver.

[6] Pure sinusoidal variations cannot be expressed as an MA(q) model.

Figure 33 shows eight examples of realizations of MA(1) processes.

As in the case of AR(p) processes, we need to estimate each $\theta_{q,j}$ and the variance σ_ϵ^2 from available data. If we attempt to use the method of matching lagged moments that led to the Yule–Walker equations in the AR(p) case, we find that

$$s_\tau = \sigma_\epsilon^2 \sum_{j=0}^{q-\tau} \theta_{q,j+\tau}\theta_{q,j}, \quad \tau = 0, 1, \ldots, q,$$

where we define $\theta_{q,0} = -1$ (cf. Equation (32b)). These are nonlinear in the MA coefficients, but an additional difficulty is that a valid solution need not exist. As an example, for an MA(1) process, we have

$$s_0 = \sigma_\epsilon^2 \left(1 + \theta_{1,1}^2\right) \quad \text{and} \quad s_1 = -\sigma_\epsilon^2 \theta_{1,1}, \tag{504a}$$

which can only be solved for a valid σ_ϵ^2 and $\theta_{1,1}$ if $|s_1/s_0| \leq 1/2$. Thus, whereas there is always an AR(p) process whose theoretical ACVS agrees out to lag p with $\hat{s}_0^{(\text{P})}, \ldots, \hat{s}_p^{(\text{P})}$ computed for any time series, the same is not true of MA processes. In this sense we can claim that the class of AR processes is "richer" than the class of MA processes.

There is, however, a second way of looking at the Yule–Walker estimation procedure for AR(p) processes as far as SDF estimation is concerned. If we fit an AR(p) process to a time series by that procedure, the ACVS of the fitted model necessarily agrees *exactly* up to lag p with the estimated ACVS. The values of the ACVS after lag p are generated recursively in accordance with an AR(p) model, namely,

$$s_\tau = \sum_{j=1}^{p} \phi_{p,j} s_{\tau-j}, \quad \tau = p+1, p+2, \ldots$$

(see Equation (450a)). The estimated SDF can thus be regarded as given by

$$\hat{S}^{(\text{YW})}(f) = \Delta_t \sum_{\tau=-\infty}^{\infty} \tilde{s}_\tau e^{-i2\pi f \tau \Delta_t}, \quad \text{where } \tilde{s}_\tau = \begin{cases} \hat{s}_\tau^{(\text{P})}, & |\tau| \leq p; \\ \sum_{j=1}^{p} \tilde{\phi}_{p,j} \tilde{s}_{|\tau|-j}, & |\tau| > p, \end{cases}$$

with $\tilde{\phi}_{p,j}$ defined to be the Yule–Walker estimator of $\phi_{p,j}$. Thus the Yule–Walker SDF estimator can be regarded as accepting $\hat{s}_\tau^{(\text{P})}$ up to lag p and then estimating the ACVS beyond that lag based upon the AR(p) model. If we follow this procedure for an MA(q) model, we would set the ACVS beyond lag q equal to 0 (see Equation (32b)). Our SDF estimate would thus be just

$$\hat{S}(f) = \Delta_t \sum_{\tau=-q}^{q} \hat{s}_\tau^{(\text{P})} e^{-i2\pi f \tau \Delta_t}. \tag{504b}$$

This is identical to the truncated periodogram of Equation (343b). Although this estimator was an early example of a lag window estimator, its smoothing window is inferior to that for other estimators of the lag window class (such as those utilizing the Parzen or Papoulis lag windows). In particular, the smoothing window is such that $\hat{S}(f)$ need *not* be nonnegative at all frequencies, thus potentially yielding an estimate that violates a fundamental property of an actual SDF (see Exercise [7.7]).

To conclude, the choice of the class of models to be used in parametric spectral analysis is obviously quite important. The class of AR models has a number of advantages when compared to MA models, not the least of which is ease of computation. Unfortunately, algorithms for explicitly fitting ARMA models are much more involved than those for fitting AR models; on the other hand, we have noted that the spectral estimator of Equation (491b) is in fact an ARMA SDF, so it is certainly possible to obtain *implicit* ARMA spectral estimates in a computationally efficient manner.

9.15 Summary of Parametric Spectral Estimators

We assume that $X_0, X_1, \ldots, X_{N-1}$ is a sample of length N from a real-valued stationary process $\{X_t\}$ with zero mean and unknown SDF $S_X(\cdot)$ defined over the interval $[-f_{\mathcal{N}}, f_{\mathcal{N}}]$, where $f_{\mathcal{N}} \stackrel{\text{def}}{=} 1/(2\Delta_t)$ is the Nyquist frequency and Δ_t is the sampling interval between observations. (If $\{X_t\}$ has an unknown mean, then we need to replace X_t with $X_t' = X_t - \overline{X}$ in all computational formulae, where $\overline{X} \stackrel{\text{def}}{=} \sum_{t=0}^{N-1} X_t/N$ is the sample mean.) The first step in parametric spectral analysis is to estimate the $p+1$ parameters – namely, $\phi_{p,1}, \ldots, \phi_{p,p}$ and σ_p^2 – of an AR(p) model (see Equation (446a)) using one of the following estimators (Section 9.11 discusses choosing the order p).

[1] *Yule–Walker estimators* $\tilde{\phi}_{p,1}, \ldots, \tilde{\phi}_{p,p}$ and $\tilde{\sigma}_p^2$

We first form the usual biased estimator $\{\hat{s}_\tau^{(\mathrm{P})}\}$ of the ACVS using $X_0, \ldots, X_{N-1}$ (see Equation (170b)). With $\tilde{\sigma}_0^2 \stackrel{\text{def}}{=} \hat{s}_0^{(\mathrm{P})}$, we obtain the Yule–Walker estimators by recursively computing, for $k = 1, \ldots, p$,

$$\tilde{\phi}_{k,k} = \frac{\hat{s}_k^{(\mathrm{P})} - \sum_{j=1}^{k-1} \tilde{\phi}_{k-1,j}\hat{s}_{k-j}^{(\mathrm{P})}}{\tilde{\sigma}_{k-1}^2}; \qquad \text{(see (458a))}$$

$$\tilde{\phi}_{k,j} = \tilde{\phi}_{k-1,j} - \tilde{\phi}_{k,k}\tilde{\phi}_{k-1,k-j}, \quad 1 \le j \le k-1; \qquad \text{(see (458b))}$$

$$\tilde{\sigma}_k^2 = \tilde{\sigma}_{k-1}^2 (1 - \tilde{\phi}_{k,k}^2) \qquad \text{(see (458c))}$$

(for $k = 1$, we obtain $\tilde{\phi}_{1,1} = \hat{s}_1^{(\mathrm{P})}/\hat{s}_0^{(\mathrm{P})}$ and $\tilde{\sigma}_1^2 = \hat{s}_0^{(\mathrm{P})}(1 - \tilde{\phi}_{1,1}^2)$). An equivalent formulation that makes direct use of $X_0, \ldots, X_{N-1}$ rather than $\{\hat{s}_k^{(\mathrm{P})}\}$ is outlined in the discussion surrounding Equation (471). A variation on the Yule–Walker method is to use the ACVS estimator corresponding to a direct spectral estimator $\hat{S}^{(\mathrm{D})}(\cdot)$ other than the periodogram (see the discussion concerning Figures 461 and 462). The Yule–Walker parameter estimators are guaranteed to correspond to a causal – and hence stationary – AR process.

[2] *Burg estimators* $\bar{\phi}_{p,1}, \ldots, \bar{\phi}_{p,p}$ and $\bar{\sigma}_p^2$

With $\vec{e}_t(0) \stackrel{\text{def}}{=} X_t$, $\overleftarrow{e}_t(0) \stackrel{\text{def}}{=} X_t$ and $\bar{\sigma}_0^2 \stackrel{\text{def}}{=} \hat{s}_0^{(\mathrm{P})}$, we obtain the Burg estimators by recursively computing, for $k = 1, \ldots, p$,

$$A_k = \sum_{t=k}^{N-1} \vec{e}_t^{\,2}(k-1) + \overleftarrow{e}_{t-k}^{\,2}(k-1) \qquad \text{(see (467d))}$$

$$B_k = 2 \sum_{t=k}^{N-1} \vec{e}_t(k-1)\overleftarrow{e}_{t-k}(k-1) \qquad \text{(see (467e))}$$

$$\bar{\phi}_{k,k} = B_k/A_k \qquad \text{(see (467f))}$$

$$\vec{e}_t(k) = \vec{e}_t(k-1) - \bar{\phi}_{k,k}\overleftarrow{e}_{t-k}(k-1), \quad k \le t \le N-1; \qquad \text{(see (467a))}$$

$$\overleftarrow{e}_{t-k}(k) = \overleftarrow{e}_{t-k}(k-1) - \bar{\phi}_{k,k}\vec{e}_t(k-1), \quad k \le t \le N-1; \qquad \text{(see (467b))}$$

$$\bar{\phi}_{k,j} = \bar{\phi}_{k-1,j} - \bar{\phi}_{k,k}\bar{\phi}_{k-1,k-j}, \quad 1 \le j \le k-1;$$

$$\bar{\sigma}_k^2 = \bar{\sigma}_{k-1}^2 (1 - \bar{\phi}_{k,k}^2). \qquad \text{(see (468b))}$$

The term A_k can be recursively computed using Equation (509b). The Burg parameter estimators in practice correspond to a causal – and hence stationary – AR process and have been found to be superior to the Yule–Walker estimators in numerous Monte Carlo experiments.

[3] *Least squares (LS) estimators*

Here we use the resemblance between an AR(p) model and a regression model to formulate LS parameter estimators. Three different formulations are detailed in Section 9.7, namely, the forward LS, backward LS and forward/backward LS estimators (FLS, BLS and FBLS, respectively). Calculation of these estimators can be done using standard least squares techniques (Golub and Van

Loan, 2013; Press et al., 2007) or specialized algorithms that exploit the structure of the AR model (Marple, 1987; Kay, 1988). None of the three LS parameter estimators is guaranteed to correspond to a causal AR process. In terms of spectrum estimation, Monte Carlo studies show that the FBLS estimator is generally superior to the Yule–Walker estimator and outperforms the Burg estimator in some cases.

[4] *Maximum likelihood estimators (MLEs)*
Under the assumption of a Gaussian process, we derived MLEs for the AR parameters in Section 9.8. Unfortunately, these estimators require the use of a nonlinear optimization routine and hence are computationally intensive compared to Yule–Walker, Burg and LS estimators (particularly for large model orders p). These computationally more efficient estimators can all be regarded as approximate MLEs.

Let $\hat{\phi}_{p,1}, \ldots, \hat{\phi}_{p,p}$ and $\hat{\sigma}_p^2$ represent any of the estimators just mentioned (Yule–Walker, Burg, LS or MLEs). There are two ways we can use these estimators to produce an estimator of the SDF $S_X(\cdot)$.

[1] *Pure parametric approach*
Here we merely substitute the AR parameter estimators for the corresponding theoretical parameters in Equation (446b) to obtain the SDF estimator

$$\hat{S}_X(f) = \frac{\hat{\sigma}_p^2 \Delta_t}{\left|1 - \sum_{j=1}^{p} \hat{\phi}_{p,j} e^{-i2\pi f j \Delta_t}\right|^2},$$

which is a periodic function of f with a period of $2f_{\mathcal{N}}$. Section 9.9 gives an approach for obtaining confidence intervals for the normalized SDF based upon this estimator.

[2] *Prewhitening approach*
Here we use the estimated coefficients $\hat{\phi}_{p,1}, \ldots, \hat{\phi}_{p,p}$ to create a prewhitening filter. Section 9.10 discusses this combined parametric/nonparametric approach in detail – its features are generally more appealing than the pure parametric approach.

9.16 Exercises

[9.1] (a) Using Equation (145b), verify the two stationary solutions stated in Equation (447) for AR(1) processes.

(b) Show that, if $\widetilde{Y}_t = \sum_{j=1}^{p} \varphi_{p,j} \widetilde{Y}_{t-j} + \varepsilon_t$ is a stationary acausal AR(p) process with SDF

$$S_{\widetilde{Y}}(f) = \frac{\sigma_\varepsilon^2}{\left|1 - \sum_{j=1}^{p} \varphi_{p,j} e^{-i2\pi f j}\right|^2},$$

there exists a causal AR(p) process, say, $Y_t = \sum_{j=1}^{p} \phi_{p,j} Y_{t-j} + \epsilon_t$, whose SDF $S_Y(\cdot)$ is given by the above upon replacing $\varphi_{p,j}$ with $\phi_{p,j}$, and $S_Y(\cdot)$ is *identically* the same as $S_{\widetilde{Y}}(\cdot)$. Here $\{\varepsilon_t\}$ and $\{\epsilon_t\}$ are both zero-mean white noise processes with finite variances given by, respectively, $\sigma_\varepsilon^2 > 0$ and $\sigma_\epsilon^2 > 0$. Hint: for a complex-valued variable z, consider the polynomial $\widetilde{\mathcal{G}}(z) = 1 - \sum_{j=1}^{p} \varphi_{p,j} z^{-j}$, rewrite it as

$$\widetilde{\mathcal{G}}(z) = \prod_{j=1}^{p} \left(1 - \frac{r_j}{z}\right),$$

where $r_1, \ldots, r_p$ are the p roots of the polynomial (at least one of which is such that $|r_j| > 1$), and then consider $z = e^{i2\pi f}$ and $|\widetilde{\mathcal{G}}(z)|^2$.

(c) For a real-valued ϕ such that $0 < |\phi| < 1$, consider the AR(2) process

$$\widetilde{Y}_t = \left(\phi + \frac{1}{\phi}\right) \widetilde{Y}_{t-1} - \widetilde{Y}_{t-2} + \varepsilon_t,$$

where $\{\varepsilon_t\}$ is a zero-mean white noise process with variance $\sigma_\varepsilon^2 > 0$. By expressing $\widetilde{Y}_t$ as a linear combination of RVs in $\{\varepsilon_t\}$, show that $\{\widetilde{Y}_t\}$ has a stationary – but acausal – solution. Use this expression to determine var $\{\widetilde{Y}_t\}$.

(d) Determine the causal AR(2) process $Y_t = \phi_{2,1} Y_{t-1} + \phi_{2,2} Y_{t-2} + \epsilon_t$ that has the same SDF as $\{\widetilde{Y}_t\}$ of part (c) (here $\{\epsilon_t\}$ is a zero-mean white noise process with variance $\sigma_\epsilon^2 > 0$). Express the stationary solution for Y_t as a linear combination of RVs in $\{\epsilon_t\}$. Use this expression to determine var $\{Y_t\}$. Show that var $\{Y_t\}$ is related to var $\{\widetilde{Y}_t\}$ of part (c) in a manner consistent with the result of part (b).

(e) Use the step-down Levinson–Durbin procedure stated in Equations (460a) and (460b) to verify the expression for var $\{Y_t\}$ obtained in part (d). Is it possible to do the same for var $\{\widetilde{Y}_t\}$ obtained in part (c)?

[9.2] In what follows, take $Y_t = \phi Y_{t-1} + \epsilon_t$ to be a causal (and hence stationary) AR(1) process with mean zero, where $\{\epsilon_t\}$ is a white noise process with mean zero and variance σ_1^2. Let $S(\cdot)$ and Δ_t denote the SDF and sampling interval for $\{Y_t\}$.

(a) Using equations developed in Section 9.3, show that the ACVS for $\{Y_t\}$ is

$$s_\tau = \frac{\phi^{|\tau|} \sigma_1^2}{1 - \phi^2}, \quad \tau \in \mathbb{Z} \tag{507a}$$

(cf. Exercise [2.17a], which tackles the above using a different approach).

(b) Use Equation (165a) to show that the variance of the sample mean $\overline{Y}_N$ of $Y_0, Y_1, \ldots, Y_{N-1}$ is

$$\text{var}\{\overline{Y}_N\} = \frac{s_0}{N}\left(\frac{1+\phi}{1-\phi} - \frac{2\phi(1-\phi^N)}{N(1-\phi)^2}\right). \tag{507b}$$

Equation (165c) suggests the approximation $S(0)/(N\Delta_t)$. Show this yields

$$\text{var}\{\overline{Y}_N\} \approx \frac{s_0}{N}\left(\frac{1+\phi}{1-\phi}\right). \tag{507c}$$

(c) For $N = 100$, plot the log of the ratio var $\{\overline{Y}_N\}/s_0$ versus $\phi = -0.99, -0.98, \ldots, 0.98, 0.99$, and then superimpose a similar plot, but with the exact variance replaced by the approximation of Equation (507c). Plot the ratio of the approximate variance of Equation (507c) to the exact variance of Equation (507b) versus ϕ. Comment on how well the approximation does.

(d) For $\phi = -0.9$, $\phi = 0$ and finally $\phi = 0.9$, plot $\log_{10}\left(\text{var}\{\overline{Y}_N\}/s_0\right)$ versus $\log_{10}(N)$ for $N = 1, 2, \ldots, 1000$. Comment briefly on the appearance of the three plots (in particular, how they differ and how they are similar).

(e) Equation (298c) expresses the degrees of freedom η in a time series of length N in terms of its ACVS. Derive an expression for η when the series is a Gaussian AR(1) process. For $N = 100$, plot η versus versus $\phi = -0.99, -0.98, \ldots, 0.98, 0.99$. Comment upon your findings.

(f) Suppose that $\{Y(t) : t \in \mathbb{R}\}$ is a continuous parameter stationary process with a Lorenzian SDF and ACVS given by

$$S_{Y(t)}(f) = \frac{2L\sigma^2}{1 + (2\pi f L)^2} \quad \text{and} \quad s_{Y(t)}(\tau) = \sigma^2 e^{-|\tau|/L}, \quad \text{for } f, \tau \in \mathbb{R},$$

where $\sigma^2 > 0$ and $L > 0$ are finite and represent the variance of $\{Y(t)\}$ and its correlation length (see Equations (131b) and (131c)). Show that the discrete parameter stationary process defined by $Y_t = Y(t\Delta_t)$ is an AR(1) process, i.e., has an SDF and ACVS in agreement with that of $\{Y_t : t \in \mathbb{Z}\}$ used in previous parts of this exercise. How does the AR(1) coefficient ϕ depend on σ^2, L and Δ_t, and what range of values can ϕ assume?

[9.3] Obtain Yule–Walker estimates of the parameters $\phi_{2,1}$, $\phi_{2,2}$ and σ_2^2 for an AR(2) process using the following biased estimates of the ACVS: $\hat{s}_0^{(P)} = 1$, $\hat{s}_1^{(P)} = -1/2$ and $\hat{s}_2^{(P)} = -1/5$. Assuming

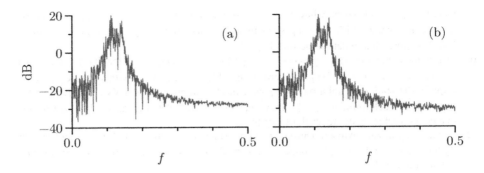

Figure 508 Periodogram (left-hand plot) and Yule–Walker SDF estimate of order $p = 1023$ (right-hand) for AR(4) time series shown in Figure 34(e) – see Exercise [9.4] for details.

$\Delta_t = 1$, state the corresponding estimated SDF in a reduced form (i.e, cosines rather than complex exponentials), and compute its value at frequencies $f = 0$, $1/4$ and $1/2$. Plot the SDF on a decibel scale versus frequency over a fine grid of frequencies, e.g., $f_k = k/256$ for $k = 0, 1, \ldots, 128$. Finally, determine the corresponding estimated ACVS at lags $\tau = 0, 1, \ldots 5$.

[9.4] Figure 508(a) shows the periodogram $\hat{S}^{(P)}(f_k)$ versus Fourier frequencies $f_k = k/N$, $k = 0,1, \ldots, N/2$, for the AR(4) time series of length $N = 1024$ displayed in Figure 34(e). Figure 508(b) shows a Yule–Walker SDF estimate $\hat{S}^{(YW)}(\cdot)$ of order $p = 1023$ over the same grid of frequencies; i.e., we have set $p = N - 1$, which is the maximum lag for which we can compute an estimate of the ACVS from $X_0, \ldots, X_{N-1}$. As noted in Section 9.3, an important property of a Yule–Walker estimator is that its corresponding ACVS is identical to the sample ACVS $\{\hat{s}^{(P)}\}$ up to lag p. Since the periodogram is the Fourier transform of $\{\hat{s}^{(P)}\}$, for this example we thus have, for $\tau = 0, \ldots, 1023$,

$$\int_{-1/2}^{1/2} \hat{S}^{(YW)}(f) e^{i2\pi f\tau}\, df = \hat{s}_\tau^{(P)} \quad \text{and} \quad \int_{-1/2}^{1/2} \hat{S}^{(P)}(f) e^{i2\pi f\tau}\, df = \hat{s}_\tau^{(P)}$$

(cf. Equation (171c)). Figure 508 indicates that the periodogram and Yule-Walker estimate resemble one another closely. Explain why they aren't identical.

[9.5] Here we prove a useful corollary to the orthogonality principle. Let $\{X_t\}$ be any stationary process with zero mean and with an ACVS $\{s_k\}$ that is positive definite. Define

$$d_t(k) = X_t - \sum_{l=1}^{k} \psi_l X_{t-l},$$

where the ψ_l terms are any set of constants. Suppose that $E\{d_t(k) X_{t-j}\} = 0$, $1 \leq j \leq k$; i.e., the $d_t(k)$ terms mimic the property stated in Equation (454a) for $\vec{\epsilon}_t(k)$. Show that we must have $d_t(k) = \vec{\epsilon}_t(k)$. Hint: note that the system of equations here is identical to that in Equation (454b).

[9.6] Equation (460d) gives a procedure for determining the ACVS for an AR(p) process given its coefficients $\phi_{p,1}, \ldots, \phi_{p,p}$ and innovation variance σ_p^2. Use this procedure to show that ACVS for the AR(2) process of Equation (34) is

$$s_0 = \tfrac{16}{9}, \quad s_1 = \tfrac{8}{9} \quad \text{and} \quad s_\tau = \tfrac{3}{4} s_{\tau-1} - \tfrac{1}{2} s_{\tau-2}, \quad \tau = 2, 3, \ldots \quad (508a)$$

and that the ACVS for the AR(4) process of Equation (35a) is

$$s_0 \doteq 1.523\,434\,580, \quad s_1 \doteq 1.091\,506\,153, \quad s_2 \doteq 0.054\,284\,646, \quad s_3 \doteq -0.975\,329\,449 \quad (508b)$$

and $s_\tau = 2.7607 s_{\tau-1} - 3.8106 s_{\tau-2} + 2.6535 s_{\tau-3} - 0.9238 s_{\tau-4}$, $\tau = 4, 5, \ldots$. Use the above to check that the top row of plots in Figure 168 is correct.

[9.7] Verify Equation (462b), which states that $\phi_{k,k}$ can be interpreted as a correlation coefficient. Hint: express $\overleftarrow{\epsilon}_{t-k}(k-1)$ in terms of $\{X_t\}$; evoke the orthogonality principle (Equation (454a)); and use Equation (456d).

[9.8] (a) Use Equations (457a) and (457b) to show that $\sigma_0^2 = s_0$; i.e., the zeroth-order mean square linear prediction error is just the process variance. Is this a reasonable interpretation for σ_0^2?

(b) Show how the Levinson–Durbin recursions can be initialized starting with $k = 0$ rather than $k = 1$ as is done in Equation (457a).

(c) Use part (a) and Equation (457b) to show that

$$\sigma_k^2 = s_0 \prod_{j=1}^{k}(1 - \phi_{j,j}^2). \tag{509a}$$

[9.9] Consider the following six AR(4) processes:

(a) $Y_t = 0.8296 Y_{t-1} + 0.8742 Y_{t-2} - 0.4277 Y_{t-3} + 0.6609 Y_{t-4} + \epsilon_t$;
(b) $Y_t = 0.2835 Y_{t-1} + 0.0382 Y_{t-2} + 0.4732 Y_{t-3} - 0.7307 Y_{t-4} + \epsilon_t$;
(c) $Y_t = 0.3140 Y_{t-1} + 0.4101 Y_{t-2} - 0.0845 Y_{t-3} + 0.4382 Y_{t-4} + \epsilon_t$;
(d) $Y_t = 0.8693 Y_{t-1} - 0.4891 Y_{t-2} - 0.0754 Y_{t-3} + 0.8800 Y_{t-4} + \epsilon_t$;
(e) $Y_t = 0.9565 Y_{t-1} - 0.7650 Y_{t-2} - 0.0500 Y_{t-3} + 0.1207 Y_{t-4} + \epsilon_t$;
(f) $Y_t = 0.8081 Y_{t-1} - 0.7226 Y_{t-2} + 0.9778 Y_{t-3} + 0.8933 Y_{t-4} + \epsilon_t$,

where, as usual, $\{\epsilon_t\}$ is a white noise process with zero mean and finite nonzero variance. Use the step-down Levinson–Durbin recursions of Equation (460a) to determine which (if any) of these processes are causal (and hence stationary).

[9.10] Suppose that we are given the following ACVS values: $s_0 = 3$, $s_1 = 2$ and $s_2 = 1$. Use the Levinson–Durbin recursions to show that $\phi_{1,1} = 2/3$, $\phi_{2,1} = 4/5$, $\phi_{2,2} = -1/5$, $\sigma_1^2 = 5/3$ and $\sigma_2^2 = 8/5$. Construct $\boldsymbol{L}_3^{-1}$ and $\boldsymbol{D}_3$, and verify that $\boldsymbol{L}_3^{-1}\boldsymbol{\Gamma}_3\boldsymbol{L}_3^{-T} = \boldsymbol{D}_3$ (see Equation (464d)). Determine $\boldsymbol{L}_3$ also, verifying that it is lower triangular. (This example is due to Therrien, 1983.)

[9.11] Any one of the following five sets of $p+1$ quantities completely characterizes the second-order properties of the AR(p) process of Equation (446a):

(i) $\phi_{p,1}, \phi_{p,2}, \ldots, \phi_{p,p}$ and σ_p^2; (ii) $\phi_{p,1}, \phi_{p,2}, \ldots, \phi_{p,p}$ and s_0;

(iii) $\phi_{1,1}, \phi_{2,2}, \ldots, \phi_{p,p}$ and s_0; (iv) $\phi_{1,1}, \phi_{2,2}, \ldots, \phi_{p,p}$ and σ_p^2; (v) $s_0, s_1, \ldots, s_p$.

Show these are equivalent by demonstrating how, given any one of them, we can get the other four.

[9.12] (a) Figure 458 shows Yule–Walker AR(4) SDF estimates based on the realization of the AR(4) process shown in Figure 34(e). The four estimates shown in the figure make use of the first 16 values of the time series, the first 64 values, the first 256 values and, finally, the entire series of 1024 values. Figure 469 shows corresponding Burg AR(4) SDF estimates. Create similar figures of Yule–Walker and Burg estimates for the time series shown in Figure 34(f), (g) and (h), which are downloadable from the website for this book.

(b) Repeat part (a), but now using the four AR(2) time series shown in Figures 34(a) to (d).

[9.13] (a) Show that Equation (470a) holds (hint: use Equation (467c)).

(b) Show that A_k of Equation (467d) obeys the recursive relationship

$$A_k = \left(1 - \bar{\phi}_{k-1,k-1}^2\right) A_{k-1} - \overrightarrow{\epsilon}_{k-1}^{\,2}(k-1) - \overleftarrow{\epsilon}_{N-k}^{\,2}(k-1), \tag{509b}$$

which is helpful in computing $\bar{\phi}_{k,k}$ in Equation (467f) (Andersen, 1978).

[9.14] (a) Figure 478 shows FBLS SDF estimates based on the realization of the AR(4) process shown in Figure 34(e). The four estimates shown in the figure make use of the first 16 values of the time series, the first 64 values, the first 256 values and, finally, the entire series of 1024 values. Create similar figures for the FLS estimates and the BLS estimates.

(b) Repeat part (a), but now using the three AR(4) time series shown in Figures 34(f), (g) and (h) and also including the FBLS estimator.

[9.15] (a) Use Burg's algorithm to fit an AR(2) model to the six-point time series $\{100, 10, 1, -1, -10, -100\}$. Report the Burg estimates $\bar{\phi}_{2,1}$, $\bar{\phi}_{2,2}$ and $\bar{\sigma}_2^2$. Assuming $\Delta_t = 1$, state the corresponding estimated SDF in a reduced form (i.e, cosines rather than complex exponentials), and plot the SDF on a decibel scale over the frequencies $f'_k = k/256$, $k = 0, 1, \ldots, 128$.

(b) In the AR(2) case, the FBLS estimator of $\boldsymbol{\Phi} = [\phi_{2,1}, \phi_{2,2}]^T$ minimizes the sum of squares $SS_{(FBLS)}(\boldsymbol{\Phi})$ of Equation (477d). Show that this estimator $\hat{\boldsymbol{\Phi}}_{(FBLS)}$ satisfies the equation

$$\begin{bmatrix} 2\sum_{t=1}^{N-2} X_t^2 & A \\ A & \sum_{t=0}^{N-3} X_t^2 + \sum_{t=2}^{N-1} X_t^2 \end{bmatrix} \boldsymbol{\Phi} = \begin{bmatrix} A \\ 2\sum_{t=0}^{N-3} X_t X_{t+2} \end{bmatrix},$$

where you are to determine what A is. For the six-point time series, find and state both elements of $\hat{\boldsymbol{\Phi}}_{(FBLS)}$ along with the estimate of the innovation variance given by Equation (477e). As done for the Burg estimate, create and plot the FBLS AR(2) SDF on a decibel scale versus the frequencies f'_k. How well do the Burg and FBLS SDF estimates agree?

(c) Compute and plot the periodogram for the six-point time series over a fine grid of frequencies. How well does the periodogram agree with the Burg and FBLS AR(2) SDF estimates?

(d) Show that the estimated coefficients $\hat{\boldsymbol{\Phi}}_{(FBLS)}$ obtained in part (b) correspond to an acausal (but stationary) AR(2) process.

(e) (Note: this part assumes familiarity with the solution to Exercise [9.1b].) Find a causal AR(2) process with parameters, say, $\check{\boldsymbol{\Phi}}_{(FBLS)}$ and $\check{\sigma}^2_{(FBLS)}$, whose SDF is in exact agreement with the one based on $\hat{\boldsymbol{\Phi}}_{(FBLS)}$ and $\hat{\sigma}^2_{(FBLS)}$. Use $\check{\boldsymbol{\Phi}}_{(FBLS)}$ in conjunction with Equation (479) to compute an alternative estimate, say $\tilde{\sigma}^2_{(FBLS)}$, for the innovation variance. Compute and plot the AR(2) SDF estimate based on $\check{\boldsymbol{\Phi}}_{(FBLS)}$ and $\tilde{\sigma}^2_{(FBLS)}$ versus f'_k. How well does this estimate agree with the Burg estimate of part (a)?

[9.16] Suppose that we wish to fit an AR(1) model to a time series that is a realization of a portion $X_0, X_1, \ldots, X_{N-1}$ of a stationary process with zero mean.

(a) Show that the Yule–Walker, Burg, FLS, and BLS estimators of the coefficient $\phi_{1,1}$ are given by, respectively,

$$\frac{\sum_{t=1}^{N-1} X_t X_{t-1}}{\sum_{t=0}^{N-1} X_t^2}, \quad \frac{\sum_{t=1}^{N-1} X_t X_{t-1}}{\frac{1}{2}X_0^2 + \sum_{t=1}^{N-2} X_t^2 + \frac{1}{2}X_{N-1}^2}, \quad \frac{\sum_{t=1}^{N-1} X_t X_{t-1}}{\sum_{t=0}^{N-2} X_t^2} \quad \text{and} \quad \frac{\sum_{t=1}^{N-1} X_t X_{t-1}}{\sum_{t=1}^{N-1} X_t^2}.$$

(b) Show that the Yule–Walker estimator of $\phi_{1,1}$ must be less than or equal to the Burg estimator in magnitude. Assuming $\{X_t\}$ is multivariate Gaussian, what is the probability that the Yule–Walker and Burg estimators are equal?

(c) Show that the FBLS estimator of $\phi_{1,1}$ is identical to that of the Burg estimator (this relationship does not hold in general when $p > 1$).

(d) Consider the case $N = 2$ (Jones, 1985, p. 226). Show that, whereas the Yule–Walker, Burg and FBLS estimators are bounded in magnitude, the FLS estimator and the BLS estimator need not be. Determine the MLE when $N = 2$ (see Equation (482d)). Assuming a bivariate Gaussian distribution for X_0 and X_1, discuss briefly which of the six estimators can or cannot yield an estimated AR(1) model that is causal (and hence stationary).

[9.17] Show that Equation (484b) is true.

[9.18] Assuming that Equation (484b) is true (the burden of Exercise [9.17]), show how the FBLS estimator of the AR(p) coefficients can be regarded as an approximation to the MLE. Hint: consider the relationship between $l(\boldsymbol{\Phi}_p, \sigma_p^2 \mid \boldsymbol{H})$ and $l(\boldsymbol{\Phi}_p, \sigma_p^2 \mid \widetilde{\boldsymbol{H}})$ and how it impacts Equation (484c); derive an equivalent of Equation (481a) that uses $\widetilde{\boldsymbol{H}}$ rather than $\boldsymbol{H}$; produce the analog of Equation (482a); and then use approximations similar to those that yielded the FLS estimator as an approximation to the MLE.

[9.19] Figure 496 shows two different AR(5) SDF estimates for the ocean wave time series (available from the website for this book): the thick smooth curve is the Yule–Walker estimate, and the dashed curve, the Burg estimate. We used the Burg estimates $\bar{\phi}_{5,1}, \ldots, \bar{\phi}_{5,5}$ to create a prewhitening filter, leading to the prediction errors shown in Figure 497(a) and the postcolored $m = 55$ Parzen lag window estimate $\hat{S}_X^{(\text{PC})}(\cdot)$ shown as the thick smooth curve in Figure 497(c). To investigate how sensitive $\hat{S}_X^{(\text{PC})}(\cdot)$ is to the chosen prewhitening filter, use the Yule–Walker estimates $\tilde{\phi}_{5,1}, \ldots, \tilde{\phi}_{5,5}$ rather than the Burg estimates to create the prewhitening filter, and produce plots similar to Figure 497(a), (b) and (c). Comment briefly on how the Yule–Walker-based postcolored $m = 55$ Parzen lag window estimate compares with the corresponding Burg-based estimate.

10

Harmonic Analysis

10.0 Introduction

If a stationary process has a purely continuous spectrum, it is natural to estimate its spectral density function (SDF) since this function is easier to interpret than the integrated spectrum. Estimation of the SDF has occupied our attention in the previous four chapters. However, if we are given a sample of a time series drawn from a process with a purely discrete spectrum (i.e., a "line" spectrum for which the integrated spectrum is a step function), our estimation problem is quite different: we must estimate the location and magnitude of the jumps in the integrated spectrum, which requires estimation techniques that differ from what we have already studied. It is more common, however, to come across processes whose spectra are a mixture of lines and an SDF stemming from a so-called "background continuum." In Section 4.4 we distinguished two cases. If the SDF for the continuum is that of white noise, we said that the process has a discrete spectrum – as opposed to a *purely* discrete spectrum, which has only a line component; on the other hand, if the SDF for the continuum differs from that of white noise (sometimes called "colored" noise), we said that the process has a mixed spectrum (see Figure 121).

In this chapter we begin with discrete parameter harmonic processes with random phases (these processes have a purely discrete spectrum). In the next section we first recall how we defined these processes in Section 2.6. We then use some standard concepts from tidal analysis to motivate and illustrate use of these processes as models.

10.1 Harmonic Processes – Purely Discrete Spectra

A real-valued discrete parameter harmonic process with random phases can be written as

$$X_t = \mu + \sum_{l=1}^{L} D_l \cos\left(2\pi f_l t \Delta_{\text{t}} + \phi_l\right), \qquad (511)$$

where μ, $D_l \geq 0$ and $f_l > 0$ are in general unknown real-valued constants; $L \geq 1$ is the number of harmonic components; and the ϕ_l terms are independent real-valued random variables (RVs) – representing random phase angles – with a rectangular (uniform) distribution on $(-\pi, \pi]$ (see Exercise [37]; Equation (511) is the same as Equation (35d) except for the insertion of the sampling interval Δ_{t}, which makes the frequencies f_l have physically meaningful units of cycles per unit time). Since $\{X_t\}$ is a discrete parameter process, we can consider each f_l to lie in the interval $(0, f_\mathcal{N}]$ due to the aliasing effect, where $f_\mathcal{N} \stackrel{\text{def}}{=} 1/(2\,\Delta_t)$ is the Nyquist frequency (see Sections 3.9 and 4.5).

This model is in one sense not really a statistical model: for each realization of the process the RVs ϕ_l are fixed quantities, so each realization is essentially deterministic and free from what is commonly thought of as random variation. For this reason the process is often dismissed as of no interest; however, its simple properties can prove useful in practice as we will demonstrate by looking at ocean tide prediction.

The theory of gravitational tidal potential predicts that at a fixed location the hourly height of the ocean tide at the hour with index t can be written as

$$X_t = \mu + \sum_{l=1}^{L'} \alpha_{t,l} H_l \cos\left(2\pi f_l t \Delta_t + v_{t,l} + u_{t,l} - g_l\right) \text{ with } \Delta_t = 1 \text{ hour,} \qquad (512a)$$

where the sum is over L' primary tidal constituent frequencies, and the harmonic constants H_l and g_l depend only upon the location of interest and must be estimated. (These primary constituent frequencies derive from consideration of a frictionless and inertia-free ocean covering the whole globe; this crude model is inadequate for the real ocean, various hydrographic effects such as the locations of land and rivers making it necessary to introduce the location-dependent amplitudes H_l and phases g_l.) The $\alpha_{t,l}$ and $u_{t,l}$ terms are the so-called nodal parameters (these are essentially corrections for the amplitudes H_l and phases g_l), and $v_{t,l}$ is the phase of the lth constituent of the potential over Greenwich at $t=0$ (these three quantities can be found from, e.g., Doodson and Warburg, 1941). For locations such as Honolulu, where the tides are virtually "linear" (Munk and Cartwright, 1966), this theory is sufficient, but at many British ports and elsewhere the distortion of tides by shallow-water effects often cannot be ignored. Nonlinear interactions are incorporated into harmonic tidal prediction by including constituents with frequencies that are sums and differences of the L' primary constituents. The total number of constituents required, say L, can exceed 100 in complicated shallow water areas such as Anchorage, Alaska, or Southampton, England.

In practice the terms of known form, namely the nodal parameters $\alpha_{t,l}$ and $u_{t,l}$ – which are continuously but slowly varying – are updated at fixed periods when computing the right-hand side of Equation (512a). Sufficient accuracy is often obtained by regarding them as fixed over a span of a year and updating them annually, although every 60 days is preferred (whichever span is used, the $v_{t,l}$ terms are also updated by "resetting" the index t to 0 at the start of each span). This implies that in Equation (512a) both the amplitudes and phases of the cosines vary slowly and slightly with time. This is a good example of what Thomson (1982) called the "convenient fiction" of pure line components. To illustrate the complicated forms that can occur, predicted heights of the tide at Southampton, England, are given in Figure 513 for two different segments of time in 1977. This uses $L = 102$ frequencies and 60 day updating of $\alpha_{t,l}$, $u_{t,l}$ and $v_{t,l}$ (see Walden, 1982). The two segments differ substantially even though they arise for the same model, which illustrates the point that lots of sinusoidal terms – when summed – can interact to give quite complicated patterns.

10.2 Harmonic Processes with Additive White Noise – Discrete Spectra

As we have noted, the model given by Equation (511) is not truly statistical in the sense that, once the random phases ϕ_l are set, there are no remaining stochastic components (the time series is just a linear combination of cosines with known frequencies, amplitudes and phases). A more realistic and useful model is given by

$$X_t = \mu + \sum_{l=1}^{L} D_l \cos\left(2\pi f_l t \Delta_t + \phi_l\right) + \epsilon_t, \qquad (512b)$$

10.2 Harmonic Processes with Additive White Noise – Discrete Spectra

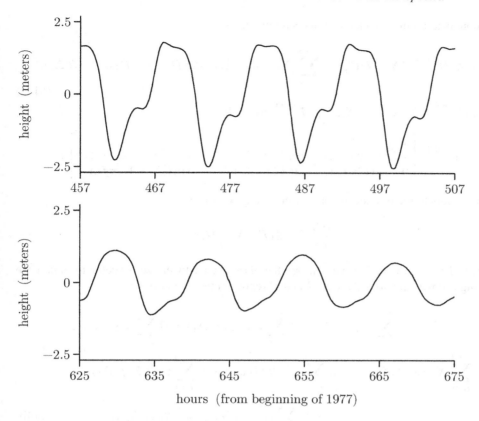

Figure 513 Predicted spring and neap tidal curves at Southampton (January, 1977). Tidal elevations relative to the mean sea level (in meters) are plotted versus an hour count starting from the beginning of 1977. The upper and lower plots show, respectively, spring and neap tides.

which differs from Equation (511) only in that ϵ_t has been added to represent observational error (always present in any sort of physical measurements). The background continuum $\{\epsilon_t\}$ is assumed to be a real-valued white noise process with zero mean and variance σ_ϵ^2. Each ϵ_t is also assumed to be independent of each ϕ_l. Whereas the model of Equation (511) has a purely discrete spectrum, that of Equation (512b) has a discrete spectrum.

Equation (512b) arises more from analytical convenience than realism – many measuring instruments are band-limited, so that $\{\epsilon_t\}$ is not truly white. We looked at predicted tidal heights in the previous section. Observed tidal heights – i.e., predictions plus noise – certainly have a mixed spectrum with colored rather than white noise (see Section 10.4). However, we shall concentrate in what follows on the model of Equation (512b) since it is commonly assumed and – if nothing else – can serve an approximation to the mixed process case.

Let us assume initially that $L = 1$ and that the frequency $f_1 = f$ is known. The unknown parameters in Equation (512b) are now just $D_1 = D$ and σ_ϵ^2. The model can then be rewritten as

$$X_t = \mu + A \cos(2\pi f t \, \Delta_\mathrm{t}) + B \sin(2\pi f t \, \Delta_\mathrm{t}) + \epsilon_t, \tag{513}$$

where $A \stackrel{\mathrm{def}}{=} D \cos(\phi_1)$ and $B \stackrel{\mathrm{def}}{=} -D \sin(\phi_1)$. Note that, for a given realization, A and B are just constants so that Equation (513) expresses a linear regression model. For a time series of length N that is assumed to be a realization of $X_0, X_1, \ldots, X_{N-1}$, we can estimate μ, A and B by least squares (see, e.g., Bloomfield, 2000, or Weisberg, 2014). For real sinusoids, this amounts to minimizing the residual sum of squares

$$\mathrm{SS}(\mu, A, B) \stackrel{\text{def}}{=} \|\boldsymbol{X} - \boldsymbol{H}\boldsymbol{\beta}\|^2 = \sum_{t=0}^{N-1} \left[X_t - \mu - A\cos\left(2\pi f t \Delta_t\right) - B\sin\left(2\pi f t \Delta_t\right)\right]^2, \tag{514a}$$

where $\boldsymbol{X}^T \stackrel{\text{def}}{=} [X_0, X_1, \ldots, X_{N-1}]$; $\boldsymbol{\beta}^T \stackrel{\text{def}}{=} [\mu, A, B]$;

$$\boldsymbol{H}^T \stackrel{\text{def}}{=} \begin{bmatrix} 1 & 1 & 1 & \cdots & 1 \\ 1 & \cos(2\pi f \Delta_t) & \cos(4\pi f \Delta_t) & \cdots & \cos([N-1]2\pi f \Delta_t) \\ 0 & \sin(2\pi f \Delta_t) & \sin(4\pi f \Delta_t) & \cdots & \sin([N-1]2\pi f \Delta_t) \end{bmatrix};$$

and $\|\boldsymbol{V}\|$ refers to the Euclidean norm of the vector $\boldsymbol{V}$. Since

$$\frac{\mathrm{dSS}}{\mathrm{d}\boldsymbol{\beta}} = -2\boldsymbol{H}^T\left(\boldsymbol{X} - \boldsymbol{H}\boldsymbol{\beta}\right)$$

(see, e.g., Rao, 1973, p. 71), we can set the above equal to the zero vector to obtain the following normal equations (here $\hat{\mu}$, $\hat{A}$ and $\hat{B}$ represent their solutions):

$$\sum_{t=0}^{N-1} X_t = N\hat{\mu} + \hat{A}\sum_{t=0}^{N-1}\cos\left(2\pi f t \Delta_t\right) + \hat{B}\sum_{t=0}^{N-1}\sin\left(2\pi f t \Delta_t\right);$$

$$\sum_{t=0}^{N-1} X_t \cos\left(2\pi f t \Delta_t\right) = \hat{\mu}\sum_{t=0}^{N-1}\cos\left(2\pi f t \Delta_t\right) + \hat{A}\sum_{t=0}^{N-1}\cos^2\left(2\pi f t \Delta_t\right)$$
$$+ \hat{B}\sum_{t=0}^{N-1}\cos\left(2\pi f t \Delta_t\right)\sin\left(2\pi f t \Delta_t\right); \tag{514b}$$

$$\sum_{t=0}^{N-1} X_t \sin\left(2\pi f t \Delta_t\right) = \hat{\mu}\sum_{t=0}^{N-1}\sin\left(2\pi f t \Delta_t\right) + \hat{B}\sum_{t=0}^{N-1}\sin^2\left(2\pi f t \Delta_t\right)$$
$$+ \hat{A}\sum_{t=0}^{N-1}\cos\left(2\pi f t \Delta_t\right)\sin\left(2\pi f t \Delta_t\right). \tag{514c}$$

We can now use Exercises [1.2c] and [1.3] to reduce these three equations somewhat. For example, Equation (514b) becomes

$$\sum_{t=0}^{N-1} X_t \cos\left(2\pi f t \Delta_t\right) = \hat{\mu}\frac{\cos\left([N-1]\pi f \Delta_t\right)\sin\left(N\pi f \Delta_t\right)}{\sin\left(\pi f \Delta_t\right)}$$
$$+ \hat{A}\left[\frac{N}{2} + \frac{\sin\left(N2\pi f \Delta_t\right)}{2\sin\left(2\pi f \Delta_t\right)}\cos\left([N-1]2\pi f \Delta_t\right)\right]$$
$$+ \hat{B}\left[\frac{\sin\left(N2\pi f \Delta_t\right)}{2\sin\left(2\pi f \Delta_t\right)}\sin\left([N-1]2\pi f \Delta_t\right)\right],$$

with similar expressions for the other two normal equations. Using these expressions, it can be argued (see Exercise [10.1]) that

$$\left|\hat{\mu} - \overline{X}\right|, \quad \left|\hat{A} - \frac{2}{N}\sum_{t=0}^{N-1} X_t \cos\left(2\pi f t \Delta_t\right)\right| \quad \text{and} \quad \left|\hat{B} - \frac{2}{N}\sum_{t=0}^{N-1} X_t \sin\left(2\pi f t \Delta_t\right)\right|$$

10.2 Harmonic Processes with Additive White Noise – Discrete Spectra

are all close to zero (at least for large N), where, as usual, $\overline{X} \stackrel{\text{def}}{=} \sum_{t=0}^{N-1} X_t/N$, the sample mean. Thus we have, to a good approximation,

$$\hat{\mu} \approx \overline{X}, \quad \hat{A} \approx \frac{2}{N} \sum_{t=0}^{N-1} X_t \cos\left(2\pi f t \, \Delta_\text{t}\right) \text{ and } \hat{B} \approx \frac{2}{N} \sum_{t=0}^{N-1} X_t \sin\left(2\pi f t \, \Delta_\text{t}\right). \quad (515\text{a})$$

If in fact $f = k/(N \, \Delta_\text{t})$, where k is an integer such that $1 \leq k < N/2$, it follows from Exercise [1.3] that the approximations in Equation (515a) are in fact equalities. As we have noted in Chapters 3 and 6, this special set of frequencies – along with 0 and, if N is even, $f_\mathcal{N}$ – is known as the *Fourier frequencies* or *standard frequencies*. Since any frequency f in the interval $(0, f_\mathcal{N}]$ is at most a distance of $1/(N \, \Delta_\text{t})$ away from a Fourier frequency not equal to 0 or $f_\mathcal{N}$, it is intuitively reasonable that the approximations in Equation (515a) are good.

Reverting back to the general case $L \geq 1$, we can rewrite the model of Equation (512b) as

$$X_t = \mu + \sum_{l=1}^{L} \left[A_l \cos\left(2\pi f_l t \, \Delta_\text{t}\right) + B_l \sin\left(2\pi f_l t \, \Delta_\text{t}\right)\right] + \epsilon_t, \quad (515\text{b})$$

where $A_l \stackrel{\text{def}}{=} D_l \cos(\phi_l)$ and $B_l \stackrel{\text{def}}{=} - D_l \sin(\phi_l)$. For the remainder of this section, let us assume that each f_l is a Fourier frequency not equal to either 0 or $f_\mathcal{N}$. The deductions already made are equally valid here, and so $\hat{\mu} = \overline{X}$,

$$\hat{A}_l = \frac{2}{N} \sum_{t=0}^{N-1} X_t \cos\left(2\pi f_l t \, \Delta_\text{t}\right) \text{ and } \hat{B}_l = \frac{2}{N} \sum_{t=0}^{N-1} X_t \sin\left(2\pi f_l t \, \Delta_\text{t}\right) \quad (515\text{c})$$

are the *exact* least squares estimators of μ, A_l and B_l, respectively. Let us consider some of the statistical properties of $\hat{A}_l$ and $\hat{B}_l$ *under the assumption that each ϕ_l is a constant* – this implies that A_l and B_l are constants. Equation (515b) is thus a standard multiple linear regression model, and the fact that the RVs ϵ_t are uncorrelated means that the RVs X_t are also uncorrelated. Note also that

$$E\{\hat{A}_l\} = \frac{2}{N} \sum_{t=0}^{N-1} E\{X_t\} \cos\left(2\pi f_l t \, \Delta_\text{t}\right)$$

$$= \frac{2}{N} \sum_{t=0}^{N-1} \sum_{k=1}^{L} \left[A_k \cos\left(2\pi f_k t \, \Delta_\text{t}\right) \cos\left(2\pi f_l t \, \Delta_\text{t}\right) \right.$$

$$\left. + B_k \sin\left(2\pi f_k t \, \Delta_\text{t}\right) \cos\left(2\pi f_l t \, \Delta_\text{t}\right)\right]$$

since $\sum_t \mu \cos\left(2\pi f_l t \, \Delta_\text{t}\right) = 0$ (see Exercise [1.2d]) and $E\{\epsilon_t\} = 0$. By interchanging the summations and using the orthogonality relationships

$$\sum_{t=0}^{N-1} \cos\left(2\pi f_k t \, \Delta_\text{t}\right) \cos\left(2\pi f_l t \, \Delta_\text{t}\right) = \sum_{t=0}^{N-1} \sin\left(2\pi f_k t \, \Delta_\text{t}\right) \cos\left(2\pi f_l t \, \Delta_\text{t}\right) = 0,$$

$$\sum_{t=0}^{N-1} \cos^2\left(2\pi f_l t \, \Delta_\text{t}\right) = N/2,$$

$k \neq l$ (see Exercise [1.3c]), we find that $E\{\hat{A}_l\} = A_l$ and, likewise, $E\{\hat{B}_l\} = B_l$. From the definitions of $\hat{A}_l$ and $\hat{B}_l$ and the fact that the RVs X_t are uncorrelated, we have

$$\text{var}\{\hat{A}_l\} = \frac{4}{N^2} \sum_{t=0}^{N-1} \text{var}\{X_t\} \cos^2(2\pi f_l t \Delta_t) = \frac{2\sigma_\epsilon^2}{N} = \text{var}\{\hat{B}_l\} \qquad (516a)$$

when each ϕ_l is fixed. It also follows from the orthogonality relationships that, for $k \neq l$,

$$\text{cov}\{\hat{A}_k, \hat{B}_l\} = \text{cov}\{\hat{A}_l, \hat{B}_l\} = \text{cov}\{\hat{A}_k, \hat{A}_l\} = \text{cov}\{\hat{B}_k, \hat{B}_l\} = 0.$$

We also note that $E\{\hat{\mu}\} = \mu$, $\text{var}\{\hat{\mu}\} = \sigma_\epsilon^2/N$ and $\text{cov}\{\hat{\mu}, \hat{A}_l\} = \text{cov}\{\hat{\mu}, \hat{B}_l\} = 0$ for all l.

We can estimate σ_ϵ^2 by the usual formula used in multiple linear regression (Weisberg, 2014), namely,

$$\hat{\sigma}_\epsilon^2 = \frac{1}{N - 2L - 1} \sum_{t=0}^{N-1} \left(X_t - \overline{X} - \sum_{l=1}^{L} \left[\hat{A}_l \cos(2\pi f_l t \Delta_t) + \hat{B}_l \sin(2\pi f_l t \Delta_t) \right] \right)^2$$

$$= \frac{1}{N - 2L - 1} \left\| \boldsymbol{X} - \boldsymbol{H}\hat{\boldsymbol{\beta}} \right\|^2, \qquad (516b)$$

where now

$$\boldsymbol{H}^T \stackrel{\text{def}}{=} \begin{bmatrix} 1 & 1 & 1 & \cdots & 1 \\ 1 & \cos(2\pi f_1 \Delta_t) & \cos(4\pi f_1 \Delta_t) & \cdots & \cos([N-1]2\pi f_1 \Delta_t) \\ 0 & \sin(2\pi f_1 \Delta_t) & \sin(4\pi f_1 \Delta_t) & \cdots & \sin([N-1]2\pi f_1 \Delta_t) \\ \vdots & \vdots & \vdots & \ddots & \vdots \\ 1 & \cos(2\pi f_L \Delta_t) & \cos(4\pi f_L \Delta_t) & \cdots & \cos([N-1]2\pi f_L \Delta_t) \\ 0 & \sin(2\pi f_L \Delta_t) & \sin(4\pi f_L \Delta_t) & \cdots & \sin([N-1]2\pi f_L \Delta_t) \end{bmatrix}$$

and $\hat{\boldsymbol{\beta}}^T \stackrel{\text{def}}{=} [\hat{\mu}, \hat{A}_1, \hat{B}_1, \ldots, \hat{A}_L, \hat{B}_L]$. The divisor $N - 2L - 1$ is due to the fact that we have estimated $2L + 1$ parameters from the data.

If the frequencies f_l are not all of the form $k/(N \Delta_t)$, the estimators $\hat{\mu}$, $\hat{A}_l$ and $\hat{B}_l$ can be regarded as approximate least squares estimates of μ, A_l and B_l. Priestley (1981, p. 394) argues that in this case

$$E\{\hat{A}_l\} = A_l + O\left(\frac{1}{N}\right) \quad \text{and} \quad E\{\hat{B}_l\} = B_l + O\left(\frac{1}{N}\right), \qquad (516c)$$

which means that, for example, there exists a constant c (independent of N) such that

$$\left| E\{\hat{A}_l\} - A_l \right| < \frac{c}{N}$$

for all N (note, however, that c need not be a small number!).

To summarize, in order to estimate A_l and B_l, we have regarded each ϕ_l as fixed and have derived some statistical properties of the corresponding estimators ($\hat{A}_l$ and $\hat{B}_l$) conditional on ϕ_l being fixed. With this conditioning, the only stochastic part of $\{X_t\}$ is $\{\epsilon_t\}$, and the process $\{X_t\}$ is uncorrelated, which makes it easy to deduce the statistical properties of $\hat{A}_l$ and $\hat{B}_l$. While this approach is reasonable for amplitude estimation, it is problematic for spectral estimation. Fixing each ϕ_l causes $\{X_t\}$ to be nonstationary since its mean value, namely,

$$E\{X_t\} = \mu + \sum_{l=1}^{L} A_l \cos(2\pi f_l t \Delta_t) + B_l \sin(2\pi f_l t \Delta_t),$$

varies with time. Hence $\{X_t\}$ is outside of the framework of the spectral representation theorem, which provides the basis for spectral estimation. Proper theoretical treatment of spectral estimation would require reintroducing the ϕ_l terms as RVs.

10.2 Harmonic Processes with Additive White Noise – Discrete Spectra

Comments and Extensions to Section 10.2

[1] Here we look at the complex-valued counterpart of Equation (512b), namely,

$$Z_t = \mu + \sum_{l=1}^{L} D_l e^{i(2\pi f_l t \Delta_t + \phi_l)} + \epsilon_t, \qquad (517a)$$

where, while the amplitude D_l is real-valued, now $\{Z_t\}$ and μ are both complex-valued, $|f_l| \in (0, f_{\mathcal{N}}]$ – hence $f_l = 0$ is prohibited – and $\{\epsilon_t\}$ is proper complex-valued white noise with zero mean and variance $\sigma_\epsilon^2 = E\{|\epsilon_t|^2\}$ (see Exercise [32] and the discussion preceding it). Although Equation (512b) is much more useful for physical applications, the complex-valued form has three advantages:

[1] it leads to more compact mathematical expressions;
[2] it brings out more clearly the connection with the periodogram; and
[3] it is the form most often used in the electrical engineering literature.

The equivalent of Equation (513) is

$$Z_t = \mu + C e^{i 2\pi f t \Delta_t} + \epsilon_t, \qquad (517b)$$

where $C \stackrel{\text{def}}{=} D \exp(i\phi)$ is the complex-valued amplitude of the complex exponential. Again, if we regard ϕ as a constant for a given realization, we can estimate the parameters μ and C by least squares:

$$\text{SS}(\mu, C) = \|\boldsymbol{Z} - \boldsymbol{H}\boldsymbol{\beta}\|^2 = \sum_{t=0}^{N-1} \left| Z_t - \mu - C e^{i 2\pi f t \Delta_t} \right|^2,$$

where $\boldsymbol{Z}^T \stackrel{\text{def}}{=} [Z_0, Z_1, \ldots, Z_{N-1}]$, $\boldsymbol{\beta}^T \stackrel{\text{def}}{=} [\mu, C]$ and

$$\boldsymbol{H}^T \stackrel{\text{def}}{=} \begin{bmatrix} 1 & 1 & 1 & \cdots & 1 \\ 1 & e^{i 2\pi f \Delta_t} & e^{i 4\pi f \Delta_t} & \cdots & e^{i(N-1) 2\pi f \Delta_t} \end{bmatrix}.$$

If we set

$$\frac{\text{dSS}}{\text{d}\boldsymbol{\beta}} = -\boldsymbol{H}^T \left(\boldsymbol{Z}^* - \boldsymbol{H}^* \boldsymbol{\beta}^* \right)$$

to zero and take the complex conjugate of the resulting expressions (the asterisk denotes this operation), we obtain the following normal equations:

$$\sum_{t=0}^{N-1} Z_t = N\hat{\mu} + \hat{C} \sum_{t=0}^{N-1} e^{i 2\pi f t \Delta_t} \quad \text{and} \quad \sum_{t=0}^{N-1} Z_t e^{-i 2\pi f t \Delta_t} = \hat{\mu} \sum_{t=0}^{N-1} e^{-i 2\pi f t \Delta_t} + N\hat{C}.$$

From Exercises [1.2c] and [1.2d], we know that

$$\sum_{t=0}^{N-1} e^{i 2\pi f_k t \Delta_t} = 0 \quad \text{for } f_k \stackrel{\text{def}}{=} \frac{k}{N \Delta_t}, \quad 1 \leq k < N.$$

If we restrict f to this set of frequencies, the following are exact least squares estimators:

$$\hat{\mu} = \frac{1}{N} \sum_{t=0}^{N-1} Z_t \quad \text{and} \quad \hat{C} = \frac{1}{N} \sum_{t=0}^{N-1} Z_t e^{-i 2\pi f t \Delta_t}. \qquad (517c)$$

Note that $\hat{C}$ would reduce to $\hat{\mu}$ if we were to allow f to be zero (recall the assumption $0 < |f_l| \leq f_{\mathcal{N}}$); i.e., C would be the amplitude associated with zero frequency and would be confounded with the mean value μ (the mean is sometimes called the DC – *direct current* – component in the electrical engineering literature).

If we introduce several frequencies, our complex-valued model now becomes

$$Z_t = \mu + \sum_{l=1}^{L} C_l e^{i2\pi f_l t \Delta_t} + \epsilon_t, \qquad (518a)$$

where we assume that each f_l is one of the Fourier frequencies not equal to 0 or $\pm f_{\mathcal{N}}$. Under the same assumptions and restrictions as for their real sinusoidal counterpart, the following results hold:

$$E\{\hat{C}_l\} = C_l, \quad \text{var}\{\hat{C}_l\} = \frac{\sigma_\epsilon^2}{N} \quad \text{and} \quad \text{cov}\{\hat{C}_k, \hat{C}_l\} = 0, \quad k \neq l,$$

where $\hat{C}_l$ is the least squares estimate of C_l (see Exercise [10.2]). In addition we have

$$E\{\hat{\mu}\} = \mu, \quad \text{var}\{\hat{\mu}\} = \frac{\sigma_\epsilon^2}{N} \quad \text{and} \quad \text{cov}\{\hat{\mu}, \hat{C}_l\} = 0.$$

In order to estimate σ_ϵ^2, we use the complex analog of the usual formula in linear regression:

$$\hat{\sigma}_\epsilon^2 = \frac{1}{N - L - 1} \left\| \mathbf{Z} - \mathbf{H}\hat{\boldsymbol{\beta}} \right\|^2, \qquad (518b)$$

where

$$\mathbf{H}^T \stackrel{\text{def}}{=} \begin{bmatrix} 1 & 1 & 1 & \cdots & 1 \\ 1 & e^{i2\pi f_1 \Delta_t} & e^{i4\pi f_1 \Delta_t} & \cdots & e^{i(N-1)2\pi f_1 \Delta_t} \\ \vdots & \vdots & \vdots & \ddots & \vdots \\ 1 & e^{i2\pi f_L \Delta_t} & e^{i4\pi f_L \Delta_t} & \cdots & e^{i(N-1)2\pi f_L \Delta_t} \end{bmatrix}$$

and $\hat{\boldsymbol{\beta}}^T \stackrel{\text{def}}{=} [\hat{\mu}, \hat{C}_1, \ldots, \hat{C}_L]$. The divisor $N - L - 1$ is due to the fact that we have estimated $L + 1$ complex-valued parameters from our complex-valued data. Under a Gaussian assumption, $\hat{\sigma}_\epsilon^2$ is an unbiased estimator of σ_ϵ^2 (Miller, 1973, theorem 7.3); however, this result depends upon $\{\epsilon_t\}$ being proper complex-valued white noise (to the best of our knowledge, the properties of the expectation of $\hat{\sigma}_\epsilon^2$ in the improper case are an open question).

10.3 Spectral Representation of Discrete and Mixed Spectra

The Lebesgue decomposition theorem for integrated spectra (Section 4.4) says that a combination of spectral lines plus white noise (a "discrete" spectrum) or a combination of spectral lines plus colored noise (a "mixed" spectrum) has an integrated spectrum $S^{(I)}(\cdot)$ given by

$$S^{(I)}(f) = S_1^{(I)}(f) + S_2^{(I)}(f),$$

where $S_1^{(I)}(\cdot)$ is absolutely continuous and $S_2^{(I)}(\cdot)$ is a step function. Consider the real-valued process

$$X_t = \sum_{l=1}^{L} D_l \cos(2\pi f_l t \Delta_t + \phi_l) + \eta_t, \qquad (518c)$$

which is the same as Equation (512b) with two substitutions: (1) the process mean μ is set to 0, and (2) the process $\{\eta_t\}$ is not necessarily white noise, but it is independent of each ϕ_l. The spectral representation theorem in combination with a decomposition of the orthogonal process $\{Z(f)\}$ states that

$$X_t = \int_{-f_{\mathcal{N}}}^{f_{\mathcal{N}}} e^{i2\pi f t \Delta_t} \, dZ(f) = \int_{-f_{\mathcal{N}}}^{f_{\mathcal{N}}} e^{i2\pi f t \Delta_t} \, dZ_1(f) + \int_{-f_{\mathcal{N}}}^{f_{\mathcal{N}}} e^{i2\pi f t \Delta_t} \, dZ_2(f),$$

10.3 Spectral Representation of Discrete and Mixed Spectra

where $\{Z_1(f)\}$ and $\{Z_2(f)\}$ are each orthogonal processes;

$$E\{|dZ_1(f)|^2\} = S_\eta(f)\,df \quad \text{and} \quad E\{|dZ_2(f)|^2\} = dS_2^{(I)}(f); \qquad (519a)$$

$S_\eta(\cdot)$ is the SDF of $\{\eta_t\}$; and $E\{dZ_1^*(f)\,dZ_2(f')\} = 0$ for all f and f' (Priestley, 1981, pp. 252–3). Recall from Section 4.1 that, by putting

$$C_l = D_l e^{i\phi_l}/2 \quad \text{and} \quad C_{-l} = C_l^*, \quad l = 1, \ldots, L,$$

we can write

$$X_t = \sum_{l=-L}^{L} C_l e^{i 2\pi f_l t\,\Delta_t} + \eta_t, \qquad (519b)$$

where $C_0 \stackrel{\text{def}}{=} 0$, $f_0 \stackrel{\text{def}}{=} 0$ and $f_{-l} = -f_l$. Furthermore, $E\{X_t\} = 0$ and

$$\operatorname{var}\{X_t\} = \sum_{l=-L}^{L} E\{|C_l|^2\} + \operatorname{var}\{\eta_t\} = \sum_{l=-L}^{L} D_l^2/4 + \sigma_\eta^2. \qquad (519c)$$

Corresponding to this result we have

$$E\{|dZ_2(f)|^2\} = \begin{cases} E\{|C_l|^2\} = D_l^2/4, & f = \pm f_l; \\ 0, & \text{otherwise.} \end{cases} \qquad (519d)$$

It follows from the argument leading to Equation (37b) that

$$\operatorname{cov}\{X_{t+\tau}, X_t\} = \sum_{l=1}^{L} D_l^2 \cos(2\pi f_l \tau\,\Delta_t)/2 + \operatorname{cov}\{\eta_{t+\tau}, \eta_t\}. \qquad (519e)$$

Comments and Extensions to Section 10.3

[1] For complex exponentials, we have

$$Z_t = \sum_{l=1}^{L} D_l e^{i(2\pi f_l t\,\Delta_t + \phi_l)} + \eta_t \qquad (519f)$$

(this complex-valued process $\{Z_t\}$ should not be confused with the complex-valued process $\{Z(f)\}$ with orthogonal increments – the notation is unfortunately similar). This is the same as Equation (517a) except that (1) the process mean $\mu = 0$ and (2) the proper complex-valued process $\{\eta_t\}$ need not be a white noise process. If we set $C_l = D_l \exp(i\phi_l)$ so that Equation (519f) becomes

$$Z_t = \sum_{l=1}^{L} C_l e^{i 2\pi f_l t\,\Delta_t} + \eta_t, \qquad (519g)$$

we see that Equation (519g) is a one-sided version of Equation (519b); i.e., l is only positive in Equation (519g), but note that f_l can be either positive or negative. Thus $E\{Z_t\} = 0$ and

$$\operatorname{var}\{Z_t\} = \sum_{l=1}^{L} E\{|C_l|^2\} + \operatorname{var}\{\eta_t\} = \sum_{l=1}^{L} D_l^2 + \sigma_\eta^2. \qquad (519h)$$

For this complex-valued model we also have

$$E\{|dZ_2(f)|^2\} = \begin{cases} E\{|C_l|^2\} = D_l^2, & f = f_l; \\ 0, & \text{otherwise,} \end{cases} \qquad (519i)$$

and

$$\operatorname{cov}\{Z_{t+\tau}, Z_t\} = \sum_{l=1}^{L} D_l^2 e^{i 2\pi f_l \tau\,\Delta_t} + \operatorname{cov}\{\eta_{t+\tau}, \eta_t\}. \qquad (519j)$$

Proof of this final result is the subject of Exercise [10.3].

10.4 An Example from Tidal Analysis

To illustrate some of the ideas introduced so far in this chapter, we will look at the observed heights of sea levels as recorded by tide gauges. It should be noted that sea level as defined here excludes the effect of individual waves – tidal gauges do *not* measure the instantaneous height of the sea at any particular time but rather the average level about which the waves are oscillating. We can thus write

$$Y_t = X_t + \eta_t,$$

where Y_t is the observed height of the sea level at time t; X_t is the predicted height as given by Equation (512a) – with L' increased to L – under the assumption that $\alpha_l \stackrel{\text{def}}{=} \alpha_{t,l}$, $v_l \stackrel{\text{def}}{=} v_{t,l}$ and $u_l \stackrel{\text{def}}{=} u_{t,l}$ are varying slowly enough that they can be assumed to be constant over the span of data under analysis; and η_t is the unexplained component. Thus (Murray, 1964, 1965)

$$\begin{aligned} Y_t &= \mu + \sum_{l=1}^{L} \alpha_l H_l \cos\left(2\pi f_l t \Delta_t + v_l + u_l - g_l\right) + \eta_t \\ &= \mu + \sum_{l=1}^{L} \alpha_l H_l \left[\cos(g_l) \cos\left(2\pi f_l t \Delta_t + v_l + u_l\right) \right. \\ &\qquad \left. + \sin(g_l) \sin\left(2\pi f_l t \Delta_t + v_l + u_l\right)\right] + \eta_t \\ &= \mu + \sum_{l=1}^{L} A_l \left[\alpha_l \cos\left(2\pi f_l t \Delta_t + v_l + u_l\right)\right] \\ &\qquad + B_l \left[\alpha_l \sin\left(2\pi f_l t \Delta_t + v_l + u_l\right)\right] + \eta_t, \end{aligned}$$

where $A_l \stackrel{\text{def}}{=} H_l \cos(g_l)$ and $B_l \stackrel{\text{def}}{=} H_l \sin(g_l)$. This is a slightly more complicated version of Equation (515b), and the process $\{\eta_t\}$ need not be white noise. The unknowns H_l and g_l can be found from A_l and B_l. The frequencies f_l are not in general related to the Fourier frequencies. The vertical lines in the top plot of Figure 521 indicate the locations of the $L = 102$ frequencies used in the model – the frequency $M_1 \doteq 0.0402557$ cycles/hour and its multiples $M_k = k \cdot M_1$ for $k = 2, 3, 4, 5, 6$ and 8 are indicated by thicker and longer lines than the other 95 frequencies.

It is convenient to write

$$Y_t = \sum_{l=0}^{2L} \beta_l \theta_{l,t} + \eta_t,$$

where, by definition, $\beta_0 = \mu$; $\beta_{2l-1} = H_l \cos(g_l)$; $\beta_{2l} = H_l \sin(g_l)$; $\theta_{0,t} = 1$;

$$\theta_{2l-1,t} = \alpha_l \cos\left(2\pi f_l t \Delta_t + v_l + u_l\right) \quad \text{and} \quad \theta_{2l,t} = \alpha_l \sin\left(2\pi f_l t \Delta_t + v_l + u_l\right).$$

Least squares then gives

$$\sum_t Y_t \theta_{m,t} = \sum_{l=0}^{2L} \beta_l \sum_t \theta_{l,t} \theta_{m,t}$$

for each of the $2L + 1$ parameters. These equations can be written in matrix form as

$$\Theta \beta = c,$$

10.4 An Example from Tidal Analysis

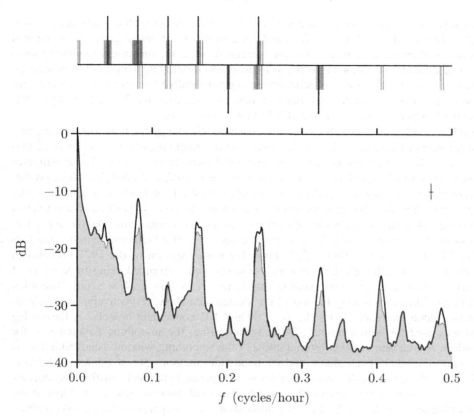

Figure 521 Tidal frequencies (upper plot) and estimated power spectra for Portsmouth residuals for 1971 (lower plot). In the upper plot, each frequency in the $L = 102$ model is indicated by a vertical line originating up or down from the thin horizontal line – there are thus 102 vertical lines in all, but their widths are such that most of them are not distinctly visible. The vertical lines going up indicate those frequencies that are in both the $L = 102$ and $L = 60$ models; the lines going down correspond to frequencies in the $L = 102$ model only. There are also seven lines that are thicker and twice as long as the others – five going up and two going down. These represent, from left to right, the tidal frequencies M_1, M_2, M_3, M_4, M_5, M_6 and M_8. In the lower plot, the thick curve is the spectrum for the 60 constituent model, while the thin curve at the top of the shaded area gives the spectrum for the 102 constituent model. A Parzen lag window with $m = 400$ was used. The height and width of the crisscross in the upper right-hand of this plot indicate, respectively, the width of a 95% confidence interval and the bandwidth measure $B_{\mathcal{U}}$ of Equation (256a).

where the matrix $\boldsymbol{\Theta}$ has $\sum_t \theta_{j-1,t}\theta_{i-1,t}$ for its (i,j)th element and

$$\boldsymbol{c} \stackrel{\text{def}}{=} \begin{bmatrix} \mu \\ \sum_t Y_t \alpha_1 \cos\left(2\pi f_1 t \Delta_t + v_l + u_1\right) \\ \vdots \\ \sum_t Y_t \alpha_1 \sin\left(2\pi f_L t \Delta_t + v_L + u_L\right) \end{bmatrix}.$$

With β estimated by $\hat{\beta}$, H_l and g_l are found from

$$\hat{H}_l = \left(\hat{\beta}_{2l}^2 + \hat{\beta}_{2l-1}^2\right)^{1/2} \quad \text{and} \quad \hat{g}_l = \tan^{-1}\left(\hat{\beta}_{2l}/\hat{\beta}_{2l-1}\right), \quad l = 1,\ldots,L.$$

In standard linear regression, the noise process is required to be uncorrelated. In our current application, even before examining estimates of η_t (formed from $Y_t - \hat{Y}_t$, where $\hat{Y}_t$ is the least squares estimate), we would expect the η_t terms to be correlated. For example, seasonal groupings of meteorological effects and, on a shorter scale, surge-generated groupings of residuals will be present. In general, for correlated noise, the maximum likelihood solution to the parameter estimation corresponds to *generalized* rather than ordinary least squares. Remarkably, ordinary least squares estimates are asymptotically efficient for a polynomial regression model such as is under consideration here (see, e.g., section 7.7 of Priestley, 1981, and also the Comments and Extensions [C&Es] for this section).

Walden (1982) calculated the harmonic constants H_l and g_l for Portsmouth, England, using three years of carefully edited consecutive hourly height data (1967–9) – over 26,000 points in all. The harmonic constants were estimated using least squares. These estimates were used to "predict" tidal elevations for each of the years 1962 and 1964–74. These predictions were in turn subtracted from the corresponding observed sea levels to produce residual series of hourly heights. This process was carried out for the so-called orthodox constituents ($L = 60$ harmonic terms comprising the primary frequencies plus some shallow-water frequencies) and for orthodox plus shallow-water terms ($L = 102$ harmonic terms) – see Figure 521. The lower plot of Figure 521 shows the power spectra of the 1971 Portsmouth residuals (note that, although the harmonic constants were computed using three years of data, each residual sequence is one year in length; i.e., $N = 365 \times 24 = 8760$). The thick curve and the thin curve tracing the top of the shaded area show, respectively, spectral estimates based upon the residuals for the $L = 60$ and 102 constituent models. A Parzen lag window was used for both estimates. After some testing, the smoothing parameter m for this window was chosen as 400. From Table 279 the smoothing window bandwidth B_W is $1.85/(m \Delta_t) \doteq 0.0046$ cycles/hour, and the bandwidth measure $B_{\mathcal{U}}$ of Equation (256a) is 0.0047. While these bandwidths are not sufficient to separate individual constituents, increasing m – i.e., decreasing the bandwidths – led to no visual improvement in the form of the spectral estimate. With $L = 102$, there is a diminution – as compared to the $L = 60$ analysis – of the contribution in the frequency bands close to integer multiples of 0.0805 cycles/hour. Since this frequency is equivalent to 1 cycle/12.42 hours (a semi-diurnal term), multiples of 2, 3 and so forth correspond to quarter-diurnal, sixth-diurnal and so forth terms. These are approximately the positions of the majority of the additional frequencies in the $L = 102$ model – see Figure 521. Note that, even for the $L = 102$ model, there is still considerable low-frequency power in the residual process so that it cannot be reasonably approximated as white noise.

How can we explain the residual power at low frequencies and in the tidal bands (i.e., the frequency ranges encompassing semi-, quarter-, eighth-diurnal, etc., tides)? The prediction model of Equation (512a) does not include nongravitational effects. For example, pressure fluctuations due to weather systems – which can induce a sea level response – have a continuous noise spectrum. Storm surges will show primarily in the band 0.03 to 0.08 cycles/hour. Land and sea breezes will cause discernible sea level changes and have associated spectral *lines* at 1 and 2 cycles/day as well as a continuous noise component (Munk and Cartwright, 1966). The overall spectrum is of course a mixed spectrum. At higher frequencies, residual power can arise through the omission of shallow-water constituents or through interaction between the tide and local mean sea level. Aliasing could also be involved. Suppose there is nonnegligible power in the *continuous time* residual series at harmonics of the main semidiurnal M_2 constituent at frequencies corresponding to M_{14} and M_{16}, i.e., approximately 0.5636 and 0.6441 cycles/hour. Sampling every hour gives a Nyquist frequency of $f_{\mathcal{N}} = 1/2$ cycle/hour. Since power from frequency $2f_{\mathcal{N}} - f$ would be aliased with f under this sampling scheme (see Equation (122b), recalling that spectra are symmetric about zero frequency), we note that 0.5636 cycles/hour aliases to 0.4364 cycles/hour, while 0.6441 aliases to 0.3559. Thus the residual power at M_{14} and M_{16} in the continuous series would appear at 0.436 and

0.356 cycles/hour in the sample spectrum. For Portsmouth there is a peak close to both of these frequencies (for other data – such as Southampton for 1955 – there is a noticeable peak about the first frequency only).

Comments and Extensions to Section 10.4

[1] If the SDF of the background continuum $\{\eta_t\}$ is colored, there is no efficiency to be gained *asymptotically* by including information on the covariance structure of the continuum. Hannan (1973), in examining a single sinusoid with frequency f_1, explained that the regression component is by nature a "signal" sent at a single frequency so that only the SDF of the background continuum at that frequency matters. Ultimately, as the sample size increases, we can gain nothing from knowledge of the covariance structure of the background continuum. He also pointed out that, for finite sample sizes and a continuum with a very irregular SDF near f_1, the influence of the covariance structure of the continuum might be appreciable even though its influence vanishes as the sample size increases.

10.5 A Special Case of Unknown Frequencies

Up to this point we have assumed that the frequencies $f_l \in (0, f_\mathcal{N}]$ in Equation (512b) are known in advance (as in the example of Section 10.4). If they are not known, then the models are no longer multiple linear regressions. The usual procedure is to somehow estimate the f_l terms and then to use these estimates in finding the other unknowns.

The classic procedure for estimating f_l – which dates from the 1890s – is based upon examining the periodogram, which we have already met in Chapter 6. The basic idea behind periodogram analysis in the current context is as follows. From Equation (512b), note that D_l^2 represents the squared amplitude of the sinusoid with frequency f_l. We also have $D_l^2 = A_l^2 + B_l^2$ from reexpressing Equation (512b) as Equation (515b). If $f_l \neq f_\mathcal{N}$, we can estimate D_l^2 by using the estimators for A_l and B_l in Equation (515c) to obtain

$$\hat{D}_l^2 = \left[\frac{2}{N}\sum_{t=0}^{N-1} X_t \cos(2\pi f_l t\, \Delta_t)\right]^2 + \left[\frac{2}{N}\sum_{t=0}^{N-1} X_t \sin(2\pi f_l t\, \Delta_t)\right]^2$$

$$= \frac{4}{N^2}\left|\sum_{t=0}^{N-1} X_t e^{-i2\pi f_l t\, \Delta_t}\right|^2 \tag{523a}$$

(a practical note: if the process mean μ is unknown, we typically need to replace X_t by $X_t - \overline{X}$ – see the discussion surrounding Equation (184)). It follows from Equation (519d) that the contribution to the power spectrum at $f = \pm f_l$ is $D_l^2/4$, which is estimated by $\hat{D}_l^2/4$. Recalling Equation (170d), the periodogram at f_l is

$$\hat{S}^{(P)}(f_l) = \frac{\Delta_t}{N}\left|\sum_{t=0}^{N-1} X_t e^{-i2\pi f_l t\, \Delta_t}\right|^2 = \frac{N\,\Delta_t}{4}\hat{D}_l^2, \tag{523b}$$

so the scheme is to plot $\hat{S}^{(P)}(f)$ versus f and look for "large" values of $\hat{S}^{(P)}(f)$. Note that, for any $f \in (0, f_\mathcal{N})$ not equal to an f_l in Equation (512b), we can imagine adding an $(L+1)$st term to the model with $f_{L+1} = f$ and $D_{L+1} = 0$. The estimator $\hat{D}_{L+1}^2$ should thus be an estimator of zero and consequently should be small.

Now let us assume that each f_l in Equation (512b) is a positive Fourier frequency not equal to $f_\mathcal{N}$ (hence each f_l can be written in the form $k_l/(N\,\Delta_t)$, where k_l is an integer – unique for each different f_l – such that $0 < k_l < N/2$).

524   Harmonic Analysis

▷ **Exercise [524]** Show that

$$E\{\hat{S}^{(P)}(f_l)\} = \frac{N\Delta_t}{4}D_l^2 + \sigma_\epsilon^2 \Delta_t. \qquad (524a)$$ ◁

The first term on the right-hand side above is $N\Delta_t$ times the contribution to the power spectrum from the line at $f = \pm f_l$, while the second is the value of the white noise contribution to the spectrum. If $f_l \in (0, f_\mathcal{N})$ is one of the Fourier frequencies *not* in Equation (512b), then

$$E\{\hat{S}^{(P)}(f_l)\} = \sigma_\epsilon^2 \Delta_t$$

by the artifact of adding a fictitious term to the model with zero amplitude (we cannot use this same trick to handle the case $f_l = f_\mathcal{N}$ – the above equation still holds then, but a special argument is needed and is left as an exercise for the reader).

Thus, in the case where each f_l is of a special nature (i.e., a Fourier frequency), plotting $\hat{S}^{(P)}(f_l)$ versus f_l on the grid of standard frequencies and searching this plot for large values is a reasonable way of finding which frequencies f_l belong in the model (Section 6.6 discusses the sampling properties for $\tilde{S}^{(P)}(f_l)$).

Comments and Extensions to Section 10.5

[1] For the complex exponential model of Equation (517a), the equivalent of Equation (524a) is

$$E\{\hat{S}^{(P)}(f_l)\} = \left(ND_l^2 + \sigma_\epsilon^2\right)\Delta_t,$$

where now $f_l \in (-f_\mathcal{N}, 0) \cup (0, f_\mathcal{N})$.

10.6 General Case of Unknown Frequencies

Assuming the general case of Equation (512b) for which f_l is not necessarily one of the Fourier frequencies, we have the following result, valid for all f and with μ known to be zero:

$$E\{\hat{S}^{(P)}(f)\} = \sigma_\epsilon^2 \Delta_t + \sum_{l=1}^{L} \frac{D_l^2}{4}\left[\mathcal{F}(f + f_l) + \mathcal{F}(f - f_l)\right], \qquad (524b)$$

where $\mathcal{F}(\cdot)$ is Fejér's kernel – see Equation (174c) for its definition and also Figure 176. The proof of Equation (524b) is Exercise [10.4], which can be most easily seen by expressing the increments of $S_2^{(I)}(\cdot)$ in Section 10.3 in terms of the Dirac delta function (see Equation (519a) along with Equation (519d)):

$$dS_2^{(I)}(f) = \sum_{l=1}^{L} \frac{D_l^2}{4}\left[\delta(f - f_l) + \delta(f + f_l)\right] df.$$

What does Equation (524b) tell us about the usefulness of the periodogram for identifying f_l in Equation (512b)? Consider the case where there is only one frequency, f_1 (i.e., $L = 1$ in Equation (512b)). Equation (524b) reduces to a constant term, $\sigma_\epsilon^2 \Delta_t$, plus the sum of $D_1^2/4$ times two Fejér kernels centered at frequencies $\pm f_1$. If $f_1 = k_1/(N\Delta_t)$ for some integer k_1 and if we evaluate $E\{\hat{S}^{(P)}(f)\}$ at just the Fourier frequencies, Equation (524b) is in agreement with our earlier analysis: a plot of $E\{\hat{S}^{(P)}(f)\}$ versus those frequencies has a single large value at f_1 against a "background" of level $\sigma_\epsilon^2 \Delta_t$. If, however, f_1 is not of the form $k_1/(N\Delta_t)$, the plot is more complicated since we are now sampling Fejér's kernel at points other than the "null points." For example, if f_1 falls exactly halfway between two of the

Fourier frequencies and if we only plot $E\{\hat{S}^{(\mathrm{P})}(\cdot)\}$ at the Fourier frequencies, we would find the two largest values at $f_1 \pm 1/(2N\Delta_t)$; moreover, these values would have approximately the same amplitude due to the symmetry of the Fejér kernel (the qualifier "approximately" is needed due to the contribution from the second Fejér kernel centered at $-f_1$). For large enough N and for f_1 not too close to zero or Nyquist frequency, the size of these two values is about 40% of the value of $E\{\hat{S}^{(\mathrm{P})}(f_1)\}$ (see Exercise [10.5]). Thus, we need to plot $\hat{S}^{(\mathrm{P})}(\cdot)$ at more than just the $\lfloor N/2 \rfloor + 1$ Fourier frequencies. In practice a grid twice as fine as that of the Fourier frequencies suffices – at least initially – to aid us in searching for periodic components. An additional rationale for this advice is that $\hat{S}^{(\mathrm{P})}(\cdot)$ is a trigonometric polynomial of degree $N-1$ and hence is uniquely determined by its values at N points. (There is, however, a potential penalty for sampling twice as finely as the Fourier frequencies – the loss of independence between the periodogram ordinates. This presents no problem when we are merely examining plots of the periodogram ordinates to assess appropriate terms for Equation (512b), but it is a problem in constructing valid statistical tests based upon periodogram ordinates.)

Let us consider the relationship between the periodogram and least squares estimation for the single-frequency model

$$X_t = D\cos\left(2\pi f_1 t\Delta_t + \phi\right) + \epsilon_t = A\cos\left(2\pi f_1 t\Delta_t\right) + B\sin\left(2\pi f_1 t\Delta_t\right) + \epsilon_t. \quad (525a)$$

This is the model described by Equations (512b) and (513) with μ again known to be zero. We can find the exact least squares estimators of A and B, say $\hat{A}$ and $\hat{B}$, from the normal equations

$$\sum_{t=0}^{N-1} X_t \cos\left(2\pi f_1 t\Delta_t\right) = \hat{A}\sum_{t=0}^{N-1}\cos^2(2\pi f_1 t\Delta_t) + \hat{B}\sum_{t=0}^{N-1}\cos\left(2\pi f_1 t\Delta_t\right)\sin\left(2\pi f_1 t\Delta_t\right)$$

$$\sum_{t=0}^{N-1} X_t \sin\left(2\pi f_1 t\Delta_t\right) = \hat{A}\sum_{t=0}^{N-1}\cos\left(2\pi f_1 t\Delta_t\right)\sin\left(2\pi f_1 t\Delta_t\right) + \hat{B}\sum_{t=0}^{N-1}\sin^2(2\pi f_1 t\Delta_t)$$

(cf. Equations (514b) and (514c)). The explicit expressions for $\hat{A}$ and $\hat{B}$ are somewhat unwieldy but simplify considerably if we restrict f_1 to be of the form $f'_k = k/(2N\Delta_t)$, i.e., a frequency on the grid of frequencies twice as fine as the Fourier frequencies. If we rule out the zero and Nyquist frequencies so that $0 < f'_k < f_\mathcal{N}$, we have

$$\sum_{t=0}^{N-1}\cos^2(2\pi f'_k t\Delta_t) = \sum_{t=0}^{N-1}\sin^2(2\pi f'_k t\Delta_t) = \frac{N}{2}$$

and

$$\sum_{t=0}^{N-1}\cos\left(2\pi f'_k t\Delta_t\right)\sin\left(2\pi f'_k t\Delta_t\right) = 0$$

(see Exercise [1.3a]). The simplified normal equations lead to

$$\hat{A} = \frac{2}{N}\sum_{t=0}^{N-1} X_t \cos\left(2\pi f'_k t\Delta_t\right) \quad \text{and} \quad \hat{B} = \frac{2}{N}\sum_{t=0}^{N-1} X_t \sin\left(2\pi f'_k t\Delta_t\right)$$

as the exact least squares estimators of A and B when $f_1 = f'_k$. The residual sum of squares is $\mathrm{SS}(f'_k)$, where

$$\mathrm{SS}(f) \stackrel{\mathrm{def}}{=} \sum_{t=0}^{N-1}\left(X_t - \left[\hat{A}\cos\left(2\pi f t\Delta_t\right) + \hat{B}\sin\left(2\pi f t\Delta_t\right)\right]\right)^2. \quad (525b)$$

If we substitute for $\hat{A}$ and $\hat{B}$ and simplify somewhat (part of the burden of Exercise [10.6a]), we obtain

$$\text{SS}(f'_k) = \sum_{t=0}^{N-1} X_t^2 - \frac{2}{\Delta_t} \hat{S}^{(\text{P})}(f'_k), \tag{526a}$$

thus establishing an interesting connection between the periodogram and the residual sum of squares from a single-frequency regression model. This connection, however, banks on the single frequency f_1 being of the special form $f'_k = k/(2N\Delta_t)$. Exercise [10.7] demonstrates that this connection breaks down when f_1 is not of this special form; however, as long as $f_1 \in (0, f_{\mathcal{N}})$ but is not too close to the zero or Nyquist frequencies, we can expect $\hat{S}^{(\text{P})}(f_1)$ to be closely approximated by

$$\widetilde{S}(f_1) \stackrel{\text{def}}{=} \frac{\Delta_t}{2} \left[\sum_{t=0}^{N-1} X_t^2 - \text{SS}(f_1) \right]. \tag{526b}$$

We assumed f_1 to be known, but in general it will not be. As N gets large, any f_1 satisfying $0 < f_1 < f_{\mathcal{N}}$ will become closer and closer to a frequency of the form $f'_k = k/(2N\Delta_t)$, and so Equation (526a) says that any large-sample results based upon *maximizing* the periodogram will be the same as those based on *minimizing* the residual sum of squares for a single-frequency regression model.

In the case where

[1] the error terms $\{\epsilon_t\}$ in the single-frequency model are an independent and identically distributed sequence of RVs with zero mean and finite variance and

[2] $\hat{f}_1$ represents the frequency of the maximum value of the periodogram,

Whittle (1952) and Walker (1971) showed that

$$E\{\hat{f}_1\} = f_1 + O\left(N^{-1}\right) \text{ and } \text{var}\{\hat{f}_1\} \approx \frac{3}{N^3 R \pi^2 \Delta_t^2}, \tag{526c}$$

where

$$R \stackrel{\text{def}}{=} \frac{A^2 + B^2}{2\sigma_\epsilon^2} = \frac{D^2}{2\sigma_\epsilon^2} \tag{526d}$$

is the signal-to-noise ratio; i.e., since

$$\text{var}\{X_t\} = \frac{A^2 + B^2}{2} + \sigma_\epsilon^2,$$

R is just the ratio of the two terms on the right-hand side, the first due to the sinusoid (the "signal"), and the second to the white noise process (the "noise"). Since the mean square error (MSE) is the variance plus the bias squared, the contribution of the variance to the MSE is $O\left(N^{-3}\right)$, whereas that of the bias is $O\left(N^{-2}\right)$. Hence the MSE of $\hat{f}_1$ is dominated by bias rather than variance.

The $O\left(N^{-3}\right)$ rate at which $\text{var}\{\hat{f}_1\}$ decreases is somewhat surprising (the rate $O\left(N^{-1}\right)$ is more common, as in Equation (164b) for example), but Bloomfield (2000, p. 32) gives the following argument to lend credence to it (see also Quinn, 2012, section 3). Suppose R is reasonably large so that, by inspecting a realization of $X_0, X_1, \ldots, X_{N-1}$, we can see that there are, say, between M and $M+1$ complete cycles of the sinusoid. Therefore we know that

$$\frac{M}{N\Delta_t} \leq f_1 \leq \frac{(M+1)}{N\Delta_t}. \tag{526e}$$

It follows that $|\tilde{f}_1 - f_1| \leq 1/(N\Delta_t)$ for any estimator $\tilde{f}_1$ satisfying the above. This implies that

$$\text{var}\{\tilde{f}_1\} \leq \frac{1}{N^2\Delta_t^2} = O\left(N^{-2}\right).$$

If we can estimate f_1 to within order $O\left(N^{-2}\right)$ with only the constraint that $\tilde{f}_1$ satisfies Equation (526e), it is plausible that the fancier periodogram estimator is $O\left(N^{-3}\right)$. Rice and Rosenblatt (1988) show that the product of the amplitude of the sinusoid and the sample size must be quite large for the asymptotic theory to be meaningful.

If there is more than just a single frequency in Equation (512b) (i.e., $L > 1$), the form of $E\{\hat{S}^{(\text{P})}(\cdot)\}$ in Equation (524b) can be quite complicated due to the superposition of $2L$ Fejér kernels centered at the points $\{\pm f_l\}$ together with the constant term $\sigma_\epsilon^2 \Delta_t$. In addition to interference due to the sidelobes, the widths of the central lobes of the superimposed kernels can contribute to the confusion. For example, if two of the f_l terms are close together and N is small enough, a plot of $E\{\hat{S}^{(\text{P})}(\cdot)\}$ might only indicate a single broad peak. Likewise, a frequency with a small D_l^2 might be effectively masked by a "sidelobe" from Fejér's kernel due to a nearby frequency with a large amplitude (see Section 10.8 and Figure 536).

In practical applications, we only observe $\hat{S}^{(\text{P})}(\cdot)$, which can differ substantially from $E\{\hat{S}^{(\text{P})}(\cdot)\}$ due to the presence of the white noise component. Our problem of identifying frequency components is much more difficult due to the distortions that arise from sampling variations. One effect of the white noise component is to introduce "spurious" peaks into the periodogram (see Figure 210b(a)). Recall that, in the null case when $L = 0$ and the white noise process is Gaussian, the $\hat{S}^{(\text{P})}(f)$ terms at the Fourier frequencies are independent χ^2 RVs with one or two degrees of freedom (see Section 6.6). Because they are independent RVs, in portions of the spectrum where the white noise component dominates, we can expect to see a rather choppy behavior with many "peaks" and "valleys." Furthermore, the χ_1^2 and χ_2^2 distributions have heavy tails (Figure 210a(a) shows the density function for the latter); in a random sample from such a distribution we can expect one or more observations that appear unusually large. These large observations could be mistaken for peaks due to a harmonic component. This fact points out the need for statistical tests to help us decide if a peak in the periodogram can be considered statistically significant (see Sections 10.9 and 10.10).

Comments and Extensions to Section 10.6

[1] Hannan (1973) looked at the $L = 1$ case of Equation (512b) and took the additive noise to be a strictly stationary process rather than just white noise. He found that

$$\lim_{N \to \infty} N\left(\hat{f}_1 - f_1\right) = 0 \text{ almost surely}$$

for $0 < f_1 < f_\mathcal{N}$. He also considered the case $f_1 = f_\mathcal{N}$ and found a stronger result: there is an integer N_0 – determined by the realization – such that $\hat{f}_1 = f_1$ for all $N \geq N_0$ with $\mathbf{P}[N_0 < \infty] = 1$ (he found this latter result also holds when f_1 is set to zero, but under the additional assumption that $\mu = 0$ in Equation (512b)).

[2] Suppose that $f'_k = k/(N'\Delta_t)$ maximizes the periodogram $\hat{S}^{(\text{P})}(\cdot)$ over a grid of equally spaced frequencies defined by $f'_j = j/(N'\Delta_t)$ for $1 \leq j < N'/2$ (the choice $N' = N$ yields the grid of Fourier frequencies, but $N' > N$ is also of interest if we pad $N' - N$ zeros to our time series for use with an FFT algorithm – see C&E [1] for Section 3.11). Suppose, however, that we are interested in obtaining the frequency maximizing $\hat{S}^{(\text{P})}(\cdot)$ over *all* frequencies. From Equation (170d), we can write

$$\hat{S}^{(\text{P})}(f) = \Delta_t \sum_{\tau=-(N-1)}^{N-1} \hat{s}_\tau^{(\text{P})} e^{-i2\pi f\tau\Delta_t} = \Delta_t \left[\hat{s}_0^{(\text{P})} + 2\sum_{\tau=1}^{N-1} \hat{s}_\tau^{(\text{P})} \cos\left(2\pi f\tau\Delta_t\right) \right],$$

where $\{\hat{s}_\tau^{(\mathrm{P})}\}$ is the biased estimator of the ACVS. If we let $\omega \stackrel{\text{def}}{=} 2\pi f \, \Delta_{\mathrm{t}}$ and

$$g(\omega) \stackrel{\text{def}}{=} \sum_{\tau=1}^{N-1} \hat{s}_\tau^{(\mathrm{P})} \cos(\omega\tau),$$

then a peak in $g(\cdot)$ at ω corresponds to a peak in $\hat{S}^{(\mathrm{P})}(\cdot)$ at $f = \omega/(2\pi \, \Delta_{\mathrm{t}})$. The first and second derivatives of $g(\cdot)$ are

$$g'(\omega) = -\sum_{\tau=1}^{N-1} \tau \hat{s}_\tau^{(\mathrm{P})} \sin(\omega\tau) \text{ and } g''(\omega) = -\sum_{\tau=1}^{N-1} \tau^2 \hat{s}_\tau^{(\mathrm{P})} \cos(\omega\tau). \tag{528a}$$

Note that a peak in $g(\cdot)$ corresponds to a root in $g'(\cdot)$. If $\omega^{(0)} \stackrel{\text{def}}{=} 2\pi f'_k \, \Delta_{\mathrm{t}}$ is not too far from the true location of the peak in $g(\cdot)$, we can apply the Newton–Raphson method to recursively compute

$$\omega^{(j)} = \omega^{(j-1)} - \frac{g'(\omega^{(j-1)})}{g''(\omega^{(j-1)})} \text{ for } j = 1, 2, \ldots, J,$$

stopping when $\left|\omega^{(J)} - \omega^{(J-1)}\right|$ is smaller than some specified tolerance. The value $\omega^{(J)}/(2\pi \, \Delta_{\mathrm{t}})$ is taken to be the location of the peak in $\hat{S}^{(\mathrm{P})}(\cdot)$ (Newton and Pagano, 1983; Newton, 1988, section 3.9.2).

In practice this scheme can fail if f'_k is not sufficiently close to the true peak location (Quinn et al., 2008; Quinn, 2012, section 9). Section 9.4 of Press et al. (2007) describes a useful method for finding the root of $g'(\cdot)$, call it $\omega^{(\mathrm{r})}$, using a Newton–Raphson scheme in combination with bisection. This method can be applied here if we can find two values of ω – call them $\omega^{(\mathrm{L})}$ and $\omega^{(\mathrm{U})}$ – that bracket the root; i.e., we have $\omega^{(\mathrm{L})} < \omega^{(\mathrm{r})} < \omega^{(\mathrm{U})}$ with $g'(\omega^{(\mathrm{L})}) > 0$ and $g'(\omega^{(\mathrm{U})}) < 0$. If N' is sufficiently large (usually $N' = 4N$ does the trick), these bracketing values can be taken to be $2\pi f'_{k-1} \, \Delta_{\mathrm{t}}$ and $2\pi f'_{k+1} \, \Delta_{\mathrm{t}}$. (There can also be problems if the equation for $g''(\omega)$ in Equation (528a) cannot be accurately computed, in which case *Brent's method* (Press et al., 2007, section 9.3) can be used to find the bracketed root of $g'(\cdot)$.)

[3] The relationship between the periodogram and the regression-based $\widetilde{S}(\cdot)$ provides motivation for the *Lomb–Scargle periodogram* (Lomb, 1976; Scargle, 1982; see also section 13.8 of Press et al., 2007, and Stoica et al., 2009). This periodogram handles an irregularly sampled time series $X(t_0)$, $X(t_1)$, $\ldots$, $X(t_{N-1})$, $t_0 < t_1 < \cdots < t_{N-1}$, in the context of a continuous parameter analog of the single-frequency model of Equation (525a):

$$X(t_n) = A\cos(2\pi f_1[t_n - c]) + B\sin(2\pi f_1[t_n - c]) + \epsilon_{t_n}, \tag{528b}$$

$n = 0, 1, \ldots, N-1$, where c is a constant (yet to be specified), and the errors $\{\epsilon_{t_n}\}$ are uncorrelated Gaussian RVs with zero mean and variance σ_ϵ^2. The least squares estimators $\hat{A}$ and $\hat{B}$ of A and B are the values minimizing the residual sum of squares

$$\begin{aligned}\mathrm{SS}_c(A, B) &\stackrel{\text{def}}{=} \|\boldsymbol{X} - \boldsymbol{H}_c\boldsymbol{\beta}\|^2 \\ &= \sum_{n=0}^{N-1} (X(t_n) - [A\cos(2\pi f_1[t_n - c]) + B\sin(2\pi f_1[t_n - c])])^2,\end{aligned} \tag{528c}$$

where $\boldsymbol{X}^T \stackrel{\text{def}}{=} [X(t_0), X(t_1), \ldots, X(t_{N-1})]$, $\boldsymbol{\beta}^T \stackrel{\text{def}}{=} [A, B]$ and

$$\boldsymbol{H}_c^T \stackrel{\text{def}}{=} \begin{bmatrix} \cos(2\pi f_1[t_0 - c]) & \cos(2\pi f_1[t_1 - c]) & \cdots & \cos(2\pi f_1[t_{N-1} - c]) \\ \sin(2\pi f_1[t_0 - c]) & \sin(2\pi f_1[t_1 - c]) & \cdots & \sin(2\pi f_1[t_{N-1} - c]) \end{bmatrix}$$

(cf. Equations (514a) and (525b)). The Lomb–Scargle periodogram is defined as

$$\hat{S}^{(\text{LS})}(f_1) = \frac{1}{2} \left[\sum_{n=0}^{N-1} X^2(t_n) - \text{SS}_c(\hat{A}, \hat{B}) \right] \quad (529\text{a})$$

(cf. Equation (526b)). The above opens up the possibility of a dependence on a shift c in the t_n's, which is undesirable since, to quote Scargle (1982), "... the power spectrum is meant to measure harmonic content of signals without regard to phase." In fact there is no such dependence: Exercise [10.8a] is to show that $\text{SS}_c(\hat{A}, \hat{B})$ is the same no matter what c is (however, in general $\hat{A}$ and $\hat{B}$ do depend on c). Setting c equal to

$$\tilde{c} = \frac{1}{4\pi f_1} \tan^{-1} \left[\frac{\sum_n \sin(4\pi f_1 t_n)}{\sum_n \cos(4\pi f_1 t_n)} \right] \quad (529\text{b})$$

allows us to express the Lomb–Scargle periodogram succinctly as

$$\hat{S}^{(\text{LS})}(f_1) = \frac{1}{2} \left(\frac{\left[\sum_n X(t_n)\cos(2\pi f_1[t_n - \tilde{c}])\right]^2}{\sum_n \cos^2(2\pi f_1[t_n - \tilde{c}])} + \frac{\left[\sum_n X(t_n)\sin(2\pi f_1[t_n - \tilde{c}])\right]^2}{\sum_n \sin^2(2\pi f_1[t_n - \tilde{c}])} \right). \quad (529\text{c})$$

Verification of the above is the burden of Exercise [10.8b], while Exercise [10.8c] shows that $\tilde{c}$ leads to the corresponding least squares estimators $\hat{A}$ and $\hat{B}$ being uncorrelated (this point is expanded upon in C&E [2] for Section 10.9). There is, however, no real need to use Equations (529c) and (529b) since the Lomb–Scargle periodogram can just as well be computed via Equation (529a) using standard linear regression techniques with c set to zero.

In the case of equally spaced data $t_n = t_0 + n\,\Delta_t$, the Lomb–Scargle periodogram – after multiplication by Δ_t – is equal to the usual periodogram if f_1 is equal to $k/(2N\,\Delta_t)$ for some k satisfying $0 < k < N$. It should also be noted that the Lomb–Scargle periodogram arises from the single-frequency model of Equation (528b) and does not in general offer a proper decomposition of the sample variance of a time series across frequencies as does the usual periodogram; i.e., while we can take the usual periodogram to be a primitive SDF estimator, we should not in general regard the Lomb–Scargle periodogram as such. Finally, we note that formulation of the Lomb–Scargle periodogram in terms of Equation (528b) presumes that the mean of the time series is zero (in practice, this assumption is handled by centering the time series with its sample mean and taking $X(t_n)$ to be the centered series). Alternatively we can add a term μ to the model of Equation (528b) to handle a nonzero mean (cf. Equations (525a) and (513)). We can then estimate μ along with A and B via least squares (Cumming et al., 1999; Zechmeister and Kürster, 2009).

[4] The relationship between the periodogram and regression analysis also provides motivation for the *Laplace periodogram* (Li, 2008; see also section 5.5 of Li, 2014). Assuming the single-frequency model of Equation (525a) with f_1 set to a frequency of the form $f'_k = k/(2N\,\Delta_t)$, $0 < k < N$, and letting $\boldsymbol{\beta}^T = [A, B]$ as before, we can express the least squares estimator of $\boldsymbol{\beta}$ as

$$\hat{\boldsymbol{\beta}}^{(\text{LS})} = \underset{A,B}{\operatorname{argmin}} \sum_{t=0}^{N-1} \left| X_t - A\cos(2\pi f'_k t\,\Delta_t) - B\sin(2\pi f'_k t\,\Delta_t) \right|^2. \quad (529\text{d})$$

The usual periodogram and $\hat{\boldsymbol{\beta}}^{(\text{LS})}$ are related by

$$\hat{S}^{(\text{P})}(f'_k) = \frac{N}{4\Delta_t} \|\hat{\boldsymbol{\beta}}^{(\text{LS})}\|^2 \quad (529\text{e})$$

(see Equation (588a), which falls out from Exercise [10.6a]). Replacing the square modulus in Equation (529d) by just the modulus leads to the least absolute deviations estimator:

$$\hat{\boldsymbol{\beta}}^{(\text{LAD})} = \underset{A,B}{\operatorname{argmin}} \sum_{t=0}^{N-1} \left| X_t - A\cos(2\pi f'_k t\,\Delta_t) - B\sin(2\pi f'_k t\,\Delta_t) \right|.$$

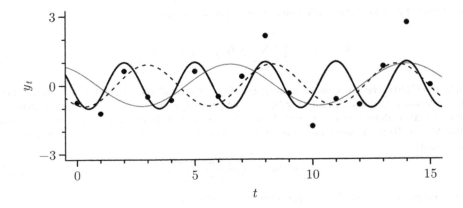

Figure 530 Sinusoids with periods 7.5 (thin solid curve), 5.3 (dashed) and 3 (thick solid) along with their summation at $t = 0, 1, \ldots, 15$ (dots).

In analogy to Equation (529e) the Laplace periodogram is defined by

$$\hat{S}^{(\mathrm{L})}(f'_k) = \frac{N}{4\Delta_t}\|\hat{\boldsymbol{\beta}}^{(\mathrm{LAD})}\|^2$$

(there is, however, no need to restrict this definition to frequencies of the form f'_k).

Li (2008) gives two compelling arguments for the Laplace periodogram. First, for regression models such as Equation (525a), least squares estimators of the parameters have well-known optimality properties when the errors are Gaussian, but these estimators can perform poorly when the errors come from a heavy-tailed distribution. This fact is the inspiration for robust regression analysis, whose goal is to find estimators that work well in non-Gaussian settings. Along with other robust estimators, least absolute deviations estimators perform better than least squares estimators for certain non-Gaussian distributions, suggesting use of the Laplace periodogram to ascertain the harmonic content of a heavy-tailed time series. Second, recall that the periodogram is a primitive estimator of SDFs that are Fourier transforms of ACVSs. The Laplace periodogram is an estimator of the Fourier transform of a sequence that depends upon zero-crossing rates as defined by $\mathbf{P}[X_{t+\tau}X_t < 0]$, $\tau \in \mathbb{Z}$. These rates provide an interpretable – and robust – characterization of stationary processes alternative to that provided by the ACVS.

10.7 An Artificial Example from Kay and Marple

Let us illustrate some of the points in the previous sections by looking at an artificial example due to Kay and Marple (1981, p. 1386). Figure 530 shows three different sinusoids and their weighted summation

$$y_t = 0.9\cos\left(2\pi[t+1]/7.5\right) + 0.9\cos\left(2\pi[t+1]/5.3 + \pi/2\right) + \cos\left(2\pi[t+1]/3\right) \quad (530)$$

at indices $t = 0, 1, \ldots, 15$ (the use of $t + 1$ in the above is merely to replicate the Kay and Marple example using our standard zero-based indices). We can regard this time series as a realization of the model of Equation (511) with $\mu = 0$, $L = 3$ and $\Delta_t = 1$. The frequencies of the three sinusoids are $1/7.5 = 0.133$ (thin solid curve), $1/5.3 = 0.189$ (dashed) and $1/3$ (thick solid). This series of length 16 has about 2+ cycles of the first sinusoid, 3 cycles of the second and 5+ cycles of the third.

In Figure 531 we study the effect of grid size on our ability to detect the presence of harmonic components when examining the periodogram $\hat{S}^{(\mathrm{P})}(\cdot)$ for $y_0, y_1, \ldots, y_{15}$. The dots in the top plot show the value of $10\log_{10}\hat{S}^{(\mathrm{P})}(\cdot)$ at the nine Fourier frequencies $f_k =$

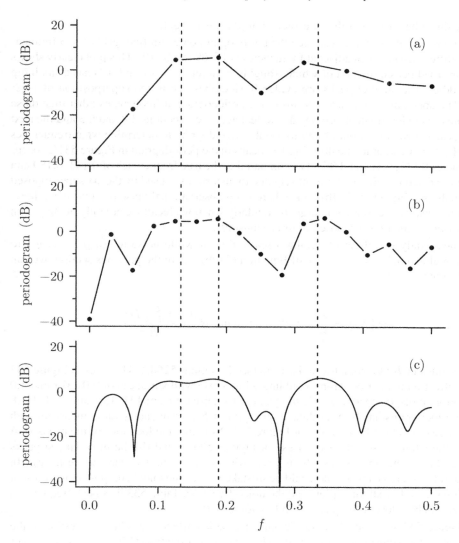

Figure 531 Effect of grid size on interpretation of the periodogram. For the time series of length 16 shown in Figure 530, the plots show the periodogram (on a decibel scale) versus frequency at the nine Fourier frequencies (dots in top plot, connected by lines), on a grid of frequencies twice as fine as the Fourier frequencies (17 dots in middle plot), and on a finely spaced grid of 513 frequencies (bottom).

$k/N = k/16$ for $k = 0, 1, \ldots, 8$. The three dashed vertical lines indicate the locations of the frequencies of the three sinusoids that form $\{y_t\}$. There are two broad peaks visible, the one with the lower frequency being larger in magnitude. Note that this plot fails to resolve the two lowest frequencies and that the apparent location of the highest frequency is shifted to the left. From this plot we might conclude that there are only two sinusoidal components in $\{y_t\}$.

Figure 531(b) shows $10 \log_{10} \hat{S}^{(P)}(\cdot)$ at the frequencies $k/32$ for $k = 0, 1, \ldots, 16$. There are still two major broad peaks visible (with maximum heights at points $k = 6$ and $k = 11$, respectively) and some minor bumps that are an order of magnitude smaller. Note now that the peak with the higher frequency is slightly larger than the one with the lower frequency. Moreover, there is a hint of some structure in the first broad bump (there is a slight dip between points $k = 4$ and 6). By going to a finer grid than that provided by the Fourier frequencies, we are getting a better idea of the sinusoidal components in $\{y_t\}$.

In Figure 531(c) we have evaluated the periodogram on a still finer grid of 513 frequencies. The difference between adjacent frequencies is $1/1024 \approx 0.001$. This plot clearly shows two major broad peaks, the first of which (roughly between $f = 0.1$ and 0.2) has a small dip in the middle. The smaller peaks are evidently sidelobes due to the superposition of Fejér kernels. The appearance of nonzero terms in the periodogram at frequencies other than those of sinusoids in the harmonic process is due to leakage – see Sections 3.8 and 6.3 and the next section. Leakage can affect our ability to use the periodogram to determine what frequencies make up $\{y_t\}$. For example, the three highest peaks in the periodogram in Figure 531(c) occur at frequencies 0.117, 0.181 and 0.338 as compared to the true frequencies of 0.133, 0.189 and 0.333. These discrepancies are here entirely due to distortions caused by the six superimposed Fejér kernels (see Equation (524b)) since there is *no* observational noise present. Note, however, that the discrepancies are not large, particularly when we recall we are only dealing with 16 points from a function composed of three sinusoids.

We next study the effect of adding a realization of white noise to $y_0, y_1, \ldots, y_{15}$ to produce a realization $x_0, x_1, \ldots, x_{15}$ of Equation (512b). As in the previous section, we can use the decomposition

$$\operatorname{var}\{X_t\} = \sum_{l=1}^{L} D_l^2/2 + \sigma_\epsilon^2 \text{ with } R \stackrel{\text{def}}{=} \frac{\sum_{l=1}^{L} D_l^2}{2\sigma_\epsilon^2},$$

where we consider R as a signal-to-noise ratio (see Equation (526d)). The plots in Figure 533 show the effect on the periodogram of adding white noise with variances of 0.01, 0.1 and 1.0 (top to bottom plots, respectively), yielding signal-to-noise ratios of 131, 13.1 and 1.31. Each of these periodograms has been evaluated on a grid of 513 frequencies for comparison with the noise-free case in the bottom plot of Figure 531. Note that the locations of the peaks shift around as the white noise variance varies. In Figures 533(a) and (b) the three largest peaks are attributable to the three sinusoids in $\{x_t\}$, whereas only the two largest are in (c). In both (b) and (c), there is a peak that could be mistaken for a sinusoidal component not present in Equation (530). For all three plots and the noise-free case, Table 533 lists peak frequencies that are reasonably close to frequencies in Equation (530).

In Figure 534 we study how the length of a series affects our ability to estimate the frequencies of the sinusoids that form it. Here we used Equation (530) to evaluate y_t for $t = 0, 1, \ldots, 127$. We added white noise with unit variance to each element of the extended series and calculated separate periodograms on a grid of 513 frequencies for the first 16, 32, 64 and 128 points in the contaminated series. These periodograms are shown in Figure 533(c) and the three plots of Figure 534. As the series length increases, it becomes easier to pick out the frequency components, and the center of the peaks in the periodogram get closer to the true frequencies of the sinusoids (again indicated by the thin vertical lines). Note also that the heights of the peaks increase directly in proportion to the number of points and that the widths of the peaks are comparable to the widths of the central lobes of the corresponding Fejér's kernels (these lobes are shown by the thick curves in the lower right-hand portion of each plot). Both of these indicate that we are dealing with a harmonic process as opposed to just a purely continuous stationary process with an SDF highly concentrated around certain frequencies.

10.7 An Artificial Example from Kay and Marple

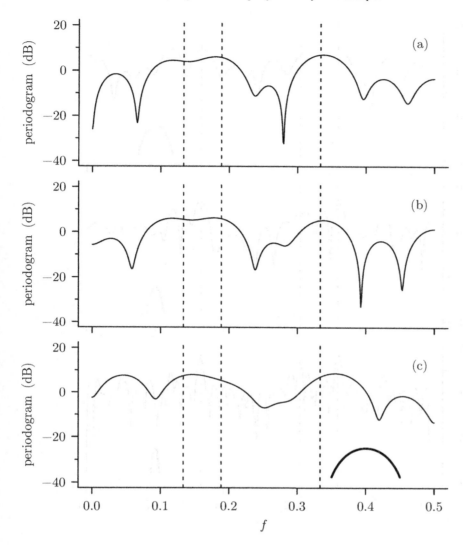

Figure 533 Effect of additive white noise on the periodogram. Here we add a realization of white noise to y_0, $y_1, \ldots, y_{15}$ of Equation (530) and compute the resulting periodogram $\hat{S}^{(P)}(\cdot)$ over a finely spaced grid of 513 frequencies. The variances of the additive white noise are 0.01, 0.1 and 1, respectively, in the top to bottom plots. The periodogram for the noise-free case is shown in the bottom plot of Figure 531. In plot (c), the thick curve centered at $f = 0.4$ depicts the central lobe of Fejér's kernel for sample size $N = 16$.

l	True f_l	531(c) $\sigma_\epsilon^2 = 0$	533(a) $\sigma_\epsilon^2 = 0.01$	533(b) $\sigma_\epsilon^2 = 0.1$	533(c) $\sigma_\epsilon^2 = 1$
1	0.133	0.117	0.117	0.117	0.147
2	0.189	0.181	0.181	0.178	—
3	0.333	0.338	0.337	0.338	0.355

Table 533 True and estimated f_l (the sample size N is fixed at 16).

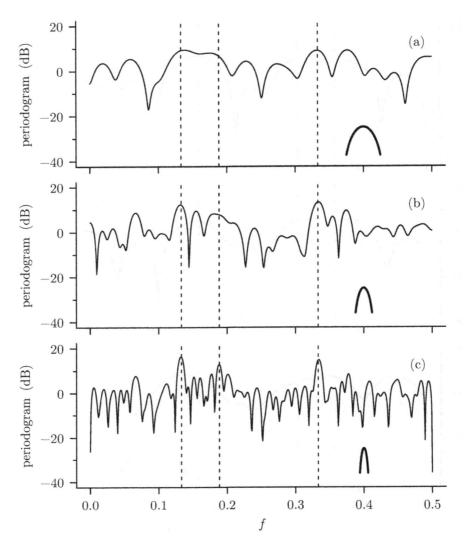

Figure 534 Effect of sample size on the periodogram. Here we add white noise with unit variance to $y_0, y_1, \ldots, y_{127}$ of Equation (530) and compute the resulting periodogram $\hat{S}^{(\text{P})}(\cdot)$ using just the first 32 and 64 values of the series (a and b), and finally all 128 values (c). The periodogram for just the first 16 values is shown in the bottom plot of Figure 533. In each plot, the thick curve centered at $f = 0.4$ depicts the central lobe of Fejér's kernel for the appropriate sample size.

Comments and Extensions to Section 10.7

[1] Since the series in our example only had 16 points, is it reasonable to expect the agreement between estimated and actual frequencies to be as close as Table 533 indicates, or is there something peculiar about our example? Let us see what the approximation of Equation (526c) gives us for the variance of the estimated frequency for the sinusoid with the highest frequency using the three noise-contaminated series. Strictly speaking, this approximation assumes that the model involves only a single frequency rather than three; however, this approximation can be taken as valid if the frequencies are spaced "sufficiently" far apart (Priestley, 1981, p. 410). If we thus ignore the other two frequencies, the signal-to-noise ratio for the $f_3 = 1/3$ component alone is $R = 1/(2\sigma_\epsilon^2)$ since $D_3^2 = 1$. This implies that

$$\text{var}\{\hat{f}_3\} \approx \frac{6\sigma_\epsilon^2}{N^3\pi^2} \doteq 0.00015 \times \sigma_\epsilon^2 \text{ for } N = 16.$$

For $\sigma_\epsilon^2 = 0.01$, 0.1 and 1, the standard deviations of $\hat{f}_3$ are 0.001, 0.004 and 0.012. From the last row of Table 533, we see that $|\hat{f}_3 - f_3|$ is 0.004, 0.005 and 0.022 for these three cases. Thus the observed deviations are of the order of magnitude expected by statistical theory.

10.8 Tapering and the Identification of Frequencies

We have already discussed the idea of data tapers and examined their properties in Section 6.4. When we taper a sample of length N of the harmonic process of Equation (512b) with a data taper $\{h_t\}$, it follows from an argument similar to that used in the derivation of Equation (524b) that

$$E\{\hat{S}^{(D)}(f)\} = \sigma_\epsilon^2 \Delta_t + \sum_{l=1}^{L} \frac{D_l^2}{4} \left[\mathcal{H}(f + f_l) + \mathcal{H}(f - f_l) \right], \tag{535a}$$

where $\mathcal{H}(\cdot)$ is the spectral window corresponding to the taper $\{h_t\}$ (see Equation (186e) and the discussion surrounding it). If the taper is the rectangular data taper, then $\hat{S}^{(D)}(\cdot) = \hat{S}^{(P)}(\cdot)$, $\mathcal{H}(\cdot) = \mathcal{F}(\cdot)$ (i.e., Fejér's kernel) and Equation (535a) reduces to Equation (524b).

It is easy to concoct an artificial example where use of a taper greatly enhances the identification of the frequencies in Equation (512b). Figure 536 shows three different direct spectral estimates $\hat{S}^{(D)}(\cdot)$ formed from one realization of length $N = 256$ from a special case of Equation (512b):

$$X_t = 0.0316 \cos(0.2943\pi t + \phi_1) + \cos(0.3333\pi t + \phi_2) + 0.0001 \cos(0.3971\pi t + \phi_3) + \epsilon_t, \tag{535b}$$

where $\{\epsilon_t\}$ is a white noise process with variance $\sigma_\epsilon^2 = 10^{-10} = -100\,\text{dB}$. For the three plots in the figure, the corresponding data tapers are, from top to bottom, a rectangular data taper (cf. Figures 190(a) and 191(a) for $N = 64$), a Slepian data taper with $NW = 2$ (Figures 190(f) and 191(f) for $N = 64$) and a Slepian data taper with $NW = 4$ (Figures 190(g) and 191(g) for $N = 64$). The thick curve in the upper left-hand corner of each plot depicts the central lobe of the associated spectral window $\mathcal{H}(\cdot)$. For the rectangular taper (plot a), the sidelobes of Fejér's kernel that is associated with the middle term in Equation (535b) completely masks the sinusoid with the smallest amplitude (the third term), and there is only a slight hint of the sinusoid with the lowest frequency (first term). Thus the periodogram only clearly reveals the presence of a single harmonic component. Use of a Slepian taper with $NW = 2$ (plot b) yields a spectral window with sidelobes reduced enough to clearly reveal the presence of a second sinusoid, but the third sinusoid is still hidden (this Slepian data taper is roughly comparable to the popular Hanning data taper). Finally, all three sinusoidal components are revealed when a Slepian taper with $NW = 4$ is used (plot c), as is the level of the background continuum (the horizontal dashed lines indicates the true level σ_ϵ^2).

Note, however, that, in comparison to the $NW = 2$ Slepian data taper, the broader central lobe of the $NW = 4$ Slepian taper somewhat degrades the resolution of the two dominant sinusoidal components. An example can be constructed in which the $NW = 4$ Slepian taper fails to resolve two closely spaced frequencies, while a taper with a narrower central lobe (say, a Slepian taper with $NW = 1$ or 2 or a 100% cosine taper) succeeds in doing so (see Exercise [10.9b]). The selection of a suitable taper is clearly dependent upon the true model – a fact that accounts for the large number of tapers suggested as being appropriate for different problems (Harris, 1978). Rife and Vincent (1970) present an interesting study on the estimation of the magnitude and frequencies of two "tones" when using different families of tapers with differing central lobe and sidelobe properties. Tseng et al. (1981) propose a family of tapers that is specified by four parameters and is effective in the detection of three-tone signals.

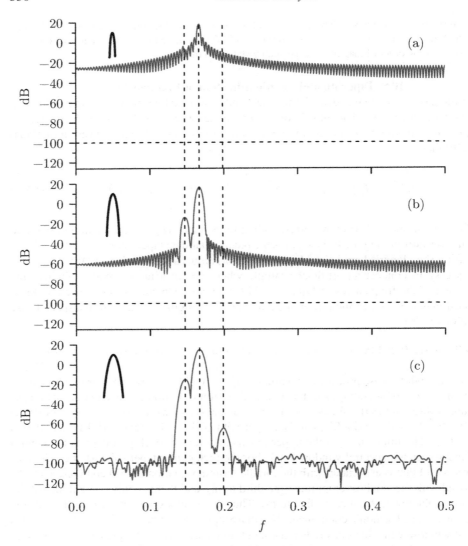

Figure 536 Effect of tapering on detection of harmonic components in process of Equation (535b). Plot (a) shows the periodogram, while the other two plots show direct spectral estimators based on (b) a Slepian taper with $NW = 2$ and (c) a Slepian taper with $NW = 4$. Each spectral estimate is plotted over a finely spaced grid of 513 frequencies. The vertical dashed lines mark the locations of the frequencies in Equation (535b), while the horizontal line shows the level σ_ϵ^2 of the background continuum. The central lobes of the associated spectral windows are shown in the upper left-hand corners.

Tapering is sometimes useful even when there is but a single sinusoidal term in a harmonic process. In this case, Equation (524b) tells us that

$$E\{\hat{S}^{(\mathrm{P})}(f)\} = \sigma_\epsilon^2 \Delta_{\mathrm{t}} + \frac{D_1^2}{4}\left[\mathcal{F}(f+f_1) + \mathcal{F}(f-f_1)\right]. \tag{536}$$

For f close to f_1, the second of the two Fejér kernels above is generally the dominant term, but the presence of the first kernel can distort $E\{\hat{S}^{(\mathrm{P})}(\cdot)\}$ such that its peak value is shifted away from f_1. The utility of tapering can also depend critically on what value the RV ϕ_1 assumes in the single-frequency model $X_t = D_1 \cos\left(2\pi f_1 t \Delta_{\mathrm{t}} + \phi_1\right) + \epsilon_t$. The burden of

10.8 Tapering and the Identification of Frequencies

Exercise [10.10a] is to show that the expected value of the periodogram conditional on ϕ_1 is

$$E\{\hat{S}^{(\text{P})}(f) \mid \phi_1\} = \sigma_\epsilon^2 \Delta_t + \frac{D_1^2}{4}[\mathcal{F}(f+f_1) + \mathcal{F}(f-f_1) + \mathcal{K}(f, f_1, \phi_1)], \quad (537a)$$

where

$$\mathcal{K}(f, f_1, \phi_1) \stackrel{\text{def}}{=} 2\Delta_t N \cos\left(2\pi[N-1]f_1\Delta_t + 2\phi_1\right)\mathcal{D}_N([f+f_1]\Delta_t)\mathcal{D}_N([f-f_1]\Delta_t),$$

and $\mathcal{D}_N(\cdot)$ is Dirichlet's kernel – see Equation (17c) for its definition and Equation (174c) for its connection to Fejér's kernel. For particular realizations of ϕ_1, the term involving Dirichlet's kernel can also push the peak value of the periodogram away from f_1. Distortions due to both kernels can potentially be lessened by tapering.

As a concrete example, consider the following single-frequency model with a fixed phase:

$$X_t = \cos\left(2\pi f_1 t + 5\pi/12\right) + \epsilon_t, \quad (537b)$$

where $f_1 = 0.0725$, and the background continuum $\{\epsilon_t\}$ is white noise with variance $\sigma_\epsilon^2 = 10^{-4}$. This process has a signal-to-noise ratio of 37 dB, which is so high that plots of individual realizations are barely distinguishable from plots of pure sinusoids. We consider series of sample size $N = 64$, for which the cosine term in Equation (537b) traces out $Nf_1 = 4.64$ oscillations over $t = 0, 1, \ldots, N-1$. The dark curve in Figure 538a shows the corresponding unconditional expectation $E\{\hat{S}^{(\text{P})}(\cdot)\}$ over an interval of frequencies centered about f_1. The peak of this curve – indicated by the dashed vertical line – is at 0.07251, which is quite close to f_1 (had the figure included a line depicting f_1, it would have been visually indistinguishable from the dashed line). The light curve shows the conditional expectation $E\{\hat{S}^{(\text{P})}(\cdot) \mid \phi_1 = 5\pi/12\}$. The solid vertical line indicates that its peak frequency occurs at 0.07215, which is distinctly below f_1. For this particular setting for ϕ_1, the distortion away from f_1 is primarily due to $\mathcal{K}(f, f_1, \phi_1)$ in Equation (537a) rather than $\mathcal{F}(f+f_1)$.

To demonstrate the effectiveness of tapering, we generated 50 realizations of length $N = 64$ from the process described by Equation (537b). For each realization, we computed the periodogram $\hat{S}^{(\text{P})}(\cdot)$ and a Hanning-based direct spectral estimate $\hat{S}^{(\text{D})}(\cdot)$. We then found the locations of the frequencies at which $\hat{S}^{(\text{P})}(\cdot)$ and $\hat{S}^{(\text{D})}(\cdot)$ attained their peak values (see C&E [2] for Section 10.6). A scatter plot of the locations of the peak frequencies for $\hat{S}^{(\text{D})}(\cdot)$ versus those for $\hat{S}^{(\text{P})}(\cdot)$ is shown in Figure 538b for the 50 realizations. The dashed horizontal and vertical lines mark the location of the true frequency $f_1 = 0.0725$ of the sinusoidal component, while the solid vertical line marks the peak frequency of $E\{\hat{S}^{(\text{P})}(\cdot) \mid \phi_1 = 5\pi/12\}$. We see that, whereas the peak frequencies derived from $\hat{S}^{(\text{D})}(\cdot)$ are clustered nicely about f_1, those from $\hat{S}^{(\text{P})}(\cdot)$ are biased low in frequency and are clustered around peak frequency for the conditional expectation of the periodogram. The mean square error in estimating f_1 via the periodogram is dominated by this bias and is considerably greater than that associated with the relatively bias-free $\hat{S}^{(\text{D})}(\cdot)$. (We will return to this example later when we discuss the parametric approach to harmonic analysis – see the discussion concerning Figure 557 in Section 10.12. Exercise [10.10] expands upon this example to demonstrate that tapering can lead to an estimator of peak frequency that is better than the periodogram-based estimator in terms of not only bias, but also variance.)

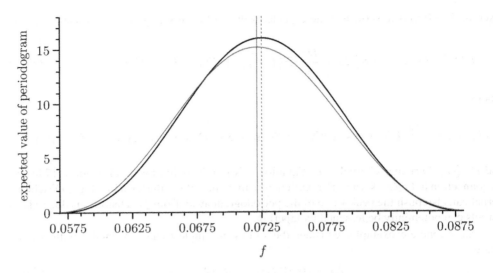

Figure 538a Unconditional expected value of periodogram (thick curve) for a harmonic regression model with a single frequency at $f_1 = 0.0725$, along with the expected value conditional on the random phase ϕ_1 being set to $5\pi/12$ (thin curve). The vertical dashed line marks the location of the peak for the unconditional expectation (this location is quite close to f_1). The solid line does the same for the conditional expectation.

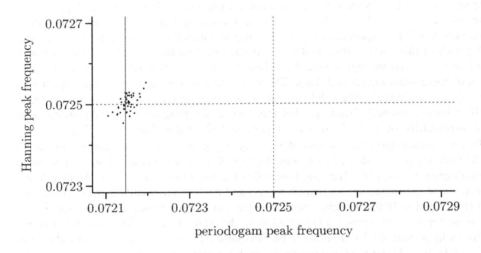

Figure 538b Scatter plot of peak frequencies computed from a Hanning-based direct spectral estimator versus those from the periodogram. The 50 plotted points are computed from 50 realizations of $\{X_t\}$ of Equation (537b), which has a single sinusoid with frequency $f_1 = 0.0725$ (dashed horizontal and vertical lines). There is a substantial downward bias in the periodogram-based peak frequencies, which are clustered around the peak frequency (solid vertical line) of the conditional expectation $E\{\hat{S}^{(P)}(\cdot) \mid \phi_1 = 5\pi/12\}$ shown in Figure 538a.

10.9 Tests for Periodicity – White Noise Case

As should be clear from the previous sections, we need statistical tests to help determine when a spectral peak is significantly larger than is likely to arise if there were no genuine periodic components. We consider two basic cases, where the decision is

[1] between white noise and spectral lines plus white noise (i.e., white noise versus a discrete spectrum) and

[2] between colored noise and spectral lines plus colored noise (i.e., colored noise versus a mixed spectrum).

Typical tests for the first case make use of the fact that, under the null hypothesis that $\{X_t\}$ is Gaussian white noise, the periodogram ordinates calculated at the Fourier frequencies (excluding the zero and Nyquist frequencies) are distributed as independent and identically distributed χ_2^2 RVs times a multiplicative constant (see Section 6.6). The tests then look for periodogram ordinates of a size incompatible with the null hypothesis. Implicitly it is assumed that any spectral lines occur at Fourier frequencies, a rather strong assumption we reconsider at the end of this section. For the remainder of this section we look at case [1] by using Equation (512b) as our model for the observed time series, where the process $\{\epsilon_t\}$ is assumed to be Gaussian white noise (we delve into case [2] in Section 10.10).

As usual, let $f_k = k/(N\,\Delta_t)$ denote the kth Fourier frequency, and let M be the number of Fourier frequencies satisfying $0 < f_k < f_\mathcal{N}$. Thus

$$M = \left\lfloor \frac{N-1}{2} \right\rfloor = \begin{cases} (N-2)/2, & N \text{ even;} \\ (N-1)/2, & N \text{ odd.} \end{cases} \tag{539a}$$

The null hypothesis is $D_1 = \cdots = D_L = 0$. We consider the periodogram terms $\hat{S}^{(\mathrm{P})}(f_k)$ for $k = 1, 2, \ldots, M$. Under the null hypothesis,

$$\frac{2\hat{S}^{(\mathrm{P})}(f_k)}{\sigma_\epsilon^2 \Delta_t} \stackrel{\mathrm{d}}{=} \chi_2^2,$$

where $\stackrel{\mathrm{d}}{=}$ means "equal in distribution" (see Equation (203a)). The probability density function for a chi-square RV with two degrees of freedom is $f(u) = \exp(-u/2)/2$, $u \geq 0$ (see Equation (204d) and Figure 210a(a)). For any $u_0 \geq 0$,

$$\mathbf{P}\left[\frac{2\hat{S}^{(\mathrm{P})}(f_k)}{\sigma_\epsilon^2 \Delta_t} \leq u_0\right] = \mathbf{P}\left[\chi_2^2 \leq u_0\right] = \int_0^{u_0} f(u)\,\mathrm{d}u = 1 - e^{-u_0/2}.$$

Let

$$\gamma \stackrel{\mathrm{def}}{=} \max_{1 \leq k \leq M} \frac{2\hat{S}^{(\mathrm{P})}(f_k)}{\sigma_\epsilon^2 \Delta_t}.$$

Under the null hypothesis, γ is the maximum of the M independent and identically distributed χ_2^2 RVs; thus, for any u_0,

$$\mathbf{P}[\gamma > u_0] = 1 - \mathbf{P}\left[\frac{2\hat{S}^{(\mathrm{P})}(f_k)}{\sigma_\epsilon^2 \Delta_t} \leq u_0 \text{ for } 1 \leq k \leq M\right] = 1 - \left(1 - e^{-u_0/2}\right)^M.$$

This equation gives us the distribution of γ under the null hypothesis. Under the alternative hypothesis that $D_l > 0$ for one or more terms in Equation (512b), γ will be large, so we use a one-sided test with a critical region of the form $\{\gamma : \gamma > u_0\}$, where u_0 is chosen so that $\mathbf{P}[\gamma > u_0] = \alpha$, the chosen level of the test.

The chief disadvantage of this test (due to Schuster, 1898) is that we must know σ_ϵ^2 in advance to be able to calculate γ. Since we usually must estimate σ_ϵ^2, Fisher (1929) derived an exact test for the null hypothesis based upon the statistic

$$g \stackrel{\mathrm{def}}{=} \max_{1 \leq k \leq M} \frac{\hat{S}^{(\mathrm{P})}(f_k)}{\sum_{j=1}^M \hat{S}^{(\mathrm{P})}(f_j)} \tag{539b}$$

(note that $\frac{1}{M} \leq g \leq 1$). The burden of Exercise [10.11] is to show that the denominator above is such that

$$\sum_{j=1}^{M} \hat{S}^{(\text{P})}(f_j) = \frac{\Delta_t}{2}\left(\sum_{t=0}^{N-1}(X_t - \overline{X})^2 - \frac{\delta_N}{N}\left[\sum_{t=0}^{N-1} X_t(-1)^t\right]^2\right), \qquad (540\text{a})$$

where $\delta_N = 1$ if N is even and is 0 otherwise. Note that, when N is odd so that the last summation vanishes, the right-hand side is proportional to the standard estimator of σ_ϵ^2 under the null hypothesis (cf. Equation (166b) with τ set to 0); when N is even, it is approximately so, with the approximation becoming better as N gets larger. We can regard g as the maximum of the sum of squares due to a single frequency f_k over the total sum of squares. Note also that σ_ϵ^2 acts as a proportionality constant in the distribution of both max $\hat{S}^{(\text{P})}(f_k)$ and $\sum \hat{S}^{(\text{P})}(f_j)$, so that σ_ϵ^2 "ratios out" of Fisher's g statistic. (In practice, we can calculate $\sum_{j=1}^{M} \hat{S}^{(\text{P})}(f_j)$ directly in the frequency domain – the time domain equivalent provided by Equation (540a) is of interest more for theory than for computations.)

Fisher (1929) found that the exact distribution of g under the null hypothesis is given by

$$\mathbf{P}[g > g_F] = \sum_{j=1}^{M'}(-1)^{j-1}\binom{M}{j}(1 - jg_F)^{M-1}, \qquad (540\text{b})$$

where M' is the largest integer satisfying both $M' < 1/g_F$ and $M' \leq M$. As a simplification, we can use just the $j = 1$ term in this summation:

$$\mathbf{P}[g > g_F] \approx (M-1)(1 - g_F)^{M-1}. \qquad (540\text{c})$$

Setting the right-hand side equal to α and solving for g_F gives us an approximation $\tilde{g}_F$ for the critical level for an α level test:

$$\tilde{g}_F = 1 - (\alpha/M)^{1/(M-1)}. \qquad (540\text{d})$$

For $\alpha = 0.05$ and $M = 5, 6, \ldots, 50$, the error in this approximation is within 0.1%; moreover, the error is in the direction of decreasing the probability; i.e., $\mathbf{P}[g > \tilde{g}_F] < 0.05$. The error in this approximation is also of about the same order for $\alpha = 0.01, 0.02$ and 0.1 (Nowroozi, 1967). Quinn and Hannan (2001, p. 15) suggest another approximation for the distribution for Fisher's g test:

$$\mathbf{P}[g > g_F] \approx 1 - \exp\left(-N\exp\left(-Ng_F/2\right)/2\right)$$

(in contrast to Equation (540c), the above is correct asymptotically as $N \to \infty$). This leads to an approximation for g_F given by

$$\hat{g}_F = -2\log\left(-2\log\left(1-\alpha\right)/N\right)/N. \qquad (540\text{e})$$

The results of Exercise [10.12] support using $\tilde{g}_F$ when N is small, but then switching to $\hat{g}_F$ when N is large (as a rule of thumb, $N = 3000$ is the boundary between small and large).

Anderson (1971) notes that Fisher's test is the uniformly most powerful symmetric invariant decision procedure against *simple* periodicities, i.e., where the alternative hypothesis is that there exists a periodicity at only one Fourier frequency. But how well does it perform when there is a *compound* periodicity, i.e., spectral lines at several frequencies? Siegel (1980) studied this problem (for another approach, see Bøviken, 1983). He suggested a test statistic based on all large values of the rescaled periodogram

$$\tilde{S}^{(P)}(f_k) \stackrel{\text{def}}{=} \frac{\hat{S}^{(P)}(f_k)}{\sum_{j=1}^{M} \hat{S}^{(P)}(f_j)} \quad (541a)$$

instead of only their maximum. For a value $0 < g_0 \leq g_F$ (where, as before, g_F is the critical value for Fisher's test) and for each $\tilde{S}^{(P)}(f_k)$ that exceeds g_0, Siegel sums the excess of $\tilde{S}^{(P)}(f_k)$ above g_0. This forms the statistic

$$\sum_{k=1}^{M} \left(\tilde{S}^{(P)}(f_k) - g_0 \right)_+,$$

where $(a)_+ = \max\{a, 0\}$ is the positive-part function. This statistic depends on the choice of g_0 and can be related to g_F by writing $g_0 = \lambda g_F$ with $0 < \lambda \leq 1$ to give

$$T_\lambda \stackrel{\text{def}}{=} \sum_{k=1}^{M} \left(\tilde{S}^{(P)}(f_k) - \lambda g_F \right)_+. \quad (541b)$$

Note that, when $\lambda = 1$, the event $g > g_F$ is identical to the event $T_1 > 0$; i.e., we have Fisher's test. Suppose that we choose g_F to correspond to the α significance level of Fisher's test so that

$$\mathbf{P}\left[T_1 > 0\right] = \mathbf{P}\left[\sum_{k=1}^{M} \left(\tilde{S}^{(P)}(f_k) - g_F \right)_+ > 0 \right] = \alpha.$$

Stevens (1939) showed that

$$\mathbf{P}\left[T_1 > 0\right] = \sum_{j=1}^{M'} (-1)^{j-1} \binom{M}{j} (1 - j g_F)^{M-1},$$

exactly as is required by Equation (540b).

We need to compute $\mathbf{P}\left[T_\lambda > t\right]$ for general λ in the range 0 to 1 so that we can find the critical points for significance tests. Siegel (1979) gives the formula

$$\mathbf{P}\left[T_\lambda > t\right] = \sum_{j=1}^{M} \sum_{k=0}^{j-1} (-1)^{j+k+1} \binom{M}{j} \binom{j-1}{k} \binom{M-1}{k} t^k (1 - j\lambda g_F - t)_+^{M-k-1}. \quad (541c)$$

Siegel (1980) gives critical values t_λ for T_λ for significance levels 0.01 and 0.05 with M ranging from 5 to 50 (corresponding to values of N ranging from 11 to 102) and with $\lambda = 0.4$, 0.6 and 0.8. For a value of $\lambda = 0.6$, Siegel found that his test statistic $T_{0.6}$ proved only slightly less powerful than Fisher's g (equivalently, T_1) against an alternative of simple periodicity, but that it outperformed Fisher's test against an alternative of *compound* periodicity (Siegel looked at periodic activity at both two and three frequencies).

Table 542 gives exact critical values for $T_{0.6}$ for $\alpha = 0.05$ and 0.01 and values of M corresponding to sample sizes N of integer powers of two ranging from 32 to 4096 (thus extending Siegel's table). For computing purposes, it is useful to rewrite Equation (541c) as

$$\mathbf{P}\left[T_\lambda > t\right] = t^{M-1} + \sum_{j=1}^{M} \sum_{k=0}^{\min\{j-1, M-2\}} (-1)^{j+k+1} e^{Q_{j,k}}, \quad (541d)$$

		$\alpha = 0.05$			$\alpha = 0.01$		
N	M	Exact	$\tilde{t}_{0.6}$	$c\chi_0^2(\beta)$	Exact	$\tilde{t}_{0.6}$	$c\chi_0^2(\beta)$
16	7	0.225	0.253	0.220	0.268	0.318	0.273
32	15	0.140	0.146	0.138	0.164	0.173	0.165
64	31	0.0861	0.0861	0.0852	0.0969	0.0971	0.0974
128	63	0.0523	0.0515	0.0520	0.0562	0.0552	0.0564
256	127	0.0315	0.0310	0.0315	0.0322	0.0316	0.0323
512	255	0.0190	0.0187	0.0190	0.0184	0.0181	0.0184
1024	511	0.0114	0.0113	0.0114	0.0104	0.0104	0.0105
2048	1023	0.00688	0.00686	0.00688	0.00595	0.00598	0.00597
4096	2047	0.00416	0.00415	0.00416	0.00341	0.00344	0.00341

Table 542 Exact critical values $t_{0.6}$ for $T_{0.6}$ and two approximations thereof.

where

$$Q_{j,k} \stackrel{\text{def}}{=} \log\binom{M}{j} + \log\binom{j-1}{k} + \log\binom{M-1}{k} + k\log(t) + (M-k-1)\log(1-j\lambda g_F - t)_+$$

(Walden, 1992). If we recall that, e.g., $\log(a!) = \log(\Gamma(a+1))$, we have

$$\log\binom{a}{b} = \log(\Gamma(a+1)) - \log(\Gamma(a-b+1)) - \log(\Gamma(b+1)),$$

and we can use standard approximations to the log of a gamma function with a large argument to evaluate $Q_{j,k}$ (see, for example, Press et al., 2007, p. 257). Use of Equation (541d) allows the construction of interpolation formulae that give accurate approximate critical values as a function of M. For $\alpha = 0.05$ the interpolation formula yields the approximation

$$t_{0.6} = 1.033 M^{-0.72356}, \tag{542a}$$

and for $\alpha = 0.01$ it is

$$t_{0.6} = 1.4987 M^{-0.79695}. \tag{542b}$$

Table 542 shows these approximations next to the exact values. The approximations are within 2% of the exact values for $31 \leq M \leq 2047$ (a range for which the formulae were designed to cover – they should be used with caution if M deviates significantly from this range).

Siegel (1979) gives an alternative way of finding the critical values for M large. He notes that

$$T_\lambda \stackrel{\text{d}}{=} c\chi_0^2(\beta) \tag{542c}$$

asymptotically, where the RV $\chi_0^2(\beta)$ obeys what Siegel calls the *noncentral chi-square distribution* with zero degrees of freedom and noncentrality parameter β. The first and second moments of the distribution in Equation (542c) are

$$E\{T_\lambda\} = (1 - \lambda g_F)^M \text{ and } E\{T_\lambda^2\} = \frac{2(1 - \lambda g_F)^{M+1} + (M-1)(1 - 2\lambda g_F)_+^{M+1}}{M+1},$$

and the parameters c and β in Equation (542c) are related by

$$c = \frac{\text{var}\{T_\lambda\}}{4E\{T_\lambda\}} = \frac{E\{T_\lambda^2\} - (E\{T_\lambda\})^2}{4E\{T_\lambda\}} \text{ and } \beta = \frac{E\{T_\lambda\}}{c}.$$

10.9 Tests for Periodicity – White Noise Case

Hence it is merely necessary to find the critical values of $c\chi_0^2(\beta)$ with c and β given by the above. This can be done using the algorithm of Farebrother (1987) or, e.g., the function qchisq in the R language. Table 542 gives the approximate critical values using this method, all of which are within 2% of the exact values.

The tests we have looked at so far assume that any spectral lines occur at Fourier frequencies. Fisher's and Siegel's tests look particularly useful in theory for the case of large samples where the Fourier frequencies are closely spaced and thus are likely to be very close to any true lines (for large enough N, it will become obvious from the periodogram that there are significant peaks in the spectrum so that in practice the tests lose some of their value). What happens, however, if a spectral line occurs midway between two Fourier frequencies? Equation (524b) can be employed to show that the expected value of the periodogram at such an intermediate frequency can be substantially less than that obtained at a Fourier frequency (cf. Equation (524a); see also Whittle, 1952). Even though there must be some degradation in performance, Priestley (1981, p. 410) argues that Fisher's test will still be reasonably powerful, provided the signal-to-noise ratio $D_l^2/(2\sigma_\epsilon^2)$ is large (he also gives a summary of some other tests for periodicity, none of which appear particularly satisfactory).

Section 10.15 gives an example of applying Fisher's and Siegel's tests to the ocean noise time series of Figure 4(d).

Comments and Extensions to Section 10.9

[1] Fisher's and Siegel's white noise tests both make use of the periodogram, as does the cumulative periodogram test for white noise described in C&E [6] for Section 6.6. Whereas all three tests should have similar performance under the null hypothesis of white noise, they are designed to reject the null hypothesis when faced with different alternative hypotheses. In the case of Fisher's and Siegel's tests, the alternatives are time series with discrete spectra, while, for the cumulative periodogram test, the alternatives are often thought of as series with purely continuous spectra. This does not mean that the tests do not have some power against alternatives for which they were not specifically designed. Exercise [10.13] asks the reader to apply all three tests to various simulated time series, some of which the tests are designed for, and others of which they are not.

[2] For an irregularly sampled Gaussian time series, the Lomb–Scargle periodogram of Equation (529a) can be used to test the null hypothesis (NH) of white noise against the alternative hypothesis (AH) of a sinusoid buried in white noise. When the frequency f_1 is known a priori, the connection between this periodogram and least squares estimation of A and B in the model of Equation (528b) suggests using the standard F-test for the regression, which evaluates the joint significance of $\hat{A}$ and $\hat{B}$. This test takes the form
$$F = \frac{(\text{SS}_{\text{NH}} - \text{SS}_{\text{AH}})/(\text{df}_{\text{NH}} - \text{df}_{\text{AH}})}{\text{SS}_{\text{AH}}/\text{df}_{\text{AH}}},$$
where, recalling Equation (528c),
$$\text{SS}_{\text{AH}} = \text{SS}_c(\hat{A}, \hat{B}) \quad \text{and} \quad \text{SS}_{\text{NH}} = \sum_{n=0}^{N-1} X^2(t_n)$$
are the residual sum of squares under the alternative and null hypotheses, while $\text{df}_{\text{AH}} = N - 2$ and $\text{df}_{\text{NH}} = N$ are the associated degrees of freedom (Weisberg, 2014). Hence
$$F = \frac{\hat{S}^{(\text{LS})}(f_1)}{\hat{\sigma}_\epsilon^2}, \quad \text{where } \hat{\sigma}_\epsilon^2 = \frac{\text{SS}_c(\hat{A}, \hat{B})}{N - 2} \qquad (543)$$
is the usual least squares estimator of σ_ϵ^2 (Exercise [10.8a] shows that $\text{SS}_c(\hat{A}, \hat{B})$ is the same for all c – hence setting c to $\tilde{c}$ of Equation (529b) to force $\hat{A}$ and $\hat{B}$ to be uncorrelated does not modify F). This F statistic is F-distributed with 2 and $N - 2$ degrees of freedom. Large values of F provide evidence

against the null hypothesis. Note that the upper $(1-\alpha) \times 100\%$ percentage point of the $F_{2,\nu}$ distribution can be surprisingly easily computed using the formula

$$\frac{\nu(1-\alpha^{2/\nu})}{2\alpha^{2/\nu}}. \tag{544a}$$

Complications arise in adapting the Lomb–Scargle approach when the frequency f_1 is unknown. One adaptation is to define a set of, say, M candidate frequencies for the sinusoids in the model of Equation (528b), compute the F-statistic of Equation (543) over the M frequencies and consider the maximum value of these to evaluate the hypothesis of white noise versus the alternative of a sinusoid with an unknown frequency buried in white noise. The distribution of this maximum under the null hypothesis is difficult to come by analytically. Zechmeister and Kürster (2009) discuss various approximations, noting that Monte Carlo simulations are a viable – but computationally intensive – choice for getting at the distribution.

10.10 Tests for Periodicity – Colored Noise Case

We now turn our attention to case [2] described at the beginning of Section 10.9 – tests for spectral lines in *colored* noise. As in the white noise case, periodogram ordinates can form the basis for a test in the mixed spectra case (see, e.g., Wen et al., 2012, and references therein), but other approaches have been explored (see, e.g., Priestley, 1981, sections 8.3–4, for a technique based upon the autocovariance function). For case [2] we turn our attention in this section to an appealing multitaper technique proposed in Thomson (1982).

We start by assuming the following version of Equation (518c) involving just a single frequency f_1:

$$X_t = D_1 \cos(2\pi f_1 t \Delta_t + \phi_1) + \eta_t, \tag{544b}$$

where $\{\eta_t\}$ (the background continuum) is a zero mean stationary process with SDF $S_\eta(\cdot)$ that is not necessarily constant; i.e., $\{\eta_t\}$ can be colored noise. We also assume in this section that $\{\eta_t\}$ has a Gaussian distribution. We would usually consider ϕ_1 in Equation (544b) to be an RV, but Thomson's approach regards it as an unknown constant. This assumption turns $\{X_t\}$ into a process with a time varying mean value:

$$E\{X_t\} = D_1 \cos(2\pi f_1 t \Delta_t + \phi_1) = C_1 e^{i2\pi f_1 t \Delta_t} + C_1^* e^{-i2\pi f_1 t \Delta_t}, \tag{544c}$$

where $C_1 \stackrel{\text{def}}{=} D_1 e^{i\phi_1}/2$. Note that we can write

$$X_t = E\{X_t\} + \eta_t. \tag{544d}$$

The multitaper test for periodicity makes use of K orthogonal data tapers $\{h_{0,t}\}$, ..., $\{h_{K-1,t}\}$, which, when used to form the basic multitaper estimator of Equation (352a), yield a spectral estimator with a standard bandwidth measure $B_{\overline{\mathcal{H}}}$ dictated by Equation (353e) (Thomson, 1982, focuses on the Slepian multitapers, but we also consider the sinusoidal tapers in what follows). Similar to Equations (186a) and (436a), let us consider

$$J_k(f) \stackrel{\text{def}}{=} \Delta_t^{1/2} \sum_{t=0}^{N-1} h_{k,t} X_t e^{-i2\pi f t \Delta_t},$$

where the kth-order data taper $\{h_{k,t}\}$ is associated with spectral window

$$\mathcal{H}_k(f) \stackrel{\text{def}}{=} \frac{1}{\Delta_t} |H_k(f)|^2, \text{ for which } H_k(f) \stackrel{\text{def}}{=} \Delta_t \sum_{t=0}^{N-1} h_{k,t} e^{-i2\pi f t \Delta_t}.$$

10.10 Tests for Periodicity – Colored Noise Case

Equation (352a) makes it clear that $|J_k(f)|^2$ is equal to the kth eigenspectrum $\hat{S}_k^{(\mathrm{MT})}(f)$ in the multitaper scheme. Using Equation (544d), the expected value of the complex-valued quantity $J_k(f)$ is

$$E\{J_k(f)\} = \Delta_t^{1/2} \sum_{t=0}^{N-1} h_{k,t} \left(C_1 e^{i2\pi f_1 t \Delta_t} + C_1^* e^{-i2\pi f_1 t \Delta_t} \right) e^{-i2\pi f t \Delta_t}$$

$$= \frac{1}{\Delta_t^{1/2}} \left[C_1 H_k(f - f_1) + C_1^* H_k(f + f_1) \right].$$

The frequency domain analog of Equation (544d) is thus

$$J_k(f) = E\{J_k(f)\} + \Delta_t^{1/2} \sum_{t=0}^{N-1} h_{k,t} \eta_t e^{-i2\pi f t \Delta_t}.$$

Note that the squared modulus of the term with η_t is a direct spectral estimator of $S_\eta(f)$ using the data taper $\{h_{k,t}\}$. At $f = f_1$, the above becomes

$$J_k(f_1) = E\{J_k(f_1)\} + \Delta_t^{1/2} \sum_{t=0}^{N-1} h_{k,t} \eta_t e^{-i2\pi f_1 t \Delta_t},$$

where

$$E\{J_k(f_1)\} = \frac{1}{\Delta_t^{1/2}} \left[C_1 H_k(0) + C_1^* H_k(2f_1) \right]. \tag{545a}$$

For a Slepian or sinusoidal taper that contributes to a basic multitaper estimator with half-bandwidth measure $B_{\overline{\mathcal{H}}}/2$, the squared modulus of $H_k(\cdot)$ is highly concentrated within the interval $[-B_{\overline{\mathcal{H}}}/2, B_{\overline{\mathcal{H}}}/2]$. Thus we have, for $2f_1 > B_{\overline{\mathcal{H}}}/2$,

$$E\{J_k(f_1)\} \approx C_1 \frac{H_k(0)}{\Delta_t^{1/2}} = \Delta_t^{1/2} C_1 \sum_{t=0}^{N-1} h_{k,t}. \tag{545b}$$

The kth-order Slepian and sinusoidal data tapers are skew-symmetric about their midpoints for odd k and symmetric for even k (see the left-hand plots of Figures 360, 362, 394 and 396), For both multitaper schemes, the above summation is in fact zero for odd k, but it is real-valued and positive for even k.

We now assume that $2f_1 > B_{\overline{\mathcal{H}}}/2$ and take the approximation in Equation (545b) to be an equality in what follows. We can write the following model for $J_k(f_1)$:

$$J_k(f_1) = C_1 \frac{H_k(0)}{\Delta_t^{1/2}} + \tilde{\epsilon}_k, \quad k = 0, \ldots, K-1, \tag{545c}$$

where

$$\tilde{\epsilon}_k \stackrel{\mathrm{def}}{=} \Delta_t^{1/2} \sum_{t=0}^{N-1} h_{k,t} \eta_t e^{-i2\pi f_1 t \Delta_t}.$$

Equation (545c) thus defines a complex-valued regression model: $J_k(f_1)$ is the observed response (dependent variable); C_1 is an unknown parameter; $H_k(0)/\Delta_t^{1/2}$ is the kth predictor (independent variable); and $\tilde{\epsilon}_k$ is the error term. We have already assumed $\{\eta_t\}$ to be a zero mean real-valued Gaussian stationary process. We now add the assumptions that its SDF

$S_\eta(\cdot)$ is slowly varying in the interval $[f_1 - B_{\overline{\mathcal{H}}}/2, f_1 + B_{\overline{\mathcal{H}}}/2]$ and that f_1 is not too close to either zero or the Nyquist frequency. Since $\tilde{\epsilon}_k$ is just the kth eigencoefficient in a multitaper estimator of $S_\eta(f_1)$, we can appeal to the arguments used in the solution to Exercise [8.3] to make two reasonable assertions, one about $\text{cov}\,\{\tilde{\epsilon}_k, \tilde{\epsilon}_l\}$, and the other concerning $\text{cov}\,\{\tilde{\epsilon}_k, \tilde{\epsilon}_l^*\}$. First,

$$\text{cov}\,\{\tilde{\epsilon}_k, \tilde{\epsilon}_l\} = E\{\tilde{\epsilon}_k \tilde{\epsilon}_l^*\} = 0, \qquad k \neq l,$$

so that the $\tilde{\epsilon}_k$ are pairwise uncorrelated, and, when $k = l$,

$$\text{var}\,\{\tilde{\epsilon}_k\} = E\{|\tilde{\epsilon}_k|^2\} \stackrel{\text{def}}{=} \sigma_{\tilde{\epsilon}}^2,$$

so that each $\tilde{\epsilon}_k$ has the same variance; in addition, we have $\sigma_{\tilde{\epsilon}}^2 = S_\eta(f_1)$. Second,

$$\text{cov}\,\{\tilde{\epsilon}_k, \tilde{\epsilon}_l^*\} = E\{\tilde{\epsilon}_k \tilde{\epsilon}_l\} = 0, \qquad 0 \leq k, l \leq K - 1.$$

These two assertions imply that the real and imaginary components of $\tilde{\epsilon}_k$, $k = 0, 1, \ldots, K-1$, are $2K$ real-valued and independent Gaussian RVs, each with the same semivariance $\sigma_{\tilde{\epsilon}}^2/2$. With all the above stipulations, $\tilde{\epsilon}_k$ matches the requirements of complex-valued regression theory as formulated in Miller (1973, p. 715).

We can now estimate C_1 by employing a version of least squares valid for complex-valued quantities. The estimator is the value, say $\hat{C}_1$, of C_1 that minimizes

$$\text{SS}(C_1) \stackrel{\text{def}}{=} \sum_{k=0}^{K-1} \left| J_k(f_1) - C_1 \frac{H_k(0)}{\Delta_t^{1/2}} \right|^2.$$

It follows from Miller (1973, equation (3.2)) that

$$\hat{C}_1 = \Delta_t^{1/2} \frac{\sum_{k=0}^{K-1} J_k(f_1) H_k(0)}{\sum_{k=0}^{K-1} H_k^2(0)} = \Delta_t^{1/2} \frac{\sum_{k=0,2,\ldots}^{2\lfloor (K-1)/2 \rfloor} J_k(f_1) H_k(0)}{\sum_{k=0,2,\ldots}^{2\lfloor (K-1)/2 \rfloor} H_k^2(0)} \qquad (546a)$$

(we can eliminate $k = 1, 3, \ldots$ from the summations because $H_k(0) = 0$ for odd k). Moreover, under the conditions on $\tilde{\epsilon}_k$ stated in the previous paragraph, we can state the following results, all based on theorem 8.1 of Miller (1973). First, $\hat{C}_1$ is a proper complex Gaussian RV (it is uncorrelated with its complex conjugate), with mean C_1 and variance $\sigma_{\tilde{\epsilon}}^2 \Delta_t / \sum_{k=0,2,\ldots}^{2\lfloor (K-1)/2 \rfloor} H_k^2(0)$. Second, an estimator of $\sigma_{\tilde{\epsilon}}^2$ is given by

$$\hat{\sigma}_{\tilde{\epsilon}}^2 = \frac{1}{K} \sum_{k=0}^{K-1} \left| J_k(f_1) - \hat{J}_k(f_1) \right|^2, \qquad (546b)$$

where $\hat{J}_k(f_1)$ is the fitted value for $J_k(f_1)$ given by

$$\hat{J}_k(f_1) = \hat{C}_1 \frac{H_k(0)}{\Delta_t^{1/2}}$$

(this is zero when k is odd). Third, the RV $2K\hat{\sigma}_{\tilde{\epsilon}}^2/\sigma_{\tilde{\epsilon}}^2$ follows a chi-square distribution with $2K - 2$ degrees of freedom. Finally, $\hat{C}_1$ and $2K\hat{\sigma}_{\tilde{\epsilon}}^2/\sigma_{\tilde{\epsilon}}^2$ are independent RVs.

Thomson's test for a periodicity in colored noise is based upon the usual F-test for the significance of a regression parameter. Under the null hypothesis that $C_1 = 0$, the RV $\hat{C}_1$ has a complex Gaussian distribution with zero mean and variance $\sigma_{\tilde{\epsilon}}^2 \Delta_t / \sum_{k=0,2,\ldots}^{2\lfloor(K-1)/2\rfloor} H_k^2(0)$; moreover, because $J_k(f_1) = \tilde{\epsilon}_k$ under the null hypothesis, it follows from our assumptions about $\tilde{\epsilon}_k$ that the real and imaginary components of $\hat{C}_1$ are uncorrelated and have the same semivariance. Since $|\hat{C}_1|^2$ is thus the sum of squares of two uncorrelated real-valued Gaussian RVs with zero means and equal semivariances, we have

$$\frac{2|\hat{C}_1|^2 \sum_{k=0,2,\ldots}^{2\lfloor(K-1)/2\rfloor} H_k^2(0)}{\sigma_{\tilde{\epsilon}}^2 \Delta_t} \stackrel{d}{=} \chi_2^2.$$

Because $\hat{C}_1$ and the χ_{2K-2}^2 RV $2K\hat{\sigma}_{\tilde{\epsilon}}^2/\sigma_{\tilde{\epsilon}}^2$ are independent, it follows that the above χ_2^2 RV is independent of $2K\hat{\sigma}_{\tilde{\epsilon}}^2/\sigma_{\tilde{\epsilon}}^2$ also. The ratio of independent χ^2 RVs with 2 and $2K-2$ degrees of freedom – each divided by their respective degrees of freedom – is F-distributed with 2 and $2K-2$ degrees of freedom. Hence, we obtain (after some reduction)

$$\frac{(K-1)|\hat{C}_1|^2 \sum_{k=0,2,\ldots}^{2\lfloor(K-1)/2\rfloor} H_k^2(0)}{\Delta_t \sum_{k=0}^{K-1} \left| J_k(f_1) - \hat{J}_k(f_1) \right|^2} \stackrel{d}{=} F_{2,2K-2} \qquad (547a)$$

(Thomson, 1982, equation (13.10)). If in fact $C_1 \neq 0$, the above statistic should give a value exceeding a high percentage point of the $F_{2,2K-2}$ distribution (the 95% point, say; Equation (544a) gives a formula for computing any desired percentage point).

If the F-test at $f = f_1$ is significant (i.e., the null hypothesis that $C_1 = 0$ is rejected), Thomson (1982) suggests reshaping the spectrum around f_1 to give a better estimate of $S_\eta(\cdot)$, the SDF of the background continuum $\{\eta_t\}$. Since $\sigma_{\tilde{\epsilon}}^2 = S_\eta(f_1)$, Equation (546b) can be rewritten as

$$\hat{S}_\eta(f_1) = \frac{1}{K} \sum_{k=0}^{K-1} \left| J_k(f_1) - \hat{J}_k(f_1) \right|^2 = \frac{1}{K} \sum_{k=0}^{K-1} \left| J_k(f_1) - \hat{C}_1 \frac{H_k(0)}{\Delta_t^{1/2}} \right|^2.$$

For f in the neighborhood of f_1, i.e., $f \in [f_1 - B_{\overline{\mathcal{H}}}/2, f_1 + B_{\overline{\mathcal{H}}}/2]$, the above generalizes to

$$\hat{S}_\eta(f) = \frac{1}{K} \sum_{k=0}^{K-1} \left| J_k(f) - \hat{C}_1 \frac{H_k(f - f_1)}{\Delta_t^{1/2}} \right|^2. \qquad (547b)$$

For f sufficiently outside of this neighborhood, we have $H_k(f - f_1) \approx 0$, and the right-hand side above reverts to the usual basic multitaper estimator of Equation (352a). To ensure a smooth transition into the basic estimator, we recommend computing the reshaped spectrum over all frequencies f such that

$$|f - f_1| \leq 1.25 B_{\overline{\mathcal{H}}}/2. \qquad (547c)$$

Finally, here are two related comments are about practical application of Thomson's method. First, since the eigencoefficients $J_k(\cdot)$ are typically computed using an FFT, f_1 would need to coincide with a standard frequency after padding with zeros to obtain a fine grid. Taking f_1 to be the standard frequency at which $\hat{C}_1$ is a maximum is one way to *define* the frequency at which to carry out the F-test. The theory we have presented assumes f_1 to be known a priori, so caution is in order in interpreting the significance of a peak if in fact f_1 is determined a posteriori using the time series under study. A quote from Thomson (1990a) is pertinent here (edited slightly to fit in with our exposition):

It is important to remember that in typical time-series problems hundreds or thousands of uncorrelated estimates are being dealt with; consequently one will encounter numerous instances of the F-test giving what would normally be considered highly significant test values that, in actuality, will only be sampling fluctuations. A good rule of thumb is not to get excited by significance levels greater than $1/N$.

Our second comment concerns multiple applications of Thomson's test. Note that the regression model in Equation (545c) is defined pointwise at each frequency. There is thus no need to assume that the eigencoefficients are uncorrelated across frequencies; however, because of the localized nature of the F-test at f_1, a test at a second frequency f_2 will be approximately independent of the first test as long as f_2 is outside of the interval $[f_1 - B_{\overline{\mathcal{H}}}, f_1 + B_{\overline{\mathcal{H}}}]$. The assumption of a slowly varying spectrum around f_1 must also now hold around f_2. This approach extends to multiple lines if analogous requirements are satisfied. (Denison and Walden, 1999, discuss the subtleties that can arise when applying Thomson's test multiple times.)

Comments and Extensions to Section 10.10

[1] So far we have restricted f_1 to be nonzero, but let now consider the special case $f_1 = 0$ so that Equation (544b) becomes $X_t = D_1 + \eta_t$ (we set $\phi_1 = 0$ since the phase of a constant term is undefined). The parameter D_1 is now the expected value of the stationary process $\{X_t\}$; i.e., $E\{X_t\} = D_1$. With C_1 defined to be $D_1/2$, Equation (545a) becomes

$$E\{J_k(0)\} = \frac{2C_1 H_k(0)}{\Delta_t^{1/2}} = \frac{D_1 H_k(0)}{\Delta_t^{1/2}},$$

and the least squares estimator of C_1 (Equation (546a)) becomes

$$\hat{C}_1 = \frac{\Delta_t^{1/2}}{2} \frac{\sum_{k=0,2,\ldots}^{2\lfloor (K-1)/2 \rfloor} J_k(0) H_k(0)}{\sum_{k=0,2,\ldots}^{2\lfloor (K-1)/2 \rfloor} H_k^2(0)}, \quad \text{and hence} \quad \hat{D}_1 = \Delta_t^{1/2} \frac{\sum_{k=0,2,\ldots}^{2\lfloor (K-1)/2 \rfloor} J_k(0) H_k(0)}{\sum_{k=0,2,\ldots}^{2\lfloor (K-1)/2 \rfloor} H_k^2(0)}.$$

If $K = 1$, the above reduces to

$$\hat{D}_1 = \Delta_t^{1/2} \frac{J_0(0) H_0(0)}{H_0^2(0)} = \frac{\sum_{t=0}^{N-1} h_{0,t} X_t}{\sum_{t=0}^{N-1} h_{0,t}}.$$

This is the alternative estimator to the sample mean given in Equation (195b). The reshaped SDF estimator appropriate for this case is

$$\hat{S}_\eta(f) = \left| J_0(f) - \hat{D}_1 \frac{H_0(f)}{\Delta_t^{1/2}} \right|^2 = \Delta_t \left| \sum_{t=0}^{N-1} h_{0,t} \left(X_t - \hat{D}_1 \right) e^{-i2\pi f t \Delta_t} \right|^2,$$

in agreement with the "mean corrected" direct spectral estimator of Equation (196a) with $\hat{D}_1$ substituted for $\tilde{\mu}$.

[2] We noted briefly that Thomson's F-test for a single spectral line is valid for multiple lines as long as the lines are well separated. Denison et al. (1999) discuss how to adapt the test to handle closely spaced spectral lines.

For well-separated lines, Wei and Craigmile (2010) propose tests that are similar in spirit to Thomson's in operating locally (i.e., the test at f_1 is based on spectral estimates just in the neighborhood of f_1) but they also consider tests operating globally (Fisher's g test of Equation (539b) is a prime example of a global test because it assesses a peak using values of the periodogram with frequencies ranging globally from just above zero to just below Nyquist). One of their tests is a local F-test like Thomson's, but it gets around the usual assumption in multitaper estimation of the SDF being locally constant by allowing the log SDF to vary linearly. They also discuss the advantages and disadvantages of local and global tests in terms of power to detect periodicities under various scenarios.

10.11 Completing a Harmonic Analysis

In Sections 10.5 and 10.6 we outlined ways of searching for spectral lines immersed in white noise by looking for peaks in the periodogram. We considered tests for periodicity – both simple and compound – against an alternative hypothesis of white noise in Section 10.9 and Thomson's multitaper-based test for detecting periodicity in a white or colored background continuum in the previous section. Suppose that, based on one of these tests, we have concluded that there are one or more line components in a time series. To complete our analysis, we must estimate the amplitudes of the line components and the SDF of the background continuum. If we have evidence that the background continuum is white noise, we need only estimate the variance σ_ϵ^2 of the white noise since its SDF is just $\sigma_\epsilon^2 \Delta_{\text{t}}$; for the case of a colored continuum, we must estimate the SDF after somehow compensating for the line components.

We begin by considering the white noise case in detail, which assumes Equation (515b) as its model (in what follows, we assume $\mu = 0$ – see C&E [1] for how to deal with an unknown process mean). Suppose we have estimates $\hat{f}_l$ of the line frequencies obtained from the periodogram (or from some other method). We can then form *approximate conditional* least squares (ACLS) estimates of A_l and B_l using Equation (515c) with f_l replaced by $\hat{f}_l$:

$$\hat{A}_l = \frac{2}{N} \sum_{t=0}^{N-1} X_t \cos(2\pi \hat{f}_l t \Delta_{\text{t}}) \quad \text{and} \quad \hat{B}_l = \frac{2}{N} \sum_{t=0}^{N-1} X_t \sin(2\pi \hat{f}_l t \Delta_{\text{t}}). \tag{549a}$$

Here "conditional" refers to the use of the estimator $\hat{f}_l$ in lieu of knowing the actual f_l, and "approximate" acknowledges that, for non-Fourier frequencies, these estimators will not be exact least squares estimators (see Section 10.2). We estimate the size of the jumps at $\pm \hat{f}_l$ in the integrated spectrum for $\{X_t\}$ by

$$\frac{\hat{A}_l^2 + \hat{B}_l^2}{4} = \frac{\hat{S}^{(\text{P})}(\hat{f}_l)}{N \Delta_{\text{t}}} \tag{549b}$$

(see the discussion surrounding Equation (523a)). The estimator of the white noise variance σ_ϵ^2 is

$$\hat{\sigma}_\epsilon^2 = \frac{\text{SS}(\{\hat{A}_l\}, \{\hat{B}_l\}, \{\hat{f}_l\})}{N - 2L}, \tag{549c}$$

where

$$\text{SS}(\{A_l\}, \{B_l\}, \{f_l\}) \stackrel{\text{def}}{=} \sum_{t=0}^{N-1} \left(X_t - \sum_{l=1}^{L} [A_l \cos(2\pi f_l t \Delta_{\text{t}}) + B_l \sin(2\pi f_l t \Delta_{\text{t}})] \right)^2. \tag{549d}$$

The divisor $N - 2L$ is motivated by standard linear regression theory, i.e., the sample size minus the number of parameters estimated (Weisberg, 2014). The estimator of the SDF $S_\epsilon(\cdot)$ of the white noise is $\hat{S}_\epsilon(f) = \hat{\sigma}_\epsilon^2 \Delta_{\text{t}}$.

A second approach would be to find *exact conditional* least squares (ECLS) estimates for A_l and B_l by locating those values of A_l and B_l actually minimizing $\text{SS}(\{A_l\}, \{B_l\}, \{\hat{f}_l\})$. These estimates can be obtained by regressing X_t on $\cos(2\pi \hat{f}_l t \Delta_{\text{t}})$ and $\sin(2\pi \hat{f}_l t \Delta_{\text{t}})$, $l = 1, \ldots, L$. If we let $\check{A}_l$ and $\check{B}_l$ denote the minimizers, the estimator of the jumps takes the form of the left-hand side of Equation (549b), while the estimator of σ_ϵ^2 is again given by Equation (549c).

A third approach would be to find *exact unconditional* least squares (EULS) estimates for A_l, B_l and f_l by locating those values of A_l, B_l and f_l that minimize $\text{SS}(\{A_l\}, \{B_l\}, \{f_l\})$.

With $\hat{A}_l$, $\hat{B}_l$ and $\hat{f}_l$ denoting the minimizers, the left-hand side of Equation (549b) would again estimate the jumps, but now linear least squares theory suggests the following estimator for the white noise variance:

$$\tilde{\sigma}_\epsilon^2 = \frac{\text{SS}(\{\hat{A}_l\}, \{\hat{B}_l\}, \{\hat{f}_l\})}{N - 3L}; \qquad (550a)$$

however, the fact that the estimators arise from nonlinear least squares dictates caution in use of the above.

In practice, we must use a nonlinear optimization routine to find EULS parameter estimates. Such routines typically require initial settings of the parameters, for which we can use the ACLS estimates. Bloomfield (2000) has an extensive discussion of EULS estimates. For most time series encountered in nature, ACLS estimates are adequate, but Bloomfield gives a pathological artificial example where EULS estimates are remarkably better than the ACLS and ECLS estimates.

Once we have obtained estimates of A_l, B_l and f_l by any of these three approaches, we can examine the adequacy of our fitted model by looking at the *residual process* $\{R_t\}$, where

$$R_t \stackrel{\text{def}}{=} X_t - \sum_{l=1}^{L} \left[\hat{A}_l \cos\left(2\pi \hat{f}_l t \Delta_\text{t}\right) + \hat{B}_l \sin\left(2\pi \hat{f}_l t \Delta_\text{t}\right)\right]. \qquad (550b)$$

If our fitted model is adequate, the series $\{R_t\}$ should resemble a realization of a white noise process because we can regard R_t as a proxy for the unknown ϵ_t. This suggests assessing $\{R_t\}$ using periodogram-based tests for white noise (e.g., Fisher's test of Equation (539b) or the normalized cumulative periodogram test that follows from Equation (215a)). Caution is in order here because $\{R_t\}$ tends to be deficient in power around the line frequencies, which is inconsistent with the hypothesis that the residuals are a realization of white noise. Figure 551 shows an example (Exercise [10.14a] expands upon it). We start with the time series whose periodogram is shown in Figure 534(c). We treat this series as if it were a realization of length $N = 128$ of the process of Equation (515b) with frequencies f_1, f_2 and f_3 dictated as per Equation (530) and with white noise variance $\sigma_\epsilon^2 = 1$. The EULS procedure yields a set of residuals $\{R_t\}$ whose periodogram is shown in Figure 551. The three dashed vertical lines indicate the EULS estimates of the line frequencies, which are in good agreement with the true frequencies (these are indicated by similar lines in Figure 534(c)). Whereas the periodogram for the original series has peaks at these frequencies, the periodogram of the residuals has pronounced troughs.

Now we consider the second possibility, namely, a colored continuum. Our model is now

$$X_t = \sum_{l=1}^{L} [A_l \cos\left(2\pi f_l t \Delta_\text{t}\right) + B_l \sin\left(2\pi f_l t \Delta_\text{t}\right)] + \eta_t,$$

where $\{\eta_t\}$ is a stationary process with zero mean and colored SDF $S_\eta(\cdot)$. As in the case of a white continuum, we could use either the ACLS, ECLS or EULS approach and then form the residual process $\{R_t\}$. If $\hat{A}_l$, $\hat{B}_l$ and $\hat{f}_l$ are good estimates, then $\{R_t\}$ should be a good approximation to $\{\eta_t\}$. We can thus estimate $S_\eta(\cdot)$ by computing an SDF estimate using these residuals, but again we note that such estimates tend to have a deficiency of power around the line frequencies.

A second approach is to reformulate the model in terms of complex exponentials, i.e., to write

$$X_t = \sum_{l=1}^{L} \left(C_l e^{i2\pi f_l t \Delta_\text{t}} + C_l^* e^{-i2\pi f_l t \Delta_\text{t}}\right) + \eta_t, \qquad (550c)$$

10.11 Completing a Harmonic Analysis

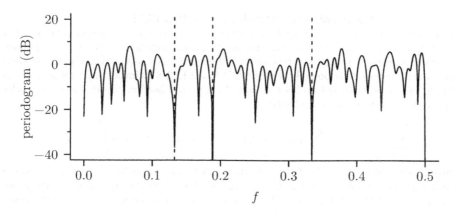

Figure 551 Periodogram for residuals from EULS estimates The residuals are derived from a time series whose periodogram (shown in Figure 534(c)) has three prominent peaks. Fitting an order $L = 3$ model of the form of Equation (515b) using the EULS method yields residuals whose periodogram, as this figure illustrates, now has pronounced troughs at the peak locations (indicated by the dashed lines).

and then to estimate C_l via the multitaper approach of the previous section. If the $\hat{f}_l$ terms are sufficiently separated and not too close to zero or the Nyquist frequency, as discussed, then the estimator of C_l is given by

$$\hat{C}_l = \Delta_t^{1/2} \frac{\sum_{k=0,2,\ldots}^{2\lfloor(K-1)/2\rfloor} J_k(\hat{f}_l) H_k(0)}{\sum_{k=0,2,\ldots}^{2\lfloor(K-1)/2\rfloor} H_k^2(0)} \qquad (551a)$$

(cf. Equation (546a)), where $J_k(\hat{f}_l)$ and $H_k(0)$ are as defined previously. The size of the jumps in the integrated spectrum at $\pm \hat{f}_l$ would be estimated by $|\hat{C}_l|^2$. The SDF $S_\eta(\cdot)$ of the background continuum would be estimated by reshaping – in the neighborhoods of each $\hat{f}_l$ – the multitaper SDF estimate based upon $\{X_t\}$ (see Equation (547b)). This estimate of $S_\eta(\cdot)$ tends to be better behaved around the line frequencies than estimates based upon the residual process.

Finally we note that the multitaper approach can also be used with a white noise continuum. Here we need only estimate the variance σ_ϵ^2 of the white noise, and the appropriate estimator is now

$$\hat{\sigma}_\epsilon^2 = \frac{1}{N - 2L} \sum_{t=0}^{N-1} \left(X_t - \sum_{l=1}^{L} \left[\hat{C}_l e^{i2\pi \hat{f}_l t \Delta_t} + \hat{C}_l^* e^{-i2\pi \hat{f}_l t \Delta_t} \right] \right)^2. \qquad (551b)$$

This expression can be reduced somewhat by noting that

$$\hat{C}_l e^{i2\pi \hat{f}_l t \Delta_t} + \hat{C}_l^* e^{-i2\pi \hat{f}_l t \Delta_t} = 2\Re(\hat{C}_l) \cos(2\pi \hat{f}_l t \Delta_t) - 2\Im(\hat{C}_l) \sin(2\pi \hat{f}_l t \Delta_t),$$

where $\Re(z)$ and $\Im(z)$ refer, respectively, to the real and imaginary components of the complex number z. Exercise [10.14b] considers an example of this approach. (Rife and Boorstyn, 1974, 1976, present an interesting discussion about the estimation of complex exponentials in white noise.)

Comments and Extensions to Section 10.11

[1] In formulating the ACLS, ECLS and EULS estimators, we assumed for convenience that $\mu = 0$ in Equation (515b), which is not a viable assumption in most practical applications. When μ is unknown, we can use the sample mean $\overline{X}$ to center the time series, thus replacing X_t by $X_t - \overline{X}$ in Equations (549a) and (549d). Another approach is possible with the ECLS and EULS estimators by treating μ as an additional parameter to be estimated via least squares. In place of Equation (549d), we now consider $\text{SS}(\mu, \{A_l\}, \{B_l\}, \{f_l\})$, which is defined to be

$$\sum_{t=0}^{N-1} \left(X_t - \mu - \sum_{l=1}^{L} [A_l \cos(2\pi f_l t \Delta_t) + B_l \sin(2\pi f_l t \Delta_t)] \right)^2.$$

The minimizing values of $\text{SS}(\mu, \{A_l\}, \{B_l\}, \{\hat{f}_l\})$ are the ECLS estimators for μ, A_l and B_l, and the ECLS estimator $\hat{\mu}$ is not in general equal to $\overline{X}$; likewise, the EULS estimators for μ, A_l, B_l and f_l are the minimizing values of $\text{SS}(\mu, \{A_l\}, \{B_l\}, \{f_l\})$, and again $\hat{\mu}$ need not be equal to the sample mean. The divisors in the estimators $\hat{\sigma}_\epsilon^2$ and $\tilde{\sigma}_\epsilon^2$ of Equations (549c) and (550a) now become, respectively, $N - 2L - 1$ and $N - 3L - 1$. C&E [1] for Section 10.15 gives an example of these estimators of μ.

[2] For certain time series, we might know the frequencies f_l a priori, in which case we can use the known f_l in place of $\hat{f}_l$ in Equations (549a), (549b) and (549c). Because we no longer need to condition on an estimator of f_l, the methods we have previously described as yielding approximate conditional and exact conditional least squares estimators now give us approximate unconditional and exact unconditional least squares estimators.

[3] We considered estimators of the jumps in the integrated spectrum in this section, but we have yet to discuss their distributions. Here we derive the distribution for a jump estimator appropriate for the model of Equation (513) with (1) $\mu = 0$, (2) f set to $f_k = k/(N \Delta_t)$ for some integer k such that $1 \leq k < N/2$ and (3) $\{\epsilon_t\}$ assumed to be Gaussian. Under these restrictive assumptions, the exact least squares estimators $\hat{A}$ and $\hat{B}$ for A and B are given by Equation (549a), and standard least square theory says that $\hat{A}$ and $\hat{B}$ are Gaussian RVs with means A and B and a covariance matrix given by $\sigma_\epsilon^2 (\boldsymbol{H}^T \boldsymbol{H})^{-1}$, where

$$\boldsymbol{H}^T = \begin{bmatrix} 1 & \cos(2\pi f_k \Delta_t) & \cos(4\pi f_k \Delta_t) & \cdots & \cos([N-1]2\pi f_k \Delta_t) \\ 0 & \sin(2\pi f_k \Delta_t) & \sin(4\pi f_k \Delta_t) & \cdots & \sin([N-1]2\pi f_k \Delta_t) \end{bmatrix}$$

(see, e.g., Weisberg, 2014). We can appeal to Exercise [1.3c] to obtain the elements of $\boldsymbol{H}^T \boldsymbol{H}$ (and hence $(\boldsymbol{H}^T \boldsymbol{H})^{-1}$), thus establishing that $\hat{A}$ and $\hat{B}$ are uncorrelated (and hence independent due to Gaussianity) and have a common variance of $2\sigma_\epsilon^2 / N$. The true jump is $J = (A^2 + B^2)/4$, and the obvious jump estimator is $\hat{J} = (\hat{A}^2 + \hat{B}^2)/4$ (cf. the left-hand side of Equation (549b)), which is a quadratic form in Gaussian RVs. Now, if U and V are independent Gaussian RVs with means μ_U and μ_V and with unit variances, then the RV $\chi_2^2(\beta) = U^2 + V^2$ is said to obey a *noncentral chi-square distribution* with two degrees of freedom and with noncentrality parameter $\beta = \mu_U^2 + \mu_V^2$ (see, e.g., Anderson, 2003; a related RV arose in Equation (542c) concerning Siegel's test statistic). After a normalization to satisfy the unit variance criterion, we can claim that the normalized jump estimator $2N\hat{J}/\sigma_\epsilon^2$ is distributed as $\chi_2^2(\beta)$ with $\beta = N(A^2 + B^2)/(2\sigma_\epsilon^2) = NR$, where R is the signal-to-noise ratio of Equation (526d). We need to know σ_ϵ^2 to formulate this normalized estimator; in the usual case where σ_ϵ^2 is unknown, we can use the standard estimator $\hat{\sigma}_\epsilon^2 = \text{SS}(\hat{A}, \hat{B}, f_k)/(N-2)$, where SS is defined as per Equation (549d), and $(N-2)\hat{\sigma}_\epsilon^2/\sigma_\epsilon^2$ has the distribution of an ordinary χ_{N-2}^2 RV. Now the ratio of a $\chi_2^2(\beta)/2$ RV to an independent $\chi_{N-2}^2/(N-2)$ RV yields an RV obeying a *noncentral F-distribution* with 2 and $N-2$ degrees of freedom and with noncentrality parameter β (Anderson, 2003). Let $F_{2,N-2,\beta}(p)$ represent the $p \times 100\%$ percentage point for this distribution. Since the statistic $N\hat{J}/\hat{\sigma}_\epsilon^2$ is distributed as such, the distribution of $\hat{J}$ is dictated by

$$\mathbf{P}[\hat{J} \leq \hat{\sigma}_\epsilon^2 F_{2,N-2,\beta}(p)/N] = p. \tag{552}$$

Unfortunately this distribution depends on the typically unknown signal-to-noise ratio R through β, a fact that complicates using the above to generate a confidence interval (CI) for the true jump J. Exercise [10.15] considers using Equation (552) with a plug-in approach for creating CIs, but, while this works reasonably well for certain settings for R, it can lead to disconcertingly conservative CIs for others.

10.12 A Parametric Approach to Harmonic Analysis

Autoregressive (AR) spectral estimation, as discussed in Chapter 9, is thought of by statisticians as a method for estimating SDFs, i.e., a spectrum with only a purely continuous component and no line components. However, this technique has been used in applied work to estimate the frequencies of line components, perhaps immersed in a background continuum (Ulrych, 1972b; Chen and Stegen, 1974; Satorius and Zeidler, 1978; Kane and de Paula, 1996; Sutcliffe et al., 2013; Tary et al., 2014). We investigate the AR approach to spectral line estimation in this section.

Consider a *deterministic* real sinusoid of the form $x_t = D\cos(2\pi f t \Delta_t + \phi)$, $t \in \mathbb{Z}$, where $0 < f \leq 1/(2\Delta_t)$, $D > 0$ and $-\pi < \phi \leq \pi$.

▷ **Exercise [553]** Show that $\{x_t\}$ gives rise to the second-order difference equation

$$x_t = 2\cos(2\pi f \Delta_t) x_{t-1} - x_{t-2}. \tag{553a}$$

Hint: write the cosines as complex exponentials. ◁

Equation (553a) does not depend on the fixed amplitude D or phase ϕ. These come into play by specifying initial conditions, which we can take to be setting any two adjacent values in $\{x_t\}$, e.g., $x_0 = D\cos(\phi)$ and $x_1 = D\cos(2\pi f \Delta_t + \phi)$. With these conditions, we can use Equation (553a) and the related equation $x_{t-2} = 2\cos(2\pi f \Delta_t) x_{t-1} - x_t$ to generate the remaining terms in $\{x_t\}$. (Exercise [10.16] explores an alternative way of creating a sinusoidal sequence satisfying Equation (553a) by specifying f, x_0 and x_1 rather than f, D and ϕ.)

Now consider a *randomly phased* sinusoid of the form $X_t = D\cos(2\pi f t \Delta_t + \phi)$, where ϕ is an RV uniformly distributed between $-\pi$ and π. The process $\{X_t\}$ satisfies

$$X_t = 2\cos(2\pi f \Delta_t) X_{t-1} - X_{t-2}, \tag{553b}$$

which is analogous to Equation (553a); here the initial conditions $X_0 = D\cos(\phi)$ and $X_1 = D\cos(2\pi f \Delta_t + \phi)$ involve the RV ϕ and the constant D. We call the above X_t a *pseudo-AR(2) process*. This terminology is meant to remind us that Equation (553b) lacks one aspect of the usual AR(2) process – the innovation term ϵ_t. If we write Equation (553b) as

$$X_t = \varphi_{2,1} X_{t-1} + \varphi_{2,2} X_{t-2},$$

it is easy to verify that the roots of the polynomial equation

$$1 - \varphi_{2,1} z^{-1} - \varphi_{2,2} z^{-2} = 0$$

are $z = \exp(\pm i 2\pi f \Delta_t)$, both of which are on the unit circle. We know that, for a randomly phased sinusoid, we hace $\text{cov}\{X_{t+\tau}, X_t\} = D^2 \cos(2\pi f \tau \Delta_t)/2$ (cf. Equation (37b) with $L = 1$ and with Δ_t inserted or Equation (519e) with no noise). Exercise [10.17] is to derive the same result directly from Equation (553b).

Just as a single sinusoid satisfies a second-order difference equation, the summation of p sinusoids satisfies a difference equation of order $2p$. If the roots $\{z_l\}$ of the polynomial equation

$$1 - \sum_{k=1}^{2p} \varphi_{2p,k} z^{-k} = 0 \tag{553c}$$

are all on the unit circle in conjugate pairs so that $z_l = \exp(\pm i 2\pi f_l \Delta_t)$, $l = 1, 2, \ldots, p$, then the pseudo-AR($2p$) equation

$$X_t = \sum_{k=1}^{2p} \varphi_{2p,k} X_{t-k} \tag{553d}$$

has a solution given by

$$X_t = \sum_{l=1}^{p} D_l \cos\left(2\pi f_l t \, \Delta_t + \phi_l\right) \tag{554a}$$

(cf. Equation (511) with μ set to zero; proof of the above constitutes Exercise [10.18]). As an example of how to obtain the $\varphi_{2p,k}$ coefficients given the f_l frequencies, consider Equation (553c) for the case of $p = 2$ sinusoids with $\Delta_t = 1$. We have

$$1 - \varphi_{4,1} e^{i 2\pi f_1} - \varphi_{4,2} e^{i 4\pi f_1} - \varphi_{4,3} e^{i 6\pi f_1} - \varphi_{4,4} e^{i 8\pi f_1} = 0$$

and

$$1 - \varphi_{4,1} e^{i 2\pi f_2} - \varphi_{4,2} e^{i 4\pi f_2} - \varphi_{4,3} e^{i 6\pi f_2} - \varphi_{4,4} e^{i 8\pi f_2} = 0.$$

The real and imaginary parts of the above lead to the matrix equation

$$\begin{bmatrix} \cos(2\pi f_1) & \cos(4\pi f_1) & \cos(6\pi f_1) & \cos(8\pi f_1) \\ \sin(2\pi f_1) & \sin(4\pi f_1) & \sin(6\pi f_1) & \sin(8\pi f_1) \\ \cos(2\pi f_2) & \cos(4\pi f_2) & \cos(6\pi f_2) & \cos(8\pi f_2) \\ \sin(2\pi f_2) & \sin(4\pi f_2) & \sin(6\pi f_2) & \sin(8\pi f_2) \end{bmatrix} \begin{bmatrix} \varphi_{4,1} \\ \varphi_{4,2} \\ \varphi_{4,3} \\ \varphi_{4,4} \end{bmatrix} = \begin{bmatrix} 1 \\ 0 \\ 1 \\ 0 \end{bmatrix}, \tag{554b}$$

from which $\varphi_{4,1}, \ldots, \varphi_{4,4}$ can be obtained (see, however, Exercise [10.19], which argues that we must have $\varphi_{4,4} = -1$ and that we can use a related three-dimensional matrix equation to solve for $\varphi_{4,1}, \varphi_{4,2}$ and $\varphi_{4,3}$).

So far we have only considered perfectly observed sinusoids. We now consider randomly phased sinusoids plus white noise. For p sinusoids observed with zero mean white noise $\{\alpha_t\}$, the observed process is the sum of a pseudo-AR($2p$) process and white noise:

$$\widetilde{X}_t = X_t + \alpha_t = \sum_{k=1}^{2p} \varphi_{2p,k} X_{t-k} + \alpha_t, \tag{554c}$$

where $\text{cov}\{\alpha_{t+\tau}, X_t\} = 0$ for all integers τ; i.e., the noise $\{\alpha_t\}$ is assumed to be uncorrelated with the "signal" $\{X_t\}$. If we substitute $X_{t-k} = \widetilde{X}_{t-k} - \alpha_{t-k}$ into the right-hand side of the above, we get

$$\widetilde{X}_t - \sum_{k=1}^{2p} \varphi_{2p,k} \widetilde{X}_{t-k} = \alpha_t - \sum_{k=1}^{2p} \varphi_{2p,k} \alpha_{t-k}, \tag{554d}$$

which is an ARMA($2p, 2p$) process whose AR and MA coefficients are identical. Such a result was first derived by Ulrych and Clayton (1976), but without the stipulation that sinusoids were randomly phased and hence stochastic. Kay and Marple (1981, p. 1403) report a method (essentially due to Pisarenko, 1973) for estimating the parameters $\varphi_{2p,k}$ in Equation (554c).

It is important to clearly distinguish between a pseudo-AR process (i.e., one without an innovation term) plus white noise and a standard AR process plus white noise. We have just seen how the former – for the case of p randomly phased sinusoids plus white noise – yields an ARMA($2p, 2p$) process with identical AR and MA coefficients. In the latter case, however, an ARMA process with an equal number of AR and MA coefficients is again obtained but now with *distinct* coefficients (Walker, 1960; Tong, 1975; Friedlander, 1982). To see this, let $\{Y_t\}$ be a standard AR(p) process with innovation variance σ_p^2 as in Equation (446a), and let $\{\xi_t\}$ be a white noise process with variance σ_ξ^2 that is uncorrelated with $\{Y_t\}$. Then the process defined by $\widetilde{Y}_t = Y_t + \xi_t$ has an SDF given by

$$S(f) = \frac{\sigma_p^2 \, \Delta_t}{\left|1 - \sum_{k=1}^{p} \phi_{p,k} e^{-i 2\pi f k \, \Delta_t}\right|^2} + \sigma_\xi^2 \, \Delta_t$$

$$= \frac{\sigma_p^2 \, \Delta_t + \sigma_\xi^2 \, \Delta_t \left|1 - \sum_{k=1}^{p} \phi_{p,k} e^{-i 2\pi f k \, \Delta_t}\right|^2}{\left|1 - \sum_{k=1}^{p} \phi_{p,k} e^{-i 2\pi f k \, \Delta_t}\right|^2}.$$

where we have evoked Equation (446b), along with the fact that the SDF for the white noise process is $\sigma_\xi^2 \Delta_t$ and the result of Exercise [4.12]. If we let

$$\sigma_p^2 + \sigma_\xi^2 \left| 1 - \sum_{k=1}^p \phi_{p,k} e^{-i2\pi f k \Delta_t} \right|^2 = \sigma_\zeta^2 \left| 1 - \sum_{k=1}^p \theta_{p,k} e^{-i2\pi f k \Delta_t} \right|^2$$

by appropriately defining σ_ζ^2 and $\theta_{p,k}$, we obtain

$$S(f) = \sigma_\zeta^2 \Delta_t \frac{\left| 1 - \sum_{k=1}^p \theta_{p,k} e^{-i2\pi f k \Delta_t} \right|^2}{\left| 1 - \sum_{k=1}^p \phi_{p,k} e^{-i2\pi f k \Delta_t} \right|^2}, \qquad (555a)$$

which is the SDF for an ARMA(p, p) process with AR parameters $\phi_{p,k}$, MA parameters $\theta_{p,k}$ and innovation variance σ_ζ^2. The $\theta_{p,k}$ terms are no longer identical to the $\phi_{p,k}$ terms, but note that the $p+1$ values $\{\theta_{p,k}\}$ and σ_ζ^2 are determined by the $p+2$ values $\{\phi_{p,k}\}$, σ_p^2 and σ_ξ^2. (Proof that Equation (555a) gives the resulting SDF is the subject of Exercise [10.20].)

Since a standard AR(p) process plus white noise is an ARMA(p, p) process, the AR parameters can be estimated from the ARMA(p, p) nature of the observed process. This is done by Tong (1975) and Friedlander (1982). These methods, however, do not make use of the fact that there are only $p+2$ free parameters, but rather they treat the model as if there were $2p+2$ free parameters ($\phi_{p,1}, \ldots, \phi_{p,p}; \theta_{p,1}, \ldots, \theta_{p,p}; \sigma_p^2$ and σ_ξ^2); i.e., the MA parameters are not considered related to the other $p+2$ parameters. Jones (1980) and Tugnait (1986) present algorithms for solving the AR parameters that involve only the $p+2$ free parameters.

As a simple illustration of the AR approach to sinusoid estimation, we consider an example due to Makhoul (1981b) of a deterministic sinusoid with a frequency of $f = 1/6\,\text{Hz}$ and a sampling interval of $\Delta_t = 1$ sec. With these values substituted into Equation (553a), the appropriate difference equation becomes

$$x_t = 2\cos(\pi/3) x_{t-1} - x_{t-2} = x_{t-1} - x_{t-2}. \qquad (555b)$$

For convenience let us assume that the amplitude of the sinusoid is $D = 1/\cos(\pi/6)$ and its phase is $\phi = -\pi/2$. The initial conditions are thus

$$x_0 = D\cos(\phi) = D\cos(-\pi/2) = 0$$

and

$$x_1 = D\cos(2\pi f + \phi) = D\cos(-\pi/6) = 1,$$

and subsequent values are

$$x_2 = x_1 - x_0 = 1, \qquad x_3 = x_2 - x_1 = 0,$$
$$x_4 = x_3 - x_2 = -1, \qquad x_5 = x_4 - x_3 = -1,$$

etc.

Suppose that we have only the samples x_0, x_1, x_2 and x_4 and that we seek to estimate the parameters $\phi_{2,1}$, $\phi_{2,2}$ and σ_2^2 of a *standard* AR(2) model. If we do so by using the forward/backward least squares (FBLS) estimator, we must find the values of $\phi_{2,1}$ and $\phi_{2,2}$ such that the forward/backward sum of squares

$$\sum_{t=2}^3 \left[(x_t - \phi_{2,1} x_{t-1} - \phi_{2,2} x_{t-2})^2 + (x_{t-2} - \phi_{2,1} x_{t-1} - \phi_{2,2} x_t)^2 \right] \qquad (555c)$$

is minimized (cf. Equation (477d)). If we call the solution points $\hat{\phi}_{2,1}$ and $\hat{\phi}_{2,2}$, we find that the minimum sum of squares is *zero* when $\hat{\phi}_{2,1} = 1$ and $\hat{\phi}_{2,2} = -1$; i.e., the fitted model is *exactly* the true model of Equation (555b). Note that the innovation term ϵ_t can be dropped from the fitted model – it has a mean of zero by assumption, and our estimate of the innovation variance would be zero, implying that $\epsilon_t = 0$ for all t.

If now we consider Burg's algorithm instead (Section 9.5), we must first fit an AR(1) model and estimate $\phi_{1,1}$ by minimizing the forward/backward sum of squares

$$\sum_{t=1}^{3} \left[(x_t - \phi_{1,1} x_{t-1})^2 + (x_{t-1} - \phi_{1,1} x_t)^2 \right] \tag{556a}$$

(see Equation (468a)). It follows from Exercise [9.16a] that the estimate is

$$\bar{\phi}_{1,1} = \frac{2 \sum_{t=1}^{3} x_t x_{t-1}}{x_0^2 + 2 \sum_{t=1}^{2} x_t^2 + x_3^2} = 1/2, \tag{556b}$$

after substitution of the sample values. We now use the Levinson–Durbin relationship

$$\phi_{2,1} = \phi_{1,1} - \phi_{2,2} \phi_{1,1} = \phi_{1,1} (1 - \phi_{2,2}).$$

If we now replace $\phi_{2,1}$ in Equation (555c) with the right-hand side of the above (with $\bar{\phi}_{1,1}$ replacing $\phi_{1,1}$), the next step in Burg's algorithm is the minimization of

$$\sum_{t=2}^{3} \left(\left[x_t - (1 - \phi_{2,2}) \bar{\phi}_{1,1} x_{t-1} - \phi_{2,2} x_{t-2} \right]^2 + \left[x_{t-2} - (1 - \phi_{2,2}) \bar{\phi}_{1,1} x_{t-1} - \phi_{2,2} x_t \right]^2 \right) \tag{556c}$$

with respect to $\phi_{2,2}$. With the data substituted we obtain $\bar{\phi}_{2,2} = -1$, and hence $\bar{\phi}_{2,1} = (1 - \bar{\phi}_{2,2}) \bar{\phi}_{1,1} = 1$. These are the same estimates as obtained previously. (Makhoul, 1981b, appears to have miscalculated $\bar{\phi}_{1,1}$ and hence derived poor estimates of $\phi_{2,1}$ and $\phi_{2,2}$ from Burg's algorithm.)

As another example of the parametric approach to harmonic analysis, let us consider the stochastic process of Equation (537b) that we discussed previously in the context of tapering (see Figure 538b). This process consists of a single sinusoid with a frequency of $f_1 = 0.0725$ plus a small amount of white noise. We use the FBLS estimator to fit AR(16) models to the same 50 realizations entertained in Figure 538b. For each realization, we determine the frequency of the peak value in the AR SDF estimate (see C&E [3] for details on how to find this peak location). Figure 538b shows a scatter plot of peak locations as determined from a Hanning-based direct spectral estimator $\hat{S}^{(D)}(\cdot)$ (vertical axis) and the periodogram $\hat{S}^{(P)}(\cdot)$ (horizontal). Figure 557 has a similar scatter plot, but now we display the AR-based peak locations on the horizontal axis (the locations for $\hat{S}^{(D)}(\cdot)$ are again shown on the vertical axis). Note that there is no indication of bias in either estimator of f_1; however, the sample variance of the estimator based upon $\hat{S}^{(D)}(\cdot)$ is larger by a factor of 1.5. This example illustrates the usefulness of the parametric approach to harmonic analysis, but we must view its apparent superiority over the nonparametric approach with caution: first, we have used the standard Hanning taper rather than seeking an optimal taper for peak estimation for the model of Equation (537b) and, second, we set $p = 16$ by optimizing over AR orders 2 to 32 via a Monte Carlo experiment whose replication is the burden of Exercise [10.21]. (Use of the FBLS method is important here – the Yule–Walker and Burg estimators perform poorly on this example, with increases in sample variances by factors of, respectively, 5.1 and 676.6 over that for the FBLS estimator.)

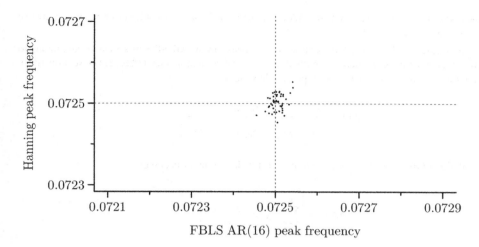

Figure 557 Scatter plot of locations of peak frequencies computed from a Hanning-based direct spectral estimator $\hat{S}^{(D)}(\cdot)$ (vertical axis) versus locations determined from an AR(16) SDF estimator with parameters estimated using the forward/backward least squares method (horizontal axis). Each point in the scatter plot corresponds to location estimators for one of 50 different realizations from the model of Equation (537b). The true location of the peak frequency is 0.0725 (indicated by the vertical and horizontal lines). (This figure should be compared with Figure 538b, which makes use of the same 50 realizations.)

Comments and Extensions to Section 10.12

[1] The discussion following Equation (446c) states that a necessary and sufficient condition for the existence of a stationary solution to the AR($2p$) process

$$Y_t = \sum_{k=1}^{2p} \phi_{2p,k} Y_{t-k} + \epsilon_t$$

(cf. Equation (446a)) is that *none* of the solutions of the polynomial equation

$$1 - \sum_{k=1}^{2p} \phi_{2p,k} z^{-k} = 0$$

lies on the unit circle in the complex plane. In this section we noted that a randomly phased harmonic process with p components can be represented as

$$X_t = \sum_{k=1}^{2p} \varphi_{2p,k} X_{t-k}$$

(Equation (553d)). This harmonic process is stationary, but its associated polynomial equation

$$1 - \sum_{k=1}^{2p} \varphi_{2p,k} z^{-k} = 0$$

(Equation (553c)) has *all* of its roots on the unit circle. This stark contrast in the conditions for stationarity is entirely due to the innovation term ϵ_t with nonzero variance and highlights the importance of carefully distinguishing between AR processes $\{Y_t\}$ and pseudo-AR processes $\{X_t\}$.

[2] For readers familiar with the notion of a backward shift operator and its use in expressing stationary nondeterministic ARMA processes (see, for example, Box et al., 2015), we point out the fact – unfortunately sometimes overlooked in the literature – that, since the roots of Equation (553c) are on the unit circle, the operator common to the AR and MA components of Equation (554d) does not have an inverse and hence cannot be canceled out.

[3] We can accurately determine the locations of the peaks in an AR SDF in a manner analogous to that used to locate the peak frequencies in the direct spectral estimators (see C&E [2] for Section 10.6). Recall that the SDF for a stationary AR(p) process has the form

$$S(f) = \frac{\sigma_p^2 \Delta_t}{\left|1 - \sum_{j=1}^{p} \phi_{p,j} e^{-i2\pi f j \Delta_t}\right|^2}$$

(this is Equation (446b)). With $\phi_{p,0}$ taken to be -1, the denominator becomes

$$D(f) = \left|\sum_{j=0}^{p} \phi_{p,j} e^{-i2\pi f j \Delta_t}\right|^2.$$

We can regard $D(\cdot)$ as proportional to the periodogram for a "time series" $X_j = \phi_{p,j}$ of length $N = p+1$. Comparing the above to Equation (170d) and then taking into consideration Equations (170c) and (170b), we can write

$$D(f) = (p+1) \sum_{\tau=-p}^{p} \hat{s}_{\phi,\tau}^{(P)} e^{-i2\pi f \tau \Delta_t}, \quad \text{where } \hat{s}_{\phi,\tau}^{(P)} = \frac{1}{p+1} \sum_{j=0}^{p-|\tau|} \phi_{p,j+|\tau|} \phi_{p,j}.$$

Since a peak in $S(\cdot)$ corresponds to a valley in $D(\cdot)$, we can use the Newton–Raphson method described following Equation (528a) – with $\hat{s}_{\phi,\tau}^{(P)}$ substituted for $\hat{s}_{\tau}^{(P)}$ – to determine the valley location. This technique requires a good initial estimate, which we can usually obtain by evaluating $D(\cdot)$ on a fine enough grid of frequencies via a DFT of $\phi_{p,0}, \phi_{p,1}, \ldots, \phi_{p,p}$ padded with lots of zeros (for details, see the discussion surrounding Equation (449b); in creating Figure 557, padding with $256 - 17 = 239$ zeros sufficed, but $128 - 17 = 111$ did not). As we noted previously in the discussion following Equation (528a), the Newton–Raphson method can fail if the initial estimate is poor; in these cases, we can use a bisection technique (combined with Newton–Raphson) or Brent's method if the location can be bracketed. Since we are now searching for a minimum rather than a maximum, we must modify the bracketing conditions to be $g'(\omega^{(L)}) < 0$ and $g'(\omega^{(U)}) > 0$.

10.13 Problems with the Parametric Approach

For a process consisting of a sinusoid in additive noise, two basic problems have been identified with the AR spectral estimates. Chen and Stegen (1974) found the location of the peak in the estimate to depend on the phase of the sinusoid, while Fougere et al. (1976) found that the estimate can sometimes contain two adjacent peaks in situations where only a single peak should appear – this is known in the literature as *spectral line splitting* or *spontaneous line splitting*. Causes and cures for these problems have been the subject of many papers – see Kay and Marple (1981, p. 1396) for a nice summary. Here we give a perspective on the historic literature.

Toman (1965) looked at the effect on the periodogram of short truncation lengths T of the *continuous* parameter deterministic sinusoid $X(t) = D\cos(2\pi f_0 t - \pi/2) = D\sin(2\pi f_0 t)$, where $f_0 = 1\,\text{Hz}$ so that the corresponding period is 1 sec. The data segment is taken to be from $t = 0$ to $t = T$, and the periodogram is here defined as

$$\frac{1}{T}\left|\int_0^T X(t) e^{-i2\pi f t}\,dt\right|^2.$$

10.13 Problems with the Parametric Approach

Frequency (Hz)	Phase	$f_\mathcal{N}$	N	p	Peak Shift	Line Splitting
Ulrych, 1972a:						
1	$-\pi/2$	10	21	11	no	no
1	0	10	21	11	no	no
1	$-\pi/2$	10	12	11	no	no
Chen and Stegen, 1974:						
1	$-\pi/2$	10	24	2,8	no	no
1	$-\pi/2$	10	24	20	yes	yes
1	$-\pi/2$	10	10–60	8	oscillates	—
1	varied	10	15	8	oscillates	—
1	varied	10	12–65	8	oscillates	—
Fougere et al., 1976:						
1	$-\pi/2\,(\pi/9)\,\pi/2$	10	21	19	no	no
5	$-\pi/2\,(\pi/90)\,\pi/2$	10	6	5	yes	yes
1.25 (2) 49.25	$-\pi/4$	50	101	24	yes (all)	yes (all)

Table 559 Summary of simulation results for a single sinusoid with additive white noise (estimation by Burg's algorithm). The first column gives the frequency of the sinusoid; the second column is phase (with respect to a cosine term); $f_\mathcal{N}$ is the Nyquist frequency; N is the sample size; and p is the order of the autoregressive model. The notation $x\,(a)\,y$ means from x to y in steps of a.

Note that, because the period is unity, the ratio of the length of the data segment to the period is just T. Toman demonstrated that, for $0 \leq T \leq 0.58$, the location of the peak value of the periodogram is at $f = 0$; as T increases from 0.58 to 0.716, the location of the peak value of the periodogram increases from 0 to 1 Hz (the latter being the true sinusoidal frequency f_0); for T greater than 0.716, the peak frequency oscillates, giving 1 Hz at the maximum of each oscillation and converging to 1 Hz as $T \to \infty$. For the special case $T = 1$ corresponding to one complete cycle of the sinusoid, the peak occurred at 0.84 Hz.

Jackson (1967) clearly and concisely explained Toman's experimental results. He noted that, with $X(t) = D\cos(2\pi f_0 t + \phi)$,

$$\int_0^T X(t)\mathrm{e}^{-\mathrm{i}2\pi ft}\,\mathrm{d}t = \frac{D}{2}\mathrm{e}^{\mathrm{i}\phi}\mathrm{e}^{-\mathrm{i}\pi(f-f_0)T}\frac{\sin(\pi(f-f_0)T)}{\pi(f-f_0)}$$
$$+ \frac{D}{2}\mathrm{e}^{-\mathrm{i}\phi}\mathrm{e}^{-\mathrm{i}\pi(f+f_0)T}\frac{\sin(\pi(f+f_0)T)}{\pi(f+f_0)}.$$

For the special case considered by Toman, namely, $\phi = -\pi/2$, $f_0 = 1$ and $T = 1$, the above reduces to

$$\mathrm{i}\frac{D}{2}\mathrm{e}^{-\mathrm{i}\pi f}\left(\operatorname{sinc}(f-1) - \operatorname{sinc}(f+1)\right),$$

where, as usual, $\operatorname{sinc}(t) = \sin(\pi t)/(\pi t)$ is the sinc function. The periodogram is thus proportional to $|\operatorname{sinc}(f-1) - \operatorname{sinc}(f+1)|^2$. Jackson noted that the interference of the two sinc functions produces extrema at $\pm 0.84f$, as observed by Toman. For a single cycle of a cosine, i.e., $\phi = 0$, the maxima occur at $\pm 1.12f$. Jackson similarly explained Toman's other observations. His explanation is obviously in agreement with our example of interference from two Fejér's kernels (see the discussion concerning Equation (536)).

Ulrych (1972a) – see also Ulrych (1972b) – thought that such spectral shifts could be prevented by the use of Burg's algorithm. In his example – see Table 559 – neither shifting nor splitting was present even with additive white noise. Chen and Stegen (1974) carried out a more detailed investigation of Burg's algorithm using as input 1 Hz sinusoids sampled at 20 samples per second with white noise superimposed. Table 559 summarizes some of their results. The particular parameter combinations used by Ulrych (1972a) happen to coincide with points in the oscillating location of the spectral peak that agree with the true value of 1 Hz; in general, however, this is not the case, so that Burg's method is indeed susceptible to these phenomena.

Fougere et al. (1976) noted that spectral line splitting is most likely to occur when

[1] the signal-to-noise ratio is high;
[2] the initial phase of sinusoidal components is some odd multiple of $45°$; and
[3] the time duration of the data sequence is such that the sinusoidal components have an odd number of quarter cycles.

Other historic papers of interest are Fougere (1977, 1985). Table 559 gives a summary of some of Fougere's results. In the 1 Hz example, an AR(19) model with 21 data points produced no anomalous results for a range of input phases; on the other hand, an AR(24) model with 101 data points produced line splitting for a range of frequencies for the input sinusoids. Fougere (1977) concluded that "... splitting does not occur as a result of using an overly large number of filter weights [AR coefficients], as is frequently claimed in the literature." Chen and Stegen (1974) had reached the *opposite* conclusion based on the second of their examples given in Table 559.

Chen and Stegen (1974) considered that the frequency shifts they observed were due to essentially the same mechanism as observed by Jackson (1967). However, Fougere (1977) notes that "Jackson's worst cases were sine waves an even number of cycles long with either $90°$ or $0°$ initial phase." For these two cases Burg's spectra are neither split nor shifted but are extremely accurate.

Fougere (1977) put the problem with line splitting in Burg's algorithm down to the substitution of estimated $\phi_{k,k}$ from lower order fitted AR models into the higher order fitted models (as we have seen, the forward/backward sums of squares from the final model order are not necessarily minimized using Burg's algorithm due to the substitution from lower order fits). Fougere suggested setting $\phi_{k,k} = U \sin(\Phi_k)$ for $k = 1, 2, \ldots, p$, with U slightly less than 1, and solving for the Φ_k terms *simultaneously* using a nonlinear optimization scheme. Once Φ_k has been estimated by $\hat{\Phi}_k$, the corresponding estimates for $\phi_{k,k}$, namely, $\hat{\phi}_{k,k} = U \sin(\hat{\Phi}_k)$, must be less than unity as required for stability. Fougere demonstrated that, at least in the single sinusoid examples examined, line splitting could be eliminated by using this procedure. (Bell and Percival, 1991, obtained similar results with a "two-step" Burg algorithm, which estimates the reflection coefficients in pairs and hence can be regarded as a compromise between Burg's algorithm and Fougere's method.)

Kay and Marple (1979) diagnosed both a different cause and cure from the suggestions of Fougere. They considered that

... spectral line splitting is a result of estimation errors and is not inherent in the autoregressive approach. In particular, the interaction between positive and negative sinusoidal frequency components [Jackson's mechanism] in the Burg reflection coefficient and Yule–Walker autocorrelation estimates and the use of the biased estimator in the Yule–Walker approach are responsible for spectral line splitting.

Their cure (for one sinusoid) was the use of *complex-valued data* and the *unbiased* ACVS estimator in the Yule–Walker case. To follow their argument, consider a noise-free randomly phased sinusoid with unit amplitude, $X_t = \cos(2\pi f t\, \Delta_t + \phi)$, and take the time series to be $X_0, X_1, \ldots, X_{N-1}$. The biased estimator of the ACVS is

$$\hat{s}_\tau^{(P)} = \frac{1}{N} \sum_{t=0}^{N-|\tau|-1} X_{t+|\tau|} X_t$$

since $E\{X_t\} = 0$ under the random phase assumption (this is Equation (170b)). Hence

$$\hat{s}_\tau^{(P)} = \frac{1}{N} \sum_{t=0}^{N-|\tau|-1} \cos([2\pi f t\, \Delta_t + \phi] + 2\pi f |\tau|\, \Delta_t) \cos(2\pi f t\, \Delta_t + \phi)$$

$$= \frac{N-|\tau|}{2N} \cos(2\pi f \tau\, \Delta_t) + \cos(2\pi [N-1]f\, \Delta_t + 2\phi) \frac{\sin(2\pi[N-|\tau|]f\, \Delta_t)}{2N \sin(2\pi f\, \Delta_t)} \quad (561\text{a})$$

(the proof of this result is Exercise [10.22]). This can be written in terms of the true ACVS $s_\tau = \cos(2\pi f \tau\, \Delta_t)/2$ as

$$\hat{s}_\tau^{(P)} = \frac{N-|\tau|}{N} s_\tau + \cos(2\pi[N-1]f\, \Delta_t + 2\phi) \frac{\sin(2\pi[N-|\tau|]f\, \Delta_t)}{2N \sin(2\pi f\, \Delta_t)}; \quad (561\text{b})$$

i.e., $\hat{s}_\tau^{(P)}$ is composed of one term that is a biased estimator of s_τ and a second term that is phase dependent. Both these terms would contribute to an inaccurate spectral estimate using the Yule–Walker equations based on $\{\hat{s}_\tau^{(P)}\}$.

Suppose now that N is large enough that the phase dependent term is close to zero. Then

$$\hat{s}_\tau^{(P)} \approx \frac{N-|\tau|}{N} s_\tau = \left(1 - \frac{|\tau|}{N}\right) s_\tau. \quad (561\text{c})$$

Kay and Marple (1979) argue that the estimator $\hat{s}_\tau^{(P)}$ "... corresponds more nearly to the ACVS of two sinusoids which beat together to approximate a linear tapering." Since an AR(p) spectral estimate fitted by the Yule–Walker method has an ACVS that is *identical* with the sample ACVS up to lag p, the spectral estimate should have two – rather than only one – spectral peaks. As a concrete example, they consider the case $f = 1/(4\,\Delta_t) = f_\mathcal{N}/2$ and $\phi = 0$ when N is odd, for which

$$\hat{s}_\tau^{(P)} = \left(1 - \frac{|\tau|-1}{N}\right) s_\tau. \quad (561\text{d})$$

Setting $N = 9$, line splitting occurs when $p = 8$ (verification of these results is part of Exercise [10.23]). Note that the model used by Kay and Marple is a noise-free sinusoid. This is reasonable for investigating line splitting, since the phenomenon has been seen to occur at *high* signal-to-noise ratios.

Kay and Marple (1979) argue that the phase-dependent term of Equation (561a), i.e., the term involving $\cos(2\pi[N-1]f\, \Delta_t + 2\phi)$, can be regarded as the interaction between complex exponentials since it can be written as

$$\frac{1}{4N} \sum_{t=0}^{N-|\tau|-1} \left(e^{i(2\pi f[2t+|\tau|]\, \Delta_t + 2\phi)} + e^{-i(2\pi f[2t+|\tau|]\, \Delta_t + 2\phi)} \right)$$

(the above falls out as part of the proof of Equation (561a) that Exercise [10.22] calls for). This fact led them to consider the *analytic series* associated with $\{X_t\}$. By definition this series is $X_t + \mathrm{i}\mathcal{HT}\{X_t\}$, where $\mathcal{HT}$ is the *Hilbert transform* of $\{X_t\}$ (Papoulis and Pillai, 2002, section 9–3). For our purposes, we can regard the Hilbert transform of a series as a phase-shifted version of the series (the shift is $\pi/2$ for $f < 0$ and $-\pi/2$ for $f > 0$); thus, if $X_t = \cos{(2\pi f t \,\Delta_{\mathrm{t}} + \phi)}$, then $\mathcal{HT}\{X_t\} = \sin{(2\pi f t \,\Delta_{\mathrm{t}} + \phi)}$ if $f > 0$ and $\mathcal{HT}\{X_t\} = -\sin{(2\pi f t \,\Delta_{\mathrm{t}} + \phi)}$ if $f < 0$. Assuming $f > 0$ in our example, the analytic series is thus

$$\cos{(2\pi f t \,\Delta_{\mathrm{t}} + \phi)} + \mathrm{i}\sin{(2\pi f t \,\Delta_{\mathrm{t}} + \phi)} = \mathrm{e}^{\mathrm{i}(2\pi f t \,\Delta_{\mathrm{t}} + \phi)} \stackrel{\text{def}}{=} Z_t,$$

say.

▷ **Exercise [562]** Show that, for $0 \leq \tau \leq N-1$,

$$\hat{s}_{Z,\tau}^{(\mathrm{P})} \stackrel{\text{def}}{=} \frac{1}{N} \sum_{t=0}^{N-\tau-1} Z_{t+\tau} Z_t^* = \frac{N-\tau}{N} \mathrm{e}^{\mathrm{i}2\pi f \tau \,\Delta_{\mathrm{t}}}.$$ ◁

The above is the biased ACVS estimator appropriate for zero-mean complex-valued data (see Equation (231)). The corresponding unbiased estimator is

$$\hat{s}_{Z,\tau}^{(\mathrm{U})} = \frac{1}{N-\tau} \sum_{t=0}^{N-\tau-1} Z_{t+\tau} Z_t^* = \frac{N}{N-\tau} \hat{s}_{Z,\tau}^{(\mathrm{P})} = \mathrm{e}^{\mathrm{i}2\pi f \tau \,\Delta_{\mathrm{t}}}. \qquad (562)$$

This is exactly the ACVS of a complex exponential with unit amplitude (cf. Equation (519j)). Since this estimate equals the exact or theoretical ACVS, the fact that the AR(p) process fitted by the Yule–Walker method has a theoretical ACVS that is identical with the estimated ACVS up to lag p will prevent the possibility of line splitting. Thus, for the Yule–Walker method, Kay and Marple (1979) recommend using complex-valued data (i.e., the analytic series) and the unbiased ACVS estimator; however, for time series other than the one considered here, this recommendation is tempered by the fact that, as opposed to $\{\hat{s}_{Z,\tau}^{(\mathrm{P})}\}$, the sequence $\{\hat{s}_{Z,\tau}^{(\mathrm{U})}\}$ in general need *not* correspond to a valid ACVS for a complex-valued process and hence might not be amenable for use with the Yule–Walker method.

Let us now turn to the case of Burg's algorithm. For the analytic series $\{Z_t\}$, we make use of the Burg estimate of $\phi_{1,1}$ appropriate for complex-valued series (see Section 9.13):

$$\bar{\phi}_{1,1} = \frac{2\sum_{t=0}^{N-2} Z_{t+1} Z_t^*}{|Z_0|^2 + 2\sum_{t=1}^{N-2} |Z_t|^2 + |Z_{N-1}|^2} = \frac{2(N-1)\mathrm{e}^{\mathrm{i}2\pi f \,\Delta_{\mathrm{t}}}}{2(N-1)} = \mathrm{e}^{\mathrm{i}2\pi f \,\Delta_{\mathrm{t}}},$$

where we have used the result in Equation (562) and the fact that $|Z_t|^2 = 1$ for all t. Kay and Marple (1979) argue that the estimates of the higher-order reflection coefficients, $\bar{\phi}_{2,2}$, $\bar{\phi}_{3,3}$, etc., are all zero. The ACVS estimator corresponding to Burg's algorithm is thus $\bar{\phi}_{1,1}^\tau = \exp{(\mathrm{i}2\pi f \tau \,\Delta_{\mathrm{t}})}$ for $\tau \geq 0$, again identical to the theoretical ACVS. Since the implied and theoretical ACVSs are identical, line splitting in Burg's algorithm is also arguably cured by using complex exponentials, i.e., by making the series "analytic."

Kay and Marple (1981) note that, for *multiple* sinusoids, the performance of Fougere's method is undocumented and that the use of complex-valued data in conjunction with Burg's algorithm can still result in line splitting. However, there is no evidence of line splitting in the FBLS method discussed in Section 9.7.

10.14 Singular Value Decomposition Approach

We stated in Equation (553d) that the summation of p *real-valued* sinusoids can be represented by a real-valued pseudo-AR($2p$) equation in which the parameters $\varphi_{2p,k}$ are the coefficients of the polynomial equation stated in Equation (553c) with roots occurring in pairs of the form $\exp\left(\pm i 2\pi f_j \Delta_t\right)$ for $j = 1, \ldots, p$. As might be anticipated, for the summation of p *complex-valued* sinusoids, i.e., complex exponentials,

$$\tilde{Z}_t = \sum_{l=1}^{p} D_l e^{i(2\pi f_l t \Delta_t + \phi_l)},$$

there is a corresponding representation in terms of a complex-valued pseudo-AR(p) equation where the parameters $\varphi_{p,k}$ are the coefficients of the polynomial equation

$$1 - \sum_{k=1}^{p} \varphi_{p,k} z^{-k} = 0 \quad \text{or, equivalently,} \quad z^p - \sum_{k=1}^{p} \varphi_{p,k} z^{p-k} = 0$$

with roots of the form $z_j = \exp\left(i 2\pi f_j \Delta_t\right)$ for $j = 1, \ldots, p$. To see that we do indeed have

$$\tilde{Z}_t = \sum_{k=1}^{p} \varphi_{p,k} \tilde{Z}_{t-k},$$

we first write

$$\tilde{Z}_t = \sum_{l=1}^{p} C_l e^{i 2\pi f_l t \Delta_t} = \sum_{l=1}^{p} C_l z_l^t, \quad \text{where} \quad C_l \stackrel{\text{def}}{=} D_l e^{i \phi_l}.$$

The relationship between the roots $\{z_j\}$ and coefficients $\{\varphi_{p,k}\}$ is given by

$$\prod_{j=1}^{p} (z - z_j) = z^p - \sum_{k=1}^{p} \varphi_{p,k} z^{p-k}. \tag{563}$$

Now $\tilde{Z}_{t-k} = \sum_{l=1}^{p} C_l z_l^{t-k}$, and hence

$$\sum_{k=1}^{p} \varphi_{p,k} \tilde{Z}_{t-k} = \sum_{k=1}^{p} \varphi_{p,k} \left(\sum_{l=1}^{p} C_l z_l^{t-k} \right) = \sum_{l=1}^{p} C_l z_l^{t-p} \sum_{k=1}^{p} \varphi_{p,k} z_l^{p-k}.$$

However, because z_l is a root, we have

$$\prod_{j=1}^{p} (z_l - z_j) = z_l^p - \sum_{k=1}^{p} \varphi_{p,k} z_l^{p-k} = 0, \quad \text{i.e.,} \quad \sum_{k=1}^{p} \varphi_{p,k} z_l^{p-k} = z_l^p,$$

so we can now write

$$\sum_{k=1}^{p} \varphi_{p,k} \tilde{Z}_{t-k} = \sum_{l=1}^{p} C_l z_l^{t-p} z_l^p = \sum_{l=1}^{p} C_l z_l^t = \tilde{Z}_t$$

as required.

We now make use of this pseudo-AR(p) representation for the summation of p complex exponentials to design an estimation scheme for the frequencies f_l for a model with additive proper complex-valued white noise $\{\epsilon_t\}$ with zero mean; i.e.,

$$Z_t = \tilde{Z}_t + \epsilon_t = \sum_{l=1}^{p} C_l e^{i2\pi f_l t \Delta_t} + \epsilon_t = \sum_{k=1}^{p} \varphi_{p,k} \tilde{Z}_{t-k} + \epsilon_t$$

(cf. Equation (517a) with $\mu = 0$). To determine the f_l's, Tufts and Kumaresan (1982) consider the following model involving a predictive filter of order p' (with $p' > p$ and p assumed unknown) in both the forward and backward directions:

$$Z_t = \sum_{k=1}^{p'} \varphi_{p',k} Z_{t-k} \text{ and } Z_t^* = \sum_{k=1}^{p'} \varphi_{p',k} Z_{t+k}^*,$$

which, given data $Z_0, Z_1, \ldots, Z_{N-1}$, yields the equations

$$Z_{p'} = \varphi_{p',1} Z_{p'-1} + \varphi_{p',2} Z_{p'-2} + \cdots + \varphi_{p',p'} Z_0$$
$$\vdots \quad \vdots \quad \vdots \quad \ddots \quad \vdots$$
$$Z_{N-1} = \varphi_{p',1} Z_{N-2} + \varphi_{p',2} Z_{N-3} + \cdots + \varphi_{p',p'} Z_{N-p'-1}$$
$$Z_0^* = \varphi_{p',1} Z_1^* + \varphi_{p',2} Z_2^* + \cdots + \varphi_{p',p'} Z_{p'}^*$$
$$\vdots \quad \vdots \quad \vdots \quad \ddots \quad \vdots$$
$$Z_{N-p'-1}^* = \varphi_{p',1} Z_{N-p'}^* + \varphi_{p',2} Z_{N-p'+1}^* + \cdots + \varphi_{p',p'} Z_{N-1}^*$$

(to see why complex conjugates are needed in the bottom $N - p'$ equations, consider the complex conjugate of Equation (502)). For the noiseless case when $Z_t = \tilde{Z}_t$, these equations would be exactly true if $p' = p$. For the moment we assume the noiseless case, but take $p' > p$. We can write the just-presented equations compactly as

$$\boldsymbol{A}\boldsymbol{\varphi} = \boldsymbol{Z}, \tag{564}$$

where $\boldsymbol{\varphi} \stackrel{\text{def}}{=} [\varphi_{p',1}, \ldots, \varphi_{p',p'}]^T$,

$$\boldsymbol{A} \stackrel{\text{def}}{=} \begin{bmatrix} Z_{p'-1} & Z_{p'-2} & \cdots & Z_0 \\ \vdots & \vdots & \ddots & \vdots \\ Z_{N-2} & Z_{N-3} & \cdots & Z_{N-p'-1} \\ Z_1^* & Z_2^* & \cdots & Z_{p'}^* \\ \vdots & \vdots & \ddots & \vdots \\ Z_{N-p'}^* & Z_{N-p'+1}^* & \cdots & Z_{N-1}^* \end{bmatrix} \text{ and } \boldsymbol{Z} \stackrel{\text{def}}{=} \begin{bmatrix} Z_{p'} \\ \vdots \\ Z_{N-1} \\ Z_0^* \\ \vdots \\ Z_{N-p'-1}^* \end{bmatrix}$$

(note that $\boldsymbol{A}$ is a $2(N - p') \times p'$ matrix).

As is true for any matrix, we can express $\boldsymbol{A}$ in terms of its singular value decomposition (see, for example, Golub and Kahan, 1965, or Businger and Golub, 1969; for real-valued matrices see Jackson, 1972, or Golub and Van Loan, 2013, section 2.4). Two sets of orthonormal vectors $\{\boldsymbol{u}_j\}$ and $\{\boldsymbol{v}_j\}$ exist such that

$$\boldsymbol{A}^H \boldsymbol{u}_j = \gamma_j \boldsymbol{v}_j \text{ and } \boldsymbol{A}\boldsymbol{v}_j = \lambda_j \boldsymbol{u}_j,$$

10.14 Singular Value Decomposition Approach

from which it follows that

$$AA^H u_j = \gamma_j \lambda_j u_j, \qquad j = 1, \ldots, 2(N-p'),$$
$$A^H A v_j = \gamma_j \lambda_j v_j, \qquad j = 1, \ldots, p', \qquad (565a)$$

(see Exercise [10.24]). Here the superscript H denotes the operation of complex-conjugate transposition. Each u_j is an eigenvector of AA^H, while v_j is an eigenvector of $A^H A$. If we rank the γ_j and λ_j terms each in decreasing order of magnitude, it can be shown that, for some integer $r \leq \min(2(N-p'), p')$,

$$\gamma_j = \begin{cases} \lambda_j > 0, & j \leq r; \\ 0, & \text{otherwise}; \end{cases} \quad \text{and } \lambda_j = 0 \text{ for } j > r;$$

i.e., there are r eigenvalues $\gamma_j \lambda_j = \lambda_j^2 > 0$ common to the two systems of equations in Equation (565a), and all the others are zero. The vectors u_j and v_j for $j = 1, \ldots, r$ are called the *left and right eigenvectors* or *singular vectors* of the matrix A; the λ_j terms are called the *eigenvalues* or *singular values* of A. The integer r is the rank of A (its importance becomes clearer when we look at the two cases of zero and nonzero additive noise). The matrix A can be written as

$$A = U\Lambda V^H, \qquad (565b)$$

where U is a $2(N-p') \times r$ matrix whose columns are the eigenvectors u_j; V is a $p' \times r$ matrix whose columns are the eigenvectors v_j; and Λ is a diagonal matrix whose diagonal elements are the eigenvalues λ_j. The right-hand side of the above is called the *singular value decomposition* (SVD) of A. By the orthonormality of the eigenvectors,

$$U^H U = V^H V = I_r, \quad UU^H = I_r \text{ if } r = 2(N-p') \text{ and } VV^H = I_r \text{ if } r = p', \quad (565c)$$

where I_r is the $r \times r$ identity matrix (Jackson, 1972).

Let us now return to Equation (564) and premultiply both sides of this equation by A^H to obtain $A^H A \varphi = A^H Z$. If we let $R = A^H A$ and $\beta = A^H Z$, then we have

$$R\varphi = \beta,$$

where the (j,k)th element of R and the jth element of β are given by, respectively,

$$\sum_{t=p'}^{N-1} Z_{t-j}^* Z_{t-k} + \sum_{t=0}^{N-p'-1} Z_{t+j} Z_{t+k}^* \quad \text{and} \quad \sum_{t=p'}^{N-1} Z_{t-j}^* Z_t + \sum_{t=0}^{N-p'-1} Z_{t+j} Z_t^*.$$

The system $R\varphi = \beta$ is the least squares system of equations corresponding to minimizing the forward/backward sum of squares

$$\sum_{t=p'}^{N-1} \left| Z_t - \sum_{k=1}^{p'} \varphi_{p',k} Z_{t-k} \right|^2 + \sum_{t=0}^{N-p'-1} \left| Z_t^* - \sum_{k=1}^{p'} \varphi_{p',k} Z_{t+k}^* \right|^2.$$

If we use the singular value decomposition given in Equation (565b) and recall that $U^H U = I_r$ (see Equation (565c)), we can write

$$R = A^H A = (V\Lambda U^H)(U\Lambda V^H) = V\Lambda I_r \Lambda V^H = V\Lambda^2 V^H,$$

where Λ^n refers to the diagonal matrix with diagonal elements λ_j^n. Also, β can be written

$$\beta = A^H Z = V\Lambda U^H Z,$$

so that $R\varphi = \beta$ becomes

$$V\Lambda^2 V^H \varphi = V\Lambda U^H Z. \tag{566a}$$

The *generalized inverse* of $R = V\Lambda^2 V^H$ is given by

$$R^\# = V\Lambda^{-2} V^H.$$

This generalized inverse always exists and gives the solution (using $V^H V = I_r$ from Equation (565c))

$$\tilde{\varphi} = \left(V\Lambda^{-2} V^H\right)\left(V\Lambda U^H\right) Z = V\Lambda^{-1} U^H Z. \tag{566b}$$

That this is indeed a solution to Equation (566a) can be checked by noting that

$$V\Lambda^2 V^H \tilde{\varphi} = V\Lambda^2 V^H V\Lambda^{-1} U^H Z = V\Lambda^2 I_r \Lambda^{-1} U^H Z = V\Lambda U^H Z$$

(again using $V^H V = I_r$). This solution minimizes $\tilde{\varphi}^H \tilde{\varphi} = |\tilde{\varphi}|^2$ (see, for example, Jackson, 1972). It is interesting to note that, if $r = p'$, then $RR^\# = R^\# R = VV^H = I_{p'}$ from Equation (565c), and in this case Equation (566b) expresses the usual least squares solution for φ.

When the additive noise $\{\epsilon_t\}$ is zero for all t, the process $\{Z_t\}$ is just the sum of p complex exponentials, and $r = p < p'$. From Equation (566b), we have

$$\tilde{\varphi} = [v_1, \ldots, v_p] \begin{bmatrix} \lambda_1^{-1} & & & \\ & \lambda_2^{-1} & & \\ & & \ddots & \\ & & & \lambda_p^{-1} \end{bmatrix} \begin{bmatrix} u_1^H \\ \vdots \\ \vdots \\ u_p^H \end{bmatrix} Z = \left(\sum_{k=1}^p \frac{1}{\lambda_k} v_k u_k^H\right) Z \tag{566c}$$

(Tufts and Kumaresan, 1982). To ensure that the rank of A is $r = p$, Hua and Sarkar (1988) point out that we should choose p' such that $p \leq p' \leq N - p$. With $\tilde{\varphi}$ determined from Equation (566c), we form the corresponding polynomial equation, namely,

$$z^{p'} - \sum_{k=1}^{p'} \tilde{\varphi}_{p',k} z^{p'-k} = 0.$$

Since $r = p$ this expression has p roots z_j on the unit circle at the locations of the sinusoidal frequencies, and the other $p' - p$ roots lie inside the unit circle (Tufts and Kumaresan, 1982). The frequency f_j is found by inverting $z_j = \exp(i2\pi f_j \Delta_t)$; i.e., $f_j = \Im(\log(z_j))/(2\pi \Delta_t)$, where $\Im(z)$ is the imaginary part of the complex number z, and we take the solution such that $|f_j| \leq f_\mathcal{N}$.

Of course, a much more realistic situation is when the additive noise $\{\epsilon_t\}$ is nonzero. In this case A has full rank, i.e., $r = \min(2(N - p'), p')$, and, by a derivation identical to that used for Equation (566c), the solution using the generalized inverse is

$$\hat{\varphi} = \left(\sum_{k=1}^{p'} \frac{1}{\hat{\lambda}_k} \hat{v}_k \hat{u}_k^H\right) Z. \tag{566d}$$

Since the matrix A and vector Z of Equation (564) here include additive noise, the left and right eigenvectors and eigenvalues are in general different from those in Equation (566c), so we have added a hat to these quantities to emphasize this. The eigenvalues $\hat{\lambda}_{p+1}, \ldots, \hat{\lambda}_{p'}$ are due to the noise and will usually be small. The solution $\hat{\varphi}$ thus becomes dominated by the noise components unless the sum in Equation (566d) is truncated at p, so we define

$$\hat{\varphi}' = \left(\sum_{k=1}^{p} \frac{1}{\hat{\lambda}_k} \hat{v}_k \hat{u}_k^H \right) Z,$$

for which again the sinusoidal terms are the dominant influence. If prior knowledge of the exact number of sinusoids p is lacking, it might be possible to ascertain the number by looking for the point k in the eigenvalue sequence

$$\hat{\lambda}_1 \geq \hat{\lambda}_2 \geq \cdots \hat{\lambda}_k \geq \hat{\lambda}_{k+1} \geq \cdots \geq \hat{\lambda}_{p'}$$

at which the value is large before rapidly decreasing at $k+1$ (sometimes such a cutoff is clear, but often it is not). The polynomial equation

$$z^{p'} - \sum_{k=1}^{p'} \tilde{\varphi}'_{p',k} z^{p'-k} = 0$$

has in general p roots z'_k close to the unit circle, and these can be used to find the frequencies $f'_k = \Im(\log(z'_k))/(2\pi \Delta_t)$.

Comments and Extensions to Section 10.14

[1] Tufts and Kumaresan (1982) consider that the SVD approach is likely to be most useful when only a short time series is available and the sinusoids are more closely spaced than the reciprocal of the observation time.

[2] If minimization of the forward/backward sum of squares alone is used for the estimation of φ, large values of p' can result in spurious spectral peaks. By using the SVD, the length of p' can be made considerably greater because of the effective gain in signal-to-noise ratio, thus increasing the resolution of the sinusoids. Tufts and Kumaresan found a good empirical choice to be $p' = 3N/4$. Since this implies that p' complex roots must be determined, it is clear that the method is only practicable for short series, as indeed was intended.

10.15 Examples of Harmonic Analysis

Willamette River Data

As a simple example of a harmonic analysis, let us consider a time series related to the flow of the Willamette River at Salem, Oregon (we used the first 128 values of this series as an example in Chapter 1). Figure 568a shows a plot of the $N = 395$ data points in the time series. Each point represents the log of the average daily water flow over a one month period from October, 1950, to August, 1983. The sampling interval is $\Delta_t = 1/12$ year, which yields a Nyquist frequency $f_\mathcal{N}$ of 6 cycles per year. There is a marked cyclical behavior of the data with a period of about 1 year. A harmonic process with one or more sinusoidal terms thus might be a reasonable model.

We start by entertaining the model of sinusoids plus white noise given by Equation (512b) and reformulated as Equation (515b). To handle the process mean μ, we center the time series by subtracting off its sample mean of 9.825 and consider Equation (515b) with $\mu = 0$ as a potential model for the centered series. To fit this model, we need first to determine

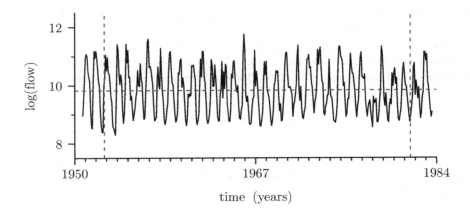

Figure 568a Willamette River data. There are $N = 395$ samples taken $\Delta_t = 1/12$ year apart (Figure 2(c) shows the first 128 values of this series). The horizontal dashed line indicates the sample mean for the series (9.825). The vertical dashed lines delineate a subseries of length 345 used in an SVD analysis.

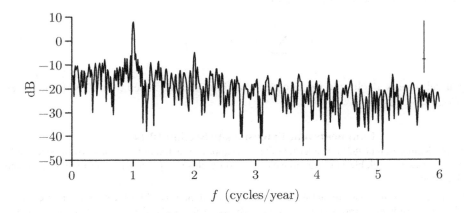

Figure 568b Periodogram for centered Willamette River data. Here the periodogram is plotted over a grid of frequencies more than twice as fine as the Fourier frequencies. The width of the crisscross in the upper right-hand corner gives the bandwidth measure $B_{\mathcal{H}}$ as per Equation (194), while its height gives the length of a 95% CI for $10 \log_{10}(S(f))$ as per Equation (205b).

the number of sinusoids L and the corresponding f_l terms. To do this we pad the centered data with 629 zeros to create a series of length 1024 and use an FFT routine to evaluate the periodogram $\hat{S}^{(P)}(\cdot)$ at the frequencies $j/(1024\,\Delta_t)$ for $j = 1, \ldots, 512$ (see C&E [1] for Section 3.11; we know from Exercise [6.15b] that $\hat{S}^{(P)}(0)$ is zero due to centering and hence the $j = 0$ term is of no interest). This grid of frequencies is more than twice as fine as that associated with the Fourier frequencies for $N = 395$, and ensures that we do not miss any important peaks in the periodogram. Figure 568b is a plot of $\hat{S}^{(P)}(\cdot)$ at these 512 frequencies. The largest peak – about 15 dB above the rest of the periodogram – occurs at $86/(1024\,\Delta_t) \doteq 1.00781$ cycles/year, which is the second closest frequency to 1 cycle/year of the form $j/(1024\,\Delta_t)$ because $85/(1024\,\Delta_t) \doteq 0.996$ cycle/year deviates only by 0.004 from 1 cycle/year.

The second largest peak in the periodogram is at $171/(1024\,\Delta_t) \doteq 2.00391$ cycles/year, which is the closest frequency to 2 cycles/year because $170/(1024\,\Delta_t) \doteq 1.992$ cycles/year. An explanation for the presence of this peak is that the river flow data have an annual variation

10.15 Examples of Harmonic Analysis

that is periodic but not purely sinusoidal. If we describe this variation by a real-valued periodic function $g_p(\cdot)$ with a period of 1 year, the results of Section 3.1 say that we can write

$$g_p(t) = \frac{a_0}{2} + \sum_{l=1}^{\infty} [a_l \cos(2\pi l t) + b_l \sin(2\pi l t)]$$

(cf. Equation (48a) with $T = 1$). The frequency that is associated with the lth cosine and sine in the summation is just $f'_l = l$ cycles/year. The lowest such frequency, namely $f'_1 = 1$ cycle/year, is called the *fundamental frequency* for $g_p(\cdot)$ and is equal to the reciprocal of its period. All the other frequencies that are involved in representing $g_p(\cdot)$ are multiples of the fundamental frequency; i.e., $f'_l = lf'_1 = l$ cycles/year. The frequency f'_{l+1} is called the lth harmonic of the fundamental frequency f'_1 so that, for example, $f'_2 = 2$ cycles/year is the first harmonic. If we now sample $g_p(\cdot)$ at points $\Delta_t = 1/12$ year apart, we obtain a periodic sequence given by

$$g_t \stackrel{\text{def}}{=} g_p(t\,\Delta_t) = \frac{a_0}{2} + \sum_{l=1}^{\infty} [a_l \cos(2\pi l t\,\Delta_t) + b_l \sin(2\pi l t\,\Delta_t)]. \quad (569a)$$

Because this choice of Δ_t produces an aliasing effect, the above can be rewritten as

$$g_t = \mu' + \sum_{l=1}^{6} [A'_l \cos(2\pi l t\,\Delta_t) + B'_l \sin(2\pi l t\,\Delta_t)], \quad (569b)$$

where μ', A'_l and B'_l are related to a_l and b_l (see Exercise [10.25]). Note that the summation in the above equation resembles the nonrandom portion of Equation (515b) if we let $L = 6$ and equate f_l with $l = f'_l$. Given the nature of this time series, it is thus not unreasonable to find a peak in the periodogram at 2 cycles/year, the first harmonic for a fundamental frequency of 1 cycle/year (recall that we also encountered harmonics in the example from tidal analysis in Section 10.4). While A'_l and B'_l for $l = 3, 4, \ldots, 6$ need not be nonzero, the visual lack of peaks in the periodogram at $f = 3, 4, \ldots, 6$ suggests that they are not significantly different from zero.

Besides the peaks corresponding to annual and semiannual periods, there are numerous other lesser peaks in the periodogram that might or might not be due to other sinusoidal components. There is a possibility that the periodogram is masking components such as the second or higher harmonics of 1 cycle/year due to leakage from the two dominant peaks. Exercise [10.26] indicates that in fact this is not the case by considering a direct spectral estimate $\hat{S}^{(D)}(\cdot)$ based upon a Slepian data taper.

As a starting point, let us assume that there is only one frequency of importance in the time series so that the following simple harmonic model is appropriate for the centered data:

$$X_t = A_1 \cos(2\pi f_1 t\,\Delta_t) + B_1 \sin(2\pi f_1 t\,\Delta_t) + \epsilon_t. \quad (569c)$$

Although it would be reasonable to consider f_1 to be 1 cycle/year based on physical arguments, let us assume – for illustrative purposes – that it is unknown and use the periodogram to estimate it. If we search the periodogram in Figure 568b for its peak value (see C&E [2] for Section 10.6), we find that it occurs at $\hat{f}_1 \doteq 1.00319$ cycles/year. Conditional on this periodogram-based estimate of f_1, we can now estimate the parameters A_1 and B_1 using the ACLS estimators of Equation (549a) or the ECLS estimators, which minimize $SS(A_1, B_1, \hat{f}_1)$ of Equation (549d). Alternatively we can find the EULS estimators A_1, B_1 and f_1 by minimizing $SS(A_1, B_1, f_1)$. The upper part of Table 570 shows the resulting estimates, along with

Method	$\hat{f}_1$	$\hat{A}_1$	$\hat{B}_1$	Jump	$\hat{f}_2$	$\hat{A}_2$	$\hat{B}_2$	Jump	SS
ACLS	1.00319	−0.3033	0.8443	0.2012					87.40164
ECLS	1.00319	−0.3029	0.8447	0.2013					87.40157
EULS	1.00326	−0.3086	0.8427	0.2013					87.39881
ACLS	1.00319	−0.3033	0.8443	0.2012	2.00339	0.0365	0.1937	0.00971	79.58391
ECLS	1.00319	−0.3030	0.8451	0.2015	2.00339	0.0364	0.1956	0.00989	79.58305
EULS	1.00333	−0.3147	0.8411	0.2016	2.00244	0.0553	0.1915	0.00994	79.54593

Table 570 Parameter estimates of f_l, A_l and B_l for single- and two-frequency models of the form of Equation (512b) with $\mu = 0$ for the centered Willamette River time series, along with estimated jumps in the integrated spectrum at $\pm \hat{f}_l$ and the corresponding residual sum of squares (SS). Three types of least squares estimates are considered: approximate conditional (ACLS), exact conditional (ECLS) and exact unconditional (EULS). The top part of the table shows estimates for the single-frequency model; the bottom, for the model with two frequencies. The ACLS and ECLS estimates are conditional on estimates of f_l determined from the periodogram of the centered series.

the corresponding estimates of the jump in the integrated spectrum at $\hat{f}_1$ (computed using the left-hand side of Equation (549b)) and the corresponding residual sum of squares (we can use the latter to estimate the white noise variance σ_ϵ^2 via Equation (549c) for the ACLS and ECLS estimates or Equation (550a) in the EULS case). The estimates of A_1, B_1, f_1 and the jump in the integrated spectrum from all three approaches are in good agreement. Theory demands that the value of the SS associated with the ACLS method cannot be lower than that for the ECLS method; likewise, the ECLS value cannot be lower than the one for EULS. The values in the table adhere to this ordering. For comparison the sum of squares of the centered time series $\sum_t X_t^2$ is 246.3938. The coefficient of determination, namely,

$$1 - \frac{\mathrm{SS}(\hat{A}_1, \hat{B}_1, \hat{f}_l)}{\sum_t X_t^2} \doteq \begin{cases} 0.6452766, & \text{for ACLS;} \\ 0.6452769, & \text{for ECLS;} \\ 0.6452881, & \text{for EULS,} \end{cases}$$

is the proportion of variability in the time series explained by the sinusoidal terms in Equation (569c) and is the same for all three methods to four significant digits.

Figure 571a(a) is a plot of the fitted model for the uncentered river flow data based upon the EULS estimates using the centered series, i.e.,

$$\overline{X} + \hat{A}_1 \cos(2\pi \hat{f}_1 t \Delta_t) + \hat{B}_1 \sin(2\pi \hat{f}_1 t \Delta_t)$$

while Figure 571a(b) shows the corresponding EULS residuals:

$$R_t = X_t - \hat{A}_1 \cos(2\pi \hat{f}_1 t \Delta_t) - \hat{B}_1 \sin(2\pi \hat{f}_1 t \Delta_t)$$

(the sum of the series in (a) and (b) is equal to the original time series shown in Figure 568a; because the ACLS and ECLS parameter estimates are so similar to the EULS estimates, figures corresponding to Figure 571a for the ACLS and ECLS cases are visually identical to the EULS case – we concentrate on the EULS case in what follows and leave the ACLS and ECLS cases as exercises for the reader). If Equation (569c) is adequate as a model for the river flow data, the residuals should resemble a white noise process. To examine whether $\{R_t\}$ can be considered to be white noise, we plot its periodogram in Figure 571b. Note that the only substantial difference between this periodogram and that of $\{X_t\}$ in Figure 568b is near $f = 1$ cycle/year, as is reasonable to expect. The peak close to $f = 2$ cycles/year is the most

10.15 *Examples of Harmonic Analysis* 571

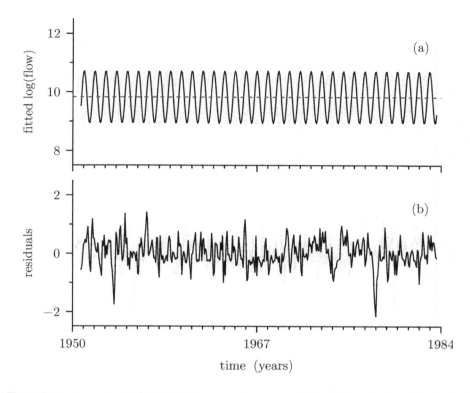

Figure 571a Mean-adjusted fitted single-frequency harmonic model for Willamette River data based on EULS parameter estimates (top plot), along with corresponding residuals (bottom).

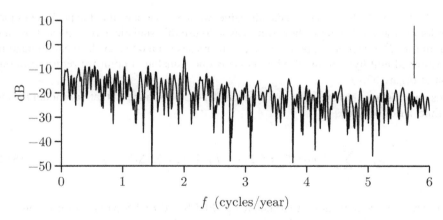

Figure 571b As in Figure 568b, but now the periodogram is based on the EULS residuals of Figure 571a(b).

prominent one in the periodogram of the residuals. Although this peak does occur close to a frequency that, given the nature of this series, is reasonable, it is only about 5 dB above the rest of the periodogram. Since we are in a situation where Fisher's g statistic (Equation (539b)) can help decide whether this peak is significant, we compute it for these residuals and obtain $g \doteq 0.088$. Under the null hypothesis of white noise, the upper 1% critical level for the test statistic is approximately 0.049 (from the approximation in Equation (540d), or using the entry for $m = 200$ in Table 1(b) of Shimshoni, 1971, which is reasonable since here

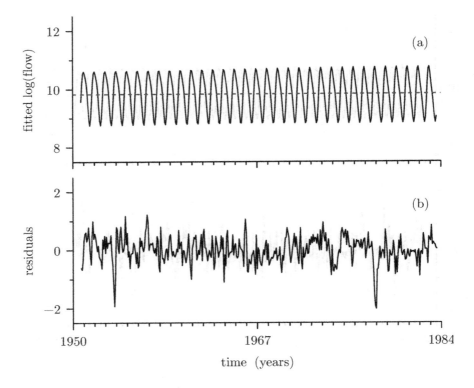

Figure 572 As in Figure 571a, but now for two-frequency harmonic model.

$m = (N-1)/2 = 197$). Since g exceeds this value, we have evidence that the peak is not just due to random variation. (There is, however, reason to use this statistic here with caution: the periodogram for $\{R_t\}$ decreases about 10 dB as f increases from 0 to 6 cycles/year, indicating that the presumed null hypothesis of white noise is questionable, a point we return to in our discussion of Figure 573.)

Let us now assume that there are two frequencies of importance so that our model for the centered time series becomes

$$X_t = \sum_{l=1}^{2} A_l \cos\left(2\pi f_l t \, \Delta_t\right) + B_l \sin\left(2\pi f_l t \, \Delta_t\right) + \epsilon_t. \tag{572}$$

The lower part of Table 570 shows that the ACLS, ECLS and EULS parameter estimates for this two-frequency model are in good agreement. For the ACLS and ECLS methods, we use the same periodogram-based estimate of f_1 as in the single-frequency model, and we estimate f_2 also from the periodogram for $\{X_t\}$, obtaining $\hat{f}_2 \doteq 2.00339$. The table shows that the estimated jumps in integrated spectrum are approximately the same for all three methods and that their SS values are similar, with the small differences amongst them obeying the pattern dictated by theory (the SS for the EULS method does not exceed the one for ECLS, and the latter does not exceed the one for ACLS). Overall there is a reduction in SS of about 10% over the single-frequency model; however, the coefficient of determination is now 0.677 for all three methods, which is a modest 3% increase in explained variability as compared to 0.645 for the single-frequency model.

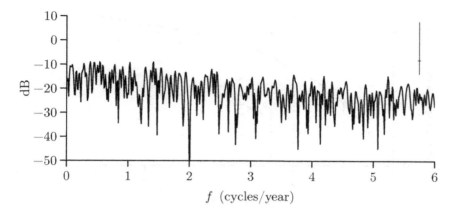

Figure 573 As in Figures 568b and 571b, but now the periodogram is based on the EULS residuals of Figure 572(b).

Figure 572 shows the mean-adjusted fitted two-frequency model

$$\overline{X} + \sum_{l=1}^{2} \left(\hat{A}_l \cos\left(2\pi \hat{f}_l t \, \Delta_t\right) + \hat{B}_l \sin\left(2\pi \hat{f}_l t \, \Delta_t\right) \right)$$

(top plot) and associated residuals

$$R_t = X_t - \sum_{l=1}^{2} \left(\hat{A}_l \cos\left(2\pi \hat{f}_l t \, \Delta_t\right) + \hat{B}_l \sin\left(2\pi \hat{f}_l t \, \Delta_t\right) \right)$$

(bottom) based upon the EULS method (plots for the ACLS and EULS methods are virtually identical). There are subtle visual differences between the two plots here and the corresponding ones in Figure 571a for the single-frequency model. Again, if the two-frequency model were adequate, $\{R_t\}$ should be close to a realization from a white noise process (we note in passing that there are two substantive downshoots in the residuals, perhaps attributable to periods of drought). Figure 573 shows the periodogram for the residuals. As is to be expected, the only substantial difference between this periodogram and that for single-frequency residuals in Figure 571b is near $f = 2$ cycles/year, where there is now a downshoot instead of a peak. There are no prominent peaks in the periodogram. Application of Fisher's g statistic now yields $g \doteq 0.039$, which fails to exceed either the 1% critical level (0.049) or the 5% critical level (0.041); however, the null hypothesis for Fisher's test is white noise, and failure to reject this hypothesis does not imply we should entertain it uncritically, particularly since the periodogram in Figure 573 visibly decreases with increasing frequency, in violation of the pattern we would expect to see from a realization of white noise.

We can test whether the residuals $\{R_t\}$ of Figure 572(b) can be considered to be drawn from a white noise process by applying the normalized cumulative periodogram test for white noise (described following Equation (215a)). Figure 574 shows the graphical equivalent of the level $\alpha = 0.05$ test. The test statistic is $D \doteq 0.334$, which, since it is greater than the critical level $\tilde{D}(0.05) \doteq 0.096$ of Equation (215b), says that we can reject the null hypothesis that the residuals come from a white noise process. This is reflected in the fact that, in Figure 574, the cumulative periodogram (the solid curve) crosses the upper slanted dashed line. The frequency associated with D is 2.46, which is indicated by a solid vertical line. To the left of this line are two dashed vertical lines. These indicate the frequencies $\hat{f}_1$ and $\hat{f}_2$ near which troughs occur

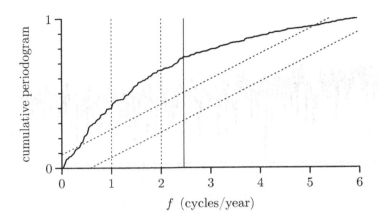

Figure 574 Normalized cumulative periodogram test for white noise as applied to EULS residuals obtained by fitting the model of Equation (572) to the centered Willamette River data.

in Figure 573 (plots similar to Figure 574 for the ACLS and ECLS residuals are visually identical to this figure).

Given that we cannot regard the background continuum as a white noise process, let us now consider a model of the form of Equation (550c), which takes the continuum to be a zero mean stationary process with an SDF $S_\eta(\cdot)$ that need not be that of white noise. Although we already have a good idea of the appropriate frequencies f_l to use in this model, let us illustrate the use of the F-test for periodicity discussed in Section 10.10. For this test, we select $K = 5$ Slepian multitapers with $NW = 4/\Delta_t$ (Exercise [10.28] is to redo what follows using $K = 5$ and $K = 6$ sinusoidal multitapers instead). We compute the F-test of Equation (547a) at frequencies $j/(1024\,\Delta_t)$ with $j = 1, 2, \ldots, 512$ and plot it in Figure 575(a). At a given fixed frequency, the F-test is F-distributed with 2 and $2K - 2 = 8$ degrees of freedom under the null hypothesis of no spectral line. We reject the null hypothesis for large values of the F-test. The two dashed horizontal lines on the plot indicate the levels of the upper 99% and 99.9% percentage points of the $F_{2,8}$ distribution (8.6 and 18.5, respectively, obtained by using Equation (544a) with $\nu = 8$ and α set to 0.01 and 0.001). The two largest excursions in the F-test occur at the same frequencies as the two largest peaks of the periodogram in Figure 568b, namely, 1.00781 and 2.00391 cycles/year. The largest excursion (at 1.00781 cycles/year) exceeds the 99.9% point and is clearly significant; the F-test at 2.00391 cycles/year exceeds the 99% point (but not the 99.9%). There are also two other frequencies at which the F-test exceeds the 99% point, namely, 2.461 and 4.195 cycles/year. The periodograms in Figures 568b, 571b and 573 do not clearly indicate a line component at either frequency, so how can we explain this result? The rule of thumb from Thomson (1990a) dictates not getting excited by significance levels greater than $1/N$ (see the discussion at the end of Section 10.10). Here $1/N = 1/395 \doteq 0.0025$, which translates into a critical value of 13.8 (indicated by the dotted horizontal line). Since only the excursions at 1.00781 and 2.00391 cycles/year exceed 13.8, the rule of thumb says that all the other peaks in the F-test are due to sampling fluctuations.

Based upon the results of the F-test, we now proceed under the assumption that there are line components at the frequencies $\hat{f}_1 \doteq 1.00781$ and $\hat{f}_2 \doteq 2.00391$ cycles/year so that $L = 2$ in Equation (550c). The associated complex-valued amplitudes for these frequencies, namely, C_1 and C_2, can be estimated using Equation (546a) to obtain

$$\hat{C}_1 \doteq -0.2912 - i0.3122 \qquad |\hat{C}_1|^2 \doteq 0.1823$$
$$\hat{C}_2 \doteq 0.0232 - i0.0984 \quad \text{yielding} \quad |\hat{C}_2|^2 \doteq 0.0102.$$

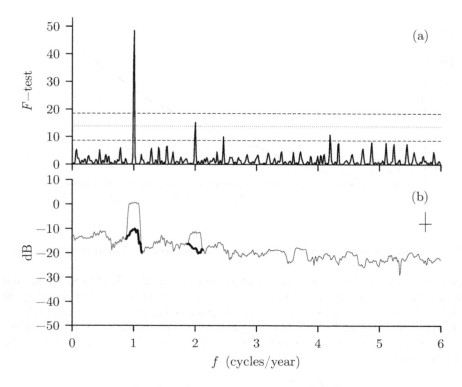

Figure 575 Thomson's F-test (top plot) and Slepian multitaper SDF estimate (bottom) for centered Willamette River data (see text for details).

The sizes of the jumps (0.1823 and 0.0102) in the integrated spectrum at $\hat{f}_1 \doteq 1.00781$ and $\hat{f}_2 \doteq 2.00391$ cycles/year are within 10% of the ACLS, ECLS and EULS estimates previously determined at frequencies close to these (see the bottom three rows of Table 570).

The thin curve in Figure 575(b) shows the basic multitaper SDF estimate $\hat{S}^{(\mathrm{MT})}(\cdot)$ of Equation (352a) for the centered Willamette River data. This estimate uses the same set of Slepian tapers as in the F-test; i.e., $K = 5$ and $NW = 4/\Delta_t$. As determined by Equation (353e), the bandwidth for the estimator is $B_{\overline{\mathcal{H}}} \doteq 0.19847$, which is the width of the crisscross on the plot (as usual, the height is the length of a 95% CI for $10\,\log_{10}(S(f))$). The peaks at 1 and 2 cycles/year are spread out by about this amount, and the distance between the two peaks is greater than $B_{\overline{\mathcal{H}}}$, indicating that the F-tests at the two frequencies are approximately independent. To obtain an estimate of $S_\eta(\cdot)$, we reshape $\hat{S}^{(\mathrm{MT})}(\cdot)$ using Equation (547b) at all frequencies dictated by Equation (547c), i.e., $|f - f_1| \leq 1.25 B_{\overline{\mathcal{H}}}/2$ or $|f - f_2| \leq 1.25 B_{\overline{\mathcal{H}}}/2$. Since f_1 corresponds to $86/(1024\,\Delta_t)$ and f_2 corresponds to $171/(1024\,\Delta_t)$, this amounts to reshaping $\hat{S}^{(\mathrm{MT})}(\cdot)$ at the frequencies $j/(1024\,\Delta_t)$ for $j = 76, 77, \ldots, 96$ and $j = 161, 162, \ldots, 181$. The two short thick curves indicate the reshaped portions of $\hat{S}^{(\mathrm{MT})}(\cdot)$. Our estimate of $S_\eta(\cdot)$ is thus given by these curves in addition to the unreshaped portions of $\hat{S}^{(\mathrm{MT})}(\cdot)$.

Let us now analyze the Willamette River data using a parametric approach. We fit AR models of orders $p = 1, 2, \ldots, 150$ using Burg's algorithm (Exercise [10.29] invites the reader to use the Yule–Walker and FBLS methods instead). Figure 576 shows a plot of the AICC order selection criterion versus p – see Section 9.11 for details. The minimum value occurs at $p = 27$ (marked by a vertical line) and suggests that an AR(27) process might be appropriate, but we also consider the case $p = 150$ for comparison. The top two plots in Figure 577 show the resulting SDF estimates over the grid of frequencies $j/(2^{16}\,\Delta_t)$, $0 \leq j \leq 2^{15} = 32{,}768$

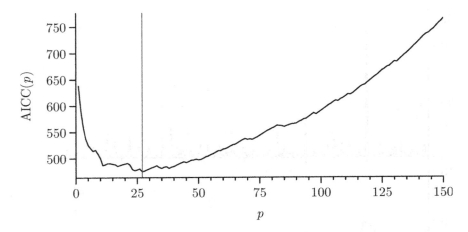

Figure 576 AICC order selection criterion for Willamette River data (parameter estimation via Burg's algorithm).

(we need a grid this fine for accurate portrayal of the peaks in the AR(150) SDF). In both estimates, the two largest peaks occur near 1 and 2 cycles/year, in agreement with periodogram shown in Figure 568b. Table 578 lists the observed peak frequencies $\hat{f}_j$, peak values $\hat{S}(\hat{f}_j)$ and peak widths for these SDFs and – for comparison – the periodogram. The peak widths at the observed peak frequency are measured by $f' + f''$, where $f' > 0$ and $f'' > 0$ are the smallest values such that

$$10 \log_{10}\left(\frac{\hat{S}(\hat{f}_j)}{\hat{S}(\hat{f}_j + f')}\right) = 10 \log_{10}\left(\frac{\hat{S}(\hat{f}_j)}{\hat{S}(\hat{f}_j - f'')}\right) = 3\,\text{dB}.$$

This table and the plots show some interesting qualitative differences and similarities amongst the SDF estimates:

[1] the peak near 2 cycles/year in the AR(27) SDF is displaced in frequency somewhat low compared to those based on the other two SDF estimates, while the peak near 1 cycle/year has about the same location for all three estimates;

[2] the widths of both AR(150) peaks are smaller than those for the corresponding AR(27) peaks by more than an order of magnitude;

[3] the widths of the peaks near 1 cycle/year in the AR SDFs are much smaller than the width of the peak in the periodogram (about 25 times smaller in the AR(150) case and two times in the AR(27)), while at 2 cycles/year the width is smaller only for $p = 150$;

[4] the widths of the peaks at 1 and 2 cycles/year for a particular AR estimate differ by about an order of magnitude, whereas the periodogram widths are similar (0.0265 and 0.0269);

[5] the heights of the two peaks in the AR(27) SDF are about an order of magnitude smaller than the corresponding peaks in the AR(150) SDF; and

[6] the AR(27) SDF is much smoother looking than the AR(150) SDF and the periodogram.

Item [1] illustrates the danger of too low an AR order (a loss of accuracy of peak location), while item [6], of too high an AR order (introduction of spurious peaks). Items [2] and [4] indicate an inherent problem with assessing peak widths in AR SDF estimates: whereas the peak widths in the periodogram can be attributed simply to the width of the central lobe of the associated spectral window, there is evidently no corresponding simple relationship for the AR estimates. If we believe that the true frequencies for this time series are at 1 and 2 cycles/year, items [1] and [3] indicate that the narrowness of the AR peaks does not translate into increased

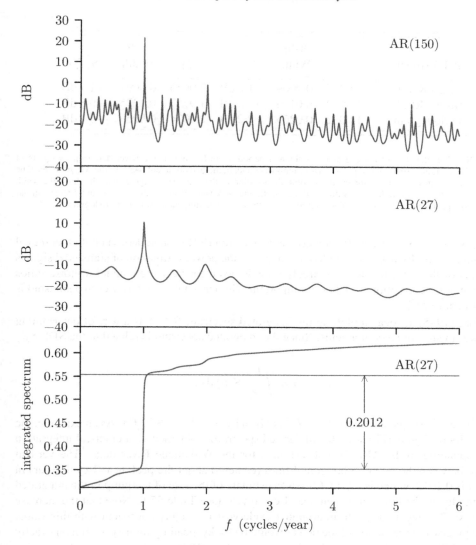

Figure 577 Burg-based AR(150) and AR(27) SDF estimates (top two plots) and integrated AR(27) spectrum (bottom plot) for Willamette River data (see text for details).

accuracy of the peak location. Indeed, as is pointed out by Burg (1975, p. 64) the narrowness of AR peaks calls for care in determining peak locations (see C&E [3] for Section 10.12).

Item [5] raises the question of how we can relate the height of a peak in an AR SDF at frequency f_l to the amplitude D_l of a sinusoidal component at that frequency in the assumed model

$$X_t = \sum_{l=1}^{L} D_l \cos\left(2\pi f_l \Delta_t + \phi_l\right) + \epsilon_t.$$

In the case of the periodogram, we have the approximate relationship

$$E\{\hat{S}^{(P)}(f_l)\} \approx \sigma_\epsilon^2 \Delta_t + N \Delta_t D_l^2 / 4$$

(since $\mathcal{F}(0) = N \Delta_t$, this follows from Equation (524b) under the assumption that f_l is well separated from the other frequencies in the model and is not too close to zero or the Nyquist

SDF Estimate	$\hat{f}_1$	3 dB Width	$\hat{S}(\hat{f}_1)$	$\hat{f}_2$	3 dB Width	$\hat{S}(\hat{f}_2)$
Burg AR(150)	1.0055	0.0009	21.6 dB	2.0055	0.0091	−1.1 dB
Burg AR(27)	1.0042	0.0126	10.3 dB	1.9757	0.1225	−9.8 dB
Periodogram	1.0032	0.0265	8.2 dB	2.0034	0.0269	−5.0 dB

Table 578 Characteristics of autoregressive SDFs and periodogram for Willamette River data (top two plots of Figure 577 and Figure 568b). The top two rows of numbers concern the two dominant peaks in the AR(p) SDFs. The first column indicates the AR estimator; the second, the location of the peak near 1 cycle/year; the third, the width of the peak 3 dB down from the peak value; and the fourth, the peak value itself. Columns 5, 6 and 7 show similar values for the peak near 2 cycles/year. The final row gives the corresponding values for the periodogram $\hat{S}^{(\mathrm{P})}(\cdot)$.

frequency; see also Figure 534). Pignari and Canavero (1991) concluded, via both theoretical calculations and Monte Carlo simulations, that "... the power estimation of sinusoidal signals by means of the Burg method is scarcely reliable, even for very high signal-to-noise ratios for which noise corruption of data is unimportant" (in our notation, the power in a sinusoid is proportional to D_l^2).

Burg (1985) advocated plotting the integrated spectrum $S^{(\mathrm{I})}(\cdot)$ as a way of determining the power in sinusoidal components. Since the integrated spectrum is related to the SDF $S(\cdot)$ by

$$S^{(\mathrm{I})}(f) = \int_{-f_{\mathcal{N}}}^{f} S(u)\,\mathrm{d}u,$$

the power in the frequency interval $[f', f'']$ is given by $S^{(\mathrm{I})}(f'') - S^{(\mathrm{I})}(f')$. As an example, the bottom plot in Figure 577 shows the integrated spectrum – obtained via numerical integration – corresponding to the AR(27) SDF estimate for the Willamette River data. The vertical distance between the two thin horizontal lines (centered about the peak near 1 cycle/year) in this plot is 0.2012, which is the ACLS-based estimate of the size of the jump in the integrated spectrum due to the line component near 1 cycle/year (see Table 570). Note that the increase in the AR(27) integrated spectrum around 1 cycle/year is roughly consistent with this value, but that precise determination of the power would be tricky, pending future research on exactly how to pick the interval $[f', f'']$ to bracket a peak. (Burg, 1975, pp. 65–6, notes that, because the integral over $f \in [-f_{\mathcal{N}}, f_{\mathcal{N}}]$ of a Burg-based AR SDF estimate must in theory be equal to the sample variance of the time series, numerical integration of the SDF over the grid of frequencies used for plotting should ideally yield a value close to the sample variance – if not, then the grid might be too coarse to accurately display the estimated SDF. Numerical integration of the AR(150) and AR(27) SDFs in Figure 577 yields, respectively, 0.6237824 and 0.6237827, both of which agree well with the sample variance for the centered Willamette River data, namely, 0.6237818. This result suggests that the frequency grid is adequate for displaying the overall patterns in the SDFs, but does not guarantee the accurate portrayal of sharp peaks.)

Let us now use the centered Willamette River time series to illustrate the SVD approach to harmonic analysis. In Section 10.14 we formulated the SVD method for complex-valued time series, whereas the Willamette River is real-valued. While a real-valued series is a special case of complex-valued series, here we render the series $\{X_t\}$ into complex-valued form by creating a corresponding "analytic" series (Exercise [10.31] explores using the real-valued series as is and also using a complex-valued series created by complex demodulation). The

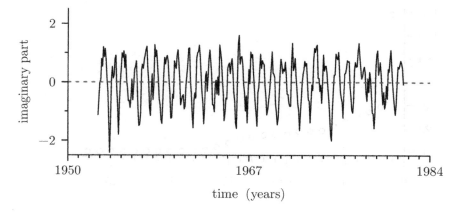

Figure 579 Imaginary part of analytic series used in SVD analysis. The series has $N = 345$ values, and the corresponding real part (prior to centering) is delineated in Figure 568a by two vertical dashed lines.

analytic series is

$$X_t + \mathrm{i}\,\mathcal{HT}\{X_t\},$$

where $\mathcal{HT}$ is the Hilbert transform of $\{X_t\}$ (see Section 10.13 above and section 9–3 of Papoulis and Pillai, 2002). By definition, the transfer function (frequency response function) of the Hilbert transform, $G(\cdot)$ say, must have unit gain, a phase angle of $-\pi/2$ for f between 0 and 1/2 and a phase angle of $\pi/2$ for f between 0 and $-1/2$. Thus $G(\cdot)$ takes the form

$$G(f) = \begin{cases} \mathrm{i}, & -1/2 < f < 0; \\ -\mathrm{i}, & 0 < f < 1/2 \end{cases} \qquad (579\mathrm{a})$$

(Oppenheim and Schafer, 2010, equation (12.62b)). The corresponding impulse response sequence $\{g_u\}$ for the Hilbert transform is

$$g_u = \int_{-1/2}^{1/2} G(f) \mathrm{e}^{\mathrm{i}2\pi f u}\,\mathrm{d}f = \begin{cases} 2/(\pi u), & u\ \text{odd}; \\ 0, & u\ \text{even} \end{cases} \qquad (579\mathrm{b})$$

(verification of the above is the subject of Exercise [10.30a]). We can obtain a realizable approximation to this ideal (infinitely long) impulse response by multiplying the g_u sequence by a set of convergence factors (see Sections 3.8 and 5.8); a reasonable choice (amongst many others) is a Hanning lag window:

$$c_u = \begin{cases} 1 - 0.5 + 0.5\cos\left(\pi u/25\right), & |u| \le 25; \\ 0, & \text{otherwise} \end{cases} \qquad (579\mathrm{c})$$

(see Equation (343c) with a set to $1/4$ and m set to 25). The convergence factors reduce the impulse response sequence smoothly to a total length of 51 nonzero coefficients (i.e., 25 each side of $u = 0$). This approximation creates no phase errors, but slight gain errors do occur, mainly near zero and Nyquist frequencies (see Exercise [10.30b]). The result of filtering the series $\{X_t\}$ of length $N = 395$ with the filter $\{c_u g_u\}$ of width 51 is a series of length 345. This series is the imaginary part of the analytic series and is plotted in Figure 579; the corresponding real part $X_{25}, X_{26}, \ldots, X_{369}$ is shown (prior to centering) in Figure 568a between the two vertical dashed lines.

In applying the methods of Section 10.14 to the complex (analytic) series $\{Z_t\}$, we need to set p', which represents the number of values prior to Z_t needed to predict it well. Our

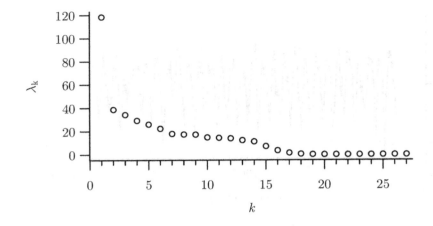

Figure 580a Eigenvalues λ_k of orders $k = 1, \ldots, 27$ in SVD of analytic series created from Willamette River data.

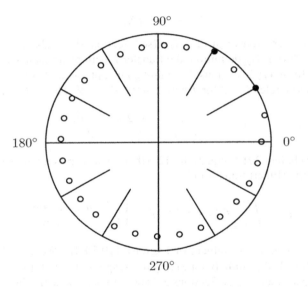

Figure 580b Complex roots z'_k (small circles) in SVD analysis of Willamette River data. The large circle is the unit circle in the complex plane. The lines indicate phase angles on the unit circle that are multiples of 30°. There are 27 roots in all, and the two filled circles indicate the two closest to the unit circle – these have phase angles close to 30° and 60°.

parametric analysis of $\{X_t\}$ suggests $p' = 27$, and experimentation with other choices indicates this to be suitable. Our previous investigations indicate two sinusoids with frequencies close to one and two cycles/year, which suggests truncating at $p = 2$ in the SVD method; on the other hand, a plot of the eigenvalues (Figure 580a) suggests $p = 1$. Neither of these choices gives results in reasonable agreement with our previous analyses, but setting $p = 3$ does. The resulting estimated roots are plotted in polar form in Figure 580b and were found using the function polyroot in the R language (the SVD itself was computed using the function svd). The two complex roots z'_k closest to the unit circle are at $(0.8635, 0.5007)$ and $(0.4997, 0.8366)$, having magnitudes 0.9981 and 0.9745, respectively, and phase angles quite close to 30° and 60°. Recalling that $\Delta t = 1/12$ year, the corresponding frequencies f'_k are

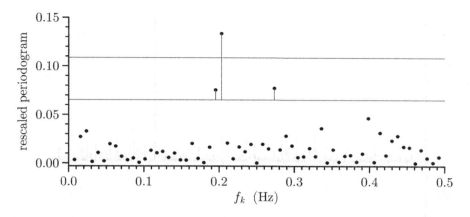

Figure 581 Application of Fisher's and Siegel's tests to ocean noise data (see text for details).

1.0035 and 1.9717 cycles/year, similar to those obtained in our other analyses. There is, however, a third root whose magnitude (0.9740) is just below that of the root with the second largest magnitude. This root has a phase angle close to 15° and is associated with 0.4876 cycles/year, a frequency that has not been flagged by our previous analyses and hence must be regarded with caution.

Ocean Noise Data
Here we give an example of the application of Fisher's and Siegel's tests using the ocean noise data in Figure 4(d). In Section 1.2 we noted that the sample ACS for this time series shows a tendency to oscillate with a period of 5 sec. What do the tests for simple or compound periodicity in white noise tell us about this series? To begin, we calculate $\hat{S}^{(\mathrm{P})}(f_k)$ over the grid of Fourier frequencies appropriate for a series of length $N = 128$, for which, as per Equation (539a), the number of Fourier frequencies satisfying $0 < f_k < f_\mathcal{N} = 1/2\,\mathrm{Hz}$ is $M = (N-2)/2 = 63$. We then form the rescaled periodogram $\tilde{S}^{(\mathrm{P})}(f_k)$ of Equation (541a). Figure 581 shows $\tilde{S}^{(\mathrm{P})}(f_k)$ versus f_k for $k = 1, 2, \ldots, M$ (solid circles). The critical value g_F for Fisher's test at $\alpha = 0.05$ obtained via Equation (540b) is marked as the upper horizontal line (its values is 0.1086, while its approximation $\tilde{g}_F$ given by Equation (540d) is 0.1088); the lower line marks $0.6 g_F$ used in Siegel's test. The maximum value of $\tilde{S}^{(\mathrm{P})}(f_k)$ is 0.1336 and substantially exceeds g_F, so that the null hypothesis of white noise is rejected at the 5% level using Fisher's test. Siegel's test statistic $T_{0.6}$ is formed from the sum of the positive excesses of the $\tilde{S}^{(\mathrm{P})}(f_k)$ terms over $0.6 g_F \doteq 0.0652$. These excesses are shown by the three vertical lines in Figure 581; their sum is 0.0909. For $\alpha = 0.05$ and $M = 63$, the critical value for Siegel's test as stated in Table 542 is 0.0523 (its approximations 0.0515 and 0.0518 via $\tilde{t}_{0.6}$ and $c\chi_0^2(\beta)$ are slightly smaller). Hence the null hypothesis of white noise is rejected at the 5% level using Siegel's test. The first two values of $\tilde{S}^{(\mathrm{P})}(f_k)$ exceeding $0.6 g_F$ are adjacent and probably are both associated with a frequency of around 0.2 Hz, which corresponds to a period of 5 sec. There appears to be an additional sinusoid present with a frequency of 0.273 Hz, corresponding to a period of about 3.7 sec.

Now consider carrying out the tests at the 1% level ($\alpha = 0.01$). The critical value g_F is now 0.1316 (its approximation $\tilde{g}_F$ is also 0.1316). The maximum value of $\tilde{S}^{(\mathrm{P})}(f_k)$ of 0.1336 just exceeds g_F, so Fisher's test rejects the null hypothesis of white noise also at the 1% level. The sum of positive excesses of $\tilde{S}^{(\mathrm{P})}(f_k)$ over $0.6 g_F \doteq 0.0789$ is 0.0547. For $\alpha = 0.01$ and $M = 63$, Table 542 says that the critical value for Siegel's test is 0.0562 (its approximations via $\tilde{t}_{0.6}$ and $c\chi_0^2(\beta)$ are quite close: 0.0552 and 0.0564). Hence we fail to reject the null hypothesis of white noise (but just barely!) at the 1% significance level using Siegel's test.

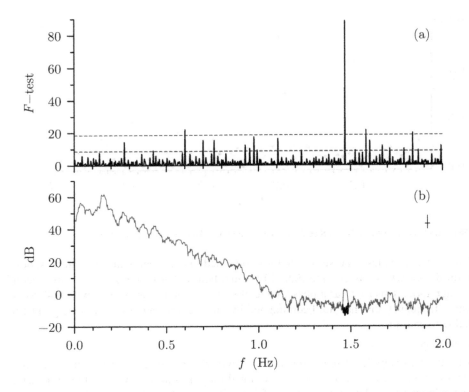

Figure 582 Thomson's F-test (top plot) and Slepian multitaper SDF estimate (bottom) for centered ocean wave data (see text for details).

The practical conclusion from these tests is that there is almost certainly one, and probably two sinusoids, of periods about 5 and 3.7 sec, present in the ocean noise data (this finding is not inconsistent with results from the normalized cumulative periodogram test for white noise discussed in C&E [6] for Section 6.8).

Ocean Wave Data
In our analysis of the ocean wave data in Section 6.8, we noted a small peak at $f = 1.469$ Hz in the direct spectral estimates shown in Figures 226(c) and (d). Here we assess the significance of that peak using Thomson's multitaper-based F-test for periodicity (see Section 10.10). As we did for the Willamette River data, we use $K = 5$ Slepian multitapers with $NW = 4/\Delta_t$ and compute the F-test of Equation (547a) over a grid of frequencies $j/(4096\,\Delta_t)$, $j = 1$, 2, ..., 2048 (since the sample size for the ocean wave data is $N = 1024$, the spacing between Fourier frequencies is $1/(1024\,\Delta_t)$, so the grid we are using has a much finer spacing). Figure 582 plots the results in a manner similar to Figure 575 for the Willamette River case. Figure 582(a) shows the F-test and has two dashed horizontal lines, which indicate the levels of the upper 99% and 99.9% percentage points of the $F_{2,8}$ distribution needed to evaluate the test (here the rule-of-thumb critical level from Thomson, 1990a, corresponds to a significance level of $1/N = 1/1024 \approx 0.001$ and hence is virtually the same as the critical level for the upper 99.9% percentage point). Recalling that we reject the null hypothesis of no spectral line for large values of the F-test, there is one obvious extreme rejection, which is associated with a level of significance of 3.5×10^{-6} and with frequency $\hat{f}_1 = 1505/(4096\,\Delta_t) \doteq 1.470$ Hz, in close agreement with the location of the peak in Figure 226 (the SDF estimates there are computed over the grid of Fourier frequencies, and the Fourier frequency closest to $\hat{f}_1$ is in

Method	Time Series	Model	μ Estimator	μ Estimate	SS
ECLS	uncentered	one frequency	least squares	9.82573	87.40153
ECLS	centered	one frequency	$\overline{X}$	9.82542	87.40157
EULS	uncentered	one frequency	least squares	9.82576	87.39877
EULS	centered	one frequency	$\overline{X}$	9.82542	87.39881
ECLS	uncentered	two frequencies	least squares	9.82564	79.58303
ECLS	centered	two frequencies	$\overline{X}$	9.82542	79.58305
EULS	uncentered	two frequencies	least squares	9.82566	79.54591
EULS	centered	two frequencies	$\overline{X}$	9.82542	79.54593

Table 583 Comparison of two ways to handle the mean μ in models of the form of Equation (512b) for the Willamette River time series. The first way is to center the series using the sample mean, set $\mu = 0$ in Equation (512b) and then estimate the remaining model parameters via least squares; the second is to use the uncentered series and estimate μ along with the other parameters. The next-to-last column in the table lists the resulting estimates of the mean for different combinations of methods and models; the last column lists the associated SSs.

fact $376/(1024\,\Delta_t) \doteq 1.469\,\mathrm{Hz}$). Figure 582(b) shows the corresponding multitaper SDF estimate (thin curve) and the reshaped estimate done using Equation (547b) over the interval of frequencies around $\hat{f}_1$ dictated by Equation (547c) (thick shorter curve). The revised estimate of the background continuum flows well with the SDF estimate before and after the interval about $\hat{f}_1$.

There are, however, three other frequencies at which the F-test exceeds the critical level associated with a level of significance 0.001, namely, $\hat{f}_2 \doteq 0.602\,\mathrm{Hz}$, $\hat{f}_3 \doteq 1.584\,\mathrm{Hz}$ and $\hat{f}_4 \doteq 1.837\,\mathrm{Hz}$. The SDF estimate in Figure 582(b) does not offer visual confirmation of peaks at these frequencies. If we consider F-tests based upon $K = 5$ and $K = 6$ sinusoidal multitapers, we find that the ones at $\hat{f}_2$ and $\hat{f}_4$ no longer exceed the critical level associated with 0.001, which suggests caution in declaring significant peaks at these two frequencies; however, both sinusoidal-based F-tests at $\hat{f}_3$ still exceed the 0.001 critical level, in agreement with the Slepian-based test. A look at the reshaped Slepian-based SDF estimate around $\hat{f}_3$ (not shown in Figure 582(b)) reveals an unappealing gouge in the estimated background continuum with a depth of about 10 dB. This result suggests considering an F-test based upon a multitapering scheme with a much wider bandwidth. If we consider the F-test based upon $K = 12$ sinusoidal multitapers, indeed the test at $\hat{f}_3$ no longer exceeds the critical level associated with 0.001. In conclusion, after considering the F-test with four different sets of multitapers, the only frequency at which we reject the null hypothesis of no spectral line in all cases is $\hat{f}_1$, and the observed critical level is always orders of magnitude smaller than 0.001.

Comments and Extensions to Section 10.15

[1] In our analysis of the Willamette River data, we centered the time series by subtracting off its sample mean $\overline{X}$ and then fit single- and two-frequency models of the form of Equation (512b) with μ set to zero using the ACLS, ECLS and EULS methods. For ECLS and EULS, an alternative to centering the series is to retain μ in the model and to estimate it using least squares along with the other model parameters. Including μ impacts the estimates of the other parameters, but almost undetectably: the new values for $\hat{f}_1$, $\hat{A}_1$, $\hat{B}_1$, $\hat{f}_2$, $\hat{A}_2$ and $\hat{B}_2$ round to the same values listed in Table 570 for estimates based on the centered data (however, for the two-frequency model with the EULS method, the estimated jump at $\hat{f}_2 \doteq 2.0024$ changes slightly from 0.00993 to 0.00994). Table 583 lists the least squares estimates of μ and the resulting SSs for the ECLS and EULS methods in combination with the single- and two-frequency models, along with corresponding results extracted from Table 570 based on centering the

time series. The least squares estimates of μ and the sample mean differ only in the third decimal place. Least squares estimation of μ must give a SS that is no greater than the one associated with setting μ equal to $\overline{X}$, a pattern that holds true in Table 583 in all four method/model combinations; however, in each case, the difference in SSs for two ways of handling μ amounts to a change in the fifth decimal place. We conclude that handling the nonzero mean for the Willamette River data by centering the series is as effective as dealing with it by including μ in the model.

10.16 Summary of Harmonic Analysis

In this section we collect together the most useful results of this chapter. It is assumed throughout that $\mu = 0$; D_l and f_l are real-valued constants; C_l is a complex-valued constant; and the ϕ_l terms are independent real-valued RVs uniformly distributed on $(-\pi, \pi]$. The spectral classifications used are as defined in Section 4.4.

[1] *Real-valued discrete parameter harmonic processes*

(a) Mixed spectrum:

$$X_t = \sum_{l=1}^{L} D_l \cos(2\pi f_l t \Delta_t + \phi_l) + \eta_t \qquad \text{(see (518c))}$$

$$= \sum_{l=-L}^{L} C_l e^{i 2\pi f_l t \Delta_t} + \eta_t, \qquad \text{(see (519b))}$$

where $\{\eta_t\}$ is a real-valued (possibly colored) noise process (independent of each ϕ_l) with zero mean and variance σ_η^2, $C_0 = 0$, $f_0 = 0$, and, for $l = 1, \ldots, L$, $C_l = D_l \exp(i\phi_l)/2$, $C_{-l} = C_l^*$ and $f_{-l} = -f_l$. Then, $E\{X_t\} = 0$,

$$\text{var}\{X_t\} = \sum_{l=-L}^{L} E\{|C_l|^2\} + \text{var}\{\eta_t\} = \sum_{l=1}^{L} D_l^2/2 + \sigma_\eta^2 \qquad \text{(see (519c))}$$

and

$$\text{cov}\{X_{t+\tau}, X_t\} = \sum_{l=1}^{L} D_l^2 \cos(2\pi f_l \tau \Delta_t)/2 + \text{cov}\{\eta_{t+\tau}, \eta_t\}. \qquad \text{(see (519e))}$$

(b) Discrete spectrum:

If the noise process $\{\eta_t\}$ is white, i.e., $\eta_t = \epsilon_t$, then the mixed spectrum reduces to a discrete spectrum, and

$$X_t = \sum_{l=1}^{L} D_l \cos(2\pi f_l t \Delta_t + \phi_l) + \epsilon_t \qquad \text{(see (512b))}$$

$$= \sum_{l=1}^{L} (A_l \cos(2\pi f_l t \Delta_t) + B_l \sin(2\pi f_l t \Delta_t)) + \epsilon_t, \qquad \text{(see (515b))}$$

where $\{\epsilon_t\}$ is a real-valued white noise process with zero mean and variance σ_ϵ^2, independent of each ϕ_l. We have $E\{X_t\} = 0$,

$$\text{var}\{X_t\} = \sum_{l=1}^{L} D_l^2/2 + \sigma_\epsilon^2 \quad \text{and} \quad \text{cov}\{X_{t+\tau}, X_t\} = \sum_{l=1}^{L} D_l^2 \cos(2\pi f_l \tau \Delta_t)/2, \quad |\tau| > 0.$$

(c) Known frequencies, discrete spectrum:
If each f_l is any of the Fourier frequencies not equal to 0 or $f_\mathcal{N}$,

$$\hat{A}_l = \frac{2}{N}\sum_{t=0}^{N-1} X_t \cos(2\pi f_l t\,\Delta_t); \quad \hat{B}_l = \frac{2}{N}\sum_{t=0}^{N-1} X_t \sin(2\pi f_l t\,\Delta_t) \qquad \text{(see (515c))}$$

give *exact* least squares estimates of A_l and B_l. If now each ϕ_l is treated as a constant so that A_l and B_l are constants and Equation (515b) is a multiple linear regression model, then

$$E\{\hat{A}_l\} = A_l, \quad E\{\hat{B}_l\} = B_l \text{ and } \operatorname{var}\{\hat{A}_l\} = \operatorname{var}\{\hat{B}_l\} = \frac{2\sigma_\epsilon^2}{N}. \qquad \text{(see (516a))}$$

Moreover, if $k \neq l$, $\operatorname{cov}\{\hat{A}_k, \hat{B}_l\}$, $\operatorname{cov}\{\hat{A}_l, \hat{B}_l\}$, $\operatorname{cov}\{\hat{A}_k, \hat{A}_l\}$ and $\operatorname{cov}\{\hat{B}_k, \hat{B}_l\}$ are all zero, and σ_ϵ^2 is estimated by Equation (516b). If the frequencies f_l are not all Fourier frequencies, then $\hat{A}_l$ and $\hat{B}_l$ give *approximate* least squares estimates, and

$$E\{\hat{A}_l\} = A_l + O\left(\frac{1}{N}\right) \text{ and } E\{\hat{B}_l\} = B_l + O\left(\frac{1}{N}\right). \qquad \text{(see (516c))}$$

(d) Unknown frequencies, discrete spectrum:
When the frequencies f_l are unknown, the standard approach is to look for peaks in the periodogram. If f_l is a Fourier frequency (not equal to 0 or $f_\mathcal{N}$) present in $\{X_t\}$, then

$$E\{\hat{S}^{(\mathrm{P})}(f_l)\} = \left(N\frac{D_l^2}{4} + \sigma_\epsilon^2\right)\Delta_t, \qquad \text{(see (524a))}$$

while, if f_l is a Fourier frequency not in $\{X_t\}$, then

$$E\{\hat{S}^{(\mathrm{P})}(f_l)\} = \sigma_\epsilon^2\,\Delta_t.$$

When f_l is not necessarily one of the Fourier frequencies, it follows that

$$E\{\hat{S}^{(\mathrm{P})}(f)\} = \sigma_\epsilon^2\,\Delta_t + \sum_{l=1}^{L} \frac{D_l^2}{4}\left[\mathcal{F}(f+f_l) + \mathcal{F}(f-f_l)\right], \qquad \text{(see (524b))}$$

where $\mathcal{F}(\cdot)$ is Fejér's kernel. If a data taper $\{h_t\}$ is applied to the process, then the expectation of the direct spectral estimator is given by

$$E\{\hat{S}^{(\mathrm{D})}(f)\} = \sigma_\epsilon^2\,\Delta_t + \sum_{l=1}^{L} \frac{D_l^2}{4}\left[\mathcal{H}(f+f_l) + \mathcal{H}(f-f_l)\right], \qquad \text{(see (535a))}$$

where $\mathcal{H}(\cdot)$ is the spectral window corresponding to $\{h_t\}$. If the taper is the default rectangular taper, then $\mathcal{H}(\cdot) = \mathcal{F}(\cdot)$.

[2] *Complex-valued discrete parameter harmonic processes*
  (a) Mixed spectrum:

$$Z_t = \sum_{l=1}^{L} D_l e^{i(2\pi f_l t\,\Delta_t + \phi_l)} + \eta_t = \sum_{l=1}^{L} C_l e^{i2\pi f_l t\,\Delta_t} + \eta_t, \qquad \text{(see (519g))}$$

where $\{\eta_t\}$ is a proper complex-valued (possibly colored) noise process with zero mean and variance σ_η^2, independent of each ϕ_l, and $C_l = D_l \exp(i\phi_l)$. Then, $E\{Z_t\} = 0$,

$$\operatorname{var}\{Z_t\} = \sum_{l=1}^{L} E\{|C_l|^2\} + \operatorname{var}\{\eta_t\} = \sum_{l=1}^{L} D_l^2 + \sigma_\eta^2 \qquad \text{(see (519h))}$$

and
$$\operatorname{cov}\{Z_{t+\tau}, Z_t\} = \sum_{l=1}^{L} D_l^2 e^{i 2\pi f_l \tau \Delta_t} + \operatorname{cov}\{\eta_{t+\tau}, \eta_t\}. \qquad \text{(see (519i))}$$

(b) Discrete spectrum:
If the noise process is white, i.e., $\eta_t = \epsilon_t$ then the mixed spectrum reduces to a discrete spectrum, and
$$Z_t = \sum_{l=1}^{L} D_l e^{i(2\pi f_l t \Delta_t + \phi_l)} + \epsilon_t, \qquad \text{(see (517a))}$$

where $\{\epsilon_t\}$ is a real-valued white noise process with zero mean and variance σ_ϵ^2, independent of each ϕ_l. We have $E\{Z_t\} = 0$,
$$\operatorname{var}\{Z_t\} = \sum_{l=1}^{L} D_l^2 + \sigma_\epsilon^2 \quad \text{and} \quad \operatorname{cov}\{Z_{t+\tau}, Z_t\} = \sum_{l=1}^{L} D_l^2 e^{i 2\pi f_l \tau \Delta_t}, \quad |\tau| > 0.$$

(c) Known frequencies, discrete spectrum:
Let us assume a process with a discrete spectrum. If each f_l is any of the Fourier frequencies not equal to 0 or $f_\mathcal{N}$,
$$\hat{C}_l = \frac{1}{N} \sum_{t=0}^{N-1} Z_t e^{-i 2\pi f_l t \Delta_t} \qquad \text{(see Exercise [10.2])}$$

is the exact least squares estimate of C_l. With each ϕ_l treated as a constant,
$$E\{\hat{C}_l\} = C_l, \quad \operatorname{var}\{\hat{C}_l\} = \frac{\sigma_\epsilon^2}{N} \quad \text{and} \quad \operatorname{cov}\{\hat{C}_k, \hat{C}_l\} = 0, \quad k \neq l.$$

(d) Unknown frequencies, discrete spectrum:
If f_l is a Fourier frequency (not equal to 0 or $f_\mathcal{N}$) present in $\{Z_t\}$, then
$$E\{\hat{S}^{(\mathrm{P})}(f_l)\} = \left(N D_l^2 + \sigma_\epsilon^2\right) \Delta_t,$$

while, if f_l is a Fourier frequency not in $\{Z_t\}$,
$$E\{\hat{S}^{(\mathrm{P})}(f_l)\} = \sigma_\epsilon^2 \Delta_t.$$

[3] *Tests for periodicity for real-valued process*
(a) White noise:
This case is discussed in detail in Section 10.9. The sample size N is usually taken to be odd so that $N = 2M + 1$ for some integer M (Section 10.15 demonstrates how to accommodate an even sample size). The null hypothesis is $D_1 = \cdots = D_L = 0$. Fisher's exact test for *simple* periodicity uses the statistic
$$g \stackrel{\text{def}}{=} \max_{1 \leq k \leq M} \frac{\hat{S}^{(\mathrm{P})}(f_k)}{\sum_{j=1}^{M} \hat{S}^{(\mathrm{P})}(f_j)}, \qquad \text{(see (539b))}$$

where $\hat{S}^{(\mathrm{P})}(f_k)$ is a periodogram term. The exact distribution of g under the null hypothesis is given by Equation (540b). Critical values g_F for Fisher's test for $\alpha = 0.01, 0.02, 0.05$

and 0.1 can be adequately approximated using Equation (540d). To test for the presence of *compound* periodicities, Siegel derived the statistic based on excesses over a threshold,

$$T_\lambda \stackrel{\text{def}}{=} \sum_{k=1}^{M} \left(\tilde{S}^{(\text{P})}(f_k) - \lambda g_F \right)_+ , \qquad \text{(see (541b))}$$

where $\tilde{S}^{(\text{P})}(f_k) = \hat{S}^{(\text{P})}(f_k) / \sum_{j=1}^{M} \hat{S}^{(\text{P})}(f_j)$, $0 < \lambda \leq 1$, and g_F is the critical value for Fisher's test. The exact distribution of T_λ under the null hypothesis is given by Equation (541c). When $\lambda = 0.6$, critical values for M from about 20 to 2000 can be approximated using Equation (542a) for $\alpha = 0.05$, and Equation (542b) for $\alpha = 0.01$.

(b) Colored noise:

This case is discussed in detail in Section 10.10. To test for periodicity at a particular frequency f_1 for a real-valued process incorporating colored *Gaussian* noise, the recommended statistic is

$$\frac{(K-1)|\hat{C}_1|^2 \sum_{k=0}^{K-1} H_k^2(0)}{\Delta_t \sum_{k=0}^{K-1} \left| J_k(f_1) - \hat{J}_k(f_1) \right|^2}, \qquad \text{(see (547a))}$$

where $H_k(0) = \Delta_t \sum_{t=0}^{N-1} h_{k,t}$,

$$J_k(f_1) = \Delta_t^{1/2} \sum_{t=0}^{N-1} h_{k,t} X_t e^{-i 2\pi f_1 t \Delta_t}, \quad \hat{J}_k(f_1) = \hat{C}_1 \frac{H_k(0)}{\Delta_t^{1/2}}$$

and $\hat{C}_1$ is given by Equation (546a). Here $\{h_{k,t}\}$ is a kth-order Slepian or sinusoidal data taper. This statistic has a simple $F_{2,2K-2}$ distribution under the null hypothesis of no periodicity at the frequency f_1.

10.17 Exercises

[10.1] Verify that the approximations in Equation (515a) are valid.

[10.2] For the complex-valued model given by Equation (518a), show that cov $\{\hat{C}_k, \hat{C}_l\} = 0$ for $k \neq l$, where $\hat{C}_k$ and $\hat{C}_l$ are the least squares estimators of, respectively, C_k and C_l; i.e.,

$$\hat{C}_k = \frac{1}{N} \sum_{t=0}^{N-1} Z_t e^{-i 2\pi f_k t \Delta_t} \quad \text{and} \quad \hat{C}_l = \frac{1}{N} \sum_{t=0}^{N-1} Z_t e^{-i 2\pi f_l t \Delta_t}.$$

[10.3] Prove that

$$\text{cov}\{Z_{t+\tau}, Z_t\} = \sum_{l=1}^{L} D_l^2 e^{i 2\pi f_l \tau \Delta_t} + \text{cov}\{\eta_{t+\tau}, \eta_t\}$$

for $\{Z_t\}$ given by Equation (519g).

[10.4] Verify Equation (524b).

[10.5] Consider a time series $X_0, X_1, \ldots, X_{N-1}$ obeying Equation (512b) specialized to the case $\mu = 0$, $L = 1$ and $\Delta_t = 1$:

$$X_t = D_1 \cos(2\pi f'_k t + \phi_1) + \epsilon_t,$$

where $f'_k = \frac{2k+1}{2N}$, and k is a nonnegative integer such that $k < (N-1)/2$; i.e., f'_k falls halfway between the Fourier frequencies k/N and $(k+1)/N$. Equation (526d) says that the signal-to-noise ratio for $\{X_t\}$ is $R = D_1^2/(2\sigma_\epsilon^2)$, where, as usual, $\sigma_\epsilon^2 = \text{var}\{\epsilon_t\}$.

(a) Assuming that N and R are sufficiently large and that f'_k is not too close to zero or Nyquist frequency, argue that the ratio

$$r(f'_k) \stackrel{\text{def}}{=} \frac{E\{\hat{S}^{(\text{P})}(f'_k + \frac{1}{2N})\}}{E\{\hat{S}^{(\text{P})}(f'_k)\}} \approx \frac{4}{\pi^2} \doteq 0.405$$

by appealing to Equation (524b) (Whittle, 1952; see also Priestley, 1981, p. 403; because $f'_k + \frac{1}{2N}$ is a Fourier frequency, the above supports the claim that evaluation of the periodogram at just the Fourier frequencies tends to underestimate its peak value when the dominant frequency in $\{X_t\}$ is halfway between Fourier frequencies).

(b) Compute and plot the normalized ratio $r(f'_k)/(4/\pi^2)$ versus f'_k, $k = 0, 1, \ldots, 511$, for the case $N = 1024$ and $R = 1$. Repeat for $R = 0.1$, 10, and 100. How well does the approximation $r(f'_k) \approx 4/\pi^2$ work here?

[10.6] (a) Verify Equation (526a) and the closely related equation

$$\hat{S}^{(P)}(f'_k) = \frac{N}{4\,\Delta_t} \left(\hat{A}^2 + \hat{B}^2\right). \tag{588a}$$

(b) Equation (526a) excludes the zero and Nyquist frequencies. For these two cases, develop expressions that relate the periodogram to a residual sum of squares from an appropriate regression model.

[10.7] For the time series $\{y_t\}$ of length $N = 16$ given by Equation (530), compute and plot the periodogram $\hat{S}^{(P)}(f'_j)$ versus $f'_j = j/1024$ for $j = 1, 2, \ldots, 511$ (note that this ignores the zero and Nyquist frequencies since $0 < f'_j < 1/2$). With f_1 replaced by f'_j in Equation (526b), do the same for the regression-based $\widetilde{S}(f'_j)$. At what frequencies do we have $\widetilde{S}(f'_j) = \hat{S}^{(P)}(f'_j)$? How does

$$\sum_{j=1}^{511} \widetilde{S}(f'_j) \text{ compare with } \sum_{j=1}^{511} \hat{S}^{(P)}(f'_j)?$$

[10.8] This exercise looks into three key properties of the Lomb–Scargle periodogram.

(a) Show that the residual sum of squares given by Equation (528c) does not depend on the constant c. Hint: recall that the standard result that the residual sum of squares can be expressed as $\boldsymbol{X}^T[\boldsymbol{I}_N - \boldsymbol{H}_c(\boldsymbol{H}_c^T\boldsymbol{H}_c)^{-1}\boldsymbol{H}_c^T]\boldsymbol{X}$ (see, e.g., Weisberg, 2014), and show that $\boldsymbol{H}_c = \boldsymbol{H}_0\boldsymbol{C}$ for a suitably defined 2×2 matrix $\boldsymbol{C}$. (As usual, $\boldsymbol{I}_N$ is the $N \times N$ identity matrix.)

(b) Verify Equation (529c).

(c) Show that the least squares estimators $\hat{A}$ and $\hat{B}$ minimizing $\text{SS}_{\tilde{c}}(A, B)$ are uncorrelated.

[10.9] (a) Reformulate Figure 536 so that it shows direct spectral estimates based upon the Hanning and Slepian tapers with $NW = 6$ and $NW = 8$ rather than the default and Slepian tapers with $NW = 2$ and $NW = 4$. Comment on how these three new estimates compare with the ones shown in Figure 536.

(b) Generate a realization of length $N = 256$ from the following process, which, like the process of Equation (535b), is a special case of Equation (512b):

$$X_t = \cos(0.311\pi t + \phi_1) + \cos(0.3167\pi t + \phi_2) + \epsilon_t, \tag{588b}$$

where $\{\epsilon_t\}$ is a white noise process with variance $\sigma_\epsilon^2 = 10^{-6} = -60\,\text{dB}$. Using both linear and decibel scales, compute and plot direct spectral estimates for the time series based upon the Hanning data taper and Slepian data tapers with $NW = 1, 2$ and 4. How well do these estimates indicate the presence of the two sinusoidal components in the process, and how well do they portray the background continuum?

[10.10] (a) Verify Equation (537a).

(b) If U and V are two RVs with a bivariate distribution, the law of total expectation says that $E\{U\} = E\{E\{U \mid V\}\}$ (for a precise statement, see, e.g., Chung, 1974, section 9.1, Priestley, 1981, p. 75, or Rao, 1973, section 2b.3). Use this law with Equation (537a) to verify that $E\{\hat{S}^{(P)}(f)\}$ is as stated in Equation (536).

(c) For $\sigma^2 = 10^{-4}$, $\Delta_t = 1$, $D_1 = 1$, $f_1 = 0.0725$ and $N = 64$, determine – via computations – the value $f \in [f_1 - \frac{1}{N}, f_1 + \frac{1}{N}]$ such that $E\{\hat{S}^{(P)}(f) \mid \phi_1\}$ of Equation (537a) is maximized when the phase is set to each of 360 different values, namely, $\phi_1 = -\pi + \frac{k\pi}{180}$, $k = 0, 1, \ldots, 359$. Plot these peak frequencies versus ϕ_1 for comparison with f_1, paying particular attention to the settings $\phi_1 = -\pi/3$, 0 and $5\pi/12$. Comment briefly.

(d) Create a figure similar to Figure 538b, but this time for the single-frequency model $X_t = \cos(2\pi f_1 t - \pi/3) + \epsilon_t$, which is the same as the model of Equation (537b), but has a different fixed phase. Do the same for the model $X_t = \cos(2\pi f_1 t) + \epsilon_t$. Comment briefly on how these two new scatter plots compare with the one in Figure 538b.

(e) Create a figure similar to Figure 538b, but this time drawing 1000 realizations from the single-frequency model $X_t = \cos(2\pi f_1 t + \phi_1) + \epsilon_t$, where each realization of X_t, $t = 0, 1, \ldots, 63$, is created using a different realization of the uniformly distributed ϕ_1. Briefly compare this new scatter plot with the one in Figure 538b and the ones considered in part (d).

[10.11] Verify Equation (540a).

[10.12] Equations (540d) and (540e) give approximations $\tilde{g}_F$ and $\hat{g}_F$ for the critical level g_F of Fisher's g test. For significance levels $\alpha = 0.01, 0.02, 0.05$ and 0.1 in combination with sample sizes $N = 21, 101, 501, 1001, 2001, 3001$ and 4001, compare these approximations by computing $\mathbf{P}[g > \tilde{g}_F]$ and $\mathbf{P}[g > \hat{g}_F]$ using Equation (540b) to see which is closer to the selected α. Comment upon your findings.

[10.13] This exercise compares the performance of three tests for white noise: the cumulative periodogram test (see Equation (215a) and the discussion following it), Fisher's test (Equation (539b)) and Siegel's test with $\lambda = 0.6$ (Equation (541b)). For each of the nine power-of-two sample sizes N used in Table 542, generate a realization of a portion $X_{j,0}, X_{j,1}, \ldots, X_{j,N-1}$ of each of the following four stationary processes:

(a) $\{X_{1,t}\}$, a Gaussian white noise process with zero mean and unit variance;

(b) $\{X_{2,t}\}$, an AR(2) process dictated by Equation (34) – see Exercise [597] for a description of how to generate a realization from this process;

(c) $\{X_{3,t}\}$, a harmonic process with additive white noise $\{\epsilon_t\}$ dictated by $X_{3,t} = 0.5\cos(\pi t/2 + \phi) + \epsilon_t$, where ϕ is uniformly distributed over $(-\pi, \pi]$, while $\{\epsilon_t\}$ is Gaussian with zero mean and unit variance (this is a special case of Equation (512b) with $\mu = 0$, $L = 1$, $D_1 = 0.1$, $f_1 = 1/4$, $\Delta_t = 1$ and $\phi_1 = \phi$); and

(d) $\{X_{4,t}\}$, another harmonic process with additive white noise $\{\epsilon_t\}$, but now given by $X_{4,t} = 0.25\cos(0.2943\pi t + \phi_1) + 0.25\cos(0.3333\pi t + \phi_2) + \epsilon_t$, where ϕ_1 and ϕ_2 are independent and uniformly distributed over $(-\pi, \pi]$, while $\{\epsilon_t\}$ is again Gaussian with zero mean and unit variance (this is a special case of Equation (512b) with $L = 2$).

For each realization of $\{X_{j,t}\}$ for the nine sample sizes, compute the cumulative periodogram test statistic D, Fisher's test statistic g and Siegel's test statistic $T_{0.6}$, and use these three statistics individually to evaluate the null hypothesis of white noise at both the $\alpha = 0.05$ and $\alpha = 0.01$ levels of significance (use Equation (215b) to determine the critical levels for D; Equation (540d), for g; and the "Exact" columns of Table 542, for $T_{0.6}$). Repeat everything a large number N_{R} of times (use different realizations for all four $\{X_{j,t}\}$ each time, and take "large" to mean something from 1000 up to 100,000). For each $\{X_{j,t}\}$ for the nine samples sizes, count the number of times out of N_{R} that each of the three white noise tests rejected the null hypothesis of white noise at the two settings for α. Comment upon your findings.

[10.14] (a) Using the time series $\{X_t\}$ whose periodogram is displayed in Figure 534(c) (downloadable in the file ts-km-noisy-128.txt from the "Data" part of the website for the book or recreatable using R code from the site), compute ACLS and ECLS estimates of A_l, B_l, the jumps in the integrated spectrum and σ_ϵ^2 for the model

$$X_t = \sum_{l=1}^{3} \left[A_l \cos(2\pi \hat{f}_l t) + B_l \sin(2\pi \hat{f}_l t) \right] + \epsilon_t, \tag{589}$$

where $\hat{f}_1 = 0.1326075$, $\hat{f}_2 = 0.1883504$ and $\hat{f}_3 = 0.334064$ (these are the maximizing values – rounded to seven decimal places – for the three largest peaks in the periodogram – see the discussion in C&E [2] for Section 10.6; as usual, σ_ϵ^2 is the variance of the white noise process $\{\epsilon_t\}$. Compute the periodogram over the grid of frequencies $k/1024$, $k = 0, 1, \ldots, 512$, for the residuals $\{R_t\}$ corresponding to the ACLS estimates (see Equation (550b)), and plot it for comparison with the periodogram shown in Figure 551, which is for the residuals for

EULS estimates. Do the same for the residuals for the ECLS estimates. For both sets of residuals, perform Fisher's test (Equation (539b)) and the normalized cumulative periodogram test (described following Equation (215a)). Comment upon your findings.

(b) For the same time series as considered in part (a), compute the jumps $|\hat{C}_l|^2$ in the integrated spectrum based upon the multitaper estimates $\hat{C}_l$ of Equation (551a) by employing the following eight sets of tapers: Slepian tapers with $NW = 2$ in combination first with $K = 1$ and then with $K = 3$; Slepian tapers with $NW = 4$ and $K = 1, 3$ and 5; and sinusoidal tapers with $K = 1, 3$ and 5. Compare the jump estimates with the ones found in part (a) and with what Equation (530) suggests. Use the multitaper approach to estimate the white noise variance via Equation (551b). Comment on how these eight estimates of σ_ϵ^2 compare with the ACLS and ECLS estimates found in part (a).

(c) The time series in the file `ts-km-noisy-16384.txt` that is accessible from the "Data" part of the website for the book (or recreatable using R code from the site) was created in a manner similar to the series of interest in parts (a) and (b), but is of length $N = 2^{14} = 16384$ rather than $N = 2^7 = 128$. Using this longer series, compute ACLS and ECLS estimates of the jumps in the integrated spectrum and σ_ϵ^2 for the model of Equation (589), but with $\hat{f}_1, \hat{f}_2$ and $\hat{f}_3$ replaced by $f_1 = 1/7.5$, $f_2 = 1/5.3$ and $f_3 = 1/3$ as suggested by Equation (530). Compute the jumps $|\hat{C}_l|^2$ in the integrated spectrum based upon the multitaper estimates $\hat{C}_l$ of Equation (551a) by employing the same eight sets of tapers as used in part (b). Compute the corresponding estimates of the white noise variance via Equation (551b). Comment on how all these estimates compare with the ones obtained in parts (a) and (b) for the shorter time series.

[10.15] Here we consider creating a 95% CI for a jump J in the integrated spectrum based upon the jump estimator $\hat{J}$ whose distribution is stated by Equation (552). Generate a large number, say M, of time series from the model

$$X_t = D\cos(2\pi f t + \phi) + \epsilon_t, \quad t = 0, 1, \ldots, N-1,$$

where $D^2 = 1$; $f = 1/8$; ϕ is an RV uniformly distributed over the interval $(-\pi, \pi]$ (a different RV independently chosen for each of the M time series); $\{\epsilon_t\}$ is Gaussian white noise with mean zero and variance $\sigma_\epsilon^2 = 1$ (independently chosen for each of the M time series); and N is taken to be first 128, then 512, 2048, 8192 and finally 32,768. Noting this model to be a special case of Equations (512b) and (513) with $\mu = 0$, $A^2 + B^2 = D^2 = 1$ and $\Delta_t = 1$,

(a) compute the least squares estimators $\hat{A}$ and $\hat{B}$ of A and B for each $\{X_t\}$;
(b) form the corresponding jump estimator $\hat{J} = (\hat{A}^2 + \hat{B}^2)/4$;
(c) count the number of cases out of M for which $\hat{J}$ is trapped by the interval

$$\left[\frac{\hat{\sigma}_\epsilon^2 F_{2,N-2,\beta}(0.025)}{N}, \frac{\hat{\sigma}_\epsilon^2 F_{2,N-2,\beta}(0.975)}{N}\right] \quad (590a)$$

with β set to $N(A^2 + B^2)/(2\sigma_\epsilon^2) = N/2$;

(d) count the number of cases out of M for which $J = (A^2 + B^2)/4 = 1/4$ is trapped by the interval whose end points are given by

$$\hat{J} \pm \frac{\hat{\sigma}_\epsilon^2 \left[F_{2,N-2,\beta}(0.975) - F_{2,N-2,\beta}(0.025)\right]}{2N}, \quad (590b)$$

with β set, as before, to $N/2$; and

(e) count the number of cases out of M for which $J = 1/4$ is trapped by the interval whose end points are given by Equation (590b) with β now set to the obvious estimator $\hat{\beta} = N(\hat{A}^2 + \hat{B}^2)/(2\hat{\sigma}_\epsilon^2)$.

Repeat (a) through (e) with $D^2 = A^2 + B^2$ changed from 1 to 0.1 and then also to 10. Finally repeat everything, but with f first changed to $1/4$ and then to $3/8$. Comment upon your findings.

[10.16] Given $f \in (0, 1/(2\Delta_t)]$, determine what conditions we need to impose on the real-valued constants x_0 and x_1 so that we can write $x_0 = D\cos(\phi)$ and $x_1 = D\cos(2\pi f \Delta_t + \phi)$ for some $D > 0$

and $\phi \in (-\pi, \pi]$. As four concrete examples, suppose $\Delta_t = 1$, $f = 1/16$, $x_0 = \pm 2$ and $x_1 = \pm 2$, and determine the corresponding D and ϕ in each case if there be such. Use Equation (553a) to form x_t, $t = 2, 3, \ldots, 16$, for the four cases. Plot x_t, $t = 0, 1, \ldots, 16$, versus t for each case. As four additional examples, repeat everything, but now with f assumed to be $1/2$.

[10.17] Using Equation (553b) and the initial conditions $X_0 = D \cos(\phi)$ and $X_1 = D \cos(2\pi f \Delta_t + \phi)$, show that $\text{cov}\{X_{t+\tau}, X_t\} = D^2 \cos(2\pi f \tau \Delta_t)/2$.

[10.18] Show that Equation (554a) provides a solution to Equation (553d) under the stipulation that the coefficients $\{\varphi_{2p,k}\}$ in the difference equation satisfy Equation (553c) when z is set to $z_l = \exp(\pm i 2\pi f_l \Delta_t)$, $l = 1, \ldots, p$.

[10.19] Given that the roots $\{z_j\}$ of polynomial Equation (553c) are all on the unit circle and occur in conjugate pairs, show that we must have $\varphi_{2p,2p} = -1$. For the case $p = 2$, merge this result with Equation (554b) to form a three-dimensional matrix equation that can be used to solve for $\varphi_{4,1}$, $\varphi_{4,2}$ and $\varphi_{4,3}$.

[10.20] Show that an AR(p) SDF plus a white noise SDF yields an ARMA(p,p) SDF. As a concrete example find the coefficients for the ARMA(2,2) process obtained by adding white noise with unit variance to the AR(2) process defined by Equation (34). Plot the SDFs for the AR(2) and ARMA(2,2) processes for comparison.

[10.21] Figure 557 shows peak frequencies estimated using 50 realizations – each of sample size $N = 64$ – from the process of Equation (537b) (the actual peak frequency is $f_1 = 0.0725$). The peak frequencies shown on the horizontal axis are estimated using the AR forward/backward least squares (FBLS) method with order $p = 16$; i.e., the coefficients from a fitted AR(16) model are used to compute the corresponding SDF estimate, and the peak frequency is the location of the largest value in this estimate. The "Data" part of the website for the book has 1000 downloadable realizations from this process (the first 50 of these are the ones used in Figure 557). For each realization, use the FBLS AR method of orders $p = 2, 3, \ldots, 32$ to obtain 31 different estimates of the peak frequency. For a given p, compute the sample variance and MSE of the 1000 estimates, and plot these quantities versus p (use a logarithm scale for the vertical axis). Comment upon your findings.

[10.22] Verify Equation (561a).

[10.23] (a) Verify Equation (561d), which assumes that the sample size N is odd. Derive a corresponding expression for $\hat{s}_\tau^{(P)}$ when N is even.
(b) Using the biased ACVS estimator $\hat{s}_\tau^{(P)}$ of Equation (561d), set $N = 9$, and compute and plot AR(p) SDF estimates based upon the Yule–Walker method for $p = 2, 3, \ldots, 8$ assuming that $\Delta_t = 1$. Comment upon your findings.
(c) Repeat part (b), but now use the expression you found for $\hat{s}_\tau^{(P)}$ in part (a) for even sample sizes with $N = 10$ and $p = 2, 3, \ldots, 9$.
(d) Repeat part (b), but now focusing on the cases $N = 15$ with $p = 14$; $N = 31$ with $p = 30$; $N = 63$ with $p = 62$; and, finally, $N = 127$ with $p = 126$.
(e) Repeat part (b), but now set $N = 127$, and consider $p = 74, 76, 78$ and 80.
(f) For the time series considered in parts (a) to (e), i.e., $x_t = \cos(\pi t/2)$, $t = 0, 1, \ldots, N-1$, with $\Delta_t = 1$, show that, for all $N \geq 3$, use of either the forward/backward least squares method or Burg's algorithm yields estimates of the spectrum that, in contrast to the Yule–Walker method, do not suffer from line splitting. Hint: show that $x_t = \varphi_{2,1} x_{t-1} + \varphi_{2,2} x_{t-2}$ (with $\varphi_{2,1}$ and $\varphi_{2,2}$ appropriately chosen), and study the conclusions drawn from Equations (555c) and (556a).

[10.24] With $m \stackrel{\text{def}}{=} 2(N - p')$ and $n = p'$, derive Equation (565b) by considering the eigenvectors and eigenvalues of the $(m + n) \times (m + n)$ matrix

$$B \stackrel{\text{def}}{=} \begin{bmatrix} 0 & A \\ A^H & 0 \end{bmatrix}.$$

[10.25] Show that Equation (569a) can be rewritten as Equation (569b). What happens if $\Delta_t = 1/5$ instead of $1/12$?

[10.26] Using the Willamette River data (downloadable from the "Data" part of the website for the book), compute a direct spectral estimate based upon an $NW = 4/\Delta_t$ Slepian data taper. Plot the estimate in a manner similar to Figure 568b to verify that the periodogram in that figure does not suffer from any apparent leakage (particularly at the second and higher harmonics of 1 cycle/year).

[10.27] Substitute appropriate estimates for the unknown quantities in the expression for the variance of a periodogram estimate of frequency in Equation (526c) to determine roughly whether the periodogram-based estimated frequencies $\hat{f}_1 \doteq 1.00319$ cycles/year and $\hat{f}_2 \doteq 2.00339$ cycles/year for the Willamette River data include – within their two standard deviation limits – the values 1 cycle/year and 2 cycles/year (see Table 570 and the discussion surrounding it). Are the corresponding AR(p) estimates in Table 578 also within these two standard deviation limits?

[10.28] Figure 575 shows Thomson's F-test and corresponding reshaped multitaper estimate for the centered Willamette River data using $K = 5$ Slepian multitapers with $NW = 4/\Delta_t$. Redo this figure using $K = 5$ sinusoidal multitapers and then using $K = 6$ tapers. Comment on how well the three F-tests agree with one other.

[10.29] The top two plots of Figure 577 show Burg-based AR(150) and AR(27) SDF estimates for the Willamette River data of Figure 568a after centering, and the top two rows of Table 578 summarize some properties of these estimates. Create similar plots and summaries using (a) the Yule–Walker method and (b) the forward/backward least squares method for estimation of the AR SDFs. Comment on how these SDF estimates compare with those based on Burg's algorithm.

[10.30] (a) Verify Equation (579b).

(b) Compute and plot the phase function and the squared gain function for the filter $\{c_u g_u\}$, where $\{g_u\}$ and $\{c_u\}$ are given by, respectively, Equations (579b) and (579c). How well do these functions match up the ones corresponding to the transfer function of Equation (579a)?

[10.31] For our SVD analysis of the Willamette River time series in Section 10.15, we used a complex-valued analytic series whose real and imaginary parts were, respectively, a portion of the original series and a corresponding filtered version of the series (the filter was an approximation to the Hilbert transform). Here we explore two alternatives to the analytic series (the Willamette River data are downloadable from the "Data" part of the website for the book).

(a) A way to create a complex-valued time series that is usefully related to a real-valued series is via complex demodulation and low-pass filtering (see Tukey, 1961, Hasan, 1983, Bloomfield, 2000 or Stoica and Moses, 2005). To do so here, take the centered Willamette River series $\{X_t\}$ and form the demodulated series $Z_t = X_t e^{-i2\pi f_0 t \Delta_t}$, $t = 0, 1, \ldots, N - 1$, where $f_0 = 1.5$ cycles/year (since $\Delta_t = 1/12$ year, the unitless demodulating frequency is $f_0 \Delta_t = 1/8$). Next filter $\{Z_t\}$ using a low-pass filter $\{c_u g_{I,u}\}$ with 51 nonzero coefficients to obtain

$$Z'_t = \sum_{u=-25}^{25} c_u g_{I,u} Z_{t-u}, \quad t = 25, 26, \ldots, N - 26,$$

where $\{g_{I,u}\}$ is the ideal low-pass filter of Equation (152b) with W set to $1/8$, while $\{c_u\}$ are the convergence factors specified by Equation (579c). Finally use the remodulated filtered series $\{Z'_t e^{i2\pi f_0 t \Delta_t}\}$ in place of the analytic series to create plots analogous to Figures 580a and 580b. How well do the two roots closest to the unit circle gotten using the analytic series agree with those from $\{Z'_t e^{i2\pi f_0 t \Delta_t}\}$? What is the rationale behind the choices $f_0 = 1.5$ cycles/year and $W = 1/8$ and behind the choice to remodulate (hint: recall Exercise [5.13a])?

(b) Since a real-valued series is a special case of a complex-valued one, there is nothing preventing the application of the theory presented in Section 10.15 to a real-valued series. Do so by taking the centered Willamette River series $\{X_t\}$ and subjecting it to the same SVD analysis as was done using the analytic series, but set $p = 6$ rather than $p = 3$. Comment upon your findings.

11

Simulation of Time Series

11.0 Introduction

Of the many reasons for estimating a hypothetical spectral density function (SDF) for a time series $\{X_t\}$, a prominent one is to use the estimated SDF as the basis for creating computer-generated artificial series with characteristics similar to those of $\{X_t\}$. Simulated series are of interest in bootstrapping, a popular computer-based method for assessing the sampling variability in certain statistics (see, e.g., Efron and Gong, 1983, or Davison and Hinkley, 1997; we touch upon this topic in Sections 11.4 to 11.6). Simulated series are also useful for assessing procedures that act upon time series. As a concrete example, consider the atomic clock data $\{X_t\}$ displayed in Figure 326(a). These data compare the time kept by two atomic clocks, both of which were part of a large ensemble of atomic clocks maintained at the US Naval Observatory and used to provide a standard of time for the United States. Since no single clock keeps perfect time, pairwise comparisons amongst clocks are used to form a nonphysical "paper" clock that in principle keeps time better than any individual physical clock. Various procedures – known as time-scale algorithms – have been proposed for manipulating pairwise comparisons to form a paper clock (see, e.g., references in Matsakis and Tavella, 2008). The relative merits of different algorithms can be assessed by applying them to simulated pairwise comparisons. One way to ensure realistic simulated series is to make them consistent with SDF estimates similar to those in Figure 329.

In what follows we first formulate procedures for generating artificial time series that can be considered to be realizations from a known stationary process. Section 11.1 describes simulation of autoregressive moving average (ARMA) processes with known parameters $\phi_{p,1}$, ..., $\phi_{p,p}$, $\theta_{q,1}$, ..., $\theta_{q,q}$ and σ_ϵ^2 (Equation (35b)) and harmonic processes with known parameters (Equations (35c) and (35d)). Section 11.2 discusses simulation of a stationary process for which we know a portion of its autocovariance sequence (ACVS), while Section 11.3 considers the case in which the SDF is fully specified and readily available. We next adjust procedures described in these sections to work with SDF estimates based on actual time series that we are interested in mimicking. Section 11.4 does this for nonparametric estimates (including direct SDF estimates, lag window estimates, multitaper estimates and Welch's overlapped segment averaging [WOSA] estimates), while Section 11.5 does it for autoregressive-based parametric SDF estimates. Section 11.6 has examples of simulated series that resemble actual time series. Up to here we assume Gaussianity for the most part. Section 11.7 comments on simulation of non-Gaussian time series. We close with a summary and exercises (Sections 11.8 and 11.9).

11.1 Simulation of ARMA Processes and Harmonic Processes

In Section 2.6 we looked at examples of realizations from several discrete parameter stationary processes including white noise (see Figure 31), moving average (MA) processes (see Equation (32a) and Figure 33), autoregressive (AR) processes (Equation (33) and Figure 34) and harmonic processes (Equations (35c) and (35d) and Figure 36). Here we discuss how to generate a realization of a contiguous portion $X_0, X_1, \ldots, X_{N-1}$ from each of these processes assuming that we have access to a suitable number of independent realizations of random variables (RVs) with certain distributions. White noise, MA and AR processes are all special cases of autoregressive moving average (ARMA) processes (Equation (35b)). We tackle these special cases first, after which we turn to ARMA processes in general and then finally to harmonic processes. For convenience we take the process mean μ for the stationary processes to be zero (a realization of a process with $\mu \neq 0$ can be easily generated by taking a realization with zero mean and adding the desired nonzero μ to each of the N values constituting the realization). We also take the sampling interval Δ_t to be unity; item [3] in the Comments and Extensions (C&Es) discusses its role in the simulations.

White Noise Process

A white noise process with zero mean is a sequence of uncorrelated RVs $\{X_t : t \in \mathbb{Z}\}$, and hence $E\{X_t\} = 0$ and $\text{var}\{X_t\} = \sigma^2$ for all t, where $0 < \sigma^2 < \infty$. Since independent RVs are also uncorrelated, we can generate a realization of length N from a white noise process by simply generating N independent RVs, each of which has mean zero and variance σ^2. Plots (a) to (f) of Figure 31 show six realizations of length $N = 100$ from white noise processes, each with zero mean and unit variance and each derived from a different selection of independent RVs (see the figure caption for details). Creation of the first four realizations requires drawing independent samples from RVs with (a) a Gaussian (normal) distribution; (b) a uniform distribution over the interval $[-\sqrt{3}, \sqrt{3}]$; (c) an exponential distribution together with a two-point distribution taking on the values ± 1 with equal probability (used to convert an exponential RV into a double exponential RV); and (d) a three-point distribution over -5, 0 and 5 with probabilities 0.02, 0.96 and 0.02. Additionally a four-point distribution over four categories with equal probability is used to create the realization in Figure 31(e) by taking a deviate from the distributions associated with the realizations in (a) to (d). In the R language the functions `rnorm`, `runif`, `rexp` and `sample` can do the random draws, but equivalent functions are available in other languages (e.g., in MATLAB the functions `normrnd`, `unifrnd` and `exprnd` give draws from normal, uniform and exponential distributions).

Moving Average Process

Given a zero mean white noise process $\{\epsilon_t : t \in \mathbb{Z}\}$ with variance σ_ϵ^2 and given a set of $q \geq 1$ constants $\theta_{q,1}, \ldots, \theta_{q,q}$ (arbitrary except for the constraint $\theta_{q,q} \neq 0$), we can construct a zero mean qth-order MA process $\{X_t : t \in \mathbb{Z}\}$ using

$$X_t = \epsilon_t - \theta_{q,1}\epsilon_{t-1} - \cdots - \theta_{q,q}\epsilon_{t-q} = \sum_{j=0}^{q} \vartheta_{q,j}\epsilon_{t-j}, \qquad (594)$$

where $\vartheta_{q,0} \stackrel{\text{def}}{=} 1$ and $\vartheta_{q,j} \stackrel{\text{def}}{=} -\theta_{q,j}$ for $j = 1, \ldots, q$. Once we have a realization of ϵ_{-q}, $\epsilon_{-q+1}, \ldots, \epsilon_{N-1}$, we can readily turn it into a realization of $X_0, X_1, \ldots, X_{N-1}$ using the formula above. The summation in the formula is a convolution, which is often most efficiently computed using the discrete Fourier transform (DFT) in the following manner. Let M be any integer such that $M \geq N + q$ (DFTs are usually computed using a fast Fourier transform (FFT) algorithm, some of which work best when M is set to a power of two). Define $\vartheta_{q,j} = 0$

for $j = q+1, q+2, \ldots, M-1$, and define

$$\varepsilon_t = \begin{cases} \epsilon_t, & 0 \leq t \leq N-1; \\ 0, & N \leq t \leq M-q-1 \text{ if } M > N+q; \\ \epsilon_{t-M}, & M-q \leq t \leq M-1. \end{cases}$$

Using the definition stated in Equation (91b), the DFTs of $\{\vartheta_{q,j}\}$ and $\{\varepsilon_t\}$ are

$$\sum_{j=0}^{M-1} \vartheta_{q,j} e^{-i2\pi kj/M} \stackrel{\text{def}}{=} \Theta_k \text{ and } \sum_{t=0}^{M-1} \varepsilon_t e^{-i2\pi kt/M} \stackrel{\text{def}}{=} E_k, \quad k = 0, 1, \ldots, M-1.$$

The desired simulated time series is the realization of

$$X_t = \frac{1}{M} \sum_{k=0}^{M-1} \Theta_k E_k e^{i2\pi kt/M}, \quad t = 0, 1, \ldots, N-1$$

(this result follows from Equations (92a), (101d) and (101e) along with an argument similar to what is needed to solve Exercise [3.20b]). Note that we require an uncorrelated series of length $N+q$ in order to simulate a correlated series of length N. Figure 33 shows eight realizations of length $N = 128$ from MA processes of order $q = 1$ defined by $X_t = \epsilon_t - \theta_{1,1}\epsilon_{t-1}$, where $\{\epsilon_t\}$ is Gaussian white noise, and $\theta_{1,1}$ is set to 1 for four realizations (top two rows) and to -1 for the remaining four (bottom two).

Autoregressive Process
Given a zero mean white noise process $\{\epsilon_t\}$ with variance σ_p^2 and given a set of $p \geq 1$ constants $\phi_{p,1}, \ldots, \phi_{p,p}$ obeying certain constraints (including $\phi_{p,p} \neq 0$ – see Section 9.2 for details), we can construct a zero mean pth-order causal (and hence stationary) AR process

$$X_t = \phi_{p,1} X_{t-1} + \phi_{p,2} X_{t-2} + \cdots + \phi_{p,p} X_{t-p} + \epsilon_t = \sum_{j=1}^{p} \phi_{p,j} X_{t-j} + \epsilon_t. \quad (595)$$

Simulation of an AR process directly from the above equation is problematic: to simulate X_0, we need realizations of $X_{-1}, \ldots, X_{-p}$, but these in turn require realizations of $X_{-p-1}, \ldots, X_{-2p}$ and so forth ad infinitum. To get around this startup problem, a commonly advocated strategy is to force the realizations of $X_{-1}, \ldots, X_{-p}$ to be zero, and then use Equation (595) with realizations of $\epsilon_0, \ldots, \epsilon_M, \epsilon_{M+1}, \ldots, \epsilon_{N+M-1}$ for some $M > 0$ to generate realizations of $X_0, \ldots, X_M, X_{M+1}, \ldots, X_{N+M-1}$ (hence X_0 is set to ϵ_0; X_1, to $\phi_{p,1}X_0 + \epsilon_1$, etc.). The desired simulated time series of length N is taken to be $X_M, X_{M+1}, \ldots, X_{N+M-1}$, with the idea that, if M is sufficiently large, the initial zero settings should have a negligible effect on the simulated series (see the discussion on asymptotic stationarity in C&E [3] for Section 9.2). The difficulty with this burn-in procedure is that it is not usually clear what M should be set to: the larger M is, the smaller the effect of the initial zeros, but a large M can lead to an unacceptable increase in computational complexity if many realizations of an AR process are needed for a Monte Carlo experiment.

Kay (1981b) describes a procedure for simulating a Gaussian AR process that solves the startup problem by generating *stationary initial conditions* (see Section 9.4 for details on what follows). The procedure falls out from unraveling the prediction errors $\vec{e}_t(t)$ associated with the best linear predictor $\vec{X}_t(t)$ of X_t given all values of the process with time indices ranging from 0 up to $t-1$, where t varies from 1 up to $N-1$; for the special case $t = 0$, we define $\vec{X}_0(0)$ to be 0 (i.e., the expected value of X_0) and take $\vec{e}_0(0)$ to be X_0. A key property of these errors is that they are pairwise uncorrelated (see Exercise [465]). In addition $E\{\vec{\epsilon}_t(t)\} = 0$ for all t, so the prediction errors would constitute a segment of length N of zero mean Gaussian white noise were it not for the fact that their variances are not the same. Assuming $N \geq p+1$ for convenience, the best linear predictor of X_t for $t = p, \ldots, N-1$ is given by the right-hand side of Equation (595) with ϵ_t replaced by its expected value of zero:

$$\vec{X}_t(t) \stackrel{\text{def}}{=} \sum_{j=1}^{p} \phi_{p,j} X_{t-j}.$$

The associated prediction error is

$$\vec{\epsilon}_t(t) \stackrel{\text{def}}{=} X_t - \vec{X}_t(t) = \epsilon_t,$$

and hence $\text{var}\{\vec{\epsilon}_t(t)\} = \sigma_p^2$. For $t = 0, \ldots, p-1$, we can't appeal to Equation (595) directly to forecast X_t since we have fewer than p values prior to X_t. The best linear predictors are given by $\vec{X}_0(0) = 0$ and, if $p \geq 2$, by

$$\vec{X}_t(t) \stackrel{\text{def}}{=} \sum_{j=1}^{t} \phi_{t,j} X_{t-j}, \quad t = 1, \ldots, p-1,$$

(see Equation (452)). Let $\sigma_0^2, \sigma_1^2, \ldots, \sigma_{p-1}^2$ denote the variances of the associated prediction errors, which, starting with σ_p^2 and using $\phi_{p,p}$ from Equation (595), can be computed recursively via

$$\sigma_{t-1}^2 = \frac{\sigma_t^2}{1 - \phi_{t,t}^2}, \quad t = p, p-1, \ldots, 1, \tag{596a}$$

(see Equation (460b)). Given a realization of a portion $Z_0, Z_1, \ldots, Z_{N-1}$ of a Gaussian white noise process with zero mean and unit variance, we can form a realization of the prediction errors $\vec{\epsilon}_t(t)$, $t = 0, 1, \ldots, N-1$, by equating them to realizations of $\sigma_0 Z_0, \ldots, \sigma_{p-1} Z_{p-1}$, $\sigma_p Z_p, \ldots, \sigma_p Z_{N-1}$. Now note that the system of N equations

$$\vec{\epsilon}_0(0) = X_0$$
$$\vec{\epsilon}_1(1) = X_1 - \phi_{1,1} X_0 \quad \text{(not needed if } p = 1\text{)}$$
$$\vec{\epsilon}_2(2) = X_2 - \sum_{j=1}^{2} \phi_{2,j} X_{2-j} \quad \text{(not needed if } p = 1 \text{ or } 2\text{)}$$
$$\vdots \tag{596b}$$
$$\vec{\epsilon}_{p-1}(p-1) = X_{p-1} - \sum_{j=1}^{p-1} \phi_{p-1,j} X_{p-1-j} \quad \text{(not needed if } p = 1, 2 \text{ or } 3\text{)}$$
$$\vec{\epsilon}_t(t) = \epsilon_t = X_t - \sum_{j=1}^{p} \phi_{p,j} X_{t-j}, \quad t = p, \ldots, N-1$$

can be reexpressed – upon replacing the prediction errors with distributionally equivalent RVs – as

$$X_0 = \sigma_0 Z_0$$
$$X_1 = \phi_{1,1} X_0 + \sigma_1 Z_1 \quad \text{(not needed if } p = 1\text{)}$$

11.1 Simulation of ARMA Processes and Harmonic Processes

$$X_2 = \sum_{j=1}^{2} \phi_{2,j} X_{2-j} + \sigma_2 Z_2 \quad \text{(not needed if } p = 1 \text{ or } 2)$$

$$\vdots \qquad (597a)$$

$$X_{p-1} = \sum_{j=1}^{p-1} \phi_{p-1,j} X_{p-1-j} + \sigma_{p-1} Z_{p-1} \quad \text{(not needed if } p = 1, 2 \text{ or } 3)$$

$$X_t = \sum_{j=1}^{p} \phi_{p,j} X_{t-j} + \sigma_p Z_t, \quad t = p, \ldots, N-1,$$

which gives us a recursive way of transforming RVs that have the same multivariate distribution as the uncorrelated prediction errors into a correlated AR series.

To implement this scheme, we assume knowledge of the $\phi_{p,j}$ coefficients, but we also need the $\phi_{t,j}$ coefficients for t less than p when $p \geq 2$. Thus, when $p = 1$, all that is required is $\phi_{1,1}$, i.e., the AR(1) coefficient. When $p \geq 2$, starting with the AR coefficients $\phi_{p,j}$ and with t set to p, we can use

$$\phi_{t-1,j} = \frac{\phi_{t,j} + \phi_{t,t}\phi_{t,t-j}}{1 - \phi_{t,t}^2}, \quad 1 \leq j \leq t-1, \qquad (597b)$$

to obtain the required coefficients for order $p - 1$ (see Equation (460a)). With these coefficients, we can use the above again if need be to obtain the coefficients for order $p - 2$ and so forth, ending the recursions with the computation of $\phi_{1,1}$.

As an example, let us consider the second-order Gaussian AR process of Equation (34), namely,

$$X_t = \tfrac{3}{4} X_{t-1} - \tfrac{1}{2} X_{t-2} + \epsilon_t \text{ with } \sigma_\epsilon^2 = 1.$$

▷ **Exercise [597]** Show that, given a realization of a portion $Z_0, Z_1, \ldots, Z_{N-1}$ of Gaussian white noise with zero mean and unit variance, we can generate a realization of length N of the AR process above via

$$X_0 = \tfrac{4}{3} Z_0$$
$$X_1 = \tfrac{1}{2} X_0 + \tfrac{2}{\sqrt{3}} Z_1$$
$$X_t = \tfrac{3}{4} X_{t-1} - \tfrac{1}{2} X_{t-2} + Z_t, \quad t = 2, \ldots, N-1.$$ ◁

For a second example, see Exercise [11.1], which considers the fourth-order Gaussian AR process of Equation (35a).

Figure 34 shows eight realizations of AR processes, each of length $N = 1024$. The top two rows show four realizations of the AR(2) process of Equation (34), and the bottom two, of the AR(4) process of Equation (35a).

Autoregressive Moving Average Process

A zero mean ARMA process $\{X_t\}$ of order $p \geq 1$ and $q \geq 1$ combines the AR process of Equation (595) with the MA process of Equation (594) by using the latter as a replacement for the white noise process driving the former:

$$X_t = \sum_{j=1}^{p} \phi_{p,j} X_{t-j} + \sum_{j=0}^{q} \vartheta_{q,j} \epsilon_{t-j}. \qquad (597c)$$

▷ **Exercise [598]** Show that an ARMA process can be decomposed as follows (Kay, 1981b):

$$X_t = \sum_{j=0}^{q} \vartheta_{q,j} Y_{t-j}, \quad \text{where } Y_t \stackrel{\text{def}}{=} \sum_{k=1}^{p} \phi_{p,k} Y_{t-k} + \epsilon_t.$$ ◁

We can thus simulate a segment of length N from a Gaussian ARMA process by first simulating a segment of length $N + q$ from the AR process $\{Y_t\}$ – to be treated as a realization of $Y_{-q}, Y_{-q+1}, \ldots, Y_{N-1}$ – and then following the recipe for simulating an MA process, but with the AR realization replacing the realization of $\epsilon_{-q}, \epsilon_{-q+1}, \ldots, \epsilon_{N-1}$.

Harmonic Process

A zero mean harmonic process $\{X_t\}$ can be written as

$$X_t = \sum_{l=1}^{L} A_l \cos{(2\pi f_l t)} + B_l \sin{(2\pi f_l t)}, \quad t \in \mathbb{Z}, \tag{598a}$$

where f_l are real-valued positive constants, L is a positive integer, and A_l and B_l are uncorrelated real-valued RVs with zero means such that $\text{var}\{A_l\} = \text{var}\{B_l\}$ (see Equation (35c)). To create a realization of $X_0, X_1, \ldots, X_{N-1}$, we need only generate uncorrelated realizations of the $2L$ RVs A_l and B_l, and then evaluate the right-hand side of the above for $t = 0, 1, \ldots, N - 1$. If each of the $2L$ RVs obeys a Gaussian distribution, then the process $\{X_t\}$ is also Gaussian with mean zero and variance $\sum_{l=1}^{L} \text{var}\{A_l\}$ (see Equation (36b)). The left-hand column of Figure 36 shows three realizations of a Gaussian harmonic process with $L = 1$ and $f_1 = 1/20$ for which the realizations of A_1 and B_1 come from a standard Gaussian distribution. Since we only need that the RVs A_l and B_l be uncorrelated, and since independence implies uncorrelatedness, we could simulate a non-Gaussian harmonic process by, e.g., generating realizations of $2L$ RVs, each with a possibly different distribution with the only restriction being that $\text{var}\{A_l\}$ and $\text{var}\{B_l\}$ must be equal for a given l.

We can reexpress the harmonic process of Equation (598a) as

$$X_t = \sum_{l=1}^{L} D_l \cos{(2\pi f_l t + \phi_l)}, \quad t \in \mathbb{Z}, \tag{598b}$$

where $D_l^2 = A_l^2 + B_l^2$, $\tan{(\phi_l)} = -B_l/A_l$, $A_l = D_l \cos{(\phi_l)}$ and $B_l = -D_l \sin{(\phi_l)}$ (see Equation (35d)). As noted following that equation, if the A_l and B_l RVs are all Gaussian, then the D_l^2 RVs are independent with exponential PDFs, but these RVs need not be identically distributed. The mean completely characterizes the exponential distribution, and the mean for the lth RV D_l^2 is $2\text{var}\{A_l\}$, which need not be the same for all l. The nonnegative random amplitude D_l is said to obey a Rayleigh distribution, so the D_l RVs are independent and Rayleigh distributed, but they need not be identically distributed. On the other hand, the phases ϕ_l are independent and identically distributed RVs. Their common distribution is uniform over the interval $(-\pi, \pi]$, and the ϕ_l RVs are independent of the D_l RVs.

To simulate a Gaussian harmonic process that obeys Equation (598b) rather than Equation (598a), we need to generate L independent realizations of RVs with exponential PDFs and with means given by $2\text{var}\{A_l\}$. The square roots of these realizations are realizations of the random amplitudes D_l. Independent of these L realizations, we also need to generate realizations of L independent and identically distributed (IID) RVs that are uniformly distribution over $(-\pi, \pi]$ – these realizations serve as realizations of the ϕ_l RVs. To generate harmonic processes that need not obey a Gaussian distribution for all t, it is sufficient to generate the L

11.1 Simulation of ARMA Processes and Harmonic Processes

random phases as before, but the L independent random amplitudes can be any set of RVs that yield positive values with probability one. In particular, the amplitudes can be fixed, i.e., have degenerate distributions. Three realizations from a harmonic process with fixed amplitudes but random phases are shown in the right-hand column of Figure 36 for the case of $L = 1$, $f_1 = 1/20$ and $D_1 = \sqrt{2}$. The ACVS for this non-Gaussian harmonic process is the same as that of the Gaussian-distributed harmonic process used to generate the realizations in the left-hand column.

Comments and Extensions to Section 11.1

[1] In defining an AR process $\{X_t\}$, we have taken the variance σ_p^2 of the associated white noise process to be one of the $p + 1$ parameters. It is sometimes convenient to use the process variance var$\{X_t\}$ instead. Noting that $\sigma_0^2 =$ var$\{\vec{\epsilon}_0(0)\} =$ var$\{X_0 - \vec{X}_0(0)\} =$ var$\{X_t\}$ since $\vec{X}_0(0) = 0$, we can readily obtain the remaining prediction error variances required for the simulation scheme by reexpressing Equation (596a) as

$$\sigma_t^2 = \sigma_{t-1}^2 \left(1 - \phi_{t,t}^2\right), \quad t = 1, 2, \ldots, p.$$

[2] The approach presented above for generating realizations of Gaussian ARMA processes with stationary initial conditions follows Kay (1981b). McLeod and Hipel (1978) describe another method for doing so that has some interesting contrasts with Kay's approach. For an MA process, their method is the same as what we have presented. For a pth-order Gaussian AR process, the first step in their method is to form the covariance matrix $\boldsymbol{\Gamma}_p$ for $X_0, \ldots, X_{p-1}$; i.e., its (j, k)th element is s_{j-k}, where $\{s_\tau\}$ is the ACVS for the AR process – cf. Equation (450c). All $p \times p$ entries of $\boldsymbol{\Gamma}_p$ are known once we know $s_0, \ldots, s_{p-1}$. Given the $p + 1$ parameters $\phi_{p,1}, \ldots, \phi_{p,p}$ and σ_p^2 for the process, part of the burden of Exercise [9.11] is to show how to obtain these ACVS values. The second step calls upon a general purpose routine to construct the Cholesky decomposition of $\boldsymbol{\Gamma}_p$ (see, e.g., Ralston and Rabinowitz, 1978, or Golub and Van Loan, 2013). This yields a lower triangular matrix $\boldsymbol{C}_p$ such that $\boldsymbol{C}_p \boldsymbol{C}_p^T = \boldsymbol{\Gamma}_p$. Let $\boldsymbol{Z}_p$ be a p-dimensional vector containing p IID standard Gaussian RVs. A standard result in the theory of Gaussian vectors says that $\boldsymbol{C}_p \boldsymbol{Z}_p$ has the same distribution as $X_0, \ldots, X_{p-1}$ (see, e.g., section A.3 of Brockwell and Davis, 2016). Hence a realization of $\boldsymbol{C}_p \boldsymbol{Z}_p$ gives us a realization of $X_0, \ldots, X_{p-1}$. Given these stationary initial conditions, we can in turn generate realizations of $X_p, \ldots, X_{N-1}$ by generating realizations of $N - p$ additional IID standard Gaussian RVs, multiplying these by σ_p so that they can be regarded as realizations of $\epsilon_p, \ldots \epsilon_{N-1}$ and then making use of the defining equation for an AR process (Equation (595)).

For the AR case, what is the difference between the methods of Kay (1981b) and of McLeod and Hipel (1978)? The two methods simulate $X_p, \ldots, X_{N-1}$ in exactly the same way once they have generated the stationary initial conditions for $\boldsymbol{X}_p \stackrel{\text{def}}{=} \left[X_0, \ldots, X_{p-1}\right]^T$. McLeod and Hipel generate the conditions via $\boldsymbol{X}_p \stackrel{\text{d}}{=} \boldsymbol{C}_p \boldsymbol{Z}_p$, which uses the Cholesky decomposition $\boldsymbol{\Gamma}_p = \boldsymbol{C}_p \boldsymbol{C}_p^T$ (recall that $\stackrel{\text{d}}{=}$ means "equal in distribution"). In contrast, Kay uses the modified Cholesky decomposition $\boldsymbol{\Gamma}_p = \boldsymbol{L}_p \boldsymbol{D}_p \boldsymbol{L}_p^T$, where $\boldsymbol{L}_p$ is a $p \times p$ lower triangular matrix whose diagonal elements are all equal to 1, and $\boldsymbol{D}_p$ is a $p \times p$ diagonal matrix whose diagonal elements are all positive (see the discussion surrounding Equation (464c)). Let $\boldsymbol{D}_p^{1/2}$ be a diagonal matrix whose diagonal elements are the square roots of those in $\boldsymbol{D}_p$, and hence $\boldsymbol{D}_p^{1/2} \boldsymbol{D}_p^{1/2} = \boldsymbol{D}_p$. It follows that the Cholesky and modified Cholesky decompositions are related by $\boldsymbol{C}_p = \boldsymbol{L}_p \boldsymbol{D}_p^{1/2}$. Let $\vec{\boldsymbol{\epsilon}}_p \stackrel{\text{def}}{=} \left[\vec{\epsilon}_0(0), \ldots, \vec{\epsilon}_{p-1}(p-1)\right]^T$. Kay's method for generating stationary initial conditions is based on $\boldsymbol{L}_p^{-1} \boldsymbol{X}_p = \vec{\boldsymbol{\epsilon}}_p$ (see Equation (465a)). Since $\vec{\boldsymbol{\epsilon}}_p \stackrel{\text{d}}{=} \boldsymbol{D}_p^{1/2} \boldsymbol{Z}_p$, we have $\boldsymbol{L}_p^{-1} \boldsymbol{X}_p \stackrel{\text{d}}{=} \boldsymbol{D}_p^{1/2} \boldsymbol{Z}_p$ and hence $\boldsymbol{X}_p \stackrel{\text{d}}{=} \boldsymbol{L}_p \boldsymbol{D}_p^{1/2} \boldsymbol{Z}_p = \boldsymbol{C}_p \boldsymbol{Z}_p$, which is the same as the basis for the McLeod and Hipel method. The two methods thus produce equivalent stationary initial conditions. The theoretical advantage of Kay's method is that the elements of $\boldsymbol{L}_p^{-1}$ and $\boldsymbol{D}_p^{1/2}$ are easily obtained from the $p + 1$ AR parameters via Equations (597b) and (596a) because

$$L_p^{-1} = \begin{bmatrix} 1 & & & & \\ -\phi_{1,1} & 1 & & & \\ -\phi_{2,2} & -\phi_{2,1} & 1 & & \\ \vdots & \vdots & & \ddots & \\ -\phi_{p-1,p-1} & \cdots & & -\phi_{p-1,1} & 1 \end{bmatrix} \text{ and } D_p^{1/2} = \begin{bmatrix} \sigma_0 & & & & \\ & \sigma_1 & & & \\ & & \sigma_2 & & \\ & & & \ddots & \\ & & & & \sigma_{p-1} \end{bmatrix}$$

(see Equation (465b) and the description following it). In addition, there is no need to get L_p explicitly by inverting L_p^{-1} since we can obtain the elements of the stationary initial conditions $L_p D_p^{1/2} Z_p$ by the unravelling procedure of Equation (597a).

Turning now to Gaussian ARMA processes, McLeod and Hipel (1978) generate stationary initial conditions based upon the joint distribution of the $p + q$ RVs $X_0, \ldots, X_{p-1}, \epsilon_{p-q}, \ldots, \epsilon_{p-1}$ (once realizations of these RVs are provided, it is an easy matter to generate realizations of $X_p, \ldots, X_{N-1}$ via Equation (597c) by generating realizations of $\epsilon_p, \ldots, \epsilon_{N-1}$). The covariance matrix for these $p + q$ RVs is given by

$$\Sigma_{p+q} = \begin{bmatrix} \Gamma_p & \sigma_\epsilon^2 \Psi_{p,q} \\ \sigma_\epsilon^2 \Psi_{p,q}^T & \sigma_\epsilon^2 I_q \end{bmatrix}, \tag{600a}$$

where the (j, k)th entry of the $p \times p$ matrix Γ_p is s_{j-k}, and s_{j-k} is the ACVS of the ARMA process; the (t, k)th entry of the $p \times q$ matrix $\sigma_\epsilon^2 \Psi_{p,q}$ is $\text{cov}\{X_t, \epsilon_{p-q+k}\}$, where $t = 0, \ldots, p-1$ and $k = 0, \ldots, q-1$; and $\sigma_\epsilon^2 I_q$ is the covariance matrix for $\epsilon_{p-q}, \ldots, \epsilon_{p-1}$, where I_q is the $q \times q$ identity matrix. Section 3.2.1 of Brockwell and Davis (2016) gives three methods for determining the ACVS for an ARMA process from its $p + q + 1$ parameters. Here is a fourth based upon the decomposition in Exercise [598] in conjunction with the method for determining the ACVS for an AR process given by Exercise [9.11]. Let $\{s_{Y,\tau}\}$ denote the ACVS for AR process $\{Y_t\}$ defined in Exercise [598]. An application of Exercise [2.1e] shows that the ACVS $\{s_\tau\}$ for the ARMA process $\{X_t\}$ is given by

$$s_\tau = \sum_{j=0}^{q} \sum_{k=0}^{q} \vartheta_{q,j} \vartheta_{q,k} s_{Y,\tau-j+k}. \tag{600b}$$

To form the elements of Γ_p, we need to evaluate $s_{Y,\tau}$ at lags $\tau = 0, \ldots, p+q-1$ in order to get s_τ at the required lags $\tau = 0, \ldots, p-1$. We next need to determine $\sigma_\epsilon^2 \Psi_{p,q}$. If we assume that the ARMA process is causal, there exists a unique set of weights $\{\psi_j\}$ such that

$$X_t = \sum_{j=0}^{\infty} \psi_j \epsilon_{t-j} = \sum_{j=-\infty}^{\infty} \psi_j \epsilon_{t-j}, \tag{600c}$$

where $\psi_0 \stackrel{\text{def}}{=} 1$ and $\psi_j \stackrel{\text{def}}{=} 0$ for $j < 0$ (as noted in section 3.1 of Brockwell and Davis, 2016, an ARMA process is causal if the AR coefficients satisfy the same condition that we need for an AR process to be causal; i.e., the roots of the polynomial $1 - \sum_{j=1}^{p} \phi_{p,j} z^{-j}$ must all lie inside the unit circle, and no constraints are needed on the MA coefficients). Hence

$$\text{cov}\{X_t, \epsilon_{p-q+k}\} = \sum_{j=-\infty}^{\infty} \psi_j E\{\epsilon_{t-j} \epsilon_{p-q+k}\} = \sigma_\epsilon^2 \psi_{t-p+q-k};$$

however, since $t = 0, \ldots, p-1$ and $k = 0, \ldots, q-1$, we actually only need to compute ψ_j for $j = 0, \ldots, q-1$ to fill in the nonzero entries of $\Psi_{p,q}$. A recursive scheme for doing so is

$$\psi_j = \sum_{k=1}^{p} \phi_{p,k} \psi_{j-k} + \vartheta_{q,j}, \quad j = 0, \ldots, q-1 \tag{600d}$$

(Brockwell and Davis, 2016, equation (3.1.7)). With Σ_{p+q} now completely determined, we form its Cholesky decomposition $\Sigma_{p+q} = C_{p+q} C_{p+q}^T$ and can then obtain a realization of $X_0, \ldots, X_{p-1}$, $\epsilon_{p-q}, \ldots, \epsilon_{p-1}$ via $C_{p+q} Z_{p+q}$, where Z_{p+q} is a vector containing $p+q$ standard Gaussian RVs. Exercise [11.2] invites the reader to work out the details of this method for the case $p=1$ and $q=1$ and to compare it to Kay's method.

[3] As noted in Section 2.2, the t in X_t is a unitless index and not an actual time. We can take the time associated with X_t to be $t_0 + t\,\Delta_t$, where t_0 is the actual time at which X_0 is observed, and $\Delta_t > 0$ is the sampling interval between adjacent values of the time series, e.g., X_t and X_{t+1}. The procedures we have described for generating ARMA processes – and special cases thereof – yield realizations of X_0, $X_1, \ldots, X_{N-1}$ that we are free to associate with times dictated by arbitrary choices for t_0 and $\Delta_t > 0$. The same is true for harmonic processes *if* we are willing to assume that the positive frequencies f_l in Equations (598a) and (598b) are unitless. If we want them to have units, e.g., Hz, then we need to replace t in the summations in these equations with $t\,\Delta_t$, where $\Delta_t = 1$ sec for this example. With this replacement, we can associate $X_0, X_1, \ldots, X_{N-1}$ with times $t_0, t_0 + \Delta_t, \ldots, t_0 + (N-1)\,\Delta_t$ for any choice of t_0 so desired.

11 Simulation of Processes with a Known Autocovariance Sequence

Here we consider how to generate a realization of a portion $X_0, X_1, \ldots, X_{N-1}$ of a Gaussian stationary process given that we know part of its ACVS $\{s_\tau\}$ (without loss of generality, we assume that the process mean for $\{X_t\}$ is zero and that its sampling interval Δ_t is unity). Under the Gaussian assumption, a particularly attractive approach is *circulant embedding*, which is also known as the Davies–Harte method since it was introduced into the statistical literature in an appendix to Davies and Harte (1987); however, the method predates this appendix, with Woods (1972) being an early reference outside the context of time series. Subject to certain conditions (stated in the next paragraph), the method is exact in the sense that it produces simulated time series whose statistical properties are *exactly* the same as those of the desired $X_0, X_1, \ldots, X_{N-1}$. The method is not entirely general because there are certain stationary processes and certain sample sizes N for which the required conditions do not hold. In what follows, we present a formulation of circulant embedding due to Dietrich and Newsam (1997), which is an appealing alternative to that of Davies and Harte (1987).

Assume that $s_0, s_1, \ldots, s_N$ are known, and compute the weights

$$S_k \stackrel{\text{def}}{=} \sum_{\tau=0}^{N} s_\tau e^{-i2\pi \tilde{f}_k \tau} + \sum_{\tau=N+1}^{2N-1} s_{2N-\tau} e^{-i2\pi \tilde{f}_k \tau}, \quad k = 0, 1, \ldots, 2N-1, \tag{601a}$$

where $\tilde{f}_k = k/(2N)$ defines a grid of frequencies twice as fine as the Fourier frequencies associated with a time series of length N. Equation (91b) tells us that $\{S_k\}$ is the DFT of

$$s_0, s_1, \ldots, s_{N-2}, s_{N-1}, s_N, s_{N-1}, s_{N-2}, \ldots, s_1. \tag{601b}$$

This "circularized" sequence of length $2N$ has "embedded" within it the $N+1$ values of the ACVS that are assumed to be known.

▷ **Exercise [601]** Show that the weights S_k are real-valued. ◁

For circulant embedding to work, the weights S_k *must* be nonnegative. This condition is known to hold for certain stationary processes, but not all (see, e.g, Gneiting, 2000, Craigmile, 2003, and references therein). Assuming that $S_k \geq 0$ for all k, let $Z_0, Z_1, \ldots, Z_{4N-1}$ be a set of $4N$ independent standard Gaussian RVs, and define

$$\mathcal{Y}_k = \left(\frac{S_k}{2N}\right)^{1/2} (Z_{2k} + i Z_{2k+1}), \quad 0 \leq k \leq 2N-1. \tag{601c}$$

Let $\{Y_t\}$ be the DFT of $\{\mathcal{Y}_k\}$:

$$Y_t = \sum_{k=0}^{2N-1} \mathcal{Y}_k e^{-i2\pi \tilde{f}_k t}. \tag{602}$$

The process $\{Y_t\}$ is complex-valued and well defined for all $t \in \mathbb{Z}$ (it is a periodic sequence with a period of $2N$). Let $Y_{\Re,t} \stackrel{\text{def}}{=} \Re\{Y_t\}$ and $Y_{\Im,t} \stackrel{\text{def}}{=} \Im\{Y_t\}$, where, as usual, $\Re\{z\}$ and $\Im\{z\}$ are the real and imaginary components of the complex-valued variable z.

▷ **Exercise [602]** Show that the real-valued processes $\{Y_{\Re,t}\}$ and $\{Y_{\Im,t}\}$ are zero mean Gaussian stationary processes, both of which have ACVSs that are in agreement with that of $\{X_t\}$ at lags $\tau = 0, 1, \ldots, N$, but not necessarily at lags $\tau > N$; moreover, the RVs in $\{Y_{\Re,t}\}$ are independent of the RVs in $\{Y_{\Im,t}\}$. ◁

We can thus regard realizations of $Y_{\Re,0}, Y_{\Re,1}, \ldots, Y_{\Re,N-1}$ and $Y_{\Im,0}, Y_{\Im,1}, \ldots, Y_{\Im,N-1}$ as two independent realizations of $X_0, X_1, \ldots, X_{N-1}$ (actually this statement holds if we replace $N-1$ with N – see C&E [1] later in this section for details).

In summary, given knowledge of a portion s_τ, $\tau = 0, 1, \ldots, N$, of the ACVS for a Gaussian stationary process $\{X_t\}$, the circulant embedding method consists of

[1] computing the DFT $\{S_k\}$ of the circularized sequence given in Equation (601b);
[2] checking that $S_k \geq 0$ for $k = 0, 1, \ldots, 2N-1$;
[3] if the S_k weights are nonnegative, using $4N$ deviates from a Gaussian white noise process $\{Z_t\}$ with zero mean and unit variance to form $\mathcal{Y}_k$ as per Equation (601c); and, finally,
[4] taking the DFT of $\{\mathcal{Y}_k\}$ to obtain $Y_t = Y_{\Re,t} + iY_{\Im,t}$ – we can consider realizations of the real-valued sequences $Y_{\Re,t}$ and $Y_{\Im,t}$, $t = 0, 1, \ldots, N-1$, to be two independent realizations of X_t, $t = 0, 1, \ldots, N-1$.

We can compute the DFTs needed in steps [1] and [4] efficiently if $2N$ is amenable to an FFT algorithm. We can obtain as many additional pairs of realizations of X_t as desired by repeating steps [3] and [4] over and over again, with a different random sample of Gaussian white noise used at each iteration of step [3].

As an example, let us generate realizations of length $N = 1024$ from the AR(2) process of Equation (34) using circulant embedding (we do this for illustrative purposes only – Exercise [597] gives the preferred method). To compute the required weights S_k using Equation (601a), we need the ACVS $\{s_\tau\}$ for the AR(2) process at lags $\tau = 0, 1, \ldots, 1024$, which we can obtain using Equation (508a). Using these ACVS values, we form the circularized sequence of Equation (601b) and take its DFT, which yields the desired weights S_k, $k = 0, 1, \ldots, 2047$. These weights are all positive (as required by the circulant embedding method) and are shown in Figure 603(a) (the smallest weight is $S_{1024} \doteq 0.19753$). Using these weights, we next create the complex values $\mathcal{Y}_k$, $k = 0, 1, \ldots, 2047$, as per Equation (601c), for which we need 4096 independent realizations from a standard Gaussian distribution. Finally we take the DFT of $\{\mathcal{Y}_k\}$ to obtain the complex-valued sequence $\{Y_t\}$ of Equation (602), the real and imaginary parts of which are shown by the curves in Figures 603(b) and (c). The first 1024 deviates in both parts are indicated by the darker portions of the curves – these are two realizations of the AR(2) process, and these realizations are independent of each other. The appearances of both realizations are qualitatively similar to the four realizations of this process shown in the top four plots of Figure 34, which were generated using the preferred procedure stated in Exercise [597]. (The lighter portions of the curves in plots (b) and (c) have a similar appearance to the darker portions – C&E [2] for this section discusses why this is to be expected; see also Exercise [11.5], which uses this AR(2) example as a follow-on to Exercise [602]).

11.2 Simulation of Processes with a Known Autocovariance Sequence

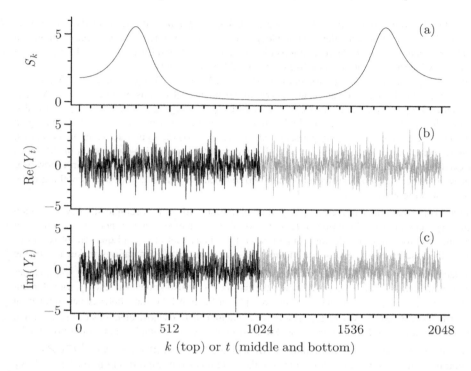

Figure 603 Creation via circulant embedding of realizations of length $N = 1024$ from the Gaussian AR(2) process of Equation (34). Plot (a) shows the weights S_k, $k = 0, 1, \ldots, 2047$, used to create the complex-valued series Y_t, $t = 0, 1, \ldots, 2047$, while (b) and (c) show, respectively, the real and imaginary parts of $\{Y_t\}$. The first 1024 values of both (the darker portions of the plots) constitute two independent realizations of the AR(2) process (cf. the top four plots of Figure 34, which were generated via the method detailed in Exercise [597]).

Comments and Extensions to Section 11.2

[1] Our goal in this section was to formulate an exact method for simulating a time series of length N, but the version of circulant embedding we presented yields simulated series of length $N + 1$ (one longer than desired). It is easy to achieve our goal by just ignoring $Y_{\Re,N}$ and $Y_{\Im,N}$, but, to avoid tossing anything away, why not just reformulate the method with N replaced throughout by $N - 1$ (in particular, the middle part of Equation (601b) would become $s_{N-3}, s_{N-2}, s_{N-1}, s_{N-2}, s_{N-3}$)? An appealing aspect of circulant embedding is that it can be implemented efficiently using an FFT algorithm. Because simple versions of this algorithm are designed to work with series whose lengths are powers of two, sample sizes of time series are often set such that $N = 2^J$ for some positive integer J. Our formulation of circulant embedding requires taking the DFT of a series of length $2N$, which is a power of two if N is a power of two. If we were to reformulate the method by replacing N with $N - 1$, we would need to take the DFT of a series of length $2N - 2$, which is in general not a power of two if N is such. The formulation we have presented is thus motivated by the desire to make the circulant embedding method computationally efficient when using a simple FFT algorithm with a desired sample size N that is a power of two (see C&E [1] for Section 6.3 for another example of the interplay between N and FFT algorithms).

One consequence of our formulation is that we need not insist on s_N being part of the circularized sequence of Equation (601b) – we are free to replace it with some other value, but keeping in mind the goal that the weights arising from the DFT of the modified circularized sequence should be nonnegative (see Exercise [11.4c] for an example).

[2] The solution to Exercise [602] indicates that both $\{Y_{\Re,t} : t \in \mathbb{Z}\}$ and $\{Y_{\Im,t} : t \in \mathbb{Z}\}$ are special cases of the harmonic process of Equation (35c), each being composed of sinusoids with frequencies $\pi k/N$, $k = 0, 1, \ldots, 2N - 1$; moreover, these frequencies are such that both processes are periodic with a period of $2N$ (an easy extension to the exercise). Knowledge of realizations of these processes over $0 \leq t \leq 2N - 1$ is thus equivalent to knowledge over $t \in \mathbb{Z}$. Turning to the example displayed in Figure 603 for which $N = 1024$, the realizations of $Y_{\Re,t}$ and $Y_{\Im,t}$ shown for $0 \leq t \leq 2047$ in plots (b) and (c) are also what the realizations look like over, e.g., $2048 \leq t \leq 4095$.

Since $\{Y_{\Re,t}\}$ and $\{Y_{\Im,t}\}$ are periodic, the series $Y_{\Re,0}, Y_{\Re,1}, \ldots, Y_{\Re,N-1}$ and $Y_{\Im,0}, Y_{\Im,1}, \ldots, Y_{\Im,N-1}$ are not the only ones we can regard as two independent realizations of $X_0, X_1, \ldots, X_{N-1}$. In fact the series $Y_{\Re,\nu_\Re}, Y_{\Re,\nu_\Re+1}, \ldots, Y_{\Re,\nu_\Re+N-1}$ and $Y_{\Im,\nu_\Im}, Y_{\Im,\nu_\Im+1}, \ldots, Y_{\Im,\nu_\Im+N-1}$ are also such for *any* choice of $\nu_\Re$ and $\nu_\Im$ (in particular these integers need *not* be the same). Looking again at Figure 603, we can take the two independent realizations to be (i) the lighter series in the two plots; (ii) the lighter series in (b) in combination with the darker series in (c); (iii) the portion of the series in (b) indexed by $t = 512, 513, \ldots, 1535$ together with the darker series in (c); and so forth; however, given that we cannot in general extract more than two independent realizations, there is nothing to be gained by not sticking with just $\nu_\Re = \nu_\Im = 0$.

[3] There are some stationary processes $\{X_t\}$ in combination with certain sample sizes N for which the weights S_k generated via Equation (601a) are not all nonnegative, hence rendering the circulant embedding method formally inapplicable (in particular, the square root of S_k in Equation (601c) poses a problem). In these cases, an alternative method for generating exact simulations makes use of the Cholesky decomposition of the $N \times N$ covariance matrix $\boldsymbol{\Gamma}_N$ for $\boldsymbol{X}_N \stackrel{\text{def}}{=} [X_0, \ldots, X_{N-1}]^T$. Under the assumption that $s_0, s_1, \ldots, s_{N-1}$ are known, the matrix $\boldsymbol{\Gamma}_N$ is also known since its (j,k)th element is s_{j-k}, and $0 \leq |j-k| \leq N - 1$. As noted in C&E [2] for Section 11.1, the Cholesky decomposition takes the form $\boldsymbol{\Gamma}_N = \boldsymbol{C}_N \boldsymbol{C}_N^T$, where $\boldsymbol{C}_N$ is a lower triangular matrix whose elements can be obtained efficiently via the Levinson–Durbin recursions. Simulation of $\boldsymbol{X}_N$ is accomplished by multiplying $\boldsymbol{C}_N$ with an N-dimensional vector $\boldsymbol{Z}_N$ containing N IID standard Gaussian RVs; i.e., $\boldsymbol{X}_N \stackrel{\text{d}}{=} \boldsymbol{C}_N \boldsymbol{Z}_N$.

In contrast to the circulant embedding method, there are no theoretical restrictions on the Cholesky decomposition method since $\boldsymbol{C}_N$ is well defined for all covariance matrices $\boldsymbol{\Gamma}_N$. There are, however, two practical problems that restrict its applicability. First, even with the help of the computationally efficient Levinson–Durbin recursions, there are certain processes for which computing $\boldsymbol{C}_N$ is challenging. One example is band-limited white noise, whose ACVS is given by $s_\tau^{(\text{BL})} = \sin(2\pi W\tau)/(\pi\tau)$, where $0 < W < 1/2$ (this is Equation (379b) with $\Delta_t = 1$). We invite readers to verify that use of the Levinson–Durbin recursions to compute $\boldsymbol{C}_N$ for the case $W = 1/8$ and $N = 1024$ leads to substantial numerical instabilities (in particular, we found that, while all 1024 diagonal elements of $\boldsymbol{C}_N$ must in theory be nonnegative, 398 of them were incorrectly computed to be negative). Second, even if $\boldsymbol{C}_N$ can be computed accurately, generating realizations via $\boldsymbol{C}_N \boldsymbol{Z}_N$ has a computational complexity of $O(N^2)$, which can render this method painfully slow even for moderate sample sizes, say $N = 1024$; by contrast, if an FFT algorithm is used, the computational complexity of the circulant embedding method is $O(N \log_2(N))$.

Since the circulant embedding method is computationally attractive, it is of interest to consider two modifications that allow its use when $S_k < 0$ for some k. First, we can generate series of longer length, say $N' > N$. If the weights for the length N' series prove to be nonnegative, we can get a series of the desired length by just extracting its first N values. Exercise [11.6a] gives an example where this works. Second, we can define a new set of weights, say $\tilde{S}_k$, by replacing any negative weights with zeros (i.e., $\tilde{S}_k = (S_k)_+ \stackrel{\text{def}}{=} \max\{S_k, 0\}$) and proceed as usual, but, while the resulting method is no longer exact, it is possible to measure how inaccurate it is. Exercise [11.6b] gives an example of this approach.

11.3 Simulation of Processes with a Known Spectral Density Function

In the previous section we considered simulation of a portion $X_0, X_1, \ldots, X_{N-1}$ of a zero mean Gaussian stationary process $\{X_t\}$ under the assumption that we know part of its ACVS $\{s_\tau\}$. In this section we consider simulation given that we know its SDF $S(\cdot)$ (for convenience, we take the sampling interval Δ_t for $\{X_t\}$ to be unity, so its Nyquist frequency $f_\mathcal{N}$ is $1/2$, and $S(\cdot)$ is a periodic function with a period of unity; C&E [1] for this section discusses adjusting for $\Delta_t \neq 1$). Under mild conditions (e.g., $S(\cdot)$ is square integrable), we have $\{s_\tau\} \longleftrightarrow S(\cdot)$, and hence knowledge of the SDF in theory gives us knowledge of the ACVS

11.3 Simulation of Processes with a Known Spectral Density Function

in its entirety. This fact would seem to say that we could just use the techniques discussed in the previous section since they only require partial knowledge of the ACVS. The motivation for this section is twofold. First, there are certain SDFs that can be described simply, but their corresponding ACVSs are hard to get at. Consider $S^{(\mathrm{PL})}(f) = |f|^{1/3}, |f| \leq 1/2$ (an example of a power-law SDF that might arise as a simplistic model for the increments of a certain type of ocean turbulence). While it is true that the ACVS can be described simply as

$$s_\tau^{(\mathrm{PL})} = \int_{-1/2}^{1/2} |f|^{1/3} e^{i 2\pi f \tau}\, df, \quad \tau \in \mathbb{Z},$$

a handy analytic expression for the integral is hard in general to come by, and use of numerical integration to evaluate the integral is tricky at best, particularly for large τ. Second, if the ACVS is readily available and if the required nonnegativity conditions hold, then circulant embedding is certainly the method of choice; however, if nonnegativity fails to hold, the ACVS-based alternatives to circulant embedding described in C&E [3] for Section 11.2 might be less appealing than the SDF-based approach described here.

In what follows we present an SDF-based method that closely mimics the approach to circulant embedding advocated by Dietrich and Newsam (1997). In particular the method generates a complex-valued time series from which we can extract two independent realizations of a real-valued stationary process whose statistical properties are potentially a close match for those of $X_0, X_1, \ldots, X_{N-1}$. We refer to this method as the *Gaussian spectral synthesis method* (GSSM), where the qualifier "Gaussian" is included to emphasize that, similar to the circulant embedding method, GSSM presumes Gaussianity. References that discuss methods similar in spirit to GSSM – but differing in details – include Thompson (1973), Mitchell and McPherson (1981), Percival (1992), Chambers (1995), Davison and Hinkley (1997) and Sun and Chaika (1997).

As motivation for GSSM, consider the following connection between the circulant embedding method and the SDF $S(\cdot)$ for $\{X_t\}$.

▷ **Exercise [605]** Suppose that the ACVS for $\{X_t\}$ is identically zero at all lags $|\tau| \geq N$. Show that the circulant embedding weights S_k as given by Equation (601a) are such that $S_k = S(\tilde{f}_k)$. Show also that

$$\{s_\tau : \tau = -(N-1), \ldots, N\} \longleftrightarrow \{S(\tilde{f}_k) : k = -(N-1), \ldots N\}. \qquad (605a) \triangleleft$$

Hence, in this special case, we can pick off the required weights directly from the SDF for $\{X_t\}$. Since it is also true that $\{s_\tau : \tau \in \mathbb{Z}\} \longleftrightarrow S(\cdot)$, the inverse Fourier transform for a finite sequence and the one appropriate for an infinite sequence say that

$$s_\tau = \frac{1}{2N} \sum_{k=-(N-1)}^{N} S(\tilde{f}_k) e^{i 2\pi \tilde{f}_k \tau} = \int_{-1/2}^{1/2} S(f) e^{i 2\pi f \tau}\, df. \qquad (605b)$$

In the above we can regard the middle term as a Riemann sum approximation to the integral, but an unusual one in that the approximation is actually exact (cf. the discussion following Equation (171f)). In the case where $s_\tau \neq 0$ for at least some $|\tau| \geq N$, the middle term is only an approximation to the integral and hence also to s_τ. We can obtain a better approximation to s_τ by decreasing the spacing between frequencies. Accordingly if we let N' be an integer greater than N, then $f'_k \stackrel{\text{def}}{=} k/(2N')$ defines a finer grid of frequencies than that of $\tilde{f}_k =$

$k/(2N)$. Assume that $S(f'_k)$ is finite for all k (this assumption does *not* hold for certain SDFs of interest – see the discussion in C&E [2]). We then have

$$s_\tau \approx \frac{1}{2N'} \sum_{k=-(N'-1)}^{N'} S(f'_k) e^{i2\pi f'_k \tau} = \frac{1}{2N'} \sum_{k=0}^{2N'-1} S(f'_k) e^{i2\pi f'_k \tau} \stackrel{\text{def}}{=} s'_\tau, \qquad (606a)$$

where changing the limits over which k varies is justified because both $\{S(f'_k) : k \in \mathbb{Z}\}$ and $\{\exp(i2\pi f'_k \tau) : k \in \mathbb{Z}\}$ are periodic sequences with a period of $2N'$. Exercise [11.7] is to verify that the approximation s'_τ is real-valued. In general, since s'_τ is a Riemann sum approximation to the integral displayed in Equation (605b), for any fixed τ, the approximation s'_τ should get closer to s_τ as N' increases.

To obtain the desired simulated series, let $Z_0, Z_1, \ldots, Z_{4N'-1}$ be a set of $4N'$ independent standard Gaussian RVs, and define

$$\mathcal{U}_k = \left(\frac{S(f'_k)}{2N'}\right)^{1/2} (Z_{2k} + i Z_{2k+1}), \quad 0 \le k \le 2N'-1, \qquad (606b)$$

and let $\{U_t\}$ be the DFT of $\{\mathcal{U}_k\}$, i.e.,

$$U_t = \sum_{k=0}^{2N'-1} \mathcal{U}_k e^{-i2\pi f'_k t} \qquad (606c)$$

(note that the two equations above closely mimic Equations (601c) and (602) used by the circulant embedding method). Let $U_{\Re,t} \stackrel{\text{def}}{=} \Re\{U_t\}$ and $U_{\Im,t} \stackrel{\text{def}}{=} \Im\{U_t\}$. The burden of Exercise [11.8] is to show that $\{U_{\Re,t} : t \in \mathbb{Z}\}$ and $\{U_{\Im,t} : t \in \mathbb{Z}\}$ are Gaussian stationary processes with zero means and ACVSs given by $\{s'_\tau\}$ of Equation (606a) and that the RVs in $\{U_{\Re,t}\}$ are independent of the RVs in $\{U_{\Im,t}\}$. To the extent that s'_τ is a good approximation to s_τ for $\tau = 0, 1, \ldots, N-1$, we can thus regard realizations of $U_{\Re,0}, U_{\Re,1}, \ldots, U_{\Re,N-1}$ and $U_{\Im,0}, U_{\Im,1}, \ldots, U_{\Im,N-1}$ as two independent realizations of processes whose ACVSs are approximately equal to that of $\{X_t\}$ at lags $|\tau| \le N-1$ (the approximation should improve if we increase N' while holding N fixed).

What remains is to set N'. To demonstrate the interplay between N' and N, consider the AR(4) process of Equation (35a). We can compute its ACVS $\{s_\tau\}$ using the procedure described surrounding Equation (508b). The dots in all three panels of Figure 607 show s_τ versus $\tau = 0, 1, \ldots, 256$. Suppose we are interested in GSSM-based simulations for sample size $N = 64$ (this is *not* the best way to generate simulations for this AR(4) process – Exercise [11.1] states the recommended procedure). If we were to set N' to $N/2 = 32$, the approximating ACVS $\{s'_\tau\}$ would be a periodic sequence with a period of 64. The solid curve in Figure 607(a) depicts this sequence. For a simulated series of length $N = 64$, we want s'_τ to be a good approximation to s_τ at lags $\tau = 0, 1, \ldots, 63$ (the vertical dashed line marks lag 63). The plot indicates that s'_τ is in general a poor approximation at the desired lags. If we increase N' by setting it to $N = 64$, the approximating ACVS – the solid curve in Figure 607(b) – is now periodic with a period of 128. At the important lags, i.e., $\tau = 0, 1, \ldots, 63$, the approximating s'_τ is now a better match to s_τ, but distortions are still visible. Finally let us consider setting N' to $2N = 128$. The approximating ACVS shown in Figure 607(c) is now periodic with a period of 256. The agreement between s'_τ and s_τ is visibly quite good at the lags that matter (the visible distortions at larger values of τ are of no concern for generating a series of length $N = 64$).

11.3 Simulation of Processes with a Known Spectral Density Function

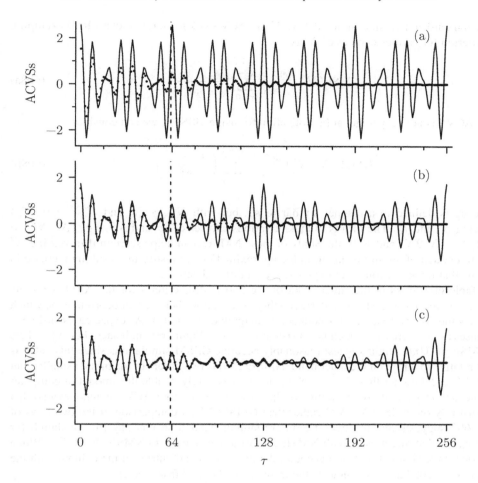

Figure 607 Actual ACVS $\{s_\tau\}$ for AR(4) process of Equation (35a) for $\tau = 0, 1, \ldots, 256$ (dots in all three plots), along with three GSSM-based approximations $\{s'_\tau\}$ (solid curves). In plot (a), the approximation s'_τ is obtained via Equation (606a) with N' set to 32 and with $S(\cdot)$ set to the AR(4) SDF; plots (b) and (c) are similar, but now with N' set to 64 in the former and to 128 in the latter. The vertical dashed line marks $\tau = 63$, and the visual good agreement between s'_τ and s_τ for $|\tau| \leq 63$ in plot (c) suggests that GSSM might provide reasonably accurate simulated AR(4) series of length $N = 64$ if we set $N' = 128$.

N'	32	64	128	256	512	1024
$\sqrt{\text{NMSE}(N')}$	0.500	0.106	0.007	3.55×10^{-5}	1.81×10^{-9}	9.21×10^{-14}
$\sqrt{\text{NMSE}(N', 2N')}$	0.349	0.099	0.007	3.55×10^{-5}	1.81×10^{-9}	3.15×10^{-14}

Table 607 Square roots of NMSE(N') and NMSE(N', N'') criteria of Equations (608a) and (608b) as applied to AR(4) process of Equation (35a) (here $N'' = 2N'$). The values in the columns marked by $N' = 32, 64$ and 128 make use of the GSSM-based approximations s'_τ to the AR(4) ACVS shown, respectively, in Figures 607(a) to (c).

Rather than relying on a visual comparison to pick N', we can assess various choices with the help of normalized mean square error (NMSE) criteria (the normalization essentially means that the assessment is actually done in terms of an autocorrelation sequence (ACS), i.e., a standardized version of an ACVS). If the ACVS is known (as in our AR(4) example), the criterion for various choices of N' is

$$\text{NMSE}(N') \stackrel{\text{def}}{=} \frac{1}{N} \sum_{\tau=0}^{N-1} \left(\frac{s'_\tau - s_\tau}{s_0} \right)^2. \tag{608a}$$

If the ACVS is not easy to get at (a motivation for using GSSM), we can consider

$$\text{NMSE}(N', N'') \stackrel{\text{def}}{=} \frac{1}{N} \sum_{\tau=0}^{N-1} \left(\frac{s'_\tau - s''_\tau}{s''_0} \right)^2, \tag{608b}$$

where s''_τ is defined in a manner similar to s'_τ, but with $N'' > N'$ and $f''_k \stackrel{\text{def}}{=} k/(2N'')$ replacing N' and f'_k. The idea is to look at various choices of N' and N'', e.g., $(N', N'') = (N/2, N)$, $(N, 2N)$ and so forth. Study of these NMSEs can help ascertain if increasing N' leads to enough of an improvement in the approximations to justify the resulting increase in the time that it takes a computer to generate a pair of realizations.

Table 607 looks at the square root of the two NMSE criteria for the AR(4) example (taking square roots recasts the criteria roughly as deviations between autocorrelations, which connects the criteria to quantities bounded in magnitude by unity). As expected, $\text{NMSE}(N')$ decreases as N' increases, which is consistent with what Figure 607 indicates. For $N' = 128$, an NMSE of 0.007 might suggest reasonably accurate GSSM simulations since this value is small relative to unity; on the other hand, NMSE(256) is two orders of magnitude smaller, but using $N' = 256$ rather than $N' = 128$ would approximately double the amount of computer time needed to generate realizations. In the case where the true ACVS is unknown so that we must rely on $\text{NMSE}(N', N'')$ rather than $\text{NMSE}(N')$, a comparison of the two rows of Table 607 suggests that the two criteria are similar enough so that any rule of thumb for selecting N' that might work with $\text{NMSE}(N')$ would also apply to $\text{NMSE}(N', 2N')$. (When numerical precision is taken into account, the two very small entries in the column with the header $N' = 1024$ are indistinguishable from each other and from zero.)

In summary, given knowledge of the SDF $S(\cdot)$ for a Gaussian stationary process $\{X_t\}$, GSSM consists of

[1] setting N' to an appropriate value relative to the desired N (setting N' to be less than N is not recommended – a reasonable default is often $N' = 2N$, but, in general, larger settings for N' imply more accurate GSSM-based simulations at the expense of longer computer times to generate the simulations);
[2] computing $S(f'_k)$, $k = 0, 1, \ldots, 2N' - 1$, where $f'_k = k/(2N')$;
[3] using the $S(f'_k)$ sequence along with $4N'$ deviates from a Gaussian white noise process $\{Z_t\}$ with zero mean and unit variance to form $\mathcal{U}_k$ as per Equation (606b); and, finally,
[4] taking the DFT of $\{\mathcal{U}_k\}$ to obtain $U_t = U_{\Re,t} + iU_{\Im,t}$ – we can consider realizations of the real-valued sequences $U_{\Re,t}$ and $U_{\Im,t}$, $t = 0, 1, \ldots, N-1$, to be independent and – at least to some degree – to have the same statistical properties as X_t, $t = 0, 1, \ldots, N-1$.

We can compute the DFT in step [4] efficiently if $2N'$ is amenable to an FFT algorithm. Repetition of steps [3] and [4] yields as many additional pairs of realizations as desired, but a different random sample of Gaussian white noise is needed at each iteration of step [3].

As an example, let us proceed as we did in illustrating the circulant embedding method by generating realizations of length $N = 1024$ from the AR(2) process of Equation (34) (see Figure 603 and the discussion surrounding it; again we emphasize that the best way to simulate this AR(2) process is *neither* GSSM *nor* the circulant embedding method, but rather the

11.3 Simulation of Processes with a Known Spectral Density Function

method described in Exercise [597]). Setting $N' = N = 1024$ allows us to directly compare the performance of GSSM and the circulant embedding method because both methods then make use of the same number of independent standard Gaussian RVs, namely, $4N = 4096$. To implement GSSM, we need to compute the SDF for the AR(2) process, namely,

$$S(f) = \frac{\sigma_2^2}{\left|1 - \sum_{j=1}^{2} \phi_{2,j} e^{-i2\pi f j}\right|^2} = \frac{1}{\left|1 - 0.75 e^{-i2\pi f} + 0.5 e^{-i4\pi f}\right|^2}, \tag{609}$$

over the grid of frequencies $f'_k = k/2048$, $k = 0, 1, \ldots, 2047$ (once we recall the assumption $\Delta_t = 1$ and once we set p to 2, the above follows from Equation (446b), which states the SDF $S(\cdot)$ for AR processes, while Equation (34) tells us that $\sigma_2^2 = 1$, $\phi_{2,1} = 0.75$ and $\phi_{2,2} = -0.5$). As noted in C&E [5] for Section 9.2, we can obtain the required sequence $\{S(f'_k)\}$ of 2048 values by taking the DFT – call it $\{G_k : k = 0, 1, \ldots, 2047\}$ – of the sequence consisting of -1, 0.75 and -0.5 followed by 2045 zeros. The desired $S(f'_k)$ is given by $1/|G_k|^2$ (see Equations (449a) and (449b), noting carefully that we need to take N' in the former to be 2048, and we need f'_k in the latter to be $k/2048$). When numerical precision is taken into account, the weight $S(f'_k)$ and the weight S_k of Equation (601a) used by the circulant embedding method are indistinguishable. As a result, a plot of $S(f'_k)$ versus k looks no different from the plot of S_k versus k shown in Figure 603(a). Additionally, if we reuse the same 4096 realizations of Gaussian white noise to form $\{U_k\}$ and then $\{U_t\}$ as we did in forming $\{\mathcal{Y}_k\}$ and then $\{Y_t\}$, plots of the real and imaginary components of $\{U_t\}$ look identical to those of $\{Y_t\}$, i.e., the bottom two plots of Figure 603. For this example, GSSM generates realizations that, while in theory not exact, are virtually identical to the exact realizations obtained via circulant embedding. The degree to which we get similar results for other stationary processes is explored in Exercises [11.10] and [11.11].

Comments and Extensions to Section 11.3

[1] We have assumed for convenience that the known SDF for the stationary process $\{X_t\}$ has an associated sampling interval Δ_t of unity. If we are supplied with an SDF that is associated with a process whose sampling interval differs from unity, we cannot use this SDF with the GSSM formulation presented here (in particular the SDF for $\{X_t\}$ is periodic with a period $1/\Delta_t$, whereas our formulation presumes an SDF with unit periodicity). We need to adjust the supplied SDF to make it work with GSSM. Fortunately the adjustment is simple (the discussion here follows that of C&E [4] for Section 4.3, but the notation differs somewhat). First, take $S(\cdot)$ to be what is required by GSSM in Equation (606b) and elsewhere. Second, take $S_{\Delta_t}(\cdot)$ to be the true SDF for $\{X_t\}$ (this SDF is thus periodic with a period of $1/\Delta_t$). Given $S_{\Delta_t}(\cdot)$, we can get what our GSSM formulation needs by setting

$$S(f) = \frac{1}{\Delta_t} S_{\Delta_t}\left(\frac{f}{\Delta_t}\right).$$

Note that

$$S(f+1) = \frac{1}{\Delta_t} S_{\Delta_t}\left(\frac{f+1}{\Delta_t}\right) = \frac{1}{\Delta_t} S_{\Delta_t}\left(\frac{f}{\Delta_t}\right) = S(f),$$

and hence a periodicity of $1/\Delta_t$ for $S_{\Delta_t}(\cdot)$ implies the required unity periodicity for $S(\cdot)$. The factor of $1/\Delta_t$ just before $S_{\Delta_t}(\cdot)$ in the above is needed so that the integral of $S(\cdot)$ over an interval of unit length is equal to $s_0 = \text{var}\{X_t\}$. Thus

$$\int_{-1/2}^{1/2} S(f')\,df' = \frac{1}{\Delta_t}\int_{-1/2}^{1/2} S_{\Delta_t}\left(\frac{f'}{\Delta_t}\right)\,df' = \int_{-f_\mathcal{N}}^{f_\mathcal{N}} S_{\Delta_t}(f)\,df = s_0$$

upon making the change of variable $f = f'/\Delta_t$, where, as usual, $f_\mathcal{N} = 1/(2\Delta_t)$ is the Nyquist frequency. Note that we have both $\{s_\tau\} \longleftrightarrow S_{\Delta_t}(\cdot)$ and $\{s_\tau\} \longleftrightarrow S(\cdot)$, which is a consequence of the way in which Δ_t is incorporated into the Fourier transform of Equation (74a) and the inverse Fourier transform of Equation (75a) (this is consistent with the fact that the definition of the ACVS, namely, $s_\tau = \text{cov}\{X_{t+\tau}, X_t\}$, does not involve Δ_t).

After taking the GSSM realizations of $U_{\Re,0}, U_{\Re,1}, \ldots, U_{\Re,N-1}$ and $U_{\Im,0}, U_{\Im,1}, \ldots, U_{\Im,N-1}$ to be surrogates for realizations of $X_0, X_1, \ldots, X_{N-1}$, we can reincorporate Δ_t by using it to associate the time index t with the actual time $t_0 + t\,\Delta_t$, where t_0 is the time associated with index $t = 0$.

[2] Our formulation of GSSM requires that $S(f_k')$ be finite for $k = 0, 1, \ldots, 2N' - 1$, where, as before, $f_k' = k/(2N')$. The requirement is not satisfied for all k for some stationary processes of interest. In particular, consider a power-law SDF defined by $S^{(\text{PL})}(f) = C|f|^\alpha$ for $|f| \leq 1/2$, where $-1 < \alpha < 0$. While $S^{(\text{PL})}(f_k')$ is finite for $1 \leq k \leq 2N' - 1$, it is infinite for $k = 0$. Fortunately we can adjust GSSM to handle this problem. First, note that

$$s_0^{(\text{PL})} = \int_{-1/2}^{1/2} S^{(\text{PL})}(f)\,\mathrm{d}f = 2C \int_0^{1/2} f^\alpha \,\mathrm{d}f = \frac{2C}{\alpha+1}\left(\frac{1}{2}\right)^{\alpha+1}. \tag{610}$$

Second, as given by Equation (606a), the approximation s_0' for $s_0^{(\text{PL})}$ is

$$s_0' = \frac{1}{2N'} \sum_{k=0}^{2N'-1} S^{(\text{PL})}(f_k'),$$

which is infinite because $S^{(\text{PL})}(f_0') = S^{(\text{PL})}(0) = \infty$. The adjustment consists of replacing $S^{(\text{PL})}(f_0')$ in the above with

$$\tilde{S}_0 \stackrel{\text{def}}{=} 2N' s_0^{(\text{PL})} - \sum_{k=1}^{2N'-1} S^{(\text{PL})}(f_k'),$$

which forces the resulting expression – call it $\tilde{s}_0'$ – to be *exactly* equal to $s_0^{(\text{PL})}$. The adjustment that is needed to generate the simulated series is simple: just redefine $\mathcal{U}_0$ in Equation (606b) to be

$$\mathcal{U}_0 = \left(\frac{\tilde{S}_0}{2N'}\right)^{1/2} (Z_0 + iZ_1),$$

but keep $\mathcal{U}_k$ for $1 \leq k \leq 2N' - 1$ as is. Note, however, that the ACVS for the simulated series is given by

$$\tilde{s}_\tau' = \frac{1}{2N'}\left(\tilde{S}_0 + \sum_{k=1}^{2N'-1} S^{(\text{PL})}(f_k') e^{i2\pi f_k' \tau}\right)$$

rather than by s_τ' of Equation (606a) (proof of this closely follows the solution to Exercise [11.8]).

An objection here is that easy access to the process variance (as Equation (610) gives for the power-law example) violates the premise motivating this section, namely, that we have easy access to the SDF, but not the ACVS. The violation is mild in that the adjustment to GSSM assumes we have access to the zero lag component of the ACVS, but nothing more; however, if we were to multiply the power-law SDF by a continuous and even function that is bounded above and away from zero and that has unit periodicity, this newly defined SDF would still diverge to infinity as $f \to 0$. We would then have no guarantee of easy access to the process variance and might have to entertain a potentially problematic numerical integration, which would distract from using the adjusted GSSM. (There are also some stationary processes for which $S(f_k')$ is potentially infinite at nonzero indices k, a prime example being Gegenbauer processes – see section 11.3 of Woodward et al., 2017, and references therein. Adjusting GSSM to handle such processes is tricky.)

11.4 Simulating Time Series from Nonparametric Spectral Estimates

Suppose we have a time series considered to be a realization of a portion $X_0, X_1, \ldots, X_{N-1}$ of a Gaussian stationary process with sampling time $\Delta_t = 1$ and with an SDF $S(\cdot)$ (we assume a sampling time of unity merely for convenience – C&E [1] for this section indicates what adjustments need to be made to the upcoming exposition when $\Delta_t \neq 1$). Suppose that, based upon this realization, we estimate the SDF using, say $\hat{S}(\cdot)$, and that this estimator satisfies two conditions: first, $\hat{S}(f) \geq 0$ for all f; and second, the ACVS estimator $\{\hat{s}_\tau\}$ corresponding to $\hat{S}(\cdot)$, i.e., $\{\hat{s}_\tau\} \longleftrightarrow \hat{S}(\cdot)$, is such that $\hat{s}_\tau = 0$ for all $|\tau| \geq N$. An argument that closely parallels the proof of Exercise [605] says that

[1] we can use the circulant embedding method to simulate a time series of length N whose *theoretical* SDF is *identical* to the estimator $\hat{S}(\cdot)$; and
[2] the weights S_k required by that method for use in Equation (601c) satisfy $S_k = \hat{S}(\tilde{f}_k)$, where, as usual, $\tilde{f}_k$ is defined to be $k/(2N)$.

With bootstrapping in mind, Percival and Constantine (2006) note that many nonparametric estimators satisfy the two conditions we imposed on $\hat{S}(\cdot)$. These include estimates resulting from direct SDF estimators $\hat{S}^{(\text{D})}(\cdot)$ (see Equation (186b)); lag window estimators $\hat{S}^{(\text{LW})}_m(\cdot)$ (Equation (248a)), but restricted to those guaranteed to be nonnegative at all frequencies; basic and weighted multitaper estimators $\hat{S}^{(\text{MT})}(\cdot)$ and $\hat{S}^{(\text{WMT})}(\cdot)$ (Equations (352a) and (352b)); and WOSA estimators $\hat{S}^{(\text{WOSA})}(\cdot)$ (Equation (414b)). Hence we can use the circulant embedding method to simulate time series of length N based on estimates from almost all the nonparametric spectral estimators considered in Chapters 6, 7 and 8 (C&E [2] for this section discusses simulating series with a length different from that of the time series used to create $\hat{S}(\cdot)$, while C&E [3] looks into some nonparametric estimators for which circulant embedding is problematic).

In what follows, we match specific nonparametric SDF estimators up with the circulant embedding method. We assume $\mu = E\{X_t\}$ to be known, which means that we can take μ to be zero without loss of generality. Define

$$\tilde{X}_t = \begin{cases} X_t, & 0 \leq t \leq N-1; \\ 0, & N \leq t \leq 2N-1 \end{cases} \qquad (611\text{a})$$

(if μ is unknown, we estimate it by the sample mean $\overline{X} \stackrel{\text{def}}{=} \sum_{t=0}^{N-1} X_t/N$ and replace X_t with $X_t - \overline{X}$ in the above). Examples of simulating time series based on nonparametric SDF estimates of actual time series are given in Section 11.6.

Direct Spectral Estimators

Given a data taper $\{h_t : t = 0, 1, \ldots, N-1\}$ satisfying the normalization $\sum_t h_t^2 = 1$, a direct spectral estimator is defined by

$$\hat{S}^{(\text{D})}(f) = \left| \sum_{t=0}^{N-1} h_t X_t e^{-i2\pi f t} \right|^2$$

(this is Equation (186b) with Δ_t set to unity). Clearly we must have $\hat{S}^{(\text{D})}(f) \geq 0$; in addition, the corresponding ACVS estimator $\{\hat{s}^{(\text{D})}_\tau\}$ is such that $\hat{s}^{(\text{D})}_\tau = 0$ for $|\tau| \geq N$ (see Equation (188b)), so the two conditions hold that we need for circulant embedding to work. Define

$$\tilde{h}_t = \begin{cases} h_t, & 0 \leq t \leq N-1; \\ 0, & N \leq t \leq 2N-1. \end{cases} \qquad (611\text{b})$$

The weights S_k needed in Equation (601c) are

$$S_k = \hat{S}^{(\mathrm{D})}(\tilde{f}_k) = \left|\sum_{t=0}^{N-1} h_t X_t \mathrm{e}^{-\mathrm{i}2\pi \tilde{f}_k t}\right|^2 = \left|\sum_{t=0}^{2N-1} \tilde{h}_t \tilde{X}_t \mathrm{e}^{-\mathrm{i}\pi k t/N}\right|^2, \quad 0 \leq k \leq 2N-1. \tag{612a}$$

The right-hand summation is the DFT of $\{\tilde{h}_t \tilde{X}_t : t = 0, 1, \ldots, 2N-1\}$ and hence can be computed efficiently if $2N$ is a length accepted by an FFT algorithm.

Lag Window Spectral Estimators

Given an ACVS estimator $\{\hat{s}^{(\mathrm{D})}_\tau\}$ corresponding to a direct spectral estimator and given a lag window $\{w_{m,\tau}\}$, a lag window spectral estimator is defined by

$$\hat{S}^{(\mathrm{LW})}_m(f) = \sum_{\tau=-(N-1)}^{N-1} w_{m,\tau} \hat{s}^{(\mathrm{D})}_\tau \mathrm{e}^{-\mathrm{i}2\pi f \tau},$$

(this is Equation (248a) with Δ_t set to unity). As noted in Section 7.5, certain lag windows ensure that $\hat{S}^{(\mathrm{LW})}_m(f) \geq 0$ for all f, including the Bartlett, Daniell, Bartlett–Priestley, Parzen, Gaussian and Papoulis lag windows. For all lag window spectral estimators, the corresponding ACVS estimator $\{\hat{s}^{(\mathrm{LW})}_\tau\}$ is such that $\hat{s}^{(\mathrm{LW})}_\tau = 0$ for $|\tau| \geq N$ (see Equation (248b)). Hence the two conditions needed for circulant embedding to work hold if $\{w_{m,\tau}\}$ is selected with care.

Consider the inverse DFT of the finite sequence $\{\hat{S}^{(\mathrm{D})}(\tilde{f}_k)\}$ of Equation (612a):

$$\frac{1}{2N} \sum_{k=0}^{2N-1} \hat{S}^{(\mathrm{D})}(\tilde{f}_k) \mathrm{e}^{\mathrm{i}\pi k \tau/N} \stackrel{\text{def}}{=} \tilde{s}^{(\mathrm{D})}_\tau, \quad \tau = 0, \ldots, 2N-1.$$

This inverse DFT has the following relationship to $\{\hat{s}^{(\mathrm{D})}_\tau\}$:

$$\tilde{s}^{(\mathrm{D})}_\tau = \begin{cases} \hat{s}^{(\mathrm{D})}_\tau, & 0 \leq \tau \leq N; \\ \hat{s}^{(\mathrm{D})}_{2N-\tau}, & N+1 \leq \tau \leq 2N-1 \end{cases} \tag{612b}$$

(deducible from computational details in Sections 6.7 and 7.11). Define

$$\tilde{w}_{m,\tau} = \begin{cases} w_{m,\tau}, & 0 \leq \tau \leq N; \\ w_{m,2N-\tau}, & N+1 \leq \tau \leq 2N-1 \end{cases} \tag{612c}$$

(an adaptation of Equation (314) once we recall that $w_{m,N}$ is always zero). The weights needed to use circulant embedding with lag window spectral estimates are

$$S_k = \hat{S}^{(\mathrm{LW})}_m(\tilde{f}_k) = \sum_{\tau=0}^{2N-1} \tilde{w}_\tau \tilde{s}^{(\mathrm{D})}_\tau \mathrm{e}^{-\mathrm{i}\pi k \tau/N}, \tag{612d}$$

i.e., the DFT of $\{\tilde{w}_\tau \tilde{s}^{(\mathrm{D})}_\tau\}$, which can be computed efficiently if $2N$ is compatible with an FFT algorithm.

Multitaper Spectral Estimators

Given a set of K data tapers $\{h_{k,t}\}$, $k = 0, 1, \ldots, K-1$, and given a set of nonnegative weights d_k that sum to unity, a weighted multitaper estimator is defined by

$$\hat{S}^{(\mathrm{WMT})}(f) = \sum_{k=0}^{K-1} d_k \left|\sum_{t=0}^{N-1} h_{k,t} X_t \mathrm{e}^{-\mathrm{i}2\pi f t}\right|^2$$

11.4 Simulating Time Series from Nonparametric Spectral Estimates

(the above follows from Equations (352b) and (352a) with Δ_t set to unity). The condition $\hat{S}^{(\text{WMT})}(f) \geq 0$ clearly holds; in addition, the corresponding ACVS estimator is a weighted sum of K individual ACVS estimators, each corresponding to the ACVS for a direct spectral estimator and hence being zero at lags $|\tau| \geq N$, from which we can conclude that the two conditions hold that we need for circulant embedding. With $\{\tilde{h}_{k,t}\}$ defined in a manner similar to $\{\tilde{h}_t\}$ in Equation (611b), the weights needed to use circulant embedding with weighted multitaper spectral estimates follow directly from Equation (612a):

$$S_k = \hat{S}^{(\text{WMT})}(\tilde{f}_k) = \sum_{k'=0}^{K-1} d_{k'} \left| \sum_{t=0}^{2N-1} \tilde{h}_{k',t} \tilde{X}_t e^{-i\pi kt/N} \right|^2, \quad 0 \leq k \leq 2N-1. \quad (613a)$$

Weights for simulating time series whose SDFs are in agreement with a basic multitaper estimate follow from the above upon setting $d_{k'} = 1/K$.

WOSA Spectral Estimators

Given a data taper $\{h_t : t = 0, 1, \ldots, N_S - 1\}$ appropriate for block size $N_S < N$, a WOSA spectral estimator is defined by

$$\hat{S}^{(\text{WOSA})}(f) = \frac{1}{N_B} \sum_{j=0}^{N_B-1} \left| \sum_{t=0}^{N_S-1} h_t X_{t+jn} e^{-i2\pi ft} \right|^2,$$

where N_B is the number of blocks to be averaged together, and n is an integer-valued shift factor satisfying the constraints of Equation (414c) (the above follows from Equations (414b) and (414a)). A minor variation on the argument used for multitaper spectral estimators says that WOSA estimators also satisfy the two conditions we need for circulant embedding to work.

Define

$$\tilde{h}'_t = \begin{cases} h_t, & 0 \leq t \leq N_S - 1; \\ 0, & N_S \leq t \leq 2N - 1. \end{cases} \quad (613b)$$

The circulant embedding weights for WOSA are

$$S_k = \hat{S}^{(\text{WOSA})}(\tilde{f}_k) = \frac{1}{N_B} \sum_{j=0}^{N_B-1} \left| \sum_{t=0}^{2N-1} \tilde{h}'_t \tilde{X}_{t+jn} e^{-i\pi kt/N} \right|^2, \quad 0 \leq k \leq 2N-1. \quad (613c)$$

Comments and Extensions to Section 11.4

[1] In generating the weights S_k needed for circulant embedding, we assumed for convenience that $\Delta_t = 1$ for the time series $\{X_t\}$ to be simulated. If not, let $\hat{S}(\cdot)$ stand for either the direct spectral estimator of Equation (186b), the lag window estimator of Equation (248a), the weighted multitaper estimator of Equation (352b) or the WOSA estimator of Equation (414b). Suppose we have computed $\hat{S}(\cdot)$ over the grid of frequencies $k/(2N \Delta_t)$, $k = 0, 1, \ldots, 2N - 1$. We can use these to form the required weights by setting

$$S_k = \frac{1}{\Delta_t} \hat{S}\left(\frac{k}{2N \Delta_t}\right)$$

(the right-hand side is in keeping with remarks made in C&E [1] for Section 11.3). The above weights will be *identical* to ones obtained using the right-hand sides of Equations (612a), (612d), (613a) or (613c), none of which have any dependence on Δ_t. Once a simulated series of length N has been generated using circulant embedding, we can take Δ_t into account by associating the time indices t for the series with actual times $t_0 + t\,\Delta_t$, $t = 0, 1, \ldots, N-1$, where t_0 is the time associated with index $t = 0$.

[2] The procedure we have described for simulating a time series yields a simulated series of the same length N as the original time series. If we desire a simulated series of shorter length, say $N' < N$, a simple solution is to use the procedure as is and just discard the last $N - N'$ values of the original simulated series. If we are basing the simulations on a lag window estimate that uses a lag window with a truncation point m (e.g., the Bartlett lag window of Equation (269a), the Parzen window of Equation (275a) or the Papoulis window of Equation (278)), and if $m < N$, we can generate series with the shorter length $N' = m$ using weights

$$S_k = \sum_{\tau=0}^{2N'-1} \tilde{w}_\tau \tilde{s}_\tau^{(D)} e^{-i\pi k\tau/N'}, \quad 0 \le k \le 2N'-1 \tag{614a}$$

(cf. Equation (612d)), for which we need to take the definitions of $\tilde{w}_\tau$ and $\tilde{s}_\tau^{(D)}$ to be

$$\tilde{w}_{m,\tau} = \begin{cases} w_{m,\tau}, & 0 \le \tau \le N'; \\ w_{m,2N'-\tau}, & N'+1 \le \tau \le 2N'-1 \end{cases} \quad \& \quad \tilde{s}_\tau^{(D)} = \begin{cases} \hat{s}_\tau^{(D)}, & 0 \le \tau \le N'; \\ \hat{s}_{2N'-\tau}^{(D)}, & N'+1 \le \tau \le 2N'-1 \end{cases}$$

(cf. Equations (612c) and (612b)). If we base the simulations on a WOSA estimate with a block size $N_S < N$, we can generate series of length $N' = N_S$ using weights

$$S_k = \frac{1}{N_B} \sum_{j=0}^{N_B-1} \left| \sum_{t=0}^{2N'-1} \tilde{h}_t' \tilde{X}_{t+jn} e^{-i\pi kt/N'} \right|^2, \quad 0 \le k \le 2N'-1$$

(cf. Equation (613c)).

We can also adjust the circulant embedding method to simulate series of length $N' > N$, but with an obvious word of warning: simulating a series of length, say, $N' = 1{,}000{,}000$ based upon an SDF estimate that uses a time series with length $N = 1000$ might fail miserably in capturing the correct low-frequency properties of the phenomenon under study, particularly if, as a preprocessing step, the series has been centered by subtracting off its sample mean. Redefine

$$\tilde{X}_t = \begin{cases} X_t, & 0 \le t \le N-1; \\ 0, & N \le t \le 2N'-1 \end{cases} \text{ and } \tilde{h}_t = \begin{cases} h_t, & 0 \le t \le N-1; \\ 0, & N \le t \le 2N'-1 \end{cases} \tag{614b}$$

(cf. Equations (611a) and (611b)). For direct spectral estimators, the weights are now

$$S_k = \left| \sum_{t=0}^{2N'-1} \tilde{h}_t \tilde{X}_t e^{-i\pi kt/N'} \right|^2, \quad 0 \le k \le 2N'-1$$

(cf. Equation (612a)). The adjustments to S_k for weighted multitaper and WOSA estimators are similar: for the former, we have

$$S_k = \sum_{k'=0}^{K-1} d_{k'} \left| \sum_{t=0}^{2N'-1} \tilde{h}_{k',t} \tilde{X}_t e^{-i\pi kt/N'} \right|^2, \quad 0 \le k \le 2N'-1$$

(cf. Equation (613a)), where $\tilde{h}_{k',t}$ and $\tilde{X}_t$ are defined in keeping with Equation (614b); for the latter, we have

$$S_k = \frac{1}{N_B} \sum_{j=0}^{N_B-1} \left| \sum_{t=0}^{2N'-1} \tilde{h}_t' \tilde{X}_{t+jn} e^{-i\pi kt/N'} \right|^2, \quad 0 \le k \le 2N'-1$$

11.4 Simulating Time Series from Nonparametric Spectral Estimates

(cf. Equation (613c)), where we need to redefine

$$\tilde{h}'_t = \begin{cases} h_t, & 0 \leq t \leq N_S - 1; \\ 0, & N_S \leq t \leq 2N' - 1 \end{cases}$$

(cf. Equation (613b)) and to again use $\tilde{X}_t$ as per Equation (614b). For lag window estimators, the weights S_k specified in Equation (614a) for the case $N' < N$ also work when $N' > N$.

[3] While the circulant embedding method is attractive for generating time series whose SDFs are in agreement with the nonparametric estimators discussed in this section, there are other nonparametric estimators for which the method is problematic. As indicated by Equation (601a), circulant embedding requires knowledge of an ACVS out to lag N in order to generate the required weights S_k. All the nonparametric estimators discussed in this section are such that the corresponding estimated ACVS, say $\{\hat{s}_\tau\}$, is known for all $\tau \in \mathbb{Z}$ (and, in particular, $\hat{s}_\tau = 0$ for $|\tau| \geq N$, which guarantees that the weights are nonnegative, as required by the circulant embedding method). For other nonparametric estimators, the corresponding estimated ACVS is either difficult to get at or ill-defined.

As an example of a nonparametric estimator whose estimated ACVS is not easily obtainable, consider nonparametric estimators that make use of a prewhitening filter such as the estimator of Equation (491b):

$$\hat{S}_X^{(\text{PC})}(f) = \frac{\hat{S}_{e,m}^{(\text{LW})}(f)}{\left|1 - \sum_{j=1}^p \bar{\phi}_{p,j} e^{-i2\pi f j \Delta_t}\right|^2}.$$

In this example the time series $\{X_t\}$ is prewhitened using a filter with coefficients $1, -\bar{\phi}_{p,1}, \ldots, -\bar{\phi}_{p,p}$; the output from the filter forms the basis for the lag window estimate $\hat{S}_{e,m}^{(\text{LW})}(\cdot)$; and this estimate is divided by the squared gain function for the filter to form the postcolored SDF estimate $\hat{S}_X^{(\text{PC})}(\cdot)$ for $\{X_t\}$. In principle, the SDF estimate $\hat{S}_X^{(\text{PC})}(\cdot)$ corresponds to the SDF for an ARMA process. Its AR coefficients are just $\bar{\phi}_{p,1}, \ldots, \bar{\phi}_{p,p}$, but its MA coefficients are not directly available – they are associated with $\hat{S}_{e,m}^{(\text{LW})}(\cdot)$, and routines to extract them are numerically unstable (if we could extract them, we would have multiple ways of computing the estimated ACVS corresponding to $\hat{S}_X^{(\text{PC})}(\cdot)$; see, e.g., section 3.2.1 of Brockwell and Davis, 2016). While the estimated ACVS corresponding to $\hat{S}_{e,m}^{(\text{LW})}(\cdot)$ is readily available, it is not easy to manipulate it to get the ACVS that goes with $\hat{S}_X^{(\text{PC})}(\cdot)$. Since circulant embedding is problematic here and since it is easy to compute $\hat{S}_X^{(\text{PC})}(\cdot)$ over any desired grid of frequencies, GSSM is an attractive option for simulating time series with statistical properties closely matching those dictated by $\hat{S}_X^{(\text{PC})}(\cdot)$.

Two examples of nonparametric estimators for which the corresponding estimated ACVS is ill-defined are the adaptive multitaper estimator $\hat{S}^{(\text{AMT})}(\cdot)$ of Equation (389b) and the discretely smoothed periodogram $\hat{S}_m^{(\text{DSP})}(\cdot)$ of Equation (307). The former is defined at any particular frequency f by an iterative scheme. The corresponding estimated ACVS requires knowing $\hat{S}^{(\text{AMT})}(f)$ for all $|f| \leq f_\mathcal{N}$, but we can only compute $\hat{S}^{(\text{AMT})}(\cdot)$ over a finite set of frequencies, which renders the ACVS ill-defined. As is also true for nonparametric estimators that make use of a prewhitening filter, GSSM is an attractive option for use with $\hat{S}^{(\text{AMT})}(\cdot)$. We noted in C&E [2] for Section 7.10 that the ACVS corresponding to $\hat{S}_m^{(\text{DSP})}(\cdot)$ does not have a standard definition; however, pending future research, circulant embedding might prove compatible with one of the ACVS definitions proposed in that C&E.

[4] Since the periodogram is a special case of a direct spectral estimator, we can use circulant embedding to simulate a Gaussian time series whose theoretical SDF is equal to the periodogram $\hat{S}^{(\text{P})}(\cdot)$ of an actual time series $X_0, X_1, \ldots, X_{N-1}$ (Equation (612a) provides the necessary weights S_k once we have set h_t in Equation (611b) to $1/\sqrt{N}$. The periodogram for the simulated series will only resemble the periodogram for the actual series to a certain extent – they will *not* be exactly equal (Exercise [11.12] demonstrates this fact). Additionally the sample mean of the simulated series in practice will differ from the sample mean of the actual time series. Keeping these two facts in mind, there is an interesting contrast between simulation via periodogram-based circulant embedding and the *method of surrogate time series* (MSTS) proposed by Theiler et al. (1992) in the context of detecting nonlinear structure in a stationary process (see also Schreiber and Schmitz, 2000; Chan and Tong, 2001; Kantz and Schreiber,

2004). Given $X_0, X_1, \ldots, X_{N-1}$, this method starts by taking its DFT:

$$\mathcal{X}_k \stackrel{\text{def}}{=} \sum_{t=0}^{N-1} X_t e^{-i2\pi kt/N} \quad \text{and hence} \quad \{X_t\} \longleftrightarrow \{\mathcal{X}_k\}$$

(see Equation (171a) with Δ_t set to unity). Note that the periodogram for $\{X_t\}$ at the kth Fourier frequency k/N is $|\mathcal{X}_k|^2/N$ (see Equation (171b)). Define ϕ_0 to be zero, and let ϕ_k, $k = 1, \ldots, \lfloor (N-1)/2 \rfloor$, be IID RVs uniformly distributed over $(-\pi, \pi]$, where $\lfloor x \rfloor$ is the greatest integer less than or equal to x. For $N/2 < k \leq N-1$, define $\phi_k = -\phi_{N-k}$. Finally, if N is even, define $\phi_{N/2}$ to be an RV that assigns a probability of $1/2$ to the points 0 and π and that is independent of the other ϕ_k RVs. The MSTS simulated series is the inverse DFT of $\{\mathcal{X}_k \exp(i\phi_k)\}$:

$$X_{\phi,t} \stackrel{\text{def}}{=} \frac{1}{N} \sum_{k=0}^{N-1} \mathcal{X}_k e^{i\phi_k} e^{i2\pi kt/N} \quad \text{and hence} \quad \{X_{\phi,t}\} \longleftrightarrow \{\mathcal{X}_k e^{i\phi_k}\}. \quad (616a)$$

The sequence $\{X_{\phi,t}\}$ is periodic with a period of N. Exercise [11.13a] is to show that it is real-valued (solving this exercise reveals that, when N is even, we need the special definition for the distribution of $\phi_{N/2}$ to ensure that $\{X_{\phi,t}\}$ is real-valued in general). Note that the periodogram for $\{X_{\phi,t}\}$ is $|\mathcal{X}_k e^{i\phi_k}|^2/N = |\mathcal{X}_k|^2/N$. Thus, for any given realization of the ϕ_k RVs, the simulated series $X_{\phi,0}, X_{\phi,1}, \ldots X_{\phi,N-1}$ has *identically* the same periodogram at the Fourier frequencies as $X_0, X_1, \ldots, X_{N-1}$; moreover, their sample means are the same since $\sum_{t=0}^{N-1} X_t = \mathcal{X}_0 = \mathcal{X}_0 e^{i\phi_0} = \sum_{t=0}^{N-1} X_{\phi,t}$. By contrast, a series simulated via circulant embedding has a periodogram and sample mean differing in practice from those of $\{X_t\}$.

As noted in the solution to Exercise [602], the stationary process from which the simulated series from circulant embedding arise is a harmonic process formulated as per Equation (35c), which involves linear combinations of sines and cosines with random amplitudes (this is the defining equation for a harmonic processs). For MSTS, with the time series $\{X_t\}$ regarded as fixed and with φ_k defined to be such that $\mathcal{X}_k = |\mathcal{X}_k| \exp(i\varphi_k)$, we have

$$X_{\phi,t} = \overline{X} + \sum_{k=1}^{\lfloor N/2 \rfloor} D_k \cos(2\pi kt/N + \phi'_k), \quad \text{where} \quad D_k \stackrel{\text{def}}{=} \begin{cases} 2|\mathcal{X}_k|/N, & 1 \leq k < N/2; \\ |\mathcal{X}_{N/2}|/N, & k = N/2, \end{cases} \quad (616b)$$

and

$$\phi'_k \stackrel{\text{def}}{=} \begin{cases} \varphi_k + \phi_k + 2\pi, & \varphi_k + \phi_k \leq -\pi; \\ \varphi_k + \phi_k - 2\pi, & \varphi_k + \phi_k > \pi; \\ \varphi_k + \phi_k, & \text{otherwise,} \end{cases} \quad (616c)$$

with the RVs ϕ'_k having the same multivariate distribution as the RVs ϕ_k. Exercise [11.13b] is to verify the above, while [11.13c] is to establish that, when N is odd, the process $\{X_{\phi,t}\}$ is a harmonic process formulated as per Equation (35d), but one involving a linear combination of cosines with random phases and fixed amplitudes; on the other hand, when N is even, $\{X_{\phi,t}\}$ is in general *not* a harmonic process. Whereas the realizations from circulant embedding are from a Gaussian distribution because we specified the random amplitudes to be Gaussian RVs, MSTS realizations do not necessarily obey a Gaussian distribution exactly, but, due to a central limit effect, the sum of $N-1$ randomly phase cosines will be close to Gaussian if the D_k amplitudes are not dominated by a few large values (Walden and Prescott, 1983; Sun and Chaika, 1997). The relative merits of circulant embedding and MSTS depend on the rationale for simulating time series, but certainly circulant embedding has an advantage over MSTS in any applications for which variations in the periodogram or in the sample mean are desirable.

11.5 Simulating Time Series from Parametric Spectral Estimates

Given a time series considered to be a realization of a portion $X_0, X_1, \ldots, X_{N-1}$ of the zero mean pth order causal (and hence stationary) AR process

$$X_t = \sum_{j=1}^{p} \phi_{p,j} X_{t-j} + \epsilon_t,$$

where $\{\epsilon_t\}$ is a zero mean white noise process, suppose we estimate the AR parameters $\phi_{p,1}$, $\ldots, \phi_{p,p}$ and $\sigma_p^2 = \text{var}\{\epsilon_t\}$ using one of the methods described in Chapter 9 (Yule–Walker, Burg, least squares or maximum likelihood). Let $\hat{\phi}_{p,1}, \ldots, \hat{\phi}_{p,p}$ and $\hat{\sigma}_p^2$ denote these parameter estimates. We can form a corresponding parametric SDF estimate, say $\hat{S}(\cdot)$, by plugging the parameter estimates into the functional form for the SDF for an AR(p) process:

$$\hat{S}(f) = \frac{\hat{\sigma}_p^2}{\left|1 - \sum_{j=1}^{p} \hat{\phi}_{p,j} e^{-i 2\pi f j}\right|^2} \tag{617}$$

(see Equation (446b) with the sampling time Δ_t taken to be unity, as assumed elsewhere in this chapter). Our task is to generate realizations from the estimated AR(p) process.

Suppose that the AR coefficient estimates $\hat{\phi}_{p,1}, \ldots, \hat{\phi}_{p,p}$ correspond to those of a causal AR process (causality is guaranteed for Yule–Walker estimates and always happens in practice for Burg estimates, but not necessarily for least squares estimates). If we are willing to assume Gaussianity, we can readily adapt the simulation scheme of Equation (597a) to get the desired realizations of the estimated AR(p) process. This scheme depends just on the $p+1$ parameters $\phi_{p,1}, \ldots, \phi_{p,p}$ and σ_p^2, and the adaptation replaces these with $\hat{\phi}_{p,1}, \ldots, \hat{\phi}_{p,p}$ and $\hat{\sigma}_p^2$. In particular, use of these estimates in conjunction with Equations (597b) and (596a) yields everything needed in Equation (597a) to generate a simulated AR series of length N based on a random sample $\{Z_t\}$ of size N from a standard Gaussian distribution (to simulate a series of length $N' \neq N$, take $\{Z_t\}$ to be a random sample of size N' instead).

We can also check the Gaussian assumption by forming the observed prediction errors in keeping with Equation (596b), but some adjustments are needed to put the errors on a common footing. Recall that the variance for the unobservable predictions error $\vec{e}_t(t)$, $t = 0, 1, \ldots, p-1$, is σ_t^2 so that $\vec{e}_t(t)/\sigma_t$ has unit variance; additionally, $\vec{e}_t(t)/\sigma_p$ has unit variance for $t = p, \ldots, N-1$. Equation (596a) allows us to express σ_t^2 for a given $t < p$ in terms of σ_p^2 and $\phi_{t+1,t+1}, \ldots, \phi_{p,p}$. In particular

$$\sigma_{p-1}^2 = \frac{\sigma_p^2}{1 - \phi_{p,p}^2}$$

$$\sigma_{p-2}^2 = \frac{\sigma_{p-1}^2}{1 - \phi_{p-1,p-1}^2} = \frac{\sigma_p^2}{\prod_{j=p-1}^{p}(1 - \phi_{j,j}^2)} \quad \text{(not needed if } p = 1\text{)}$$

$$\vdots$$

$$\sigma_0^2 = \frac{\sigma_1^2}{1 - \phi_{1,1}^2} = \frac{\sigma_p^2}{\prod_{j=1}^{p}(1 - \phi_{j,j}^2)} \quad \text{(not needed if } p = 1 \text{ or } 2\text{)}.$$

For $t = 0, \ldots, p-1$, we thus have

$$\sigma_t^2 = \frac{\sigma_p^2}{\prod_{j=t+1}^{p}(1 - \phi_{j,j}^2)}.$$

The observed prediction errors are defined in a manner similar to the unobservable prediction errors of Equation (596b):

$$\vec{e}_0(0) = X_0$$

$$\vec{e}_t(t) = X_t - \sum_{j=1}^{t} \hat{\phi}_{t,j} X_{t-j}, \quad t = 1, \ldots, p-1 \text{ (not needed if } p=1\text{)} \quad (618a)$$

$$\vec{e}_t(t) = X_t - \sum_{j=1}^{p} \hat{\phi}_{p,j} X_{t-j}, \quad t = p, \ldots, N-1.$$

The observed prediction errors normalized so that their variances are on a common footing are

$$\vec{Z}_t(t) \stackrel{\text{def}}{=} \frac{\vec{e}_t(t)}{\hat{\sigma}_t}, \quad \text{where } \hat{\sigma}_t^2 \stackrel{\text{def}}{=} \begin{cases} \hat{\sigma}_p^2 / \prod_{j=t+1}^{p}(1 - \hat{\phi}_{j,j}^2), & 0 \leq t \leq p-1; \\ \hat{\sigma}_p^2, & p \leq t \leq N-1. \end{cases} \quad (618b)$$

If $\{X_t\}$ is indeed a Gaussian AR(p) process, these normalized errors $\{\vec{Z}_t(t)\}$ should resemble a random sample from a standard Gaussian distribution (they will not be exactly so due to the use of estimated AR parameters). In the spirit of bootstrapping (see, e.g., Efron and Gong, 1983; Swanepoel and Van Wyk, 1986; and Davison and Hinkley, 1997), rather than creating simulated series via Equation (597a) using a random sample of size N from a standard Gaussian distribution, we can use a random sample (with replacement) from $\{\vec{Z}_t(t) : t = 0, 1, \ldots, N-1\}$ (this scheme is used with the ocean wave data in Section 11.6).

If the AR coefficient estimates do *not* correspond to those of a causal AR process (as is a possibility with least squares estimators), this fact will become apparent either because $|\hat{\phi}_{p,p}| \geq 1$ or because, when attempting to use the coefficients estimates $\hat{\phi}_{p,j}$ to get corresponding coefficients for the best linear predictors of orders less than p via recursive use of Equation (597b), the estimate $\hat{\phi}_{j,j}$ for some $\phi_{j,j}$, $j < p$, is such that $|\hat{\phi}_{j,j}| \geq 1$ (see C&E [5] for Section 9.4). Since we cannot estimate all the coefficients required by the simulation scheme of Equation (597a), we cannot use that method to generate realizations exactly consistent with the statistical properties of the estimated AR(p) process. If, however, the coefficients $\hat{\phi}_{p,j}$ correspond to a stationary acausal AR(p) process, then the corresponding SDF estimate $\hat{S}(\cdot)$ of Equation (617) has all the properties of an SDF (see C&E [1] for Section 9.2; note that, if the noncausal coefficients $\hat{\phi}_{p,j}$ were to correspond to a nonstationarity process, then $\hat{S}(f) = \infty$ for some f since one of the roots of $1 - \sum_{j=1}^{p} \hat{\phi}_{p,j} z^{-j}$ would lie on the unit circle). In lieu of an exact simulation method and in view of the fact that we can readily compute $\hat{S}(\cdot)$ over various grids of frequencies (see C&E [5] for Section 9.2), a good candidate for an approximate simulation method is GSSM – we just need to substitute $\hat{S}(f'_k)$ for $S(f'_k)$ in Equation (606b).

Comments and Extensions to Section 11.5

[1] We can regard the observed prediction errors $\{\vec{e}_t(t)\}$ of Equation (618a) as forward errors in the same spirit as ones encountered in Burg's algorithm (see Equation (467a), which is the same as Equation (618a) when $p \leq t \leq N-1$). Burg's algorithm also makes use of observed backward prediction errors. We can also entertain these here – denote them by $\{\overleftarrow{e}_t(t) : t = 0, 1, \ldots, N-1\}$. A simple way to define the backward errors is take the right-hand sides of the equalities in Equation (618a) and replace $X_0, X_1, \ldots, X_{N-1}$ with its time reversal, i.e., $X_{N-1}, X_{N-2}, \ldots, X_0$. The observed forward and backward prediction errors do not constitute a set of $2N$ uncorrelated deviates, but, nonetheless, we can entertain bootstrapping schemes in which we randomly sample (with replacement) from $\vec{Z}_0(0), \overleftarrow{Z}_0(0), \vec{Z}_1(1), \overleftarrow{Z}_1(1), \ldots, \vec{Z}_{N-1}(N-1), \overleftarrow{Z}_{N-1}(N-1)$, where $\overleftarrow{Z}_t(t)$ is defined by replacing $\vec{e}_t(t)$ in the definition for $\vec{Z}_t(t)$ in Equation (618b) with $\overleftarrow{e}_t(t)$.

11.6 Examples of Simulation of Time Series

Ocean Wave Data

We have previously considered the ocean wave data in Sections 6.8, 7.12, 8.9, 9.12 and 10.15. This series $\{X_t\}$ is shown in Figure 620(a), which is a copy of Figure 225. The series is of length $N = 1024$, has a sampling interval of $\Delta_t = 1/4$ sec and hence spans a total of 256 sec. Let us first use these data as an example of simulating a time series whose properties are in keeping with a lag window spectral estimate. The dark curves in Figures 318 and 319 show several lag window SDF estimates for this series, all of which are smoothed versions of a direct spectral estimate $\hat{S}^{(D)}(\cdot)$ using an $NW = 2/\Delta_t$ Slepian data taper (the light curves in the figures show $\hat{S}^{(D)}(\cdot)$). Tapering is advised because, as demonstrated in Figure 226, the periodogram shows evidence of bias due to leakage. The $m = 150$ Parzen lag window estimate in Figure 318 tracks $\hat{S}^{(D)}(\cdot)$ nicely and is considerably smoother, but arguably too bumpy over $0.2\,\text{Hz} \le f \le 1.0\,\text{Hz}$ in light of a hypothesized monotonic rolloff. The heavier degree of smoothing achieved by the four lag window estimates shown in Figure 319 depicts the rolloff better, but at the price of smearing out the dominant peak centered near $f = 0.16$ Hz (the $m = 150$ Parzen estimate does a better job of capturing this peak). These four lag window estimates have comparable smoothing window bandwidths, and, in our discussion of that figure, we settled on the $m = 23.666$ Gaussian lag window estimate (the dark curve in Figure 319(d)) as the most attractive of the four. Under the assumption that the ocean wave series is a realization of a stationary process, the $m = 150$ Parzen and Gaussian lag window SDF estimates are good candidates for simulating time series whose statistical properties are in agreement with those of the actual series.

Plots (b) to (e) in Figure 620 show four simulations of $\{X_t\}$ via the circulant embedding method described in Section 11.2 and adapted in Section 11.4 for use with lag window estimates. Plots (b) and (d) show simulations based on the $m = 150$ Parzen estimate. The weights S_k needed by circulant embedding are calculated using Equation (612d), which, in forming $\tilde{w}_{m,\tau}$ as per Equation (612c), makes uses of the Parzen lag window $w_{m,\tau}$ of Equation (275a) with m set to 150. The weights also need $\tilde{s}_\tau^{(D)}$, which is formed as per Equation (612b) and which in turn depends on the ACVS estimate $\hat{s}_\tau^{(D)}$ corresponding to the direct spectral estimate. Equation (188b) defines $\hat{s}_\tau^{(D)}$, but, prior to using that equation, we must center the time series by subtracting off its sample mean $\overline{X} \doteq 209.1$ because the ocean wave data are not compatible with the assumed stationary process having a mean of zero; in addition, the data taper used in Equation (188b) is the Slepian taper as approximated by Equation (196b) with W set to $2/N$. As must be true, all the weights S_k are nonnegative, with the smallest and largest being 0.4141 and 4.261×10^6 (these are associated with indices $k = 810$ and $k = 81$ and with frequencies 1.582 Hz and 0.1582 Hz). The simulations in Figures 620(b) and (d) are first $N = 1024$ parts of the real and imaginary components of the complex-valued series $\{Y_t\}$ of length $2N = 2048$ – see Equation (602), which, in view of Equation (601c), depends upon the weights S_k and a random sample of size $4N = 4096$ from a standard Gaussian distribution. The actual series has a prominent nonzero sample mean, whereas the simulated series are drawn from a zero mean stationary process (but their sample means are different from zero). To compensate for this mismatch, we have added the sample mean $\overline{X} \doteq 209.1$ to both simulated series. This addition results in the sample mean for the series in plot (b) being 230.6, and it is 205.8 in plot (d). Visually comparing the two simulated series with the actual series in Figure 620(a) does not suggest a serious mismatch in their statistical properties.

Plots (c) and (e) in Figure 620 show simulations created in a manner similar to those in (b) and (d), but now using the Gaussian lag window estimate rather than the $m = 150$ Parzen. Some details are necessarily different. We now use the Gaussian lag window of Equation (276) with m set to 23.666. The smallest and largest weights S_k are 0.5272 and 2.039×10^6 (these are associated with indices $k = 809$ and $k = 81$ and with frequencies

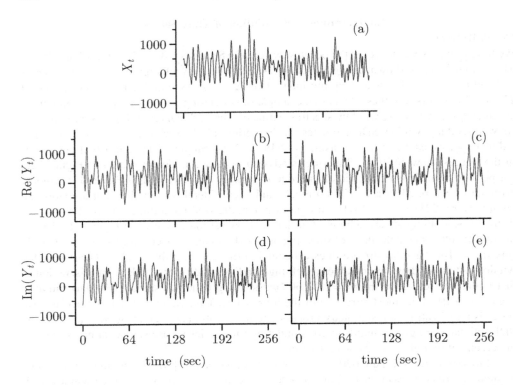

Figure 620 Ocean wave data $\{X_t : t = 0, 1, \ldots, N - 1\}$ (plot (a) – also shown in Figure 225) and four simulations of this series based upon circular embedding (plots (b) to (e)). Here $N = 1024$, and the series has a sampling interval of $\Delta_t = 1/4$ sec, resulting in a time span of 256 sec for it and its simulations. For plots (b) and (d), the weights S_k, $k = 0, 1, \ldots, 2N - 1$, used to create the complex-valued series $\{Y_t : t = 0, 1, \ldots, 2N - 1\}$ of Equation (602) are based on the $m = 150$ Parzen lag window estimate shown as the dark curve in Figure 318; (b) and (d) show the real and imaginary parts of Y_t, $t = 0, 1, \ldots, N - 1$. Plots (c) and (e) are similar, but now use the $m = 23.666$ Gaussian lag window estimate shown as the dark curve in Figure 319(d). The same sample of $4N$ deviates from a standard Gaussian distribution is used to create the Y_t series from both lag window estimates.

1.580 Hz and 0.1582 Hz). The sample means for the series in plots (c) and (e) are 241.7 and 199.8. In addition we used *exactly* the same sample of size $4N = 4096$ from a standard Gaussian distribution as in the Parzen case. The simulations in plots (b) and (c) are thus *not* independent (in fact they are highly correlated – their sample correlation coefficient $\hat{\rho}$ is 0.96). The same is true for plots (d) and (e) (now $\hat{\rho} \doteq 0.97$). Using the same random sample allows us to visualize changes in the simulated series due entirely to the different weighting schemes (Parzen and Gaussian). Simulations (b) and (c) track each other well, as do (d) and (e). The two weighting schemes thus do not yield simulated series with much visual difference.

In view of Figure 620, should we use the $m = 150$ Parzen or the Gaussian lag window estimate as the basis for simulating the ocean wave data? If our interest is in simulating series whose spectral properties over $0.2 \text{ Hz} \leq f \leq 1.0 \text{ Hz}$ are estimated as best as possible under the working hypothesis of a monotonic rolloff, the discussion surrounding Figures 318 and 319 suggests using the Gaussian lag window estimate; on the other hand, oceanographic considerations might support the hypothesis that, due to the mechanism generating ocean waves, the high-power part of the SDF has a peak centered near 0.16 Hz and that the peak has a narrower bandwidth than what the Gaussian lag window estimate suggests. This hypothesis would favor the $m = 150$ Parzen estimate if we want the simulated series to be faithful to the high-power part of the SDF. The choice between the Parzen or the Gaussian estimate thus

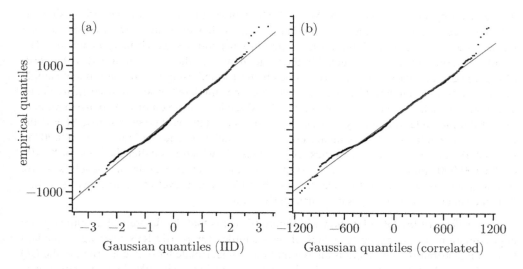

Figure 621 Gaussian QQ plots for ocean wave data (Figure 620(a) shows the series itself). The empirical quantiles (order statistics) for the time series are plotted as small dots versus the expected value of quantiles under an IID Gaussian assumption (left-hand plot) and under an assumption of a Gaussian stationary process with an SDF given by an $m = 150$ Parzen lag window estimate $\hat{S}_m^{(LW)}(\cdot)$ (right-hand plot; the dark curve in Figure 318 shows $\hat{S}_m^{(LW)}(\cdot)$). The lines are simple linear least squares fits of the empirical versus theoretical quantiles.

depends upon what we want to achieve with the simulated series in terms of delving into various hypotheses about the ocean wave data. (If we were willing to entertain the hypothesis that suggests the Gaussian estimate in addition to the hypothesis that suggests Parzen, we could entertain a scheme that would somehow merge the weights generated by the two estimates to yield simulations in support of both hypotheses).

Theoretical justification for all four simulations in Figure 620 follows if the ocean wave data obey a Gaussian distribution. To investigate the univariate distribution for $\{X_t\}$, Figure 621(a) shows a Gaussian QQ plot for this series constructed in the following manner (section 1.6, Brockwell and Davis, 2016; see also Chambers et al., 1983). Let $X_{(t)}$, $t = 0, 1, \ldots, N-1$, denote the order statistics for the series, i.e., the X_t values rearranged from smallest to largest so that $X_{(0)} \leq X_{(1)} \leq \cdots \leq X_{(N-1)}$. The small dots in Figure 621(a) show $X_{(t)}$ versus q_t, where $q_t \stackrel{\text{def}}{=} \Phi^{-1}([t+0.5]/N)$, and $\Phi^{-1}(p)$ is the $p \times 100\%$ percentage point of the standard Gaussian distribution; i.e., we are plotting empirical quantiles versus corresponding theoretical quantiles. Visual evidence against Gaussianity consists of $X_{(t)}$ not appearing to vary linearly with q_t. The line in Figure 621(a) is a least squares fit of the linear model $E\{X_{(t)}\} = \alpha + \beta q_t$, where α and β are relatable to μ and σ, the theoretical mean and standard deviation of the ocean wave series under the null hypothesis of Gaussianity. The bulk of the dots seem to have a linear alignment, but there are notable deviations in the lower and upper tails. To quantify the degree of linearity, consider the coefficient of determination

$$R^2 \stackrel{\text{def}}{=} \frac{\left(\sum_{t=0}^{N-1} \left(X_{(t)} - \overline{X}\right) q_t\right)^2}{\sum_{t=0}^{N-1} \left(X_{(t)} - \overline{X}\right)^2 \sum_{t=0}^{N-1} q_t^2}, \quad (621)$$

a standard measure for assessing the efficacy of a simple linear regression (see, e.g., Weisberg, 2014). This coefficient can range from 0 to 1. If R^2 is too small, there is evidence against the null hypothesis. To determine what constitutes "too small," we need the distribution of R^2 under the null hypothesis. This distribution is analytically intractable, but we can approximate it by generating a large number, say $N_\text{R} = 100{,}000$, of simulated ocean wave series using circulant embedding based on the $m = 150$ Parzen lag window estimate (if we use the Gaussian lag window estimate instead, the results stated in what follows are virtually identical). The simulated series are each constructed in the same manner as those shown in parts (b) and (d) of Figure 620, but with each set of two series being formed using a different random sample of size $4N$ from a standard Gaussian distribution. The simulated series are necessarily Gaussian, and, for each of them, we can compute R^2, which leads to an empirical approximation to its distribution under the null hypothesis of a zero mean stationary process whose SDF is the $m = 150$ Parzen lag window estimate. For the N_R simulated series, R^2 ranges from 0.96349 to 0.99948. For the ocean wave series, $R^2 \doteq 0.99227$, which is smaller than that of $88{,}406$ of the $100{,}000$ simulated series. The p-value, i.e., the observed significance level, is 0.116, which means we would fail to reject the null hypothesis at a level of significance $\alpha = 0.1$. Evidence against Gaussianity is not particularly strong.

A critique of the QQ plot shown in Figure 621(a) is that the theoretical quantiles q_t are approximations to the expected values of the quantiles for N IID RVs obeying a standard Gaussian distribution; i.e., there is an assumption of uncorrelatedness in constructing the QQ plot, whereas the time series we are dealing with is associated with RVs having a covariance structure that should be in keeping with the $m = 150$ Parzen lag window estimate $\hat{S}_m^{(\text{LW})}(\cdot)$. This suggests plotting the empirical quantiles against the expected values of the quantiles from a time series of length N whose SDF is given by $\hat{S}_m^{(\text{LW})}(\cdot)$. We can approximate these expected values by considering again the N_R simulated series. For each pair $\{Y_{\Re,t}\}$ and $\{Y_{\Im,t}\}$ of simulated series of length N, we form their order statistics $\{Y_{\Re,(t)}\}$ and $\{Y_{\Im,(t)}\}$. Separately for each t, we average the N_R order statistics collected from the $N_\text{R}/2$ individual realizations of $Y_{\Re,(t)}$ and $Y_{\Im,(t)}$ – let q'_t denote this average, and regard it as an approximation to a theoretical replacement for q_t that takes into account the statistical properties dictated by $\hat{S}_m^{(\text{LW})}(\cdot)$. Figure 621(b) shows another Gaussian QQ plot for the ocean wave series, but this time consisting of the empirical quantiles $X_{(t)}$ plotted against the theoretical quantiles q'_t. The degree of linearity in the two QQ plots in Figure 621 appears remarkably similar. With q'_t replacing q_t in Equation (621), we now have $R^2 \doteq 0.99144$ in comparison to 0.99227 for the QQ plot in Figure 621(a). To assess if R^2 is too small, we use a procedure analogous to the one previously described. For the N_R simulated series, R^2 now ranges from 0.96119 to 0.99965. For the ocean wave series, this R^2 is now smaller than that for $92{,}288$ of the N_R simulated series, yielding a p-value of 0.077, which means we would fail to reject the null hypothesis at a level of significance $\alpha = 0.05$, but would reject for $\alpha = 0.1$. There is thus some evidence against Gaussianity, but not particularly strong. Since the evidence against the ocean wave data being Gaussian is somewhat weak, there is no compelling reason to be dissatisfied with the simulations shown in Figure 620.

Let us now consider using simulated series to assess the variability in the Gaussian lag window estimate $\hat{S}_m^{(\text{LW})}(\cdot)$ shown in Figure 319(d) and redisplayed as the solid curve in Figure 623a. The assessment involves a bootstrapping scheme advocated in Percival and Constantine (2006), which is based on the same circulant embedding method that led to the simulations of the ocean wave series shown in Figures 620(c) and (e). The scheme starts by generating a large number N_R of simulated series based upon $\hat{S}_m^{(\text{LW})}(\cdot)$ (as before, we take "large" to be $100{,}000$). For each such series, we compute a Gaussian lag window estimate formed in exactly the same manner as we formed $\hat{S}_m^{(\text{LW})}(\cdot)$ (in particular, we use an $NW = 2/\Delta_\text{t}$ Slepian data taper after centering the simulated series by subtracting off its sample mean, and we set the lag window parameter m to 23.666). Let $\tilde{S}_n^{(\text{LW})}(\tilde{f}_k)$ be the lag window estimate at frequency $k/(2N\Delta_\text{t})$ for the nth simulated series, $n = 1, 2, \ldots, N_\text{R}$. For each frequency, we sort the N_R estimates $\tilde{S}_n^{(\text{LW})}(\tilde{f}_k)$ – call these $\tilde{S}_{(n)}^{(\text{LW})}(\tilde{f}_k)$ so that

11.6 Examples of Simulation of Time Series

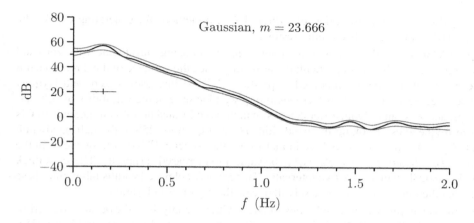

Figure 623a Gaussian lag window spectral estimate $\hat{S}_m^{(\mathrm{LW})}(\cdot)$ with $m = 23.666$ for ocean wave data (dark curve, reproduced from Figure 319(d)). The width and height of the crisscross gives the bandwidth measure $B_\mathcal{U}$ (Equation (256a)) and the length of a 95% CI for $10\log_{10} S(f)$ (Equation (266a)). The thin curves surrounding the dark curve show bootstrapped 95% CIs based upon circulant embedding.

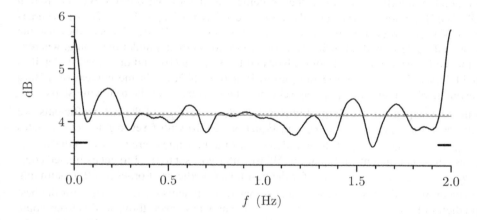

Figure 623b Widths of bootstrapped 95% CIs shown as thin curves in Figure 623a (dark curve). The solid and dashed thin horizontal lines show, respectively, the average of these dB widths (4.14 dB) and the length of a 95% CI as calculated by Equation (266a) (4.16 dB; the height of the crisscross in Figure 623a also depicts this length). The widths of the short thick lines emanating from $f = 0$ Hz and $f = 2$ Hz depict the half-bandwidth measure $B_\mathcal{U}/2$ as calculated by Equation (256a) (the width of the crisscross in Figure 623a depicts $B_\mathcal{U}$).

$\tilde{S}_{(n)}^{(\mathrm{LW})}(\tilde{f}_k) \leq \tilde{S}_{(n+1)}^{(\mathrm{LW})}(\tilde{f}_k)$ for $n = 1, 2, \ldots, N_\mathrm{R} - 1$. We then pick out $\tilde{S}_{(\lfloor 0.975 N_\mathrm{R} \rfloor)}^{(\mathrm{LW})}(\tilde{f}_k)$ and $\tilde{S}_{(\lfloor 0.025 N_\mathrm{R} \rfloor)}^{(\mathrm{LW})}(\tilde{f}_k)$ to serve as upper and lower bounds for a bootstrapped pointwise 95% CI for the unknown $S(\tilde{f}_k)$. These bounds are shown as the two thin curves in Figure 623a, where they depict a 95% CI for $10 \log_{10} S(\tilde{f}_k)$. The SDF estimate $\hat{S}_m^{(\mathrm{LW})}(\cdot)$ is always trapped between the upper and lower bounds of the bootstrapped CIs, but there are some frequencies where $\hat{S}_m^{(\mathrm{LW})}(\tilde{f}_k)$ is just barely below the upper bound, and others where it is barely above the lower. The crisscross in Figure 623a is redisplayed from Figure 319(d). Its height gives the length of a 95% CI based upon Equation (266a). This height appears to be comparable to the typical widths of the bootstrapped CIs. Figure 623b compares the bootstrapped CIs with the theory-based CIs in more detail. The dark curve shows the widths of the bootstrapped CIs versus frequency. The average of these dB widths is 4.14 dB, which is indicated by a solid thin line. The length of the theory-based 95% CI is 4.16 dB and is indicated by a dashed thin line (barely distinguishable from the solid line). The two methods for generating CIs for the unknown SDF yield remarkably similar results.

Figure 623b shows the widths of the bootstrapped CIs being elevated near the zero and Nyquist frequencies. This is not particularly surprising since the variance of the direct spectral estimate $\hat{S}^{(D)}(\cdot)$ that is being smoothed to produce $\hat{S}_m^{(LW)}(\cdot)$ increases at zero and Nyquist frequencies (see Section 6.6). This fact would imply an increase in the variability of $\hat{S}_m^{(LW)}(\cdot)$ as we approach zero and Nyquist frequencies, which would translate into an increase in CIs near these frequencies. Evoking the bandwidth $B_\mathcal{U}$ of Equation (256a), the half-bandwidth $B_\mathcal{U}/2$ of $\hat{S}_m^{(LW)}(\cdot)$ comes into play here in that we might expect $\hat{S}_m^{(LW)}(\cdot)$ at frequencies in the intervals $[0, B_\mathcal{U}/2]$ and $[f_\mathcal{N} - B_\mathcal{U}/2, f_\mathcal{N}]$ to be subject to increased variability. The short thick lines in Figure 623b depict these two intervals. The regions of elevated widths fall within these intervals, but the intervals fail to precisely delineate the regions of elevation.

As a final example of how simulations can help with the analysis of the ocean wave series, let us reconsider its Burg AR(25) SDF estimate $\hat{S}(\cdot)$ (shown as the thick curve in Figure 625a, which is the same as the thick curves in both panels of Figure 500). Our goal is to assess the variability in $\hat{S}(\cdot)$. To do so, we adapt the simulation scheme discussed in Section 11.5. The adaptation yields the bootstrapping scheme proposed by Swanepoel and Van Wyk (1986), but with two notable modifications. First, their scheme uses the normalized observed prediction errors $\vec{Z}_t(t)$ of Equation (618b), but only those indexed by $t = p, p+1, \ldots, N-1$, whereas we use all N such errors (for this example, $p = 25$ and $N = 1024$). Second, they use the burn-in procedure described following Equation (595) to get their simulations going, whereas we use the stationary initial conditions advocated by Kay (1981b) and described in detail in Section 11.1 (see the discussion leading up to Equation (597a)). In the context of a Burg AR(p) estimate $\hat{S}(\cdot)$, their bootstrapping scheme scheme consists of the following steps.

In the first step, given the Burg estimates $\hat{\phi}_{p,1}, \hat{\phi}_{p,2}, \ldots, \hat{\phi}_{p,p}$ of the AR coefficients, we substitute these for $\phi_{p,1}, \phi_{p,2}, \ldots, \phi_{p,p}$ in Equation (597b) with t set to p to get estimates $\hat{\phi}_{p-1,1}, \hat{\phi}_{p-1,2}, \ldots, \hat{\phi}_{p-1,p-1}$ of the coefficients for the best linear predictor of order $p-1$. We then use these coefficients in Equation (597b) with t now set to $p-1$ to get estimated coefficients $\hat{\phi}_{p-2,1}, \hat{\phi}_{p-2,2}, \ldots, \hat{\phi}_{p-2,p-2}$ for the best linear predictor of order $p-2$. Continuing in this manner, we end up with Burg-based estimated coefficients $\hat{\phi}_{k,1}, \ldots, \hat{\phi}_{k,k}$ for the best linear predictor of orders $k = p-1, p-2, \ldots, 1$. We now have everything needed to compute the observed prediction errors $\vec{e}_t(t), t = 0, 1, \ldots, N-1$ as per Equation (618a). Given the Burg estimate $\hat{\sigma}_p^2$ of σ_p^2, we can also compute the normalized observed prediction errors $\vec{Z}_t(t)$ as per Equation (618b). (As noted in Section 9.5, Burg's algorithm is recursive. A look at how its recursions work reveals that code for it could extract all the required prediction errors along with the $\hat{\phi}_{j,j}$ coefficients on the fly. If the code were to return these along with the $p+1$ Burg parameter estimates, it would be an easy task to compute the normalized observed prediction errors $\vec{Z}_t(t)$.)

In the second step, given the Burg coefficient estimates and the normalized observed prediction errors $\{\vec{Z}_t(t) : t = 0, 1, \ldots, N-1\}$, we now consider Equation (597a), which tells us how to simulate an AR(p) time series given (i) the coefficients of the best linear predictors of orders unity up to p; (ii) the square roots of $\sigma_t^2, t = 0, 1, \ldots, p$; and (iii) a random sample $\{Z_t\}$ of size N from a standard Gaussian distribution. In the bootstrapping scheme, for (i), we substitute the estimated coefficients we just formed in the first step; for (ii), we use square roots of the estimates $\hat{\sigma}_t^2$ defined in Equation (618b); and, most importantly, for (iii), we substitute a random sample of size N with replacement from the normalized observed prediction errors $\{\vec{Z}_t(t)\}$. Since we are using $\{X_t\}$ to denote the ocean wave data, let $X'_t, t = 0, 1, \ldots, N-1$, denote the simulated series we get from the right-hand sides of Equation (597a).

In the third step, we use $\{X'_t\}$ in exactly the same manner in which we used $\{X_t\}$ to

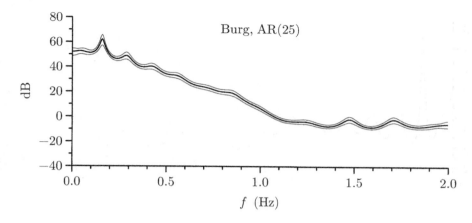

Figure 625a Burg AR(p) spectral estimate $\hat{S}(\cdot)$ of order $p = 25$ for ocean wave data (dark curve, reproduced from Figures 500(a) and (b)). The thin curves surrounding the dark curve show 95% CIs based upon the bootstrapping scheme of Swanepoel and Van Wyk (1986).

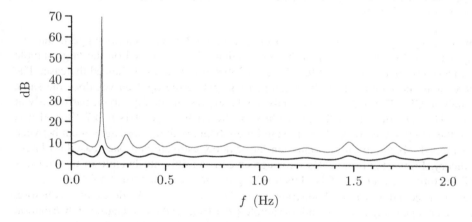

Figure 625b Widths of bootstrapped 95% CIs shown as thin curves in Figure 625a (dark curve), along with widths of no less than 95% CIs shown as thin curves in Figure 500(a) (light curve).

form an AR(25) Burg SDF estimate (in particular, we subtract off the sample mean of $\{X'_t\}$ as a preprocessing step). Denote this estimate by $\hat{S}_1(\cdot)$ (for comparison with the actual SDF estimate $\hat{S}(\cdot)$ for the ocean wave series, we compute $\hat{S}_1(\cdot)$ over all Fourier frequencies $f_k = k/(N \, \Delta_t)$ satisfying $0 < f_k < f_{\mathcal{N}}$). We then generate a second simulated series in the same manner as the first, but using a different random sample with replacement from $\{\vec{Z}_t(t)\}$. We use this to create another AR(25) SDF estimate, namely, $\hat{S}_2(\cdot)$. We repeat this procedure over and over again, ending up with a large number of AR(25) SDF estimates $\hat{S}_n(\cdot)$, $n = 1, 2, \ldots,$ N_R, where here we set $N_R = 100{,}000$.

In the fourth and final step, we form a bootstrapped pointwise 95% CI for the unknown $S(f_k)$ by sorting the $\hat{S}_n(f_k)$ estimates – call these $\hat{S}_{(n)}(f_k)$ so that $\hat{S}_{(n)}(f_k) \leq \hat{S}_{(n+1)}(f_k)$. We then use $\hat{S}_{(\lfloor 0.975 N_R \rfloor)}(f_k)$ and $\hat{S}_{(\lfloor 0.025 N_R \rfloor)}(f_k)$ to define the upper and lower bounds of the 95% CI.

Figure 625a shows the bootstrapped 95% CIs for $S(f_k)$ versus f_k as two light curves above and below a dark curve depicting the Burg AR(25) estimate for the ocean wave series. The Burg estimate $\hat{S}(f_k)$ is trapped between the upper and lower CI at all Fourier frequencies

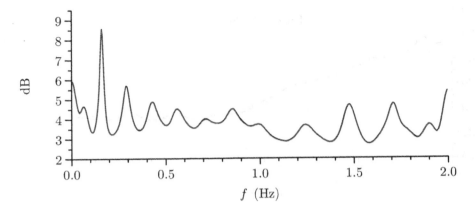

Figure 626 Widths of bootstrapped 95% CIs shown as thin curves in Figure 625a (dark curve, replicated from Figure 625b). These CIs make use of a random sample with replacement from the normalized observed prediction errors $\{\vec{Z}_t(t)\}$. There is a light curve showing corresponding widths when using a random sample $\{Z_t\}$ from a standard Gaussian distribution, but this curve is virtually identical to the dark curve (the maximum absolute deviation between the two curves is 0.06 dB).

satisfying $0 < f_k < f_{\mathcal{N}}$. A visual comparison of Figure 625a with 500(a) suggests that the bootstrapped 95% CIs are tighter than the no less than 95% CIs based on the large-sample analytic theory presented in Section 9.9. Figure 625b shows that this is indeed the case. The dark curve plots the widths of the bootstrapped CIs, while the light curves does the same for the analytic CIs. The widths of the analytic CIs are systematically larger, particularly at $f_{41} \doteq 0.16$ Hz. The median of the dB widths for the bootstrapped CIs is 3.7 dB, and it is 8.3 dB for the analytic CIs. In stark contrast to Figure 623b, with its close agreement between the bootstrapped CIs based on the Gaussian lag window estimate and ones based on large-sample theory, Figure 625b indicates poor agreement, possibly attributable to the "no less than 95%" nature of the analytic CIs (thus resulting in wider CIs than true 95% CIs).

There is a second way to do bootstrapping in the AR context. We do exactly as before, with the sole change being to use a random sample $\{Z_t\}$ from a standard Gaussian distribution rather than a random sample with replacement from the normalized observed prediction errors $\{\vec{Z}_t(t)\}$. This second way is closer in spirit to the circulant embedding approach used with the Gaussian lag window estimate in that both use $\{Z_t\}$. Figure 626 shows that this second way yields CI widths that are remarkably similar to those of the first. One explanation for this close match is that $\{\vec{Z}_t(t)\}$ is indistinguishable from deviates drawn from a standard Gaussian distribution, so it does not matter if we sample from $\{\vec{Z}_t(t)\}$ or from $\{Z_t\}$.

In conclusion, simulated ocean wave series are useful not only for mimicking potentially important aspects of the original series, but also in analyzing the series. On the analysis side, simulated series help assess the assumption of Gaussianity and the variability of nonparametric and parametric SDF estimates. In the case of a nonparametric lag window SDF estimate, the assessment of variability via simulated series agrees remarkably well with that provided by the large-sample theory of Section 7.4. In the case of a parametric AR estimate, assessments from simulated series indicate much less variability than what the theory presented in Section 9.9 suggests. As noted in that section, this theory is inherently conservative. Simulated series verify this conservativeness and provide a second – arguably more realistic – take on the variability in a parametric AR estimate.

Ship Altitude Data
As an interesting contrast to the ocean wave data, let us consider the ship altitude time series

11.6 Examples of Simulation of Time Series 627

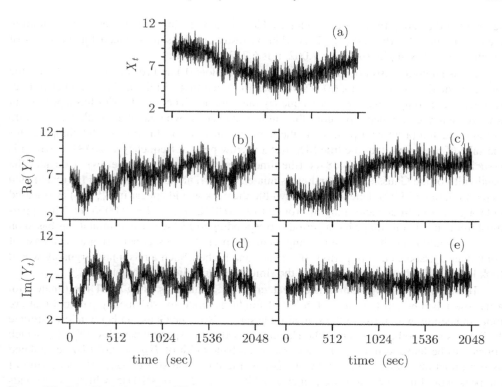

Figure 627 Ship altitude time series $\{X_t : t = 0, 1, \ldots, N - 1\}$ (plot (a) – also shown in Figure 330) and four simulations of this series based upon circulant embedding (plots (b) to (e)). Here $N = 2048$, and the data have a sampling interval of $\Delta_t = 1$ sec. For plots (b) and (d), the weights S_k, $k = 0, 1, \ldots, 2N - 1$, used to create the complex-valued series $\{Y_t : t = 0, 1, \ldots, 2N - 1\}$ of Equation (602) are based on the $m = 80$ Gaussian lag window estimate $\hat{S}_m^{(\text{LW})}(\cdot)$ shown as the dark curve in Figure 331(c); (b) and (d) show, respectively, the real and imaginary parts of Y_t, $t = 0, 1, \ldots, N - 1$. Plots (c) and (e) are similar, but now use the Hanning-based direct spectral estimate $\hat{S}^{(\text{D})}(\cdot)$ shown as the dark curve in Figure 331(b). The same sample of $4N$ deviates from a standard Gaussian distribution was used in creating the Y_t series from both $\hat{S}_m^{(\text{LW})}(\cdot)$ and $\hat{S}^{(\text{D})}(\cdot)$.

$\{X_t\}$ as a second example of simulating a time series using circulant embedding. This series has a sampling interval of $\Delta_t = 1$ sec and is shown in Figure 627(a), which is a replication of Figure 330. Figure 331 shows several nonparametric SDF estimates for this series. Leakage is a concern for some of these estimates. The most attractive estimate is the $m = 80$ Gaussian lag window estimate $\hat{S}_m^{(\text{LW})}(\cdot)$ shown as the dark curve in Figure 331(c). This estimate is a smoothed version of the Hanning-based direct spectral estimate $\hat{S}^{(\text{D})}(\cdot)$ shown in 331(b). The Hanning data taper is useful here because the periodogram shows evidence of bias due to leakage, and the Gaussian lag window is also beneficial because the resulting $\hat{S}_m^{(\text{LW})}(\cdot)$ shows no evidence of bias due to smoothing window leakage. Under the assumption that the ship altitude series is a realization of a stationary process, the nonparametric SDF estimate $\hat{S}_m^{(\text{LW})}(\cdot)$ is potentially a good basis for simulating time series whose statistical properties are in agreement with those of the actual series.

Plots (b) and (d) of Figure 627 shows two simulations of $\{X_t\}$ using circulant embedding in conjunction with the Gaussian lag window estimate $\hat{S}_m^{(\text{LW})}(\cdot)$. The weights S_k are formulated in a manner similar to what we described for the preceding ocean wave example; however, here the ACVS estimate $\hat{s}_\tau^{(\text{D})}$ of Equation (188b) uses the Hanning taper of Equation (189b), but, prior to forming the estimate, we again center the time series by subtracting off its sample mean, which here is $\overline{X} \doteq 6.943$. To facilitate comparison of the simulated series with the actual time series, we have added $\overline{X}$ to both simulated series prior to plotting them in Figures 627(b) and (d). This addition results in the sample mean for the series of length $N = 2048$ in plot (b) being 7.002; it is 6.733 in plot (d).

A visual comparison of the simulated series in plots (b) and (d) of Figure 627 with the ship altitude time series in plot (a) is somewhat disconcerting. While the high-frequency fluctuations seem comparable, the low-frequency ones are not. The actual series has a prominent low-frequency component spanning the entire time series, whereas the simulated series do not have such a prominent pattern. Is there an explanation for this mismatch? Suppose, for the sake of argument, that the true SDF has a narrow peak at frequency $1/2048$ Hz, which is associated with a period of 2048 sec (the time span of the entire time series). Consider the bandwidth of the Gaussian lag window estimate $\hat{S}_m^{(\text{LW})}(\cdot)$, namely, $B_{\mathcal{U}} \doteq 0.01$ Hz (calculated as per Equation (256a)). Due to this bandwidth, the peak would appear in $\hat{S}_m^{(\text{LW})}(\cdot)$ as smeared out roughly over frequencies ranging from 0 to $1/2048$ Hz $+ B_{\mathcal{U}}/2 \doteq 0.0055$ Hz. The uppermost frequency in this range has a period corresponding to 182 sec. Thus simulations based on $\hat{S}_m^{(\text{LW})}(\cdot)$ would not exhibit a single component of a period 2048 sec, but rather a mishmash of components associated with periods of 182 sec and higher. Such a mishmash arguably would look like the low-frequency patterns in the simulated series in plots (b) and (d).

If the mishmash explanation is correct, creating simulations using an SDF estimate with a narrower bandwidth than that of $\hat{S}_m^{(\text{LW})}(\cdot)$ might generate low-frequency patterns that are more in keeping with those of the ship altitude series. Now the Gaussian lag window estimate $\hat{S}_m^{(\text{LW})}(\cdot)$ is a smoothed version of the Hanning-based direct spectral estimate $\hat{S}_m^{(\text{LW})}(\cdot)$, which appears to be leakage free and has an associated bandwidth of $B_{\mathcal{H}} \doteq 0.001$ Hz (calculated as per Equation (194)). The bandwidth $B_{\mathcal{H}}$ is a tenth of the size of $B_{\mathcal{U}}$. The smearing of the peak would now go out to frequency $1/2048 + B_{\mathcal{H}}/2 \doteq 0.0010$ Hz, which corresponds to a period of 1004 sec, i.e., about half the span of the entire time series. Plots (c) and (e) of Figure 627 shows two simulations of $\{X_t\}$ using circulant embedding in conjunction with the Hanning-based estimate $\hat{S}^{(\text{D})}(\cdot)$ shown in Figure 331(b), with the weights S_k now being dictated by Equation (612a). The random sample of size $4N = 8192$ from a standard Gaussian distribution used to create these two simulated series is the same as was used to create the lag window-based simulations in plots (b) and (d) (reusing the random sample makes the simulated series such that any differences between them are solely due to their weighing schemes). The low-frequency patterns in the simulated series in plots (c) and (e) seem to match reasonably well with that of the ship altitude time series in plot (a). This result lends credence to the hypothesis that the reason why the series based on $\hat{S}_m^{(\text{LW})}(\cdot)$ fail to replicate the ship altitude data at the very lowest frequencies is due to the bandwidth associated with $\hat{S}_m^{(\text{LW})}(\cdot)$.

To conclude, while the Gaussian lag window estimate $\hat{S}_m^{(\text{LW})}(\cdot)$ does a credible job for the most part, its bandwidth is potentially a detriment to reliable estimation of the SDF at the very lowest frequencies. Visual inspection of simulations using circulant embedding based on $\hat{S}_m^{(\text{LW})}(\cdot)$ brings out this mismatch between the time series and the SDF estimate. Caution must be exercised in using simulated series based on $\hat{S}_m^{(\text{LW})}(\cdot)$ if the purpose of the simulation includes matching the properties of the ship altitude series at the very lowest frequencies. Of course, at higher frequencies than these, there are regions of potential interest in the SDF. For example, there is a curious dip centered near 0.05 Hz that is evident in all the estimated SDFs in Figure 331. We could use simulations based on $\hat{S}_m^{(\text{LW})}(\cdot)$ to investigate hypotheses about this dip.

Atomic Clock Data

In Sections 8.9 and 9.12 we looked at estimating the innovation variance for the atomic clock fractional frequency deviates $\{Y_t\}$ using, respectively, multitaper and AR approaches (see Figure 326(b) for a plot of $\{Y_t\}$ itself). The bottom row of Table 432 lists the multitaper-

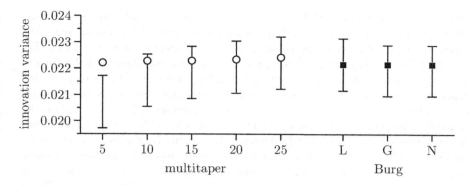

Figure 629 Multitaper-based estimates (circles) and Burg-based AR estimates (squares) of innovation variance for atomic clock fractional frequency deviates $\{Y_t\}$. The five displayed multitaper-based estimates differ in the number K of sinusoidal tapers used to create the estimate (5, 10, ..., 25). The three displayed AR estimates are identical – each is based on Burg's algorithm with $p = 42$. The vertical lines (all but one interrupted by a circle or square) depict 95% CIs for the unknown theoretical innovation variance. The CIs associated with multitapering employ simulations based on circulant embedding. The first ("L") of the three Burg-based CIs makes use of large-sample theory; the second ("G"), of simulated AR series driven by samples from a standard Gaussian distribution; and the third, also of simulated series, but now driven by normalized observed prediction errors.

based estimates $\hat{\sigma}^2_{(\mathrm{MT})}$ using sinusoidal multitapering with $K = 5$, 10, 15, 20 and 25 tapers; these are also shown by circles in Figure 629. These five estimates agree well with one another (the ratio of the largest to the smallest is 1.01). The bottom row of Table 500 lists AR estimates based on the Yule–Walker method and Burg's algorithm with p set to 42 in both cases. Both estimates agree well with the multitaper estimates; for simplicity, we focus here just on Burg. Each square in Figure 629 portrays the Burg estimate and has a common vertical displacement of $\bar{\sigma}^2_p \doteq 0.02216$. Large-sample statistical theory suggests that, based on the Burg estimate, an approximate 95% CI for the unknown theoretical innovation variance is $\bar{\sigma}^2_p \pm 1.96 \times \bar{\sigma}^2_p \sqrt{(2/N)}$, where $N = 3999$ for the fractional frequency series. The resulting CI is $[0.02119, 0.02313]$, which is portrayed by the vertical line intersecting the left-most square in Figure 629.

A large-sample statistical theory has yet to be formulated that would allow construction of similar CIs for multitapering. In lieu of suitable theory, let us explore CIs based on simulated series with statistical properties in agreement with the multitaper SDF estimate used to form a particular $\hat{\sigma}^2_{(\mathrm{MT})}$. In addition, we use simulated series to check the reasonableness of the CI based on the Burg estimate $\bar{\sigma}^2_p$ and large-sample theory.

Turning first to multitapering, let $\hat{S}^{(\mathrm{MT})}(\cdot)$ denote the sinusoidal multitaper SDF estimate associated with one of the five estimates $\hat{\sigma}^2_{(\mathrm{MT})}$ of the innovation variance shown as circles in Figure 629 (thus $\hat{S}^{(\mathrm{MT})}(\cdot)$ employs either $K = 5$, 10, 15, 20 or 25 tapers). We can use circulant embedding to simulate series whose theoretical SDF is the same as $\hat{S}^{(\mathrm{MT})}(\cdot)$. Since $\Delta_t = 1$ minute for $\{Y_t\}$, we can just ignore the sampling time and take the weights S_k needed by circulant embedding to be $\hat{S}^{(\mathrm{MT})}(\tilde{f}_k)$, $0 \leq k \leq 2N - 1$, where $\tilde{f}_k = k/(2N)$. This is in agreement with Equation (613a) if we set $d_{k'} = 1/K$ and take $\tilde{X}_t$ of Equation (611a) to be $Y_t - \overline{Y}$, i.e., the fractional frequency deviates centered about their sample mean. The weights, along with a random sample of size $4N = 15{,}996$ from a standard Gaussian distribution, go into the construction of the $\mathcal{Y}_k$ series of Equation (601c). The Fourier transform of $\{\mathcal{Y}_k\}$ yields the complex-valued series $Y_t = Y_{\Re,t} + iY_{\Im,t}$ of Equation (602) (note carefully that, while Equation (602) uses Y_t to denote this complex-valued series, it is *not* the same as the real-valued fractional frequency deviates, which unfortunately are also denoted by Y_t). The series $Y_{\Re,t}$ and $Y_{\Im,t}$, $t = 0, 1, \ldots, N - 1$, are two independent realizations of a zero mean

Gaussian process whose theoretical SDF is $\hat{S}^{(\mathrm{MT})}(\cdot)$. After centering each series by subtracting off their respective sample means, we then compute multitaper SDF estimates using the same K sinusoidal tapers that went into creating $\hat{S}^{(\mathrm{MT})}(\cdot)$. These two new multitaper SDF estimates are then fed into Equation (405b) to produce two new multitaper-based estimates of the innovation variance. We repeat this procedure a large number of times (each with a different random sample of size $4N$), ending up with N_{R} estimates of the innovation variance (we set $N_{\mathrm{R}} = 100{,}000$). We then sort these estimates, index them by $1, 2, \ldots, N_{\mathrm{R}}$ and use the values indexed by 2,500 and 97,500 as lower and upper limits for a 95% CI for the unknown theoretical innovation variance. Figure 629 depicted these limits for the five settings of K. The heights of these CIs and the one based upon large-sample theory for the Burg estimate (surrounding the left-most square) are quite similar, differing at most by 4%; however, the lower and upper CI limits are markedly different, with the $K = 5$ multitaper-based CI not even trapping the corresponding estimate! The downward displacements of the CIs relative to their corresponding estimates for small settings of K flag these CIs as questionable.

Turning now to the AR case, we consider two schemes for creating a 95% CI for the unknown theoretical innovation variance using the $p = 42$ Burg-based parameter estimates $\bar{\phi}_{p,1}$, $\bar{\phi}_{p,2}, \ldots, \bar{\phi}_{p,p}$ and $\bar{\sigma}_p^2$ for $\{Y_t\}$ after centering. For both schemes, we take the p coefficient estimates and recursively use Equation (597b) to obtain corresponding estimates $\bar{\phi}_{j,k}$ for the coefficients of the best linear predictors of orders $j = 1, 2, \ldots, p-1$. We then substitute $\bar{\phi}_{j,j}$ and $\bar{\sigma}_p^2$ for $\hat{\phi}_{j,j}$ and $\hat{\sigma}_p^2$ in Equation (618b) to obtain an estimate $\bar{\sigma}_t^2$ for σ_t^2, $0 \leq t \leq p - 1$. The first scheme is based upon Equation (597a) with the unknown $\phi_{j,k}$ and σ_t^2 terms replaced by estimates $\bar{\phi}_{j,k}$ and $\bar{\sigma}_t^2$. Using these substitutions, we use a sample $\{Z_t\}$ of length N from a standard Gaussian distribution to create a simulated series $\{X_t\}$ to be regarded as a realization of a Gaussian AR(42) process with theoretical parameters identical to the Burg parameter estimates. We then take $\{X_t\}$, center it using its sample mean and feed the centered series into Burg's algorithm to get a new estimate of the innovation variance σ_p^2. We repeat this procedure N_{R} times (each time using a different random sample of size N), ending up with $N_{\mathrm{R}} = 100{,}000$ Burg-based estimates of the innovation variance. We then sort these estimates and use those indexed by 2,500 and 97,500 as the lower and upper limits of a 95% CI for the unknown theoretical σ_p^2. Figure 629 shows these limits emanating from the middle square. The height of this CI is similar to the one to its immediate left, which was obtained from large-sample theory; however, the latter CI is symmetric about $\bar{\sigma}_p^2$, but the former is not.

The second AR scheme for creating a 95% CI is *exactly* the same as the first, with one exception: in the spirit of bootstrapping, instead of using a random sample $\{Z_t\}$ from a standard Gaussian distribution in Equation (597a), we take instead a random sample with replacement of size N from the N normalized observed prediction errors $\{\vec{Z}_t(t)\}$ formed as per Equation (618b) (we interpret $\hat{\phi}_{j,j}$ and $\hat{\sigma}_p^2$ as the Burg estimates $\bar{\phi}_{j,j}$ and $\bar{\sigma}_p^2$ obtained from $\{Y_t\}$ after centering). The resulting CI is the right-most one depicted in Figure 629. This CI is virtually identical to the one to its immediate left, which comes from the first AR scheme. The fact that the two schemes yield such similar results is presumably attributable to $\{\vec{Z}_t(t)\}$ being a good approximation to a random sample from a standard Gaussian distribution.

Thus, as noted in Section 9.12, the nonparametric multitaper-based estimates of the innovation variance and the parametric Burg-based AR estimates agree reasonably well. In addition, simulation-based CIs for the unknown theoretical innovation variance have remarkably similar heights in the nonparametric and parametric cases (and all agree well with the height of the CI determined by large-sample theory for the Burg-based estimate of σ_p^2); however, while the CIs themselves are in reasonable agreement for the $K = 20$ and 25 multitaper-based cases and the three Burg-based cases, the $K = 5$ and 10 multitaper-based CIs are markedly out of sync and intuitively unappealing.

Comments and Extensions to Section 11.6

[1] In the examples presented in this section, we looked at two different methods for simulating time series that fit in with the notion of bootstrapping: a parametric method based on the AR model (Swanepoel and Van Wyk, 1986) and a nonparametric method based on circulant embedding (Percival and Constantine, 2006). While these methods allow for bootstrapping certain statistics associated with time series, they are certainly not the only methods that do so. Comprehensive reviews of the literature on bootstrapping for time series are Politis (2003) and Kreiss and Paparoditis (2011). The books by Davison and Hinkley (1997) and Lahiri (2003b) discuss both parametric and nonparametric methods. A critique of some of the nonparametric methods discussed in these references is an over-reliance on asymptotic properties of the periodogram, particularly regarding unbiasedness and independence across the grid of Fourier frequencies (as we have emphasized throughout this book, these are dicey assumptions in some practical applications). Use of circulant embedding with nonparametric SDF estimates other than the periodogram can potentially overcome the problems associated with the periodogram. Circulant embedding worked credibly in some – but not all – of the examples we presented (see the discussion surrounding Figure 629), illustrating that more work is needed to establish its limitations.

11.7 Comments on Simulation of Non-Gaussian Time Series

Simulation of a non-Gaussian time series having a specified ACVS or SDF structure is generally considerably more involved than the Gaussian case discussed extensively in this chapter. One approach for ARMA processes (Davies et al., 1980) is to generate an IID noise sequence having a distribution such that, when it is subjected to the filter required to obtain the specified ARMA form, the marginal distribution of the filter output has the desired first four moments. An alternative is to start with IID Gaussian noise and pass it through both a linear filter and a nonlinear transform; the correct design of both these components yields the desired second-order characteristics (ACVS, SDF) and marginal distribution. Note that in the first approach the aim is to produce a *constrained* marginal distribution having specified first four moments, and in the second case, a *full* marginal distribution, but in both cases only first-order PDF modeling is attempted.

For the first approach, the desire is to simulate a zero mean non-Gaussian, causal, ARMA process. Let $\{\epsilon_t\}$ denotes a zero mean IID sequence with variance σ_ϵ^2 (a stronger assumption than white noise, but IID implies that the RVs are uncorrelated). We can generate such a sequence in practice by simply sampling independent RVs from a suitable zero mean distribution and rescaling if necessary. Then for the causal ARMA process there exists a unique set of weights $\{\psi_j\}$ such that

$$X_t = \sum_{j=0}^{\infty} \psi_j \epsilon_{t-j} \text{ with } \psi_0 \stackrel{\text{def}}{=} 1$$

(this is Equation (600c)). The IID sequence is passed through the linear time-invariant (LTI) filter $\{\psi_j\}$ to produce the zero mean linear process $\{X_t\}$. Since $\{\epsilon_t\}$ is strictly stationary, the output $\{X_t\}$ of the LTI filter is also strictly stationary (Papoulis and Pillai, 2002, section 9.2). In particular, each RV ϵ_t in the sequence $\{\epsilon_t\}$ has the same marginal distribution, and likewise each X_t in $\{X_t\}$ has the same marginal distribution, which is in general different from that of ϵ_t and is intended to be non-Gaussian.

Now $\{\epsilon_t\}$ is IID with variance σ_ϵ^2, so $\sigma^2 \stackrel{\text{def}}{=} \text{var}\{X_t\} = \sigma_\epsilon^2 \sum_{j=0}^{\infty} \psi_j^2$. For the zero mean RV X_t, the moment coefficient of skewness $\sqrt{\delta_1(X_t)}$ and moment coefficient of kurtosis $\delta_2(X_t)$ are given by

$$\sqrt{\delta_1(X_t)} \stackrel{\text{def}}{=} \frac{E\{X_t^3\}}{(E\{X_t^2\})^{3/2}} = \frac{E\{X_t^3\}}{\sigma^3} \text{ and } \delta_2(X_t) \stackrel{\text{def}}{=} \frac{E\{X_t^4\}}{(E\{X_t^2\})^2} = \frac{E\{X_t^4\}}{\sigma^4},$$

which leads to

$$\sqrt{\delta_1(X_t)} = \frac{\sum_{j=0}^{\infty} \psi_j^3}{(\sum_{j=0}^{\infty} \psi_j^2)^{3/2}} \sqrt{\delta_1(\epsilon_t)} \qquad (632a)$$

and

$$\delta_2(X_t) = \frac{\sum_{j=0}^{\infty} \psi_j^4}{(\sum_{j=0}^{\infty} \psi_j^2)^2} \delta_2(\epsilon_t) + 6 \frac{\sum_{j=0}^{\infty} \sum_{i>j}^{\infty} \psi_i^2 \psi_j^2}{(\sum_{j=0}^{\infty} \psi_j^2)^2} \qquad (632b)$$

(Davies et al., 1980). Since X_t has zero mean, its first four cumulants are given by $\kappa_1(X_t) = 0$, $\kappa_2(X_t) = E\{X_t^2\} = \sigma^2$, $\kappa_3(X_t) = E\{X_t^3\}$ and $\kappa_4(X_t) = E\{X_t^4\} - 3\sigma^4$.

▷ **Exercise [632a]** Show that

$$\delta_2(X_t) - 3 = \frac{\sum_{j=0}^{\infty} \psi_j^4}{(\sum_{j=0}^{\infty} \psi_j^2)^2} [\delta_2(\epsilon_t) - 3] \text{ and also } \frac{\kappa_4(X_t)}{\sigma^4} = \frac{\sum_{j=0}^{\infty} \psi_j^4}{(\sum_{j=0}^{\infty} \psi_j^2)^2} \frac{\kappa_4(\epsilon_t)}{\sigma_\epsilon^4}. \quad (632c) \triangleleft$$

Hence the scaled fourth-order cumulant of X_t is equal to the scaled fourth-order cumulant of ϵ_t multiplied by the factor $\sum_{j=0}^{\infty} \psi_j^4 / (\sum_{j=0}^{\infty} \psi_j^2)^2$, a quantity called the kurtosis norm of the filter $\{\psi_j\}$ in Longbottom et al. (1988). Here are some notes:

[1] If either $\sqrt{\delta_1(\epsilon_t)} = 0$ or $\sum_{j=0}^{\infty} \psi_j^3 / (\sum_{j=0}^{\infty} \psi_j^2)^{3/2} = 0$, we see from Equation (632a) that $\sqrt{\delta_1(X_t)} = 0$. So zero-skewness noise gives rise to a zero-skewness process, but this is not the only way that the latter can occur.

[2] Equation (632c) shows that $\delta_2(X_t) = 3$ if and only if $\delta_2(\epsilon_t) = 3$. So a kurtosis of 3 for ϵ_t is a necessary and sufficient condition for a kurtosis of 3 for X_t.

[3] Since $(\sum_{j=0}^{\infty} \psi_j^2)^2 \geq \sum_{j=0}^{\infty} \psi_j^4$, the kurtosis norm of the filter is bounded by unity, and so the scaled fourth cumulant of X_t never exceeds the scaled fourth cumulant of ϵ_t (the solution to Exercise [632a] makes this inequality obvious, and shows the inequality is strict unless all $\psi_j = 0$ for $j > 0$, i.e., unless $\{X_t\}$ is IID noise).

Properties [1] and [2] are consistent with filtered independent Gaussian noise being itself Gaussian. Property [3] tells us that, in the sense of the scaled fourth cumulant, the distribution of the filtered IID noise X_t is closer to Gaussian than that of the IID noise itself (recall that $\kappa_4(X_t)$ would be zero if X_t were Gaussian).

As an example, in the case of an ARMA(1,1) process, $X_t = \phi X_{t-1} + \epsilon_t + \vartheta \epsilon_{t-1}$, the forms of the sums in Equations (632a) and (632c) can be readily evaluated analytically, leading to the following results.

▷ **Exercise [632b]** Show that, for an ARMA(1,1) process,

$$\text{var}\{X_t\} = \sigma^2 = \sigma_\epsilon^2 [1 - \phi^2 + (\phi + \vartheta)^2]/(1 - \phi^2), \qquad (632d)$$

while Equation (632a) becomes

$$\sqrt{\delta_1(X_t)} = \frac{[1 - \phi^3 + (\phi + \vartheta)^3](1 - \phi^2)^{3/2}}{[1 - \phi^2 + (\phi + \vartheta)^2]^{3/2}(1 - \phi^3)} \sqrt{\delta_1(\epsilon_t)} \qquad (632e)$$

and Equation (632c) leads to

$$\delta_2(X_t) - 3 = \frac{[1 - \phi^4 + (\phi + \vartheta)^4](1 - \phi^2)}{[1 - \phi^2 + (\phi + \vartheta)^2]^2(1 + \phi^2)} [\delta_2(\epsilon_t) - 3]. \qquad (632f) \triangleleft$$

11.7 Comments on Simulation of Non-Gaussian Time Series

If we wish to simulate a sequence $\{X_t\}$ having mean zero, variance σ^2, skewness $\sqrt{\delta_1(X_t)}$ and kurtosis $\delta_2(X_t)$ and having a causal ARMA(1,1) structure, then we need to generate IID noise $\{\epsilon_t\}$ with mean zero, variance satisfying Equation (632d) and skewness and kurtosis satisfying Equations (632e) and (632f), respectively. This can be accomplished by generating an IID sequence from a distribution with these specified moment constraints. The Johnson system of distributions covers many combinations of skewness and kurtosis; see, e.g., Figure 2 of Johnson (1949). Hill (1976) and Hill et al. (1976) provide FORTRAN routines to calculate the Johnson parameters corresponding to a required mean, variance, skewness and kurtosis, which can be used to simulate Johnson RVs from Gaussian RVs (translations are also available in R).

Exercise [11.14] establishes that, in the ARMA(1,1) case, $|\sqrt{\delta_1(X_t)}| < |\sqrt{\delta_1(\epsilon_t)}|$ under the assumption $\phi + \vartheta \neq 0$; i.e., the size of the skewness of X_t is less than the size of the skewness of ϵ_t (when $\phi + \vartheta = 0$, the ARMA(1,1) process degenerates into IID noise; we then have $X_t = \epsilon_t$, and Equation (632e) says $|\sqrt{\delta_1(X_t)}| = |\sqrt{\delta_1(\epsilon_t)}|$). Thus, since the skewness of the Gaussian distribution is zero, if we only consider skewness, the distribution of the filtered IID noise X_t is closer to Gaussian than that of the IID noise itself (we reached a similar conclusion when we considered just the scaled fourth cumulant). The relationship between the PDFs for X_t and ϵ_t in cases other than ARMA(1,1) has been studied theoretically (Mallows, 1967; Granger, 1979).

Walden (1993) uses this general approach to synthesize sequences of primary reflection coefficients in the layered-earth model. Drilled wells at three locations enabled statistical model structures to be derived: ARMA(1,1) models with marginal distributions for $\{X_t\}$ having zero mean, zero skewness and kurtosis values of either 3.8, 4.7 or 8.1. Instead of the Johnson class, he uses the generalized Gaussian (GG) distribution as a model for the IID noise. The zero mean and symmetric GG has a PDF for ϵ_t given by

$$f(\epsilon) = \frac{\alpha A(\alpha, \sigma_\epsilon)}{2\Gamma(1/\alpha)} \exp\left(-[A(\alpha, \sigma_\epsilon)|\epsilon|]^\alpha\right), \quad |\epsilon| < \infty,$$

with $A(\alpha, \sigma_\epsilon) \stackrel{\text{def}}{=} \{\Gamma(3/\alpha)/[\sigma_\epsilon^2 \Gamma(1/\alpha)]\}^{1/2}$, where α is the shape parameter, and $\Gamma(\cdot)$ is the gamma function. The shape parameter determines the peakedness at the origin and long-tailedness of the distribution relative to the Gaussian. The Laplace (double-sided exponential) distribution corresponds to $\alpha = 1$; the Gaussian, to $\alpha = 2$; and the uniform distribution is obtained as $\alpha \to \infty$. The zero mean and symmetric GG has

$$E\{\epsilon_t^{2n}\} = [\Gamma^{n-1}(1/\alpha)\Gamma([2n+1]/\alpha)\sigma_\epsilon^{2n}]/\Gamma^n(3/\alpha),$$

so that the value of α that gives the required kurtosis can be found. Simulation of GG variates can be accomplished using R code such as pgnorm. In Walden (1993) values of the four parameters input to the simulation, namely, variance and kurtosis, and ARMA parameters ϕ and ϑ, agree well with the corresponding estimated values from the simulated sequences.

We now turn to an alternative approach for the simulation of a non-Gaussian time series having a specified ACVS or SDF structure. This methodology is discussed in Gujar and Kavanagh (1968), Liu and Munson (1982) and Sondhi (1983). The approach can be illustrated thus:

$$\{\epsilon_t\} \longrightarrow \boxed{G(\cdot)} \longrightarrow \{X_t\} \longrightarrow \boxed{\mathcal{Z}(\cdot)} \longrightarrow \{Y_t\} \tag{633}$$

In the above $\{\epsilon_t\}$ is now taken to be a zero mean, unit variance, Gaussian IID sequence. The Gaussian IID sequence is passed through a LTI filter with transfer function $G(\cdot)$ to produce a linear Gaussian process $\{X_t\}$ with zero mean and ACVS $\{s_{X,\tau}\}$. The filter coefficients are normalized so that the output has a unit variance, i.e., $s_{X,0} = 1$. The sequence $\{X_t\}$ is strictly stationary with mean zero and variance unity and is jointly Gaussian. This linear process is then subjected to a zero-memory (instantaneous) nonlinearity (ZMNL) denoted $\mathcal{Z}(\cdot)$ which is chosen so that the output process $\{\mathcal{Z}(X_t)\} \stackrel{\text{def}}{=} \{Y_t\}$ has a desired ACVS and marginal non-Gaussian PDF. Since $\{X_t\}$ is strictly stationary, $\{Y_t\}$, the output of the ZMNL, is itself strictly stationary (Papoulis and Pillai, 2002, section 9.2). Each RV X_t in the process $\{X_t\}$ has the same Gaussian marginal distribution, and each Y_t in $\{Y_t\}$ has the same non-Gaussian marginal distribution. Denote the Gaussian distribution function as usual by $\Phi(\cdot)$, and the distribution function of Y_t by $F_Y(\cdot)$. By the probability integral transform, the RV U resulting from transformation of the RVs X_t and Y_t via $U = \Phi(X_t)$ and $U = F_Y(Y_t)$ is uniformly distributed on $[0, 1]$. Then

$$\mathcal{Z}(X_t) \stackrel{\text{d}}{=} Y_t \stackrel{\text{d}}{=} F_Y^{-1}(U) \stackrel{\text{d}}{=} F_Y^{-1}(\Phi(X_t)).$$

This result shows how the form of the ZMNL is determined by the choice of $F_Y(\cdot)$. As an example suppose it is desired that the output $\{Y_t\}$ should have a marginal uniform distribution with mean zero and variance of unity, with PDF $f(y) = 1/(2\sqrt{3})$, $|y| \leq \sqrt{3}$. Then

$$F_Y(y) = \int_{-\sqrt{3}}^{y} [1/(2\sqrt{3})] \, \mathrm{d}z = \frac{y + \sqrt{3}}{2\sqrt{3}},$$

so $U \stackrel{\text{d}}{=} (Y_t + \sqrt{3})/(2\sqrt{3})$ and therefore

$$\mathcal{Z}(X_t) \stackrel{\text{d}}{=} Y_t \stackrel{\text{d}}{=} [2U - 1]\sqrt{3} \stackrel{\text{d}}{=} [2\Phi(X_t) - 1]\sqrt{3}. \tag{634a}$$

Having derived the form of the nonlinearity, the next step is to find the relationship between $s_{X,\tau}$ and $s_{Y,\tau}$. Since $X_{t+\tau}$ and X_t both have mean zero, $s_{X,\tau} = E\{X_{t+\tau} X_t\}$, and since both RVs also have a variance of unity, $s_{X,\tau} = \rho_{X,\tau}$; i.e., the ACVS and ACS are identical for $\{X_t\}$. The bivariate standard Gaussian PDF for $(X_{t+\tau}, X_t)$ is

$$f(v, w; \rho_{X,\tau}) = \frac{1}{2\pi(1 - \rho_{X,\tau}^2)^{1/2}} \exp\left(-\frac{v^2 + w^2 - 2\rho_{X,\tau} vw}{2(1 - \rho_{X,\tau}^2)} \right).$$

If we choose to standardize $\{Y_t\}$ such that $s_{Y,0} = 1$ so that the ACVS and ACS are identical for $\{Y_t\}$ also, then $s_{Y,\tau} = \rho_{Y,\tau}$ and

$$\rho_{Y,\tau} \stackrel{\text{def}}{=} E\{Y_{t+\tau} Y_t\} - E\{Y_{t+\tau}\}E\{Y_t\} = E\{\mathcal{Z}(X_{t+\tau})\mathcal{Z}(X_t)\} - E\{\mathcal{Z}(X_{t+\tau})\}E\{\mathcal{Z}(X_t)\}.$$

Sondhi (1983) gives three examples where $\rho_{Y,\tau}$ can be found analytically. For Y_t having the uniform distribution with mean zero and variance of unity, $\mathcal{Z}(X_t)$ is given by Equation (634a), and Sondhi obtains

$$\rho_{Y,\tau} = (6/\pi)\sin^{-1}(\rho_{X,\tau}/2). \tag{634b}$$

A general approach for finding $\rho_{Y,\tau}$, developed in Liu and Munson (1982) and Sondhi (1983), makes use of Hermite polynomials. First,

$$E\{\mathcal{Z}(X_t)\} = \frac{1}{(2\pi)^{1/2}} \int_{-\infty}^{\infty} \mathcal{Z}(v) e^{-v^2/2} \, \mathrm{d}v.$$

11.7 Comments on Simulation of Non-Gaussian Time Series

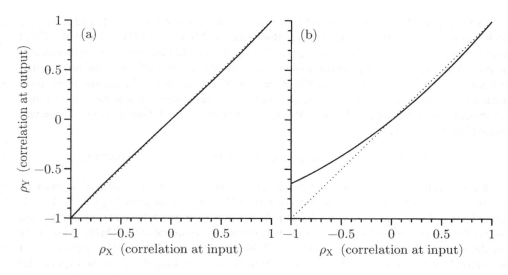

Figure 635 The relationship between the autocorrelation $\rho_{Y,\tau}$ at the output of the ZMNL and $\rho_{X,\tau}$ at the input. The dotted line marks equality. For (a) the distribution for Y_t is the uniform with mean zero and variance unity. The exact mapping given by Equation (634b) is shown by a dash-dot line, and that given by Equation (635d), using coefficients $\{d_k\}$ up to $k = 7$, is shown by a solid line; the two are indistinguishable. For (b) the distribution for Y_t is the exponential with variance unity. The mapping given by Equation (635d), using coefficients $\{d_k\}$ up to $k = 7$, is again shown by a solid line. Since in (a) the distribution is symmetric for Y_t, the relationship shows the expected odd symmetry. In (b) the distribution for Y_t is asymmetric, and the relationship does not show any recognizable symmetries.

Then $\mathcal{Z}(v)$ is expanded in terms of Hermite polynomials $\mathrm{He}_k(v)$ with coefficients $\{d_k\}$,

$$\mathcal{Z}(v) = \sum_{k=0}^{\infty} d_k \frac{\mathrm{He}_k(v)}{\sqrt{(k!)}}. \tag{635a}$$

The orthogonality properties of Hermite polynomials say that $E\{\mathcal{Z}(X_t)\} = d_0$, leading to

$$E\{\mathcal{Z}(X_{t+\tau})\}E\{\mathcal{Z}(X_t)\} = d_0^2.$$

Next,

$$E\{\mathcal{Z}(X_{t+\tau})\mathcal{Z}(X_t)\} = \int_{-\infty}^{\infty} \int_{-\infty}^{\infty} \mathcal{Z}(v)\mathcal{Z}(w) f(v,w;\rho_{X,\tau})\,dv\,dw. \tag{635b}$$

But the bivariate standard Gaussian PDF can also be expanded in terms of Hermite polynomials using Mehler's expansion (Barrett and Lampard, 1955):

$$f(v,w;\rho_{X,\tau}) = \frac{1}{2\pi} e^{-(v^2+w^2)/2} \sum_{k=0}^{\infty} \rho_{X,\tau} \frac{\mathrm{He}_k(v)\mathrm{He}_k(w)}{k!}. \tag{635c}$$

Equations (635a), (635b) and (635c) give $E\{\mathcal{Z}(X_{t+\tau})\mathcal{Z}(X_t)\} = \sum_{k=0}^{\infty} d_k^2 \rho_{X,\tau}^k$ (Wise et al., 1977). Finally,

$$\rho_{Y,\tau} = \sum_{k=0}^{\infty} d_k^2 \rho_\tau^k - d_0^2 = \sum_{k=1}^{\infty} d_k^2 \rho_{X,\tau}^k, \tag{635d}$$

and by setting $\tau = 0$ we see that $\sum_{k=1}^{\infty} d_k^2 = 1$.

Hence, given $\{\rho_{Y,\tau}\}$, Equation (635d) can be used to solve for $\{\rho_{X,\tau}\}$, and then $\{\epsilon_t\}$ can be correctly colored to produce $\{X_t\}$. Both Liu and Munson (1982) and Sondhi (1983) observe that, if the PDF of Y_t is symmetric, $\mathcal{Z}(\cdot)$ will be an odd function of its argument, and the even coefficients in $\{d_k\}$ vanish, making $\rho_{Y,\tau}$ an odd function of $\rho_{X,\tau}$; the relationship between $\rho_{Y,\tau}$ and $\rho_{X,\tau}$ is then invertible. Returning to the example of Y_t having the uniform distribution with mean zero and variance of unity, the invertibility can also be seen in Equation (634b). For this example, Liu and Munson (1982) obtain the following first few nonzero squared Hermite coefficients:

$$d_1^2 \doteq 0.9550627; \quad d_3^2 \doteq 0.0397943; \quad d_5^2 \doteq 0.0044768; \quad d_7^2 \doteq 0.0006662.$$

Using these values, we can compare the exact mapping between $\rho_{X,\tau}$ and $\rho_{Y,\tau}$ given by Equation (634b), with that given by Equation (635d), using coefficients $\{d_k\}$ up to $k=7$. Figure 635(a) shows that there is no discernible difference between these two mappings. We also see that the choice to obtain a uniform distribution for the output of the ZMNL causes only a moderate distortion between $\rho_{X,\tau}$ and $\rho_{Y,\tau}$. Table 1 of Liu and Munson (1982) gives sets of (squared) Hermite coefficients for other distributions for Y_t. Taking those for the exponential distribution with variance unity, a highly asymmetric distribution, the relationship between $\rho_{X,\tau}$ and $\rho_{Y,\tau}$ can again be found from Equation (635d) and is given in Figure 635(b), which indicates a more substantial distortion between $\rho_{X,\tau}$ and $\rho_{Y,\tau}$. Our use of the term "distortion" is in line with Johnson (1994) calling this approach the *correlation distortion method*.

Liu and Munson (1982) discuss the condition for Equation (635d) to be solved for $\rho_{X,\tau}$ for arbitrary distributions for Y_t. They also point out that under some circumstances the derived $\{\rho_{X,\tau}\}$ may not be positive semidefinite as required (see Equation (28b)).

Suppose a solution to Equation (635d) has been found. Express it as $\hat{\rho}_{X,\tau}(\rho_{Y,\tau})$. (For the example of Y_t having the uniform distribution with mean zero and variance of unity, Equation (634b) gives the exact result $\hat{\rho}_{X,\tau}(\rho_{Y,\tau}) = 2\sin(\pi\rho_{Y,\tau}/6)$.) The sequence $\{\hat{\rho}_{X,\tau}(\rho_{Y,\tau})\}$ is symmetric and its Fourier transform $\hat{S}_X(f)$, say, is real. For many choices of $\{\rho_{Y,\tau}\}$, it happens that $\hat{S}_X(f) \geq 0$ for all f, so that the required filter $\{g_u\}$ can be found by finding a transfer function $G(\cdot)$ such that $|G(f)|^2 = \hat{S}_X(f)$ (the procedure of finding a suitable "square root" for $\hat{S}_X(\cdot)$ is known as spectral factorization). If $\hat{S}_X(f) < 0$ at some frequency f, Sondhi (1983) suggests proceeding with $(\hat{S}_X(f))_+$ instead; practical algorithmic steps are provided. Liu and Munson (1982, p. 976) show that such an approach is often nearly optimal in minimizing distortion between the realized and desired output sequence $\{\rho_{Y,\tau}\}$.

Serroukh et al. (2000) use the simulation scheme in Equation (633) with three different non-Gaussian distributions for Y_t and a variety of ACS sequences $\{\rho_{Y,\tau}\}$ to test the theory associated with the wavelet variance estimator. Good agreement between simulation results and the expected theory was obtained.

One drawback with the scheme in Equation (633) is that the marginal PDF and ACS/SDF are linked together, so that changing one requires a change to the other. A novel method for both representing and generating realizations of a stationary non-Gaussian process is described by Kay (2010). Here specifications of the PDF and ACS/SDF are uncoupled, allowing independent specification of the first-order PDF of the random process and the autocorrelation structure; however, the PDF must be symmetric and infinitely divisible (this means that the desired PDF should correspond to that of a sum of an arbitrary number of IID random variables). By way of contrast, the scheme in Equation (633), when it can be implemented, has no such restrictions on the marginal PDF. Finally, Picinbono (2010) shows that, by using a structure with appropriate random coefficients, it is possible to construct time series with ARMA second-order properties, while the marginal PDF is symmetric, but otherwise unconstrained, and can be specified in advance.

11.8 Summary of Simulation of Time Series

Suppose we have either (i) a theoretical stationary process $\{X_t\}$ that is specified to some degree or (ii) a nonparametric or parametric estimate $\hat{S}(\cdot)$ of the spectral density function (SDF) based upon a realization of a portion $X_0, X_1, \ldots, X_{N-1}$ of a stationary process whose theoretical SDF is unknown. Suppose that, in case (i), we are interested in computer-generated artificial series (simulations) that can be regarded as realizations from $\{X_t\}$ and, in case (ii), in simulations whose statistical properties are in agreement with those of the estimate $\hat{S}(\cdot)$. There are a number of ways in which we can accomplish our goals, but these depend on exactly what we have to work with.

Turning first to case (i), suppose that $\{X_t\}$ is a zero mean autoregressive moving average (ARMA) process of order (p, q) (see Equation (35b)). We presume that its $p + q + 1$ parameters $\phi_{p,j}$, $\theta_{q,j}$ and σ_ϵ^2 are known so that $\{X_t\}$ is fully specified. If $\{X_t\}$ is in fact just white noise ($p = 0$, $q = 0$), it is an easy matter to generate simulations using computer routines purported to give random samples from a particular distribution (Gaussian is a popular choice, but white noise can be simulated easily with certain non-Gaussian distributions, e.g., uniform or exponential). If $\{X_t\}$ is a moving average (MA) process ($p = 0$, $q \geq 1$), the fact that it is a finite linear combination of white noise (not necessarily Gaussian) points to easy methods for generating simulations (see the discussion following Equation (594)). If $\{X_t\}$ is an autoregressive (AR) process ($p \geq 1$, $q = 0$), we face a challenge. By definition X_t involves a linear combination of $X_{t-1}, \ldots, X_{t-p}$, so simulation of X_t would seem to require a simulation of X_{t-p}, which in turn needs a simulation of X_{t-2p} and so forth ad infinitum. This startup problem can be elegantly circumvented by generating stationary initial conditions (Kay, 1981b; see the discussion surrounding Equation (597a)). Using these conditions, we can simulate an AR series of length N based upon a random sample of size N, typically from a standard Gaussian distribution. If $\{X_t\}$ is an ARMA process with $p \geq 1$ and $q \geq 1$, we can merge the techniques for simulating AR and MA processes to come up with simulations of $\{X_t\}$ (see the discussion following Exercise [598]). If, on the other hand, the process of interest is a fully specified harmonic process, its formulations in terms of either random amplitudes (Equation (598a)) or random phases (Equation (598b)) suggest easy pathways to generate simulations.

Consider now another formulation of case (i), in which $\{X_t\}$ is specified either in terms of its autocovariance sequence (ACVS) $\{s_\tau\}$ or its SDF $S(\cdot)$. If $s_0, s_1, \ldots, s_N$ are known, we can use circulant embedding to simulate $X_0, X_1, \ldots, X_{N-1}$ subject to a set of weights S_k being nonnegative. The weights are obtained from the Fourier transform of a circularized version of the $N + 1$ known elements of the ACVS (see Equation (601a)). Unfortunately there are some stationary processes and some sample sizes N for which some of the weights are negative, which formally renders circulant embedding inapplicable (however, setting the offending weights to zero permits simulation of time series whose statistical properties are approximately consistent with the portion of the ACVS we know, with the degree of approximation being quantifiable). When the weights are nonnegative, circulant embedding is a computationally efficient method for generating exact simulations (this method is associated with a harmonic process, and here "exact" means that the ACVS for the harmonic process agrees perfectly with s_τ at the known lags). Its efficiency is due to its use of a Fourier transform involving the weights and a random sample of size $2N$, typically from a standard Gaussian distribution. (If circulant embedding cannot provide exact simulations because of negative weights, an alternative approach involves a Cholesky decomposition of a covariance matrix containing the known ACVS values, but this approach is computationally unappealing and is sometimes subject to numerical instabilities.)

If, on the other hand, $\{X_t\}$ is specified by its SDF $S(\cdot)$ rather than its ACVS, the Gaussian spectral synthesis method (GSSM) is worth exploring. Circulant embedding and GSSM use weights and random samples in exactly the same way to generate simulations. The salient difference between the two methods is that GSSM uses weights extracted directly from $S(\cdot)$, which are guaranteed to be nonnegative. Like circulant embedding, simulation via GSSM is computationally appealing; however, in general it does *not* yield exact simulations. Better approximations to exact simulations can be achieved at the expensive of larger random samples and additional time to generate each simulation, which might compromise its appeal.

Turning now to case (ii), suppose that $\hat{S}(\cdot)$ is a nonparametric SDF estimate based upon a realization of $X_0, X_1, \ldots, X_{N-1}$. If $\hat{S}(\cdot)$ is either (a) a direct spectral estimate, (b) a lag window estimate that uses a lag window such as the Parzen or Gaussian, (c) a basic or weighted multitaper estimate or

(d) a WOSA estimate, we can *always* use circulant embedding to generate simulations whose statistical properties are in accordance with $\hat{S}(\cdot)$. The weights required for circulant embedding are gotten by evaluating $\hat{S}(\cdot)$ over a grid of frequencies twice as finely spaced as that of the Fourier frequencies (an adjustment might be needed to take into account the sampling interval Δ_t; see Section 11.4 for details). On the other hand, if $\hat{S}(\cdot)$ is a parametric AR SDF estimate and if the estimated AR coefficients used to formulate $\hat{S}(\cdot)$ correspond to those of a causal AR process (the Yule–Walker method guarantees this, as does Burg's algorithm in practice), then we can generate simulations using the same procedure used in case (i) (we just use the estimated AR parameters as substitutes for the parameters that would be available for a fully specified AR process). This simulation procedure uses a sample of size N drawn typically from a standard Gaussian distribution. Motivated by bootstrapping, we can alternatively use samples drawn (with replacement) from normalized observed prediction errors associated with the time series behind $\hat{S}(\cdot)$ (Swanepoel and Van Wyk, 1986; see the discussion leading up to Equation (618b) for details). If the estimated AR coefficients do not correspond to those of a causal AR process, we can revert to GSSM to generate simulations that are approximately consistent with $\hat{S}(\cdot)$.

Finally, we note that the simulation techniques that address cases (i) and (ii) typically work best when our goal is to generate realizations from a Gaussian stationary process (if we deviate from this goal, in general we no longer have control over the marginal distribution of the simulations). If we want to simultaneously control the ACVS and the marginal probability density function (PDF) of the simulated series, we face challenges that can be addressed in two ways. First, under certain conditions, we can generate an independent and identically distributed (IID) noise sequence and manipulate it to produce a simulated series that both obeys a desired ARMA structure and matches at least certain aspects of a desired PDF (mean, variance, skewness and kurtosis). Second, again under certain conditions, we can filter IID Gaussian noise and then subject it to a zero-memory nonlinear (ZMNL) transform to yield a simulated series with a tractable ACVS and a tractable (and fully specified) PDF (details are in Section 11.7).

11.9 Exercises

[11.1] Consider a Gaussian AR(4) process defined by Equation (35a), namely,

$$X_t = 2.7607 X_{t-1} - 3.8106 X_{t-2} + 2.6535 X_{t-3} - 0.9238 X_{t-4} + \epsilon_t = \sum_{j=1}^{4} \phi_{4,j} X_{t-j} + \epsilon_t,$$

where $\{\epsilon_t\}$ is Gaussian white noise with zero mean and variance $\sigma_4^2 = 0.002$ (hence $\sigma_4 \doteq 0.04472136$). Assuming for convenience that $N \geq 5$, an adaptation of Equation (597a) for the case $p = 4$ yields

$$X_0 = \sigma_0 Z_0$$
$$X_1 = \phi_{1,1} X_0 + \sigma_1 Z_1$$
$$X_2 = \sum_{j=1}^{2} \phi_{2,j} X_{2-j} + \sigma_2 Z_2$$
$$X_3 = \sum_{j=1}^{3} \phi_{3,j} X_{3-j} + \sigma_3 Z_3$$
$$X_t = \sum_{j=1}^{4} \phi_{4,j} X_{t-j} + \sigma_4 Z_t, \quad t = 4, \ldots, N-1,$$

which gives us a recipe for manipulating realizations of $Z_0, Z_1, \ldots, Z_{N-1}$ (a random sample of Gaussian RVs with zero means and unit variances) to produce what can be regarded as a realization of $X_0, X_1, \ldots, X_{N-1}$. Show that, to make this scheme work for the AR(4) process of interest here, we need to set

$$\sigma_0 \doteq 1.23427492$$
$$\sigma_1 \doteq 0.86104315, \quad \phi_{1,1} \doteq 0.71647721$$
$$\sigma_2 \doteq 0.16440913, \quad \phi_{2,1} \doteq 1.41977221, \quad \phi_{2,2} \doteq -0.98160136$$
$$\sigma_3 \doteq 0.11680396, \quad \phi_{3,1} \doteq 2.11057498, \quad \phi_{3,2} \doteq -1.98076723, \quad \phi_{3,3} \doteq 0.70375083.$$

[11.2] (a) In the last part of C&E [2] for Section 11.1, we discussed a procedure described in McLeod and Hipel (1978) for generating realizations of a zero mean Gaussian ARMA(p,q) process $\{X_t\}$ with stationary initial conditions. Work out the details of their procedure for an ARMA(1,1) process by determining the entries of the covariance matrix Σ_{p+q} of Equation (600a). As noted in the C&E, generating the stationary initial conditions requires forming the Cholesky decomposition of Σ_{p+q}, i.e., finding a lower triangular matrix C_{p+q} such that $C_{p+q}C_{p+q}^T = \Sigma_{p+q}$. Determine C_{p+q} and describe how to use it to generate the required stationary initial conditions. Finally state how these conditions can be used in conjunction with the defining equation for an ARMA(1,1) process to generate a realization of X_0, X_1, ..., X_{N-1}. Hint: assuming $a \neq 0$, the Cholesky decomposition of a symmetric positive semidefinite matrix of the form

$$A = \begin{bmatrix} a & b \\ b & c \end{bmatrix} \text{ is } A = \begin{bmatrix} \sqrt{a} & 0 \\ \frac{b}{\sqrt{a}} & \sqrt{(c-\frac{b^2}{a})} \end{bmatrix} \begin{bmatrix} \sqrt{a} & \frac{b}{\sqrt{a}} \\ 0 & \sqrt{(c-\frac{b^2}{a})} \end{bmatrix}.$$

(b) Following Exercise [598], we discussed a different procedure – as described in Kay (1981b) – for generating realizations of an ARMA(p,q) process with stationary initial conditions. Work out the details of this procedure for the ARMA(1,1) case. Comment briefly on how it compares with the procedure formulated in part (a).

(c) As one check to verify that the procedures of parts (a) and (b) yield equivalent simulations, show that expressions for var$\{X_0\}$ that can be deduced from the two procedures are in agreement.

[11.3] Equation (597a) describes a procedure advocated by Kay (1981b) for simulating a portion X_0, X_1, ..., X_{N-1} of a zero mean Gaussian AR(p) process. The procedure involves unraveling prediction errors associated with best linear predictors, with the coefficients for the predictors being provided by the Levinson–Durbin recursions described in Section 9.4. No matter what the model order p is, each simulated series requires a random sample of length N from a standard Gaussian distribution. By contrast, the procedures described in Section 11.1 for simulating a portion of length N from a zero mean MA(q) process require a random sample of length $N+q$. Focusing for simplicity on the MA(1) case, explore an approach similar to Kay's to simulate a portion X_0, X_1, ..., X_{N-1} of a zero mean Gaussian MA(1) process, with the goal being to use just N realizations of a standard Gaussian RV as opposed to $N+1$ (the discussion at the beginning of Section 9.4 notes that the Levinson–Durbin recursions are not restricted to AR(p) processes, but rather can be used to form the coefficients for the best linear predictors of other stationary processes). With this same goal in mind, explore also adaptation of the approach of McLeod and Hipel (1978) that is discussed in C&E [2] for Section 11.1. Comment briefly on how the two adapted approaches compare with the procedures described in Section 11.1 for simulating an MA(1) process.

[11.4] Consider an AR(1) process $\{X_t\}$ dictated by the equation

$$X_t = \phi_{1,1} X_{t-1} + \epsilon_t,$$

where $0 < |\phi_{1,1}| < 1$, and $\{\epsilon_t\}$ is Gaussian white noise with mean zero and variance σ_1^2 (we take the sampling interval Δ_t to be unity). As noted by Equation (507a), the ACVS for this process is given by $s_\tau = \phi_{1,1}^{|\tau|} \sigma_1^2 / (1 - \phi_{1,1}^2)$.

(a) Given a realization of a portion Z_0, Z_1, ..., Z_{N-1} of Gaussian white noise with zero mean and unit variance, show how to create a realization of X_0, X_1, ..., X_{N-1} using the exact method described by Equation (597a) (i.e., develop a recipe similar to the one given in Exercise [597] for the AR(2) process of Equation (34)).

(b) Show that the weights $\{S_k\}$ given by Equation (601a) and used by the circulant embedding method to create a realization of length N from $\{X_t\}$ take the form

$$S_k = \frac{\sigma_1^2 \left(1 - (-1)^k \phi_{1,1}^N\right)}{1 - 2\phi_{1,1} \cos(\pi k/N) + \phi_{1,1}^2} \qquad k = 0, 1, \ldots, 2N - 1. \qquad (640a)$$

Argue that $S_k > 0$ (thus the conditions for the circulant embedding method to work hold for $\{X_t\}$).

(c) Consider the case $\phi_{1,1} = 0.9$, $\sigma_1^2 = 1 - \phi_{1,1}^2$ and $N = 10$, for which the ACVS at the maximum lag in the circularized sequence of Equation (601b) is $s_{10} = \phi_{1,1}^{10} \doteq 0.3487$. Suppose that we replace s_{10} in the sequence with

$$\gamma \stackrel{\text{def}}{=} \frac{2\phi_{1,1}^N + \phi_{1,1} - 1}{\phi_{1,1} + 1} \doteq 0.3144 \qquad (640b)$$

and then generate weights, say $\tilde{S}_k$, in keeping with Equation (601a); i.e.,

$$\tilde{S}_k = \sum_{\tau=0}^{N-1} s_\tau e^{-i2\pi \tilde{f}_k \tau} + \gamma e^{-i2\pi \tilde{f}_k N} + \sum_{\tau=N+1}^{2N-1} s_{2N-\tau} e^{-i2\pi \tilde{f}_k \tau}.$$

How do the $\tilde{S}_k$ weights compare with the S_k weights? (Once this question is answered, an extension to this exercise is to explain how γ came about!)

[11.5] Exercise [602] notes that that the simulations $\{Y_{\Re,t}\}$ and $\{Y_{\Im,t}\}$ generated by the circular embedding method have ACVSs that agree with that of the desired time series $\{X_t\}$ for lags satisfying $0 \leq \tau \leq N$, but not necessarily for lags $\tau > N$. For the AR(2) process that is the subject of Figure 603, but now setting $N = 20$ rather than 1024, determine how well the ACVSs for $\{Y_{\Re,t}\}$ and $\{Y_{\Im,t}\}$ agree with that for $\{X_t\}$ at lags $\tau > 20$.

[11.6] Consider the AR(4) process of Equation (35a) (its ACVS $\{s_\tau\}$ can be computed using the procedure described surrounding Equation (508b)). Suppose we want to generate simulations of length $N = 128$ using the circulant embedding method.

(a) Calculate the required $2N = 256$ weights S_k as per Equation (601a), noting that some of the weights are negative. Next calculate the $2N' = 512$ weights associated with a desired length of 256, i.e., $N' = 2N$ rather than N. Are all 512 of these weights nonnegative?

(b) Reconsider the 256 weights S_k computed in part (a) for the desired length $N = 128$, and define $\tilde{S}_k = (S_k)_+$ (i.e., the maximum of S_k and 0). If we use $\tilde{S}_k$ as a substitute for S_k in Equation (601c) and then proceed as usual with the circulant embedding method, show that the resulting simulations $Y_{\Re,0}, Y_{\Re,1}, \ldots, Y_{\Re,127}$ and $Y_{\Im,0}, Y_{\Im,1}, \ldots, Y_{\Im,127}$ are two independent realizations from stationary processes with a common ACVS, say $\{\tilde{s}_\tau\}$, but that this ACVS is not in perfect agreement with $\{s_\tau\}$ at lags $\tau = 0, 1, \ldots, 128$. Use $\left[\frac{1}{128} \sum_{\tau=1}^{128} (\tilde{s}_\tau/\tilde{s}_0 - s_\tau/s_0)^2\right]^{1/2}$ to ascertain the degree of imperfection. Comment upon your findings.

[11.7] Show that s'_τ as defined in Equation (606a) is real-valued.

[11.8] Let U_t be as defined by Equation (606c), and let $U_{\Re,t}$ and $U_{\Im,t}$ be its real and imaginary components. Show that both $\{U_{\Re,t} : t \in \mathbb{Z}\}$ and $\{U_{\Im,t} : t \in \mathbb{Z}\}$ are Gaussian stationary processes with zero mean and ACVS $\{s'_\tau\}$, where s'_τ is defined by Equation (606a). Also show that each RV in $\{U_{\Re,t}\}$ is independent of each RV in $\{U_{\Im,t}\}$. Hint: study the solution to Exercise [602].

[11.9] Verify the contents of Figure 607 and Table 607.

[11.10] Consider again the AR(1) process $\{X_t\}$ that is the focus of Exercise [11.4].

(a) Assuming that $N' = N$ so that f'_k is set to $k/(2N)$, derive an expression for the weights $\{S(f'_k)\}$ needed for GSSM (these are used in Equation (606b)). Compare these weights with the weights $\{S_k\}$ of Equation (640a) that are used in circulant embedding by deriving an expression for $\max_k |S(f'_k) - S_k|$. Finally, comment upon the following question: while circulant embedding is an exact simulation method for simulating AR(1) time series, can the same be said for GSSM?

(b) In the paragraph containing Equation (609), we illustrated GSSM by generating simulated series of length $N = 1024$ from the AR(2) process of Equation (34). We found that, although GSSM is in theory not an exact method, when numerical accuracy comes into play, it is computationally indistinguishable from circulant embedding (an exact method). To lend credence to this finding, consider the following. As noted in the discussion surrounding Equation (177a), the dynamic range $10 \log_{10} \left(\max_f S(f) / \min_f S(f) \right)$ is a crude measure of the complexity of an SDF. The AR(2) SDF is such that $10 \log_{10} \left(\max_f S(f) \right) \doteq 7.5$ dB and $10 \log_{10} \left(\min_f S(f) \right) \doteq -7.0$ dB, so its dynamic range is 14.5 dB. Determine the values of $\phi_{1,1} > 0$ and σ_1^2 that are needed so that the AR(1) SDF has the same maximum and minimum values as the AR(2) and hence the same dynamic range. For this AR(1) process and with N set to 1024, use the expression you derived in part (a) to compute $\max_k |S(f'_k) - S_k|$. Does this computation lend credence to the apparent numerical agreement between the circulant embedding weights and the GSSM weights we found in the AR(2) case?

[11.11] At the end of Section 11.2, we illustrated the circulant embedding method by generating simulated series of length $N = 1024$ from the AR(2) process of Equation (34) – see Figure 603 and the discussion surrounding it. Replicate what was done there, but now using the AR(4) process of Equation (35a); in particular, generate the equivalent of Figure 603. We also used the AR(2) process to illustrate GSSM (see the paragraph containing Equation (609)), but did not show the equivalent of Figure 603 (such a figure was not needed because it turned out to be visually identical to Figure 603). Replicate what was done there, but now using the AR(4) process; if need be, generate the equivalent of Figure 603. Comment upon your findings.

[11.12] For each of the four AR(2) time series of length $N = 1024$ shown in Figure 34, compute the periodogram over the grid of frequencies $\tilde{f}_k = k/(2N)$, $k = 0, 1, \ldots, 2N - 1$ (the series are downloadable from the "Data" part of the website for the book). Based upon each of the four periodograms, use the circulant embedding method to generate a simulated series whose theoretical SDF is equal to the periodogram (this method gives two simulated series, but just pick one of them for subsequent use). Compute the periodograms for each of the four simulated series. Compare these periodograms to the corresponding periodograms for the four actual AR(2) time series. Repeat this exercise using the four AR(4) series in Figure 34. Comment upon your findings.

[11.13] (a) Show that $X_{\phi,t}$ of Equation (616a) is real-valued.

(b) Verify that $X_{\phi,t}$ of Equation (616a) can be reexpressed as in Equation (616b).

(c) Argue that, when N is odd, $\{X_{\phi,t}\}$ is a harmonic process as defined by Equation (35c) (but formulated as per Equation (35d)), whereas, when N is even, $\{X_{\phi,t}\}$ is *not* a harmonic process in general. Hint: study Exercise [37] and its solution.

[11.14] Consider the causal ARMA(1,1) process $X_t = \phi X_{t-1} + \epsilon_t + \vartheta \epsilon_{t-1}$, which is the focus of Exercise [632b]. Assume that $\phi + \vartheta \neq 0$, and note that causality implies $|\phi| < 1$. Starting from Equation (632a), show that $|\sqrt{\delta_1(X_t)}| < |\sqrt{\delta_1(\epsilon_t)}|$ for this process. Hint: study the proof of Equation (632e) in Exercise [632b].

References

Acheson, D. (1997) *From Calculus to Chaos: An Introduction to Dynamics.* Oxford: Oxford University Press.

Adams, J. W. (1991) A New Optimal Window. *IEEE Transactions on Signal Processing*, 39, 1753–69.

Ahmed, N., Natarajan, T. and Rao, K. R. (1974) Discrete Cosine Transform. *IEEE Transactions on Computers*, 23, 90–3.

Akaike, H. (1969) Fitting Autoregressive Models for Prediction. *Annals of the Institute of Statistical Mathematics*, 21, 243–7.

——— (1970) Statistical Predictor Identification. *Annals of the Institute of Statistical Mathematics*, 22, 203–17.

——— (1974) A New Look at the Statistical Model Identification. *IEEE Transactions on Automatic Control*, 19, 716–22.

Alavi, A. S. and Jenkins, G. M. (1965) An Example of Digital Filtering. *Applied Statistics*, 14, 70–4.

Amrein, M. and Künsch, H. R. (2011) Approximate Variances for Tapered Spectral Estimates. *Signal Processing*, 91, 2685–9.

Andersen, N. O. (1978) Comments on the Performance of Maximum Entropy Algorithms. *Proceedings of the IEEE*, 66, 1581–2.

Anderson, T. W. (1971) *The Statistical Analysis of Time Series.* New York: John Wiley & Sons.

——— (2003) *An Introduction to Multivariate Statistical Analysis* (Third Edition). Hoboken, NJ: John Wiley & Sons.

Babadi, B. and Brown, E. N. (2014) A Review of Multitaper Spectral Analysis. *IEEE Transactions on Biomedical Engineering*, 61, 1555–64.

Baggeroer, A. B. (1976) Confidence Intervals for Regression (MEM) Spectral Estimates. *IEEE Transactions on Information Theory*, 22, 534–45 (also in Childers, 1978).

Barbé, K., Pintelon, R. and Schoukens, J. (2010) Welch Method Revisited: Nonparametric Power Spectrum Estimation via Circular Overlap. *IEEE Transactions on Signal Processing*, 58, 553–65.

Barnes, J. A., Chi, A. R., Cutler, L. S., Healey, D. J., Leeson, D. B., McGunigal, T. E., Mullen, J. A., Jr., Smith, W. L., Sydnor, R. L., Vessot, R. F. C. and Winkler, G. M. R. (1971) Characterization of Frequency Stability. *IEEE Transactions on Instrumentation and Measurement*, 20, 105–20.

Barrett, J. F. and Lampard, D. G. (1955) An Expansion for Some Second-Order Probability Distributions and Its Application to Noise Problems. *IRE Transactions on Information Theory*, 1, 10–15.

Bartlett, M. S. (1950) Periodogram Analysis and Continuous Spectra. *Biometrika*, 37, 1–16.

——— (1955) *An Introduction to Stochastic Processes.* Cambridge: Cambridge University Press.

——— (1963) Statistical Estimation of Density Functions. *Sankhyā: The Indian Journal of Statistics, Series A*, 25, 245–54.

Bartlett, M. S. and Kendall, D. G. (1946) The Statistical Analysis of Variance-Heterogeneity and the Logarithmic Transformation. *Supplement to the Journal of the Royal Statistical Society*, 8, 128–38.

Beauchamp, K. G. (1984) *Applications of Walsh and Related Functions*. London: Academic Press.

Bell, B. M. and Percival, D. B. (1991) A Two Step Burg Algorithm. *IEEE Transactions on Signal Processing*, 39, 185–9.

Beltrão, K. I. and Bloomfield, P. (1987) Determining the Bandwidth of a Kernel Spectrum Estimate. *Journal of Time Series Analysis*, 8, 21–38.

Beran, J. (1994) *Statistics for Long-Memory Processes*. New York: Chapman & Hall.

Berk, K. N. (1974) Consistent Autoregressive Spectral Estimates. *Annals of Statistics*, 2, 489–502.

Blackman, R. B. and Tukey, J. W. (1958) *The Measurement of Power Spectra*. New York: Dover Publications.

Bloomfield, P. (2000) *Fourier Analysis of Time Series: An Introduction* (Second Edition). New York: John Wiley & Sons.

Bogert, B. P., Healy, M. J. and Tukey, J. W. (1963) The Quefrency Alanysis of Time Series for Echoes: Cepstrum, Pseudo-Autocovariance, Cross-Cepstrum and Saphe Cracking. In *Proceedings of the Symposium on Time Series Analysis*, edited by M. Rosenblatt, New York: John Wiley & Sons, 209–43 (also in Brillinger, 1984a).

Bohman, H. (1961) Approximate Fourier Analysis of Distribution Functions. *Arkiv för Matematik*, 4, 99–157.

Bolt, B. A. and Brillinger, D. R. (1979) Estimation of Uncertainties in Eigenspectral Estimates from Decaying Geophysical Time Series. *Geophysical Journal of the Royal Astronomical Society*, 59, 593–603.

Bøviken, E. (1983) New Tests of Significance in Periodogram Analysis. *Scandinavian Journal of Statistics*, 10, 1–9.

Box, G. E. P. (1954) Some Theorems on Quadratic Forms Applied in the Study of Analysis of Variance Problems, I. Effect of Inequality of Variance in the One-Way Classification. *Annals of Mathematical Statistics*, 25, 290–302.

Box, G. E. P., Jenkins, G. M., Reinsel, G. C. and Ljung, G. M. (2015) *Time Series Analysis: Forecasting and Control* (Fifth Edition). Hoboken, NJ: John Wiley & Sons.

Bracewell, R. N. (2000) *The Fourier Transform and Its Applications* (Third Edition). Boston: McGraw-Hill.

Briggs, W. L. and Henson, V. E. (1995) *The DFT: An Owner's Manual for the Discrete Fourier Transform*. Philadelphia: SIAM.

Brillinger, D. R. (1981a) *Time Series: Data Analysis and Theory* (Expanded Edition). San Francisco: Holden-Day.

——— (1981b) The Key Role of Tapering in Spectrum Estimation. *IEEE Transactions on Acoustics, Speech, and Signal Processing*, 29, 1075–6.

——— (1987) Fitting Cosines: Some Procedures and Some Physical Examples. In *Applied Probability, Stochastics Processes, and Sampling Theory*, edited by I. B. MacNeill and G. J. Umphrey, Dordrecht: D. Reidel Publishing Company, 75–100.

Brillinger, D. R., editor (1984a) *The Collected Works of John W. Tukey, Volume I, Time Series: 1949–1964*. Belmont, CA: Wadsworth.

——— (1984b) *The Collected Works of John W. Tukey, Volume II, Time Series: 1965–1984*. Belmont, CA: Wadsworth.

Brockwell, P. J. and Davis, R. A. (1991) *Time Series: Theory and Methods* (Second Edition). New York: Springer-Verlag.

——— (2016) *Introduction to Time Series and Forecasting* (Third Edition). New York: Springer.

Broersen, P. M. T. (2000) Finite Sample Criteria for Autoregressive Order Selection. *IEEE Transactions on Signal Processing*, 48, 3550–8.

——— (2002) Automatic Spectral Analysis with Time Series Models. *IEEE Transactions on Instrumentation and Measurement*, 51, 211–16.

——— (2006) *Automatic Autocorrelation and Spectral Analysis*. London: Springer.

Bronez, T. P. (1985) Nonparametric Spectral Estimation of Irregularly-Sampled Multidimensional Random Processes. Ph.D. dissertation, Department of Electrical Engineering, Arizona State University.

———— (1986) Nonparametric Spectral Estimation with Irregularly Sampled Data. In *Proceedings of the Third IEEE ASSP Workshop on Spectrum Estimation and Modeling*, Boston, 133–6.

———— (1988) Spectral Estimation of Irregularly Sampled Multidimensional Processes by Generalized Prolate Spheroidal Sequences. *IEEE Transactions on Acoustics, Speech, and Signal Processing*, 36, 1862–73.

———— (1992) On the Performance Advantage of Multitaper Spectral Analysis. *IEEE Transactions on Signal Processing*, 40, 2941–6.

Bruce, A. G. and Martin, R. D. (1989) Leave-k-out Diagnostics for Time Series. *Journal of the Royal Statistical Society, Series B*, 51, 363–424.

Bunch, J. R. (1985) Stability of Methods for Solving Toeplitz Systems of Equations. *SIAM Journal on Scientific and Statistical Computing*, 6, 349–64.

Burg, J. P. (1967) Maximum Entropy Spectral Analysis. In *Proceedings of the 37th Meeting of the Society of Exploration Geophysicists* (also in Childers, 1978).

———— (1968) A New Analysis Technique for Time Series Data. In *NATO Advanced Study Institute on Signal Processing with Emphasis on Underwater Acoustics* (also in Childers, 1978).

———— (1975) Maximum Entropy Spectral Analysis. Ph.D. dissertation, Department of Geophysics, Stanford University.

———— (1985) Absolute Power Density Spectra. In *Maximum-Entropy and Bayesian Methods in Inverse Problems*, edited by C. R. Smith and W. T. Grandy, Jr., Dordrecht: D. Reidel Publishing Company, 273–86.

Burshtein, D. and Weinstein, E. (1987) Confidence Intervals for the Maximum Entropy Spectrum. *IEEE Transactions on Acoustics, Speech, and Signal Processing*, 35, 504–10.

———— (1988) Corrections to "Confidence Intervals for the Maximum Entropy Spectrum". *IEEE Transactions on Acoustics, Speech, and Signal Processing*, 36, 826.

Businger, P. A. and Golub, G. H. (1969) Algorithm 358: Singular Value Decomposition of a Complex Matrix. *Communications of the ACM*, 12, 564–5.

Carter, G. C. (1987) Coherence and Time Delay Estimation. *Proceedings of the IEEE*, 75, 236–55.

Chambers, J. M., Cleveland, W. S., Kleiner, B. and Tukey, P. A. (1983) *Graphical Methods for Data Analysis*. Boston: Duxbury Press.

Chambers, M. J. (1995) The Simulation of Random Vector Time Series with Given Spectrum. *Mathematical and Computer Modelling*, 22, 1–6.

Champeney, D. C. (1987) *A Handbook of Fourier Theorems*. Cambridge: Cambridge University Press.

Chan, G., Hall, P. and Poskitt, D. S. (1995) Periodogram-Based Estimators of Fractal Properties. *Annals of Statistics*, 23, 1684–711.

Chan, K.-S. and Tong, H. (2001) *Chaos: A Statistical Perspective*. New York: Springer.

Chandna, S. and Walden, A. T. (2011) Statistical Properties of the Estimator of the Rotary Coefficient. *IEEE Transactions on Signal Processing*, 59, 1298–1303.

Chatfield, C. (2004) *The Analysis of Time Series: An Introduction* (Sixth Edition). London: Chapman & Hall/CRC.

Chave, A. D., Thomson, D. J. and Ander, M. E. (1987) On the Robust Estimation of Power Spectra, Coherences, and Transfer Functions. *Journal of Geophysical Research*, 92, 633–48.

Chen, W. Y. and Stegen, G. R. (1974) Experiments with Maximum Entropy Power Spectra of Sinusoids. *Journal of Geophysical Research*, 79, 3019–22.

Childers, D. G., editor (1978) *Modern Spectrum Analysis*. New York: IEEE Press.

Chilès, J.-P. and Delfiner, P. (2012) *Geostatistics: Modeling Spatial Uncertainty* (Second Edition). Hoboken, NJ: John Wiley & Sons.

Choi, B. S. (1992) *ARMA Model Identification*. New York: Springer-Verlag.

Chonavel, T. (2002) *Statistical Signal Processing*. London: Springer-Verlag.

Chung, K. L. (1974) *A Course in Probability Theory* (Second Edition). New York: Academic Press.

Cleveland, W. S. (1979) Robust Locally Weighted Regression and Smoothing Scatterplots. *Journal of the American Statistical Association*, 74, 829–36.

Cleveland, W. S. and Parzen, E. (1975) The Estimation of Coherence, Frequency Response, and Envelope Delay. *Technometrics*, 17, 167–72.

Conover, W. J. (1999) *Practical Nonparametric Statistics* (Third Edition). New York: John Wiley & Sons.

Conradsen, K. and Spliid, H. (1981) A Seasonal Adjustment Filter for Use in Box–Jenkins Analysis of Seasonal Time Series. *Applied Statistics*, 30, 172–7.

Constantine, W. L. B. (1999) Wavelet Techniques for Chaotic and Fractal Dynamics. Ph.D. dissertation, Department of Mechanical Engineering, University of Washington.

Courant, R. and Hilbert, D. (1953) *Methods of Mathematical Physics, Volume I*. New York: Interscience Publishers.

Craigmile, P. F. (2003) Simulating a Class of Stationary Gaussian Processes using the Davies–Harte Algorithm, with Application to Long Memory Processes. *Journal of Time Series Analysis*, 24, 505–11.

Cramér, H. (1942) On Harmonic Analysis in Certain Functional Spaces. *Arkiv för Matematik, Astronomi och Fysik*, 28B, 1–7.

Cumming, A., Marcy, G. W. and Butler, R. P. (1999) The Lick Planet Search: Detectability and Mass Thresholds. *Astrophysical Journal*, 526, 890–915.

Daniell, P. J. (1946) Discussion on the Papers by Bartlett, Foster, Cunningham and Hynd. *Supplement to the Journal of the Royal Statistical Society*, 8, 88–90.

Davies, N., Spedding, T. and Watson, W. (1980) Autoregressive Moving Average Processes With Non-Normal Residuals. *Journal of Time Series Analysis*, 1, 103–9.

Davies, R. B. (2001) Integrated Processes and the Discrete Cosine Transform. *Journal of Applied Probability*, 38A, 701–17.

Davies, R. B. and Harte, D. S. (1987) Tests for Hurst Effect. *Biometrika*, 74, 95–101.

Davis, H. T. and Jones, R. H. (1968) Estimation of the Innovation Variance of a Stationary Time Series. *Journal of the American Statistical Association*, 63, 141–9.

Davison, A. C. and Hinkley, D. V. (1997) *Bootstrap Methods and Their Application*. Cambridge: Cambridge University Press.

de Gooijer, J. G., Abraham, B., Gould, A. and Robinson, L. (1985) Methods for Determining the Order of an Autoregressive-Moving Average Process: A Survey. *International Statistical Review*, 53, 301–29.

Denison, D. G. T. and Walden, A. T. (1999) The Search for Solar Gravity-Mode Oscillations: An Analysis Using *Ulysses* Magnetic Field Data. *Astrophysical Journal*, 514, 972–78.

Denison, D. G. T., Walden, A. T., Balogh, A. and Forsyth, R. J. (1999) Multitaper Testing of Spectral Lines and the Detection of the Solar Rotation Frequency and Its Harmonics. *Applied Statistics*, 48, 427–39.

Dietrich, C. R. and Newsam, G. N. (1997) Fast and Exact Simulation of Stationary Gaussian Processes through Circulant Embedding of the Covariance Matrix. *SIAM Journal on Scientific Computing*, 18, 1088–107.

Diggle, P. J. and al Wasel, I. (1997) Spectral Analysis of Replicated Biomedical Time Series. *Applied Statistics*, 46, 31–71.

Doetsch, G. (1943) *Theorie und Anwendung der Laplace-Transformation*. New York: Dover Publications.

Doodson, A. T. and Warburg, H. D. (1941) *Admiralty Manual of Tides*. London: H. M. Stationery Office.

Dzhaparidze, K. O. and Yaglom, A. M. (1983) Spectrum Parameter Estimation in Time Series Analysis. *Developments in Statistics*, 4, 1–96.

Efron, B. and Gong, G. (1983) A Leisurely Look at the Bootstrap, the Jackknife, and Cross-Validation. *The American Statistician*, 37, 36–48.

Einstein, A. (1914) Méthode pour la Détermination des Valeurs Statistiques d'Observations concernant des Grandeurs Soumises à des Fluctuations Irrégulières. *Archives de Sciences Physiques et Naturalles*, 37, 254–6 (translated as "Method for the Determinination of the Statistical Values of Observations Concerning Quantities Subject to Irregular Fluctuations" in *IEEE ASSP Magazine*, 4(4), 6).

Emery, W. J. and Thomson, R. E. (2001) *Data Analysis Methods in Physical Oceanography* (Second and Revised Edition). Amsterdam: Elsevier.

Epanechnikov, V. A. (1969) Non-Parametric Estimation of a Multivariate Probability Density. *Theory of Probability & Its Applications*, 14, 153–8.

Fairfield Smith, H. (1936) The Problem of Comparing the Result of Two Experiments with Unequal Errors. *Journal of the Council for Scientific and Industrial Research*, 9, 211–12.

Falk, R. and Well, A. D. (1997) Many Faces of the Correlation Coefficient. *Journal of Statistics Education*, 5(3).

Fan, J. and Kreutzberger, E. (1998) Automatic Local Smoothing for Spectral Density Estimation. *Scandinavian Journal of Statistics*, 25, 359–69.

Farebrother, R. W. (1987) The Distribution of a Noncentral χ^2 Variable with Nonnegative Degrees of Freedom. *Applied Statistics*, 36, 402–5.

Fay, G. and Soulier, P. (2001) The Periodogram of an I.I.D. Sequence. *Stochastic Processes and their Applications*, 92, 315–43.

Ferguson, T. S. (1995) A Class of Bivariate Uniform Distributions. *Statistical Papers*, 36, 31–40.

Fisher, R. A. (1929) Tests of Significance in Harmonic Analysis. *Proceedings of the Royal Society of London, Series A*, 125, 54–9.

Fodor, I. K. and Stark, P. B. (2000) Multitaper Spectrum Estimation for Time Series with Gaps. *IEEE Transactions on Signal Processing*, 48, 3472–83.

Folland, G. B. and Sitaram, A. (1997) The Uncertainty Principle: A Mathematical Survey. *Journal of Fourier Analysis and Applications*, 3, 207–38.

Fougere, P. F. (1977) A Solution to the Problem of Spontaneous Line Splitting in Maximum Entropy Power Spectrum Analysis. *Journal of Geophysical Research*, 82, 1051–4.

———— (1985a) A Review of the Problem of Spontaneous Line Splitting in Maximum Entropy Power Spectral Analysis. In *Maximum-Entropy and Bayesian Methods in Inverse Problems*, edited by C. R. Smith and W. T. Grandy, Jr., Dordrecht: D. Reidel Publishing Company, 303–15.

———— (1985b) On the Accuracy of Spectrum Analysis of Red Noise Processes Using Maximum Entropy and Periodogram Methods: Simulation Studies and Application to Geophysical Data. *Journal of Geophysical Research*, 90, 4355–66.

Fougere, P. F., Zawalick, E. J. and Radoski, H. R. (1976) Spontaneous Line Splitting in Maximum Entropy Power Spectrum Analysis. *Physics of the Earth and Planetary Interiors*, 12, 201–7.

Franke, J. (1985) ARMA Processes Have Maximal Entropy Among Time Series with Prescribed Autocovariances and Impulse Responses. *Advances in Applied Probability*, 17, 810–40.

Friedlander, B. (1982) System Identification Techniques for Adaptive Signal Processing. *IEEE Transactions on Acoustics, Speech, and Signal Processing*, 30, 240–6.

Fuller, W. A. (1996) *Introduction to Statistical Time Series* (Second Edition). New York: John Wiley & Sons.

Gao, H.-Y. (1993) Wavelet Estimation of Spectral Densities in Time Series Analysis. Ph.D. dissertation, Department of Statistics, University of California, Berkeley.

———— (1997) Choice of Thresholds for Wavelet Shrinkage Estimate of the Spectrum. *Journal of Time Series Analysis*, 18, 231–51.

Geçkinli, N. C. and Yavuz, D. (1978) Some Novel Windows and a Concise Tutorial Comparison of Window Families. *IEEE Transactions on Acoustics, Speech, and Signal Processing*, 26, 501–7.

———— (1983) *Discrete Fourier Transformation and Its Applications to Power Spectra Estimation* (Studies in Electrical and Electronic Engineering 8). Amsterdam: Elsevier Scientific Publishing Company.

Geweke, J. and Porter-Hudak, S. (1983) The Estimation and Application of Long Memory Time Series Models. *Journal of Time Series Analysis*, 4, 221–38.

Gneiting, T. (2000) Power-Law Correlations, Related Models for Long-Range Dependence and Their Simulation. *Journal of Applied Probability*, 37, 1104–9.

Godolphin, E. J. and Unwin, J. M. (1983) Evaluation of the Covariance Matrix for the Maximum Likelihood Estimator of a Gaussian Autoregressive-Moving Average Process. *Biometrika*, 70, 279–84.

Golub, G. H. and Kahan, W. (1965) Calculating the Singular Values and Pseudo-Inverse of a Matrix. *SIAM Journal on Numerical Analysis*, 2, 205–24.

Golub, G. H. and Van Loan, C. F. (2013) *Matrix Computations* (Fourth Edition). Baltimore: Johns Hopkins University Press.

Gonzalez, R. C. and Woods, R. E. (2007) *Digital Image Processing* (Third Edition). Upper Saddle River, NJ: Prentice-Hall.

Gradshteyn, I. S. and Ryzhik, I. M. (1980) *Table of Integrals, Series, and Products* (Corrected and Enlarged Edition). New York: Academic Press.

Granger, C. W. J. (1966) The Typical Spectral Shape of an Economic Variable. *Econometrica*, 34, 150–61.

——— (1979) Nearer-Normality and Some Econometric Models. *Econometrica*, 47, 781–4.

Granger, C. W. J. and Joyeux, R. (1980) An Introduction to Long-Memory Time Series Models and Fractional Differencing. *Journal of Time Series Analysis*, 1, 15–29.

Graybill, F. A. (1983) *Matrices with Applications in Statistics* (Second Edition). Belmont, CA: Wadsworth.

Greene, D. H. and Knuth, D. E. (1990) *Mathematics for the Analysis of Algorithms* (Third Edition). Boston: Birkhäuser.

Greenhall, C. A. (2006) Decorrelation Bandwidth of Multitaper Spectral Estimators. Unpublished technical report, personal communication.

Grenander, U. (1951) On Empirical Spectral Analysis of Stochastic Processes. *Arkiv för Matematik*, 1, 503–31.

Grenander, U. and Rosenblatt, M. (1984) *Statistical Analysis of Stationary Time Series* (Second Edition). New York: Chelsea Publishing Company.

Grenander, U. and Szegő, G. (1984) *Toeplitz Forms and Their Applications* (Second Edition). New York: Chelsea Publishing Company.

Grünbaum, F. A. (1981) Eigenvectors of a Toeplitz Matrix: Discrete Version of the Prolate Spheroidal Wave Functions. *SIAM Journal on Algebraic and Discrete Methods*, 2, 136–41.

Gujar, U. G. and Kavanagh, R. J. (1968) Generation of Random Signals with Specified Probability Density Functions and Power Density Spectra. *IEEE Transactions on Automatic Control*, 13, 716–9.

Hamming, R. W. (1983) *Digital Filters* (Second Edition). Englewood Cliffs, NJ: Prentice-Hall.

Hannan, E. J. (1970) *Multiple Time Series*. New York: John Wiley & Sons.

——— (1973) The Estimation of Frequency. *Journal of Applied Probability*, 10, 510–19.

Hannan, E. J. and Nicholls, D. F. (1977) The Estimation of the Prediction Error Variance. *Journal of the American Statistical Association*, 72, 834–40.

Hannig, J. and Lee, T. C. M. (2004) Kernel Smoothing of Periodograms Under Kullback–Leibler Discrepancy. *Signal Processing*, 84, 1255–66.

Hanssen, A. (1997) Multidimensional Multitaper Spectral Estimation. *Signal Processing*, 58, 327–32.

Hansson, M. (1999) Optimized Weighted Averaging of Peak Matched Multiple Window Spectrum Estimates. *IEEE Transactions on Signal Processing*, 47, 1141–6.

Hansson, M. and Salomonsson, G. (1997) A Multiple Window Method for Estimation of Peaked Spectra. *IEEE Transactions on Signal Processing*, 45, 778–81.

Hansson-Sandsten, M. (2012) A Welch Method Approximation of the Thomson Multitaper Spectrum Estimator. In *Proceedings of 20th European Signal Processing Conference (EUSIPCO)*, Bucharest, 440–4.

Hardin, J. C. (1986) An Additional Source of Uncertainty and Bias in Digital Spectral Estimates Near the Nyquist Frequency. *Journal of Sound and Vibration*, 110, 533–7.

Harris, F. J. (1978) On the Use of Windows for Harmonic Analysis with the Discrete Fourier Transform. *Proceedings of the IEEE*, 66, 51–83 (also in Kesler, 1986).

Hasan, T. (1983) Complex Demodulation: Some Theory and Applications. In *Handbook of Statistics 3: Time Series in the Frequency Domain*, edited by D. R. Brillinger and P. R. Krishnaiah, Amsterdam: North-Holland, 125–56.

Hastie, T. J. and Tibshirani, R. J. (1990) *Generalized Additive Models*. London: Chapman & Hall.

Hayashi, Y. (1979) Space-Time Spectral Analysis of Rotary Vector Series. *Journal of the Atmospheric Sciences*, 36, 757–66.

Heideman, M. T., Johnson, D. H. and Burrus, C. S. (1987) Gauss and the History of the Fast Fourier Transform. *IEEE ASSP Magazine*, 1, 14–21.

Hill, I. D. (1976) Algorithm AS 100: Normal–Johnson and Johnson–Normal Transformations. *Applied Statistics*, 25, 190–2.

Hill, I. D., Hill, R. and Holder, R. L. (1976) Algorithm AS 99: Fitting Johnson Curves by Moments. *Applied Statistics*, 25, 180–9.

Hogan, J. A. and Lakey, J. D. (2012) *Duration and Bandwidth Limiting: Prolate Functions, Sampling, and Applications*. New York: Springer.

Hosking, J. R. M. (1981) Fractional Differencing. *Biometrika*, 68, 165–76.

Hua, Y. and Sarkar, T. K. (1988) Perturbation Analysis of TK Method for Harmonic Retrieval Problems. *IEEE Transactions on Acoustics, Speech, and Signal Processing*, 36, 228–40.

Hurvich, C. M. (1985) Data-Driven Choice of a Spectrum Estimate: Extending the Applicability of Cross-Validation Methods. *Journal of American Statistical Association*, 80, 933–40.

Hurvich, C. M. and Beltrão, K. I. (1990) Cross-Validatory Choice of a Spectrum Estimate and its Connections with AIC. *Journal of Time Series Analysis*, 11, 121–37.

Hurvich, C. M. and Chen, W. W. (2000) An Efficient Taper for Potentially Overdifferenced Long-Memory Time Series. *Journal of Time Series Analysis*, 21, 155–80.

Hurvich, C. M. and Tsai, C.-L. (1989) Regression and Time Series Model Selection in Small Samples. *Biometrika*, 76, 297–307.

Ibragimov, I. A. and Linnik, Yu. V. (1971) *Independent and Stationary Sequences of Random Variables*. Gröningen, The Netherlands: Wolters–Noordhoff.

Ihara, S. (1984) Maximum Entropy Spectral Analysis and ARMA Processes. *IEEE Transactions on Information Theory*, 30, 377–80.

Isserlis, L. (1918) On a Formula for the Product-Moment Coefficient of Any Order of a Normal Frequency Distribution in Any Number of Variables. *Biometrika*, 12, 134–9.

Jackson, D. D. (1972) Interpretation of Inaccurate, Insufficient and Inconsistent Data. *Geophysical Journal of the Royal Astronomical Society*, 28, 97–109.

Jackson, P. L. (1967) Truncations and Phase Relationships of Sinusoids. *Journal of Geophysical Research*, 72, 1400–3.

Jacovitti, G. and Scarano, G. (1987) On a Property of the PARCOR Coefficients of Stationary Processes Having Gaussian-Shaped ACF. *Proceedings of the IEEE*, 75, 960–1.

Jaynes, E. T. (1982) On the Rationale of Maximum-Entropy Methods. *Proceedings of the IEEE*, 70, 939–52.

Jenkins, G. M. (1961) General Considerations in the Analysis of Spectra. *Technometrics*, 3, 133–66.

Jenkins, G. M. and Watts, D. G. (1968) *Spectral Analysis and Its Applications*. San Francisco: Holden-Day.

Johnson, G. E. (1994) Constructions of Particular Random Processes. *Proceedings of the IEEE*, 82, 270–85.

Johnson, N. L. (1949) Systems of Frequency Curves Generated by Methods of Translation. *Biometrika*, 36, 149–76.

Jones, N. B., Lago, P. J. and Parekh, A. (1987) Principal Component Analysis of the Spectra of Point Processes – An Application in Electromyography. In *Mathematics in Signal Processing*, edited by T. S. Durrani, J. B. Abbiss, J. E. Hudson, R. N. Madan, J. G. McWhirter and T. A. Moore, Oxford: Clarendon Press, 147–64.

Jones, R. H. (1971) Spectrum Estimation with Missing Observations. *Annals of the Institute of Statistical Mathematics*, 23, 387–98.

——— (1976a) Estimation of the Innovation Generalized Variance of a Multivariate Stationary Time Series. *Journal of the American Statistical Association*, 71, 386–8.

——— (1976b) Autoregression Order Selection. *Geophysics*, 41, 771–3 (also in Childers, 1978).

——— (1980) Maximum Likelihood Fitting of ARMA Models to Time Series with Missing Observations. *Technometrics*, 22, 389–95.

——— (1985) Time Series Analysis – Time Domain. In *Probability, Statistics, and Decision Making in the Atmospheric Sciences*, edited by A. H. Murphy and R. W. Katz, Boulder, CO: Westview Press, 223–59.

Jones, R. H., Crowell, D. H., Nakagawa, J. K. and Kapuniai, L. (1972) Statistical Comparisons of EEG Spectra Before and During Stimulation in Human Neonates. In *Computers in Biomedicine, a*

Supplement to the Proceedings of the Fifth Hawaii International Conference on System Sciences, North Hollywood, CA: Western Periodicals Company.

Joughin, I. (1995) Estimation of Ice-Sheet Topography and Motion Using Interferometric Synthetic Aperture Radar. Ph.D. dissertation, Department of Electrical Engineering, University of Washington.

Kaiser, J. F. (1966) Digital Filters. In *System Analysis by Digital Computer*, edited by F. F. Kuo and J. F. Kaiser, New York: John Wiley & Sons, 218–85.

——— (1974) Nonrecursive Digital Filter Design Using the $I_0 - SINH$ Window Function. In *Proceedings – 1974 IEEE International Symposium on Circuits and Systems*, 20–3.

Kane, R. P. and de Paula, E. R. (1996) Atmospheric CO_2 Changes at Mauna Loa, Hawaii. *Journal of Atmospheric and Terrestrial Physics*, 58, 1673–81.

Kantz, H. and Schreiber, T. (2004) *Nonlinear Time Series Analysis* (Second Edition). Cambridge: Cambridge University Press.

Kasyap, R. L. and Eom, K.-B. (1988) Estimation in Long-Memory Time Series Model. *Journal of Time Series Analysis*, 9, 35–41.

Kaveh, M. and Cooper, G. R. (1976) An Empirical Investigation of the Properties of the Autoregressive Spectral Estimator. *IEEE Transactions on Information Theory*, 22, 313–23 (also in Childers, 1978).

Kay, S. M. (1981a) The Effect of Sampling Rate on Autocorrelation Estimation. *IEEE Transactions on Acoustics, Speech, and Signal Processing*, 29, 859–67.

——— (1981b) Efficient Generation of Colored Noise. *Proceedings of the IEEE*, 69, 480–1.

——— (1983) Recursive Maximum Likelihood Estimation. *IEEE Transactions on Acoustics, Speech, and Signal Processing*, 31, 56–65.

——— (1988) *Modern Spectral Estimation: Theory and Application*. Englewood Cliffs, NJ: Prentice-Hall.

——— (2010) Representation and Generation of Non-Gaussian Wide-Sense Stationary Random Processes With Arbitrary PSDs and a Class of PDFs. *IEEE Transactions on Signal Processing*, 58, 3448–58.

Kay, S. M. and Makhoul, J. (1983) On the Statistics of the Estimated Reflection Coefficients of an Autoregressive Process. *IEEE Transactions on Acoustics, Speech, and Signal Processing*, 31, 1447–55 (also in Kesler, 1986).

Kay, S. M. and Marple, S. L., Jr. (1979) Sources of and Remedies for Spectral Line Splitting in Autoregressive Spectrum Analysis. In *Proceedings of the 1979 IEEE International Conference on Acoustics, Speech, and Signal Processing*, 151–4.

——— (1981) Spectrum Analysis – A Modern Perspective. *Proceedings of the IEEE*, 69, 1380–419 (also in Kesler, 1986).

Keeling, R. F., Piper, S. C., Bollenbacher, A. F. and Walker, J. S. (2009) Atmospheric CO2 Records from Sites in the SIO Air Sampling Network. In *Trends: A Compendium of Data on Global Change*, Oak Ridge, Tennessee: Carbon Dioxide Information Analysis Center, Oak Ridge National Laboratory, US Department of Energy (doi: 10.3334/CDIAC/atg.035).

Kerner, C. and Harris, P. E. (1994) Scattering Attenuation in Sediments Modeled by ARMA Processes—Validation of Simple Q Models. *Geophysics*, 59, 1813–26.

Kesler, S. B., editor (1986) *Modern Spectrum Analysis, II*. New York: IEEE Press.

Kikkawa, S. and Ishida, M. (1988) Number of Degrees of Freedom, Correlation Times, and Equivalent Bandwidths of a Random Process. *IEEE Transactions on Information Theory*, 34, 151–5.

King, M. E. (1990) Multiple Taper Spectral Analysis of Earth Rotation Data. Ph.D. dissertation, Scripps Institute of Oceanography, University of California, San Diego.

Kokoszka, P. and Mikosch, T. (2000) The Periodogram at the Fourier Frequencies. *Stochastic Processes and their Applications*, 86, 49–79.

Kooperberg, C., Stone, C. J. and Truong, Y. K. (1995) Logspline Estimation of a Possibly Mixed Spectral Distribution. *Journal of Time Series Analysis*, 16, 359–88.

Koopmans, L. H. (1974) *The Spectral Analysis of Time Series*. New York: Academic Press.

Koslov, J. W. and Jones, R. H. (1985) A Unified Approach to Confidence Bounds for the Autoregressive Spectral Estimator. *Journal of Time Series Analysis*, 6, 141–51.

Kowalski, A., Musial, F., Enck, P. and Kalveram, K.-T. (2000) Spectral Analysis of Binary Time Series: Square Waves vs. Sinusoidal Functions. *Biological Rhythm Research*, 31, 481–98.

Kreiss, J.-P. and Paparoditis, E. (2011) Bootstrap Methods for Dependent Data: A Review. *Journal of the Korean Statistical Society*, 40, 357–78.

Kromer, R. E. (1969) Asymptotic Properties of the Autoregressive Spectral Estimator. Ph.D. dissertation, Department of Statistics, Stanford University.

Kullback, S. and Leibler, R. A. (1951) On Information and Sufficiency. *Annals of Mathematical Statistics*, 22, 79–86.

Kung, S. Y. and Arun, K. S. (1987) Singular-Value-Decomposition Algorithms for Linear System Approximation and Spectrum Estimation. In *Advances in Statistical Signal Processing* (Volume 1), edited by H. V. Poor, Greenwich, CT: JAI Press, 203–50.

Kuo, C., Lindberg, C. R. and Thomson, D. J. (1990) Coherence Established Between Atmospheric Carbon Dioxide and Global Temperature. *Nature*, 343, 709–14.

Lagunas-Hernández, M. A., Santamaría-Perez, M. E. and Figueiras-Vidal, A. R. (1984) ARMA Model Maximum Entropy Power Spectral Estimation. *IEEE Transactions on Acoustics, Speech, and Signal Processing*, 32, 984–90.

Lahiri, S. N. (2003a) A Necessary and Sufficient Condition for Asymptotic Independence of Discrete Fourier Transforms under Short- and Long-Range Dependence. *Annals of Statistics*, 31, 613–41.

——— (2003b) *Resampling Methods for Dependent Data.* New York: Springer.

Landers, T. E. and Lacoss, R. T. (1977) Some Geophysical Applications of Autoregressive Spectral Estimates. *IEEE Transactions on Geoscience Electronics*, 15, 26–32 (also in Childers, 1978).

Lang, S. W. and McClellan, J. H. (1980) Frequency Estimation with Maximum Entropy Spectral Estimators. *IEEE Transactions on Acoustics, Speech, and Signal Processing*, 28, 716–24 (also in Kesler, 1986).

Lanning, E. N. and Johnson, D. M. (1983) Automated Identification of Rock Boundaries: An Application of the Walsh Transform to Geophysical Well-Log Analysis. *Geophysics*, 48, 197–205.

Lawrance, A. J. (1991) Directionality and Reversibility in Time Series. *International Statistical Review*, 59, 67–79.

Lee, T. C. M. (1997) A Simple Span Selector for Periodogram Smoothing. *Biometrika*, 84, 965–9.

——— (2001) A Stabilized Bandwidth Selection Method for Kernel Smoothing of the Periodogram. *Signal Processing*, 81, 419–30.

Li, T.-H. (2008) Laplace Periodogram for Time Series Analysis. *Journal of the American Statistical Association*, 103, 757–68.

——— (2014) *Time Series with Mixed Spectra.* Boca Raton, FL: CRC Press.

Lii, K. S. and Rosenblatt, M. (2008) Prolate Spheroidal Spectral Estimates. *Statistics and Probability Letters*, 78, 1339–48.

Lindberg, C. R. and Park, J. (1987) Multiple-Taper Spectral Analysis of Terrestrial Free Oscillations: Part II. *Geophysical Journal of the Royal Astronomical Society*, 91, 795–836.

Liu, B. and Munson, D. C., Jr. (1982) Generation of a Random Sequence Having a Jointly Specified Marginal Distribution and Autocovariance. *IEEE Transactions on Acoustics, Speech, and Signal Processing*, 30, 973–83.

Liu, T.-C. and Van Veen, B. D. (1992) Multiple Window Based Minimum Variance Spectrum Estimation for Multidimensional Random Fields. *IEEE Transactions on Signal Processing*, 40, 578–89.

Lomb, N. R. (1976) Least-Squares Frequency Analysis of Unequally Spaced Data. *Astrophysics and Space Science*, 39, 447–62.

Longbottom, J., Walden, A. T. and White, R. E. (1988) Principles and Application of Maximum Kurtosis Phase Estimation. *Geophysical Prospecting*, 36, 115–38.

Lopes, A., Lopes, S. and Souza, R. R. (1997) On the Spectral Density of a Class of Chaotic Time Series. *Journal of Time Series Analysis*, 18, 465–74.

Loupas, T. and McDicken, W. N. (1990) Low-Order Complex AR Models for Mean and Maximum Frequency Estimation in the Context of Doppler Color Flow Mapping. *IEEE Transactions on Ultrasonics, Ferroelectrics, and Frequency Control*, 37, 590–601.

Lysne, D. and Tjøstheim, D. (1987) Loss of Spectral Peaks in Autoregressive Spectral Estimation. *Biometrika*, 74, 200–6.

Maitani, T. (1983) Statistics of Wind Direction Fluctuations in the Surface Layer over Plant Canopies. *Boundary-Layer Meteorology*, 26, 15–24.

Makhoul, J. (1981a) On the Eigenvectors of Symmetric Toeplitz Matrices. *IEEE Transactions on Acoustics, Speech, and Signal Processing*, 29, 868–72.

——— (1981b) Lattice Methods in Spectral Estimation. In *Applied Time Series Analysis II*, edited by D. F. Findley, New York: Academic Press, 301–25.

——— (1986) Maximum Confusion Spectral Analysis. In *Proceedings of the Third IEEE ASSP Workshop on Spectrum Estimation and Modeling*, Boston, 6–9.

——— (1990) Volume of the Space of Positive Definite Sequences. *IEEE Transactions on Acoustics, Speech, and Signal Processing*, 38, 506–11.

Mallows, C. L. (1967) Linear Processes are Nearly Gaussian. *Journal of Applied Probability*, 4, 313–29.

Mann, H. B. and Wald, A. (1943) On the Statistical Treatment of Linear Stochastic Difference Equations. *Econometrica*, 11, 173–220.

Marple, S. L., Jr. (1980) A New Autoregressive Spectrum Analysis Algorithm. *IEEE Transactions on Acoustics, Speech, and Signal Processing*, 28, 441–54 (also in Kesler, 1986).

——— (1987) *Digital Spectral Analysis with Applications*. Englewood Cliffs, NJ: Prentice-Hall.

Martin, R. D. and Thomson, D. J. (1982) Robust-Resistant Spectrum Estimation. *Proceedings of the IEEE*, 70, 1097–115.

Matsakis, D. and Tavella, P. (2008) Special Issue on Time Scale Algorithms. *Metrologia*, 45, iii–iv.

McCloud, M. L., Scharf, L. L. and Mullis, C. T. (1999) Lag-Windowing and Multiple-Data-Windowing Are Roughly Equivalent for Smooth Spectrum Estimation. *IEEE Transactions on Signal Processing*, 47, 839–43.

McCoy, E. J., Walden, A. T. and Percival, D. B. (1998) Multitaper Spectral Estimation of Power Law Processes. *IEEE Transactions on Signal Processing*, 46, 655–68.

McCullagh, P. and Nelder, J. A. (1989) *Generalized Linear Models* (Second Edition). London: Chapman & Hall.

McHardy, I. and Czerny, B. (1987) Fractal X-ray Time Variability and Spectral Invariance of the Seyfert Galaxy NGC5506. *Nature*, 325, 696–8.

McLeod, A. I. and Hipel, K. W. (1978) Simulation Procedures for Box–Jenkins Models. *Water Resources Research*, 14, 969–75.

McLeod, A. I. and Jiménez, C. (1984) Nonnegative Definiteness of the Sample Autocovariance Function. *The American Statistician*, 38, 297–8.

——— (1985) Reply to Discussion by Arcese and Newton. *The American Statistician*, 39, 237–8.

McQuarrie, A. D. R. and Tsai, C.-L. (1998) *Regression and Time Series Model Selection*. Singapore: World Scientific.

Mercer, J. A., Colosi, J. A., Howe, B. M., Dzieciuch, M. A., Stephen, R. and Worcester, P. F. (2009) LOAPEX: The Long-Range Ocean Acoustic Propagation EXperiment. *IEEE Journal of Oceanic Engineering*, 34, 1–11.

Miller, K. S. (1973) Complex Linear Least Squares. *SIAM Review*, 15, 706–26.

——— (1974a) *Complex Stochastic Processes: An Introduction to Theory and Application*. Reading, MA: Addison-Wesley.

Miller, R. G. (1974b) The Jackknife – A Review. *Biometrika*, 61, 1–15.

Mitchell, R. L. and McPherson, D. A. (1981) Generating Nonstationary Random Sequences. *IEEE Transactions on Aerospace and Electronic Systems*, AES–17, 553–60.

Mohr, D. L. (1981) Modeling Data as a Fractional Gaussian Noise. Ph.D. dissertation, Department of Statistics, Princeton University.

Mombeni, H., Rezaei, S. and Nadarajah, S. (2017) Linex Discrepancy for Bandwidth Selection. *Communications in Statistics – Simulation and Computation*, 46, 5054–69.

Monro, D. M. and Branch, J. L. (1977) Algorithm AS 117: The Chirp Discrete Fourier Transform of General Length. *Applied Statistics*, 26, 351–61.

Moon, F. C. (1992) *Chaotic and Fractal Dynamics: An Introduction for Applied Scientists and Engineers*. New York: John Wiley & Sons.

Morettin, P. A. (1981) Walsh Spectral Analysis. *SIAM Review*, 23, 279–91.

——— (1984) The Levinson Algorithm and its Applications in Time Series Analysis. *International Statistical Review*, 52, 83–92.

Moulin, P. (1994) Wavelet Thresholding Techniques for Power Spectrum Estimation. *IEEE Transactions on Signal Processing*, 42, 3126–36.

Mullis, C. T. and Scharf, L. L. (1991) Quadratic Estimators of the Power Spectrum. In *Advances in Spectrum Analysis and Array Processing*, Volume I, edited by S. Haykin, Englewood Cliffs, NJ: Prentice-Hall, 1–57.

Munk, W. H. and Cartwright, D. E. (1966) Tidal Spectroscopy and Prediction. *Philosophical Transactions of the Royal Society of London, Series A*, 259, 533–81.

Munk, W. H. and MacDonald, G. J. F. (1975) *The Rotation of the Earth: A Geophysical Discussion.* Cambridge: Cambridge University Press.

Murray, M. T. (1964) A General Method for the Analysis of Hourly Heights of Tide. *International Hydrographic Review*, 41, 91–102.

——— (1965) Optimization Processes in Tidal Analysis. *International Hydrographic Review*, 42, 73–82.

Narasimhan, S. V. and Harish, M. (2006) Spectral Estimation Based on Discrete Cosine Transform and Modified Group Delay. *Signal Processing*, 86, 1586–96.

Neave, H. R. (1972) A Comparison of Lag Window Generators. *Journal of the American Statistical Association*, 67, 152–8.

Nelsen, R. B. (1998) Correlation, Regression Lines, and Moments of Inertia. *The American Statistician*, 52, 343–45.

Newton, H. J. (1988) *TIMESLAB: A Time Series Analysis Laboratory.* Pacific Grove, CA: Wadsworth & Brooks/Cole.

Newton, H. J. and Pagano, M. (1983) A Method for Determining Periods in Time Series. *Journal of the American Statistical Association*, 78, 152–7.

——— (1984) Simultaneous Confidence Bands for Autoregressive Spectra. *Biometrika*, 71, 197–202.

Nitzberg, R. (1979) Spectral Estimation: An Impossibility? *Proceedings of the IEEE*, 67, 437–8.

Nowroozi, A. A. (1967) Table for Fisher's Test of Significance in Harmonic Analysis. *Geophysical Journal of the Royal Astronomical Society*, 12, 517–20.

Nuttall, A. H. (1981) Some Windows with Very Good Sidelobe Behavior. *IEEE Transactions on Acoustics, Speech, and Signal Processing*, 29, 84–91.

Nuttall, A. H. and Carter, G. C. (1982) Spectral Estimation Using Combined Time and Lag Weighting. *Proceedings of the IEEE*, 70, 1115–25.

Ombao, H. C., Raz, J. A., Strawderman, R. L. and von Sachs, R. (2001) A Simple Generalised Cross-validation Method of Span Selection for Periodogram Smoothing. *Biometrika*, 88, 1186–92.

Oppenheim, A. V. and Schafer, R. W. (2010) *Discrete-Time Signal Processing* (Third Edition). Upper Saddle River, NJ: Pearson.

Pagano, M. (1973) When is an Autoregressive Scheme Stationary? *Communications in Statistics*, 1, 533–44.

Papoulis, A. (1973) Minimum-Bias Windows for High-Resolution Spectral Estimates. *IEEE Transactions on Information Theory*, 19, 9–12.

——— (1985) Levinson's Algorithm, Wold's Decomposition, and Spectral Estimation. *SIAM Review*, 27, 405–41.

Papoulis, A. and Pillai, S. U. (2002) *Probability, Random Variables, and Stochastic Processes* (Fourth Edition). New York: McGraw-Hill.

Park, J., Lindberg, C. R. and Thomson, D. J. (1987a) Multiple-Taper Spectral Analysis of Terrestrial Free Oscillations: Part I. *Geophysical Journal of the Royal Astronomical Society*, 91, 755–94.

Park, J., Lindberg, C. R. and Vernon, F. L., III. (1987b) Multitaper Spectral Analysis of High-Frequency Seismograms. *Journal of Geophysical Research*, 92, 12675–84.

Parzen, E. (1957) On Choosing an Estimate of the Spectral Density Function of a Stationary Time Series. *Annals of Mathematical Statistics*, 28, 921–32.

——— (1961) Mathematical Considerations in the Estimation of Spectra. *Technometrics*, 3, 167–90.

Pawitan, Y. and O'Sullivan, F. (1994) Nonparametric Spectral Density Estimation Using Penalized Whittle Likelihood. *Journal of the American Statistical Association*, 89, 600–10.

Percival, D. B. (1983) The Statistics of Long Memory Processes. Ph.D. dissertation, Department of Statistics, University of Washington.

——— (1991) Characterization of Frequency Stability: Frequency-Domain Estimation of Stability Measures. *Proceedings of the IEEE*, 79, 961–72.

——— (1992) Simulating Gaussian Random Processes with Specified Spectra. *Computing Science and Statistics*, 24, 534–8.

Percival, D. B. and Constantine, W. L. B. (2006) Exact Simulation of Gaussian Time Series from Nonparametric Spectral Estimates with Application to Bootstrapping. *Statistics and Computing*, 16, 25–35.

Percival, D. B. and Walden, A. T. (2000) *Wavelet Methods for Time Series Analysis*. Cambridge: Cambridge University Press.

Picinbono, B. (2010) ARMA Signals With Specified Symmetric Marginal Probability Distribution. *IEEE Transactions on Signal Processing*, 58, 1542–52.

Picinbono, B. and Bondon, P. (1997) Second-Order Statistics of Complex Signals. *IEEE Transactions on Signal Processing*, 45, 411–20.

Pignari, S. and Canavero, F. G. (1991) Amplitude Errors in the Burg Spectrum Estimation of Sinusoidal Signals. *Signal Processing*, 22, 107–12.

Pisarenko, V. F. (1973) The Retrieval of Harmonics from a Covariance Function. *Geophysical Journal of the Royal Astronomical Society*, 33, 347–66 (also in Kesler, 1986).

Pisias, N. G. and Mix, A. C. (1988) Aliasing of the Geologic Record and the Search for Long-Period Milankovitch Cycles. *Paleoceanography*, 3, 613–9.

Politis, D. N. (2003) The Impact of Bootstrap Methods on Time Series Analysis. *Statistical Science*, 18, 219–30.

Press, H. and Tukey, J. W. (1956) Power Spectral Methods of Analysis and Application in Airplane Dynamics. In *AGARD Flight Test Manual, Vol. IV, Instrumentation*, edited by E. J. Durbin, Paris: North Atlantic Treaty Organization, Advisory Group for Aeronautical Research and Development, C:1–C:41 (also in Brillinger, 1984a).

Press, W. H., Teukolsky, S. A., Vetterling, W. T. and Flannery, B. P. (2007) *Numerical Recipes: The Art of Scientific Computing* (Third Edition). Cambridge: Cambridge University Press.

Priestley, M. B. (1962) Basic Considerations in the Estimation of Spectra. *Technometrics*, 4, 551–64.

——— (1981) *Spectral Analysis and Time Series*. London: Academic Press.

Proakis, J. G. and Manolakis, D. G. (2007) *Digital Signal Processing: Principles, Algorithms and Applications* (4th Edition). Upper Saddle River, NJ: Pearson Prentice Hall.

Pukkila, T. and Nyquist, H. (1985) On the Frequency Domain Estimation of the Innovation Variance of a Stationary Univariate Time Series. *Biometrika*, 72, 317–23.

Quinn, B. G. (2012) The Estimation of Frequency. In *Handbook of Statistics 30: Time Series Analysis: Methods and Applications*, edited by T. Subba Rao, S. Subba Rao and C. R. Rao, Amsterdam: Elsevier, 585–621.

Quinn, B. G. and Hannan, E. J. (2001) *The Estimation and Tracking of Frequency*. Cambridge: Cambridge University Press.

Quinn, B. G., McKilliam, R. G. and Clarkson, I. V. L. (2008) Maximizing the Periodogram. In *IEEE GLOBECOM 2008*, 1–5.

Rabiner, L. R. and Gold, B. (1975) *Theory and Application of Digital Signal Processing*. Englewood Cliffs, NJ: Prentice-Hall.

Rabiner, L. R. and Schafer, R. W. (1978) *Digital Processing of Speech Signals*. Englewood Cliffs, NJ: Prentice-Hall.

Ralston, A. and Rabinowitz, P. (1978) *A First Course in Numerical Analysis* (Second Edition). New York: McGraw-Hill (reprinted in 2001 by Dover Publications, New York).

Ramsey, F. L. (1974) Characterization of the Partial Autocorrelation Function. *Annals of Statistics*, 2, 1296–301.

Rao, C. R. (1973) *Linear Statistical Inference and Its Applications* (Second Edition). New York: John Wiley & Sons.

Ray, B. K. and Tsay, R. S. (2000) Long-Range Dependence in Daily Stock Volatilities. *Journal of Business & Economic Statistics*, 18, 254–62.

Reid, J. S. (1979) Confidence Limits and Maximum Entropy Spectra. *Journal of Geophysical Research*, 84, 5289–301.

Rice, J. A. and Rosenblatt, M. (1988) On Frequency Estimation. *Biometrika*, 75, 477–84.

Rice, S. O. (1945) Mathematical Analysis of Random Noise, Part III: Statistical Properties of Random Noise Currents. *Bell System Technical Journal*, 24, 46–156.

Riedel, K. S. and Sidorenko, A. (1995) Minimum Bias Multiple Taper Spectral Estimation. *IEEE Transactions on Signal Processing*, 43, 188–95.

——— (1996) Adaptive Smoothing of the Log-Spectrum with Multiple Tapering. *IEEE Transactions on Signal Processing*, 44, 1794–800.

Rife, D. C. and Boorstyn, R. R. (1974) Single-Tone Parameter Estimation from Discrete-Time Observations. *IEEE Transactions on Information Theory*, 20, 591–8.

——— (1976) Multiple Tone Parameter Estimation from Discrete-Time Observations. *Bell System Technical Journal*, 55, 1389–410.

Rife, D. C. and Vincent, G. A. (1970) Use of the Discrete Fourier Transform in the Measurement of Frequencies and Levels of Tones. *Bell System Technical Journal*, 49, 197–228.

Roberts, R. A. and Mullis, C. T. (1987) *Digital Signal Processing*. Reading, MA: Addison-Wesley.

Rogers, J. L. and Nicewander, W. A. (1988) Thirteen Ways to Look at the Correlation Coefficient. *The American Statistician*, 42, 59–66.

Rosenblatt, M. (1985) *Stationary Sequences and Random Fields*. Boston: Birkhäuser.

Rovine, M. J. and von Eye, A. (1997) A 14th Way to Look at a Correlation Coefficient: Correlation as the Proportion of Matches. *The American Statistician*, 51, 42–6.

Rowe, D. B. (2005) Modeling Both the Magnitude and Phase of Complex-Valued fMRI Data. *NeuroImage*, 25, 1310–24.

Rutman, J. (1978) Characterization of Phase and Frequency Instabilities in Precision Frequency Sources: Fifteen Years of Progress. *Proceedings of the IEEE*, 66, 1048–75.

Sakai, H. (1979) Statistical Properties of AR Spectral Analysis. *IEEE Transactions on Acoustics, Speech, and Signal Processing*, 27, 402–9.

Sakai, H., Soeda, T. and Tokumaru, H. (1979) On the Relation Between Fitting Autoregression and Periodogram with Applications. *Annals of Statistics*, 7, 96–107.

Samarov, A. and Taqqu, M. S. (1988) On the Efficiency of the Sample Mean in Long-Memory Noise. *Journal of Time Series Analysis*, 9, 191–200.

Satorius, E. H. and Zeidler, J. R. (1978) Maximum Entropy Spectral Analysis of Multiple Sinusoids in Noise. *Geophysics*, 43, 1111–18 (also in Kesler, 1986).

Satterthwaite, F. E. (1941) Synthesis of Variance. *Psychometrika*, 6, 309–16.

——— (1946) An Approximate Distribution of Estimates of Variance Components. *Biometrics Bulletin*, 2, 110–14.

Scargle, J. D. (1982) Studies in Astronomical Time Series Analysis. II. Statistical Aspects of Spectral Analysis of Unequally Spaced Data. *Astrophysical Journal*, 263, 835–53.

Scheffé, H. (1959) *The Analysis of Variance*. New York: John Wiley & Sons.

Schreiber, T. and Schmitz, A. (2000) Surrogate Time Series. *Physica D*, 142, 346–82.

Schreier, P. J. and Scharf, L. L. (2003) Second-Order Analysis of Improper Complex Random Vectors and Processes. *IEEE Transactions on Signal Processing*, 51, 714–25.

——— (2010) *Statistical Signal Processing of Complex-Valued Data: The Theory of Improper and Noncircular Signals*. Cambridge: Cambridge University Press.

Schuster, A. (1898) On the Investigation of Hidden Periodicities with Application to a Supposed 26 Day Period of Meteorological Phenomena. *Terrestrial Magnetism*, 3, 13–41.

Serroukh, A., Walden, A. T. and Percival, D. B. (2000) Statistical Properties and Uses of the Wavelet Variance Estimator for the Scale Analysis of Time Series. *Journal of the American Statistical Association*, 95, 184–96.

Shimshoni, M. (1971) On Fisher's Test of Significance in Harmonic Analysis. *Geophysical Journal of the Royal Astronomical Society*, 23, 373–7.

Shumway, R. H. and Stoffer, D. S. (2017) *Time Series Analysis and Its Applications: With R Examples* (4th Edition). New York: Springer.

Siegel, A. F. (1979) The Noncentral Chi-Squared Distribution with Zero Degrees of Freedom and Testing for Uniformity. *Biometrika*, 66, 381–6.

——— (1980) Testing for Periodicity in a Time Series. *Journal of the American Statistical Association*, 75, 345–8.

Singleton, R. C. (1969) An Algorithm for Computing the Mixed Radix Fast Fourier Transform. *IEEE Transactions on Audio and Electroacoustics*, 17, 93–103.

Sjoholm, P. F. (1989) Statistical Optimization of the Log Spectral Density Estimate. In *Twenty-third Asilomar Conference on Signals, Systems & Computers, Volume 1*, San Jose, CA: Maple Press, 355–9.

Slepian, D. (1976) On Bandwidth. *Proceedings of the IEEE*, 64, 292–300.

——— (1978) Prolate Spheroidal Wave Functions, Fourier Analysis, and Uncertainty – V: The Discrete Case. *Bell System Technical Journal*, 57, 1371–430.

——— (1983) Some Comments on Fourier Analysis, Uncertainty and Modeling. *SIAM Review*, 25, 379–93.

Slepian, D. and Pollak, H. O. (1961) Prolate Spheroidal Wave Functions, Fourier Analysis and Uncertainty – I. *Bell System Technical Journal*, 40, 43–63.

Sloane, E. A. (1969) Comparison of Linearly and Quadratically Modified Spectral Estimates of Gaussian Signals. *IEEE Transactions on Audio and Electroacoustics*, 17, 133–7.

Sondhi, M. M. (1983) Random Processes with Specified Spectral Density and First-Order Probability Density. *Bell System Technical Journal*, 62, 679–701.

Spencer-Smith, J. L. and Todd, H. A. C. (1941) A Time Series Met With in Textile Research. *Supplement to the Journal of the Royal Statistical Society*, 7, 131–45.

Stein, M. L. (1999) *Interpolation of Spatial Data.* New York: Springer-Verlag.

Stephens, M. A. (1974) EDF Statistics for Goodness of Fit and Some Comparisons. *Journal of the American Statistical Association*, 69, 730–7.

Stevens, W. L. (1939) Solution to a Geometrical Problem in Probability. *Annals of Eugenics*, 9, 315–20.

Stoffer, D. S. (1991) Walsh–Fourier Analysis and Its Statistical Applications. *Journal of the American Statistical Association*, 86, 461–79.

Stoffer, D. S., Scher, M. S., Richardson, G. A., Day, N. L. and Coble, P. A. (1988) A Walsh–Fourier Analysis of the Effects of Moderate Maternal Alcohol Consumption on Neonatal Sleep-State Cycling. *Journal of the American Statistical Association*, 83, 954–63.

Stoica, P., Li, J. and He, H. (2009) Spectral Analysis of Nonuniformly Sampled Data: A New Approach Versus the Periodogram. *IEEE Transactions on Signal Processing*, 57, 843–58.

Stoica, P. and Moses, R. (2005) *Spectral Analysis of Signals.* Upper Saddle River, NJ: Pearson Prentice-Hall.

Stoica, P. and Sandgren, N. (2006) Smoothed Nonparametric Spectral Estimation via Cepstrum Thresholding: Introduction of a Method for Smoothed Nonparametric Spectral Estimation. *IEEE Signal Processing Magazine*, 23, 34–45.

Stoica, P. and Selén, Y. (2004) Model-Order Selection: A Review of Information Criterion Rules. *IEEE Signal Processing Magazine*, 21, 36–47.

Strang, G. (1999) The Discrete Cosine Transform. *SIAM Review*, 41, 135–47.

Sun, T. C. and Chaika, M. (1997) On Simulation of a Gaussian Stationary Process. *Journal of Time Series Analysis*, 18, 79–93.

Sutcliffe, P. R., Heilig, B. and Lotz, S. (2013) Spectral Structure of Pc3–4 Pulsations: Possible Signature of Cavity Modes. *Annales Geophysicae*, 31, 725–43.

Swanepoel, J. W. H. and Van Wyk, J. W. J. (1986) The Bootstrap Applied to Power Spectral Density Function Estimation. *Biometrika*, 73, 135–41.

Sykulski, A. M., Olhede, S. C., Lilly, J. M. and Danioux, E. (2016) Lagrangian Time Series Models for Ocean Surface Drifter Trajectories. *Journal of the Royal Statistical Society, Series C (Applied Statistics)*, 65, 29–50.

Tary, J. B., Herrera, R. H., Han, J. and van der Baan, M. (2014) Spectral Estimation–What Is New? What Is Next? *Reviews of Geophysics*, 52, 723–49.

Taylor, A. E. and Mann, W. R. (1972) *Advanced Calculus* (Second Edition). Lexington, MA: Xerox College Publishing.

Theiler, J., Eubank, S., Longtin, A., Galdrikian, B. and Farmer, J. D. (1992) Testing for Nonlinearity in Time Series: The Method of Surrogate Data. *Physica D*, 58, 77–94.

Therrien, C. W. (1983) On the Relation Between Triangular Matrix Decomposition and Linear Prediction. *Proceedings of the IEEE*, 71, 1459–60.

Thompson, R. (1973) Generation of Stochastic Processes with Given Spectrum. *Utilitas Mathematica*, 3, 127–37.

Thomson, D. J. (1977) Spectrum Estimation Techniques for Characterization and Development of WT4 Waveguide – I. *Bell System Technical Journal*, 56, 1769–815.

——— (1982) Spectrum Estimation and Harmonic Analysis. *Proceedings of the IEEE*, 70, 1055–96.

——— (1990a) Time Series Analysis of Holocene Climate Data. *Philosophical Transactions of the Royal Society of London, Series A*, 330, 601–16.

——— (1990b) Quadratic-Inverse Spectrum Estimates: Applications to Palaeoclimatology. *Philosophical Transactions of the Royal Society of London, Series A*, 332, 539–97.

——— (2007) Jackknifing Multitaper Spectrum Estimates. *IEEE Signal Processing Magazine*, 24, 20–30.

Thomson, D. J. and Chave, A. D. (1991) Jackknifed Error Estimates for Spectra, Coherences, and Transfer Functions. In *Advances in Spectrum Analysis and Array Processing*, Volume I, edited by S. Haykin, Englewood Cliffs, NJ: Prentice-Hall, 58–113.

Titchmarsh, E. C. (1939) *The Theory of Functions* (Second Edition). Oxford: Oxford University Press.

Todoeschuck, J. P. and Jensen, O. G. (1988) Joseph Geology and Seismic Deconvolution. *Geophysics*, 53, 1410–14.

Toman, K. (1965) The Spectral Shifts of Truncated Sinusoids. *Journal of Geophysical Research*, 70, 1749–50.

Tong, H. (1975) Autoregressive Model Fitting with Noisy Data by Akaike's Information Criterion. *IEEE Transactions on Information Theory*, 21, 476–80 (also in Childers, 1978).

Tsakiroglou, E. and Walden, A. T. (2002) From Blackman–Tukey Pilot Estimators to Wavelet Packet Estimators: A Modern Perspective on an Old Spectrum Estimation Idea. *Signal Processing*, 82, 1425–41.

Tseng, F. I., Sarkar, T. K. and Weiner, D. D. (1981) A Novel Window for Harmonic Analysis. *IEEE Transactions on Acoustics, Speech, and Signal Processing*, 29, 177–88.

Tufts, D. W. and Kumaresan, R. (1982) Estimation of Frequencies of Multiple Sinusoids: Making Linear Prediction Perform Like Maximum Likelihood. *Proceedings of the IEEE*, 70, 975–89.

Tugnait, J. K. (1986) Recursive Parameter Estimation for Noisy Autoregressive Signals. *IEEE Transactions on Information Theory*, 32, 426–30.

Tukey, J. W. (1961) Discussion, Emphasizing the Connection Between Analysis of Variance and Spectrum Analysis. *Technometrics*, 3, 191–219.

——— (1967) An Introduction to the Calculations of Numerical Spectrum Analysis. In *Spectral Analysis of Time Series*, edited by B. Harris, New York: John Wiley & Sons, 25–46 (also in Brillinger, 1984b).

——— (1980) Can We Predict Where "Time Series" Should Go Next? In *Directions in Time Series Analysis*, edited by D. R. Brillinger and G. C. Tiao, Hayward, CA: Institute of Mathematical Statistics, 1–31 (also in Brillinger, 1984b).

Ulrych, T. J. (1972a) Maximum Entropy Power Spectrum of Truncated Sinusoids. *Journal of Geophysical Research*, 77, 1396–400.

——— (1972b) Maximum Entropy Power Spectrum of Long Period Geomagnetic Reversals. *Nature*, 235, 218–9.

Ulrych, T. J. and Bishop, T. N. (1975) Maximum Entropy Spectral Analysis and Autoregressive Decomposition. *Reviews of Geophysics and Space Physics*, 13, 183–200 (also in Childers, 1978).

Ulrych, T. J. and Clayton, R. W. (1976) Time Series Modelling and Maximum Entropy. *Physics of the Earth and Planetary Interiors*, 12, 188–200.

Ulrych, T. J. and Ooe, M. (1983) Autoregressive and Mixed Autoregressive-Moving Average Models and Spectra. In *Nonlinear Methods of Spectral Analysis* (Second Edition), edited by S. Haykin,

Berlin: Springer-Verlag, 73–125.

Van Schooneveld, C. and Frijling, D. J. (1981) Spectral Analysis: On the Usefulness of Linear Tapering for Leakage Suppression. *IEEE Transactions on Acoustics, Speech, and Signal Processing*, 29, 323–9.

Van Veen, B. D. and Scharf, L. L. (1990) Estimation of Structured Covariance Matrices and Multiple Window Spectrum Analysis. *IEEE Transactions on Acoustics, Speech, and Signal Processing*, 38, 1467–72.

von Storch, H. and Zwiers, F. W. (1999) *Statistical Analysis in Climate Research*. Cambridge: Cambridge University Press.

Wahba, G. (1980) Automatic Smoothing of the Log Periodogram. *Journal of American Statistical Association*, 75, 122–32.

Walden, A. T. (1982) The Statistical Analysis of Extreme High Sea Levels Utilizing Data from the Solent Area. Ph.D. dissertation, University of Southampton.

——— (1989) Accurate Approximation of a 0th Order Discrete Prolate Spheroidal Sequence for Filtering and Data Tapering. *Signal Processing*, 18, 341–8.

——— (1990a) Variance and Degrees of Freedom of a Spectral Estimator Following Data Tapering and Spectral Smoothing. *Signal Processing*, 20, 67–79.

——— (1990b) Improved Low-Frequency Decay Estimation Using the Multitaper Spectral Analysis Method. *Geophysical Prospecting*, 38, 61–86.

——— (1992) Asymptotic Percentage Points for Siegel's Test Statistic for Compound Periodicities. *Biometrika*, 79, 438–40.

——— (1993) Simulation of Realistic Synthetic Reflection Sequences. *Geophysical Prospecting*, 41, 313–21.

——— (1995) Multitaper Estimation of the Innovation Variance of a Stationary Time Series. *IEEE Transactions on Signal Processing*, 43, 181–7.

——— (2013) Rotary Components, Random Ellipses and Polarization: A Statistical Perspective. *Philosophical Transactions of the Royal Society of London. Series A, Mathematical, Physical and Engineering Sciences*, 371, 20110554.

Walden, A. T. and Hosken, J. W. J. (1985) An Investigation of the Spectral Properties of Primary Reflection Coefficients. *Geophysical Prospecting*, 33, 400–35.

Walden, A. T., McCoy, E. J. and Percival, D. B. (1995) The Effective Bandwidth of a Multitaper Spectral Estimator. *Biometrika*, 82, 201–14.

Walden, A. T., Percival, D. B. and McCoy, E. J. (1998) Spectrum Estimation by Wavelet Thresholding of Multitaper Estimators. *IEEE Transactions on Signal Processing*, 46, 3153–65.

Walden, A. T. and Prescott, P. (1983) Statistical Distributions for Tidal Elevations. *Geophysical Journal of the Royal Astronomical Society*, 72, 223–36.

Walden, A. T. and White, R. E. (1984) On Errors of Fit and Accuracy in Matching Synthetic Seismograms and Seismic Traces. *Geophysical Prospecting*, 32, 871–91.

——— (1990) Estimating the Statistical Bandwidth of a Time Series. *Biometrika*, 77, 699–707.

Walker, A. M. (1960) Some Consequences of Superimposed Error in Time Series Analysis. *Biometrika*, 47, 33–43.

——— (1971) On the Estimation of a Harmonic Component in a Time Series with Stationary Independent Residuals. *Biometrika*, 58, 21–36.

Walker, J. (1985) Searching for Patterns of Rainfall in a Storm. *Scientific American*, 252 (1), 112–19.

Wei, L. and Craigmile, P. F. (2010) Global and Local Spectral-Based Tests for Periodicities. *Biometrika*, 97, 223–30.

Weisberg, S. (2014) *Applied Linear Regression* (Fourth Edition). Hoboken, NJ: John Wiley & Sons.

Weiss, G. (1975) Time-Reversibility of Linear Stochastic Processes. *Journal of Applied Probability*, 12, 831–6.

Welch, B. L. (1936) The Specification of Rules for Rejecting Too Variable a Product, with Particular Reference to an Electric Lamp Problem. *Supplement to the Journal of the Royal Statistical Society*, 3, 29–48.

——— (1938) The Significance of the Difference Between Two Means when the Population Variances are Unequal. *Biometrika*, 29, 350–62.

Welch, P. D. (1967) The Use of Fast Fourier Transform for the Estimation of Power Spectra: A Method Based on Time Averaging Over Short, Modified Periodograms. *IEEE Transactions on Audio and Electroacoustics*, 15, 70–3 (also in Childers, 1978).

Wen, Q. H., Wong, A. and Wang, X. L. (2012) Overlapped Grouping Periodogram Test for Detecting Multiple Hidden Periodicities in Mixed Spectra. *Journal of Time Series Analysis*, 33, 255–68.

Whittle, P. (1952) The Simultaneous Estimation of a Time Series' Harmonic Components and Covariance Structure. *Trabajos de Estadistica y de Investigacion Operativa*, 3, 43–57.

——— (1953) Estimation and Information in Stationary Time Series. *Arkiv för Matematik*, 2, 423–34.

Wiener, N. (1949) *Extrapolation, Interpolation, and Smoothing of Stationary Time Series*. Cambridge, MA: MIT Press.

Wilson, R. (1987) Finite Prolate Spheroidal Sequences and Their Applications I: Generation and Properties. *IEEE Transactions on Pattern Analysis and Machine Intelligence*, 9, 787–95.

Wilson, R. and Spann, M. (1988) Finite Prolate Spheroidal Sequences and Their Applications II: Image Feature Description and Segmentation. *IEEE Transactions on Pattern Analysis and Machine Intelligence*, 10, 193–203.

Wise, G. L., Traganitis, A. P. and Thomas, J. B. (1977) The Effect of a Memoryless Nonlinearity on the Spectrum of a Random Process. *IEEE Transactions on Information Theory*, 23, 84–9.

Woods, J. W. (1972) Two-Dimensional Discrete Markovian Fields. *IEEE Transactions on Information Theory*, 18, 232–40.

Woodward, W. A., Gray, H. L. and Elliott, A. C. (2017) *Applied Time Series Analysis With R* (2nd Edition). Boca Raton, FL: CRC Press.

Wright, J. H. (2002) Log-Periodogram Estimation of Long Memory Volatility Dependencies with Conditionally Heavy Tailed Returns. *Econometric Reviews*, 21, 397–417.

Wunsch, C. (2000) On Sharp Spectral Lines in the Climate Record and the Millennial Peak. *Paleoceanography*, 15, 417–24.

Wunsch, C. and Gunn, D. E. (2003) A Densely Sampled Core and Climate Variable Aliasing. *Geo-Marine Letters*, 23, 64–71.

Yajima, Y. (1989) A Central Limit Theorem of Fourier Transforms of Strongly Dependent Stationary Processes. *Journal of Time Series Analysis*, 10, 375–83.

Yaglom, A. M. (1958) Correlation Theory of Processes with Random Stationary nth Increments. *American Mathematical Society Translations* (Series 2), 8, 87–141.

——— (1987a) *Correlation Theory of Stationary and Related Random Functions, Volume I: Basic Results*. New York: Springer-Verlag.

——— (1987b) Einstein's 1914 Paper on the Theory of Irregularly Fluctuating Series of Observations. *IEEE ASSP Magazine*, 4(4), 7–11.

Yuen, C. K. (1979) Comments on Modern Methods for Spectrum Estimation. *IEEE Transactions on Acoustics, Speech, and Signal Processing*, 27, 298–9.

Zechmeister, M. and Kürster, M. (2009) The Generalised Lomb–Scargle Periodogram: A New Formalism for the Floating-Mean and Keplerian Periodograms. *Astronomy & Astrophysics*, 496, 577–584.

Zhang, H.-C. (1992) Reduction of the Asymptotic Bias of Autoregressive and Spectral Estimators by Tapering. *Journal of Time Series Analysis*, 13, 451–69.

Author Index

Abbiss, J. B., 648
Abraham, B., 645
Acheson, D., 228, 642
Adams, J. W., 197, 642
Ahmed, N., 217, 642
Akaike, H., 493–4, 642
Alavi, A. S., 151, 201, 642
al Wasel, I., 215, 645
Amrein, M., 262–3, 642
Ander, M. E., 644
Andersen, N. O., 509, 642
Anderson, T. W., 35, 37, 166, 215, 447, 503, 540, 552, 642
Arun, K. S., 16, 650

Babadi, B., 359, 642
Baggeroer, A. B., 485, 642
Balogh, A., 645
Barbé, K., 424, 642
Barnes, J. A., 162, 327, 329, 642
Barrett, J. F., 635, 642
Bartlett, M. S., 215, 269–70, 275, 302, 344, 414, 642–3
Beauchamp, K. G., 138–40, 643
Bell, B. M., 560, 643
Beltrão, K. I., 307, 643, 648
Beran, J., 407, 643
Berk, K. N., 452, 485, 643
Bishop, T. N., 35, 494, 656
Blackman, R. B., 185, 201, 264, 343, 643
Bloomfield, P., 92, 154, 160, 232, 237, 252, 280, 307, 513, 526, 550, 592, 643
Bluestein, L. I., 94
Bogert, B. P., 227, 301, 473, 643
Bohman, H., 278, 643
Bollenbacher, A. F., 649
Bolt, B. A., 204, 643
Bondon, P., 31, 232, 653

Boorstyn, R. R., 551, 654
Bøviken, E., 541, 643
Box, G. E. P., 16, 34–5, 40, 264, 333, 431, 446, 557, 643
Bracewell, R. N., 55, 61, 69, 72, 137, 643
Branch, J. L., 94, 651
Briggs, W. L., 92, 643
Brillinger, D. R., 175–6, 195, 203–4, 206, 215, 262, 643, 647, 653, 656
Brockwell, P. J., 21, 28, 40, 43, 158, 203, 333, 404, 448, 465, 471, 473, 482–3, 485, 488, 599–601, 615, 621, 643
Broersen, P. M. T., 174, 492, 643
Bronez, T. P., 355, 359, 374, 382, 424, 643–4
Brown, E. N., 359, 642
Bruce, A. G., 486, 644
Bunch, J. R., 462, 644
Burg, J. P., 457, 466–7, 471–5, 577–8, 644
Burrus, C. S., 647
Burshtein, D., 485, 488, 491, 644
Businger, P. A., 564, 644
Butler, R. P., 645

Canavero, F. G., 578, 653
Carter, G. C., 257, 414, 420, 644, 652
Cartwright, D. E., 512, 522, 652
Chaika, M., 605, 616, 655
Chambers, J. M., 621, 644
Chambers, M. J., 605, 644
Champeney, D. C., 49, 53, 55, 62, 69, 84, 137, 165, 170, 644
Chan, G., 217, 644
Chan, K.-S., 228, 615, 644
Chandna, S., 30, 644
Chatfield, C., 151, 644
Chave, A. D., 196, 356, 419, 644, 656
Chen, W. W., 197, 648
Chen, W. Y., 553, 558–60, 644

Chi, A. R., 642
Childers, D. G., 642, 644, 648–50, 656, 658
Chilès, J.-P., 165, 340, 644
Choi, B. S., 492, 644
Chonavel, T., 110, 644
Chung, K. L., 120, 588, 644
Clarkson, I. V. L., 653
Clayton, R. W., 494–5, 554, 656
Cleveland, W. S., 181, 280, 644
Coble, P. A., 655
Colosi, J. A., 651
Conover, W. J., 215–6, 645
Conradsen, K., 40, 645
Constantine, W. L. B., 228, 230, 611, 622, 631, 645, 653
Cooper, G. R., 485, 649
Courant, R., 63, 645
Craigmile, P. F., 440, 548, 601, 645, 657
Cramér, H., 108, 645
Crowell, D. H., 648
Cumming, A., 529, 645
Cutler, L. S., 642
Czerny, B., 407, 651

Daniell, P. J., 271, 645
Danioux, E., 655
Davies, N., 631–2, 645
Davies, R. B., 219, 601, 645
Davis, H. T., 404, 645
Davis, R. A., 21, 28, 40, 43, 158, 203, 333, 404, 448, 465, 471, 473, 482–3, 485, 488, 599–601, 615, 621, 643
Davison, A. C., 593, 605, 618, 631, 645
Day, N. L., 655
de Gooijer, J. G., 492, 494, 645
Delfiner, P., 165, 340, 644
Denison, D. G. T., 548, 645
de Paula, E. R., 553, 649
Dietrich, C. R., 601, 605, 645
Diggle, P. J., 215, 645
Doetsch, G., 55, 645
Doodson, A. T., 512, 645
Durbin, E. J., 653
Durrani, T. S., 648
Dzhaparidze, K. O., 307, 645
Dzieciuch, M. A., 651
Dziuba, S., 40

Efron, B., 356, 593, 618, 645
Einstein, A., 219, 645
Elliott, A. C., 658
Emery, W. J., 337, 645
Enck, P., 650
Eom, K.-B., 408, 649
Epanechnikov, V. A., 275, 646
Eubank, S., 656

Fairfield Smith, H., 264, 646
Falk, R., 3, 646
Fan, J., 307, 646
Farebrother, R. W., 543, 646

Farmer, J. D., 656
Fay, G., 215, 646
Ferguson, T. S., 45, 646
Figueiras-Vidal, A. R., 650
Filliben, J., 39–40
Findley, D. F., 651
Fisher, R. A., 1, 539–40, 646
Flannery, B. P., 653
Fodor, I. K., 355, 646
Folland, G. B., 61, 646
Forsyth, R. J., 645
Fougere, P. F., 181, 183, 194, 558–60, 646
Fox, W., 321–2
Franke, J., 474, 646
Friedlander, B., 554–5, 646
Frijling, D. J., 221, 315, 657
Fuller, W. A., 164, 166, 203, 214–5, 646

Galdrikian, B., 656
Gao, H.-Y., 307, 412, 646
Geçkinli, N. C., 197, 646
Geweke, J., 408, 646
Gneiting, T., 601, 646
Godolphin, E. J., 484, 646
Gold, B., 80, 145, 156, 653
Golub, G. H., 89, 375, 464, 505, 564, 599, 644, 646
Gong, G., 356, 593, 618, 645
Gonzalez, R. C., 217, 647
Gould, A., 645
Gradshteyn, I. S., 50, 345, 647
Grandy, W. T., Jr., 644, 646
Granger, C. W. J., 151, 407, 633, 647
Gray, H. L., 658
Graybill, F. A., 375, 647
Greene, D. H., 25, 647
Greenhall, C. A., 355, 369, 402, 437, 441, 647
Grenander, U., 145, 166, 185, 251, 257, 647
Grünbaum, F. A., 94–5, 647
Gujar, U. G., 633, 647
Gunn, D. E., 123, 658

Hall, P., 644
Hamming, R. W., 145, 647
Han, J., 655
Hannan, E. J., 262, 404, 523, 527, 540, 647, 653
Hannig, J., 307, 312, 647
Hanssen, A., 355, 647
Hansson, M., 382, 647
Hansson-Sandsten, M., 424–5, 647
Hardin, J. C., 123, 647
Harish, M., 219, 652
Harris, B., 656
Harris, F. J., 196, 277, 535, 647
Harris, P. E., 407, 649
Harte, D. S., 601, 645
Hasan, T., 160, 232, 592, 647
Hastie, T. J., 309, 647
Hayashi, Y., 30, 647
Haykin, S., 652, 656–7
He, H., 655
Healey, D. J., 642

Healy, M. J., 643
Heideman, M. T., 92, 647
Heilig, B., 655
Henson, V. E., 92, 643
Herrera, R. H., 655
Hilbert, D., 63, 645
Hill, I. D., 633, 648
Hill, R., 648
Hinkley, D. V., 593, 605, 618, 631, 645
Hipel, K. W., 599–600, 639, 651
Hogan, J. A., 359, 648
Holder, R. L., 648
Hosken, J. W. J., 407, 657
Hosking, J. R. M., 407, 648
Howe, B. M., 651
Hua, Y., 566, 648
Hudson, J. E., 648
Hurvich, C. M., 197, 307, 495, 648

Ibragimov, I. A., 165, 648
Ihara, S., 474, 648
Ishida, M., 298, 300, 649
Isserlis, L., 30, 648

Jackson, D. D., 564–6, 648
Jackson, P. L., 559–60, 648
Jacovitti, G., 131, 648
Jaynes, E. T., 473, 648
Jenkins, G. M., 151, 160, 169, 201, 251, 264, 290, 642–3, 648
Jensen, O. G., 407, 656
Jessup, A., 225
Jiménez, C., 169–70, 185, 651
Johnson, D. H., 647
Johnson, D. M., 140, 650
Johnson, G. E., 636, 648
Johnson, N. L., 633, 648
Jones, N. B., 16, 648
Jones, R. H., 16, 210, 404, 482, 485, 494, 510, 555, 645, 648–9
Joughin, I., 443, 649
Joyeux, R., 407, 647

Kahan, W., 564, 646
Kaiser, J. F., 196, 649
Kalveram, K.-T., 650
Kane, R. P., 553, 649
Kantz, H., 615, 649
Kapuniai, L., 648
Kasyap, R. L., 408, 649
Katz, R. W., 648
Kavanagh, R. J., 633, 647
Kaveh, M., 485, 649
Kay, S. M., 123, 478, 480, 485, 493–5, 502, 506, 530, 554, 558, 560–2, 595, 598–9, 624, 636–7, 639, 649
Keeling, R. F., 332, 649
Kendall, D. G., 302, 643
Kerner, C., 407, 649
Kesler, S. B., 647, 649, 650–1, 653–4
Kikkawa, S., 298, 300, 649

King, M. E., 366, 649
Kleiner, B., 644
Knuth, D. E., 25, 647
Kokoszka, P., 215, 649
Kooperberg, C., 307, 649
Koopmans, L. H., 108, 134, 210, 649
Koslov, J. W., 485, 649
Kowalski, A., 140, 650
Kreiss, J.-P., 631, 650
Kreutzberger, E., 307, 646
Krishnaiah, P. R., 647
Kromer, R. E., 485, 490, 650
Kubrick, S., 215, 219, 228, 297, 307–9, 311–12, 337, 540, 615, 644–6, 650, 652–3
Kullback, S., 297, 345, 650
Kumaresan, R., 564, 566–7, 656
Kung, S. Y., 16, 650
Künsch, H. R., 262–3, 642
Kuo, C., 366, 650
Kuo, F. F., 649
Kürster, M., 529, 544, 658

Lacoss, R. T., 494–5, 650
Lago, P. J., 648
Lagunas-Hernández, M. A., 474, 650
Lahiri, S. N., 215, 631, 650
Lakey, J. D., 359, 648
Lampard, D. G., 635, 642
Landers, T. E., 494–5, 650
Lang, S. W., 32, 650
Lanning, E. N., 140, 650
Lawrance, A. J., 485, 650
Lee, T. C. M., 296, 307, 312, 647, 650
Leeson, D. B., 642
Leibler, R. A., 297, 345, 650
Li, J., 655
Li, T.-H., 529–30, 650
Lii, K. S., 370, 650
Lilly, J. M., 655
Lindberg, C. R., 366, 650, 652
Linnik, Yu. V., 165, 648
Liu, B., 633–4, 636, 650
Liu, T.-C., 355, 650
Ljung, G. M., 643
Lomb, N. R., 528, 650
Longbottom, J., 632, 650
Longtin, A., 656
Lopes, A., 230, 650
Lopes, S., 650
Lotz, S., 655
Loupas, T., 30, 650
Lynch, D., 76–8, 178–9, 192, 200–1, 218–9, 294, 300–1, 306, 337, 382, 458–9, 489
Lysne, D., 466, 484, 650

MacDonald, G. J. F., 407, 652
MacNeill, I. B., 643
Madan, R. N., 648
Maitani, T., 30, 651
Makhoul, J., 28, 156, 473–4, 493, 555–6, 649, 651
Mallows, C. L., 633, 651

Mann, H. B., 486, 651
Mann, W. R., 25, 656
Manolakis, D. G., 241, 653
Marcy, G. W., 645
Marple, S. L., Jr., 478, 480, 494–5, 506, 530, 554, 558, 560–2, 649, 651
Martin, R. D., 196, 227, 486, 492, 644, 651
Matsakis, D., 326, 593, 651
McClellan, J. H., 32, 650
McCloud, M. L., 423, 651
McCoy, E. J., 408–9, 412, 651, 657
McCullagh, P., 308, 651
McDicken, W. N., 30, 650
McGunigal, T. E., 642
McHardy, I., 407, 651
McKilliam, R. G., 653
McLeod, A. I., 169–70, 185, 599–600, 639, 651
McPherson, D. A., 605, 651
McQuarrie, A. D. R., 477, 651
McWhirter, J. G., 648
Mercer, J. A., 330, 651
Mikosch, T., 215, 649
Miller, K. S., 30, 518, 546, 651
Miller, R. G., 356, 651
Mitchell, R. L., 605, 651
Mix, A. C., 124, 653
Mohr, D. L., 408, 651
Mombeni, H., 307, 651
Monro, D. M., 94, 651
Moon, F. C., 228, 651
Moore, T. A., 648
Morettin, P. A., 140, 462, 651–2
Moses, R., 160, 232, 492, 592, 655
Moulin, P., 307, 412, 652
Mullen, J. A., Jr., 642
Mullis, C. T., 359, 444, 651–2, 654
Munk, W. H., 407, 512, 522, 652
Munson, D. C., Jr., 633–4, 636, 650
Murphy, A. H., 648
Murray, M. T., 520, 652
Musial, F., 650

Nadarajah, S., 651
Nakagawa, J. K., 648
Narasimhan, S. V., 219, 652
Natarajan, T., 642
Neave, H. R., 277, 652
Nelder, J. A., 308, 651
Nelsen, R. B., 3, 652
Newsam, G. N., 601, 605, 645
Newton, H. J., 35, 170, 451, 463–4, 470, 480, 483–5, 528, 652
Nicewander, W. A., 3, 654
Nicholls, D. F., 404, 647
Nitzberg, R., 474, 652
Nowroozi, A. A., 540, 652
Nuttall, A. H., 197, 257, 414, 420, 652
Nyquist, H., 405, 440, 653

O'Sullivan, F., 307, 653
Olhede, S. C., 655
Ombao, H. C., 297, 307–9, 311–12, 652

Ooe, M., 494, 656
Oppenheim, A. V., 84, 92, 94, 579, 652

Pagano, M., 484–5, 528, 652
Paparoditis, E., 631, 650
Papoulis, A., 21, 114, 278, 452, 457, 463, 562, 579, 631, 634, 652
Parekh, A., 648
Park, J., 359, 366, 650, 652
Parzen, E., 251, 275, 280, 644, 652
Pawitan, Y., 307, 653
Percival, D. B., 307, 408, 412, 560, 605, 611, 622, 631, 643, 651, 653–4, 657
Picinbono, B., 31, 232, 636, 653
Pignari, S., 578, 653
Pillai, S. U., 21, 114, 562, 579, 631, 634, 652
Pintelon, R., 642
Piper, S. C., 649
Pisarenko, V. F., 554, 653
Pisias, N. G., 124, 653
Politis, D. N., 631, 653
Pollak, H. O., 66, 655
Poor, H. V., 650
Porter-Hudak, S., 408, 646
Poskitt, D. S., 644
Prescott, P., 38, 616, 657
Press, H., 197, 653
Press, W. H., 506, 528, 542, 653
Priestley, M. B., 21, 37, 108, 110, 113–5, 117–8, 213, 240, 250–2, 256, 269, 275, 292–3, 404, 412, 446, 448–9, 463, 477, 516, 519, 522, 534, 543–4, 588, 653
Proakis, J. G., 241, 653
Pukkila, T., 405, 440, 653

Quinn, B. G., 526, 528, 540, 653

Rabiner, L. R., 80, 145, 156, 447, 653
Rabinowitz, P., 599, 653
Radoski, H. R., 646
Ralston, A., 599, 653
Ramsey, F. L., 462, 653
Rao, C. R., 514, 588, 653
Rao, K. R., 642
Ray, B. K., 407, 654
Raz, J. A., 652
Reid, J. S., 485, 654
Reinsel, G. C., 643
Rezaei, S., 651
Rice, J. A., 527, 654
Rice, S. O., 298, 654
Richardson, G. A., 655
Riedel, K. S., 355, 391–2, 400, 412, 434, 654
Rife, D. C., 535, 551, 654
Roberts, R. A., 444, 654
Robinson, L., 645
Rogers, J. L., 3, 654
Rosenblatt, M., 145, 166, 257, 370, 463, 494, 527, 643, 647, 650, 654
Rovine, M. J., 3, 654
Rowe, D. B., 30, 654
Rutman, J., 431, 654
Ryzhik, I. M., 50, 345, 647

Sakai, H., 451, 485, 654
Salomonsson, G., 382, 647
Samarov, A., 166, 654
Sandgren, N., 307, 655
Santamaría-Perez, M. E., 650
Sarkar, T. K., 566, 648, 656
Satorius, E. H., 553, 654
Satterthwaite, F. E., 264, 654
Scarano, G., 131, 648
Scargle, J. D., 528–9, 654
Schafer, R. W., 84, 92, 94, 447, 579, 652–3
Scharf, L. L., 30–1, 232, 359, 366, 651–2, 654, 657
Scheffé, H., 486, 654
Scher, M. S., 655
Schmidt, L., 326
Schmitz, A., 615, 654
Schoukens, J., 642
Schreiber, T., 615, 649, 654
Schreier, P. J., 30–1, 232, 654
Schuster, A., 539, 654
Selén, Y., 492, 655
Serroukh, A., 636, 654
Shimshoni, M., 571, 654
Shumway, R. H., 21, 203, 404, 655
Sidorenko, A., 355, 391–2, 400, 412, 434, 654
Siegel, A. F., 541–2, 655
Singleton, R. C., 94, 655
Sitaram, A., 61, 646
Sjoholm, P. F., 301, 304, 655
Slepian, D., 57–8, 61–3, 66, 85, 87–8, 105, 384, 655
Sloane, E. A., 195, 371, 655
Smith, C. R., 644, 646
Smith, W. L., 642
Soeda, T., 654
Sondhi, M. M., 633–4, 636, 655
Soulier, P., 215, 646
Souza, R. R., 650
Spann, M., 95, 658
Spedding, T., 645
Spencer-Smith, J. L., 33, 503, 655
Spliid, H., 40, 645
Stark, P. B., 355, 646
Stegen, G. R., 553, 558–60, 644
Stein, M. L., 340, 655
Stephen, R., 651
Stephens, M. A., 216, 655
Stevens, W. L., 541, 655
Stoffer, D. S., 21, 140, 203, 404, 655
Stoica, P., 160, 232, 307, 492, 528, 592, 655
Stone, C. J., 649
Strang, G., 217, 655
Strawderman, R. L., 652
Subba Rao, S., 653
Subba Rao, T., 653
Sun, T. C., 605, 616, 655
Sutcliffe, P. R., 553, 655
Swanepoel, J. W. H., 618, 624–5, 631, 638, 655
Sydnor, R. L., 642
Sykulski, A. M., 30, 655
Szegő, G., 185, 647

Taqqu, M. S., 166, 654
Tary, J. B., 553, 655
Tavella, P., 593, 651
Taylor, A. E., 25, 656
Teukolsky, S. A., 653
Theiler, J., 615, 656
Therrien, C. W., 464, 509, 656
Thomas, J. B., 658
Thompson, R., 605, 656
Thomson, D. J., 176, 196, 211–12, 227, 351–2, 355–7, 359, 366, 374, 381–3, 386, 390–1, 415–6, 424, 434, 492, 512, 544, 547–9, 574, 582, 644, 650–2, 656
Thomson, R. E., 337, 645
Tiao, G. C., 656
Tibshirani, R. J., 309, 647
Titchmarsh, E. C., 78, 120, 167, 656
Tjøstheim, D., 466, 484, 650
Todd, H. A. C., 33, 503, 655
Todoeschuck, J. P., 407, 656
Tokumaru, H., 654
Toman, K., 558, 656
Tong, H., 228, 554–5, 615, 644, 656
Traganitis, A. P., 658
Truong, Y. K., 649
Tsai, C.-L., 477, 495, 648, 651
Tsakiroglou, E., 201, 656
Tsay, R. S., 407, 654
Tseng, F. I., 535, 656
Tufts, D. W., 564, 566–7, 656
Tugnait, J. K., 555, 656
Tukey, J. W., 160, 174, 185, 197–8, 201, 264, 343, 592, 643, 653, 656
Tukey, P. A., 644

Ulrych, T. J., 35, 494–5, 553–4, 559–60, 656
Umphrey, G. J., 643
Unwin, J. M., 484, 646

van der Baan, M., 655
Van Loan, C. F., 89, 375, 464, 506, 564, 599, 646
Van Schooneveld, C., 221, 315, 657
Van Veen, B. D., 355, 366, 650, 657
Van Wyk, J. W. J., 618, 624–5, 631, 638, 655
Vernon, F. L., III., 652
Vessot, R. F. C., 642
Vetterling, W. T., 653
Vincent, G. A., 535, 654
von Eye, A., 3, 654
von Sachs, R., 652
von Storch, H., 215, 335, 657

Wahba, G., 301, 306, 657
Wald, A., 486, 651
Walden, A. T., 30, 38, 194, 196, 201, 210, 213, 232, 262–3, 280, 290, 297, 300, 307, 353, 359, 366, 400, 402, 405, 407, 410, 412, 440, 512, 522, 542, 548, 616, 633, 644–5, 650–1, 653–4, 656–7
Walker, A. M., 526, 554, 657
Walker, J., 16, 657

Walker, J. S., 649
Wang, X. L., 658
Warburg, H. D., 512, 645
Watson, W., 645
Watts, D. G., 160, 169, 290, 648
Wei, L., 548, 657
Weiner, D. D., 656
Weinstein, E., 485, 488, 491, 644
Weisberg, S., 408, 440, 513, 516, 543, 549, 552, 588, 621, 657
Weiss, G., 484–5, 657
Welch, B. L., 264, 657
Welch, P. D., 351, 413, 415, 658
Well, A. D., 3, 646
Wen, Q. H., 544, 658
White, R. E., 262, 290, 297, 300, 650, 657
Whittle, P., 307, 526, 543, 588, 658
Wiener, N., 78, 658
Wilson, R., 95, 658
Winkler, G. M. R., 642
Wise, G. L., 635, 658
Wong, A., 658

Wood, S., 39
Woods, J. W., 601, 658
Woods, R. E., 217, 647
Woodward, W. A., 610, 658
Worcester, P. F., 651
Wright, J. H., 407, 658
Wunsch, C., 123–4, 658

Yaglom, A. M., 21, 165, 214, 219, 307, 340, 645, 658
Yajima, Y., 408, 658
Yavuz, D., 197, 646
Yuen, C. K., 194–5, 658

Zawalick, E. J., 646
Zechmeister, M., 529, 544, 658
Zeidler, J. R., 553, 654
Zhang, H.-C., 461, 658
Zwiers, F. W., 215, 335, 657

Subject Index

100% cosine data taper, 189

absolute deviations sinusoidal amplitude estimator (Laplace method), 529
absolutely
  continuous, 120
  summable, 165
acausal
  autoregressive (AR) process, 446, 506
  filter, 147–8, 152, 155
  purely
    autoregressive (AR) process, 447–8
ACF (*see* autocorrelation function)
acoustic properties of human speech
  AR model, 447
ACLS (*see* approximate conditional least squares)
ACS (*see* autocorrelation sequence)
ACVF (*see* autocovariance function)
ACVS (*see* autocovariance sequence)
adaptive multitaper spectral estimator, 389
  ACVS issue, 615
  distribution of, 390
  EDOFs, 390
  ocean wave data, 427–8
  Parseval's theorem in expectation, 390, 439
aggregated processes, 42
AIC (*see* Akaike's information criterion)
AICC (*see* Akaike's information criterion corrected for bias)
Akaike's information criterion, 494
  corrected for bias (AICC), 495
  order selection, 494
    centered time series, 495
Akaike's information criterion corrected for bias (AICC), 495
  order selection, 495
    centered time series, 495
    Willamette River data, 575–6

aliases, 82
aliasing, 83, 108, 113, 122–3, 569
  and averaging, 161
  antialiasing filter, 57, 224
  different sampling intervals example, 131
  discrete time/continuous frequency, 81
  sea level residuals, 522
amplitude spectrum, 54
analytic series, 114, 561, 578–9
  imaginary part of, 579
  Willamette River data, 578, 592
angular frequency, 8
antialiasing filter, 57, 224
applications of spectral analysis, 16–7
approximate conditional least squares (ACLS)
  conditions needed for ACLS to become ECLS
    complex-valued case, 517
    real-valued case, 515
  sinusoidal amplitude estimator, 515, 525, 549, 569–70, 572–3, 585, 589
  white noise variance estimator, 549, 570, 589
approximate least squares, 516, 585
approximate least squares sinusoidal amplitude estimator
  expected value of, 516, 585
AR (autoregressive) process (*see* autoregressive (AR) process)
AR(1) process
  asymptotic stationarity and causality, 448–9
  autocovariance sequence, 44
  backward least squares autoregressive parameter estimator, 510
  Burg autoregressive spectral estimator, 510
  forward least squares autoregressive parameter estimator, 510
  from sampling a continuous parameter/time process, 507
  infinite-order moving average form, 44
  maximum likelihood estimator, 482

model for colored noise, 335
parameter estimation from estimated ACVS, 335–6
simulation of Gaussian
  via circulant embedding, 639–40
  via Gaussian spectral synthesis method (GSSM), 639–40
  via Kay's method, 639
Yule–Walker autoregressive parameter estimator, 510
AR(2) process, 34, 168, 172–3
  alternative models, 440
  asymptotic time series bandwidth, 347
  bias in periodogram, 177
  Burg autoregressive spectral estimator, 509
  cumulative periodogram test, 216
  direct spectral estimator, 186
    and variability, 208
  discretely smoothed periodogram, 309–10, 348–9
  form of FBLS parameter estimator, 510
  innovation variance estimators, 406–7
  lag window spectral estimator, 294, 348–9
  log SDF estimator, 304–5
  maximum likelihood autoregressive spectral estimator, 489
  numerical examples
    Burg's algorithm, 509
    periodogram, 172–3, 510
    Yule–Walker estimator, 507–8
  simulated examples, 34
    direct spectral estimator properties, 242
    periodogram properties, 242
  simulation of Gaussian
    via circulant embedding, 602–3
    via Gaussian spectral synthesis method (GSSM), 608–9
    via Kay's method, 597
  tapering, 195
  unimodality of spectrum and bandwidth of time series, 300
  WOSA estimator, 416–7
  Yule–Walker autoregressive spectral estimator, 509
AR(4) process, 35, 168, 172–3
  and end-point matching, 239
  asymptotic time series bandwidth, 347
  bias in periodogram, 178
  Burg autoregressive spectral estimator, 468–9, 509
  DCT-based periodograms, 218
  direct spectral estimator, 186
  discretely smoothed direct spectral estimator, 310–11, 348–9
  eigenspectra, 361, 363, 368, 395, 397, 437
  FBLS autoregressive spectral estimator, 478, 509
  innovation variance estimators, 406–7
  lag window spectral estimator, 295, 301, 349
  log SDF estimator, 305–6
  low-rank weighted multitaper spectral estimator, 422–3
  maximum likelihood autoregressive spectral estimator, 482–3, 490

multitaper spectral estimator, 358–9, 364–5, 398–9, 400, 419
multitaper versus WOSA spectral estimation, 418–9
nonunimodality of spectrum and bandwidth of time series, 300–1
numerical examples
  Burg's algorithm, 468–9, 509
  periodogram, 172–3
  Yule–Walker estimator, 458–9, 461–2, 509
prewhitening, 198–201
SDF and expectation of direct spectral estimator, 193
simulated examples, 34
  direct spectral estimator properties, 242
  periodogram properties, 242
simulation of Gaussian
  via circulant embedding, 640–1
  via Gaussian spectral synthesis method (GSSM), 606, 641
  via Kay's method, 638–9
smoothing window
  leakage, 289
  parameter choice, 292
tapering, 185
WOSA estimator, 417–8
Yule–Walker autoregressive spectral estimator, 458–9, 461–2, 509
Yule–Walker versus periodogram spectral estimators, 508
ARIMA (autoregressive, integrated, moving average) process, 16
ARMA (see autoregressive moving average (ARMA) process)
ARMA(1,1) process, 44
  autocovariance sequence (ACVS), 44
  infinite-order moving average form, 44
  kurtosis, 632
  skewness, 632
    inequality, 633, 641
  variance, 632
asymptotic equivalence of autoregressive estimators, 484
asymptotic relative efficiency, 165
asymptotic stationarity and causality, 448–9
asymptotically efficient estimator, 166, 522–3
atmospheric CO_2 data, 332–3
atomic clock data, 1–2
  ACS for
    sample, 4
    theoretical, 10
  lag 1 scatter plot for, 3
  simulation of, 11
  spectrum for
    estimated, 12, 14
    theoretical, 10, 12, 14
atomic clock fractional frequency deviates, 326, 430
  associated time differences, 325–6
  direct spectral estimator, 327
  periodogram, 327
  Burg autoregressive spectral estimator, 501
  innovation variance
    estimation, 431–2, 499–500

Subject Index 669

simulation, 628–9
  lag window spectral estimator, 328, 501
  multitaper spectral estimator, 501
  periodogram 328
atomic clock timekeeping, 162, 593
augmented Yule–Walker equations, 452
autocorrelated cosine lag window, 284, 345
autocorrelation, 72, 96–8, 100, 102, 112
autocorrelation function (ACF), 27
autocorrelation method, 480
autocorrelation sequence (ACS), 4, 9, 27
  conjugate symmetry, 30
  estimation, 5, 19–20
  MA(2) process, 159
  sample, 3
  white noise, 19
autocorrelation width, 73, 97–8, 100, 102
  smoothing window, 251
  spectral window, 191–2, 194
autocovariance function (ACVF), 27, 38
  and the integrated spectrum, 114
  and the SDF, 114
  conjugate symmetry, 30
  Gaussian-shaped, 131
  positive semidefinite, 30
  symmetry, 27
autocovariance sequence (ACVS), 27, 29
  and integrated spectrum, 111
  ARMA process, 44, 600
  AR process, 44, 460, 508
  band-limited white noise, 379, 604
  biased estimator of, 2-7, 166, 168–9, 232, 295, 505, 561, 591
    for complex-valued process, 231, 562
    positive definite, 451
  classification examples, 121
  complex-valued process, 29, 42
  corresponding to
    direct spectral estimator, 188, 611–12
    lag window spectral estimator, 248, 612
    multitaper estimator, 433
    periodogram, 170–1
    WOSA estimator, 420, 435
    Yule–Walker autoregressive spectral estimator, 504, 508
  decay to zero, 120
  estimation of, 166, 231
  for process squared, 43
  fractionally differenced (FD) process, 440
  GSSM-based approximation, 606–7, 640
  harmonic process, 36–7, 117, 591
    complex-valued, 519, 562
  inverse Fourier transform of SDF, 111, 126
  moving average process, 32
    MA(2), 159
  piecewise-constant SDF, 348
  positive semidefinite, 28, 30, 43
  power-law process, 391, 605
  randomly phased sinusoid, 35, 511, 553, 598
  subsampling, 130
  symmetry and conjugate symmetry, 27, 29
  unbiased estimator of, 166, 168–9, 236–7, 271
    for complex-valued process, 562

versus spectral density function, 124
white noise, 31, 117
Wold's theorem, 117, 126
autoregressive (AR) process, 33, 445–6
  acausal, 446, 506
  acoustic properties of human speech, 447
  ACVS, 460, 508
  asymptotic stationarity and causality, 448–9
  causality, 446, 448, 506, 618, 638
    and parameter estimators, 448, 617
    assessment, 463
    conditions for, 34, 446
  characterization, 509
  order, 446
  order selection (*see* order selection for autoregressive process)
  plus white noise is an ARMA process, 554, 591
  pseudo-, 553
  purely acausal, 447
  simulation, 595
    zero initial conditions, 595
  simulation of Gaussian
    stationary initial conditions, 595, 599, 624, 637
    via Kay's method, 595–7
    via McLeod and Hipel's method, 599
  sinusoidal variations, 447
  spectral density function (SDF), 144–5, 446, 485, 609, 617–8
  stationarity condition, 34, 145, 446, 448, 557
autoregressive estimator
  and prewhitening, 491, 506
  asymptotic equivalences, 484
  Burg's algorithm, 466, 505
    complex-valued process, 502
  least squares estimators, 475, 478, 505
  maximum entropy method, 471
  maximum likelihood estimators, 480, 506
  pure parametric approach, 506
  variability assessment
    via large-sample theory, 485
    via simulation, 626
  Yule–Walker method, 449–51, 505
    complex-valued process, 502
autoregressive moving average (ARMA) process, 35, 492, 597–8
  ACVS, 600
  AR process plus white noise, 159, 554, 591
  causal, 600, 631
  maximum entropy, 473–4
  pseudo-AR process plus white noise, 554
  SDF, 145, 159
  simulation of Gaussian, 593–4, 597–8
    stationary initial conditions, 599–600, 637, 639
    via Kay's method, 598, 639
    via McLeod and Hipel's method, 600, 639
  simulation of non-Gaussian, 631, 638
  simulation via step-down scheme, 461
  with distinct AR and MA coefficients, 554, 591
  with identical AR and MA coefficients, 554

autoregressive moving average (ARMA) spectral estimator, 492, 504
autorelation sequence, 29, 31
average value of a function, 71

background continuum, 511, 513, 523, 544, 547, 549, 551, 553, 583, 588
backward least squares (BLS)
   autoregressive parameter estimator, 477, 484
   AR(1) example, 510
backward linear prediction, 455
   complex-valued process, 502, 564
backward linear prediction error, 455
band-limited function, 57–8, 93, 139
   extrapolation of, 66
   recovering the continuous from the sampled, 84
band-limited white noise, 128, 131, 379, 604
band-pass filter, 146
band-pass process, 299, 347
bandwidth
   of AR(2) and AR(4) time series, 347
   of spectrum, 292, 296, 346
      estimation of, 297
   of time series, 300–1, 321–2, 347
      estimation of, 300
   regularization, 357, 377, 425, 428, 434
   smoothing window, 245, 251, 268, 281, 292, 296, 300–1, 318, 322, 341, 346–7
      Bartlett–Priestley design, 344
      choice of, 252
      finer frequency grid, 255
      for sea level residuals, 522
      Grenander, 251, 268, 279, 281
      lag window spectral estimator, 279
      Parzen, 251
      repeated lag window estimator, 349–50
   spectral window, 206, 226–7, 229, 235, 268, 294, 296, 317–8, 323, 325, 328, 331, 335, 341, 347, 568, 623–4, 628
      approximation for sinusoidal multitaper spectral estimator, 400, 409
      basic multitaper spectral estimator, 437, 575
      definition for basic multitaper spectral estimator, 353
      definition for direct spectral estimator, 194
      definition for eigenspectrum, 353
      definition for lag window estimator, 256
      definition for weighted multitaper spectral estimator, 353
      definition for WOSA, 414–5
      for sea level residuals, 521–2
      frequency range for statistical results, 213, 244
      versus equivalent width, 214, 355, 369, 402
      versus regularization bandwidth, 402
      WOSA, 429
   spectral window versus smoothing window, 257
Bartlett
   lag window, 268–9, 281, 344, 412
   smoothing window, 269–70
   spectral window, 269–70

Bartlett–Priestley
   design window, 273–4, 344
   lag window, 268, 274, 281, 320, 344
   smoothing window, 274–5
   spectral window, 274–5
basic multitaper spectral estimator, 352–5, 432–4, 611, 613
   ACVS estimator for, 433
   and Gaussian white noise, 370–4, 402–3
   and improved estimation of
      innovation variance, 403–7
      exponent of power-law process, 407–12, 440
      log SDF, 412
   and periodicity testing, 544–8
   approximated by WOSA spectral estimator, 424
   AR(4) data, 358–65, 368–70, 393–400, 405–7, 419
   atomic clock fractional frequency deviates, 430–2, 501, 628–30
   bias of, 359, 392
   distribution of, 354, 356–7, 381–2, 409, 434
   energy preservation of, 366–7
   equivalent degrees of freedom (EDOFs) of, 354–5, 434
   equivalent width, 355
   expected value of, 353, 433
   in circulant embedding simulation method, 611
   jackknifing of, 356–7, 369–70
   ocean wave data, 425–9
   spectral window for, 353, 433
      bandwidth of, 353, 359, 433
         approximation (sinusoidal), 400–1, 434
      sinusoidal, 398–400
      Slepian, 358–9, 364–5
   variance of, 353–4, 359
      Gaussian white noise, 372–4, 402–3
   via quadratic spectral estimator, 374
basis for the class of band-limited functions, 66
Bessel function, 196, 234
best (backward) linear predictor, 455
   complex-valued process, 502, 564
best (forward) linear predictor, 452–3
   complex-valued process, 502, 564
best linear predictor, 404, 455, 460–1, 463, 472
best predictor, 463
bias
   and dynamic range, 177
   broad-band, 359, 380–2, 387–8, 390
      defined, 378
      indicator, 379–81, 439
      minimizing indicator of, 380
   due to leakage, 185, 192, 351
   due to smoothing window, 256, 290
   indicator for minimum-bias multitaper spectral estimator, 392, 441
   local, 290, 359, 381–2
      defined, 378
      magnitude indicator, 378–9, 381
   of direct spectral estimator, 192
   of multitaper spectral estimator, 359
   of periodogram, 163

of weighted multitaper spectral estimator, 391, 441
bivariate standard Gaussian PDF, 634
  Mehler's expansion, 635
Blackman–Tukey estimator, 252
block averaging, 270–1, 412–14
Bluestein's substitution, 94
blurring function, 93
bootstrapping, 593, 611, 618, 631, 638
  confidence intervals for innovation variance, 630
    second approach in AR case, 630
  confidence intervals for log of SDF, 623–5
    second approach in AR case, 626
boundary conditions (see circular or reflecting)
Brent's method, 528, 558
broad-band bias, 359, 380–2, 387–8, 390
  defined, 378
  indicator, 379–81, 439
  minimizing indicator of, 380
Burg autoregressive estimator, 466–8, 505
  AR(1) process, 510
  AR(2) data, 509–10
  AR(4) data, 468–9, 509
  assessing variability (SDF), 485, 624–6
  atomic clock fractional frequency deviates, 501
  causality, 466, 505
  complex-valued process, 502, 562
  FPE criterion, 494
  ocean wave data, 496–9
    order selection via PACS, 498–9
  of innovation variance, 629
  peak in SDF
    height, 576
    shifting and splitting, 560, 562, 576, 591
    width, 576
  Willamette River data, 575–7
Burg's algorithm, 466, 468, 470, 473, 478, 482, 491, 493, 505, 560, 575–6
  analytic series, 562
  and Yule–Walker method, 471
  AR(2) numerical examples, 509–10, 556
  observed prediction error, 466
  two-step, 560

carbon dioxide and global temperature, 366
cascaded filter, 145
Cauchy inequality, 257, 262, 371, 442
causal process
  autoregressive (AR), 34, 446, 448, 506, 618, 638
  autoregressive moving average (ARMA), 600, 631
causal filter, 147, 161
causality, 446, 448, 506, 618, 638
  and parameter estimators, 448, 617
  assessment, 463
  conditions for, 34, 446
centering
  AIC and AICC, 494–5
  and direct spectral estimator, 195, 241, 244, 548

  and periodogram, 184, 213, 240
  and simulations, 619, 627
  and tapering of a time series, 195
  and WOSA spectral estimator, 419–20
  of data for harmonic analysis, 583
central peak (lobe), 175
cepstrum, 301, 473, 503
  estimator, 302–3
Cesàro summability and sums, 78, 165
chaotic beam data, 228
  direct spectral estimator, 228–30
  periodogram, 228–9
Chebyshev's inequality, 164
chi-square distribution, 37, 204, 209–10, 539
  noncentral, 542, 552
chi-square RV, 37, 203, 209–10, 235, 408, 539, 546
  and confidence intervals, 438, 486, 490
  logarithm of, 209, 412
  expected value and variance, 302
  scaled, 264, 298, 343
  sum of chi-square RVs, 299
  weighted sum of chi-square RVs, 264
chirp transform algorithm, 94, 216
Cholesky decomposition, 599, 604, 637, 639
  modified, 464, 480, 599
circulant embedding simulation method, 440, 601–4, 611, 620, 627, 629, 631
  ACVS properties, 602, 640
  and direct SDF estimator, 611
  and lag window SDF estimator, 611–12
  and multitaper SDF estimator, 611, 613
  and periodogram, 615, 641
  and WOSA SDF estimator, 611, 613
  compared to method of surrogate time series (MSTS), 616
  computational efficiency, 603
  issues with ACVS, 615
  overcoming negative weights, 604, 637, 640
  weights, 601–2, 605, 611, 629, 637, 640
    AR(1) process, 640–1
    direct SDF estimator, 612
    lag window SDF estimator, 612
    multitaper SDF estimator, 613
    WOSA SDF estimator, 613
circular boundary conditions, 424
circularized sequence, 601–2, 640
classification of spectra, 120–2
CO_2 data, 332
  lag window spectral estimator for filtered, 335
  periodogram for filtered, 335
coefficient of determination, 570, 572, 621
colored noise, 120, 126, 219, 241, 377–8, 381, 389, 511, 513, 518, 520, 523, 539, 544, 550, 574, 584
  proper complex-valued, 519, 585
complementary covariance, 31
complete stationarity, 215
complex demodulation, 160, 232, 592
complex exponentials, 17, 48, 134
  linear combinations of, 132, 137
complex-valued colored noise (proper), 519, 585
complex-valued stochastic process
  continuous parameter/time, 23

discrete parameter/time, 23
complex-valued Gaussian or normal RV (proper), 30, 546–7
complex-valued process
  by complex demodulation, 160, 374
    resulting ACVS, 160
    resulting SDF, 160, 374
  example of, 127
complex-valued roots, 566–7, 580, 592
complex-valued stationary process, 29, 231, 374, 501–2, 519, 585
  time reverse of conjugate, 502
complex-valued white noise (proper), 517–8, 564
compound periodicity, 540–1
computational pathways, 221, 224, 315
concentration measure
  continuous time/continuous frequency, 62, 65
  discrete time/continuous frequency, 85, 105, 155, 386
  discrete time/discrete frequency, 94
  for eigenspectrum spectral window, 368
condition number of covariance matrix, 185
  and dynamic range, 185
confidence interval for
  innovation variance, 629
    bootstrapping, 630
    via simulation, 630
  jump in the integrated spectrum, 590
  log of SDF, 205, 227, 229, 266–7, 323, 328, 331, 335, 342, 354, 359, 425, 521
    bootstrapping, 623–5
  SDF, 205, 265, 267, 342, 357, 485, 488–90
    chi-square versus jackknifing, 438
conjugate symmetry, 29–30, 48, 54, 75, 92
consistent estimator of
  mean, 164
  SDF
    discretely smoothed direct spectral estimator, 246
    lag window spectral estimator, 260, 342
constructed multitaper spectral estimator, 355, 376
continuous parameter deterministic sinusoid, 558
continuous parameter white noise, 38
continuous parameter/time stochastic process, 22, 38, 113–4, 131, 507
  sampling and aliasing, 122–3
  sampling and averaging, 161
convergence
  factors, 80, 152–3, 157, 579, 592
  in mean square, 49, 53
  pointwise, 50, 53, 84
convolution, 67, 96, 98–9, 101
  as an LTI filter, 137, 157
    continuous parameter, 142, 247
    discrete parameter, 143, 247
  as smoothing operation, 69–70, 247, 385
  cyclic, 101, 106
  Fourier transform of, 67, 96, 98–9, 101
  illustration of, 68–9, 104
  in frequency domain, 96, 98–9, 101, 105
  of periodic functions, 96, 105
copper wires, 57

correlation distortion due to zero-memory nonlinearity (ZMNL), 635–6
correlation length, 131
correlation of direct spectral estimator, 212, 243
correlation time, 214
correlogram method power spectral density estimator, 252
cosine data taper, 189–90, 260
cosine lag window, 290
covariance
  complex-valued case, 25, 41
  real-valued case, 24
covariance matrix, 28, 42, 44, 184, 376, 379, 382, 391, 440, 450, 457, 465, 485
  Cholesky decomposition, 599, 604, 637, 639
  modified, 464, 480, 599
  condition number, 185
covariance method, 480
covariances of
  direct spectral estimator, 205, 209–13, 243
  in WOSA blocks, 415, 442
  eigenspectra, 354, 372, 436, 438
  lag window spectral estimator, 260–1, 342
  periodogram, 203–4, 235
    subtle issues, 214–5
  weighted multitaper spectral estimator, 437
  weighted sums of time series values, 242
CPDF (see cumulative probability distribution function)
cross-correlation (complex), 72, 96, 98–9, 101–2
cumulant, 215, 632
  inequality for fourth, 632–3
cumulative periodogram test, 215–6
  examples
    AR(2) data, 216
    ocean noise, 230–1
    Willamette River data, 550, 573–4
cumulative probability distribution function (CPDF), 23
  N-dimensional, 23
cyclic
  autocorrelation, 220
  convolution, 220, 273
  convolution theorem, 101, 106
  convolutions via FFTs, 106
cyclical frequency, 8

Daniell
  design window, 272
  lag window, 268, 273, 281, 290
  smoothing window, 250, 272–3, 320
  spectral window, 272–3
data taper (see taper)
data window, 186
DC (direct current), 518
DCT-based periodogram, 217, 244
  AR(4) process, 218
  asymptotic unbiasedness of, 218
  bias of, 218–9
  statistical properties, 219
decay
  of envelope of sidelobes, 269
  rate of sidelobes

Bartlett smoothing window, 269, 276
Bartlett–Priestley smoothing window, 275
Daniell smoothing window, 273
Gaussian smoothing window, 277
Papoulis smoothing window, 278
Parzen smoothing window, 276
spectral windows, 281, 328
decibel (dB) scale, 13, 209, 225–7, 266
deconstructed multitaper spectral estimator, 355, 376, 420–3
decorrelation time, 214
default data taper, 188, 190
degenerate kernel, 86
degrees of freedom
  for chi-square RV, 37, 202
  one, 13, 203
  two, 13, 203
  zero, 542
  equivalent (EDOFs), 264, 266
    discretely smoothed direct spectral estimator, 343
    discretely smoothed periodogram, 347
    lag window spectral estimator, 264, 279, 342
    multitaper spectral estimator, 354–5, 390, 434
    WOSA spectral estimator, 415–6, 442–3
  in a time series, 298, 507
  of order 2, 298
derivative as an LTI filter, 157
design window, 247
  for Bartlett–Priestley, 273–4, 319, 344
  for Daniell, 272, 319
determinant of a matrix, 480
deterministic
  function, 48
  sequence, 48
  sinusoid, 553
detrending (see trend removal and quadratic trend removal)
deviance, 308
  generalized cross-validated (GCV), 309, 313, 324, 339
DFT (see discrete Fourier transform)
diagonal matrix, 464, 565–6, 599
differencing
  for linear trend removal, 40
  for seasonality/periodicity removal, 40, 42, 333
  higher-order, 340
  to achieve stationarity, 340, 431
digamma function, 210, 296, 302, 345, 405, 409
Dirac delta function, 74, 113, 120, 157, 250, 524
direct current (DC), 518
direct spectral estimator, 164, 186, 201, 233, 340, 375, 535–6, 545, 611
  and centering a time series, 195–6, 241, 244, 548
  atomic clock data, 327–9
  average value of, 241
  averaging of uncorrelated, 443
  bias, 192
  chaotic beam data, 228–9
  computation of, 219–24
  correlations, 212, 243

corresponding ACVS estimator, 188, 452, 611
covariances, 205, 209–13, 243
discretely smoothed, 246–7
  equivalent degrees of freedom (EDOFs), 347
  spectral window bandwidth, 347
distribution of, 204, 235
  for complex-valued process, 232
eigenspectrum, 352
expected value of, 186, 234
  for real-valued harmonic process, 535, 585
exponential averaging, 444
for blocks in WOSA, 414
harmonic process examples, 536, 588
in circulant embedding simulation method, 611–12, 627–8
inconsistent estimator of SDF, 207, 235
linear combination of independent, 441–2
ocean wave data, 225–6, 496
relationship to quadratic spectral estimator, 376
ship altitude data, 330–1
simulating properties of, 242
smoothing of, 245
spectrum of, 213
use with AR order selection, 493
use with Yule–Walker autoregressive estimator, 452, 461–2, 505
variability, 208
variance of, 204, 235, 624
  for complex-valued process, 232
white noise, 189, 371
Dirichlet's kernel, 17–8, 76–7, 80, 85, 93, 106, 127, 233, 383, 537
discrete cosine transform, 184, 217
discrete Fourier transform (DFT), 74–5, 92, 219, 594, 602, 608
  as a convolution, 93–4
  chirp algorithm, 94
  conjugate symmetry, 75, 92
discrete parameter/time stochastic process, 22
discrete power spectrum, 50, 139–40
discrete prolate spheroidal sequence (DPSS), 87, 155, 189, 357
discrete prolate spheroidal wave function (DP-SWF), 87, 105–6, 383–5, 439
discrete spectrum, 120, 511–13, 518, 538, 584–6
discrete time sequence, 74
  autocorrelation width, 100
  convolution theorem, 99
  equivalent width, 100
  Fourier representation, 75, 99
  Parseval's theorem, 75, 99
  segment of, 91
    autocorrelation width, 102
    convolution theorem, 101
    equivalent width, 102
    Fourier representation, 92, 100
    Parseval's theorem, 92, 101
    spectral properties, 101
  spectral properties, 99
discretely smoothed direct spectral estimator, 246–7
  computation of, 316
  consistent estimator of SDF, 246

equivalence to lag window spectral estimator, 249–50
  finer frequency grid, 253
discretely smoothed periodogram, 307
  bias/variance trade-off, 309
  computational scheme, 347
  corresponding ACVS estimator, 312–13, 348, 615
  equivalent degrees of freedom (EDOFs), 347
  evaluating SDF estimates, 349
  ice profile data, 324
  parameter choice, 307, 309–10, 312, 324, 339, 348
  spectral window bandwidth, 347
distribution of
  adaptive multitaper spectral estimator, 390
  direct spectral estimator, 204, 235
    complex-valued process, 232
  lag window spectral estimator, 264, 342, 490
  maximum likelihood AR parameter estimator, 483
    other AR parameter estimators, 484
  multitaper spectral estimator, 354, 356, 381, 409, 434
  periodogram, 203, 235, 408
  WOSA spectral estimator, 436
Doppler measurements of blood flow, 131
double orthogonality
  of eigenfunctions, 65, 87
  of Slepian sequences (DPSSs), 106
double-sided verus single-sided SDF, 118
DPSS (see discrete prolate spheroidal sequence)
DPSWF (see discrete prolate spheroidal wave function)
duration–bandwidth product, 66, 357
dynamic range, 177, 184, 197, 492, 641
  and condition number, 185

echoes, 227
ECLS (see exact conditional least squares)
EDOFs (see equivalent degrees of freedom)
effective bandwidth (see spectral window bandwidth)
eigenfunction, 64
  double orthogonality, 65, 87
  for a linear filter, 135
eigenspectra, 352, 385, 432, 545
  bandwidth of spectral windows, 353, 358
  covariances of, 354, 372, 436, 438
  plots of for AR(4) data, 361, 363, 368, 395, 397
  regularization bandwidth, 357
  weights associated with, 352, 388–9
eigenvalue, 64, 86–7, 95, 185, 375, 380, 392, 565, 567, 580, 591
  for a linear filter, 135
eigenvector, 87, 95, 380, 392, 591
  left and right, 565, 567
end-point mismatch, 181, 183
energy
  of a nonperiodic function, 54–5
  of a periodic function, 49
energy spectral density function, 55, 97, 99, 132

ensemble, 21–2
entire function, 57
entropy, 471–3
envelope of sidelobes, 269, 273
Epanechnikov lag window (see Bartlett–Priestley lag window)
equal
  by definition, 8, 23
  in distribution, 203, 485, 599
  in the mean square sense, 49, 53
equivalence of time and frequency domains, 55
equivalent degrees of freedom (EDOFs), 264, 266
  adaptive multitaper spectral estimator, 390
    example (ocean wave data), 427–8
  basic multitaper spectral estimator, 354–5, 434
    example (ocean wave data), 427
  discretely smoothed direct spectral estimator, 343
  discretely smoothed periodogram, 347
  lag window spectral estimator, 264, 279, 342
    example (ocean wave data), 318–20
  via eigenvalues, 423–4
  weighted multitaper spectral estimator, 354
  WOSA spectral estimator, 415–6, 442–3
    example (ocean wave data), 429–30
equivalent width, 58–60, 73, 97–8, 100, 102, 213–4
  multitaper spectral estimator, 355, 437
  sinusoidal multitaper spectral estimator, 401, 441
  smoothing window, 251
  versus spectral window bandwidth, 214, 355, 369
  sinusoidal multitapers, 402
  Slepian multitapers, 369
ergodic theorems, 165
Euler relationship, 17
Euler's constant, 210, 302, 404, 408
exact conditional least squares (ECLS)
  including mean, 552
  sinusoidal amplitude estimator, 549, 569–70, 572, 589–90
  white noise variance estimator, 549, 552, 570, 589–90
exact unconditional least squares (EULS)
  including mean, 552
  sinusoidal amplitude estimator, 549, 569–70, 572
  sinusoidal frequency estimator, 549
  white noise variance estimator, 550, 552, 570
expected value (expectation), 4
expected value of
  direct spectral estimator, 186, 234
    real-valued harmonic process, 535, 585
  lag window spectral estimator, 255, 341, 343
  multitaper spectral estimator
    adaptive, 389–90, 439
    basic, 353, 433
    weighted, 352
  periodogram, 174, 233, 240
    complex-valued harmonic process, 524, 586
    real-valued harmonic process, 524, 585
    conditional on phase value, 537–8
  quadratic spectral estimator, 376–7, 439

sinusoidal frequency estimator using periodogram maximum, 526
  WOSA spectral estimator, 414, 435
exponential averaging, 444
exponential distribution, 202, 209, 215, 598
exponential RV, 209
extrapolation of a band-limited function, 66

fader, 186
fast Fourier transform (FFT), 92, 94, 237, 602, 608
FBLS (see forward and backward least squares)
F-distribution, 547, 574, 582, 587
  non-central, 552
  percentage points (special case), 544
Fejér's kernel, 80, 174–6, 180, 233, 240, 524, 532, 534, 537
  different expressions for, 236
  sampling of, 524
  sidelobes, 178–80, 328, 527, 532, 535–6
FFT (see fast Fourier transform)
filter
  acausal, 147–8, 152, 155
  analog, 133
  antialias, 57, 224
  band-pass, 146
  cascaded, 145
  causal, 147
  demodulation, 160
  digital, 140
  effect on integrated spectrum (filtering theorem)
    analog, 136, 157
    digital, 141
  example of an LTI digital, 148
  finite impulse response (FIR), 147–8, 152, 155, 161, 503
  first difference, 333
  for pilot analysis, 160
  high-pass, 146
    as trend remover, 151
    formed by subtracting output of low-pass filter from its input, 149
  infinite impulse response (IIR), 147, 503
  Kalman, 482
  kurtosis norm of, 632
  linear time-invariant (LTI), 132, 631, 633
    analog, 133
    and convolution, 137
    digital, 140
  low-pass, 146, 592
    cutoff frequency, 151
    design by least squares, 152
    design using Slepian sequences, 155
  prediction error, 198
  prewhitening, 197, 234, 241, 447, 482, 491–3, 496, 510, 615
  seasonal differencing, 333
  transfer function, 136, 141
  transformation or operator, 133
final prediction error (FPE), 493
finite impulse response (FIR) filter, 147–8, 152, 155, 161, 503

FIR (see finite impulse response)
first difference filter, 40, 333
  effect of filter on SDF, 158, 326
Fisher's test for simple periodicity, 539–40, 550, 586
  critical value, 581
    approximation, 540, 589
  ocean noise data, 581
  Willamette River data residuals, 571, 573
flicker frequency noise, 431
folding frequency (Nyquist frequency), 82
forward and backward least squares (FBLS)
  and SVD approach, 565
  autoregressive parameter estimator, 477, 484
    AR(2) examples, 510, 555–6
  autoregressive spectral estimator
    AR(4) example, 478, 482
    Willamette River data, 592
  same as modified covariance method, 480
forward least squares (FLS)
  autoregressive parameter estimator, 477, 484
    AR(1) example, 510
  same as covariance method, 480
forward linear prediction, 452–3
  complex-valued process, 502, 564
forward linear prediction error, 453, 596
Fourier coefficient, 49
Fourier frequencies, 8, 91, 171, 515, 523–4
  fundamental, 357
  grid of, 171
    alternative grid for uncorrelated SDF estimators, 206, 208, 212
Fourier integral representation, 54
Fourier representation of a periodic function, 49, 95
Fourier synthesis, 54
Fourier theory
  continuous time/continuous frequency, 53, 97
  continuous time/discrete frequency, 48, 95
  discrete time/continuous frequency, 74, 99
  discrete time/discrete frequency, 91, 100
Fourier transform pair, 54, 96–7, 99–100
  examples of, 171, 181, 188, 212, 220–2, 237, 247–8, 301–2
  for convolution of two functions, 67
  for derivative of a function, 61
FPE (see final prediction error)
fractionally differenced (FD) process, 166, 407–8, 440
  autocovariance sequence (ACVS), 440
  SDF, 407–8
  simulation by circulant embedding, 440
Fredholm integral equation
  of the first kind, 383
  of the second kind, 63, 86
frequency grid, 171, 527, 531, 558, 578, 605
  for circulant embedding, 601
  for uncorrelated SDF estimators, 206, 208, 212
frequency response function, 136
  Hilbert transform, 579
function, special type of
  blurring, 93
  digamma, 210

Dirac delta, 74
gamma, 440
modified Bessel, 196
periodic, 48
positive-part, 541
trigamma, 296
function width
  autocorrelation, 73, 97–8, 105, 191, 194, 288
  equivalent, 58, 73, 97–8, 105, 213
  half-power, 191–2, 288
  variance, 60, 73, 105, 191–2, 194, 288
fundamental Fourier frequency, 357
fundamental frequency, 569

gain function, 136
  squared
    for differencing, 334
    for Hilbert transform approximation, 592
    for ideal low-pass, high-pass and band-pass filters, 146
gamma function, 297, 345, 440, 542, 633
Gaussian distribution, 265
  generalized, 633
Gaussian kernel, 69
Gaussian lag window, 268, 276, 278, 281, 320, 330, 430
Gaussian or normal process, 29, 445, 463, 480, 490, 544
  autoregressive, 34, 445, 595
  complex-valued, 30
  entropy, 472
  moving average, 33, 595
  white noise, 31, 201, 210, 539, 594
Gaussian or normal RV
  complex-valued, 30, 210
  PDF and its Fourier transform, 55
Gaussian smoothing window, 276–8
Gaussian spectral synthesis method (GSSM), 605–8, 637, 640–1
  ACVS approximation, 607
  adjustment for power-law process, 610
  and adaptive multitaper spectral estimator, 615
  and prewhitening, 615
  role of sampling interval, 609
  weights for AR(1) process, 640–1
Gaussian spectral window, 278
Gaussian white noise, 31, 201, 210, 539, 594
  multitaper spectral estimator, 370
Gaussian-shaped ACVF and SDF, 131
Gegenbauer process, 610
generalized cross-validated (GCV) deviance, 309, 313, 324, 339
generalized Gaussian (GG) distribution, 633
generalized inverse, 566
generalized least squares, 412, 522
Gibbs phenomenon, 77, 79, 81, 152
goodness-of-fit test
  Kolmogorov, 215
grid size, 179, 527, 531, 558, 568, 578, 605
group delay, 136
GSSM (*see* Gaussian spectral synthesis method)

half-power width of function, 191–2, 288
Hamming lag window, 343
Hanning
  data taper, 189–90, 222, 230, 234, 260, 305, 327, 331, 345, 415–8, 535
    alternative form, 240, 285
  lag window, 343, 579
harmonic analysis, 511
  ocean noise data, 581
  ocean wave data, 582
  parametric approach, 553
    AR examples, 555–6
    problems with, 558
  singular value decomposition (SVD) approach, 563
  summary, 584
  Willamette River data, 567
    centering of time series, 583
harmonic process, 35, 108, 511
  ACVS for, 36–7, 117
  expected value of, 36
  periodicity, 603–4
  random phase, 37, 45
  simulation, 593–4, 598–9, 616, 637, 641
    examples, 36
  variance of, 36, 108, 110
  variance spectrum of, 108
harmonic process with additive noise
  complex-valued, 519, 585–6
    ACVS for, 519, 562, 585–7
    expected value of, 519, 585–6
    known frequencies, 517, 586
    unknown frequencies, 586
    variance of, 519, 585–6
  real-valued, 512, 544, 584–5
    ACVS for, 519, 584, 591
    expected value of, 516, 584
    known frequencies, 513, 515, 585
    random phase, 516
    signal-to-noise ratio, 526, 532
    tapered, 535–6, 588
    unknown frequencies, 523, 585
    variance of, 519, 526, 532, 584
harmonic tidal prediction, 512
  at Southampton, England, 513
harmonics
  in tidal prediction, 522
  of fundamental frequency, 569
Heisenberg's uncertainty principle, 61–2, 104
Hermite polynomials, 634–5
Hermitian Toeplitz matrix, 30
high-cut frequency, 57
high-pass
  filter, 146
  process, 299, 347
  creation of from a low-pass filter, 160
Hilbert transform, 114, 562, 579
  frequency response function of, 579, 592
  impulse response sequence of, 579, 592
  transfer function of, 579, 592

ice profile data, 321
  lag window spectral estimator, 322–3
  periodogram, 321, 323

Subject Index 677

plot of, 322
IID (independent and identically distributed)
  noise, 32, 631
  process, 32
IIR (see infinite impulse response)
imaginary part of
  analytic series, 579
  complex-valued number, 551
  complex-valued process, 602, 606, 640
impulse response
  function, 142
  sequence, 143
    for Hilbert transform, 579, 592
    normalization of, 150
inconsistency of
  direct spectral estimator, 235
  periodogram, 163–4, 207–8
index-limited sequence, 85
indirect nonparametric estimator, 252
infinite impulse response (IIR) filter, 147, 503
initial conditions
  AR process simulation, 595
  Gaussian AR process simulation, 595, 599
  Gaussian ARMA process simulation, 600, 639
inner product, 470
innovation process, 446
  estimation of, 17
innovation variance, 403–4, 446, 449, 455, 468,
    471–2, 510
  Burg autoregressive estimator, 499–500, 629
  confidence interval for, 629
  different estimators compared, 406, 440
  direct spectral estimator, 405
  mean square linear prediction error for AR(p),
    455
  multitaper spectral estimator, 405, 432, 629
    variability assessment via simulation, 630
  periodogram-based estimator, 404
  simulation, 628–9
    variability assessment via, 629
  smoothed periodogram estimator, 404
  Yule–Walker autoregressive estimator, 499–
    500
integrable (in square sense), 49, 53–4
integral equation, 63, 86
integral time scale, 131, 214
integrated magnitude square error, 52, 76
integrated spectrum (spectral distribution func-
    tion), 110, 117, 163, 511
  absolutely continuous, 120
  and AR SDF (parametric sinusoidal power
    estimation), 578
  and LTI filters, 135–6, 141
    cascade of filters, 146
  basic properties, 116
  classification examples, 121
  continuous parameter stationary process, 114
  continuous singular function, 120
  differentiability, 120
  discrete parameter stationary process, 110
  jumps (steps), 120, 511, 549, 551, 589–90
    confidence interval for, 590
    distribution of estimator, 552, 590
    Willamette River example, 570, 572, 575

intrinsically stationary process
  order d, 340
  order unity, 431
inverse Fourier transform (see Fourier transform
    pair)
invertibility of moving average process, 43
irregularly sampled observations, 355, 528, 543
Isserlis theorem, 30, 43, 210, 236, 241, 298, 371

jackknifing
  and multitaper spectral estimator, 356, 369–70
  for confidence interval calculation, 438
Johnson system of distributions, 633
joint moment, 27
joint stationarity, 29
jump process, 109
jumps in the integrated spectrum, 120, 511, 549,
    551, 589–90
  confidence interval for, 590
  distribution of estimator, 552, 590
  Willamette River example, 570, 572, 575

Kalman filter, 482
kernel
  degenerate, 86
  Dirichlet's, 17–8, 76, 93
  Fejér's, 80, 174–5
  Gaussian, 69, 290
  of an integral equation, 63, 86
  rectangular, 71, 290
Kolmogorov goodness-of-fit test, 215
Kronecker's delta function, 44
kth-order Slepian data taper, 357, 382, 384
  plots of, 360, 362
Kullback–Leibler (KL) discrepancy, 297, 312,
    346, 349
  expected value, 297
  for scaled chi-square RVs, 345
  symmetrized form, 345–6, 349
  with swapped arguments, 345–6, 349
kurtosis, 631
  ARMA(1,1) process, 632
  norm of a filter, 632

lag, 2, 27
  lag 1 scatter plot, 2
  lag p differencing, 40
  lag τ scatter plot, 2
lag window, 248, 341
  choice of, 287
  compared to data taper, 282–3, 285
  constructed from data taper, 283, 344–5
  parameter m (controls smoothing), 247, 287
    choice of, 291, 304, 313–4
    examples of, 318, 322, 328, 330, 339
  properties of, 250
  reshaped, 257, 281–2, 375
  type
    autocorrelated 100% cosine, 283–5, 344–5
    Bartlett, 268–9, 281, 344, 412
    Bartlett–Priestley, 268, 273, 281, 320, 344
    cosine (100%), 283–4, 290

678 Subject Index

Daniell, 268, 271, 281, 290, 320
Gaussian, 268, 276, 278, 281, 320, 331, 430
Hamming, 343
Hanning, 343, 579
Papoulis, 268, 278–9, 281, 345
Parzen, 248, 268, 275, 281, 301, 320–1, 323, 331, 344, 346, 496, 521–2
Slepian, 285, 345
lag window spectral estimator, 245, 248, 340–1, 492, 612
  alterative names for, 252
  as a quadratic spectral estimator, 375
  as deconstructed multitaper estimator, 420–2
  asymptotic distribution of, 264–5, 297, 342, 490
  asymptotic unbiasedness of, 256, 341
  asymptotic variance of, 258–9, 262, 265, 279, 341
    verification of validity by simulation, 622–3
  atomic clock data, 328–9, 501
  computation of, 314–7
  consistent estimator of SDF, 260, 342
  corresponding ACVS estimator, 248, 341, 612
  covariances, 260, 342
  equivalence to discretely smoothed direct spectral estimator, 249–50
  equivalent degrees of freedom (EDOFs), 264, 279, 342
  expected value of, 255, 341, 343
  filtered CO_2 data, 335
  gamma PDF, 297
  ice profile data, 321–4
  in circulant embedding simulation method, 611–12, 619–20, 627
  ocean wave data, 316–9, 349–50, 430, 496
  parameter m (controls smoothing), 247, 287
    choice of, 291, 304, 313–4
      examples of, 318, 322, 328, 330, 339
    repeated, 349–50
  sea level residuals, 521–2
  ship altitude data, 330–2
  smoothing window bandwidth B_W, 251, 279
    alternative measure β_W, 251, 279
  use in prewhitening, 491
Laplace periodogram, 529–30
law of total expectation, 588
leakage, 77–8, 178, 180, 196, 233, 287, 358, 532, 592
  and end-point mismatch, 181, 239
  bias in spectra due to, 225, 227, 233, 351
  reduction of, 78, 185
  smoothing window, 268, 287–9, 294–5, 316, 318–9, 498
least squares (LS), 40, 50, 332, 409, 430–1, 475–8, 513–6
  approximate, 516, 585
  approximate conditional (*see* approximate conditional least squares)
  complex-valued, 517, 546, 565
  exact conditional (*see* exact conditional least squares)
  exact unconditional (*see* exact unconditional least squares)
  filter design, 152, 155

generalized (GLS), 412, 522
ordinary (OLS), 409, 430, 440, 522
least squares (LS) autoregressive estimators, 475–9, 505–6
  AR(1) process, 510
  assessing variability (SDF), 485
  backward, 477, 484, 510
  causality not guaranteed, 479, 506, 510, 617
  forward, 475–7, 484, 510
  forward and backward, 477–8, 484, 509, 555–6, 565
  relation to Yule–Walker scheme, 479–80
Lebesgue decomposition theorem, 120, 518
Levinson–Durbin recursions, 452, 456–7, 461–2, 464–6, 509, 556, 604, 639
  for complex-valued time series, 502
  linear algebra derivation, 463–4
  step-down scheme, 460–1, 507–8, 597
    causality assessment, 461, 463, 509
    generation of ACVS for an AR process, 460, 508
    simulation of ARMA process, 461, 595–8
likelihood function, 308, 480
  and order selection criteria, 494–5
  log of, 480–1
  reduced (profile), 482–4
  Whittle, 307
line spectrum, 120, 511, 522
linear and log scales for SDFs, 336–9
linear model, 2, 39, 46, 621
linear regression, 151, 408–10, 412, 440, 446, 501
  complex-valued, 517, 545
  real-valued, 513–5, 529
  robust, 530
linear taper, 186
linear time-invariant (LTI) filter, 132
  analog, 133
  digital, 140
  linearity of, 133, 140
linear trend removal
  by differencing, 40
  by linear least squares estimation, 40
linear window, 186
local bias, 290, 359, 381–2
  defined, 378
  indicator, 378–9, 381
  minimizing indicator of, 381
log and linear scales for SDFs, 336–9
log spectral density function, 301
  estimation of
    log direct spectral estimator, 301–2
    with smoothing, 303–6, 324
Lomb–Scargle periodogram, 528–9, 543–4, 588
long memory process, 407
long-range dependence, 407–8, 411
Lorenzian SDF, 131, 507
loss of resolution, 76, 78, 271, 290, 357, 359, 400, 434, 535
low-cut frequency, 57
low-pass
  filter, 146, 592
  process, 299, 347
  use of to create a high-pass filter, 160

low-rank approximation to SDF estimator, 420–3
lower triangular matrix, 464, 484, 509, 599, 639
LTI (*see* linear time-invariant)

MA (moving average) process (*see* moving average (MA) process)
male speech, 57
Mann–Wald theorem, 486
marginal distribution, 45, 631, 634
matching condition, 135–6, 141–3
matrix
   Cholesky decomposition of, 599, 604, 637, 639
     modified, 464, 480, 599
   covariance, 28, 42, 44, 184, 376, 379, 382, 391, 440, 450, 457, 465, 485
   determinant of, 480
   diagonal, 464, 565–6, 599
   generalized inverse of, 566
   lower triangular, 464, 484, 509, 599, 639
   rank of, 375–6, 380–1, 420, 422–3, 490, 565–6
   Schur, 484
   singular value decomposition (SVD) of, 425, 564–5
     and short time series, 567
     approach to harmonic analysis, 563
       Willamette River data, 578–80, 592
     singular value, 565
     singular vector, 565
   Toeplitz, 29, 156, 457, 462, 480, 484
     Hermitian, 30
   trace of, 376, 378, 439
   tridiagonal, 88, 95
maximum confusion spectral analysis, 473
maximum entropy
   principle of, 471, 473
     for ARMA process, 473–4
   spectral analysis (MESA), 194, 447, 466, 471–4, 503
     example for MA processes, 475–6
maximum likelihood autoregressive estimator (MLE), 480–4, 506
   approximations to, 484, 510
   AR(1) process, 482
   AR(2) data, 489
   AR(4) data, 482–3, 490
   assessing variability (SDF), 485
   distribution of, 483
   recursive, 485
mean, 4, 24
   complex-valued case, 25
   estimation of, 3, 164–6, 184, 195–6, 213, 548
mean correction (*see* centering)
mean integrated square error (MISE), 303
mean square convergence, 49, 52–3
mean square error (MSE), 167, 236, 293, 346, 349, 386–8, 404, 463, 526, 537
   normalized (NMSE), 296, 314, 346, 349
mean square linear prediction error, 453–5, 457, 460, 468, 493, 509
   same as AR(p) innovation variance, 455
mean square log error (MSLE), 296, 346, 349

mean square prediction error, 463
Mehler's expansion, 635
method of surrogate time series (MSTS), 615–6, 641
   and harmonic processes, 641
   compared to circulant embedding, 616
minimum-bias multitapers, 392–3, 441
   bias indicator of, 392, 441
   plots of, 393
   relationship to sinusoidal multitapers, 392–3
mixed spectrum, 120–1, 359, 511, 513, 518–9, 521–2, 539, 584–5
modified
   Cholesky decomposition, 464, 480, 599
   covariance method, 480
   Bessel function, 196
   Daniell spectral estimator and windows, 280
   periodogram, 186
moment
   coefficient of kurtosis, 631
   coefficient of skewness, 631
   first and second, 27
   joint, 27
moving average (MA) process, 32, 43, 503–4, 594–5
   autocovariance sequence (ACVS), 32, 128
   invertibility, 43
   noninvertibility, 474
   MA(1)
     ACVS, 504
     as infinite-order AR process, 43
     simulated examples, 33
   MA(2), 159–60
   simulation of, 594–5, 637, 639
   spectral density function (SDF), 130, 144, 503
   textile slivers, 503
moving average (MA) spectral estimator, 503
   and truncated periodogram, 504
MSTS (*see* method of surrogate time series)
multitaper spectral estimator, 351–5, 432–4
   adaptive, 389–90
   ACVS (issue of being ill-defined), 615
   distribution of, 390
   versus weighted multitaper estimator, 390
   and improved estimation of
     innovation variance, 403–7
     exponent of power-law process, 407–12, 440
     log SDF, 412
   and periodicity testing, 544–8
   approximated by WOSA spectral estimator, 424
   AR(4) data, 358–65, 368–70, 393–400, 405–7, 419
   atomic clock fractional frequency deviates, 430–2, 501, 628–30
   basic, 352–5, 432–4, 611, 613
     ACVS estimator for, 433
     and Gaussian white noise, 370–4, 402–3
     bias of, 359, 392
     distribution of, 354, 356–7, 381–2, 409, 434
     energy preservation of, 366–7
     EDOFs of, 354–5, 434
     equivalent width, 355

expected value of, 353, 433
jackknifing of, 356–7, 369–70
spectral window for, 353, 433
  bandwidth of, 353, 359, 433
    approximation (sinusoidal), 400–1, 434
  sinusoidal, 398–400
  Slepian, 358–9, 364–5
  variance of, 353–4, 359
    Gaussian white noise, 372–4, 402–3
constructed, 355, 376
deconstructed, 355, 376
eigenspectra (see basic entry for eigenspectra)
in circulant embedding simulation method, 611–13
information recovery, 195, 366, 372
minimum-bias, 391–3
ocean wave data, 425–9
peak-matched multitapers, 382
sinusoidal multitapers, 355, 391–4, 396, 434
Slepian multitapers, 355, 357–8, 360, 362, 368, 434
versus prewhitening, 424
versus WOSA spectral estimator, 414, 417–9, 424
via quadratic spectral estimator, 374
via regularization, 382
weighted, 352–4, 388, 433, 611–13
  and quadratic spectral estimator, 376
  bias of, 391
  covariances of, 437
  distribution of, 354–5
  EDOFs of, 354–5
  equivalent width, 355, 437
  expected value of, 352, 389
  low-rank approximation, 422–3
  Parseval's theorem in expectation, 390
  spectral window for, 352–3, 391
    bandwidth of, 353
  variance of, 354
multitaper-based test for periodicity, 544–8
  sinusoidal amplitude estimator (complex-valued), 546, 551, 590
  Willamette River data, 574–5
  white noise variance estimator, 546, 551, 590
multivariate Gaussian (normal), 4, 29–30, 483, 485, 510
mutually (pairwise) uncorrelated random variables, 8, 13, 108, 203–4, 206, 235, 246, 258, 302, 306, 381, 410, 443, 465, 546, 595–6

Newton–Raphson method, 528, 558
noise
  colored, (see entry for colored noise)
  flicker frequency, 431
  IID, 32, 631
  in signal-to-noise ratio, 526
  random walk frequency, 431
  white, (see entry for white noise)
  with heavy-tailed distribution, 530
noncentral
  F-distribution, 552

chi-square distribution, 542, 552
nonergodic stationary process, 128
non-Gaussian (non-normal) process, 36, 210, 412, 463, 530, 631–6
  plots with examples of, 31
nonindependent but uncorrelated random variables, 45–6, 108
nonlinear lag relationship, 5–6
nonnegative definite (see positive semi-definite)
nonperiodic function, 53–4, 107, 143
nonstationary process
  complex-valued, 42
  differencing for stationarity, 333, 340, 431
  examples of, 39–41, 44–6, 332
nonwhite noise (see colored noise)
normal equations, 514, 517
normal process (see Gaussian process)
normalized mean square error (NMSE), 296, 314, 346, 349, 607
normalized observed prediction error
  backward, 618
  forward, 618, 624, 626, 630, 638
normalized root mean-square error, 406, 440
Nyquist
  frequency, 82, 84, 108, 122–3, 142
  rate, 84, 123

observed prediction error
  backward, 466, 492, 618
    normalized, 618
  forward, 466, 491, 496, 617–8
    normalized, 618, 624, 626, 630, 638
ocean noise data, 1–2
  ACS for
    sample, 4
    theoretical, 10
  lag 1 scatter plot for, 3
  periodogram for, 231
  simulation of, 11
  spectrum for
    estimated, 12, 14
    theoretical, 10, 12, 14
  testing for white noise, 11, 230–1, 581–2
ocean tide prediction, 512
ocean wave data, 224–8, 316–21, 425–30, 496–9, 582–3, 619–26
  adaptive multitaper spectral estimator, 428
  equivalent degrees of freedom (EDOFs), 427–8
  Burg autoregressive spectral estimator, 496, 500, 625
  direct spectral estimator, 226
  estimated PACS, 498–9
  harmonic analysis of, 582–3
  lag window spectral estimator, 318–9, 349–50, 430, 496, 500, 623
  multitaper spectral estimator, 426, 582
  periodogram, 226–7
  plot of, 225, 620
  prewhitening, 496–7
  reshaped spectrum, 582
  simulated series, 620
  uses for, 626
  Thomson's F-test for periodicity, 582

WOSA spectral estimator, 430
Yule–Walker autoregressive spectral estimator, 496
octave, 269, 273, 276–7, 282, 288–9
one-sided versus two-sided SDF, 118
orchestral music, 57
order
  of autoregressive process, 33, 446
  of autoregressive moving average process, 35
  of autoregressive spectral estimator, 458
  of linear predictor, 452–3, 455
  of moving average process, 32, 503
  of taper in multitapering scheme, 357, 392–3
order selection for autoregressive process, 492–5
  AIC method, 494
    centered time series, 495
  AICC method, 495
    centered time series, 495, 575–6
  direct spectral estimator criterion, 493
  examples of use of, 496, 498–500
  FPE method, 493
  PACS method, 493, 498
ordinary least squares (OLS), 409, 430, 440, 522
orthogonal
  data tapers, 354
  increments, 109
  process, 109
orthogonality
  of Hermite polynomials, 635
  principle, 454
    corollary to, 508
  relationships (trigonometric), 19, 202, 516
orthonormality
  basis exhibiting, 138, 384
    nonuniqueness of, 138
  of data tapers, 432
  of eigenvectors, 565
oscilloscope, 21–2
outliers, 196, 227
oversmoothing, 71
oxygen isotope ratios, 366

PACS (see partial autocorrelation sequence)
pairwise uncorrelated random variables (see mutually (pairwise) uncorrelated random variables)
Papoulis
  lag window, 268, 278–9, 281, 290, 345
  smoothing window, 278–9
  spectral window, 279
parametric approach to harmonic analysis, 553–8
  AR(2) example, 555–6
  AR(16) example, 556–7
  problems with, 558–62
  Willamette River data, 575–8
parametric spectral estimation, 14, 445, 482, 491–2, 503, 506
  AR versus MA, 503
  summary, 505
Parseval's theorem, 49, 55, 75, 92
  in expectation

adaptive multitaper spectral estimator, 390, 439
weighted multitaper spectral estimator, 390
one discrete time sequence, 99
  example of usage, 85, 111, 115, 188, 194, 251, 260, 303, 385, 390
one nonperiodic function, 97
  example of usage, 61
one periodic function, 96
  example of usage, 49, 52, 137–8
one segment of discrete time sequence, 101
two discrete time sequences, 99
two nonperiodic functions, 97
two periodic functions, 96
two segments of discrete time sequences, 101
partial autocorrelation
  coefficient, 457, 466, 469, 509
  sequence (PACS), 462–3
    complex-valued process, 502
    estimated for ocean wave data, 498–9
    for AR order selection, 493, 498
Parzen
  lag window, 248, 268, 275–6, 281–2, 292, 316, 320–1, 331, 344, 346, 496, 521–2
  smoothing window, 275, 280–1, 344
  spectral window, 275
PDF (see probability density function)
peak-matched multitapering, 382
peak shifting, 559–60, 576
  and splitting, 560
    analytic series, 561
    multiple sinusoids, 562
    with Burg's algorithm, 560, 562, 591
    with Yule–Walker approach, 560–2, 591
peak width, 576
Pearson product moment correlation coefficient, 3
percentage of block overlap in WOSA, 415
periodic extension, 54, 92
periodic function, 48–52, 74, 95–7
  convolution theorems, 96–7
  Fourier representation, 49, 95–6
  Parseval's theorem, 49, 96
  spectral properties, 50, 96
periodicity
  compound, 540–1, 587
  elimination by differencing, 40
  simple, 540–1, 586
  testing with white noise, 538–44, 586–7
    Fisher's test for (simple periodicity), 539, 586
      example of use, 571, 573, 581
    irregular sampling (Lomb–Scargle), 543–4
    Schuster's test for (simple periodicity), 539
    Siegel's test for (compound periodicity), 541, 587
      example of use, 581
    Thomson's F-test for (colored noise), 544–8, 587
      example of use, 574–5, 582–3, 592
periodogram, 170–185, 232–3
  and DFT of time series, 116, 171
  and real-valued harmonic process plus white noise, 523

average value of, 241
averaging of independent, 441
bias, 163, 176–7, 218
  AR(2) process (low dynamic range), 177
  AR(4) process (high dynamic range), 178–80
  asymptotic (unbiasedness), 175
  average value of, 239
  due to use of log periodogram, 304–5, 307
bootstrapping and, 615–6, 631
computation of, 219–24
covariances, 203–4, 235
  verification via simulation, 242
  subtle issues, 214
DCT-based, 217–9, 244
deficiency of power after removal of lines, 550–1
definition, 170, 232
  for complex-valued time series, 231
discretely smoothed (see discretely smoothed periodogram)
distribution of, 203, 235, 307–8, 408
  verification via simulation, 242
effect on periodogram of
  centering of time series, 184, 213, 240
    at nonzero Fourier frequencies, 184
    at zero frequency, 184
  circular shifting of time series, 238–9
  scaling of time series, 238
  time indexing of time series, 184
examples of
  AR(2) data, 172–3, 182, 208, 216, 246, 249, 294, 305, 310
  AR(4) data, 172–3, 182, 199, 508
  atmospheric CO_2 data, 335
  atomic clock data, 327–8
  chaotic beam data, 228
  ice profile data, 323
  ocean noise data, 231
  ocean wave data, 226–7, 497
  rotation of earth data, 224
  harmonic process plus white noise, 533–4, 536, 551
  ship altitude data, 331
  three sinusoids (Kay and Marple), 531
  white noise data, 207, 210
  Willamette River data, 568, 571, 573
expected value of, 174–5, 203, 233, 240
  complex-valued harmonic process plus white noise, 524, 586
  real-valued harmonic process plus white noise, 524, 536, 585
    conditional on phase value, 537–8
  verification via simulation, 242
in circulant embedding simulation method, 615
inconsistent estimator of SDF, 163–4, 207, 235
Laplace, 529
log likelihood, 308
Lomb–Scargle, 528–9, 543–4, 588
modified, 186
power-law exponent estimation, 408
reflecting boundary conditions, 244
rescaled, 541, 581

spurious peaks, 527
sum over
  Fourier frequencies, 237, 540
  non-Fourier frequencies, 233, 237
symmetry, 170
three-point example to verify properties, 238
truncated, 343, 504
utility compared to multitaper spectral estimator, 403–12
variance of, 203, 235, 241
  for AR(1) process, 243
  verification via simulation, 242
phase function (phase shift function), 136
phase-shifted Fourier transform, 383
phases (see random phases)
pilot analysis (filters for), 160
pointwise convergence, 50, 53, 84
Poisson's formula, 84
polynomial regression model, 522
positive
  definite, 28, 170, 375, 380, 422, 450–2, 454, 457, 461, 464, 480
  semidefinite, 28, 30, 167, 170, 374–5, 457
positive-part function, 541, 636, 640
postcoloring, 198–201, 234, 496, 615
  AR model approach, 491–2, 510
  atomic clock data, 329, 339–40
  ocean wave data, 497–500
power
  of a nonperiodic function, 55
  of a periodic function, 50
power spectral density function (PSDF), 113, 116
power-law exponent, 403, 407–12
  estimation using linear regression and multitaper estimator, 409
    atomic clock data, 430
  periodogram, 408
  periodogram versus multitaper estimation, 410–11
power-law process, 327–30, 336–7, 403, 407, 430–1
  ACVS, 391, 434, 605
  spectral density function (SDF), 391, 434, 605, 610
predictability of a time series, 17, 403–4, 471–2
prediction error
  backward, 455
  forward, 453, 596
prediction error filter, 198
prewhitening, 179, 197–201, 447, 491–2, 503, 506, 615
  versus multitaper spectral estimator, 424
  ocean wave data, 496–8, 510
prewhitening filter, 197, 234, 241, 447, 482, 491–3, 496, 510, 615
principle of maximum entropy, 447, 471, 473
probability density function (PDF), 24
  bivariate standard Gaussian, 634–5
  chi-square with two degrees of freedom, 37
  exponential (scaled chi-square with two degrees of freedom), 37, 209
  gamma (scaled chi-square with arbitrary degrees of freedom), 297

generalized Gaussian (GG), 633
Gaussian (normal) with zero mean, 55
modeling, 631
spectral window as an example of, 192
uniform
  over interval $(-\pi, \pi]$, 37, 45, 598
  over interval $(-\sqrt{3}, \sqrt{3}]$, 634
probability integral transform, 634
probability of an event, 23
process
  band-pass, 299, 347
  high-pass, 299, 347
  low-pass, 299, 347
  residual, 550
  with stationary differences of order d, 340
profile likelihood, 482–4
projections, 384, 439
prolate spheroidal wave functions (PSWFs), 64
proper
  complex-valued Gaussian or normal RV, 546–7
  complex-valued white noise, 517–8, 564
  process, 30, 32
PSDF (*see* power spectral density function)
pseudo-AR process, 553
  complex-valued, 563
  finding the coefficients, 591
  plus white noise is an ARMA process, 554
  solution to defining equation, 554, 591
  stationarity condition, 557
  SVD
    parameter estimator, 566
    sinusoidal frequency estimator, 567
purely
  acausal, 447–8
  continuous spectrum, 120, 163, 165, 167, 511
  continuous stationary process, 457, 532
  discrete spectrum, 120, 511, 513
  nondeterministic process, 404, 471

QQ plot
  Gaussian stationary process assumption, 621–2
  IID Gaussian assumption, 621
quadratic lag window (*see* Bartlett–Priestley lag window)
quadratic spectral estimator, 374–82, 439
  bias of
    broad-band, 378
      indicator, 379–81
      minimizing indicator of, 380
    for white noise, 377–8
    local, 378
    magnitude indicator, 378–9, 381
  deconstructed multitaper spectral estimator, 376
  examples of
    direct spectral estimator, 375
    lag window spectral estimator, 375
    weighted multitaper spectral estimator, 376
      justification for, 351
    WOSA spectral estimator, 423
  expected value of, 376–7, 439

regularization bandwidth, 377
  variance of, 380
quadratic taper, 248
quadratic trend removal
  by ordinary least squares (OLS), 332, 443
  by differencing, 46, 333–4
quadratic window, 248
  estimator, 252

radar clutter, 131
random amplitudes, 35–7, 111
random phases, 37–8, 111
  nonuniformly distributed, 38, 45
  uniformly distributed, 37, 511, 553, 598, 616
random variable, 3, 21–3
  complex-valued, 21, 23
  vector-valued, 21–2
random walk, 43
  frequency noise, 431
randomly phased sinusoid, 35, 511, 553, 598
  ACVS for, 553, 591
  plus white noise, 554
rank of matrix, 375–6, 380–1, 420, 422–3, 490, 565–6
Rayleigh
  distribution, 37, 598
  quotient, 89, 105
  theorem, 49
real part of complex-valued
  process, 602, 606, 640
  variable, 61, 486
real-valued stochastic process
  continuous parameter/time, 22
    stationary, 26
  discrete parameter/time, 22
    stationary, 26
realizations, 3, 21–2
reciprocity relationships, 58–62
  equivalent width, 58–9
  fundamental uncertainty relationship, 60–2
  similarity theorem, 58–60
rectangular
  data taper, 188–90, 535
  kernel, 71, 290
recursive maximum likelihood autoregressive parameter estimator (RMLE), 485
reduced (profile) likelihood, 482–4
reflecting boundary conditions, 181, 183, 216, 219, 244, 424
reflection coefficient (*see* partial autocorrelation coefficient)
regularization, 377, 386
  bandwidth, 357, 377, 425, 428, 434
  versus spectral window bandwidth, 402
rejection filtration, 201
relation function, 31
relative efficiency, 165
remodulated filtered series, 592
repeated lag window spectral estimator, 349–50
rescaled periodogram, 541, 581
reshaped
  lag window, 257, 281–2, 375
  spectrum, 547, 551, 575, 582, 592
    Willamette River data, 575

residual process (EULS), 550
  periodogram of
    harmonic process plus white noise, 551
    Willamette River data, 570
resistor, 21, 40–1
resolution
  loss of, 76, 78, 271, 290, 357, 359, 400, 434, 535
restoration of power, 195
Riemann integral, 24, 26, 112–13, 258, 261
  approximation by a Riemann sum, 174, 254, 258–9, 404, 605–6
Riemann sum, 174, 254, 258–9, 404, 605–6
Riemann–Lebesgue lemma, 120, 167
Riemann–Stieltjes integral
  ordinary, 24–6, 109, 112, 127
  stochastic, 110, 112, 127
ringing, 196, 225, 227, 277, 320, 331
ripples, 72, 77, 79, 152–3, 189, 253, 275, 290, 294
robust estimation of SDF via WOSA, 419
rotation of the earth, 148, 222, 366, 407
rule of thumb for significance levels, 548, 574, 582

sample autocorrelation, 3
sample autocorrelation sequence (sample ACS), 3
  examples, 4
  correlation of, 4
  variance of, 4
sample mean, 3, 164–6, 184, 195–6, 213
  variance for AR(1) process, 507
sampled function, 74
sampling and aliasing, 122–3, 569
sampling interval, 84–5, 113, 119, 122–5, 142–3, 163, 207, 232, 352, 374, 432, 445, 505, 511
  computational details, 220–4, 314–5
  irregular, 382, 528–9, 543
  jitter, 124
    effect on oxygen isotope time series, 124
  oversampling, 123
  role in simulation of time series, 594, 601, 609, 613
  subsampling, 124
    example, 130–1
scale preservation (LTI filters), 133, 140
scatter plot
  lag 1 and four examples, 2–3
  lag τ, 2, 5
  $\tau = 6$ and 9 examples, 6
  locations of peak frequencies from estimated spectra, 537–8, 556–7, 589, 591
Schur matrix, 484
Schuster's test for simple periodicity, 539
Schwarz inequality, 55, 57, 61, 104, 257
SDF (see spectral density function)
sea levels, 520
seasonality/periodicity, 40, 332
  removal by differencing, 40, 42, 333
second differencing for quadratic trend removal, 46

seismic trace of earthquake, 58
semiperiodogram, 217
semivariance, 32, 546–7
sequency (Walsh functions), 138
sequency-limited and time-limited, 140
shading sequence, 186
Shannon number, 66–7, 90, 357, 372–4, 381
ship altitude data, 330–2, 626–8
  analysis of longer series, 350
  direct spectral estimator, 331
  lag window spectral estimator, 331
  periodogram, 331
  plot of, 330, 627
  simulation of, 626–8
sidelobe leakage, 77, 185, 386, 388, 390
sidelobes
  of smoothing windows, 268–9, 273, 275–8
  of spectral windows, 281–2
    Fejér's kernel, 178–80, 328, 527, 532, 535–6
Siegel's test for compound periodicity, 540–3, 581, 587
  asymptotic critical value, 542–3, 581
  exact critical value, 541, 581
  interpolated critical value, 542, 581
sign of i convention, 55
signal-to-noise ratio, 526, 532, 543, 552, 560–1, 567, 578, 587
  signal in, 526
similarity theorem, 58–60
simple periodicity, 540–1, 586
simulation of
  atomic clock data, 628–30
  autoregressive (AR) process, 595
    zero initial conditions, 595
  fractionally differenced (FD) process, 440
  Gaussian AR(1) process
    via Kay's method, 639
  Gaussian AR(2) process, 597, 602
    via circulant embedding, 603
    via Gaussian spectral synthesis method (GSSM), 608
    via Kay's method, 597
    examples, 34
  Gaussian AR(4) process, 597
    via circulant embedding, 640–1
    via Gaussian spectral synthesis method (GSSM), 606, 641
    via Kay's method, 597, 638
    examples, 34
  Gaussian ARMA process, 593–4
    via Kay's method, 598, 639
    via McLeod and Hipel's method, 600, 639
  Gaussian autoregressive (AR) process
    stationary initial conditions, 595, 599, 624, 637
    via Kay's method, 597
    via McLeod and Hipel's method, 599
  harmonic process, 593–4, 637
  moving average (MA) process, 594, 637, 639
    MA(1) examples, 33
  non-Gaussian time series
    ARMA, 631, 636, 638
    with specified ACVS and PDF, 631, 633, 638

ocean wave data, 619–26
  uses, 626
reduced-length time series
  based on nonparametric SDF estimate, 614
ship altitude data, 626–8
time series, 593
  based on
    nonparametric SDF estimate, 611, 619–20, 627, 629, 637
      with non-unity sampling interval, 613, 638
      method of surrogate time series (MSTS), 615–6
    parametric SDF estimate, 617, 638
    periodogram, 615
  centering, 619, 627–8
  Gaussianity testing, 621
  given a portion of its ACVS, 593, 601, 637
  given its fully specified SDF, 593, 604, 637
  role of sampling interval, 594, 609, 613
  summary, 637–8
  white noise, 594, 637
    examples, 31
simulation via
  circulant embedding, 601–4, 619–20, 627, 629, 631
    ACVS properties, 602, 640
    computational efficiency, 603
    issues with ACVS, 615
    overcoming negative weights, 604, 640
    weights, 601–2, 605, 629, 640–1
  Gaussian spectral synthesis method (GSSM), 605–6, 608
    adjustment for power-law process, 610
    and adaptive multitaper spectral estimator, 615
    and prewhitening, 615
    role of sampling interval, 609
    weights, 640–1
  method of surrogate time series (MSTS), 615, 641
    relation of simulated process to harmonic process, 641
sinc function, 63, 559
single-sided versus double-sided SDF, 118
singular value decomposition (SVD), 425, 564–5
  and short time series, 567
  approach to harmonic analysis, 563
  Willamette River data, 578–80, 592
  singular value, 565
  singular vector, 565
sinusoidal amplitude estimator
  approximate conditional least squares (ACLS), 515, 525, 549, 569–70, 572–3, 585, 589
  approximate least squares, 516, 585
  exact conditional least squares (ECLS), 549, 569–70, 572, 589–90
    including mean, 552
  exact least squares
    complex-valued, 517, 586
    real-valued, 515, 525, 585
  exact unconditional least squares (EULS), 549, 569–70, 572
    including mean, 552
  Laplace method, 529

Lomb–Scargle method, 528
multitaper (complex-valued), 546, 551, 574
sinusoidal frequency estimator
  exact unconditional least squares (EULS), 549, 569–70, 572
    including mean, 552
  location of maximum (global or local) of AR SDF, 556–8, 591
    forward and backward least squares, 556
  direct spectral estimator, 537–8, 556–7, 585, 589
    rationale for tapering, 535–7
  periodogram, 526–8, 537–8, 589
    at Fourier frequencies, 523–4, 585–6
    at general frequencies, 524–5, 585
    expected value of estimator, 526
    convergence of estimator, 527
    MSE of estimator, 526, 537
    variance of estimator, 526, 534
  Thomson's F-test, 574, 582–3
  parametric approach, 553–8
    problems with, 558–63
  three-sinusoids example
    noise-free (Kay and Marple), 530
      effect of grid size, 530–2
    with noise, 532
      effect of noise variance, 533–4
      effect of series length, 532, 534
sinusoidal multitapers, 355, 372, 391, 393–4, 396, 434, 544, 574, 583, 590, 592
  compared to Slepian multitapers, 393
  defined, 392
  equivalent width, 401, 441
    versus spectral window bandwidth, 402
  relationship to minimum-bias multitapers, 392
sinusoidal power estimation (AR SDF), 578
sinusoidal variations
  deterministic, 553–4, 558, 590
  expressible as AR model, 447
  inexpressible as MA model, 503
  randomly phased, 35, 511, 553, 561
    ACVS for, 37, 553, 591
    examples of realizations, 36
    plus white noise, 512, 554
    as ARMA process, 554
  sum of, 6–7
skewness, 631
  ARMA(1,1) process, 632
  inequality, 633, 641
Slepian
  data taper, 190–1, 206, 208, 225, 228, 234, 260, 263, 282, 287, 416, 461, 496, 535–6, 569, 591
    approximation of zeroth-order, 196
  lag window, 285, 345
  multitapers, 355, 360, 362, 372, 391, 434, 544, 574, 582, 590, 592
    compared to sinusoidal, 393
    decomposition of energy, 367
    equivalent width versus spectral window bandwidth, 369
  sequence, 87, 105, 145, 189, 357
    double orthogonality, 106
    zeroth-order as a low-pass filter, 155

smoothing of
  direct spectral estimator, 245
  log direct spectral estimator, 303
  periodogram estimator, 246, 307
smoothing window, 245, 248, 341
  autocorrelation width of, 251
  bandwidth, 245, 251, 268, 281, 292, 296, 300–1, 318, 322, 341, 346–7
  Bartlett–Priestley design, 344
  choice of, 252
  finer frequency grid, 255
  for sea level residuals, 522
  Grenander, 251, 268, 279, 281
  lag window spectral estimator, 279
  Parzen, 251
  repeated lag window estimator, 349–50
  bias due to, 256, 290
  equivalent width of, 251
  leakage, 268, 287–9, 294, 316, 318, 331, 498
  parameter m (controls smoothing), 247, 287
    choice of, 291, 304, 313–4
    examples of, 318, 322, 328, 330, 339
  properties of, 250
  type
    Bartlett, 269–70
    Bartlett–Priestley, 274–5
    Daniell (rectangular), 250, 272–3
    Gaussian, 277–8
    modified Daniell, 280
    Papoulis, 278–9
    Parzen, 275–6, 280–2, 344
    rectangular, 271
spectral bandwidth, 292, 296, 346
spectral density function (SDF), 111
  after first difference and seasonal filtering, 333–4
  aliasing effect, 122–3
  alternative definitions, 114–6
  and LTI digital filters, 143
  and subsampling, 130
  asymmetry for complex-valued process, 118
  basic properties, 116–7
  classification examples, 121
  confidence interval for, 205, 265–7, 342
    for log of, 205, 227, 229, 266–7, 342
  double-sided versus single-sided, 118
  dynamic range, 177, 184, 197, 492, 641
  for first difference of a process, 158, 326
  for particular stationary processes
    AR, 145, 446, 485, 609, 617–8
    AR(1), 336
    AR(2), 172–3, 609
    AR(4), 172–3
    computation using zero padding, 449
    ARMA, 145, 159
    band-limited white noise, 379
    fractionally differenced (FD), 166, 407–8
    MA, 130, 144
    MA(1), 130
    MA(2), 159
    power-law, 605, 610
    white noise, 117
  Fourier transform of ACVS, 111, 114, 126
  imputed from noise model, 336
  linear and log scales, 336–8
  log of, 301
  one-sided versus two-sided, 118
  nonparametric estimators of
    direct spectral estimator, 186
      discretely smoothed, 246–7
    lag window estimator, 248
    multitaper, 352
    periodogram, 170
      discretely smoothed, 307
    WOSA, 414
    robust estimation via, 419
  parametric estimators of
    AR, 446–7
    Burg, 466
    least squares, 475
    maximum likelihood, 480
    Yule–Walker, 451
    MA, 503
  single-sided versus double-sided, 118
  symmetry for real-valued process, 111–12, 118, 126
  two-sided versus one-sided, 118
  type
    Gaussian-shaped, 131
    Lorenzian, 131
    piecewise constant, 348
    rectangular, 128
    triangular, 128
  unimodal, 297
  units of (physically meaningful), 374
  variance-preserving, 337–9
  versus autocovariance sequence, 124–5
  weighted sum of uncorrelated processes with SDFs, 130
spectral distribution function, 117
spectral estimator of the Grenander–Rosenblatt type, 252
spectral factorization, 636
spectral lines, 518, 538–40, 543–4, 548–9
  splitting of, 558, 560
spectral representation
  interpretation, 110–11
  of stationary process, 108
    complex-valued, 114
    continuous parameter, 113–4, 135
    discrete parameter, 110, 125
    nonzero mean, 113
    with discrete spectra, 518–9
    with mixed spectra, 518–9
  theorem, 7, 108–10, 382, 518–9
  uniqueness, 110
spectral reshaping, 547, 551, 575, 582
spectral window, 175, 186–8, 255, 353, 414
  bandwidth, 192, 194, 213–4, 355
  definitions of
    autocorrelation width (standard), 194
    correlation time, 213–4, 355
    half-power width, 192
    standard (autocorrelation width), 194
    variance width, 192, 194
  definitions (standard) for spectral estimators
    direct, 194, 235
    discretely smoothed periodogram, 347

lag window, 256, 341
multitaper, 353
WOSA, 415
frequency range for statistical results, 206, 213, 244, 268
standard measure of, 194
  examples of use, 226–7, 229, 294, 317–8, 323, 325, 328, 331, 335, 568, 623–4
  versus regularization bandwidth, 358, 402
  versus smoothing window bandwidth, 296
  versus correlation time, 214, 368–9, 401–2
bias, 192, 391
  and minimum-bias multitapers, 391–2
  due to central lobe (spectral window bias), 194, 535
  due to sidelobes (leakage), 194, 535
decay rate, 281, 328
definitions of for spectral estimators
  direct, 186–8, 234
  lag window, 255, 341
  multitaper, 353
  periodogram, 175
  WOSA, 414
type
  autocorrelated 100% cosine lag window, 285
  Bartlett lag window, 268, 270
  Bartlett–Priestley lag window, 268, 274
  Bartlett–Priestley taper, 288
  Daniell (rectangular) lag window, 268, 272
  Daniell taper, 288
  Gaussian lag window, 268, 278
  Gaussian taper, 288
  Hanning taper, 191
  modified Daniell window, 280
  100% cosine lag window, 284
  Papoulis lag window, 268, 279
  Papoulis taper, 288
  Parzen lag window, 268, 276
  $p \times 100\%$ taper, 191
  periodogram, 176
  Slepian lag window, 286
  Slepian multitapers, 358, 364–5
    kth-order taper, 361, 363, 368
  Slepian taper (zeroth-order), 191, 288
  sinusoidal multitapers, 398–401
    kth-order taper, 395, 397
    standard bandwidth approximation, 400–1
spectrograph estimator, 252
spectrum, 107–8
classification of, 120–2
discrete, 120, 511, 513, 518, 538, 584, 586
for simple time series model, 5–11
  constant, 11
  estimated
    nonparametric, 11–13
    parametric, 14–5
  high frequency, 9–10
  low frequency, 9–10
  theoretical, 9–11
integrated, 120
interpretation via band-pass filtering, 147–8
line, 120, 511, 522
mixed, 120, 511, 513, 518, 521–2, 539, 584–5

purely
  continuous, 120, 165, 167
  discrete, 120, 511
spectrum analyzers, 419, 444
spinning rotor, 39
spontaneous line splitting (*see* spectral lines)
spurious peaks in
  FBLS pseudo-AR sinusoidal frequency estimator, 567
  periodogram, 527
square integrable, 48, 53–4, 75
squared norm, 470
St. Paul temperature data, 5–6
standard frequencies (Fourier frequencies), 8, 91, 515
stationarity
conditions for
  autoregressive (AR) process, 145, 446–9, 557
  pseudo-AR process, 557
second-order/weak/covariance, 26–7
strict/complete/strong, 26, 631, 634
stationary differences of order d
  process with, 340
stationary process, 4, 7, 21, 26–31, 107
  and modeling data, 38–41, 45
  examples of
    autoregressive (AR), 33
    autoregressive moving average (ARMA), 35
    bandlimited white noise, 379
    harmonic, 35
    fractionally differenced, 166
    moving average (MA), 32
    white noise, 31
  intrinsically of order d, 340
  nonergodic, 128
  reverse of conjugate of complex-valued process, 502
  reverse of real-valued process, 455, 477, 484
step function, 24, 26, 120
step-down Levinson–Durbin recursions, 460–1
  and causality assessment, 453, 509
  and generation of ACVS for an AR process, 460–1, 508
  and simulation of ARMA process, 594–8
stochastic continuity, 114, 118
stochastic process, 21–2, 23–5
  jump, 109
  orthogonal, 109
storm surges, 522
strict stationarity, 26, 631, 634
sum of squares (SS), 467, 476, 514
summability
  absolute, 165
  Cesàro, 78, 165
superposition (LTI filters), 133, 140
SVD (*see* singular value decomposition)
symmetry, 27
  conjugate, 29–30, 48, 54
Szegő–Kolmogorov theorem, 404

tables
  AR-based estimates of innovation variance, 500

Burg's algorithm and sinusoidal frequency estimation, 559
characteristics of AR SDFs and periodogram for Willamette River data, 578
comparison of
 spectral bandwidth measures for direct spectral estimator, 214
 spectral bandwidth measures for sinusoidal multitaper spectral estimator, 402
 spectral bandwidth measures for Slepian multitaper spectral estimator, 369
 window bandwidth measures for discretely smoothed direct spectral estimator, 255
 Yule–Walker AR(4) and AR(8) coefficient estimates, 459
exact and approximate critical values for Siegel's test, 542
harmonic analysis and centering for Willamette River data, 583
horizontal and vertical components of crisscrosses (ice profile data), 325
multitaper-based estimates of innovation variance, 432
NMSE criteria for GSSM-based approximations for AR(4) process, 607
parameter estimates in harmonic analysis for Willamette River data, 570
properties of lag windows, 279
ratio of variances for estimators of power-law exponent, 411
reduced likelihoods for estimated AR(4) coefficients, 483
true and estimated sinusoidal frequencies, 533
variance inflation factor for tapers, 260
taper, 186
 compared to lag window, 282–3
 advantages/disadvantages in use of
  better performance of Yule–Walker autoregressive spectral estimator, 452, 461
  decrease in bias due to sidelobes, 179, 185, 189, 234, 535–6
  decrease in effective sample size, 194–5
  decrease in pairwise correlation, 207
  increase in bias due to central lobe, 192, 535
 converting
  taper into lag window, 283
  lag window into taper, 285
 linear, 186
 normalization
  unit energy, 188
  data-dependent, 195
 orthogonality, 354, 357, 392, 432
 plethora of types, 196–7
 quadratic, 248
 type
  Bartlett–Priestley, 287
  Daniell, 287
  default, 188, 190
  Gaussian, 287
  Hanning, 189–90, 230, 234, 327, 331, 415–7
  Hanning (alternative definition), 240
  kth-order minimum-bias, 392–3
  kth-order sinusoidal, 392–4, 396
  kth-order Slepian, 357, 360, 362, 368
  minimum-bias multitapers, 392–3
  $p \times 100\%$ cosine, 189–90, 260
  Papoulis, 287
  Parzen, 287
  rectangular, 188–90, 535–6
  sinusoidal multitapers, 355, 392–3, 402–3
  Slepian, 190, 196, 212, 225, 228, 230, 234, 260, 263, 287, 416, 461, 496, 535–6
  Slepian multitapers, 355, 357, 372–3
  variance inflation factor, 260, 262, 342, 417
temporal window, 186
terrestrial free oscillations, 359, 366
test for
 periodicity
  Fisher's, 540–1, 586–7
  in colored noise, 544, 587
  in white noise, 538, 586–7
  Schuster's, 539
  Siegel's, 541, 587
  Thomson's F-test (multitaper approach), 228, 359, 544–8, 574–5, 582–3, 587, 592
  with irregular sampling, 543–4
  white noise (cumulative periodogram), 215–6, 230–1, 573–4
textile slivers, 33, 503
Thomson's F-test for periodicity, 228, 359, 544–8, 574–5, 582–3, 587, 592
tidal
 analysis, 40, 520–3
 frequencies, 521
time indexing of a series
 and periodogram, 184
time invariance (LTI filters), 133, 140
time series simulation, 593
 based on
  nonparametric SDF estimate, 611, 619–20, 627, 629, 637
   with non-unity sampling interval, 613, 638
  method of surrogate time series (MSTS), 615–6
  parametric SDF estimate, 617, 638
  periodogram, 615
 centering, 619, 627–8
 examples of,
  atomic clock data, 628–30
  ocean wave data, 619–26
  uses, 626
  ship altitude data, 626–8
 for testing of Gaussianity, 621
 given a portion of its ACVS, 593, 601, 637
 given its fully specified SDF, 593, 604, 637
 non-Gaussian
  ARMA process, 631, 636, 638
  with specified ACVS and PDF, 631, 633, 638
 reduced-length
  based on nonparametric SDF estimate, 614
 role of sampling interval, 594, 609, 613
 summary, 637–8
time-limited function, 58
time reverse of
 conjugate of complex-valued stationary process is stationary with same ACVS, 502

real-valued stationary process is stationary with same ACVS, 455, 477, 484
time-reversibility, 484
Toeplitz
   matrix, 29, 156, 457, 462, 480, 484
      Hermitian, 30
   system of equations, 452, 461
trace of a matrix, 376, 378, 439
trade-off
   between bias and variance, 80, 296, 309
   between fidelity and resolution, 80
transfer function, 136, 142–3, 290
   and
      gain function, 136
      phase function, 136
   as Fourier transform of impulse response sequence, 143
   for
      AR filter, 144–5, 487
      Hilbert transform, 579, 592
      MA filter, 144
transpose of vector, 28
trend, 40, 151, 332
   removal, 39, 334, 443
   as high-pass filtering, 151
tridiagonal matrix, 88, 95
trigamma function, 296, 302, 405, 410
truncated periodogram, 343, 504
twin peaks, 76–8, 178–9, 192, 200–1, 218–9, 291, 294, 300–1, 306, 337, 382, 458–9, 489
two-sided versus one-sided SDF, 118

uncorrelated random variables
   definition, 31
   relationship to independent random variables, 31–2
undersmoothing, 71
uniform distribution
   over interval $(-\pi, \pi]$, 37, 45, 598
   over interval $(-\sqrt{3}, \sqrt{3}]$, 634
uniqueness of
   Fourier transform, 55, 82
   integrated spectrum, 135
   spectral representation, 110
   stationary solution to AR equation, 144–5, 446–8
unit circle, 446, 448–9, 463, 553, 557–8, 566–7, 580, 591–2, 600, 618

variability of SDF estimators
   assessment via
      asymptotic theory, 204–5, 265–7, 342, 354, 485–91
      simulation, 622–6
variance
   definition of
      complex-valued case, 25
      real-valued case, 24
variance inflation factor due to tapering, 260, 262, 342, 417
variance of
   linear combination of random variables, 41

sample mean for AR(1) process, 507
sinusoidal frequency estimator using periodogram maximum, 526–7, 534, 592
spectral estimators
   basic multitaper, 353–4, 359
      for Gaussian white noise, 372–4, 402–3
   direct, 204, 235, 624
      for complex-valued process, 232
      verification via simulation, 242
   lag window, 258–9, 262, 265, 279, 341
   periodogram, 203, 235, 241
      for AR(1) process, 243
      verification via simulation, 242
   quadratic, 380
   weighted multitaper, 354
   WOSA, 415, 435
stationary process, 27–8
   bias in estimator (unknown mean), 168, 237
variance-preserving SDF, 337–9
variance spectrum for
   harmonic process, 108
   simple time series model, 8, 47
variance-stabilizing transform, 13, 209
variance width of function, 60, 73, 105, 191–2, 194, 288
vector transposition, 28

Walsh functions, 138–40
weighted covariance estimator, 252
weighted multitaper spectral estimator, 352–4, 388, 433, 611–13
   and quadratic spectral estimator, 376
   bias of, 391
   covariances of, 437
   distribution of, 354–5
   equivalent degrees of freedom (EDOFs) of, 354–5
   equivalent width, 355, 437
   expected value of, 352, 389
   in circulant embedding simulation method, 611–13
   low-rank approximation, 422–3
   Parseval's theorem in expectation, 390
   spectral window for, 352–3, 391
      bandwidth of, 353
   variance of, 354
   via regularization, 382
weighted overlapped segment averaging (*see* WOSA spectral estimator)
weights for circulant embedding simulation method (*see* circulant embedding simulation method)
Welch's overlapped segment averaging spectral estimator (*see* WOSA spectral estimator)
white noise, 31–2, 120, 169, 184, 197, 236, 381, 386–8, 446, 450, 511, 513, 524, 532, 534, 538, 586
   and direct spectral estimator, 189, 371
   and periodogram, 177, 207, 371
   and quadratic spectral estimator, 377
   and tapering, 195
   autocovariance sequence (ACVS), 31
   band-limited, 379
   complex-valued, 32

and proper, 517–8, 564
continuous parameter, 38
examples of, 31
Gaussian, 201, 539
spectral density function (SDF), 117
simulation of, 594, 637
test for (cumulative periodogram), 215, 230–1
versus colored noise, 120, 126, 241, 372
white noise variance estimator (harmonic analysis)
  approximate conditional least squares (ACLS), 549, 570
  exact conditional least squares (ECLS), 549, 552, 570
  exact least squares
    complex-valued, 518
    real-valued, 516
  exact unconditional least squares (EULS), 550, 552, 570
  multitaper, 551
Whittle likelihood, 307
Wiener–Khintchine theorem, 118
Willamette River data, 1–2, 567–8
  AICC order selection, 575–6
  analytic series version, 578–81, 592
  Burg autoregressive spectral estimator, 575–7
  cumulative periodogram test of residuals, 573–4
  direct spectral estimator, 569, 591–2
  FBLS autoregressive spectral estimator, 592
  fitted values, 571–2
  harmonic analysis of, 567–81
    effect of centering, 583
  integrated autoregressive spectrum, 577
  multitaper spectral estimator, 575
  parametric approach to harmonic analysis of, 575–8
  periodogram, 568
    of residuals, 570–1, 573
  plot of
    complex roots for, 580
    fitted model, 571–2
    residuals, 571–2
  remodulated filtered series, 592
  reshaped spectrum, 575, 592
  residuals, 571–2
  small portion of, 1–2
    ACS for
      sample, 4
      theoretical, 10
    lag 1 scatter plot for, 3
    sample ACVS for, 169
    simulation of, 11
    spectrum for
      estimated, 12, 14
      theoretical, 10, 12, 14
  SVD approach to harmonic analysis of, 578, 592
  table of characteristics of AR SDFs and periodogram, 578
  Thomson's F-test for periodicity, 575, 592
  Yule–Walker autoregressive spectral estimator, 592
wind speed data, 1–2

ACS for
  sample, 4
  theoretical, 10
lag 1 scatter plot for, 3
sample ACVS for, 169
simulation of, 11
spectrum for
  estimated, 12, 14
  theoretical, 10, 12, 14
window
  data, 186
  design, 247
    for Bartlett–Priestley, 273–4, 319, 344
    for Daniell, 272, 319
  lag (see entry for lag window)
  linear, 186
  quadratic, 248
  smoothing (see entry for smoothing window)
  spectral (see entry for spectral window)
  temporal, 186
window closing, 293
  and WOSA, 416
  examples, 293–5
windowed periodogram estimator, 252
windows and convergence factors, 80
Wold's decomposition theorem, 404, 471
Wold's theorem, 117–8, 126, 185
WOSA spectral estimator, 351, 412–19, 432, 434–6
  ACVS estimator for, 420, 435
  AR(2) data, 416–7
  AR(4) data, 417–8
  at zero frequency, 420
  boundary conditions
    circular, 424
    reflecting, 424
  centering for unknown mean, 419–20
  computational efficiency, 419
  connection to Bartlett lag window spectral estimator, 270
  deconstructed multitaper spectral estimator, 420–1, 423
  EDOFs of, 415–6, 423–4, 442–3
  expected value of, 414, 435
  for approximating a multitaper spectral estimator, 424–5
  in circulant embedding simulation method, 611, 613
  ocean wave data, 425, 429–30
  robust SDF estimation, 419
  shift factor restriction, 420
  variance of, 415, 435
  versus multitaper spectral estimator, 414, 417–9, 424

Yule–Walker autoregressive estimator, 449–52, 505
  and ACVS for direct spectral estimator, 452, 461–2, 505
  AR(1) process, 510
  AR(2) numerical example, 507–8
  AR(4) data, 458–9, 461–2, 508–9
  assessing variability (SDF), 485
  causality, 448, 505

complex-valued process, 502
   corresponding ACVS, 451, 504, 508
   FPE criterion, 494
   ocean wave data, 496–9
     order selection via PACS, 498–9
   order of, 458–9
   peak shifting and splitting, 560–2, 591
   relation to
     autocorrelation method, 480
     Burg's algorithm, 471
     forward least squares, 479
   Willamette River data, 592
Yule–Walker equations, 450, 561
   arising from two related problems, 454–5
   augmented, 452
   Levinson–Durbin recursions and, 457

zero padding, 93–4
   use in computing AR SDF, 449
   finer frequency grid, 179, 527, 558, 568
   for noncyclic convolutions, 106, 169–70, 220
zero-memory nonlinearity (ZMNL), 634
   correlation distortion, 635–6